26
34

8
9

34

24
34 | 80
68
120
6

17
9
26
8
34.

Merrill

Algebra 1

Applications and Connections

GLENCOE

McGraw-Hill

New York, New York
Columbus, Ohio
Mission Hills, California
Peoria, Illinois

Send all inquiries to:
Glencoe/McGraw-Hill
936 Eastwind Drive
Westerville, OH 43081

ISBN: 0-02-824178-9 (Student Edition)
ISBN: 0-02-824179-7 (Teacher's Wraparound Edition)

5 6 7 8 9 10 VH/LP 03 02 01 00 99 98 97 96 95

AUTHORS

Alan G. Foster is chairperson of the mathematics department at Addison Trail High School, Addison, Illinois. He has taught mathematics courses at every level of the high school curriculum. Mr. Foster obtained his B.S. degree from Illinois State University and M.A. in mathematics from the University of Illinois. He is active in professional organizations at the local, state, and national levels. He is a past president of the Illinois Council of Teachers of Mathematics and a recipient of the Illinois Council of Teachers of Mathematics T.E. Rine Award for excellence in the teaching of mathematics. He also was a recipient of the 1987 Presidential Award for Excellence in the Teaching of Mathematics in the state of Illinois. Mr. Foster is a coauthor of *Merrill Geometry, Merrill Algebra Essentials,* and *Merrill Algebra 2.*

Joan M. Gell is a mathematics teacher and department chairperson at Palos Verdes Peninsula High School in Palos Verdes Estates, California. Ms. Gell has taught mathematics at every level from junior high school to college. She received her B.S. degree in mathematics education from the State University of New York-Cortland, and M.A. degree in mathematics from Bowdoin College in Brunswick, Maine. Ms. Gell has developed and conducted in-service classes in mathematics and computer science and is past president of the California Mathematics Council. She serves as the chairperson of the 1992 MATHCOUNTS Problem Writing Committee. Ms. Gell was a finalist for the 1984 Presidential Award for Excellence in the Teaching of Mathematics in the state of California.

Leslie J. Winters is the Secondary Mathematics Specialist for the Los Angeles Unified School District. He has thirty years of classroom experience in teaching mathematics at every level from junior high school to college. Mr. Winters received bachelor's degrees in mathematics and secondary education from Pepperdine University and the University of Dayton, and master's degrees from the University of Southern California and Boston College. He is a past president of the California Mathematics Council-Southern Section and was a recipient of the 1983 Presidential Award for Excellence in the Teaching of Mathematics in the state of California. Mr. Winters is a coauthor of *Merrill Algebra Essentials* and *Merrill Algebra 2.*

James N. Rath has 30 years of classroom experience in teaching mathematics at every level of the high school curriculum. He is a mathematics teacher and former head of the mathematics department at Darien High School, Darien, Connecticut. Mr. Rath earned his B.A. degree in philosophy from the Catholic University of America and his M.Ed. and M.A. degrees in mathematics from Boston College. He is active in professional organizations at the local, state, and national levels. Mr. Rath is a coauthor of *Merrill Pre-Algebra, Merrill Algebra Essentials,* and *Merrill Algebra 2.*

Berchie W. Gordon is a Mathematics/Computer coordinator for the Northwest Local School District in Cincinnati, Ohio. Dr. Gordon has taught mathematics at every level from junior high school to college. She received her B.S. degree in Mathematics from Emory University in Atlanta, Georgia and M.A.T. in education from Northwestern University, Evanston, Illinois. She has done further study at the University of Illinois and the University of Cincinnati where she received her doctorate in curriculum and instruction. Dr. Gordon has developed and conducted numerous in-service workshops in mathematics and computer applications. She has served as a consultant for IBM. She has traveled nationally to make presentations on the graphing calculator to teacher groups.

CONSULTANTS

Donald W. Collins
Department of Mathematics and Informational Sciences
Sam Houston State University
Huntsville, Texas

Gail F. Burrill
Chairperson, Department of Mathematics
Whitnall High School
Greenfield, Wisconsin

EDITORIAL ADVISORS

William Collins
Chairperson, Department of Mathematics
James Lick High School
San Jose, California

Jo Helen Williams
Supervisor, Secondary Mathematics
Dayton Public Schools
Dayton, Ohio

REVIEWERS

Ron Ashby
Mathematics Supervisor
Vigo County Secondary Schools
Terre Haute, IN

Paul Benz
Mathematics Department Chairman
Clay Junior High School
Carmel, IN

Grace Berlin
Mathematics Department Chairman/Teacher
Stillwater Junior High School
Stillwater, OK

Esta Rae Carelli
Mathematics Teacher
Staten Island Technical High School
Staten Island, NY

Dr. Marva Fears
Mathematics Supervisor
Dekalb County School District
Decatur, GA

Terry Hacker
Mathematics Teacher
Oliver Springs High School
Oliver Springs, TN

Marietta Harris
Curriculum Coordinator
Memphis City Schools
Memphis, TN

Cathy Jahr
Mathematics Department Chairman
Martin Westview High School
Martin, TN

Joe Lee
Mathematics Department Chairman/Teacher
Parkway West Junior High School
Chesterfield, MO

Betty Mayberry
Mathematics Department Head
Riverdale High School
Murfreesboro, TN

Dr. James Reney
Supervisor of Mathematics
Altoona Area School District
Altoona, PA

Lowell Wilson
Mathematics Teacher
George Washington High School
Greenfield, IN

Table of Contents

CHAPTER 1 An Introduction to Algebra 6

1-1 Variables and Expressions 8
1-2 Evaluating Expressions 12
Reading Algebra 17
1-3 Open Sentences 18
1-4 Identity and Equality Properties 22
Mid-Chapter Review 25
1-5 The Distributive Property 26
1-6 Commutative and Associative Properties 31
History Connection: Sir William Rowan Hamilton 35
1-7 Formulas 36
Technology: Formulas 40
1-8 Problem-Solving: Explore Verbal Problems 41
Cooperative Learning Activity 43
Chapter Summary and Review 44
Chapter Test 47

CHAPTER 2 Rational Numbers 48

2-1 Integers and the Number Line 50
Application: Meteorology 54
2-2 Adding and Subtracting Integers 55
Puzzle 59
2-3 Inequalities and the Number Line 60
2-4 Comparing and Ordering Rational Numbers 65
2-5 Adding and Subtracting Rational Numbers 69
Mid-Chapter Review 73
2-6 Multiplying Rational Numbers 74
History Connection: Sonya Kovalevskaya 78
Technology: Adding Fractions 79
2-7 Dividing Rational Numbers 80
2-8 Problem-Solving Strategy: Write an Equation 85
Cooperative Learning Activity 87
Chapter Summary and Review 88
Chapter Test 91

APPLICATIONS AND CONNECTIONS

APPLICATIONS
Basketball 26
Golf 56
Business 62
Consumerism 66
Stock Market 69

CONNECTIONS
Geometry 10, 14, 32, 36, 37
Logic 18
Mental Math 27
Set Theory 51
Sequences 76

Technology

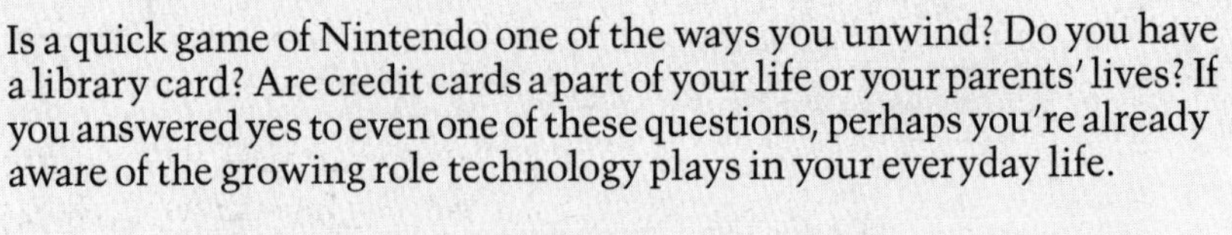

Is a quick game of Nintendo one of the ways you unwind? Do you have a library card? Are credit cards a part of your life or your parents' lives? If you answered yes to even one of these questions, perhaps you're already aware of the growing role technology plays in your everyday life.

In fact, calculators and computers have become so essential that it would be rare for your life not to be affected by these amazingly versatile machines. Schools, banks, department stores, hospitals, government offices – even police departments – rely heavily on technology to perform many important tasks.

The Technology pages in this text let you use technology to explore patterns, make conjectures, and discover mathematics. You will learn to use programs written in the BASIC computer language as well as computer software. You will also investigate mathematical concepts using scientific calculators and graphing calculators.

BASIC and Spreadsheets: Formulas 40
BASIC: Adding Fractions 79
MET: Solving Equations 121
BASIC: Successive Discounts 150
MET: Solving Inequalities 191
Spreadsheets: Volume and Surface Area 247
MET: Factoring 285
MET: Rational Expressions 315
Graphing Calculator: Graphing Relations 384
MET: Graphing Linear Equations 427
Graphing Calculator: Solving Systems of Equations 462
MET: Solving Radical Equations 505
Graphing Calculator: Solving Quadratic Equations 530
Graphing Calculator: Regression Lines 588
BASIC: Trigonometric Functions 637

MET (Mathematics Exploration Toolkit) was developed for IBM by Wicat Systems, Inc.

CHAPTER 3 Equations 92

3-1 Solving Equations by Using Addition 94
3-2 Solving Equations by Using Subtraction 99
3-3 Solving Equations by Using Multiplication and Division ... 103
Mid-Chapter Review 107
3-4 Problem-Solving Strategy: Work Backwards 108
Cooperative Learning Activity 110
3-5 Solving Equations Using More Than One Operation 111
Puzzle 115
3-6 Solving Equations with the Variable on Both Sides 116
Puzzle 120
Technology: Solving Equations 121
3-7 More Equations 122
History Connection 125
Chapter Summary and Review 126
Chapter Test 129
College Entrance Exam Preview 130

CHAPTER 4 Applications of Rational Numbers 132

4-1 Ratios and Proportions 134
4-2 Percent 138
4-3 Application: Simple Interest 142
Application: Compound Interest 145
4-4 Percent of Change 146
Technology: Successive Discounts 150
4-5 Problem-Solving Strategy: Make a Table or Chart 151
Cooperative Learning Activity 153
4-6 Application: Mixtures 154
Mid-Chapter Review 157
4-7 Application: Uniform Motion 158
Application: Travel Agents 161
4-8 Direct Variation 162
History Connection: A'h-Mose 165
4-9 Inverse Variation 166
Chapter Summary and Review 170
Chapter Test 173

APPLICATIONS AND CONNECTIONS

APPLICATIONS

Aviation 95
Weightlifting 99
Finance 111, 142, 143
Biology 134
Models 135
Travel 136, 158
Sales ... 140, 146, 147, 148, 154
Chemistry 155
Uniform Motion 159
Space 162
Electronics 163
Music 166
Physics 167,168

CONNECTIONS

Geometry 105, 117
Number Theory 113

CHAPTER 5 Inequalities 174

5-1 Solving Inequalities Using Addition and Subtraction 176
History Connection: Recorde, Harriot, and Oughtred 180
5-2 Solving Inequalities Using Multiplication and Division ... 181
Reading Algebra 185
5-3 Inequalities with More Than One Operation 186
Mid-Chapter Review 190
Technology: Solving Inequalities 191
5-4 Problem-Solving Strategy: Make a Diagram 192
Cooperative Learning Activity 193
5-5 Compound Inequalities 194
5-6 Open Sentences Involving Absolute Value 199
Chapter Summary and Review 204
Chapter Test 207

POLLS OPEN 7 A.M. CLOSE 8 P.M.

POLLING PLACE

CHAPTER 6 Polynomials 208

6-1 Problem-Solving Strategy: Look for a Pattern 210
Cooperative Learning Activity 212
6-2 Multiplying Monomials 213
6-3 Dividing Monomials 217
Reading Algebra 220
6-4 Scientific Notation 221
Application: Black Holes 225
6-5 Polynomials 226
Mid-Chapter Review 229
6-6 Adding and Subtracting Polynomials 230
History Connection: Emmy Noether 233
6-7 Multiplying a Polynomial by a Monomial 234
6-8 Multiplying Polynomials 238
Extra 242
6-9 Some Special Products 243
Application: Punnett Squares 246
Technology: Volume and Surface Area 247
Chapter Summary and Review 248
Chapter Test 251
College Entrance Exam Preview 252

APPLICATIONS AND CONNECTIONS

APPLICATIONS
Sports 178
Sales 183, 196
Machinery 201
Biology 222
Economics 222
Finance 227
Gardening 240

CONNECTIONS
Geometry 182, 213, 214, 231, 235
Statistics 188

CHAPTER 7

Factoring 254

7-1	Factors and Greatest Common Factor	256
	History Connection: Eratosthenes	260
7-2	Factoring Using the Distributive Property	261
7-3	Factoring by Grouping	265
7-4	Problem-Solving Strategy: Guess and Check	269
	Cooperative Learning Activity	270
7-5	Factoring Trinomials	271
7-6	Factoring Differences of Squares	276
	Mid-Chapter Review	280
7-7	Perfect Squares and Factoring	281
	Technology: Factoring	285
7-8	Summary of Factoring	286
7-9	Solving Equations by Factoring	290
7-10	More Solving Equations by Factoring	295
	Chapter Summary and Review	300
	Chapter Test	303

CHAPTER 8

Rational Expressions 304

8-1	Simplifying Rational Expressions	306
	History Connection: Fractions	310
8-2	Multiplying Rational Expressions	311
	Technology: Rational Expressions	315
8-3	Dividing Rational Expressions	316
	Extra	318
8-4	Dividing Polynomials	319
8-5	Rational Expressions with Like Denominators	322
	Mid-Chapter Review	325
8-6	Problem-Solving Strategy: List Possibilities	326
	Cooperative Learning Activity	328
8-7	Rational Expressions with Unlike Denominators	329
8-8	Mixed Expressions and Complex Fractions	334
8-9	Solving Rational Equations	338
8-10	Application: Formulas	343
	Chapter Summary and Review	348
	Chapter Test	351

APPLICATIONS AND CONNECTIONS

APPLICATIONS

Work 338
Fishing 339
Photography 343
Electronics 344, 345

CONNECTIONS

Geometry .. 262, 272, 278, 282, 287, 312, 323, 343
Number Theory 297

Special Features

How did algebra get its name? Read the **History Connection** on page 549 to find out. These features contain information about real people from the past and present and from many different cultures who have had a great influence on what you study in algebra today.

What does algebra have to do with genetics? The **Application** on page 246 can answer this question. In these features, you'll discover how algebra is used in most of the subjects you study in school as well as in your everyday life.

How can working together with my classmates help me solve problems? The fun, but challenging, problem presented in each **Cooperative Learning Activity** gives you an opportunity to cooperate, not compete, with other students.

History Connections

Sir William Rowan Hamilton 35
Sonya Kovalevskaya 78
Diophantus 125
A'h-Mose 165
Recorde, Harriot, and Oughtred 180
Emmy Noether 233
Eratosthenes 260
Fractions 310
René Descartes 383
Benjamin Banneker 409
The K'iu-ch'ang Suan-shu 467
Jaime Escalante 481
Muhammed ibn Musa al Khwarizmi 549
Graunt and Bernoulli 593
Pythagoras 636

Applications

Meteorology 54
Compound Interest 145
Travel Agents 161
Black Holes 225
Punnett Squares 246
Escape Velocity 486

Cooperative Learning Activities

Cooperative Learning Activities 43, 87, 110, 153, 193, 212, 270, 328, 391, 433, 441, 476, 529, 599, 616

CHAPTER 9 Functions and Graphs 352

9-1 Ordered Pairs 354
9-2 Relations 359
9-3 Equations as Relations 364
9-4 Graphing Linear Relations 369
Mid-Chapter Review 373
9-5 Functions 374
9-6 Graphing Inequalities in Two Variables 379
History Connection: René Descartes 383
Technology: Graphing Relations 384
9-7 Finding Equations from Relations 385
9-8 Problem-Solving Strategy: Use a Graph 389
Cooperative Learning Activity 391
Chapter Summary and Review 392
Chapter Test 395
College Entrance Exam Preview 396

CHAPTER 10 Graphing Linear Equations 398

10-1 Slope of a Line 400
10-2 Point-Slope and Standard Forms of Linear Equations 405
History Connection: Benjamin Banneker 409
10-3 Slope-Intercept Form of Linear Equations 410
10-4 Graphing Linear Equations 415
Mid-Chapter Review 418
10-5 Writing Slope-Intercept Equations of Lines 419
10-6 Parallel and Perpendicular Lines 423
Technology: Graphing Linear Equations 427
10-7 Midpoint of a Line Segment 428
10-8 Problem-Solving Strategy: Use a Graph 432
Cooperative Learning Activity 433
Chapter Summary and Review 434
Chapter Test 437

APPLICATIONS AND CONNECTIONS

APPLICATIONS

Cartography 356
Career 366
Meteorology 371
Physics 376
Sales 380
Business 386
Statistics 390
Driving 400
Travel 411
Communication 416
Manufacturing 419

CONNECTIONS

Probability 361
Geometry 423, 428

CHAPTER 11

Systems of Open Sentences 438

11-1	Problem-Solving Strategy: Hidden Assumptions	440
	Cooperative Learning Activity	441
11-2	Graphing Systems of Equations	442
11-3	Substitution	447
11-4	Elimination Using Addition and Subtraction	452
	Mid-Chapter Review	456
11-5	Elimination Using Multiplication	457
	Technology: Solving Systems of Equations	462
11-6	Graphing Systems of Inequalities	463
	History Connection: The *K'iu-ch'ang Suan-shu*	467
	Chapter Summary and Review	468
	Chapter Test	471

CHAPTER 12

Radical Expressions 472

12-1	Problem-Solving Strategy: Use a Table	474
	Cooperative Learning Activity	476
12-2	Square Roots	477
	History Connection: Jaime Escalante	481
12-3	The Pythagorean Theorem	482
	Application: Escape Velocity	486
12-4	Real Numbers	487
	Mid-Chapter Review	491
12-5	Simplifying Square Roots	492
12-6	Adding and Subtracting Radical Expressions	497
12-7	Radical Equations	501
	Technology: Solving Radical Equations	505
12-8	The Distance Formula	506
	Chapter Summary and Review	510
	Chapter Test	513
	College Entrance Exam Preview	514

APPLICATIONS AND CONNECTIONS

APPLICATIONS

Metallurgy ... 448
Uniform Motion ... 454, 459
Banking ... 459
Agriculture ... 459
Plumbing ... 464
Construction ... 479
Electricity ... 484

CONNECTIONS

Geometry ... 444, 494, 497, 498, 507
Number Theory ... 449, 453, 502
Statistics ... 474

CHAPTER 13 Quadratics 516

13-1 Graphing Quadratic Functions 518
13-2 Solving Quadratic Equations by Graphing 523
13-3 Problem-Solving Strategy: Identify Subgoals 527
Cooperative Learning Activity 529
Technology: Solving Quadratic Equations 530
13-4 Solving Quadratic Equations by Completing the Square .. 531
13-5 Solving Quadratic Equations Using the Quadratic Formula 536
Mid-Chapter Review 540
13-6 Using the Discriminant 541
13-7 Application: Solving Quadratic Equations 546
History Connection: Muhammed ibn Musa al Khwarizmi 549
13-8 The Sum and Product of Roots 550
Chapter Summary and Review 554
Chapter Test 557

CHAPTER 14 Statistics and Probability 558

14-1 Statistics and Line Plots 560
14-2 Stem-and-Leaf Plots 565
14-3 Measures of Central Tendency 570
14-4 Measures of Variation 575
14-5 Box-and-Whisker Plots 579
14-6 Scatter Plots 583
Mid-Chapter Review 587
Technology: Regression Lines 588
14-7 Probability and Odds 589
History Connection: Graunt and Bernoulli 593
14-8 Empirical Probability 594
14-9 Problem-Solving Strategy: Solve a Simpler Problem 598
Cooperative Learning Activity 599
14-10 Compound Events 600
Chapter Summary and Review 604
Chapter Test 607

APPLICATIONS AND CONNECTIONS

APPLICATIONS

Finance 520, 576
Construction 533
Physics 538, 551
Manufacturing 547
Transportation 561
Income 566
Meteorology 570
Baseball 571
Sports 572
Fundraising 580
Weather 590
Civics 595

CONNECTIONS

Number Theory 524
Geometry 543, 546

College Entrance Exam Preview

What will you do after you graduate from high school? Go to college? Get a job? Join the armed forces? If college is for you, you need to get ready **now.** That's right – NOW!

To get into most colleges and universities, you need to take the SAT (Scholastic Aptitude Test) or the ACT (American College Test) in your junior or senior year of high school. To help you prepare for these tests and other similar tests, you can use the College Entrance Exam Previews given after every third chapter in this text. They are on pages 130-131, 252-253, 396-397, 514-515, and 642-643.

CHAPTER 15 Trigonometry 608

15-1 Angles and Triangles 610
15-2 Problem-Solving Strategy: Make a Model 615
Cooperative Learning Activity 616
15-3 30°-60° Right Triangles 617
Mid-Chapter Review 621
15-4 Similar Triangles 622
15-5 Trigonometric Ratios 626
15-6 Solving Right Triangles 632
History Connection: Pythagoras 636
Technology: Trigonometric Functions 637
Chapter Summary and Review 638
Chapter Test 641
College Entrance Exam Preview 642

APPLICATIONS

Recreation 618
Construction 619, 633
Transportation 628
Physics 628
Forestry 633
Fishing 633

Using Tables 644
Algebraic Skills Review 647
Glossary 664
Selected Answers 674
Index 698
Photo Credits 710

Manipulative Activities A1-A24

Extended Projects B1

Project 1: The Principle of Balance B2
Project 2: Mathematics and Cryptography B6
Project 3: Extending the Pythagorean Theorem B10
Project 4: What's on TV? B14

Manipulative Activities

How can you discover the rules of algebra and know that they really work? Many of the concepts in algebra can be modeled by using manipulatives, such as counters, algebra tiles, and graphing calculators. You will use yellow and red counters to represent positive and negative integers and a cup to represent a variable. You will use these models to add and subtract integers, as well as to solve equations. You will use area models called algebra tiles to add, subtract, multiply, and factor algebraic expressions like $x^2 + 2x - 1$ and $4x + 2$.

In addition to modeling algebraic expressions, you will explore the Pythagorean theorem by using a geoboard. Base-ten models can help you estimate the square root of a number. You can use the graphing calculator to graph functions quickly and accurately. You can graph several functions on the same screen so you can compare their similarities and differences.

The Labs found in the **Manipulative Activities** section at the back of this book can help you understand many of the concepts you will study in this course.

Lab 1 The Distributive Property A2
Lab 2 Adding and Subtracting Integers A3
Lab 3 Solving One-Step Equations A5
Lab 4 Solving Multi-Step Equations A7
Lab 5 Polynomials A8
Lab 6 Adding and Subtracting Polynomials A10
Lab 7 Multiplying a Polynomial by a Monomial A12
Lab 8 Multiplying Polynomials A13
Lab 9 Factoring Using the Distributive Property A15
Lab 10 Factoring Trinomials A16
Lab 11 **Graphing Calculator:** Slope A18
Lab 12 **Graphing Calculator:** Intercepts A19
Lab 13 The Pythagorean Theorem A20
Lab 14 Estimating Square Roots A21
Lab 15 **Graphing Calculator:** Quadratic Functions A22
Lab 16 **Graphing Calculator:** Solving Quadratic Equations A23
Lab 17 Completing the Square A24

SYMBOLS

$=$	is equal to
$\neq$	is not equal to
$>$	is greater than
$<$	is less than
$\geq$	is greater than or equal to
$\leq$	is less than or equal to
$\approx$	is approximately equal to
$\cdot$	times
$-$	negative
$+$	positive
$\pm$	positive or negative
$-a$	opposite or additive inverse of a
$\lvert a \rvert$	absolute value of a
$a \stackrel{?}{=} b$	Does a equal b?
π	pi
$\{\ \}$	set
$\%$	percent
$^\circ$	degrees
$a:b$	ratio of a to b
$f(x)$	f of x, the value of f at x
(a, b)	ordered pair a, b
$\overline{AB}$	line segment AB
AB	measure of $\overline{AB}$
$\sqrt{\ }$	principal square root
$\cos A$	cosine of A
$\sin A$	sine of A
$\tan A$	tangent of A

Metric System

mm	millimeter
cm	centimeter
m	meter
km	kilometer
g	gram
kg	kilogram
mL	milliliter
L	liter
h	hour
min	minute
s	second
km/h	kilometer per hour
m/s	meters per second
°C	degrees Celsius

INSIDE YOUR BOOK

Understanding the Lesson

Each chapter is organized into lessons to make learning manageable. The basic plan of the lesson is easy to follow, beginning with a relevant application, followed by the development of the mathematical concept with plenty of examples, and ending with various types of exercises for you to complete.

Objectives clarify what concepts and skills you are expected to know after studying the lesson and completing the exercises.

Just as a journalist opens a story with a compelling "hook," nearly every lesson in this book opens with a relevant **application** that connects the mathematics to the real world.

Interesting math-related trivia and historical facts, presented in **FYI**—"for your information"—enhance the relevance of the mathematics content.

To help ensure your success in *Merrill Algebra 1*, completely worked out **examples** are provided for each type of practice exercise.

Annotations, printed in blue, provide hints as to the reasoning or property required to complete each step of the solution to a problem.

Connections, in both examples and exercises, highlight ways in which the study of algebra is related to other areas of mathematics like geometry and statistics.

6-6 Adding and Subtracting Polynomials

Objective After studying this lesson, you should be able to:
- add and subtract polynomials.

Application The standard m... is the *united inch*... measurement of a... sum of the length... window. If the le... right is $2x + 8$ in... inches, what is th... united inches?

FYI... The largest windows in the world are in the Palace of Industry and Technology in France. They measure 715.2 feet wide by 164 feet high.

$x - 3$ in.

The size of the ... $(x - 3)$ inches. To add two polynomials, add the like terms.

... term with its

$$(2x + 8) + (x - 3) = 2x + 8 + x - 3$$
$$= (2x + x) + (8 - 3) \quad \text{Commutative and associative properties}$$
$$= 3x + 5$$

The size of the window in united inches is $3x + 5$ inches.

You can add polynomials by grouping the like terms together and then finding the sum (as in the example above), or by writing them in column form.

Example 1 **Find $(3y^2 + 5y - 6) + (7y^2 - 9)$.**

Method 1 Group the like terms together.

$$(3y^2 + 5y - 6) + (7y^2 - 9) = (3y^2 + 7y^2) + 5y + [-6 + (-9)]$$
$$= (3 + 7)y^2 + 5y + (-15)$$
$$= 10y^2 + 5y - 15$$

Method 2 Add in column form.

$$\begin{array}{rl} 3y^2 + 5y - 6 & \text{Notice that like terms are aligned.} \\ (+)\ 7y^2 - 9 & \text{There is no term in the y column.} \\ \hline 10y^2 + 5y - 15 & \end{array}$$

Recall that you can subtract a rational number by adding its ... inverse or opposite. Similarly, you can subtract a polynomial b... its additive inverse.

230 CHAPTER 6 POLYNOMIALS

Example 3 CONNECTION Geometry

Find the measure of the third side of the triangle at the right. P is the measure of the perimeter.

The perimeter is the sum of the measures of the three sides of the triangle. Let s represent the measure of the third side.

$$(12x^2 - 7x + 9) = (3x^2 + 2x - 1) + (8x^2 - 8x + 5) + s \quad \text{Substitution}$$
$$12x^2 - 7x + 9 - (3x^2 + 2x - 1) - (8x^2 - 8x + 5) = s \quad \text{Solve for } s.$$
$$12x^2 - 7x + 9 - 3x^2 - 2x + 1 - 8x^2 + 8x - 5 = s$$
$$(12x^2 - 3x^2 - 8x^2) + (-7x - 2x + 8x) + (9 + 1 - 5) = s \quad \text{Group the like terms.}$$
$$x^2 - x + 5 = s$$

The measure of the third side is $x^2 - x + 5$.

LESSON 6-6 ADDING AND SUBTRACTING POLYNOMIALS 231

Communicating Mathematics exercises provide you with an opportunity to check your understanding of the material you have just read using verbal, written, pictorial, graphical, and algebraic methods.

CHECKING FOR UNDERSTANDING

Communicating Mathematics — Read and study the lesson to answer each question.

1. What is the first step when adding or subtracting in column form?
2. What is the best way to check the subtraction of two polynomials?

Guided Practice — Find the additive inverse of each polynomial.

3. $3x + 2y$
4. $-8m + 7n$
5. $x^2 + 3x + 7$
6. $-4h^2 - 5hk - k^2$
7. $-3ab^2 + 5a^2b - b^3$
8. $x^3 + 5x^2 - 3x - 11$

Name the like terms in each group.

9. $5m, 4mn, -3m, 2n, -mn, 8n$
10. $2x^3, 5xy, -x^2y, 14xy, 12xy$
11. $-7ab^2, 8a^2b, 11b^2, 16a^2b, -2b^2$
12. $3p^3q, -2p, 10p^3q, 15pq, -p$

EXERCISES

Practice — Find each sum or difference.

13. $5ax^2 + 3a^2x - 7a^3$ / $(+)\ 2ax^2 - 8a^2x + 4$
14. $a^3 - b^3$ / $(+)\ 3a^3 + 2a^2b - b^2 + 2b^3$
15. $4a + 5b - 6c + d$ / $3a - 7b + 2c + 8d$ / $(+)\ 2a - b + 7d$
16. $2x^2 - 5x + 7$ / $5x^2 + 7x - 3$ / $(+)\ x^2 - x + 11$
17. $6x^2y^2 - 3xy - 7$ / $(-)\ 5x^2y^2 + 2xy + 3$
18. $5x^2 - 4$ / $(-)\ 3x^2 + 8x + 4$
19. $11m^2n^2 + 2mn - 11$ / $(-)\ 5m^2n^2 - 6mn + 17$
20. $2a - 7$ / $(-)\ 5a^2 + 8a - 11$

n)

$8n + 8)$

of the

Sharpening your reasoning skills is a major goal of challenging critical thinking exercises. Exercises of this type frequently appear on scholastic aptitude tests.

Critical Thinking — The sum of the degree measures of the three angles of a triangle is 180. Find the degree measure of the third angle of each triangle given the degree measures of the other two angles.

34. $5 - 2x;\ 7 + 8x$
35. $3x^2 - 5;\ 4x^2 + 2x + 1$
36. $4 - 2x;\ x^2 - 1$
37. $x^2 - 8x + 2;\ x^2 - 3x - 1$

Applications

38. **Travel** Joan Be... average rate of 4... miles per hour. ... average speed is...
39. **Basketball** On... Detroit Pistons... game. The two t... scored 2 less po... score?

Word problems are easier to solve because applications relate to your own experiences or to things you have read or heard about.

Mixed Review exercises help you retain the concepts and skills you have learned. The lesson reference given with each exercise makes it easy for you to re-study the concept.

Mixed Review

40. Evaluate $|x|$ if $x = -2.1$. **(Lesson 2-2)**
41. **Sports** There were 3 times as many sports in the 1976 Summer Olympics as there were in the 1976 Winter Olympics. There were 14 more sports in the Summer Olympics than there were in the Winter Olympics. How many sports were there in the 1976 Summer Olympics? **(Lesson 3-6)**
42. **Bicycling** Adita rode his bicycle 72 kilometers. How long did it take him if his rate was: **a.** 9 km/h? **b.** 18 km/h? **(Lesson 4-7)**
43. Solve $3m < 2m - 7$. **(Lesson 5-1)**
44. Find the degree of $x + 2y^2 + 3z^3$. **(Lesson 6-5)**

HISTORY CONNECTION

Emmy Noether

Emmy Noether (1882–1935) was a German mathematician whose strength was in abstract algebra. In her hands the axiomatic method (using axioms or properties) became a powerful tool of mathematical research. Noether did much of her research while living in Göttingen, Germany, which was then the principal center for mathematics research in Europe. Because she was female, Noether was unable to secure a teaching position at the University of Göttingen. However, her influence there was vast. During World War II, Noether was forced to leave Germany. She became a professor at Bryn Mawr College, in the United States, where she remained until her death. Albert Einstein paid her a great tribute in 1935: "In the judgment of the most competent living mathematicians, (Emmy) Noether was the most significant creative mathematical genius thus far produced since the higher education of women began."

History Connections highlight the mathematics contributions of individuals from many cultures and times.

Getting into the Chapter

Every chapter begins with a two-page application including a large full-color photograph to help you connect the mathematics you will learn in the chapter with your real world experiences.

CHAPTER OBJECTIVES

In this chapter, you will:
- Arrange the terms of a polynomial in order.
- Add, subtract, and multiply polynomials.
- Solve polynomial equations.
- Solve problems by looking for a pattern.

The list of **chapter objectives** lets you know what you can expect to learn in the chapter.

CHAPTER 6

Polynomials

The **Application** describes a real-life situation where the algebra in the chapter can be used. It will help you answer the question "When am I ever going to use this?"

Population control for deer? This might seem like an unwarranted interference with nature, but this problem is much more complex.

APPLICATION IN BIOLOGY

These days, many cities support herds of deer in their metropolitan parks, supplying them with food, medical care, shelter in winter, and protection from predators.

But uncontrolled, an animal population develops *polynomially.* That is, if one animal has x offspring, then it will have an average of x^2 grandoffspring, x^3 great-grandoffspring, x^4 great-great-grandoffspring, and so on, assuming that each animal has an average of x babies in its lifetime. We can represent the number of descendants of one animal with the polynomial expression $x + x^2 + x^3 + x^4 + \ldots$, continuing for any number of generations, assuming that none of the animals dies early. So, if an animal has an average of 13 offspring in its lifetime, then it will probably have 30,940 descendants in four generations.

So this is the problem caused by cities' support of deer in parks. Any natural ecosystem has a delicate balance. In the wild, deer populations are controlled naturally—by predators, disease, starvation, and other forms of natural competition. Since we have eliminated nearly all of the population controls for the deer living in city parks, we now have a responsibility to provide them with new ones or face the consequences of an uncontrolled population.

In **Algebra in Action**, you can begin to actually apply the algebra that is presented in the chapter.

ALGEBRA IN ACTION

A deer has an average of 4 offspring during its lifetime. Complete the chart below to determine the probable number of deer descendants in five generations.

Generation	Polynomial	Number of Deer Descendants
1	x	4
2	$x + x^2$	4 + 16 = 20
3	$x + x^2 + x^3$	4 + 16 + 64 = ?
4	$x + x^2 + x^3 + x^4$	4 + 16 + 64 + ? = ?
5	$x + x^2 + x^3 + x^4 +$?	4 + 16 + 64 + ? + ? = ?

209

Wrapping Up the Chapter

Review pages at the end of each chapter allow you to complete your mastery of the material. The vocabulary, objectives, examples, and exercises help you make sure you understand the skills and concepts presented in the chapter.

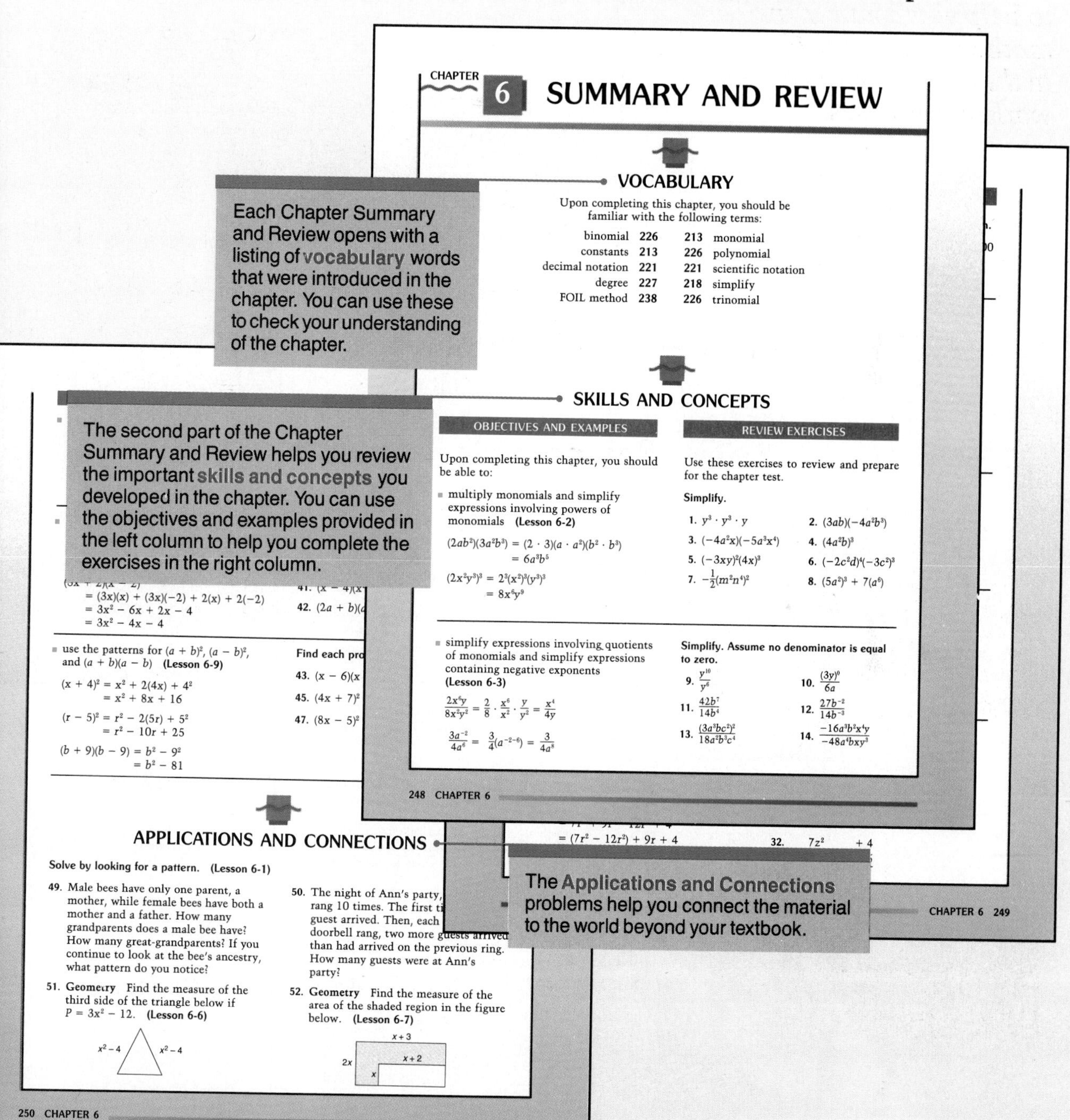

CHAPTER 6 SUMMARY AND REVIEW

VOCABULARY

Upon completing this chapter, you should be familiar with the following terms:

binomial	226	213	monomial
constants	213	226	polynomial
decimal notation	221	221	scientific notation
degree	227	218	simplify
FOIL method	238	226	trinomial

SKILLS AND CONCEPTS

OBJECTIVES AND EXAMPLES

Upon completing this chapter, you should be able to:

- multiply monomials and simplify expressions involving powers of monomials **(Lesson 6-2)**

$(2ab^2)(3a^2b^3) = (2 \cdot 3)(a \cdot a^2)(b^2 \cdot b^3) = 6a^3b^5$

$(2x^2y^3)^3 = 2^3(x^2)^3(y^3)^3 = 8x^6y^9$

- simplify expressions involving quotients of monomials and simplify expressions containing negative exponents **(Lesson 6-3)**

$\frac{2x^6y}{8x^2y^2} = \frac{2}{8} \cdot \frac{x^6}{x^2} \cdot \frac{y}{y^2} = \frac{x^4}{4y}$

$\frac{3a^{-2}}{4a^6} = \frac{3}{4}(a^{-2-6}) = \frac{3}{4a^8}$

REVIEW EXERCISES

Use these exercises to review and prepare for the chapter test.

Simplify.

1. $y^3 \cdot y^3 \cdot y$
2. $(3ab)(-4a^2b^3)$
3. $(-4a^2x)(-5a^3x^4)$
4. $(4a^2b)^3$
5. $(-3xy)^2(4x)^3$
6. $(-2c^2d)^4(-3c^2)^3$
7. $-\frac{1}{2}(m^2n^4)^2$
8. $(5a^2)^3 + 7(a^6)$

Simplify. Assume no denominator is equal to zero.

9. $\frac{y^{10}}{y^6}$
10. $\frac{(3y)^0}{6a}$
11. $\frac{42b^7}{14b^4}$
12. $\frac{27b^{-2}}{14b^{-3}}$
13. $\frac{(3a^3bc^2)^2}{18a^2b^3c^4}$
14. $\frac{-16a^3b^2x^4y}{-48a^4bxy^3}$

248 CHAPTER 6

$= (3x)(x) + (3x)(-2) + 2(x) + 2(-2)$
$= 3x^2 - 6x + 2x - 4$
$= 3x^2 - 4x - 4$

42. $(2a + b)(a$

- use the patterns for $(a + b)^2$, $(a - b)^2$, and $(a + b)(a - b)$ **(Lesson 6-9)**

$(x + 4)^2 = x^2 + 2(4x) + 4^2 = x^2 + 8x + 16$

$(r - 5)^2 = r^2 - 2(5r) + 5^2 = r^2 - 10r + 25$

$(b + 9)(b - 9) = b^2 - 9^2 = b^2 - 81$

Find each pro

43. $(x - 6)(x$
45. $(4x + 7)^2$
47. $(8x - 5)^2$

$= (7r^2 - 12r^2) + 9r + 4$ 32. $7z^2$ $+ 4$

CHAPTER 6 249

APPLICATIONS AND CONNECTIONS

Solve by looking for a pattern. (Lesson 6-1)

49. Male bees have only one parent, a mother, while female bees have both a mother and a father. How many grandparents does a male bee have? How many great-grandparents? If you continue to look at the bee's ancestry, what pattern do you notice?

50. The night of Ann's party, rang 10 times. The first t guest arrived. Then, each doorbell rang, two more guests arrived than had arrived on the previous ring. How many guests were at Ann's party?

51. **Geometry** Find the measure of the third side of the triangle below if $P = 3x^2 - 12$. **(Lesson 6-6)**

$x^2 - 4$ $x^2 - 4$

52. **Geometry** Find the measure of the area of the shaded region in the figure below. **(Lesson 6-7)**

$x + 3$ $2x$ $x + 2$ x

250 CHAPTER 6

CHAPTER OBJECTIVES

In this chapter, you will:

- Evaluate expressions and formulas.
- Use mathematical properties to simplify expressions.
- Solve open sentences.
- Translate verbal expressions into mathematical expressions and equations.

An Introduction to Algebra

APPLICATION IN SPORTS

Most of our sports developed from skills that we had to use to survive: running, dodging, kicking, throwing, climbing, and so on. Because we were intelligent and aggressive, staying alive became easier. With more free time, we began to find ways to enjoy the use of our survival skills. By combining our natural love of action with vivid imagination, we not only created a wide variety of sports, but we also invented a means to measure our sports achievement: numbers.

"Who won?" "What was the score?" "How did my favorite player do?" "Has my team moved up in the standings?"

Without numbers, we could not evaluate the performances of individuals and teams. Every baseball game is full of numbers used to compare the players and their teams. Which team scored the most runs? Who pitched the fastest? How far was the ball hit? Who has the most home runs? stolen bases? runs batted in? What is a player's batting average? earned run average?

As the volume of numbers used to describe sports increased, we developed tools to measure, calculate, and store them. Radar guns measure the speed of a fastball. Cameras film the swing of the bat and the flight of the ball. Computers compile, calculate, and store numbers for later evaluation.

ALGEBRA IN ACTION

A *batting average* is calculated by dividing the number of hits by the number of at bats. The result is rounded so that the average has three digits after the decimal point. The chart below shows the players with the most hits in the American League in 1992. Copy and complete the chart. Who had the best average?

Player	Hits	At Bats	Batting Average
Kirby Puckett	210	639	$210 \div 639 = 0.329$
Carlos Baerga	205	657	$205 \div 657 = ?$
Paul Molitor	195	609	?
Shane Mack	189	600	?
Frank Thomas	185	573	?

1-1 Variables and Expressions

Objectives

After studying this lesson, you should be able to:

- translate verbal expressions into mathematical expressions, and
- write an expression containing identical factors as an expression using exponents.

Application

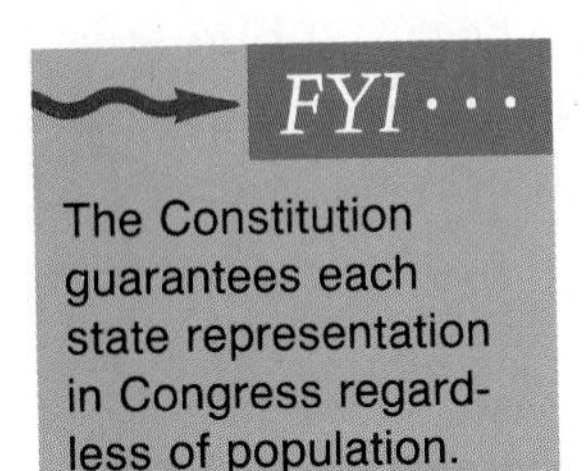

The United States House of Representatives has 435 members. The number of members from each state is determined by the state's population. Each state is divided into congressional districts, which have an average of about 575,000 people. There is one representative from each district.

Alaska, Delaware, North Dakota, South Dakota, Vermont, and Wyoming each have one representative in the House. You can find the number of representatives from other states if you know the populations of those states.

Since you cannot have a fraction of a representative, round to the nearest whole number.

If a state has a population of 2,910,000, it will have 2,910,000 ÷ 575,000 or 5 representatives.

If a state has a population of 6,280,000, it will have 6,280,000 ÷ 575,000 or 11 representatives.

If a state has a population of p, it will have $p \div 575{,}000$ representatives.

$6 + 3$ and $4 \times 5 - 8$ are numerical expressions.

The letter p is called a **variable,** and $p \div 575{,}000$ is an **algebraic expression.** In algebra, variables are symbols that are used to represent unspecified numbers. Any letter may be used as a variable. We chose the letter p since it is the first letter in the word "population."

An algebraic expression consists of one or more numbers and variables along with one or more arithmetic operations. Here are some other examples of algebraic expressions.

$$b + 4 \qquad \frac{s}{t} - 1 \qquad a \times 4n \qquad 8rs \div 3k$$

In a multiplication expression, the quantities being multiplied are called **factors** and the result is called the **product.**

$$4 \times 5 \times 8 = 160$$

factors (4, 5, 8) *product* (160)

In algebraic expressions, a raised dot or parentheses are often used to indicate multiplication. When variables are used to represent factors, the multiplication sign is usually omitted. Here are five ways to represent the product of a and b.

To indicate the product of 3 and a, write $3a$.

$ab \qquad a \cdot b \qquad a(b) \qquad a \times b \qquad (a)(b)$

Fraction bars are used to indicate division.

$\frac{x}{3}$ means $x \div 3$

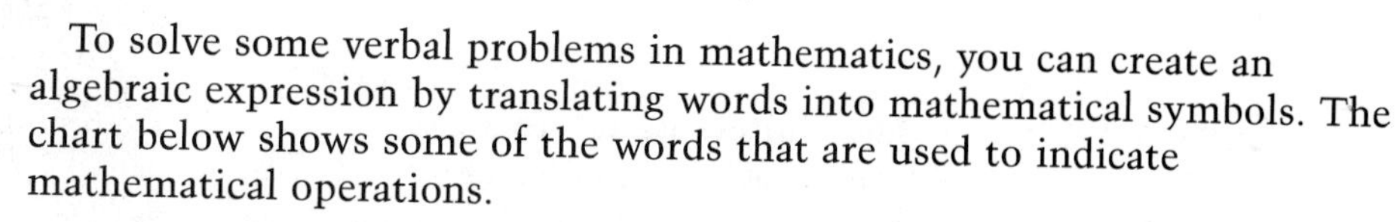

To solve some verbal problems in mathematics, you can create an algebraic expression by translating words into mathematical symbols. The chart below shows some of the words that are used to indicate mathematical operations.

Addition	Subtraction	Multiplication	Division
■ the sum of ■ increased by ■ plus ■ more than ■ added to ■ the total of	■ the difference of ■ decreased by ■ minus ■ less than ■ subtracted from	■ the product of ■ multiplied by ■ times	■ the quotient of ■ divided by ■ the ratio of

Example 1

Write an algebraic expression for *the sum of a and b.*

An algebraic expression is $a + b$.

Example 2

Write an algebraic expression for *a number k divided by a number n.*

An algebraic expression is $\frac{k}{n}$.

Example 3

Write a verbal expression for $4x$.

A verbal expression is the product of 4 and x, or 4 times x.

Example 4

Write a verbal expression for $z - 6$.

One verbal expression is six less than z. Another is the difference of z and 6. *Can you think of another?*

In an expression like 10^3, 10 is the **base** and 3 is the **exponent.** The exponent indicates the number of times the base is used as a factor.

10^3 means $10 \cdot 10 \cdot 10$ $\qquad$ a^2 means $a \cdot a$

An expression in the form x^n is called a power.

Symbols	Words	Meaning
8^1	8 to the first power	8
8^2	8 to the second power or 8 squared	$8 \cdot 8$
8^3	8 to the third power or 8 cubed	$8 \cdot 8 \cdot 8$
8^4	8 to the fourth power	$8 \cdot 8 \cdot 8 \cdot 8$
$6n^5$	6 times *n* to the fifth power	$6 \cdot n \cdot n \cdot n \cdot n \cdot n$

Example 5

Find the area of the square at the right.

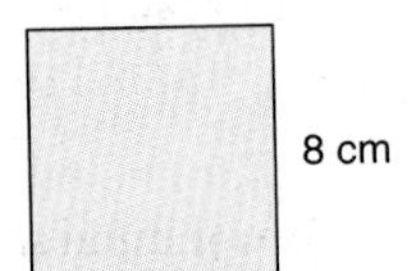

The area of a square is the length of the side squared.
The area of the square is $(8)^2$ or 64 cm^2.
Area is expressed in units squared.

Example 6

Find the volume of the cube at the right.

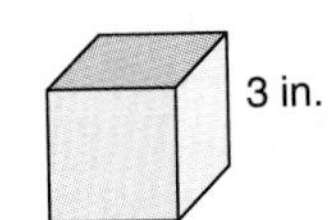

The volume of a cube is the length of the side cubed.
The volume of the cube is $(3)^3$ or 27 in^3.
Volume is expressed in units cubed.

Example 7

Evaluate 2^5.

$$2^5 = 2 \cdot 2 \cdot 2 \cdot 2 \cdot 2$$
$$= 32$$

You can use the y^x key on a calculator to raise a number to a power.

Enter:	2	y^x	5	=
Display:	2		5	32

Example 8

Write an algebraic expression for the expression *the cube of k increased by seven*.

An algebraic expression is $k^3 + 7$.

CHECKING FOR UNDERSTANDING

Communicating Mathematics

Read and study the lesson to answer each question.

1. If a state has a population of 3,431,000, how would you determine the number of representatives it has?
2. What is the difference between numerical expressions and algebraic expressions?
3. How would you represent the sum of 3 and a? the product of 3 and a?
4. Can you find the volume of a cube if you only know the length of a side? How?

Guided Practice

Write an algebraic expression for each verbal expression.

5. the product of x and 7
6. the quotient of r and s
7. the sum of a and 19
8. a number n decreased by 4
9. a number b to the third power
10. 25 squared

Write as expressions that use exponents.

11. $5 \cdot 5 \cdot 5$
12. $7 \cdot a \cdot a \cdot a \cdot a$
13. $2(m)(m)(m)$
14. $5 \cdot 5 \cdot 5 \cdot x \cdot x \cdot y$

Evaluate each expression.

15. 2^4
16. 5^3
17. 10^4
18. 3^6

EXERCISES

Practice

Write a verbal expression for each algebraic expression.

19. $m - 1$
20. xy
21. n^4
22. 5^3
23. $8y^2$
24. $z^7 + 2$

Write an algebraic expression for each verbal expression.

25. a number increased by 17
26. seven times a number
27. twice the cube of a number
28. a number to the sixth power

Write an algebraic expression for each verbal expression.

29. one-half the square of a number
30. six times a number decreased by 17
31. 94 increased by twice a number
32. three fourths of the square of a number

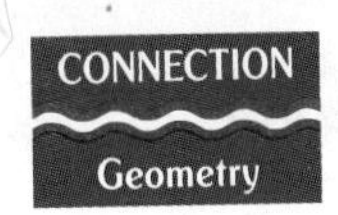

33. Find the area of the figure at the right.

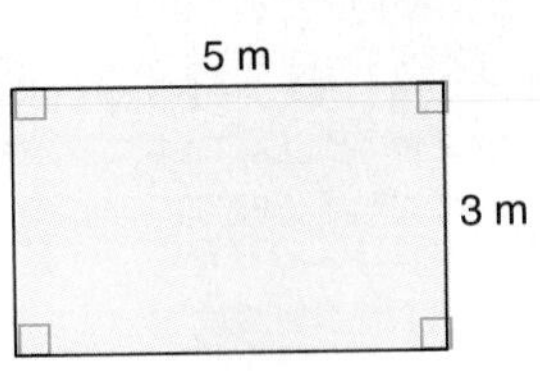

34. Find the volume of the figure at the right.

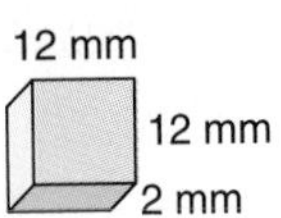

Use your calculator to evaluate each expression to the nearest thousandth.

35. $(6.2)^6$ 36. $(4.8)^5$ 37. $5^5 + 6^6 + 7^7$ 38. $2^2 \cdot 3^3 \cdot 4^4 \cdot 5^5$

Write an algebraic expression for each verbal expression.

39. the sum of a and b, decreased by the product of a and b
40. the difference of a and b, increased by the quotient of a and b

Critical Thinking

41. Evaluate 4^2 and 2^4. What do you notice? From this example, can you say that the same relationship exists between 2^5 and 5^2? What about between 3^4 and 4^3?

Applications

42. **Electronics** For a round conductor, the diameter in mils squared (d^2) equals the cross-sectional area in circular mils (C.M.). Find the cross-sectional area in C.M. for the wire shown at the right.

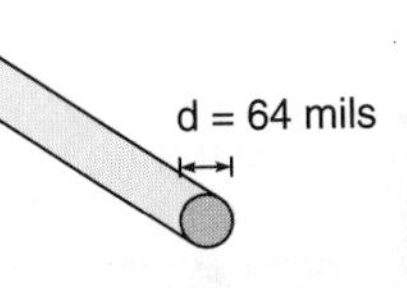

43. **Government** The White House has 132 rooms, including a barber shop and a bowling alley. The main building is shaped like a rectangle that is 168 feet long and 85 feet wide. What is the area of the main building of the White House?

Computer

44. BASIC is a computer language. The symbols used in BASIC are similar to those used in algebra.

+ means add.
– means subtract.
* means multiply.
/ means divide.
↑ means exponent.

Numeric variables in BASIC are represented by capital letters. Input the following sets of values for a and b into the program below to find $a + b$, $a - b$, $a \times b$, $a \div b$, and a^b: 2, 4; 10, 3; −9, 2; 5.2, 2.

```
10 INPUT A,B
20 PRINT "A + B = ";A + B
30 PRINT "A - B = ";A - B
40 PRINT "A * B = ";A * B
50 PRINT "A / B = ";A / B
60 PRINT "A ↑ B = ";A ↑ B
```

1-2 Evaluating Expressions

Objective

After studying this lesson, you should be able to:

- use the order of operations to evaluate expressions.

Application

Janet Graves scored 7 points in each of the first four volleyball games. In the fifth game, she scored 3 points. The following expression represents the number of points she scored in the five games.

$$7 \cdot 4 + 3$$

To find the number of points she scored, evaluate the expression $7 \cdot 4 + 3$. Which of the following methods is correct?

$7 \cdot 4 + 3 = 28 + 3$	*Multiply first.*	$7 \cdot 4 + 3 = 7 \cdot 7$	*Add first.*
$= 31$	*Then add.*	$= 49$	*Then multiply.*

The answers are not equal because a different order of operations was used in each method. Since a numerical expression must have only one value, we use the following order of operations in algebra.

Order of Operations

1. **Simplify the expressions inside grouping symbols, such as parentheses and brackets, and as indicated by fraction bars.**
2. **Evaluate all powers.**
3. **Do all multiplications and divisions from left to right.**
4. **Do all additions and subtractions from left to right.**

Therefore, Janet scored $7 \cdot 4 + 3$ or 31 points in the five games.

Let's evaluate another numerical expression.

Example 1

Evaluate $12 \div 3 \cdot 5 - 4^2$.

$12 \div 3 \cdot 5 - 4^2 = 12 \div 3 \cdot 5 - 16$ — *Evaluate 4^2.*

$= 4 \cdot 5 - 16$ — *Divide 12 by 3.*

$= 20 - 16$ — *Multiply 5 by 4.*

$= 4$ — *Subtract 16 from 20.*

Algebraic expressions can be evaluated when the value of the variables is known. First, the variables are replaced by their values. Then, the value of the numerical expression is calculated.

Example 2 Find the perimeter of the figure when $x = 9$ and $y = 4$.

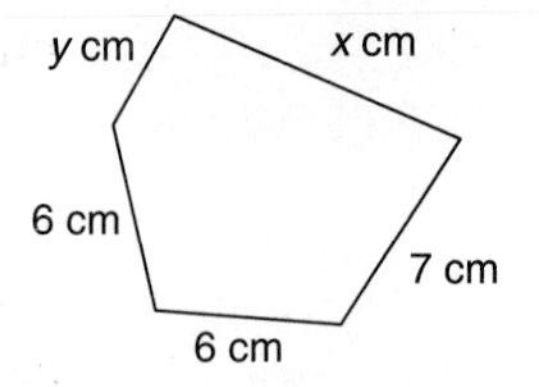

CONNECTION Geometry

The perimeter (P) is the sum of the lengths of the sides.

$P = x + y + 6 + 6 + 7$

$P = x + y + 19$

Now replace each variable with its given value.

$P = 9 + 4 + 19$

$P = 32$

The perimeter is 32 cm.

In mathematics, grouping symbols such as parentheses ( ) and brackets [ ] are used to clarify or change the order of operations. They indicate that the expression within is to be evaluated first.

$$50 - 5 \cdot 6 + 3 = 50 - 30 + 3 \text{ or } 23$$

$$50 - 5(6 + 3) = 50 - 5 \cdot 9 \text{ or } 5$$

Example 3 Evaluate $4(1 + 5)^2 \div 8$.

$4(1 + 5)^2 \div 8 = 4(6)^2 \div 8$	*Add 1 and 5.*
$= 4(36) \div 8$	*Evaluate 6^2.*
$= 144 \div 8$	*Multiply 36 by 4.*
$= 18$	*Divide 144 by 8.*

When more than one grouping symbol is used, start evaluating within the innermost grouping symbols.

$$50 - [5(6 + 3)] = 50 - (5 \cdot 9) = 50 - 45 \text{ or } 5$$

Example 4 Evaluate $\frac{2}{3}[8(a - b)^2 + 3b]$ if $a = 5$ and $b = 2$.

$\frac{2}{3}[8(a - b)^2 + 3b] = \frac{2}{3}[8(5 - 2)^2 + 3 \cdot 2]$	*Substitute 5 for a and 2 for each b.*
$= \frac{2}{3}[8(3)^2 + 3 \cdot 2]$	*Subtract 2 from 5.*
$= \frac{2}{3}[8 \cdot 9 + 3 \cdot 2]$	*Evaluate 3^2.*
$= \frac{2}{3}[72 + 6]$	*Multiply 9 by 8 and 2 by 3.*
$= \frac{2}{3}[78]$	*Add 72 and 6.*
$= 52$	*Multiply 78 by $\frac{2}{3}$ and simplify.*

The fraction bar is another grouping symbol. It indicates that the numerator and denominator should each be treated as a single value.

$$\frac{3 \times 8}{2 \times 4} \quad \textbf{means} \quad (3 \times 8) \div (2 \times 4) \text{ or } 3$$

When evaluating an expression with a scientific calculator, it may be necessary to use grouping symbols.

Example 5 **Use a calculator to evaluate $\frac{b^7 - 5}{b^2 + c^4}$ if $b = 6$ and $c = 8$.**

You can use the x^2 key on a calculator to square a number like 6.

Enter:   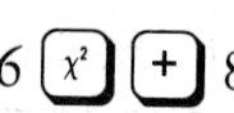 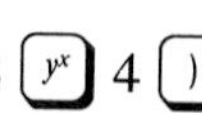 

Display: 67.747096

Why is it necessary to have parentheses around the numerator and denominator?

CHECKING FOR UNDERSTANDING

Communicating Mathematics

Read and study the lesson to answer each question.

1. When evaluating the expression $2 + 5 \cdot 3$, what do you do first?
2. Name two types of grouping symbols.

Explain how to evaluate each expression. Do *not* evaluate.

3. $(12 - 6) \cdot 2$
4. $9 - 3^2$
5. $4(5 - 3)^2$
6. $8 + 6 \div (2 + 1)$

Guided Practice

Evaluate each expression.

7. $3 + 8 \div 2 - 5$
8. $5(9 + 3) - 3 \cdot 4$
9. $5^3 + 6^3 - 5^2$
10. $\frac{38 - 12}{2 \cdot 13}$

Evaluate each expression if $a = 6$, $b = 4$, and $c = 3$.

11. $a + b^2 + c^2$
12. $3ab - c^2$
13. $8(a - c)^2 + 3$
14. $\frac{2ab - c^3}{7}$

EXERCISES

Practice

Evaluate each expression.

15. $4 + 7 \cdot 2 + 8$
16. $12 \div 4 + 15 \cdot 3$
17. $29 - 3(9 - 4)$
18. $4(11 + 7) - 9 \cdot 8$
19. $16 \div 2 \cdot 5 \cdot 3 \div 6$
20. $288 \div [3(9 + 3)]$

Evaluate each expression.

21. $6(4^3 + 2^2)$

22. $\dfrac{9 \cdot 4 + 2 \cdot 6}{7 \cdot 7}$

23. $\dfrac{2 \cdot 8^2 - 2^2 \cdot 8}{2 \cdot 8}$

24. $\frac{3}{4}(6) + \frac{1}{3}(12)$

25. $25 - \frac{1}{3}(18 + 9)$

26. $7(0.2 + 0.5) - 0.6$

A formula for the perimeter or circumference of each figure is given. Find the perimeter or circumference when $x = 3$, $y = 4$, and $z = 5.5$. Use 3.14 for π.

27. triangle:
$P = x + y + z$

x mm, y mm, z mm

28. square:
$P = 4z$

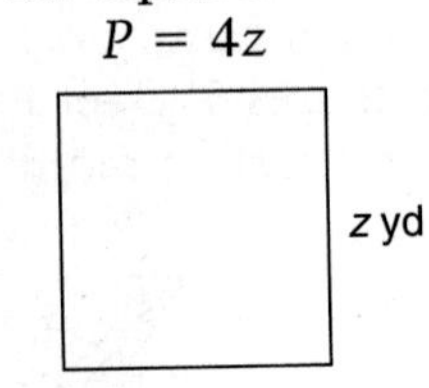

29. parallelogram:
$P = 2(x + y)$

y in., x in., x in., y in.

See Exercise 32. On a regulation track, x is 75 and y is 101.

30. isosceles trapezoid:
$P = x + 2y + z$

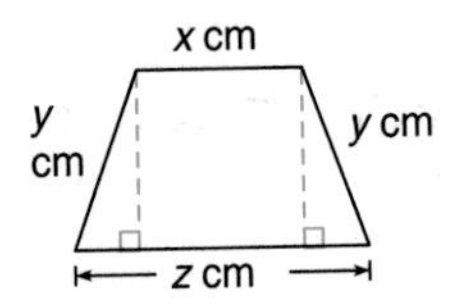

31. circle:
$C = \pi y$

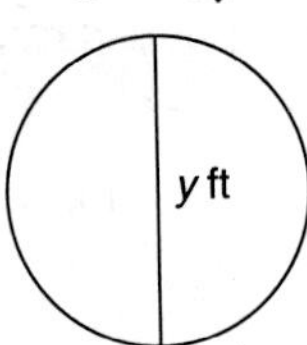

32. oval:
$P = \pi x + 2y$

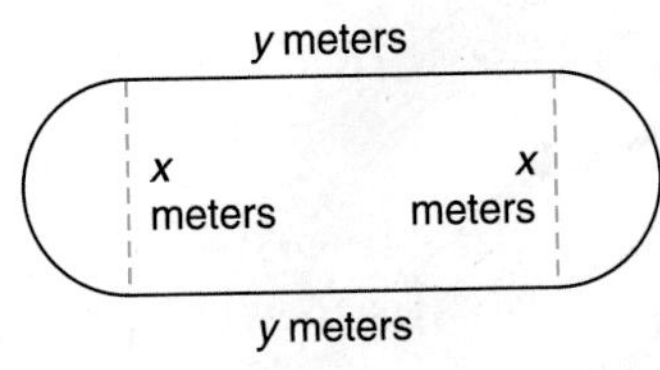

Evaluate each expression if $a = 6$, $b = 4$, $c = 3$, $d = \frac{1}{2}$, $n = \frac{2}{3}$, $x = 0.2$, and $y = 1.3$.

33. $12d + bc$

34. $a(8 - 3n) + 4d$

35. $ax + bc$

36. $x^2 + 6d$

37. $(100x)^2 + 10y$

38. $\dfrac{8b^2c}{x}$

39. $\dfrac{a + b^2}{3bc}$

40. $\dfrac{a^2 - b^2}{2 + d^3}$

Write an algebraic expression for each verbal expression. Then, evaluate the expression if $a = 3$, $b = \frac{1}{2}$, and $c = 0$.

41. twice the sum of a and b

42. twice the product of a and b

43. the square of b increased by c

44. the cube of a decreased by b

Critical Thinking

45. If you change one sign in the expression below, the value of the expression will double. Find the sign. Why does this work?

$$61 - 13 - 12 - 8 - 7 - 6 - 5$$

Applications

46. Merchandising If 12 items are in a dozen, 12 dozen are in a gross, and 12 gross are in a great-gross, how many items are in a great-gross?

47. Health Your optimum exercise heart rate per minute is given by the expression $0.7(220 - a)$, where a represents your age. Find your optimum exercise heart rate.

48. Carpentry Ana Martinez is putting molding around the ceiling of a paneled family room. If the room measures 12 feet by 16 feet, how many feet of molding are needed?

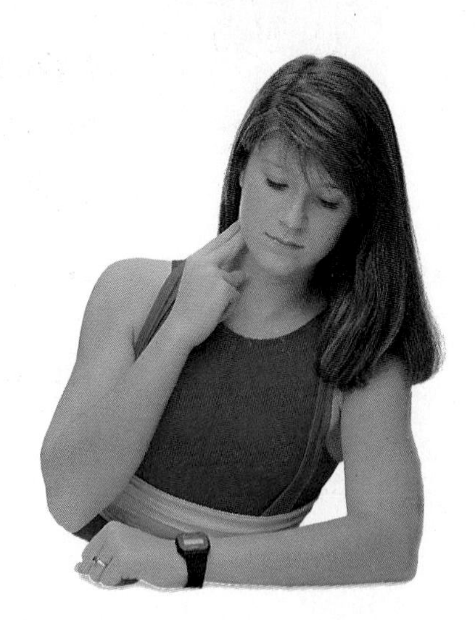

Mixed Review

49. Write an algebraic expression for *7 less than w*. **(Lesson 1-1)**

50. Write a verbal expression for y^5. **(Lesson 1-1)**

51. Write $4 \cdot 4 \cdot 4 \cdot 4$ as an expression using exponents. **(Lesson 1-1)**

52. Write an algebraic expression for *twice a number decreased by 25.* **(Lesson 1-1)**

53. Find the area of the rectangle. **(Lesson 1-1)**

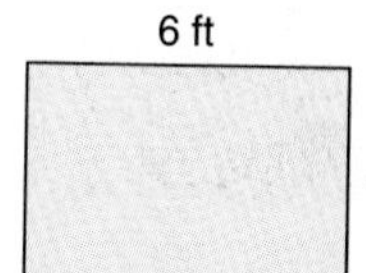

READING ALGEBRA

Suppose you are asked to read each expression below.

$$3x + y \qquad\qquad 3(x + y)$$

In the expression $3(x + y)$, parentheses are used to show that the sum $x + y$ is multiplied by three. In algebraic expressions, terms enclosed by parentheses are treated as one quantity. The expression $3(x + y)$ is read *three times the quantity x plus y*. The expression $3x + y$ is read *three x plus y*.

Try this experiment. Have a classmate close his or her book. Read the expressions below to your classmate. Then have him or her write each one as an algebraic expression.

a. seven times the cube of the quantity b minus 4 — $7(b - 4)^3$
b. twice the quantity a plus 3 times the quantity b minus 9 — $2(a + 3)(b - 9)$
c. four divided by the difference of a number and 6 — $4 \div (n - 6)$

Compare your classmate's answers with the correct answers shown above. How do they compare?

1-3 Open Sentences

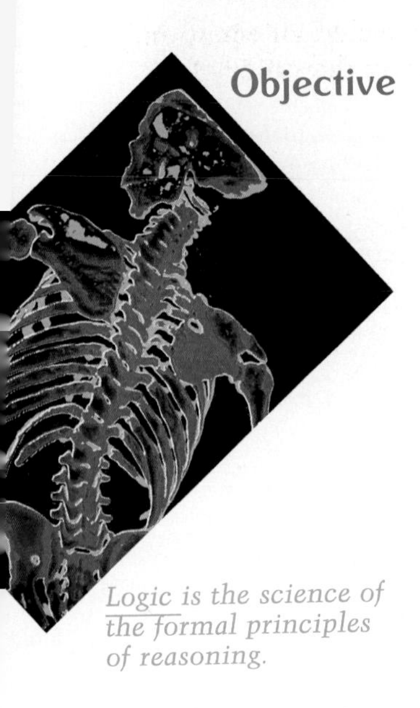

Objective

After studying this lesson, you should be able to:

- solve open sentences by performing arithmetic operations.

Which of the following sentences are true? Which are false?

There are over 200 bones in the human body. *true*
The name Mary is a verb. *false*
California is larger than New Jersey. *true*
Twenty-seven divided by 3 is less than 7. *false*

Logic is the science of the formal principles of reasoning.

Note that each of the previous sentences is either true or false. A **statement** is any sentence that is either true or false, but not both. Statements are the building blocks in a system of *logic.* One way of referring to a specific statement is to represent it with a letter such as p or q. Let p be the statement *France is a country in Europe.* This statement is true. To find the *negation* of statement p, write $\sim p$ (read "not p"). This represents the statement *France is not a country in Europe.* This statement is false.

Example 1

Let p represent the statement $5 + 2 = 8$. Let q represent the statement *The President of the United States must be born in the U.S.* State whether each statement is *true* or *false.* Then state whether its negation is *true* or *false.*

a. p $\quad 5 + 2 = 8 \quad$ *This statement is false.*

$\sim p$ $\quad 5 + 2 \neq 8 \quad$ *This statement is true.*

b. q $\quad$ The President of the United States must be born in the U.S. $\quad$ *This statement is true.*

$\sim q$ $\quad$ The President of the United States can be born outside of the U.S. $\quad$ *This statement is false.*

Are the statements below true or false?

Words	Symbols
A number x plus six is equal to eight.	$x + 6 = 8$
A number n is less than ten.	$n < 10$
Seven is greater than twice a number c.	$7 > 2c$

Open sentences are neither true nor false.

Mathematical sentences like these are called **open sentences.** Before you can determine whether these sentences are true or false, you must know what numbers will replace the variables x, n, and c. Finding the replacements for the variable that make a sentence true is called **solving the open sentence.** Each replacement is called a **solution** of the open sentence.

An open sentence that contains an equals sign, =, is called an **equation.** Sometimes you can solve an equation by simply applying the order of operations.

Example 2

Solve $\frac{5 \cdot 3 + 3}{4 \cdot 2 - 2} = t$.

$$\frac{5 \cdot 3 + 3}{4 \cdot 2 - 2} = t$$

$$\frac{15 + 3}{8 - 2} = t$$ *Evaluate the numerator and denominator.*

$$\frac{18}{6} = t$$ *Divide.*

$$3 = t$$

The solution is 3.

A **set** is a collection of objects or numbers. Sets are often shown by using braces. Each object or number in a set is called an **element.**

$1 \in \{1, 2, 3\}$ is read *One is an element of the set containing 1, 2, 3.*

Sets are usually named by capital letters.

$$A = \{1, 2, 3\} \qquad B = \{1, 2\}$$

Since every element of set B is an element of set A, then set B is a **subset** of set A. Other subsets of set A are $\{1\}$, $\{2\}$, $\{3\}$, $\{1, 3\}$, $\{2, 3\}$, and $\{1, 2, 3\}$. A set with no elements is called the **null** or **empty set,** shown by $\{\ \}$ or $\emptyset$. The empty set is a subset of every set.

The set of numbers from which replacements for a variable may be chosen is called a **replacement set.** The set of all replacements for the variable that make an open sentence true is called the solution set for the sentence.

Example 3

Find the solution set for the open sentence $n - 1 < 2$ if the replacement set is $\{1, 2, 3\}$.

If $n = 1$, then the sentence becomes $1 - 1 < 2$. This sentence is true.
If $n = 2$, then the sentence becomes $2 - 1 < 2$. This sentence is true.
If $n = 3$, then the sentence becomes $3 - 1 < 2$. This sentence is false.

Therefore, $\{1, 2\}$ is the solution set.

CHECKING FOR UNDERSTANDING

Communicating Mathematics

Complete.

1. A __?__ is a sentence that is either true or false.
2. $\sim p$ refers to the __?__ of statement p.
3. $17 = n - 4$ is called an __?__.
4. $\{2, 3, 4, 5, 6\}$ is a __?__ for the sentence $p + 2 > 4$.

Write in symbols.

5. the set containing 3, 4, 5, 6
6. Three is an element of the set containing 3, 4, 5, 6.

Guided Practice

State whether each sentence is true or false.

7. $3(11 - 5) > 18$
8. $0.01 + 0.01 = 0.0002$
9. $\frac{3 + 15}{6} = \frac{1}{2}(6)$
10. $\frac{1}{2} + \frac{3}{4} = \frac{3}{2} + \frac{1}{4}$

Replace the variable to make each open sentence true.

11. $18 + y = 20$
12. $3 \cdot x = 24$
13. a is an author of this book.
14. There are s states in the U.S.

Solve each equation.

15. $x = 8 + 3$
16. $y = 12 - 0.03$
17. $a = \frac{3}{4} \cdot 12$
18. $8.2 - 6.75 = m$

EXERCISES

Practice

Write the negation of each statement. Then state whether the negation is *true* or *false*.

19. The capital of the U.S. is Houston.
20. George Bush is President of the United States.
21. Birds have wings.
22. Oranges are not a citrus fruit.

List all subsets of each set to show that each statement is true.

23. $\{1, 2\}$ has 4 subsets.
24. $\{5, 6, 7, 8\}$ has 16 subsets.

Solve each equation.

25. $a = \frac{12 + 8}{4}$

26. $\frac{21 - 3}{12 - 3} = x$

27. $14.8 - 3.75 = t$

28. $n = \frac{84 \div 7}{18 \div 9}$

29. $\frac{2}{13} + \frac{5}{13} = p$

30. $\frac{5}{8} + \frac{1}{4} = y$

31. $d = 3\frac{1}{2} \div 2$

32. $r = 5\frac{1}{2} + \frac{1}{3}$

Find the solution set for each open sentence if the replacement set is {4, 5, 6, 7, 8}.

33. $x + 2 > 7$

34. $x - 3 > \frac{x + 1}{2}$

35. $\frac{2(x - 2)}{3} = \frac{4}{7 - 5}$

36. $9x - 20 = x^2$

37. $0.3(x + 4) \leq 0.4(2x + 3)$

38. $1.3x - 12 < 0.9x + 4$

Critical Thinking

39. Write five open sentences that have 2 as a solution.

Applications

40. **Music** The Oakbrook School chorus has eight voices in each section. The sections are as follows: soprano, alto, tenor, and bass. Write an equation to represent the total number of voices in the chorus. Then, solve the equation.

41. **Banking** In the beginning of the month, the balance in Marisa Fuentes's checking account was \$428.79. After writing checks totaling \$1097.31, depositing 2 checks of \$691.53 each, and withdrawing \$100 from a bank machine, what is the new balance in Marisa's account?

42. **Health** Two out of every three people in the United States wear eyeglasses. If the population of the U.S. was about 250,000,000 in 1990, about how many people were wearing eyeglasses?

Mixed Review

43. Write an algebraic expression for *a number h cubed.* **(Lesson 1-1)**

44. Write $\frac{1}{2} \cdot a \cdot a \cdot b \cdot b \cdot b$ as an expression using exponents. **(Lesson 1-1)**

Evaluate each expression. (Lesson 1-2)

45. $3 \cdot 6 - 12 \div 4$

46. $\frac{9 \cdot 3 - 4^2}{3^2 + 2^2}$

47. Evaluate $(15x)^3 - y$ if $x = 0.2$ and $y = 1.3$. **(Lesson 1-2)**

1-4 Identity and Equality Properties

Objective After studying this lesson, you should be able to:

- recognize and use the properties of identity and equality.

In algebra, there are certain statements, or *properties*, that are true for any number. Solve the equations below.

$$b + 15 = 15 \qquad 4780 + c = 4780$$

What value makes each statement true?

The solution of each equation is 0. Since the sum of any number and 0 is equal to the number, zero is called the **additive identity.**

Additive Identity Property	**For any number a, $a + 0 = 0 + a = a$.**

Solve the equations below.

$$5x = 5 \qquad z \cdot 655 = 655$$

What value makes each statement true?

The solution of each equation is 1. Since the product of any number and 1 is equal to the number, one is called the **multiplicative identity.**

Multiplicative Identity Property	**For any number a, $a \cdot 1 = 1 \cdot a = a$.**

Zero is a factor in each of these statements.

$$0(11) = 0 \qquad 2 \cdot 5 \cdot 0 = 0 \qquad 4 \cdot 0 \cdot 27 \cdot 8 = 0$$

When 0 is a factor, what can you say about the product?

Multiplicative Property of Zero

For any number a, $a \cdot 0 = 0 \cdot a = 0$.

Take a look at the statements below.

1. If $10 - 2 = 8$ and $8 = 5 + 3$, then $10 - 2 = 5 + 3$.
2. $5 = 5$
3. If $3 + 9 = 12$, then $8 + (3 + 9) = 8 + 12$.
4. If $7 + 3 = 10$, then $10 = 7 + 3$.

These are examples of the **properties of equality.**

Properties of Equality

The following properties are true for any numbers a, b, and c.

Reflexive Property: $a = a$
Symmetric Property: If $a = b$, then $b = a$.
Transitive Property: If $a = b$ and $b = c$, then $a = c$.
Substitution Property: If $a = b$, then a may be replaced by b.

Which property of equality corresponds to each of the four numbered statements above?

Example

Evaluate $6(12 - 48 \div 4) + 9 \cdot 1$. Indicate the property used in each step.

$6(12 - 48 \div 4) + 9 \cdot 1 = 6(12 - 12) + 9 \cdot 1$ *Substitution (=)*
$= 6(0) + 9 \cdot 1$ *Substitution (=)*
$= 0 + 9 \cdot 1$ *Multiplicative property of zero*
$= 0 + 9$ *Multiplicative identity*
$= 9$ *Additive identity*

CHECKING FOR UNDERSTANDING

Communicating Mathematics

Read and study the lesson to answer each question.

1. Can a property ever be false? Why or why not?
2. Can 1 be an additive identity? Why or why not?
3. Can 0 be a multiplicative identity? Why or why not?

Guided Practice

Name the property illustrated by each statement.

4. $7 + 6 = 7 + 6$
5. If $6 = 3 + 3$, then $3 + 3 = 6$.
6. $3 + (2 + 1) = 3 + 3$
7. $9 + 5 = (6 + 3) + 5$
8. If $8 = 6 + 2$ and $6 + 2 = 5 + 3$, then $8 = 5 + 3$.

Evaluate each expression. Indicate the property used in each step.

9. $1(8 - 2^3)$
10. $3 + 5(4 - 2^2) - 1$

EXERCISES

Practice

Solve each equation.

11. $0 + x = 7$
12. $a \cdot 1 = 5$
13. $7b = 7$
14. $0(18) = n$

Name the property illustrated by each statement.

15. If $8 + 1 = 9$, then $9 = 8 + 1$.
16. $0 \cdot 36 = 0$
17. $1(87) = 87$
18. $9 + (2 + 10) = 9 + 12$
19. $14 + 16 = 14 + 16$
20. $0 + 17 = 17$
21. $(9 - 7)(5) = 2(5)$
22. $abc = 1abc$
23. $7(0) = 0$
24. If $3 = 4 - 1$, then $4 - 1 = 3$.
25. If $9 + 1 = 10$ and $10 = 5(2)$, then $9 + 1 = 5(2)$.

Evaluate each expression. Indicate the property used in each step.

26. $5(9 \div 3^2)$
27. $3 + 18(12 \div 6 - 2)$
28. $(19 - 12) \div 7 \cdot 23$
29. $(2^5 - 5^2) + (4^2 - 2^4)$
30. $(9 - 2 \cdot 3)^3 - 27 + 9 \cdot 2$
31. $(13 + \frac{2}{5} \cdot 5)(3^2 - 2^3)$

Critical Thinking

32. Think about the relationship "is greater than," represented by the symbol >. Does > work with the **a)** reflexive property, **b)** symmetric property, or **c)** the transitive property? Why or why not?

Applications

33. **History** Lincoln's Gettysburg address began "Four score and seven years ago, ..." Write an expression to represent four score and seven. How many years is that?

34. **Culture** Each year in the Chinese calendar is named for one of 12 animals. Every 12 years, the same animal is named. If 1992 is the Year of the Monkey, how many years in the 20th century have been Years of the Monkey?

35. **Money** Geraldina makes a call to a teen talkline. The cost of the call is \$3.00 for the first minute plus 85¢ for each additional minute. If Geraldina talks for 40 minutes, how much did her call cost?

Mixed Review

36. Write a verbal expression for x^2. **(Lesson 1-1)**

37. Evaluate the expression $7(8 - 4) + 11$. **(Lesson 1-2)**

38. *True* or *false:* $15 \div 3 + 7 < 13$. **(Lesson 1-3)**

39. Replace the variable to make the sentence $\frac{1}{2}(m) = 10$ true. **(Lesson 1-3)**

MID-CHAPTER REVIEW

Write an algebraic expression for each verbal expression. Use *n* for the variable. (Lesson 1-1)

1. the cube of a number
2. the square of a number increased by seven

Explain how to evaluate each expression. Do not evaluate. (Lesson 1-2)

3. $8 \div 2 + 6 \cdot 2$
4. $3(3^2 - 3)$
5. $(8 + 6) \div 2 + 2$

Evaluate each expression if $a = 6$, $b = 4$, $c = 3$, and $d = \frac{1}{2}$. (Lesson 1-2)

6. $\frac{2}{3}(b^2) - \frac{1}{3}(a)$
7. $\frac{1}{2}(2b + 10c) - 4$
8. $d(a + b) - c$

Solve each equation. (Lesson 1-3)

9. $x = 6 + 0.28$
10. $m = \frac{64 + 4}{17}$
11. $y = \frac{96 \div 6}{8 \div 2}$

Find the solution set for each open sentence if the replacement set is {4, 5, 6, 7, 8}. (Lesson 1-3)

12. $10 - x < 7$
13. $\frac{x + 3}{2} < 5$
14. $\frac{2x + 1}{7} \geq \frac{x + 4}{5}$

Name the property illustrated by each statement. (Lesson 1-4)

15. $3abc \cdot 0 = 0$
16. If $12xy - 3 = 4$, then $4 = 12xy - 3$.

1-5 The Distributive Property

Objective After studying this lesson, you should be able to:

use the distributive property to simplify expressions.

Application There are four children in the Walker family: Renita, Randy, Robert, and Rochelle. Renita and Robert play basketball; Randy and Rochelle prefer tennis. Mr. and Mrs. Walker buy 2 pairs of basketball shoes for \$79.95 each and 2 pairs of tennis shoes for \$69.97 each. How much do they spend on athletic shoes?

This problem can be solved in two ways.

1. **Total cost:**

   $2 \times$ (cost of basketball shoes plus the cost of tennis shoes)

   $2(\$79.95 + \$69.97) = 2(\$149.92) = \$299.84$

2. **Total cost:**

   $(2 \times$ cost of basketball shoes$) + (2 \times$ cost of tennis shoes$)$

   $2(\$79.95) + 2(\$69.97) = \$159.90 + \$139.94 = \$299.84$

Either way you look at it, the Walkers are spending the same amount of money. That's because the following is true.

$$2(\$79.95 + \$69.97) = 2(\$79.95) + 2(\$69.97)$$

This is an example of the **distributive property.**

Distributive Property	**For any numbers a, b, and c,** **$a(b + c) = ab + ac$ and $(b + c)a = ba + ca$;** **$a(b - c) = ab - ac$ and $(b - c)a = ba - ca$.**

Notice that the distributive property works both ways; that is, it doesn't matter whether a is placed on the right or the left of the expression in parentheses.

Example 1 **Find the area of the basketball court in two different ways.**

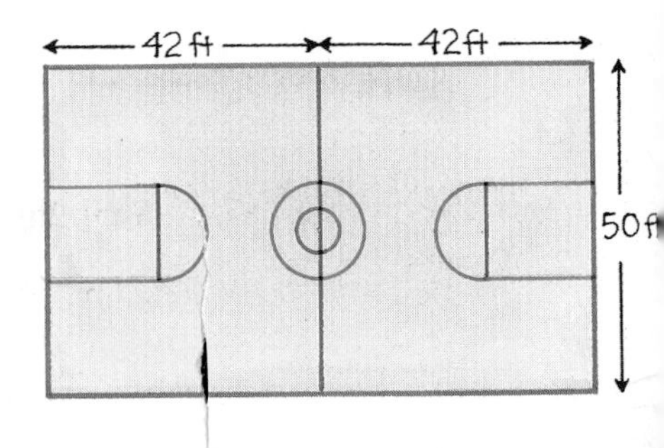

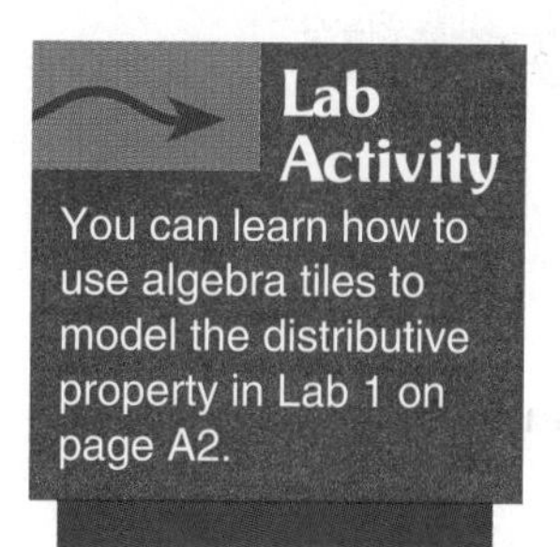

Method 1: Multiply the width by the length.

$$A = 50(42 + 42)$$
$$= 50(84)$$
$$= 4200$$

Method 2: Add the areas of the smaller rectangles.

$$A = 50(42) + 50(42)$$
$$= 2100 + 2100$$
$$= 4200$$

The area is 4200 ft^2.

You can use the distributive property to multiply in your head.

Example 2

Use the distributive property to find each product.

a. $5 \cdot 97 = 5(100 - 3)$
$= 5 \cdot 100 - 5 \cdot 3$
$= 500 - 15$
$= 485$

b. $12(8.5) = 12(8 + 0.5)$
$= (12)(8) + (12)(0.5)$
$= 96 + 6$
$= 102$

The symmetric property of equality allows the distributive property to be rewritten as follows.

$$\text{If } a(b + c) = ab + ac, \text{ then } ab + ac = a(b + c).$$

A **term** is a number, a variable, or a product or quotient of numbers and variables. Some examples of terms are $5x^2$, $\frac{ab}{4}$, and $7k$. **Like terms** are terms that contain the same variables, with corresponding variables raised to the same power. Some pairs of like terms are $5x$ and $3x$, $2xy$ and $7xy$, and $7ax^2$ and $11ax^2$.

Note that x and x^2 are not like terms.

An expression in **simplest form** has no like terms and no parentheses.

Example 3

Simplify $5a + 7a$.

$5a + 7a = (5 + 7)a$ *Distributive property*
$= 12a$ *Substitution property of equality*

Example 4 **Simplify $\frac{4}{5}x^2y - \frac{2}{5}x^2y$.**

$\frac{4}{5}x^2y - \frac{2}{5}x^2y = \left(\frac{4}{5} - \frac{2}{5}\right) x^2y$ *Distributive property*

$= \frac{2}{5} x^2y$ *Substitution property of equality*

The **coefficient** is the numerical part of a term. For example, in $6ab$, the coefficient is 6. In rs, the coefficient is 1 since, by the multiplicative identity property, $1 \cdot rs = rs$. Like terms may also be defined as terms that are the same or that differ only in their coefficients.

Example 5 **Name the coefficient in each term.**

a. $19g^2h$ The coefficient is 19.

b. xy^2z^2 The coefficient is 1 since $xy^2z^2 = 1xy^2z^2$.

c. $\frac{2x^3}{5}$ The coefficient is $\frac{2}{5}$ since $\frac{2x^3}{5}$ can be written as $\frac{2}{5}(x^3)$.

Example 6 **Simplify $8n^2 + n^2 + 7n + 3n$.**

$8n^2 + n^2 + 7n + 3n = 8n^2 + 1n^2 + 7n + 3n$ *Multiplicative identity property*

$= (8 + 1)n^2 + (7 + 3)n$ *Distributive property*

$= 9n^2 + 10n$ *Substitution property of equality*

Example 7 **Simplify $\frac{b}{2} + b$.**

$\frac{b}{2} + b = \frac{1}{2}b + 1b$ *Multiplicative identity property*

$= \left(\frac{1}{2} + 1\right) b$ *Distributive property*

$= \left(\frac{1}{2} + \frac{2}{2}\right) b$ *Substitution property of equality*

$= \frac{3}{2}b$ *Substitution property of equality*

Example 8 **Use a calculator to simplify $374.2n + 582.6n + 52.38y + 38.02y$.**

By the distributive property,

$$374.2n + 582.6n + 52.38y + 38.02y = (374.2 + 582.6)n + (52.38 + 38.02)y$$
$$= 956.8n + 90.4y.$$

Since $956.8n$ and $90.4y$ are not like terms, the expression $956.8n + 90.4y$ is in simplest form.

CHECKING FOR UNDERSTANDING

Communicating Mathematics

Read and study the lesson to answer each question.

1. Write the distributive property in four different ways.
2. The distributive property is sometimes called the distributive property of multiplication over addition (or subtraction). Why do you think it is called that?

Guided Practice

Name the coefficient in each term.

3. $5m$
4. a^2q
5. $\frac{am}{3}$
6. $0.5bm$

Name the like terms in each list of terms.

7. $6b$, $6bc$, bc
8. $7a$, $7a^2$, $29a$, $7a^3$
9. $4xy$, $5xy$, $6xy^2$, $6x^2y$
10. $2rs^2$, r^2s, rs^2

Simplify each expression, if possible. If not possible, write *in simplest form.*

11. $x + x + x + x + x$
12. yyy
13. $x + \frac{1}{x}$
14. $0(x + y)$

EXERCISES

Practice

Mental Math: Find each product mentally.

15. $3 \cdot 215$
16. $4 \cdot 98$

Use the distributive property to rewrite each expression.

17. $8(3x + 7)$
18. $4m - 4n$

Simplify each expression, if possible. If not possible, write *in simplest form.*

19. $13a + 5a$
20. $21x - 10x$
21. $3(5am - 4)$
22. $15x^2 + 7x^2$
23. $9y^2 + 13y^2 + 3$
24. $11a^2 - 11a^2 + 12a^2$
25. $14a^2 + 13b^2 + 27$
26. $5a + 7a + 10b + 5b$
27. $3(x + 2y) - 2y$
28. $5x + 3(x - y)$
29. $6(5a + 3b - 2b)$
30. $5ab^2 + 2a^2b + a^2b^2$
31. $4(3x + 2) + 2(x + 3)$
32. $x^2 + \frac{7}{8}x - \frac{x}{8}$

Use a calculator to simplify each expression.

33. $3.047xy^3 - 0.012y^3 + 5.78xy^3$

34. $1.042a^2 + 8.0879a + 5.265a$

35. $1436x^2 - 789x^2 + 5689x^2$

36. $5.8rs^3 - 4.06rs^3 + 0.92r^2s$

Critical Thinking

37. If $2(b + c) = 2b + 2c$, does $2 + (b \cdot c) = (2 + b)(2 + c)$? Choose values for b and c to show that these may be true or find *counterexamples* to show that they are not.

Applications

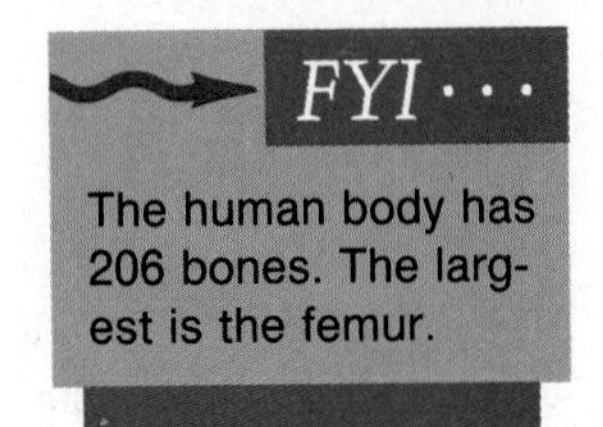

38. **Retailing** Maria and Mark are salesclerks at a local department store. Each earns $4.95 per hour. Maria works 24 hours per week and Mark works 32 hours per week. How much do the two of them earn together each week?

39. **Archaeology** To calculate the height of a person, archaeologists measure the lengths of certain bones. The bones measured are the femur or thigh bone (F), the tibia or leg bone (T), the humerus or upper arm bone (H), and the radius or lower arm bone (R). When the length of one of these bones is known, scientists can use one of the formulas below to determine the person's height (h), in centimeters.

Males	Females
$h = 69.089 + 2.238F$	$h = 61.412 + 2.317F$
$h = 81.688 + 2.392T$	$h = 72.572 + 2.533T$
$h = 73.570 + 2.970H$	$h = 64.977 + 3.144H$
$h = 80.405 + 3.650R$	$h = 73.502 + 3.876R$

a. The femur of a 37-year-old woman measured 47.9 cm. Use a calculator to find the height of the woman to the nearest tenth of a centimeter.

b. The humerus of a 49-year-old man measured 35.7 cm. Use a calculator to find the height of the man to the nearest tenth of a centimeter.

Computers

40. Use the BASIC program at the right to test whether the distributive property extends to division and exponents. Input different values for A, B, and C.

```
10 INPUT A, B, C
20 PRINT "A = ";A, "B = ";B,"C = ";C
30 PRINT "(A + B)/C = ";(A+B)/C
40 PRINT "A/C + B/C = ";A/C+B/C
50 PRINT
60 PRINT "(A + B)↑C = ";(A+B)↑C
70 PRINT "A↑C + B↑C = ";A↑C+B↑C
```

a. Does $\frac{a + b}{c} = \frac{a}{c} + \frac{b}{c}$?

b. Does $(a + b)^c = a^c + b^c$?

Mixed Review

41. Write $3 \cdot 3 \cdot a \cdot a \cdot a$ as an expression using exponents. **(Lesson 1-1)**

42. Evaluate $196 \div [4(11 - 4)]$. **(Lesson 1-2)**

43. Solve the equation $9.6 + 4.53 = b$. **(Lesson 1-3)**

44. Name the property shown by the following: If $19 - 3 = 16$, then $16 = 19 - 3$. **(Lesson 1-4)**

45. Solve the equation $4 + 5 \cdot 0 = y$. **(Lesson 1-4)**

1-6 Commutative and Associative Properties

Objective After studying this lesson, you should be able to:

- recognize and use the commutative and associative properties when simplifying expressions.

Application

John Mattox owns a roofing business. He has a contract to roof three houses for Tradition Homes. These houses require 800, 1050, and 1200 square feet of shingles respectively. How many square feet of shingles are needed to roof the three houses?

To find the sum of the three numbers easily, you can use the **commutative property** and the **associative property.**

Commutative Properties	**For any numbers a and b, $a + b = b + a$ and $a \cdot b = b \cdot a$.**

The commutative property says that the order in which you add or multiply two numbers does not change their sum or product.

$800 + 1050 + 1200 = 800 + 1200 + 1050$ *Commutative property (+)*

One easy way to evaluate $800 + 1200 + 1050$ is to group, or *associate,* the numbers. The associative property allows you to do this.

Associative Properties	**For any numbers a, b, and c, $(a + b) + c = a + (b + c)$ and $(ab)c = a(bc)$.**

The associative property says that the way you group three numbers when adding or multiplying does not change their sum or product. Since you can add 800 and 1200 mentally, group them together.

$800 + (1200 + 1050) = (800 + 1200) + 1050$ *Associative property (+)*
$= 2000 + 1050$ *Substitution property (=)*
$= 3050$

So John needs 3050 square feet of shingles to roof the three houses.

Example 1 Find the volume of a rectangular prism 4 inches long, $3\frac{1}{2}$ inches wide, and $\frac{1}{2}$ inch high.

Recall that the volume (V) of a rectangular prism is the length (ℓ) times the width (w) times the height (h).

$$
\begin{aligned}
V &= \ell wh \\
&= (4)\left(3\tfrac{1}{2}\right)\left(\tfrac{1}{2}\right) \\
&= (4)\left(\tfrac{1}{2}\right)\left(3\tfrac{1}{2}\right) && \text{Commutative property } (\times) \\
&= (2)\left(3\tfrac{1}{2}\right) \\
&= 7
\end{aligned}
$$

The volume of the prism is 7 in^3.

The commutative and associative properties can be used with the other properties you have studied when simplifying expressions.

Example 2 Simplify $3x + (5 + 6x)$, indicating all of the properties used.

$$
\begin{aligned}
3x + (5 + 6x) &= 3x + (6x + 5) && \text{Commutative property } (+) \\
&= (3x + 6x) + 5 && \text{Associative property } (+) \\
&= (3 + 6)x + 5 && \text{Distributive property} \\
&= 9x + 5 && \text{Substitution property } (=)
\end{aligned}
$$

The chart below summarizes the properties that are used when simplifying expressions.

The following properties are true for any numbers *a*, *b*, and *c*.		
	Addition	**Multiplication**
Commutative	$a + b = b + a$	$ab = ba$
Associative	$(a + b) + c = a + (b + c)$	$(ab)c = a(bc)$
Identity	0 is the identity. $a + 0 = 0 + a = a$	1 is the identity. $a \cdot 1 = 1 \cdot a = a$
Zero		$a \cdot 0 = 0 \cdot a = 0$
Distributive	$a(b + c) = ab + ac$ and $(b + c)a = ba + ca$	
Substitution	If $a = b$, then a may be substituted for b.	

Example 3 Simplify $5n + 2(n^2 + 4n) + n^2$, indicating all of the properties used.

$$
\begin{aligned}
5n + 2(n^2 + 4n) + n^2 &= 5n + (2n^2 + 8n) + n^2 && \text{Distributive property} \\
&= 5n + (8n + 2n^2) + n^2 && \text{Commutative property } (+) \\
&= (5n + 8n) + (2n^2 + 1n^2) && \text{Associative property } (+) \\
&= (5 + 8)n + (2 + 1)n^2 && \text{Distributive property} \\
&= 13n + 3n^2 && \text{Substitution property } (=)
\end{aligned}
$$

CHECKING FOR UNDERSTANDING

Communicating Mathematics

Name the property illustrated by each statement.

1. The way you group numbers when adding doesn't change their sum.
2. The order that you multiply numbers doesn't change their product.

Guided Practice

Name the property illustrated by each statement.

3. $(8 + 4) + 2 = 8 + (4 + 2)$
4. $5 + 3 = 3 + 5$
5. $8(a + b) = 8a + 8b$
6. $(3 + 8) + x = 11 + x$
7. $3x = x \cdot 3$
8. $(a + b) + 3 = a + (b + 3)$
9. $5(ab) = (5a)b$
10. $cb + ab = (c + a)b$
11. $10(x + y) = 10(y + x)$
12. $3(a + 2b) = (a + 2b)3$

Name the property that justifies each step.

13. Simplify $6a + (8b + 2a)$.
    a. $6a + (8b + 2a) = 6a + (2a + 8b)$
    b. $= (6a + 2a) + 8b$
    c. $= (6 + 2)a + 8b$
    d. $= 8a + 8b$

14. Simplify $8a^2 + (8a + a^2) + 7a$.
    a. $8a^2 + (8a + a^2) + 7a = 8a^2 + (a^2 + 8a) + 7a$
    b. $= (8a^2 + a^2) + (8a + 7a)$
    c. $= (8a^2 + 1a^2) + (8a + 7a)$
    d. $= (8 + 1)a^2 + (8 + 7)a$
    e. $= 9a^2 + 15a$

EXERCISES

Practice

Name the property illustrated by each statement.

15. $5a + 2b = 2b + 5a$
16. $1 \cdot a^2 = a^2$
17. $(a + 3b) + 2c = a + (3b + 2c)$
18. $x^2 + (y + z) = x^2 + (z + y)$
19. $ax + 2b = xa + 2b$
20. $(3 \cdot x) \cdot y = 3 \cdot (x \cdot y)$
21. $29 + 0 = 29$
22. $5(a + 3b) = 5a + 15b$
23. $5a + 3b = 3b + 5a$
24. $5a + \left(\frac{1}{2}b + c\right) = \left(5a + \frac{1}{2}b\right) + c$

Name the property illustrated by each statement.

25. The quantity m plus n times a equals m times a plus n times a.

26. Zero plus 7 equals 7.

27. The product of one and the quantity a plus b equals a plus b.

28. Three times m plus n times q equals 3 times m plus q times n.

Simplify.

29. $5a + 6b + 7a$

30. $8x + 2y + x$

31. $3x + 2y + 2x + 8y$

32. $\frac{2}{3}x^2 + 5x + x^2$

33. $3a + 5b + 2c + 8b$

34. $5 + 7(ac + 2b) + 2ac$

35. $3(4x + y) + 2x$

36. $3(x + 2y) + 4(3x + y)$

37. $\frac{3}{4} + \frac{2}{3}\left(x + 2y\right) + x$

38. $\frac{3}{5}\left(\frac{1}{2}x + 2y\right) + 2x$

39. $0.2(3x + 0.2) + 0.5(5x + 3)$

40. $3[4 + 5(2x + 3y)]$

Critical Thinking

41. Is $24 \div 6 = 6 \div 24$? Is $36 \div 9 = 9 \div 36$? Write a sentence about division and the commutative property.

42. Is $12 - 8 = 8 - 12$? Is $27 - 10 = 10 - 27$? Write a sentence about subtraction and the commutative property.

Applications

43. **Marketing** Betty Castillo is the product manager for Toasty Oatsies cereal. Part of her job is to determine in what size box Toasty Oatsies should be packaged. Betty can choose from among the following sizes: 8″ by 11″ by $2\frac{1}{2}$″, $8\frac{1}{2}$″ × 10″ × 2″, or $7\frac{7}{8}$″ by 11″ by 3″.
   a. Name the size that has the greatest volume.
   b. Write an expression to represent the volume of the largest box in three different ways.
   c. In how many different ways can you write this expression?

44. **America** The United States flag has either 5 stars or 6 stars in each row. Rows of 5 stars alternate with rows of 6 stars. How many rows are there with 6 stars?

45. **Volleyball** In the game of volleyball, the top of the net is 7 feet 11 inches from the floor. The bottom of the net is 4 feet 8 inches from the floor. How wide is a volleyball net?

46. **Weather** Meteorologists can predict when a storm will hit their area by examining the travel time of the storm system. To do this, they use the following formula.

distance of system (in miles) ÷ speed of system (in miles per hour) = travel time of system (in hours)

It is 4:00 P.M., and a storm is heading towards the coast at a speed of 30 miles per hour. The storm is about 150 miles from the coast. What time will the storm hit?

Mixed Review

47. Write an algebraic expression for *twice the square of a number.* **(Lesson 1-1)**

Evaluate each expression if $c = 3$, $x = 0.2$, and $y = 1.3$. (Lesson 1-2)

48. $c^2 + y^2$

49. $(15x)^3 - y$

50. Solve the equation $\frac{5}{6} \cdot 18 = m$. **(Lesson 1-3)**

51. Name the coefficient of $\frac{4am^2}{5}$. **(Lesson 1-5)**

52. Simplify the expression $9a + 14(a + 3)$. **(Lesson 1-5)**

Journal

Make a list of all of the properties you have learned in this chapter. Give an example of each.

Name the property illustrated by each statement.

53. If $a + b = c$ and $a = b$, then $a + a = c$. **(Lesson 1-4)**

54. $3(4 + 12) = 3(4) + 3(12)$ **(Lesson 1-5)**

HISTORY CONNECTION

Sir William Rowan Hamilton

The use of the term *associative* is often credited to Sir William Rowan Hamilton (1805–1865). Hamilton was a mathematician who studied and taught at Trinity College in Dublin, Ireland. A child prodigy, he was appointed Royal Astronomer of Ireland, Director of the Dunsink Observatory, and Professor of Astronomy at Trinity College at age 22, and was knighted at age 30. Hamilton also holds the distinction of being the first foreign associate named to the National Academy of Sciences in the United States.

1-7 Formulas

Objective

After studying this lesson, you should be able to:

- translate verbal sentences into equations or formulas.

A **formula** is an equation that states a rule for the relationship between certain quantities.

Application

The First Community Bank shows the time and temperature on an electronic billboard. The temperature is given in degrees Celsius. If the temperature shown is 25°C, what is the temperature in degrees Fahrenheit?

The highest temperature ever recorded was 136.4°F at El Azizia, Libya on Sept. 13, 1922.

Use the formula $F = \frac{9}{5}C + 32$. The variables F and C represent the degree measures in Fahrenheit and Celsius, respectively.

$F = \frac{9}{5}C + 32$

$F = \frac{9}{5}(25) + 32$ *Replace C with 25.*

$F = 45 + 32$

$F = 77$

The temperature is 77°F.

Example 1

The formula for the area of a triangle is $A = \frac{1}{2}bh$. Find the area of a triangle with a base (b) of 6 feet and a height (h) of 9 feet.

$A = \frac{1}{2}bh$

$= \frac{1}{2}(6)(9)$

$= 3 \cdot 9$

$= 27$

9 ft

6 ft

The area of the triangle is 27 ft^2.

Many sentences can be written as equations or formulas. Use variables to represent the unspecified numbers or measures referred to in the sentence. Then write the verbal expressions as algebraic expressions. Some verbal expressions that suggest the *equals sign* are listed below.

- is
- equals
- is equal to
- is the same as
- is as much as
- is identical to

Example 2

Translate the sentence below into an equation.
The number z is equal to twice the sum of x and y.

The number z is equal to twice the sum of x and y.

$z \quad = \quad 2$ times add x and y

The equation is $z = 2(x + y)$.

Example 3

Translate the sentence below into a formula.
The area of a circle equals the product of π and the square of the radius (r).

The area of a circle equals the product of π and the square of the radius (r).

$A \quad = \quad$ multiply π and r^2

The formula is $A = \pi r^2$.

Sometimes you can discover a formula by making a model.

Example 4

Find the surface area of a rectangular solid.

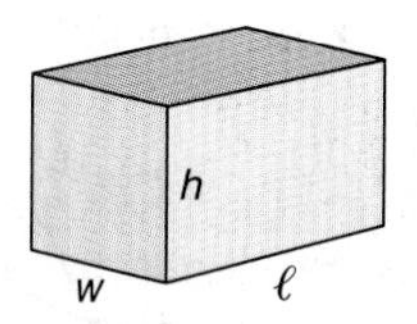

Take a rectangular box apart. It could look like the model at the right.

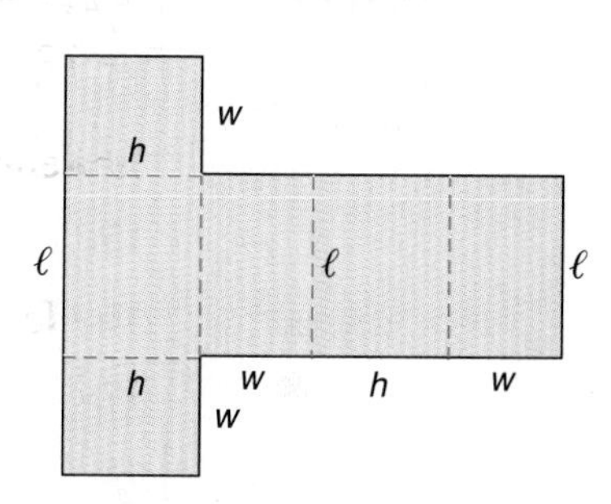

You can find the surface area of the solid by finding the area of each side. Each side is a rectangular surface.

$$S = wh + wh + \ell h + \ell h + \ell w + \ell w$$
$$= 2wh + 2\ell h + 2\ell w$$
$$= 2(wh + \ell h + \ell w)$$

CHECKING FOR UNDERSTANDING

Communicating Mathematics

Read and study the lesson to answer each question.

1. Is an equation a type of formula, or is a formula a type of equation?
2. In the formula $C = \frac{5}{9}(F - 32)$, what do the C and F represent?

Guided Practice

Write a formula for each sentence. Use the variables indicated.

3. The area (A) of a square is the square of the length of one of its sides (s).
4. The perimeter (P) of a parallelogram is twice the sum of the lengths of two adjacent sides (a and b).
5. The perimeter (P) of a square is 4 times the length of a side (s).
6. The circumference (C) of a circle is the product of 2, π, and the radius (r).

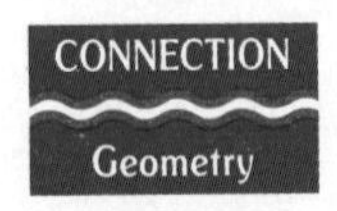

Copy each chart. Then use the formula for the surface area (SA) of a rectangular solid to complete the chart.

	ℓ	w	h	SA
7.	5	8	6	
9.	18	10	4	
11.	$20\frac{1}{2}$	3	8	

	ℓ	w	h	SA
8.	$5\frac{1}{2}$	12	$3\frac{1}{2}$	
10.	12.9	11	4.6	
12.	21.8	6.5	9.7	

EXERCISES

Practice

Write an equation for each sentence.

13. Twice x increased by the square of y is equal to z.
14. The square of a decreased by the cube of b is equal to c.
15. The square of the sum of x and a is equal to m.
16. The number r equals the cube of the difference of a and b.
17. The square of the product of a, b, and c is equal to k.
18. The product of a, b, and the square of c is f.
19. Twenty-nine decreased by the product of x and y is equal to z.

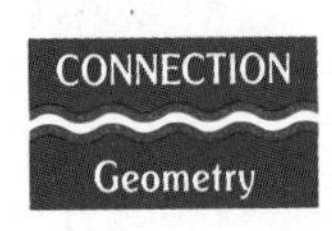

The formula for the area (A) of a trapezoid as shown at the right is $A = \frac{1}{2}h(a + b)$. Copy and complete each chart.

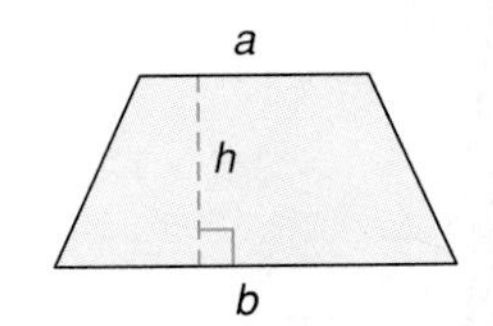

	h	a	b	A
20.	6	24	19	
21.	11	37	23	
22.	12	24	40	
24.	10	19	54	
26.	$\frac{5}{8}$	$\frac{3}{4}$	$\frac{1}{2}$	

	h	a	b	A
23.	$3\frac{1}{3}$	12	$8\frac{1}{4}$	
25.	4	18.9	12.7	
27.	2.4	8.25	3.15	

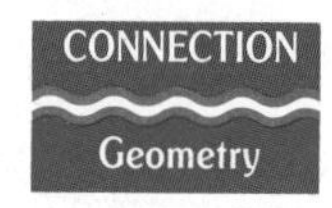

The formula for the area (A) of the shaded region of the figure below is $A = \frac{1}{2}\pi a^2 - a^2$. Find the area of the shaded region for each value of a to the nearest whole number. Use 3.14 for π.

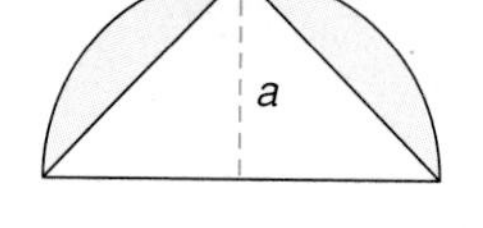

28. 6 **29.** 4 **30.** 81 **31.** 64

32. 3.8 **33.** 5.6 **34.** 18.3 **35.** 27.4

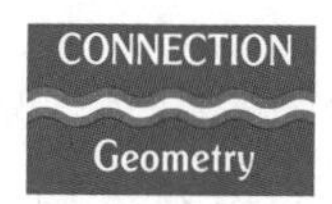

Write a formula for the area of each figure.

36.

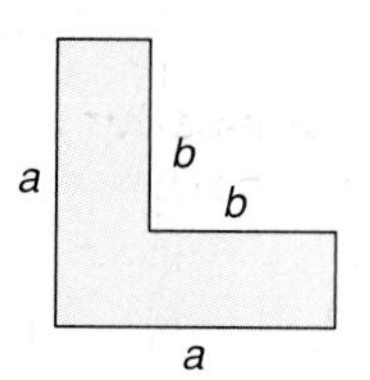

37.

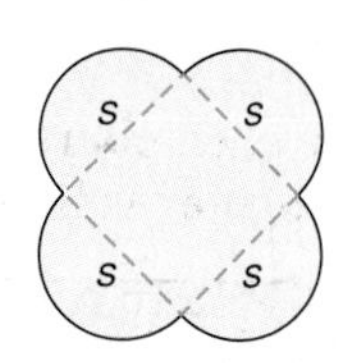

38.

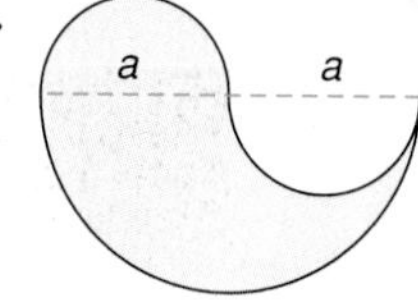

Critical Thinking

39. The area of a triangle is 420 cm². Its base is 30 cm. Use the formula $A = \frac{1}{2}bh$ to find its height.

Applications

The distance traveled (d) is equal to the rate (r) times the time (t). **Write this sentence as a formula and use it to solve each problem.**

40. Travel Find the distance from Danville to the beach if it takes 3 hours to drive there at an average rate of 50 miles per hour.

41. Fitness Kimi runs for 30 minutes each day. Find the distance she runs if she averages 660 feet per minute.

42. Physics The speed of sound through air is about 330 meters per second. Find the distance between Carlos and an explosion if it takes 10 seconds for the sound to reach him.

43. Astronomy The speed of light is about 186,000 miles per second. If the sun is about 93 million miles from Earth, about how long does it take for the rays of the sun to reach Earth?

Mixed Review

44. Evaluate $12 \cdot 6 \div 3 \cdot 2 \div 8$. **(Lesson 1-2)**

45. Solve the equation $m = 3\frac{1}{3} \div 3$. **(Lesson 1-3)**

46. Name the like terms in the following list: a^3, b^3, $3a^2$, $4b^3$. **(Lesson 1-5)**

47. Simplify the expression $16a + 21a + 30b - 7b$. **(Lesson 1-5)**

48. State the property shown by $8(2 \cdot 6) = (2 \cdot 6)8$. **(Lesson 1-6)**

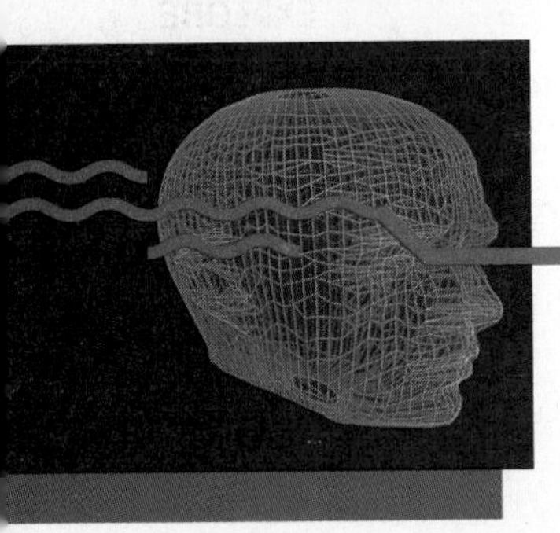

Technology

Formulas

▶ **BASIC**
Graphing calculator
Graphing software
▶ **Spreadsheets**

In BASIC, formulas used with READ and DATA statements are helpful when creating charts involving many calculations.

This program makes a chart that finds and records the areas of triangles using the formula $A = \frac{1}{2}bh$.

```
10 PRINT "B","H","A"
20 PRINT
30 READ B, H
40 IF B = 0 THEN 100
50 DATA 4,6,12,10,6.2,5.4
60 DATA 120,91,10.6,15.3,0.0
70 LET A = 0.5*B*H
80 PRINT B,H,A
90 GOTO 30
100 END
```

The READ and DATA statements assign values to the variable b for base and h for height. Enter and run the program on your computer.

Spreadsheets are computer programs especially designed to create charts involving many calculations. These charts are composed of cells named by column letters and row numbers. In the spreadsheet below, cells A1, B1, and C1 hold the labels B, H, and A for base, height, and area. Cells A3 to A7 hold the values for the base of each triangle. Cells B3 to B7 hold the values for the height of each triangle. Cells C3 to C7 contain the values for the area of each triangle.

```
=====A=======B=======C=======D=======E=======F=====
1:   B       H       A
2:-------------------------
3:     4       6        12
4:   12      10        60
5:     6.2     5.4      16.74
6: 120       91      5460
7:   10.6    15.3      81.09
```

Each cell in column C holds a formula to compute the area. Cell C3 holds the formula 0.5 * A3 * B3. The formula refers to the cells containing the measures of the base and height of the first triangle.

EXERCISES

1. Write a BASIC program to make a chart for the perimeters of rectangles. Use the formula $P = 2(\ell + w)$. Choose values for the lengths and widths of five rectangles.
2. Set up a spreadsheet to make a chart like the chart in Exercise 1.
3. Write a BASIC program to make a chart for the areas of trapezoids. Use the formula $A = \frac{1}{2}h(a + b)$. Choose values for the heights and bases of five trapezoids.
4. Set up a spreadsheet to make a chart like the chart in Exercise 3.

EXPLORE
PLAN
SOLVE
EXAMINE

1-8 Problem-Solving: Explore Verbal Problems

Objective

After studying this lesson, you should be able to:

- explore problem situations by asking and answering questions.

The four steps that can be used to solve verbal problems are listed below.

Problem-Solving Plan

1. **Explore the problem.**
2. **Plan the solution.**
3. **Solve the problem.**
4. **Examine the solution.**

Let's examine the first step of this plan, exploring the problem. To solve a verbal problem, first read the problem carefully and explore what the problem is about.

- Identify what information is given.
- Identify what you are asked to find.
- Choose a variable to represent one of the unspecified numbers in the problem. This is called *defining the variable.*
- Use the variable to write expressions for the other unspecified numbers in the problem.

Example

One day John and Inger picked peaches for 4 hours. In all they picked 30 baskets of peaches. Inger picked 6 baskets fewer than John.

Questions	Answers
a. How many baskets were picked in all?	a. 30
b. How long did John and Inger work?	b. 4 hours
c. Who worked longer?	c. They worked the same.
d. Who picked more?	d. John
e. If Inger picked n baskets, how many did John pick?	e. $n + 6$
f. If John picked r baskets, how many did Inger pick?	f. $r - 6$

CHECKING FOR UNDERSTANDING

Communicating Mathematics

Read and study the lesson to answer each question.

1. What is the emphasis of this lesson?
2. What happens if you do not understand the problem?

Guided Practice

For each situation, answer the related questions.

3. Louis Limotta checked his cash register at the end of the day. He found that he had 7 fewer \$5 bills than \$1 bills. He had eleven \$10 bills and no larger bills. In all he had \$267 in bills.

   a. Did he have more \$5 bills or \$1 bills?
   b. How many more?
   c. How much money did he have in all?
   d. How many \$20 bills did he have?
   e. How much money did he have in \$1 and \$5 bills together?
   f. When did he check his cash register?
   g. If he had n \$5 bills, how much money did he have in \$5 bills?

4. Luisa has 20 books on crafts and cooking. She also has 21 novels. She has 6 more cookbooks than craft books.

   a. Does she have more cookbooks or craft books?
   b. How many more?
   c. Does she have more novels than craft books?
   d. How many books does she have in all?
   e. If she has n cookbooks, how many craft books has she?
   f. Of what kind of book does she have the least?

EXERCISES

Practice

For each situation, answer the related questions.

5. Two breakfast cereals, Kornies and Krispies, together cost \$3.59. One of them costs 7¢ more than the other.

   a. Which one costs more?
   b. What is the difference in their prices?
   c. How much would two boxes of each cost?
   d. Would two boxes of the more expensive cereal cost more or less than \$3.59?
   e. If the more expensive cereal costs n cents, what is the cost of the other cereal?

6. The Vegetable Mart offers corn at 18¢ per ear, cucumbers at 12¢ each, and tomatoes at 59¢ per basket. Mirna has only \$3 to spend and wants one basket of tomatoes and 5 fewer ears of corn than cucumbers.

   a. How many baskets of tomatoes does she want?
   b. How much more is a basket of tomatoes than an ear of corn?
   c. After buying the tomatoes, how much money does she have remaining?
   d. Will she buy more cucumbers or ears of corn?
   e. If she buys n cucumbers, how many ears of corn will she buy?

7. Phoebe goes to Pluto's Platters to buy cassette tapes. She bought 3 more rock tapes than classical and 2 fewer western tapes than rock. Including 5 jazz tapes, she bought 18.
   a. Did she buy more classical than rock?
   b. Did she buy more rock than western?
   c. Which did she buy the most of?
   d. How many rock, classical, and western tapes did she buy?
   e. If she bought n classical tapes, how many rock tapes did she buy?

8. Craig said, "I am 24 years younger than my mom and the sum of our ages is 68 years."
   a. How old was Craig's mother when Craig was born?
   b. How much older than Craig is his mother now?
   c. What will be the sum of their ages in 5 years?
   d. How old was Craig's mother when Craig was 10 years old?
   e. In ten years, how much younger than his mother will Craig be?

9. Selam was working some math problems. She thought that she could finish 3 pages of 24 problems per page in 2 hours. After $1\frac{1}{2}$ hours, she had finished 2 pages.
   a. How many problems had Selam completed in $1\frac{1}{2}$ hours?
   b. How many problems are there in all?
   c. At the rate she is working, will she finish the three pages in 2 hours?

10. Lorena can paint the house in 25 hours. Mia can paint the same house in 30 hours.
   a. Who paints faster, Mia or Lorena?
   b. Working alone, how much of the house can Lorena paint in 20 hours?
   c. Working alone, how much of the house can Mia paint in x hours?
   d. If Lorena and Mia work together, how many hours will it take to paint the house?

Portfolio

A portfolio is representative samples of your work, collected over a period of time. Begin your portfolio by selecting an item that shows something new you learned in this chapter.

COOPERATIVE LEARNING ACTIVITY

Work in groups of three. Each person in the group must understand your solution and be able to explain it to any person in the class.

In the final round of the game of *Jeopardy*, the contestants have to decide how much money to wager to win the game. During a recent game, the scores were as follows.

Roberta Jackson	Felicia Gonzalez	Peter Thomas
\$4800	\$5200	\$2400

Decide how much each person should wager. Keep in mind that (1) only the person with the most money wins the money, (2) some of the contestants may not get the question right, and (3) you don't know how the other contestants will wager.

CHAPTER 1 SUMMARY AND REVIEW

VOCABULARY

Upon completing this chapter, you should be familiar with the following terms:

additive identity	22	18	open sentence
algebraic expression	8	13	order of operations
base	9	10	power
coefficient	28	8	product
element	19	19	replacement set
empty set	19	19	set
equation	19	27	simplest form
exponent	9	19	solution
factor	8	19	subset
formula	36	27	term
like terms	27	8	variable
multiplicative identity	22		

SKILLS AND CONCEPTS

OBJECTIVES AND EXAMPLES

Upon completing this chapter, you should be able to:

- translate verbal expressions into mathematical expressions **(Lesson 1-1)**

Write an algebraic expression for the verbal expression *twice a number decreased by 17.* $2n - 17$

REVIEW EXERCISES

Use these exercises to review and prepare for the chapter test.

Write an algebraic expression for each verbal expression. Use y as the variable.

1. the product of 8 and a number
2. twice the cube of a number

- write an expression containing identical factors as an expression using exponents **(Lesson 1-1)**

Write $m \cdot m \cdot m$ as an expression using exponents. m^3

Write as an expression using exponents.

3. $a \cdot a \cdot a \cdot a$
4. $15 \cdot x \cdot x \cdot x \cdot y \cdot y$

OBJECTIVES AND EXAMPLES

REVIEW EXERCISES

- use the order of operations to evaluate expressions **(Lesson 1-2)**

Evaluate $2x + 2y + z^2$ if $x = 7$, $y = 3$, and $z = 1.5$.

$$2x + 2y + z^2 = 2 \cdot 7 + 2 \cdot 3 + (1.5)^2 = 14 + 6 + 2.25 = 22.25$$

Evaluate each expression if $a = 5$, $b = 8$, $c = \frac{2}{3}$, $d = \frac{1}{2}$, and $e = 0.3$.

5. ab^2
6. $3ac - bd$

- solve open sentences by performing arithmetic operations **(Lesson 1-3)**

$$x = 6(7) + 2^4 = 42 + 16 = 58$$

Solve each equation.

7. $a = 29 - 5^2$
8. $5(6) - 3(5) = y$
9. $w = (0.2)(8 + 3)$
10. $m = \frac{2}{3}\left(3 - \frac{1}{2}\right)$

- recognize and use the properties of identity and equality **(Lesson 1-4)**

Name the property illustrated by $(3 + 4) + 2 = 7 + 2$.

substitution property of equality

Name the property illustrated by each statement.

11. $7 + 0 = 7$
12. $2(1) = 2$
13. If $a + b = 5$, then $5 = a + b$.

- use the distributive property to simplify expressions **(Lesson 1-5)**

$$5b + 3(b + 2) = 5b + 3 \cdot b + 3 \cdot 2 = 5b + 3b + 6 = 8b + 6$$

Simplify.

14. $10x + x$
15. $2(a + b) + b$
16. $2(3a + 2a + b)$
17. $9(r + s) - 2s$

- recognize and use the commutative and associative properties when simplifying expressions **(Lesson 1-6)**

Name the property illustrated by $2b + (3c + 1) = (2b + 3c) + 1$.

associative property of addition

$$2x + 4xy + 5x = 2x + 5x + 4xy = (2x + 5x) + 4xy = 7x + 4xy$$

Name the property illustrated by each statement.

18. $5(a + c) = 5(c + a)$
19. $10(ab) = (10a)b$
20. $4 + (x + y) = (4 + x) + y$
21. $a(b + c) = (b + c)a$

Simplify.

22. $6a + 7b + 8a + 2b$
23. $\frac{3a^2}{4} + \frac{2ab}{3} + ab$

OBJECTIVES AND EXAMPLES	REVIEW EXERCISES
▪ translate verbal sentences into equations or formulas **(Lesson 1-7)** Write an equation for the sentence *The sum of the square of x and b is equal to a.* $x^2 + b = a$	**Write an equation for each sentence.** 24. Eighteen decreased by the square of d is equal to f. 25. The number c equals the cube of the product of 2 and x.

APPLICATIONS AND CONNECTIONS

Geometry: Find the perimeter of each figure when $x = 7$, $y = 3$, and $z = 1\frac{1}{2}$. (Lesson 1-2)

26.

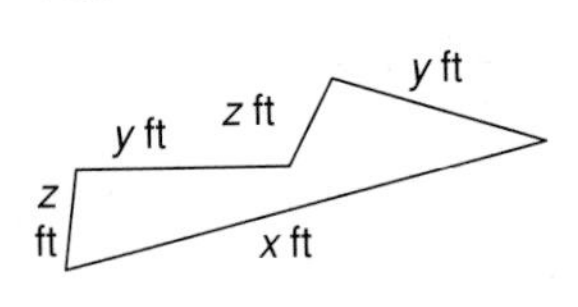

27.

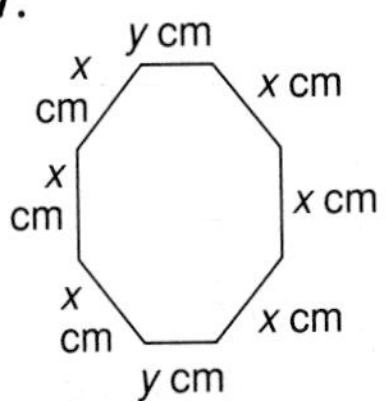

28.

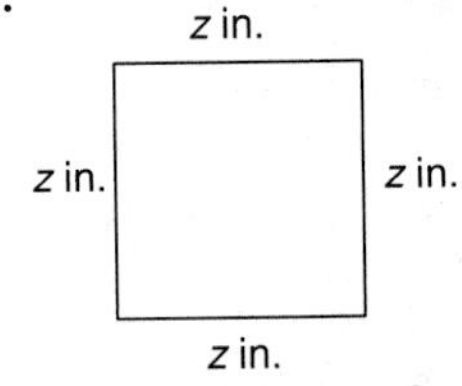

Logic: Write the negation of each statement. Then state whether the negation is *true* or *false*. (Lesson 1-3)

29. Ostriches cannot fly.

30. Delaware is the largest state in the U.S.

Two cans of vegetables together cost \$1.08. One of them costs 10¢ more than the other. (Lesson 1-8)

31. Would 2 cans of the less expensive vegetable cost more or less than \$1.08?
32. How much would 3 cans of each cost?
33. How much is the more expensive vegetable?

34. **Time** Write an expression to represent the number of seconds in a day. Then evaluate the expression. **(Lessons 1-1, 1-2)**

35. **Foods** There are 111 calories in one cup of orange juice. Write an equation to represent the number of calories in 3 cups of orange juice. Then solve the equation. **(Lesson 1-7)**

36. **Sports** A pitcher's earned-run average is given by the formula $ERA = 9\left(\frac{a}{b}\right)$, where a is the number of earned runs the pitcher has allowed and b is the number of innings pitched. Find a pitcher's ERA if he has allowed 23 earned runs in 180 innings. **(Lesson 1-7)**

37. **Masonry** What is the maximum number of whole bricks in a wall 25 inches wide and 7 inches high if each brick is 8 inches wide by 2 inches high and there is $\frac{1}{2}$ inch of mortar between each brick? **(Lesson 1-8)**

CHAPTER 1 TEST

Write an algebraic expression for each verbal expression. Use x as the variable.

1. a number increased by 17
2. twice the square of a number
3. the sum of a number and its cube
4. twice Nica's age decreased by 23 years

Evaluate.

5. $(12 - 10)^4$
6. $13 + 4 \cdot 5^2$
7. $0.7(1.4 + 0.6)$
8. $\frac{3}{4}(8 + 28)$
9. $23 - 12(1.5)$
10. $6 + 3(3.4)$

Evaluate if $m = 8$, $n = 3$, $p = \frac{3}{4}$, $q = \frac{2}{3}$, and $r = 0.5$.

11. $(mn)^2$
12. pq^2
13. $n + r^2$

Solve each equation.

14. $v = \frac{6^2 - 2^3}{7}$
15. $8(0.03) - 0.05 = y$
16. $\frac{3}{4} - \left(\frac{1}{2}\right)^2 = k$

Name the property illustrated by each statement.

17. $t + 0 = t$
18. $7(m + 2n) = 7(2n + m)$
19. $34 \cdot 1 = 34$
20. $3(2a + b) = 6a + 3b$
21. If $m = a + b$ then $a + b = m$.
22. $3(2) = 2(3)$

Simplify.

23. $n + 5n$
24. $2.5x - x + y + 3.5y$
25. $3(a + 2) + 5a$
26. $4an + \frac{2}{3}am + 8an + \frac{1}{3}am$

Write an equation for each sentence.

27. The product of π and the square of r is A.
28. The sum of a, b, and c is equal to P.
29. If a car travels at an average speed of 60 miles per hour, how far can the car travel in 4 hours?
30. Mr. Greenpeas is planning a garden 6 feet wide by 12 feet long. If each plant takes up 1 square foot of space, how many plants can Mr. Greenpeas plant in his garden?
31. Kenny has 4 pairs of socks in each of the following colors: black, brown, and navy. Write an equation to represent the number of socks Kenny has. Then, solve the equation.

Bonus **Simplify.**

$$x = \frac{4 + (2 + 6)^2 - \frac{2}{3} + 6 \div 2 + 9^2}{\frac{3}{4} - \frac{2}{3} + 7^2 - 4^2 + 3^2}$$

CHAPTER OBJECTIVES

In this chapter, you will:

- Simplify expressions.
- Add, subtract, multiply, divide, and compare rational numbers.
- Graph inequalities.
- Write equations for verbal problems.

Rational Numbers

APPLICATION IN RECREATION

At one time or another, almost everyone in the U.S. has played Monopoly®. It's no wonder that the game has become so popular since its invention by Charles Darrow in the 1930s. Playing Monopoly® lets a person pretend to direct a long business career in just a few hours. The players all start out even, and each one gets the same number of turns. Their success at the game will depend upon their strategy, a good memory, *and* a bit of luck.

When you play Monopoly®, what is your favorite playing piece? the top hat? the car? Or do you have your own playing piece, like a lucky coin?

In order to play Monopoly®, we must know how to move the playing pieces around the board. To do this, we need to add the numbers on the dice and then move our playing piece the corresponding number of spaces forward. We also need to know how to move our playing piece *backwards* when directed to do so. Thus, we can only play and enjoy Monopoly® by understanding how to add and subtract numbers.

ALGEBRA IN ACTION

Elyse, Tom, Emilio, and Jean have just begun a game of Monopoly®. The chart below shows each of their first three turns. Who is farther around the board after three turns? How can you represent each player's moves using integers?

Player	Turn 1	Turn 2	Turn 3
Elyse	Rolls double 5s; rolls 9.	Rolls 6.	Rolls 11; Chance - go back 3.
Tom	Rolls 3.	Rolls 4; Chance - Advance to Illinois Ave. (forward 17).	Rolls 6 - Go directly to jail (back 20).
Emilio	Rolls 6.	Rolls double 3s; rolls 7.	Rolls 8.
Jean	Rolls 8.	Rolls 9; Community Chest - Advance to next utility (forward 11).	Rolls double 2s; rolls 3.

2-1 Integers and the Number Line

Objectives

After studying this lesson, you should be able to:

- state the coordinate of a point on a number line,
- graph integers on a number line, and
- add integers by using a number line.

Application

The high temperature in Milwaukee on January 23 was 6°F. The following day, the high temperature was 9° lower. What was the high temperature on January 24?

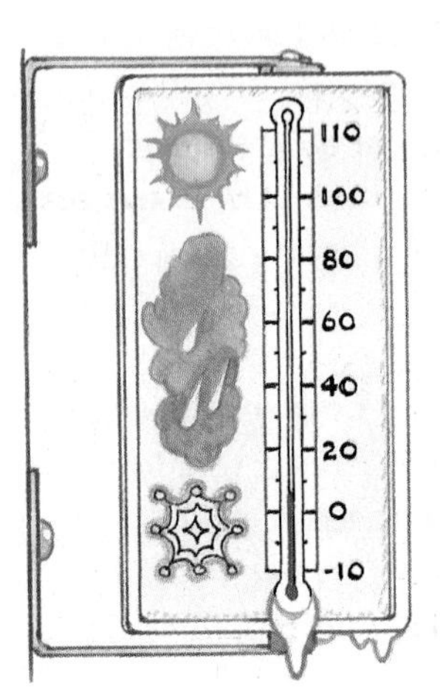

You can solve problems like the one above by using a **number line.** A number line is drawn by choosing a starting position on a line, usually 0, and marking off equal distances from that point. The set of **whole numbers** is often represented on a number line. This set can be written {0, 1, 2, 3, . . . }, where " . . . " means that the set continues indefinitely.

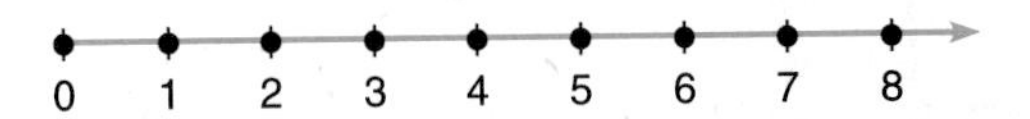

Although only a portion of the number line is shown, the arrowhead indicates that the line and the set of numbers continue.

We can use the expression $6 - 9$ to represent the problem given above. The number line below shows that the value of $6 - 9$ should be 3 less than 0. However, there is no whole number that corresponds to 3 less than 0. You can write the number *3 less than 0* as -3. This is an example of a **negative** number.

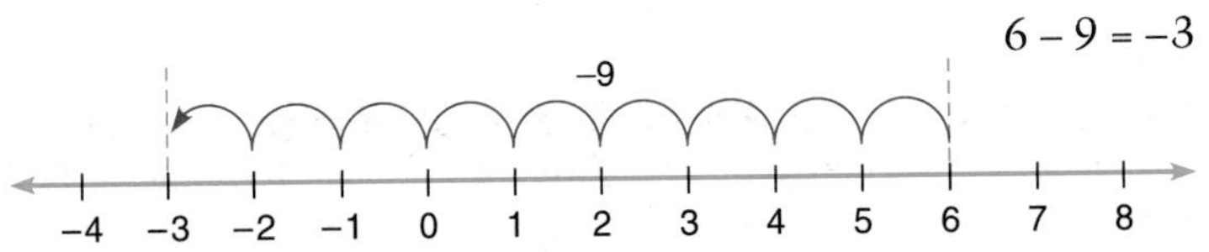

Any nonzero number written without a sign is understood to be positive.

To include negative numbers on a number line, extend the line to the left of zero and mark off equal distances. To avoid confusion, name the points to the right of zero using the *positive sign* (+) and to the left of zero using the *negative sign* (−). Zero is neither positive nor negative.

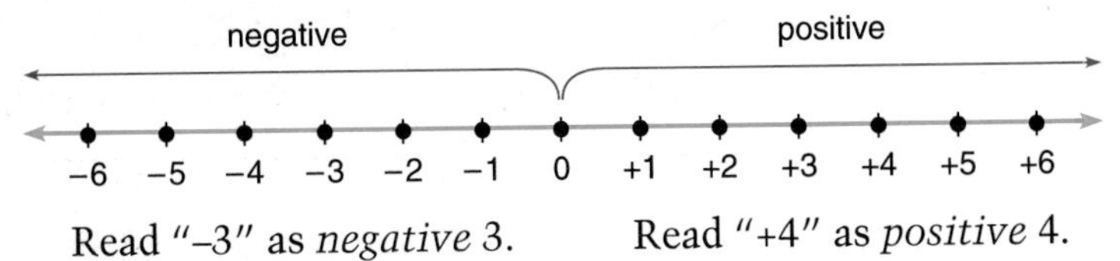

Read "–3" as *negative* 3. Read "+4" as *positive* 4.

The set of numbers used to name the points on the number line above is called the set of **integers.** This set can be written { . . . , −3, −2, −1, 0, 1, 2, 3, . . . }.

Sets of numbers are often represented by figures called **Venn diagrams.** The size and shape of a Venn diagram are unimportant. However, usually a rectangle is used to represent the *universal set,* and circles or ovals are used to represent subsets of the universal set.

Sets	Examples
Natural numbers	1, 2, 3, 4, 5, . . .
Whole numbers	0, 1, 2, 3, 4, . . .
Integers	. . . , −2, −1, 0, 1, 2, . . .

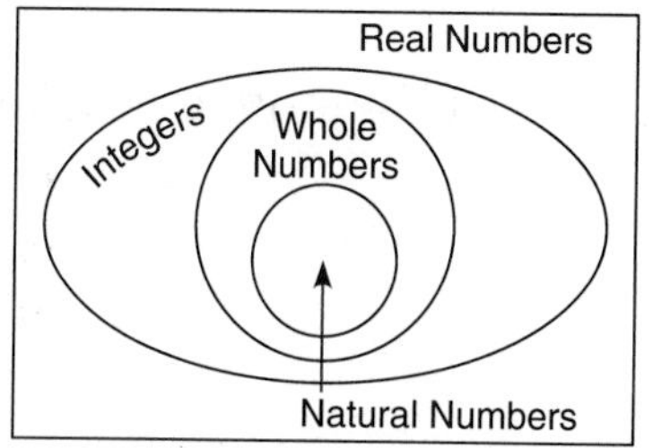

Notice that the natural numbers are a subset of the whole numbers, the whole numbers are a subset of the integers, and the integers are a subset of the universal set, the **real numbers.**

Venn diagrams can also show how various sets are related. Let A and B be two sets. The **intersection** of A and B ($A \cap B$) is the set of elements common to *both A and B.* The **union** of two sets A and B ($A \cup B$) is the set of all elements contained *either in A or in B or in both.*

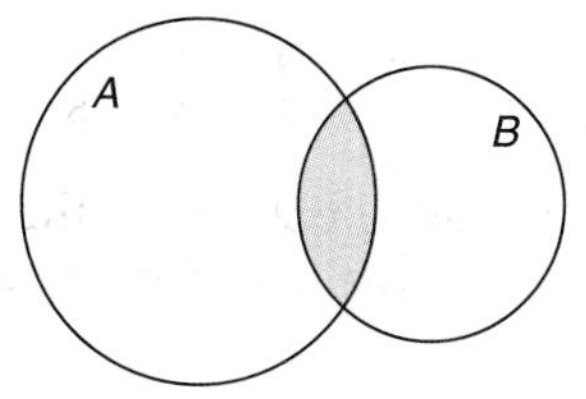

The shaded area is $A \cap B$.

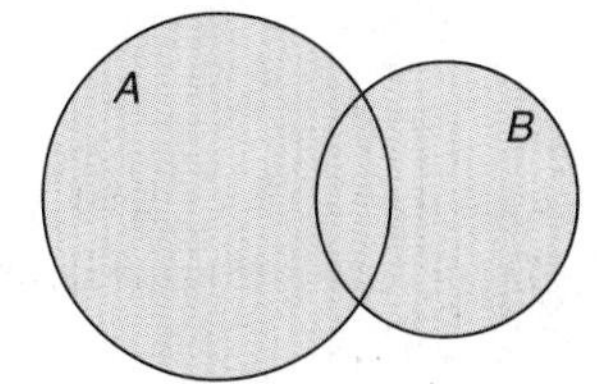

The shaded area is $A \cup B$.

Two sets that have no members in common are said to be **disjoint.** Their intersection is the **empty set, Ø.**

Example 1

Let $A = \{1, 2, 3, 4, 5\}$, $B = \{2, 4, 6, 8\}$, and $C = \{7, 8, 9\}$. Use a Venn diagram to show $A \cap B$, $A \cup B$, and $A \cap C$.

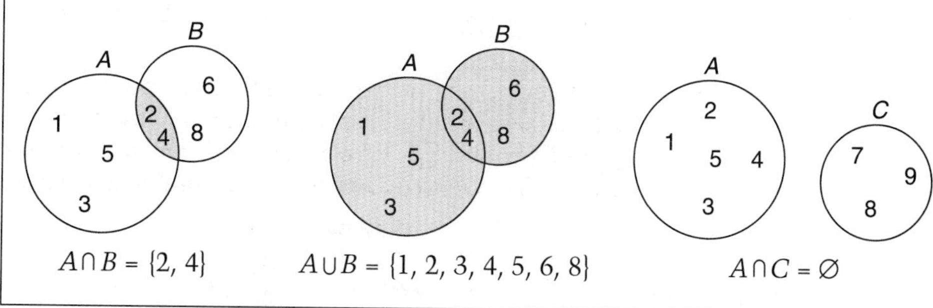

$A \cap B = \{2, 4\}$ $A \cup B = \{1, 2, 3, 4, 5, 6, 8\}$ $A \cap C = \varnothing$

To **graph** a set of numbers means to locate the points named by those numbers on a number line. The number that corresponds to a point on a number line is called the **coordinate** of that point.

Example 2

Name the set of numbers graphed.

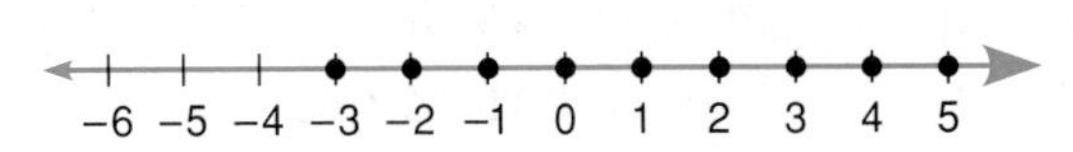

The bold arrow means that the graph continues indefinitely in that direction.

The set is $\{-3, -2, -1, 0, 1, 2, 3, 4, 5, \ldots\}$.

You can use a number line to add integers. To find the sum of -4 and 6, follow the steps below.

Step 1 Draw an arrow starting at 0 and going to -4, the first addend.

Step 2 Starting at -4, draw an arrow to the right 6 units long, the second addend.

Step 3 The second arrow ends at the sum, 2.

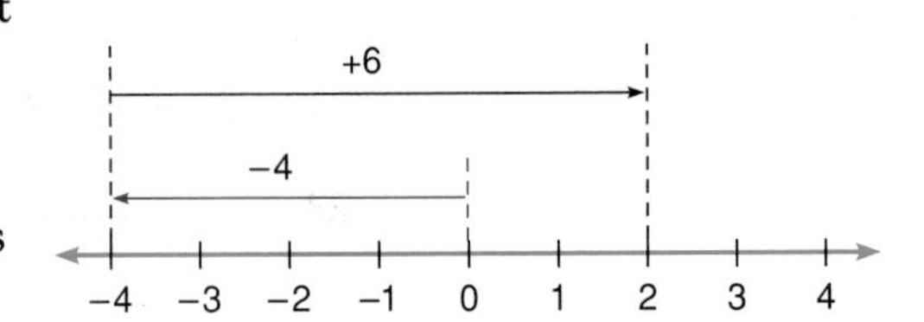

Example 3

Find $-2 + (-5)$ on a number line.

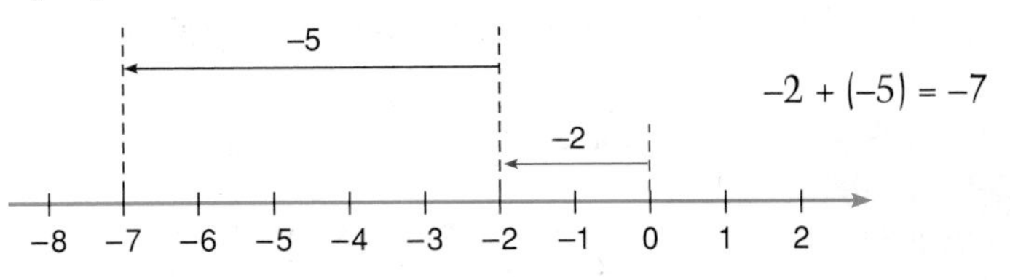

$-2 + (-5) = -7$

Notice that parentheses are used in the expression $-2 + (-5)$ so that the sign of the number is not confused with the operation symbol.

CHECKING FOR UNDERSTANDING

Communicating Mathematics

Read and study the lesson to answer each question.

1. What are three real-life situations where you would use negative integers?
2. Compare and contrast the intersection and union of sets.
3. How does the set of integers differ from the set of whole numbers?
4. Explain how to use a number line and arrows to find $4 + (-3)$. What is the sum?

Guided Practice

Name the coordinate of each point.

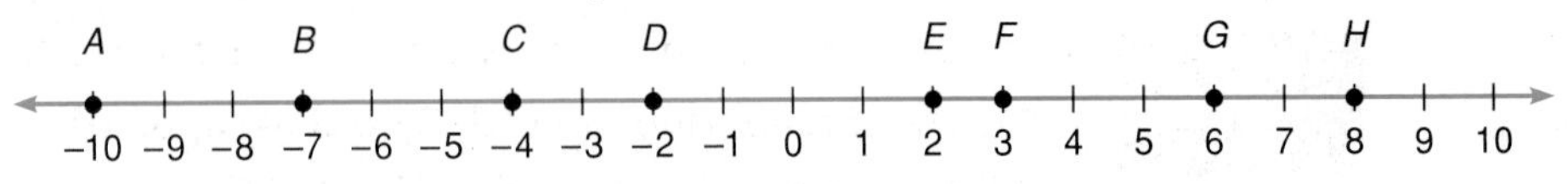

5. *B*
6. *H*
7. *A*
8. *D*
9. *E*
10. *G*
11. *C*
12. *F*

Use the Venn diagram to list the members of each set.

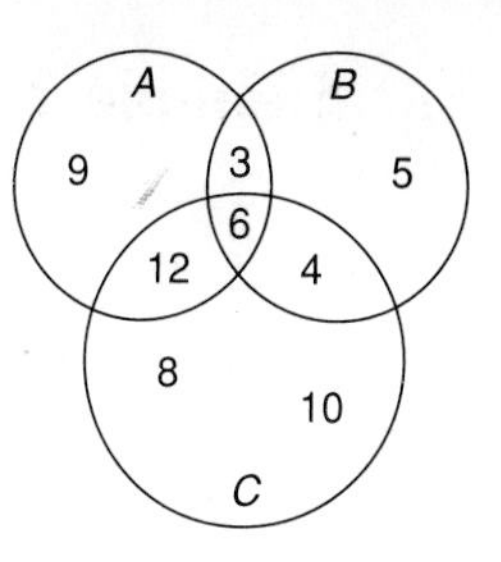

13. A
14. B
15. C
16. $A \cap B$
17. $A \cup B$
18. $B \cap C$
19. $B \cup C$
20. $A \cup B \cup C$

Write a corresponding addition sentence for each diagram.

21.

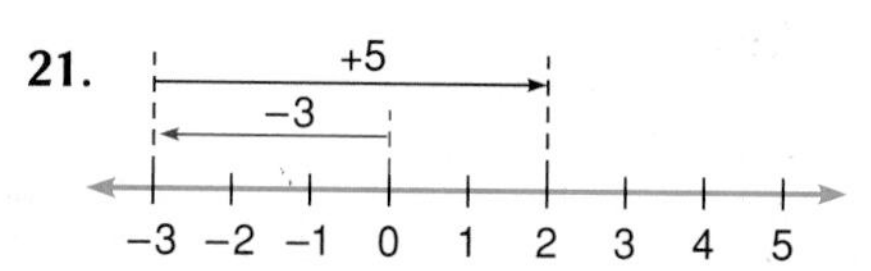

22.

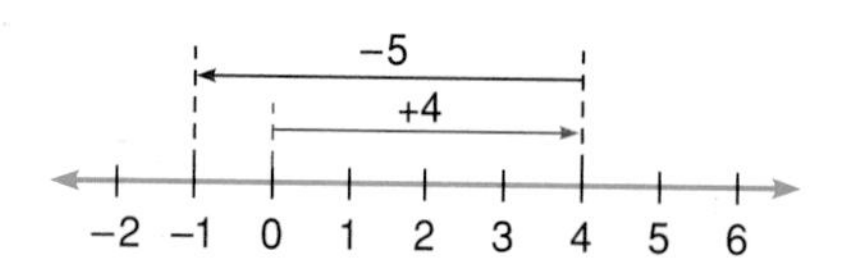

23.

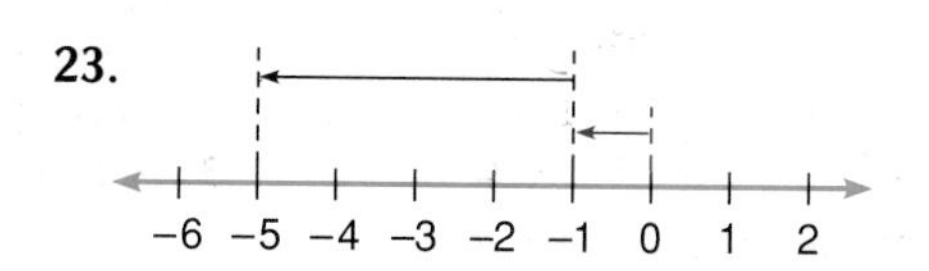

24.

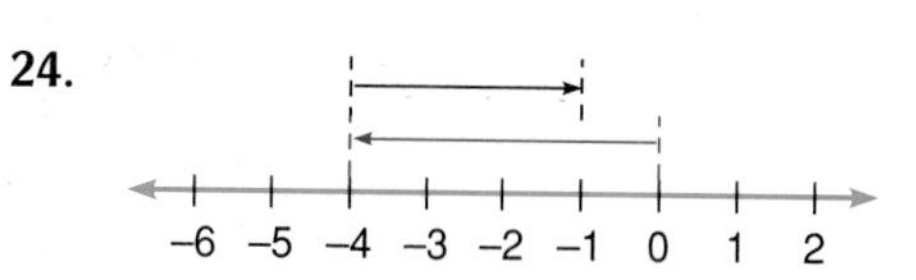

EXERCISES

Practice

Name the set of numbers graphed.

25.

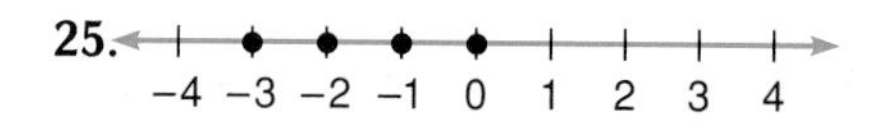

26. −2 −1 0 1 2 3 4 5 6

27. −4 −3 −2 −1 0 1 2 3 4

28. −4 −3 −2 −1 0 1 2 3 4

Graph each set of numbers on a number line.

29. $\{-1, 1, 3, 5\}$
30. $\{-2, -4, -6\}$
31. $\{-3, 0, 2\}$
32. $\{\ldots, -3, -2, -1, 0\}$
33. {integers greater than -2}
34. {integers less than 4 and greater than or equal to -2}

Find each sum. If necessary, use a number line.

35. $7 + 6$
36. $-9 + (-7)$
37. $-8 + (-12)$
38. $-5 + 0$
39. $-9 + 4$
40. $6 + (-13)$
41. $4 + (-4)$
42. $9 + (-5)$
43. $-3 + 9$
44. $-11 + 11$
45. $2 + (-7)$
46. $-11 + 10$
47. $9 + (-4)$
48. $-6 + (-14)$
49. $-13 + (-9)$

Draw a Venn diagram to find each of the following.

50. the intersection of {vowels} and {first seven letters of the alphabet}
51. the intersection of {1, 3, 5} and {4, 6, 8, 10}
52. the union of {a, b, c, d} and {14, 15}
53. the union of {letters in the word *dictionary*} and {letters in the word *nation*} and {letters in the word *tidal*}

Critical Thinking

Find the next three numbers in each pattern.

54. 33, 25, 17, ?, ?, ?

55. −21, −16, −11, ?, ?, ?

Applications

For Exercises 56 and 57, write an open sentence using addition. Then solve each problem.

56. **Oceanography** A scuba diver was exploring mineral deposits at a depth of 75 meters. She went up 30 meters. At what depth was she then?

57. **Football** A football team gains 4 yards on the first down and loses 20 yards on the second down. What was the net gain or loss after the two plays?

58. **School** Thirty students are in the French club. Twenty-one students are in the drama club. Six are in both. Make a Venn diagram of this situation. How many are in at least one club?

Mixed Review

59. Write $\frac{3}{4} \cdot x \cdot y \cdot y \cdot y \cdot y \cdot y$ as an expression using exponents. **(Lesson 1-1)**

60. Solve $y = 7 - 4$. **(Lesson 1-3)**

61. Simplify $23y^2 + 32y^2$. **(Lesson 1-5)**

62. Write as an equation: *the sum of x and the square of a is equal to n.* **(Lesson 1-7)**

APPLICATION

Meteorology

Windchill factor is an estimate of the cooling effect the wind has on a person in cold weather. The chilling effect of cold increases as the speed of the wind increases. Meteorologists use a chart like the one at the right to predict the windchill factor.

Wind speed in mph	Actual temperature (°Fahrenheit) 30	20	10	0	−10	−20	−30
	Equivalent temperature (°Fahrenheit)						
0	30	20	10	0	−10	−20	−30
5	27	16	6	−5	−15	−26	−36
10	16	4	−9	−21	−33	−46	−58
15	9	−4	−18	−31	−45	−58	−72

Use the windchill chart to find each windchill factor.

1. 30°F, 10 mph
2. 20°F, 5 mph
3. 0°F, 15 mph

2-2 Adding and Subtracting Integers

Objectives

After studying this lesson, you should be able to:

- find the absolute value of a number,
- add integers without using a number line, and
- subtract integers.

Application

In the first half of 1992, Skillplex Incorporated's expenses were \$713,000 under budget. In the second half, their expenses were \$425,000. How close to the 1992 budget were Skillplex's expenses?

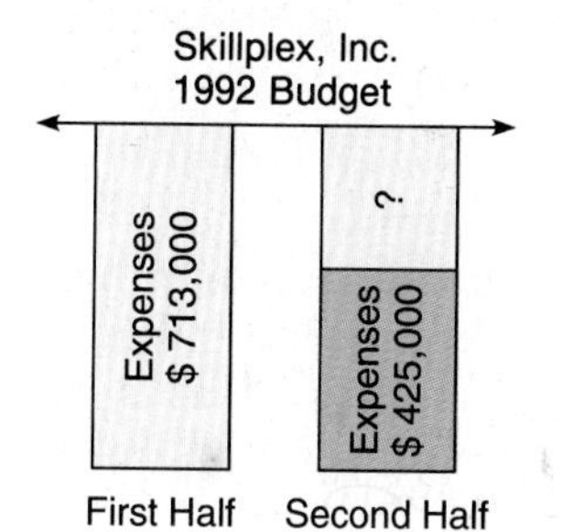

To find how close Skillplex was to the budget, we would add −\$713,000 and +\$425,000. But it would be really hard to draw a number line big enough to find sums like $-713{,}000 + 425{,}000$! Some concepts that we can illustrate on a number line will help us find easier ways to add and subtract integers.

Looking at 4 and −4 on a number line, you can see that they are the same number of units from 0. Two numbers that are the same distance from 0 on a number line are said to have the same **absolute value.**

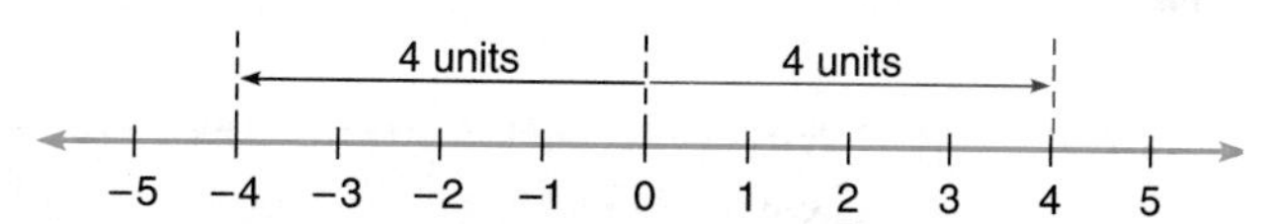

Definition of Absolute Value

The absolute value of a number is the number of units it is from 0 on a number line.

The symbol for the absolute value of a number is two vertical bars around the number.

$|-14| = 14$ is read *The absolute value of −14 is 14.*

$|-8 + 7| = 1$ is read *The absolute value of the quantity −8 + 7 is 1.*

Note that absolute value bars can serve as grouping symbols.

Example 1

Evaluate the expression $-|m + 4|$ if $m = -6$.

$$\begin{aligned} -|m + 4| &= -|-6 + 4| && \textit{Substitution property of equality} \\ &= -|-2| \\ &= -(2) && \textit{The absolute value of } -2 \textit{ is 2.} \\ &= -2 \end{aligned}$$

You can use absolute value to find sums of integers.

$3 + 2 = 5$

Notice that the sign of each addend is positive. The sum is positive.

$-3 + (-2) = -5$

Notice that the sign of each addend is negative. The sum is negative.

These examples suggest the following rule.

Adding Integers with the Same Sign	**To add integers with the same sign, add their absolute values. Give the result the same sign as the integers.**

Example 2

Find the sum: $-5 + (-7)$.

$-5 + (-7) = -(|-5| + |-7|)$ *Both numbers are negative, so the sum is negative.*
$= -(5 + 7)$ *Add the absolute values.*
$= -12$

You can use absolute value to add integers with different signs.

$4 + (-9) = -5$

Notice that $9 - 4$ is 5. Which integer, 4 or -9, has the greater absolute value? What is its sign? What is the sign of the sum?

$-5 + 8 = 3$

Notice that $8 - 5$ is 3. Which integer, -5 or 8, has the greater absolute value? What is its sign? What is the sign of the sum?

These examples suggest the following rule.

Adding Integers with Different Signs	**To add integers with different signs, subtract the lesser absolute value from the greater absolute value. Give the result the same sign as the integer with the greater absolute value.**

In golf, a score of 0 is called *even par.* A score of 4 under par is written as -4. A score of 1 over par is written as $+1$ or 1.

Example 3

In two consecutive rounds of golf, Nancy Lopez shot 3 over par and 4 under par. Calvin Peete shot 5 over par and 1 under par. Which player had the better score at the end of the second round?

Lopez: $+3 + (-4) = -(|-4| - |3|)$ *-4 has the greater absolute value, so the sum is negative.*
$= -(4 - 3)$
$= -1$ Lopez is 1 under par.

Peete: $5 + (-1) = +(|5| - |-1|)$ *5 has the greater absolute value, so the sum is positive.*
$= +(5 - 1)$
$= 4$ Peete is 4 over par. So, Lopez had the better score.

FYI . . .

Before 1850, golf balls were made of leather and stuffed with feathers.

What is the result when you add two numbers like 3 and -3? If the sum of two numbers is 0, the numbers are called **additive inverses** or **opposites.**

Zero is its own opposite.

-3 is the additive inverse, or opposite, of 3. $\quad -3 + 3 = 0$

7 is the additive inverse, or opposite, of -7. $\quad 7 + (-7) = 0$

These examples suggest the following rule.

Additive Inverse Property

For every number a, $a + (-a) = 0$.

Lab Activity

You can learn how to use counters to add and subtract integers in Lab 2 on pages A3-A4.

Additive inverses are used when you subtract numbers.

Subtraction	*Addition*	*Subtraction*	*Addition*
$7 - 3 = 4$	$7 + (-3) = 4$	$8 - (-2) = 10$	$8 + 2 = 10$

additive inverses — *same result*

It appears that subtracting a number is equivalent to adding its additive inverse.

Subtracting Integers

To subtract a number, add its additive inverse.
For any numbers a and b, $a - b = a + (-b)$.

The [+/-] key, called the **change-sign key**, changes the sign of the number on the display.

Example 4

Find the difference: $-4 - (-7)$.

Method 1

To subtract -7, add $+7$.

$$-4 - (-7) = -4 + (+7) = 3$$

Method 2

Use a calculator.

Enter: 4 [+/-] [−] 7 [+/-] [=] 3

Example 5

Simplify $7x - 12x$.

$7x - 12x = 7x + (-12x)$ *To subtract 12x, add −12x.*

$= [7 + (-12)]x$ *Distributive property*

$= -5x$

CHECKING FOR UNDERSTANDING

Communicating Mathematics

Read and study the lesson to answer each question.

1. Use an integer to show how far under budget Skillplex's expenses were for 1992.
2. In Example 3, by how many strokes is Lopez leading?
3. Have a friend write an integer on a piece of paper. Without knowing what it is, explain how to find its absolute value without a number line.
4. If you add a positive integer and a negative integer, how do you determine whether the sum is positive or negative?

Guided Practice

State the additive inverse and absolute value of each integer.

5. $+8$
6. -24
7. 0

State the sign of each sum.

8. $-6 + (-11)$
9. $-8 + (+16)$
10. $-27 + 31$

State the number named.

11. $-(-5)$
12. $-(+7)$
13. $-(4)$

Find each sum or difference.

14. $-13 + (-8)$
15. $-10 + 4$
16. $-21 - (-14)$

EXERCISES

Practice

Find each sum or difference.

17. $18 + 22$
18. $-6 + (-13)$
19. $-3 + (+16)$
20. $27 - 19$
21. $8 - 13$
22. $17 - (-23)$
23. $14 + (-9)$
24. $-5 + 31$
25. $-18 + (-11)$
26. $0 - 21$
27. $19m + 12m$
28. $8h - 23h$
29. $-25 + 47$
30. $-104 + 16$
31. $97 + (-79)$
32. $-18p - 4p$
33. $24b - (-9b)$
34. $41y - (-41y)$
35. $4 + (-12) + (-18)$
36. $7 + (-11) + 32$

Evaluate each expression if $a = -5$, $k = 3$, and $m = -6$.

37. $a + 13$
38. $k - 7$
39. $15 + m$
40. $k + (-18)$
41. $m + 6$
42. $|m|$
43. $|7 + a|$
44. $|m - 4|$
45. $|a + k|$
46. $|k| + |m|$
47. $-|-24 + m|$
48. $-|a + (-11)|$

Find each sum or difference.

49. $|-285 + (-641)|$
50. $|-931| - (-643)$
51. $-|-423 - (-148)|$
52. $-||-843| + |-231||$

Critical Thinking

Complete.

53. If $n > 0$, then $|n| = \underline{\ ?\ }$.
54. If $n < 0$, then $|-n| = \underline{\ ?\ }$.

Applications

For each problem, write an open sentence using addition or subtraction. Then solve the problem.

55. **Business** Ken Martin purchased a share of stock for $30. The next day the price of the stock dropped $6. On the third day, the price rose by $15. How much was Ken's stock worth at the end of the third day?

56. **Mining** The entrance to a silver mine is 4275 feet above sea level. The mine is 6324 feet deep. What is the elevation at the bottom of the mine?

Computer

57. This program creates patterns of addition. Use the patterns to study and review addition of integers. Enter each addend and the steps between the addends from one sentence to the next. For example, to create the sentences $3 + (-4) = -1$, $3 + (-2) = 1$, and $3 + 0 = 3$, enter 3, 0 (because 3 is the first addend and it will not change) and then enter -4, 2 (because -4 is the second addend and it will increase by 2 each time).

```
10 PRINT "ENTER THE FIRST
   ADDEND AND STEP"
20 INPUT A1, S1
30 PRINT "ENTER THE SECOND
   ADDEND AND STEP"
40 INPUT A2, S2
50 FOR I = 1 TO 10
60 LET X1=A1+(I-1)*S1
70 LET X2=A2+(I-1)*S2
80 PRINT X1;"+";X2;"=";
   X1+X2
90 NEXT I
100 END
```

Use the program to find each set of sums.

a. $-1 + 3$
$-3 + 0$
$-5 + (-3)$
etc.

b. $-20 + 5$
$-15 + 10$
$-10 + 15$
etc.

c. $-12 + 7$
$-11 + 4$
$-10 + 1$
etc.

d. How are the steps between the addends related to the steps between the sums?

Mixed Review

58. Evaluate $0.2(0.6) + 3(0.4)$. **(Lesson 1-2)**

59. Solve $\frac{14 + 28}{4 + 3} = y$. **(Lesson 1-3)**

60. Simplify $\frac{3}{4}y + \frac{x}{4} + 3x$. **(Lesson 1-5)**

61. Name the property illustrated by $a + (2b + 5c) = (a + 2b) + 5c$. **(Lesson 1-6)**

62. Graph $\{-3, -2, 2, 3\}$ on a number line. **(Lesson 2-1)**

63. Find $-3 + 12$ on a number line. **(Lesson 2-1)**

PUZZLE

Magic squares are square arrays of numbers. The sum of the numbers in each row, column, and diagonal is the same.

1	2	−3
−4	0	4
3	−2	−1

The sum of the numbers in each row, column, and diagonal of this magic square is 0.

New magic squares can be formed by adding the same number to each entry in the magic square above. In the squares below, how was each entry obtained? Find each new magic square.

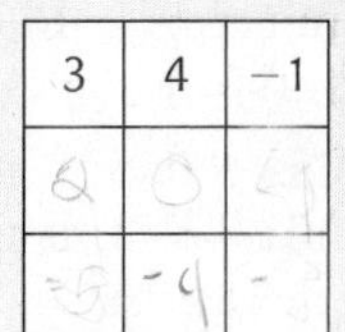

−2		
	−3	
		−4

	−7	

2-3 Inequalities and the Number Line

Objectives

After studying this lesson, you should be able to:

- compare numbers,
- write inequalities for graphs on number lines, and
- graph inequalities on number lines.

Application

The Lincoln High School hockey team played ten games during its regular season. During the season, each win is worth 2 points, each tie is worth 1 point, and each loss is 0 points. To qualify for the playoffs, a team must have more than 14 points. If Lincoln's team finished the regular season with 6 wins, 3 ties, and 1 loss, did they qualify for the playoffs?

You can solve this problem by determining if Lincoln's team received more than 14 points. Since $6(2) + 3(1) + 1(0) = 12 + 3 + 0$ or 15, and 15 is *more than* 14, Lincoln's team qualified for the playoffs.

The statement *15 is more than 14* can be symbolized in two ways.

$15 > 14$, which means 15 *is greater than* 14.
$14 < 15$, which means 14 *is less than* 15.

Note that the $<$ and $>$ point to the lesser number.

A mathematical sentence that uses $>$ or $<$ to compare two expressions is called an **inequality**. If $>$, $<$, and $=$ are used to compare the same two numbers, then only one of the three sentences is true.

Comparison Property

For any two numbers a and b, exactly one of the following sentences is true.

$$a < b \qquad a = b \qquad a > b$$

Example 1

Replace ? with <, >, or = to make -14 ? $3 + (-20)$ a true sentence.

$-14 \ \underline{?}\ 3 + (-20)$
$-14 \ \underline{?}\ -17$ *Simplify.*

Since -14 is greater than -17, the true sentence is $-14 > 3 + (-20)$.

In Chapter 1 you learned that the reflexive, symmetric, and transitive properties hold for equality. Do these properties hold for inequalities? Check each property using specific numbers.

Reflexive:	Is $6 > 6$? *no*	Is $6 < 6$? *no*
Symmetric:	If $8 > 3$, is $3 > 8$? *no*	If $3 < 8$, is $8 < 3$? *no*
Transitive:	If $7 > 1$ and $1 > -4$, is $7 > -4$? *yes*	

Check to see if the transitive property is true for other examples.

From these and other examples, we can conclude that only the transitive property holds for $<$ and $>$.

Transitive Property of Order	**For all numbers *a*, *b*, and *c*,** **1. if $a < b$ and $b < c$, then $a < c$, and** **2. if $a > b$ and $b > c$, then $a > c$.**

The symbols $\neq$, $\leq$, and $\geq$ can also be used to compare numbers. The chart below shows several inequality symbols and their meanings.

Symbol	Meaning
$<$	is less than
$>$	is greater than
$\neq$	is not equal to
$\leq$	is less than or equal to
$\geq$	is greater than or equal to

Example 2

Is $5.1 \leq -2.4 + 8.3$ *true* or *false*?

$5.1 \overset{?}{\leq} -2.4 + 8.3$ *Is 5.1 less than or equal to −2.4 + 8.3?*

$5.1 \leq 5.9$ *Simplify.*

Since 5.1 is less than 5.9, the sentence is true.

Consider the graphs of -3, -1, $2\frac{1}{2}$, and 4.5 shown on the number line below.

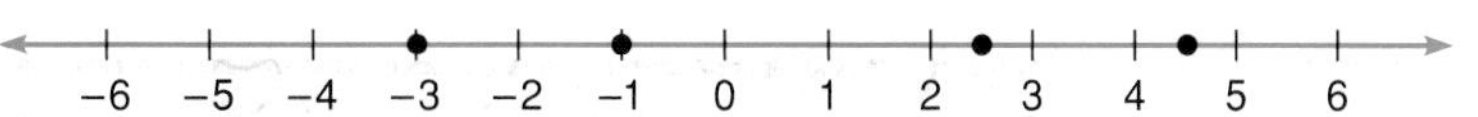

The following statements can be made about the numbers and their graphs.

The graph of -3 is to the left of the graph of -1.	$-3 < -1$
The graph of -1 is to the left of the graph of $2\frac{1}{2}$.	$-1 < 2\frac{1}{2}$
The graph of $2\frac{1}{2}$ is to the right of the graph of -3.	$2\frac{1}{2} > -3$
The graph of 4.5 is to the right of the graph of $2\frac{1}{2}$.	$4.5 > 2\frac{1}{2}$

Comparing Numbers on the Number Line	**If *a* and *b* represent any numbers and the graph of *a* is to the left of the graph of *b*, then $a < b$. If the graph of *a* is to the right of the graph of *b*, then $a > b$.**

Some points on a number line cannot be named by integers. The number line below is separated into fourths to show the graphs of some common fractions and decimals. The numbers graphed on this number line are examples of **rational numbers.**

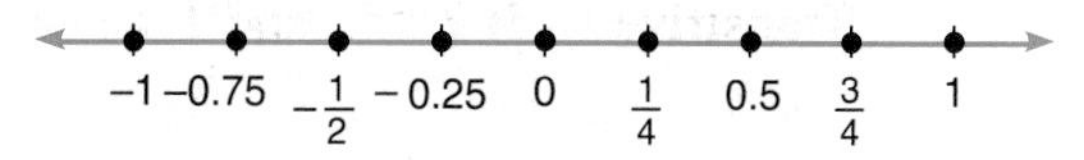

Definition of a Rational Number	**A rational number is a number that can be expressed in the form $\frac{a}{b}$, where a and b are integers and b is not equal to zero.**

Examples of rational numbers expressed in the form $\frac{a}{b}$ are shown below.

Notice that all integers are rational numbers.

Rational Number	3	$-2\frac{3}{4}$	0.125	0	$0.33\overline{3}$
Form $\frac{a}{b}$	$\frac{3}{1}$	$-\frac{11}{4}$	$\frac{1}{8}$	$\frac{0}{1}$	$\frac{1}{3}$

Recall that equations like $x - 4 = 9$ are open sentences. Inequalities like $x < 5$ are also considered to be open sentences. To find the solution set of $x < 5$, determine what replacements for x make $x < 5$ true. All numbers less than 5 make the inequality true. This can be shown by the solution set {all numbers less than 5}. Not only does this include integers like 4, 0, and -3, but it also includes all of the rational numbers less than 5. Examples of rational numbers less than 5 are $\frac{1}{3}$, $-4\frac{2}{5}$, and -2.

Example 3

Graph the solution set of $x < 5$.

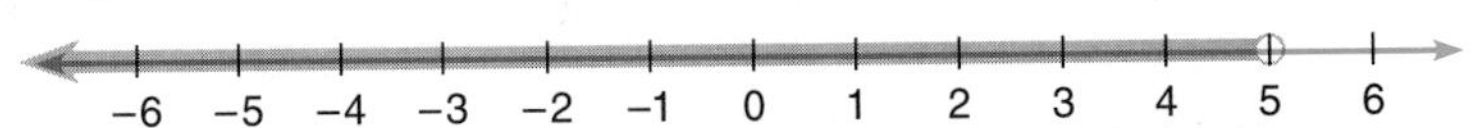

The heavy arrow indicates that all numbers to the left of 5 are included. The *circle* indicates that the point corresponding to 5 is *not* included in the graph of the solution set.

Example 4

Graph the solution set of $n \geq -\frac{3}{2}$.

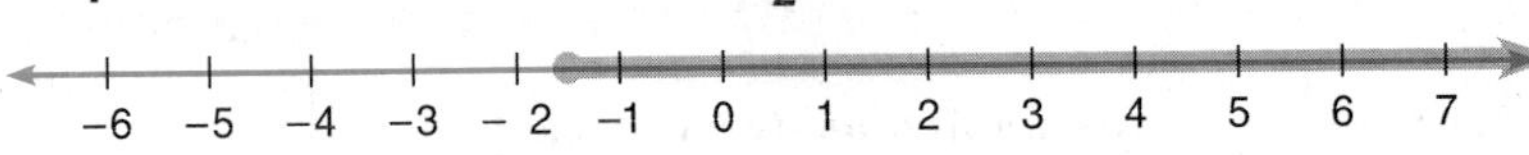

The solution set is $\left\{-\frac{3}{2}\text{ and all numbers greater than }-\frac{3}{2}\right\}$. The *dot* indicates that the point corresponding to $-\frac{3}{2}$ *is* included in the graph of the solution set.

Example 5

When cars are ordered from a certain distributor, the order must be for no more than 10 cars. Determine the number of cars that could be in an order from this distributor. Then graph the solution set.

An order from this distributor must be for no more than 10 cars. Since it is not possible to order a fractional number of cars or a negative number of cars, the number of cars in an order must be a whole number that is less than or equal to 10. Since a dealer would not place an order for 0 cars, the solution set is {1, 2, 3, 4, 5, 6, 7, 8, 9, 10}.

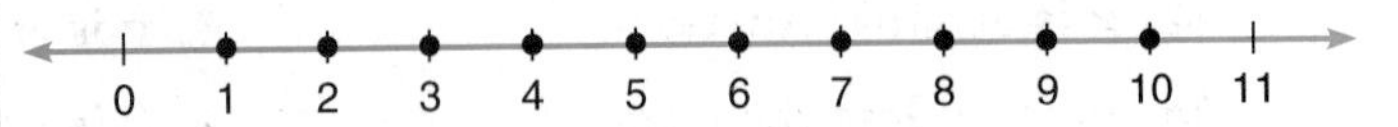

CHECKING FOR UNDERSTANDING

Communicating Mathematics

Complete.

1. If $a \leq b$, then a is less than b or a __?__ b.
2. A dot is used to indicate that a point is __?__ in a graph on a number line.
3. Numbers of the form $\frac{a}{b}$, where a and b are integers and $b \neq 0$, are called __?__.
4. The definition of a rational number states that b cannot be zero. Why must this condition be included in the definition?

Guided Practice

Determine whether each sentence is *true* or *false*.

5. $7 < 4$
6. $-3 \leq 3$
7. $-9 > -4$
8. $-5\frac{2}{3} \neq -6\frac{2}{3}$
9. $-4 < 2 - 8$
10. If $1 < 3$, then $1 > 3$.

If n is a whole number, write the solution set for each inequality.

11. $n > 3$
12. $n \leq 4$
13. $n \neq 6$

Determine whether each number is included on the graph below.

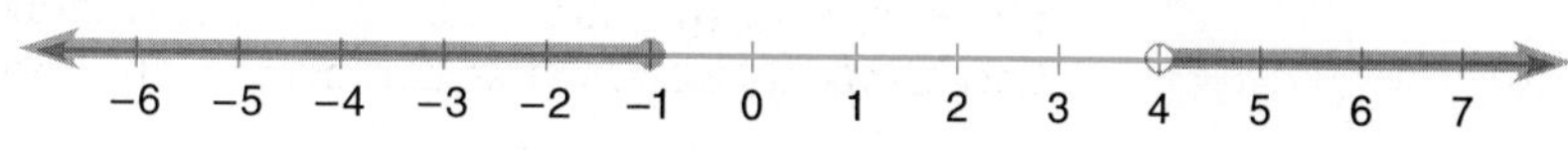

14. -1
15. 4
16. 4.1
17. $-1\frac{1}{2}$

EXERCISES

Practice

Write an inequality for each graph.

18. −4 −3 −2 −1 0 1 2 3 4 5

19. −6 −5 −4 −3 −2 −1 0 1 2 3

20. −5 −4 −3 −2 −1 0 1 2 3 4

21. −5 −4 −3 −2 −1 0 1 2 3 4

22. −4 −3 −2 −1 0 1 2 3 4 5

23.

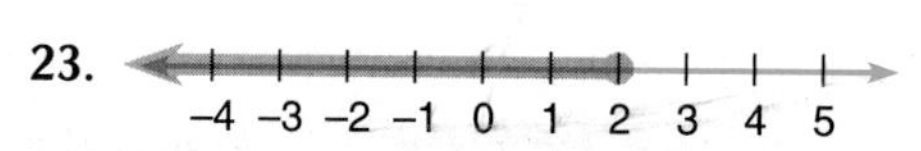

Graph the solution set of each inequality on a number line.

24. $n > 5$
25. $y < -2$
26. $x \neq -1$
27. $y \leq 6$
28. $x \geq -2$
29. $x < -10$
30. $a \geq -3$
31. $y \neq 3$

Write an algebraic expression for each verbal expression.

32. x is at most thirty.
33. y is at least negative 5.
34. m is no less than ten.
35. b is negative.

Replace each ___?___ with <, >, or = to make each sentence true.

36. $-5 \ \underline{?} \ 7$

37. $-2 \ \underline{?} \ -3$

38. $-7 - 2 \ \underline{?} \ -9$

39. $-5 \ \underline{?} \ 0 - 3$

40. $3 \ \underline{?} \ \frac{15}{3}$

41. $8 \ \underline{?} \ 4.1 + 3.9$

42. $5 \ \underline{?} \ 8.4 - 1.5$

43. $\frac{4}{3}(6) \ \underline{?} \ 4\left(\frac{3}{2}\right)$

44. $\frac{27.155}{3} \ \underline{?} \ 2(4.459)$

45. Draw a Venn diagram to show the relationship among the following sets of numbers: natural numbers, whole numbers, integers, rational numbers, and real numbers.

Critical Thinking

46. Three numbers x, y, and z satisfy the following conditions: $y - z < 0$, $x - y > 0$, and $z - x < 0$. Which one is the greatest?

47. If $x < y$ and $x < z$, what conclusion can be made about y and z?

Applications

48. **School** Less than half the students in Mrs. Chen's science class are boys. If there are 34 students in the class, what is the greatest number of boys that could be in the class?

49. **Population** At least one-fourth of the residents of Riverdale are over the age of 60. If Riverdale has 19,200 residents, what is the least number of residents that could be over the age of 60?

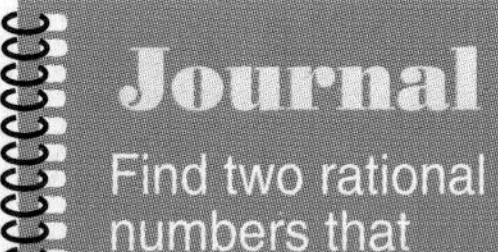

Find two rational numbers that have been printed in a newspaper or magazine. Write an inequality using the numbers. Explain how you know your inequality is correct.

50. **Armed Forces** A battalion of 880 people is made up of four companies. Each company is made up of five platoons. Each platoon contains four squads. If each company has the same number of people, as do each platoon and each squad, how many people are in each squad?

51. **History** In 1883, the original Metropolitan Opera House opened in New York City. Seven decades and two years later, Marian Anderson became the first African-American soloist to sing with the Metropolitan Opera. When did this event take place?

Mixed Review

52. Evaluate $5y + 3$ if $y = 1.3$. **(Lesson 1-2)**

53. Name the like terms: $\frac{m^2n}{2}$, $3mn$, $5m^2n$, $\frac{mn^2}{4}$. **(Lesson 1-5)**

54. Find the next three numbers in the pattern 1, 3, 9, 27, ___?___, ___?___, ___?___. **(Lesson 2-1)**

55. Write a corresponding addition sentence for the diagram shown at the right. **(Lesson 2-1)**

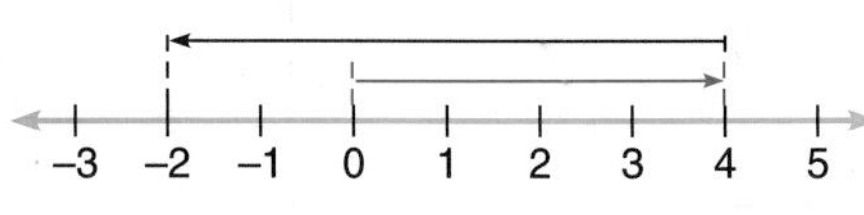

56. Find the sum: $82 + (-78)$. **(Lesson 2-2)**

57. Evaluate $a - 12$ if $a = -8$. **(Lesson 2-2)**

2-4 Comparing and Ordering Rational Numbers

Objectives

After studying this lesson, you should be able to:

- compare rational numbers,
- write rational numbers in increasing or decreasing order, and
- find a number between two rational numbers.

Application

Paul and Vivian work at a neighborhood fast-food restaurant after school. They each make $4.95 per hour and work 20 hours per week. Paul saves $\frac{3}{8}$ of his wages and Vivian saves $\frac{4}{11}$ of her wages. Who saves more money?

When two fractions are compared, the cross products are the products of the terms on the diagonals.

You know that $\frac{6}{9}$ is less than $\frac{7}{9}$ because the denominators are the same and 6 is less than 7. But how can you compare $\frac{3}{8}$ and $\frac{4}{11}$? You can use *cross products* to compare rational numbers.

Comparison Property for Rational Numbers

For any rational numbers $\frac{a}{b}$ and $\frac{c}{d}$, with $b > 0$ and $d > 0$:

1. if $\frac{a}{b} < \frac{c}{d}$, then $ad < bc$, and

2. if $ad < bc$, then $\frac{a}{b} < \frac{c}{d}$.

This property also holds if $<$ is replaced by $>$, $\leq$, $\geq$, or $=$.

Use this property to compare $\frac{3}{8}$ and $\frac{4}{11}$.

Find the product of 3 and 11. *Find the product of 8 and 4.*

$$\frac{3}{8} \;?\; \frac{4}{11}$$

$$3(11) \;?\; 8(4)$$

$$33 > 32$$

Since $33 > 32$, we can conclude that $\frac{3}{8} > \frac{4}{11}$. Thus, Paul is saving more money.

Example 1

Replace the ? with <, >, or = to make $\frac{6}{11}$? $\frac{7}{12}$ a true sentence.

Find the cross products.

$$\frac{6}{11} \;?\; \frac{7}{12}$$

$$6(12) \;?\; 11(7)$$

$$72 < 77$$

The true sentence is $\frac{6}{11} < \frac{7}{12}$.

Every rational number can be expressed as a terminating or repeating decimal. You can use a calculator to write rational numbers as decimals so they can be easily compared.

Example 2

Use a calculator to write the fractions $\frac{3}{5}$, $\frac{2}{3}$, and $\frac{6}{11}$ as decimals. Then write the fractions in order from greatest to least.

$\frac{3}{5} = 0.6$ *This is a terminating decimal.*

$\frac{2}{3} = 0.66666\ldots$ or $0.\overline{6}$ *This is a repeating decimal.*

$\frac{6}{11} = 0.545454\ldots$ or $0.\overline{54}$ *This is also a repeating decimal.*

Since $0.\overline{6} > 0.6$ and $0.6 > 0.\overline{54}$, $\frac{2}{3} > \frac{3}{5} > \frac{6}{11}$.

Example 3

Super Saver Mart advertised a 9.4-ounce tube of toothpaste for \$1.61. Is this a better buy than another brand's 6-ounce tube for 95¢?

Compare the cost per ounce, or **unit cost** of each tube. Let's assume that the quality of the two items is the same. Then the item with the lesser unit cost is the better buy.

$$\text{unit cost} = \text{total cost} \div \text{number of units}$$

Use a calculator to find the unit cost of each brand. In each case, the unit cost is expressed in cents per ounce.

unit cost of first brand: 1.61 [÷] 9.4 [=] 0.17127659

unit cost of second brand: 0.95 [÷] 6 [=] 0.15833333

Since 0.16¢ < 0.17¢, the 6-ounce tube for 95¢ is the better buy.

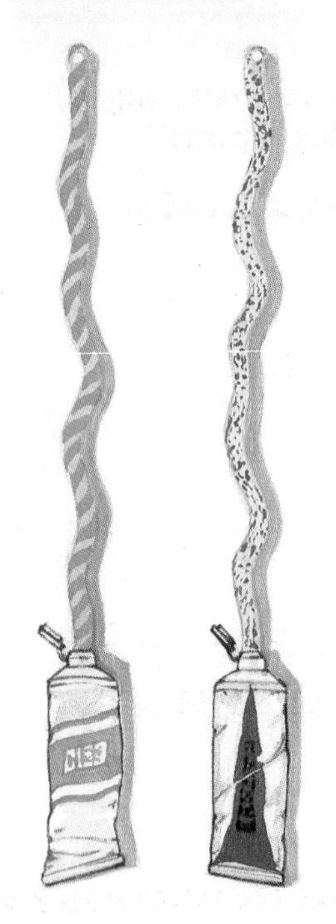

Given any two integers, can you always find another integer between them? Consider the integers 3 and 4. There is no integer between them. What about any two rational numbers? Can you always find another rational number between them? One point that lies between any two numbers is the point midway between them. The coordinates of the points on the number line below are $\frac{1}{3}$ and $\frac{1}{2}$.

To find the coordinate of the point midway between $\frac{1}{3}$ and $\frac{1}{2}$, find the average of the two numbers.

$$\text{average: } \frac{1}{2}\left(\frac{1}{3} + \frac{1}{2}\right) = \frac{1}{2}\left(\frac{5}{6}\right) \text{ or } \frac{5}{12}$$

This process can be continued indefinitely. The property described above is called the **density property.**

Density Property for Rational Numbers	**Between every pair of distinct rational numbers, there are infinitely many rational numbers.**

Example 4

Find a rational number between $\frac{3}{4}$ and $\frac{6}{7}$.

Find the average of the two rational numbers.

$$\frac{1}{2}\left(\frac{3}{4} + \frac{6}{7}\right) = \frac{1}{2}\left(\frac{21}{28} + \frac{24}{28}\right) = \frac{1}{2}\left(\frac{45}{28}\right) = \frac{45}{56}$$

Enter: 3 [÷] 4 [+] 6 [÷] 7

Enter: [=] [÷] 2 [=]

Display: 0.8035714

A rational number between $\frac{3}{4}$ and $\frac{6}{7}$ is $\frac{45}{56}$ or 0.804.

CHECKING FOR UNDERSTANDING

Communicating Mathematics

Read and study the lesson to answer each question.

1. On page 66, we discussed terminating decimals and repeating decimals. What is a nonterminating, nonrepeating decimal?
2. Explain how to find three rational numbers between $\frac{1}{5}$ and $\frac{1}{4}$.

Guided Practice

Name the greater rational number.

3. $\frac{3}{4}, \frac{4}{5}$
4. $\frac{11}{12}, \frac{7}{8}$
5. $\frac{9}{10}, \frac{10}{11}$
6. $\frac{7}{8}, \frac{8}{9}$
7. $\frac{6}{5}, \frac{7}{6}$
8. $\frac{11}{9}, \frac{12}{10}$

Which is the better buy?

9. a 21-ounce can of baked beans for 79¢ or a 28-ounce can for 97¢
10. a 10-ounce jar of coffee for $4.27 or an 8-ounce jar for $3.64

EXERCISES

Practice

Replace each ___?___ with <, >, or = to make each sentence true.

11. $\frac{6}{7}$ ___?___ $\frac{7}{8}$
12. $\frac{8}{7}$ ___?___ $\frac{9}{8}$
13. $\frac{7}{19}$ ___?___ $\frac{6}{17}$
14. $\frac{8}{15}$ ___?___ $\frac{9}{16}$
15. $\frac{5}{14}$ ___?___ $\frac{25}{70}$
16. $\frac{0.4}{3}$ ___?___ $\frac{1.2}{8}$

Write the fractions in order from least to greatest.

17. $\frac{17}{21}, \frac{20}{27}, \frac{19}{24}$

18. $\frac{17}{19}, \frac{32}{35}, \frac{45}{49}$

19. $\frac{3}{14}, \frac{5}{23}, \frac{9}{43}$

Find a number between the given numbers.

20. $\frac{1}{2}$ and $\frac{6}{7}$

21. $\frac{4}{7}$ and $\frac{9}{4}$

22. $\frac{2}{9}$ and $\frac{8}{11}$

23. $\frac{19}{30}$ and $\frac{31}{45}$

24. $\frac{7}{6}$ and $\frac{21}{18}$

25. $-\frac{3}{8}$ and $\frac{9}{10}$

Critical Thinking

26. Find the coordinates of B and C if the distances between each pair of points are equal.

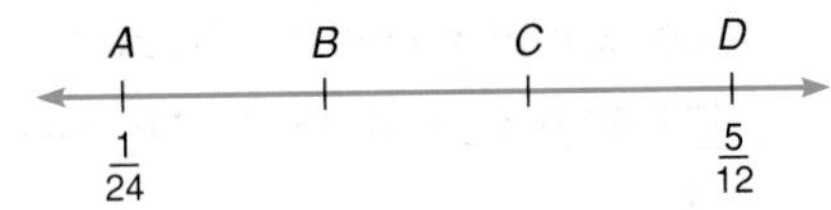

27. To find a number between $\frac{3}{4}$ and $\frac{6}{7}$, John added as follows.

$$\frac{3+6}{4+7} = \frac{9}{11}$$ *$\frac{9}{11}$ is between $\frac{3}{4}$ and $\frac{6}{7}$.*

He claimed that this method will always work. Do you agree?

Applications

Which is the better buy?

28. a half-pound bag of cashews for \$2.93 or a $\frac{3}{4}$-pound bag for \$4.19

29. three liters of soda for \$2.25 or two liters for \$1.69

30. a 27-ounce loaf of bread for 93¢ or a 20-ounce loaf for 79¢

31. a dozen oranges for \$1.59 or half a dozen oranges for 85¢

32. five pounds of green beans for \$3.50 or 2 pounds for \$1.38

33. a 48-ounce bottle of dishwashing liquid for \$2.39 or a 22-ounce bottle for \$1.09

Mixed Review

34. Write a mathematical expression for *seven more than r*. **(Lesson 1-1)**

Name the property illustrated by each statement.

35. $A(1) = A$ **(Lesson 1-4)**

36. $3 + 4 = 4 + 3$ **(Lesson 1-6)**

Replace the $\underline{\ ?\ }$ with <, >, or = to make each sentence true. (Lesson 2-3)

37. $12 \underline{\ ?\ } 15 - 27$

38. $(-7.502)(0.511) \underline{\ ?\ } -3.115$

2-5 Adding and Subtracting Rational Numbers

Objectives

After studying this lesson, you should be able to:

- add two or more rational numbers,
- subtract rational numbers, and
- simplify expressions that contain rational numbers.

Application

On Monday, stock in Paxtel Corporation rose $3\frac{1}{8}$ points. The next day, because of a rumor about a merger with another company, the stock dropped $1\frac{3}{4}$ points. What was the net change in the price of the Paxtel stock?

Rational numbers can be added using a number line. Consider the sum of $\frac{1}{4}$ and $-\frac{3}{8}$. First replace $\frac{1}{4}$ with $\frac{2}{8}$, because the number line is separated into eighths. Then add $\frac{2}{8}$ and $-\frac{3}{8}$.

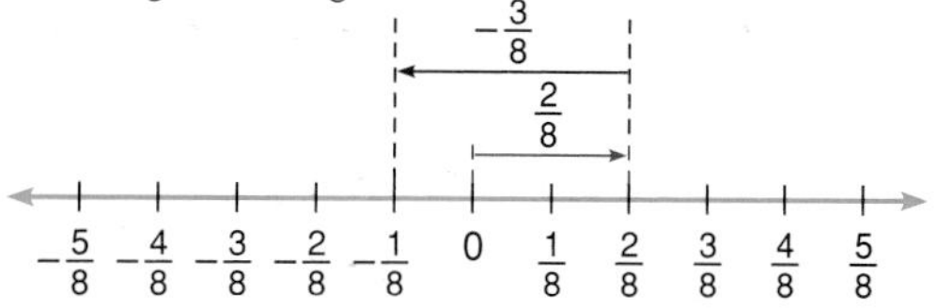

The sum of $\frac{1}{4}$ and $-\frac{3}{8}$ is $-\frac{1}{8}$.

Using a number line to show the addition of rational numbers is very inconvenient. However, the rules used to add integers can also be used to add rational numbers. If two rational numbers have the same sign, add their absolute values. Give the result the same sign as the addends. If two rational numbers have different signs, subtract the lesser absolute value from the greater absolute value. Give the result the same sign as the addend with the greater absolute value.

Example 1

Find the net change in the price of the Paxtel stock.

Find the net change by finding the sum of $3\frac{1}{8}$ and $-1\frac{3}{4}$.

$3\frac{1}{8} + \left(-1\frac{3}{4}\right) = 3\frac{1}{8} + \left(-1\frac{6}{8}\right)$ *Because the least common denominator is 8, replace $-1\frac{3}{4}$ with $-1\frac{6}{8}$.*

$= +\left(\left|3\frac{1}{8}\right| - \left|-1\frac{6}{8}\right|\right)$ *$3\frac{1}{8}$ has the greater absolute value. So, the sign of the sum is positive.*

$= +\left(3\frac{1}{8} - 1\frac{6}{8}\right)$ *Since the addends have different signs, find the difference of their absolute values.*

$= +\left(2\frac{9}{8} - 1\frac{6}{8}\right)$ *Replace $3\frac{1}{8}$ with $2\frac{9}{8}$. Then subtract.*

$= 1\frac{3}{8}$ *The net change is up $1\frac{3}{8}$ points.*

Example 2 Find the sum: $-1.354 + (-0.765)$.

$$-1.354 + (-0.765) = -(|-1.354| + |-0.765|)$$ *The numbers have the same sign.*
$$= -(1.354 + 0.765)$$ *Add their absolute values.*
$$= -2.119$$

When adding three or more rational numbers, you can use the commutative and associative properties to rearrange the addends.

Example 3 Find the sum: $-\frac{4}{3} + \frac{5}{8} + \left(-\frac{7}{3}\right)$.

$$-\frac{4}{3} + \frac{5}{8} + \left(-\frac{7}{3}\right) = \left[-\frac{4}{3} + \left(-\frac{7}{3}\right)\right] + \frac{5}{8}$$ *Commutative and associative properties of addition*
$$= -\frac{11}{3} + \frac{5}{8}$$ *The least common denominator is 24.*
$$= -\frac{88}{24} + \frac{15}{24}$$ $-\frac{11}{3} = -\frac{88}{24}$ *and* $\frac{5}{8} = \frac{15}{24}$
$$= -\frac{73}{24}$$
$$= -3\frac{1}{24}$$

Example 4 Find the sum: $28.32 + (-56.17) + 32.41 + (-75.13)$.

Group the positive numbers together and the negative numbers together.

$$28.32 + (-56.17) + 32.41 + (-75.13)$$
$$= 28.32 + 32.41 + (-56.17) + (-75.13)$$
$$= (28.32 + 32.41) + [(-56.17) + (-75.13)]$$
$$= 60.73 + (-131.30)$$
$$= -70.57$$

To subtract rational numbers, use the same process you used to subtract integers.

Example 5 Find the difference: $-\frac{7}{8} - \left(-\frac{3}{16}\right)$.

$$-\frac{7}{8} - \left(-\frac{3}{16}\right) = -\frac{7}{8} + \frac{3}{16}$$ *To subtract* $-\frac{3}{16}$, *add* $+\frac{3}{16}$.
$$= -\frac{14}{16} + \frac{3}{16}$$ *The least common denominator is 16.*
$$= -\frac{11}{16}$$

Example 6 Evaluate $y - 0.5$, if $y = -0.8$.

$y - 0.5 = -0.8 - 0.5$ *Replace y with −0.8.*

$= -1.3$

CHECKING FOR UNDERSTANDING

Communicating Mathematics

Describe how you could use the commutative and associative properties to find each sum mentally.

1. $1.6x - 4y + (-3y) + 0.4x$
2. $-4\frac{1}{4} + 6\frac{1}{3} + 2\frac{2}{3} + \left(-3\frac{3}{4}\right)$

Guided Practice

Find the least common denominator of each pair of numbers.

3. $\frac{1}{2}, \frac{3}{4}$
4. $\frac{3}{4}, \frac{5}{6}$
5. $\frac{2}{9}, \frac{5}{6}$
6. $\frac{4}{7}, \frac{3}{5}$

Find each sum or difference.

7. $\frac{17}{21} + \left(-\frac{13}{21}\right)$
8. $4.57 + (-3.69)$
9. $-72.5 - 81.3$
10. $\frac{1}{6} - \frac{2}{3}$
11. $-\frac{2}{7} + \frac{3}{14} + \frac{3}{7}$
12. $-3a + 12a + (-14a)$

EXERCISES

Practice

Find each sum or difference.

13. $-\frac{11}{9} + \left(-\frac{7}{9}\right)$
14. $\frac{5}{11} - \frac{6}{11}$
15. $-\frac{7}{12} + \frac{5}{12}$
16. $-4.8 + 3.2$
17. $-38.9 + 24.2$
18. $-1.7 - 3.9$
19. $\frac{2}{7} - \frac{3}{14}$
20. $-\frac{1}{8} + \left(-\frac{5}{2}\right)$
21. $\frac{2}{3} + \left(-\frac{2}{9}\right)$
22. $-0.007 + 0.06$
23. $-0.0005 + (-0.3)$
24. $-\frac{3}{5} + \frac{5}{6}$
25. $\frac{3}{8} + \left(-\frac{7}{12}\right)$
26. $-\frac{7}{15} + \left(-\frac{5}{12}\right)$
27. $-4.5 - 8.6$
28. $89.3 - (-14.2)$
29. $-5\frac{7}{8} - 2\frac{3}{4}$
30. $7\frac{3}{10} - \left(-4\frac{1}{5}\right)$
31. $\begin{array}{r} -15m \\ +\quad 6m \\ \hline \end{array}$
32. $\begin{array}{r} -13c \\ -\quad 28c \\ \hline \end{array}$
33. $\begin{array}{r} 5.8k \\ +(-3.6k) \\ \hline \end{array}$
34. $\begin{array}{r} -0.23x \\ +(-0.5\ x) \\ \hline \end{array}$

Evaluate each expression.

35. $a - (-7)$, if $a = 1.9$

36. $h - (-1.3)$, if $h = -18$

37. $w - 3.7$, if $w = -1.8$

38. $\frac{11}{2} - m$, if $m = -\frac{5}{2}$

39. $\frac{11}{4} - x$, if $x = \frac{27}{8}$

40. $-\frac{12}{7} - z$, if $z = \frac{16}{21}$

Journal

Write a paragraph to explain how adding and subtracting rational numbers is like adding and subtracting integers.

Find each sum.

41. $5y + (-12y) + (-21y)$

42. $-3z + (-17z) + (-18z)$

43. $\frac{7}{3} + \left(-\frac{5}{6}\right) + \left(-\frac{2}{3}\right)$

44. $-\frac{3}{5} + \frac{6}{7} + \left(-\frac{2}{35}\right)$

45. $\frac{3}{4} + \left(-\frac{5}{8}\right) + \frac{3}{32}$

46. $6.7 + (-8.1) + (-7.3)$

47. $-4.13 + (-5.18) + 9.63$

48. $-14a + 36k + 12k + (-83a)$

Critical Thinking

Use a calculator to find each sum. Express each result in decimal form.

49. $-0.37 + \left(-\frac{21}{8}\right)$

50. $-8.66 + 6\frac{7}{8}$

51. $7.43 + \left(-\frac{9}{4}\right)$

52. $-11\frac{7}{16} + 7.225$

Applications

53. **Stock Market** Ted Burton invested his savings in 50 shares of stock. In 3 days, the stock gained $3\frac{1}{8}$ points, lost $\frac{3}{4}$ of a point, and gained another $\frac{1}{8}$ of a point. What was the net gain of the value of one share of Ted's stock?

54. **Golf** In four rounds of a recent golf tournament, Julie shot 3 under par, 2 over par, 4 under par, and 1 under par. What was her score for the tournament?

55. **Football** On *Monday Night Football,* the Philadelphia Eagles made the following yardage on their first seven plays: +3, −7, +15, −5, +32, +6, and −14. What was their net yardage?

56. **Personal Finance** On Tuesday, Delia Puente wrote checks for \$35.76 and \$41.32. On Wednesday, she deposited \$135.59 into her checking account. On Friday, she wrote a check for \$63.17. What was the net increase or decrease in Delia's account for the week?

Mixed Review

57. Evaluate $\frac{4d^2}{4ce}$ if $c = 3$, $d = \frac{1}{2}$, and $e = \frac{2}{3}$. **(Lesson 1-2)**

Name an integer that describes each situation. (Lesson 2-1)

58. 650 meters above sea level

59. 25°F below zero

60. Write an open sentence using addition and solve. **(Lesson 2-2)**
An elevator in the Empire State Building went up to the 36th floor. Then it came down 11 floors. At what floor was it then?

61. Graph the solution set of $m < 1$ on a number line. **(Lesson 2-3)**

62. Which is the better buy: a 184-gram can of peanuts for 91¢ or a 340-gram can for $1.89? **(Lesson 2-4)**

MID-CHAPTER REVIEW

Graph each set of numbers on a number line. (Lesson 2-1)

1. $\{-2, 2, 4\}$
2. $\{-4, -3, -2, -1\}$
3. $\{4, 5, 6, \ldots\}$

Find each sum or difference. (Lesson 2-2)

4. $43 + (-67)$
5. $-23 + (-47)$
6. $-93 + 39$
7. $13 - 31$
8. $-23 - (-12)$
9. $-47 - 35$
10. $-\frac{8}{13} + \left(-\frac{11}{13}\right)$
11. $-\frac{3}{8} + \frac{5}{24}$
12. $-3.948 - 4.826$

Replace each $\underline{?}$ with <, >, or = to make each sentence true. (Lesson 2-3, 2-4)

13. $6 \underline{\ ?\ } 4 + 2$
14. $10 \underline{\ ?\ } \frac{27}{3}$
15. $8\left(\frac{3}{4}\right) \underline{\ ?\ } 6\left(\frac{2}{3}\right)$
16. $\frac{2}{3} \underline{\ ?\ } \frac{3}{5}$
17. $-\frac{7}{6} \underline{\ ?\ } -\frac{21}{18}$
18. $\frac{1.1}{4} \underline{\ ?\ } \frac{2.2}{5}$

Solve. (Lesson 2-4, 2-5)

19. **Stock Market** Last week, the following day-to-day changes were recorded in the Dow Jones stock exchange: Monday, $+5\frac{3}{8}$; Tuesday, $-6\frac{1}{4}$; Wednesday, $+11\frac{1}{8}$; Thursday, $+3\frac{5}{8}$; Friday, $-7\frac{1}{2}$. What was the net change for the week?

20. **Consumerism** Haloke wants to buy soda for a party. Eight 1-liter bottles of Brand X cost $3.92. Six 1-liter bottles of Brand Y cost $3.08. A 2-liter bottle of Brand Z costs $1.09. Which soda is the least expensive per liter?

2-6 Multiplying Rational Numbers

Objective After studying this lesson, you should be able to:
- multiply rational numbers.

Application Lourdes Ortiz wants to buy a washer and dryer from Bargain TV and Appliances. If she buys on the "90 days same as cash" plan, she will have to make 4 equal payments of \$237.11 each. What is the net effect on her bank account?

FYI . . .

Before the electric washing machine was invented in 1906 by Alva J. Fisher, clothes were washed in wringer machines.

One way to solve this problem is to use repeated addition. Think of a bank withdrawal of \$237.11 as −237.11. Then 4(−237.11) can be expressed as follows.

$$(-237.11) + (-237.11) + (-237.11) + (-237.11) = -948.44$$

Therefore, the net effect on Ms. Ortiz's bank account would be −\$948.44.

Since this method would not work if we wanted to find the product of $\frac{2}{3}$ and $-\frac{4}{5}$, we can use the following procedure to show that this product is negative.

$0 = \frac{2}{3}(0)$ *Multiplication property of zero*

$0 = \frac{2}{3}\left[\frac{4}{5} + \left(-\frac{4}{5}\right)\right]$ *Substitution and additive inverse properties*

$0 = \frac{2}{3}\left(\frac{4}{5}\right) + \frac{2}{3}\left(-\frac{4}{5}\right)$ *Distributive property*

$0 = \frac{8}{15} + \frac{2}{3}\left(-\frac{4}{5}\right)$ *Substitution property*

By the additive inverse property, 0 is equal to $\frac{8}{15} + \left(-\frac{8}{15}\right)$. Therefore, $\frac{2}{3}\left(-\frac{4}{5}\right)$ must equal $-\frac{8}{15}$.

This example suggests the following rule.

Multiplying Two Numbers with Different Signs	**The product of two numbers having different signs is negative.**

Example 1 **Multiply (−0.6)5.**

$(-0.6)5 = -3$ *Since the factors have different signs, the product is negative.*

Example 2 **Simplify $(-3a)(4b)$.**

$$(-3a)(4b) = (-3)(4)ab \quad \textit{Commutative and associative properties } (\times)$$
$$= -12ab$$

You already know that the product of two positive numbers is positive. What is the sign of the product of two negative numbers? To find the product of -6 and -8.5, you can use a procedure similar to the one used on the previous page.

$0 = -6(0)$	*Multiplicative property of zero*
$0 = -6[8.5 + (-8.5)]$	*Substitution and additive inverse properties*
$0 = -6(8.5) + (-6)(-8.5)$	*Distributive property*
$0 = -51 + (-6)(-8.5)$	*Substitution property (=)*

By the additive inverse property, 0 is equal to $-51 + 51$. Therefore, $(-6)(-8.5)$ must be equal to $+51$.

This example suggests the following rule.

Multiplying Two Numbers with the Same Sign

The product of two numbers having the same sign is positive.

Example 3 **Multiply $\left(-\frac{2}{3}\right)\left(-\frac{1}{5}\right)$.**

$$\left(-\frac{2}{3}\right)\left(-\frac{1}{5}\right) = \frac{2}{15} \quad \textit{The sign of the product is positive. Why?}$$

Example 4 **Simplify $6x(-7y) + (-3x)(-5y)$.**

$$6x(-7y) + (-3x)(-5y) = -42xy + 15xy$$
$$= -27xy$$

Notice that multiplying a number by -1 results in the opposite of the number.

$-1(3) = -3$ $\qquad$ $(1.7)(-1) = -1.7$ $\qquad$ $(-1)(-4n) = 4n$

Multiplicative Property of -1

The product of any number and -1 is its additive inverse.
$-1(a) = -a$ and $a(-1) = -a$

To find the product of three or more numbers, first group the numbers in pairs.

Example 5 **Multiply $\left(-\frac{1}{2}\right)(-6)(5)\left(-\frac{2}{5}\right)(-2)$.**

$$\left(-\frac{1}{2}\right)(-6)(5)\left(-\frac{2}{5}\right)(-2) = \left[\left(-\frac{1}{2}\right)(-6)\right]\left[(5)\left(-\frac{2}{5}\right)\right](-2) \quad \textit{Associative property } (\times)$$
$$= (3)(-2)(-2) \quad \textit{Substitution property } (=)$$
$$= [(3)(-2)](-2) \quad \textit{Associative property } (\times)$$
$$= (-6)(-2) \quad \textit{Substitution property } (=)$$
$$= 12$$

A **sequence** is a set of numbers in a specific order. The **terms** of a sequence are the numbers in it. The first term of a sequence is represented by a, the second term a_2, and so on up to the nth term, a_n.

Example 6 **Find the twenty-fifth term in the sequence 8, 5, 2, −1, ...**

Notice that each term in the sequence is found by adding −3 to the term preceding it.

$8 + (-3) = 5 \quad \Rightarrow \quad 5 + (-3) = 2 \quad \Rightarrow \quad 2 + (-3) = -1$

To find the twenty-fifth term, you can add −3 to 8 twenty-four times, or you can use the formula $a_n = a + (n - 1)d$, where n is the number of the term you want to find and d is the difference between terms.

$$a_n = a + (n - 1)d$$
$$= 8 + (25 - 1)(-3) \quad \textit{Replace n with 25 and d with } -3.$$
$$= 8 + (24)(-3)$$
$$= 8 + (-72)$$
$$= -64$$

So the twenty-fifth term is −64.

CHECKING FOR UNDERSTANDING

Communicating Mathematics

Read and study the lesson to answer each question.

1. Under what conditions is ab positive?
2. Under what conditions is ab negative?
3. Under what conditions is ab equal to zero?

Guided Practice Determine whether each product is positive or negative. Then find the product.

4. $3(-2)$
5. $(-2)(-8)$
6. $\left(\frac{7}{3}\right)\left(\frac{7}{3}\right)$
7. $(-3)(4)(-2)$
8. $\left(-\frac{4}{5}\right)\left(-\frac{1}{5}\right)(-5)$

EXERCISES

Practice Find each product.

9. $5(12)$
10. $(-6)(11)$
11. $(-6)(-5)$
12. $\left(\frac{3}{5}\right)\left(-\frac{4}{7}\right)$
13. $\left(-\frac{7}{8}\right)\left(-\frac{1}{3}\right)$
14. $(-5)\left(-\frac{2}{5}\right)$
15. $(-2.93)(-0.003)$
16. $(-6)(4)(-3)$
17. $\left(\frac{3}{5}\right)\left(\frac{2}{3}\right)(-3)$
18. $\left(-\frac{7}{12}\right)\left(\frac{6}{7}\right)\left(-\frac{3}{4}\right)$
19. $(-6.8)(-5.415)(3.1)$
20. $(-4)(0)(-2)(-3)$
21. $(4)(-2)(-1)(-3)$
22. $\frac{3}{5}(5)(-2)\left(-\frac{1}{2}\right)$
23. $\frac{2}{11}(-11)(-4)\left(-\frac{3}{4}\right)$

Complete.

24. Find the 19th term of the sequence for which $a = 11$ and $d = -2$.
25. Find the 16th term of the sequence for which $a = 1.5$ and $d = 0.5$.
26. Find the 43rd term of the sequence $-19, -15, -11, \ldots$.
27. Find the 58th term of the sequence $10, 4, -2, \ldots$.

Simplify.

28. $5x(-6y) + 2x(-8y)$
29. $-4(4) + (-2)(-3)$
30. $5(-2) - 3(-8)$
31. $(-2x)(-4y) - (-9x)(10y)$
32. $4(7) - 3(11)$
33. $(-1.9)(3.04) - (-5.4)(-11.6)$
34. $\left(-\frac{1}{2}\right)\left(-\frac{1}{2}\right) - \frac{2}{3}\left(\frac{3}{2}\right)$
35. $\frac{5}{6}\left(\frac{6}{7}\right) - \left(\frac{5}{6}\right)\left(-\frac{6}{7}\right)$
36. $5(3t - 2t) + 2(4t - 3t)$
37. $4[3x + (-2x)] - 5(3x + 2x)$
38. $\frac{5}{6}(-24a + 36b) + \left(-\frac{1}{3}\right)(60a - 42b)$
39. $1.2(4x - 5y) - 0.2(-1.5x + 8y)$
40. $9(2ab - 3c) - (4ab - 6c)$

Critical Thinking

41. If a^2 is positive, what can you conclude about a?
42. If a^3 is positive, what can you conclude about a?
43. If a^3 is negative, what can you conclude about a?
44. If a multiplication involves an even number of negative factors, what must be true of the product?
45. If a multiplication involves an odd number of negative factors, what must be true of the product?

Applications

46. **Statistics** Neil's algebra average dropped three fourths of a point for each of eight consecutive months. If his average was 82 originally, what was his average at the end of eight months?

47. **Sales** Mirna bought a pair of shoes that cost \$32.87 with tax. If she gave the clerk two twenty-dollar bills, how much should Mirna receive in change?

Computer

48. Temperature is commonly measured in degrees Celsius or degrees Fahrenheit. In this program C represents Celsius temperature and F represents Fahrenheit temperature. The formula in line 20 converts the Fahrenheit temperature to Celsius. There is one temperature at which the value of F is the same as C. Use this program to find that temperature.

```
10 FOR F = -50 TO 50
20 LET C = 5/9*(F-32)
30 IF C<> F THEN 50
40 PRINT "F = ";F,"C = ";C
50 NEXT F
60 END
```

Mixed Review

49. Simplify $x^2 + 3x + 2x + 5x^2$. **(Lesson 1-6)**
50. Find the next three numbers in the sequence $-6, -3, 0, \underline{?}, \underline{?}, \underline{?}$. **(Lesson 2-1)**
51. Find the sum: $-197 + |-483|$. **(Lesson 2-2)**
52. Write the solution set of $y \geq -4$ if y is an integer. **(Lesson 2-3)**
53. Which number is greater, $\frac{1}{3}$ or $\frac{1}{4}$? **(Lesson 2-4)**
54. Find the difference: $41y - (-41y)$. **(Lesson 2-5)**

HISTORY CONNECTION

Sonya Kovalevskaya

Sonya Kovalevskaya (1850–1891) was born in Russia in the midst of sweeping reforms. However, the situation for women was very oppressive. Kovalevskaya was inspired by a paternal uncle who instilled in her a reverence for mathematics. As a child, an error caused her room to be wallpapered by notes from the calculus lectures of a prominent Russian mathematician. Later, she astonished her professors by the speed with which she grasped mathematical ideas. In 1883, Sonya Kovalevskaya became a professor at the University of Stockholm. A Swedish newspaper wrote of her appointment, "Today we do not herald the arrival of some . . . prince of noble blood. No, the Princess of Science, Madam Kovalevskaya, has honoured our city with her arrival. She is to be the first woman lecturer in all Sweden."

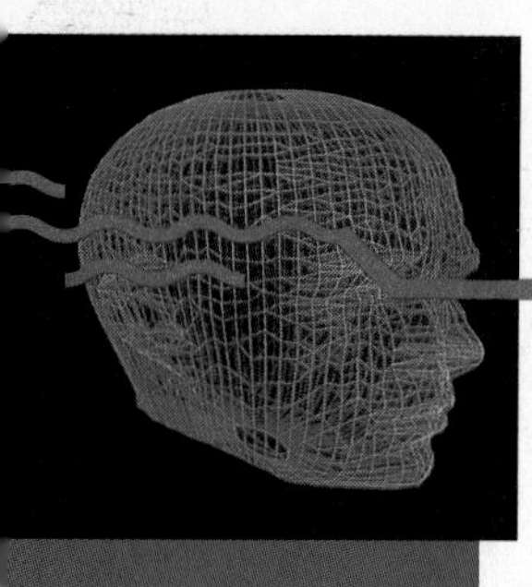

Technology

Adding Fractions

▶ **BASIC**
Graphing calculator
Graphing software
Spreadsheets

Computers add integers and decimals automatically. However, you have to write a special program for a computer to add fractions. The BASIC program below instructs the computer to add two fractions based on the following definition.

$$\frac{a}{b} + \frac{c}{d} = \frac{ad + bc}{bd}$$

```
10 PRINT "ENTER THE NUMERATOR AND THE
   DENOMINATOR OF THE FIRST FRACTION."
20 INPUT A,B
30 PRINT "ENTER THE NUMERATOR AND THE
   DENOMINATOR OF THE SECOND FRACTION."
40 INPUT C,D
50 LET N = A * D + B * C
60 LET X = B * D
70 PRINT "THE SUM IS ";N;"/";X
80 END
```

This program does not always give the sum in simplest form. You can add the following lines to the program so the sum will always be given in simplest form.

```
70 IF N < X THEN 100
80 LET Y = X
90 GOTO 110
100 LET Y = N
110 FOR I = Y TO 1 STEP - 1
120 IF INT (N/I) < > N/I THEN 140
130 IF INT (X/I) = X/I THEN 150
140 NEXT I
150 LET N = N/I
160 LET X = X/I
170 PRINT "THE SUM IS ";N;"/";X
180 END
```

EXERCISES

Use the program to find each sum in simplest form.

1. $\frac{2}{3} + \frac{4}{9}$
2. $\frac{8}{11} + \frac{1}{22}$
3. $\frac{4}{6} + \frac{1}{4}$
4. $\frac{7}{16} + \frac{9}{10}$
5. $\frac{5}{12} + \frac{7}{18}$
6. Write a program similar to the one above to subtract two fractions.

2-7 Dividing Rational Numbers

Objective After studying this lesson, you should be able to:

- divide rational numbers.

Other kinds of averages include median and mode, which are presented in Chapter 14.

The **mean** is a number that represents a set of data. The mean is found by adding the elements in the set and then dividing that sum by the number of elements in the set.

Application Norm Taylor and Sandy Kim play golf every Tuesday. During the month of July, their scores were as shown.

Scores	
Norm	Sandy
−1	−3
+1	−2
−3	−1
−2	−3
+5	−1

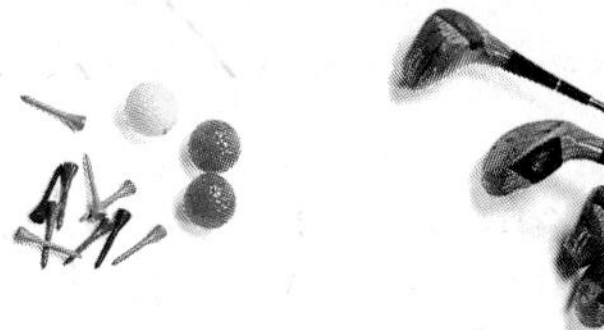

Who had the better average?

Find each person's mean score.

Norm: $\frac{-1 + 1 + (-3) + (-2) + 5}{5} = \frac{0}{5}$, or 0

So Norm's mean score is 0, or even par.

Sandy: $\frac{-3 + (-2) + (-1) + (-3) + (-1)}{5} = -10 \div 5$

We can divide 10 by 5 to find the quotient. But how do we know if the quotient is positive or negative?

You have already learned that addition and subtraction are inverse operations. In the same way, multiplication and division are inverse operations. So the rule for finding the sign of the quotient of two numbers is the same as the rule for finding the sign of the product.

Dividing Rational Numbers

The quotient of two numbers is positive if the numbers have the same sign. The quotient of two numbers is negative if the numbers have different signs.

So Sandy's average is $-10 \div 5 = -2$ or 2 under par, and she has the better average.

Example 1 **Find the quotient of −45 and −9.**

This division problem may be written as $-45 \div (-9)$ or $\frac{-45}{-9}$.

$\frac{-45}{-9} = 5$ *The fraction bar indicates division. The quotient is positive.*

Recall that it is possible to subtract a number by adding its additive inverse. In a similar manner, it is possible to divide a number by multiplying by its multiplicative inverse. Two numbers whose product is 1 are called **multiplicative inverses** or **reciprocals.**

The reciprocal of $\frac{4}{3}$ is $\frac{3}{4}$ because $\frac{4}{3} \cdot \frac{3}{4}$ is 1.

The reciprocal of -5 is $-\frac{1}{5}$ because $-5\left(-\frac{1}{5}\right)$ is 1.

The reciprocal of x is $\frac{1}{x}$ because $x \cdot \frac{1}{x}$ is 1.

Zero has no reciprocal because any number times 0 is 0, not 1.

These examples can be summarized in the multiplicative inverse property.

Multiplicative Inverse Property

For every nonzero number *a*, there is exactly one number $\frac{1}{a}$ such that

$$a\left(\frac{1}{a}\right) = \frac{1}{a}(a) = 1.$$

Recall that you can change any division expression to an equivalent multiplication expression. To divide by any nonzero number, multiply by the reciprocal of that number.

Division Rule

For any numbers *a* and *b*, with $b \neq 0$,

$$a \div b = \frac{a}{b} = a\left(\frac{1}{b}\right) = \frac{1}{b}(a).$$

Example 2

Divide $-\frac{3}{4} \div 8$.

$-\frac{3}{4} \div 8 = -\frac{3}{4} \cdot \frac{1}{8}$ *Multiply by $\frac{1}{8}$, the reciprocal of 8.*

$= -\frac{3}{32}$ *Since one factor is negative, the product is negative.*

If a fraction has one or more fractions in the numerator or denominator, it is called a **complex fraction.** To simplify a complex fraction, rewrite it as a division sentence.

Example 3

Simplify $\frac{\frac{3}{11}}{6}$.

Rewrite the fraction as $\frac{3}{11} \div 6$, since fractions indicate division.

$\frac{3}{11} \div 6 = \frac{3}{11} \cdot \frac{1}{6}$ *Multiply by $\frac{1}{6}$, the reciprocal of 6.*

$= \frac{3}{66}$ or $\frac{1}{22}$

Example 4 Simplify $\frac{4a + 32}{4}$.

Method 1

$$\frac{4a+32}{4} = (4a + 32) \div 4$$

$$= (4a + 32)\left(\frac{1}{4}\right)$$ *To divide by 4, multiply by $\frac{1}{4}$.*

$$= 4a\left(\frac{1}{4}\right) + 32\left(\frac{1}{4}\right)$$ *Distributive property*

$$= a + 8$$

Method 2

$$\frac{4a+32}{4} = \frac{4a}{4} + \frac{32}{4}$$

$$= a + 8$$

You can use the comparison property for rational numbers to compare two negative fractions, or a positive fraction and a negative fraction. Note that the fractions below are equivalent.

Note that $-\frac{3}{4} \neq \frac{-3}{-4}$.

$$\frac{-3}{4} \qquad \frac{3}{-4} \qquad -\frac{3}{4}$$

Since the comparison property requires that the denominators must be positive, rewrite each negative fraction with a negative numerator.

Example 5 **Replace the ___?___ with <, >, or = to make $-\frac{1}{4}$ ___?___ $-\frac{1}{5}$ a true sentence.**

Rewrite $-\frac{1}{4}$ as $\frac{-1}{4}$ and $-\frac{1}{5}$ as $\frac{-1}{5}$.

$$\frac{-1}{4} \ ? \ \frac{-1}{5}$$ *Find the cross products.*

$$-5 < -4$$

The true sentence is $-\frac{1}{4} < -\frac{1}{5}$.

CHECKING FOR UNDERSTANDING

Communicating Mathematics

Read and study the lesson to answer each question.

1. Why are multiplication and division called inverse operations?
2. What is the reciprocal of $\frac{1}{5}$?
3. Explain in one sentence how to divide rational numbers.
4. Words that have the same meaning are called *synonyms*. Write a synonym for each term.
   a. additive inverse
   b. multiplicative inverse
   c. mean

Guided Practice

Name the reciprocal of each number.

5. 3
6. -5
7. 0
8. -1
9. $\frac{2}{3}$
10. $-\frac{1}{15}$
11. $3\frac{1}{4}$
12. $-2\frac{3}{7}$
13. $-6\frac{5}{11}$

Simplify.

14. $\frac{-30}{-5}$
15. $\frac{-48}{8}$
16. $\frac{-200x}{50}$
17. $\frac{45b}{9}$
18. $\frac{-\frac{5}{6}}{8}$
19. $\frac{3a + 9}{3}$

EXERCISES

Practice

Simplify.

20. $\frac{30}{-6}$
21. $\frac{-55}{11}$
22. $\frac{-96}{-16}$
23. $\frac{-450n}{10}$
24. $\frac{-36a}{-6}$
25. $\frac{63a}{-9}$
26. $-49 \div (-7)$
27. $-16 \div 8$
28. $65 \div (-13)$
29. $-\frac{3}{4} \div 9$
30. $\frac{-1}{3} \div (-4)$
31. $-9 \div \left(-\frac{10}{17}\right)$
32. $\frac{\frac{7}{8}}{-10}$
33. $\frac{7}{-\frac{2}{5}}$
34. $\frac{-5}{\frac{2}{7}}$
35. $\frac{6a + 24}{6}$
36. $\frac{20a + 30b}{-2}$
37. $\frac{-5x + (-10y)}{-5}$
38. $\frac{70x - 30y}{-5}$

Find the mean for each set of data.

39. 4, 6, 9, 12, 5
40. 10, 3, 8, 15
41. 10, 4, −21, 6, −3, 8, 5, 5, 2, −2
42. 2.5°, 6°, 18.5°, 29.5°, 32.5°, 28°, 24.5°, 20°, 16.5°, 5°, −2°, −1°

Replace each __?__ with <, >, or = to make each sentence true.

43. $-\frac{6}{5}$ __?__ $-\frac{7}{6}$
44. $-\frac{9}{7}$ __?__ $-\frac{7}{5}$
45. $-\frac{3}{4}$ __?__ $-\frac{2}{3}$
46. $-\frac{13}{11}$ __?__ $-\frac{15}{13}$
47. $-\frac{1}{3}$ __?__ $-\frac{2}{7}$
48. $-\frac{7}{6}$ __?__ $-\frac{21}{18}$

Critical Thinking

49. The $\boxed{1/x}$ key on a calculator is called the *reciprocal key.* When this key is pressed, the calculator replaces the number on the display with its reciprocal. Enter 0 and then press the reciprocal key. What happened? Why?

50. Enter a number and then press the reciprocal key twice. What happened? Predict what will happen if you press the key n times.

51. Enter 7.328 $\boxed{1/x}$ $\boxed{\times}$ 7.328 $\boxed{=}$. What is the result? Why?

Applications

While Hartsfield Airport has more takeoffs and landings, more passengers fly through Chicago's O'Hare International Airport than any other.

52. **Basketball** According to NBA standards, a basketball should bounce back $\frac{2}{3}$ of the distance from which it is dropped. How high should a basketball bounce when it is dropped from a height of $3\frac{3}{4}$ yards?

53. **Aviation** The busiest airport in the world is Atlanta's Hartsfield Airport. On average, an airplane takes off or lands every 39.66 seconds. At this rate, how many airplanes take off or land in 17 hours? Round to the nearest whole number.

54. **Space** Halley's Comet flashes through the sky every 76.3 years. It last appeared in 1986. In what year of the 23rd century is Halley's Comet expected to reappear?

55. **World Records** The longest loaf of bread ever baked was 2132 feet $2\frac{1}{2}$ inches. If this loaf were cut into $\frac{1}{2}$-inch slices, how many slices of bread would there have been?

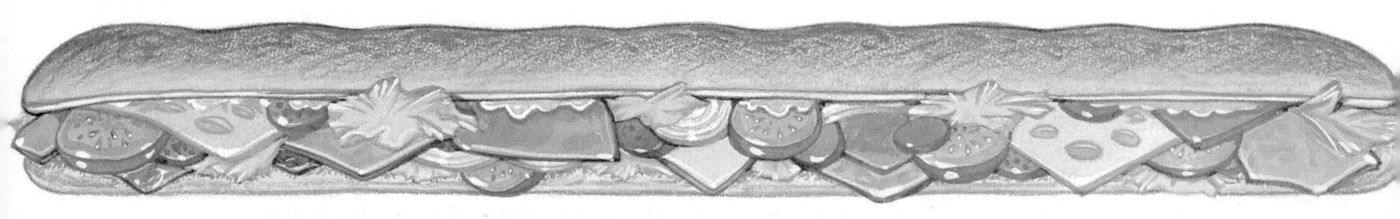

Mixed Review

56. Write an equation for the sentence *The number b equals x decreased by the cube of m.* **(Lesson 1-7)**

57. Evaluate $|-8 + k|$ if $k = 3$. **(Lesson 2-2)**

58. *True* or *false:* $0.6 \leq 1 - 0.4$. **(Lesson 2-3)**

59. Which is the better buy: a 1-pound package of lunch meat for $1.98 or a 12-ounce package for $1.80? **(Lesson 2-4)**

60. Find the sum: $9m + 43m + (-16m)$. **(Lesson 2-5)**

61. Find the product: $4\left(-\frac{7}{8}\right)$. **(Lesson 2-6)**

2-8 Problem-Solving Strategy: Write an Equation

EXPLORE
PLAN
SOLVE
EXAMINE

Objectives

After studying this lesson, you should be able to:

- define variables and write equations for verbal problems, and
- write verbal problems for equations.

In Chapter 1, you learned how to explore verbal problems before solving them. In this lesson, you will study how to plan the solution to a verbal problem by writing an equation.

First, explore the problem. Then plan the solution by doing the following.

- Read the problem again. Decide how the unspecified numbers relate to each other and to the other given information.
- Write an equation to represent the relationship.

Example 1

Write an equation for the following problem.

Six years ago, twice Jennifer's age was 16 years. How old is she now?

Let j = Jennifer's age now.

Then, $j - 6$ = Jennifer's age six years ago.

twice Jennifer's age six years ago is 16 years

$2 \cdot \quad (j - 6) \quad = \quad 16$

So, the equation is $2(j - 6) = 16$.

You can make up your own verbal problem if you are given an equation or two. First, you need to know what the variable in the equation represents. Then you can use the equation to establish the conditions of the problem.

Example 2

Write a problem based on the given information.

Let w = Steve's weight in pounds.
Then, $w - 12$ = Toshiro's weight in pounds.
$w + (w - 12) = 118$

Here's a sample problem.

Toshiro weighs 12 pounds less than Steve. The sum of their weights is 118 pounds. How much does Steve weigh? How much does Toshiro weigh?

CHECKING FOR UNDERSTANDING

Communicating Mathematics

Read and study the lesson to answer each question.

1. What are the four steps in the problem-solving plan?
2. Why should you read a verbal problem twice before trying to solve it?
3. Write an equation for Example 1 given the following information.
   Let $j + 6$ = Jennifer's age now.
   Then, j = Jennifer's age 6 years ago.

Guided Practice

Write an expression for each problem.

4. This year's senior class has 117 fewer students than last year's class. This year's class has 947 students. How many were in last year's class?
5. Buzz Jackson is now 49 years old. How old was he n years ago?
6. The length of a rectangle is 4 feet more than the width. The width is w feet. Find the length.

7. Sheldon is 8 years older than Atepa. If Sheldon is n years old, how old is Atepa?
8. Tess types 42 words per minute. How many minutes would it take for her to type a 3000-word paper?
9. Anthony makes $5.65 per hour as a sales clerk at Tom's Sporting Goods. How much does he make for working n hours?
10. Alonso has 8 more than twice as many red pens as blue pens. He has b blue pens. How many red pens does he have?

EXERCISES

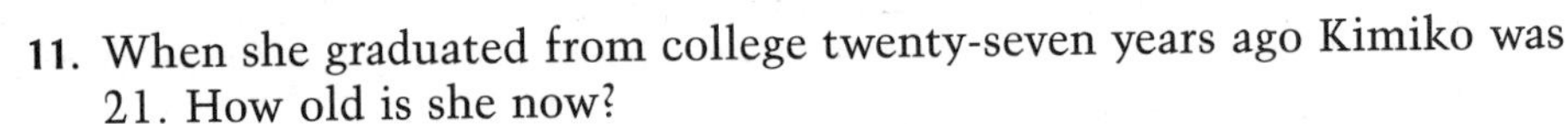

Practice

Define a variable, then write an equation for each problem. Do *not* try to solve.

11. When she graduated from college twenty-seven years ago Kimiko was 21. How old is she now?
12. For 6 consecutive weeks, Connie lost the same amount of weight. Six weeks ago she weighed 145 pounds. She now weighs 125 pounds. How many pounds did Connie lose each week?
13. Ponderosa pines grow about $1\frac{1}{2}$ feet each year. If a pine tree is now 17 feet tall, about how long will it take the tree to become $33\frac{1}{2}$ feet tall?
14. Four years ago, three times Cecile's age was 42, her father's age now. How old is Cecile now?
15. Kerri has 31 CDs and tapes altogether. If she has 4 more than half as many CDs as tapes, how many tapes does Kerri have?
16. Sonia is 3 years older than Melissa. The sum of their ages in 4 years will be 59 years. How old is Sonia now?

Strategies

Look for a pattern.
Solve a simpler problem.
Act it out.
Guess and check.
Draw a diagram.
Make a chart.
Work backwards.

17. Shannon has 4 more dimes than quarters and 7 fewer nickels than dimes. She has 28 coins in all. How many quarters does Shannon have?

18. The sum of Mrs. Blakely's age and her daughter Karen's age is 53 years. In 8 years, Mrs. Blakely will be twice as old as Karen. How old is Karen now?

Write a problem based on the given information.

19. Let a = Quincy's age now.
$2(a - 7) = 58$

20. Let w = Willie's age now.
$w + 4$ = Elena's age now.
$(w + 10) + (w + 4 + 10) = 54$

21. Let s = weight of Seth's car in pounds.
$s + 250$ = weight of Ramón's car in pounds.
$s + (s + 250) = 7140$

22. Let m = Manuel's height in inches.
$m + 7$ = Jason's height in inches.
$2m + (m + 7) = 193$

23. Let s = Soto's height in centimeters.
$s - 31$ = Reggie's height in centimeters.
$s + 2(s - 31) = 502$

Portfolio

Select one of the assignments from this chapter that you found especially challenging and place it in your portfolio.

COOPERATIVE LEARNING ACTIVITY

Work in groups of four. Each person in the group must understand the solution and be able to explain it to any person in the class.

Leah has promised to make at least 15 pounds of chocolate fudge for a graduation party. She will have to buy baking chocolate, milk, butter, and sugar. The grocery carries only one size of each ingredient: a box of baking chocolate is \$2.69, a carton of milk is 69¢, a pound of butter is \$1.09, and a box of sugar is 75¢.

With one box of chocolate and one carton of milk, Leah can make up to 6 pounds of fudge. With one more box of chocolate, she can make up to 9 pounds. With another carton of milk, she can make up to 12 pounds. There are 4 sticks of butter to a pound, and each stick makes exactly $4\frac{1}{2}$ pounds of fudge. Each box of sugar makes $1\frac{1}{2}$ pounds of fudge.

1. If she buys just enough of everything to come out even, can Leah buy all that she needs for \$20?
2. What will her grocery bill be?
3. How many pounds of fudge can she make?

CHAPTER 2

SUMMARY AND REVIEW

VOCABULARY

Upon completing this chapter, you should be familiar with the following terms:

absolute value	**55**	**81**	multiplicative inverses
additive inverses	**56**	**50**	negative
complex fraction	**81**	**50**	number line
coordinate	**51**	**56**	opposites
cross products	**65**	**61**	rational numbers
disjoint	**51**	**50**	whole numbers
graph	**51**	**51**	real numbers
inequality	**60**	**81**	reciprocals
integers	**50**	**51**	union
intersection	**51**	**66**	unit cost
mean	**80**	**51**	Venn diagrams

SKILLS AND CONCEPTS

OBJECTIVES AND EXAMPLES

Upon completing this chapter, you should be able to:

- graph integers on a number line **(Lesson 2-1)**

Graph $\{\ldots, -1, 0, 1, 2\}$.

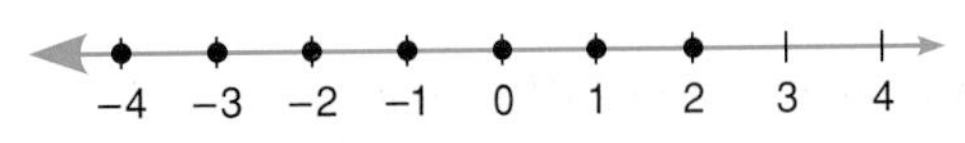

REVIEW EXERCISES

Use these exercises to review and prepare for the chapter test.

Graph each set of numbers on a number line.

1. $\{5, 3, -1, -3\}$

2. $\{-3, -2, -1, 0, \ldots\}$

- add integers **(Lessons 2-1, 2-2)**

$-14 + (-9) = -23$
$20 + (-4) = 16$
$-7 + 3 = -4$

Find each sum.

3. $17 + (-9)$

4. $-9 + (-12)$

5. $-12 + 8$

6. $-17 + (-31)$

- subtract integers **(Lesson 2-2)**

$7 - 9 = -2$
$-6 - 5 = -11$
$-4 - (-8) = 4$

Find each difference.

7. $14 - 36$

8. $8 - (-5)$

9. $-7 - (-11)$

10. $-13x - (-7x)$

OBJECTIVES AND EXAMPLES	REVIEW EXERCISES

- find the absolute value of a number **(Lesson 2-2)**

Evaluate $|m - 2|$ if $m = -3$.

$|m - 2| = |-3 - 2|$
$= |-5|$
$= 5$

Evaluate each expression.

11. $|4 - x|$, if $x = -2$
12. $|x| - 2.6$, if $x = -5$
13. $-|a + (-12)|$, if $a = -3$

- graph inequalities on number lines **(Lesson 2-3)**

Graph the solution set of $x \geq -3$.

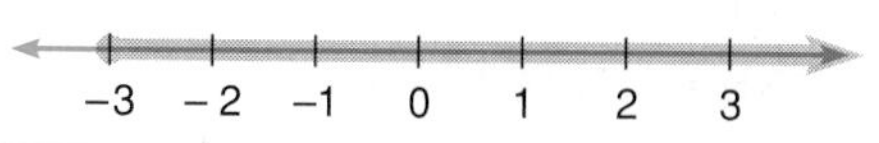

Graph the solution set of each inequality on a number line.

14. $x < 2$
15. $x \neq -4$

- compare numbers **(Lessons 2-3, 2-4)**

$\frac{2}{3} \underline{\ ?\ } \frac{4}{5}$

$5 \cdot 2 \underline{\ ?\ } 3 \cdot 4$

$10 < 12$

$\frac{2}{3} < \frac{4}{5}$

Replace each $\underline{\ ?\ }$ with <, >, or = to make each sentence true.

16. $-9 \underline{\ ?\ } -11$
17. $-13 \underline{\ ?\ } 13$
18. $-7 \underline{\ ?\ } \frac{-3.6}{0.6}$
19. $\frac{3}{8} \underline{\ ?\ } \frac{4}{11}$
20. $-\frac{10}{11} \underline{\ ?\ } \frac{11}{12}$
21. $-\frac{9}{11} \underline{\ ?\ } -\frac{7}{8}$

- find a number between two rational numbers **(Lesson 2-4)**

The average of $\frac{3}{5}$ and $\frac{7}{12}$ is as follows.

$\frac{1}{2}\left(\frac{3}{5} + \frac{7}{12}\right) = \frac{1}{2}\left(\frac{71}{60}\right)$
$= \frac{71}{120}$

Find a number between the given numbers.

22. $\frac{2}{9}$ and $\frac{5}{8}$
23. $\frac{3}{5}$ and $\frac{7}{12}$
24. $-\frac{1}{2}$ and $\frac{7}{11}$

- add or subtract rational numbers **(Lesson 2-5)**

$-0.37 + (-0.812) = -1.182$

Find each sum or difference.

25. $\frac{6}{7} + \left(-\frac{13}{7}\right)$
26. $-\frac{4}{3} + \frac{5}{6} + \left(-\frac{7}{3}\right)$
27. $3.72 - (-8.65)$
28. $-4.5y - 8.1y$

- multiply rational numbers **(Lesson 2-6)**

$-4(-2) + 6(-3) = 8 + (-18)$
$= -10$

Simplify.

29. $(-11)(9)$
30. $(-8)(-12)$
31. $\frac{3}{5}\left(-\frac{5}{7}\right)$
32. $-3(7) + (-8)(-9)$
33. $\frac{1}{2}(6a + 8b) - \frac{2}{3}(12a + 24b)$

OBJECTIVES AND EXAMPLES	REVIEW EXERCISES

- divide rational numbers **(Lesson 2-7)**

$$\frac{-12}{-\frac{2}{3}} = -12 \div \left(-\frac{2}{3}\right)$$
$$= -12\left(-\frac{3}{2}\right)$$
$$= 18$$

Simplify.

34. $\frac{-54}{6}$

35. $\frac{63b}{-7}$

36. $\frac{\frac{4}{5}}{-7}$

37. $\frac{33a + 66}{-11}$

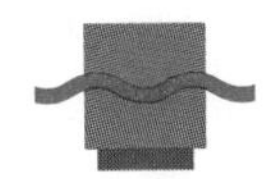

APPLICATIONS AND CONNECTIONS

Set Theory: Draw a Venn diagram to find each of the following. (Lesson 2-1)

38. the union of {vowels} and {last ten letters of the alphabet}

39. the intersection of {1, 3, 5, 7} and {1, 2, 3, 4, 5}

40. Aquatics A submarine descended to a depth of 432 meters and then rose 189 meters. How far below the surface of the water was the submarine? **(Lesson 2-1)**

41. Savings Pam was trying to save money for a ten-speed bicycle. She opened a savings account with a deposit of $75. Then she withdrew $37 for a new pair of shoes. What is her new balance? **(Lesson 2-2)**

Sequences: Complete. (Lesson 2-6)

42. Find the next five terms of the sequence 3, 4.5, 6, ...

43. Find the 20th term of the sequence for which $a = 7$ and $d = -4$.

44. Find the 24th term of the sequence $-6, -1, 4, \ldots$.

45. Consumerism Which is the better buy: 0.75 liter of soda for 89¢ or 1.25 liters of soda for $1.31? **(Lesson 2-4)**

46. Statistics Find the mean for the following set of data: 4, 7.2, 4, 9, 21, 15, 6, 6.3, 29, 0. **(Lesson 2-7)**

47. Metallurgy The gold content of jewelry is given in karats. For example, 24-karat gold is pure gold, but 18-karat gold is $\frac{18}{24}$ or 0.75 gold. **(Lessons 2-6, 2-7)**

a. Ten-karat gold is what fraction gold? What fraction is *not* gold?

b. If a piece of jewelry is $\frac{2}{3}$ gold, how would you describe it in karats?

Define a variable, then write an equation for each problem. Do *not* try to solve. (Lesson 2-8)

48. Minal weighs 8 pounds less than Claudia. Together they weigh 182 pounds. How much does Minal weigh?

49. Three times a number decreased by 21 is 57. Find the number.

CHAPTER 2 TEST

Graph each set of numbers on a number line.

1. $\{1, 2, 5\}$
2. {all numbers less than 2}

Find each sum or difference.

3. $-11 + (-13)$
4. $12 - 19$
5. $1.654 + (-2.367)$
6. $-41 - (-52)$
7. $6.32 - (-7.41)$
8. $12x + (-21x)$
9. $\frac{3}{7} + \left(-\frac{9}{7}\right)$
10. $-\frac{7}{16} - \frac{3}{8}$
11. $18b + 13xy - 46b$
12. $[4 + (-13)] - 12$
13. $\frac{5}{8} + \left(-\frac{3}{16}\right) + \left(-\frac{3}{4}\right)$

Evaluate each expression.

14. $-x - 38$, if $x = -2$
15. $\left|-\frac{1}{2} + z\right|$, if $z = \frac{1}{4}$
16. $-|a| + |b|$, if $a = 6$ and $b = -2$
17. $d - (-3.8)$, if $d = 0$

Replace each $\underline{?}$ with <, >, or = to make each sentence true.

18. $2 \underline{\ ?\ } -7$
19. $-4 \underline{\ ?\ } -3$
20. $\frac{5.4}{18} \underline{\ ?\ } -4 + 1$
21. $(4.1)(0.2) \underline{\ ?\ } 8.2$
22. $\frac{7}{6} \underline{\ ?\ } \frac{13}{12}$
23. $-\frac{12}{17} \underline{\ ?\ } -\frac{9}{14}$

Find a number between the given numbers.

24. $\frac{5}{11}$ and $\frac{13}{7}$
25. $-\frac{2}{3}$ and $-\frac{9}{14}$
26. $-\frac{13}{2}$ and $\frac{12}{7}$

Simplify.

27. $\frac{8(-3)}{2}$
28. $(-5)(-2)(-2) - (-6)(-3)$
29. $\frac{2}{3}\left(\frac{1}{2}\right) - \left(-\frac{3}{2}\right)\left(-\frac{2}{3}\right)$
30. $\frac{70a - 42b}{-14}$
31. $\frac{3}{4}(8x + 12y) - \frac{5}{7}(21x - 35y)$
32. $\frac{\frac{11}{5}}{-6}$

33. Define a variable and write an equation for the following problem.

    Each week for several weeks, Save-a-Buck stores reduced the price of a sofa by \$18.25. The original price was \$380.25. The final reduced price was \$252.50. For how many weeks was the sofa on sale?

Bonus Find the sum: $4|(-5 + 2)|^2 + (-72)$.

CHAPTER OBJECTIVES

In this chapter, you will:

- Solve equations using one or more operations.
- Solve equations containing fractions or decimals.
- Solve problems that can be represented by equations.
- Work backwards to solve problems.

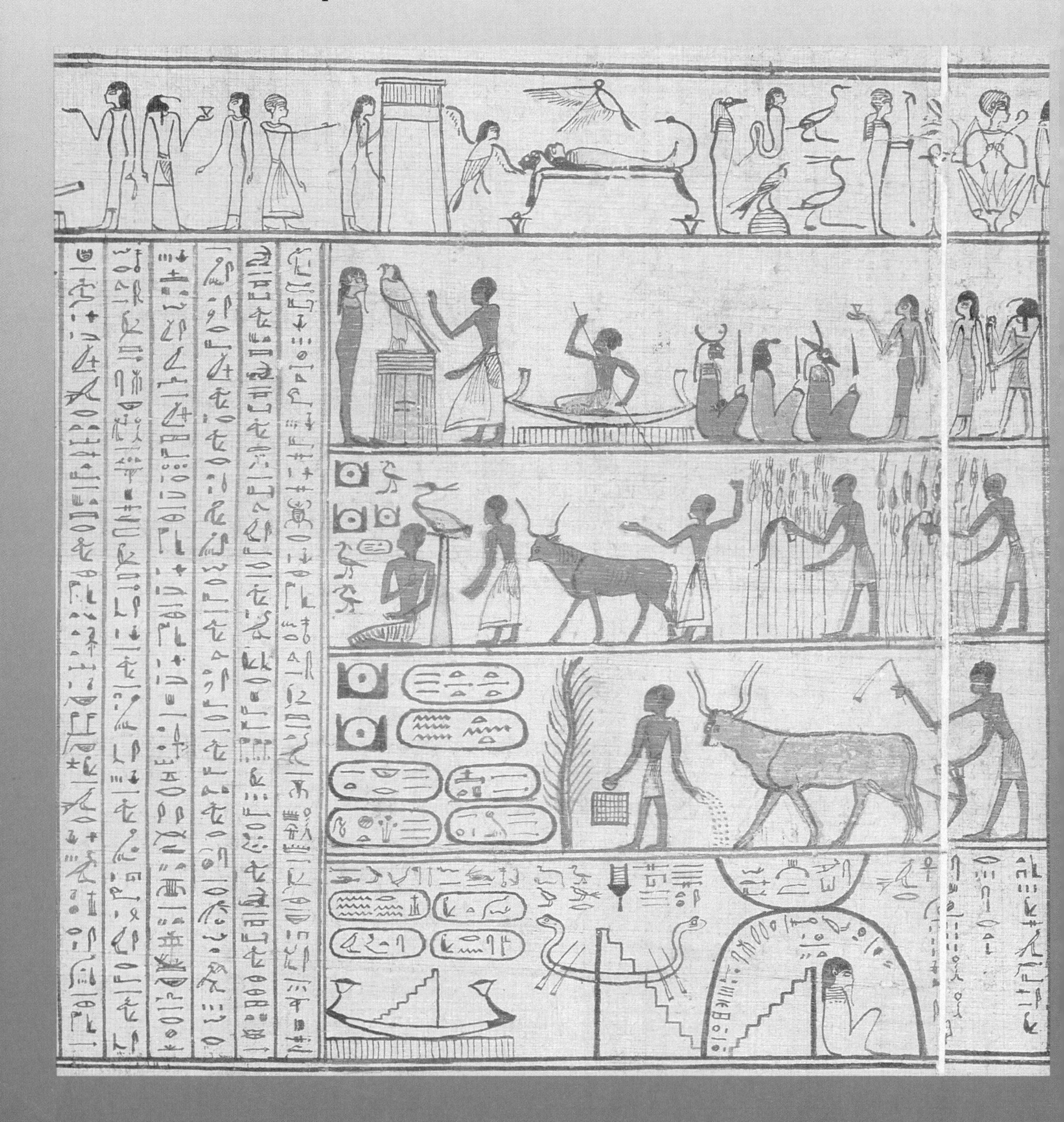

CHAPTER 3

Equations

APPLICATION IN HISTORY

Just like you, students in ancient Babylon had algebra homework, although they did not write their assignments in notebooks filled with lined paper. Instead, they wrote on clay tablets using the ends of little sticks to make wedge-shaped marks. About one thousand years later, in Egypt, students wrote their algebra assignments on papyrus, a parchment-like material that was easier to write on than a clay tablet.

If you could read either of these assignments today, you might not recognize them as algebra assignments. Neither the ancient Babylonians nor the ancient Egyptians used letters for unknown values in their algebraic equations. One reason for this is that the alphabet of today had not yet been invented! So, instead of letters, they made little pictures that stood for the unknowns. This would be the same as if you wrote entire words instead of letters like x and y in your equations. Algebraic symbolism using letters did not gain wide acceptance until the Greek mathematician, Diophantus, introduced the *syncopated* style of writing equations.

They may not have had skateboards, compact disks, or VCRs, but 4500 years ago, Babylonian teenagers did have algebra homework!

ALGEBRA IN ACTION

Diophantus made numerous contributions in the area of algebraic symbolism. The chart below shows examples of his syncopated equations along with their modern algebraic form. How would Diophantus have written the equation $2x + 9 = x - 3$?

Diophantus' Equation	Modern Meaning
$\zeta\iota\sigma\beta$	$x = 2$
$\zeta\gamma\iota\sigma\theta$	$x + 3 = 9$
$\zeta\gamma\beta\iota\sigma\theta$	$3x + 2 = 9$
$\zeta\theta\Lambda\gamma\iota\sigma\beta$	$9x - 3 = 2$
$\zeta\beta\Lambda\theta\iota\sigma\zeta\gamma$	$2x - 9 = x + 3$

3-1 Solving Equations by Using Addition

Objective After studying this lesson, you should be able to:

- solve equations by using addition.

Application Record Universe and Music Madhouse each sell CDs for \$10.95. If each store raises its price by \$2, they will still be selling CDs for the same price.

$$10.95 = 10.95$$
$$10.95 + 2 = 10.95 + 2$$

This example illustrates the **addition property of equality.**

Addition Property of Equality	**For any numbers a, b, and c, if $a = b$, then $a + c = b + c$.**

Note that c can be positive or negative.

$$15 + 3 = 15 + 3 \qquad 15 + (-3) = 15 + (-3)$$

Think of an equation as a scale in balance. A scale balances when each side holds equal weight. If you add weight to only one side, as shown in the center illustration below, then the scale is no longer in balance. However, if you add the same weight to each side, the scale will balance.

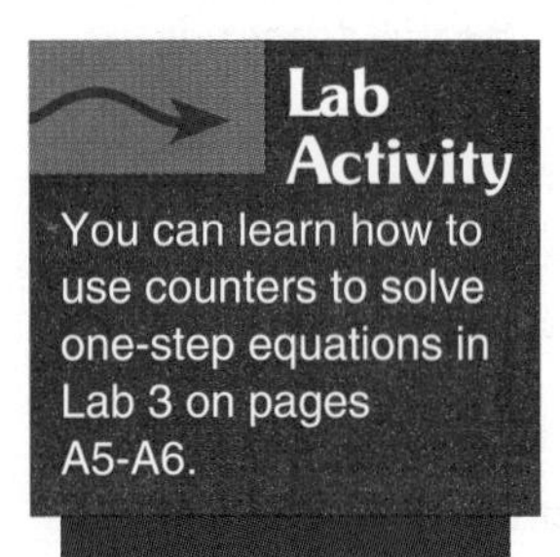

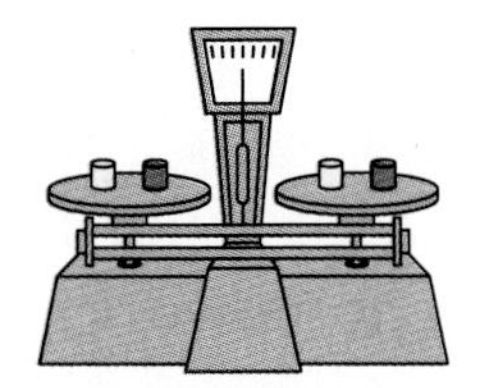

If the same number is added to each side of an equation, then the result is an **equivalent equation.** Equivalent equations are equations that have the same solution.

$11 = x + 3$ *The solution to this equation is 8.*

$11 + 5 = x + 3 + 5$ *Using the addition property of equality, add 5 to each side.*

$16 = x + 8$ *The solution to this equation is also 8.*

Remember, x means $1 \cdot x$. The coefficient of x is 1.

To **solve an equation** means to isolate the variable having a coefficient of 1 on one side of the equation. You can do this by using the addition property of equality.

Example 1 **Solve $r + 16 = -7$.**

$$r + 16 = -7$$
$$r + 16 + (-16) = -7 + (-16) \quad \textit{Add } -16 \textit{ to each side.}$$
$$r + 0 = -23 \quad \textit{The sum of } -16 \textit{ and } 16 \textit{ is } 0.$$
$$r = -23$$

To check that -23 is the solution, substitute -23 for r in the original equation.

Check: $r + 16 = -7$

$$-23 + 16 \stackrel{?}{=} -7$$
$$-7 = -7 \ \checkmark$$

The solution is -23.

Recall from Lesson 2-7 how to solve a problem by writing and solving an equation.

Example 2

A traffic helicopter descended 160 meters to observe road conditions. It leveled off at 225 meters. What was its original altitude?

EXPLORE Read the problem to find out what is asked. Then define a variable.

The problem asks for the helicopter's original altitude.
Let a = the helicopter's original altitude.
Then $a - 160$ = the helicopter's altitude after it descends 160 meters.

PLAN Write an equation.

The problem states that the helicopter leveled at 225 meters. So, the equation is $a - 160 = 225$.

SOLVE Solve the equation and answer the problem.

$$a - 160 = 225$$
$$a - 160 + 160 = 225 + 160 \quad \textit{Add 160 to each side.}$$
$$a + 0 = 385 \quad \textit{The sum of } -160 \textit{ and 160 is 0.}$$
$$a = 385$$

The original altitude was 385 meters.

EXAMINE Check to see if the answer makes sense.

If the original altitude was 385 meters, then the new altitude is $385 - 160$ or 225 meters.

Solving some equations requires additional steps.

Example 3

Solve $-8 - y = 13$.

$$-8 - y = 13$$

$$-8 + 8 - y = 13 + 8 \quad \textit{Add 8 to each side.}$$

$$-y = 21 \quad \textit{The opposite of y is positive 21.}$$

$$y = -21 \quad \textit{Therefore, y is negative 21.}$$

Check:

$$-8 - y = 13$$

$$-8 - (-21) \stackrel{?}{=} 13$$

$$-8 + 21 \stackrel{?}{=} 13$$

$$13 = 13 \ \checkmark$$

The solution is -21.

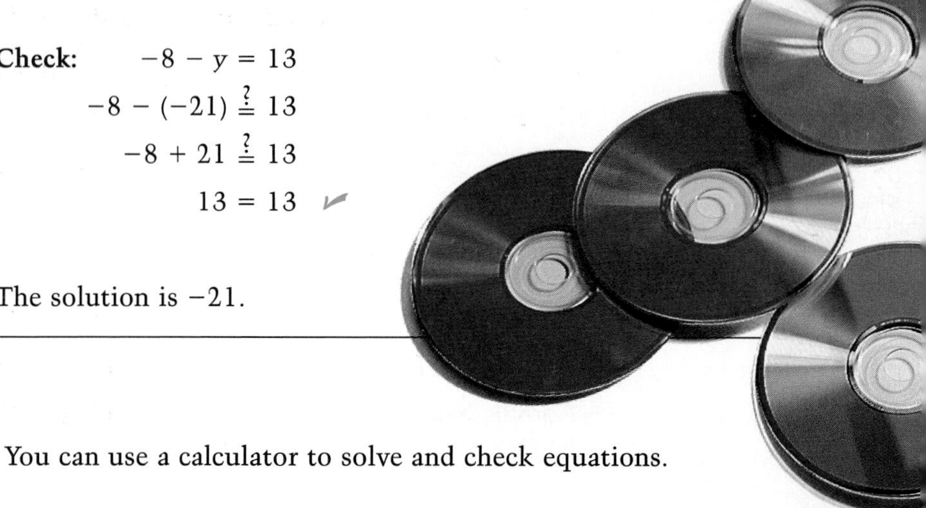

You can use a calculator to solve and check equations.

Example 4

Use a calculator to solve $t + (-3.28) = -17.56$.

$$t + (-3.28) = -17.56$$

$$t + (-3.28) + 3.28 = -17.56 + 3.28$$

$$t = -17.56 + 3.28 \quad \textit{Use a calculator to simplify.}$$

Enter: 17.56 [+/-] [+] 3.28 

Display: −14.28

Check:

$$t + (-3.28) = -17.56$$

$$-14.28 + (-3.28) \stackrel{?}{=} -17.56$$

Enter: 14.28 [+/-] [+] 3.28  [=]

Display: −17.56 ✓

The solution is -14.28.

CHECKING FOR UNDERSTANDING

Communicating Mathematics

Read and study the lesson to answer each question.

1. What will happen if Record Universe and Music Madhouse each lower their prices on CDs by $3?
2. Write three equivalent equations.
3. What is the [+/-] key on your calculator called? What does it do?

Guided Practice

State the number you would add to each side of the equation to solve it.

4. $y + 21 = -7$
5. $13 + x = -16$
6. $y - 5 = 11$
7. $z + (-9) = 34$
8. $-10 + k = 34$
9. $y - 13 = 45$

Solve and check each equation.

10. $m + 10 = 7$
11. $a - 15 = -32$
12. $5 + a = -14$
13. $y + (-7) = -19$
14. $9 = x + 13$
15. $b + (-14) = 6$

EXERCISES

Practice

Solve and check each equation.

16. $k + 11 = -21$
17. $0 = t + (-1.4)$
18. $-11 = a + 8$
19. $-12 + z = -36$
20. $14 + c = -5$
21. $x - 13 = 45$
22. $p + 12 = -4$
23. $r + (-8) = 7$
24. $-12 + b = 12$
25. $r + (-11) = -21$
26. $h + (-13) = -5$
27. $-11 = k + (-5)$
28. $-7 = -16 - k$
29. $-27 = -6 - p$
30. $-14 - a = -21$
31. $-23 = -19 + n$
32. $-4.1 = m + (-0.5)$
33. $r - 6.5 = -9.3$
34. $x + 4.2 = 1.5$
35. $-1.43 + w = 0.89$

Write an equation and solve. Then check each solution.

36. A number increased by 5 is equal to 34. Find the number.
37. A number decreased by 14 is -46. Find the number.

Write an equation and solve. Then check each solution.

38. Thirteen subtracted from a number is -5. Find the number.

39. A number increased by -45 is 77. Find the number.

40. Twenty-three minus a number is 42. Find the number.

41. The sum of two numbers is -23. One of the numbers is 9. What is the other number?

Critical Thinking

42. What value of x makes the statement $x + x = x$ true?

Applications

Define a variable, write an equation, and solve each problem.

FYI . . .

Spelunking is the hobby of exploring caves. The greatest recorded depth to which a team of spelunkers has descended is 5036 feet, recorded in October 1983.

43. **Spelunking** Four cave explorers descended to a depth of 112 meters below the cave entrance. They discovered a large cavern whose ceiling was 27 meters above them. At what depth below the cave entrance was the cavern ceiling?

44. **Gardening** The area of Mr. Hooper's triangular courtyard is 1520.2 square feet. The area occupied by a circular fountain in the middle of the courtyard is 132.7 square feet. A walkway covers 253.6 square feet. If the remaining area is used for gardens, how much area will Mr. Hooper have for his gardens?

45. **Sales** Jeff Simons sold 27 cars last month. This is 36 fewer cars than he sold during the same time period one year ago. What were his sales one year ago?

46. **Weather** The temperature at mid-afternoon was 12°C. By early evening, the temperature was -7°C. What was the temperature change?

Mixed Review

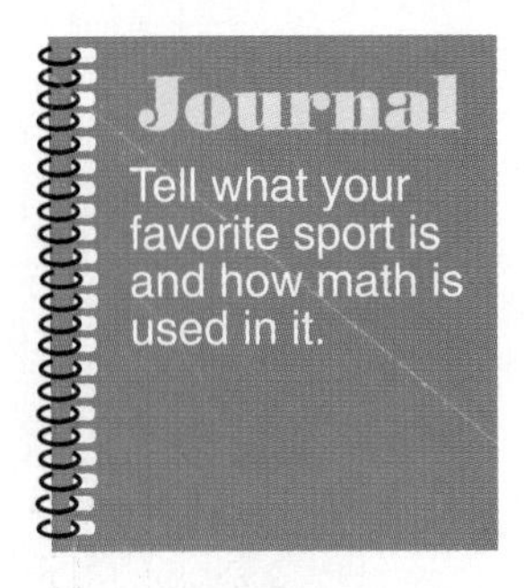

47. Evaluate $\frac{6ab^2}{x^3 + y^2}$ if $a = 6$, $b = 4$, $x = 0.2$, and $y = 1.3$. **(Lesson 1-2)**

48. Find the sum: $-21 + 52$. **(Lesson 2-2)**

49. Find the difference: $-67.1 - (-38.2)$. **(Lesson 2-5)**

50. Simplify $\frac{7a + 35}{-7}$. **(Lesson 2-7)**

51. Define the variable, then write an equation.
Juan has 15 pennies and 8 more nickels than dimes. He has 51 coins in all. How many dimes does he have? **(Lesson 2-8)**

3-2 Solving Equations by Using Subtraction

Objective

After studying this lesson, you should be able to:

- solve equations by using subtraction.

You are now familiar with the addition property of equality that states that when the same number is added to each side of an equation, an equivalent equation results. If the same number is subtracted from each side of an equation, the result is also an equivalent equation. Consider the following example.

Example 1

Nolan Young has bench pressed 190 pounds. He is working toward a goal of pressing 215 pounds. How many more pounds does he need to bench press to reach his goal?

Let x = the number of pounds still needed. Since an amount must be added to 190 pounds to reach the goal of 215 pounds, the equation is $190 + x = 215$.

$$190 + x = 215$$
$$190 - 190 + x = 215 - 190$$
$$x + 0 = 25$$
$$x = 25$$

Nolan needs to press 25 pounds more.

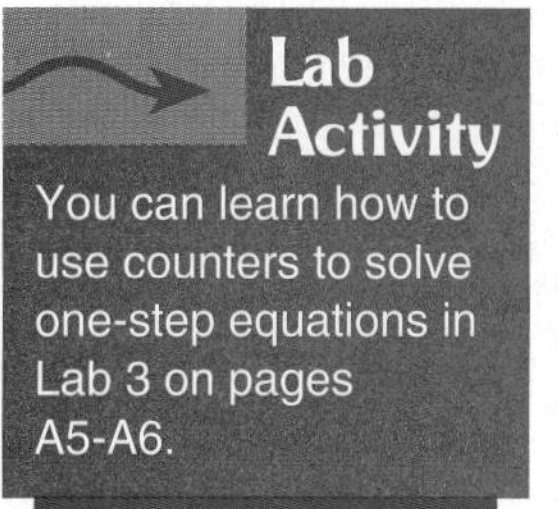

The property that is used to subtract the same number from each side of an equation is called the **subtraction property of equality.**

Subtraction Property of Equality	**For any numbers a, b, and c, if $a = b$, then $a - c = b - c$.**

Most equations can be solved in two ways. Recall that subtracting a number is the same as adding its inverse.

Example 2

Solve $x + 15 = -6$ in two ways.

Method 1:

$x + 15 = -6$ *Use the subtraction property of equality.*

$x + 15 - 15 = -6 - 15$ *Subtract 15 from each side.*

$x = -21$

(continued on the next page)

Check: $x + 15 = -6$

$$-21 + 15 \stackrel{?}{=} -6$$

$$-6 = -6 \checkmark$$

Method 2: $x + 15 = -6$ *Use the addition property of equality.*

$$x + 15 + (-15) = -6 + (-15)$$ *Add −15 to each side.*

$$x = -21$$ *The solutions are the same.*

The solution is −21.

Sometimes an equation can be solved more easily if it is first rewritten in a different form.

Example 3

Solve $b - (-8) = 23$.

This equation is equivalent to $b + 8 = 23$. *Why?*

$$b + 8 = 23$$

$$b + 8 - 8 = 23 - 8$$ *Subtract 8 from each side.*

$$b = 15$$

Check: $b - (-8) = 23$

$$15 - (-8) \stackrel{?}{=} 23$$

$$15 + 8 \stackrel{?}{=} 23$$

$$23 = 23 \checkmark$$

The solution is 15.

Example 4

Solve $y + (-7.5) = -12.2$.

This equation is equivalent to $y - 7.5 = -12.2$. *Why?*

$$y - 7.5 = -12.2$$

$$y - 7.5 + 7.5 = -12.2 + 7.5$$ *Add 7.5 to each side.*

$$y = -4.7$$

Check: $y + (-7.5) = -12.2$

$$-4.7 + (-7.5) \stackrel{?}{=} -12.2$$

$$-12.2 = -12.2 \checkmark$$

The solution is −4.7.

CHECKING FOR UNDERSTANDING

Communicating Mathematics

Read and study the lesson to answer each question.

1. Explain why it was not really necessary to state a subtraction property of equality.
2. In Example 1, if Nolan's goal was 225 pounds, how many more pounds would he need to bench press to reach his goal?

Guided Practice

State the number you would subtract from each side of the equation to solve it.

3. $m + 16 = 14$
4. $k + 9 = -16$
5. $t + 5 = 8$
6. $y + 9 = -53$
7. $z + (-3) = -8$
8. $x + (-4) = -37$

Rename each expression by using its inverse operation.

9. $m + (-8)$
10. $y - (-11)$
11. $z + (-31)$
12. $p - (-47)$

Solve and check each equation.

13. $y + 16 = 7$
14. $b + 15 = -32$
15. $x + (-8) = -31$
16. $d - (-27) = 13$

EXERCISES

Practice

Solve and check each equation.

17. $18 + m = -57$
18. $y + 3 = -15$
19. $y + 2.3 = 1.5$
20. $2.4 = m + 3.7$
21. $h - 26 = -29$
22. $-15 + d = 13$
23. $16 - y = 37$
24. $41 = 32 - r$
25. $k + (-13) = 21$
26. $z + (-17) = 0$
27. $m - (-13) = 37$
28. $-27 - b = -7$
29. $t - (-16) = 9$
30. $y + (-13) = -27$
31. $-\frac{5}{8} + w = \frac{5}{8}$
32. $x - \left(-\frac{5}{6}\right) = \frac{2}{3}$

Write an equation and solve.

33. The sum of two integers is -23. One of the integers is 9. What is the other integer?

34. Eighty-two increased by some number is -34. Find the number.

35. A number increased by -56 is -82. Find the number.

36. What number decreased by 45 is -78?

37. What number decreased by -67 is -34?

38. The difference of a number and -23 is 35. Find the number.

Critical Thinking

39. Suppose the solution of an equation is a number n. Is it possible for $-n$ to also be a solution of this equation? If so, name an example.

Applications

Define a variable, write an equation, and solve each problem.

40. **Baseball** In a mid-season slump, the Yankees scored only 17 runs in 9 games. Their opponents scored 41 runs. By how many runs did their opponents outscore them?

41. **Skiing** Lisa Thorson skied down the slalom run in 139.8 seconds. This was 13.7 seconds slower than her best time. What was her best time?

42. **Finance** Shares of stock in Olympia Motors were listed at $37\frac{3}{4}$ per share when the market opened. When the market closed, the shares had dropped $2\frac{1}{8}$ points. What was the new listing?

43. **Farming** A rancher lost 47 cattle because of the summer drought. His herd now numbers 396. How large was the herd before the drought?

Mixed Review

44. Is the sentence $0.101 > 0.110$ *true* or *false*? **(Lesson 1-3)**

45. Graph $\{0, 2, 6\}$ on a number line. **(Lesson 2-1)**

46. Find a number between $-\frac{8}{17}$ and $\frac{1}{9}$. **(Lesson 2-4)**

47. Simplify $3(-4) + 2(-7)$. **(Lesson 2-6)**

48. Solve $x + (-7) = 36$. **(Lesson 3-1)**

49. Solve $r - 21 = -37$. **(Lesson 3-1)**

3-3 Solving Equations by Using Multiplication and Division

Objectives

After studying this lesson, you should be able to:

- solve equations by using multiplication, and
- solve equations by using division.

Application

Sheet metal is often used as a roofing material. The pitch of a roof is important for the sheet metal technician to know. The pitch is equal to the rise, or height of the roof, divided by the span, or width of the roof.

$$\text{pitch} = \frac{\text{rise}}{\text{span}}$$

Hector Reyes is a sheet metal technician for Reliable Roofing Company. He wants to find the pitch of a roof with a rise of 11 feet and a span of 22 feet.

Lab Activity

You can learn how to use counters to solve one-step equations in Lab 3 on pages A5-A6.

Let p = the pitch of the roof. Then, $p = \text{rise} \div \text{span}$ or $11 \div 22$.

$p = \frac{11}{22}$

$p = \frac{1}{2}$

The pitch is $\frac{1}{2}$.

Suppose Hector knew that the pitch of a certain roof was $\frac{5}{12}$ and the span was 24 feet. How would he find the rise? Let r = the rise. Then solve the equation $\frac{5}{12} = \frac{r}{24}$.

To solve this equation, you would use the **multiplication property of equality.**

Multiplication Property of Equality	**For any numbers a, b, and c, if $a = b$, then $ac = bc$.**

Example 1

Solve $\frac{5}{12} = \frac{r}{24}$.

$$\frac{5}{12} = \frac{r}{24}$$

$$24\left(\frac{5}{12}\right) = 24\left(\frac{r}{24}\right) \quad \textit{Multiply each side by 24.}$$

$$10 = r$$

Check: $\frac{5}{12} = \frac{r}{24}$

$$\frac{5}{12} \stackrel{?}{=} \frac{10}{24} \quad \textit{Replace r with 10.}$$

$$\frac{5}{12} = \frac{5}{12} \ \checkmark$$

The solution is 10, so the rise of the roof is 10 feet.

Example 2

Solve $\left(2\frac{1}{3}\right)m = 3\frac{1}{9}$.

$$\left(2\frac{1}{3}\right)m = 3\frac{1}{9}$$

$$\frac{7}{3}m = \frac{28}{9} \quad \textit{Rewrite the mixed numbers as improper fractions.}$$

$$\frac{3}{7}\left(\frac{7}{3}m\right) = \frac{3}{7}\left(\frac{28}{9}\right) \quad \textit{Multiply each side by } \tfrac{3}{7}\textit{, the reciprocal of } \tfrac{7}{3}.$$

$$m = \frac{4}{3} \quad \textit{Check this result.}$$

The solution is $\frac{4}{3}$.

Example 3

Solve $24 = -2a$.

$$24 = -2a$$

$$-\frac{1}{2}(24) = -\frac{1}{2}(-2a) \quad \textit{Multiply each side by } -\tfrac{1}{2}.$$

$$-12 = a \quad \textit{Check this result.}$$

The solution is -12.

The equation $24 = -2a$ was solved by multiplying each side by $-\frac{1}{2}$. The same result could have been obtained by dividing each side by -2. This method uses the **division property of equality.**

Division Property of Equality	**For any numbers a, b, and c, with $c \neq 0$, if $a = b$, then $\frac{a}{c} = \frac{b}{c}$.**

The division property of equality is often easier to use than the multiplication property of equality.

Example 4 **Solve $-6x = 11$.**

$$-6x = 11$$
$$\frac{-6x}{-6} = \frac{11}{-6} \quad \textit{Divide each side by } -6.$$
$$x = -\frac{11}{6}$$

Check:
$$-6x = 11$$
$$-6\left(-\frac{11}{6}\right) \stackrel{?}{=} 11$$
$$11 = 11 \checkmark$$

The solution is $-\frac{11}{6}$.

Example 5

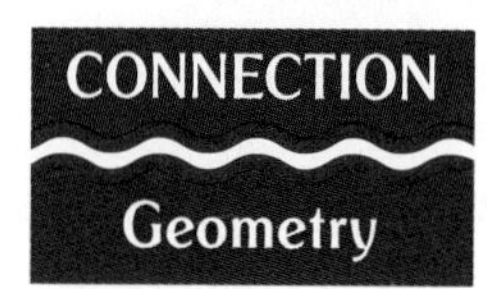

The area of the rectangle at the right is 28 cm². Find the length.

$$A = \ell w$$
$$28 = \ell(3)$$
$$\frac{28}{3} = \frac{3\ell}{3}$$
$$9.\overline{3} = \ell$$

The length of the rectangle is $9.\overline{3}$ cm.

CHECKING FOR UNDERSTANDING

Communicating Mathematics

Read and study the lesson to answer each question.

1. Write a formula for the pitch of a roof. Let p = the pitch, r = the rise, and s = the span.
2. Can the division property of equality ever be used in place of the multiplication property of equality? When?
3. Write the multiplication property of equality in your own words.
4. Write the division property of equality in your own words.

Guided Practice

State the number by which you would multiply each side to solve each equation.

5. $\frac{b}{3} = -6$
6. $\frac{x}{5} = 10$
7. $\frac{3}{4}n = 30$
8. $-\frac{5}{9}x = 15$
9. $-8n = 24$
10. $1 = \frac{k}{9}$

State the number by which you would divide each side to solve each equation.

11. $4x = 24$
12. $35 = 4y$
13. $-36 = 4z$
14. $-5x = 14$
15. $-8x = -9$
16. $-6x = -36$

EXERCISES

Practice

Solve and check each equation.

17. $-4r = -28$
18. $-8t = 56$
19. $5x = -45$
20. $-5s = -85$
21. $9x = 40$
22. $-3y = 52$
23. $3w = -11$
24. $434 = -31y$
25. $42.51x = 8$
26. $5c = 8$
27. $17b = -391$
28. $0.49x = 6.277$
29. $\frac{k}{8} = 6$
30. $11 = \frac{x}{5}$
31. $-10 = \frac{b}{-7}$
32. $\frac{h}{11} = -25$
33. $-65 = \frac{f}{29}$
34. $\frac{c}{-8} = -14$
35. $\frac{2}{5}t = -10$
36. $\frac{4}{9}t = 72$
37. $-\frac{3}{5}y = -50$
38. $-\frac{11}{8}x = 42$
39. $-\frac{13}{5}y = -22$
40. $\frac{5}{2}x = -25$
41. $3x = 4\frac{2}{3}$
42. $-5x = -3\frac{2}{3}$
43. $\left(-4\frac{1}{2}\right)x = 36$

Write an equation and solve.

44. Eight times a number is 216. What is the number?
45. Negative twelve times a number is -156. What is the number?
46. Negative seven times a number is 1.476. What is the number?
47. One fourth of a number is -16.325. What is the number?
48. Four thirds of a number is 4.82. What is the number?

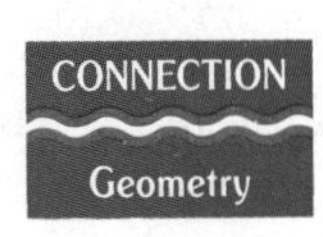

Find the missing measure.

49. $A = ?\ in^2$

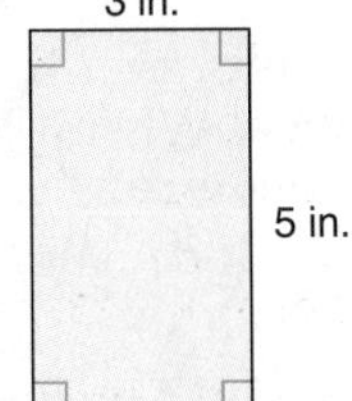

50. $A = 49\ cm^2$

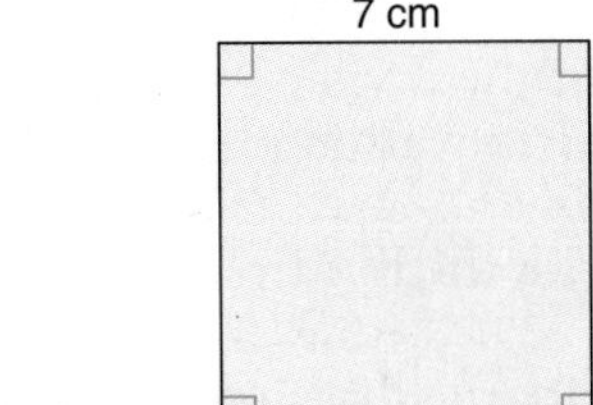

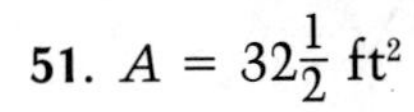
51. $A = 32\frac{1}{2}\ ft^2$

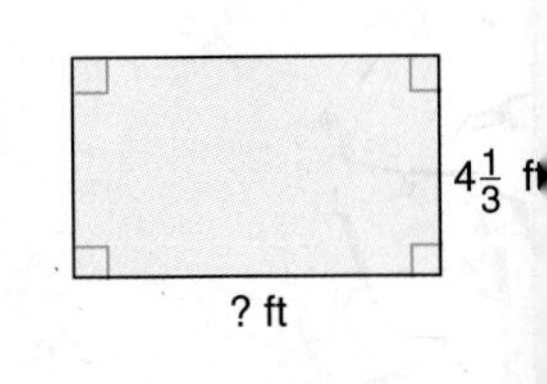

Journal

Draw a model and explain how to find the length of a rectangle if its area is 51 m^2 and its length is 17 m.

Complete.

52. If $3x = 15$, then $9x = \underline{?}$.
53. If $10y = 46$, then $5y = \underline{?}$.
54. If $2a = -10$, then $-6a = \underline{?}$.
55. If $12b = -1$, then $4b = \underline{?}$.
56. If $7k - 5 = 4$, then $21k - 15 = \underline{?}$.

Critical Thinking

57. A number is multiplied by two, then squared, then divided by the multiplicative identity. The result is zero. What was the number?

Applications

Define a variable, write an equation, and solve each problem.

58. **Sports** Joyce Conners paid $47.50 for five football tickets. What was the cost per ticket?
59. **Sales** A store sells a six-pack of ginger ale for $2.28. Each time Mika buys ginger ale, he sells the empty cans to a recycler at a rate of 1¢ per can. How many cans of ginger ale did Mika buy if the net amount he spent after recycling the cans was $7.40?

Mixed Review

60. Simplify $4[1 + 4(5x + 2y)]$. **(Lesson 1-6)**
61. Find the next three numbers in the pattern 37, 26, 15, $\underline{?}$, $\underline{?}$, $\underline{?}$. **(Lesson 2-1)**

Solve. (Lessons 3-1 and 3-2)

62. $d + (-6) = -9$
63. $x - (-33) = 14$

MID-CHAPTER REVIEW

Solve and check each equation. (Lessons 3-1, 3-2, 3-3)

1. $4.4 = b + 6.3$
2. $z + (-18) = 34$
3. $y - 7 = -32$
4. $r - (-31) = 16$
5. $-19 - s = 41$
6. $6x = -42$
7. $-13 = \frac{b}{-8}$
8. $\frac{3}{4}x = -12$
9. $\left(5\frac{1}{2}\right)x = 33$
10. **Budgeting** The total of Jon Young's gas bill and electric bill was $210.87. His electric bill was $95.25. How much was his gas bill? **(Lesson 3-2)**
11. **Plumbing** Two meters of copper tubing weighs 0.25 kilograms. How much does 50 meters of the same tubing weigh? **(Lesson 3-3)**

3-4 Problem-Solving Strategy: Work Backwards

Objective After studying this lesson, you should be able to:

- solve problems by working backwards.

Problem

Four families went to a baseball game. A vendor selling bags of popcorn came by. The Wilson family bought half of the bags of popcorn plus one. The Martin family bought half of the remaining bags of popcorn plus one. The Perez family bought half of the remaining bags of popcorn plus one. And the Royster family bought half of the remaining bags of popcorn plus one, leaving the vendor with no bags of popcorn. If the Roysters bought 2 bags of popcorn, how many bags did the four families buy?

Sometimes you can work backwards to solve problems. Make a table to show what happened.

Family	Bags Left	+	Bags Bought	=	Original Number of Bags
Royster	0		1 + 1		2
Perez	2		1 + 3		6
Martin	6		1 + 7		14
Wilson	14		1 + 15		30

So the families bought 30 bags of popcorn.

Most problems can be solved using one of several strategies. You may find the following strategies helpful as you solve the problems in this lesson.

- work backwards
- make a table
- guess and check
- act it out
- solve a simpler (or a similar) problem
- look for a pattern
- make a diagram
- eliminate possibilities

CHECKING FOR UNDERSTANDING

Communicating Mathematics

Read and study the lesson to answer each question.

1. How could you use the guess-and-check strategy to solve the problem?
2. How can you check the solution to the problem?

Guided Practice

Solve by working backwards.

3. An ice sculpture is melting at a rate of half its weight every hour. After 8 hours, it weighs $\frac{5}{16}$ of a pound. How much did it weigh in the beginning?

4. A number is decreased by 35, then multiplied by 6, then added to 87, then divided by 3. The result is 67. What is the number?

5. Kristin spent one fifth of her money for gasoline. Then she spent half of what was left for a haircut. She bought lunch for $7. When she got home, she had $13 left. How much did Kristin have originally?

EXERCISES

Practice

Solve. Use any strategy.

6. The digits below are in a special order. What order is it?

   0, 2, 3, 6, 7, 1, 9, 4, 5, 8

7. Four dominoes are shown below. Arrange the dominoes into a domino donut so that all sides equal the same sum.

Example:

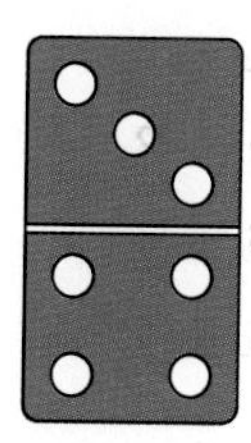 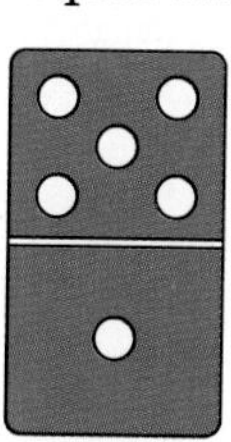 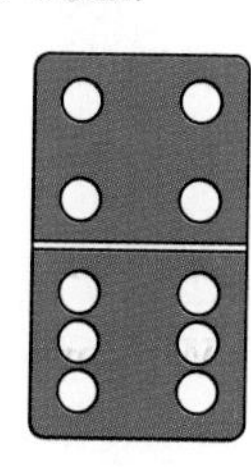 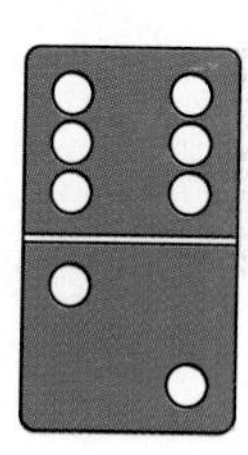

8. A bacteria population triples in number each day. If there are 2,187,000 bacteria on the seventh day, how many bacteria were there on the first day?

FYI . . .

The most dominoes set up and toppled by one person is 281,581. It took about 13 minutes for all of them to fall.

9. **Number Theory** A number that is *balanced* has exactly one digit that is the sum of all of the other digits. For example, 1236 is balanced because 1 + 2 + 3 = 6. What is the greatest three-digit balanced number?

10. Paper plates can be purchased in packages of 15 or 25. Joe Spanato bought 7 packages and got 125 plates. How many packages of 25 did he buy?

11. Explain the placement of numbers in the grid at the right. (*Hint:* Think of the words that the numbers represent.)

1	6	2
5	4	9
8	7	3

12. In Mary Ann's garden, all of the flowers are either pink, yellow, or white. Given any three of the flowers, at least one of them is pink. Given any three of the flowers, at least one of them is white. Can you say that given any three of the flowers, at least one of them must be yellow? Why?

Strategies

Look for a pattern.
Solve a simpler problem.
Act it out.
Guess and check.
Draw a diagram.
Make a chart.
Work backwards.

13. Patty Lee invited 17 people to her party. She gave each person a number from 2 to 18, and kept the number 1 for herself. After a while, Patty noticed that the sum of the numbers on each couple dancing was a perfect square. What was the number of Patty's partner?

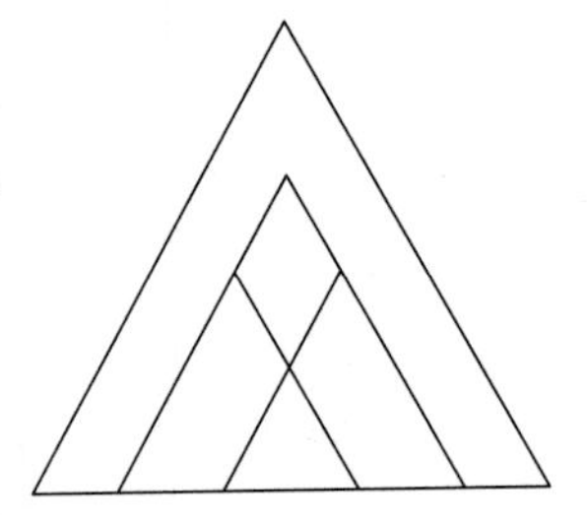

14. What is the least number of colors needed to paint the regions in the picture? No two regions that share a border can be colored the same color.

15. Mr. McCutcheon has been asked to direct the school play. After holding auditions, he finds four boys and five girls capable of being in the play. Unfortunately, Paul won't join the cast unless Kim is in it; Kim won't join if either Carlos or John is in it; neither Carlos nor John will join unless Marquita is in it; if either Marquita or Kim is in the cast, Lori will not be in it; unless Lori is in it, Renee won't be in it. Maria and Kevin agree to be in the play no matter what cast is chosen. If the cast has two boys and three girls, who should Mr. McCutcheon choose?

COOPERATIVE LEARNING ACTIVITY

Work in groups of three. Each person in the group must understand the solution and be able to explain it to any person in the class.

The names of eight colors—*blue, brown, green, maroon, orange, red, white,* and *yellow*—can be placed in sections of the wheel at the right. Use the following clues to fill each section.

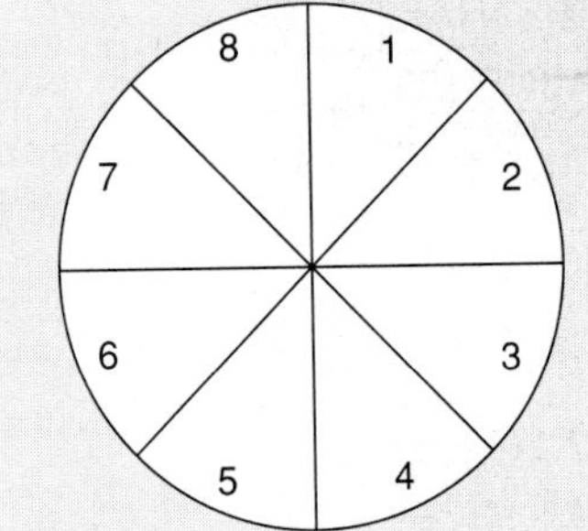

1. No two words next to each other are the same length.
2. No two words with double letters are next to each other.
3. The word in Section 1 has no letters in common with the words in either section next to it.
4. The word in Section 3 has the same number of letters as the word in Section 7; the same is true for Sections 4 and 8.
5. The words *red, yellow,* and *blue* do not all appear in the same half of the wheel.

3-5 Solving Equations Using More Than One Operation

Objective

After studying this lesson, you should be able to:

- solve equations involving more than one operation.

Application

If you are playing in the outfield during a baseball game and a fly ball is hit in your direction, there are several steps that you must take in order to catch the ball. Some of these steps include the following.

1. Watch the direction that the ball is going in.
2. Begin running in that direction.
3. Estimate where the ball will land.
4. Gauge your running speed to be there in time.
5. Have your glove in the correct position to catch the ball.

Solving some types of equations requires more than one step. To solve an equation with more than one operation, undo the operations in reverse order. In other words, work backwards.

Example 1

APPLICATION
Finance

Bonnie Huston sold some stock for $42 per share. This was $10 per share more than twice what she paid for it. What was the price when she bought the stock?

EXPLORE Read the problem and define a variable.

Let p = the price Bonnie paid.

PLAN Write an equation.

$42 equals $10 more than twice what she paid

$42 = 10 + 2p$

SOLVE Solve the equation and answer the problem.

Work backwards to solve the equation.

$42 = 10 + 2p$

$42 - 10 = 10 - 10 + 2p$ *Undo the addition first. Use the subtraction property of equality.*

$32 = 2p$

$\frac{32}{2} = \frac{2p}{2}$ *Then undo the multiplication. Use the division property of equality.*

$16 = p$

Bonnie originally paid $16 per share.

EXAMINE Check to see if the answer makes sense.

Twice $16 is $32. Ten more than twice $16 is $42.

Example 2

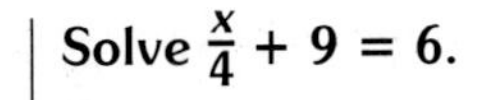

Solve $\frac{x}{4} + 9 = 6$.

$$\frac{x}{4} + 9 = 6$$

$$\frac{x}{4} + 9 - 9 = 6 - 9 \qquad \textit{First, subtract 9 from each side. Why?}$$

$$\frac{x}{4} = -3$$

$$4\left(\frac{x}{4}\right) = 4(-3) \qquad \textit{Then, multiply each side by 4.}$$

$$x = -12$$

Check: $\frac{x}{4} + 9 = 6$

$$\frac{-12}{4} + 9 \stackrel{?}{=} 6$$

$$-3 + 9 \stackrel{?}{=} 6$$

$$6 = 6 \ \checkmark$$

The solution is -12.

Lab Activity

You can learn how to use counters to solve equations in Lab 4 on page A7.

Example 3

Solve $\frac{d - 4}{3} = 5$.

$$\frac{d - 4}{3} = 5$$

$$3\left(\frac{d - 4}{3}\right) = 3(5) \qquad \textit{Multiply each side by 3. Why?}$$

$$d - 4 = 15$$

$$d - 4 + 4 = 15 + 4 \qquad \textit{Add 4 to each side.}$$

$$d = 19$$

Check: $\frac{d - 4}{3} = 5$

$$\frac{19 - 4}{3} \stackrel{?}{=} 5$$

$$\frac{15}{3} \stackrel{?}{=} 5$$

$$5 = 5 \ \checkmark$$

The solution is 19.

Consecutive numbers are numbers in counting order such as 3, 4, 5. Beginning with an even integer and counting by two will result in *consecutive even integers.* For example, -6, -4, -2, 0, and 2 are consecutive even integers. Beginning with an odd integer and counting by two will result in *consecutive odd integers.* For example, -1, 1, 3, and 5 are consecutive odd integers.

The study of odd and even numbers as well as the study of numbers in general is called **number theory**.

Example 4

Find three consecutive even integers whose sum is -12.

Let x = the least even integer.
Then $x + 2$ = the next greater even integer,
and $x + 4$ = the greatest of the three even integers.

$$x + (x + 2) + (x + 4) = -12$$
$$3x + 6 = -12$$
$$3x + 6 - 6 = -12 - 6$$
$$3x = -18$$
$$\frac{3x}{3} = \frac{-18}{3}$$
$$x = -6$$

$$x + 2 = -6 + 2 \qquad x + 4 = -6 + 4$$
$$x = -4 \qquad x = -2$$

The integers are -6, -4, and -2. *Does the answer make sense?*

CHECKING FOR UNDERSTANDING

Communicating Mathematics

Read and study the lesson to complete the following.

1. How do you undo addition? division?
2. Write an example of an equation requiring more than one operation to solve.
3. If n is an even integer, explain how to find the even integer just before it.

Guided Practice

Explain how to solve each equation. Then solve.

4. $3x - 7 = 2$
5. $8 + 3x = 5$
6. $\frac{a + 2}{5} = 10$
7. $-\frac{4}{13}y - 7 = 6$

List three consecutive integers that satisfy each condition.

8. the least one is -2
9. even, the greatest one is 10
10. odd, the least one is -7

For each sentence, define a variable. Then write an equation.

11. The sum of two consecutive integers is 17.
12. The sum of three consecutive even integers is 48.
13. The sum of two consecutive odd integers is -36.
14. Seventeen decreased by twice a number is 5.

ERCISES

Practice

Solve.

15. $4t - 7 = 5$
16. $6 = 4n + 2$
17. $4 + 7x = 39$
18. $34 = 8 - 2t$
19. $-3x - 7 = 18$
20. $0.2n + 3 = 8.6$
21. $\frac{3}{4}n - 3 = 9$
22. $7 = 3 - \frac{n}{3}$
23. $7 = \frac{x}{2} + 5$
24. $\frac{y}{3} + 6 = -45$
25. $\frac{c}{-4} - 8 = -42$
26. $\frac{d + 5}{3} = -9$
27. $\frac{3 + n}{7} = -5$
28. $5 = \frac{m - 5}{4}$
29. $16 = \frac{s - 8}{-7}$
30. $\frac{4d + 5}{7} = 7$
31. $\frac{7n + (-1)}{8} = 8$
32. $\frac{-3n - (-4)}{-6} = -9$

Define a variable, write an equation, and solve each problem. Some problems may have no solution.

33. Find three consecutive integers whose sum is 87.
34. Find four consecutive integers whose sum is 130.
35. Find two consecutive even integers whose sum is 115.
36. The lengths of the sides of a triangle are consecutive odd integers. The perimeter is 27 meters. What are the lengths of the sides?
37. Find four consecutive even integers such that twice the least increased by the greatest is 96.

Critical Thinking

38. Write an expression for the sum of three consecutive odd integers where $2n - 1$ is the smallest integer.

Applications

Define a variable, write an equation, and solve each problem.

39. **Sales** Karen has 6 more than twice as many newspaper customers as when she started selling newspapers. She now has 98 customers. How many did she have when she started?
40. **Baseball** One season, Reggie Walker scored 9 more than twice the number of runs he batted in. He scored 117 runs that season. How many runs did he bat in?
41. **Statistics** Namid has an average of 76 on four tests. What score does he have to get on the 100-point final test if it counts double and he wants to have an average of 80 or better? Is it possible for Namid to have an average of 85?

42. **Running** In cross-country, a team's score is the sum of the place numbers of the first five finishers on the team. The captain of a team placed second in a meet. The next four finishers on the team placed in consecutive order. The team score was 40. In what places did the other members finish?

Computer

43. The BASIC program at the right finds three consecutive odd or even integers for a given sum. Not all numbers can be expressed as the sum of three consecutive odd or even integers. Line 40 checks for these numbers. Use the program to answer the following questions.

```
10 PRINT "ENTER THE SUM."
20 INPUT S
30 LET N = (S-6)/3
40 IF N = INT(N) THEN 70
50 PRINT "NO SOLUTION."
60 GOTO 80
70 PRINT "THE NUMBERS ARE ";
   N;", "; N+2;", AND "; N+4
80 END
```

a. Find five sums of three consecutive odd integers. What do these sums have in common?

b. Find five sums of three consecutive even integers. What do these sums have in common?

c. Find five numbers that are not the sum of three consecutive odd or even integers. What do these numbers have in common?

Mixed Review

44. Name the property illustrated by the following statement. If $6 = 2a$ and $a = 3$, then $6 = 2 \cdot 3$. **(Lesson 1-4)**

45. Evaluate $m + 17$ if $m = -6$. **(Lesson 2-2)**

46. Which is greater, $\frac{0.06}{0.4}$ or $\frac{0.9}{5}$? **(Lesson 2-4)**

Solve. (Lessons 3-1, 3-3)

47. $d - 27 = -63$

48. $-7w = -49$

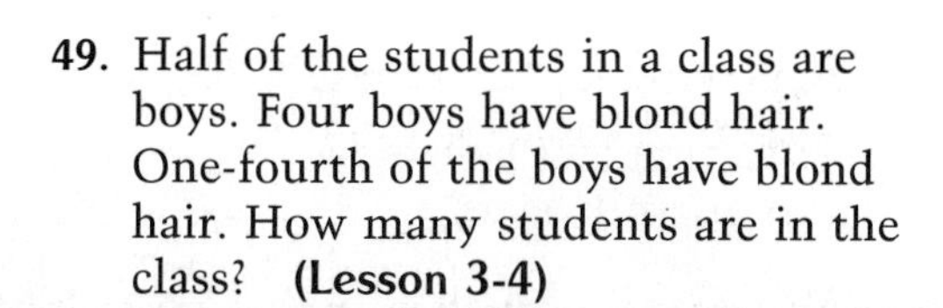

49. Half of the students in a class are boys. Four boys have blond hair. One-fourth of the boys have blond hair. How many students are in the class? **(Lesson 3-4)**

PUZZLE

Copy.

Arrange the digits 1 through 8 in the grid at the right so that no two consecutive integers occupy neighboring squares horizontally, vertically, or diagonally.

<table>
<tr><td></td><td>?</td><td>?</td><td></td></tr>
<tr><td>?</td><td>?</td><td>?</td><td>?</td></tr>
<tr><td></td><td>?</td><td>?</td><td></td></tr>
</table>

3-6 Solving Equations with the Variable on Both Sides

Objectives

After studying this lesson, you should be able to:

- solve equations with the variable on both sides, and
- solve equations containing grouping symbols.

Application

FYI . . .

Dallas serves as the headquarters of more oil companies than any other U.S. city. More cars are produced in the Detroit area than anywhere else in the U.S.

Many equations contain variables on each side. To solve these types of equations, first use the addition or subtraction property of equality to write an equivalent equation that has all of the variables on one side. Then solve the equation.

In 1980, the population of Detroit, Michigan was 1,203,000. During the 1970s, the population decreased at an average rate of 31,100 people per year. In 1980, the population of Dallas, Texas was 905,000. During the 1970s, the population increased at an average rate of 6100 people per year. Suppose the population of each city continued to increase or decrease at these rates. In how many years would the populations of the two cities be the same?

After y years, the population of Detroit would be $1{,}203{,}000 - 31{,}100y$. After y years, the population of Dallas would be $905{,}000 + 6100y$. The populations would be the same when the two expressions are equal.

Use a calculator.

$$1{,}203{,}000 - 31{,}100y = 905{,}000 + 6100y$$

Add 31,100y to each side.

$$1{,}203{,}000 - 31{,}100y + 31{,}100y = 905{,}000 + 6100y + 31{,}100y$$

$$1{,}203{,}000 = 905{,}000 + 37{,}200y$$

Subtract 905,000 from each side.

$$1{,}203{,}000 - 905{,}000 = 905{,}000 - 905{,}000 + 37{,}200y$$

$$298{,}000 = 37{,}200y$$

Divide each side by 37,200.

$$\frac{298{,}000}{37{,}200} = \frac{37{,}200y}{37{,}200}$$

$$8 \approx y$$

At these rates, the populations would be the same after about 8 years.

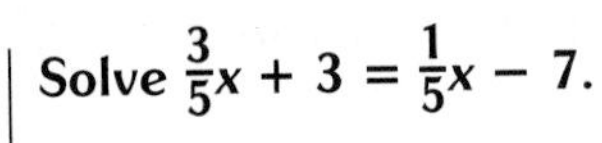

Example 1 **Solve $\frac{3}{5}x + 3 = \frac{1}{5}x - 7$.**

$$\frac{3}{5}x + 3 = \frac{1}{5}x - 7$$

$$\frac{3}{5}x - \frac{1}{5}x + 3 = \frac{1}{5}x - \frac{1}{5}x - 7 \quad \textit{Subtract } \tfrac{1}{5}x \textit{ from each side.}$$

$$\frac{2}{5}x + 3 = -7$$

$$\frac{2}{5}x + 3 - 3 = -7 - 3 \quad \textit{Subtract 3 from each side.}$$

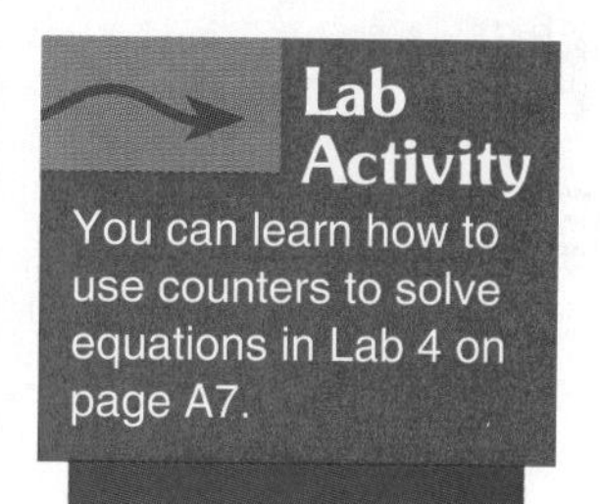

$$\frac{2}{5}x = -10$$

$$\frac{5}{2}\left(\frac{2}{5}x\right) = \frac{5}{2}(-10) \quad \textit{Multiply each side by } \tfrac{5}{2}.$$

$$x = -25 \quad \textit{Check this result.}$$

The solution is -25.

Many equations also contain grouping symbols. When solving equations of this type, first use the distributive property to remove the grouping symbols.

Example 2

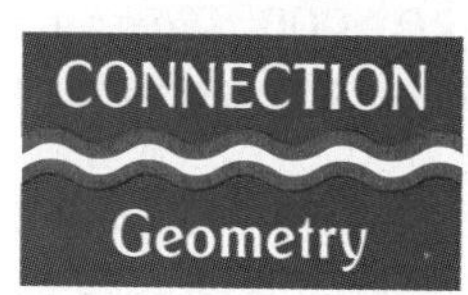

The perimeter of a rectangle is 148 inches. Find its dimensions if the length is 17 inches greater than three times the width.

Let w = the width in inches.
Then $3w + 17$ = the length in inches.
Recall that the formula for perimeter can be expressed as $2w + 2\ell = P$.

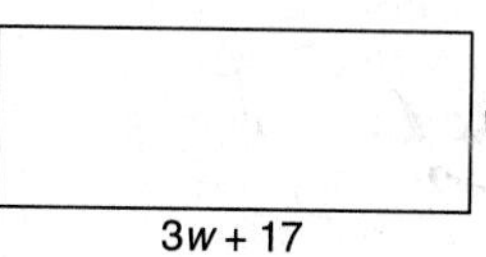

$$2w + 2(3w + 17) = 148$$

$$2w + 6w + 34 = 148 \quad \textit{Use the distributive property.}$$

$$8w + 34 = 148 \quad \textit{Simplify.}$$

$$8w + 34 - 34 = 148 - 34$$

$$8w = 114$$

$$\frac{8w}{8} = \frac{114}{8}$$

$$w = \frac{114}{8} \text{ or } 14\frac{1}{4} \quad \textit{Check this result.}$$

$$3w + 17 = 3\left(\frac{114}{8}\right) + 17 \text{ or } 59\frac{3}{4}$$

The width is $14\frac{1}{4}$ inches and the length is $59\frac{3}{4}$ inches.

Some equations may have no solution. That is, there is no value of the variable that will result in a true equation.

Example 3

Solve $2x + 5 = 2x - 3$.

$$2x + 5 = 2x - 3$$

$$2x - 2x + 5 = 2x - 2x - 3$$

$$5 = -3$$

Since $5 = -3$ is a false statement, this equation has no solution.

Some equations may have *every number* in their solution sets. An equation that is true for every value of the variable is called an **identity**.

Example 4

Solve $3(x + 1) - 5 = 3x - 2$.

$$3(x + 1) - 5 = 3x - 2$$
$$3x + 3 - 5 = 3x - 2$$
$$3x - 2 = 3x - 2 \quad \textit{Reflexive property of equality}$$

Since the expressions on each side of the equation are the same, this equation is an identity. The statement $3(x + 1) - 5 = 3x - 2$ is true for all values of x.

CHECKING FOR UNDERSTANDING

Communicating Mathematics

Read and study the lesson to answer each question.

1. Do some research: In 1988, what were the populations of Detroit and Dallas? Were they nearly the same?
2. What do you think are the reasons why the populations of Detroit and Dallas were or were not the same in 1988?
3. What is the difference between an identity and an equation with no solution?

Guided Practice

Explain how to solve each equation. Do *not* solve.

4. $3x + 2 = 4x - 1$
5. $8y - 10 = -3y + 2$
6. $4(3 + 5w) = -11$
7. $-7(x - 3) = -4$

Solve and check each equation.

8. $6x + 7 = 8x - 13$
9. $3(h + 2) = 12$
10. $7 - 3x = x - 4(2 + x)$
11. $-3(x + 5) = 3(x - 1)$

EXERCISES

Practice

Find the dimensions of each rectangle. The perimeter is given.

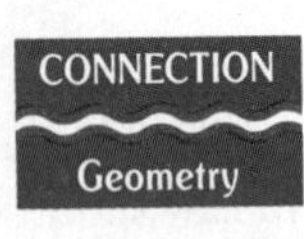

12. $P = 920$ m

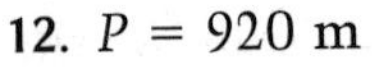

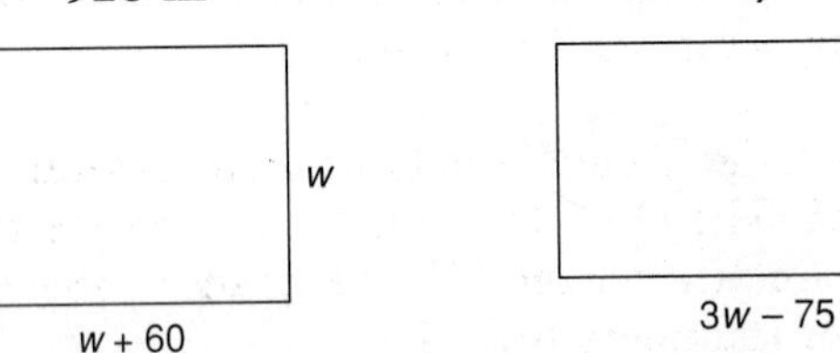

13. $P = 370$ yd

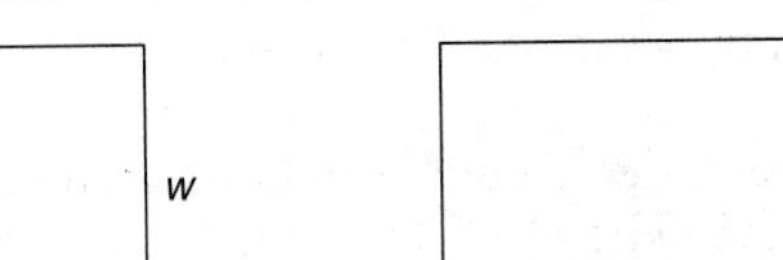

14. $P = 220$ ft

2w – 40

w

Solve and check each equation.

15. $3 - 4x = 10x + 10$

16. $17 + 2n = 21 + 2n$

17. $14b - 6 = -2b + 8$

18. $\frac{2}{3}n + 8 = \frac{1}{3}n - 2$

19. $18 - 3.8x = 7.36 - 1.9x$

20. $\frac{3}{4}n + 16 = 2 - \frac{1}{8}n$

21. $6(y + 2) - 4 = -10$

22. $3x - 2(x + 3) = x$

23. $7 + 2(x + 1) = 2x + 9$

24. $6 = 3 + 5(y - 2)$

25. $4(x - 2) = 4x$

26. $5x - 7 = 5(x - 2) + 3$

27. $5 - \frac{1}{2}(b - 6) = 4$

28. $5n + 4 = 7(n + 1) - 2n$

29. $4(2x - 1) = -10(x - 5)$

30. $-8(4 + 9x) = 7(-2 - 11x)$

31. $4(2a - 8) = \frac{1}{7}(49a + 70)$

32. $2(x - 3) + 5 = 3(x - 1)$

33. $-3(2n - 5) = \frac{1}{2}(-12n + 30)$

34. $2[x + 3(x - 1)] = 18$

Define a variable, write an equation, and solve each problem.

35. Twice a number increased by 12 is 31 less than three times the number. Find the number.

36. Twice the greater of two consecutive odd integers is 13 less than three times the lesser. Find the integers.

37. Three times the greatest of three consecutive even integers exceeds twice the least by 38. Find the integers.

Critical Thinking

38. Is the inequality $x > -x$ sometimes true, never true, or always true? Is the inequality $x < -x$ sometimes true, never true, or always true?

Applications

Define a variable, write an equation, and solve each problem.

39. **Sales** Last year, Marla Ames sold 7 sedans more than twice the number of vans Toshio Kanazawa sold. Marla sold 83 sedans. How many vans did Toshio sell?

40. **Running** One member of a cross-country team placed fourth in a meet. The next four finishers on the team placed in consecutive order, but farther behind. The team score was 70. In what places did the other members finish?

41. **Soccer** A soccer field is 75 yards shorter than 3 times its width. Its perimeter is 370 yards. Find its dimensions.

42. **Travel** Paloma Rey drove to work on Wednesday at 40 miles per hour and arrived one minute late. She left home at the same time on Thursday, drove 45 miles per hour, and arrived one minute early. How far does Ms. Rey drive to work? (*Hint:* Convert hours to minutes.)

Mixed Review

43. Write the following sentence as an equation.

    A is equal to the sum of *m* and the square of *n*. **(Lesson 1-7)**

44. Find the sum: $\frac{3}{4} + \left(-\frac{7}{12}\right)$. **(Lesson 2-5)**

45. Find the product: $(-2.93)(-0.003)$. **(Lesson 2-6)**

Solve. (Lessons 3-1 through 3-3 and 3-5)

46. $y - 7.3 = 5.1$

47. $w - (-37) = 28$

48. $\frac{9}{2}r = -30$

49. $7n - 4 = 17$

PUZZLE

"This thermometer's no good," complained Mrs. Weatherby. "It's in Celsius, and I want to know the temperature in Fahrenheit."

"You're so old-fashioned, Mother," chided Mercuria, with a quick, silvery laugh. "Just use the formula $\frac{9}{5}C + 32 = F$ (where *C* is the temperature in Celsius and *F* is the temperature in Fahrenheit)."

"My daughter, the science major!" Mrs. Weatherby rolled her eyes.

"If that's too hard for you, then just double the number on the thermometer and add 30. Of course," Mercuria added sadly, "you won't get a precise answer."

The two women each computed the temperature in degrees Fahrenheit, Mercuria with her formula and Mrs. Weatherby by the simpler approximation. Surprisingly, they got exactly the same answer.

Just how hot was it?

—reprinted from *Games* magazine

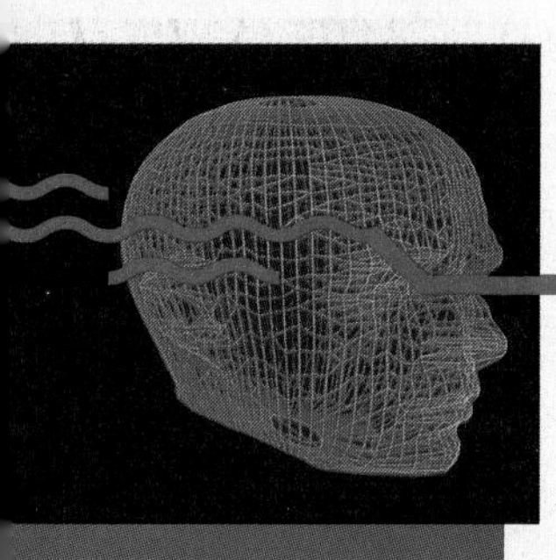

Technology

Solving Equations

BASIC
Graphing calculator
▶ Graphing software
Spreadsheets

The *Mathematics Exploration Toolkit* can help you practice the steps for solving equations. Since the computer performs all calculations and simplifying steps, you can concentrate on deciding which steps to choose. You will use the CALC commands listed below. You may enter the abbreviations in parentheses instead of typing the entire command.

Command	Purpose
ADD	Add the same expression to each side of the equation.
SUBTRACT (sub)	Subtract the same expression from each side of the equation.
MULTIPLY (mul)	Multiply each side of the equation by the same expression.
DIVIDE (div)	Divide each side of the equation by the same expression.
SIMPLIFY (simp) or [F5]	Simplify the equation.

Example 1 **Solve $6x = 3x + 18$.**

Enter	Result
6x=3x+18	$6x = 3x + 18$
sub 3x	$6x - 3x = 3x + 18 - 3x$
simp	$3x = 18$
div 3	$\frac{3x}{3} = \frac{18}{3}$
simp	$x = 6$

Example 2 **Solve $2(x - 2) = 3x - (4x - 5)$.**

Enter	Result
2(x−2)=3x−(4x−5)	$2(x - 2) = 3x - (4x - 5)$
simp	$2x - 4 = -x + 5$
add x	$2x - 4 + x = -x + 5 + x$
simp	$3x - 4 = 5$
add 4	$3x - 4 + 4 = 5 + 4$
simp	$3x = 9$
div 3	$\frac{3x}{3} = \frac{9}{3}$
simp	$x = 3$

EXERCISES

Use CALC to solve each equation. Record each step and solution.

1. $5x = x - 12$
2. $7 - x + 5 = -3 + 4x$
3. $3(1 - 6x) = 2x + 1$
4. $2 - (3x - 1) = 2(1 - 2x)$
5. $x = 4 - 5(x - 1) + 6x$
6. $3(x - 2) - 1 = x - (7 - 2x)$

3-7 More Equations

Objectives

After studying this lesson, you should be able to:

- solve equations containing fractions or decimals, and
- solve equations containing more than one variable.

Application

FYI . . .

Studies show that boys spend most of their money on food (43%) while girls spend most of their money (41%) on clothes.

Kelly and Todd Washington went shopping with their parents for back-to-school clothes. At the end of the day, the Washingtons had bought the same number of skirts, shirts, slacks, and jeans. They spent \$311.67. How many of each item did they buy?

Use the price list at the right to write an equation.

Price List

Skirts	\$29.99
Shirts	19.99
Slacks	24.95
Jeans	28.96

Let x = the number of each item that the Washingtons bought.

$$29.99x + 19.99x + 24.95x + 28.96x = 311.67$$

Since each decimal involves hundredths, multiply each side by 100 to clear the decimals.

$$100(29.99x + 19.99x + 24.95x + 28.96x) = 100(311.67)$$
$$100(29.99x) + 100(19.99x) + 100\ (24.95x) + 100(28.96x) = 100(311.67)$$
$$2999x + 1999x + 2495x + 2896x = 31{,}167$$
$$10{,}389x = 31{,}167$$
$$x = 3$$

So the Washingtons bought 3 skirts, 3 shirts, 3 pairs of slacks, and 3 pairs of jeans. *Check this result.*

You can also solve equations containing fractions by first using the multiplication property of equality to eliminate the fractions.

Example 1

Solve $\frac{2x}{5} + \frac{x}{4} = \frac{26}{5}$.

$$\frac{2x}{5} + \frac{x}{4} = \frac{26}{5}$$ *The least common denominator is 20.*

$$20\left(\frac{2x}{5} + \frac{x}{4}\right) = 20\left(\frac{26}{5}\right)$$ *Multiply each side of the equation by 20.*

$$20\left(\frac{2x}{5}\right) + 20\left(\frac{x}{4}\right) = 20\left(\frac{26}{5}\right)$$ *Use the distributive property.*

$$8x + 5x = 104$$ *The fractions are eliminated.*

$$13x = 104$$
$$x = 8$$

Check: $\frac{2x}{5} + \frac{x}{4} = \frac{26}{5}$

$\frac{2(8)}{5} + \frac{8}{4} \stackrel{?}{=} \frac{26}{5}$

$\frac{16}{5} + \frac{8}{4} \stackrel{?}{=} \frac{26}{5}$

$\frac{64}{20} + \frac{40}{20} \stackrel{?}{=} \frac{104}{20}$

$\frac{104}{20} = \frac{104}{20}$ ✓

The solution is 8.

Some equations contain more than one variable. You will often be asked to solve such equations for a specific variable.

Example 2 **Solve for x in $ax + b = dx + c$.**

$$ax + b = dx + c$$
$$ax + b - b = dx + c - b$$
$$ax = dx + c - b$$
$$ax - dx = dx - dx + c - b$$
$$(a - d)x = c - b$$
$$\frac{(a - d)x}{a - d} = \frac{c - b}{a - d}$$

Division by zero is undefined. Therefore $a - d \neq 0$, or $a \neq d$.

$$x = \frac{c - b}{a - d}$$

CHECKING FOR UNDERSTANDING

Communicating Mathematics

Read and study the lesson to answer each question.

1. What happens when you multiply each side of the equation $\frac{a}{16} + \frac{1}{2} = \frac{1}{8}$ by 16?
2. Can the same result be achieved by multiplying each side by 32?
3. How would you change the equation in the Application on p. 122 if the number of shirts was 2 times the number of jeans?

Guided Practice

State the number by which you can multiply each side to eliminate the fractions or decimals. Then rewrite each equation.

4. $1.2s + 8.1 = 3.5 - 2s$
5. $8.17y = 4.2 - 3.7y$
6. $\frac{3}{4}x - 7 = 8 + \frac{2}{3}x$
7. $\frac{2}{5}x = 7 - \frac{3}{4}x$

Solve and check each equation.

8. $0.2x + 1.7 = 3.9$
9. $5.3 - 0.3x = -9.4$
10. $\frac{4 - x}{5} = \frac{1}{5}x$
11. $\frac{5}{8}x + \frac{3}{5} = x$

:RCISES

Practice

Solve and check each equation.

12. $\frac{y+5}{3} = 7$

13. $\frac{3n-2}{5} = \frac{7}{10}$

14. $1.9s + 6 = 3.1 - s$

15. $28 - 2.2y = 11.6y + 262.6$

16. $\frac{3}{4}x - 4 = 7 + \frac{1}{2}x$

17. $\frac{3}{8} - \frac{1}{4}x = \frac{1}{2}x - \frac{3}{4}$

18. $5.4y + 8.2 = 9.8y - 2.8$

19. $1.03x - 4 = -2.15x + 8.72$

20. $3y - \frac{4}{5} = \frac{1}{3}y$

21. $\frac{7+3t}{4} = -\frac{1t}{8}$

22. $\frac{3}{2}y - y = 4 + \frac{1}{2}y$

23. $\frac{x}{2} - \frac{1}{3} = \frac{x}{3} - \frac{1}{2}$

Solve for x.

24. $5x = y$

25. $\frac{x+a}{3} = c$

26. $ax + b = cx$

27. $ex - 2y = 3z$

Solve for y.

28. $ay - b = c$

29. $ay + z = am - ny$

30. $a(y + 1) = b$

31. $\frac{3}{5}y + a = b$

Define a variable, write an equation, and solve each problem.

32. Five eighths of a number is three more than half the number. Find the number.

33. One half of a number increased by 16 is four less than two thirds of the number. Find the number.

34. Five more than two thirds of a number is the same as three less than one half of the number. Find the number.

35. One fifth of a number plus five times that number is equal to seven times the number less 18. Find the number.

36. The sum of two numbers is 25. Twelve less than four times one of the numbers is 16 more than twice the other number. Find both numbers.

37. The difference of two numbers is 12. Two fifths of one of the numbers is six more than one third of the other number. Find both numbers.

Critical Thinking

38. Explain how to solve the following equation.

$$\frac{3}{x+1} = \frac{1}{x+1} - 7$$

Applications

Define a variable, write an equation, and solve each problem.

39. **Construction** A rectangular playground is 60 meters longer than it is wide. It is enclosed by 920 meters of fencing. Find its length.

40. **Sales** Luisa stopped by the grocery store to buy 2 cans of peas at $.69 each, 1 carton of orange juice at $2.49, and some tomato sauce at $.95 each. The grocery was having "Double Coupon Day," and Luisa had a 25¢ coupon for the orange juice. If Luisa spent $6.22, how many cans of tomato sauce did she buy?

Computer

41. The program at the right solves equations of the form $ax + b = cx + d$.

Write each equation below in the form $ax + b = cx + d$. Then run the program to find the solution.

a. $2(2x + 3) = 4x + 6$

b. $5x - 7 = x + 3$

c. $6 - 3x = 3x - 6$

d. $6.8 + 5.4x = 4.6x + 2.8$

e. $5x - 8 - 3x = 2(x - 3)$

```
10 PRINT "ENTER A, B, C, D:"
20 INPUT A, B, C, D
30 PRINT A;"X + ";B;" = ";
40 PRINT C;"X + ";D
50 IF A-C <> 0 THEN 110
60 IF D-B <> 0 THEN 90
70 PRINT "THIS IS AN
   IDENTITY."
80 GOTO 120
90 PRINT "NO SOLUTION."
100 GOTO 120
110 PRINT "X = ";
    (D-B)/(A-C)
120 END
```

Mixed Review

42. Replace the $\underline{?}$ with <, >, or = to make the sentence true.
$(-6.01)(-4.122) \underline{?} \frac{9.624}{2.2}$ **(Lesson 2-3)**

43. Simplify $\frac{\frac{3}{11}}{-6}$. **(Lesson 2-7)**

44. Write a problem based on the following information.
Let n = number of nickels that Yvette has.
$n - 17$ = number of pennies that Yvette has.
$n + (n - 17) = 63$ **(Lesson 2-8)**

Portfolio

Select one of the assignments from this chapter that you found especially challenging and place it in your portfolio.

Solve. (Lessons 3-5 and 3-6)

45. $2 - 7s = -19$

46. $3x - 5 = 7x + 7$

HISTORY CONNECTION

DIOPHANTUS

Diophantus of Alexandria was one of the greatest mathematicians of the Greek civilization. It is believed that he lived in the third century, but no one knows for sure. Diophantus is best known for writing a book called *Arithmetica,* in which some algebraic symbols were used for what may have been the first time.

Little is known of Diophantus' life except for a problem printed in a work called the *Greek Anthology.* Although the problem was not written by Diophantus, it is believed to accurately describe his life.

> Diophantus passed one sixth of his life in childhood, one twelfth in youth, and one seventh more as a bachelor. Five years after his marriage, there was born a son who died four years before his father, at half his father's (final) age. How old was Diophantus when he died?

PTER 3

SUMMARY AND REVIEW

VOCABULARY

Upon completing this chapter, you should be familiar with the following terms:

consecutive numbers	112	118	identity
equivalent equation	94	112	number theory

SKILLS AND CONCEPTS

OBJECTIVES AND EXAMPLES

Upon completing this chapter, you should be able to:

- solve equations by using addition **(Lesson 3-1)**

$$x - 3 = 5$$
$$x - 3 + 3 = 5 + 3$$
$$x = 8$$

$$4 = 8 - y$$
$$4 - 8 = 8 - 8 - y$$
$$-4 = -y$$
$$4 = y$$

REVIEW EXERCISES

Use these exercises to review and prepare for the chapter test.

Solve and check each equation.

1. $x - 16 = 37$
2. $k + 13 = 5$
3. $15 - y = 9$
4. $-13 = 6 - k$
5. $19 = -8 + d$
6. $m + (-5) = -17$
7. Thirteen less than some number is 64. What is the number?
8. The sum of a number and -35 is 98. Find the number.

- solve equations by using subtraction **(Lesson 3-2)**

$$m + 16 = 8$$
$$m + 16 - 16 = 8 - 16$$
$$m = -8$$

$$q - (-2) = -10$$
$$q + 2 = -10$$
$$q + 2 - 2 = -10 - 2$$
$$q = -12$$

Solve and check each equation.

9. $z + 15 = -9$
10. $19 = y + 7$
11. $p + (-7) = 31$
12. $r - (-5) = -8$
13. $y + (-9) = -35$
14. $m - (-4) = 21$
15. Some number added to -16 is equal to 39. What is the number?
16. A number decreased by -11 is -176. Find the number.

OBJECTIVES AND EXAMPLES	REVIEW EXERCISES
■ solve equations by using multiplication or division **(Lesson 3-3)** $-14x = 42$ $\frac{-14x}{-14} = \frac{42}{-14}$ $x = -3$ $\frac{m}{6} = -8$ $6\left(\frac{m}{6}\right) = (-8)6$ $m = -48$	**Solve and check each equation.** 17. $-7r = -56$ 18. $23y = 1035$ 19. $534 = -89r$ 20. $\frac{x}{5} = 7$ 21. $\frac{3}{4}x = -12$ 22. $1\frac{2}{3}n = 1\frac{1}{2}$ 23. Six times a number is -96. Find the number. 24. Seven eighths of a number is 14. What is the number?
■ solve equations involving more than one operation **(Lesson 3-5)** $2x + 16 = 18$ $2x + 16 - 16 = 18 - 16$ $\frac{2x}{2} = \frac{2}{2}$ $x = 1$	**Solve and check each equation.** 25. $3x - 8 = 22$ 26. $-4y + 2 = 32$ 27. $0.5n + 3 = -6$ 28. $-6 = 3.1t + 6.4$ 29. $\frac{x}{-3} + 2 = -21$ 30. $\frac{r - 8}{-6} = 7$
■ solve equations with the variable on both sides and equations containing grouping symbols **(Lesson 3-6)** $2(c + 1) = 8c - 22$ $2c + 2 = 8c - 22$ $2c - 8c + 2 = 8c - 8c - 22$ $-6c + 2 = -22$ $-6c + 2 - 2 = -22 - 2$ $\frac{-6c}{-6} = \frac{-24}{-6}$ $c = 4$	**Solve and check each equation.** 31. $5a - 5 = 7a - 19$ 32. $-3(x + 2) = -18$ 33. $4(2y - 1) = -10(y - 5)$ 34. $11.2n + 6 = 5.2n$ 35. Twice a number increased by 12 is 31 less than three times the number. Find the number.
■ solve equations containing fractions or decimals **(Lesson 3-7)** $\frac{2}{5}y + \frac{y}{2} = 9$ $10\left(\frac{2}{5}y + \frac{y}{2}\right) = 10(9)$ $4y + 5y = 90$ $\frac{9y}{9} = \frac{90}{9}$ $y = 10$	**Solve and check each equation.** 36. $\frac{2}{3}x + 5 = \frac{1}{2}x + 4$ 37. $2.9m + 1.7 = 3.5 + 2.3m$ 38. $\frac{3t + 1}{4} = \frac{3}{4}t - 5$ 39. $2.85y - 7 = 12.85y - 2$

OBJECTIVES AND EXAMPLES

- solve equations containing more than one variable **(Lesson 3-7)**

Solve for x in $\frac{ax + 1}{2} = b$.

$$2\left(\frac{ax + 1}{2}\right) = 2(b)$$
$$ax + 1 = 2b$$
$$ax + 1 - 1 = 2b - 1$$
$$\frac{ax}{a} = \frac{2b - 1}{a}$$
$$x = \frac{2b - 1}{a}$$

REVIEW EXERCISES

Solve for x.

40. $\frac{x + y}{c} = d$

41. $5(2a + x) = 3b$

42. $\frac{2x - a}{3} = \frac{a + 3b}{4}$

43. $\frac{2}{3}x + a = a + b$

APPLICATIONS AND CONNECTIONS

Geometry: Find the missing measure. (Lesson 3-3)

44. $A = 42\frac{1}{4}$ ft^2

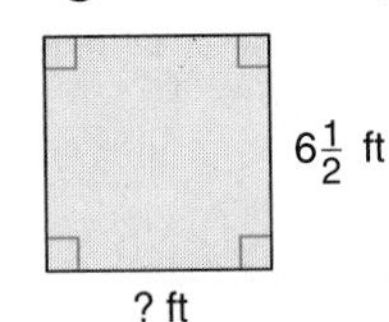

45. $A = 17.85$ cm^2

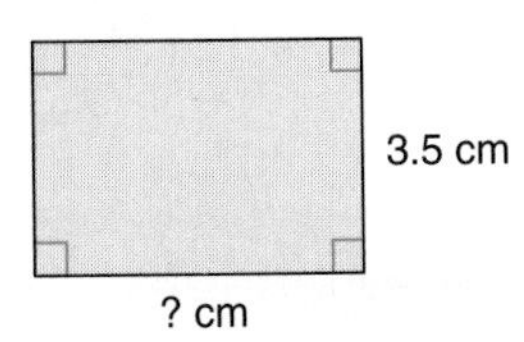

Solve by working backwards. (Lesson 3-4)

46. The Broadview Library charges fines for overdue books as follows: 10¢ per day for each of the first three days, and 5¢ per day thereafter. Bill Rogers paid a fine of $1.50. How many days was his book overdue?

47. Pia has a beaker filled with water. She uses half of the water and gives half of the remainder to Kwon. She has 225 mL left in the beaker. How much water was originally in the beaker?

Number Theory: Define a variable, write an equation, and solve each problem. (Lesson 3-5)

48. Find two consecutive even integers whose sum is 94.

49. Find three consecutive odd integers whose sum is 81.

Geometry: Find the dimensions of each rectangle. The perimeter is given. (Lesson 3-6)

50. $P = 70$ in.

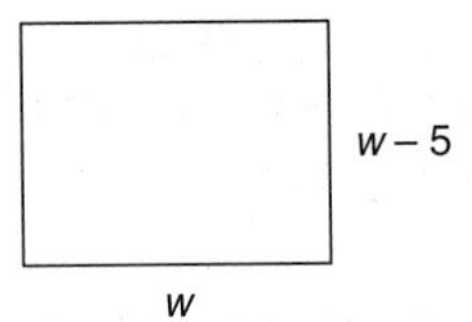

51. $P = 188$ m

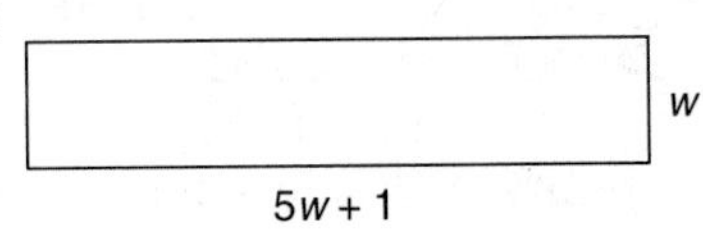

52. **Games** Darlene's score at the end of the first game was −7.8. At the end of the second game, her score was 19.2. How many points did Darlene score during the second game? **(Lesson 3-1)**

53. **Gardening** The lengths of the sides of Mrs. Garcia's garden are consecutive even integers. The perimeter is 156 feet. What are the lengths of the four sides? **(Lesson 3-5)**

CHAPTER 3 TEST

Solve and check each equation.

1. $m + 13 = -9$
2. $k + 16 = -4$
3. $y + (-3) = 14$
4. $x + (-6) = 13$
5. $k - (-3) = 28$
6. $-5 - k = 14$
7. $r - (-1.2) = -7.3$
8. $b - \frac{2}{3} = -\frac{5}{6}$
9. $-3y = 63$
10. $\frac{3}{4}y = -27$
11. $3x + 1 = 16$
12. $5.2n + 0.7 = 2.8 + 2.2n$
13. $5(8 - 2n) = 4n - 2$
14. $3(n + 5) - 6 = 3n + 9$
15. $7x + 9 = 3(x + 3)$
16. $-2(3n - 5) + 3n = 2 - n$
17. $\frac{3}{4}n - \frac{2}{3}n = 5$
18. $\frac{t - 7}{4} = 11$
19. $\frac{2r - 3}{-7} = 5$
20. $8r - \frac{r}{3} = 46$

Define a variable, write an equation, and solve each problem.

21. The sum of two integers is −11. One integer is 8. Find the other integer.
22. The difference of two integers is 26. The lesser integer is −11. What is the greater integer?
23. Four times a number decreased by twice the number is 100. What is the number?
24. Find two consecutive integers such that twice the lesser integer increased by the greater integer is 50.

Solve for *x*.

25. $x + r = q$
26. $\frac{x + y}{b} = c$
27. $yx - a = cx$

Solve.

28. **Skiing** Gary Carson skied down the slalom run in 131.3 seconds. This was 21.7 seconds faster than his sister Jenny. What was Jenny's time?
29. **Golfing** Joe's golf score was 68. This was 4 less than Maria's golf score. What was Maria's score?
30. **Riding** Alma rides her bicycle for three fourths of an hour every day. Find the distance she rides if she averages 13.65 miles per hour.

Bonus A sporting goods store sells T-shirts for one price and sweatshirts for a different price. Editon spent the same amount on 5 T-shirts as he spent on 2 sweatshirts. If he spent a total of $115 on the 5 T-shirts and 2 sweatshirts, what is the cost of each T-shirt?

College Entrance Exam Preview

The test questions on these pages deal with number concepts and basic operations.

Directions: Choose the best answer. Write A, B, C, or D.

1. Which of the following numbers is *not* a prime number?

   (A) 17 (B) 23 (C) 37 (D) 87

2. How many integers between 325 and 400, inclusive, are divisible by 4?

   (A) 18 (B) 19 (C) 20 (D) 24

3. How many integers between 99 and 201 are divisible by 2 or 5?

   (A) 60 (B) 61 (C) 70 (D) 71

4. How many integers are between, but not including, 5 and 1995?

   (A) 1988 (B) 1989
   (C) 1990 (D) 2000

5. A person is standing in line, thirteenth from the front and eleventh from the back. How many people are in the line?

   (A) 22 (B) 23 (C) 24 (D) 25

6. What number is missing from the sequence 2, 5, 15, 18, 54, 171, 174, 522?

   (A) 36 (B) 56 (C) 57 (D) 25

7. Which of the following is the least?

   (A) 0.77 (B) $\frac{7}{9}$ (C) $\frac{8}{11}$ (D) $\frac{3}{4}$

8. Which fraction is greater than $\frac{1}{4}$ but less than $\frac{1}{3}$?

   (A) $\frac{4}{13}$ (B) $\frac{1}{5}$ (C) $\frac{5}{12}$ (D) $\frac{3}{4}$

9. Which of the following fractions is less than $\frac{1}{5}$?

   (A) $\frac{3}{14}$ (B) $\frac{21}{100}$ (C) $\frac{2}{11}$ (D) $\frac{101}{501}$

10. Which of the following is the greatest?

    (A) $\frac{1}{2}$ (B) $\frac{5}{11}$ (C) $\frac{4}{9}$ (D) $\frac{7}{13}$

11. What digit is represented by ■ in this subtraction problem?

$$\begin{array}{r} 80\blacktriangle \\ -\ 602 \\ \hline \blacksquare 98 \end{array}$$

    (A) 1 (B) 2 (C) 3 (D) 4

12. The difference between $42\frac{3}{8}$ minutes and $41\frac{2}{3}$ minutes is approximately how many seconds?

    (A) 18 (B) 63 (C) $22\frac{1}{2}$ (D) 43

13. A patient must be given medication every 5 hours, starting at 10:00 A.M. Thursday. What is the first day on which the patient will receive medication at noon?

    (A) Thursday (B) Friday
    (C) Saturday (D) Sunday

14. Which group of numbers is arranged from greatest to least?

(A) $3, \frac{1}{4}, -1, -0.5$
(B) $-1, -0.5, \frac{1}{4}, 3$
(C) $3, \frac{1}{4}, -0.5, -1$
(D) $-0.5, \frac{1}{4}, -1, 3$

15. Find the smallest positive integer which gives a remainder of 3 when divided by 4, 6, or 8.

(A) 51 (B) 27 (C) 15 (D) 26

16. $8(916) + 916 =$

(A) $4(916) + 3(916)$
(B) $5(916) + 4(916)$
(C) $6(916) + 4(916)$
(D) $3(916) + 4(916)$

17. Alan owes Benito $4, Benito owes Carl $12, and Carl in turn owes Alan some money. If all three debts could be settled by having Benito pay $3 to Alan and $5 to Carl, how much does Carl owe Alan?

(A) $4 (B) $5 (C) $6 (D) $7

18. A person has 100 green, 100 orange, and 100 yellow jelly beans. How many jars can be filled if each jar must contain 8 green, 5 orange, and 6 yellow jelly beans?

(A) 12 (B) 15 (C) 16 (D) 25

19. Which is the difference of two consecutive prime numbers less than 30?

(A) 5 (B) 6 (C) 7 (D) 8

20. If $(-9)(-9)(-9) = (-9)(-9)(-9)n$, then $n =$

(A) 1 (B) 0 (C) -9 (D) 9

TEST TAKING TIP

As part of your preparation for a standardized entrance exam, review basic definitions such as the ones below.

1. A number is *divisible* by another number if it can be divided exactly with no remainder. A number is divisible by each of its factors.
2. A number is *prime* if it has no factors except itself and 1.
   7 is a prime number.
   8 is not prime since it has factors other than itself and 1. 2 and 4 are also factors of 8.

21. The sum of five consecutive integers is always divisible by

(A) 2 (B) 4 (C) 5 (D) 10

22. $16 - 12 \div 2^2 \times 3 =$

(A) $\frac{1}{3}$ (B) 3 (C) 7 (D) 15

23. Which group of numbers is arranged in descending order?

(A) $\frac{5}{7}, \frac{7}{12}, \frac{6}{11}, \frac{3}{13}$
(B) $\frac{7}{12}, \frac{5}{7}, \frac{6}{11}, \frac{3}{13}$
(C) $\frac{5}{7}, \frac{6}{11}, \frac{7}{12}, \frac{3}{13}$
(D) $\frac{3}{13}, \frac{7}{12}, \frac{6}{11}, \frac{5}{7}$

PTER OBJECTIVES

n this chapter, you will:

- Solve proportions.
- Solve problems that can be represented by proportions.
- Solve uniform motion problems.
- Solve problems involving direct and inverse variation.
- Use a table or chart to solve problems.

Applications of Rational Numbers

How do you know which is the best deal? Just being able to read an advertisement is not enough anymore. Now, you must also understand the mathematics in the advertisement!

APPLICATION IN CONSUMER AWARENESS

The job of every ad writer is to create an advertisement that catches your eye. Ads using large type and flashy photos of exciting and glamorous locations promise you almost anything if you will just buy the product. Somehow, you must read through all of the glitz to find real bargains on things that you actually need. That is the only way that *you* can control your money instead of letting the ad writers do it for you.

How can you tell what discounts and financing actually mean in dollars and cents? Is a 20% discount on an item selling for \$12.95 more than $\frac{1}{4}$ off the price of an item selling for \$15.98? When is a discount more appealing than a rebate? What is a.p.r. financing? What is the finance charge on your parents' charge card, if they have one? How much money will it *really* cost you to buy an item using that charge card if it will take you more than one month to pay off the balance?

To find answers to these questions, you must understand the language of mathematics and become *math literate.* This will help you to actually read advertisements, and let you, not the advertiser, determine the things that you really need.

ALGEBRA IN ACTION

Three electronic stores are selling the same car stereo system at a discount. Their advertisements are shown below. From which store should you buy the system to get the best price?

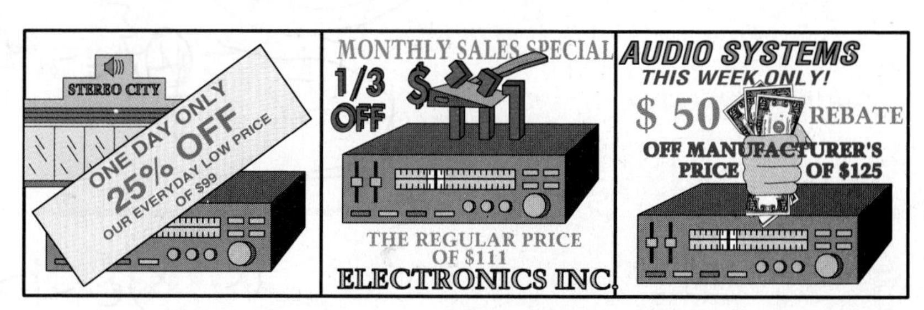

4-1 Ratios and Proportions

Objective After studying this lesson, you should be able to:

- solve proportions.

Application

With a single throw of the net, 120 million fish were caught in the Barents Sea in August of 1982.

The title of the newspaper article implies that the fish in the creek were counted one by one. In fact, counting each fish individually would be difficult and might endanger the fish. The *capture-recapture* method uses **ratios** to determine the fish population.

A **ratio** is a comparison of two numbers by division. Ratios can be expressed in the following ways.

$$a \text{ to } b \qquad x{:}y \qquad \frac{3}{7}$$

Ratios are most often expressed as fractions in simplest form. A ratio that is equivalent to a whole number is written with a denominator of 1.

Example 1 **What is the ratio of 20 inches to 4 feet?**

The units must be the same, so change feet to inches: 4 feet = 48 inches. The ratio is $\frac{20}{48}$ or $\frac{5}{12}$.

An equation of the form $\frac{a}{b} = \frac{c}{d}$ stating that two ratios are equal is called a **proportion.** Every proportion consists of four terms.

first ↓ $\frac{a}{b}$ ↑ *second* = *third* ↓ $\frac{c}{d}$ ↑ *fourth*

The first and fourth terms, a and d, are called the **extremes.**

The second and third terms, b and c, are called the **means.**

Means-Extremes Property of Proportions

In a proportion, the product of the extremes is equal to the product of the means.

$$\text{If } \frac{a}{b} = \frac{c}{d}, \text{ then } ad = bc.$$

To solve a proportion, cross-multiply.

Example 2

Twenty fish were captured from Black Creek, tagged, and returned to the creek. Later, 29 fish were captured. Of these, 3 had tags. Estimate the number of fish in the creek.

Let f = the approximate number of fish in the creek.

$$\frac{\text{tagged fish}}{\text{approximate number of fish in creek}} = \frac{\text{captured fish that were tagged}}{\text{captured fish}}$$

$$\frac{20}{f} = \frac{3}{29}$$

$$20(29) = 3f \quad \textit{Cross multiply.}$$

$$580 = 3f$$

$$193\tfrac{1}{3} = f$$

A good estimate would be that there are about 193 fish in the creek.

Models are often made to **scale.** A scale is a ratio that compares the size of the model to the actual size of the object being modeled.

Example 3

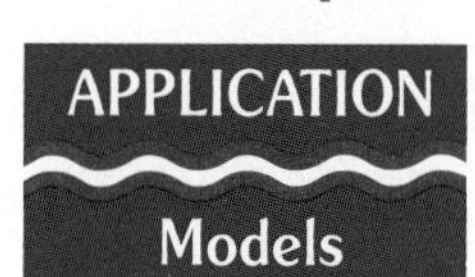

A model car is made to the following scale: 1 inch to 10 inches. If the door of the actual car is 33 inches long, what is the door length of the model?

$$\frac{1}{10} = \frac{x}{33}$$

$$33 = 10x \quad \textit{Means-extremes property}$$

$$3\tfrac{3}{10} = x \quad \textit{Check this result.}$$

The door length of the model car is $3\frac{3}{10}$ inches.

Example 4

Use a calculator to solve the proportion $\frac{6}{2.56} = \frac{9.32}{m}$.

By the means-extremes property, $6m = (2.56)(9.32)$. Multiply the means. Then divide by 6.

Enter: 2.56 [×] 9.32 [÷] 6 [=]

Display: 3.9765333

Rounded to the nearest hundredth, the solution is 3.98.

Example 5

Solve $\frac{x}{5} = \frac{x + 3}{10}$.

$$\frac{x}{5} = \frac{x+3}{10}$$

$$10x = 5(x + 3) \quad \textit{Means-extremes property}$$

$$10x = 5x + 15 \quad \textit{Distributive property}$$

$$5x = 15$$

$$x = 3 \quad \textit{Check this result.}$$

The solution is 3.

A ratio of two measurements having different units of measure is called a **rate**. For example, 20 miles per gallon is a rate. Proportions are often used to solve problems involving rates.

Example 6

A 96-mile trip requires 6 gallons of gasoline. At that rate, how many gallons would be required for a 152-mile trip?

EXPLORE Let n = number of gallons required for a 152-mile trip.

PLAN Write a proportion for the problem.

$$\frac{96 \text{ miles}}{6 \text{ gallons}} = \frac{152 \text{ miles}}{n \text{ gallons}}$$ *Notice that both ratios compare miles to gallons.*

SOLVE

$$\frac{96}{6} = \frac{152}{n}$$

$$96n = 912$$

$$n = 9.5$$

A 152-mile trip would require 9.5 gallons of gasoline.

EXAMINE A trip of about 100 miles requires 6 gallons of gasoline. A trip of about 150 miles would require about $1\frac{1}{2}$ times the amount of gasoline or 9 gallons. Therefore, for a 152-mile trip a solution of 9.5 gallons is reasonable.

CHECKING FOR UNDERSTANDING

Communicating Mathematics

Read and study the lesson to complete the following.

1. A ratio of two measurements having different units of measure is called a __?__.
2. In a proportion, the product of the means equals the product of the __?__.
3. A __?__ is a comparison of two numbers by division.
4. In a proportion, the second and third terms are called the __?__.

Guided Practice

Write each ratio as a fraction in simplest form.

5. 3 grams to 11 grams
6. 21 meters to 16 meters
7. 12 ounces to 6 ounces
8. 8 feet to 28 inches
9. 16 cm to 40 mm
10. 35 minutes to 2 hours

Solve each proportion.

11. $\frac{3}{4} = \frac{x}{8}$
12. $\frac{2}{10} = \frac{1}{y}$
13. $\frac{10}{a} = \frac{20}{28}$

EXERCISES

Practice

Solve each proportion.

14. $\frac{3}{15} = \frac{b}{45}$

15. $\frac{6}{8} = \frac{7}{x}$

16. $\frac{x}{9} = \frac{-7}{16}$

17. $\frac{5}{2n} = \frac{-2}{1.6}$

18. $\frac{x+2}{5} = \frac{7}{5}$

19. $\frac{6}{14} = \frac{7}{x-3}$

20. $\frac{3}{5} = \frac{x+2}{6}$

21. $\frac{14}{10} = \frac{5+x}{x-3}$

22. $\frac{9}{x-8} = \frac{4}{5}$

23. $\frac{4-x}{3+x} = \frac{16}{25}$

24. $\frac{x-12}{6} = \frac{x+7}{-4}$

25. $\frac{x+9}{5} = \frac{x-10}{11}$

Use a calculator to solve each proportion.

26. $\frac{4.3}{25.8} = \frac{1}{2n}$

27. $\frac{2}{0.19} = \frac{12}{0.5n}$

28. $\frac{n}{8} = \frac{0.21}{2}$

29. $\frac{x}{4.085} = \frac{5}{16.33}$

30. $\frac{2.405}{3.67} = \frac{g}{1.88}$

31. $\frac{3s}{9.65} = \frac{21}{1.066}$

Critical Thinking

Find the ratio of *a* to *b*.

32. $10a = 25b$

33. $4a + 2b = 16a$

34. $\frac{3a}{5} = \frac{12b}{7}$

Applications

35. Biology In a game preserve, 239 deer are caught, marked, and then released. Later, out of 198 deer caught, 42 are marked. Estimate the total deer population in the preserve.

36. Drafting The scale on the blueprint for a house is 1 inch to 3 feet. If the living room on the blueprint is $5\frac{1}{2}$ inches by 7 inches, what are the dimensions of the actual room?

37. Cartography The scale on a map is 1 centimeter to 57 kilometers. Fargo and Bismarck are 4.7 centimeters apart on the map. What is the actual distance between these cities?

38. Machinery In the belt drive shown below, pulley A is 30 inches in diameter and pulley B is 15 inches.

a. What is the ratio of the diameter of A to the diameter of B?

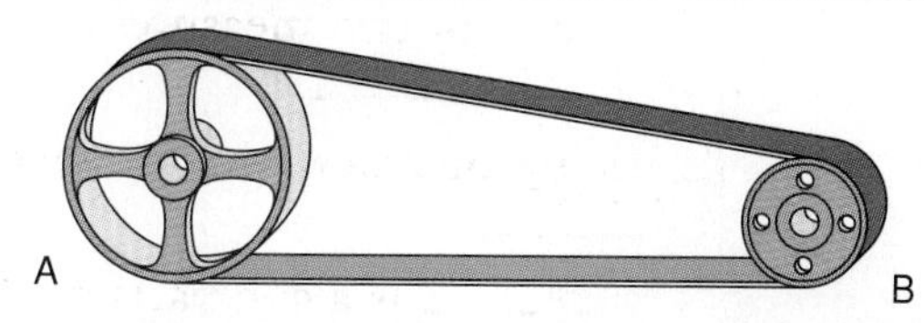

b. How many times does pulley B turn while pulley A turns once?

Mixed Review

39. Evaluate $3^2 + 4(8 - 2) - 6(3) + 4 \div 2 + (8 - 2)^2$. **(Lesson 1-2)**

40. Simplify $6(2x + y) + 2(x + 4y)$. **(Lesson 1-6)**

41. Find the sum: $-17px + 22bg + 35px + (-37bg)$. **(Lesson 2-5)**

42. Cooking If there are four sticks of butter in a pound of butter, and each stick of butter is $\frac{1}{2}$ cup, how many cups of butter are there in a pound of butter? **(Lesson 2-6)**

43. Solve: $-\frac{7}{6} + k = \frac{5}{6}$. **(Lesson 3-1)**

4-2 Percent

Objective After studying this lesson, you should be able to:
- solve percent problems.

Application

The chart at the right illustrates how 50 people responded to a poll that asked the question, "How did you spend your vacation?" What is the rate of people staying at home per 100 people?

How People Spend Their Vacation	
Visit family/friends	15
Travel/sightsee	13
Stay at home	11
Summer/winter resort	8
No vacation/don't know	3

The Epcot Poll, taken at Disney World's Epcot Center, is one of the world's largest ongoing opinion polls. From 1982 to 1990, more than 500,000 people were polled.

You can solve this problem by using a proportion. The ratio of people who stay at home to the total polled is $\frac{11}{50}$. Write a proportion that sets $\frac{11}{50}$ equal to a ratio with a denominator of 100.

$$\frac{11}{50} = \frac{n}{100}$$
$$22 = n$$

People stayed at home at a rate of 22 per 100, or 22 percent. The word **percent** means *per hundred,* or *hundredths.* The symbol for percent is %.

$$22\% = \frac{22}{100} = 0.22$$

Example 1 **Change $\frac{3}{5}$ to a percent.**

Use a proportion.

$$\frac{3}{5} = \frac{n}{100}$$
$$300 = 5n$$
$$60 = n$$

Use a calculator.

$$\frac{3}{5} = \frac{n}{100}$$
$$100\left(\frac{3}{5}\right) = n$$

Enter: 3 [÷] 5 [×] 100 [=]

Display: 60

Thus, $\frac{3}{5}$ is equal to $\frac{60}{100}$ or 60%.

Proportions are often used to solve percent problems. One of the ratios in these proportions is always a comparison of two numbers called the **percentage** and the **base.** The other ratio, called the **rate,** is a fraction with a denominator of 100.

percentage → 3, *base* → 5; $\frac{3}{5} = \frac{60}{100}$ ← *rate*

Percent Proportion	$\frac{\text{Percentage}}{\text{Base}} = \text{Rate or } \frac{\text{Percentage}}{\text{Base}} = \frac{r}{100}$

The $\frac{percentage}{base}$ represents $\frac{part}{whole}$.

Example 2

50 is what percent of 60?

Use the percent proportion.

$$\frac{\text{Percentage}}{\text{Base}} = \frac{r}{100}$$

$$\frac{50}{60} = \frac{r}{100}$$ *The percentage is 50. The base is 60.*

$$5000 = 60r$$

$$83\frac{1}{3} = r$$

Thus, 50 is $83\frac{1}{3}\%$ of 60. The decimal solution is $83.\overline{3}\%$.

You can also write an equation to solve problems with percents.

Example 3

a. What number is 36% of 150?

$$x = \frac{36}{100} \cdot 150$$

$$x = 54$$

Thus, 54 is 36% of 150.

b. 40% of what number is 30?

$$\frac{40}{100} \cdot x = 30$$

$$x = 75$$

Thus, 40% of 75 is 30.

Most calculators have a key labeled [%]. This key can be used when solving problems involving percents. The keying sequences vary for different types of calculators. One possible keying sequence is shown in the next example.

Example 4

Find 18% of 46.3.

Enter: 18 [%] [×] 46.3 [=]

You may not need to press the [=] key.

Display: 8.334

Thus, 18% of 46.3 is 8.334.

Example 5

APPLICATION
Sales

Jim Byars earns 3% commission on his sales of new cars. If he earned $861 in commissions last week, what was the dollar amount of his total sales?

EXPLORE Let x = total sales in dollars.

PLAN *3% of total sales equals $861.*

$$\frac{3}{100} \cdot x = 861$$

How could you solve this by using a decimal instead of a fraction?

SOLVE

$$\frac{3}{100}x = 861$$

$$3x = 86{,}100$$

$$x = 28{,}700$$

The total sales amount was $28,700.

EXAMINE **Enter:** 3 [%] [×] 28,700 [=] 861

CHECKING FOR UNDERSTANDING

Communicating Mathematics

Read and study the lesson to answer each question.

1. In the percent proportion, what is the rate?
2. Name three areas in your everyday life where percent is used.

Guided Practice

Write each ratio as a percent.

3. $\frac{31}{100}$
4. $\frac{3}{10}$
5. $\frac{1}{25}$
6. $\frac{7}{20}$
7. $\frac{3}{8}$
8. $\frac{9}{5}$

Solve.

9. What is 40% of 60?
10. Seventy-five is what percent of 250?
11. Twenty-one is 35% of what number?
12. Fifty-two is what percent of 80?

EXERCISES

Practice

Use a proportion to answer each question.

13. Six is what percent of 15?
14. What percent of 50 is 35?
15. Five is what percent of 40?
16. What percent of 75 is 225?

Use an equation to answer each question.

17. What number is 40% of 80?
18. 17.65 is 25% of what number?
19. Fourteen is what percent of 56?
20. What percent of 72 is 12?

Solve.

21. Thirty-six is 45% of what number?
22. Find 81% of 32.
23. Find 4% of $6070.
24. Fifty-five is what percent of 88?
25. $54,000 is 108% of what amount?
26. Find 112% of $500.
27. Find 0.12% of $5200.
28. What is 98.5% of $140.32?

Critical Thinking

29. If a is 225% of b, then b is what percent of a?

Applications

30. **Metallurgy** In a 180-kilogram sample of ore, there was 3.2% metal. How many kilograms of metal were in the sample?
31. **Education** Janice scored 85% on the last test. She answered 34 questions correctly. How many questions were on the test?
32. **Sales** The sales tax on a $20 purchase was $0.90. What was the rate of sales tax?
33. **Research** Suppose 6% of 8000 people polled regarding an election expressed no opinion. How many people had an opinion?
34. **Statistics** Circle graphs are often used to describe sets of data. They show the relationship between parts of the data and the whole. The circle graph at the right was made using the information in the chart on page 138.
    a. What do most people do on their vacation?
    b. What do people do least often on their vacation?
    c. What is the sum of the percents in the circle graph?

Journal

Find three examples of how percents are used in real life. Explain each use.

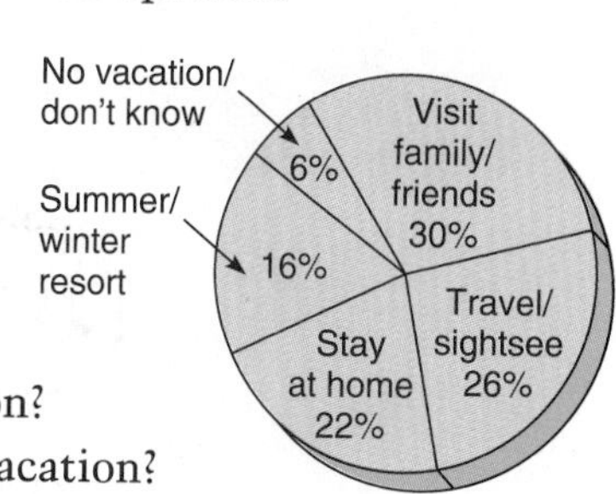

Mixed Review

35. Write the following sentence as an equation.
    The sum of twice m and the square of n is y. **(Lesson 1-7)**
36. Find the difference: $18 - (-34)$. **(Lesson 2-2)**
37. Solve $y + \frac{7}{16} = -\frac{5}{8}$. **(Lesson 3-2)**
38. **Personal Finance** Isabel saves $18 in 4 weeks. How long will it take her to save $81 at the same rate? **(Lesson 4-1)**

4-3 Application: Simple Interest

Objective After studying this lesson, you should be able to:

- solve problems involving simple interest.

The formula $I = prt$ is used to solve problems involving **simple interest.** In the formula, I represents the interest, p represents the amount of money invested, called the *principal*, r represents the annual interest rate, and t represents time in years.

Example 1

Kiko Murimoto opened a LifeBank savings account that earns 7% annual interest. After 6 months, she received \$52.50 in interest. How much money had Kiko deposited when she opened the account?

EXPLORE Let p = the amount of money Kiko deposited.

PLAN

$I = prt$

$52.50 = p(0.07)(0.5)$ *Change 7% to 0.07 and 6 months to 0.5 year.*

SOLVE

$52.50 = 0.035p$ *Multiply 0.07 by 0.5.*

$1500 = p$ *Divide each side by 0.035.*

Kiko had deposited \$1500 in the bank.

EXAMINE When p is 1500, $I = (1500)(0.07)\left(\frac{1}{2}\right)$ or \$52.50.

Example 2

Marilyn Mallinson invested \$30,000, part at 6% annual interest and the rest at 7.5% annual interest. Last year, she earned \$1995 in interest. How much money did she invest at each rate?

EXPLORE Let n = the amount of money invested at 6%.
Then $30{,}000 - n$ = the amount of principal invested at 7.5%.

PLAN

$\underbrace{\text{interest on 6\% investment}}$

$n(0.06)(1)$ or $0.06n$

$\underbrace{\text{interest on 7.5\% investment}}$

$(30{,}000 - n)(0.075)(1)$ or $2250 - 0.075n$

SOLVE

$\underbrace{\text{interest at 6\%}} + \underbrace{\text{interest at 7.5\%}} = \underbrace{\text{total interest}}$

$0.06n + (2250 - 0.075n) = 1995$

$-0.015n + 2250 = 1995$ *Add 0.06n and −0.075n.*

$-0.015n = -255$

$n = 17{,}000$ *Check this solution.*

Marilyn invested \$17,000 at 6% and 30,000 − 17,000 or \$13,000 at 7.5%.

Example 3

Joe Parker invested \$7625, part at 8% annual interest and the rest at 6.5% annual interest. In the same amount of time, he earned three times as much interest from the 6.5% investment as he did from the 8% investment. How much money did he have invested at 6.5%?

EXPLORE Let n = the amount invested at 6.5%.
Then $7625 - n$ = the amount invested at 8%.

PLAN *interest at 6.5% = 3 times interest at 8%*

$$0.065n = 3[0.08(7625 - n)]$$

SOLVE

$$0.065n = 3(610 - 0.08n)$$
$$0.065n = 1830 - 0.24n$$
$$0.305n = 1830$$
$$n = 6000$$

Joe invested \$6000 at 6.5%.

EXAMINE Since (0.065)(6000) = \$390 and (0.08)(7625 − 6000) = \$130, and \$390 = 3(\$130), the investment at 6.5% earned three times as much as the one at 8%.

CHECKING FOR UNDERSTANDING

Communicating Mathematics

Read and study the lesson to answer each question.

1. In Example 1, suppose Kiko adds the \$52.50 in interest to the principal. What will the interest be the next year?
2. Suppose an investor invested \$5000 at 8% and \$2000 at 12%. Why do you think they invested the greater amount at the lower rate?

Guided Practice

Use $I = prt$ to find the missing quantity.

3. Find I if p = \$8000, r = 6%, and t = 1 year.
4. Find I if p = \$5000, $r = 12\frac{1}{2}\%$, and t = 5 years.
5. Find t if I = \$1890, p = \$6000, and r = 9%.
6. Find t if I = \$2160, p = \$6000, and r = 8%.
7. Find r if I = \$2430, p = \$9000, and t = 2 years, 3 months.
8. Find r if I = \$780, p = \$6500, and t = 1 year.
9. Find p if I = \$756, r = 9%, and $t = 3\frac{1}{2}$ years.
10. Find p if I = \$196, r = 10%, and t = 7 years.

EXERCISES

Applications: Finance

11. Michelle Limotta invested \$10,000 for one year, part at 8% annual interest and the rest at 12% annual interest. Her total interest for the year was \$944. How much money did she invest at each rate?
12. Steve Devine invested \$7200 for one year, part at 10% annual interest and the rest at 14% annual interest. His total interest for the year was \$960. How much money did he invest at each rate?

13. Fred Ferguson invested \$5000 for one year, part at 9% annual interest and the rest at 12% annual interest. The interest from the investment at 9% was \$198 more than the interest from the investment at 12%. How much money did he invest at 9%?
14. Angela Raimondi wants to invest \$8500, part at 14% annual interest and part at 12% annual interest. If she wants to earn the same amount of interest from each investment, how much should she invest at 14%? (Round the answer to the nearest cent.)
15. In one year, Cholena Youngblood earned the same amount of interest from an investment at 8% annual interest as an investment at 12% annual interest. She had invested \$1500 more at 8%. How much money did she have invested at 12%?
16. John and Iris Johnson have \$8000 to invest. They want to earn \$850 interest for the year. Their money can be invested at annual interest rates of 10% or 11%. What is the minimum amount at 11%, with the rest at 10%, that will earn them \$850 in interest?
17. Carlos and Bonita Díaz have invested \$2500 at 10% annual interest. They have \$6000 more to invest. At what rate must they invest the \$6000 to have \$9440 at the end of the year?
18. Ken Bauman invested \$9450, part at 8% annual interest and the rest at 11% annual interest. He earned twice as much interest at 11% as he did at 8%. How much money did he have invested at 11%?
19. If Li Fong had earned one fourth of a percent more in annual interest on an investment, the interest for one year would have been \$45 greater. How much did he invest at the beginning of the year?

Computer

20. The program at the right calculates a percentage increase and then a percentage decrease of the new total at the same rate. First, enter the original amount. Then, enter the percent of increase and decrease. The program will print the amount after the increase and the amount after both the increase and decrease.

```
10 PRINT "ENTER THE AMOUNT."
20 INPUT A
30 PRINT "ENTER THE PERCENT."
40 INPUT R
50 LET A1 = A + A * (R/100)
60 LET A2 = A1 - A1 * (R/100)
70 PRINT A;" INCREASES TO ";A1
80 PRINT "THEN DECREASES TO ";A2
90 END
```

a. Copy and complete the chart.

Original Amount	Percent	After Increase	After Decrease
\$100	10%		
\$40	20%		
\$500	15%		
\$6500	5%		

b. Compare the original amounts and the final amounts. What do you notice?

Critical Thinking

21. Polly Crawford deposited \$1000 in a savings account at 8% annual interest computed semiannually. If no withdrawals or deposits were made, what was the amount in the account at the end of eighteen months?

Mixed Review

22. Solve $\frac{5}{6} \cdot 18 = m$. **(Lesson 1-3)**
23. Replace the $\underline{?}$ in $(-6.01)(-4.122) \underline{?} 25.005$ with <, >, or = to make the sentence true. **(Lesson 2-3)**
24. Find a number between $-\frac{11}{19}$ and $\frac{5}{14}$. **(Lesson 2-4)**

Work backwards to solve the problem. (Lesson 3-4)

25. Two barrels, X and Y, contain water. X contains more water than Y. John pours from X into Y as much water as Y already contains. Then he pours from Y into X as much water as X already contains. Finally, John pours from X into Y as much water as Y already contains. Both barrels now contain 64 gallons of water. How many gallons of water were in each barrel at the beginning?
26. Twenty-eight is 20% of what number? **(Lesson 4-2)**

APPLICATION

Compound Interest

Compound interest is the amount of interest paid or earned on the original principal *plus* the accumulated interest. Thus, the interest becomes part of the principal. Suppose you deposited $500 in a savings account for 3 years at a rate of 6% annual interest.

	amount deposited +	*interest*	*= new principal*	
End of first year	$500	+ $500(0.06) =	$530	
End of second year	$530	+ $530(0.06) =	$561.80	
End of third year	$561.80	+ $561.80(0.06) =	$595.51	*Rounded to the nearest cent.*

A formula that can be used to find compound interest is $T = p(1 + r)^t$, where T is the total amount in the account (principal plus interest), p is the principal, r is the annual interest rate divided by the number of times interest is compounded yearly, and t is the number of times interest is compounded.

Example

Alonso invested $750 for 1 year at 8% annual interest. If interest is compounded semiannually, how much will he have at the end of the year?

$T = p(1 + r)^t$

$= 750\left(1 + \frac{0.08}{2}\right)^{2 \cdot 1}$ *interest compounded twice per year for 1 year*

$= 750(1.04)^2$ *Use a calculator to evaluate.*

$= 811.20$ Alonso will have $811.20 at the end of the year.

Now try this.

Ana invests $600 in a savings account at 12% compounded semiannually. Jim invests $600 in an account at 12% compounded quarterly. At the end of two years, who will have more money? How much more?

4-4 Percent of Change

Objectives

After studying this lesson, you should be able to:

- solve problems involving percent of increase or decrease, and
- solve problems involving discount or sales tax.

Application

A pair of jeans that originally cost \$40 is on sale for \$30. You can write a ratio that compares the amount of decrease to the original price. This ratio can be written as a percent.

$$\frac{\text{amount of decrease}}{\text{original price}} \rightarrow \frac{10}{40} = \frac{r}{100} \rightarrow r = 25$$

The amount of decrease is $\frac{25}{100}$ or 25% of the original price. So, we can say that the **percent of decrease** is 25%.

The **percent of increase** can be found in a similar manner, as shown in the following example.

Example 1

The base price of the 1991 Buyer's Car of the Year was \$15,925. The base price of the same car in 1994 is \$19,705. Find the percent of increase.

EXPLORE The price increased from \$15,925 to \$19,705. The amount of increase was \$3780.

PLAN You need to write a ratio that compares the amount of increase with the original price. Then, express the ratio as a percent.

$$\frac{\text{amount of increase}}{\text{original price}} \quad \frac{3780}{15{,}925} = \frac{r}{100}$$

SOLVE

$$\frac{3780}{15{,}925} = \frac{r}{100}$$

$$378{,}000 = 15{,}925r \quad \text{Use a calculator to divide each side by 15,925.}$$

$$23.736263 = r$$

The percent of increase is about 24%.

EXAMINE Use a calculator to find 24% of \$15,925.

Enter: 24 [%] [×] 15,925 [=] 3822

\$15,925 + \$3822 = \$19,747

Since 24% > 23.736%, the answer is reasonable.

Sometimes an increase or decrease is given as a percent, rather than an amount. Two applications of percent of change are discounts and sales tax. Keep in mind that a discount of 35% means that the price is decreased by 35%. Sales tax of 5% means that the price is increased by 5%.

Example 2

Amy bought a television set that had an original price of \$295.95. Because she worked at the store, she received a 20% discount. What was the discounted price?

EXPLORE The original price was \$295.95, and the discount was 20%.

PLAN You want to find the amount of discount, then subtract that amount from \$295.95. The result is the discounted price.

SOLVE 20% of \$295.95 = 0.20(295.95) *Note that $20\% = \frac{20}{100} = 0.20$.*
= 59.19

Subtract this amount from the original price.
\$295.95 − \$59.19 = \$236.76

The discounted price was \$236.76.

EXAMINE Here's another way to solve the problem. The discount was 20%, so the discounted price was 80% (100% − 20%) of the original price.

Find 80% of \$295.95. 0.80(295.95) = 236.76

This method produces the same discount price, \$236.76.

Example 3

Frank bought a new tennis racket for \$79.99. He also had to pay sales tax of $5\frac{1}{2}$%. What was the total price?

EXPLORE The price was \$79.99 and the tax rate was $5\frac{1}{2}$%.

PLAN First find $5\frac{1}{2}$% of \$79.99. Then add the result to \$79.99.

SOLVE $5\frac{1}{2}$% of \$79.99 = 0.055(79.99) *Note that $5\frac{1}{2}\% = 0.055$.*
= 4.39945 *Round 4.39945 to 4.40.*

Add this amount to the original price.
\$79.99 + \$4.40 = \$84.39

The total price was \$84.39.

EXAMINE The tax rate is $5\frac{1}{2}$%, so the total price was 100% + $5\frac{1}{2}$% or 105.5% of the purchase price. Find 105.5% of \$79.99.

(1.055)(79.99) = 84.38945

Thus, the total price of \$84.39 is correct.

Example 4

Connie's Clothing prices their goods 25% above the wholesale price. If the retail price of a jacket is $79, what was the wholesale price?

The *wholesale price* is the price the store pays for their goods. The *retail price* is the price at which the goods are sold in the store.

EXPLORE Let j = wholesale price of jacket.
Then $0.25j$ = amount of markup.
Note that 25% = 0.25.

PLAN $\underbrace{\text{wholesale price}} + \underbrace{\text{markup}} = \underbrace{\text{retail price}}$

$$j \quad + \quad 0.25j \quad = \quad 79$$

SOLVE

$$(1 + 0.25)j = 79 \quad \textit{Distributive property}$$
$$1.25j = 79$$
$$j = 63.2$$

The wholesale price was $63.20.

EXAMINE If the wholesale price is $63.20 and the retail price is $79, the markup is $15.80. Since $15.80 is 25% of $63.20, the solution is correct.

CHECKING FOR UNDERSTANDING

Communicating Mathematics

Read and study the lesson. Then explain how to solve each problem.

1. percent of increase, if you know the old and new prices
2. discount, if you know the percent of discount
3. sales tax, if you know the rate of tax
4. wholesale price, if you know the retail price and the percent of markup

Guided Practice

Copy and complete the chart. The first line is given as a sample.

	Original Amount	Later Amount	Did the amount increase (I) or decrease (D)?	Amount of Increase or Decrease	Percent of Increase or Decrease
	$50	$70	I	$20	40%
5.	$100	$94			
6.	$100	$108			
7.	$200		D		14%
8.	$313.49		I		14.5%
9.		$60.10	I	$12.21	
10.		$36	D	$36	

Notice that 20 is 40% of 50.

EXERCISES

Practice

11. What number increased by 40% equals 14?
12. What number decreased by $66\frac{2}{3}\%$ is 18?
13. Fourteen is 50% less than what number?
14. Twenty is 20% more than what number?
15. What is 30% more than 30?
16. What is 75% less than 80?
17. A price decreased from \$50 to \$40. Find the percent of decrease.
18. A price increased from \$40 to \$50. Find the percent of increase.
19. A price plus 5% tax is equal to \$3.15. Find the original price.
20. An item sells for \$36 after a 25% discount. Find the original price.

Find the final price of each item. When there is a discount and sales tax, compute the discount price first.

21. stereo: \$345.00
    discount: 12%
22. clothing: \$74.00
    sales tax: 6.5%
23. shoes: \$44.00
    discount: 10%
    sales tax: 4%
24. tires: \$154.00
    discount: 20%
    sales tax: 5%

Critical Thinking

25. An amount is increased by 10%. The result is decreased by 10%. Is the final result equal to the original amount?

Applications: Sales

26. Zoe got a discount of \$4.50 on a new radio. The discounted price was \$24.65. What was the percent of discount to the nearest percent?
27. Chapa paid \$92.04 for new school clothes. This included 4% sales tax. What was the cost of the clothes before taxes?
28. The wholesale price of a stove was \$550. The retail price was 20% greater than the wholesale price. The retail price was then discounted by 10% during a sale. What was the sale price?
29. Tires-4-Less allows a 10% discount if a purchase is paid for within 30 days. An additional 5% discount is given if the purchase is paid for within 15 days. Bob Cordoba buys a set of tires that originally cost \$360. If he pays the entire amount at the time of purchase, how much does he pay for his tires? (*Hint:* Compute the discounts separately.)

FYI ···

The largest tires are manufactured by the Goodyear Co. for giant dump trucks. They are 12 feet in diameter and weigh 12,500 pounds.

Mixed Review

30. Simplify $9a + 16b + 14a$. **(Lesson 1-5)**
31. Simplify $\frac{77b}{-11}$. **(Lesson 2-7)**
32. Solve $-5 = 4 - 2(x - 5)$. **(Lesson 3-6)**
33. Eighteen is what percent of 60? **(Lesson 4-2)**
34. Use $I = prt$ to find t if $I = \$3528$, $P = \$8400$, and $r = 10\frac{1}{2}\%$. **(Lesson 4-3)**

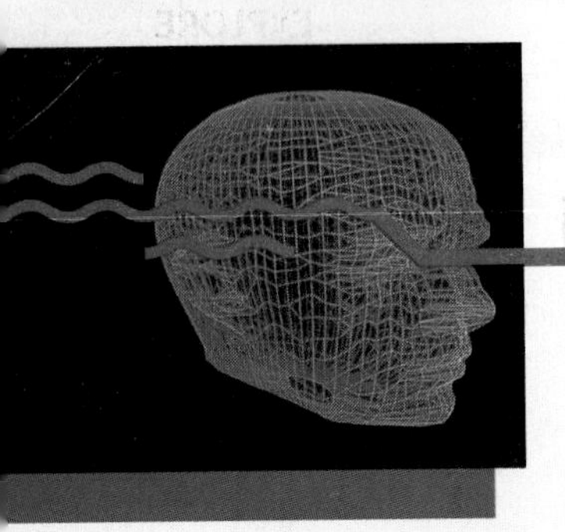

Technology

Successive Discounts

▶ **BASIC**
Graphing Calculator
Graphing Software
Spreadsheets

Application

Bonnie's Appliances sells microwave ovens for $250. One oven is discounted 20% because it is scratched. Then the store puts all microwave ovens on sale at a discount of 15%. What is the final sale price of the scratched microwave?

The regular price of microwave ovens is $250. The scratched microwave oven is $250 − $250(0.20) or $250(0.80) or $200. On sale, the scratched microwave oven is $200(0.85) or $170.

The two successive discounts are summarized as follows.

$$(\$250 \times 0.80) \times 0.85 = \$170$$

A combined discount would be 20% + 15% or 35%. This is 65% of the original amount.

$$\$250 \times 0.65 = \$162.50$$

The program below can help you determine whether two successive discounts or one combined discount will produce the lower price.

```
10 PRINT "ENTER ORIGINAL PRICE AND TWO DISCOUNTS AS
   DECIMALS."
20 INPUT P,X1,X2
30 LET S1 = P * (1 - X1) * (1 - X2)      successive discounts
40 LET S2 = P * (1 - (X1 + X2))          combined discount
50 PRINT "SUCCESSIVE DISCOUNTS: $ ";S1
60 PRINT "COMBINED DISCOUNT: $" ;S2
70 IF S2 < S1 THEN PRINT "A COMBINED DISCOUNT HAS A
   LOWER SALE PRICE."
80 IF S1 < S2 THEN PRINT "SUCCESSIVE DISCOUNTS HAVE A
   LOWER SALE PRICE."
90 IF S1 = S2 THEN PRINT "THERE IS NO DIFFERENCE."
100 END
```

EXERCISES

Find the sale price of each item using successive discounts and then one combined discount.

1. price, $49; discounts, 20% and 10%
2. price, $185; discounts, 25% and 10%
3. price, $12.50; discounts, 30% and 12.5%
4. price, $156.95; discounts, 30% and 15%
5. What is the relationship between the sale price using successive discounts and the sale price using combined discounts?

4-5 Problem-Solving Strategy: Make a Table or Chart

EXPLORE
PLAN
SOLVE
EXAMINE

Objective After studying this lesson, you should be able to:

- solve problems by making a table or chart.

Many problems can be solved more easily if you organize the information in a table or chart.

Example 1

After college, Mark Brandauer is offered two jobs. Job A has a starting salary of $17,500 per year with a guaranteed raise of $3000 per year. Job B has a starting salary of $20,000 with a guaranteed raise of $2000 per year. Which job would pay more after 5 years?

Make a chart to show the effect of each option over a 5-year period.

Year	1	2	3	4	5
Job A	$17,500	$20,500	$23,500	$26,500	$29,500
Job B	$20,000	$22,000	$24,000	$26,000	$28,000

Job A would pay more after 5 years.

When solving logic problems in which several clues are given, it is often helpful to use a table called a *matrix*.

Example 2

Five students each have a favorite food. No two of them have the same favorite. Using the clues below, which food goes with each person?

1. **Adrienne's favorite food is not pizza.**
2. **Benito hates pizza and hamburgers.**
3. **One of the students' favorite foods is chicken wings.**
4. **Cindy's favorite is french fries.**
5. **Darryl is allergic to onion rings.**
6. **Eddie's favorite is onion rings.**

Set up a table to solve the problem. Write all of the names and categories in the table. Place a Y (yes) or N (no) wherever you can. No two have the same favorite food, so once you write a Y in a box, you can write an N in all of the other boxes in that row and column. Then complete the table.

(Continued on page 152)

Name	Pizza	Hamburgers	Chicken Wings	French Fries	Onion Rings
Adrienne	N	Y	N	N	N
Benito	N	N	Y	N	N
Cindy	N	N	N	Y	N
Darryl	Y	N	N	N	N
Eddie	N	N	N	N	Y

Adrienne's favorite is hamburgers, Benito's favorite is chicken wings, Cindy's favorite is french fries, Darryl's favorite is pizza, and Eddie's favorite is onion rings.

CHECKING FOR UNDERSTANDING

Communicating Mathematics

Read and study the lesson to answer each question.

1. Over the 5-year period described in Example 1, which job would pay the most total money?
2. Would the matrix solution in Example 2 work if more than one person had the same favorite food? Why or why not?

Guided Practice

Use a table or chart to solve each problem.

3. Carla Agosto and Ken Takamura are both electricians. Ms. Agosto charges \$15 for a service call, plus \$15 per hour. Mr. Takamura charges \$35 for a service call, plus \$10 per hour. If they work the same number of hours and are paid the same amount of money, how many hours did they work?
4. Four friends have birthdays in January, July, September, and December. Amelia was not born in the winter. Brandon celebrates his birthday near Independence Day. Lianna's is the month after Jason's. In what month was each one born?
5. Fly Boy Aviation Products uses two machines to make 33,000 airplane bolts very quickly. Machine A can make 3000 bolts per hour and Machine B can make 1800 bolts per hour. Machine A starts at 7:30 A.M. and Machine B starts at 10:30 A.M. At what time is the job completed?

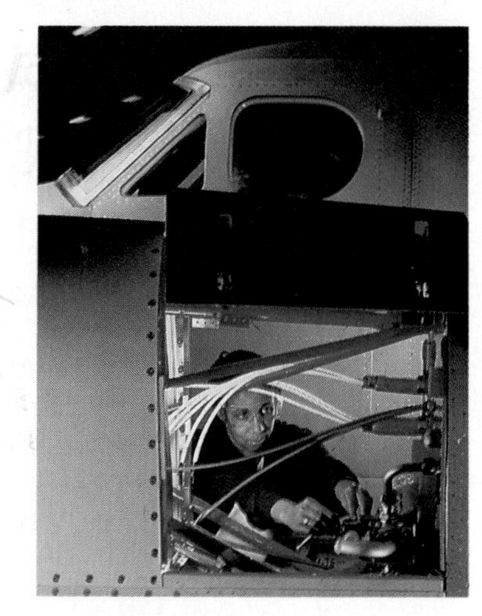

6. Mr. Smith, Mr. Jones, and Mr. Chang are a butcher, a baker, and a candlestick maker, but not in that order. The baker is married to Mr. Smith's sister. Mr. Chang is not the candlestick maker. Mr. Jones is a bachelor. Mr. Smith is a regular customer of the butcher's. Who does what?

EXERCISES

Practice

Solve. Use any strategy.

Strategies

Look for a pattern.
Solve a simpler problem.
Act it out.
Guess and check.
Draw a diagram.
Make a chart.
Work backwards.

7. Place the numbers 1 through 7 in the seven compartments formed by the three circles at the right so that the sums of the numbers in each circle are the same. The numbers 1, 4, and 6 have already been placed for you.

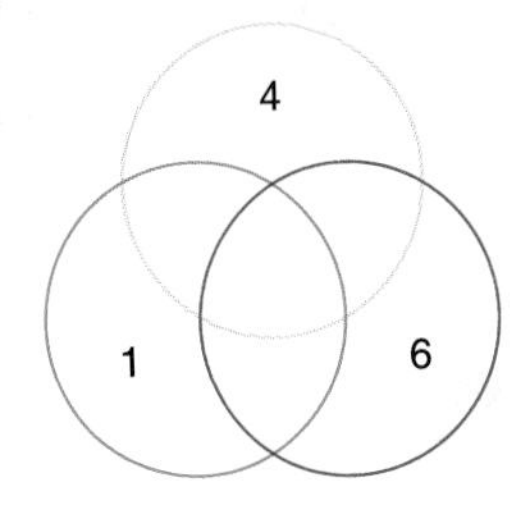

8. Lindsay, Paul, and Kelly each have hair of a different color. One has blonde hair, one has red hair, and one has brown hair. If Lindsay is a blonde, Paul has red hair. If Lindsay has red hair, Paul has brown hair. If Paul is not a blonde, Kelly has red hair. What color hair does each person have?

FYI . . .

The hair of Diane Witt of Worcester, Massachusetts measured 10 feet 9 inches in 1989.

9. Imagine you are playing a game of tic-tac-toe. What is the greatest number of squares that can be left empty when the game is won?

10. Numbers that can be represented by geometric figures are called *figurate numbers.* One type of figurate number is a *triangular number.* Triangular numbers are those that can be represented by a triangular array of dots, with n on each side.

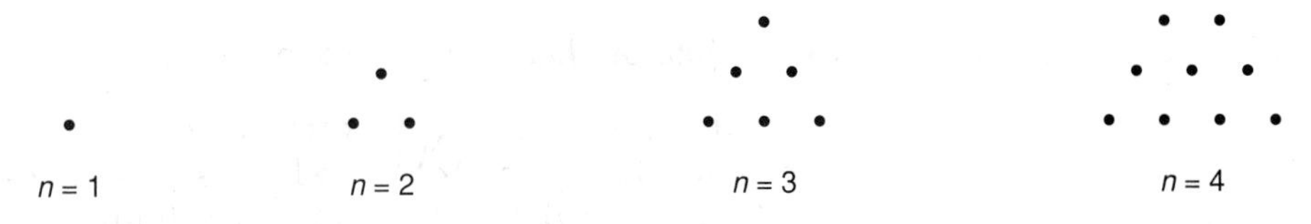

a. Make a table that compares n with the number of dots in the array. What pattern do you notice?

b. What is the 10th triangular number?

COOPERATIVE LEARNING ACTIVITY

Work in groups of four. Each person in the group must understand the solution and be able to explain it to any person in the class.

Eight squares cover the large square at the right. All of the eight squares are the same size. The square marked 1 is completely shown; however, you can only see a part of the other seven squares. Number the squares in order from the top layer to the bottom.

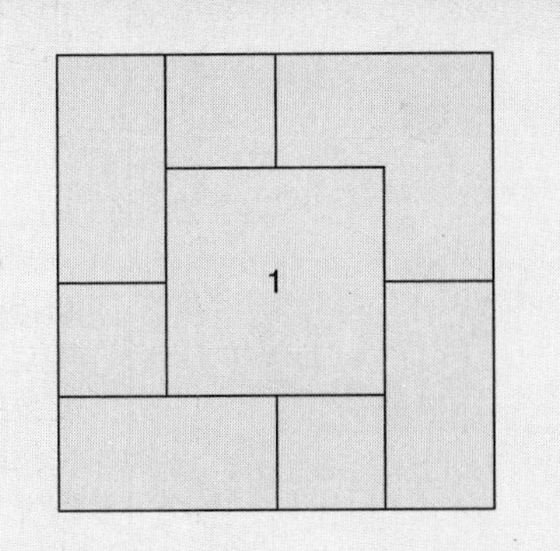

4-6 Application: Mixtures

Objective

After studying this lesson, you should be able to:

- solve mixture problems.

It is often helpful to make a chart to organize the information.

Mixture problems involve a combination of two or more parts into a whole. While studying mixture problems, you will be asked to find information about one or more of these parts when information about the whole is given. You can use equations to solve problems like this.

Example 1

Delectable Dan's Cookie Company sells two kinds of cookies daily: chocolate chip at $6.50 per dozen and white chocolate macadamia at $9.00 per dozen. On Thursday, Dan sold 85 dozen more chocolate chip than white chocolate macadamia cookies. The total sales for both were $4055.50. How many dozen of each were sold?

EXPLORE Let w = the number of dozens of white chocolate macadamia cookies sold. Then $w + 85$ = the number of dozens of chocolate chip cookies sold.

PLAN

	Number of Dozens	Price per Dozen	Total Price
White Chocolate	w	$9.00	9w
Chocolate Chip	w + 85	$6.50	6.5(w + 85)

$$\underbrace{\text{total sales of white chocolate macadamia}}_{9w} + \underbrace{\text{total sales of chocolate chip}}_{6.5(w+85)} = \underbrace{\text{total sales}}_{4055.5}$$

$$9w + 6.5(w + 85) = 4055.5$$

SOLVE

$$9w + 6.5w + 552.5 = 4055.5$$
$$15.5w + 552.5 = 4055.5$$
$$15.5w = 3503$$
$$w = 226$$

There were 226 dozen white chocolate macadamia cookies sold
There were 226 + 85, or 311 dozen chocolate chip cookies sold

EXAMINE If 226 dozen white chocolate macadamia cookies were sold, the total sales of those cookies would be 226($9) or $2034. If 311 dozen chocolate chip cookies were sold, the total sales of those cookies would be 311($6.50) or $2021.50.

Since $2034 + $2021.50 = $4055.50, the solution is correct.

Sometimes mixture problems are expressed in terms of percents.

Kendra is doing a chemistry experiment that calls for a 30% solution of copper sulfate. She has 40 mL of 25% solution. How many milliliters of 60% solution should Kendra add to obtain the required 30% solution?

EXPLORE Let x = the amount of 60% solution to be added.

PLAN

	Amount of Solution (mL)	Amount of Copper Sulfate
25% solution	40	0.25(40)
60% solution	x	$0.60x$
30% solution	$40 + x$	$0.30(40 + x)$

amount of copper sulfate in 25% solution + amount of copper sulfate in 60% solution = amount of copper sulfate in mixture

$$0.25(40) + 0.60x = 0.30(40 + x)$$

SOLVE

$$10 + 0.6x = 12 + 0.3x$$
$$0.3x = 2$$
$$x \approx 6.7 \quad \textit{Round to the nearest tenth.}$$

Kendra should add 6.7 mL of the 60% solution to the 40 mL of 25% solution. *Examine this solution.*

CHECKING FOR UNDERSTANDING

Communicating Mathematics

Read and study the lesson to answer each question.

1. If Example 1 had stated that Dan sold 85 dozen less white chocolate cookies than chocolate chip, how would you define the variable?
2. In Example 2, suppose Kendra had added an 80% solution of copper sulfate. Write an equation to represent this situation.

Guided Practice

Copy and complete each chart. Write an equation to represent the situation.

3. **Money** Rodolfo has \$2.55 in dimes and quarters. He has eight more dimes than quarters. How many quarters does he have?

	Number	Total Value
Quarters	x	
Dimes		

4. **Sales** Peanuts sell for \$3.00 per pound. Cashews sell for \$6.00 per pound. How many pounds of cashews should be mixed with 12 pounds of peanuts to obtain a mixture that sells for \$4.20 per pound?

	Pounds	Total Price
\$3.00 peanuts	12	
\$6.00 cashews		
\$4.20 mixture		

5. **Advertising** An advertisement for an orange drink claims that the drink contains 10% orange juice. How much pure orange juice would have to be added to 5 quarts of the drink to obtain a mixture containing 40% orange juice?

	Quarts	Amount of Orange Juice
10% juice	5	
Pure juice		
40% juice		

6. **Finance** Paul Yu is investing $6000 in two accounts, part at 4.5% and the remainder at 6%. If the total annual interest earned from the two accounts is $279, how much did Paul deposit at each rate?

	Amount	Yearly Interest
At 4.5%	x	
At 6%		

EXERCISES

Applications

7. **Entertainment** At the Golden Oldies Theater, tickets for adults cost $5.50 and tickets for children cost $3.50. How many of each kind of ticket was purchased if 21 tickets were bought for $83.50?

8. **Food** A liter of cream has 9.2% butterfat. How much skim milk containing 2% butterfat should be added to the cream to obtain a mixture with 6.4% butterfat?

9. **Sales** Anne Leibowitz owns "The Coffee Pot," a specialty coffee store. She wants to create a special mix using two coffees, one priced at $6.40 per pound and the other at $7.28 per pound. How many pounds of the $7.28 coffee should she mix with 9 pounds of the $6.40 coffee to sell the mixture for $6.95 per pound?

10. **Pharmacy** A pharmacist has 150 dL of a 25% solution of peroxide in water. How many deciliters of pure peroxide should be added to obtain a 40% solution?

11. **Sales** Ground chuck sells for $1.75 per pound. How many pounds of ground round selling for $2.45 per pound should be mixed with 20 pounds of ground chuck to obtain a mixture that sells for $2.05 per pound?

12. **Entertainment** The Martins are going to Funtasticland, an amusement park. The total cost of tickets for a family of two adults and three children is $79.50. If an adult ticket costs $6.00 more than a child's ticket, find the cost of each.

13. **Automotives** A car radiator has a capacity of 16 quarts and is filled with a 25% antifreeze solution. How much must be drained off and replaced with pure antifreeze to obtain a 40% antifreeze solution?

14. **Finance** Editon Longwell invested $33,600, part at 5% interest and the remainder at 8% interest. If he earned twice as much from his 5% investment as his 8% investment, how much did he invest at each rate?

15. **Chemistry** A solution is 50% alcohol. If 10 liters are removed from the solution and replaced with 10 liters of pure alcohol, the resulting solution is 75% alcohol. How many liters of the solution are there?

Critical Thinking

16. Write a mixture problem for the equation $1.00x + 0.28(40) = 0.40(x + 40)$.

Mixed Review

The largest brain ever recorded was that of a 50-year-old male. It weighed 41 pounds 8 ounces.

17. Define a variable, then write an equation for the following problem. In a football game, Diego gained 134 yards running. This was 17 yards more than the previous game. How many yards did he gain in both games? **(Lesson 2-8)**

18. Solve $-7h = -91$. **(Lesson 3-3)**

19. **Physiology** The brain of the average adult weighs about 2% of that person's total body weight. What is the approximate weight of the brain of a 160-pound person? **(Lesson 4-2)**

20. **Sales** Charlene paid $45.10, including $7\frac{1}{2}$% tax, for a pair of jeans. What was the cost of the jeans before tax? **(Lesson 4-4)**

MID-CHAPTER REVIEW

Solve each proportion. (Lesson 4-1)

1. $\frac{6}{5} = \frac{18}{t}$
2. $\frac{21}{27} = \frac{x}{18}$
3. $\frac{19.25}{a} = \frac{5.5}{2.94}$

Answer the following. (Lesson 4-2)

4. Find 37.5% of 80.
5. Sixteen is 40% of what number?
6. Thirty-seven is what percent of 296?

Solve. (Lesson 4-3)

7. **Finance** Isaac deposited an amount of money in the bank at 7% interest. After 3 months he received $14 interest. How much money had Isaac deposited in the bank?

Find the final price of each item. (Lesson 4-4)

8. 10-speed bike: $148.00
discount: 18%
9. books: $38.50
sales tax: 6%

Use a table to solve the problem. (Lesson 4-5)

10. **Consumer Awareness** The telephone company charges a base rate of $5.12 for a 20-minute long-distance telephone call. If Bob Hartschorn was charged $5.32 for a 22-minute call, how much should he be charged for an hour call?

4-7 Application: Uniform Motion

Objective

After studying this lesson, you should be able to:

- solve problems involving uniform motion by using the formula $d = rt$.

When an object moves at a constant speed, or rate, it is said to be in **uniform motion.** The formula $d = rt$ is used to solve uniform motion problems. In the formula, d represents distance, r represents rate, and t represents time.

Example 1

An airplane flies 1000 miles due east in 2 hours and 1000 miles due south in 3 hours. What is the average speed of the airplane?

EXPLORE Let r = the average rate of speed at which the airplane travels. The airplane flies a total of 2000 miles in 5 hours.

PLAN $d = rt$

$2000 = 5r$ *Replace d with 2000 and t with 5.*

SOLVE $400 = r$ The average speed is 400 miles per hour.

EXAMINE When r is 400, $2000 = 5(400)$. So, the answer is correct.

When solving uniform motion problems, it is often helpful to draw a diagram and to organize relevant information in a chart.

Example 2

Don and Donna Wyatt leave their home at the same time, traveling in opposite directions. Don travels at 80 kilometers per hour and Donna travels at 72 kilometers per hour. In how many hours will they be 760 kilometers apart?

EXPLORE Draw a diagram to help analyze the problem.

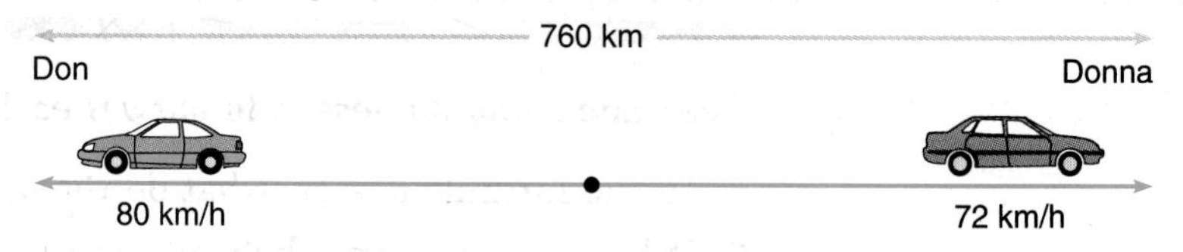

PLAN Organize the information in a chart. Let t represent the number of hours.

	r	$\cdot\ t$	$= d$	
Don	80	t	$80t$	*Don travels 80t km.*
Donna	72	t	$72t$	*Donna travels 72t km.*

Don's distance + Donna's distance = total distance

$$80t + 72t = 760$$

SOLVE

$$152t = 760$$
$$t = 5 \quad \text{Check this solution.}$$

In 5 hours, Don and Donna will be 760 kilometers apart.

Example 3

At 8:00 A.M. Felicia leaves home on a business trip driving 35 miles per hour. A half hour later, José discovers that Felicia forgot her briefcase. He drives 50 miles per hour to catch up with her. If José is delayed 15 minutes with a flat tire, when will he catch up with Felicia?

EXPLORE Let x = the time Felicia drives until José catches up with her.

PLAN

	r	t	$r \cdot t = d$
Felicia	35	x	$35x$
José	50	$\left(x - \frac{3}{4}\right)$	$50\left(x - \frac{3}{4}\right)$

Felicia travels for x hours.

José starts $\frac{1}{2}$ hour later and is delayed $\frac{1}{4}$ hour (flat tire).

José travels for $\left(x - \frac{3}{4}\right)$ hours.

José and Felicia travel the same distance.

Felicia's distance = José's distance

$$35x = 50\left(x - \frac{3}{4}\right)$$

SOLVE

$$35x = 50x - \frac{75}{2}$$
$$-15x = -\frac{75}{2}$$
$$x = 2\frac{1}{2} \quad \text{Check this solution.}$$

Felicia has been traveling for $2\frac{1}{2}$ hours when José catches up to her. José reaches Felicia at 8 A.M. + $2\frac{1}{2}$ hours or 10:30 A.M.

CHECKING FOR UNDERSTANDING

Communicating Mathematics

Read and study the lesson to answer each question.

1. In the formula $d = rt$, what do the d, r, and t represent?
2. Why is it sometimes helpful to use charts and diagrams?

Guided Practice

Use $d = rt$ to answer each question.

3. Makya is driving 40 miles per hour. How far will she travel:
   a. in 3 hours? **b.** in $4\frac{1}{2}$ hours? **c.** in 15 minutes? **d.** in k minutes?
4. Juan traveled 270 kilometers. What was his rate if he made the trip:
   a. in 5 hours? **b.** in x hours?
5. Patsy traveled 360 miles. How long did it take her if her rate was:
   a. 40 miles per hour? **b.** 30 miles per hour? **c.** x miles per hour?

EXERCISES

Applications

6. At 1:30 P.M., an airplane leaves Tucson for Baltimore, a distance of 2240 miles. The plane flies at 280 miles per hour. A second airplane leaves Tucson at 2:15 P.M., and is scheduled to land in Baltimore 15 minutes before the first airplane. At what rate must the second airplane travel to arrive on schedule?

7. Two trains leave York at the same time, one traveling north, the other south. The first train travels at 40 miles per hour and the second at 30 miles per hour. In how many hours will the trains be 245 miles apart?

8. Rosita drove from Boston to Cleveland, a distance of 616 miles, to visit her grandparents. Her rest, gasoline, and food stops took 2 hours. What was her average speed if the trip took 16 hours altogether?

9. Two cyclists are traveling in the same direction on the same bike path. One travels at 20 miles per hour and the other at 14 miles per hour. After how many hours will they be 15 miles apart?

10. At the same time Kris leaves Washington, D.C. for Detroit, Amy leaves Detroit for Washington, D.C. The distance between the cities is 510 miles. Amy's average speed is 5 miles per hour faster than Kris's. How fast is Kris driving if they pass each other in 6 hours?

11. *The Harvest Moon* leaves the pier at 9:00 A.M. at 8 knots (nautical miles per hour). A half hour later, *The River Nymph* leaves the same pier in the same direction traveling at 10 knots. At what time will *The Nymph* overtake *The Moon*?

12. Art leaves at 10:00 A.M., traveling at 50 miles per hour. At 11:30 A.M., Jennifer starts in the same direction at 45 miles per hour. When will they be 100 miles apart?

13. Guillermo is driving 40 miles per hour. After he has driven 30 miles, his brother Jorge starts driving in the same direction. At what rate must Jorge drive to catch up with Guillermo in 5 hours?

14. Two airplanes leave Dallas at the same time and fly in opposite directions. One airplane travels 80 miles per hour faster than the other. After three hours, they are 2940 miles apart. What is the rate of each airplane?

15. An express train travels 80 kilometers per hour from Wheaton to Ward. A local train, traveling at 48 kilometers per hour, takes 2 hours longer for the same trip. How far apart are Wheaton and Ward?

16. Marcus Jackson runs a 440-yard run in 55 seconds and Alfonso Rey runs it in 88 seconds. To have Marcus and Alfonso finish at the same time, how much of a headstart should Marcus give Alfonso? State your answer in yards.

Critical Thinking

17. Two trains are 240 miles apart traveling toward each other on parallel tracks. One travels at 35 miles per hour and the other travels at 45 miles per hour. At the front of the faster train is a bee that can fly at 75 miles per hour. The bee flies from one train to the other until the trains pass each other. When the trains pass, how far has the bee flown?

18. Mary Baylor drove to work on Thursday at 40 miles per hour and arrived one minute late. She left at the same time on Friday, drove at 45 miles per hour, and arrived one minute early. How far does Ms. Baylor drive to work?

Journal

Make up a word problem involving uniform motion and solve it.

Mixed Review

19. Write an algebraic expression for *five less than w*. **(Lesson 1-1)**

20. Name the property illustrated by $5(3 - 3) = 0$. **(Lesson 1-4)**

21. Find the sum: $-15 + 23$. **(Lesson 2-1)**

22. **Finance** A stockbroker has invested part of $25,000 at 10.5% interest and the rest at $12\frac{1}{4}$% interest. If the annual income earned from these investments is $2,843.75, find the amount invested at each rate. **(Lesson 4-6)**

APPLICATION

Travel Agents

Many travelers rely on travel agents to make the best arrangements for them. Travel agents arrange for transportation, hotel accomodations, car rentals, and vacation packages. They may also give advice on weather conditions, time-zone changes, tourist attractions, and restaurants. For the international traveler, travel agents provide information on customs regulations, required papers (like passports and visas), and monetary exchange rates.

Suppose you wanted to fly from Atlanta, Georgia, to Frankfurt, Germany. The timetable at the right shows that Lufthansa Airlines has a number of nonstop flights from Atlanta to Frankfurt. One flight leaves Atlanta at 5:00 P.M. and arrives in Frankfurt at 7:25 A.M. Your travel agent would point out that the time in Frankfurt is 6 hours ahead of the time in Atlanta. With this information, you can determine the length of the flight. The time from 5:00 P.M. to 7:25 A.M. is about $14\frac{1}{2}$ hours. So the flight is about $14\frac{1}{2} - 6$ or $8\frac{1}{2}$ hours long.

From: **ATLANTA, GA.**

Frankfurt, West Germany
4615 mi

LUFTHANSA GERMAN AIRLINES FROM ATLANTA TO FRANKFURT

56	5:00p	7:25a	447	*Nonstop*
24	6:25p	8:45a	445	*Nonstop*
7	7:10p	11:35a	441	*Nonstop*
3	9:15p	11:35a	443	*Nonstop*
1	9:15p	11:35a	439	*Nonstop*

The timetable also shows that the air distance between the two cities is 4615 miles. Using the formula for uniform motion, you can find the average speed of the airplane.

$$d = rt$$
$$4615 = 8.5r$$
$$540 = r$$

The average speed would be 540 miles per hour.

4-8 Direct Variation

Objective After studying this lesson, you should be able to:

- solve problems involving direct variations.

Application Yana is taking part in a work-study program offered by Jackson High School. After school, he works as an assistant manager at Victory Sporting Goods, where he earns \$5.95 per hour. The table below relates the number of hours that he works (x) and his pay (y).

x	5	10	15	20	25
y	\$29.75	\$59.50	\$89.25	\$119	\$148.75

Yana's income depends *directly* on the number of hours that he works. The relationship between the number of hours worked and his income is shown by the equation $y = 5.95x$. This type of equation is called a **direct variation.** We say that *y varies directly as* x.

Definition of Direct Variation

A direct variation is described by an equation of the form $y = kx$, where $k \neq 0$.

In the equation $y = kx$, k is called the **constant of variation.** To find the constant of variation, divide each side by x.

$$\frac{y}{x} = k$$

Example 1

APPLICATION — Space

The weight of an object on the moon varies directly as its weight on Earth. With all his gear on, Neil Armstrong weighed 360 pounds on Earth, but when he stepped on the moon on July 20, 1969, he weighed 60 pounds. Kristina weighs 108 pounds on Earth. What would she weigh on the moon?

Let x = weight on Earth, and let y = weight on the moon. Find the value of k in the equation $y = kx$.

$k = \frac{y}{x}$

$k = \frac{60}{360}$ *Substitute the astronaut's weights for x and y.*

$k = \frac{1}{6}$

Next find Kristina's weight on the moon. Find y when x is 108.

$y = kx$

$y = \frac{1}{6}(108)$ *Substitute 108 for x and $\frac{1}{6}$ for k.*

$y = 18$ Thus, Kristina would weigh 18 pounds on the moon.

Direct variations are also related to proportions. From the table on Yana's income, many proportions can be formed. Two examples are shown.

number of hours (numerators); *income* (denominators)

$$\frac{5}{2975} = \frac{15}{8925}$$

number of hours $\frac{10}{25} = \frac{5950}{14,875}$ *income*

x_1 is read "x sub 1."

Two general forms for proportions like these can be derived from the equation $y = kx$. Let (x_1, y_1) be a solution for $y = kx$. Let a second solution be (x_2, y_2). Then $y_1 = kx_1$ and $y_2 = kx_2$.

$y_1 = kx_1$ — *This equation describes a direct variation.*

$\frac{y_1}{y_2} = \frac{kx_1}{kx_2}$ — *Use the division property of equality. Since y_2 and kx_2 are equivalent, you can divide the left side by y_2 and the right side by kx_2.*

$\frac{y_1}{y_2} = \frac{x_1}{x_2}$ — *Simplify.*

Let's see how another proportion can be derived from this proportion.

$x_2y_1 = x_1y_2$ — *Find the cross products of the proportion above.*

$\frac{x_2y_1}{y_1y_2} = \frac{x_1y_2}{y_1y_2}$ — *Divide each side by y_1y_2.*

$\frac{x_2}{y_2} = \frac{x_1}{y_1}$ — *Simplify.*

Example 2

If y varies directly as x, and $y = 27$ when $x = 6$, find x when $y = 45$.

Use $\frac{y_1}{y_2} = \frac{x_1}{x_2}$ to solve the problem.

$\frac{27}{45} = \frac{6}{x_2}$ — *Let $y_1 = 27$, $x_1 = 6$, and $y_2 = 45$.*

$27x_2 = 45(6)$ — *Find the cross products.*

$27x_2 = 270$

$x_2 = 10$

Thus, $x = 10$ when $y = 45$.

Example 3

In an electrical transformer, voltage is directly proportional to the number of turns on the coil. If 110 volts comes from 55 turns, what would be the voltage produced by 66 turns?

Write an equation of the form $y = kx$ for the variation.

$110 = 55k$

$2 = k$

The equation is $y = 2x$. Thus, the voltage produced by 66 turns would be 2(66) or 132 volts.

CHECKING FOR UNDERSTANDING

Communicating Mathematics

Read and study the lesson to answer each question.

1. How do you find the constant of variation in a direct variation?
2. *True* or *false:* A person's height varies directly as his or her age. Why or why not?

Guided Practice

Find the constant of variation for each direct variation.

3. $y = 3x$
4. $-3a = b$
5. $d = 7t$
6. $n = \frac{1}{3}m$

Solve. Assume that y varies directly as x.

7. If $y = 12$ when $x = 3$, find y when $x = 7$.
8. If $y = -8$ when $x = 2$, find y when $x = 10$.

EXERCISES

Practice

Solve. Assume that y varies directly as x.

9. If $y = 3$ when $x = 15$, find y when $x = -25$.
10. If $y = -7$ when $x = -14$, find y when $x = 20$.
11. If $y = -6$ when $x = 9$, find x when $y = -4$.
12. If $y = -8$ when $x = -3$, find x when $y = 6$.
13. If $y = 12$ when $x = 15$, find x when $y = 21$.
14. If $y = 2\frac{2}{3}$ when $x = \frac{1}{4}$, find y when $x = 1\frac{1}{8}$.

15-20. For Exercises 9-14, find the constant of variation. Then write an equation of the form $y = kx$ for each variation.

21. y varies directly as the square of x. If $y = 14$ when $x = 4$, find y when $x = 9$.
22. y varies directly as $x - 10$. If $y = -12$ when $x = -3$, find x when $y = 7$.

Critical Thinking

23. Look back at the proportions on page 163. Does it follow that $\frac{y_1}{x_2} = \frac{x_1}{y_2}$? Why or why not?

Applications

24. **Electronics** If 2.5 m of copper wire weighs 0.325 kg, how much will 75 m of wire weigh?
25. **Consumerism** Six pounds of sugar costs $2.00. At that rate, how much will 40 pounds of sugar cost?
26. **Travel** A car uses 5 gallons of gasoline to travel 143 miles. At that rate, how much gasoline will the car use to travel 200 miles?
27. **Chemistry** Charles' Law says that the volume of a gas is directly proportional to its temperature. If the volume of a gas is 2.5 cubic feet at 150° (absolute temperature), what is the volume of the same gas at 200° (absolute temperature)?

Mixed Review

28. Solve $5n + 3 = 9$. **(Lesson 3-5)**

29. Sales tax of 6% is added to a purchase of $11. Find the total price. **(Lesson 4-4)**

30. Make a table to solve the following problem. **(Lesson 4-5)**

    Mei tore a sheet of paper in half. Then she tore each of the resulting pieces in half. If she continued this process 15 more times, how many pieces of paper would she end up with?

31. **Sales** How much coffee that costs $3 a pound should be mixed with 5 pounds of coffee that costs $3.50 a pound to obtain a mixture that costs $3.25 a pound? **(Lesson 4-6)**

32. **Travel** Brad and Scott leave home at the same time, riding on their bicycles in opposite directions. Scott rides at 10 kilometers per hour and Brad rides at 12 kilometers per hour. In how many hours will they be 110 kilometers apart? **(Lesson 4-7)**

HISTORY CONNECTION

A'h-Mose

A'h-mose (more commonly referred to as Ahmes) was an Egyptian scribe who copied an earlier mathematics text on papyrus in 1650 B.C. Papyrus was a writing material made from the papyrus plant that was used by the ancient Egyptians. A'h-mose's papyrus was purchased by the English Egyptologist A. Henry Rhind. As a result, it has come to be called the *Rhind Papyrus.*

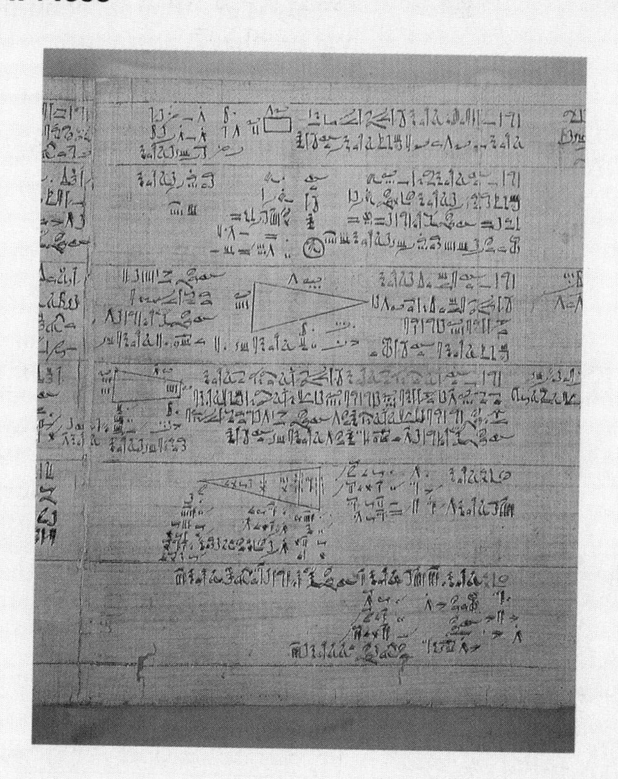

A'h-mose began his text with a table of fractions and an explanation of the Egyptian method of using fractions. The ancient Egyptians defined all fractions, with the single exception of the fraction $\frac{2}{3}$, in terms of *unit fractions.* That is, they only used fractions with a numerator of one. For example, instead of writing $\frac{3}{4}$, they wrote $\frac{1}{2} + \frac{1}{4}$. Instead of writing $\frac{5}{6}$, they wrote $\frac{1}{2} + \frac{1}{3}$.

Write each fraction in terms of unit fractions only.

1. $\frac{2}{5}$
2. $\frac{3}{8}$
3. $\frac{5}{12}$
4. $\frac{7}{24}$

4-9 Inverse Variation

Objective

After studying this lesson, you should be able to:

- solve problems involving inverse variations.

Connection

The three rectangles at the right have the same area, 4 square units. Notice that as the length increases, the width decreases. As the length decreases, the width increases. However, their product stays the same.

8 units

$\frac{1}{2}$ units

4 units

1 units

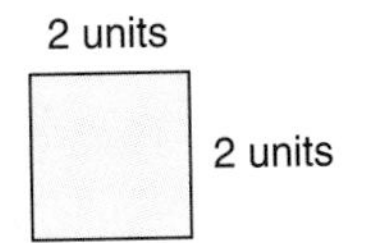

ℓ	2	4	8
w	2	1	$\frac{1}{2}$
A	4	4	4

This is an example of an **inverse variation.** We say that *ℓ varies inversely as w.*

Definition of Inverse Variation	**An inverse variation is described by an equation of the form $xy = k$, where $k \neq 0$.**

Sometimes inverse variations are written in the form $y = \frac{k}{x}$.

Example 1

The pitch of a musical tone varies inversely as its wavelength. If one tone has a pitch of 440 vibrations per second and a wavelength of 2.4 feet, find the wavelength of a tone that has a pitch of 660 vibrations per second.

Let p = pitch and w = wavelength. Find the value of k.

$pw = k$

$(440)(2.4) = k$ *Substitute 440 for p and 2.4 for w.*

$1056 = k$ *The constant of variation is 1056.*

Next, find the wavelength of the second tone.

$w = \frac{k}{p}$ *Divide each side of $pw = k$ by p.*

$w = \frac{1056}{660}$ *Substitute 1056 for k and 660 for p.*

$w = 1.6$ The wavelength is 1.6 feet.

Let (x_1, y_1) be a solution of an inverse variation, $xy = k$. Let (x_2, y_2) be a second solution. Then $x_1y_1 = k$ and $x_2y_2 = k$.

$x_1y_1 = k$

$x_1y_1 = x_2y_2$ *You can substitute x_2y_2 for k because $x_2y_2 = k$.*

The equation $x_1y_1 = x_2y_2$ is called the *product rule for inverse variations.* Study how it can be used to form a proportion.

$$x_1y_1 = x_2y_2$$

$$\frac{x_1y_1}{x_2y_1} = \frac{x_2y_2}{x_2y_1} \quad \text{Divide each side by } x_2y_1.$$

$$\frac{x_1}{x_2} = \frac{y_2}{y_1}$$

Notice that this proportion is different from the proportions for direct variation on page 163.

Example 2

If y varies inversely as x, and $y = 3$ when $x = 12$, find x when $y = 4$.

Let $x_1 = 12$, $y_1 = 3$, and $y_2 = 4$. Solve for x_2.

Method 1

Use the product rule.

$$x_1y_1 = x_2y_2$$
$$12 \cdot 3 = x_2 \cdot 4$$
$$\frac{36}{4} = x_2$$
$$9 = x_2$$

Method 2

Use the proportion.

$$\frac{x_1}{x_2} = \frac{y_2}{y_1}$$
$$\frac{12}{x_2} = \frac{4}{3}$$
$$36 = 4x_2$$
$$9 = x_2$$

Thus, $x = 9$ when $y = 4$.

If you have observed people on a seesaw, you may have noticed that the heavier person must sit closer to the fulcrum (pivot point) to balance the seesaw. A seesaw is a type of *lever,* and all lever problems involve inverse variation.

Property of Levers

Suppose weights w_1 and w_2 are placed on a lever at distances d_1 and d_2, respectively, from the fulcrum. The lever is balanced when $w_1d_1 = w_2d_2$.

The property of levers is illustrated at the right.

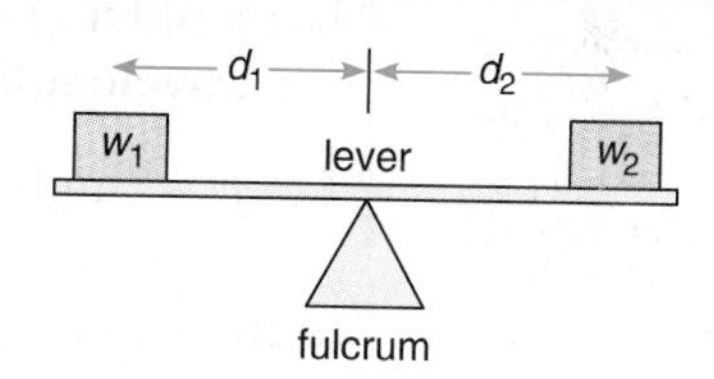

Example 3

An 8-ounce weight is placed at one end of a yardstick. A 10-ounce weight is placed at the other end. Where should the fulcrum be placed to balance the yardstick?

Let d = distance from fulcrum to 8-ounce weight, in inches.
Then $36 - d$ = distance from fulcrum to 10-ounce weight, in inches.

$$w_1d_1 = w_2d_2$$
$$8d = 10(36 - d)$$
$$8d = 360 - 10d \quad \text{Distributive property}$$
$$18d = 360$$
$$d = 20$$

The fulcrum should be placed 20 inches from the 8-ounce weight.

Example 4

APPLICATION

Physics

Pat and Ann are seated on the same side of a seesaw. Pat is 6 feet from the fulcrum and weighs 115 pounds. Ann is 8 feet from the fulcrum and weighs 120 pounds. Kai is seated on the other side of the seesaw, 10 feet from the fulcrum. If the seesaw is balanced, how much does Kai weigh?

Draw a diagram.

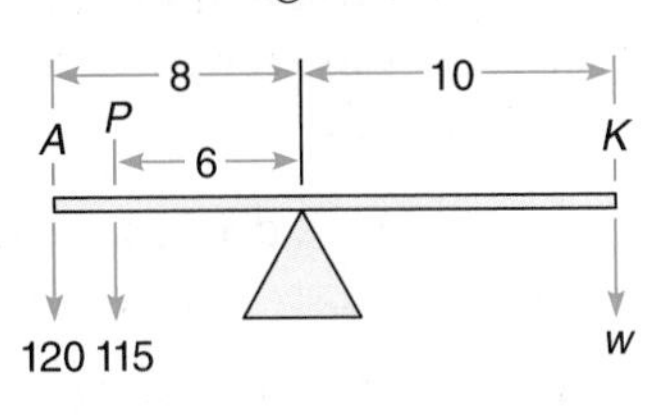

Let w = Kai's weight.

$$120(8) + 115(6) = 10w$$
$$960 + 690 = 10w$$
$$1650 = 10w$$
$$165 = w$$

Thus, Kai weighs 165 pounds.

CHECKING FOR UNDERSTANDING

Communicating Mathematics

Read and study the lesson to fill in the blanks.

1. A(n) $\underline{\ ?\ }$ variation is described by an equation of the form $xy = k$, where $k \neq 0$.
2. A(n) $\underline{\ ?\ }$ variation is described by an equation of the form $y = kx$, where $k \neq 0$.
3. The equation $x_1y_1 = x_2y_2$ is called the $\underline{\ ?\ }$.

Guided Practice

Determine which equations are inverse variations and which are direct variations. Then find the constant of variation.

4. $ab = 6$
5. $c = 3.14d$
6. $\frac{50}{y} = x$
7. $\frac{1}{5}a = d$
8. $s = 3t$
9. $14 = ab$
10. $xy = 1$
11. $a = \frac{7}{b}$
12. $2x = y$

Suppose two people are on a seesaw. For the seesaw to balance, which person must sit closer to the fulcrum?

13. Jorge, 168 pounds
    Emilio, 220 pounds
14. Shawn, 114 pounds
    Shannon, 97 pounds

EXERCISES

Practice

Solve. Assume that y varies inversely as x.

15. If $y = 24$ when $x = 8$, find y when $x = 4$.
16. If $y = -6$ when $x = -2$, find y when $x = 5$.
17. If $y = 99$ when $x = 11$, find x when $y = 11$.
18. If $y = 7$ when $x = \frac{2}{3}$, find y when $x = 7$.
19. If $x = 2.7$ when $y = 8.1$, find y when $x = 3.6$.
20. If $x = 6.1$ when $y = 4.4$, find x when $y = 3.2$.

Critical Thinking

21. Assume y varies inversely as x. If the value of x is doubled, what happens to the value of y? If the value of y is tripled, what happens to the value of x?

Applications

Chemistry: Boyle's Law states that the volume of a gas (V) varies inversely with applied pressure (P). This is shown by the formula $P_1V_1 = P_2V_2$. Use the formula to solve Exercises 22 and 23.

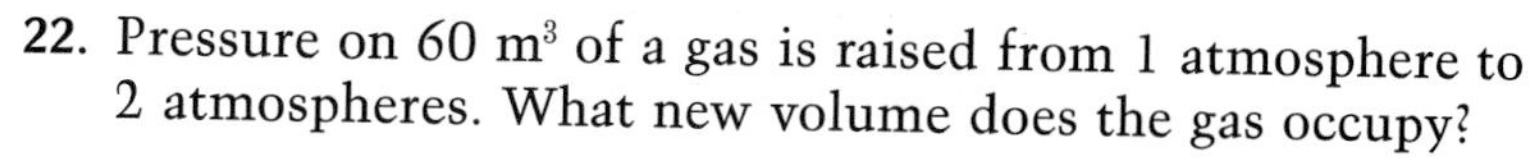

22. Pressure on 60 m^3 of a gas is raised from 1 atmosphere to 2 atmospheres. What new volume does the gas occupy?

23. A helium-filled balloon has a volume of 16 m^3 at sea level. The pressure at sea level is 1 atmosphere. The balloon rises to a point in the air where the pressure is 0.75 atmosphere. What is its volume?

Harmonics: The frequency of a vibrating string is inversely proportional to its length. Use this information to solve Exercises 24 and 25.

24. A violin string 10 inches long vibrates at a frequency of 512 cycles per second. Find the frequency of an 8-inch string.

25. A piano string 36 inches long vibrates at a frequency of 480 cycles per second. Find the frequency of the string if it were shortened to 24 inches.

Physics: Use the property of levers to solve Exercises 26-28.

26. Grant, who weighs 150 pounds, is seated 8 feet from the fulcrum of a seesaw. Mariel is seated 10 feet from the fulcrum. If the seesaw is balanced, how much does Mariel weigh?

27. Weights of 100 pounds and 115 pounds are placed on a lever. The two weights are 15 feet apart, and the lever is balanced. How far from the fulcrum is the 100-pound weight?

28. Macawi, who weighs 108 pounds, is seated 5 feet from the fulcrum of a seesaw. Barbara is seated on the same side of the seesaw, two feet farther from the fulcrum than Macawi. Barbara weighs 96 pounds. The seesaw is balanced when Sue, who weighs 101 pounds, sits on the other side. How far is Sue from the fulcrum?

Mixed Review

29. Evaluate $5 - y$, if $y = 20$. **(Lesson 2-2)**

30. Solve $\frac{x - 7}{5} = 12$. **(Lesson 3-7)**

31. Use $I = prt$ to find r if $I = \$3487.50$, $p = \$6000$, and $t = 3\frac{1}{2}$ years. **(Lesson 4-3)**

32. **Chemistry** A chemist has 2.5 liters of a solution that is 70% acid. How much water should be added to obtain a 50% acid solution? **(Lesson 4-6)**

33. **Employment** Hugo's wages vary directly as the time he works. If his wages for 4 days are \$110, how much will they be for 17 days? **(Lesson 4-8)**

Select an item from this chapter that you feel shows your best work and place it in your portfolio. Explain why you selected it.

CHAPTER 4

SUMMARY AND REVIEW

VOCABULARY

Upon completing this chapter, you should be familiar with the following terms:

base	138	146	percent of decrease
constant of variation	162	146	percent of increase
direct variation	162	134	proportion
extremes	134	136, 138	rate
inverse variation	166	134	ratio
means	134	142	simple interest
percent	138	158	uniform motion
percentage	138		

SKILLS AND CONCEPTS

OBJECTIVES AND EXAMPLES

Upon completing this chapter, you should be able to:

- solve proportions **(Lesson 4-1)**

$$\frac{x}{3} = \frac{x+1}{2}$$
$$2(x) = 3(x+1)$$
$$2x = 3x + 3$$
$$-x = 3$$
$$x = -3$$

REVIEW EXERCISES

Use these exercises to review and prepare for the chapter test.

Solve each proportion.

1. $\frac{6}{15} = \frac{n}{45}$
2. $\frac{4}{8} = \frac{11}{t}$
3. $\frac{x}{11} = \frac{35}{55}$
4. $\frac{5}{6} = \frac{a-2}{4}$
5. $\frac{y+4}{y-1} = \frac{4}{3}$
6. $\frac{z-7}{6} = \frac{z+3}{7}$

- solve percent problems **(Lesson 4-2)**

Nine is what percent of 15?

$$\frac{9}{15} = \frac{r}{100}$$
$$9(100) = 15r$$
$$\frac{900}{15} = \frac{15r}{15}$$
$$60 = r$$

Solve.

7. What number is 60% of 80?
8. Twenty-one is 35% of what number?
9. Eighty-four is what percent of 96?
10. What percent of 17 is 34?
11. What number is 0.3% of 62.7?

- solve problems involving simple interest **(Lesson 4-3)**

Use the formula $I = prt$, where I = interest, p = principal, r = annual interest rate, and t = time in years.

Solve.

12. Maria Cruz invested \$8000 for one year, part at 8% and the rest at 12%. Her total interest for the year was \$744. How much money did Maria invest at each rate?

OBJECTIVES AND EXAMPLES

- solve problems involving percent of increase or decrease **(Lesson 4-4)**

$$\frac{\text{amount of decrease}}{\text{original price}} = \frac{r}{100}$$

$$\frac{\text{amount of increase}}{\text{original price}} = \frac{r}{100}$$

REVIEW EXERCISES

Solve.

13. A price decreased from \$40 to \$35. Find the percent of decrease.

14. A price increased from \$35 to \$37.10. Find the percent of increase.

- solve problems involving discount or sales tax **(Lesson 4-4)**

At Debbie's Dresses, a dress priced at \$45 has a 5% discount. Find the discount price before sales tax.

$$5\% \text{ of } \$45 = 0.05(45) = 2.25$$

$$\$45 - \$2.25 = \$42.75$$

Solve.

15. Inez Alonso made purchases at Martin's Department Store totaling \$179.96. If she has to pay 5.5% sales tax, what should be the dollar amount she writes on her check?

16. Larry is buying a stereo that costs \$399. Since he is an employee of the store, he receives a 15% discount. He also has to pay 6% sales tax. If the discount is computed first, what is the total cost of Larry's stereo?

- solve mixture problems **(Lesson 4-6)**

Juana bought 16 used paperback books for \$10.95. Some cost 60¢ and the rest cost 75¢ each. How many of each did she buy?

	Number	Total Price
60¢ books	x	$0.60x$
75¢ books	$16 - x$	$0.75(16 - x)$

$$0.60x + 0.75(16 - x) = 10.95$$

7 at 60¢ and 16 − 7 or 9 at 75¢

Copy and complete the chart. Then write an equation and use it to solve the problem.

17. How much whipping cream (9% butterfat) should be added to 1 gallon of milk (4% butterfat) to obtain a 6% butterfat mixture?

	Gallons	Amt. of Butterfat
9% butterfat	x	
4% butterfat		
6% butterfat		

- solve problems involving uniform motion by using the formula $d = rt$ **(Lesson 4-7)**

d = distance

r = rate

t = time

Use $d = rt$ to answer each question.

18. Pat is driving at 80 kilometers per hour. How far will she travel:
a. in 2 hours? b. in 6 hours?
c. in h hours?

19. Marilyn traveled 240 miles. What was her rate if she made the trip:
a. in 6 hours? b. in t hours?

OBJECTIVES AND EXAMPLES	REVIEW EXERCISES

▪ solve problems involving direct variation **(Lesson 4-8)**

If x varies directly as y, and x = 15 when y = 1.5, find x when y = 9.

$$\frac{1.5}{9} = \frac{15}{x}$$
$$\frac{1.5x}{1.5} = \frac{135}{1.5}$$
$$x = 90$$

Solve. Assume that *y* varies directly as *x*.

20. If y = 15 when x = 5, find y when x = 7.

21. If y = 35 when x = 175, find y when x = 75.

22. If y = 3 when x = 99.9, find y when x = 522.81.

▪ solve problems involving inverse variation **(Lesson 4-9)**

If y varies inversely as x, and y = 24 when x = 30, find x when y = 10.

$$\frac{30}{x} = \frac{10}{24}$$
$$\frac{10x}{10} = \frac{720}{10}$$
$$x = 72$$

Solve. Assume that *y* varies inversely as *x*.

23. If y = 15 when x = 5, find y when x = 7.

24. If y = 35 when x = 175, find y when x = 75.

25. If y = 28 when x = 42, find y when x = 56.

APPLICATIONS AND CONNECTIONS

26. Entertainment The television series with the largest audience to date was the final episode of *M*A*S*H.* 77% of the 162 million people watching TV that night saw the program. How many people were watching? **(Lesson 4-2)**

27. Cooking George is in charge of preparing Thanksgiving dinner. The menu is shown below, along with the preparation time and cooking time for each dish. Dinner must be ready by 5:00 P.M. Make a timetable to determine when George should begin preparing each dish. **(Lesson 4-5)**

Thanksgiving Dinner Menu

Item	Preparation Time	Cooking Time
15-pound turkey		4 hours (plus 30 minutes to cool)
Mashed potatoes	30 minutes to peel, mash	45 minutes to boil
Candied yams	1 hour to boil and prepare	30 minutes to bake
Green beans	30 minutes to cut and wash	20 minutes to cook
Hot cranberry sauce		5 minutes to heat
Gravy		10 minutes to cook
Dinner rolls		15 minutes to bake
Jello mold	15 minutes to make	4 hours to set

28. Physics Lee and Rena sit on opposite ends of a 12-foot teetertotter. Lee weighs 180 pounds and Rena weighs 108 pounds. Where should the fulcrum be placed in order for the teetertotter to be balanced? **(Lesson 4-9)**

CHAPTER 4 TEST

Solve each proportion.

1. $\frac{7}{8} = \frac{5}{t}$

2. $\frac{n}{4} = \frac{3.25}{52}$

3. $\frac{y+2}{8} = \frac{7}{5}$

4. $\frac{2}{5} = \frac{x-3}{-2}$

5. $\frac{9}{11} = \frac{x-3}{x+5}$

6. $\frac{x+1}{-3} = \frac{x-4}{5}$

Solve.

7. Find 6.5% of 80.

8. Forty-two is what percent of 126?

9. Eighty-four is 60% of what number?

10. What number decreased by 20% is 16?

11. What is 50% of 17?

12. Twenty-four is what percent of 8?

13. What number increased by $33\frac{1}{3}\%$ is 52?

14. 54 is 20% more than what number?

15. A price decreased from \$60 to \$45. Find the percent of decrease.

16. The price in dollars (p) minus a 15% discount is \$3.40. Find p.

17. The price in dollars (p) plus 5.75% sales tax is \$10.52. Find p.

18. **Sales** Jason paid \$21.96 for a new shirt that was marked 20% off. What was the price before the discount?

19. **Employment** Ahmed's wages vary directly as the days he works. If his wages for 5 days are \$26, how much would they be for 12 days?

20. **Chemistry** How many ounces of a 6% iodine solution needs to be added to 12 ounces of a 10% iodine solution to create a 7% iodine solution?

21. **Cartography** The scale on a map is 2 centimeters to 5 kilometers. Doe Creek and Kent are 15.75 kilometers apart. How far apart are they on the map?

22. **Travel** Susan drove 3.25 hours at a rate of 50 miles per hour. To the nearest tenth, how long would it take her to drive the same distance at 45 miles per hour?

23. **Physics** A person weighs 0.5% less at the Equator than at the North or South Poles. How much would a person weigh at the Equator if the person weighed 148 pounds at the North Pole?

24. **Finance** Gladys deposited an amount of money in the bank at 6.5% annual interest. After 6 months, she received \$7.80 interest. How much money had Gladys deposited in the bank?

25. **Physics** Marcie weighs 133 pounds and sits 6 feet from the fulcrum of a seesaw. She balances with Rosalinda if Rosalinda sits 7 feet from the fulcrum. How much does Rosalinda weigh?

Bonus

Travel Ben, who is bicycling at the rate of x miles per hour, has a head start of 2 miles on Rachel, who is bicycling at the rate of y miles per hour ($y > x$). In how many hours will Rachel overtake Ben?

CHAPTER OBJECTIVES

In this chapter, you will:

- Solve inequalities.
- Solve compound inequalities.
- Graph solutions of open sentences that involve absolute value.
- Solve problems that can be represented by inequalities.
- Solve problems by making a diagram.

Inequalities

APPLICATION IN STATISTICS

During election campaigns, candidates need to know how people in their area will vote. Since it is not possible to question all of the voters, candidates use *polls* to find this information. In a poll, a sampling of voters is asked how they will vote in the election. Then, based on the results, a projection is made on how the actual vote will turn out.

Since the results of a poll are based on projections, their accuracy in predicting the actual election results depend greatly on the number of people used in the sampling. One measure of the accuracy of a poll is its *tolerance.* Tolerance is based in part on the size of the poll's sampling. The tolerance, along with the poll's results, define a range of values within which the actual election results should fall. This range of values can be described using a *compound inequality.*

Suppose that a poll projects that a candidate will receive 48% of the votes for mayor, and that the poll's results are accurate to within ±4%. This means the poll has a 4% tolerance and in the actual election, the candidate should receive somewhere between 44% (48% − 4%) and 52% (48% + 4%) of the votes. If x represents the percent of the vote received by this candidate, then we write $44 \leq x \leq 52$.

According to the latest poll on Issue 3, 51% of the voters plan to vote yes, 42% plan to vote no, and 7% are undecided. If these results are off by as much as 3%, could Issue 3 lose? what about 5%?

ALGEBRA IN ACTION

The chart below shows the results of polls taken to determine the voters' preference for governor. The results of the poll are accurate to within ±3%. During what months do the results indicate that *either* of the candidates could win the election?

Candidate	April	June	August	October
Alvarez	44%	48%	49%	49%
Taylor	40%	41%	42%	44%
Undecided	16%	11%	9%	7%

5-1 Solving Inequalities Using Addition and Subtraction

Objective

After studying this lesson, you should be able to:

- solve inequalities by using addition and subtraction.

Application

Eric Stevens had bowling scores of 183 and 164 for the first two games of a three-game series. What is the lowest score he can bowl in his third game and have a three-game total that is more than 500?

Let t = Eric's score for the third game. Write an inequality to represent the situation.

$$\underbrace{\text{The sum of the scores}}_{183 + 164 + t} \quad \underbrace{\text{is more than}}_{>} \quad \underbrace{500.}_{500}$$

$$347 + t > 500$$

If we were solving an equation instead of an inequality, at this point we would subtract 347 from each side, or add -347 to each side. Can these same procedures be applied to inequalities? Consider an inequality, such as $5 > -2$, that we know is true. If 3 is added to each side of $5 > -2$, the resulting inequality is $8 > 1$, which is also true. If 4 is subtracted from each side of $5 > -2$, the resulting inequality is $1 > -6$, which is also true.

These examples illustrate the following properties.

Addition and Subtraction Properties for Inequalities

For all numbers a, b, and c,
1. if $a > b$, then $a + c > b + c$ and $a - c > b - c$, and
2. if $a < b$, then $a + c < b + c$ and $a - c < b - c$.

These properties are also true when $<$ and $>$ are replaced by $\leq$ and $\geq$.

Example 1

Solve $347 + t > 500$ to answer the bowling problem presented above.

$$347 + t > 500$$
$$347 - 347 + t > 500 - 347 \qquad \textit{Subtract 347 from each side.}$$
$$t > 153$$

The solution set is {all numbers greater than 153}.

To check this solution, substitute a number greater than 153 and a number less than 153 in the original inequality. The inequality should be true only for whole numbers greater than 153 and less than or equal to 300, the highest score in bowling.

Check:

$347 + t \overset{?}{>} 500$
$347 + 155 \overset{?}{>} 500$ *Try 155.*
$502 > 500$ *True*

$347 + t \overset{?}{>} 500$
$347 + 150 \overset{?}{>} 500$ *Try 150.*
$497 > 500$ *False*

FYI ...

Thousands of years ago, the Egyptians bowled on alleys not unlike ours.

Since the solution set for $347 + t > 500$ is {all whole numbers greater than 153 and less than or equal to 300}, the lowest score Eric can bowl in his third game and have a three-game total that is more than 500 is 154.

The solution to the inequality in Example 1 was expressed as a set. A more concise way of writing this set is to use **set-builder notation.** The solution in set-builder notation is $\{t \mid t$ is a whole number and $153 < t \leq 300\}$. This is read *the set of all whole numbers t such that t is greater than 153 and less than or equal to 300.*

Example 2

Solve $8y + 3 > 9y - 14$.

$$8y + 3 > 9y - 14$$
$$8y - 8y + 3 > 9y - 8y - 14 \quad \text{Subtract } 8y \text{ from each side.}$$
$$3 > y - 14$$
$$3 + 14 > y - 14 + 14 \quad \text{Add 14 to each side.}$$
$$17 > y$$

Since $17 > y$ is the same as $y < 17$, the solution set is usually written $\{y \mid y < 17\}$.

Verbal problems containing phrases like *greater than* or *less than* can often be solved by using inequalities. The following chart shows some other phrases that indicate inequalities.

Inequalities			
$<$	$>$	$\leq$	$\geq$
■ less than ■ fewer than	■ greater than ■ more than	■ at most ■ no more than ■ less than or equal to	■ at least ■ no less than ■ greater than or equal to

Example 3

APPLICATION

Budgeting

Jessie's budget allows her to spend at most \$17.50 on new equipment for her model railroad. She has chosen a new railroad car that costs \$9.98. How much can Jessie spend on other equipment?

EXPLORE *At most \$17.50* means less than or equal to ($\leq$) \$17.50. Let x = the amount that Jessie can spend on other equipment.

PLAN *Total to spend is at most \$17.50.*

$$9.98 + x \leq 17.50$$

SOLVE $9.98 - 9.98 + x \leq 17.50 - 9.98$ *Subtract 9.98 from each side.*

Jessie can spend \$7.52 or less on other equipment.

EXAMINE Since \$9.98 + \$7.52 = \$17.50, Jessie can spend \$7.52 or less on other equipment.

Example 4

APPLICATION

Sports

Roberto has scores of 63.2, 59.8, 61.5, and 71.1 for his first four dives in a springboard diving competition. He has one more dive remaining. If Roberto's total score must be greater than 322.7 to win the competition, what score must he receive on his fifth dive?

EXPLORE You want to find the score for Roberto's fifth dive that will make his total score, which is the sum of his scores on the five dives, greater than 322.7. Let s = the score on Roberto's fifth dive.

PLAN $\underbrace{\text{Roberto's total score}}$ $\underbrace{\text{is greater than}}$ $\underbrace{322.7.}$

$$63.2 + 59.8 + 61.5 + 71.1 + s \quad > \quad 322.7$$

SOLVE

$$255.6 + s > 322.7$$
$$255.6 - 255.6 + s > 322.7 - 255.6$$
$$s > 67.1$$

Roberto will win if the score on his fifth dive is greater than 67.1.

EXAMINE Since $63.2 + 59.8 + 61.5 + 71.1 + 67.1 = 322.7$, if Roberto scores more than 67.1 on his fifth dive, his total score will be more than 322.7.

CHECKING FOR UNDERSTANDING

Communicating Mathematics

Read and study the lesson to answer each question.

1. Explain why it was not really necessary to state the subtraction property for inequalities.
2. Write three inequalities that are equivalent to $x < 5$.
3. Write two inequalities that each have $\{y|y > -4\}$ as their solution set.
4. Refer to the situation at the beginning of the lesson. If Eric wanted to have a three-game total that was greater than 525, what is the lowest score he could bowl in his third game?

Guided Practice

State the number you would add to each side to solve each inequality.

5. $x - 17 > 43$
6. $-8 + a \geq -7$
7. $11 \leq m - 14.5$
8. $y + 4 < -4$

State the number you would subtract from each side to solve each inequality.

9. $z + 1 \leq 5$
10. $-3 \geq b + 11$
11. $n + (-8) > 22$
12. $9x < 8x - 2$

Solve each inequality. Check the solution.

13. $r + 11 < 6$
14. $y - 18 > -3$
15. $10 \geq -3 + x$
16. $4a - 3 \leq 5a$

EXERCISES

Practice

Solve each inequality. Check the solution.

17. $a + 4 < 13$
18. $x - 3 < -17$
19. $r - 19 \geq 23$
20. $y + 15 \geq -2$
21. $9 + w \leq 9$
22. $-9 + c > 9$
23. $-11 > d - 4$
24. $-11 \leq k - (-4)$
25. $2x > x - 3$
26. $5y + 4 < 6y$
27. $-7 < 16 - z$
28. $-p - 11 \geq 23$
29. $2x - 3 \geq x$
30. $7h - 1 \leq 6h$
31. $-5 + 14b \leq -4 + 15b$
32. $6r + 4 \geq 5r + 4$
33. $2s - 6.5 < -11.4 + s$
34. $1.1v - 1 > 2.1v - 3$
35. $3x + \frac{4}{5} \leq 4x + \frac{3}{5}$
36. $\frac{1}{2}t + \frac{1}{4} \geq \frac{3}{2}t - \frac{2}{3}$
37. $17.42 - 7.029z \geq 15.766 - 8.029z$

Write an inequality and solve. Then check each solution.

38. A number decreased by 17 is less than -13.
39. A number increased by 4 is at least 3.
40. A number subtracted from 21 is no less than -2.
41. Twice a number added to 8 is more than three times the number.
42. Four more than 6 times a number is at most 5 times the number decreased by 3.
43. The sum of two numbers is less than 25. One of the numbers is 18. What is the other number?

If $2x \geq x - 5$, then complete each inequality.

44. $2x + 3 \geq x - \underline{\ ?\ }$
45. $2x - 7 \geq x - \underline{\ ?\ }$
46. $2x + \underline{\ ?\ } \geq x + 3$
47. $\underline{\ ?\ } \leq x$

Critical Thinking

48. Is it possible for the solution set of an inequality to be the empty set ($\emptyset$)? If so, name an example.

Applications

Define a variable, write an inequality, and solve each problem.

49. **School** Eva has earned 453 points prior to the 200-point semester test in her biology class. To get an A for the semester, she must earn at least 630 points. What is the minimum number of points she can score on the test and still get an A?
50. **Banking** City Bank requires a minimum balance of \$1200 for free checking. If Mr. Findlay still has free checking after making withdrawals of \$2000 and \$1454, how much was in his account before the withdrawals?

51. **Sports** Janine has a total score of 39.35 in the four events of a gymnastics competition. Her closest competitor, Nyoko, has scores of 9.8, 9.75, and 9.9 for three events and has one event left. What must Nyoko's score be in the last event for Janine to win the competition?

Computer

52. This BASIC program compares two rational numbers. The IF-THEN statements direct the flow of the program. When the condition following IF is true, the program goes to the line number after THEN. When the condition is false, the program goes to the next line. By the comparison property, either $\frac{A}{B} = \frac{C}{D}$ or $\frac{A}{B} > \frac{C}{D}$ or $\frac{A}{B} < \frac{C}{D}$.

```
10 PRINT "ENTER THE FIRST
   NUMERATOR AND DENOMINATOR."
20 INPUT A, B
30 PRINT "ENTER THE SECOND
   NUMERATOR AND DENOMINATOR."
40 INPUT C, D
50 IF A/B > C/D THEN 90
60 IF A/B < C/D THEN 110
70 PRINT A;"/";B;" = ";C;"/";D
80 GOTO 120
90 PRINT A;"/";B;" > ";C;"/";D
100 GOTO 120
110 PRINT A;"/";B;" < ";C;"/";D
120 END
```

Compare each pair of rational numbers. If a number is negative, enter the negative sign with the numerator.

a. $\frac{9}{11}, \frac{15}{19}$ b. $\frac{7}{8}, \frac{13}{15}$ c. $-\frac{7}{8}, -\frac{13}{15}$ d. $\frac{16}{17}, \frac{17}{18}$ e. $-\frac{7}{6}, -\frac{21}{18}$

Mixed Review

53. Simplify $0.3(0.2 + 3y) + 0.21y$. **(Lesson 1-6)**
54. Graph the solution set of $n \leq -2$. **(Lesson 2-3)**
55. Solve $\frac{4}{9}x = -9$. **(Lesson 3-3)**
56. **Sales** Denise bought a new dress for \$32.86. This included 6% sales tax. What was the cost of the dress before tax? **(Lesson 4-4)**
57. If y varies inversely as x, and $y = 8$ when $x = 24$, find y when $x = 6$. **(Lesson 4-8)**

HISTORY CONNECTION

Recorde, Harriot, and Oughtred

Three British mathematicians had a great impact on the way mathematics is written today. In 1557, in his *The Whetstone of Witte,* Robert Recorde (1510–1558) introduced the modern equals sign (=). He used a pair of equal parallel line segments for the symbol "bicause noe 2 thynges can be moare equalle." Thomas Harriot (1560–1621), in his *Artis analyticae praxis,* introduced the inequality signs, > and <. William Oughtred (1574–1660) placed great emphasis on mathematical symbols. It was he who introduced the cross (×) for multiplication.

5-2 Solving Inequalities Using Multiplication and Division

Objective

After studying this lesson, you should be able to:

- solve inequalities by using multiplication and division.

Application

Jerry Hamilton is an architect for a commercial architecture firm. He needs to design a roof that has a span of 32 feet and a pitch that is at most $\frac{3}{8}$. What is the largest possible rise the roof can have?

Let r = the rise of the roof. In Lesson 3-3, you learned that the pitch of a roof is equal to its rise divided by its span. Thus, the inequality to be solved is $\frac{r}{32} \leq \frac{3}{8}$.

If you were solving an equation instead of an inequality, at this point you would multiply each side by 32. Can the same procedure be applied to inequalities? Consider an inequality, such as $5 > -2$, that we know is true.

If each side of $5 > -2$ is multiplied by 3, the resulting inequality is $15 > -6$, which is also true.

If each side of $5 > -2$ is multiplied by -4, the resulting inequality is $-20 > 8$, which is *false*. To be true, the resulting inequality would need to be $-20 < 8$.

The results of these examples can be summarized as follows.

If each side of a true inequality is multiplied by the same positive number, the resulting inequality is also true.

If each side of a true inequality is multiplied by the same negative number, the direction of the inequality symbol must be *reversed* so that the resulting inequality is also true.

Multiplication Property for Inequalities

For all numbers a, b, and c,
1. **if c is positive and $a < b$, then $ac < bc$, and if c is positive and $a > b$, then $ac > bc$.**
2. **If c is negative and $a < b$, then $ac > bc$, and if c is negative and $a > b$, then $ac < bc$.**

These properties also hold for $\leq$ and $\geq$.

Example 1

Solve $\frac{r}{32} \leq \frac{3}{8}$ to answer the roofing problem presented above.

$$\frac{r}{32} \leq \frac{3}{8}$$

$$32\left(\frac{r}{32}\right) \leq 32\left(\frac{3}{8}\right) \quad \textit{Multiply each side by 32.}$$

$$r \leq 12$$

The solution set is $\{r | r \leq 12\}$. The largest rise for the roof is 12 feet.

There is a house in Rockport, Massachusetts, made entirely of newspaper.

Example 2

a. **Solve $-\frac{a}{7} > 2$.**

$$-\frac{a}{7} > 2$$

$$-7\left(\frac{-a}{7}\right) < -7(2)$$ *Multiply each side by −7 and change > to <.*

$$a < -14$$

The solution set is $\{a|a < -14\}$.

b. **Solve $\frac{4x}{3} \leq -12$.**

$$\frac{4x}{3} \leq -12$$

$$\frac{4}{3}x \leq -12$$ *The reciprocal of $\frac{4}{3}$ is $\frac{3}{4}$.*

$$\frac{3}{4}\left(\frac{4}{3}x\right) \leq \frac{3}{4}(-12)$$ *Multiply each side by $\frac{3}{4}$.*

$$x \leq -9$$

The solution set is $\{x|x \leq -9\}$.

Example 3

For what values of d is an angle with measure $(5d)°$ an acute angle?

The measure of an acute angle is less than 90°. Since the measure of an angle must be positive, d must also be greater than zero.

$$5d < 90$$

$$5d\left(\frac{1}{5}\right) < 90\left(\frac{1}{5}\right)$$ *The reciprocal of 5 is $\frac{1}{5}$. Multiply each side by $\frac{1}{5}$.*

$$d < 18$$

The angle is acute when d is a number less than 18 but greater than 0.

The inequality $5d < 90$ was solved by multiplying each side by $\frac{1}{5}$. The same result could have been obtained by *dividing* each side by 5. This method uses the **division property for inequalities.**

Division Property for Inequalities

For all numbers a, b, and c, with $c \neq 0$,

1. if c is positive and $a < b$, then $\frac{a}{c} < \frac{b}{c}$, and
if c is positive and $a > b$, then $\frac{a}{c} > \frac{b}{c}$.

2. If c is negative and $a < b$, then $\frac{a}{c} > \frac{b}{c}$, and
if c is negative and $a > b$, then $\frac{a}{c} < \frac{b}{c}$.

These properties also hold for ≤ and ≥.

Example 4

Solve $-72 \geq -6n$.

The inequality can be solved using either the multiplication or division properties for inequalities. *Choose the method that is easier for you.*

a.

$$-72 \geq -6n$$

$$\left(-\frac{1}{6}\right)(-72) \leq \left(-\frac{1}{6}\right)(-6n)$$ *Multiply each side by $-\frac{1}{6}$ and change ≥ to ≤.*

$$12 \leq n$$

b.

$$-72 \geq -6n$$

$$\frac{-72}{-6} \leq \frac{-6n}{-6}$$ *Divide each side by −6 and change ≥ to ≤.*

$$12 \leq n$$

The solution set is $\{n|12 \leq n\}$. *This can also be written $\{n|n \geq 12\}$.*

Example 5

APPLICATION

Sales

Readalot Bookstore makes a profit of $4.48 on the sale of each two-volume set of the book *Modern Astronomy*. How many sets must be sold for Readalot Bookstore to make a profit of at least $175?

EXPLORE Let s = the number of two-volume sets sold.
At least $175 means greater than or equal to $175.

PLAN The profit on the sales of *Modern Astronomy* is the profit per set sold times the number of sets sold.

Profit per set sold times number of sets sold is at least $175.

$$4.48 \times s \geq 175$$

SOLVE

$$\frac{4.48s}{4.48} \geq \frac{175}{4.48} \quad \text{Divide each side by 4.48.}$$

$$s \geq 39.0625$$

Since the number of two-volume sets sold must be a whole number, Readalot Bookstore must sell 40 or more sets to make a profit of at least $175.

EXAMINE The profit on sales of 39 two-volume sets is ($4.48)(39) or $174.72. The profit on sales of 40 two-volume sets is ($4.48)(40) or $179.20. Thus, the solution is correct.

CHECKING FOR UNDERSTANDING

Communicating Mathematics

Complete.

1. Explain why it was not really necessary to state the division property for inequalities.
2. If each side of a true inequality is multiplied by the same __?__ number, the direction of the inequality symbol must be reversed so that the resulting inequality is also true.
3. Multiplying by __?__ is the same as dividing by -6.
4. The measure of an __?__ angle is less than 90°.

Guided Practice

State the number by which you multiply each side to solve each inequality. Then indicate if the direction of the inequality symbol reverses.

5. $\frac{x}{4} < -5$
6. $\frac{-2z}{7} \geq -12$
7. $10x > 20$

State the number by which you divide each side to solve the inequality. Then indicate if the direction of the inequality symbol reverses.

8. $-6y \geq -24$
9. $-10a > 13$
10. $3.3m < -33$

Solve and check each inequality.

11. $8x < -48$
12. $-10a < -30$
13. $-12d \leq 30$
14. $0.1t \geq 3$

EXERCISES

Practice

Solve each inequality. Check the solution.

15. $16x < 96$
16. $-y \leq 44$
17. $-8z \geq -72$
18. $-102 > 17r$
19. $396 > -11t$
20. $-15a < -28$
21. $4c \geq -6$
22. $6 \leq 0.8n$
23. $-4.3x < -2.58$
24. $\frac{b}{-12} \leq 3$
25. $-25 > \frac{a}{-6}$
26. $13 \geq \frac{t}{13}$
27. $\frac{2}{3}m \geq -22$
28. $-\frac{7}{9}x < 42$
29. $\frac{3y}{8} \leq 32$
30. $\frac{-5x}{6} < \frac{2}{9}$
31. $-\frac{2}{5} > \frac{4z}{7}$
32. $\frac{3b}{4} \leq \frac{2}{3}$

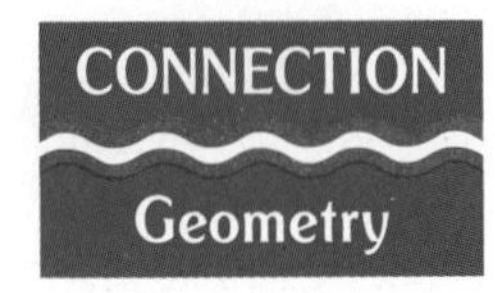

For what values of *d* is an angle with the given measure an acute angle? Assume that *d* is greater than zero.

33. $(6d)°$
34. $(24d)°$
35. $\left(\frac{5d}{8}\right)°$
36. $(3.6d)°$

Write an inequality and solve. Then check each solution.

37. Six times a number is less than 216.
38. Negative eight times a number is at most 112.
39. A number divided by negative 4 is no less than -2.
40. Five thirds of a number is no more than -15.
41. Eighty percent of a number is greater than 24.
42. The product of two numbers is no less than 144. One of the numbers is -18. What is the other number?

Complete.

43. If $5x < 12$, then $20x <$ __?__.
44. If $24k \geq 16$, then __?__ ≥ 12.
45. If $-10a > 21$, then $30a$ __?__ -63.
46. If $-9 \leq 15t$, then $25t$ __?__ -15.

Critical Thinking

Determine the conditions under which each sentence is true.

47. If $\frac{2}{x} < 5$, then $2 < 5x$.
48. If $x > y$, then $x^2 > y^2$.

Applications

Define a variable, write an inequality, and solve each problem.

49. **Business** Goodearth Recycling Center pays \$0.40 for every pound of glass that is turned in. If Sheila wants to make at least \$15, how many pounds of glass must she turn in?
50. **Sales** Julio's budget allows him to spend at most \$17 on gasoline for his car this week. If gasoline sells for \$1.36 a gallon, how many gallons can he buy for his car this week?
51. **Finance** Michael uses at most 60% of his annual Merricoe stock dividend to purchase more shares of Merricoe stock. If his dividend last year was \$885 and Merricoe stock is selling for \$14 a share, what is the greatest number of shares he can purchase?

Computer

52. For three line segments to form a triangle, the sum of the lengths of any two sides must exceed the length of the third side. Let the lengths be represented by a, b, and c. Then these three inequalities must be true: $a + b > c$, $a + c > b$, and $b + c > a$. Lines 30 through 50 of the BASIC program determine if the segments form a triangle.

```
10 PRINT "ENTER THREE LENGTHS."
20 INPUT A,B,C
30 IF C >= A + B THEN 80
40 IF B >= A + C THEN 80
50 IF A >= B + C THEN 80
60 PRINT "THIS IS A TRIANGLE."
70 GOTO 90
80 PRINT "NOT A TRIANGLE."
90 END
```

Use the program to determine whether segments with the given lengths form a triangle.

a. 11 in., 14 in., 26 in.
b. 5 ft, 12 ft, 13 ft
c. 75 cm, 87 cm, 110 cm
d. 1.5 m, 2.0 m, 2.5 m

Mixed Review

53. **Consumerism** Which is a better buy: 12 ounces of orange juice at \$1.69 or 16 ounces at \$2.29? **(Lesson 2-4)**

Solve each equation or inequality.

54. $5 - 3x = 32$ **(Lesson 3-5)**
55. $6 - 9y < -10y$ **(Lesson 5-1)**
56. $\frac{4r + 8}{16} = 7$ **(Lesson 3-5)**
57. $\frac{12 - x}{6} = \frac{x + 7}{4}$ **(Lesson 4-1)**
58. $3(4a - 9) = -7(2a - 3)$ **(Lesson 3-6)**

READING ALGEBRA

The two sentences below are examples of **compound sentences.**

A square has 4 sides *and* 3 < 3.　　A square has 4 sides *or* 3 < 3.

A compound sentence consists of two sentences that are connected by the words *and* or *or.* A **conjunction** is a compound sentence where the sentences are connected using *and.* A **disjunction** is a compound sentence where the sentences are connected using *or.* Thus, the sentence at the left above is a conjunction, while the sentence on the right is a disjunction.

To determine if a compound sentence is true, determine whether the two sentences are true or false. A conjunction is true only if *both* sentences are true. A disjunction is true if *one* or *both* sentences are true.

Determine whether each compound sentence is *true* or *false.*

1. A square has 4 sides and 3 < 3.
2. A square has 4 sides or 3 < 3.
3. 0.2 is an integer or September has 31 days.
4. Chicago is in Illinois and every whole number is a rational number.

5-3 Inequalities with More Than One Operation

Objective

After studying this lesson, you should be able to:

- solve inequalities involving more than one operation.

Application

Manuel has to choose between two after-school job offers at local video stores. The job at Videotyme pays a weekly salary of \$90. The job at Videobarn pays a weekly salary of \$70 *plus* a 10% commission on his weekly sales of videotapes. To help him choose the better job, Manuel needs to know how many dollars worth of videotapes he must sell in order to earn more per week at Videobarn.

You can write and solve an inequality to determine how much Manuel needs to sell. Let s = Manuel's weekly sales of videotapes.

Weekly salary plus 10% of weekly sales is more than \$90.

$$70 + 0.10s > 90$$

This inequality involves more than one operation. It can be solved by undoing the operations in reverse of the order of operations.

$$70 + 0.10s > 90$$
$$70 - 70 + 0.10s > 90 - 70$$ *Undo the addition first. Use the subtraction property for inequalities.*
$$0.10s > 20$$
$$10(0.10s) > 10(20)$$ *Then undo the multiplication. Use the multiplication property for inequalities.*
$$s > 200$$ *You could also divide each side by 0.10.*

Manuel must sell more than \$200 worth of videotapes each week to earn more money per week at Videobarn. Since Manuel considers himself a good salesperson, he takes the job at Videobarn.

Inequalities involving more than one operation can be solved by applying methods similar to those used for solving equations involving more than one operation.

Example 1

Solve $16 - 5b \geq 29$.

$$16 - 5b \geq 29$$

$$16 - 16 - 5b \geq 29 - 16 \quad \text{Subtract 16 from each side.}$$

$$-5b \geq 13$$

$$\frac{-5b}{-5} \leq \frac{13}{-5} \quad \text{Divide each side by } -5 \text{ and change} \geq \text{to} \leq.$$

$$b \leq -\frac{13}{5}$$

The solution set is $\left\{b \middle| b \leq \frac{-13}{5}\right\}$.

Example 2

Solve $9x + 4 < 7 - 13x$.

$$9x + 4 < 7 - 13x$$

$$9x + 13x + 4 < 7 - 13x + 13x \quad \text{Add 13x to each side.}$$

$$22x + 4 < 7$$

$$22x + 4 - 4 < 7 - 4 \quad \text{Subtract 4 from each side.}$$

$$22x < 3$$

$$\frac{22x}{22} < \frac{3}{22} \quad \text{Divide each side by 22.}$$

$$x < \frac{3}{22}$$

The solution set is $\left\{x \middle| x < \frac{3}{22}\right\}$.

When solving some inequalities that contain grouping symbols, remember to first use the distributive property to remove the grouping symbols.

Example 3

Solve $0.7(n - 3) \leq n - 0.6(n + 5)$.

$$0.7(n - 3) \leq n - 0.6(n + 5)$$

$$0.7n - 2.1 \leq n - 0.6n - 3.0 \quad \text{Use the distributive property.}$$

$$0.7n - 2.1 \leq 0.4n - 3.0$$

$$0.7n - 0.4n - 2.1 \leq 0.4n - 0.4n - 3.0 \quad \text{Subtract 0.4n from each side.}$$

$$0.3n - 2.1 \leq -3.0$$

$$0.3n - 2.1 + 2.1 \leq -3.0 + 2.1 \quad \text{Add 2.1 to each side.}$$

$$0.3n \leq -0.9$$

$$\frac{0.3n}{0.3} \leq \frac{-0.9}{0.3} \quad \text{Divide each side by 0.3.}$$

$$n \leq -3$$

The solution set is $\{n | n \leq -3\}$.

Example 4

CONNECTION Statistics

Owen's scores on the first three of four 100-point tests were 89, 92, and 82. What score must he receive on the fourth test to have an average (mean) of at least 90 for all the tests?

EXPLORE Let s = Owen's score on the fourth test.

PLAN The sum of Owen's scores on the four tests, divided by 4, must be at least 90.

$$\frac{89 + 92 + 82 + s}{4} \geq 90$$

SOLVE

$$4\left(\frac{89 + 92 + 82 + s}{4}\right) \geq 4(90) \quad \textit{Multiply each side by 4.}$$

$$89 + 92 + 82 + s \geq 360$$

$$263 + s \geq 360$$

$$263 - 263 + s \geq 360 - 263 \quad \textit{Subtract 263 from each side.}$$

$$s \geq 97$$

Owen's score on the fourth test must be at least 97.

EXAMINE Here's another way to solve the problem. Look at Owen's scores on the first three tests. 89 is 1 less than 90, 92 is 2 more than 90, and 82 is 8 less than 90. Since the sum $-1 + 2 + (-8)$ is -7, it follows that Owen needs a score at least 7 more points than 90, or 97, on the fourth test to have an average of 90 on all the tests.

CHECKING FOR UNDERSTANDING

Communicating Mathematics

Read and study the lesson to answer each question.

1. Write an example of an inequality requiring more than one operation to solve.
2. How could you solve $16 - 5b > 29$ *without* dividing by -5?
3. Refer to the situation at the beginning of the lesson. If the commission on videotape sales at Videobarn is 16% instead of 10%, how much does Manuel need to sell to earn more per week at Videobarn?
4. In Example 4, can Owen have an average of at least 91 points on all four 100-point tests? Explain.

Guided Practice

Explain how to solve each inequality. Then solve.

5. $3x - 1 > 14$
6. $-20 \geq 8 + 7a$
7. $\frac{n - 11}{-2} \leq -6$
8. $12 - \frac{5z}{4} < 37$
9. $y + 1 \geq 5y + 5$
10. $2k + 7 > 11 - k$

EXERCISES

even 12-30, 31 32 34

Practice

Solve each inequality. Check the solution.

11. $9x + 2 > 20$
12. $4y - 7 < 21$
13. $-7a + 6 \leq 48$
14. $-5 - 8b \geq 59$
15. $-12 + 11m \leq 54$
16. $5 - 6n > -19$
17. $\frac{z}{4} + 7 \geq -5$
18. $\frac{2x}{3} - 3 \leq 7$
19. $-2 - \frac{d}{5} < 23$
20. $\frac{2t + 5}{3} < -9$
21. $\frac{11 - 6w}{5} > 10$
22. $7y - 27 \geq 4y$
23. $13r - 11 > 7r + 37$
24. $6a + 9 < -4a + 29$
25. $0.1y - 2 \leq 0.3y - 5$
26. $1.3x + 6.7 \geq 3.1x - 1.4$
27. $7(g + 8) < 3(g + 12)$
28. $-5(k + 4) \geq 3(k - 4)$
29. $8c - (c - 5) > c + 17$
30. $3d - 2(8d - 9) < 3 - (2d + 7)$

Write an inequality and solve. Then check each solution.

31. The sum of twice a number and 17 is no greater than 41.
32. Two-thirds of a number decreased by 27 is at least 9.
33. Three times the sum of a number and 7 is greater than five times the number less 13.
34. Twice a number increased by 32 is no less than three times the number subtracted from 2.

Define a variable, write an inequality, and solve each problem.

35. The sum of two consecutive even integers is greater than 75. Find the pair with the least sum.
36. The sum of two consecutive odd integers is at most 123. Find the pair with the greatest sum.
37. Find all sets of two consecutive positive odd integers whose sum is no greater than 18.
38. Find all sets of three consecutive positive even integers whose sum is less than 30.

Solve each inequality. Check the solution.

39. $\frac{5y - 4}{3} > \frac{y + 5}{3}$
40. $\frac{2n + 1}{7} \geq \frac{n + 4}{5}$
41. $\frac{c + 8}{4} \leq \frac{5 - c}{9}$

Critical Thinking

42. Write an inequality that has no solution.

Applications

Define a variable, write an inequality, and solve each problem.

43. **Statistics** Abeytu's scores on the first four of five 100-point tests were 85, 89, 90, and 81. What score must she receive on the fifth test to have an average of at least 87 points for all the tests?

Journal

Write a paragraph comparing and contrasting how you solve an inequality with how you solve an equation.

44. **Sales** Don earns \$24,000 per year in salary and 8% commission on his sales. If he must have a total annual income of at least \$30,000 to pay all of his expenses, how much sales per year must he have?

45. **Employment** Maria earns \$4.76 an hour at the local supermarket. Each week, 25% of her total pay is deducted for taxes. In addition, she pays \$25 a week, after taxes, for union dues. If she wants her take-home pay to be at least \$100 a week, how many hours must she work?

Mixed Review

46. Define the variable, then write an equation.
Sports Megan scored 12 points in a basketball game. This was 4 points less than the previous game. How many points did Megan score in both games? **(Lesson 2-8)**

47. Solve $\frac{2}{5}x - 1 = \frac{1}{4}x + 2$. **(Lesson 3-6)**

48. **Employment** Deidre's wages vary directly as the time she works. If her wages for 6 days are \$121, what are her wages for 20 days? **(Lesson 4-7)**

49. Solve $-13z > -1.04$. **(Lesson 5-2)**

MID-CHAPTER REVIEW

Solve each inequality. Check the solution. (Lessons 5-1, 5-2, and 5-3)

1. $a + 2 < 10$
2. $-12 \leq s - (-3)$
3. $-3 + 13z > 14z$
4. $23b \geq 276$
5. $-\frac{1}{3}k > \frac{6}{7}$
6. $\frac{4a}{5} \leq -\frac{13}{8}$
7. $-6x + 9 < 4x + 29$
8. $9d - 5 \geq 8 - d$
9. $3(7 - 2n) < 2(3n + 13)$

Write an inequality and solve. Then check each solution.

10. The sum of a number and 11 is at least 23. **(Lesson 5-1)**

11. Three fourths of a number is less than 90. **(Lesson 5-2)**

12. Twice a number less 9 is no more than 75. **(Lesson 5-3)**

13. Three times a number subtracted from 10 is at least 6 more than the number. **(Lesson 5-3)**

14. **Business** *The Gahanna Tribune* pays its carriers 8¢ for each newspaper delivered. How many newspapers must a newspaper carrier deliver to earn at least \$5.00 a day? **(Lesson 5-2)**

15. **Communications** City Phone charges \$13 a month plus \$0.21 per call for local phone service. If Jay can spend no more than \$20 a month on this service, how many calls can he make? **(Lesson 5-3)**

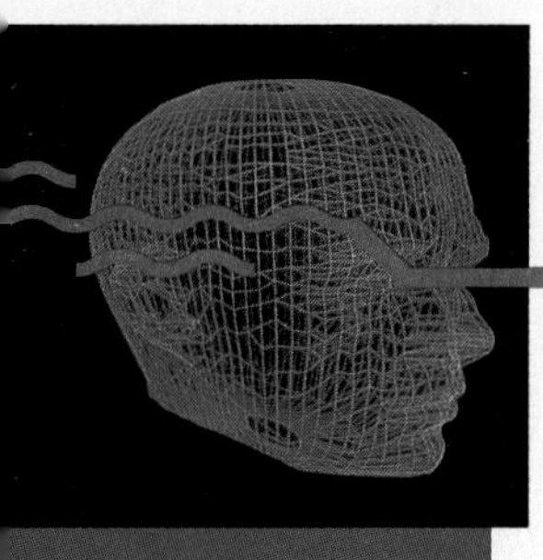

Technology

Solving Inequalities

BASIC
Graphing Calculators
▶ Graphing Software
Spreadsheets

The examples below show how to use the *Mathematics Exploration Toolkit* to practice the steps for solving inequalities. You will use the same CALC commands that were used in Chapter 3. Remember that you may enter the abbreviations shown instead of typing the entire command.

ADD (add) SUBTRACT (sub) MULTIPLY (mult)
DIVIDE (div) SIMPLIFY (simp) or [F5]

To enter the symbol ≤ type the left brace, {. To enter the symbol ≥ type the right brace, }.

Example 1

Solve $3x - 4 < 7x + 20$.

Enter	Result
3x − 4 < 7x + 20	$3x - 4 < 7x + 20$
sub 7x	$3x - 4 - 7x < 7x + 20 - 7x$
simp	$-4x - 4 < 20$
add 4	$-4x - 4 + 4 < 20 + 4$
simp	$-4x < 24$
div −4	$\frac{-4x}{-4} > \frac{24}{-4}$
simp	$x > -6$

Example 2

Solve $\left(\frac{x-2}{3}\right) \geq 6$.

Enter	Result
(x−2)/3}6	$\frac{x-2}{3} \geq 6$
mult 3	$3\left(\frac{x-2}{3}\right) \geq 3(6)$
simp	$x - 2 \geq 18$
add 2	$x - 2 + 2 \geq 18 + 2$
simp	$x \geq 20$

EXERCISES

Use CALC to solve each inequality. Record each step and solution.

1. $2x + 8 < 5x + 11$
2. $5 - x > x - 7$
3. $6(x + 4) \geq 5x - 3 - 2x$
4. $8.6x - (6.4 + 4.2x) \leq 0.2$
5. $\frac{2x - 6}{5} > \frac{3x + 2}{5}$
6. $\frac{x}{3} \leq \frac{x + 1}{4}$

EXPLORE
PLAN
SOLVE
EXAMINE

5-4 Problem-Solving Strategy: Make a Diagram

Objective

After studying this lesson, you should be able to:

- solve problems by making a diagram.

You can solve many problems more easily if you draw a picture or diagram. Sometimes a picture will help you decide how to work the problem. At other times the picture will show the answer to the problem.

Application

Let's listen in on a meeting of the Brookhaven High School Chess Club.

"Mei, would you please read the minutes of the last meeting?"

"Sure, John. At the last meeting it was decided that each member should play each of the other nine members so we can figure out who are the five best players. These five players will then play in interschool competitions. Then the meeting was adjourned."

How many games must be played to determine the five best players?

FYI...

Chess originated in ancient India. It was originally called *Chaturanga.*

Since each member plays every other member, you might think that there would be a total of 90 games (9 games $\times$ 10 members). But let's try looking at this problem another way. Draw a diagram to represent the 10 players. An easy way is to use dots.

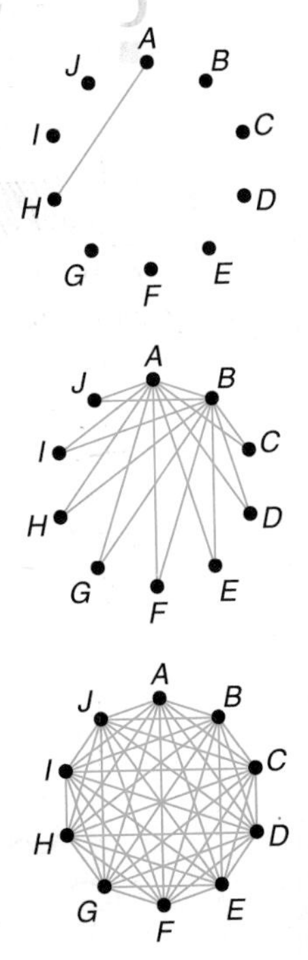

Label the points from A to J. Each point represents a player. Use a line segment joining two points to represent a game between two players. How many line segments are required to join the ten dots?

First, draw all possible segments from A. There are nine. Then draw all possible segments from B. There are eight new segments.

To check this solution, solve the problem using a different strategy.

Continue until all of the dots are connected. If you keep careful count, you will find that there are $9 + 8 + 7 + 6 + 5 + 4 + 3 + 2 + 1$, or 45 segments. Thus, 45 games would be played.

CHECKING FOR UNDERSTANDING

Communicating Mathematics

Read and study the lesson to answer each question.

1. Name two ways a picture or diagram can help you solve a problem more easily.
2. Suppose two students join the chess club. How many games will have to be played now?

Guided Practice

Solve by making a drawing or diagram.

3. Kathy Velasquez is a geography teacher. She wants to put 16 rectangular pictures of "Countries Around the World" on the wall of her classroom, but she only has a few thumbtacks. What is the least number of thumbtacks that she can use to tack up these pictures so they can all be seen? (Assume that each corner of each picture must be tacked and that the pictures are the same size.)

4. You can cut a pizza into 7 pieces with only 3 straight cuts. What is the greatest number of pieces you can make with 5 straight cuts?

5. There are eight houses on Elm Street, all in a row. These houses are numbered from 1 to 8. Allison, whose house number is greater than 2, lives next door to her best friend, Adrienne. Belinda, whose house number is greater than 5, lives two doors away from her boyfriend, Benito. Chumani, whose house number is greater than Benito's, lives three doors away from her piano teacher, Mr. Crawford. Darryl, whose house number is less than 4, lives four doors away from his teammate, Don. Who lives in each house?

EXERCISES

Practice

Solve. Use any strategy.

Strategies

Look for a pattern.
Solve a simpler problem.
Act it out.
Guess and check.
Draw a diagram.
Make a chart.
Work backwards.

6. What is the next letter in the sequence below?

   F, S, T, F, F, S, S, $\underline{?}$

7. **Number Theory** *Automorphic numbers* are whole numbers whose square ends in the same number. Two examples are $5^2 = 25$ and $6^2 = 36$. 10 is not an automorphic number, since the last two digits of 10^2 are not 1 and 0. Find another automorphic number.

8. On Saturday, Kevin challenged Luis to a 10-mile bicycle race. When Luis reached the finish line, Kevin was two miles behind him. On Sunday, Kevin challenged Luis to another race. To even up the contest, Luis started two miles behind Kevin. Assuming they cycled the same distance at the same speeds as before, which cyclist won the race?

COOPERATIVE LEARNING ACTIVITY

Work in groups of three. Each person in the group must understand the solution and be able to explain it to any person in the class.

When changing a dollar bill, you can give one coin (silver dollar), two coins (2 half-dollars), three coins (2 quarters and 1 half-dollar), and so forth. What is the least positive number of coins that is *impossible* to give as change for a dollar bill?

5-5 Compound Inequalities

Objectives

After studying this lesson, you should be able to:

- solve compound inequalities and graph their solution sets, and
- solve problems that involve compound inequalities.

Application

FYI . . .

In 1914, only 1% of Americans paid income tax, at an average of 41¢ each.

Melody Harrison works part-time while attending college. Last year she paid \$629 in federal income tax. The tax table that she used is shown at the right. Since she is single, her taxable income must have been at least \$4650 but less than \$4700.

If 1040A, line 17, OR 1040EZ, line 7 is—		And you are—			
At least	But less than	Single (and 1040EZ filers)	Married filing jointly	Married filing separately	Head of a household
		Your tax is—			
4,000	4,050	532	484	544	504
4,050	4,100	539	491	551	511
4,100	4,150	547	499	559	519
4,150	4,200	554	506	566	526
4,200	4,250	562	514	574	534
4,250	4,300	569	521	581	541
4,300	4,350	577	529	589	549
4,350	4,400	584	536	596	556
4,400	4,450	592	544	604	564
4,450	4,500	599	551	611	571
4,500	4,550	607	559	619	579
4,550	4,600	614	566	626	586
4,600	4,650	622	574	634	594
4,650	4,700	629	581	641	601
4,700	4,750	637	589	649	609
4,750	4,800	644	596	656	616
4,800	4,850	652	604	664	624
4,850	4,900	659	611	671	631
4,900	4,950	667	619	679	639
4,950	5,000	674	626	686	646

Let I = her taxable income. Then the two inequalities below describe her income.

$$I \geq 4650 \text{ and } I < 4700$$

When considered together, these two inequalities form a **compound inequality.** The compound inequality $I \geq 4650$ and $I < 4700$ can also be written without using *and* in either of the following ways.

$$4650 \leq I < 4700 \text{ or } 4700 > I \geq 4650$$

The statement $4650 \leq I < 4700$ can be read *4650 is less than or equal to I, which is less than 4700.* The statement $4700 > I \geq 4650$ can be read *4700 is greater than I, which is greater than or equal to 4650.*

Example 1

Write $x > -5$ and $x \leq 1$ as a compound inequality without using *and*.

$x > -5$ and $x \leq 1$ can be written as $-5 < x \leq 1$ or $1 \geq x > -5$.

Example 2

Write *y is positive and y is less than 10* as a compound inequality without using *and*.

y is positive means $y > 0$. *y is less than 10* means $y < 10$. Thus, the compound inequality can be written as $0 < y < 10$ or $10 > y > 0$.

Recall that the graph of an inequality is the graph of its solution set.

A compound inequality containing *and* is true only if *both* inequalities are true. Thus, the graph of a compound inequality containing *and* is the *intersection* of the graphs of the two inequalities. The intersection can be found by graphing the two inequalities and then determining where these graphs overlap. In other words, make a diagram.

Example 3 **Graph the solution set of $x > -5$ and $x \leq 1$.**

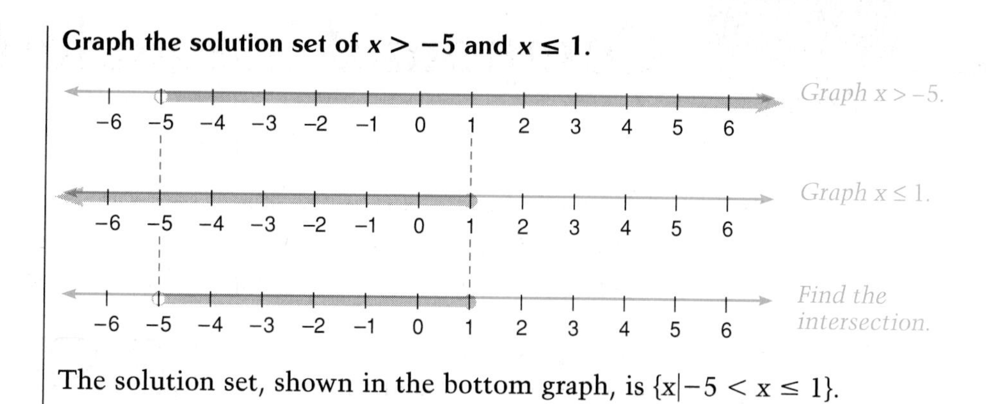

The solution set, shown in the bottom graph, is $\{x | -5 < x \leq 1\}$.

Example 4 **Solve $x - 4 < -1$ and $x - 4 > 1$. Then graph the solution set.**

$x - 4 < -1$ and $x - 4 > 1$

$x - 4 + 4 < -1 + 4$ $\qquad$ $x - 4 + 4 > 1 + 4$

$x < 3$ $\qquad$ $x > 5$

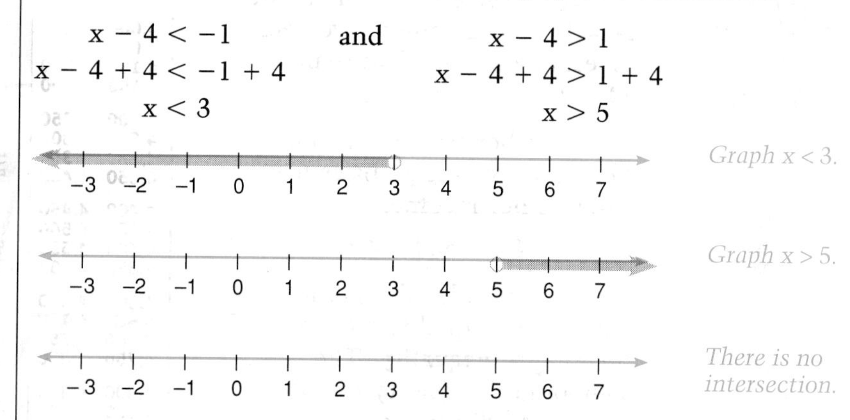

There are *no* points in the intersection of the graphs. That is, there are no numbers that are *both* less than 3 and greater than 5. Therefore, the solution set, shown in the bottom graph, is the empty set (Ø).

Another type of compound inequality contains *or* instead of *and.* A compound inequality containing *or* is true if *either* of the inequalities or *both* of the inequalities are true. The graph of a compound inequality containing *or* is the *union* of the graphs of the two inequalities.

Example 5 **Graph the solution set of $x \geq 3$ or $x < -2$.**

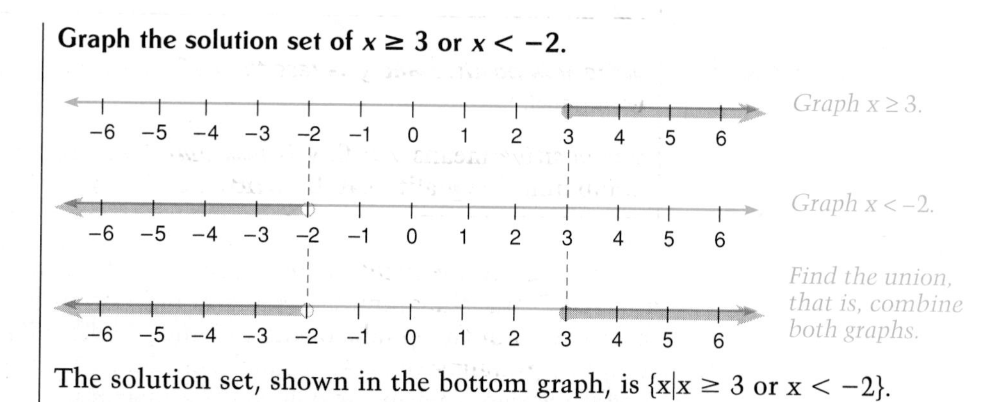

The solution set, shown in the bottom graph, is $\{x | x \geq 3 \text{ or } x < -2\}$.

Example 6

Solve $2y > y - 3$ or $3y < y + 6$. Then graph the solution set.

$$2y > y - 3 \quad \text{or} \quad 3y < y + 6$$
$$2y - y > y - y - 3 \qquad 3y - y < y - y + 6$$
$$y > -3 \qquad 2y < 6$$
$$\frac{2y}{2} < \frac{6}{2}$$
$$y < 3$$

Graph the solution sets for $y > -3$ and $y < 3$.

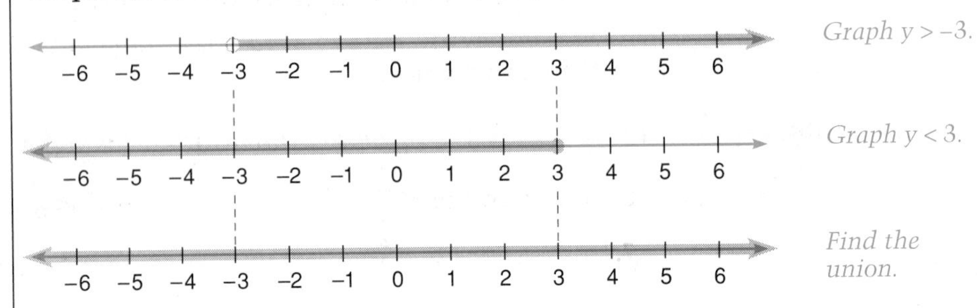

The solution set, shown in the bottom graph, is {all numbers}.

Example 7

Dwayne earns $15,600 per year in salary and 6% commission on his sales at Stride City Shoes. What would his sales have to be if his total income falls between $21,000 and $25,500?

EXPLORE Dwayne's total income is $15,600 plus 6% of his sales. Let s = Dwayne's sales. Then $15{,}600 + 0.06s$ = his total income.

PLAN If Dwayne's total income falls between $21,000 and $25,500, then it was greater than $21,000 and it was less than $25,500.

$$21{,}000 < 15{,}600 + 0.06s \quad \text{and} \quad 15{,}600 + 0.06s < 25{,}500$$

SOLVE

$$21{,}000 < 15{,}600 + 0.06s \quad \text{and} \quad 15{,}600 + 0.06s < 25{,}500$$
$$5400 < 0.06s \qquad 0.06s < 9900$$
$$\frac{5400}{0.06} < \frac{0.06s}{0.06} \qquad \frac{0.06s}{0.06} < \frac{9900}{0.06}$$
$$90{,}000 < s \qquad s < 165{,}000$$

Dwayne's sales are between $90,000 and $165,000.

EXAMINE If Dwayne's sales are $90,000, then his total income would be $15,600 + 0.06($90,000) or $21,000. If his sales are $165,000, then his total income would be $15,600 + 0.06($165,000) or $25,500. Any amount of sales between $90,000 and $165,000 would make his total income fall between $21,000 and $25,500. Thus, the solution is correct.

CHECKING FOR UNDERSTANDING

Communicating Mathematics

Read and study the lesson to answer each question.

1. What is a compound inequality?
2. Refer to the application at the beginning of the lesson. If Melody had paid \$607 in federal income tax last year, what would be the range of her taxable income, I?
3. What is the difference between a compound inequality containing *and* and a compound inequality containing *or*?
4. Can you write a compound inequality containing *or* without using *or*? Why or why not?

Guided Practice

Write each compound inequality without using *and*.

5. $0 \leq m$ and $m < 9$
6. $0 < y$ and $y \leq 12$
7. $z > -\frac{4}{5}$ and $z < \frac{2}{3}$
8. $r \geq -\frac{3}{4}$ and $r \leq \frac{1}{10}$

Graph the solution set of each compound inequality.

9. $m < -7$ or $m \geq 0$
10. $x \geq -2$ and $x \leq 5$
11. $y > -5$ and $y < 0$
12. $r > 2$ or $r \leq -2$
13. $w > -3$ or $w < 1$
14. $n \leq -5$ and $n \geq -1$

EXERCISES

Practice

Graph the solution set of each compound inequality.

15. $b > 5$ or $b \leq 0$
16. $d > 0$ or $d < 4$
17. $d \geq -6$ and $d \leq -3$
18. $1 > q \geq -5$
19. $s \geq 8$ or $s < 5$
20. $-1 < p \leq 6$
21. $t < -3$ and $t > 3$
22. $a \leq 8$ or $a \geq 3$
23. $r > -4$ or $r \leq 0$

Solve each compound inequality and graph the solution set.

24. $3 + x < -4$ or $3 + x > 4$
25. $-1 + d > -4$ or $-1 + d < 3$
26. $5n < 2n + 9$ and $9 - 2n > 11$
27. $2 > 3z + 2 > 14$
28. $-2 \leq x + 3 < 4$
29. $8 + 3t < 2$ or $-12 < 11t - 1$
30. $a \neq 6$ and $3a + 1 > 10$
31. $3x + 11 \leq 13$ or $2x \geq 5x - 12$
32. $5(x - 3) + 2 < 7$ and $5x > 4(2x - 3)$
33. $2 - 5(2x - 3) > 2$ or $3x < 2(x - 8)$

Write a compound inequality for each solution set shown below.

34. −4 −3 −2 −1 0 1 2 3 4 5

35. −5 −4 −3 −2 −1 0 1 2 3 4

36. −5 −4 −3 −2 −1 0 1 2 3 4

37. 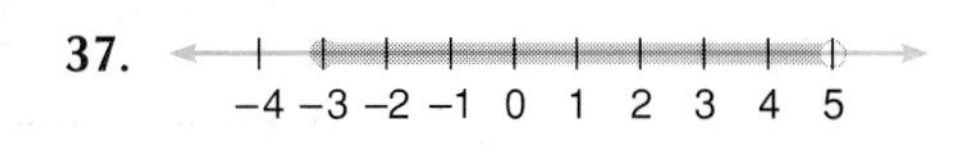

Write a compound inequality and solve. Then check each solution.

38. The sum of a number and 2 is no more than 6 or no less than 10.

39. The sum of twice a number and 5 lies between 7 and 11.

40. Five less than 6 times a number is at most 37 and at least 31.

Solve each inequality and graph the solution set.

41. $\frac{5}{x} + 3 > 0$
42. $-3 - x < 2x < 3 + x$
43. $x - 5 < 2x - 1 < x$

Critical Thinking

44. Under what conditions will the compound sentence $x < a$ and $-a < x$ have no solutions?

45. **Probability** A certain event has a probability of 1. An impossible event has a probability of 0. Write a compound inequality about possible probabilities.

Applications

Define a variable, write a compound inequality, and solve each problem.

46. **Consumerism** Consuela estimates that between 14 and 18 gallons of paint are needed to paint her house. The paint costs \$9.75 per gallon. What will the total cost of the paint be?

47. **Career** Wanita earns \$17,500 per year in salary and 8% commission on her sales at Fitright Shoes. What would her sales have to be if her total income falls between \$28,000 and \$32,000?

48. **Running** One member of a cross-country team placed first in a meet. The next four finishers on the team placed in consecutive order. To find the team score, the finishing places of team members are added. If the team score was between 35 and 45, in what places did the other members finish?

Journal

Explain the differences in graphing an inequality that contains the word *and* with one that contains the word *or*. Include graphs in your explanation.

Mixed Review

Solve. (Lessons 3-6 and 3-7)

49. $4(1 - 2x) = 10(x + 13)$
50. $5.1n + 8.6 = 9.5n - 2.4$

51. **Entertainment** At Backintime Cinema, tickets for adults cost \$5.75 and tickets for children cost \$3.75. How many of each kind of ticket were purchased if 16 tickets were bought for \$68? **(Lesson 4-5)**

Solve each inequality. (Lessons 5-2 and 5-3)

52. $2r - 2.1 < -8.7 + r$
53. $10x - 2 \geq 4(x - 2)$

5-6 Open Sentences Involving Absolute Value

Objective

After studying this lesson, you should be able to:

- solve open sentences involving absolute value and graph the solutions.

Application

Suppose you have a container full of water at 50°C. If the temperature of the water is raised to 100°C, it boils and becomes a gas (steam). If the temperature of the water drops to 0°C, it freezes and becomes a solid (ice). Thus, the water changes to a gas or a solid if its temperature is raised or lowered by 50°C.

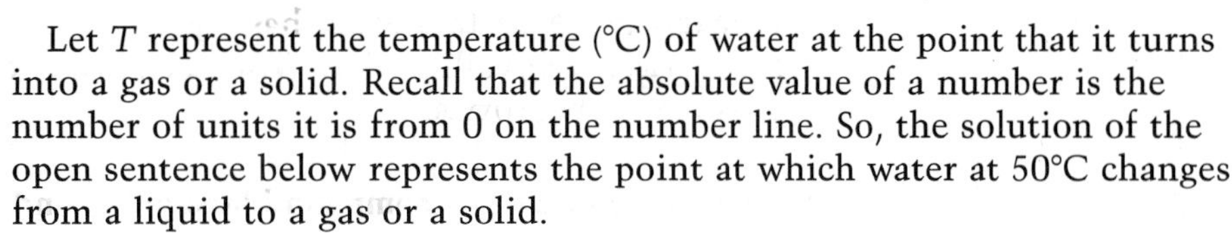

Let T represent the temperature (°C) of water at the point that it turns into a gas or a solid. Recall that the absolute value of a number is the number of units it is from 0 on the number line. So, the solution of the open sentence below represents the point at which water at 50°C changes from a liquid to a gas or a solid.

$|T - 50| = 50$ *The difference between the water's temperature and 50°C is 50°C.*

Open sentences that involve absolute value must be interpreted carefully. There are 3 cases: $|x| = n$, $|x| < n$, and $|x| > n$.

$$|x| = 2$$

The distance from 0 to x is 2 units.

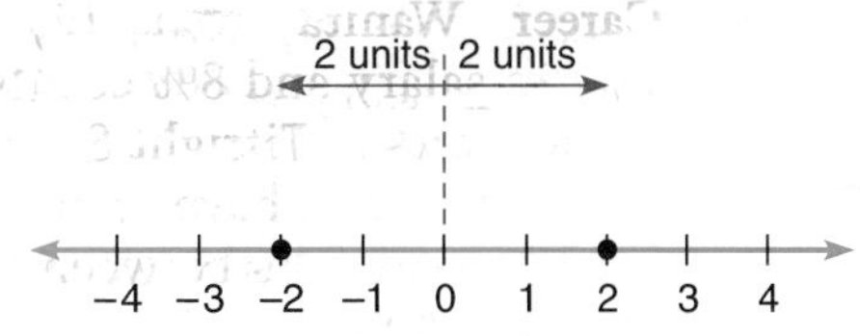

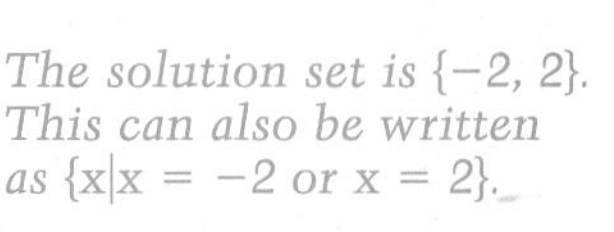

The solution set is {−2, 2}. This can also be written as {x|x = −2 or x = 2}.

Therefore, $x = -2$ or $x = 2$.

Equations involving absolute value can be solved by graphing them on a number line or by writing them as a compound sentence and solving it.

Example 1

Solve $|x - 2| = 4$ in two ways.

Graphing: $|x - 2| = 4$ means the distance between x and 2 is 4 units. So to find x on the number line, start at 2 and move 4 units in either direction.

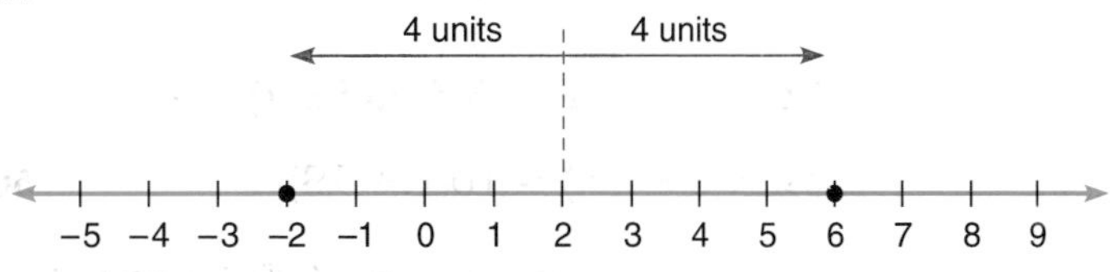

The solution set is {−2, 6}. *This can be written {x|x = −2 or x = 6}.*

Compound Sentence: $|x - 2| = 4$ also means $x - 2 = 4$ *or* $x - 2 = -4$.

$x - 2 = 4$ or $x - 2 = -4$

$x = 6$ $\quad$ $x = -2$ $\quad$ This verifies the solution set.

Inequalities involving absolute value can also be represented on a number line or as compound inequalities.

$|x| < 2$

The distance from 0 to x is less than 2 units.

The solution set is $\{x|-2 < x < 2\}$.

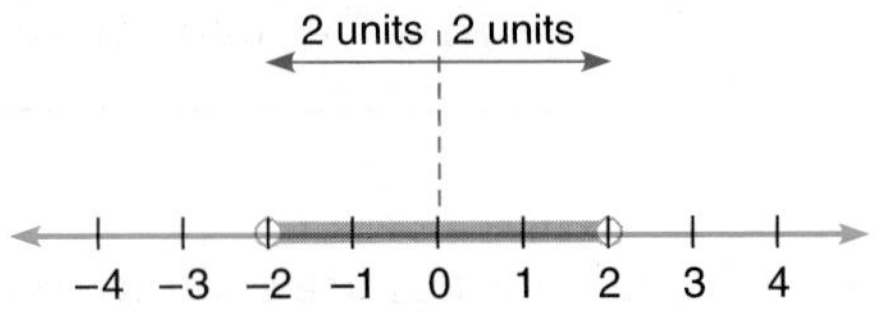

Therefore, $x > -2$ and $x < 2$.

$|x| > 2$

The distance from 0 to x is more than 2 units.

The solution set is $\{x|x < -2 \text{ or } x > 2\}$.

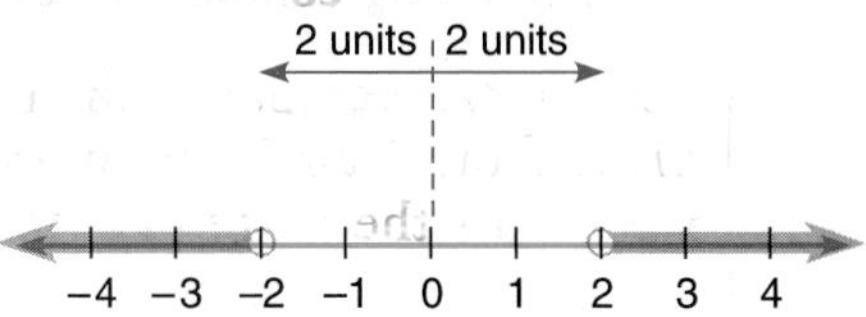

Therefore, $x < -2$ or $x > 2$.

Example 2

Solve $|3x + 4| < 8$ and graph the solution set.

$|3x + 4| < 8$ means $3x + 4 > -8$ and $3x + 4 < 8$.

$3x + 4 > -8$	and	$3x + 4 < 8$	
$3x > -12$		$3x < 4$	*Subtract 4 from each side.*
$\frac{3x}{3} > \frac{-12}{3}$		$\frac{3x}{3} < \frac{4}{3}$	*Divide each side by 3.*
$x > -4$		$x < \frac{4}{3}$	

The solution set is $\left\{x|x > -4 \text{ and } x < \frac{4}{3}\right\}$ or $\left\{x|-4 < x < \frac{4}{3}\right\}$.

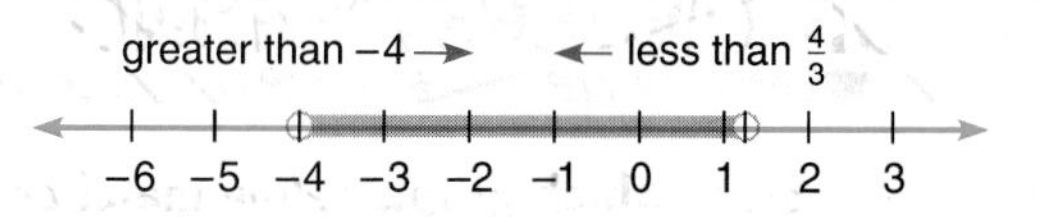

Example 3

Solve $|t + 4| \geq 3$ in two ways.

Graphing: $|t + 4| \geq 3$ means the distance between t and -4 is greater than or equal to 3 units. So to find t on the number line, start at -4 and move 3 units in either direction.

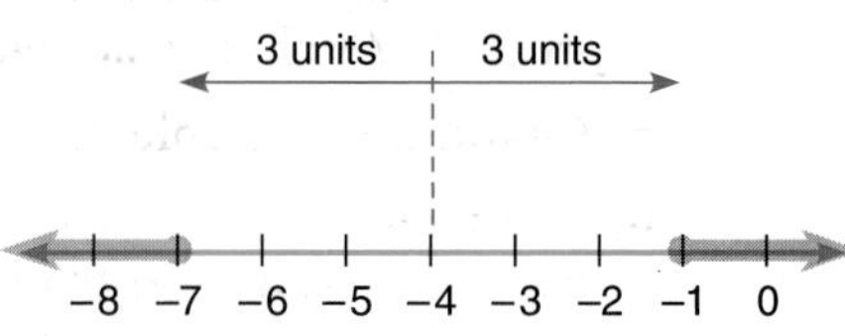

The solution set is $\{t|t \leq -7 \text{ or } t \geq -1\}$.

Compound Inequality: $|t + 4| \geq 3$ also means $t + 4 \leq -3$ or $t + 4 \geq 3$.

$t + 4 \leq -3$ or $t + 4 \geq 3$

$t \leq -7$ $\quad$ $t \geq -1$ *Subtract 4 from each side.*

This verifies the solution set found graphically.

Example 4

Ball bearings are used to connect moving parts in such a way that friction between the parts is minimized. A certain ball bearing used in automobiles will work properly only if its diameter differs from 1 cm by no more than 0.001 cm. Write an open sentence involving absolute value to represent the range of acceptable diameters for this ball bearing. Then find and graph the corresponding compound sentence.

Let d = the acceptable diameter for this type of ball bearing. Then d can differ from 1 cm by no more than 0.001 cm. Write an open sentence to represents the range of acceptable diameters.

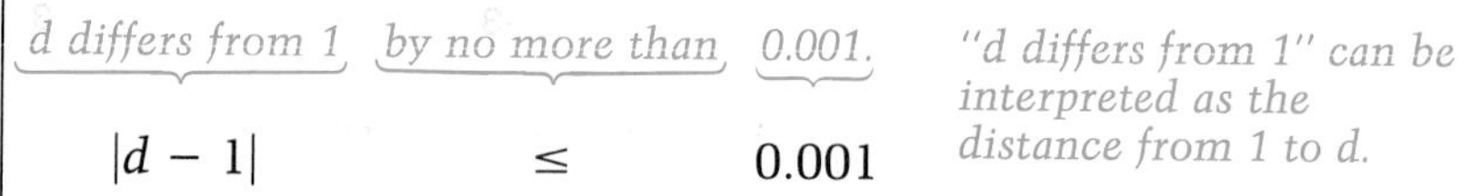

"d differs from 1" can be interpreted as the distance from 1 to d.

Now, solve $|d - 1| \leq 0.001$ to find the corresponding compound sentence.

$d - 1 \geq -0.001$ and $d - 1 \leq 0.001$

$d \geq 0.999$ $\quad$ $d \leq 1.001$ *Add 1 to each side.*

The corresponding compound sentence is $0.999 \leq d \leq 1.001$.

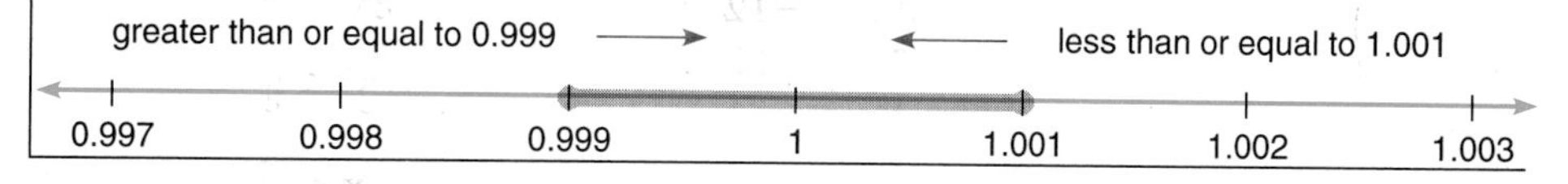

CHECKING FOR UNDERSTANDING

Communicating Mathematics

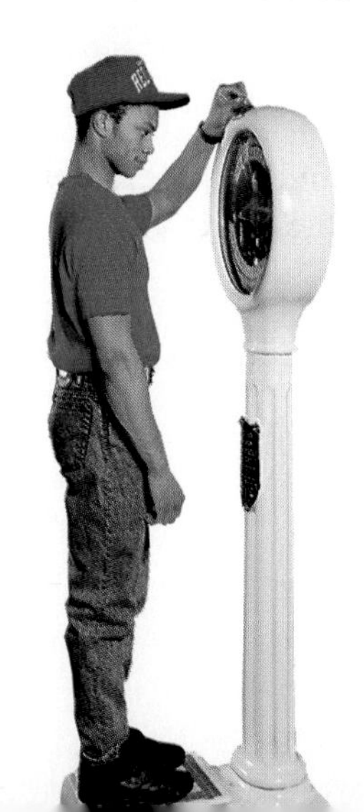

State which graph below matches each open sentence in Exercises 1–4.

1. $|x| = 4$
2. $|x| < 4$
3. $|x| > 4$
4. $|x| \leq 4$

a. 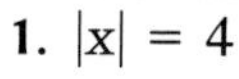

b. 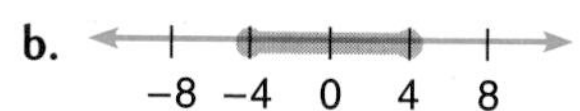

c. 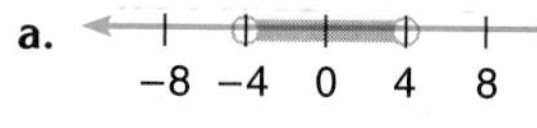

d. −8 −4 0 4 8

5. When Michael weighed himself, the scale read 122 pounds. However, the scale may have been off by as much as 3 pounds. Suppose that w represents Michael's actual weight in pounds.
   a. Graph the possible values of w on a number line.
   b. Write an open sentence involving absolute value for the variation between w and the reading on the scale.
   c. Use the open sentence in part b to write a compound inequality for the possible values of Michael's weight.

Guided Practice

Complete.

6. If the distance from 0 to x is less than 3 units, then $|x|$ ___?___ 3.
7. $|2x - 5| = 7$ has ___?___ solution(s).
8. $|2x - 5| = -1$ has ___?___ solution(s).
9. $|7 - 3y| < 10$ means $7 - 3y < 10$ ___?___ $7 - 3y > -10$.

Describe what each open sentence means in terms of distance on the number line. Then write the sentence as a compound sentence and solve.

10. $|x| = 4$
11. $|y| > 3$
12. $|t - 1| \leq 5$
13. $|x - 12| < 9$
14. $|x + 2| \geq 9$
15. $|7 - r| = 4$

EXERCISES

Practice

Write an open sentence involving absolute value for each problem.

16. The temperature was more than 5°F from 0°F.
17. Karen's golf score was within 6 strokes of her average score of 90.
18. The board needs to be 1.5 meters long, plus or minus 0.005 meters.

Solve each open sentence and graph the solution set.

19. $|x - 1| < 4$
20. $|y - 7| < 2$
21. $|a + 8| \geq 1$
22. $|2 - t| \leq 1$
23. $|9 - y| \geq 13$
24. $|3 - 3x| = 0$
25. $|14 - 2z| = 16$
26. $|2b - 11| \geq 7$
27. $|3x - 12| < 12$
28. $|4k + 2| \leq 14$
29. $|10w + 10| > 90$
30. $|x + 1| > -2$
31. $|2y - 7| \geq -6$
32. $|a - 5| = -3$
33. $|5b + 6| < 0$
34. $|13 - 5y| = 8$
35. $\left|3t - \frac{1}{2}\right| \geq \frac{11}{2}$
36. $\left|\frac{1}{2} - 3n\right| < \frac{7}{2}$

For each graph, write an open sentence involving absolute value.

37. -5 -4 -3 -2 -1 0 1 2 3 4

38. 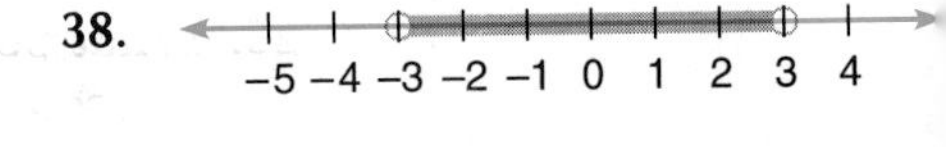
-5 -4 -3 -2 -1 0 1 2 3 4

39. 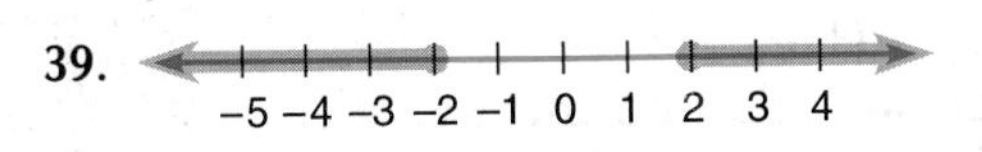
-5 -4 -3 -2 -1 0 1 2 3 4

40. 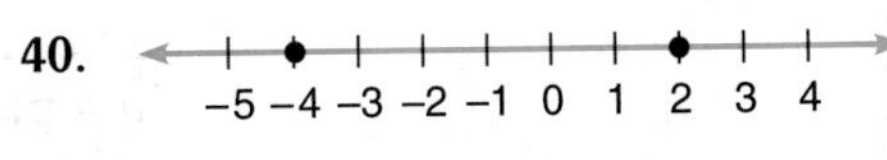
-5 -4 -3 -2 -1 0 1 2 3 4

41. 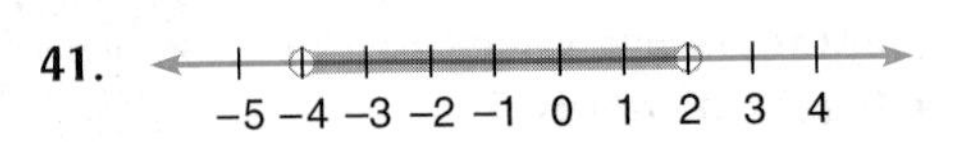
-5 -4 -3 -2 -1 0 1 2 3 4

42. 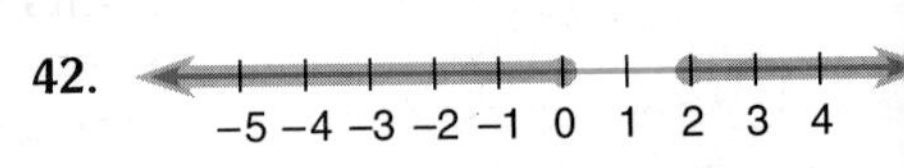
-5 -4 -3 -2 -1 0 1 2 3 4

Solve each problem.

43. Find all integer solutions of $|x| \leq 2$.
44. Find all integer solutions of $|x| < 4$.

45. If $a \geq 0$, how many integer solutions exist for $|x| \leq a$?

46. If $a \geq 0$, how many integer solutions exist for $|x| < a$?

Critical Thinking

47. Suppose $x < 0$, $y > 0$, and $x + y = 0$. Is $|x| > |y|$, $|x| = |y|$, or $|x| < |y|$?

48. Under what conditions is $-|a|$ negative? positive?

Applications

Define a variable, write a open sentence involving absolute value, and solve each problem.

49. **Chemistry** For hydrogen to be a liquid, its temperature must be within 2°C of −257°C. What is the range of temperatures for this substance to remain a liquid?

50. **Travel** Ben's car gets between 18 and 21 miles per gallon of gasoline. If his car's tank holds 15 gallons, what is the range of distance that Ben can drive his car on one tank of gasoline?

51. **Manufacturing** A certain bolt used in lawn mowers will work properly only if its diameter differs from 2 cm by no more than 0.04 cm. What is the range of acceptable diameters for this bolt?

52. **Shipping** A crate weighs 6 kg when empty. The weight of a certain book is about 0.8 kg. For shipping, a crate of books must weigh between 40 and 50 kg. What is an acceptable number of books that can be packed in the crate?

Mixed Review

53. Solve $3(2n - x) = 1 - n$ for x. **(Lesson 3-7)**

54. Twelve is 15% of what number? **(Lesson 4-2)**

55. **Finance** Marco Flores invested \$10,000 for one year, part at 8% annual interest and the balance at 10% annual interest. His total interest for the year was \$873. How much did he invest at each rate? **(Lesson 4-3)**

56. What percent of 87 is 290? **(Lesson 4-4)**

57. **Geometry** There are seven points in a plane such that no three of the points lie on the same line. How many line segments are needed to connect each pair of points? **(Lesson 5-4)**

58. **Statistics** Darlene's scores on the first three of four 100-point biology tests were 88, 90, and 91. To get an A− in the class, she must have an average between 88 and 92, inclusive, on all tests. What score must she receive on the fourth test to get an A− in biology? **(Lesson 5-5)**

Portfolio

Select some of your work from this chapter that shows how you used a calculator or computer. Place it in your portfolio.

CHAPTER 5 SUMMARY AND REVIEW

VOCABULARY

Upon completing this chapter, you should be familiar with the following terms:

compound inequality **194**
compound sentence **185**
conjunction **185**
185 disjunction
177 set-builder notation

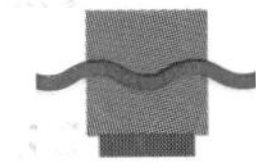

SKILLS AND CONCEPTS

OBJECTIVES AND EXAMPLES

Upon completing this chapter, you should be able to:

- solve inequalities by using addition and subtraction **(Lesson 5-1)**

$$3x - 2 < 4x$$
$$3x - 3x - 2 < 4x - 3x$$
$$-2 < x$$
$$\{x|x > -2\}$$

REVIEW EXERCISES

Use these exercises to review and prepare for the chapter test.

Solve each inequality. Check the solution.

1. $n - 4 < 9$
2. $r + 8 \leq -3$
3. $a - 2.3 \geq -7.8$
4. $5z - 6 > 4z$
5. $2x + 7 < 3x$
6. $y + \frac{5}{8} > \frac{11}{24}$

Write an inequality and solve. Then check each solution.

7. A number added to 7 is at least 12.

8. Three times a number is greater than four times the number less 8.

OBJECTIVES AND EXAMPLES

- solve inequalities by using multiplication and division **(Lesson 5-2)**

$$-\frac{2}{3}m \leq 10$$
$$-\frac{3}{2}\left(-\frac{2}{3}m\right) \geq -\frac{3}{2}(10)$$
$$m \geq -15$$

$$\{m | m \geq -15\}$$

REVIEW EXERCISES

Solve each inequality. Check the solution.

9. $6x \leq -24$
10. $-7y \geq 91$
11. $0.8t > 0.96$
12. $-\frac{4}{3}m < -16$
13. $\frac{2}{3}k \geq \frac{2}{15}$
14. $\frac{4}{7}z \leq -\frac{2}{5}$

Write an inequality and solve. Then check each solution.

15. Seven times a number is less than -154.
16. Negative three fourths of a number is no more than 30.

- solve inequalities involving more than one operation **(Lesson 5-3)**

$$14c - 11 \geq 6c + 37$$
$$14c - 6c - 11 \geq 6c - 6c + 37$$
$$8c - 11 \geq 37$$
$$8c - 11 + 11 \geq 37 + 11$$
$$8c \geq 48$$
$$\frac{8c}{8} \geq \frac{48}{8}$$
$$c \geq 6$$

$$\{c | c \geq 6\}$$

Solve each inequality. Check the solution.

17. $7x - 12 < 30$
18. $2r - 3.1 > 0.5$
19. $4y - 11 \geq 8y + 7$
20. $4(n - 1) < 7n + 8$
21. $0.3(z - 4) \leq 0.8(0.2z + 2)$

Write an inequality and solve. Then check each solution.

22. The sum of three times a number and 11 is at most 47.
23. Twice a number subtracted from 12 is no less than the number increased by 27.

- solve compound inequalities and graph the solution sets **(Lesson 5-5)**

$2a > a - 3$	and	$3a < a + 6$
$2a - a > a - a - 3$		$3a - a < a - a + 6$
$a > -3$		$2a < 6$
		$\frac{2a}{2} < \frac{6}{2}$
		$a < 3$

$$\{a | -3 < a < 3\}$$

−5 −4 −3 −2 −1 0 1 2 3 4 5

Graph the solution set of each compound inequality.

24. $x > -1$ and $x \leq 3$
25. $y \leq -3$ or $y > 0$
26. $2a + 5 \leq 7$ or $2a \geq a - 3$
27. $4r \geq 3r + 7$ and $3r + 7 < r + 29$

Write the compound inequality whose solution set is graphed.

28. −5 −4 −3 −2 −1 0 1 2 3 4

29. 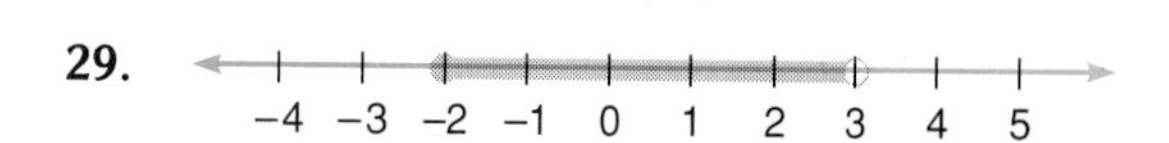

OBJECTIVES AND EXAMPLES

- solve open sentences involving absolute value and graph the solution sets **(Lesson 5-6)**

$$|2x + 1| > 1$$

$$2x + 1 > 1 \quad \text{or} \quad 2x + 1 < -1$$

$$2x + 1 - 1 > 1 - 1 \qquad 2x + 1 - 1 < -1 - 1$$

$$\frac{2x}{2} > \frac{0}{2} \qquad \frac{2x}{2} < \frac{-2}{2}$$

$$x > 0 \qquad x < -1$$

$$\{x|x > 0 \text{ or } x < -1\}$$

−4 −3 −2 −1 0 1 2 3 4

REVIEW EXERCISES

Solve each open sentence and graph the solution set.

30. $|y - 2| > 0$

31. $|1 - n| \leq 5$

32. $|7x - 10| < 0$

33. $|2p - \frac{1}{2}| > \frac{9}{2}$

APPLICATIONS AND CONNECTIONS

34. **Geometry** For what values of d is an angle with measure $(3.6d)°$ an acute angle? **(Lesson 5-2)**

35. **Number Theory** The sum of three consecutive integers is less than 100. Find the three integers with the greatest sum. **(Lesson 5-3)**

Solve by drawing a diagram. (Lesson 5-4)

36. Six softball teams qualified for the city softball playoffs. To determine the two representatives to the state playoffs, each team plays every other team exactly once. How many playoff games will be played?

37. A spider begins climbing a 10-foot tree on Monday. Each day it climbs 4 feet. Each night it slips back 1 foot. On which day will the spider reach the top of the tree?

Solve.

38. **Sales** Linda plans to spend at most \$85 on jeans and shirts. She bought 2 shirts for \$15.30 each. How much can she spend on jeans? **(Lesson 5-3)**

39. **Work** At most, Sung and Jill can earn a total of \$312 in a week. If Jill earns twice as much as Sung, what is the greatest amount each one can earn? **(Lesson 5-3)**

40. **Banking** Sean budgets between \$3 and \$4 a month to spend on bank checking charges. His bank charges \$1.75 a month plus \$0.08 per check. How many checks can Sean write each month and still meet his budget? **(Lesson 5-5)**

41. **Shipping** An empty book crate weighs 30 pounds. The weight of a book is 1.5 pounds. For shipping, the crate can weight between 55 and 60 pounds. What is the acceptable number of books that can be packed in the crate? **(Lesson 5-5)**

CHAPTER 5 TEST

Solve each inequality.

1. $a - 2 > 11$
2. $-12 \leq d + 7$
3. $7x < 6x - 11$
4. $z - 1 \geq 2z - 3$
5. $3y > 63$
6. $-\frac{2}{3}r \leq \frac{7}{12}$
7. $3x + 1 \geq 16$
8. $5 - 4b > -23$
9. $\frac{2n - 3}{-7} \leq 5$
10. $8y + 3 < 13y - 9$
11. $8(1 - 2z) \leq 25 + z$
12. $0.3(m + 4) > 0.5(m - 4)$

Solve each inequality and graph the solution set.

13. $x + 1 > -2$ and $3x < 6$
14. $2n + 1 \geq 15$ or $2n + 1 \leq -1$
15. $8 + 3t > 2$ and $-12 > 11t - 1$
16. $|n| > 3$
17. $|2x - 1| < 5$
18. $|5 - 3b| \geq 1$

Write an inequality and solve.

19. Four times a number less 8 is at least five times the number.
20. Seven less than twice a number is between 71 and 83.
21. The product of two integers is no less than 30. One of the integers is 6. What is the other integer?
22. The average of four consecutive odd integers is less than 20. What are the greatest integers that satisfy this condition?

Solve.

23. **Gymnastics** Montega has scores of 9.1, 9.3, 9.6, and 8.7 in a pommel horse competition. He has one more trial. If Montega's total score must be greater than 46.1 points to win the competition, what score must he receive on his final trial?
24. **Statistics** Jean's scores on the first three of four 100-point tests were 82, 86, and 91. What score must she receive on the fourth test to have an average of at least 87 points for all the tests?
25. **Sales** Kathy earns \$18,000 per year in salary and 5% commission on her sales at Autorama Motors. What would her sales have to be if her total income falls between \$37,000 and \$40,000?

Bonus For what values of d is an angle with measure $\left(30 - \frac{1}{2}d\right)^\circ$ an acute angle?

CHAPTER OBJECTIVES

In this chapter, you will:

- Arrange the terms of a polynomial in order.
- Add, subtract, and multiply polynomials.
- Solve polynomial equations.
- Solve problems by looking for a pattern.

CHAPTER 6

Polynomials

APPLICATION IN BIOLOGY

These days, many cities support herds of deer in their metropolitan parks, supplying them with food, medical care, shelter in winter, and protection from predators.

Population control for deer? This might seem like an unwarranted interference with nature, but this problem is much more complex.

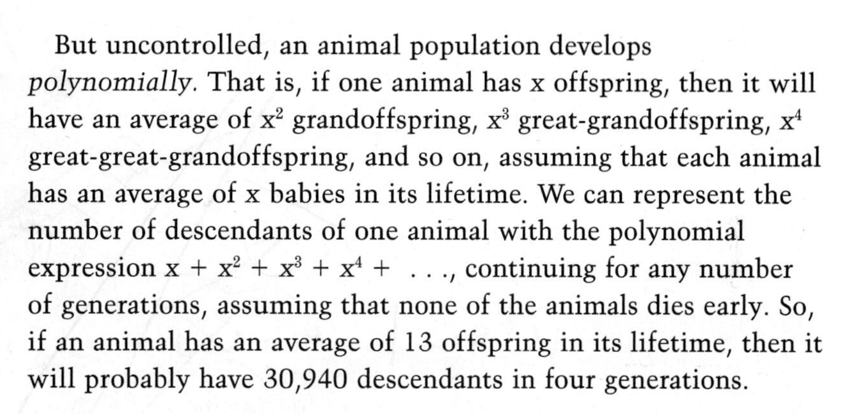

But uncontrolled, an animal population develops *polynomially.* That is, if one animal has x offspring, then it will have an average of x^2 grandoffspring, x^3 great-grandoffspring, x^4 great-great-grandoffspring, and so on, assuming that each animal has an average of x babies in its lifetime. We can represent the number of descendants of one animal with the polynomial expression $x + x^2 + x^3 + x^4 + \ldots$, continuing for any number of generations, assuming that none of the animals dies early. So, if an animal has an average of 13 offspring in its lifetime, then it will probably have 30,940 descendants in four generations.

So this is the problem caused by cities' support of deer in parks. Any natural ecosystem has a delicate balance. In the wild, deer populations are controlled naturally—by predators, disease, starvation, and other forms of natural competition. Since we have eliminated nearly all of the population controls for the deer living in city parks, we now have a responsibility to provide them with new ones or face the consequences of an uncontrolled population.

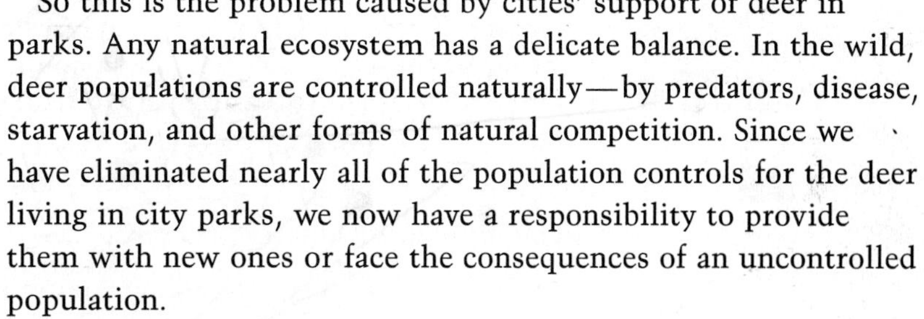

ALGEBRA IN ACTION

A deer has an average of 4 offspring during its lifetime. Complete the chart below to determine the probable number of deer descendants in five generations.

Generation	Polynomial	Number of Deer Descendants
1	x	4
2	$x + x^2$	4 + 16 = 20
3	$x + x^2 + x^3$	4 + 16 + 64 = ?
4	$x + x^2 + x^3 + x^4$	4 + 16 + 64 + ? = ?
5	$x + x^2 + x^3 + x^4 +$?	4 + 16 + 64 + ? + ? = ?

EXPLORE
PLAN
SOLVE
EXAMINE

6-1 Problem-Solving Strategy: Look for a Pattern

Objective After studying this lesson, you should be able to:

- solve problems by looking for a pattern.

One of the most-used strategies in problem solving is *look for a pattern.* When using this strategy, you will often need to make a table to organize the information.

Example How many squares are there on a checkerboard?

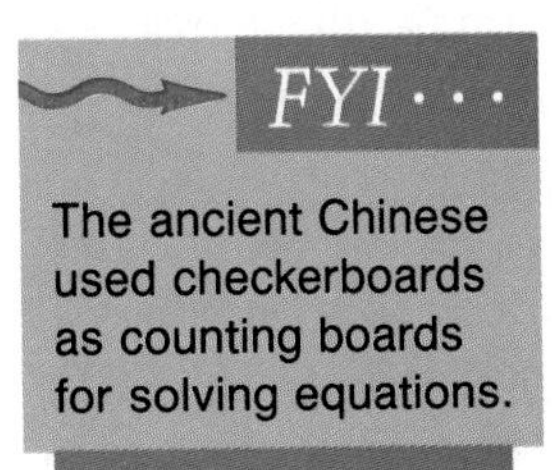

FYI . . .

The ancient Chinese used checkerboards as counting boards for solving equations.

You might say that there are 64 squares on an 8×8 checkerboard. But the problem does not ask for only 1×1 squares. There are also 2×2 squares, 3×3 squares, and so on. Since the checkerboard is 8×8, you know that there are 64 of the 1×1 squares. Count the number of 2×2 squares. There are seven 2×2 squares on each row, and 7 rows, so there are 49 of the 2×2 squares. Then count the number of 3×3 squares. There are 36. Do you notice a pattern in the number of squares of each size?

$$64 = 8 \cdot 8 \qquad 49 = 7 \cdot 7 \qquad 36 = 6 \cdot 6$$

The number of squares of a given size will always be a perfect square.

Make a table to see the pattern more clearly.

Square Size	1×1	2×2	3×3	4×4	5×5	6×6	7×7	8×8
Number of Squares	64	49	36	25	16	9	4	1

The total number of squares on an 8×8 checkerboard is $64 + 49 + 36 + 25 + 16 + 9 + 4 + 1$ or 204 squares.

Another strategy you could use to solve this problem is to solve a simpler problem and then look for the pattern. Think about the number of squares on a 1×1 checkerboard, a 2×2 checkerboard, a 3×3 checkerboard, and so on. Make a table to see the pattern more clearly.

Checkerboard Size	1×1	2×2	3×3	4×4	5×5	6×6	7×7	8×8
Number of Squares	1	5	14	30	55	?	?	?

Can you see a pattern in the total number of squares? Try finding the difference between each pair of numbers.

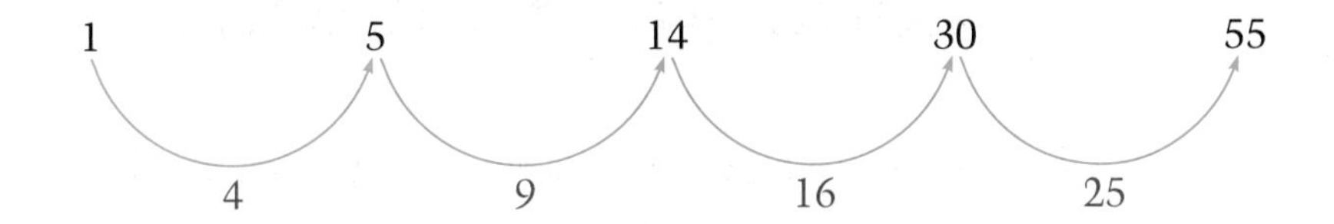

Now do you see the pattern? The numbers 4, 9, 16, and 25 are all perfect squares. If you continued adding squares to each entry, you would eventually verify that there are 204 squares on an 8×8 checkerboard.

CHECKING FOR UNDERSTANDING

Communicating Mathematics

Read and study the lesson to answer each question.

1. How many squares are there on a 10×10 checkerboard?
2. Look at the first table on the previous page. If $n \times n$ represents the size of the squares on the checkerboard, write an expression to represent the total number of squares of that size.

Guided Practice

Solve by extending the pattern.

3. $4 \times 6 = 24$
   $14 \times 16 = 224$
   $24 \times 26 = 624$
   $34 \times 36 = 1224$
   $124 \times 126 = ?$

4. $1 \times \frac{1}{2} = 1 - \frac{1}{2}$
   $2 \times \frac{2}{3} = 2 - \frac{2}{3}$
   $3 \times \frac{3}{4} = 3 - \frac{3}{4}$
   $4 \times \frac{4}{5} = 4 - \frac{4}{5}$
   $12 \times \frac{?}{?} = 12 - \frac{?}{?}$

5. $1^3 = 1^2 - 0^2$
   $2^3 = 3^2 - 1^2$
   $3^3 = 6^2 - 3^2$
   $4^3 = 10^2 - 6^2$
   $6^3 = ? - ?$

Solve by looking for a pattern.

6. How many triangles are shown at the right? Count only the triangles pointing upward.

7. At the City Center Mall, there are 25 lockers. Suppose a shopper opens every locker. Then a second shopper closes every second locker. Next a third shopper changes the state of every third locker. (If it's open, the shopper closes it. If it's closed, the shopper opens it.) Suppose this process continues until the 25th person changes the state of the 25th locker. Which lockers will still be open?

8. **Number Theory** The sequence 1, 1, 2, 3, 5, 8, . . . is called the *Fibonacci sequence.* The terms of this sequence are called *Fibonacci numbers.*

   a. List the first 10 Fibonacci numbers.

   b. Find the following quotients: $\frac{\text{2nd term}}{\text{1st term}}, \frac{\text{3rd term}}{\text{2nd term}}, \frac{\text{4th term}}{\text{3rd term}}, \ldots, \frac{\text{10th term}}{\text{9th term}}$. What do you notice?

EXERCISES

Practice

Solve. Use any strategy.

Strategies

Look for a pattern.
Solve a simpler problem.
Act it out.
Guess and check.
Draw a diagram.
Make a chart.
Work backwards.

9. Fred the frog loves to sit on lily pads and sun himself. When Fred came to his newest spot, there was only one lily pad on the surface of the pond. However, the number of lily pads doubled each day. After 30 days, the lily pads completely covered the surface of the pond. How long did it take for the pond to be half-covered?

10. At Chicken Little's, you can order Little Bits in boxes of 6, 9, or 20. If you order two boxes of 6, you can get 12 bits. But you cannot order 13, since no combination of 6, 9, and 20 adds up to 13. What is the greatest number of Little Bits that you *cannot* order?

11. In the town of Verityville, there are two types of residents. The Factites make only true statements, and the Falsites make only false statements. On a visit to Verityville, you meet a resident who says, "This is not the first time I have said what I am now saying." Is this a Factite or a Falsite? How do you know?

12. The symbol used for a U.S. dollar is a capital S with a vertical line through it. The line separates the S into 4 parts, as shown at the right. How many parts would there be if the S had 100 vertical lines through it?

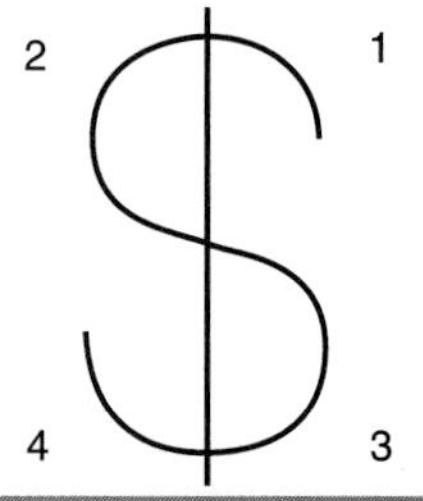

COOPERATIVE LEARNING ACTIVITY

Work in groups of three. Each person must understand the solution and be able to explain it to any person in the class.

Place a digit from 0 to 9 in each box below. The digit you place in Box #0 should indicate the number of 0s in all of the boxes. The digit you place in Box #1 should indicate the number of 1s in all of the boxes. Continue this process until Box #9 indicates the number of 9s in all of the boxes. (*Hint:* You may use a digit more than once, and you may not need to use every digit.) Good luck!

0	1	2	3	4	5	6	7	8	9
?	?	?	?	?	?	?	?	?	?

6-2 Multiplying Monomials

Objectives

After studying this lesson, you should be able to:

- multiply monomials, and
- simplify expressions involving powers of monomials.

A **monomial** is a number, a variable, or a product of a number and one or more variables. Monomials that are real numbers are called **constants.**

These are monomials.

$$-9 \qquad y \qquad 7a \qquad 3y^3 \qquad \frac{1}{2}abc^5$$

These are not monomials. *Why not?*

$$m + n \qquad \frac{x}{y} \qquad 3 - 4ab \qquad \frac{1}{x^2} \qquad \frac{7y}{9z}$$

Recall that an expression of the form x^n is a power. The base is x and the exponent is n. A table of powers of 2 is shown below.

2^1	2^2	2^3	2^4	2^5	2^6	2^7	2^8	2^9	2^{10}
2	4	8	16	32	64	128	256	512	1024

Notice that each of the following is true.

$$4 \cdot 16 = 64 \qquad 8 \cdot 16 = 128 \qquad 8 \cdot 32 = 256$$

$$\downarrow \qquad \downarrow \qquad \downarrow$$

$$2^2 \cdot 2^4 = 2^6 \qquad 2^3 \cdot 2^4 = 2^7 \qquad 2^3 \cdot 2^5 = 2^8$$

Look for a pattern in the products shown. If you consider only the exponents, you will find that $2 + 4 = 6$, $3 + 4 = 7$, and $3 + 5 = 8$.

These examples suggest that you can multiply powers that have the same base by adding exponents.

Product of Powers

For any number a and all integers m and n,
$$a^m \cdot a^n = a^{m+n}.$$

Example 1

Find the measure of the area of the rectangle.

$A = \ell w$
$= x^3 \cdot x^4$
$= x^{3+4}$
$= x^7$

Since $x^3 > x^4$, what can you say about the value of x?

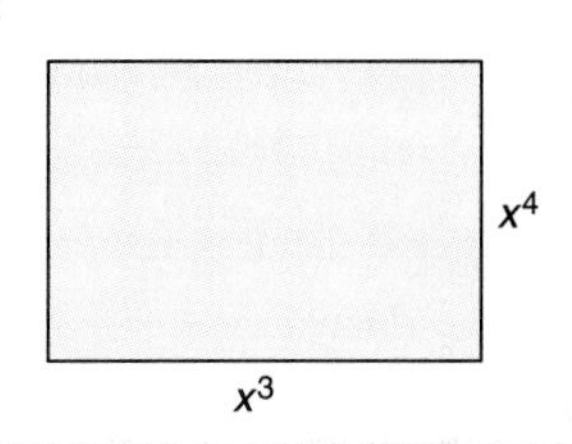

Example 2

Simplify $(-5x^2)(3x^3y^2)\left(\frac{2}{5}xy^4\right)$.

$$(-5x^2)(3x^3y^2)\left(\frac{2}{5}xy^4\right) = \left(-5 \cdot 3 \cdot \frac{2}{5}\right)(x^2 \cdot x^3 \cdot x)(y^2 \cdot y^4)$$ *Commutative and associative properties*

$$= -6x^{2+3+1}y^{2+4}$$ *Product of powers property*

$$= -6x^6y^6$$

Take a look at the examples below.

$$(5^2)^4 = (5^2)(5^2)(5^2)(5^2) = 5^{2+2+2+2} = 5^8$$

$$(x^6)^2 = (x^6)(x^6) = x^{6+6} = x^{12}$$

Since $(5^2)^4 = 5^8$ and $(x^6)^2 = x^{12}$, these examples suggest that you can find the power of a power by multiplying exponents.

Power of a Power	**For any number a and all integers *m* and *n*, $(a^m)^n = a^{mn}$.**

Here are a few more examples.

$$(xy)^3 = (xy)(xy)(xy) = (x \cdot x \cdot x)(y \cdot y \cdot y) = x^3y^3$$

$$(4ab)^4 = (4ab)(4ab)(4ab)(4ab) = (4 \cdot 4 \cdot 4 \cdot 4)(a \cdot a \cdot a \cdot a)(b \cdot b \cdot b \cdot b) = 4^4a^4b^4 \text{ or } 256a^4b^4$$

These examples suggest that the power of a product is the product of the powers.

Power of a Product	**For all numbers *a* and *b* and any integer *m*, $(ab)^m = a^mb^m$.**

Example 3

Find the measure of the volume of the cube.

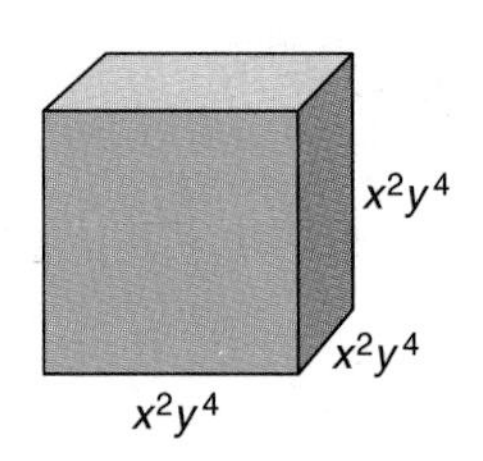

$$V = s^3$$
$$= (x^2y^4)^3$$
$$= (x^2)^3 \cdot (y^4)^3$$
$$= x^{2 \cdot 3}y^{4 \cdot 3}$$
$$= x^6y^{12}$$

In Example 4, the two power properties on this page are used together. This can be stated as follows.

Power of a Monomial	**For all numbers *a* and *b* and any integers *m, n,* and *p*, $(a^mb^n)^p = a^{mp}b^{np}$.**

Example 4 Simplify $(9b^4y)^2[(-b)^2]^3$.

$$(9b^4y)^2[(-b)^2]^3 = 9^2(b^4)^2y^2(b^2)^3 \quad (-b)^2 = b^2$$
$$= 81b^8y^2b^6$$
$$= 81b^{14}y^2$$

The variables in a monomial are usually arranged in alphabetical order.

Some calculators have a *power key* labeled y^x. You can use it to find powers of numbers.

Example 5 Evaluate $(0.14)^3$ to the nearest thousandth.

Enter: 0.14 [y^x] 3 [=]

Display: 0.002744 So, $(0.14)^3$ is about 0.003.

CHECKING FOR UNDERSTANDING

Communicating Mathematics

Read and study the lesson to answer each question.

1. Why does the product of powers property *not* apply to the expression $x^3 \cdot y^7$?
2. Write the product of powers property in your own words.
3. Write the power of a power property in your own words.
4. Write the power of a product property in your own words.

Guided Practice

Are the expressions equivalent? Write *yes* or *no*.

5. $(4y)^2$ and $4y^2$
6. $3xy^5$ and $3(xy)^5$
7. $(2a)^3$ and $8a^3$
8. $(ab)^2$ and $a^2 \cdot b^2$
9. xy^3 and x^3y^3
10. $4(x^3)^2$ and $16x^6$

EXERCISES

Practice

Find the measure of the area of each rectangle and the measure of the volume of each rectangular solid.

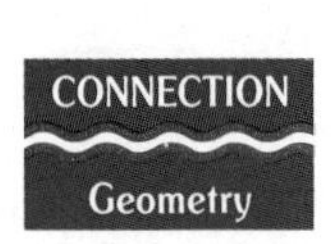

11.
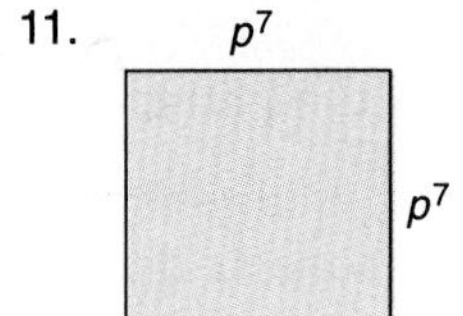

12.
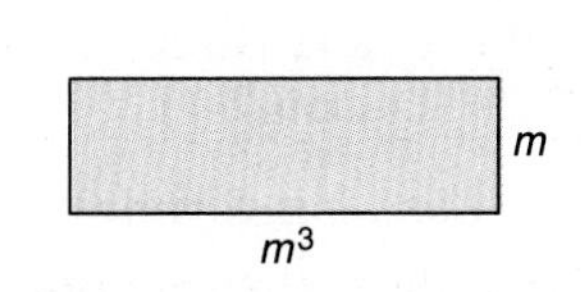

13.
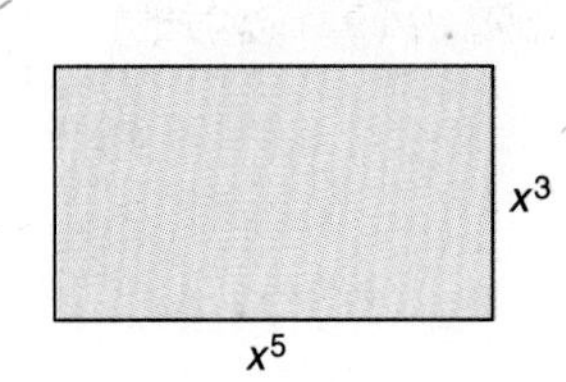

14. r^3, r^2, r^3

15.
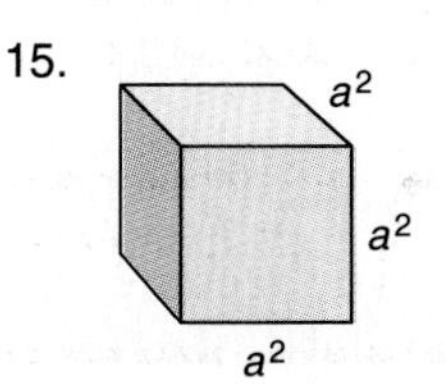

16.
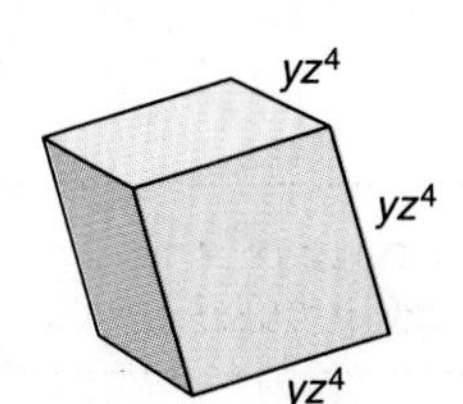

Journal

Make a list of all of the rules of powers you learned in this lesson. Then write an example to demonstrate each rule.

Simplify.

17. $a^5(a)(a^7)$
18. $(a^2b)(ab^4)$
19. $(m^3n)(mn^2)$
20. $(10^2)^2$
21. $[(-4)^2]^2$
22. $[(3^2)^4]^2$
23. $(r^3t^4)(r^4t^4)$
24. $(3a^2)(4a^3)$
25. $(-10x^3y)(2x^2)$
26. $(m^2)^5$
27. $(-7z)^3$
28. $\left(\frac{2}{5}d\right)^2$
29. $(3y^3z)(7y^4)$
30. $m^4(m^3b^2)$
31. $(3x^2y^2z)(2x^2y^2z^3)$
32. $(0.6d)^3$
33. $(a^3x^2)^4$
34. $(2a^2b)^2$
35. $(ab)(ac)(bc)$
36. $-\frac{5}{6}c(12a^3)$
37. $(-27ay^3)\left(-\frac{1}{3}ay^3\right)$
38. $-3(ax^3y)^2$
39. $\left(\frac{1}{2}xy^2\right)^3$
40. $(0.3x^3y^2)^2$
41. $(-3ab)^3(2b^3)$
42. $(2x^2)^2\left(\frac{1}{2}y^2\right)^2$
43. $\left(\frac{3}{10}y^2\right)^2(10y^2)^3$
44. $\left(-\frac{1}{8}a\right)\left(-\frac{1}{6}\right)(b)(48c)$
45. $\left(-\frac{1}{3}c^2b^3a\right)(18a^2b^2c^3)$
46. $(3a^2)^3 + (5a^2)^3$
47. $(-3x^3y)^3 - 3(x^2y)^2(x^5y)$

Critical Thinking

Answer each question. You may wish to substitute numbers and evaluate.

48. For all numbers a and b and any integer m, is $(a + b)^m = a^m + b^m$ a true sentence?

49. For all numbers a and b and any integer m, is $\left(\frac{a}{b}\right)^m = \frac{a^m}{b^m}$ a true sentence?

Applications

50. **Finance** To determine how your money will grow in an account that is compounded annually, you can use the formula $T = p(1 + r)^t$, where T is the total amount, p is the principal (amount invested), r is the annual interest rate, and t is the time in years. Use a calculator to find the final amount if you invest \$5000 at 7% for 10 years.

51. **Finance** To determine how your money will grow if you make regular deposits to an account that is compounded annually, you can use the formula $T = P\left[\frac{(1 + r)^t - 1}{r}\right]$, where T is the total amount, P is the regular payment, r is the annual interest rate, and t is the time in years. Will you have more money if you deposit \$500 per year for 10 years at 7% than you would if you invested \$5000, as in Exercise 50?

Mixed Review

52. Find the next three numbers in the pattern: $-8, -4, -2,$ ___?___, ___?___, ___?___. **(Lesson 2-1)**

53. Mrs. Lazo has 5 times as many sons as daughters. She has 6 children altogether. How many daughters does she have? **(Lesson 3-5)**

54. Solve: $2x + 4 \leq 6$ or $x \geq 2x - 4$. **(Lesson 5-5)**

55. Solve $|6 - x| \leq x$. Then graph the solution set. **(Lesson 5-6)**

56. How many rectangles are shown at the right? **(Lesson 6-1)**

6-3 Dividing Monomials

Objectives

After studying this lesson, you should be able to:

- simplify expressions involving quotients of monomials, and
- simplify expressions containing negative exponents.

Consider each of the following quotients. Each number can be expressed as a power of 3.

$$\frac{81}{27} = 3 \qquad \frac{27}{3} = 9 \qquad \frac{243}{9} = 27$$

$$\downarrow \qquad \downarrow \qquad \downarrow$$

$$\frac{3^4}{3^3} = 3^1 \qquad \frac{3^3}{3^1} = 3^2 \qquad \frac{3^5}{3^2} = 3^3$$

Once again, look for a pattern in the quotients shown. If you consider only the exponents, you may notice that $4 - 3 = 1$, $3 - 1 = 2$, and $5 - 2 = 3$.

Now simplify $\frac{b^5}{b^2}$, $b \neq 0$.

$\frac{b^5}{b^2} = \frac{\cancel{b} \cdot \cancel{b} \cdot b \cdot b \cdot b}{\cancel{b} \cdot \cancel{b}}$ *Notice that $\frac{b \cdot b}{b \cdot b}$ is equal to 1.*

$= b \cdot b \cdot b$ *The quotient has b as a factor 5 − 2 or 3 times.*

$= b^3$

These examples suggest that to divide powers with the same base, you can subtract the exponents.

Quotient of Powers

For all integers *m* and *n*, and any nonzero number *a*,

$$\frac{a^m}{a^n} = a^{m-n}.$$

Example 1

Simplify $\frac{a^4b^3}{ab^2}$.

$\frac{a^4b^3}{ab^2} = \left(\frac{a^4}{a^1}\right)\left(\frac{b^3}{b^2}\right)$ *Group the powers that have the same base.*

$= a^{4-1}b^{3-2}$ *Subtract the exponents by the quotient of powers property.*

$= a^3b$ *Recall that $b^1 = b$.*

Study the two ways shown below to simplify $\frac{a^3}{a^3}$, $a \neq 0$.

$$\frac{a^3}{a^3} = \frac{a \cdot a \cdot a}{a \cdot a \cdot a} = 1 \qquad \frac{a^3}{a^3} = a^{3-3} = a^0$$

0^0 is not defined.

Since $\frac{a^3}{a^3}$ cannot have two different values, we can conclude that $a^0 = 1$. In general, any nonzero number raised to the zero power is equal to 1.

Zero Exponent	**For any nonzero number a, $a^0 = 1$.**

Study the two ways shown below to simplify $\frac{k^2}{k^7}$, $k \neq 0$.

$$\frac{k^2}{k^7} = \frac{\not{k} \cdot \not{k}}{\not{k} \cdot \not{k} \cdot k \cdot k \cdot k \cdot k \cdot k} = \frac{1}{k \cdot k \cdot k \cdot k \cdot k} = \frac{1}{k^5}$$

$$\frac{k^2}{k^7} = k^{2-7} = k^{-5}$$

Since $\frac{k^2}{k^7}$ cannot have two different values, we can conclude that $k^{-5} = \frac{1}{k^5}$. This example suggests the following definition.

Negative Exponents	**For any nonzero number a and any integer n, $a^{-n} = \frac{1}{a^n}$.**

To **simplify** an expression involving monomials, write an equivalent expression that has positive exponents and no powers of powers. Also, each base should appear only once and all fractions should be in simplest form.

Example 2

Simplify $\frac{-6r^3s^5}{18r^{-7}s^5t^{-2}}$.

$$\frac{-6r^3s^5}{18r^{-7}s^5t^{-2}} = \left(\frac{-6}{18}\right)\left(\frac{r^3}{r^{-7}}\right)\left(\frac{s^5}{s^5}\right)\left(\frac{1}{t^{-2}}\right)$$

$$= -\frac{1}{3}r^{3-(-7)}s^{5-5}t^2 \qquad \textit{$\frac{1}{t^{-2}} = t^2$}$$

$$= -\frac{1}{3}r^{10}s^0t^2 \qquad \textit{Subtract the exponents.}$$

$$= -\frac{r^{10}t^2}{3} \qquad \textit{Remember that $s^0 = 1$.}$$

Example 3

Simplify $\frac{(4a^{-1})^{-2}}{(2a^4)^2}$.

$$\frac{(4a^{-1})^{-2}}{(2a^4)^2} = \frac{4^{-2}a^2}{2^2a^8} \qquad \textit{Power of a product property}$$

$$= \frac{4^{-2}a^2}{4a^8}$$

$$= (4^{-2-1})(a^{2-8}) \qquad \textit{Subtract the exponents.}$$

$$= 4^{-3}a^{-6}$$

$$= \frac{1}{4^3a^6} \qquad \textit{Definition of negative exponents}$$

$$= \frac{1}{64a^6}$$

CHECKING FOR UNDERSTANDING

Communicating Mathematics

Read and study the lesson to answer each question.

1. In the quotient of powers property, why must a be nonzero?
2. Use the fact that $2^4 = 16$, $2^3 = 8$, $2^2 = 4$, and $2^1 = 2$ to make a convincing argument that $2^0 = 1$.
3. Why is 0^0 undefined?

Guided Practice

Evaluate.

4. 5^0
5. $(-8)^{-1}$
6. 10^{-2}
7. $(-2)^{-3}$
8. $(5^{-1})^2$
9. $\frac{4^{-2}}{4}$
10. $\left(\frac{1}{3} \cdot \frac{1}{6}\right)^{-1}$
11. $(2^0 3^{-2})^{-2}$

Simplify. Assume no denominator is equal to zero. (Remember to express the results with positive exponents.)

12. $m^{-5}n^0$
13. $a^0b^{-2}c^{-1}$
14. $x^5y^0z^{-5}$
15. $\frac{5n^5}{n^8}$
16. $\frac{b^9}{b^4c^3}$
17. $\frac{1}{r^{-4}}$
18. $\frac{a^{-4}}{b^{-3}}$
19. $\frac{an^6}{n^5}$

EXERCISES

Practice

Simplify. Assume no denominator is equal to zero.

20. $\frac{a^0}{a^{-2}}$
21. $\frac{1}{r^{-3}}$
22. $\frac{k^{-2}}{k^4}$
23. $\frac{m^2}{m^{-4}}$
24. $\frac{b^6c^5}{b^3c^2}$
25. $\frac{(-a)^4b^8}{a^4b^7}$
26. $\frac{(-x)^3y^3}{x^3y^6}$
27. $\frac{12b^5}{4b^4}$
28. $\frac{10m^4}{30m}$
29. $\frac{x^3y^6}{x^3y^3}$
30. $\frac{b^6c^5}{b^{14}c^2}$
31. $\frac{(-r)^5s^8}{r^5s^2}$
32. $\frac{30x^4y^7}{-6x^{13}y^2}$
33. $\frac{16b^4c}{-4bc^3}$
34. $\frac{22a^2b^5c^7}{-11abc^2}$
35. $\frac{24x^2y^7z^3}{-6x^2y^3z}$
36. $\frac{7x^3z^5}{4z^{15}}$
37. $\frac{27a^4b^6c^9}{15a^3c^{15}}$
38. $\frac{(a^7b^2)^2}{(a^{-2}b)^{-2}}$
39. $\frac{r^{-5}s^{-2}}{(r^2s^5)^{-1}}$
40. $\frac{(r^{-4}k^2)^2}{(5k^2)^2}$
41. $\left(\frac{3m^2n^2}{6m^{-1}k}\right)^0$
42. $\frac{(-b^{-1}c)^0}{4a^{-1}c^2}$
43. $\left(\frac{7m^{-1}n^3}{n^2r^{-2}}\right)^{-1}$
44. $\left(\frac{2xy^{-2}z^4}{3xyz^{-1}}\right)^{-2}$

Critical Thinking

Simplify.

45. $y^2 \cdot y^b$
46. $x^{2a} \cdot x^{4a}$
47. $(2^{7x+6})(2^{3x-4})$
48. $\frac{x^{y+2}}{x^{y-3}}$
49. $\frac{(a^{x+2})^2}{(a^{x-3})^2}$
50. $\frac{a^b}{a^{a-b}}$

Applications

51. **Finance** To determine what is still owed on a car loan, you can use the formula $B = P\left[\frac{1 - (1 + i)^{k-n}}{i}\right]$, where B is the balance due (or payoff), P is the current monthly payment, i is the monthly interest rate (or the annual rate ÷ 12), k is the number of payments already made, and n is the total number of monthly payments. Use a calculator to find the payoff on a 48-month loan, after 24 payments of \$213.87 have been made, at an annual interest rate of 10.8%.

52. **Finance** To determine the monthly payment on a home, you can use the formula $P = A\left[\frac{i}{1 - (1 + i)^{-n}}\right]$, where P is the monthly payment, A is the price of the home less down payment, i is the *monthly* interest rate (or the annual rate ÷ 12), and n is the total number of monthly payments. Use a calculator to find the monthly payment on an \$80,000 home, with 5% down, over 30 years at an *annual* interest rate of 12%.

Mixed Review

53. **Consumerism** Which is the better buy: 12 extra-large eggs weighing 27 ounces for \$1.09, or 12 large eggs weighing 24 ounces for 99¢? **(Lesson 2-4)**

54. Solve $-\frac{7}{6} + k = \frac{5}{6}$. **(Lesson 3-1)**

55. **Government** The shape of the U.S. flag is determined by federal standards set by Executive Order of President Eisenhower in 1959. The width-to-length ratio must be 1:1.9. How long would a U.S. flag be that is 3 feet wide? **(Lesson 4-1)**

56. Three spiders are on a 9-foot-high wall. Susie is 4 feet from the top. Sam is 7 feet from the bottom. Shirley is 3 feet below Sam. Which spider is nearest the top of the wall? **(Lesson 5-4)**

57. Simplify $a^2(a^5)$. **(Lesson 6-2)**

READING ALGEBRA

You can read powers containing the exponent 2 or 3 in two ways.

x^2 is read "x squared" or "x to the second power."
x^3 is read "x cubed" or "x to the third power."

Powers containing other numerical exponents are usually read as follows.

x^6 is read "x to the sixth power."

The phrase *the quantity* is used to indicate parentheses when reading expressions.

$3x^2$ is read "three x squared."
$(3x)^2$ is read "three x *the quantity* squared."
$(2a + b)^4$ is often read "*the quantity* $2a$ plus b all to the fourth power."

Read each expression to a classmate while they write the expression in words.

1. 4^2
2. 3^3
3. a^5
4. $5b^2$
5. $(12r)^5$
6. $9x^3y$
7. $(x + 2y)^2$
8. $4m^2n^4$
9. $(6a^2b)^4$
10. $a - b^3$

6-4 Scientific Notation

Objectives

After studying this lesson, you should be able to:

- express numbers in scientific and decimal notation, and
- find products and quotients of numbers expressed in scientific notation.

Application

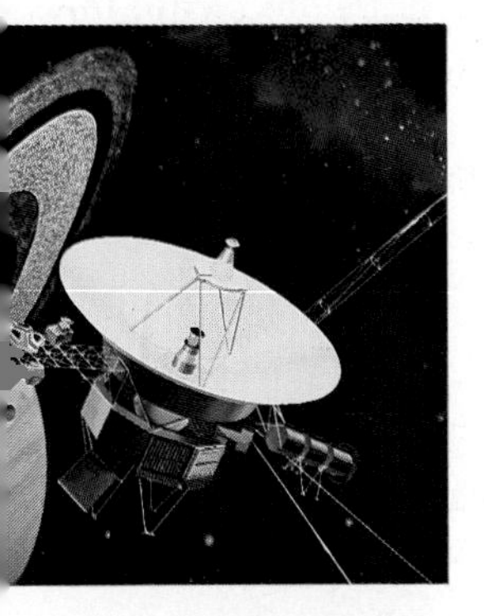

What can fly through hundreds of thousands of miles of space in a single day? What can photograph objects billions of miles from Earth? What can send messages at the speed of light—186,282 miles per second? A space probe! Space probes allow scientists to explore other planets without ever leaving Earth.

The chart at the right shows how far nine space probes have traveled. These distances are NASA estimates for travel as of late 1989 and have been rounded to the nearest million miles.

Space Probe	Miles Traveled
Magellan	40,000,000
Pioneer 6	62,000,000
Pioneer 7	78,000,000
Pioneer 8	142,000,000
Pioneer 10	4,382,000,000
Pioneer 11	2,705,000,000
Pioneer 12	100,000,000
Voyager 1	3,641,000,000
Voyager 2	2,793,000,000

FYI . . .

Our universe is estimated to be 2.1×10^{23} miles wide.

Astronomers use large numbers like these all of the time when measuring distances in space. Sometimes it is not desirable to express these numbers in **decimal notation,** as they are shown in the chart. So scientists use **scientific notation** to express very large numbers.

Definition of Scientific Notation

A number is expressed in scientific notation when it is in the form

$$a \times 10^n,$$

where $1 \leq a < 10$ and n is an integer.

For example, the average distance between the sun and Earth is about 93,000,000 miles. To write this number in scientific notation, express it as the product of a number greater than or equal to 1, but less than 10, and a power of 10.

$93{,}000{,}000 = 9.3 \times 10{,}000{,}000$ *Move the decimal point 7 places to the left.*

$= 9.3 \times 10^7$ *The exponent is 7.*

Example 1

Express 5093.4 in scientific notation.

$5093.4 = 5.0934 \times 1000$ *Move the decimal point 3 places to the left.*

$= 5.0934 \times 10^3$ *The exponent is 3.*

Example 2

Express 2.6×10^5 in decimal notation.

$$2.6 \times 10^5 = 2.6 \times 100{,}000$$
$$= 260{,}000$$

Because the exponent on 10 was 5, the decimal point was moved 5 places to the right.

Scientific notation is also used to express very small numbers. When numbers between zero and one are written in scientific notation, the exponent of 10 is negative.

Example 3

An organism called the *Mycoplasma laidlawii* has a diameter of only 100 millimicrons, or 0.000004 inches, during the early part of its life. Write this number in scientific notation.

$0.000004 = 4 \times 0.000001$ *Move the decimal point 6 places to the right.*

$= 4 \times \frac{1}{1{,}000{,}000}$

$= 4 \times \frac{1}{10^6}$ *The exponent is 6.*

$= 4 \times 10^{-6}$ *Remember that $\frac{1}{a^n} = a^{-n}$.*

So the Mycoplasma laidlawii has a diameter of 4×10^{-6} inches.

Example 4

Express 3.2×10^{-7} in decimal notation.

$$3.2 \times 10^{-7} = 3.2 \times 0.0000001$$
$$= 0.00000032$$

Because the exponent on 10 was −7, the decimal point was moved 7 places to the left.

You can find products or quotients of numbers that are expressed in scientific notation.

Example 5

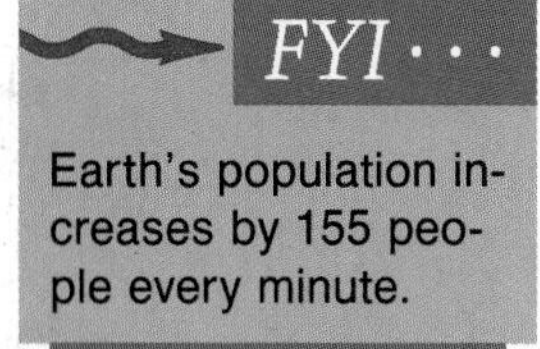

Earth's population increases by 155 people every minute.

Suppose you have to feed all of the people on Earth using the 4.325×10^{11} kilograms of food produced by the United States and Canada each year. You need to divide this food among the world population so that each person receives the same amount of food each day. How would you do this, if Earth has about 4.8×10^9 inhabitants?

If the food is to be equally divided among all inhabitants, each person would receive $(4.325 \times 10^{11}) \div (4.8 \times 10^9)$ kilograms of food.

$$\frac{4.325 \times 10^{11}}{4.8 \times 10^9} = \frac{4.325}{4.8} \times \frac{10^{11}}{10^9}$$
$$\approx 0.9 \times 10^2$$
$$\approx 90$$

Thus, each person would receive about 90 kilograms of food each year, or about $90 \div 365$ or 0.25 kilograms of food each day.

Example 6 **Use scientific notation to find the product of 0.000008 and 3,500,000,000. Express the result in scientific notation and decimal notation.**

$(0.000008)(3,500,000,000) = (8 \times 10^{-6})(3.5 \times 10^{9})$ *Express in scientific notation.*

$= (8 \times 3.5)(10^{-6} \times 10^{9})$ *Use the commutative and associative properties.*

$= 28 \times 10^{3}$ *Recall that $1 \leq a < 10$.*

$= 2.8 \times 10^{4}$ or 28,000

Many calculators use scientific notation to display answers that are either too large or too small to display on the calculator screen in decimal notation. You can enter numbers in scientific notation using keys often labeled [EE] or [EXP].

Example 7 **Use a calculator to express 230×10^{15} in scientific notation.**

Enter: 230 [EE] 15 [=] 2.3 17

The calculator has expressed 230×10^{15} as 2.3 17. That means 2.3×10^{17}.

CHECKING FOR UNDERSTANDING

Communicating Mathematics

Read and study the lesson to answer each question.

1. When do you use positive exponents in scientific notation?
2. When do you use negative exponents in scientific notation?

Complete.

3. The number 876,200 is written in __?__ notation.
4. The number 8.762×10^{5} is written in __?__ notation.

Guided Practice

Express each number in the second column in decimal notation. Express each number in the third column in scientific notation.

	Planet	Diameter (km)	Distance From Sun (km)
5.	Mercury	5.0×10^{3}	57,900,000
6.	Venus	1.208×10^{4}	108,230,000
7.	Earth	1.276×10^{4}	149,590,000
8.	Mars	6.79×10^{3}	227,920,000
9.	Jupiter	1.432×10^{5}	778,320,000
10.	Saturn	1.21×10^{5}	1,427,000,000
11.	Uranus	5.18×10^{4}	2,870,000,000
12.	Neptune	4.95×10^{4}	4,497,000,000
13.	Pluto	3×10^{3}	5,900,000,000

EXERCISES

Practice

Express each number in scientific notation.

14. 4293
15. 240,000
16. 0.000319
17. 0.004296
18. 0.000000092
19. 0.00000000317
20. 32×10^5
21. 284×10^3
22. 0.0031×10^3

Evaluate. Express each result in scientific and decimal notation.

23. $\frac{4.8 \times 10^3}{1.6 \times 10^1}$
24. $\frac{5.2 \times 10^5}{1.3 \times 10^2}$
25. $\frac{7.8 \times 10^{-5}}{1.3 \times 10^{-7}}$
26. $\frac{8.1 \times 10^2}{2.7 \times 10^{-3}}$
27. $\frac{1.32 \times 10^{-6}}{2.4 \times 10^2}$
28. $\frac{2.31 \times 10^{-2}}{3.3 \times 10^{-3}}$
29. $(2 \times 10^5)(3 \times 10^{-8})$
30. $(4 \times 10^2)(1.5 \times 10^6)$
31. $(3.1 \times 10^{-2})(2.1 \times 10^5)$
32. $(3.1 \times 10^4)(4.2 \times 10^{-3})$
33. $(78 \times 10^6)(0.01 \times 10^3)$
34. $(0.2 \times 10^5)(31 \times 10^{-6})$

Use a calculator to evaluate. Express each result in scientific and decimal notation.

35. (0.000003)(70,000)
36. (86,000,000)(0.005)
37. $24{,}000 \div 0.00006$
38. $0.0000039 \div 650{,}000$

Critical Thinking

Use a calculator to evaluate. Express each result in scientific and decimal notation.

39. $\frac{(35{,}987{,}000)(58 \times 10^3)}{42.5 \times 10^4}$
40. $\frac{(3 \times 10^8)(43 \times 10^{-4})}{23{,}000{,}000}$
41. $\frac{8.9 \times 10^5}{(98{,}000)(14 \times 10^3)}$
42. $\frac{57{,}800{,}000{,}000}{(2.3 \times 10^6)(38 \times 10^{-5})}$

Applications

Use scientific notation to solve each problem.

43. **Health** A radio station advertised the Columbus Marathon by saying that about 12,000,000 calories would be burned in one day. If there were 4000 runners, about how many calories did each runner burn?

44. **Census** In 1990, the population of the United States was 248,200,000. The area of the United States is 3,540,000 square miles. If the population were equally spaced over the land, how many people would there be for each square mile?

45. **Government** In 1990, the U.S. federal budget deficit was \$220,000,000,000. Using the U.S. population figures given in Exercise 44, find how much each American would have had to pay in 1990 to erase the deficit.

46. **Space** We learned in the lesson introduction that one of the space probes to travel space was *Pioneer 10.*
    a. How long did it take *Pioneer 10*'s radio signals, traveling at the speed of light (about 1.86×10^5 miles per second), to reach Earth from a distance of 2.85×10^9 miles?
    b. What was *Pioneer 10*'s average speed (in miles per hour) if it traveled about 2.85×10^9 miles in 4070 days?

Computer

47. In the BASIC language, powers are written using the symbols ↑ or ^. The program at the right evaluates the expression $(2ab^2)^3$ when $a = 9$ and $b = 10$. Notice that the output for the program is in E notation. This is the computer equivalent of scientific notation. Thus, 5.832E + 09 means 5.832×10^9. The computer used E notation because there were more than six significant digits in the output.

```
10 READ A, B
20 DATA 9, 10
30 PRINT (2*A*B↑2)↑3
40 END

RUN
5.832E + 09
```

Write a program that will evaluate each expression if $a = 4$, $b = 6$, and $c = 8$. Then use the program to evaluate each expression.

a. $a^2b^3c^4$
b. $(-2a)^2(4b)^3$
c. $(4a^2b^4)^3$
d. $(ac)^3 + (3b)^2$

Mixed Review

48. Simplify $\frac{60a - 30b}{-6}$. **(Lesson 2-7)**

49. **Biology** A crab has 2 more legs than a scorpion. Together they have 18 legs. How many legs does a scorpion have? **(Lesson 3-5)**

50. Thirty-five is 50% of what number? **(Lesson 4-2)**

51. Solve $-16q > -128$. **(Lesson 5-2)**

Simplify.

52. $4(a^2b^3)^3$ **(Lesson 6-2)**

53. $\frac{48a^8}{12a}$ **(Lesson 6-3)**

APPLICATION

Black Holes

A *black hole* is a region in space where matter seems to disappear. Scientists believe that when a large star runs out of nuclear fuel, it begins to break down under the force of its own gravity. The gravity eventually becomes so intense that even light cannot escape, making the star look black.

How do we know when a star has become a black hole? A star is a black hole when the radius of the star reaches a certain critical value called the *Schwarzschild radius*. This value is given by the equation $R_s = \frac{2GM}{c^2}$, where R_s is the Schwarzschild radius in meters, M is the mass in kilograms, G is the gravitational constant (6.7×10^{-11}), and c is the speed of light $(3 \times 10^8$ meters per second).

Our sun is a star, but it is not large enough to become a black hole. However, we can use it as an example of how a black hole is formed. The mass of the sun is about 2×10^{30} kilograms.

$$R_s = \frac{2(6.7 \times 10^{-11})(2 \times 10^{30})}{(3 \times 10^8)^2} \approx 3000 \text{ meters}$$

The radius of the sun is about 700,000 kilometers. If the radius were reduced to about 3000 meters, in theory, it would become a black hole.

6-5 Polynomials

Objectives

After studying this lesson, you should be able to:

- find the degree of a polynomial, and
- arrange the terms of a polynomial so that the powers of a certain variable are in ascending or descending order.

Application

Each summer, Li Chiang works at Burger World. She earns about \$1600 and saves \$1000, which she invests in a savings account at a local bank. If Li works at Burger World each summer that she is in high school, will she have enough money to pay for her first year of college, a cost of \$6000?

Banks and savings institutions pay interest on savings account deposits. To calculate how Li's money will grow, use the formula for compound interest, $T = p(1 + r)^t$, where T is the total amount, p is the principal, r is the annual interest rate, and t is the time in years that the money has been in the account. To simplify the formula, let $x = 1 + r$. When Li goes to college three years later, the money she earned the first summer will be worth $1000x^3$. The money she earned the second summer will be worth $1000x^2$, and the money she earned the third summer will be worth $1000x$. The money she earns the last summer will not have a chance to earn interest, since Li will enter college at the end of that summer. The **polynomial** below represents the amount Li will have saved.

$$1000x^3 + 1000x^2 + 1000x + 1000$$

Polynomials with more than three terms have no special name.

A polynomial is a monomial or a sum of monomials. Recall that a monomial is a number, a variable, or a product of numbers and variables. A monomial is one term with a positive exponent. A **binomial** is the sum of two monomials, and a **trinomial** is the sum of three monomials.

Monomial	Binomial	Trinomial
$5x^2$	$3x + 2$	$5x^2 - 2x + 7$
$4abc$	$4x + 5y$	$a^2 + 2ab + b^2$
-7	$3x^2 - 8xy$	$4a + 2b^2 - 3c$

Recall that $5x^2 - 2x + 7$ is equivalent to $5x^2 + (-2x) + 7$.

Example 1

State whether each expression is a polynomial. If the expression is a polynomial, identify it as either a monomial, binomial, or trinomial.

a. $8x^2 - 3xy$

The expression $8x^2 - 3xy$ can be written as $8x^2 + (-3xy)$. Therefore, $8x^2 - 3xy$ is a polynomial. Since $8x^2 - 3xy$ can be written as the sum of two monomials, $8x^2$ and $-3xy$, it is also a binomial.

b. $\frac{5}{2y^2} + 7y + 6$

The expression $\frac{5}{2y^2} + 7y + 6$ is not a polynomial because $\frac{5}{2y^2}$ is not a monomial. It is not the product of a number and a variable.

Example 2

If Li's savings account pays 8% annually, will Li have enough money for her first year of college?

If $r = 8\%$, the polynomial below represents the amount Li will have saved after working for four summers.

$$1000(1.08)^3 + 1000(1.08)^2 + 1000(1.08) + 1000$$

Use a calculator to evaluate this sum.

Enter: 1.08 [y^x] 3 [×] 1000 [+] 1.08 [y^x] 2 [×] 1000 [+] 1.08 [×] 1000 [+] 1000 [=]

Display: 4506.112

So Li will have only \$4506.11, not enough to pay for her first year.

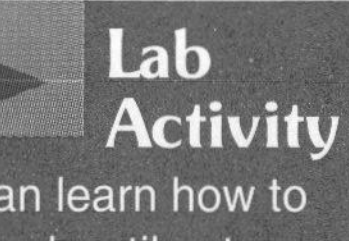

Lab Activity

You can learn how to use algebra tiles to model polynomials in Lab 5 on pages A8-A9.

The **degree** of a monomial is the sum of the exponents of its variables.

Monomial	Degree
$5x^2$	2
$4ab^3c^4$	$1 + 3 + 4 = 8$
-9	0

Remember that $a = a^1$, $x^0 = 1$, and $-9 = -9x^0$.

To find the degree of a polynomial, find the degree of each of its terms. The degree of the polynomial is the greatest of the degrees of its terms.

Example 3

Find the degree of each polynomial.

a. $8x^3 - 2x^2 + 7$

This polynomial has three terms, $8x^3$, $-2x^2$, and 7. Their degrees are 3, 2, and 0, respectively. The greatest degree is 3. Therefore, the degree of $8x^3 - 2x^2 + 7$ is 3.

b. $6x^2y + 5x^3y^2z - x + x^2y^2$

This polynomial has four terms, $6x^2y$, $5x^3y^2z$, $-x$, and x^2y^2. Their degrees are 3, 6, 1, and 4, respectively. Therefore, the degree of $6x^2y + 5x^3y^2z - x + x^2y^2$ is 6.

The terms of a polynomial are usually arranged so that the powers of one variable are in either ascending or descending order.

Ascending Order	Descending Order
$3 + 5a - 8a^2 + a^3$	$a^3 - 8a^2 + 5a + 3$
(in x) $5xy + x^3y^2 - x^4 + x^5y^2$	(in x) $x^5y^2 - x^4 + x^3y^2 + 5xy$
(in y) $x^3 - 3x^2y + 4x^2y^2 - y^3$	(in y) $-y^3 + 4x^2y^2 - 3x^2y + x^3$

CHECKING FOR UNDERSTANDING

Communicating Mathematics

Read and study the lesson to answer each question.

1. Why is $\frac{17}{5x}$ *not* a monomial?
2. Why is $a + \frac{3}{b}$ *not* a binomial?
3. Why is the degree of 12 zero?

Guided Practice

State whether each expression is a polynomial. If the expression is a polynomial, identify it as a monomial, binomial, or trinomial.

4. $5x^2y + 3xy + 7$
5. $\frac{5}{k} - k^2y$
6. 0
7. $\frac{a^3}{3}$
8. $3a^2x - 5a$
9. $x^2 - \frac{x}{2} + \frac{1}{3}$

Find the degree of each polynomial.

10. $100x$
11. -18
12. $29xyz$
13. $8x^2$
14. $-14x^3z$
15. 0
16. $12s + 21t$
17. $22x + 5y^4$
18. $-2a^2b^{10} + 3a^9b$
19. $29x + x^{29}$
20. $27x^4 - 3x^3yz$
21. $17x^3y^4 - 11xy^5$

EXERCISES

Practice

Find the degree of each polynomial.

22. $5x^2 - 2x^5$
23. $11wxyz - 9w^4$
24. $12x^4y^5 + xy^6$
25. $7x^3 + 4xy + 3xz^3$
26. $5r - 3s + 7t$
27. $17r^2t + 3r + t^2$
28. $13s^2t^2 + 4st^2 - 5s^5t$
29. $32xyz + 11x^2y + 17xz^2$
30. $2x^3yz - 5xy^3z + 11z^5$
31. $-4yzw^4 + 10x^4z^2w - 2z^2w^3$

Arrange the terms of each polynomial so that the powers of *x* are in ascending order.

32. $3 + x^4 + 2x^2$
33. $2x^2 + 5ax + a^3$
34. $1 + x^3 + x^5 + x^2$
35. $-2x^2y + 3xy^3 + x^3$
36. $17bx^2 + 11b^2x - x^3$
37. $5b + b^3x^2 + \frac{2}{3}bx$

Arrange the terms of each polynomial so that the powers of *x* are in descending order.

38. $-6x + x^5 + 4x^3 - 20$
39. $5x^2 - 3x^3 + 7 + 2x$
40. $4x^3y + 3xy^4 - x^2y^3 + y^4$
41. $11x^2 + 7ax^3 - 3x + 2a$
42. $\frac{3}{4}x^3y - x^2 + 4 + \frac{2}{3}x$
43. $7a^3x - 8a^3x^3 + \frac{1}{5}x^5 + \frac{2}{3}x^2$

Critical Thinking

44. A numeral in base ten can be written in polynomial form. For example, $2137 = 2(10)^3 + 1(10)^2 + 3(10) + 7$. Suppose 3254 is a numeral in base *b*. Write 3254 in polynomial form.

Applications

45. **Finance** Look back at the application at the beginning of this lesson. Suppose that upon her graduation from high school, Li receives a \$2000 inheritance from her grandfather. Would she then have enough money to finance her first year in college?

46. **Finance** Upon his graduation from college, Mark Price inherited \$10,000. If he invests this money in an account with an annual interest rate of 6% and adds \$1000 of his own money to the account at the end of each year, will his money have doubled after 5 years? If not, when?

47. **Biology** The average number of eggs carried by a certain type of female moth is given by the polynomial $14x^3 - 17x^2 - 16x + 34$, where x represents the width of the abdomen. About how many eggs should you expect this type of moth to produce if her abdomen measures 3 millimeters?

Mixed Review

48. Graph {all integers greater than 3} on a number line. **(Lesson 2-3)**

49. Solve $11y = -77$. **(Lesson 3-3)**

50. **Finance** Delores Delgado invested some of \$12,500 at 6.2% interest and the rest at 8.6% interest. How much did she invest at each rate if her total income from both investments is \$967? **(Lesson 4-5)**

51. Solve $10p - 14 < 8p - 17$. **(Lesson 5-3)**

Express each number in scientific notation. (Lesson 6-4)

52. 42,350

53. 0.00000628

MID-CHAPTER REVIEW

1. Solve by looking for a pattern. What is the ones digit of 3^{999}? **(Lesson 6-1)**

Simplify. Assume no denominator is equal to zero. (Lessons 6-2, 6-3)

2. $(b^4)(b^4)b$
3. $(x^3y)(xy^3)$
4. $(-2n^4y^3)(3ny^4)$
5. $[(-5)^2]^3$
6. $(4xy)^2(-3x)$
7. $(-3a^2b^5)^2$
8. $\frac{n^8}{n^5}$
9. $\frac{24a^3b^6}{-2a^2b^2}$
10. $\frac{(5r^{-1}s)^3}{(s^2)^3}$

Express each number in scientific notation. (Lesson 6-4)

11. 28.5×10^6

12. 0.005×10^{-3}

Arrange the terms of each polynomial so that the powers of x are in ascending order. (Lesson 6-5)

13. $\frac{1}{3}s^2x^3 + 4x^4 - \frac{2}{5}s^4x^2 + \frac{1}{4}x$

14. $21p^2x + 3px^3 + p^4$

6-6 Adding and Subtracting Polynomials

Objective

After studying this lesson, you should be able to:

- add and subtract polynomials.

Application

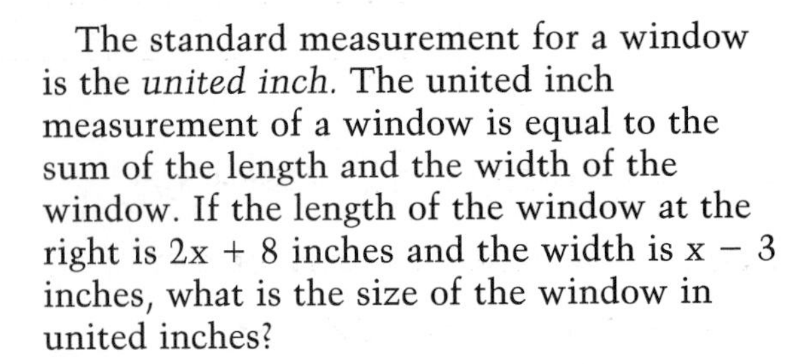

The standard measurement for a window is the *united inch.* The united inch measurement of a window is equal to the sum of the length and the width of the window. If the length of the window at the right is $2x + 8$ inches and the width is $x - 3$ inches, what is the size of the window in united inches?

> **FYI · · ·**
> The largest windows in the world are in the Palace of Industry and Technology in France. They measure 715.2 feet wide by 164 feet high.

The size of the window is $(2x + 8) + (x - 3)$ inches. To add two polynomials, add the like terms.

$$
\begin{aligned}
(2x + 8) + (x - 3) &= 2x + 8 + x - 3 \\
&= (2x + x) + (8 - 3) && \textit{Commutative and associative properties} \\
&= 3x + 5
\end{aligned}
$$

The size of the window in united inches is $3x + 5$ inches.

You can add polynomials by grouping the like terms together and then finding the sum (as in the example above), or by writing them in column form.

Example 1

Find $(3y^2 + 5y - 6) + (7y^2 - 9)$.

Method 1 Group the like terms together.

$$
\begin{aligned}
(3y^2 + 5y - 6) + (7y^2 - 9) &= (3y^2 + 7y^2) + 5y + [-6 + (-9)] \\
&= (3 + 7)y^2 + 5y + (-15) \\
&= 10y^2 + 5y - 15
\end{aligned}
$$

Method 2 Add in column form.

$$
\begin{array}{rrrr}
 & 3y^2 & + 5y & - 6 \\
(+) & 7y^2 & & - 9 \\
\hline
 & 10y^2 & + 5y & - 15
\end{array}
$$

Notice that like terms are aligned.

There is no term in the y column.

> **Lab Activity**
> You can learn how to use algebra tiles to add and subtract polynomials in Lab 6 on pages A10-A11.

Recall that you can subtract a rational number by adding its additive inverse or opposite. Similarly, you can subtract a polynomial by adding its additive inverse.

To find the additive inverse of a polynomial, replace each term with its additive inverse.

Polynomial	Additive Inverse
$x + 2y$	$-x - 2y$
$2x^2 - 3x + 5$	$-2x^2 + 3x - 5$
$-8x + 5y - 7z$	$8x - 5y + 7z$
$3x^3 - 2x^2 - 5x$	$-3x^3 + 2x^2 + 5x$

Example 2

Find $(4x^2 - 3y^2 + 5xy) - (8xy + 6x^2 + 3y^2)$.

Method 1 Group the like terms together.

$$\begin{aligned}(4x^2 - 3y^2 + 5xy) &- (8xy + 6x^2 + 3y^2)\\ &= (4x^2 - 3y^2 + 5xy) + (-8xy - 6x^2 - 3y^2)\\ &= 4x^2 - 3y^2 + 5xy - 8xy - 6x^2 - 3y^2\\ &= (4x^2 - 6x^2) + (5xy - 8xy) + (-3y^2 - 3y^2)\\ &= -2x^2 + (-3xy) + (-6y^2)\\ &= -2x^2 - 3xy - 6y^2\end{aligned}$$

Method 2 Subtract in column form.

First, reorder the terms so that the powers of x are in descending order.

$$(4x^2 + 5xy - 3y^2) - (6x^2 + 8xy + 3y^2)$$

Then, subtract.

$$\begin{array}{r} 4x^2 + 5xy - 3y^2 \\ (-)\ 6x^2 + 8xy + 3y^2 \\ \hline \end{array} \quad \text{Add the additive inverse.} \rightarrow \quad \begin{array}{r} 4x^2 + 5xy - 3y^2 \\ (+)\ -6x^2 - 8xy - 3y^2 \\ \hline -2x^2 - 3xy - 6y^2 \end{array}$$

To check this result, add $-2x^2 - 3xy - 6y^2$ and $6x^2 + 8xy + 3y^2$.

Example 3

Find the measure of the third side of the triangle at the right. *P* is the measure of the perimeter.

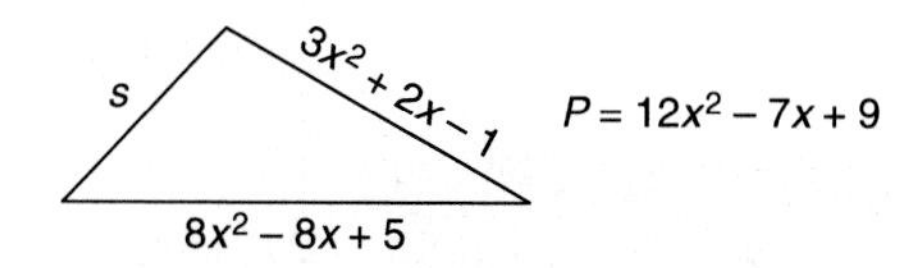

The perimeter is the sum of the measures of the three sides of the triangle. Let *s* represent the measure of the third side.

$$\begin{aligned}(12x^2 - 7x + 9) &= (3x^2 + 2x - 1) + (8x^2 - 8x + 5) + s && \textit{Substitution}\\ 12x^2 - 7x + 9 - (3x^2 + 2x - 1) - (8x^2 - 8x + 5) &= s && \textit{Solve for s.}\\ 12x^2 - 7x + 9 - 3x^2 - 2x + 1 - 8x^2 + 8x - 5 &= s\\ (12x^2 - 3x^2 - 8x^2) + (-7x - 2x + 8x) + (9 + 1 - 5) &= s && \textit{Group the like terms.}\\ x^2 - x + 5 &= s\end{aligned}$$

The measure of the third side is $x^2 - x + 5$.

CHECKING FOR UNDERSTANDING

Communicating Mathematics

Read and study the lesson to answer each question.

1. What is the first step when adding or subtracting in column form?
2. What is the best way to check the subtraction of two polynomials?

Guided Practice

Find the additive inverse of each polynomial.

3. $3x + 2y$
4. $-8m + 7n$
5. $x^2 + 3x + 7$
6. $-4h^2 - 5hk - k^2$
7. $-3ab^2 + 5a^2b - b^3$
8. $x^3 + 5x^2 - 3x - 11$

Name the like terms in each group.

9. $5m, 4mn, -3m, 2n, -mn, 8n$
10. $2x^3, 5xy, -x^2y, 14xy, 12xy$
11. $-7ab^2, 8a^2b, 11b^2, 16a^2b, -2b^2$
12. $3p^3q, -2p, 10p^3q, 15pq, -p$

EXERCISES

Practice

Find each sum or difference.

13.
$$\begin{array}{rrrr} & 5ax^2 + 3a^2x - 7a^3 & \\ (+) & 2ax^2 - 8a^2x + 4 \\ \hline \end{array}$$

14.
$$\begin{array}{rr} & a^3 - b^3 \\ (+) & 3a^3 + 2a^2b - b^2 + 2b^3 \\ \hline \end{array}$$

15.
$$\begin{array}{rr} & 4a + 5b - 6c + d \\ & 3a - 7b + 2c + 8d \\ (+) & 2a - b + 7d \\ \hline \end{array}$$

16.
$$\begin{array}{rr} & 2x^2 - 5x + 7 \\ & 5x^2 + 7x - 3 \\ (+) & x^2 - x + 11 \\ \hline \end{array}$$

17.
$$\begin{array}{rr} & 6x^2y^2 - 3xy - 7 \\ (-) & 5x^2y^2 + 2xy + 3 \\ \hline \end{array}$$

18.
$$\begin{array}{rr} & 5x^2 - 4 \\ (-) & 3x^2 + 8x + 4 \\ \hline \end{array}$$

19.
$$\begin{array}{rr} & 11m^2n^2 + 2mn - 11 \\ (-) & 5m^2n^2 - 6mn + 17 \\ \hline \end{array}$$

20.
$$\begin{array}{rr} & 2a - 7 \\ (-) & 5a^2 + 8a - 11 \\ \hline \end{array}$$

21. $(5x + 6y) + (2x + 8y)$
22. $(7n + 11m) - (4m + 2n)$
23. $(3x - 7y) + (3y + 4x)$
24. $(5a - 6m) - (2a + 5m)$
25. $(5m + 3n) + 8m$
26. $(13x + 9y) - 11y$
27. $(3 + 2a + a^2) - (5 + 8a + a^2)$
28. $(n^2 + 5n + 3) + (2n^2 + 8n + 8)$
29. $(5ax^2 + 3a^2x - 5x) + (2ax^2 - 5ax + 7x)$
30. $(x^3 - 3x^2y + 4xy^2 + y^3) - (7x^3 - 9xy^2 + x^2y + y^3)$

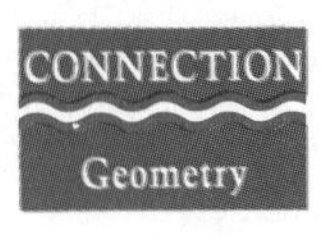

Find the measure of the third side of each triangle. *P* is the measure of the perimeter.

31. $P = 3x + 3y$

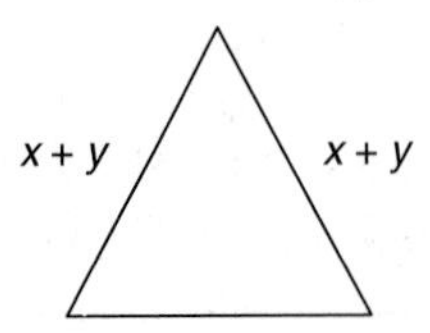

32. $P = 7x + 2y$

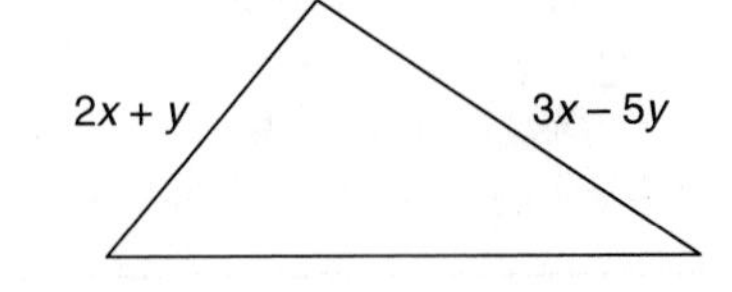

33. $P = 11x^2 - 29x + 10$

$5x^2 - 13x + 24$

$x^2 + 7x + 9$

Critical Thinking

The sum of the degree measures of the three angles of a triangle is 180. Find the degree measure of the third angle of each triangle given the degree measures of the other two angles.

34. $5 - 2x; 7 + 8x$

35. $3x^2 - 5; 4x^2 + 2x + 1$

36. $4 - 2x; x^2 - 1$

37. $x^2 - 8x + 2; x^2 - 3x - 1$

Applications

38. Travel Joan Bedney travels from Cincinnati to Winston-Salem at an average rate of 40 miles per hour and returns at an average rate of 60 miles per hour. What is her average speed for the trip? (*Hint:* The average speed is *not* 50 miles per hour.)

39. Basketball On December 13, 1983, the Denver Nuggets and the Detroit Pistons broke the record for the highest score in a basketball game. The two teams scored a total of 370 points. If the Nuggets scored 2 less points than the Pistons, what was the Nuggets' final score?

Mixed Review

40. Evaluate $|x|$ if $x = -2.1$. **(Lesson 2-2)**

41. Sports There were 3 times as many sports in the 1976 Summer Olympics as there were in the 1976 Winter Olympics. There were 14 more sports in the Summer Olympics than there were in the Winter Olympics. How many sports were there in the 1976 Summer Olympics? **(Lesson 3-6)**

42. Bicycling Adita rode his bicycle 72 kilometers. How long did it take him if his rate was: **a.** 9 km/h? **b.** 18 km/h? **(Lesson 4-7)**

43. Solve $3m < 2m - 7$. **(Lesson 5-1)**

44. Find the degree of $x + 2y^2 + 3z^3$. **(Lesson 6-5)**

Journal

Tell whether you prefer to group terms or use columns to add or subtract polynomials and why you prefer that method.

HISTORY CONNECTION

Emmy Noether

Emmy Noether (1882–1935) was a German mathematician whose strength was in abstract algebra. In her hands the axiomatic method (using axioms or properties) became a powerful tool of mathematical research. Noether did much of her research while living in Göttingen, Germany, which was then the principal center for mathematics research in Europe. Because she was female, Noether was unable to secure a teaching position at the University of Göttingen. However, her influence there was vast. During the thirties, Noether was forced to leave Germany. She became a professor at Bryn Mawr College, in the United States, where she remained until her death. Albert Einstein paid her a great tribute in 1935: "In the judgment of the most competent living mathematicians, (Emmy) Noether was the most significant creative mathematical genius thus far produced since the higher education of women began."

6-7 Multiplying a Polynomial by a Monomial

Objectives

After studying this lesson, you should be able to:

- multiply a polynomial by a monomial, and
- simplify expressions involving polynomials.

Application

The world's largest swimming pool is the Orthlieb Pool in Casablanca, Morocco. It is 30 meters longer than 6 times its width. Express the area of the swimming pool algebraically.

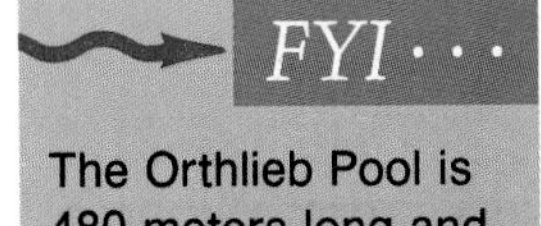

The Orthlieb Pool is 480 meters long and 75 meters wide. Its area is 36,000 square meters.

To find the area of the swimming pool, multiply the length by the width. Let w represent the width. Then $6w + 30$ represents the length.

This diagram of the swimming pool shows that the area is $w(6w + 30)$ square meters.

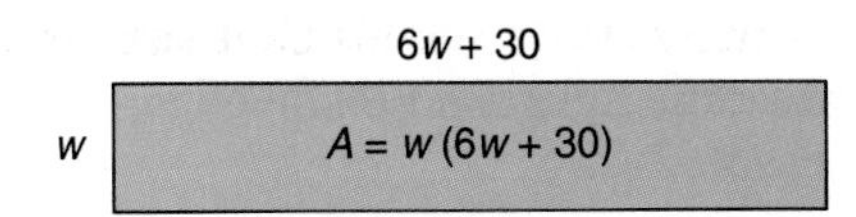

This diagram of the same swimming pool shows that the area is $(6w^2 + 30w)$ square meters.

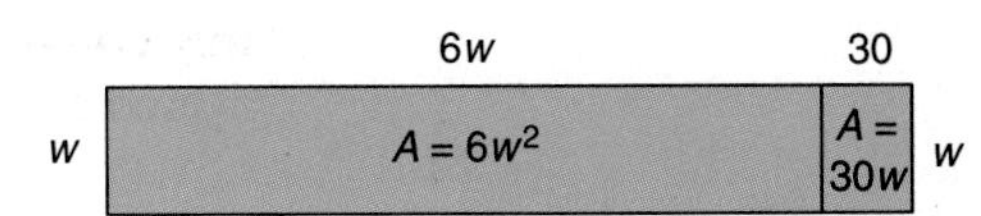

Since the areas are equal, $w(6w + 30) = 6w^2 + 30w$.

The application above shows how the distributive property can be used to multiply a polynomial by a monomial.

Example 1

Find $5a(3a^2 + 4)$.

You can multiply either horizontally or vertically.

a. Use the distributive property.

$$5a(3a^2 + 4) = 5a(3a^2) + 5a(4)$$
$$= 15a^3 + 20a$$

b. Multiply each term by $5a$.

$$\begin{array}{r} 3a^2 + 4 \\ (\times) \qquad 5a \\ \hline 15a^3 + 20a \end{array}$$

Example 2

Find $2m^2(5m^2 - 7m + 8)$.

$$2m^2(5m^2 - 7m + 8) = 2m^2(5m^2) + 2m^2(-7m) + 2m^2(8)$$
$$= 10m^4 - 14m^3 + 16m^2 \quad \textit{Product of powers property}$$

Sometimes you can simplify expressions by using the distributive property and then combining like terms.

Example 3

Find $-3xy(2x^2y + 3xy^2 - 7y^3)$.

$$-3xy(2x^2y + 3xy^2 - 7y^3) = -3xy(2x^2y) + (-3xy)(3xy^2) + (-3xy)(-7y^3)$$
$$= -6x^3y^2 - 9x^2y^3 + 21xy^4$$

Example 4

Find the measure of the area of the shaded region in simplest terms.

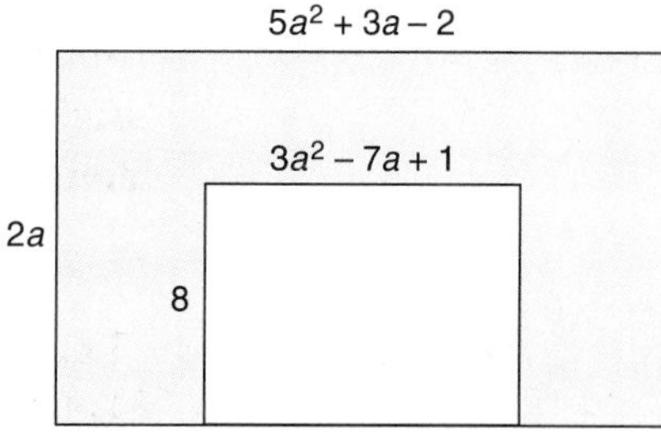

Subtract the measure of the area of the smaller rectangle from the measure of the area of the larger rectangle.

$$2a(5a^2 + 3a - 2) - 8(3a^2 - 7a + 1)$$
$$= 2a(5a^2) + 2a(3a) + 2a(-2) + (-8)(3a^2) + (-8)(-7a) + (-8)(1)$$
$$= 10a^3 + 6a^2 - 4a - 24a^2 + 56a - 8$$
$$= 10a^3 - 18a^2 + 52a - 8 \quad \textit{Combine like terms.}$$

The measure of the area of the shaded region is $10a^3 - 18a^2 + 52a - 8$.

Many equations contain polynomials that must be added, subtracted, or multiplied before the equation can be solved.

Example 5

Lab Activity

You can learn how to use algebra tiles to multiply a polynomial by a monomial in Lab 7 on page A12.

Solve $x(x - 3) + 4x - 3 = 8x + 4 + x(3 + x)$.

$x(x - 3) + 4x - 3 = 8x + 4 + x(3 + x)$	
$x^2 - 3x + 4x - 3 = 8x + 4 + 3x + x^2$	*Multiply.*
$x^2 + x - 3 = x^2 + 11x + 4$	*Combine like terms.*
$x - 3 = 11x + 4$	*Subtract x^2 from each side.*
$-3 = 10x + 4$	*Subtract x from each side.*
$-7 = 10x$	*Subtract 4 from each side.*
$-\frac{7}{10} = x$	*Check this result.*

The solution is $-\frac{7}{10}$.

CHECKING FOR UNDERSTANDING

Communicating Mathematics

Read and study the lesson to answer each question.

1. When you simplify $4y(2y + 1)$, what property is used?
2. Draw a model of a rectangular garden $5a + 6$ units long and $3a$ units wide.
3. Write an expression for the area of the garden described in Exercise 2 in the following two ways.
   a. a product of a monomial and a polynomial
   b. in simplified form

Guided Practice

Find each product.

4. $-5a(12a^2)$
5. $3m(8m + 7)$
6. $-4m^3(5m^2 + 2m)$
7. $\begin{array}{r} 5x - 3 \\ (\times) \quad 2 \\ \hline \end{array}$
8. $\begin{array}{r} m - 7 \\ (\times) \quad 2mn \\ \hline \end{array}$
9. $\begin{array}{r} 5ab^2 + b^2 \\ (\times) \quad 7ab \\ \hline \end{array}$

Simplify.

10. $b(4b - 1) + 10b$
11. $2a(a^3 - 2a^2 + 7) + 5(a^4 + 5a^3 - 3a + 5)$

Solve.

12. $11(a - 3) + 5 = 2a + 44$
13. $x(x + 2) + 3x = x(x - 3)$

EXERCISES

Practice

Find each product.

14. $5(3a + 7)$
15. $-3(8x + 5)$
16. $\frac{1}{2}x(8x + 6)$
17. $3b(5b + 8)$
18. $-2x(5x + 11)$
19. $1.1a(2a + 7)$
20. $7a(3a^2 - 2a)$
21. $3st(5s^2 + 2st)$
22. $7xy(5x^2 - y^2)$
23. $2a(5a^3 - 7a^2 + 2)$
24. $7x^2y(5x^2 - 3xy + y)$
25. $5y(8y^3 + 7y^2 - 3y)$
26. $-4x(7x^2 - 4x + 3)$
27. $5x^2y(3x^2 - 7xy + y^2)$
28. $4m^2(9m^2n + mn - 5n^2)$
29. $-8xy(4xy + 7x - 14y^2)$
30. $-\frac{1}{3}x(9x^2 + x - 5)$
31. $-\frac{3}{4}ab^2\left(\frac{1}{3}b^2 - \frac{4}{9}b + 1\right)$
32. $-2mn(8m^2 - 3mn + n^2)$

Find the measure of the area of each shaded region in simplest terms.

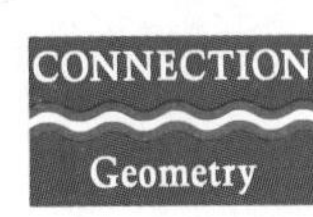

33.

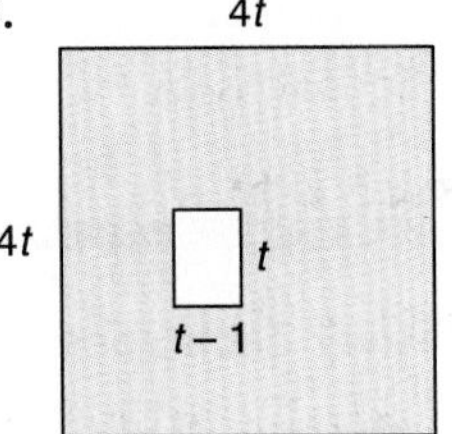

34.

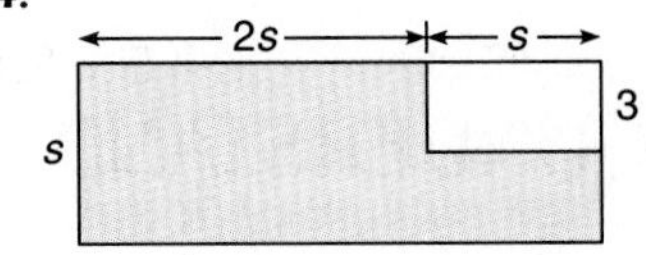

35.

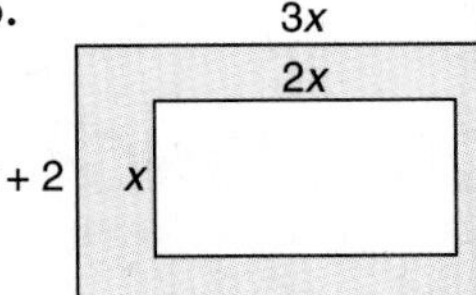

Simplify.

36. $6m(m^2 - 11m + 4) - 7(m^3 + 8m - 11)$
37. $2.5t(8t - 12) + 5.1(6t^2 + 10t - 20)$
38. $\frac{3}{4}m(8m^2 + 12m - 4) + \frac{3}{2}(8m^2 - 9m)$

39. $5m^2(m + 7) - 2m(5m^2 - 3m + 7) + 2(m^3 - 8)$

40. $6a^2(3a - 4) + 5a(7a^2 - 6a + 5) - 3(a^2 + 6a)$

41. $3a^2(a - 4) + 6a(3a^2 + a - 7) - 4(a - 7)$

42. $8r^2(r + 8) - 3r(5r^2 - 11) - 9(3r^2 - 8r + 1)$

Solve.

43. $-3(2a - 12) + 48 = 3a - 3$

44. $2(5w - 12) = 6(-2w + 3) + 2$

45. $-6(12 - 2w) = 7(-2 - 3w)$

46. $7(x - 12) = 13 + 5(3x - 4)$

47. $\frac{1}{2}(2x - 34) = \frac{2}{3}(6x - 27)$

48. $w(w + 12) = w(w + 14) + 12$

49. $a(a - 6) + 2a = 3 + a(a - 2)$

50. $q(2q + 3) + 20 = 2q(q - 3)$

51. $x(x + 8) - x(x + 3) - 23 = 3x + 11$

52. $y(y - 12) + y(y + 2) + 25 = 2y(y + 5) - 15$

Critical Thinking

53. **Geometry** A trapezoid has an area of 162 m^2 and a height of 12 m. The lower base is 6 m more than twice the upper base. Find the length of the upper base. Use $A = \frac{1}{2}h(a + b)$.

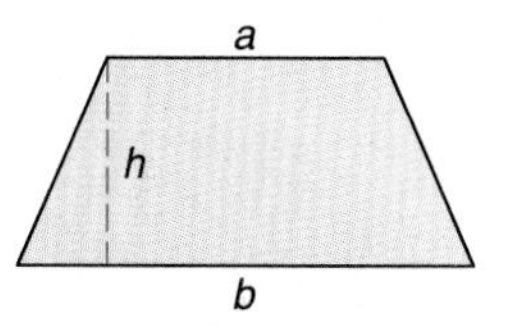

Applications

54. **Sports** The perimeter of a football field is 1040 feet. The length of the field is 120 feet less than 3 times the width. What are the dimensions of the field?

55. **Construction** Mr. Herrera had a concrete sidewalk built on three sides of his yard as shown at the right. The yard measures 24 feet by 42 feet. The longer walk is 3 feet wide. The price of the concrete was $22 per square yard, and the total bill was $902. What is the width of the walk on the remaining two sides?

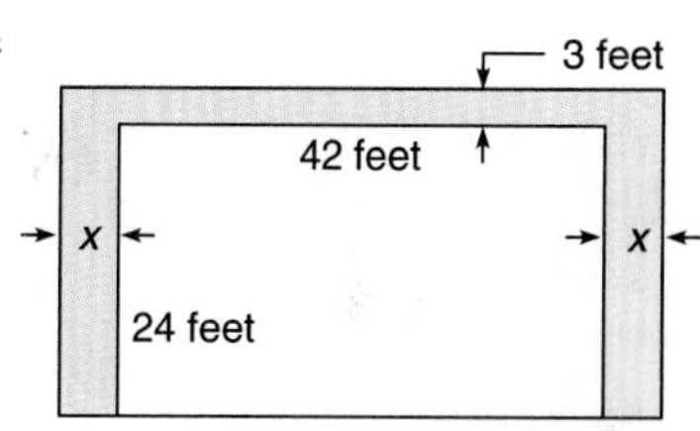

Mixed Review

56. Find $-7.9 + 3.5 + 2.4$. **(Lesson 2-5)**

57. Find $\left(-\frac{1}{3}\right)\left(-\frac{3}{4}\right)\left(-\frac{4}{5}\right)$. **(Lesson 2-6)**

58. Solve $x + \frac{4}{9} = -\frac{2}{27}$. **(Lesson 3-2)**

59. **Finance** Patricia invested $5000 for one year. Martin also invested $5000 for one year. Martin's account earned interest at a rate of 10% per year. At the end of the year, Martin's account earned $125 more than Patricia's account. What was the annual interest rate on Patricia's account? **(Lesson 4-3)**

60. Solve $|7 - x| \geq 4$. Then graph the solution set. **(Lesson 5-6)**

61. Find $(4a + 6b) + (2a + 3b)$. **(Lesson 6-6)**

6-8 Multiplying Polynomials

Objectives

After studying this lesson, you should be able to:

- use the FOIL method to multiply two binomials, and
- multiply any two polynomials by using the distributive property.

Connection

You know that the area of a rectangle is the product of its length and width. You can multiply $2x + 3$ and $5x + 8$ to find the area of the large rectangle.

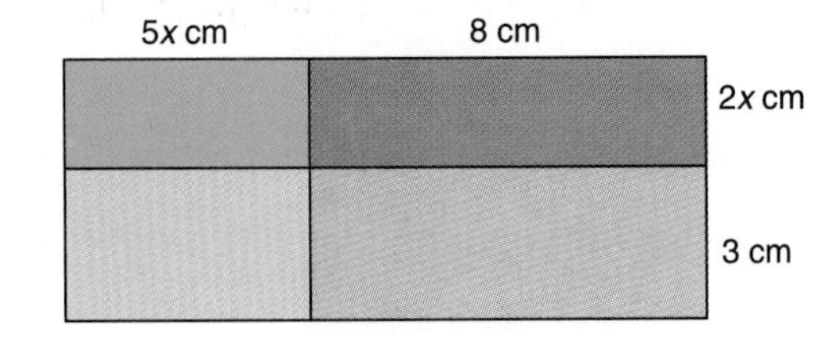

$$\begin{aligned}(2x + 3)(5x + 8) &= 2x(5x + 8) + 3(5x + 8) \quad \textit{Distributive property}\\ &= (2x)(5x) + (2x)(8) + (3)(5x) + (3)(8)\\ &= 10x^2 + 16x + 15x + 24\\ &= 10x^2 + 31x + 24\end{aligned}$$

But you also know that the area of the large rectangle equals the sum of the areas of the four smaller rectangles.

$$\begin{aligned}(2x + 3)(5x + 8) &= 2x \cdot 5x + 2x \cdot 8 + 3 \cdot 5x + 3 \cdot 8\\ &= 10x^2 + 16x + 15x + 24\\ &= 10x^2 + 31x + 24\end{aligned}$$

Lab Activity

You can learn how to use algebra tiles to multiply polynomials in Lab 8 on pages A13-A14.

This example illustrates a shortcut of the distributive property called the **FOIL method.**

F L I O

$$(2x + 3)(5x + 8) = \underset{\substack{\uparrow\\ \textbf{F}}}{(2x)(5x)} + \underset{\substack{\uparrow\\ \textbf{O}}}{(2x)(8)} + \underset{\substack{\uparrow\\ \textbf{I}}}{(3)(5x)} + \underset{\substack{\uparrow\\ \textbf{L}}}{(3)(8)}$$

F	O	I	L
product of FIRST terms	product of OUTER terms	product of INNER terms	product of LAST terms

$$\begin{aligned}&= 10x^2 + 16x + 15x + 24\\ &= 10x^2 + 31x + 24\end{aligned}$$

FOIL Method for Multiplying Two Binomials

To multiply two binomials, find the sum of the products of

F the first terms,
O the outer terms,
I the inner terms, and
L the last terms.

Example 1

Find $(y + 5)(y + 7)$.

$$\begin{aligned}
&\qquad\qquad\quad F \qquad\quad O \qquad\quad I \qquad\quad L\\
(y + 5)(y + 7) &= y \cdot y + y \cdot 7 + y \cdot 5 + 5 \cdot 7\\
&= y^2 + 7y + 5y + 35\\
&= y^2 + 12y + 35 \quad \textit{Combine like terms.}
\end{aligned}$$

Example 2

Find $(3x - 5)(5x + 2)$.

$$\begin{aligned}
&\qquad\qquad\quad F \qquad\qquad O \qquad\qquad I \qquad\qquad L\\
(3x - 5)(5x + 2) &= (3x)(5x) + (3x)(2) + (-5)(5x) + (-5)(2)\\
&= 15x^2 + 6x - 25x - 10\\
&= 15x^2 - 19x - 10
\end{aligned}$$

The distributive property can be used to multiply any two polynomials.

Example 3

Find $(2x - 5)(3x^2 - 5x + 4)$.

$$\begin{aligned}
&(2x - 5)(3x^2 - 5x + 4)\\
&\quad = 2x(3x^2 - 5x + 4) - 5(3x^2 - 5x + 4) && \textit{Distributive property}\\
&\quad = (6x^3 - 10x^2 + 8x) - (15x^2 - 25x + 20)\\
&\quad = 6x^3 - 10x^2 + 8x + (-15x^2 + 25x - 20) && \textit{Additive inverse}\\
&\quad = 6x^3 + (-10x^2 - 15x^2) + (8x + 25x) - 20\\
&\quad = 6x^3 - 25x^2 + 33x - 20
\end{aligned}$$

Example 4

Find $(x^2 - 5x + 4)(2x^2 + x - 7)$.

$$\begin{aligned}
&(x^2 - 5x + 4)(2x^2 + x - 7)\\
&\quad = x^2(2x^2 + x - 7) - 5x(2x^2 + x - 7) + 4(2x^2 + x - 7)\\
&\quad = 2x^4 + x^3 - 7x^2 - 10x^3 - 5x^2 + 35x + 8x^2 + 4x - 28\\
&\quad = 2x^4 - 9x^3 - 4x^2 + 39x - 28
\end{aligned}$$

Polynomials can also be multiplied in column form. Be careful to align the like terms.

Example 5

Find $(x^3 + 5x - 6)(2x - 9)$ in column form.

Since there is no x^2 term in $x^3 + 5x - 6$, $0x^2$ is used as a placeholder.

$$\begin{array}{rrrrrl}
 & x^3 & + 0x^2 & + 5x & - 6 & \\
(\times) & & & 2x & - 9 & \\
\hline
 & -9x^3 & - 0x^2 & - 45x & + 54 & \textit{This is the product of } x^3 + 5x - 6 \textit{ and } -9.\\
2x^4 & + 0x^3 & + 10x^2 & - 12x & & \textit{This is the product of } x^3 + 5x - 6 \textit{ and } 2x.\\
\hline
2x^4 & - 9x^3 & + 10x^2 & - 57x & + 54 & \textit{This is the sum of the partial products.}
\end{array}$$

Example 6

Maria Orozco has a rectangular garden that is 10 feet longer than it is wide. A brick path 3 feet in width surrounds the garden. The total area of the path is 396 square feet. What are the dimensions of the garden?

EXPLORE Draw a diagram to represent the situation. Let w = the width of the garden, $w + 10$ = the length of the garden, $w + 6$ = the width of the garden and path, and $w + 16$ = the length of the garden and path.

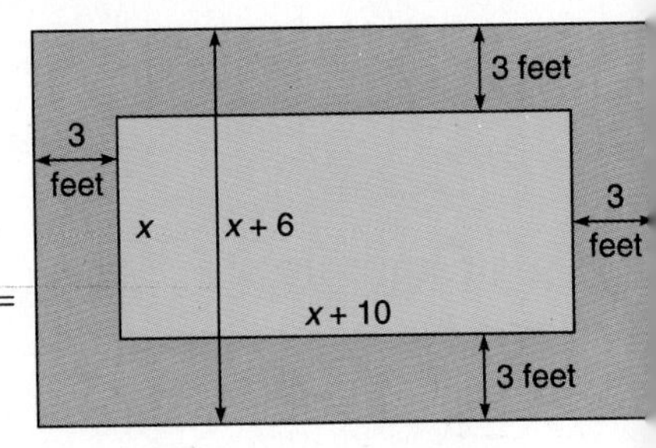

Then $w(w + 10)$ = the area of the garden, and $(w + 6)(w + 16)$ = the area of the garden and path.

PLAN

$$\underbrace{\text{area of garden and path}} - \underbrace{\text{area of garden}} = \underbrace{\text{area of path}}$$

$$(w + 6)(w + 16) - w(w + 10) = 396$$

SOLVE

$$w^2 + 16w + 6w + 96 - w^2 - 10w = 396 \quad \textit{Use the FOIL method.}$$
$$(w^2 - w^2) + (16w + 6w - 10w) + 96 = 396$$
$$12w + 96 = 396 \quad \textit{Combine like terms.}$$
$$12w = 300$$
$$w = 25$$

The width is 25 feet and the length is $w + 10$ or 35 feet.

EXAMINE To examine the solution, compute the total area of the garden and path in two ways and compare the results.

$$\underbrace{\text{area of garden}} + \underbrace{\text{area of path}} = \underbrace{\text{total area}}$$

$$25 \times 35 + 396 = 1271$$

$$\underbrace{\text{total length}} \times \underbrace{\text{total width}} = \underbrace{\text{total area}}$$

$$(35 + 6) \times (25 + 6) = 41 \times 31 \text{ or } 1271$$

Thus, the solution checks.

CHECKING FOR UNDERSTANDING

Communicating Mathematics

Read and study the lesson to answer each question.

1. In the FOIL method, what do the letters F, O, I, and L stand for?
2. Use the FOIL method to evaluate $4\frac{1}{2} \cdot 6\frac{3}{4}$. (*Hint:* Rewrite as $\left(4 + \frac{1}{2}\right)\left(6 + \frac{3}{4}\right)$.)

Guided Practice

Find the sum of the product of the inner terms and the product of the outer terms.

3. $(x + 5)(x + 3)$
4. $(2a + 1)(a + 5)$
5. $(x + 5)(5x - 3)$
6. $(3x + 2)(2x + 3)$
7. $(5b - 3)(2b + 1)$
8. $(2m + 4)(m + 5)$

Find each product.

9. $(a + 3)(a + 7)$
10. $(m - 5)(m - 11)$
11. $(x + 11)(x - 4)$
12. $(2x + 1)(x + 8)$

EXERCISES

Practice

Find each product.

13. $(c + 2)(c + 8)$
14. $(x - 4)(x - 8)$
15. $(y + 3)(y - 7)$
16. $(5y - 3)(y + 2)$
17. $(4a + 3)(2a - 1)$
18. $(7y - 1)(2y - 3)$
19. $(2x + 3y)(5x + 2y)$
20. $(2a + 3b)(5a - 2b)$
21. $(5q + 2r)(8q - 3r)$
22. $(5r - 7s)(4r + 3s)$
23. $(2r + 0.1)(5r - 0.3)$
24. $(0.7x + 2y)(0.9x + 3y)$
25. $\left(3x + \frac{1}{4}\right)\left(6x - \frac{1}{2}\right)$
26. $(x - 2)(x^2 + 2x + 4)$
27. $\begin{array}{r} x^2 + 7x - 9 \\ (\times) \quad 2x + 1 \\ \hline \end{array}$
28. $\begin{array}{r} a^2 - 3a + 11 \\ (\times) \quad 5a + 2 \\ \hline \end{array}$
29. $\begin{array}{r} 3x^2 + 5xy + y^2 \\ (\times) \quad 4x - 3y \\ \hline \end{array}$
30. $\begin{array}{r} 6x^2 - 5xy + 9y^2 \\ (\times) \quad 5x - 2y \\ \hline \end{array}$
31. $(3x + 5)(2x^2 - 5x + 11)$
32. $(4s + 5)(3s^2 + 8s - 9)$
33. $(3a + 5)(2a - 8a^2 + 3)$
34. $(5x - 2)(7 - 5x^2 + 2x)$
35. $\begin{array}{r} 5x^2 - 6x + 9 \\ (\times) \; 4x^2 + 3x + 11 \\ \hline \end{array}$
36. $\begin{array}{r} 5x^4 + 0x^3 - 2x^2 + 1 \\ (\times) \quad x^2 - 5x + 3 \\ \hline \end{array}$
37. $(x^2 - 7x + 4)(2x^2 - 3x - 6)$
38. $(a^2 + 2a + 5)(a^2 - 3a - 7)$
39. $(-2x^2 + 3x - 8)(3x^2 + 7x - 5)$
40. $(-7b^3 + 2b - 3)(5b^2 - 2b + 4)$

Critical Thinking

If $R = 2x - 1$, $S = 3x + 2$, and $T = -3x^2$, find each of the following.

41. $R \cdot S$
42. $TS - R$
43. $R(S + T)$
44. $3R(2S + 5T)$

Applications

45. **Gardening** A rectangular garden is 5 feet longer than twice its width. It has a sidewalk 3 feet wide on two of its sides, as shown at the right. The area of the sidewalk is 213 square feet. Find the dimensions of the garden.

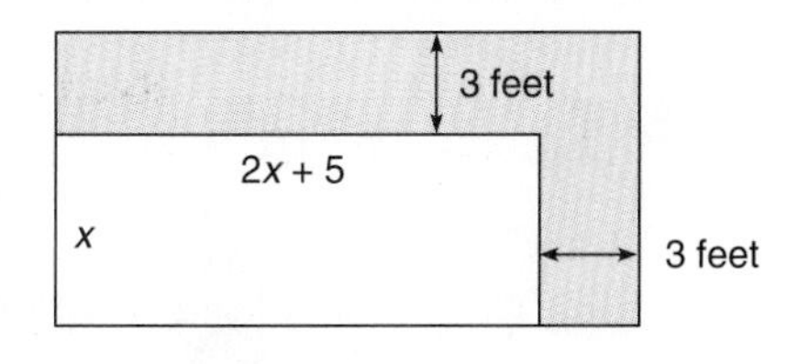

46. **Construction** The length of a rectangular lot is 7 yards less than twice its width. If the length was increased by 11 yards and the width decreased by 6 yards, the area would be decreased by 40 square yards. Find the original dimensions of the lot.

Mixed Review

47. Solve by working backwards. **(Lesson 3-4)**

Jack, Jared, and Jason are playing a game in which two of them win and one of them loses on each play. The one who loses takes some of his own points and gives them to each winner so that their points are doubled. The boys play the game three times; each wins twice and loses once. At the end of three plays, each boy has 40 points. How many points did each player have at the beginning of the game?

Portfolio

Select an item from this chapter that you feel shows your best work and place it in your portfolio. Explain why you selected it.

48. **Sales** Sales tax of 6% is added to a purchase of \$11. Find the total price. **(Lesson 4-4)**

49. Solve $14 + 7x > 8x$. **(Lesson 5-1)**

50. Simplify $\frac{ab^5c}{ac}$. Assume that the denominator is not equal to zero. **(Lesson 6-3)**

51. Multiply $\frac{2}{3}a(6a + 15)$. **(Lesson 6-7)**

EXTRA

Another way to multiply two polynomials is to use an array. Write the coefficients of the first polynomial from left to right as the headings of the columns. Then write the coefficients of the second polynomial as the headings of the rows. Next fill in the array with the products of the column headings and the row headings. Finally, add diagonally as shown. These sums will be the coefficients of the product. Study the example below.

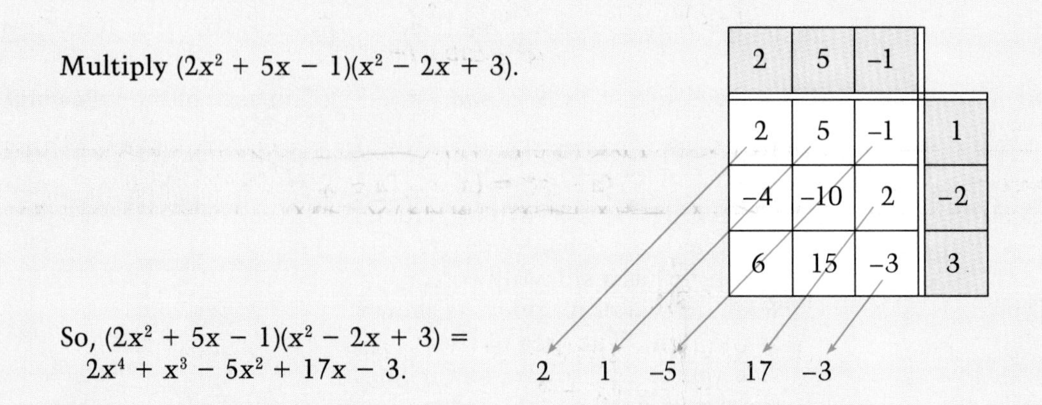

6-9 Some Special Products

Objective After studying this lesson, you should be able to:

- use the patterns for $(a + b)^2$, $(a - b)^2$, and $(a + b)(a - b)$.

Connection Study the diagram below. There are two ways to find the area of the large square.

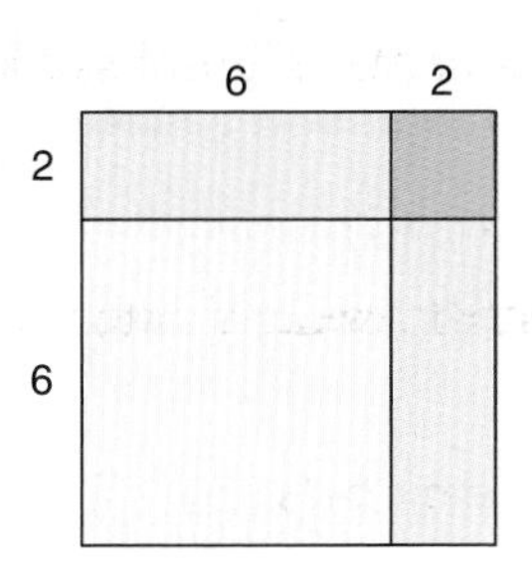

Method 1: The length of each side is 6 + 2 units. The area is the product of (6 + 2) and (6 + 2) or $(6 + 2)^2$.

$$(6 + 2)^2 = 8^2 = 64$$

Method 2: The area of the large square is the sum of the areas of the four smaller rectangles.

$$6^2 + 6 \cdot 2 + 6 \cdot 2 + 2^2 = 36 + 12 + 12 + 4 = 64$$

Thus, $(6 + 2)^2 = 6^2 + 6 \cdot 2 + 6 \cdot 2 + 2^2$.

Using a procedure similar to the one described above, we can derive a general form for the expression $(a + b)^2$.

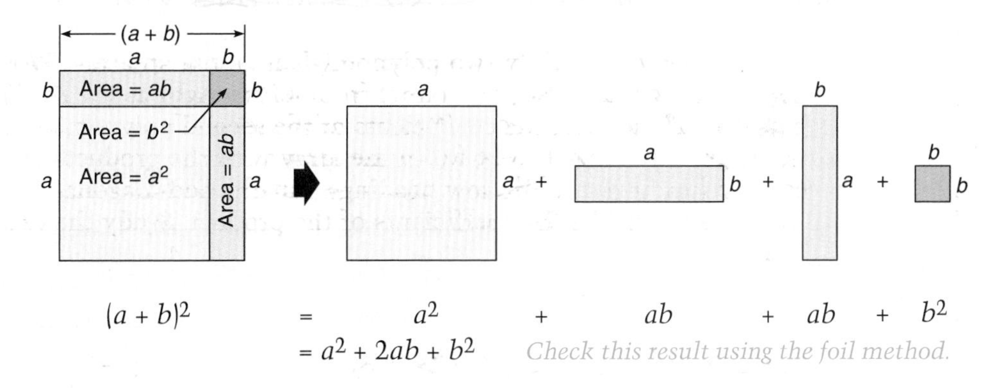

$$(a + b)^2 = a^2 + ab + ab + b^2$$
$$= a^2 + 2ab + b^2$$

Check this result using the foil method.

Square of a Sum	$(a + b)^2 = (a + b)(a + b) = a^2 + 2ab + b^2$

Example 1 **Find $(x + 3)^2$.**

$(x + 3)^2 = x^2 + 2(x)(3) + 3^2$ *Replace a with x and b with 3.*

$= x^2 + 6x + 9$

Example 2 **Find $(5m + 3n)^2$.**

$$(5m + 3n)^2 = (5m)^2 + 2(5m)(3n) + (3n)^2 \quad \text{Replace } a \text{ with } 5m \text{ and } b \text{ with } 3n.$$
$$= 25m^2 + 30mn + 9n^2$$

To find $(a - b)^2$, express $(a - b)$ as $[a + (-b)]$ and square it.

$$\begin{aligned}(a - b)^2 &= [a + (-b)]^2 \\ &= a^2 + (2)(a)(-b) + (-b)^2 \\ &= a^2 - 2ab + b^2\end{aligned}$$

Square of a Difference	$(a - b)^2 = (a - b)(a - b) = a^2 - 2ab + b^2$

Example 3 **Find $(c - 2)^2$.**

$$(c - 2)^2 = c^2 - 2(c)(2) + (2)^2 \quad \text{Replace } a \text{ with } c \text{ and } b \text{ with } 2.$$
$$= c^2 - 4c + 4$$

Example 4 **Find $(3x - 2y)^2$.**

$$(3x - 2y)^2 = (3x)^2 - 2(3x)(2y) + (2y)^2 \quad \text{Replace } a \text{ with } 3x \text{ and } b \text{ with } 2y.$$
$$= 9x^2 - 12xy + 4y^2$$

You can use the FOIL method to find the product of a sum and a difference of the same two numbers.

$$\begin{aligned}(a + b)(a - b) &= (a)(a) + (a)(-b) + (a)(b) + (b)(-b) \\ &= a^2 - ab + ab - b^2 \\ &= a^2 - b^2\end{aligned}$$

This product is called the **difference of squares.**

Product of a Sum and a Difference	$(a + b)(a - b) = (a - b)(a + b) = a^2 - b^2$

Example 5 **Find $(x + 5)(x - 5)$.**

$$(x + 5)(x - 5) = x^2 - 5^2 \quad \text{Replace } a \text{ with } x \text{ and } b \text{ with } 5.$$
$$= x^2 - 25$$

Example 6 **Find $(5a + 6b)(5a - 6b)$.**

$$(5a + 6b)(5a - 6b) = (5a)^2 - (6b)^2 \quad \text{Replace } a \text{ with } 5a \text{ and } b \text{ with } 6b.$$
$$= 25a^2 - 36b^2$$

CHECKING FOR UNDERSTANDING

Communicating Mathematics

Read and study the lesson to answer each question.

1. Compare the square of a sum and the square of a difference. How do they vary?
2. Compare the square of a difference and the difference of two squares.

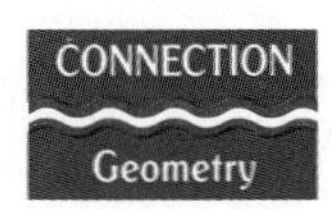

3. What does the diagram at the right represent if the shading represents regions to be removed or subtracted?

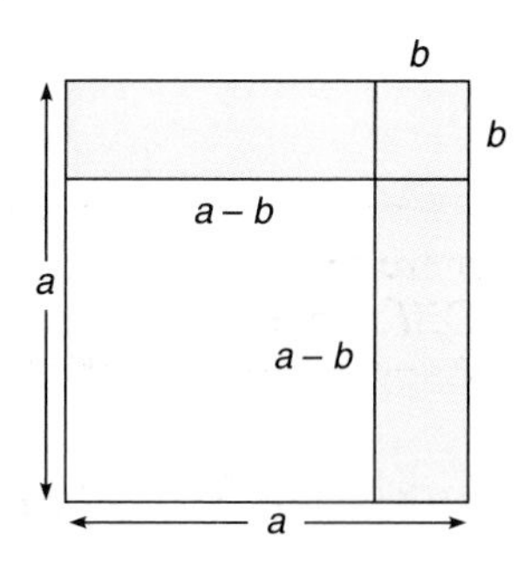

Guided Practice

Find each product.

4. $(a + 2b)^2$
5. $(a - 3b)^2$
6. $(2x + y)^2$
7. $(3x - 2y)^2$
8. $(2a + 3)(2a - 3)$
9. $(5a - 3b)(5a + 3b)$

EXERCISES

Practice

Find each product.

10. $(4x + y)^2$
11. $(2a - b)^2$
12. $(6m + 2n)^2$
13. $(4x - 9y)^2$
14. $(5a - 12b)^2$
15. $(5x + 6y)^2$
16. $\left(\frac{1}{2}a + b\right)^2$
17. $(5 - x)^2$
18. $(1 + x)^2$
19. $(1.1x + y)^2$
20. $(a^2 - 3b^2)^2$
21. $(x^3 - 5y^2)^2$
22. $(3x + 5)(3x - 5)$
23. $(8a - 2b)(8a + 2b)$
24. $(7a^2 + b)(7a^2 - b)$
25. $(8x^2 - 3y)(8x^2 + 3y)$
26. $\left(\frac{4}{3}x^2 - y\right)\left(\frac{4}{3}x^2 + y\right)$
27. $(x + 2)(x - 2)(2x + 5)$
28. $(4x - 1)(4x + 1)(x - 4)$
29. $(x - 3)(x + 4)(x + 3)(x - 4)$
30. $(x - 2y)^3$
31. $(2x - 3y)^3$
32. $(a + b)^4$
33. $(2x - y)^4$
34. $(3m + 2n)^4$
35. $(a - b)^5$

Critical Thinking

36. The diagram at the right represents $(a + b + c)^2$. Use the diagram to write the measure of the area as a polynomial.

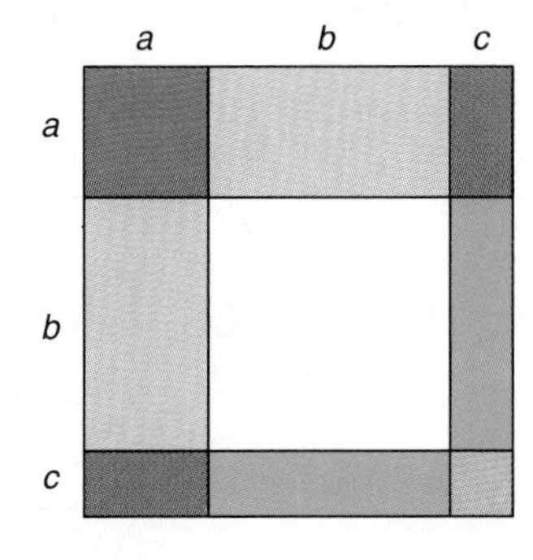

Applications

37. **City Planning** A certain section of town is shaped like a trapezoid with an area of 81 square miles. The distance between Union St. and Lee St. is 9 miles. The length of Union St. is 14 miles less than 3 times the length of Lee St. Find the length of Lee St. Use $A = \frac{1}{2}h(a + b)$.

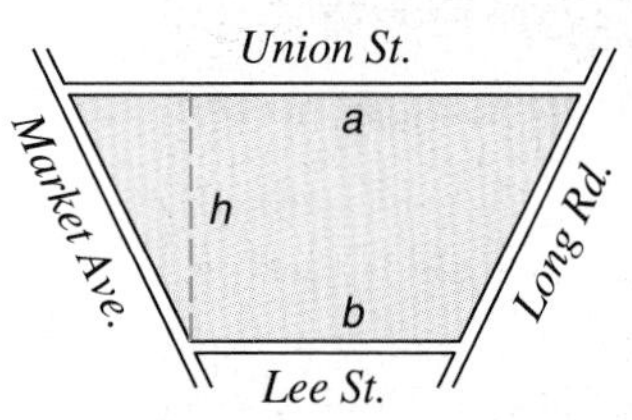

38. **Photography** To get a square photograph to fit into a square frame, Linda LaGuardia had to trim a 1-inch strip from each side of the photo, as shown at the right. In all, she trimmed off 40 square inches. What were the original dimensions of the photograph?

Mixed Review

39. Solve $\frac{4y + 3}{7} = \frac{9}{14}$. **(Lesson 3-7)**

40. **Sales** On the first day of school, 264 school notebooks were sold. Some sold for 95¢ each and the rest sold for \$1.25 each. How many of each were sold if the total sales were \$297? **(Lesson 4-6)**

41. Is the statement $3 - 1\frac{1}{2} = \frac{1}{2}$ *true* or *false*? **(Lesson 1-3)**

APPLICATION

Punnett Squares

Punnett squares are used to show possible ways that genes can combine at fertilization. In a Punnett square, *dominant* genes are shown with capital letters. *Recessive* genes are shown with lowercase letters. Letters representing the parents' genes are placed on the outer sides of the Punnett square. Letters inside the boxes of the square show the possible gene combinations for their offspring.

The Punnett square at the right represents a cross between tall pea plants and short pea plants. Let T represent the dominant gene for tallness. Let t represent the recessive gene for shortness. Since the parents have one of each type of gene, they are called *hybrids*.

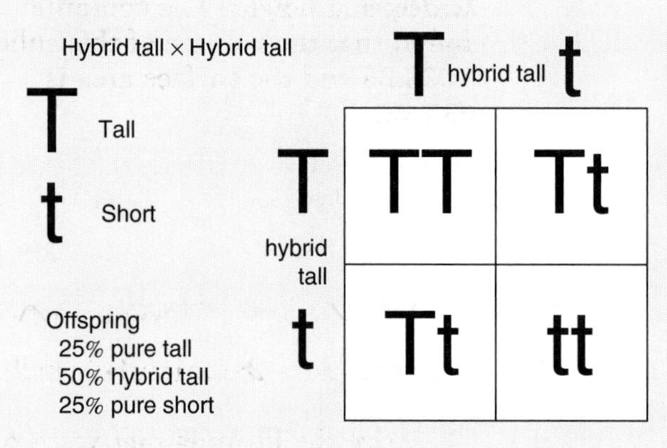

Because the parent plants have both a dominant tall gene and a recessive short gene, their offspring can be predicted as follows: 1 TT (tall), 2 Tt (tall with a recessive short gene), and 1 tt (short). Do you notice the similarity between this Punnett square and the square of a polynomial $(a + b)^2 = a \cdot a + 2ab + b \cdot b$?

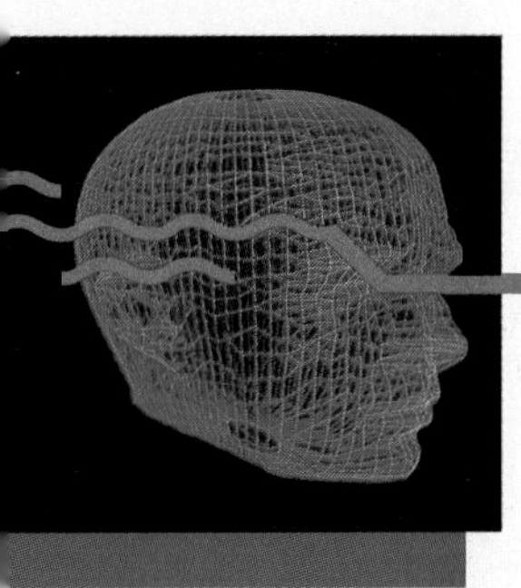

Technology

Volume and Surface Area

BASIC
Graphing calculator
Graphing software
▶ Spreadsheets

As you learned in Chapter 1, a spreadsheet is a computer program that allows the user to easily prepare tables and charts. You can use a spreadsheet to project results, make calculations, and print anything that can be organized in a table.

The spreadsheet below is set up to find the volume and surface area of a rectangular solid. To use the spreadsheet, you would enter the length, width, and height and the computer would find the volume and the surface area of the solid. The formula in cell B4 tells the computer to multiply the values in cells B1, B2, and B3 together.

```
======A==========B======================C
1: LENGTH        B1
2: WIDTH         B2
3: HEIGHT        B3
4: VOLUME        B1*B2*B3
5: SURFACE AREA  2B1*B2+2B1*B3+2B2*B3
```

A printout of the spreadsheet for finding the volume and surface area of a rectangular solid is shown at the right. The values 14.2, 7.5, and 9.7 have been entered for the length, width, and height. The computer has found that the volume of this solid is 1033.05 and the surface area is 633.98.

VOLUME AND SURFACE AREA OF A RECTANGULAR SOLID		
	A	B
1	LENGTH	14.2
2	WIDTH	7.5
3	HEIGHT	9.7
4	VOLUME	1033.05
5	SURFACE AREA	633.98

EXERCISES

1. What does the formula in cell B5 tell the computer to do?

2. Write the formula that you would enter to have the computer find the volume and the surface area of a cube if the length of a side is entered in cell B5.

3. How could you modify the program to find the area and perimeter of a rectangle?

CHAPTER 6

SUMMARY AND REVIEW

VOCABULARY

Upon completing this chapter, you should be familiar with the following terms:

binomial	**226**	**213**	monomial
constants	**213**	**226**	polynomial
decimal notation	**221**	**221**	scientific notation
degree	**227**	**218**	simplify
FOIL method	**238**	**226**	trinomial

SKILLS AND CONCEPTS

OBJECTIVES AND EXAMPLES

Upon completing this chapter, you should be able to:

- multiply monomials and simplify expressions involving powers of monomials **(Lesson 6-2)**

$$(2ab^2)(3a^2b^3) = (2 \cdot 3)(a \cdot a^2)(b^2 \cdot b^3) = 6a^3b^5$$

$$(2x^2y^3)^3 = 2^3(x^2)^3(y^3)^3 = 8x^6y^9$$

REVIEW EXERCISES

Use these exercises to review and prepare for the chapter test.

Simplify.

1. $y^3 \cdot y^3 \cdot y$
2. $(3ab)(-4a^2b^3)$
3. $(-4a^2x)(-5a^3x^4)$
4. $(4a^2b)^3$
5. $(-3xy)^2(4x)^3$
6. $(-2c^2d)^4(-3c^2)^3$
7. $-\frac{1}{2}(m^2n^4)^2$
8. $(5a^2)^3 + 7(a^6)$

- simplify expressions involving quotients of monomials and simplify expressions containing negative exponents **(Lesson 6-3)**

$$\frac{2x^6y}{8x^2y^2} = \frac{2}{8} \cdot \frac{x^6}{x^2} \cdot \frac{y}{y^2} = \frac{x^4}{4y}$$

$$\frac{3a^{-2}}{4a^6} = \frac{3}{4}(a^{-2-6}) = \frac{3}{4a^8}$$

Simplify. Assume no denominator is equal to zero.

9. $\frac{y^{10}}{y^6}$
10. $\frac{(3y)^0}{6a}$
11. $\frac{42b^7}{14b^4}$
12. $\frac{27b^{-2}}{14b^{-3}}$
13. $\frac{(3a^3bc^2)^2}{18a^2b^3c^4}$
14. $\frac{-16a^3b^2x^4y}{-48a^4bxy^3}$

OBJECTIVES AND EXAMPLES	REVIEW EXERCISES

■ express numbers in scientific and decimal notation **(Lesson 6-4)**

$3{,}600{,}000 = 3.6 \times 10^6$

$0.0021 = 2.1 \times 10^{-3}$

Express each number in scientific notation.

15. 240,000
16. 4,880,000,000
17. 0.000314
18. 0.00000187

■ find products and quotients of numbers expressed in scientific notation **(Lesson 6-4)**

$$\begin{aligned}(2 \times 10^2)(5.2 \times 10^6) &= (2 \times 5.2)(10^2 \times 10^6)\\ &= 10.4 \times 10^8\\ &= 1.04 \times 10^9\end{aligned}$$

$$\begin{aligned}\frac{1.2 \times 10^{-2}}{0.6 \times 10^3} &= \frac{1.2}{0.6} \times \frac{10^{-2}}{10^3}\\ &= 2 \times 10^{-5}\end{aligned}$$

Evaluate. Express each result in scientific notation.

19. $(2 \times 10^5)(3 \times 10^6)$
20. $(3 \times 10^3)(1.5 \times 10^6)$
21. $\dfrac{5.4 \times 10^3}{0.9 \times 10^4}$
22. $\dfrac{8.4 \times 10^{-6}}{1.4 \times 10^{-9}}$

■ find the degree of a polynomial **(Lesson 6-5)**

Find the degree of $2xy^3 + x^2y$.

degree of $2xy^3$: $1 + 3$ or 4
degree of x^2y: $2 + 1$ or 3
degree of $2xy^3 + x^2y$: 4

Find the degree of each polynomial.

23. $n - 2p^2$
24. $29n^2 + 17n^2t^2$
25. $4xy + 9xz^2 + 17rs^3$
26. $-6y - 2y^4 + 4 - 8y^2$

■ arrange the terms of a polynomial so that the powers of a certain variable are in ascending or descending order **(Lesson 6-5)**

Arrange the terms of $4x^2 + 9x^3 - 2 - x$ in descending order.

$9x^3 + 4x^2 - x - 2$

Arrange the terms of each polynomial so that the powers of x are in descending order.

27. $3x^4 - x + x^2 - 5$
28. $ax^2 - 5x^3 + a^2x - a^3$

■ add and subtract polynomials **(Lesson 6-6)**

$$\begin{array}{r} 4x^2 - 3x + 7 \\ (+)\ 2x^2 + 4x \phantom{{}+7} \\ \hline 6x^2 + x + 7 \end{array}$$

$$\begin{aligned}(7r^2 + 9r) - (12r^2 - 4) &= 7r^2 + 9r - 12r^2 + 4\\ &= (7r^2 - 12r^2) + 9r + 4\\ &= -5r^2 + 9r + 4\end{aligned}$$

Find each sum or difference.

29. $(2x^2 - 5x + 7) - (3x^3 + x^2 + 2)$
30. $(x^2 - 6xy + 7y^2) + (3x^2 + xy - y^2)$
31. $$\begin{array}{r} 11m^2n^2 + 4mn - 6 \\ (+)\ \ 5m^2n^2 - 6mn + 17 \\ \hline \end{array}$$
32. $$\begin{array}{r} 7z^2 \phantom{{}+ 2z} + 4 \\ (-)\ 3z^2 + 2z - 6 \\ \hline \end{array}$$

OBJECTIVES AND EXAMPLES	REVIEW EXERCISES

- multiply a polynomial by a monomial and simplify expressions involving polynomials **(Lesson 6-7)**

$x^2(x + 2) + 3(x^3 + 4x^2)$
$= x^3 + 2x^2 + 3x^3 + 12x^2$
$= 4x^3 + 14x^2$

Simplify.

33. $4ab(3a^2 - 7b^2)$
34. $7xy(x^2 + 4xy - 8y^2)$
35. $x(3x - 5) + 7(x^2 - 2x + 9)$
36. $4x^2(x + 8) - 3x(2x^2 - 8x + 3)$

- use the FOIL method to multiply two binomials and multiply any two polynomials by using the distributive property **(Lesson 6-8)**

$(3x + 2)(x - 2)$
$= (3x)(x) + (3x)(-2) + 2(x) + 2(-2)$
$= 3x^2 - 6x + 2x - 4$
$= 3x^2 - 4x - 4$

Find each product.

37. $(r - 3)(r + 7)$
38. $(x + 5)(3x - 2)$
39. $(4x - 3)(x + 4)$
40. $(2x + 5y)(3x - y)$
41. $(x - 4)(x^2 + 5x - 7)$
42. $(2a + b)(a^2 - 17ab - 3b^2)$

- use the patterns for $(a + b)^2$, $(a - b)^2$, and $(a + b)(a - b)$ **(Lesson 6-9)**

$(x + 4)^2 = x^2 + 2(4x) + 4^2$
$= x^2 + 8x + 16$

$(r - 5)^2 = r^2 - 2(5r) + 5^2$
$= r^2 - 10r + 25$

$(b + 9)(b - 9) = b^2 - 9^2$
$= b^2 - 81$

Find each product.

43. $(x - 6)(x + 6)$
44. $(5x - 3y)(5x + 3y)$
45. $(4x + 7)^2$
46. $(a^2 + b)^2$
47. $(8x - 5)^2$
48. $(6a - 5b)^2$

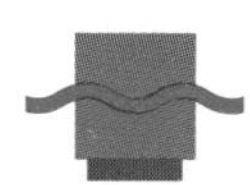

APPLICATIONS AND CONNECTIONS

Solve by looking for a pattern. (Lesson 6-1)

49. Male bees have only one parent, a mother, while female bees have both a mother and a father. How many grandparents does a male bee have? How many great-grandparents? If you continue to look at the bee's ancestry, what pattern do you notice?

50. The night of Ann's party, her doorbell rang 10 times. The first time, one guest arrived. Then, each time the doorbell rang, two more guests arrived than had arrived on the previous ring. How many guests were at Ann's party?

51. **Geometry** Find the measure of the third side of the triangle below if $P = 3x^2 - 12$. **(Lesson 6-6)**

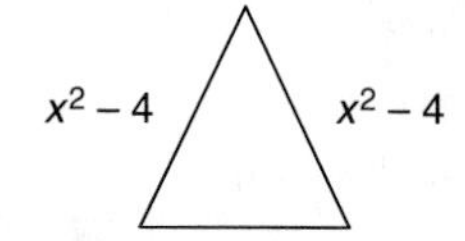

52. **Geometry** Find the measure of the area of the shaded region in the figure below. **(Lesson 6-7)**

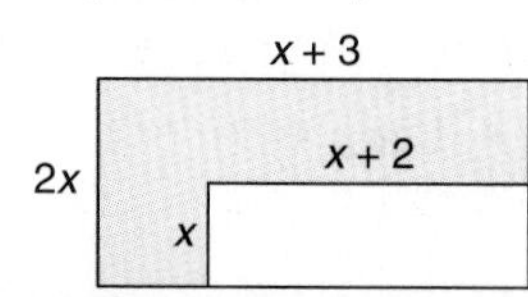

CHAPTER 6 TEST

Simplify. Assume no denominator is equal to zero.

1. $a^2 \cdot a^3 \cdot b^4 \cdot b^5$
2. $(-12abc)(4a^2b^3)$
3. $(9a)^2$
4. $(-3a)^4(a^5b)$
5. $(-5a^2)(-6b)^2$
6. $(5a)^2b + 7a^2b$
7. $\frac{y^{11}}{y^6}$
8. $\frac{y^3x}{yx}$
9. $\frac{63a^2bc}{9abc}$
10. $\frac{48a^2bc^5}{(3ab^3c^2)^2}$
11. $\frac{14ab^{-3}}{21a^2b^{-5}}$
12. $\frac{10a^2bc}{20a^{-1}b^{-1}c}$

Express in scientific notation.

13. 52,800
14. 0.00378

Evaluate. Express each result in scientific notation.

15. $(3 \times 10^3)(2 \times 10^4)$
16. $(4 \times 10^{-3})(3 \times 10^{16})$
17. $\frac{2.5 \times 10^3}{5 \times 10^{-3}}$
18. $\frac{91 \times 10^{18}}{13 \times 10^{14}}$

Arrange the terms of each polynomial so that the powers of *x* are in descending order.

19. $5x^2 - 3 + x^3 + 5x$
20. $5 - xy^3 + x^3y^2 - x^2$

Simplify.

21. $(a + 5)^2$
22. $(2x - 5)(7x + 3)$
23. $(3a^2 + 3)[2a - (-6)]$
24. $3x^2y^3(2x - xy^2)$
25. $-4xy(5x^2 - 6xy^3 + 2y^2)$
26. $(4x^2 - y^2)(4x^2 + y^2)$
27. $0.3b(0.4b^2 - 0.7b + 4)$
28. $(2a^2b + b^2)^2$
29. $x^2(x - 8) - 3x(x^2 - 7x + 3) + 5(x^3 - 6x^2)$
30. $a^2(a + 5) + 7a(a^2 + 8) - 7(a^3 - a + 2)$

Solve each equation.

31. $5y - 8 - 13y = 12y + 6$
32. $2(a + 2) + 3a = 13 - (2a + 2)$

33. **Geometry** The length of a rectangle is eight times its width. If the length was decreased by 10 meters and the width was decreased by 2 meters, the area would be decreased by 162 square meters. Find the original dimensions.

Bonus Find n if $2^{n+3} \cdot 2^{3n-2} \cdot 2^{5n+1} = 32$. (*Hint:* Write 32 as a power of 2.)

CHAPTER 6

College Entrance Exam Preview

The test questions on these pages deal with fraction concepts, ratios, proportions, and percents.

Directions: Choose the best answer. Write A, B, C, or D.

1. In a class of 27 students, six are honor students. What part of the class is *not* honor students?

   (A) $\frac{7}{11}$ (B) $\frac{7}{9}$ (C) $\frac{2}{7}$ (D) $\frac{2}{9}$

2. The months from April to December, inclusive, are what fractional part of a year?

   (A) $\frac{7}{12}$ (B) $\frac{3}{4}$ (C) $\frac{5}{12}$ (D) $\frac{2}{3}$

3. What fractional part of a week is one sixth of a day?

   (A) $\frac{1}{42}$ (B) $\frac{1}{30}$ (C) $\frac{4}{7}$ (D) $\frac{7}{168}$

4. 0.7% is the ratio of 7 to

   (A) 100 (B) $\frac{1}{10}$
   (C) $\frac{1}{1000}$ (D) 1000

5. 37.5% of a pound is equivalent to what fractional part of a pound?

   (A) $\frac{37}{100}$ (B) $\frac{3}{8}$ (C) $\frac{37}{99}$ (D) $37\frac{1}{2}$

6. Five sixths is how many sevenths?

   (A) $5\frac{5}{7}$ (B) $8\frac{2}{5}$ (C) $5\frac{5}{6}$ (D) $4\frac{2}{7}$

7. How many elevenths is 75%?

   (A) $8\frac{1}{4}$ (B) $\frac{3}{4}$ (C) $6\frac{9}{11}$ (D) $\frac{12}{11}$

8. 90% of 270 is 2.7% of

   (A) 729 (B) 900
   (C) 2000 (D) 9000

9. A box contains 60 red, blue, and green pens. If 35% are red and 9 pens are blue, what percent are green?

   (A) 40% (B) 50% (C) 60% (D) 65%

10. Doralina wishes to run a certain distance in 20% less time than she usually takes. By what percent must she increase her overall average speed?

    (A) 5% (B) 20% (C) 25% (D) 50%

11. 150% of $5c$ is b. What percent of $2b$ is c?

    (A) $5\frac{1}{2}$% (B) $6\frac{2}{3}$% (C) 15% (D) 75%

12. Two fifths times five sevenths is equal to what number times six elevenths?

    (A) $\frac{11}{21}$ (B) $\frac{55}{63}$ (C) $\frac{20}{77}$ (D) $\frac{22}{43}$

13. A woman owns $\frac{3}{4}$ of a business. She sells half of her share for \$30,000. What is the total value of the business?

    (A) \$45,000 (B) \$75,000
    (C) \$80,000 (D) \$112,000

14. On a map, a line segment 1.75 inches long represents 21 miles. What distance in miles is represented by a segment 0.6 inch long?

(A) 6 (B) 7.2

(C) 12.6 (D) 24.15

15. A cake recipe that serves 6 people calls for 2 cups of flour. How many cups of flour would be needed to bake a smaller cake that serves 4 people?

(A) $\frac{3}{4}$ (B) 1 (C) $1\frac{1}{4}$ (D) $1\frac{1}{3}$

16. The price of an item was reduced by 10%, then later reduced by 20%. The two reductions were equivalent to a single reduction of

(A) 15% (B) 28% (C) 30% (D) 70%

17. The discount price for a \$50 item is \$40. What is the discount price for a \$160 item if the rate of discount is $1\frac{1}{2}$ times the rate of the discount for the \$50 item?

(A) \$20 (B) \$48 (C) \$96 (D) \$112

18. A car was driven twice as many miles in July as in each of the other 11 months of the year. What fraction of the total mileage for the year occurred in July?

(A) $\frac{2}{11}$ (B) $\frac{2}{13}$ (C) $\frac{1}{6}$ (D) $\frac{1}{7}$

19. $\frac{68 + 68 + 68 + 68}{4} =$

(A) 17 (B) 68 (C) 136 (D) 272

20. The sum of twelve numbers is 4965. If each number is increased by 16, what is the sum of the new numbers?

(A) 192 (B) 5157 (C) 4773 (D) 4981

TEST TAKING TIP

Most standardized tests have a time limit, so you must budget your time carefully. Some questions will be much easier than others. If you cannot answer a question within a few minutes, go on to the next one. If there is still time left when you get to the end of the test, go back to the ones that you skipped.

21. Which of the following *cannot* be the average of 10, 7, 13, 2, and x if $x > 4$?

(A) 7 (B) 9 (C) 13 (D) 258

22. The average of eight integers *can* be

(A) 127.6 (B) 130.8

(C) 131.3 (D) 135.5

23. If the product of a number and b is increased by y, the result is t. Find the number in terms of b, y, and t.

(A) $\frac{t - y}{b}$ (B) $\frac{y - t}{b}$

(C) $\frac{b + y}{t}$ (D) $t - by$

24. Mark can type 60 words per minute and there is an average of 360 words per page. At this rate, how many *hours* would it take him to type k pages?

(A) $\frac{k}{6}$ (B) $\frac{k}{60}$ (C) $\frac{k}{10}$ (D) $\frac{10}{k}$

25. Eight people can sit at a square table. If three such tables are placed end-to-end to form a rectangle, how many people can be seated?

(A) 12 (B) 16 (C) 20 (D) 24

CHAPTER OBJECTIVES

In this chapter, you will:

- Find prime factorizations of integers.
- Factor polynomials.
- Solve quadratic equations using factoring and the zero product property.
- Solve problems that can be represented by quadratic equations.
- Guess and check to solve problems.

Factoring

APPLICATION IN MARINE BIOLOGY

Have you ever wondered how long a dolphin can stay in the air after jumping out of the water? In a standard aquatic park pool, dolphins can reach speeds in excess of 24 ft/s just before breaking the surface of the water. If a dolphin leaves the water traveling at 24 ft/s, then its height h, in feet, above the water after t seconds is given by the formula $h = 24t - 16t^2$.

When the dolphin returns to the water, it is 0 feet above the water. Thus, the amount of time that the dolphin stays in the air can be found by solving $0 = 24t - 16t^2$ for t. An obvious solution of the equation is $t = 0$, since after 0 seconds, the dolphin has not jumped! But how do we find the real solution?

Notice that this is a polynomial equation. In this chapter, you will learn a technique called *factoring* that may be used to solve equations like the one above.

Any trip to an aquatic park is just not complete until you have seen a dolphin leap high above the water to grab a fish or a whale splash the visitors who are brave enough to sit in the front row!

ALGEBRA IN ACTION

You can solve the problem by finding the height of the dolphin at various times using the formula $h = 24t - 16t^2$. Copy and complete the chart below. Does $h = 0$ for any value of t? If so, that value is the amount of time that our dolphin stays in the air.

t	$24t - 16t^2$	h
0.25	$24(0.25) - 16(0.25)^2$	5
0.5	$24(0.5) - 16(0.5)^2$	8
0.75	$24(0.75) - 16(0.75)^2$	?
1.0	$24(1.0) - 16(1.0)^2$	?
1.25	?	?
1.5	?	?
1.75	?	?
2.0	?	?

7-1

Factors and Greatest Common Factors

Objectives

After studying this lesson, you should be able to:

- find the prime factorization of an integer, and
- find the greatest common factor (GCF) for a set of monomials.

Connection

In mathematics, there are many situations where there is more than one correct answer. Suppose three students are asked to draw and label a rectangle that has an area of 18 square inches. As shown below, each student can draw a different rectangle, and each drawing is correct.

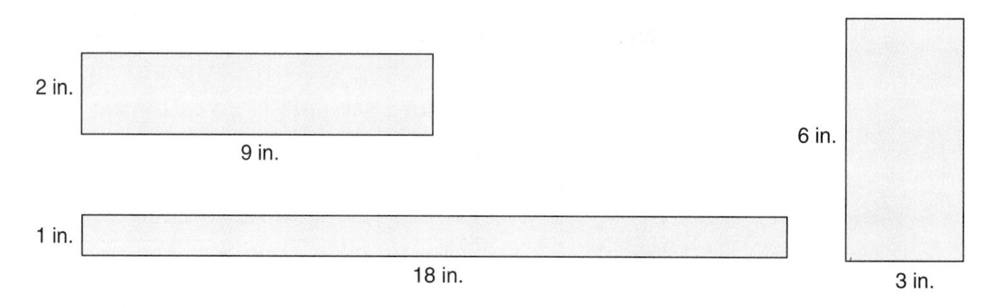

Since 2×9, 3×6, and 1×18 all equal 18, each rectangle above has an area of 18 square inches.

FYI . . .
The Ishango bone of Zaire is thought to show that Africans knew of prime numbers thousands of years ago.

When two or more numbers are multiplied, each number is a **factor** of the product. In the example above, 18 is expressed as the product of different pairs of whole numbers.

$$18 = 2 \cdot 9 \qquad 18 = 3 \cdot 6 \qquad 18 = 1 \cdot 18$$

The whole numbers 1, 18, 2, 9, 3, and 6 are factors of 18. Some whole numbers have exactly two factors, the number itself and 1. These numbers are called **prime numbers.** Whole numbers that have more than two factors are called **composite numbers.**

Definition of Prime and Composite Numbers

A prime number is a whole number, greater than 1, whose only factors are 1 and itself. A composite number is a whole number, greater than 1, that is not prime.

0 and 1 are neither prime nor composite.

The number 9 is a factor of 18, but not a **prime factor** of 18, since 9 is not a prime number. When a whole number is expressed as a product of factors that are all prime, the expression is called the **prime factorization** of the number. Thus, the prime factorization of 18 is $2 \cdot 3 \cdot 3$ or $2 \cdot 3^2$.

The prime factorization of every number is unique except for the order in which the factors are written. For example, $3 \cdot 2 \cdot 3$ is also a prime factorization of 18, but it is the same as $2 \cdot 3 \cdot 3$. This property of numbers is called the *Unique Factorization Theorem.*

Example 1 **Find the prime factorization of 84.**

You can begin by dividing 84 by its least prime factor. Continue dividing by least prime factors until all the factors are prime.

$84 = 2 \cdot 42$ *The least prime factor of 84 is 2.*

$= 2 \cdot 2 \cdot 21$ *The least prime factor of 42 is 2.*

$= 2 \cdot 2 \cdot 3 \cdot 7$ *The least prime factor of 21 is 3.*

All of the factors in the last row are prime. Thus, the prime factorization of 84 is $2 \cdot 2 \cdot 3 \cdot 7$ or $2^2 \cdot 3 \cdot 7$.

Example 2 **Factor −525.**

To factor a negative integer, first express it as the product of a whole number and −1. Then find the prime factorization.

$$\begin{aligned} -525 &= -1 \cdot 525 \\ &= -1 \cdot 3 \cdot 175 \\ &= -1 \cdot 3 \cdot 5 \cdot 35 \\ &= -1 \cdot 3 \cdot 5 \cdot 5 \cdot 7 \text{ or } -1 \cdot 3 \cdot 5^2 \cdot 7 \end{aligned}$$

A monomial is written in *factored form* when it is expressed as the product of prime numbers and variables where no variable has an exponent greater than 1.

Example 3 **Factor $20a^2b$.**

$$\begin{aligned} 20a^2b &= 2 \cdot 10 \cdot a \cdot a \cdot b \\ &= 2 \cdot 2 \cdot 5 \cdot a \cdot a \cdot b \end{aligned}$$

Two or more numbers may have some common factors. Consider the prime factorizations of 90 and 105 shown below.

$90 = 2 \cdot 3 \cdot 3 \cdot 5$ $\qquad$ $105 = 3 \cdot 5 \cdot 7$

1 is always a common factor.

The integers 90 and 105 have 3 and 5 as common prime factors. The product of these prime factors, $3 \cdot 5$ or 15, is called the **greatest common factor (GCF)** of 90 and 105.

Definition of Greatest Common Factor

The greatest common factor of two or more integers is the product of the prime factors common to the integers.

The GCF of two or more monomials is the product of their common factors, when each monomial is expressed as a product of prime factors.

Example 4 Find the GCF of 54, 63, and 180.

$54 = 2 \cdot 3 \cdot 3 \cdot 3$ *Factor each number.*

$63 = 3 \cdot 3 \cdot 7$

$180 = 2 \cdot 2 \cdot 3 \cdot 3 \cdot 5$ *Then circle the common factors.*

The GCF of 54, 63, and 180 is $3 \cdot 3$ or 9.

Example 5 Find the GCF of $8a^2b$ and $18a^2b^2c$.

$8a^2b = 2 \cdot 2 \cdot 2 \cdot a \cdot a \cdot b$

$18a^2b^2c = 2 \cdot 3 \cdot 3 \cdot a \cdot a \cdot b \cdot b \cdot c$

The GCF of $8a^2b$ and $18a^2b^2c$ is $2a^2b$.

CHECKING FOR UNDERSTANDING

Communicating Mathematics

Read and study the lesson to answer each question.

1. Draw and label as many rectangles as possible with sides of integral length that have an area of 50 square inches.
2. Name the first ten prime numbers.
3. Is $2 \cdot 3^2 \cdot 4$ the prime factorization of 72? Why or why not?
4. What is the difference between a common factor of two numbers and a greatest common factor?

Guided Practice

State whether each number is prime or composite. If the number is composite, find its prime factorization.

5. 89
6. 39
7. 24
8. 91

Find the GCF of each pair of numbers.

9. 4, 12
10. 9, 36
11. 15, 5
12. 11, 22
13. 10, 15
14. 20, 30
15. 18, 35
16. 16, 18

EXERCISES

Practice

Find the prime factorization of each number.

17. 21
18. 28
19. 60
20. 51
21. 63
22. 72
23. 112
24. 150
25. 304
26. 216
27. 300
28. 1540

Factor each expression. Do not use exponents.

29. -64 30. -26 31. -240 32. -231

33. $98a^2b$ 34. $44rs^2t^3$ 35. $756(mn)^3$ 36. $-102x^3y$

Find the GCF of the given monomials.

37. 16, 60 38. 15, 50 39. -80, 45

40. 29, -58 41. 305, 55 42. 252, 126

43. 128, 245 44. 95, 304 45. $7y^2$, $14y^2$

46. $17a$, $34a^2$ 47. $-12ab$, $4a^2b^2$ 48. $4xy$, $-6x$

49. $50n^4$, $40n^2p^2$ 50. $60x^2y^2$, $35xz^3$ 51. $12an^2$, $40a^4$

52. $56x^3y$, $49ax^2$ 53. 5, 15, 10 54. 16, 24, 28

55. 18, 30, 54 56. 24, 84, 168 57. 16, 24, 30

58. $12mn$, $10mn$, $15mn$ 59. $6a^2$, $18b^2$, $9b^3$

60. $8b^4$, $5c$, $3a^2b$ 61. $15abc$, $35a^2c$, $105a$

62. $14a^2b^2$, $18ab$, $2a^3b^3$ 63. $18x^2y^2$, $6y^2$, $42x^2y^3$

Find the missing factor.

64. $48a^5b^4 = 3a^3b^2(\underline{\ ?\ })$ 65. $-28x^4y^3 = 7x^2(\underline{\ ?\ })$

66. $36m^5n^7 = 2m^3n(6n^5)(\underline{\ ?\ })$ 67. $48a^5b^5 = 2ab^2(4ab)(\underline{\ ?\ })$

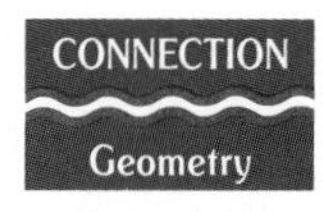

68. The area of a rectangle is 1363 square inches. If the measures of the length and width of this rectangle are both prime numbers, what are its dimensions?

Critical Thinking

69. **Geometry** Draw and label all the rectangles that satisfy the following conditions. The area of a rectangle is $8b^2$ square centimeters and the measure of each of its sides can be expressed as a monomial with integral coefficients. (*Hint:* There are 6.)

70. *Twin primes* are two consecutive odd prime numbers, such as 11 and 13. How many pairs of twin primes are there where both primes are less than 100? Find them.

Applications

71. **Gardening** Anita is planning to have 100 tomato plants in her garden. In what ways can she arrange them so that she has the same number of plants in each row and at least 5 rows of plants?

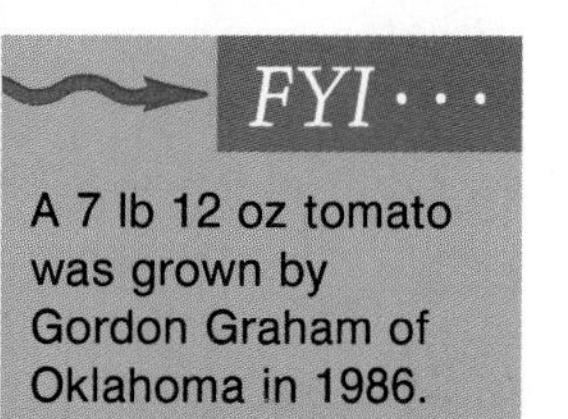

72. **Travel** Jackie and Jaime left at the same time on a 300-mile trip. Jaime's average speed for the trip was 10 miles per hour slower than Jackie's. If it took Jaime one hour longer than Jackie to complete the trip, what was his average speed for the trip?

73. **Sports** A new athletic field is being sodded at Beck High School using 2-yard-by-2-yard squares of sod. If the width of the field is 70 yards less than its length and its area is 6000 square yards, how many squares of sod will be needed?

Computer

The BASIC program at the right finds the GCF of two numbers. The INT or greatest integer function is used to determine factors. Line 40 determines if F is a factor of the first number. Line 60 determines if F is a common factor. Since values for F are checked from greatest to least, the first common factor found is the GCF.

```
10 PRINT "ENTER TWO
   NUMBERS."
20 INPUT A, B
30 FOR F = A TO 1 STEP -1
40 IF INT(A/F) = A/F THEN 60
50 NEXT F
60 IF INT(B/F) = B/F THEN 80
70 NEXT F
80 PRINT "THE GCF OF ";A;
90 PRINT " AND ";B;" IS ";F
100 END
```

Journal

Tell how you use the GCF of the terms in an expression to factor the expression.

74. Use the program to find the GCF of each pair of numbers.
a. 30, 42 **b.** 27, 81 **c.** 76, 133 **d.** 29, 37

75. Change the program so it will find the GCF of three numbers. Use the new program to find the GCF of each set of numbers below.
a. 16, 48, 128 **b.** 27, 32, 42 **c.** 60, 84, 132

Mixed Review

76. Solve $6(x - 2) = 5(x - 11) - 21$. **(Lesson 3-6)**

77. Solve $x + r = 2d$ for x. **(Lesson 3-7)**

78. Metallurgy An aluminum alloy that is 48% aluminum is to be made by combining 30% and 60% alloys. How many pounds of 60% alloy must be added to 24 pounds of 30% alloy to produce the alloy? **(Lesson 4-6)**

Solve each inequality. Check the solution. (Lessons 5-3, 5-6)

79. $5x \leq 10(3x + 4)$ **80.** $|2y - 5| \geq 4$

Simplify. (Lessons 6-7, 6-9)

81. $-8a(5a^2 + 8a - 3)$ **82.** $(5r - 7s)^2$

HISTORY CONNECTION

Eratosthenes

Eratosthenes was a Greek astronomer and poet (c. 200 B.C.). He is known for the "sieve of Eratosthenes," a method for finding prime numbers. To use this method, first list the positive integers beginning with 2. Then cross out every second number after 2, every third number after 3, every fifth number after 5, and so on. When the process is completed, the numbers *not* crossed out are prime.

Copy and complete the sieve of Eratosthenes below to find all of the prime numbers from 2 to 100. *The multiples of 2 have been crossed out for you.*

2	3	~~4~~	5	~~6~~	7	~~8~~	9	~~10~~	11	~~12~~	13	~~14~~	15	~~16~~
17	~~18~~	19	~~20~~	21	~~22~~	23	~~24~~	25	~~26~~	27	~~28~~	29	~~30~~	31
~~32~~	33	~~34~~	35	~~36~~	37	~~38~~	39	~~40~~	41	~~42~~	43	~~44~~	45	~~46~~
47	~~48~~	49	~~50~~	51	~~52~~	53	~~54~~	55	~~56~~	57	~~58~~	59	~~60~~	61
~~62~~	63	~~64~~	65	~~66~~	67	~~68~~	69	~~70~~	71	~~72~~	73	~~74~~	75	~~76~~
77	~~78~~	79	~~80~~	81	~~82~~	83	~~84~~	85	~~86~~	87	~~88~~	89	~~90~~	91
~~92~~	93	~~94~~	95	~~96~~	97	~~98~~	99	100						

7-2 Factoring Using the Distributive Property

Objective

After studying this lesson, you should be able to:

- use the GCF and the distributive property to factor polynomials.

Connection

Since $A = \ell w$, $w = \frac{A}{\ell}$.

To find the width of the rectangle at the right, divide the area by the length.

$$w = \frac{3x^2 - 8x}{x}$$
$$= \frac{3x^2}{x} - \frac{8x}{x}$$
$$= 3x - 8$$

x

A = $3x^2 - 8x$

w = ?

Since the area of the rectangle is the product of its length and width, $3x^2 - 8x = x(3x - 8)$. The expression $x(3x - 8)$ is called the *factored form* of $3x^2 - 8x$. A polynomial is in factored form, or *factored,* when it is expressed as the product of monomials and polynomials.

Lab Activity

You can learn how to use algebra tiles to factor by using the distributive property in Lab 9 on page A15.

In Chapter 6, you multiplied a polynomial by a monomial by using the distributive property. You can also reverse this process and express a polynomial in factored form by using the distributive property.

Multiplying Polynomials	Factoring Polynomials
$3(a + b) = 3a + 3b$	$3a + 3b = 3(a + b)$
$x(y - z) = xy - xz$	$xy - xz = x(y - z)$
$3y(4x + 2) = 3y(4x) + 3y(2)$	$12xy + 6y = 3y(4x) + 3y(2)$
$= 12xy + 6y$	$= 3y(4x + 2)$

The expression $3y(4x + 2)$ at the right above is not considered completely factored since the polynomial $4x + 2$ can be factored as $2(2x + 1)$. The completely factored form of $12xy + 6y$ is $6y(2x + 1)$. Factoring a polynomial or finding the factored form of a polynomial means to find its *completely* factored form.

Example 1

Use the distributive property to factor $10y^2 + 15y$.

First, find the greatest common factor for $10y^2$ and $15y$.

$10y^2 = 2 \cdot (5) \cdot (y) \cdot y$

$15y = 3 \cdot (5) \cdot (y)$ *The GCF is 5y.*

Then, express each term as the product of the GCF and its remaining factors.

$10y^2 + 15y = 5y(2y) + 5y(3)$

$= 5y(2y + 3)$ *Distributive property*

Example 2

Factor $21ab^2 - 33a^2bc$.

$$21ab^2 = 3 \cdot 7 \cdot a \cdot b \cdot b$$
$$33a^2bc = 3 \cdot 11 \cdot a \cdot a \cdot b \cdot c \quad \textit{The GCF is } 3ab.$$

$$21ab^2 - 33a^2bc = 3ab(7b) - 3ab(11ac)$$
$$= 3ab(7b - 11ac) \quad \textit{Distributive property}$$

Example 3

Factor $6x^3y^2 + 14x^2y + 2x^2$.

$$6x^3y^2 = 2 \cdot 3 \cdot x \cdot x \cdot x \cdot y \cdot y$$
$$14x^2y = 2 \cdot 7 \cdot x \cdot x \cdot y$$
$$2x^2 = 2 \cdot x \cdot x \quad \textit{The GCF is } 2x^2.$$

$$6x^3y^2 + 14x^2y + 2x^2 = 2x^2(3xy^2) + 2x^2(7y) + 2x^2(1)$$
$$= 2x^2(3xy^2 + 7y + 1)$$

Example 4

A deck 3 meters wide is to be built around a swimming pool like the one shown in the figure below. In order to determine the amount of material that is needed to build the deck, a carpenter needs to calculate the area of the deck. Write an equation that represents this area.

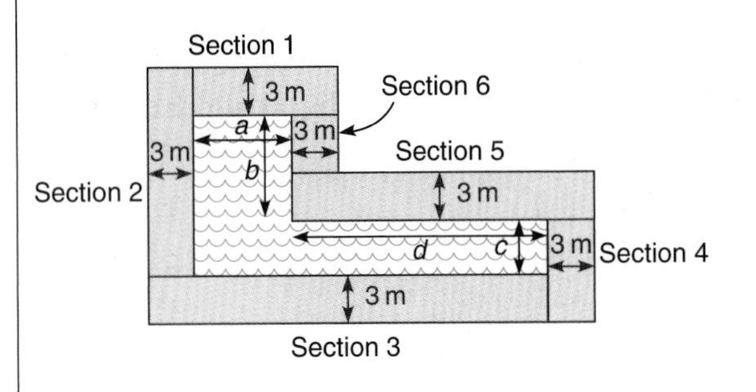

You can find the area of the deck by finding the sum of the areas of the 6 rectangular sections shown in the figure. The resulting expression can be simplified by first using the distributive property and then factoring.

Section 1   Section 2   Section 3   Section 4   Section 5   Section 6

$$A = 3(a + 3) + 3(b + c + 3) + 3(a + d + 3) + 3(c + 3) + 3(d + 3) + 3(b - 3)$$
$$= 3a + 9 + 3b + 3c + 9 + 3a + 3d + 9 + 3c + 9 + 3d + 9 + 3b - 9$$
$$= 6a + 6b + 6c + 6d + 36 \quad \textit{Combine like terms.}$$
$$= 6(a + b + c + d + 6) \quad \textit{The GCF is 6.}$$

The area of the deck is $6(a + b + c + d + 6)$ square units.

CHECKING FOR UNDERSTANDING

Communicating Mathematics

Read and study the lesson to answer each question.

1. When you factor $15a + 12a^2$, what property do you use?
2. Express $15a + 12a^2$ as a product in three different ways.
3. Which of the three answers in Exercise 2 is the completely factored form of $15a + 12a^2$? Why?

Guided Practice

Find the GCF of the terms in each expression.

4. $3y^2 + 12$
5. $4a + 2b$
6. $5y - 9y^2$
7. $9b + 5c$
8. $9a^2 + 3a$
9. $7mn - 21m^3$

Complete.

10. $6x + 3y = 3(\underline{?} + y)$
11. $8x^2 - 4x = 4x(2x - \underline{?})$
12. $12a^2b + 6a = 6a(\underline{?} + 1)$
13. $14r^2t - 42t = 14t(\underline{?} - 3)$

Factor each expression.

14. $29xy - 3x$
15. $x^5y - x$
16. $27a^2b + 9b^3$
17. $3c^2d - 6c^2d^2$

EXERCISES

Practice

Complete.

18. $16x + 4y = 4(4x + \underline{?})$
19. $24x^2 + 12y^2 = 12(\underline{?} + y^2)$
20. $12xy + 12x^2 = \underline{?}(y + x)$
21. $5a^2b + 10ab = \underline{?}(a + 2)$

Factor each expression.

22. $11x + 44x^2y$
23. $16y^2 + 8y$
24. $14xz - 18xz^2$
25. $14mn^2 + 2mn$
26. $18xy^2 - 24x^2y$
27. $15xy^3 + y^4$
28. $25a^2b^2 + 30ab^3$
29. $36p^2q^2 - 12pq$
30. $17a - 41a^3b$
31. $2m^3n^2 - 16m^2n^3 + 8mn$
32. $3x^3y + 9xy^2 + 36xy$
33. $28a^2b^2c^2 + 21a^2bc^2 - 14abc$
34. $12ax + 20bx + 32cx$
35. $a + a^2b + a^3b^3$

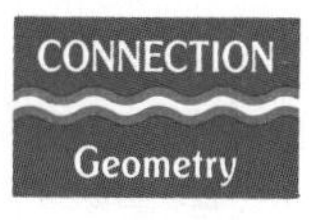

Write an equation that represents the measure of the area of the shaded region in factored form.

36.

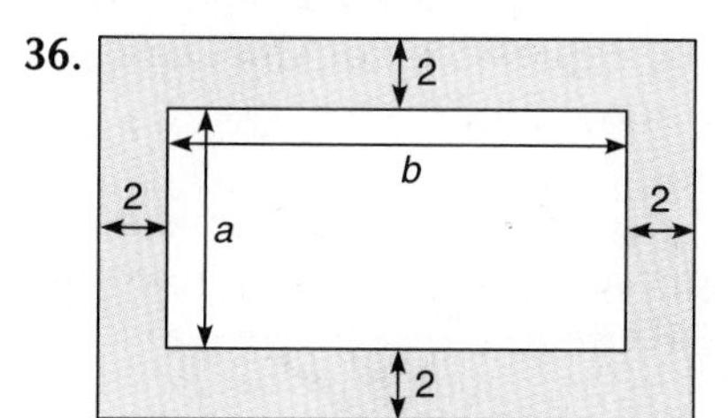

37.

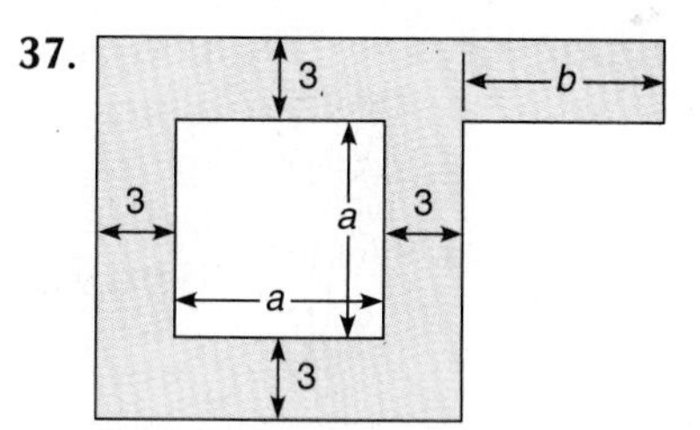

38.

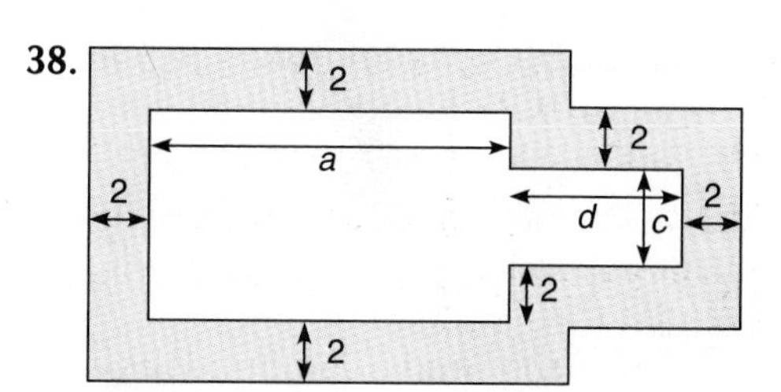

39.

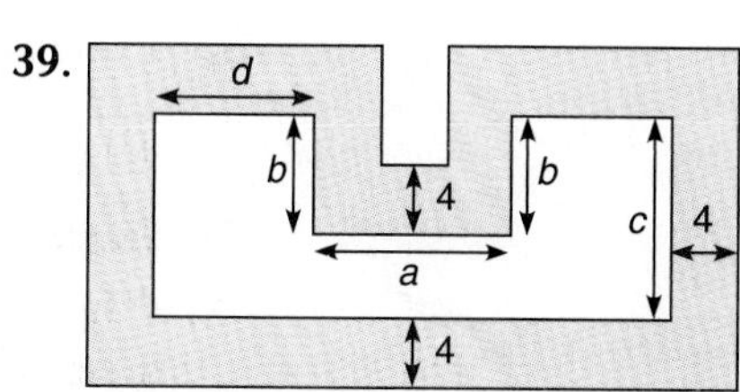

Factor each expression.

40. $ax^3 + 5bx^3 + 9cx^3$

41. $y^5 + 5y^4 + 3y^2 + 2y$

42. $\frac{2}{3}x + \frac{1}{3}x^2 - \frac{4}{3}xy^2$

43. $\frac{4}{5}a^2b - \frac{3}{5}ab^2 - \frac{1}{5}ab$

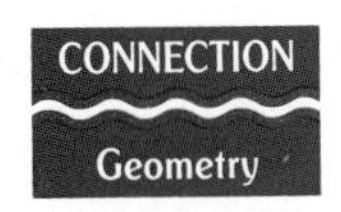

44. The measure of the perimeter of a square is $16a + 20b$. Find the measure of the area of the square.

Critical Thinking

45. **Geometry** The measures of the length and width of a rectangle are $(2x + 3)$ and $(9 - 4x)$. If x must be an integer, what are the possible measures for the area of this rectangle?

Applications

46. **Gardening** The length of Jeremy's garden is 5 feet more than twice its width, w. This year, Jeremy decided to make the garden 4 feet longer and double its width. How much additional area did Jeremy add to his garden?

47. **Stocks** During the first hour of trading, Sheila sold x shares of a stock that cost \$4 per share. During the next hour, she sold stock that cost \$8 per share. She sold 5 more shares during the first hour than the second hour. If she had sold only the \$4-per-share stock during the first two hours, how many shares would she have needed to sell to have the same amount of total sales?

Mixed Review

48. Write an algebraic expression for the expression x *is no more than negative 3.* **(Lesson 2-3)**

49. **Sales** Jerome bought a used computer for \$8 less than one-third its original price. Jerome paid \$525 for the computer. What was the original price? **(Lesson 3-5)**

50. Solve $\frac{2}{11} = \frac{x - 5}{x + 3}$. **(Lesson 4-1)**

51. Evaluate $\frac{7.8 \times 10^4}{2.6 \times 10^2}$. Express the result in scientific notation and decimal notation. **(Lesson 6-4)**

52. Arrange the terms of $2mx^4 - 3x^5 + 4m^5 + 6x^3$ so that the powers of x are in descending order. **(Lesson 6-5)**

53. Find the GCF of $38ab^3$ and $-74a^3b$. **(Lesson 7-1)**

7-3 Factoring by Grouping

Objective

After studying this lesson, you should be able to:

- use grouping techniques to factor polynomials with four or more terms.

Connection

Suppose you know that the area of a certain rectangle is $3xy - 21y + 5x - 35$ square feet. Is it possible for this rectangle to have dimensions that can be represented by binomials with integral coefficients? If so, what are these dimensions?

Since the area of a rectangle is the product of its length and width, you can determine an answer to the questions above if you can find two binomials with integral coefficients whose product is $3xy - 21y + 5x - 35$. You can do this by factoring.

Polynomials with four or more terms, like $3xy - 21y + 5x - 35$, can sometimes be factored by grouping terms of the polynomials. One method is to group the terms into binomials that can each be factored using the distributive property. Then use the distributive property again with a binomial as the common factor.

Example 1

Factor $3xy - 21y + 5x - 35$ to answer the questions presented above.

$3xy - 21y + 5x - 35 = (3xy - 21y) + (5x - 35)$ *Group terms that have a common monomial factor.*

$= 3y(x - 7) + 5(x - 7)$ *Factor. Notice that $(x - 7)$ is a common factor.*

$= (3y + 5)(x - 7)$ *Distributive property*

Check by using FOIL.

$$(3y + 5)(x - 7) = \overset{F}{3y(x)} + \overset{O}{3y(-7)} + \overset{I}{5(x)} + \overset{L}{5(-7)}$$
$$= 3xy - 21y + 5x - 35 \checkmark$$

Thus, the dimensions of this rectangle can be represented by binomials with integral coefficients. These dimensions would be $3y + 5$ feet and $x - 7$ feet.

Example 2

Factor $8m^2n - 5m - 24mn + 15$.

$$8m^2n - 5m - 24mn + 15 = (8m^2n - 5m) + (-24mn + 15)$$
$$= m(8mn - 5) + (-3)(8mn - 5)$$
$$= (m - 3)(8mn - 5)$$

Check: $(m - 3)(8mn - 5) = m(8mn) + m(-5) + (-3)(8mn) + (-3)(-5)$
$= 8m^2n - 5m - 24mn + 15 \checkmark$

Sometimes you can group the terms in more than one way when factoring a polynomial. For example, the polynomial in Example 2 could have been factored in the following way.

$$\begin{aligned} 8m^2n - 5m - 24mn + 15 &= (8m^2n - 24mn) + (-5m + 15) \\ &= 8mn(m - 3) + (-5)(m - 3) \\ &= (8mn - 5)(m - 3) \end{aligned}$$

The result is the same as in Example 2.

Example 3

Factor $2a^2 + 2bc + ab + 4ac$ in two different ways.

Method 1:
$$\begin{aligned} 2a^2 + 2bc + ab + 4ac &= (2a^2 + ab) + (4ac + 2bc) \\ &= a(2a + b) + 2c(2a + b) \\ &= (a + 2c)(2a + b) \end{aligned}$$

Method 2:
$$\begin{aligned} 2a^2 + 2bc + ab + 4ac &= (2a^2 + 4ac) + (ab + 2bc) \\ &= 2a(a + 2c) + b(a + 2c) \\ &= (2a + b)(a + 2c) \end{aligned}$$

The result is the same

Check:
$$\begin{aligned} (2a + b)(a + 2c) &= 2a(a) + 2a(2c) + b(a) + b(2c) \\ &= 2a^2 + 4ac + ba + 2bc \\ &= 2a^2 + 2bc + ab + 4ac \ \checkmark \end{aligned}$$

$-1(a - 3) = -a + 3$
$= 3 - a$

Recognizing binomials that are additive inverses is often helpful in factoring. For example, the binomials $3 - a$ and $a - 3$ are additive inverses since the sum of $3 - a$ and $a - 3$ is 0. Thus, $3 - a$ and $-a + 3$ are equivalent. What is the additive inverse of $5 - y$?

Example 4

Factor $15x - 3xy + 4y - 20$.

$$\begin{aligned} 15x - 3xy + 4y - 20 &= (15x - 3xy) + (4y - 20) \\ &= 3x(5 - y) + 4(y - 5) \\ &= -3x(y - 5) + 4(y - 5) \\ &= (-3x + 4)(y - 5) \end{aligned}$$

$(5 - y)$ and $(y - 5)$ are additive inverses.
$(5 - y) = -1(y - 5)$

Check:
$$\begin{aligned} (-3x + 4)(y - 5) &= (-3x)(y) + (-3x)(-5) + 4(y) + 4(-5) \\ &= -3xy + 15x + 4y - 20 \\ &= 15x - 3xy + 4y - 20 \ \checkmark \end{aligned}$$

CHECKING FOR UNDERSTANDING

Communicating Mathematics

Read and study the lesson to answer each question.

1. When you factor $7mn + 14n + 5m + 10$ by grouping, what properties do you use?
2. Group the terms of $7mn + 14n + 5m + 10$ in pairs in two different ways so that the pairs of terms have a common monomial factor.
3. What is the additive inverse of $3a^2 - b$?

Guided Practice

Express each polynomial in factored form.

4. $k(r + s) - m(r + s)$
5. $t(t - s) + s(t - s)$
6. $3ab(a - 4) - 8(a - 4)$
7. $8m(x + y) + (x + y)$

Complete. In each exercise, both $\underline{\quad ? \quad}$s represent the same expression.

8. $(bx + by) + (3ax + 3ay) = b(\underline{\quad ? \quad}) + 3a(\underline{\quad ? \quad})$
9. $(3mx + 2my) + (3kx + 2ky) = m(\underline{\quad ? \quad}) + k(\underline{\quad ? \quad})$
10. $(a^2 + 3ab) + (2ac + 6bc) = a(\underline{\quad ? \quad}) + 2c(\underline{\quad ? \quad})$
11. $(6xy - 15x) + (-8y + 20) = 3x(\underline{\quad ? \quad}) - 4(\underline{\quad ? \quad})$

Factor each polynomial. Check by using FOIL.

12. $rx + 2ry + kx + 2ky$
13. $ay - ab + cb - cy$
14. $a^2 - 4ac + ab - 4bc$
15. $5a + 10a^2 + 2b + 4ab$

EXERCISES

Practice

Complete. In each exercise, both $\underline{\quad ? \quad}$s represent the same expression.

16. $(10x^2 - 6xy) + (15x - 9y) = 2x(\underline{\quad ? \quad}) + 3(\underline{\quad ? \quad})$
17. $(6x^3 + 6x) + (7x^2y + 7y) = 6x(\underline{\quad ? \quad}) + 7y(\underline{\quad ? \quad})$
18. $(4m^2 - 3mp) + (6p - 8m) = m(\underline{\quad ? \quad}) - 2(\underline{\quad ? \quad})$
19. $(20k^2 - 28kp) + (7p^2 - 5kp) = 4k(\underline{\quad ? \quad}) - p(\underline{\quad ? \quad})$

Factor each polynomial. Check by using FOIL.

20. $2ax + 6xc + ba + 3bc$
21. $6mx - 4m + 3rx - 2r$
22. $2my + 7x + 7m + 2xy$
23. $3my - ab + am - 3by$
24. $3ax - 6bx + 8b - 4a$
25. $a^2 - 2ab + a - 2b$
26. $2ab + 2am - b - m$
27. $3m^2 - 5m^2p + 3p^2 - 5p^3$
28. $5a^2 - 4ab + 12b^3 - 15ab^2$
29. $4ax - 14bx + 35by - 10ay$
30. $6a^2 - 6ab + 3cb - 3ca$
31. $ax + a^2x - a - a^2$
32. $a^3 - a^2b + ab^2 - b^3$
33. $2x^3 - 5xy^2 - 2x^2y + 5y^3$

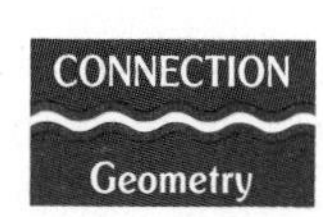

Find the dimensions of a rectangle having the given area if its dimensions can be represented by binomials with integral coefficients.

34. $5xy + 15x - 6y - 18$ cm^2
35. $4z^2 - 24z - 18m + 3mz$ ft^2

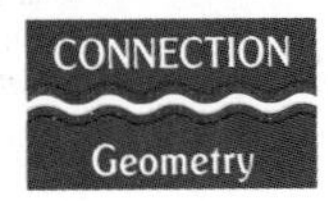

Write an expression in factored form for the measure of the perimeter of each quadrilateral shown below.

36.

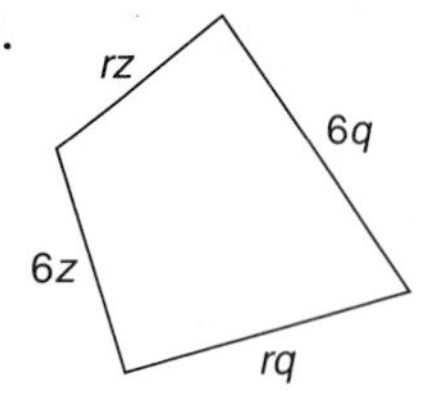

37.

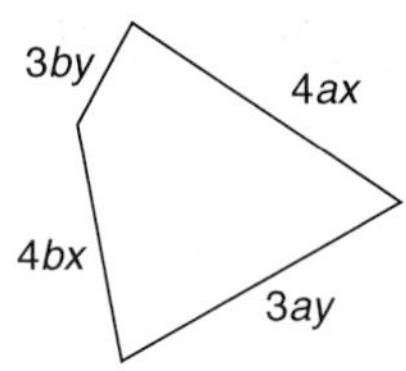

38.

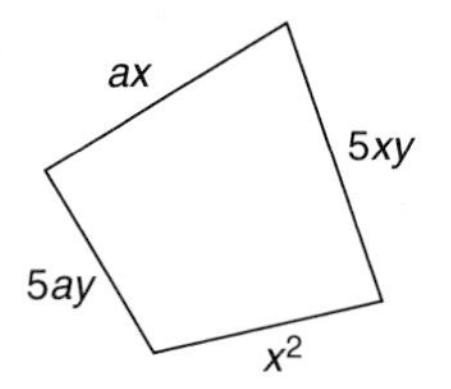

Factor each polynomial. Check by using FOIL.

39. $7xa + 7xb + 3ma + 3mb - 4b - 4a$

40. $ax - ay - 4yb + 4xb + 5x - 5y$

41. $2ax + bx - 6ay - 3by - bz - 2az$

42. $ar - 3ya + br - 3by + 3cy - rc$

Critical Thinking

43. **Geometry** The measure of the perimeter of a rectangle is $2x + 4y + 8xy + 2$. Find all the possible expressions, in factored form, for the measure of its area. (*Hint:* There are three.)

Applications

44. **Conference** A North American fellowship conference is being attended by 96 students from the United States, 56 from Mexico, and 72 from Canada.

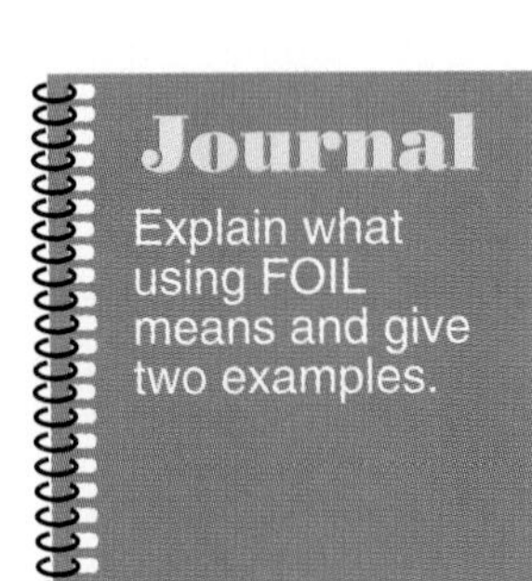

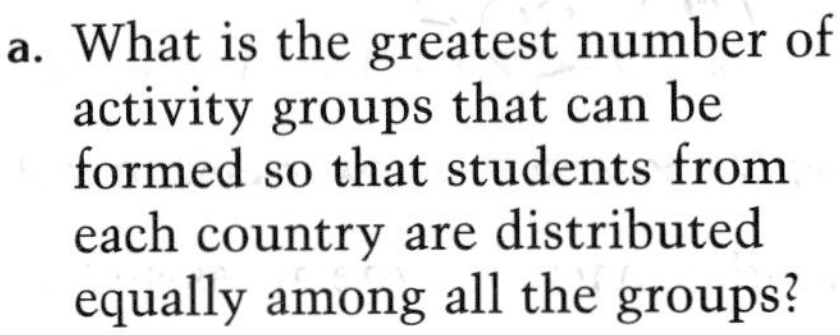

a. What is the greatest number of activity groups that can be formed so that students from each country are distributed equally among all the groups?

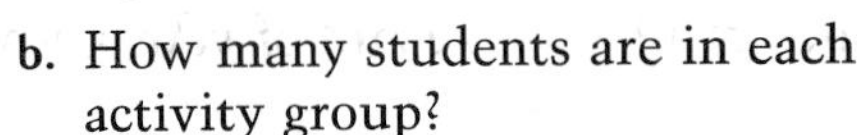

b. How many students are in each activity group?

45. **Construction** A 4-foot wide stone path is to be built along each of the longer sides of a rectangular flower garden. The length of the longer side of the garden is 3 feet less than twice the length of the shorter side, *s*. Write an expression, in factored form, to represent the total area of the garden and path.

Mixed Review

46. Find the solution set for $5 - 2n < 3n$ if the replacement set is $\{-1, 0, 1, 2, 3\}$. **(Lesson 1-3)**

47. **Sales** Blaine bought a stereo for \$476.79. This price was 20% less than the original price. What was the original price? **(Lesson 4-4)**

48. What is the next number in the sequence below? **(Lesson 6-1)**

$$-2, -1, 1, 5, 13, 29, \ldots$$

49. Multiply $(2x + 5)(3x - 8)$. **(Lesson 6-8)**

50. Factor $10x^4 - 6x^3y - 8x^2y^2$. **(Lesson 7-2)**

7-4 Problem-Solving Strategy: Guess and Check

EXPLORE
PLAN
SOLVE
EXAMINE

Objective After studying this lesson, you should be able to:

- solve problems by using guess and check.

An important problem-solving strategy is guess and check, or trial and error. To use this strategy, guess the answer to the problem, then check whether the guess is correct. If the first guess is incorrect, guess and check again until you find the right answer. Often, the results of one guess can help you make a better guess. Always keep an organized record of your guesses so you don't make the same guess twice.

Example

Insert one set of parentheses so that a true equation results.

$$10 - 4 \cdot 2 - 1 = 3$$

$10 - 4 \cdot (2 - 1) = 6$ *Try again since the value does not equal 3.*
$(10 - 4) \cdot 2 - 1 = 11$ *Try again.*
$10 - (4 \cdot 2 - 1) = 3$ *Correct!*

CHECKING FOR UNDERSTANDING

Communicating Mathematics

Read and study the lesson to answer each question.

1. What is another name for the guess-and-check strategy?
2. Why should you keep a record of your guesses?

Guided Practice

Strategies

Look for a pattern.
Solve a simpler problem.
Act it out.
Guess and check.
Draw a diagram.
Make a chart.
Work backwards.

Insert one set of parentheses so that a true equation results.

3. $4 \cdot 5 - 2 + 7 = 19$
4. $25 - 4 \cdot 2 + 3 = 5$

Solve by using guess and check.

5. Place the digits 1, 2, 3, 4, 5, 6, 8, 9, 10, and 12 at the dots at the right so that the sum of the integers on any line equals the sum on any other line.

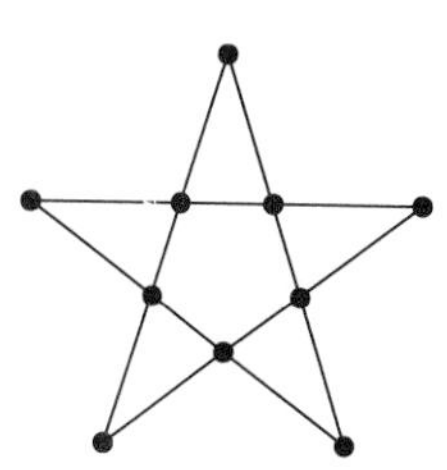

6. Numero Uno says, "I am thinking of a three-digit number. If you multiply the digits together and then multiply the result by 4, the answer is the number I'm thinking of. What is my number?"
7. Using the digits 1, 2, 3, 4, 5, and 6 only once, find two whole numbers whose product is as large as possible.

EXERCISES

Practice

Solve. Use any strategy.

8. On a recent safari, a group of people and elephants contained 100 knees and 100 trunks. If each person took 3 trunks, how many people and elephants went on safari?

9. Fill in each box at the right with a digit from 1 to 6 to make this multiplication work. Use each digit exactly once.

$$\begin{array}{r} \square \\ \times \square\square \\ \hline \square\square\square \end{array}$$

10. **Number Theory** The *word* "six" has three letters. Since the *number* 6 is divisible by 3, we say that 6 is *word-factorable.* In the same way, 36 is word-factorable because "thirty-six" has 9 letters and 36 is divisible by 9. Find five other word-factorable numbers less than 100.

11. The Canadian Can Company sent a bill to the American Aluminum Association. The payment deadline on the bill was given in the form day/month/year. AAA read this as month/day/year and paid the bill *exactly* 60 days after the due date. What two dates were involved?

COOPERATIVE LEARNING ACTIVITY

Work in groups of three. Each person in the group must understand the solution and be able to explain it to any person in the class.

In Lesson 4-5, you learned about figurate numbers. The table below shows several figurate number patterns. The letter *n* refers to the number of dots on each side.

Figurate Number	Shape of Figure	*n*									
		1	2	3	4	5	6	7	8	9	10
triangular	3-sided	1	3	6	10	15	21	28	36	45	
square	4-sided	1	4	9	16	25	36	49	64	81	
pentagonal	5-sided	1	5	12	22	35	51	70	92		
hexagonal	6-sided	1	6	15	28	45	66				
heptagonal	7-sided	1	7	18	34						
octagonal	8-sided	1	8								

1. Draw an example of each type of figurate number, with 3 on each side.
2. Copy and complete the table.
3. Describe how the numbers increase in each column of the table.
4. Write an expression to describe the number of dots in a square number.

7-5 Factoring Trinomials

Objective After studying this lesson, you should be able to:

- factor quadratic trinomials.

In Lesson 7-1, you learned that when two numbers are multiplied, each number is a *factor* of the product. Similarly, if two binomials are multiplied, each binomial is a *factor* of the product.

Lab Activity
You can learn how to use algebra tiles to factor trinomials in Lab 10 on pages A16-A17.

Consider the binomials $5x + 2$ and $3x + 7$. You can use the FOIL method to find their product.

$$\begin{aligned}(5x + 2)(3x + 7) &= \overset{F}{(5x)(3x)} + \overset{O}{(5x)(7)} + \overset{I}{(2)(3x)} + \overset{L}{(2)(7)}\\ &= 15x^2 + 35x + 6x + 14\\ &= 15x^2 + (35 + 6)x + 14\\ &= 15x^2 + 41x + 14\end{aligned}$$

Notice that $15 \cdot 14 = 210$ and $35 \cdot 6 = 210$.

The binomials $5x + 2$ and $3x + 7$ are factors of $15x^2 + 41x + 14$.

When using the FOIL method above, look at the product of the coefficients of the first and last terms, 15 and 14. Notice that it is the same as the product of the two terms, 35 and 6, whose sum is the coefficient of the middle term. You can use this pattern to factor quadratic trinomials, such as $2y^2 + 7y + 6$.

$2y^2 + 7y + 6$ — The product of 2 and 6 is 12.

$2y^2 + (\underline{\ ?\ } + \underline{\ ?\ })y + 6$ — You need to find two integers whose *product is 12* and whose *sum is 7*.

You can use the guess-and-check strategy to find these numbers.

Factors of 12	Sum of Factors	
1, 12	1 + 12 = 13	*no*
2, 6	2 + 6 = 8	*no*
3, 4	3 + 4 = 7	*yes*

$2y^2 + (3 + 4)y + 6$ — *Select the factors 3 and 4.*

$2y^2 + 3y + 4y + 6$

$(2y^2 + 3y) + (4y + 6)$ — *Group terms that have a common monomial factor.*

$y(2y + 3) + 2(2y + 3)$ — *Factor.*

$(y + 2)(2y + 3)$ — *Use the distributive property.*

Therefore, $2y^2 + 7y + 6 = (y + 2)(2y + 3)$. *Check this by using FOIL.*

Example 1 **Factor $5x^2 - 17x + 14$.**

$5x^2 - 17x + 14$

$5x^2 + (\underline{\ ?\ } + \underline{\ ?\ })x + 14$

The product of 5 and 14 is 70. Since the product is positive and the sum is negative, both factors of 70 must be negative. *Why?*

Factors of 70	Sum of Factors	
$-1, -70$	$-1 + (-70) = -71$	*no*
$-2, -35$	$-2 + (-35) = -37$	*no*
$-5, -14$	$-5 + (-14) = -19$	*no*
$-7, -10$	$-7 + (-10) = -17$	*yes*

You can stop listing factors when you find a pair that works.

$5x^2 + [-10 + (-7)]x + 14$

$5x^2 - 10x - 7x + 14$

$(5x^2 - 10x) + (-7x + 14)$

$5x(x - 2) + (-7)(x - 2)$ *Factor the GCF from each group.*

$(5x - 7)(x - 2)$ *Use the distributive property.*

Therefore, $5x^2 - 17x + 14 = (5x - 7)(x - 2)$. *Check this by using FOIL.*

Example 2

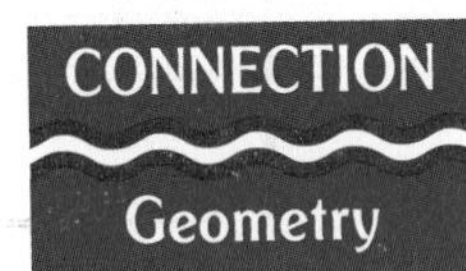

The area of a rectangle is $a^2 - 9a - 36$ square centimeters. This area is increased by increasing both the length and width by 4 centimeters. If the dimensions of the original rectangle are represented by binomials with integral coefficients, find the area of the new rectangle.

To determine the area of the new rectangle, you must first find the dimensions of the original rectangle by factoring $a^2 - 9a - 36$. The coefficient of a^2 is 1. Thus, you must find two numbers whose product is $1 \cdot -36$ or -36 and whose sum is -9.

Factors of -36	Sum of Factors	
$1, -36$	$1 + (-36) = -35$	*no*
$2, -18$	$2 + (-18) = -16$	*no*
$3, -12$	$3 + (-12) = -9$	*yes*

The factors of -36 should be chosen so that exactly one factor in each pair is negative and that factor has the greater absolute value. Why?

$$\begin{aligned} a^2 - 9a - 36 &= a^2 + [3 + (-12)]a - 36 \\ &= a^2 + 3a - 12a - 36 \\ &= (a^2 + 3a) + (-12a - 36) \\ &= a(a + 3) + (-12)(a + 3) \\ &= (a - 12)(a + 3) \end{aligned}$$

Check this by using FOIL.

The dimensions of the original rectangle are $a - 12$ cm and $a + 3$ cm. Therefore, the dimensions of the new rectangle are $(a - 12) + 4$ or $a - 8$ cm and $(a + 3) + 4$ or $a + 7$ cm. Now, find an expression for the area of the new rectangle.

$(a - 8)(a + 7) = a^2 + 7a - 8a - 56$ or $a^2 - a - 56$

The area of the new rectangle is $a^2 - a - 56$ cm^2.

Let us study the factorization of $a^2 - 9a - 36$ from Example 2 more closely.

$$a^2 - 9a - 36 = (a - 12)(a + 3)$$

Notice that the sum of the constant terms -12 and 3 is equal to 9, the coefficient of a in the trinomial. Also, the product of these terms is equal to 36, the constant term of the trinomial. This pattern holds for all trinomials whose quadratic term has a coefficient of 1.

Example 3

Factor $5x - 6 + x^2$.

The trinomial $5x - 6 + x^2$ can be written as $x^2 + 5x - 6$. For this trinomial, the constant term is -6 and the coefficient of x is 5. Thus, we need to find two factors of -6 whose sum is 5.

Factors of -6	Sum of Factors	
$1, -6$	$1 + (-6) = -5$	*no*
$-1, 6$	$-1 + 6 = 5$	*yes*

Select the factors -1 and 6.
Therefore, $x^2 + 5x - 6 = (x - 1)(x + 6)$.

Check by using FOIL and the method in Examples 1 and 2.

Example 4

Factor $2n^2 - 11n + 7$.

You must find two numbers whose product is $2 \cdot 7$ or 14 and whose sum is -11.

Factors of 14	Sum of Factors
$-1, -14$	$-1 + (-14) = -15$
$-2, -7$	$-2 + (-7) = -9$

Why should you test only negative factors of 14?

There are no factors of 14 whose sum is -11.
Therefore, $2n^2 - 11n + 7$ cannot be factored using integers.

A polynomial that *cannot* be written as a product of two polynomials with integral coefficients is called a **prime polynomial.** Thus, the trinomial $2n^2 - 11n + 7$ from Example 4 is a prime polynomial.

Example 5

Find all values of k so that the trinomial $x^2 + kx + 8$ can be factored using integers.

For $x^2 + kx + 8$, the value of k is the sum of the factors of $1 \cdot 8$ or 8.

Factors of 8	Sum of Factors
$1, 8$	$1 + 8 = 9$
$-1, -8$	$-1 + (-8) = -9$
$2, 4$	$2 + 4 = 6$
$-2, -4$	$-2 + (-4) = -6$

Therefore, the values of k are 9, -9, 6, and -6.

CHECKING FOR UNDERSTANDING

Communicating Mathematics

Complete.

1. You factor a trinomial by writing it as the ___?___ of two binomials.
2. When you factor $3x^2 + 11x + 6$, you want to find two numbers whose product is ___?___ and whose sum is 11.
3. When you factor $a^2 - 24a + 12$, you know that both factors of 12 must be ___?___.
4. A polynomial that *cannot* be written as the product of two polynomials is called a ___?___ polynomial.

Guided Practice

For each trinomial of the form $ax^2 + bx + c$, find two integers whose product is equal to *ac* and whose sum is equal to *b*.

5. $3x^2 + 11x + 6$
6. $3x^2 + 14x + 8$
7. $x^2 + 9x + 14$
8. $x^2 + 5x - 36$
9. $4x^2 - 8x + 3$
10. $5x^2 - 13x - 6$

Complete.

11. $p^2 + 9p - 10 = (p + \underline{?})(p - 1)$
12. $y^2 - 2y - 35 = (y + 5)(y - \underline{?})$
13. $4a^2 + 4a - 63 = (2a - 7)(2a + \underline{?})$
14. $4r^2 - 25r + 6 = (r - 6)(\underline{?}\ \underline{?}\ 1)$

Factor each trinomial.

15. $x^2 + 2x - 15$
16. $n^2 - 8n + 15$
17. $b^2 + 12b + 35$
18. $2x^2 + x - 21$

EXERCISES

Practice

Complete.

19. $3a^2 - 2a - 21 = (a\ \underline{?}\ \underline{?})(3a + 7)$
20. $4y^2 + 11y + 6 = (\underline{?}\ \underline{?}\ 3)(y + 2)$
21. $2z^2 - 11z + 15 = (\underline{?} - 5)(z - \underline{?})$
22. $6n^2 + 7n - 3 = (2n + \underline{?})(\underline{?} - 1)$

Factor each trinomial, if possible. If the trinomial cannot be factored using integers, write *prime*.

23. $y^2 + 12y + 27$
24. $a^2 + 22a + 21$
25. $c^2 + 2c - 3$
26. $h^2 + 5h - 8$
27. $x^2 - 5x - 24$
28. $3y^2 + 8y + 5$
29. $7a^2 + 22a + 3$
30. $8m^2 - 10m + 3$
31. $6y^2 - 11y + 4$
32. $2r^2 + 3r - 14$
33. $2x^2 - 5x - 12$
34. $2q^2 - 9q - 18$
35. $12r^2 - 11r + 3$
36. $10 + 19m + 6m^2$
37. $36 - 13y + y^2$
38. $x^2 - 4xy - 5y^2$
39. $a^2 + 2ab - 3b^2$
40. $15x^2 - 13xy + 2y^2$
41. $3s^2 - 10st - 8t^2$
42. $9k^2 + 30km + 25m^2$
43. $10a^2 - 34ab + 27b$

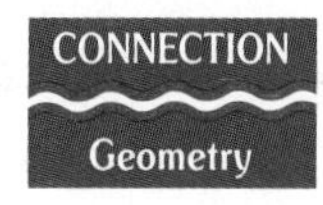

44. The area of a rectangle is $3x^2 + 14x + 15$ square meters. This area is reduced by decreasing both the length and width by 3 meters. If the dimensions of the original rectangle are represented by binomials with integral coefficients, find the area of the new rectangle.

Find all values of *k* so that each trinomial can be factored using integers.

45. $x^2 + kx + 10$
46. $m^2 + km + 6$
47. $r^2 + kr - 13$
48. $2c^2 + kc + 12$
49. $3s^2 + ks - 14$
50. $4y^2 + ky - 15$

Factor.

51. $40x^4 - 116x^3 + 84x^2$
52. $20a^4b - 59a^3b^2 + 42a^2b^3$
53. $2a^2x + 3a^2y - 14ax - 21ay + 24x + 36y$
54. $4ax^2 - 12bx^2 - ax + 3bx - 18a + 54b$

Critical Thinking

55. **Geometry** The measure of the area of a triangle is $7.5x^2 + 15.5x - 12$ where x is a positive integer. If the measure of the height of the triangle is $3x + 8$, what is the least possible measure of its base?

Applications

56. **Shipping** A shipping crate is to be built in the shape of a rectangular solid. The volume of the crate is $45x^2 - 174x + 144$ cubic feet where x is a positive integer. If the height of the crate is 3 feet, what is the minimum possible volume for this crate?

57. **Agriculture** The length of a rectangular vineyard is 20 meters greater than its width. Due to land rezoning, the length of the vineyard was increased by 16 meters and the width was decreased by 10 meters. If the area remained the same, find the original dimensions of the vineyard.

Mixed Review

58. **Finance** Hiroko invested \$11,700, part at 5% interest and the balance at 7% interest. If her annual earnings from both investments is \$733, how much is invested at each rate? **(Lesson 4-3)**
59. Solve $2n - 6 > n + 8.2$. **(Lesson 5-1)**
60. Simplify $(-2x^2y)(-6x^4y^7)$. **(Lesson 6-2)**
61. Subtract $(3x^2 - 7x + 4) - (2x^2 + 8x - 6)$. **(Lesson 6-6)**
62. Factor $15a + 6b + 10a^2 + 4ab$. **(Lesson 7-3)**
63. Solve by using guess and check. **(Lesson 7-4)**

    At a banquet, every 3 guests shared a dish of rice, every 2 guests shared a dish of fish, and every 4 guests shared a dish of corn. There were 65 dishes in all. How many guests attended the banquet?

7-6 Factoring Differences of Squares

Objective

After studying this lesson, you should be able to:

- identify and factor polynomials that are the differences of squares.

Application

Every Friday in Ms. Leshlock's algebra class, each student is given a problem that must be solved without using paper and pencil. This week, Justin's problem was to find the product $93 \cdot 87$. How did he determine that the answer was 8091 without doing the multiplication on paper?

Justin noticed that this product could be written as the product of a sum and a difference.

You can refer to Lesson 6-9 to review the rule for the product of a sum and a difference.

$$93 \cdot 87 = (90 + 3)(90 - 3)$$

He then did the calculation mentally using the rule for this special product.

$$\begin{aligned}(90 + 3)(90 - 3) &= 90^2 - 3^2 \\ &= 8100 - 9 \\ &= 8091\end{aligned}$$

The product of the sum and difference of two expressions is called the *difference of squares.* The process for finding this product can be reversed in order to factor the difference of squares. Factoring the difference of squares can also be modeled geometrically. Consider the two squares shown below. The area of the larger square is a^2 and the area of the smaller square is b^2.

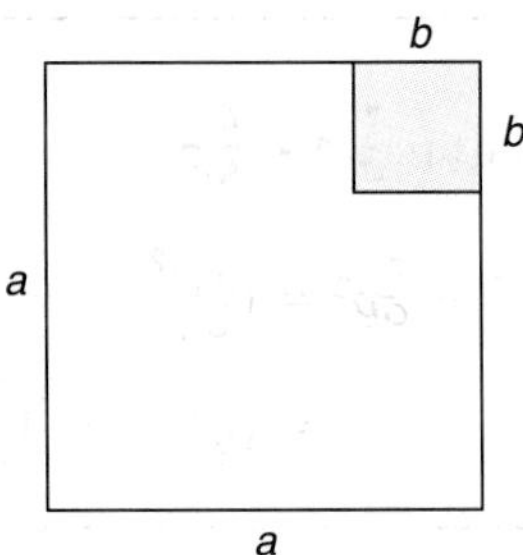

The area $a^2 - b^2$ can be found by subtracting the area of the smaller square from the area of the larger square.

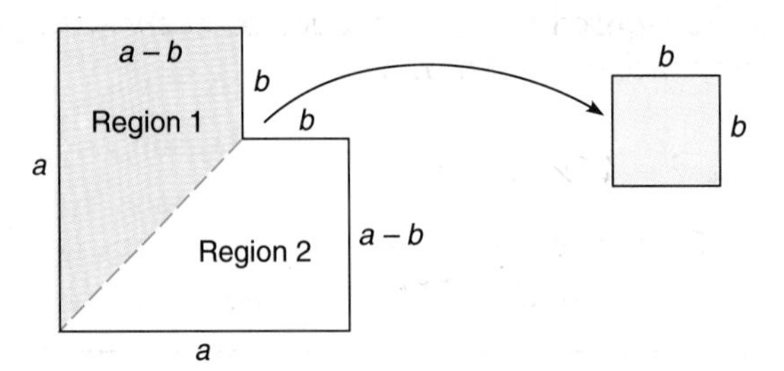

Notice that the square with area b^2 has been removed.

By rearranging these regions as shown below, you can see that $a^2 - b^2$ is equal to the product of $(a - b)$ and $(a + b)$.

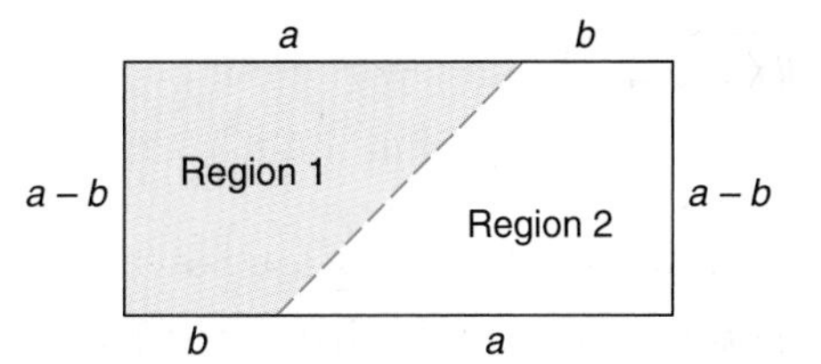

The area of the rectangle is $(a - b)(a + b)$. This is the same as $a^2 - b^2$.

Both the geometric model and the rule for the product of a sum and a difference lead us to the following rule for factoring the difference of squares.

Difference of Squares

$$a^2 - b^2 = (a - b)(a + b) = (a + b)(a - b)$$

You can use this rule to factor binomials that can be written in the form $a^2 - b^2$.

Example 1

Factor $a^2 - 64$.

$a^2 - 64 = (a)^2 - (8)^2$ *$a \cdot a = a^2$ and $8 \cdot 8 = 64$*

$= (a - 8)(a + 8)$ *Use the difference of squares.*

Example 2

Factor $9x^2 - 100y^2$.

$9x^2 - 100y^2 = (3x)^2 - (10y)^2$ *$3x \cdot 3x = 9x^2$ and $10y \cdot 10y = 100y^2$*

$= (3x - 10y)(3x + 10y)$ *Use the difference of squares.*

Example 3

Factor $\frac{1}{4}t^2 - \frac{4}{9}p^2$.

$\frac{1}{4}t^2 - \frac{4}{9}p^2 = \left(\frac{1}{2}t\right)^2 - \left(\frac{2}{3}p\right)^2$ *Why?*

$= \left(\frac{1}{2}t - \frac{2}{3}p\right)\left(\frac{1}{2}t + \frac{2}{3}p\right)$ *Check this result by using FOIL.*

Sometimes the terms of a binomial have common factors. If so, the GCF should always be factored out first. Occasionally, the difference of squares needs to be applied more than once or along with grouping in order to completely factor a polynomial.

Example 4

Factor $12x^3 - 27xy^2$.

$12x^3 - 27xy^2 = 3x(4x^2 - 9y^2)$ *The GCF of $12x^3$ and $27xy^2$ is $3x$.*

$= 3x(2x - 3y)(2x + 3y)$ *$2x \cdot 2x = 4x^2$ and $3y \cdot 3y = 9y^2$*

Example 5

Factor $162m^4 - 32n^8$.

$$162m^4 - 32n^8 = 2(81m^4 - 16n^8) \quad \textit{Why?}$$
$$= 2(9m^2 - 4n^4)(9m^2 + 4n^4)$$
$$= 2(3m - 2n^2)(3m + 2n^2)(9m^2 + 4n^4)$$

$9m^2 + 4n^4$ cannot be factored. Why not?

Example 6

The measure of the volume of a rectangular solid is $5x^3 - 20x + 2x^2 - 8$. Find the measures of the dimensions of the solid if each one can be written as a binomial with integral coefficients.

The volume of a rectangular solid is the product of its length, width, and height. To find these dimensions, you must factor $5x^3 - 20x + 2x^2 - 8$ as the product of three binomials.

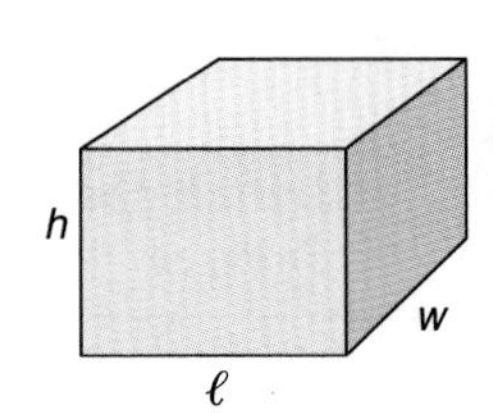

$$5x^3 - 20x + 2x^2 - 8 = (5x^3 - 20x) + (2x^2 - 8)$$
$$= 5x(x^2 - 4) + 2(x^2 - 4)$$
$$= (5x + 2)(x^2 - 4)$$
$$= (5x + 2)(x - 2)(x + 2)$$

The measures of the dimensions are $(5x + 2)$, $(x - 2)$, and $(x + 2)$.

CHECKING FOR UNDERSTANDING

Communicating Mathematics

Read and study the lesson to answer each question.

1. Under what conditions can a binomial be classified as the difference of two squares?
2. Can a difference of squares be factored using the method for factoring trinomials presented in Lesson 7-5? Explain.
3. Use the difference of squares to evaluate $\frac{15}{16} \cdot \frac{17}{16}$.

Guided Practice

State whether each binomial can be factored as a difference of squares.

4. $x^2 - y^2$
5. $a^2 - b^2$
6. $4c^2 - 7$
7. $a^2 - 4b^2$
8. $a - 9$
9. $0.04m^2 - 0.09n^2$

Match each binomial with its factored form.

10. $4x^2 - 25$
11. $16x^2 - 4$
12. $25x^2 - 4$
13. $25x^2 - 25$

a. $25(x - 1)(x + 1)$
b. $(5x - 2)(5x + 2)$
c. $(2x - 5)(2x + 5)$
d. $4(2x - 1)(2x + 1)$

EXERCISES

Practice

Factor each polynomial, if possible. If the polynomial cannot be factored, write *prime*.

14. $a^2 - 9$
15. $x^2 - 49$
16. $4x^2 - 9y^2$
17. $x^2 - 36y^2$
18. $1 - 9y^2$
19. $16a^2 - 9b^2$
20. $49 - a^2b^2$
21. $2a^2 - 25$
22. $8x^2 - 12y^2$
23. $2z^2 - 98$
24. $12a^2 - 48$
25. $8x^2 - 18$
26. $45x^2 - 20y^2z^2$
27. $25y^2 - 49z^4$
28. $17 - 68a^2$
29. $0.01n^2 - 1.69r^2$
30. $a^2x^2 - 0.64y^2$
31. $36x^2 - 125y^2$
32. $9x^4 - 16y^2$
33. $-16 + 49x^2$
34. $-9x^2 + 81$
35. $\frac{1}{16}x^2 - 25z^2$
36. $\frac{9}{2}a^2 - \frac{49}{2}b^2$
37. $\frac{1}{4}n^2 - 16$
38. $(a + b)^2 - m^2$
39. $(r - t)^2 + t^2$
40. $(x - y)^2 - y^2$

41. The difference of two numbers is 3. If the difference of their squares is 15, what is the sum of the numbers?

Find the dimensions of each rectangular solid having the given volume measure if each dimension can be written as a binomial with integral coefficients.

42. $x^3 + x^2 - x - 1$
43. $9a^3 + 18a^2 - 4a - 8$
44. $7mp^2 + 2np^2 - 7mr^2 - 2nr^2$
45. $5a^3 - 125ab^2 - 75b^3 + 3a^2b$

Factor each polynomial.

46. $x^4 - y^4$
47. $16 - a^4$
48. $9x^4 - 25y^4$
49. $48x^5 - 3xy^4$
50. $48s^5t - 243st^5$
51. $x^8 - 1$

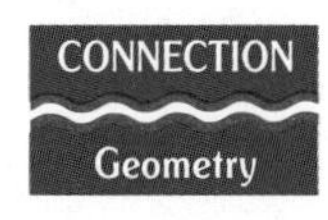

52. The side of a square is x centimeters long. The length of a rectangle is 5 centimeters longer than a side of this square and the width is 5 centimeters shorter.
    a. Which has the greater area, the square or the rectangle?
    b. How much greater is that area?

Critical Thinking

Find positive integers *m* and *n* such that $m^2 - n^2$ has each value.

53. 11
54. 15
55. 45
56. 105

Applications

57. **Photography** To get a square photograph to fit into a rectangular frame, Li-Chih had to trim a 1-inch strip from one pair of opposite sides of the photo and a 2-inch strip from the other two sides. In all, he trimmed off 64 square inches. What were the original dimensions of the photograph?

58. **Finance** Maria invested some money at 8.5% annual interest. Juanita invested $800 more than Maria at 7.5% annual interest. If Maria and Juanita received the same amount of interest after one year, how much money did each invest?

Mixed Review

59. Solve $\frac{8}{3}x = \frac{4}{9}$. **(Lesson 3-3)**

60. **Travel** Bob and Vicki took a trip to Zuma Beach. On the way there, their average speed was 42 miles per hour. On the way home, their average speed was 56 miles per hour. If their total travel time was 7 hours, find the distance to the beach. **(Lesson 4-7)**

61. Solve $|2x + 7| < 11$. **(Lesson 5-6)**

62. Simplify $\frac{-8a^3b^7}{a^2b^6}$. **(Lesson 6-3)**

63. Find the degree of $6mn^4t - 3m^4nt^2 + mn^6$. **(Lesson 6-5)**

64. Factor $16x^2 + 14x - 15$. **(Lesson 7-5)**

MID-CHAPTER REVIEW

Find the GCF of the given monomials. (Lesson 7-1)

1. $39a^2, 13a^3$
2. $64a^2c^2, 4ab^2c^4$
3. $55m^3n, 275m^5n^2, 165m^2n^3$

Factor each polynomial, if possible. If the polynomial cannot be factored using integers, write *prime*. (Lessons 7-2, 7-3, 7-5, and 7-6)

4. $y^2 - 8y + 7$
5. $25m^3n + 15m^2n^2$
6. $r^2 + 3r - 18$
7. $3a^2b - 8a - 12ab - 32$
8. $5a^3 + 45a^2 - 15a$
9. $6p^2 + 7p - 3$
10. $8k^2 - 72z^2$
11. $-7a^2b^2 - 77a^3b^2 + 77ab^3$
12. $5x^2 - 19x + 14$
13. $y^2m + y^2n - 4m - 4n$
14. Solve by using guess and check. **(Lesson 7-4)**
Write an eight-digit number using the digits 1, 2, 3, and 4 each twice so that the 1s are separated by 1 digit, the 2s are separated by 2 digits, the 3s are separated by 3 digits, and the 4s are separated by 4 digits.
15. **Geometry** The area of a rectangle is $4x^2 - 196$ square meters. This area was increased by increasing both the length and width by 9 meters. If the dimensions of the original rectangle are binomials with integral coefficients, find the area of the new rectangle. **(Lesson 7-6)**

7-7 Perfect Squares and Factoring

Objective After studying this lesson, you should be able to:

- identify and factor perfect square trinomials.

Numbers such as 1, 4, 9, and 16 are called *perfect squares* since they can be expressed as the square of an integer. Products of the form $(a + b)^2$ and $(a - b)^2$ are also called perfect squares, and the expansions of these products are called **perfect square trinomials.**

Perfect Square Trinomials

$$(a + b)^2 = a^2 + 2ab + b^2$$
$$(a - b)^2 = a^2 - 2ab + b^2$$

These patterns can be used to help you factor trinomials.

Finding a Product

$$(y + 8)^2 = y^2 + 2(y)(8) + 8^2$$
$$= y^2 + 16y + 64$$

$$(2x - 5z)^2$$
$$= (2x)^2 - 2(2x)(5z) + (5z)^2$$
$$= 4x^2 - 20xz + 25z^2$$

Factoring

$$y^2 + 16y + 64 = (y)^2 + 2(y)(8) + (8)^2$$
$$= (y + 8)^2$$

$$4x^2 - 20xz + 25z^2$$
$$= (2x)^2 - 2(2x)(5z) + (5z)^2$$
$$= (2x - 5z)^2$$

To determine whether a trinomial can be factored by using these patterns, you must first decide if it is a perfect square trinomial. In other words, you must determine if it can be written in the form $a^2 + 2ab + b^2$ or in the form $a^2 - 2ab + b^2$.

Example 1

Determine whether $x^2 + 22x + 121$ is a perfect square trinomial. If so, factor it.

To determine whether $x^2 + 22x + 121$ is a perfect square trinomial, answer each question.

a. Is the first term a perfect square? $x^2 \stackrel{?}{=} (x)^2$ *yes*

b. Is the last term a perfect square? $121 \stackrel{?}{=} (11)^2$ *yes*

c. Is the middle term twice the product of x and 11? $22x \stackrel{?}{=} 2(x)(11)$ *yes*

So, $x^2 + 22x + 121$ is a perfect square trinomial. It can be factored as follows.

$$x^2 + 22x + 121 = (x)^2 + 2(x)(11) + (11)^2$$
$$= (x + 11)^2$$

Example 2

Determine whether $16a^2 + 81 - 72a$ is a perfect square trinomial. If so, factor it.

First arrange the terms of $16a^2 + 81 - 72a$ so that the powers of a are in descending order.

$16a^2 + 81 - 72a = 16a^2 - 72a + 81$

a. Is the first term a perfect square? $16a^2 \stackrel{?}{=} (4a)^2$ *yes*

b. Is the last term a perfect square? $81 \stackrel{?}{=} (9)^2$ *yes*

c. Is the middle term twice the product of $4a$ and 9? $72a \stackrel{?}{=} 2(4a)(9)$ *yes*

$16a^2 - 72a + 81$ is a perfect square trinomial.

$$16a^2 - 72a + 81 = (4a)^2 - 2(4a)(9) + (9)^2$$
$$= (4a - 9)^2$$

Example 3

Determine whether $9p^2 - 56p + 49$ is a perfect square trinomial. If so, factor it.

a. Is the first term a perfect square? $9p^2 \stackrel{?}{=} (3p)^2$ *yes*

b. Is the last term a perfect square? $49 \stackrel{?}{=} (7)^2$ *yes*

c. Is the middle term twice the product of $3p$ and 7? $56p \stackrel{?}{=} 2(3p)(7)$ *no*

$9p^2 - 56p + 49$ is *not* a perfect square trinomial.

Example 4

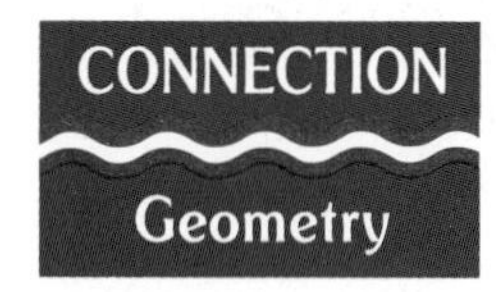

Is it possible for $9x^2 + 12xy + 4y^2$ to be the measure of the area of a square? If so, what is the measure of each side of the square?

For $9x^2 + 12xy + 4y^2$ to be the measure of the area of a square, it must be a perfect square trinomial. *Why?*

Since $9x^2 = (3x)^2$, $4y^2 = (2y)^2$, and $12xy = 2(3x)(2y)$, the trinomial $9x^2 + 12xy + 4y^2$ is a perfect square trinomial.

$$9x^2 + 12xy + 4y^2 = (3x)^2 + 2(3x)(2y) + (2y)^2$$
$$= (3x + 2y)^2$$

Thus, $9x^2 + 12xy + 4y^2$ can be the measure of the area of a square. The measure of each side of this square is $3x + 2y$.

CHECKING FOR UNDERSTANDING

Communicating Mathematics

Read and study the lesson to answer each question.

1. If an integer is a perfect square, what are the possible values for its units digit?
2. In a perfect square trinomial with a constant term, can the constant term ever be a negative integer? Why or why not?
3. *True* or *False:* The middle term of a perfect square trinomial is twice the product of the first and last term of the trinomial.

Guided Practice

Complete.

4. $b^2 + 10b + 25 = (b + \underline{\ ?\ })^2$
5. $y^2 + 14y + 49 = (y + \underline{\ ?\ })^2$
6. $m^2 - 24m + 144 = (m - \underline{\ ?\ })^2$
7. $64b^2 - 16b + 1 = (\underline{\ ?\ } - 1)^2$
8. $81n^2 + 36n + 4 = (\underline{\ ?\ } + 2)^2$
9. $1 - 12x + 36x^2 = (1 - \underline{\ ?\ })^2$

Determine whether each trinomial is a perfect square trinomial. If so, factor it.

10. $a^2 + 4a + 4$
11. $x^2 + 20x - 100$
12. $y^2 - 10y + 100$
13. $b^2 - 14b + 49$
14. $4x^2 + 4x + 1$
15. $2n^2 + 17n + 21$

EXERCISES

Practice

Determine whether each trinomial is a perfect square trinomial. If so, factor it.

16. $n^2 - 13n + 36$
17. $p^2 + 12p + 36$
18. $9b^2 - 6b + 1$
19. $9y^2 + 8y - 16$
20. $9x^2 - 10x + 4$
21. $4a^2 - 20a + 25$

Factor each trinomial, if possible. If the trinomial cannot be factored, write *prime*.

22. $x^2 + 16x + 64$
23. $n^2 - 8n + 16$
24. $a^2 + 22a + 121$
25. $4k^2 - 4k + 1$
26. $100x^2 + 20x + 1$
27. $x^2 + 6x - 9$
28. $9a^2 + 12a - 4$
29. $1 - 10z + 25z^2$
30. $4 - 28r + 49r^2$
31. $50x^2 + 40x + 8$
32. $18a^2 - 48a + 32$
33. $49m^2 - 126m + 81$
34. $64b^2 - 72b + 81$
35. $25x^2 - 120x + 144$
36. $9x^2 + 24xy + 16y^2$
37. $m^2 + 16mn + 64n^2$
38. $16p^2 - 40pr + 25r^2$
39. $4x^2 + 4xz^2 + z^4$
40. $16m^4 - 72m^2n^2 + 81n^4$
41. $\frac{1}{4}a^2 + 3a + 9$
42. $\frac{4}{9}x^2 - \frac{16}{3}x + 16$

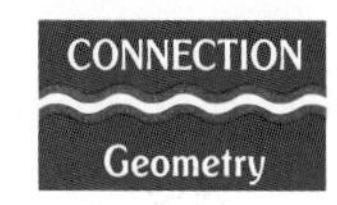

Determine whether each trinomial could be the measure of the area of a square. If so, find the measure of the side of the square.

43. $121y^2 + 22y + 1$
44. $64x^2 - 80x + 25$
45. $49m^2 + 150m + 100$
46. $4b^2 - 24bc + 36c^2$

Find all values of c so that each trinomial is a perfect square.

47. $cx^2 + 28x + 49$
48. $cn^2 - 220n + 121$
49. $9a^2 - 12ab + c$
50. $9x^2 + cxy + 49y^2$
51. $16m^2 + cmp + 25p^2$
52. $64x^2 - 16xy + c$

Factor each polynomial.

53. $a^2 + 4a + 4 - 9b^2$
54. $x^2 + 2xy + y^2 - r^2$
55. $m^2 - k^2 + 6k - 9$
56. $16 - 9x^2 - 12xz - 4z^2$
57. $a^2m - 2a^2 + 6am - 12a + 9m - 18$
58. $8ay^2 + 12y^2 + 40ay + 60y + 50a + 75$

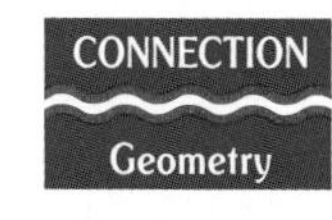

59. The measure of the area of a circle is $(9y^2 + 78y + 169)\pi$. What is the measure of its diameter?

Critical Thinking

60. **Geometry** The measure of the area of a square is $81 - 90x + 25x^2$, where x is a positive integer. What is the smallest possible perimeter measure for this square?

Applications

61. **Photography** A square photograph is placed in a frame that is 2 inches wide on all sides. If the measure of the area of the frame is 184 square inches, what are the dimensions of the photograph?

62. **Construction** The length of a rectangular lot is 60 yards greater than its width, w. To change the lot into a square lot, 900 square yards were added. Find the length of a side of this square lot.

Mixed Review

63. Fifteen is 40% of what number? **(Lesson 4-2)**

64. **Travel** A car uses 8 gallons of gasoline to travel 264 miles. How much gasoline will the car use to travel 500 miles? **(Lesson 4-8)**

65. **Number Theory** If 6.5 times an integer is increased by 11, the result is between 55 and 75. What is the integer? **(Lesson 5-5)**

Simplify. (Lessons 6-8, 6-9)

66. $(3x - 2)(3x + 2)$
67. $(8t - 3)(2t + 5)$

Factor each polynomial. (Lessons 7-2, 7-6)

68. $5x^2 - 80y^4$
69. $15x^3y^4 - 30x^2yz$

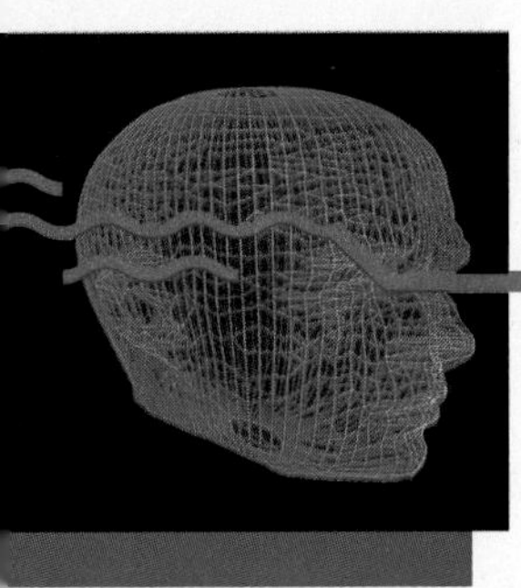

Technology

Factoring

BASIC
Graphing calculators
▶ **Graphing software**
Spreadsheets

The examples below show how to use the *Mathematics Exploration Toolkit* to factor polynomials. You will need the following CALC commands.

FACTOR (fac) FSTEPS (fst) GFACTOR (gcf) SIMPLIFY (simp)

GFACTOR factors out the greatest common monomial factor of a polynomial. If the polynomial has no common monomial factors, the original polynomial will be displayed again. Sometimes SIMPLIFY must be used first, as in Example 1.

Example 1

Factor $2a(3b + 6) + 3ab(4a - 2)$.

Enter	Result
$2a(3b + 6) + 3ab(4a - 2)$	$2a(3b + 6) + 3ab(4a - 2)$
simp	$12a^2b + 12a$
gcf	$12a(ab + 1)$

FACTOR finds the factors of polynomials with one variable. FSTEPS displays the steps used to execute the FACTOR command.

Example 2

Factor $x^5 - 16x$.

Enter	Result
x^5 − 16x	$x^5 - 16x$
gcf	$x(x^4 - 16)$
fac	$x((x - 2)(x + 2)(x^2 + 4))$
fst	$x^4 - 16$
	$(x^2 - 4)(x^2 + 4)$
	$((x - 2)(x + 2))(x^2 + 4)$
	$x((x - 2)(x + 2)(x^2 + 4))$

EXERCISES

Use GFACTOR to find the greatest common factor of each polynomial.

1. $96x^3y^3 - 80x^2y + 64xy^2$

2. $5x(y^2 + 10y) - 20x^2$

Use FACTOR and FSTEPS to factor each polynomial. Record each step.

3. $36x^2 - 64$

4. $25y^2 + 10y + 1$

5. $81z^4 - 625$

6. $36a^3 - 5a^2 - a$

7-8 Summary of Factoring

Objective After studying this lesson, you should be able to:

- factor polynomials by applying the various methods of factoring.

In this chapter, you have used various methods to factor different types of polynomials. The following chart summarizes these methods and can help you decide when to use a specific method.

Check for:	Number of Terms		
	Two	Three	Four or More
greatest common factor	✓	✓	✓
difference of squares	✓		✓
perfect square trinomial		✓	
trinomial that has two binomial factors		✓	
pairs of terms that have a common monomial factor			✓

Whenever there is a GCF other than 1, always factor it out first. Then, check the appropriate factoring methods in the order shown in the table. Use these methods to factor until all of the factors are prime.

Example 1 **Factor $3x^2 - 27$.**

First check for a GCF. Then, since there are two terms, check for the difference of squares.

$3x^2 - 27 = 3(x^2 - 9)$ *3 is the GCF.*

$= 3(x - 3)(x + 3)$ *$x^2 - 9$ is the difference of squares since $x \cdot x = x^2$ and $3 \cdot 3 = 9$.*

Thus, $3x^2 - 27$ is completely factored as $3(x - 3)(x + 3)$.

Example 2 **Factor $9y^2 - 58y + 49$.**

The polynomial has three terms. So, check for the following.

GCF: The GCF is 1.

Perfect square trinomial: Although $9y^2 = (3y)^2$ and $49 = (7)^2$, $58y \neq 2(3y)(7)$.

Trinomial with two binomial factors: Are there two numbers whose product is $9 \cdot 49$ or 441 and whose sum is -58? Yes, the product of -9 and -49 is 441 and their sum is -58.

$$9y^2 - 58y + 49 = 9y^2 + [-9 + (-49)]y + 49$$
$$= (9y^2 - 9y) + (-49y + 49)$$
$$= 9y(y - 1) + (-49)(y - 1)$$
$$= (9y - 49)(y - 1)$$ *Check this by using FOIL.*

Thus, $9y^2 - 58y + 49$ is completely factored as $(9y - 49)(y - 1)$.

Example 3

Factor $4m^4n + 6m^3n^2 - 16m^2n - 24mn^2$.

Since $4m^4n + 6m^3n^2 - 16m^2n - 24mn^2$ has four terms, first check for the GCF. Then, check for pairs of terms that have a common monomial factor.

$$4m^4n + 6m^3n^2 - 16m^2n - 24mn^2 = 2mn(2m^3 + 3m^2n - 8m - 12n)$$
$$= 2mn[(2m^3 + 3m^2n) + (-8m - 12n)]$$
$$= 2mn[m^2(2m + 3n) + (-4)(2m + 3n)]$$
$$= 2mn[(m^2 - 4)(2m + 3n)]$$
$$= 2mn(m - 2)(m + 2)(2m + 3n)$$

Thus, $4m^4n + 6m^3n^2 - 16m^2n - 24mn^2$ is completely factored as $2mn(m - 2)(m + 2)(2m + 3n)$. *Check this by multiplying the factors.*

Example 4

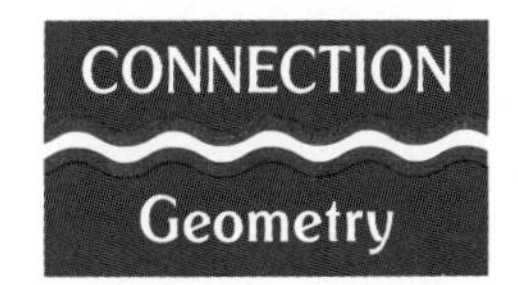

The length of a rectangle is 3 centimeters more than the length of the side of a square. The width of this rectangle is one-half the length of the side of the square. If the area of the square is $4x^2 + 12x + 9$ square centimeters, what is the area of the rectangle?

You need to factor $4x^2 + 12x + 9$ in order to determine the length of the side of the square. $4x^2 + 12x + 9$ is a perfect square trinomial. *Why?*

$$4x^2 + 12x + 9 = (2x)^2 + 2(2x)(3) + (3)^2$$
$$= (2x + 3)^2$$

The length of the side of the square is $2x + 3$ cm. Thus, the length of the rectangle is $(2x + 3) + 3$ or $2x + 6$ cm and the width is $\frac{1}{2}(2x + 3)$ cm. Now multiply to determine the area of the rectangle.

$$(2x + 6)\left[\frac{1}{2}(2x + 3)\right] = \frac{1}{2}[2x(2x) + (2x)(3) + 6(2x) + 6(3)]$$ *Use FOIL.*
$$= \frac{1}{2}(4x^2 + 18x + 18)$$
$$= 2x^2 + 9x + 9$$

The area of the rectangle is $2x^2 + 9x + 9$ cm^2.

CHECKING FOR UNDERSTANDING

Communicating Mathematics

Complete.

1. When factoring, the ___?___ of the terms of a polynomial should always be factored out first.
2. To factor a ___?___, you check for the GCF, for a perfect square, and for two binomial factors.
3. You should check for pairs of terms that have a common monomial factor when you factor a polynomial with ___?___ terms.
4. A polynomial is factored completely when each of its factors is ___?___.

Guided Practice

Indicate which method of factoring you would apply first to each polynomial.

5. $x^2 - 5x$
6. $64c^2 - 25p^2$
7. $9x^2 - 12xy + 4y^2$
8. $16m^2 - 64n^2$
9. $35z^2 + 13z - 12$
10. $2x^2 + 4xy - x - 2$

Factor. Check by using FOIL or the distributive property.

11. $3x^2 + 15$
12. $8mn^2 - 13m^2n$
13. $a^2 - 9b^2$
14. $x^2 - 5x + 6$
15. $a^2 + 8a + 16$
16. $3a^2b + 6ab + 9ab^2$

EXERCISES

Practice

Factor. Check by using FOIL or the distributive property.

17. $12a^2 + 18ay^2$
18. $2a^2 - 72$
19. $3y^2 - 147$
20. $2k^2 + 3k + 1$
21. $m^3 + 6m^2 + 9m$
22. $18y + 12y^2 + 2y^3$
23. $6r^2 + 13r + 6$
24. $4x^3 - 3x^2 - 12x + 9$
25. $m^4 - p^2$
26. $4a^3 - 36a$
27. $3x^3 - 27x$
28. $3y^2 + 21y - 24$
29. $20n^2 + 34n + 6$
30. $m^2 + 8mn + 16n^2$
31. $4a^2 + 12ab + 9b^2$
32. $4y^3 - 12y^2 + 8y$
33. $9t^3 + 66t^2 - 48t$
34. $a^2b^3 - 25b$
35. $m^3n^2 - 49m$
36. $0.4r^2 + 1.6r + 1.6$
37. $0.7y^2 - 3.5y + 4.2$
38. $\frac{1}{3}b^2 + 2b + 3$
39. $m^2 + \frac{5}{12}m - \frac{1}{6}$
40. $\frac{1}{4}x^2 + \frac{3}{2}x + 2$
41. $x^3y + 2x^2 + 8xy + 16$
42. $4a^3 + 3a^2b^2 + 8a + 6b^2$
43. $x^2y^2 - z^2 - y^2 + x^2z^2$
44. $20a^2x - 4a^2y - 45xb^2 + 9yb^2$
45. $(x + y)^2 - (a - b)^2$
46. $x^2 + 6x + 9 - y^2$
47. $(x + 1)^2 - 3(x + 1) + 2$
48. $(2x - 3)^2 - 4(2x - 3) - 5$
49. $x^4 + 6x^3 + 9x^2 - 3x^2y - 18xy - 27y$
50. $12mp^2 - 15np^2 - 16m + 20np - 16mp + 20n$

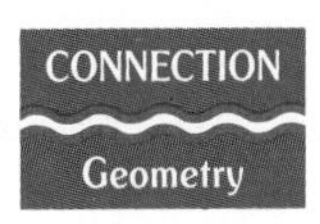

51. The measure of the volume of a rectangular solid is $x^3y - 63y^2 + 7x^2 - 9xy^3$. What are the measures of its dimensions?

Critical Thinking

52. A *nasty* number is a positive integer with at least four different factors such that the difference between the numbers in one pair of factors equals the sum of the numbers in another pair. The first nasty number is 6 since $6 = 6 \cdot 1 = 2 \cdot 3$ and $6 - 1 = 2 + 3$. Find the next five nasty numbers. (*Hint:* They are all multiples of 6.)

Applications

53. **Gardening** The length of Jillian's rectangular garden is 12 feet greater than its width. Because she wants to plant some flowers next to the garden, she decreased its length by 5 feet and increased its width by 2 feet. If the area of the garden was decreased by 55 square feet, what were its original dimensions?

54. **Conferences** A one-day Pacific Coast Land Usage Conference is being attended by 32 delegates from California, 24 from Washington, and 20 from Oregon.
    a. What is the greatest number of meetings that can be held so that the delegates from each state are distributed equally among all the meetings?
    b. If each meeting room can seat 20 people, how many extra seats will there be at each meeting?

Mixed Review

55. A price was increased from \$85 to \$110. Find the percent of increase, to the nearest tenth of a percent. **(Lesson 4-4)**

56. **Sales** Carita plans to spend at most \$50 on shorts and blouses. She bought 2 pair of shorts for \$14.20 each. How much can she spend on blouses? **(Lesson 5-1)**

57. Solve by drawing a diagram. **(Lesson 5-4)**

    Eight soccer teams play in the Central League. Each team is required to play every other team exactly once. How many games will be played in this league?

58. Simplify $(-0.2x^3y)^3$. **(Lesson 6-2)**

59. Simplify $\frac{(2a^{-1}b^2)^2}{4a^{-3}b}$. **(Lesson 6-3)**

60. **Geometry** The measure of the area of a square is $9s^2 - 42s + 49$. What is the measure of its perimeter? **(Lesson 7-7)**

7-9 Solving Equations by Factoring

Objective

After studying this lesson, you should be able to:

- use the zero product property to solve equations.

Application

The Assyrians first used life rafts made of inflated goat skins in 880 B.C.

A flare is launched from a life raft with an initial upward velocity of 144 feet per second. How many seconds will it take for the flare to return to the sea?

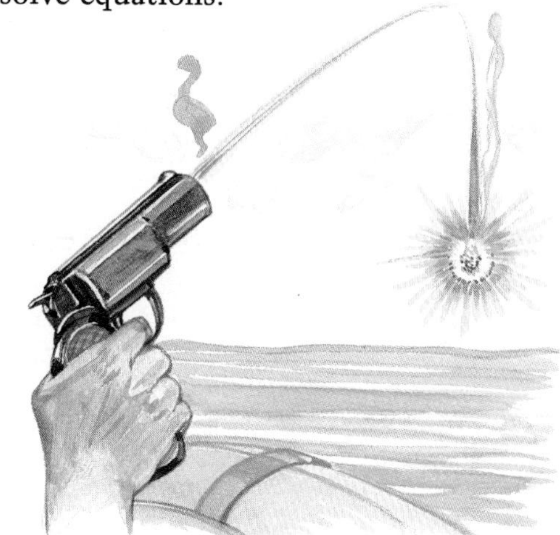

If an object is launched from ground level, it reaches its maximum height in the air at a time halfway between the launch and impact times. Its height above the ground after t seconds is given by the formula $h = vt - 16t^2$. In this formula, h represents the height of the object in feet, and v represents the object's initial velocity in feet per second.

Since the flare's height when it returns to the sea is 0 feet, $h = 0$. Since the flare's initial velocity is 144 feet per second, $v = 144$. Thus, this problem can be represented by the equation $0 = 144t - 16t^2$.

$$0 = 144t - 16t^2 \quad \text{The GCF is } 16t.$$
$$0 = 16t(9 - t)$$

To solve this equation, you need to find the values of t that make the product $16t(9 - t)$ equal to 0. Consider the following products.

$$3(0) = 0 \qquad 0(-8) = 0 \qquad \left(\frac{1}{2} - \frac{1}{2}\right)(0) = 0 \qquad 0(x + 2) = 0$$

Notice that in each case, *at least one* of the factors is zero. These examples illustrate the **zero product property.**

Zero Product Property	**For all numbers a and b, if $ab = 0$, then $a = 0$, $b = 0$, or both a and b equal 0.**

Thus, if an equation can be written in the form $ab = 0$, then the zero product property can be applied to solve that equation.

Example 1

Solve $16t(9 - t) = 0$ to find how long it took the flare to return to the sea.

If $16t(9 - t) = 0$, then $16t = 0$ or $9 - t = 0$. *Zero product property*

$16t = 0$ or $9 - t = 0$ *Solve each equation.*

$t = 0$ $\qquad 9 = t$

Check: Substitute 0 and 9 for t in the original equation.

$$16t(9 - t) = 0$$

$$16(0)(9 - 0) \stackrel{?}{=} 0 \quad \text{or} \quad 16(9)(9 - 9) \stackrel{?}{=} 0$$
$$0(9) \stackrel{?}{=} 0 \qquad\qquad 144(0) \stackrel{?}{=} 0$$
$$0 = 0 \ \checkmark \qquad\qquad 0 = 0 \ \checkmark$$

The solutions of $16t(9 - t) = 0$ are 0 and 9. So, the flare returns to the sea in 9 seconds. The answer 0 seconds is not reasonable since it represents the time when the flare was launched from the life raft.

Example 2

Solve $(y + 2)(3y + 5) = 0$. Then check the solution.

If $(y + 2)(3y + 5) = 0$, then $y + 2 = 0$ or $3y + 5 = 0$. *Zero product property*

$$y + 2 = 0 \quad \text{or} \quad 3y + 5 = 0$$
$$y = -2 \qquad\qquad 3y = -5$$
$$y = -\frac{5}{3}$$

Check: $(y + 2)(3y + 5) = 0$

$$(-2 + 2)[3(-2) + 5] \stackrel{?}{=} 0 \quad \text{or} \quad \left(-\frac{5}{3} + 2\right)\left[3\left(-\frac{5}{3}\right) + 5\right] \stackrel{?}{=} 0$$
$$0(-1) \stackrel{?}{=} 0 \qquad\qquad \frac{1}{3}(0) \stackrel{?}{=} 0$$
$$0 = 0 \ \checkmark \qquad\qquad 0 = 0 \ \checkmark$$

The solution set is $\left\{-2, -\frac{5}{3}\right\}$.

Example 3

Solve $x^2 = 7x$. Then check the solution.

To use the zero product property, the equation must be written in the form $ab = 0$.

$$x^2 = 7x$$
$$x^2 - 7x = 0$$
$$x(x - 7) = 0 \qquad \textit{Factor out the GCF, x.}$$
$$x = 0 \quad \text{or} \quad x - 7 = 0 \qquad \textit{Zero product property}$$
$$x = 7$$

Check: $x^2 = 7x$

$$(0)^2 \stackrel{?}{=} 7(0) \quad \text{or} \quad (7)^2 \stackrel{?}{=} 7(7)$$
$$0 = 0 \ \checkmark \qquad\qquad 49 = 49 \ \checkmark$$

The solution set is $\{0, 7\}$.

In Example 3, if you divide each side of $x^2 = 7x$ by x, the result, $x = 7$, would *not* be an equivalent equation since zero is not a solution of $x = 7$. For this reason, when you are solving an equation, you should not divide by expressions that contain a variable.

Example 4

Maria told this puzzle to her friends. "The product of four times my age and 45 less than three times my age is zero. How old am I?" Find Maria's age now.

EXPLORE Let m = Maria's age now.

PLAN *Four times her age times 45 less than three times her age is 0.*

$$\underbrace{4m}_{\text{Four times her age}} \underbrace{(3m - 45)}_{\text{45 less than three times her age}} = 0$$

SOLVE $4m(3m - 45) = 0$

$4m = 0$ or $3m - 45 = 0$ *Zero product property*

$m = 0$ $\quad 3m = 45$

$m = 15$

Maria is 15 years old now. The solution 0 is not reasonable since Maria cannot be 0 years old.

EXAMINE If $m = 15$, then $3m - 45$ is 0, and the product will be 0. Thus, the answer is correct.

Example 5

The product of twice a number increased by five and three times the number decreased by two is zero. Find the number.

EXPLORE Let n = the number.

PLAN *Twice a number increased by 5 times three times the number decreased by 2 is 0.*

$$(2n + 5) \quad (3n - 2) = 0$$

SOLVE $(2n + 5)(3n - 2) = 0$

$2n + 5 = 0$ or $3n - 2 = 0$ *Zero product property*

$2n = -5$ $\quad 3n = 2$

$n = -\frac{5}{2}$ $\quad n = \frac{2}{3}$

The number is either $-\frac{5}{2}$ or $\frac{2}{3}$.

EXAMINE If n is $-\frac{5}{2}$, then $2n + 5$ is 0, and the product will be 0. If n is $\frac{2}{3}$, then $3n - 2$ is 0 and the product will be 0. Thus, the answer is correct.

CHECKING FOR UNDERSTANDING

Communicating Mathematics

Read and study the lesson to answer each question.

1. Write the zero product property in your own words.
2. If a product of more than two factors is equal to 0, what must be true of the factors?
3. Can you solve $(x + 3)(x + 1) = 0$ by dividing each side of the equation by $x + 3$? Why or why not?

Guided Practice

State the conditions under which each equation will be true.
Example: **$x(x + 7) = 0$ will be true if $x = 0$ or $x + 7 = 0$.**

4. $x(x + 3) = 0$
5. $3r(r - 4) = 0$
6. $3t(4t - 32) = 0$
7. $(x - 6)(x + 4) = 0$
8. $(2y + 8)(3y + 24) = 0$
9. $(4x - 7)(3x + 5) = 0$

Solve. Check each solution.

10. $n(n - 3) = 0$
11. $8c(c + 4) = 0$
12. $3x^2 - \frac{3}{4}x = 0$
13. $7y^2 = 14y$
14. $8a^2 = -4a$
15. $-13x = -26x^2$

Practice

Solve. Check each solution.

16. $y(y - 12) = 0$
17. $7a(a + 6) = 0$
18. $2x(5x - 10) = 0$
19. $(b - 3)(b - 5) = 0$
20. $(t - 5)(t + 5) = 0$
21. $(4x + 4)(2x + 6) = 0$
22. $(p - 8)(2p + 7) = 0$
23. $(3x - 5)^2 = 0$
24. $x^2 - 6x = 0$
25. $m^2 + 36m = 0$
26. $2x^2 + 4x = 0$
27. $4s^2 = -36s$
28. $y^2 - 3y = 4y$
29. $2x^2 = x^2 - 8x$
30. $7y - 1 = -3y^2 + y - 1$
31. $z^2 - 8z + 2 = 2 - 13z$
32. $\frac{1}{2}y^2 - \frac{1}{4}y = 0$
33. $\frac{2}{3}x = \frac{1}{3}x^2$
34. $\frac{5}{6}x^2 - \frac{1}{3}x = \frac{1}{3}x$
35. $\frac{3}{4}a^2 + \frac{7}{8}a = a$

For each problem, define a variable. Then use an equation to solve the problem. Disregard any unreasonable solutions.

36. The product of a certain negative number decreased by 5 and the same number increased by 7 is 0. What is the number?
37. Randy told Honovi, "The product of twice my age decreased by 32 and 5 times my age increased by 6 is 0. How old am I?" Find Randy's age.

For each problem, define a variable. Then use an equation to solve the problem. Disregard any unreasonable solutions.

38. The square of a number subtracted from 8 times the number is equal to twice the number. Find the number.

39. When one integer is added to the square of the next consecutive integer, the sum is 1. Find the integers.

Critical Thinking

40. Write an equation with integral coefficients that has $\{0, 7, -3\}$ as its solution set.

Applications

41. **Gardening** Mr. Steinborn tripled the area of his square garden by increasing its length by 18 feet. What were the dimensions of his original garden?

42. **Construction** The length of a rectangular playground is 10 yards less than 3 times its width. In order to add a baseball diamond next to it, the width of the playground is decreased by 15 yards and the length is decreased by 35 yards. The area of this reduced playground is 675 square yards. What are the dimensions of the original playground?

Use the formula $h = vt - 16t^2$ to solve each problem.

43. **Physics** A flare is launched from a life raft with an initial upward velocity of 192 feet per second. How many seconds will it take for the flare to return to the sea?

44. **Physics** A ball is tossed directly upward with an initial velocity of 120 feet per second. How many seconds will it take for the ball to hit the ground?

Mixed Review

45. **Sales** During a special promotion, Southern Air issued round-trip first-class tickets from Orlando to Dallas at $255 and round-trip coach tickets at $198. If 167 tickets were purchased for a total price of $40,191, how many first-class tickets were sold? **(Lesson 4-6)**

46. Solve $-\frac{5s}{8} \geq \frac{15}{4}$. **(Lesson 5-2)**

47. Find $(3mn^2 + 3mn - n^3) + (5mn^2 - n - 2n^3)$. **(Lesson 6-6)**

48. Find the prime factorization of 2700. **(Lesson 7-1)**

49. Find all values of k so that $4y^2 + ky - 4$ can be factored using integers. **(Lesson 7-5)**

50. Factor $12c^2 + 10cd - 42d^2$. **(Lesson 7-8)**

7-10 More Solving Equations by Factoring

Objective

After studying this lesson, you should be able to:

- solve equations by using various factoring methods and applying the zero product property.

Application

FYI ...

In 1985, Malcolm Forbes, Jr. paid \$104,500 for a signed photograph of Abraham Lincoln and his son Tad.

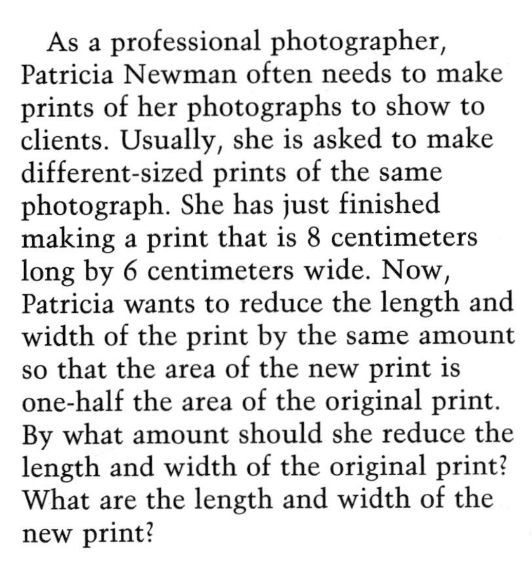

As a professional photographer, Patricia Newman often needs to make prints of her photographs to show to clients. Usually, she is asked to make different-sized prints of the same photograph. She has just finished making a print that is 8 centimeters long by 6 centimeters wide. Now, Patricia wants to reduce the length and width of the print by the same amount so that the area of the new print is one-half the area of the original print. By what amount should she reduce the length and width of the original print? What are the length and width of the new print?

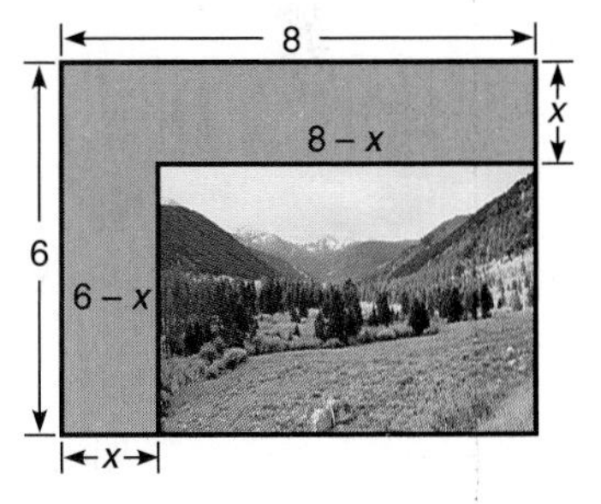

To solve this problem, you need to write an equation for the area of the new print. Let x = the amount by which the length and width should be reduced. Then, the new print is $8 - x$ cm long and $6 - x$ cm wide. Since the area of the original print is $8 \cdot 6$ or 48 cm^2, the area of the new print should be $\frac{1}{2}(48)$ or 24 cm^2.

Therefore, the area of the new print of the photograph can be represented by the following equation.

$$24 = (8 - x)(6 - x) \qquad \textit{area = length} \cdot \textit{width}$$

You can solve equations like this one by first writing them as a product of factors equal to zero and then applying the zero product property.

$$24 = (8 - x)(6 - x)$$
$$24 = 48 - 8x - 6x + x^2$$
$$x^2 - 14x + 24 = 0 \qquad \textit{Rewrite the equation.}$$
$$(x - 12)(x - 2) = 0 \qquad \textit{Factor } x^2 - 14x + 24.$$
$$x - 12 = 0 \text{ or } x - 2 = 0 \qquad \textit{Zero product property}$$
$$x = 12 \qquad x = 2$$

The solution set is {2, 12}.

Patricia cannot reduce the length and width of the original print by 12 cm each since the length is only 8 cm and the width is only 6 cm. Thus, 12 cm is not a reasonable solution.

Therefore, Patricia should reduce the length and width by 2 cm each. The length of the new print will be 8 − 2 or 6 cm and the width will be 6 − 2 or 4 cm.

Check:
$$(8 - x)(6 - x) = 24$$
$$(8 - 2)(6 - 2) \stackrel{?}{=} 24$$
$$6(4) \stackrel{?}{=} 24$$
$$24 = 24 \checkmark$$

Some equations have two solutions that are equal.

Example 1

Solve $a^2 + 64 = -16a$.

$$a^2 + 64 = -16a$$
$$a^2 + 16a + 64 = 0 \quad \textit{Rewrite the equation.}$$
$$(a + 8)^2 = 0 \quad \textit{Factor } a^2 + 16a + 64 \textit{ as a perfect square trinomial.}$$
$$(a + 8)(a + 8) = 0$$

$a + 8 = 0$ or $a + 8 = 0$ *Zero product property*
$a = -8$ $\quad$ $a = -8$

Check:
$$a^2 + 64 = -16a$$
$$(-8)^2 + 64 \stackrel{?}{=} -16(-8)$$
$$64 + 64 = 128 \checkmark$$

The solution set is $\{-8\}$.

You can apply the zero product property to an equation that is written as the product of *any* number of factors equal to zero.

Example 2

Solve $2y^3 - 24y = 13y^2$.

$$2y^3 - 24y = 13y^2$$
$$2y^3 - 13y^2 - 24y = 0 \quad \textit{Arrange the terms so that the powers of y are in descending order.}$$
$$y(2y^2 - 13y - 24) = 0 \quad \textit{Factor out the GCF, y.}$$
$$y(y - 8)(2y + 3) = 0 \quad \textit{Factor } 2y^2 - 13y - 24.$$

$y = 0$ or $y - 8 = 0$ or $2y + 3 = 0$
$y = 8$ $\quad$ $2y = -3$
$y = -\frac{3}{2}$

The solution set is $\left\{0, 8, -\frac{3}{2}\right\}$. *Check this result.*

Example 3

CONNECTION

Number Theory

Find two consecutive integers whose product is 72.

Let n = the least integer.
Then, $n + 1$ = the next greater integer.

$$n(n + 1) = 72$$
$$n^2 + n = 72$$
$$n^2 + n - 72 = 0$$
$$(n + 9)(n - 8) = 0$$

$n + 9 = 0$ or $n - 8 = 0$

$n = -9$ $\quad$ $n = 8$

If $n = -9$, then $n + 1 = -8$.
If $n = 8$, then $n + 1 = 9$.

Check: $(-9)(-8) = 72$ ✓
$8(9) = 72$ ✓

Therefore, the two consecutive integers are -9 and -8 or 8 and 9.

CHECKING FOR UNDERSTANDING

Communicating Mathematics

Read and study the lesson to answer each question.

1. When can the zero product property be used to solve an equation?
2. When solving a problem by using an equation, why must you check each solution with the original problem and not the equation?
3. Refer to the photography problem at the beginning of the lesson. If Patricia wants the new photo to have an area of 35 square centimeters, by what amount should she reduce the length and width of the original photo?

Guided Practice

Solve. Check each solution.

4. $a^2 + 4a - 21 = 0$
5. $2y^2 + 5y + 2 = 0$
6. $10m^2 + m - 3 = 0$
7. $y^2 - 16 = 0$
8. $7x^2 = 70x - 175$
9. $x^3 + 29x^2 = -28x$

Define a variable and write an equation for each problem.

10. The product of two consecutive integers is 110.
11. The sum of two integers is 15 and their product is 44.
12. The area of a living room is 40 square meters. The length of the room is 3 meters more than its width.
13. A rectangle is 4 inches wide and 7 inches long. When the length and width are increased by the same amount, the area is increased by 26 square inches.

EXERCISES

Practice

Solve. Check each solution.

14. $x^2 + 13x + 36 = 0$
15. $a^2 + a - 56 = 0$
16. $b^2 - 8b - 33 = 0$
17. $y^2 - 64 = 0$
18. $c^2 - 17c + 60 = 0$
19. $m^2 - 24m + 144 = 0$
20. $p^2 = 5p + 24$
21. $r^2 = 18 + 7r$
22. $6z^2 + 5 = -17z$
23. $3y^2 + 16y = 35$
24. $\frac{x^2}{12} - \frac{2x}{3} - 4 = 0$
25. $x^2 - \frac{x}{6} - \frac{35}{6} = 0$
26. $m^3 - 81m = 0$
27. $5b^3 + 34b^2 = 7b$
28. $81n^3 + 36n^2 = -4n$
29. $(x + 8)(x + 1) = -12$
30. $(r - 1)(r - 1) = 36$
31. $(3y + 2)(y + 3) = y + 14$

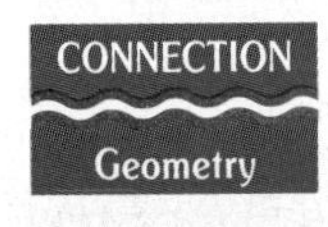

32. The area of a square is $4s^2 + 28s + 49$ square inches. Find the value of s if the perimeter of the square is 60 inches.

For each problem, define a variable. Then use an equation to solve the problem.

33. Find two consecutive even integers whose product is 120.
34. Find two integers whose sum is 11 and whose product is 24.
35. Find two integers whose difference is 3 and whose product is 88.
36. The sum of the squares of two consecutive positive odd integers is 202. Find the integers.
37. When the square of the second of two consecutive even integers is added to twice the first integer, the sum is 76. Find the integers.

Solve. Check each solution.

38. $h^3 + h^2 - 4h - 4 = 0$
39. $9m^3 - 18m^2 - m + 2 = 0$
40. $xy + 4x - 3y - 12 = 0$
41. $4pz - z + 12p - 3 = 0$

Critical Thinking

Write an equation with integral coefficients that has the given solutions.

42. $\{-3, 5\}$
43. $\left\{\frac{2}{3}, -1\right\}$
44. $\{-2, 2, 5\}$

Applications

45. **Gardening** The length of Rachel's rectangular garden is 5 yards more than its width. The area of the garden is 234 square yards. What are its dimensions?

46. **Photography** A rectangular photograph is 8 centimeters wide and 12 centimeters long. The photograph is enlarged by increasing the length and the width by an equal amount. If the area of the new photograph is 69 square centimeters greater than the area of the original photograph, what are the dimensions of the new photograph?

47. **Agriculture** A strip of uniform width is plowed along all four sides of a 12-kilometer by 9-kilometer rectangular cornfield. How wide is the strip if the cornfield is half plowed?

Use the formula $h = vt - 16t^2$ to solve each problem.

48. **Physics** A certain fireworks rocket is set off at an initial upward velocity of 440 feet per second. This type of fireworks is designed to explode at a height of 3000 feet. How many seconds after it is set off will the rocket reach 3000 feet and explode?

49. **Physics** A missile is fired with an initial upward velocity of 2320 feet per second. When will it reach an altitude of 40,000 feet?

Mixed Review

50. **Basketball** In 1962, Wilt Chamberlain set an NBA record by averaging 50.4 points per game. If he maintained this average over an 82-game season, how many total points would he score? Round to the nearest whole number. **(Lesson 2-7)**

51. Solve $\frac{x + 2}{x - 3} = \frac{4}{9}$. **(Lesson 4-1)**

52. If y varies inversely as x, and $y = 24$ when $x = 20$, find y when $x = 30$. **(Lesson 4-9)**

53. Solve $5t - (t - 3) < 6t + 7$. **(Lesson 5-3)**

54. Multiply $8cd^2(7c^2d - cd + d^2)$. **(Lesson 6-7)**

55. Find the value of c that would make $25x^2 + 5cxy + 64y^2$ a perfect square trinomial. **(Lesson 7-7)**

56. **Physics** A flare is launched with an initial upward velocity of 128 feet per second. Use the formula $h = vt - 16t^2$ to determine how long it will take for the flare to hit the ground. **(Lesson 7-9)**

Portfolio

Review items in your portfolio. Make a table of contents of the items, noting why each item was chosen. Replace any items that are no longer appropriate.

CHAPTER 7

SUMMARY AND REVIEW

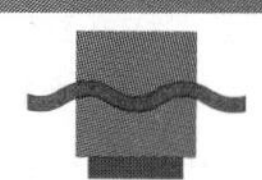

VOCABULARY

Upon completing this chapter, you should be familiar with the following terms:

composite number	**256**	**256**	prime factor
factor	**256**	**256**	prime factorization
greatest common factor, GCF	**257**	**256**	prime number
perfect square trinomial	**281**	**273**	prime polynomial

SKILLS AND CONCEPTS

OBJECTIVES AND EXAMPLES

Upon completing this chapter, you should be able to:

- find the greatest common factor (GCF) for a set of monomials **(Lesson 7-1)**

Find the GCF of $15x^2y$ and $45xy^2$.

$15x^2y = (3) \cdot (5) \cdot x \cdot (x) \cdot (y)$

$45xy^2 = (3) \cdot 3 \cdot (5) \cdot (x) \cdot (y) \cdot y$

The GCF is $3 \cdot 5 \cdot x \cdot y$ or $15xy$.

REVIEW EXERCISES

Use these exercises to review and prepare for the chapter test.

Find the GCF of the given monomials.

1. 35, 30

2. $12ab$, $-4a^2b^2$

3. 12, 18, 40

4. $16mrt$, $30m^2r$

5. $20n^2$, $24np^5$

6. $60x^2y^2$, $35xz^3$

7. $2m^2n^3p$, $8mn^2p^3$, $5m^2np^3$

- use the GCF and the distributive property to factor polynomials **(Lesson 7-2)**

Factor $12a^2 - 8ab$.

$12a^2 - 8ab = 4a(3a - 2b)$

Factor each polynomial.

8. $13x + 26y$

9. $6x^2y + 12xy + 6$

10. $24a^2b^2 - 18ab$

11. $26ab + 18ac + 32a^2$

12. $m + m^2n + m^3n^3$

13. $\frac{3}{5}a - \frac{3}{5}b + \frac{6}{5}c$

- use grouping techniques to factor polynomials with four or more terms **(Lesson 7-3)**

Factor $2x^2 - 3xz - 2xy + 3yz$.

$2x^2 - 3xz - 2xy + 3yz$
$= x(2x - 3z) - y(2x - 3z)$
$= (x - y)(2x - 3z)$

Factor each polynomial. Check with FOIL.

14. $a^2 - 4ac + ab - 4bc$

15. $24am - 9an + 40bm - 15bn$

16. $2rs + 6ps + mr + 3mp$

17. $16k^3 - 4k^2p^2 - 28kp + 7p^3$

18. $dm + 7r + mr + 7d$

OBJECTIVES AND EXAMPLES	REVIEW EXERCISES

- factor quadratic trinomials **(Lesson 7-5)**

$a^2 - 3a - 4 = (a + 1)(a - 4)$

$$\begin{aligned} 4x^2 - 4xy - 15y^2 &= 4x^2 + (-10 + 6)xy - 15y^2 \\ &= (4x^2 - 10xy) + (6xy - 15y^2) \\ &= 2x(2x - 5y) + 3y(2x - 5y) \\ &= (2x + 3y)(2x - 5y) \end{aligned}$$

Factor each trinomial, if possible. If the trinomial cannot be factored using integers, write *prime*.

19. $y^2 + 7y + 12$
20. $x^2 - 9x - 36$
21. $b^2 + 5b - 6$
22. $2r^2 - 3r - 20$
23. $3a^2 - 13a + 14$
24. $6z^2 + 7z + 3$
25. $a^2 - 10ab + 9b^2$
26. $r^2 - 8rs - 65s^2$
27. $56m^2 - 93mn + 27n^2$

- identify and factor polynomials that are the differences of squares **(Lesson 7-6)**

$a^2 - 9 = (a)^2 - (3)^2 = (a - 3)(a + 3)$

$$\begin{aligned} 3x^3 - 75x &= 3x(x^2 - 25) \\ &= 3x(x - 5)(x + 5) \end{aligned}$$

Factor each binomial, if possible. If the binomial cannot be factored using integers, write *prime*.

28. $b^2 - 16$
29. $25 - 9y^2$
30. $16a^2 - 81b^4$
31. $2y^3 - 128y$
32. $\frac{1}{4}n^2 - \frac{9}{16}r^2$
33. $81x^4 - 16$

- identify and factor perfect square trinomials **(Lesson 7-7)**

$$\begin{aligned} &16z^2 - 8z + 1 \\ &= (4z)^2 - 2(4z)(1) + (1)^2 \\ &= (4z - 1)^2 \end{aligned}$$

Factor each trinomial.

34. $a^2 + 18a + 81$
35. $16x^2 - 8x + 1$
36. $9k^2 - 12k + 4$
37. $32n^2 - 80n + 50$
38. $6b^3 - 24b^2g + 24bg^2$
39. $y^2 - \frac{3}{2}yz^2 + \frac{9}{16}z^4$

- factor polynomials by applying the various methods of factoring **(Lesson 7-8)**

Check for:	Number of Terms		
	2	3	4+
greatest common factor	✓	✓	✓
difference of squares	✓		✓
perfect square trinomial		✓	
trinomial with two binomial factors		✓	
pairs of terms with a common monomial factor			✓

Factor each polynomial.

40. $3x^2 - 12$
41. $28y^2 - 13y - 6$
42. $56a^2 - 93a + 27$
43. $6m^3 + m^2 - 15m$
44. $15ay^2 + 37ay + 20a$
45. $2r^3 - 18r^2 + 30r$
46. $12mx + 3xb + 4my + by$
47. $mx^2 + bx^2 - 49m - 49b$

OBJECTIVES AND EXAMPLES	REVIEW EXERCISES

- use the zero product property to solve equations **(Lesson 7-9)**

$$x^2 = -4x$$
$$x^2 + 4x = -4x + 4x$$
$$x(x + 4) = 0$$
$$x = 0 \quad \text{or} \quad x + 4 = 0$$
$$x = -4$$

The solution set is $\{0, -4\}$.

Solve. Check each solution.

48. $y(y + 11) = 0$

49. $4t(2t - 10) = 0$

50. $(3x - 2)(4x + 7) = 0$

51. $2a^2 - 9a = 0$

52. $n^2 = -17n$

53. $\frac{3}{4}y = \frac{1}{2}y^2$

- solve equations by using various factoring methods and applying the zero product property **(Lesson 7-10)**

$$m^2 - m = 2$$
$$m^2 - m - 2 = 2 - 2$$
$$(m + 1)(m - 2) = 0$$
$$m + 1 = 0 \quad \text{or} \quad m - 2 = 0$$
$$m = -1 \qquad m = 2$$

Solve. Check each solution.

54. $y^2 + 13y + 40 = 0$

55. $2a^2 - 98 = 0$

56. $2m^2 + 13m = 24$

57. $25r^2 + 4 = -20r$

58. $(x + 6)(x - 1) = 78$

59. $6x^3 + 29x^2 + 28x = 0$

APPLICATIONS AND CONNECTIONS

Solve by using guess and check. (Lesson 7-4)

60. Teresa wants to give \$250 to each of her favorite charities, but she needs another \$187 to be able to do so. If she gives each charity \$180, she will have \$163 left over. How much money does Teresa have?

61. Richard needs 228 balloons to use for games at his party. Balloons came in packages of 24 or 36. If Richard bought 8 packages of balloons, how many of each size did he buy?

62. Geometry The measure of the perimeter of a square is $20m + 32p$. Find the measure of its area. **(Lesson 7-2)**

63. Geometry The measure of the area of a rectangle is $16x^2 - 9$. Find the measure of its perimeter. **(Lesson 7-6)**

64. Geometry The measure of the area of a square is $4m^2 - 3mp + 3p - 4m$. What are its dimensions? **(Lesson 7-8)**

65. Number Theory The product of two consecutive odd integers is 99. Find the integers. **(Lesson 7-10)**

66. Construction A tinsmith has a rectangular piece of tin with a 3-inch square cut from each corner so that the sides can be folded up to form a box. She has designed the box so that it is twice as long as it is wide. What are the dimensions of the box if it has a volume of 1350 cubic inches? **(Lesson 7-10)**

CHAPTER 7 TEST

Find the GCF of the given monomials.

1. $18a^2b$, $28a^3b^2$
2. $6x^2y^3$, $12x^2y^2z$, $15x^2y$

Factor each polynomial, if possible. If the polynomial cannot be factored using integers, write *prime*.

3. $25y^2 - 49w^2$
4. $t^2 - 16t + 64$
5. $x^2 + 14x + 24$
6. $28m^2 + 18m$
7. $12x^2 + 23x - 24$
8. $a^2 - 11ab + 18b^2$
9. $2h^2 - 3h - 18$
10. $6x^3 + 15x^2 - 9x$
11. $36m^2 + 60mn + 25n^2$
12. $36a^2b^3 - 45ab^4$
13. $4my - 20m + 15p - 3py$
14. $x^3 - 5x^2 - 9x + 45$
15. **Geometry** The measure of the area of a rectangle is $2x^2 + 3x - 20$. What is the measure of its perimeter?

Solve. Check each solution.

16. $18s^2 + 72s = 0$
17. $4x^2 = 36$
18. $t^2 + 25 = 10t$
19. $a^2 - 9a - 52 = 0$
20. $x + 6 = 12x^2$
21. $x^3 - 5x^2 - 66x = 0$

For each problem, define a variable. Then use an equation to solve the problem.

22. **Number Theory** Find two integers whose sum is 21 and whose product is 104.
23. **Geometry** A rectangle is 4 inches wide by 7 inches long. When the length and width are increased by the same amount, the area is increased by 26 square inches. What are the dimensions of the new rectangle?
24. **Physics** A rocket is fired upward at an initial velocity of 2240 feet per second. Use the formula $h = vt - 16t^2$ to determine when the rocket will have an altitude of 78,400 feet.
25. **Construction** A rectangular lawn is 24 feet wide by 32 feet long. A sidewalk will be built along the inside edges of all four sides. The remaining lawn will have an area of 425 square feet. How wide will the walk be?

Bonus The measure of the area of a rectangle is $12x^2 + x - 20$ where x is a positive integer. What is the smallest possible perimeter measure for this rectangle?

CHAPTER OBJECTIVES

In this chapter, you will:

- Simplify rational expressions.
- Add, subtract, multiply, and divide rational expressions.
- Solve rational equations.
- Solve problems involving formulas that contain rational expressions.
- Solve problems by listing possibilities.

Chapter 8 Rational Expressions

APPLICATION IN PHYSICS

THE FAR SIDE By GARY LARSON

Why do some rearview and sideview mirrors on cars and trucks carry the warning OBJECTS IN MIRROR ARE CLOSER THAN THEY APPEAR?

Most of the mirrors that you see each day are flat mirrors, which simply means that the surface of the mirror is flat. Some mirrors, like the ones found in fun houses, have curved surfaces. A mirror whose surface curves inward is called a *concave mirror* while one whose surface curves outward is called a *convex mirror.* Convex mirrors are often used in sideview and rearview mirrors for cars and trucks since they provide a wider field of vision than a flat mirror. The problem is that objects that are the same distance away from a convex mirror and a flat mirror appear smaller in the convex mirror.

Suppose you are driving on a highway and want to change lanes. If you see a car in your sideview mirror, then you can use its size to estimate how far away it is from your car. At this point, it is important that you know whether the sideview mirror is convex or flat. If the mirror is convex, then the car you see is actually closer to your car than one with the same size in a flat mirror. Thus, if you think your sideview mirror is flat when it is really convex, you might believe there is enough room to change lanes when there actually is not! It is for this reason that manufacturers place a warning on convex mirrors stating that objects in the mirror are closer than they appear.

ALGEBRA IN ACTION

Is the actual size of the object shown in the mirror at the left smaller or larger than the image shown in the mirror?

8-1 Simplifying Rational Expressions

Objective After studying this lesson, you should be able to:

- simplify rational expressions.

A **rational expression** is an algebraic fraction whose numerator and denominator are polynomials. The expressions $\frac{36a^2bc}{24z^2}$, $\frac{2x}{x-5}$, and $\frac{p^2+25}{p^2+4p+1}$ are examples of rational expressions.

Recall that division by zero is undefined.

Because algebraic fractions indicate division, zero cannot be used as a denominator. Therefore, any value assigned to a variable that results in a denominator of zero must be excluded from the domain of the variable. These are called **excluded values.**

For $\frac{5}{x}$, exclude $x = 0$.

For $\frac{6b-3}{b+4}$, exclude $b = -4$. $-4 + 4 = 0$

For $\frac{y^2-5}{x^2-5x+6}$, exclude $x = 2$ and $x = 3$. *Factor $x^2 - 5x + 6$ to see why.*

Example 1 **For each fraction, state the values of the variable that must be excluded.**

a. $\frac{6n}{n+7}$

Exclude the values for which $n + 7 = 0$.

$n + 7 = 0$

$n = -7$

Therefore, n cannot equal -7.

b. $\frac{2a-3}{a^2-a-12}$

Exclude the values for which $a^2 - a - 12 = 0$.

$a^2 - a - 12 = 0$

$(a-4)(a+3) = 0$ *Factor $a^2 - a - 12$.*

$a = 4$ or $a = -3$ *Zero product property*

Therefore, a cannot equal 4 or -3.

To simplify a rational expression such as $\frac{14a^2bc}{42abc^2}$, first factor the numerator and denominator. Then divide each by the greatest common factor.

$$\frac{14a^2bc}{42abc^2} = \frac{2 \cdot 7 \cdot a \cdot a \cdot b \cdot c}{2 \cdot 3 \cdot 7 \cdot a \cdot b \cdot c \cdot c}$$

Notice that $a \neq 0$, $b \neq 0$, and $c \neq 0$.

$$\frac{\overset{1}{\cancel{2}} \cdot \overset{1}{\cancel{7}} \cdot \overset{1}{\cancel{a}} \cdot a \cdot \overset{1}{\cancel{b}} \cdot \overset{1}{\cancel{c}}}{\underset{1}{\cancel{2}} \cdot 3 \cdot \underset{1}{\cancel{7}} \cdot \underset{1}{\cancel{a}} \cdot \underset{1}{\cancel{b}} \cdot \underset{1}{\cancel{c}} \cdot c} \text{ or } \frac{a}{3c}$$

The GCF is 14abc.

To simplify a rational expression in which the numerator is a polynomial and the denominator is a monomial, rewrite the expression as a sum of rational expressions. Then simplify each expression.

Example 2

Simplify $\frac{3x^3 - 4x^2 + 2x - 16}{2x}$. State the excluded values of x.

$$\frac{3x^3 - 4x^2 + 2x - 16}{2x} = \frac{3x^3}{2x} - \frac{4x^2}{2x} + \frac{2x}{2x} - \frac{16}{2x}$$

Rewrite the expression.

$$= \frac{3 \cdot x^2 \cdot \overset{1}{\cancel{x}}}{2 \cdot \underset{1}{\cancel{x}}} - \frac{2x \cdot \overset{1}{\cancel{2x}}}{\underset{1}{\cancel{2x}}} + \frac{\overset{1}{\cancel{2x}}}{\underset{1}{\cancel{2x}}} - \frac{\overset{1}{\cancel{2}} \cdot 8}{\underset{1}{\cancel{2}} \cdot x}$$

Divide by each GCF.

$$= \frac{3}{2}x^2 - 2x + 1 - \frac{8}{x}$$

Exclude values for which $2x = 0$. Therefore, $x \neq 0$.

You can simplify a rational expression that has a polynomial in the numerator and denominator by dividing each by the greatest common factor.

Example 3

Simplify $\frac{3a^2 + a - 2}{a^2 + 7a + 6}$. State the excluded values of a.

$$\frac{3a^2 + a - 2}{a^2 + 7a + 6} = \frac{(a + 1)(3a - 2)}{(a + 6)(a + 1)}$$

Factor $3a^2 + a - 2$.
Factor $a^2 + 7a + 6$.

$$= \frac{\overset{1}{\cancel{(a + 1)}}(3a - 2)}{(a + 6)\underset{1}{\cancel{(a + 1)}}}$$

The GCF is $a + 1$.

$$= \frac{3a - 2}{a + 6}$$

The excluded values of a are any values for which $a^2 + 7a + 6 = 0$.

$a^2 + 7a + 6 = 0$

$(a + 6)(a + 1) = 0$

$a = -6$ or $a = -1$ *Zero product property*

Therefore, a cannot equal -6 or -1.

Example 4 Simplify $\frac{2x - 2y}{y^2 - x^2}$. State the excluded values of x and y.

$$\frac{2x - 2y}{y^2 - x^2} = \frac{2(x - y)}{(y - x)(y + x)}$$ *Distributive property. Factor $y^2 - x^2$.*

$$= \frac{2(-1)(y - x)}{(y - x)(y + x)}$$ *$x - y = -1(y - x)$*

$$= \frac{-2\overset{1}{\cancel{(y - x)}}}{\underset{1}{\cancel{(y - x)}}(y + x)}$$ *The GCF is $(y - x)$.*

$$= \frac{-2}{y + x} \text{ or } -\frac{2}{y + x}$$

The excluded values of x and y are any values for which $y^2 - x^2 = 0$.

$$y^2 - x^2 = 0$$

$$(y - x)(y + x) = 0$$ *Zero product property*

$$y = x \text{ or } y = -x$$

Therefore, y cannot equal x or $-x$.

CHECKING FOR UNDERSTANDING

Communicating Mathematics

Read and study the lesson to answer each question.

1. What is the first step when trying to find the GCF for a rational expression?
2. How would you determine the values to be excluded from $\frac{x + 1}{x^2 + 5x + 4}$?
3. Tell simply by looking at the denominator what values should be excluded in the expression $\frac{(a - 9)}{(a + 4)(a - 9)}$.
4. Write a rational expression that has -2, 3, and 7 as excluded values.

Guided Practice

For each expression, find the greatest common factor (GCF) of the numerator and denominator. Then simplify and state the excluded values of the variables.

5. $\frac{42y}{18xy}$
6. $\frac{-3x^2y^5}{18x^5y^2}$
7. $\frac{x(y + 1)}{x(y - 2)}$
8. $\frac{-6a^3 + 8a^2 + 12a}{2a}$
9. $\frac{(a + b)(a - b)}{(a - b)(a - b)}$
10. $\frac{y - 4}{y^2 - 16}$

Practice

Simplify. State the excluded values of the variables.

11. $\frac{13x}{39x^2}$
12. $\frac{14y^2z}{49yz^3}$
13. $\frac{38a^2}{42ab}$
14. $\frac{79a^2b}{158a^3bc}$
15. $\frac{m + 5}{2(m + 5)}$
16. $\frac{9z^4 - 6z^3 + 4z^2 - 15}{3z}$
17. $\frac{y + 4}{(y - 4)(y + 4)}$
18. $\frac{(a - 4)(a + 4)}{(a - 2)(a - 4)}$
19. $\frac{-1(3w - 2)}{(3w - 2)(w + 4)}$

20. $\dfrac{a + b}{a^2 - b^2}$

21. $\dfrac{c^2 - 4}{(c + 2)^2}$

22. $\dfrac{a^2 - a}{a - 1}$

23. $\dfrac{m^2 - 2m}{m - 2}$

24. $\dfrac{x^2 + 4}{x^4 - 16}$

25. $\dfrac{r^3 - r^2}{r - 1}$

26. $\dfrac{4n^2 - 8}{4n - 4}$

27. $\dfrac{3m^3}{6m^2 - 3m}$

28. $\dfrac{6y^3 - 12y^2}{12y - 18}$

29. $\dfrac{-4y^2}{2y^2 - 4y^3}$

30. $\dfrac{3a^3}{3a^3 + 6a^2b}$

31. $\dfrac{7a^3b^2}{21a^2b + 49ab^3}$

32. $\dfrac{12s^5 + 15s^4 + 20s^2 - 7s}{-5s^3}$

33. $\dfrac{x + y}{x^2 + 2xy + y^2}$

34. $\dfrac{x - 3}{x^2 + x - 12}$

35. $\dfrac{6x^2 + 24x}{x^2 + 8x + 16}$

36. $\dfrac{3 - x}{6 - 17x + 5x^2}$

37. $\dfrac{2x - 14}{x^2 - 4x - 21}$

38. $\dfrac{x^2 - x^2y}{x^3 - x^3y}$

39. $\dfrac{5x^2 + 10x + 5}{3x^2 + 6x + 3}$

40. $\dfrac{6x^4y + 8x^3y^2 + 3x^2y^3}{2xy}$

41. $\dfrac{4k^2 - 25}{4k^2 - 20k + 25}$

42. $\dfrac{2x^2 - 5x + 3}{3x^2 - 5x + 2}$

43. $\dfrac{b^2 - 5b + 6}{b^4 - 13b^2 + 36}$

44. $\dfrac{25 - x^2}{x^2 + x - 30}$

45. $\dfrac{n^2 - 8n + 12}{n^3 - 12n^2 + 36n}$

46. $\dfrac{16a^3 - 24a^2 - 160a}{8a^4 - 36a^3 + 16a^2}$

47. $\dfrac{-x^2 + 6x - 9}{x^2 - 6x + 9}$

48. $\dfrac{x^3y^3 + 5x^3y^2 + 6x^3y}{xy^5 + 5xy^4 + 6xy^3}$

49. $\dfrac{x^4 - 16}{x^4 - 8x^2 + 16}$

Critical Thinking

50. Explain why $\dfrac{x^2 - 4}{x + 2}$ is not the same as $x - 2$.

Applications

51. **Motion** If it takes 90 seconds to walk up a broken escalator and 60 seconds to ride up the same escalator (without walking) when it is moving, how long would it take to walk up the moving escalator?

52. **Physics** Chico needs to move a rock that weighs 1050 pounds. If he has a 6-foot long pinchbar and uses a fulcrum placed one foot from the rock, will he be able to lift the rock if he weighs 205 pounds?

53. **Chemistry** At sea level, water boils at 212° F. At the top of Mt. Everest, water boils at 159.8° F. If Mt. Everest is 29,002 *feet* above sea level, how many degrees does the boiling point drop for every *mile* up from sea level?

54. **Statistics** A crowd watching the Blueberry Festival Parade fills the sidewalks along Elm Street for about 1 mile on each side. If the sidewalks are 10 feet wide and an average person uses 4 square feet of sidewalk, about how many people are watching the parade?

Mixed Review

55. Solve $\frac{5}{12} = \frac{2 - x}{3 + x}$. **(Lesson 4-1)**
56. **Government** The term of office for a United States Senator is 150% as long as the term of office for the President. How long is the term of office for a U.S. Senator? **(Lesson 4-2)**
57. Solve $9q + 2 \leq 7q - 25$. Express the solution in set-builder notation. **(Lesson 5-3)**
58. Evaluate $a^3 + b^2 - (c^2 + abcde) - a \div b + d \div e + c$ if $a = 2$, $b = \frac{1}{3}$, $c = 2.2$, $d = 4$, and $e = \frac{2}{5}$. **(Lesson 1-2)**
59. Simplify $(-5x^3)(4x^4)$. **(Lesson 6-2)**
60. Find the common factor for $(3a^2b + 2ab^3)$ and $(6ab + 4b^3)$. **(Lesson 7-3)**
61. Solve $x^2 + 22 = 58$. **(Lesson 7-9)**

HISTORY CONNECTION

Fractions

The Romans used a somewhat cumbersome system of fractions based on the number twelve, even though their counting system was based on the number ten. They used a unit of measure called the *as*, one-twelfth of which was called the *uncia.* All of the fractions that the Romans needed could be expressed in terms of uncia. Notice from the table that uncia could also be divided, allowing the Romans a fraction as small as $\frac{1}{2304}$.

In seventh-century India, people began writing fractions as we write them today, but without the fraction bar. Arab mathematicians later added the bar. However, it presented such a problem to the printers of the day that the use of the fraction bar did not become common practice until the sixteenth century.

Roman Fractions

	As = 1	Uncia = 1
As	1	12
Deunx	$\frac{11}{12}$	11
Dextans	$\frac{5}{6}$	10
Dodrans	$\frac{3}{4}$	9
Bes	$\frac{2}{3}$	8
Septunx	$\frac{7}{12}$	7
Semis	$\frac{1}{2}$	6
Quincunx	$\frac{5}{12}$	5
Triens	$\frac{1}{3}$	4
Quadrans	$\frac{1}{4}$	3
Sextans	$\frac{1}{6}$	2
Sescuncia	$\frac{1}{8}$	$1\frac{1}{2}$
Uncia	$\frac{1}{12}$	1
Semuncia	$\frac{1}{24}$	$\frac{1}{2}$
Duella	$\frac{1}{36}$	$\frac{1}{3}$
Sicilicus	$\frac{1}{48}$	$\frac{1}{4}$
Sextula	$\frac{1}{72}$	$\frac{1}{6}$
Drachma	$\frac{1}{96}$	$\frac{1}{8}$
Dimidio sextula	$\frac{1}{144}$	$\frac{1}{12}$

8-2 Multiplying Rational Expressions

Objective After studying this lesson, you should be able to:

- multiply rational expressions.

To multiply fractions, you multiply the numerators and multiply the denominators.

$$\frac{3}{5} \cdot \frac{4}{7} = \frac{3 \cdot 4}{5 \cdot 7}$$
$$= \frac{12}{35}$$

You can use this same method to multiply rational expressions.

Example 1 **Find $\frac{5}{a} \cdot \frac{b}{7}$. State any excluded values.**

$$\frac{5}{a} \cdot \frac{b}{7} = \frac{5 \cdot b}{a \cdot 7}$$ *Multiply the numerators. Multiply the denominators.*

$$= \frac{5b}{7a}$$

Since $7a$ cannot equal 0, $a \neq 0$.

Example 2 **Find $\frac{2a^2d}{3bc} \cdot \frac{9b^2c}{16ad^2}$. State any excluded values.**

$$\frac{2a^2d}{3bc} \cdot \frac{9b^2c}{16ad^2} = \frac{18a^2b^2cd}{48abcd^2}$$ *Divide the numerator and denominator by the GCF, 6abcd.*

$$= \frac{3ab}{8d}$$ *Write the fraction in simplest form.*

Since neither bc nor ad^2 can equal 0, $a \neq 0$, $b \neq 0$, $c \neq 0$, and $d \neq 0$.

From this point on, it will be assumed that all replacements for variables in rational expressions that result in denominators equal to zero will be excluded.

You may find it easier to use a shortcut before multiplying fractions, as shown below.

$$\frac{3}{4} \cdot \frac{16}{21} = \frac{\overset{1}{\cancel{3}}}{\underset{1}{\cancel{4}}} \cdot \frac{\overset{4}{\cancel{16}}}{\underset{7}{\cancel{21}}} = \frac{4}{7}$$

You can use the same shortcut when multiplying rational expressions.

Example 3

Find $\frac{x + 5}{3x} \cdot \frac{12x^2}{x^2 + 7x + 10}$.

$$\frac{x + 5}{3x} \cdot \frac{12x^2}{x^2 + 7x + 10} = \frac{x + 5}{3 \cdot x} \cdot \frac{2 \cdot 2 \cdot 3 \cdot x \cdot x}{(x + 5)(x + 2)}$$ *$x \neq 0$, $x \neq -5$, and $x \neq -2$*

$$= \frac{\overset{1}{\cancel{x + 5}}}{\underset{1}{\cancel{3}} \cdot \underset{1}{\cancel{x}}} \cdot \frac{2 \cdot 2 \cdot \overset{1}{\cancel{3}} \cdot \overset{1}{\cancel{x}} \cdot x}{\underset{1}{\cancel{(x + 5)}}(x + 2)}$$ *Simplify.*

$$= \frac{4x}{x + 2}$$

Example 4

Find $\frac{4a + 8}{a^2 - 25} \cdot \frac{a - 5}{5a + 10}$.

$$\frac{4a + 8}{a^2 - 25} \cdot \frac{a - 5}{5a + 10} = \frac{4(a + 2)}{(a - 5)(a + 5)} \cdot \frac{a - 5}{5(a + 2)}$$ *$a \neq 5$, $a \neq -5$, and $a \neq -2$*

$$= \frac{4\overset{1}{\cancel{(a + 2)}}}{\underset{1}{\cancel{(a - 5)}}(a + 5)} \cdot \frac{\overset{1}{\cancel{a - 5}}}{5\underset{1}{\cancel{(a + 2)}}}$$ *Simplify.*

$$= \frac{4}{5(a + 5)}$$

$$= \frac{4}{5a + 25}$$

Example 5

Find $\frac{x^2 - x - 6}{9 - x^2} \cdot \frac{x^2 + 7x + 12}{x^2 + 4x + 4}$.

$$\frac{x^2 - x - 6}{9 - x^2} \cdot \frac{x^2 + 7x + 12}{x^2 + 4x + 4} = \frac{(x - 3)(x + 2)}{(3 - x)(3 + x)} \cdot \frac{(x + 3)(x + 4)}{(x + 2)(x + 2)}$$ *$x \neq 3$, $x \neq -3$, and $x \neq -2$*

$$= \frac{(x - 3)(x + 2)}{-1(x - 3)(x + 3)} \cdot \frac{(x + 3)(x + 4)}{(x + 2)(x + 2)}$$ *Notice that $3 - x = -1(x -$*

$$= \frac{\overset{1}{\cancel{(x - 3)}}\overset{1}{\cancel{(x + 2)}}}{-1\underset{1}{\cancel{(x - 3)}}\underset{1}{\cancel{(x + 3)}}} \cdot \frac{\overset{1}{\cancel{(x + 3)}}(x + 4)}{\underset{1}{\cancel{(x + 2)}}(x + 2)}$$ *Simplify.*

$$= \frac{x + 4}{-1(x + 2)}$$

$$= -\frac{x + 4}{x + 2}$$

Example 6

Find the measure of the area of the rectangle in simplest form.

$$A = \ell w$$

$$= \frac{2x + 4}{x} \cdot \frac{x^3 - 4x}{x^2 + 4x + 4}$$

$$= \frac{2\overset{1}{\cancel{(x + 2)}}\overset{1}{\cancel{(x)}}\overset{1}{\cancel{(x + 2)}}(x - 2)}{\underset{1}{\cancel{x}}\underset{1}{\cancel{(x + 2)}}\underset{1}{\cancel{(x + 2)}}}$$

$$= 2(x - 2)$$

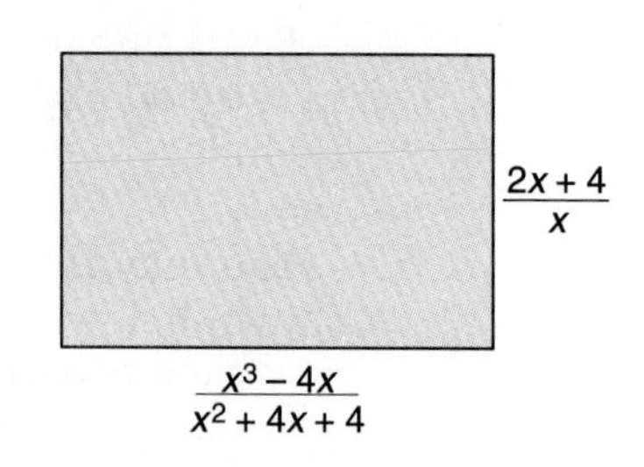

The measure of the area of the rectangle is $2(x - 2)$.

CHECKING FOR UNDERSTANDING

Communicating Mathematics

Read and study the lesson to answer each question.

1. Describe the value(s) for b such that $\frac{ab}{bc} = \frac{a}{c}$.
2. Find the error: $\frac{\overset{2}{\not{4}x}}{x + \underset{1}{\not{2}}} = \frac{2}{2} = 1$.

Guided Practice

Find each product. Assume that no denominator is equal to zero.

3. $\frac{3}{a} \cdot \frac{b}{4}$
4. $\frac{3a}{5} \cdot \frac{2x}{y}$
5. $\frac{ab}{ac} \cdot \frac{c}{d}$
6. $\frac{6a^2n}{8n^2} \cdot \frac{12n}{9a}$
7. $\frac{y-3}{7} \cdot \frac{14}{y-3}$
8. $\frac{9}{m-3} \cdot \frac{m^2-9}{12}$

EXERCISES

Practice

Find each product. Assume that no denominator is equal to zero.

9. $\frac{a^2b}{b^2c} \cdot \frac{c}{d}$
10. $\frac{10n^3}{6x^3} \cdot \frac{12n^2x^4}{25n^2x^2}$
11. $\left(\frac{2a}{b}\right)^2 \frac{5c}{6a}$
12. $\frac{8}{m^2}\left(\frac{m^2}{2c}\right)^2$
13. $\frac{6m^3n}{10a^2} \cdot \frac{4a^2m}{9n^3}$
14. $\frac{7xy^3}{11z^2} \cdot \frac{44z^3}{21x^2y}$
15. $\frac{5n-5}{3} \cdot \frac{9}{n-1}$
16. $\frac{3a-3b}{a} \cdot \frac{a^2}{a-b}$
17. $\frac{-(2a+7c)}{6} \cdot \frac{36}{-7c-2a}$
18. $\frac{2a+4b}{5} \cdot \frac{25}{6a+8b}$
19. $\frac{3x+30}{2x} \cdot \frac{4x}{4x+40}$
20. $\frac{3}{x-y} \cdot \frac{(x-y)^2}{6}$
21. $\frac{a^2-b^2}{4} \cdot \frac{16}{a+b}$
22. $\frac{r^2}{r-s} \cdot \frac{r^2-s^2}{s^2}$
23. $\frac{a^2-b^2}{a-b} \cdot \frac{7}{a+b}$
24. $\frac{x^2-16}{9} \cdot \frac{x+4}{x-4}$
25. $\frac{x^2-y^2}{x^2-1} \cdot \frac{x-1}{x-y}$
26. $\frac{r^2+s^2}{r^2-s^2} \cdot \frac{r-s}{r+s}$
27. $\frac{3k+9}{k} \cdot \frac{k^2}{k^2-9}$
28. $\frac{3a-6}{a^2-9} \cdot \frac{a+3}{a^2-2a}$
29. $\frac{y^2-x^2}{y} \cdot \frac{x}{x-y}$
30. $\frac{b+a}{b-a} \cdot \frac{a^2-b^2}{a}$
31. $\frac{3mn^2-3m}{n} \cdot \frac{3m}{n^2-1}$
32. $\frac{x}{x^2+8x+15} \cdot \frac{2x+10}{x^2}$
33. $\frac{x-5}{x^2-7x+10} \cdot \frac{x-2}{3}$
34. $\frac{b^2+20b+99}{b+9} \cdot \frac{b+7}{b^2+12b+11}$

Find the measure of the area of each rectangle in simplest form.

35.

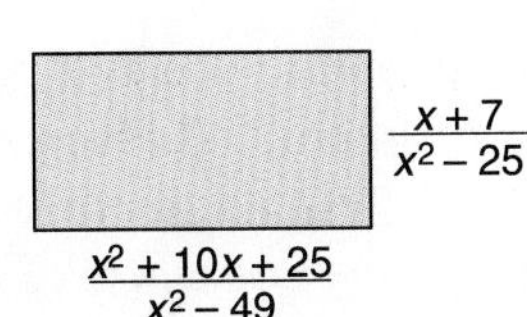

36. 37.

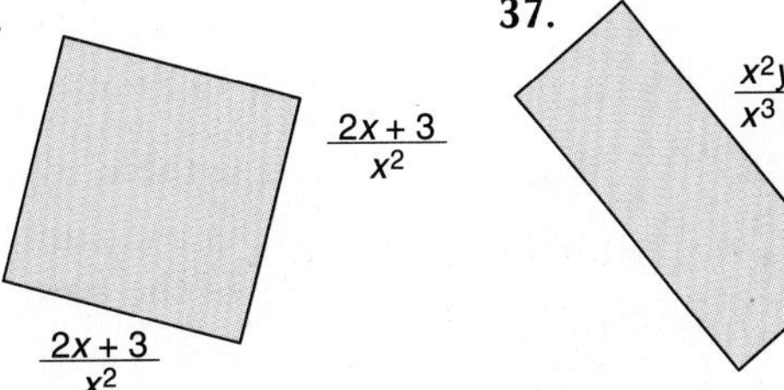

Find each product. Assume that no denominator is equal to zero.

38. $\frac{z^2-15z+50}{z^2-9z+20} \cdot \frac{z^2-11z+24}{z^2-18z+80}$
39. $\frac{y^2+3y^3}{y^2-4} \cdot \frac{2y+y^2}{y+4y^2+3y^3}$

Find each product. Assume that no denominator is equal to zero.

40. $\frac{3t^3 - 14t^2 + 8t}{2t^2 - 3t - 20} \cdot \frac{16t^2 + 34t - 15}{24t^2 - 25t + 6}$

41. $\frac{6y^2 - 5y - 6}{3y^2 - 20y - 7} \cdot \frac{y^2 - 49}{12y^3 + 23y^2 + 10y}$

42. $\frac{2m^2 - 9m + 9}{3m^2 + 19m - 14} \cdot \frac{m^2 + 14m + 49}{9 - 6m + m^2}$

43. $\frac{a^2x - b^2x}{y} \cdot \frac{y^2 + y}{a - 2} \cdot \frac{4 - 2a}{axy - bxy}$

44. $\frac{x^2y}{x^2 + 4xy + 4y^2} \cdot \frac{x^2 + 2xy}{xy} \cdot \frac{y}{x^4 - 9x^2}$

Critical Thinking

45. Find two different pairs of rational expressions whose product is $\frac{6x^2 - 6x - 36}{x^2 + 3x - 28}$.

Applications

46. **Consumer Awareness** The price of eggs goes up 10% the first month, goes down 10% the second month, and then goes up 10% again the third month. By what percentage did the price of eggs go up from the initial price?

47. **Construction** A construction supervisor needs to determine how many truckloads of earth will have to be removed from a site before the foundation can be poured. The bed of the truck has the shape shown at the right. Use the formula $V = \frac{d(a + b)}{2} \times \ell$ to find the volume of the truck bed if $d = 5$ ft, $a = 18$ ft, $b = 15$ ft, and $\ell = 9$ ft.

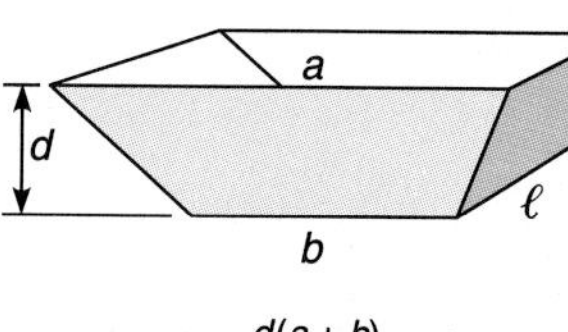

48. **Car Racing** The Indianapolis 500 is a 500-mile automobile race. If each lap is $2\frac{1}{2}$ miles long, how many laps does each driver complete to finish the race?

Mixed Review

49. Make a table to solve the problem. **(Lesson 4-5)**
What is the sum of the first three consecutive odd whole numbers?
What is the sum of the first five consecutive odd whole numbers?
What is the sum of the first fifty consecutive odd whole numbers?

50. Solve $\frac{h}{-18} > -25$. **(Lesson 5-2)**

51. Express 0.76×10^7 in scientific notation. **(Lesson 6-4)**

52. Simplify $(3x + 7)(5x - 1)$. **(Lesson 6-8)**

53. Factor $6y^2 - 24x^2$. **(Lesson 7-10)**

54. Simplify $\frac{2x^3}{2x^2(x^2 - 4)}$. **(Lesson 8-1)**

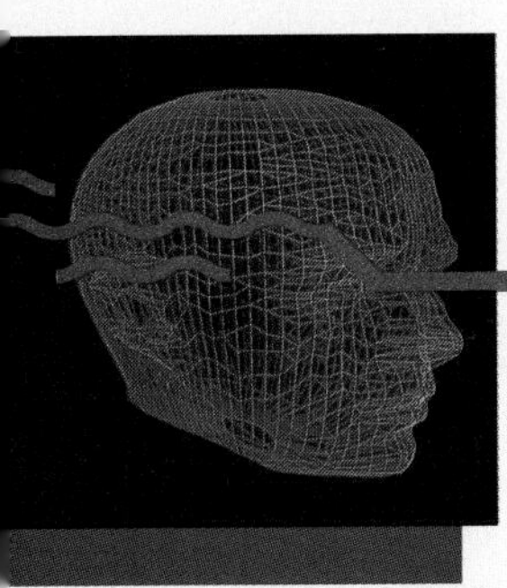

Technology

Rational Expressions

BASIC
Graphing calculators
▶ Graphing software
Spreadsheets

The examples below show how to use the *Mathematics Exploration Toolkit* to simplify rational expressions. The following CALC commands will be used.

FACTOR (fac) REDUCE (red) SIMPLIFY (simp)

First, use the FACTOR command to factor the numerator and denominator. Then, use the REDUCE command to divide the numerator and denominator by the greatest common factor. Use the SIMPLIFY command to express monomials with exponents.

Example 1 Simplify $\frac{30a^2bc^4}{25a^3b^2c}$. *Exponents are entered with the ^ symbol.*

Enter	Result
(30a^2bc^4)/(25a^3b^2c)	$\frac{30a^2bc^4}{25a^3b^2c}$
fac	$\frac{5 \cdot 3 \cdot 2(aa)b(cccc)}{5 \cdot 5(aaa)(bb)c}$
red	$\frac{ccc \cdot 2 \cdot 3}{ba \cdot 5}$
simp	$\frac{6c^3}{5ab}$

Example 2 Simplify $\frac{3x^2 - 4x + 1}{9x^2 - 1}$.

Enter	Result
(3x^2 −4x + 1)/(9x^2 − 1)	$\frac{3x^2 - 4x + 1}{9x^2 - 1}$
fac	$\frac{(x - 1)(3x - 1)}{(3x - 1)(3x + 1)}$
red	$\frac{x - 1}{3x + 1}$

EXERCISES

Use CALC to simplify each rational expression.

1. $\frac{12p^3q^5}{18p^2q^7}$
2. $\frac{y^2 - 9}{y^2 - 6y + 9}$
3. $\frac{x^3 - 2x^2}{x^4 - x^2}$
4. $\frac{2y^2 - y - 1}{2y^2 + y}$
5. $\frac{x^2 + 19x + 60}{x^2 - 225}$

8-3 Dividing Rational Expressions

Objective After studying this lesson, you should be able to:

- divide rational expressions.

You should recall that two numbers whose product is 1 are called **multiplicative inverses** or **reciprocals.**

To find the quotient of two fractions, you multiply by the reciprocal of the divisor.

$$\frac{2}{3} \div \frac{3}{4} = \frac{2}{3} \cdot \frac{4}{3} \qquad \text{The reciprocal of } \tfrac{3}{4} \text{ is } \tfrac{4}{3}.$$
$$= \frac{8}{9}$$

You can use this same method to divide rational expressions.

Example 1

Find $\frac{5}{x} \div \frac{y}{z}$.

$$\frac{5}{x} \div \frac{y}{z} = \frac{5}{x} \cdot \frac{z}{y} \qquad \text{The reciprocal of } \tfrac{y}{z} \text{ is } \tfrac{z}{y}.$$
$$= \frac{5z}{xy}$$

Example 2

Find $\frac{2x}{x+1} \div (x - 1)$.

$$\frac{2x}{x+1} \div (x-1) = \frac{2x}{x+1} \cdot \frac{1}{x-1} \qquad \text{The reciprocal of } x - 1 \text{ is } \tfrac{1}{x-1}. \text{ Why?}$$
$$= \frac{2x}{(x+1)(x-1)}$$
$$= \frac{2x}{x^2 - 1}$$

Example 3

Find $\frac{x-y}{x^2 - y^2} \div \frac{x+y}{x^2 + 2xy + y^2}$.

$$\frac{x-y}{x^2-y^2} \div \frac{x+y}{x^2+2xy+y^2} = \frac{x-y}{x^2-y^2} \cdot \frac{x^2+2xy+y^2}{x+y} \qquad \text{Multiply by the reciprocal.}$$
$$= \frac{x-y}{(x+y)(x-y)} \cdot \frac{(x+y)(x+y)}{x+y} \qquad \text{Factor } x^2 - y^2 \text{ and } x^2 + 2xy + y^2.$$
$$= \frac{\overset{1}{\cancel{x-y}}}{\underset{1}{\cancel{(x+y)}}\underset{1}{\cancel{(x-y)}}} \cdot \frac{\overset{1}{\cancel{(x+y)}}\overset{1}{\cancel{(x+y)}}}{\underset{1}{\cancel{x+y}}}$$
$$= 1$$

CHECKING FOR UNDERSTANDING

Communicating Mathematics

Read and study the lesson to answer each question.

1. *True* or *false:* Every real number other than zero has a reciprocal.
2. Why does zero *not* have a reciprocal?

Guided Practice

Find the reciprocal of each expression.

3. $\frac{m}{2}$
4. $\frac{x^2}{4}$
5. $\frac{-8}{3n}$
6. x
7. $\frac{2m^2}{5}$
8. $2bc$
9. $\frac{x + y}{x - y}$
10. $x - 2$

Find each quotient. Assume that no denominator is equal to zero.

11. $\frac{y}{5} \div \frac{y^2 - 25}{5 - y}$
12. $\frac{m^2 + 2m + 1}{2} \div \frac{m + 1}{m - 1}$

EXERCISES

Practice

Find each quotient. Assume that no denominator is equal to zero.

13. $\frac{a^2}{b^2} \div \frac{b^2}{a^2}$
14. $\frac{a^2}{b} \div \frac{a^2}{b^2}$
15. $\frac{(-a)^2}{b} \div \frac{a}{b}$
16. $\frac{3m}{m + 1} \div (m - 2)$
17. $\frac{b^2 - 9}{4b} \div (b - 3)$
18. $\frac{y^2 + 8y + 16}{y^2} \div (y + 4)$
19. $\frac{2a^3}{a + 1} \div \frac{a^2}{a + 1}$
20. $\frac{p^2}{y^2 - 4} \div \frac{p}{2 - y}$
21. $\frac{x^2 - 16}{16 - x^2} \div \frac{7}{x}$
22. $\frac{x^2 - 4x + 4}{3} \div \frac{x^2 - 4}{2x}$
23. $\frac{a^2 + 2ab + b^2}{2x} \div \frac{a + b}{x^2}$
24. $\frac{k^2 - 81}{k^2 - 36} \div \frac{k - 9}{k + 6}$
25. $\frac{t^2 + 8t + 16}{w^2 - 6w + 9} \div \frac{2t + 8}{3w - 9}$
26. $\frac{k + 2}{m^2 + 4m + 4} \div \frac{4k + 8}{m + 4}$
27. $\frac{x}{x + 2} \div \frac{x^2}{x^2 + 5x + 6}$
28. $\frac{x^2 + x - 2}{x^2 + 5x + 6} \div \frac{x^2 + 2x - 3}{x^2 + 7x + 12}$
29. $\frac{2m^2 + 7m - 15}{m + 5} \div \frac{9m^2 - 4}{3m + 2}$
30. $\frac{2x^2 - x - 15}{x^2 - 2x - 3} \div \frac{2x^2 + 3x - 5}{1 - x^2}$
31. $\frac{x^2 + 5x + 6}{x^2 - x - 12} \cdot \frac{x - 4}{x^2 + 11x + 18} \div \frac{x + 7}{x^2 + 14x + 45}$
32. $\frac{2x - 3}{2x^2 - 7x + 6} \cdot \frac{x^2 + 3x - 10}{5x + 1} \div \frac{3x^2 + 14x - 5}{3x^2 + 2x - 1}$
33. $\frac{x^2 - 1}{2x^2 + 14x + 12} \div \frac{2x^2 - 3x + 1}{8x^2 + 36x - 72} \cdot \frac{x^2 + 3x - 4}{x^2 + 5x - 6}$

Critical Thinking

34. **Geometry** The measure of the area of a rectangle is $\frac{x^2 - y^2}{2}$. If the measure of the length is $2x + 2y$, find the measure of the width.

Applications

35. **Personal Finance** Brice had his allowance doubled. Next, he received a \$3 increase. Then his allowance was cut in half. If Brice's allowance is now \$10 per week, what was his original allowance?

Journal

Write a paragraph to describe the procedure for dividing two rational expressions.

36. **Work** JoAnn Young made $40,000 last year. In recognition of her good performance, JoAnn's boss offered her a 15% raise, to be taken in one of two ways: either she can take the 15% raise this year and receive no raise next year, or she can take a 7.5% raise this year and another 7.5% raise next year. Which option should JoAnn choose?

37. **Oceanography** The part of an iceberg that is below water is about seven times the size of the part that is above water. What percent of the iceberg is above water?

38. **Travel** A one-mile-long train is traveling at 60 miles per hour. If the train enters a one-mile-long tunnel, how long does it take the train to go completely through?

Mixed Review

39. **Finance** Luis invested $5000 for one year at 9% interest. Marta invested some money at the same time at 8% interest. At the end of the year, Luis and Marta together had earned $810 in interest. How much money did Marta invest? **(Lesson 4-3)**

40. Solve $x + 3 < -17$. **(Lesson 5-1)**

41. Find $(3m^2 + 2n)^2$. **(Lesson 6-9)**

42. Factor $\frac{1}{2}x^2 - \frac{1}{4}ax$. **(Lesson 7-2)**

43. Find $\frac{m^2 - 4}{2} \cdot \frac{4}{m - 2}$. **(Lesson 8-2)**

EXTRA

Does 2 = 1?

Find the fallacy in the following "proof" of "2 = 1."

$a = b$	
$a \cdot a = b \cdot a$	*Multiply each side by a.*
$a^2 = ab$	
$a^2 - b^2 = ab - b^2$	*Subtract b^2 from each side.*
$(a - b)(a + b) = b(a - b)$	*Factor.*
$\frac{(a - b)(a + b)}{(a - b)} = \frac{b(a - b)}{(a - b)}$	*Divide each side by $(a - b)$.*
$a + b = b$	
$b + b = b$	*Substitute b for a.*
$2b = b$	
$\frac{2b}{b} = \frac{b}{b}$	*Divide each side by b.*
$2 = 1$	

8-4 Dividing Polynomials

Objective

After studying this lesson, you should be able to:

- divide polynomials by binomials.

To divide a polynomial by a polynomial, you can use a long division process similar to that used in arithmetic. For example, you can divide $x^2 + 8x + 15$ by $x + 5$ as shown below.

Step 1

To find the first term of the quotient, divide the first term of the dividend (x^2) by the first term of the divisor (x).

$$\begin{array}{r|l} x & \\ x + 5 \overline{) x^2 + 8x + 15} & \textit{When dividing } x^2 \textit{ by } x, \textit{ the result is } x. \\ x^2 + 5x & \textit{Multiply } x(x + 5). \\ \hline 3x & \textit{Subtract.} \end{array}$$

Step 2

To find the next term of the quotient, divide the first term of the partial dividend ($3x$) by the first term of the divisor (x).

$$\begin{array}{r|l} x + 3 & \\ x + 5 \overline{) x^2 + 8x + 15} & \textit{When dividing } 3x \textit{ by } x, \textit{ the result is } 3. \\ x^2 + 5x & \\ \hline 3x + 15 & \\ 3x + 15 & \textit{Multiply } 3(x + 5). \\ \hline 0 & \textit{Subtract.} \end{array}$$

Therefore, $x^2 + 8x + 15$ divided by $x + 5$ is $x + 3$. Since the remainder is 0, the divisor is a factor of the dividend. This means that $(x + 5)(x + 3) = x^2 + 8x + 15$.

If the divisor is *not* a factor of the dividend, there will be a nonzero remainder. The quotient can be expressed as follows.

$$\text{quotient} = \text{partial quotient} + \frac{\text{remainder}}{\text{divisor}}$$

Example 1

Find $(2x^2 - 11x - 20) \div (2x + 3)$.

$$\begin{array}{r|l} x - 7 & \\ 2x + 3 \overline{) 2x^2 - 11x - 20} & \\ 2x^2 + 3x & \leftarrow \textit{Multiply } x(2x + 3). \\ \hline -14x - 20 & \leftarrow \textit{Subtract, then bring down } -20. \\ -14x - 21 & \leftarrow \textit{Multiply } -7(2x + 3). \\ \hline 1 & \leftarrow \textit{Subtract. The remainder is } 1. \end{array}$$

The quotient is $x - 7$ with remainder 1.

Thus, $(2x^2 - 11x - 20) \div (2x + 3) = x - 7 + \frac{1}{2x + 3}$. ← *remainder* ← *divisor*

In an expression like $s^3 + 9$, there is no s^2 term and no s term. In such situations, rename the expression using 0 as the coefficient of these terms.

$$s^3 + 9 = s^3 + 0s^2 + 0s + 9$$

Example 2 Find $\frac{s^3 + 9}{s - 3}$.

$$
\begin{array}{r l}
 & s^2 + 3s + 9 \\
s - 3 \overline{)} & s^3 + 0s^2 + 0s + 9 \\
 & \underline{s^3 - 3s^2} \\
 & 3s^2 + 0s \\
 & \underline{3s^2 - 9s} \\
 & 9s + 9 \\
 & \underline{9s - 27} \\
 & 36
\end{array}
$$

Insert $0s^2$ and $0s$. Why?

Therefore, $\frac{s^3 + 9}{s - 3} = s^2 + 3s + 9 + \frac{36}{s - 3}$.

CHECKING FOR UNDERSTANDING

Communicating Mathematics

Read and study the lesson to answer each question.

1. What relationship is there between the degree of the divisor, the degree of the dividend, and the degree of the quotient?
2. What is a factor? Is 3 a factor of 7? Is 2x a factor of $4x^3$?

Guided Practice

State the first term of each quotient.

3. $\frac{a^2 + 3a + 2}{a + 1}$
4. $\frac{b^2 + 8b - 20}{b - 2}$
5. $\frac{8m^3 + 27}{2m + 3}$
6. $\frac{2x^2 + 3x - 2}{2x - 1}$
7. $\frac{x^3 + 2x^2 - 5x + 12}{x + 4}$
8. $\frac{2x^3 - 5x^2 + 22x + 51}{2x + 3}$

Find each quotient.

9. $(x^2 + 7x + 12) \div (x + 3)$
10. $(x^2 + 9x + 20) \div (x + 5)$

EXERCISES

Practice

Find each quotient.

11. $(a^2 - 2a - 35) \div (a - 7)$
12. $(x^2 + 6x - 16) \div (x - 2)$
13. $(c^2 + 12c + 36) \div (c + 9)$
14. $(y^2 - 2y - 30) \div (y + 7)$
15. $(2r^2 - 3r - 35) \div (2r + 7)$
16. $(3t^2 - 14t - 24) \div (3t + 4)$
17. $\frac{10x^2 + 29x + 21}{5x + 7}$
18. $\frac{12n^2 + 36n + 15}{6n + 3}$
19. $\frac{x^3 - 7x + 6}{x - 2}$
20. $\frac{4m^3 + 5m - 21}{2m - 3}$
21. $\frac{4t^3 + 17t^2 - 1}{4t + 1}$
22. $\frac{2a^3 + 9a^2 + 5a - 12}{a + 3}$
23. $\frac{27c^2 - 24c + 8}{9c - 2}$
24. $\frac{48b^2 + 8b + 7}{12b - 1}$
25. $\frac{6n^3 + 5n^2 + 12}{2n + 3}$
26. $\frac{t^3 - 19t + 9}{t - 4}$
27. $\frac{3s^3 + 8s^2 + s - 7}{s + 2}$
28. $\frac{9d^3 + 5d - 8}{3d - 2}$
29. $\frac{20t^3 - 27t^2 + t + 6}{4t - 3}$
30. $\frac{6x^3 - 9x^2 - 4x + 6}{2x - 3}$
31. $\frac{56x^3 + 32x^2 - 63x - 36}{7x + 4}$

Critical Thinking

32. Find the value of k if $(x + 2)$ is a factor of $x^3 + 7x^2 + 7x + k$.
33. Find the value of k if $(2m - 5)$ is a factor of $2m^3 - 5m^2 + 8m + k$.
34. Find the value of k if the remainder is 15 when $x^3 - 7x^2 + 4x + k$ is divided by $(x - 2)$.

Applications

35. **Sports** A football field, not including endzones, is 160 feet wide by 100 yards long. A polo field is 160 yards wide and 300 yards long. How many football fields would it take to fill one polo field?

36. **Statistics** A recent survey found that American teenagers between the ages of 12 and 17 watch television an average of 22 hours per week.

    a. How much television does the average American teenager watch in one year?

    b. If about 20% of all television time is for commercials, how much time does the average American teenager spend watching commercials each year?

Computer

37. The BASIC program at the right finds the quotient and the remainder when a polynomial is divided by a binomial of the form $(x - r)$. First, enter the degree of the polynomial. Next, enter the coefficients of the polynomial. Finally, enter the value of R in the divisor $(x - r)$. For example, in Exercise **c** below, enter 3 for the degree, 1, 2, −4, and −8 for the coefficients, and 2 for the value of r.

```
10 INPUT "DEGREE OF POLYNOMIAL:
   ";N
20 PRINT "ENTER COEFFICIENTS:"
30 FOR X = 1 TO N+1
40 INPUT A(X)
50 NEXT X
60 INPUT "CONSTANT R: ";R
70 LET B(1) = A(1)
80 PRINT "COEFFICIENTS OF
   QUOTIENT ARE:"
90 PRINT B(1);" ";
100 FOR X = 1 TO N-1
110 B(X+1)=A(X+1)+R*B(X)
120 PRINT B(X+1);" ";
130 NEXT X
140 PRINT
150 PRINT "REMAINDER: ";
    A(N+1)+R*B(N)
160 END
```

Use the program to find each quotient.

a. $(3x^2 + 5x + 2) \div (x - 2)$

b. $(3x^2 + 7x + 2) \div (x + 2)$

c. $(x^3 + 2x^2 - 4x - 8) \div (x - 2)$

d. $(x^3 + 2x^2 + 4x + 8) \div (x + 2)$

e. $(2x^4 + x^3 - 2x^2 + 3x + 1) \div (x - 1)$

Mixed Review

38. **Physics** Weights of 50 pounds and 75 pounds are placed on a lever. The two weights are 16 feet apart, and the lever is balanced. How far from the fulcrum is the 50-pound weight? **(Lesson 4-8)**

39. Make a diagram to solve the problem. **(Lesson 5-4)**

    An ant is climbing a 30-foot flagpole. Each day it climbs up 7 feet. Each night it slides back 4 feet. How many days will it take it to reach the top of the flagpole?

40. Simplify $\frac{9xyz^5}{x^4}$. **(Lesson 6-3)**

41. Factor $p^2 + 10p + 25$. **(Lesson 7-5)**

42. Find $\frac{y^2}{x^2} \div \frac{a^2}{x^2}$. **(Lesson 8-3)**

8-5 Rational Expressions with Like Denominators

Objective After studying this lesson, you should be able to:

- add and subtract rational expressions with like denominators.

To add or subtract fractions with like denominators, you add or subtract the numerators. Then you can write the sum or difference over the common denominator.

$$\frac{3}{7} + \frac{2}{7} = \frac{5}{7} \qquad \frac{7}{9} - \frac{2}{9} = \frac{5}{9}$$

You can use these same methods to add or subtract rational expressions.

Example 1 **Find $\frac{3}{x+2} + \frac{1}{x+2}$.**

$$\frac{3}{x+2} + \frac{1}{x+2} = \frac{3+1}{x+2}$$ *Since x + 2 is the common denominator, add the numerators.*

$$= \frac{4}{x+2}$$

When subtracting rational expressions, remember to add the additive inverse of the second expression.

Example 2 **Find $\frac{3a+2}{a-7} - \frac{a-3}{a-7}$.**

$$\frac{3a+2}{a-7} - \frac{a-3}{a-7} = \frac{(3a+2) + [-(a-3)]}{a-7}$$ *Add the additive inverse of $a - 3$.*

$$= \frac{3a + 2 - a + 3}{a-7}$$ *Distributive property*

$$= \frac{2a+5}{a-7}$$

When adding or subtracting fractions, sometimes the result can be expressed in simplest form.

$$\frac{3}{8} + \frac{1}{8} = \frac{4}{8}$$ *The GCF of 4 and 8 is 4.*

$$= \frac{1}{2}$$

$$\frac{9}{16} - \frac{3}{16} = \frac{6}{16}$$ *The GCF of 6 and 16 is 2.*

$$= \frac{3}{8}$$

This process may also be used when adding or subtracting rational expressions.

Example 3 **Find $\frac{8n + 3}{3n + 4} - \frac{2n - 5}{3n + 4}$.**

$$\frac{8n + 3}{3n + 4} - \frac{2n - 5}{3n + 4} = \frac{(8n + 3) - (2n - 5)}{3n + 4}$$ *Since $3n + 4$ is the common denominator, subtract the numerators.*

$$= \frac{8n + 3 - 2n + 5}{3n + 4}$$

$$= \frac{6n + 8}{3n + 4}$$ *Combine like terms.*

$$= \frac{2\overset{1}{\cancel{(3n + 4)}}}{\underset{1}{\cancel{(3n + 4)}}}$$ *Factor and simplify.*

$$= 2$$

Example 4 **Find $\frac{x}{x - 2} - \frac{x + 1}{2 - x}$.**

$$\frac{x}{x - 2} - \frac{x + 1}{2 - x} = \frac{x}{x - 2} - \frac{x + 1}{-(x - 2)}$$ *Rewrite $2 - x$ as $-(x - 2)$.*

$$= \frac{x}{x - 2} - \left(-\frac{x + 1}{x - 2}\right)$$ *Remember that $\frac{1}{-x} = -\frac{1}{x}$.*

$$= \frac{x}{x - 2} + \frac{x + 1}{x - 2}$$

$$= \frac{2x + 1}{x - 2}$$

Example 5

Find the measure of the perimeter of the rectangle.

(Rectangle with length $\frac{r}{r^2 - s^2}$ and width $\frac{s}{r^2 - s^2}$.)

$$P = 2\ell + 2w$$

$$= 2\left(\frac{r}{r^2 - s^2}\right) + 2\left(\frac{s}{r^2 - s^2}\right)$$

$$= \frac{2r}{r^2 - s^2} + \frac{2s}{r^2 - s^2}$$

$$= \frac{2r + 2s}{r^2 - s^2}$$

$$= \frac{2\overset{1}{\cancel{(r + s)}}}{\underset{1}{\cancel{(r + s)}}(r - s)}$$ *$(r + s)$ is the GCF.*

$$= \frac{2}{r - s}$$

The measure of the perimeter is $\frac{2}{r - s}$.

CHECKING FOR UNDERSTANDING

Communicating Mathematics

Read and study the lesson to answer each question.

1. If the sum of two fractions with the same denominator is zero, what can you say about the two fractions?
2. Can you find the sum of $\frac{1}{0}$ and $\frac{2}{0}$? Why or why not?
3. Write two fractions with a denominator of 5 that have a sum of 1.

Guided Practice

Find each sum or difference in simplest form.

4. $\frac{5}{8} + \frac{2}{8}$
5. $\frac{4}{a} + \frac{3}{a}$
6. $\frac{b}{x} + \frac{2}{x}$
7. $\frac{5}{2z} + \frac{-7}{2z}$
8. $\frac{3}{11} - \frac{2}{11}$
9. $\frac{14}{16} - \frac{15}{16}$
10. $\frac{a}{5} - \frac{b}{5}$
11. $\frac{8k}{5m} - \frac{3k}{5m}$

EXERCISES

Practice

Find each sum or difference in simplest form.

12. $\frac{y}{2} + \frac{y}{2}$
13. $\frac{a}{12} + \frac{2a}{12}$
14. $\frac{5x}{24} - \frac{3x}{24}$
15. $\frac{7t}{t} - \frac{8t}{t}$
16. $\frac{y}{2} + \frac{y-6}{2}$
17. $\frac{m+4}{5} + \frac{m-1}{5}$
18. $\frac{a+2}{6} - \frac{a+3}{6}$
19. $\frac{x}{x+1} + \frac{1}{x+1}$
20. $\frac{8}{y-2} - \frac{6}{y-2}$
21. $\frac{y}{b+6} - \frac{2y}{b+6}$
22. $\frac{2n}{2n-5} + \frac{5}{5-2n}$
23. $\frac{x+y}{y-2} + \frac{x-y}{2-y}$
24. $\frac{y}{a+1} - \frac{y}{a+1}$
25. $\frac{a+b}{x-3} + \frac{a+b}{3-x}$
26. $\frac{a+b}{x-3} - \frac{a+b}{3-x}$
27. $\frac{r^2}{r-s} + \frac{s^2}{r-s}$
28. $\frac{x^2}{x-y} - \frac{y^2}{x-y}$
29. $\frac{m^2}{m+n} + \frac{2mn+n^2}{m+n}$
30. $\frac{x^2}{x-y} - \frac{2xy+y^2}{x-y}$
31. $\frac{12n}{3n+2} + \frac{8}{3n+2}$
32. $\frac{6x}{x+y} + \frac{6y}{x+y}$
33. $\frac{a^2}{a-b} + \frac{-b^2}{a-b}$
34. $\frac{r^2}{r-3} + \frac{9}{3-r}$
35. $\frac{x^2}{x^2-1} + \frac{2x+1}{x^2-1}$
36. $\frac{2x+1}{(x+1)^2} + \frac{x^2}{(x+1)^2}$
37. $\frac{x-1}{(x+1)^2} - \frac{x-1}{(x+1)^2}$
38. $\frac{25}{k+5} - \frac{k^2}{k+5}$
39. $\frac{x}{x^2+2x+1} + \frac{1}{x^2+2x+1}$
40. $\frac{8y}{4y^2+12y+9} + \frac{12}{4y^2+12y+9}$
41. $\frac{2}{t^2-t-2} - \frac{t}{t^2-t-2}$

Find the measure of the perimeter of each rectangle.

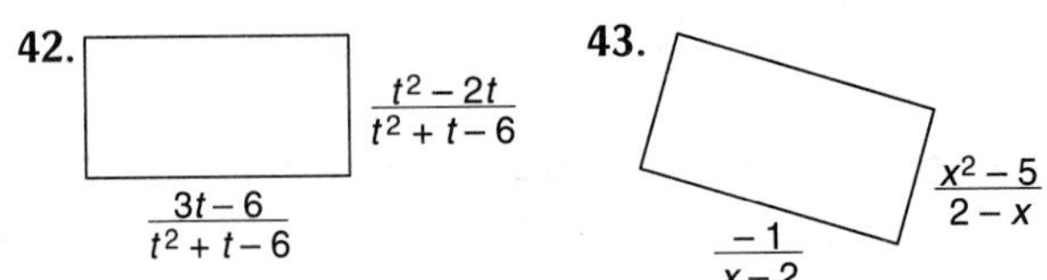

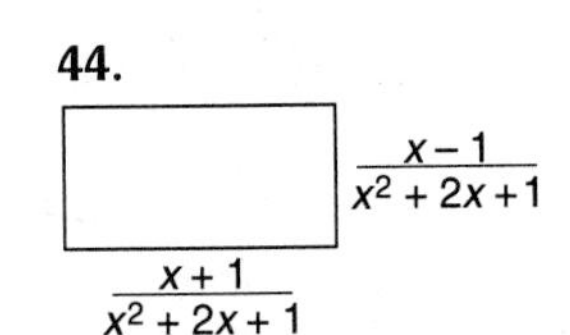

Critical Thinking

45. Is $\frac{-4}{9} = -\frac{4}{9}$? Is $\frac{4}{-9} = -\frac{4}{9}$? Is $\frac{-4}{-9} = -\frac{4}{9}$?

Applications

46. Entertainment The rock group MAXXX receives \$12,500 for each concert performance. If there are five members who receive equal shares of the money and $\frac{1}{3}$ of the money goes to pay for other salaries and expenses, how much money does each member of MAXXX actually receive for a tour of 20 concert dates?

47. History In 1624, Peter Minuit paid 60 guilders, or about \$24, to the Indians for Manhattan Island. What would be the value of that \$24 in 1992 if it had been invested at 5% annual interest compounded annually since then? Use $T = p(1 + r)^t$.

Mixed Review

48. Travel At 8:00 A.M., Alma drove west at 35 miles per hour. At 9:00 A.M., Reiko drove east from the same point at 42 miles per hour. When will they be 266 miles apart? **(Lesson 4-7)**

49. Write $r > -\frac{1}{2}$ and $r < \frac{8}{3}$ without using *and*. **(Lesson 5-5)**

50. Multiply $8(7m + 2)$. **(Lesson 6-7)**

51. Find the prime factorization of 1000. **(Lesson 7-1)**

52. Find $(m^3 - 6m^2 - m + 32) \div (m - 4)$. **(Lesson 8-4)**

MID-CHAPTER REVIEW

Simplify. State the excluded values of the variables. (Lesson 8-1)

1. $\frac{(a - 7)}{(a + 1)(a - 7)}$
2. $\frac{y - 3}{y^2 - 9}$
3. $\frac{4y^2 + 7y - 2}{8y^2 + 15y - 2}$

Find each product. Assume that no denominator is equal to zero. (Lesson 8-2)

4. $\frac{y^2 - 4}{y^2 - 1} \cdot \frac{y + 1}{y + 2}$
5. $\frac{m^2 + 16}{m^2 - 16} \cdot \frac{m - 4}{m + 4}$
6. $\frac{x + 3}{x + 4} \cdot \frac{x}{x^2 + 7x + 12}$

Find each quotient. Assume that no denominator is equal to zero. (Lessons 8-3, 8-4)

7. $\frac{a^2}{b} \div \frac{a^2}{b^2}$
8. $\frac{q}{y^2 - 4} \div \frac{q^2}{y + 2}$
9. $\frac{m^2 + 2mn + n^2}{3m} \div \frac{m^2 - n^2}{2}$
10. $(2m^2 + 5m - 3) \div (m + 3)$
11. $(3t^3 - 11t^2 - 31t + 7) \div (3t + 1)$

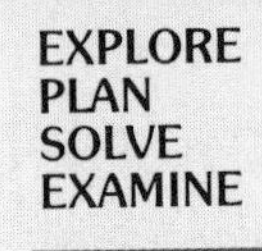

8-6 Problem-Solving Strategy: List Possibilities

Objective After studying this lesson, you should be able to:

- solve problems by making an organized list of the possibilities.

You can solve problems by listing the possibilities in an organized way. Use a systematic approach so you do not omit any important items.

Example 1 **Which whole numbers less than 30 are divisible by both 3 and 4?**

List the numbers that are divisible by 3.

0, 3, 6, 9, 12, 15, 18, 21, 24, 27

List the numbers that are divisible by 4.

0, 4, 8, 12, 16, 20, 24, 28

Since 0, 12, and 24 are in both lists, they are divisible by both 3 and 4.

Example 2 **Wacky Wheels carries bicycles, tricycles, and wagons. They have an equal number of tricycles and wagons in stock. If there are 60 pedals and 180 wheels, how many bicycles, tricycles, and wagons are there in stock?**

Make a table to show the possibilities. *Recall that wagons do not have pedals and bicycles and tricycles each have 2 pedals.*

Bicycles (2 wheels)	10	20	18	10	8	6
Tricycles (3 wheels)	10	20	20	22	24	24
Wagons (4 wheels)	10	20	21	22	24	24
Total Wheels	90	180	180	174	184	180
Total Pedals	40	80	76	64	64	60
Solution?	no	no	no	no	no	yes

So there are 6 bicycles, 24 tricycles, and 24 wagons in stock.

A **tree diagram** is a special kind of organized list. Some people prefer to use tree diagrams because they show all of the possibilities.

Example 3 **Ye Olde Ice Cream Shoppe makes chocolate, butterscotch, or strawberry sundaes. Any sundae can be served with whipped cream, nuts, neither or both. In how many ways can a sundae be served?**

The tree diagram on the next page illustrates the various combinations.

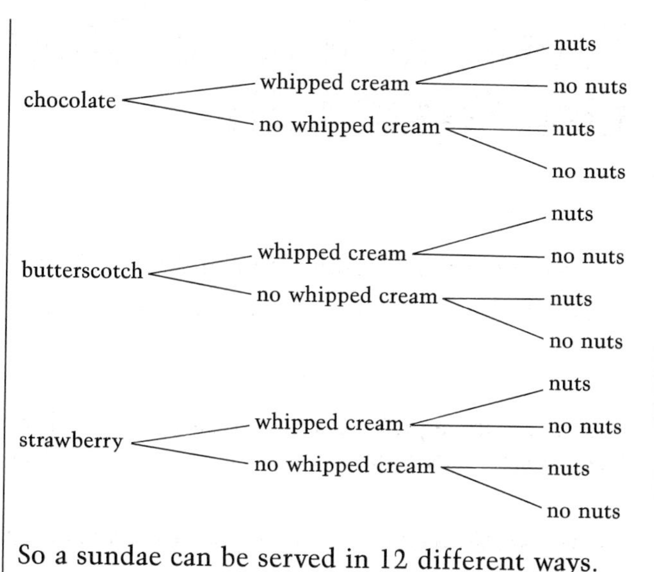

So a sundae can be served in 12 different ways.

CHECKING FOR UNDERSTANDING

Communicating Mathematics

Read and study the lesson to answer each question.

1. What do you notice about the numbers that are divisible by both 3 and 4?
2. Name another strategy that can be used with the strategy described in this lesson.

Guided Practice

Solve each problem by listing the possibilities.

3. Which whole numbers less than 50 are divisible by both 8 and 12?
4. Three darts are thrown at the target shown at the right. If we assume that each of the darts lands within one of the rings, how many different point totals are possible?

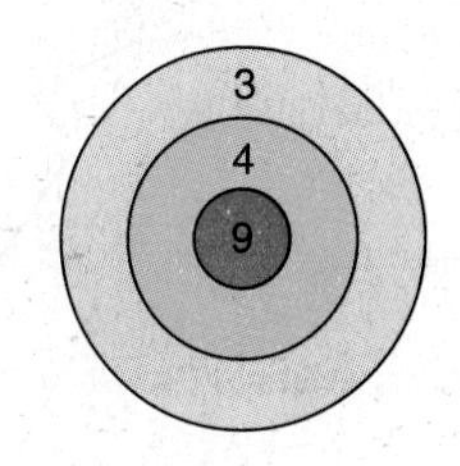

5. How many different three-digit security codes can you make using the numbers 1, 2, and 3? You can use the numbers more than once.

EXERCISES

Practice

Solve. Use any strategy.

6. In a twelve-hour period of time, how many times is the sum of the digits on a digital clock greater than 15?
7. Frugal Fred Fudd left half of his estate to his wife, Flora; $50,000 to his daughter, Fawn; half of what remained to his butler, Franklin; half of the rest for the care and feeding of his dog, Fluffy; and the remaining $10,000 to his favorite charity. What was the value of Frugal Fred's estate?

Strategies

Look for a pattern.
Solve a simpler problem.
Act it out.
Guess and check.
Draw a diagram.
Make a chart.
Work backwards.

8. **Probability** The president, vice president, secretary, and treasurer of the Drama Club are to be seated in four chairs for a yearbook picture. How many different seating arrangements are possible?
9. A visitor to Nowheresville decided to get a haircut. The town had only two barbers, each with his own shop. The visitor glanced into Floyd's Barber Shop and saw that the shop was a mess and Floyd needed a shave and a haircut. When the visitor glanced into Ed's Hair Happening, he saw that Ed was freshly shaved and his hair neatly trimmed. The visitor went to Floyd's for his haircut. Why?
10. Can the design at the right be traced without going over any line segment more than once?

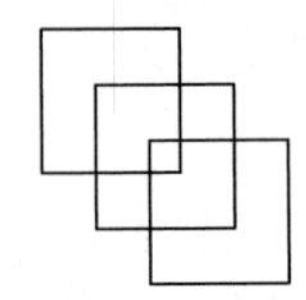

11. As Felicia waited in line at the ticket office, she noticed that two more people were ahead of her than were behind her. There were three times as many people in line as there were people behind Felicia. How many people are ahead of Felicia in line?
12. A domino has two square spaces on its face. Each of the two spaces is marked with 1, 2, 3, 4, 5, or 6 dots, or it is blank. A complete set of dominoes includes one domino for each possible combination of dots. Doubles, such as 3 dots in each space, are included. How many dominoes are in a complete set?
13. **Number Theory** A positive integer is *happy* if the sum of the squares of its digits is 1, or if this process is continued, the result is 1.

$$23 \rightarrow 2^2 + 3^2 = 13$$
$$13 \rightarrow 1^2 + 3^2 = 10$$
$$10 \rightarrow 1^2 + 0^2 = 1$$

23 is a happy number.

   a. If 23 is happy, what can you say about 32?

   b. Find all happy numbers less than 100.

COOPERATIVE LEARNING ACTIVITY

Work in groups of four. Each person in the group should understand the solution and be able to explain it to any person in the class.

Throughout the text, you have learned about different types of numbers. This is part of the study of *number theory.* The Pythagoreans, a group of scholars in ancient Greece, studied the abstract relationships among numbers. They called some numbers *perfect.* A number is perfect if it is the sum of its *proper* divisors. The proper divisors of a number are all of its factors except itself. For example, 496 is a perfect number.

$$496 = 1 + 2 + 4 + 8 + 16 + 31 + 62 + 124 + 248$$

1. Find two other perfect numbers.
2. *True* or *false:* If $2^n - 1$ is prime, then $(2^n - 1) \cdot 2^{n-1}$ is perfect.

8-7 Rational Expressions with Unlike Denominators

Objective After studying this lesson, you should be able to:

- add or subtract rational expressions with unlike denominators.

Application

The longest-running musical to date is *The Fantasticks*. So far, there have been over 12,000 performances.

Judith Paulsen, the choreographer of a big Broadway musical, has requested that the producer of the show hire enough dancers so that they can be arranged in groups of 4, 6, or 9 with no one sitting out. What is the least number of dancers that can be hired?

If the choreographer wanted to arrange the dancers in groups of 6, then the producer should hire a number of dancers that is a multiple of 6. But since she wants to arrange the dancers in groups of 4, 6, or 9, the producer must be concerned with multiples of all three numbers. The least number of dancers is given by the **least common multiple (LCM)** of the three numbers. The least common multiple is the smallest number that is a common multiple of two or more numbers.

The multiples of 4, 6, and 9 can be found by multiplying 4, 6, and 9 by each whole number.

You can find the LCM by making an organized list.

multiples of 4: $4 \cdot 0, 4 \cdot 1, 4 \cdot 2, 4 \cdot 3, 4 \cdot 4, \ldots, 4 \cdot 9, \ldots$
$0, \quad 4, \quad 8, \quad 12, \quad 16, \quad \quad 36, \ldots$

multiples of 6: $6 \cdot 0, 6 \cdot 1, 6 \cdot 2, 6 \cdot 3, 6 \cdot 4, 6 \cdot 5, 6 \cdot 6, \ldots$
$0, \quad 6, \quad 12, \quad 18, \quad 24, \quad 30, \quad 36, \ldots$

multiples of 9: $9 \cdot 0, 9 \cdot 1, 9 \cdot 2, 9 \cdot 3, 9 \cdot 4, 9 \cdot 5, \ldots$
$0, \quad 9, \quad 18, \quad 27, \quad 36, \quad 45, \ldots$

Compare these multiples. Aside from 0, the least number that is common to all three sets of multiples is 36. Thus, the LCM of 4, 6, and 9 is 36, and the producer should hire 36 dancers.

You can also use prime factorization to find the LCM.

1. Find the prime factorization of each number.

$4 = 2 \cdot 2$
$6 = 2 \cdot 3$
$9 = 3 \cdot 3$

2. Use each prime factor the greatest number of times it appears in any of the factorizations.

2 appears two times in 4.
3 appears two times in 9.

Thus, the LCM of 4, 6, and 9 is $2 \cdot 2 \cdot 3 \cdot 3$, or 36.

Example 1

Find the LCM of $12x^2y$ and $15x^2y^2$.

$12x^2y = 2 \cdot 2 \cdot 3 \cdot x \cdot x \cdot y$ *Factor each expression.*

$15x^2y^2 = 3 \cdot 5 \cdot x \cdot x \cdot y \cdot y$

$\text{LCM} = 2 \cdot 2 \cdot 3 \cdot 5 \cdot x \cdot x \cdot y \cdot y$ *Use each factor the greater number of times it appears in either factorization.*

$= 60x^2y^2$

Example 2

Find the LCM of $x^2 + x - 2$ and $x^2 + 5x - 6$.

$x^2 + x - 2 = (x - 1)(x + 2)$ *Factor each expression.*

$x^2 + 5x - 6 = (x + 6)(x - 1)$

$\text{LCM} = (x - 1)(x + 2)(x + 6)$ *Use each factor the greater number of times it appears in either factorization.*

To add or subtract fractions with unlike denominators, first rename the fractions so the denominators are alike. Any common denominator could be used. However, the computation is usually easier if you use the **least common denominator (LCD).** Recall that the least common denominator is the LCM of the denominators.

You will usually use the following steps to add or subtract rational expressions.

1. Find the LCD.
2. Change each rational expression into an equivalent expression with the LCD as the denominator.
3. Add or subtract as with rational expressions with like denominators.
4. Simplify if necessary.

Example 3

Find $\frac{6}{5x} + \frac{7}{10x^2}$.

$5x = 5 \cdot x$ *Use each factor the greater number of times it appears.*

$10x^2 = 2 \cdot 5 \cdot x \cdot x$

The LCD for $\frac{6}{5x}$ and $\frac{7}{10x^2}$ is $2 \cdot 5 \cdot x \cdot x$ or $10x^2$. Since the denominator of $\frac{7}{10x^2}$ is already $10x^2$, only $\frac{6}{5x}$ needs to be renamed.

$$\frac{6}{5x} + \frac{7}{10x^2} = \frac{6}{5x} \cdot \frac{2x}{2x} + \frac{7}{10x^2}$$ *Why do you multiply $\frac{6}{5x}$ by $\frac{2x}{2x}$?*

$$= \frac{12x}{10x^2} + \frac{7}{10x^2}$$

$$= \frac{12x + 7}{10x^2}$$

Example 4

Find $\frac{a}{a^2 - 4} - \frac{4}{a + 2}$.

Since $a^2 - 4 = (a - 2)(a + 2)$, the LCD for $\frac{a}{a^2 - 4}$ and $\frac{4}{a + 2}$ is $(a - 2)(a + 2)$, or $a^2 - 4$.

$$\frac{a}{a^2 - 4} - \frac{4}{a + 2} = \frac{a}{(a - 2)(a + 2)} - \frac{4}{(a + 2)} \cdot \frac{(a - 2)}{(a - 2)} \quad \text{Multiply } \frac{4}{a + 2} \text{ by } \frac{a - 2}{a - 2}.$$
$$= \frac{a - 4(a - 2)}{(a - 2)(a + 2)}$$
$$= \frac{a - 4a + 8}{a^2 - 4}$$
$$= \frac{-3a + 8}{a^2 - 4}$$

Example 5

Find $\frac{x + 4}{(2 - x)(x + 3)} + \frac{x - 5}{(x - 2)^2}$.

Multiply the first fraction by $\frac{-1}{-1}$ to change $(2 - x)$ to $(x - 2)$.

$$\frac{x + 4}{(2 - x)(x + 3)} + \frac{x - 5}{(x - 2)^2} = \frac{-1}{-1} \cdot \frac{(x + 4)}{(2 - x)(x + 3)} + \frac{x - 5}{(x - 2)^2}$$
$$= \frac{-(x + 4)}{(x - 2)(x + 3)} + \frac{x - 5}{(x - 2)^2} \quad -1(2 - x) = x - 2$$

The LCD for $(x - 2)(x + 3)$ and $(x - 2)^2$ is $(x + 3)(x - 2)(x - 2)$. *Why?*

$$\frac{-(x + 4)}{(x - 2)(x + 3)} + \frac{x - 5}{(x - 2)^2} = \frac{-(x + 4)}{(x - 2)(x + 3)} \cdot \frac{(x - 2)}{(x - 2)} + \frac{(x - 5)}{(x - 2)^2} \cdot \frac{(x + 3)}{(x + 3)}$$
$$= \frac{-x^2 - 2x + 8 + x^2 - 2x - 15}{(x + 3)(x - 2)(x - 2)}$$
$$= \frac{-4x - 7}{(x + 3)(x - 2)^2}$$

CHECKING FOR UNDERSTANDING

Communicating Mathematics

Read and study the lesson to answer each question.

1. Can the LCD of two rational expressions be equal to the denominator of one of the rational expressions? When?
2. Why can't $\frac{4}{x - 1} + \frac{5}{11}$ equal $\frac{9}{x + 10}$?

Guided Practice

Find the LCD for each pair of rational expressions.

3. $\frac{4}{a^2}, \frac{5}{a}$
4. $\frac{6}{b^3}, \frac{7}{ab}$
5. $\frac{3}{20a^2}, \frac{1}{24ab^3}$
6. $\frac{1}{12an^2}, \frac{3}{40a^4}$
7. $\frac{11}{56x^3y}, \frac{10}{49ax^2}$
8. $\frac{7}{a + 5}, \frac{a}{a - 3}$
9. $\frac{m}{m + n}, \frac{6}{n}$
10. $\frac{x + 5}{3x - 6}, \frac{x - 3}{x - 2}$

Find each sum or difference.

11. $\frac{m - n}{m + n} - \frac{1}{m^2 - n^2}$
12. $\frac{a}{a - b} + \frac{b}{2b + 3a}$

EXERCISES

Practice **Find each sum or difference.**

13. $\frac{t}{3} + \frac{2t}{7}$

14. $\frac{2n}{5} - \frac{3m}{4}$

15. $\frac{5}{2a} + \frac{-3}{6a}$

16. $\frac{7}{3a} - \frac{3}{6a^2}$

17. $\frac{5}{xy} + \frac{6}{yz}$

18. $\frac{3z}{7w^2} - \frac{2z}{w}$

19. $\frac{2}{t} + \frac{t+3}{s}$

20. $\frac{m}{1(m-n)} - \frac{5}{m}$

21. $\frac{4a}{2a+6} + \frac{3}{a+3}$

22. $\frac{-3}{a-5} + \frac{-6}{a^2-5a}$

23. $\frac{2y}{y^2-25} + \frac{y+5}{y-5}$

24. $\frac{3a+2}{3a-6} - \frac{a+2}{a^2-4}$

25. $\frac{x^2-1}{x+1} + \frac{x^2+1}{x-1}$

26. $\frac{k}{2k+1} - \frac{2}{k+2}$

27. $\frac{-18}{y^2-9} + \frac{7}{3-y}$

28. $\frac{a-1}{4ab} + \frac{a^2-a}{16b^2}$

29. $\frac{x-1}{3xy} - \frac{x^2-x}{9y^2}$

30. $\frac{a-2}{a^2+4a+4} + \frac{a+2}{a-2}$

31. $\frac{x^2}{4x^2-9} + \frac{x}{(2x+3)^2}$

32. $\frac{y}{y^2-2y+1} - \frac{1}{y-1}$

33. $\frac{x^2+4x-5}{x^2-2x-3} + \frac{2}{x+1}$

34. $\frac{a+2}{a^2-9} - \frac{2a}{6a^2-17a-3}$

35. $\frac{3m}{m^2+3m+2} - \frac{3m-6}{m^2+4m+4}$

36. $\frac{4a}{6a^2-a-2} - \frac{5a+1}{2-3a}$

37. $\frac{2x+1}{(x-1)^2} + \frac{x-2}{(1-x)(x+4)}$

38. $\frac{a+3}{3a^2-10a-8} + \frac{2a}{a^2-8a+16}$

Critical Thinking 39. Complete.

a. $18 \cdot 20 = ?$
GCF of 18 and 20 = ?
LCM of 18 and 20 = ?
GCF · LCM = ?

b. $16 \cdot 48 = ?$
GCF of 16 and 48 = ?
LCM of 16 and 48 = ?
GCF · LCM = ?

c. Write a rule that describes the relationship between the product of two numbers and their GCF and LCM.

d. Use your rule to describe how to find the GCF of two numbers if you already know the LCM.

Applications 40. **Astronomy** Earth, Jupiter, and Saturn revolve around the Sun about once every 1, 12, and 30 years, respectively. The last time Jupiter and Saturn appeared close to each other in Earth's night sky was in 1982. When will this happen again?

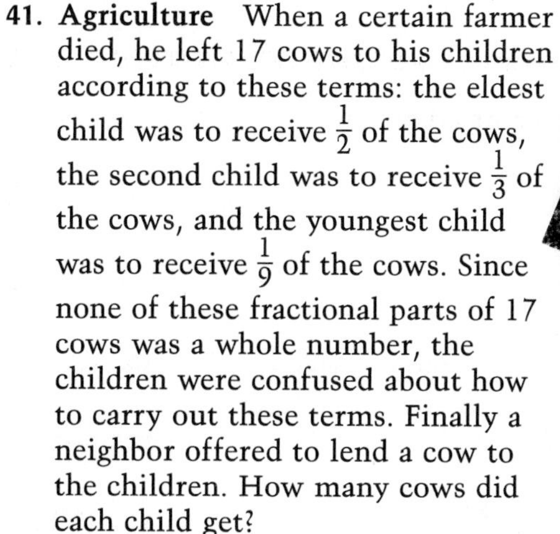

41. Agriculture When a certain farmer died, he left 17 cows to his children according to these terms: the eldest child was to receive $\frac{1}{2}$ of the cows, the second child was to receive $\frac{1}{3}$ of the cows, and the youngest child was to receive $\frac{1}{9}$ of the cows. Since none of these fractional parts of 17 cows was a whole number, the children were confused about how to carry out these terms. Finally a neighbor offered to lend a cow to the children. How many cows did each child get?

42. Parades At the Veteran's Day parade, the local members of the Veterans of Foreign Wars (VFW) found that they could arrange themselves in rows of 6, 7, or 8, with no one left over. What is the least number of VFW members in the parade?

43. Work At Migdalia Montenegro's retirement dinner, there were two sizes of banquet tables. One size seated 5 people and the other size seated 8 people. Seventy-nine people attended Mrs. Montengro's dinner. If there were no empty places, how many tables of each size were there?

44. Travel George Farquar spent, in this order, a third of his life in the United States, a sixth of his life in England, twelve years in Mexico, half the remainder in Australia, and as long in Canada as he spent in Spain. How many years did George live if he spent his 45th birthday in Mexico?

Mixed Review

45. 50% more than what number is 25% less than 60% more than 20? **(Lesson 4-4)**

46. Ping-pong Each ping-pong ball weighs about $\frac{1}{10}$ of an ounce. How many ping-pong balls weigh 1 pound altogether? **(Lesson 3-3)**

47. Find the degree of $51x^5yz$. **(Lesson 6-5)**

48. Paula Robinson went to the corner store to buy four items. The clerk at the store had to add the four prices by hand since her cash register was broken. When the clerk figured Paula's bill, she mistakenly multiplied the four figures instead of adding them. Paula had already mentally computed her sum, so, realizing that the total was correct, she paid the clerk $7.11. How much was each item that Paula bought? **(Lesson 7-4)**

49. Find $\frac{y}{y-1} - \frac{1}{y-1}$. **(Lesson 8-5)**

50. Solve by listing the possibilities. **(Lesson 8-6)**

How many ways can you receive change for a half-dollar if you receive at least one dime and one quarter?

Journal

Make up a word problem that uses rational expressions. Solve the problem and explain your solution.

8-8 Mixed Expressions and Complex Fractions

Objective After studying this lesson, you should be able to:

- simplify mixed expressions and complex fractions.

Mixed expressions contain monomials and algebraic fractions.

Algebraic expressions such as $a + \frac{b}{c}$ and $5 + \frac{x - y}{x + 3}$ are called **mixed expressions.** Changing mixed expressions to rational expressions is similar to changing mixed numbers to improper fractions.

Mixed Number to Improper Fraction

$$3\tfrac{2}{5} \text{ or } 3 + \frac{2}{5} = \frac{3(5)}{5} + \frac{2}{5} = \frac{3(5) + 2}{5} = \frac{15 + 2}{5} = \frac{17}{5}$$

Mixed Expression to Rational Expression

$$a + \frac{a^2 + b}{a - b} = \frac{a(a - b)}{a - b} + \frac{a^2 + b}{a - b} = \frac{a(a - b) + (a^2 + b)}{a - b} = \frac{a^2 - ab + a^2 + b}{a - b} = \frac{2a^2 - ab + b}{a - b}$$

Example 1 Find $8 + \frac{x^2 - y^2}{x^2 + y^2}$.

$$8 + \frac{x^2 - y^2}{x^2 + y^2} = \frac{8(x^2 + y^2)}{x^2 + y^2} + \frac{x^2 - y^2}{x^2 + y^2}$$
$$= \frac{8(x^2 + y^2) + (x^2 - y^2)}{x^2 + y^2}$$
$$= \frac{8x^2 + 8y^2 + x^2 - y^2}{x^2 + y^2}$$
$$= \frac{9x^2 + 7y^2}{x^2 + y^2}$$

If a fraction has one or more fractions in the numerator or denominator, it is called a **complex fraction.** Some complex fractions are shown below.

$$\frac{3\frac{1}{2}}{5\frac{2}{3}} \qquad \frac{8}{\frac{a}{b}} \qquad \frac{\frac{a+b}{a}}{\frac{a-b}{b}} \qquad \frac{\frac{1}{x} - \frac{1}{y}}{\frac{1}{x} + \frac{1}{y}}$$

Consider the complex fraction $\frac{\frac{3}{5}}{\frac{7}{8}}$. To simplify this fraction, rewrite it as $\frac{3}{5} \div \frac{7}{8}$ and proceed as follows.

$$\frac{3}{5} \div \frac{7}{8} = \frac{3}{5} \cdot \frac{8}{7} \text{ or } \frac{24}{35}$$

Recall that to find the quotient, you multiply by $\frac{8}{7}$, the reciprocal of $\frac{7}{8}$.

Similarly, to simplify $\frac{\frac{a}{b}}{\frac{c}{d}}$, rewrite it as $\frac{a}{b} \div \frac{c}{d}$ and proceed as follows.

$$\frac{a}{b} \div \frac{c}{d} = \frac{a}{b} \cdot \frac{d}{c} \text{ or } \frac{ad}{bc}$$ *The reciprocal of $\frac{c}{d}$ is $\frac{d}{c}$.*

Simplifying Complex Fractions

Any complex fraction $\frac{\frac{a}{b}}{\frac{c}{d}}$, where $b \neq 0$, $c \neq 0$, and $d \neq 0$, may be expressed as $\frac{ad}{bc}$.

Example 2

Simplify $\frac{\frac{1}{x} + \frac{1}{y}}{\frac{1}{x} - \frac{1}{y}}$.

$$\frac{\frac{1}{x} + \frac{1}{y}}{\frac{1}{x} - \frac{1}{y}} = \frac{\frac{y}{xy} + \frac{x}{xy}}{\frac{y}{xy} - \frac{x}{xy}}$$ *The LCD of the numerator $\frac{1}{x} + \frac{1}{y}$ and the denominator $\frac{1}{x} - \frac{1}{y}$ is xy.*

$$= \frac{\frac{y + x}{xy}}{\frac{y - x}{xy}}$$ *Add to simplify the numerator. Subtract to simplify the denominator.*

$$= \frac{y + x}{xy} \cdot \frac{xy}{y - x}$$ *The reciprocal of $\frac{y - x}{xy}$ is $\frac{xy}{y - x}$.*

$$= \frac{y + x}{\cancel{xy}_1} \cdot \frac{\cancel{xy}^1}{y - x}$$ *Eliminate common factors.*

$$= \frac{y + x}{y - x}$$

Example 3

Simplify $\frac{x + 4 - \frac{1}{x + 4}}{x + 11 + \frac{48}{x - 3}}$.

$$\frac{x + 4 - \frac{1}{x + 4}}{x + 11 + \frac{48}{x - 3}} = \frac{\frac{(x + 4)(x + 4) - 1}{x + 4}}{\frac{(x + 11)(x - 3) + 48}{x - 3}}$$ *The LCD of the numerator is x + 4. The LCD of the denominator is x − 3.*

$$= \frac{\frac{x^2 + 8x + 16 - 1}{x + 4}}{\frac{x^2 + 8x - 33 + 48}{x - 3}}$$ *Subtract to simplify the numerator. Add to simplify the denominator.*

$$= \frac{\frac{x^2 + 8x + 15}{x + 4}}{\frac{x^2 + 8x + 15}{x - 3}}$$ *Simplify.*

$$= \frac{\cancel{x^2 + 8x + 15}^1}{x + 4} \cdot \frac{x - 3}{\cancel{x^2 + 8x + 15}_1}$$ *Multiply by the reciprocal. $x^2 + 8x + 15$ is a common factor.*

$$= \frac{x - 3}{x + 4}$$

Example 4 Simplify $\dfrac{x - \dfrac{x+4}{x+1}}{x-2}$.

$$\frac{x - \frac{x+4}{x+1}}{x-2} = \frac{\frac{x(x+1)-(x+4)}{x+1}}{x-2}$$ *The LCD of the numerator is x + 1. The LCD of the denominator is x − 2.*

$$= \frac{\frac{x^2+x-x-4}{x+1}}{x-2}$$ *Subtract to simplify the numerator.*

$$= \frac{\frac{x^2-4}{x+1}}{\frac{x-2}{1}}$$ *Simplify.*

$$= \frac{x^2-4}{x+1} \cdot \frac{1}{x-2}$$ *Multiply by the reciprocal.*

$$= \frac{(x+2)\overset{1}{\cancel{(x-2)}}}{(x+1)\underset{1}{\cancel{(x-2)}}}$$ *x − 2 is a common factor.*

$$= \frac{x+2}{x+1}$$

CHECKING FOR UNDERSTANDING

Communicating Mathematics **Read and study the lesson to answer each question.**

1. Is it possible for a complex fraction to be its own reciprocal?
2. In the mixed expression $8 + \frac{x^2 - y^2}{x^2 + y^2}$, what is the denominator of 8?

Guided Practice **Find each sum.**

3. $4 + \frac{2}{x}$
4. $8 + \frac{5}{3y}$
5. $2m + \frac{4+m}{m}$
6. $3a + \frac{a+1}{2a}$
7. $b^2 + \frac{2}{b-2}$
8. $3r^2 + \frac{4}{2r+1}$

EXERCISES

Practice **Simplify.**

9. $\dfrac{3\frac{1}{2}}{4\frac{3}{4}}$
10. $\dfrac{\frac{x^2}{y}}{\frac{y}{x^3}}$
11. $\dfrac{\frac{x+4}{y-2}}{\frac{x^2}{y^3}}$
12. $\dfrac{\frac{x^3}{y^2}}{\frac{x+y}{x-y}}$
13. $\dfrac{\frac{x+y}{a+b}}{\frac{x^2-y^2}{a^2-b^2}}$
14. $\dfrac{\frac{x-y}{x+y}}{\frac{x+y}{x-y}}$

15. $\dfrac{\frac{1}{x}+\frac{1}{y}}{\frac{1}{y}-\frac{1}{x}}$

16. $\dfrac{\frac{a+b}{x}}{\frac{a-b}{y}}$

17. $\dfrac{\frac{x^2+8x+15}{x^2+x-6}}{\frac{x^2+2x-15}{x^2-2x-3}}$

18. $\dfrac{\frac{a^2-6a+5}{a^2+13a+42}}{\frac{a^2-4a+3}{a^2+3a-18}}$

19. $\dfrac{\frac{y^2-1}{y^2+3y-4}}{y+1}$

20. $\dfrac{\frac{a^2-2a-3}{a^2-1}}{a-3}$

21. $\dfrac{\frac{a^2+2a}{a^2+9a+18}}{\frac{a^2-5a}{a^2+a-30}}$

22. $\dfrac{\frac{x^2+4x-21}{x^2-9x+18}}{\frac{x^2+3x-28}{x^2-10x+24}}$

23. $\dfrac{x-\frac{15}{x-2}}{x-\frac{20}{x-1}}$

24. $\dfrac{m+\frac{35}{m+12}}{m-\frac{63}{m-2}}$

25. $7+\dfrac{x^2+y^2}{x^2-4y^2}$

26. $5+\dfrac{a^2+11}{a^2-1}$

27. $\dfrac{x+2+\frac{2}{x+5}}{x+6+\frac{6}{x+1}}$

28. $\dfrac{x+5+\frac{3}{x+1}}{x-1-\frac{3}{x+1}}$

29. $\dfrac{\frac{a^2-a-1}{a-1}}{a-\frac{1}{a-1}}$

Critical Thinking

30. Simplify $\dfrac{1}{1-\frac{1}{1+a}}-\dfrac{1}{\frac{1}{1-a}-1}$.

Applications

31. **History** Construction of the Washington Monument began in 1848 and was put on hold in 1854. Construction resumed 26 years later and continued until the monument was completed 4 years later. The monument was finally opened to the public 4 years after it was completed. When was the Washington Monument opened to the public?

32. **Statistics** In 1988, New Jersey was the most densely populated state, and Alaska was the least densely populated. The population of New Jersey was 7,721,000, and the population of Alaska was 524,000. If the land area of New Jersey is 7,468 square miles and the land area of Alaska is 570,833 square miles, how many more people were there per square mile in New Jersey than in Alaska in 1988?

Mixed Review

33. **Sales** How many pounds of apples costing 64¢ per pound must be added to 30 pounds of apples costing 49¢ per pound to create a mixture that would sell for 58¢ per pound? **(Lesson 4-6)**

34. If $\frac{4}{9}=0.\overline{4}$, $\frac{43}{99}=0.\overline{43}$, and $\frac{817}{999}=0.\overline{817}$, find fractions that are equivalent to the following decimals: $0.\overline{6}$, $0.\overline{57}$, $0.\overline{253}$, and $0.\overline{6001}$. **(Lesson 6-1)**

35. Factor $12a^2-12$. **(Lesson 7-6)**

36. Find $\frac{5b}{7x}+\frac{3a}{21x^2}$. **(Lesson 8-7)**

8-9 Solving Rational Equations

Objectives

After studying this lesson, you should be able to:

- solve rational equations, and
- solve problems involving work and uniform motion.

Rational equations are equations containing rational expressions.

Example 1

The Blendonville 4-H Club is holding their annual car wash to raise funds for club projects. Olivia can wash and wax 1 car in 3 hours. George can wash and wax 1 car in 4 hours. If Olivia and George work together, how long will it take them to wash and wax one car?

EXPLORE In one hour, Olivia can complete $\frac{1}{3}$ of the job. In 2 hours, she can complete $\frac{1}{3} \cdot 2$ or $\frac{2}{3}$ of the job. In t hours, she can complete $\frac{1}{3} \cdot t$ or $\frac{t}{3}$ of the job. Use the following formula to solve the problem.

rate of work · time = work done

$$r \cdot t = w$$

Let t = time in hours of them to wash and wax one car.

PLAN In t hours, Olivia can do $\frac{t}{3}$ of the job. In t hours, George can do $\frac{t}{4}$ of the job.

	r	t	w
Olivia	$\frac{1}{3}$	t	$\frac{t}{3}$
George	$\frac{1}{4}$	t	$\frac{t}{4}$

$$\frac{t}{3} + \frac{t}{4} = 1$$ *Together they wash and wax 1 car.*

SOLVE

$$12\left(\frac{t}{3} + \frac{t}{4}\right) = 12(1)$$

$$4t + 3t = 12$$ *Multiply each side of the equation by the LCD,*

$$7t = 12$$

$$t = \frac{12}{7}$$

Olivia and George can wash and wax 1 car in $\frac{12}{7}$ hours or about one hour and 43 minutes.

EXAMINE Olivia does $\frac{1}{3}t$ or $\frac{1}{3} \cdot \frac{12}{7}$ of the job. $\frac{1}{3} \cdot \frac{12}{7} = \frac{4}{7}$

George does $\frac{1}{4}t$ or $\frac{1}{4} \cdot \frac{12}{7}$ of the job. $\frac{1}{4} \cdot \frac{12}{7} = \frac{3}{7}$

Since $\frac{3}{7} + \frac{4}{7} = \frac{7}{7}$ or 1, the solution checks.

Example 2

Solve $\frac{11}{2x} - \frac{2}{3x} = \frac{1}{6}$.

$\frac{11}{2x} - \frac{2}{3x} = \frac{1}{6}$ *The LCD is 6x.*

$6x\left(\frac{11}{2x} - \frac{2}{3x}\right) = 6x\left(\frac{1}{6}\right)$ *Multiply each side of the equation by the LCD.*

$3(11) - 2(2) = x$ *The fractions are eliminated.*

$29 = x$ *Simplify.*

Check: $\frac{11}{2x} - \frac{2}{3x} = \frac{1}{6}$

$\frac{11}{2(29)} - \frac{2}{3(29)} \stackrel{?}{=} \frac{1}{6}$

$\frac{11}{58} - \frac{2}{87} \stackrel{?}{=} \frac{1}{6}$ *Use a calculator to check.*

$\frac{1}{6} = \frac{1}{6}$ ✓ The solution is 29.

Recall that uniform motion problems can be solved by using the formula below.

$$\textit{distance} = \textit{rate} \cdot \textit{time}$$
$$d = r \cdot t$$

Example 3

Sally and her brother rented a boat to fish in Jones Creek. The maximum speed of the boat in still water was 3 miles per hour. At this rate, a 9-mile trip downstream with the current took the same amount of time as a 3-mile trip upstream against the current. What was the rate of the current?

EXPLORE Let c = the rate of the current. The rate of the boat when traveling downstream, or with the current, is 3 miles per hour *plus* the rate of the current. That is, $3 + c$. The rate when traveling upstream, or against the current, is 3 miles per hour *minus* the rate of the current. That is $3 - c$.

(continued on the next page)

PLAN To represent the time, t, solve $d = rt$ for t. Thus, $t = \frac{d}{r}$.

	d	r	t
Downstream	9	$3 + c$	$\frac{9}{3+c}$
Upstream	3	$3 - c$	$\frac{3}{3-c}$

SOLVE $\frac{9}{3+c} = \frac{3}{3-c}$ *time downstream = time upstream. Note that $c \neq 3$ or -3. Why?*

$$(3+c)(3-c)\left(\frac{9}{3+c}\right) = (3+c)(3-c)\left(\frac{3}{3-c}\right)$$ *Multiply each side of the equation by the LCD.*

$$(3-c)9 = (3+c)3$$ *The fractions are eliminated.*

$$27 - 9c = 9 + 3c$$
$$-12c = -18$$
$$c = \frac{3}{2}$$

The rate of the current was $\frac{3}{2}$ or $1\frac{1}{2}$ miles per hour. *Examine this solution.*

Example 4

Solve $\frac{2m}{1-m} + \frac{m+3}{m^2-1} = 1$.

$$\frac{2m}{1-m} + \frac{m+3}{m^2-1} = 1$$ *What values are excluded?*

$$\frac{2m}{1-m} + \frac{m+3}{(m+1)(m-1)} = 1$$ *Note that $1 - m = -(m - 1)$.*

$$-\frac{2m}{(m-1)} + \frac{m+3}{(m+1)(m-1)} = 1$$ *The LCD is $(m + 1)(m - 1)$.*

$$(m+1)(m-1)\left(-\frac{2m}{(m-1)} + \frac{m+3}{(m+1)(m-1)}\right) = (m+1)(m-1)1$$
$$-2m(m+1) + (m+3) = m^2 - 1$$
$$-2m^2 - 2m + m + 3 = m^2 - 1$$
$$-3m^2 - m + 4 = 0$$
$$3m^2 + m - 4 = 0$$ *Why?*
$$(3m+4)(m-1) = 0$$ *Factor.*

$$3m + 4 = 0 \quad \text{or} \quad m - 1 = 0$$ *Zero product property*
$$m = -\frac{4}{3} \qquad m = 1$$

The solution is $-\frac{4}{3}$.

CHECKING FOR UNDERSTANDING

Communicating Mathematics

Read and study the lesson to answer each question.

1. In Example 3, what would happen if the maximum speed of the boat in still water was 1 mile per hour?
2. In Example 4, why is 1 not a solution?

Guided Practice

Find the LCD for each set of rational expressions.

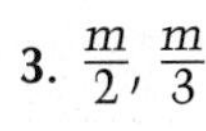

3. $\frac{m}{2}, \frac{m}{3}$
4. $\frac{x}{5}, \frac{2x}{3}$
5. $\frac{3b}{4}, \frac{5b}{8}$
6. $\frac{1}{x}, \frac{5x}{x+1}$
7. $\frac{4}{r^2-1}, \frac{5}{r-1}$
8. $\frac{m}{2m^2+3m-35}, \frac{8}{2m-7}$
9. $\frac{5k}{k+5}, \frac{k^2}{k+3}, \frac{1}{k+3}$
10. $\frac{7}{h+1}, \frac{1}{2}, \frac{2h+5}{h-1}$
11. $\frac{2x+1}{4x^2-1}, \frac{1}{3}, \frac{1}{2x+1}$
12. Luisa can paint her house in 8 days. What part of it can she paint:
    a. in 1 day? b. in 3 days? c. in x days?
13. Drew can build a garage in n days. What part of it can he build:
    a. in 1 day? b. in 4 days? c. in x days?
14. Dimas can landscape a yard in 8 days. Stephanie can do the same job in 10 days.
    a. What part of the job can Dimas do: in 1 day? in x days?
    b. What part of the job can Stephanie do: in 1 day? in x days?
    c. What part of the job can they do together: in 1 day? in x days?

EXERCISES

Practice

Solve each equation.

15. $\frac{2a-3}{6} = \frac{2a}{3} + \frac{1}{2}$
16. $\frac{3x}{5} + \frac{3}{2} = \frac{7x}{10}$
17. $\frac{2b-3}{7} - \frac{b}{2} = \frac{b+3}{14}$
18. $\frac{x+1}{x} + \frac{x+4}{x} = 6$
19. $\frac{18}{b} = \frac{3}{b} + 3$
20. $\frac{3}{5x} + \frac{7}{2x} = 1$
21. $\frac{5x}{x+1} + \frac{1}{x} = 5$
22. $\frac{2}{3r} - \frac{3r}{r-2} = -3$
23. $\frac{m}{m+1} + \frac{5}{m-1} = 1$
24. $\frac{r-1}{r+1} - \frac{2r}{r-1} = -1$
25. $\frac{4x}{2x+3} - \frac{2x}{2x-3} = 1$
26. $\frac{5}{5-p} - \frac{p^2}{5-p} = -2$
27. $\frac{14}{b-6} = \frac{1}{2} + \frac{6}{b-8}$
28. $\frac{2a-3}{a-3} - 2 = \frac{12}{a+3}$
29. $\frac{r}{3r+6} - \frac{r}{5r+10} = \frac{2}{5}$
30. $\frac{x-2}{x} - \frac{x-3}{x-6} = \frac{1}{x}$
31. $\frac{z+3}{z-1} + \frac{z+1}{z-3} = 2$
32. $\frac{x+2}{x-2} - \frac{2}{x+2} = \frac{-7}{3}$
33. $\frac{7}{x^2-5x} + \frac{3}{5-x} = \frac{4}{x}$
34. $\frac{6}{z+2} + \frac{3}{z^2-4} = \frac{2z-7}{z-2}$
35. $\frac{3w}{w^2-5w+4} = \frac{2}{w-4} + \frac{3}{w-1}$
36. $\frac{4}{k^2-8k+12} = \frac{k}{k-2} + \frac{1}{k-6}$
37. $\frac{m+3}{m+5} + \frac{2}{m-9} = \frac{-20}{m^2-4m-45}$
38. $\frac{h^2-7h-8}{3h^2+2h-8} + \frac{1}{h+2} = 0$

Critical Thinking

39. What number would you add to the numerator and denominator of $\frac{2}{11}$ to make a fraction equivalent to $\frac{1}{2}$?

Applications

Work: Use $rt = w$ to solve each problem.

40. Jane can wash the windows of a building in 4 hours. Jaime can do the same job in 6 hours. If they work together, how long will it take them to wash the windows?

41. A swimming pool can be filled by one pipe in 10 hours. The drain pipe can empty the pool in 15 hours. If both pipes are open, how long will it take to fill the pool?

42. Kiko and Marcus can clean the garage together in $3\frac{3}{5}$ hours. Kiko can do the job alone in 6 hours. How many hours would it take Marcus to do the job alone?

Physics: Use $d = rt$ to solve each problem.

43. A long-distance cyclist pedaling at a steady rate travels 30 miles with the wind. He can travel only 18 miles against the wind in the same amount of time. If the rate of the wind is 3 miles per hour, what is the cyclist's rate without the wind?

44. A tugboat pushing a barge up the Ohio River takes 1 hour longer to travel 36 miles up the river than to travel the same distance down the river. If the rate of the current is 3 miles per hour, find the speed of the tugboat and barge in still water.

45. An airplane can fly at a rate of 600 miles per hour in calm air. It can fly 2520 miles with the wind in the same time it can fly 2280 miles against the wind. Find the speed of the wind.

46. A motorboat takes $\frac{2}{3}$ as much time to travel 10 miles downstream as it takes to travel the same distance upstream. If the rate of the current is 5 miles per hour, find the speed of the motorboat in still water.

Mixed Review

47. **World Records** The Huey P. Long Bridge, in Metairie, Louisiana, is the longest railroad bridge in the world. If you were traveling on a train going 60 miles per hour across the 22,996-foot bridge, how long would it take you to cross it? **(Lesson 4-7)**

48. **Physics** Shannon weighs 126 pounds and Minal weighs 154 pounds. They are seated at opposite ends of a seesaw. Shannon and Minal are 16 feet apart, and the seesaw is balanced. How far is Shannon from the fulcrum? **(Lesson 4-9)**

49. Find $(4a + 6b) + (2a + 3b)$. **(Lesson 6-6)**

50. Factor $a^2 + 12a + 36$. **(Lesson 7-7)**

51. Find $x + \frac{x}{y}$. **(Lesson 8-8)**

8-10 Application: Formulas

Objectives After studying this lesson, you should be able to:
- solve formulas for a specified variable, and
- use formulas that involve rational expressions.

Rational expressions and rational equations often contain more than one variable. Sometimes it is useful to solve for one of the variables. Then you can find the values of any of the variables.

Example 1

CONNECTION Geometry

Solve for h in $A = \frac{1}{2}h(a + b)$. *This is the formula for the area of a trapezoid.*

$A = \frac{1}{2}h(a + b)$ *The LCD is 2.*

$2A = 2\left(\frac{1}{2}h(a + b)\right)$ *Multiply each side by 2.*

$2A = h(a + b)$

$\frac{2A}{a + b} = h$ *Divide each side by $(a + b)$.*

FYI···

The first photograph was taken in 1826 by Joseph Niépce, a French physicist. The photo of the view from his window took eight hours to develop.

The formula below applies to camera and lens systems.

$$\frac{1}{f} = \frac{1}{a} + \frac{1}{b}$$

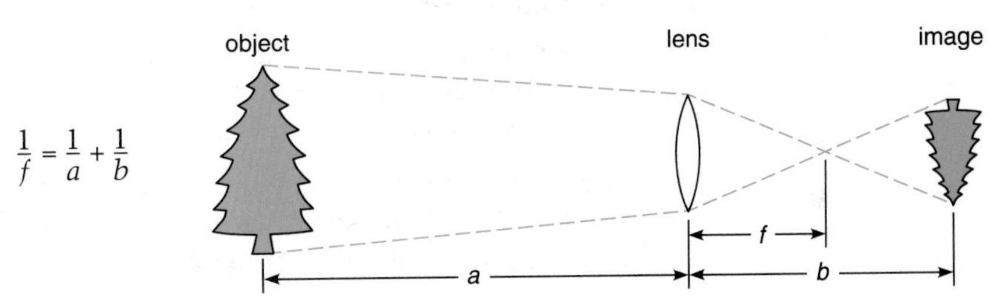

In the formula, f is the focal length of the lens, a is the distance from the object to the lens, and b is the distance from the image to the lens.

Example 2

APPLICATION Photography

Solve the formula above for f.

$\frac{1}{f} = \frac{1}{a} + \frac{1}{b}$

$abf\left(\frac{1}{f}\right) = abf\left(\frac{1}{a} + \frac{1}{b}\right)$ *Multiply each side by the LCD, abf.*

$ab = bf + af$

$ab = (b + a)f$ *Factor $bf + af$.*

$\frac{ab}{b + a} = f$ *Divide each side by $b + a$.*

The *ohm* is named for German physicist Georg Simon Ohm. The unit of conductance, the reciprocal of resistance, is the *mho*—Ohm's name spelled backward.

Electricity can be described as the flow of electrons through a conductor, such as a copper wire. Electricity flows more freely through some conductors than others. The force opposing the flow is called *resistance.* The unit of resistance commonly used is the *ohm.*

Resistances can occur one after another, that is, *in series.* Resistances can also occur in branches of the conductor going in the same direction, or *in parallel.* Look at the diagrams below. Formulas for the total resistance, R_T, are given under the diagrams.

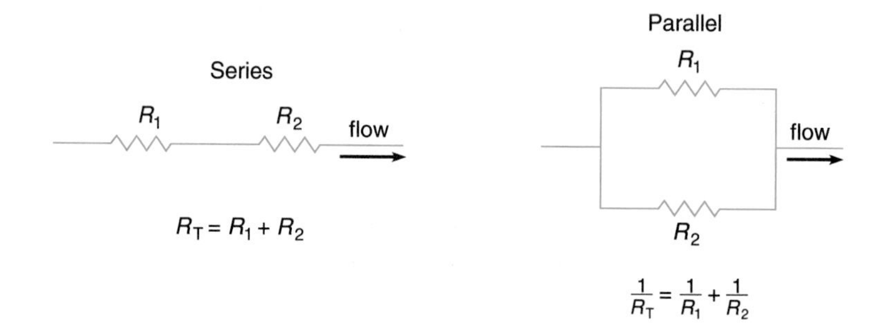

Example 3

Assume that R_1 = 4 ohms and R_2 = 3 ohms. Compute the total resistance of the conductor when the resistances are in series.

$$\begin{aligned} R_T &= R_1 + R_2 \\ &= 4 + 3 \\ &= 7 \end{aligned}$$

Thus, the total resistance is 7 ohms.

Example 4

Assume that R_1 = 5 ohms and R_2 = 6 ohms. Compute the total resistance of the conductor when the resistances are in parallel.

$$\frac{1}{R_T} = \frac{1}{R_1} + \frac{1}{R_2}$$

$$\frac{1}{R_T} = \frac{1}{5} + \frac{1}{6} \qquad R_1 = 5 \text{ and } R_2 = 6$$

$$\frac{1}{R_T} = \frac{11}{30} \qquad \frac{1}{5} + \frac{1}{6} = \frac{6}{30} + \frac{5}{30}$$

$$1 \cdot 30 = R_T \cdot 11 \qquad \textit{Find the cross products.}$$

$$\frac{30}{11} = R_T \qquad \textit{Divide each side by 11.}$$

Thus, the total resistance is $\frac{30}{11}$ or 2.727 ohms.

A *circuit,* or path for the flow of electrons, often has some resistances connected in series and others in parallel.

Example 5

A parallel circuit has one branch in series as shown at the right. Given that the total resistance is 2.25 ohms, R_1 = 3 ohms, and R_2 = 4 ohms, find R_3.

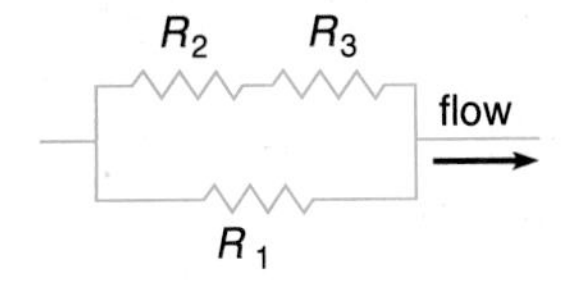

$$\frac{1}{R_T} = \frac{1}{R_1} + \frac{1}{R_2 + R_3}$$ *The total resistance of the branch in series is $R_2 + R_3$.*

$$\frac{1}{2.25} = \frac{1}{3} + \frac{1}{4 + R_3}$$ *$R_T = 2.25$ and $R_1 = 3$*

$$\frac{1}{2.25} - \frac{1}{3} = \frac{1}{4 + R_3}$$ $\frac{1}{2.25} - \frac{1}{3} = \frac{4}{9} - \frac{3}{9}$

$$\frac{1}{9} = \frac{1}{4 + R_3}$$

$$4 + R_3 = 9$$ *Find the cross products.*

$$R_3 = 5$$

Thus, R_3 is 5 ohms.

CHECKING FOR UNDERSTANDING

Communicating Mathematics

Read and study the lesson to answer each question.

1. In Example 1, what steps would you take to solve for *a*?
2. In Example 1, what steps would you take to solve for *b*?
3. In Example 2, what steps would you take to solve for *a*?
4. In Example 2, what steps would you take to solve for *b*?

Guided Practice

Exercises 5–10 refer to the diagram at the right.

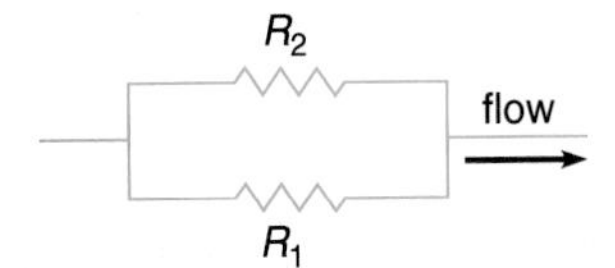

5. Find the total resistance, R_T, given that R_1 = 8 ohms and R_2 = 6 ohms.
6. Find the total resistance, R_T, given that R_1 = 4.5 ohms and R_2 = 3.5 ohms.
7. Find R_1, given that R_T is $2.\overline{2}$ ohms and R_2 is 5 ohms.
8. Find R_1, given that R_T is $3\frac{3}{7}$ ohms and R_2 is 8 ohms.
9. Find R_1 and R_2, given that the total resistance is $2.\overline{6}$ ohms and R_1 is two times as great as R_2.
10. Find R_1 and R_2, given that the total resistance is 2.25 ohms and R_1 is three times as great as R_2.

EXERCISES

Applications

Physics: Solve each formula for the variable indicated.

11. $a = \frac{v}{t}$, for t
12. $v = r + at$, for a
13. $s = vt + \frac{1}{2}at^2$, for v
14. $s = vt + \frac{1}{2}at^2$, for a
15. $F = G\left(\frac{Mm}{d^2}\right)$, for M
16. $f = \frac{W}{g} \cdot \frac{V^2}{R}$, for R

Business: Solve each formula for the variable indicated.

17. $A = p + prt$, for p
18. $I = prt$, for r
19. $I = \left(\frac{100 - P}{P}\right)\frac{365}{R}$, for P
20. $I = \frac{365d}{360 - dr}$, for d
21. $a = \frac{r}{2y} - 0.25$, for y
22. $c = \frac{P - 100}{P}$, for P

Electronics: Solve each formula for the variable indicated.

23. $H = (0.24)I^2Rt$, for R
24. $P = \frac{E^2}{R}$, for R
25. $\frac{1}{R_T} = \frac{1}{R_1} + \frac{1}{R_2}$, for R_1
26. $I = \frac{E}{r + R}$, for R
27. $I = \frac{nE}{nr + R}$, for n
28. $I = \frac{E}{\frac{r}{n} + R}$, for r

Mathematics: Solve each formula for the variable indicated.

29. $y = mx + b$, for m
30. $S = \frac{n}{2}(A + t)$, for n
31. $m = \frac{y_2 - y_1}{x_2 - x_1}$, for y_2
32. $m = \frac{y_2 - y_1}{x_2 - x_1}$, for x_1
33. $\frac{P}{D} = Q + \frac{R}{D}$, for R
34. $\frac{P}{D} = Q + \frac{R}{D}$, for D

Electronics: Exercises 35–37 refer to the diagram below.

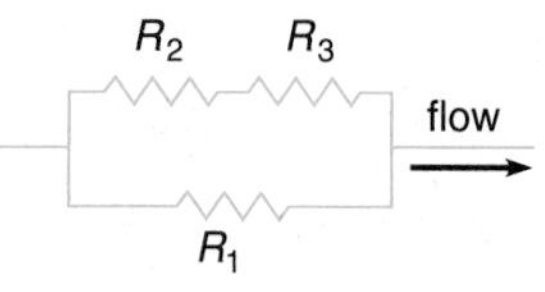

35. Find R_T, given that $R_1 = 5$ ohms, $R_2 = 4$ ohms, and $R_3 = 3$ ohms.
36. Find R_1, given that $R_T = 2\frac{10}{13}$ ohms, $R_2 = 3$ ohms, and $R_3 = 6$ ohms.
37. Find R_2, given that $R_T = 3.5$ ohms, $R_1 = 5$ ohms, and $R_3 = 4$ ohms.

Electronics: Solve each problem.

38. Resistances of 3 ohms, 6 ohms, and 9 ohms are connected in series. What is the total resistance?

39. Eight lights on a decorated tree are connected in series. Each has a resistance of 12 ohms. What is the total resistance?

40. Three coils with resistances of 3 ohms, 4 ohms, and 6 ohms are connected in parallel. What is the total resistance?

41. Three appliances are connected in parallel: a lamp with a resistance of 60 ohms, an iron with a resistance of 20 ohms, and a heating coil with a resistance of 80 ohms. Find the total resistance.

Critical Thinking

Electronics: Exercises 42–43 refer to the diagram below.

42. Write an equation for the total resistance for the diagram at the right.

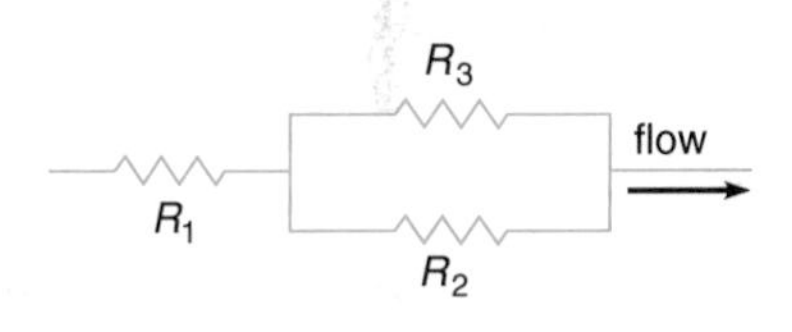

43. Find the total resistance, given that $R_1 = 5$ ohms, $R_2 = 4$ ohms, and $R_3 = 6$ ohms.

Mixed Review

44. **Temperature** The formula for finding the Celsius temperature (C) when you know the Fahrenheit temperature (F) is $C = \frac{5}{9}(F - 32)$. Find the Celsius temperature when the Fahrenheit temperature is 59°. **(Lesson 1-7)**

45. Solve: $|4x + 4| \leq 14$. **(Lesson 5-6)**

46. **Finance** The current balance on a car loan can be found by evaluating the expression $P\left[\frac{1 - (1 + r)^{k-n}}{r}\right]$, where P is the monthly payment, r is the monthly interest rate, k is the number of payments already made, and n is the total number of monthly payments. Find the current balance if $P = \$256$, $r = 0.01$, $k = 20$, and $n = 60$. **(Lesson 6-2)**

47. **Astronomy** When a solar flare occurs on the sun, it can send out a blast wave that travels at 3×10^6 kilometers per hour. How long would it take for a blast wave to reach Earth where it could be detected by a satellite in orbit? (Assume that the distance from the sun to the satellite would be 1.5×10^8 kilometers.) **(Lesson 6-4)**

48. Factor $2m^2 - 32n^2$. **(Lesson 7-8)**

49. **Work** Hugo and Denise can mow a lawn together in 24 minutes. It takes Hugo 40 minutes to do the job alone. How long would it take Denise to do the job alone? **(Lesson 8-9)**

Portfolio

Select an item that shows something new you learned in this chapter and place it in your portfolio.

CHAPTER 8

SUMMARY AND REVIEW

VOCABULARY

Upon completing this chapter, you should be familiar with the following terms:

complex fractions	334	316	multiplicative inverse
excluded values	306	338	rational equations
least common denominator (LCD)	330	306	rational expressions
least common multiple (LCM)	329	316	reciprocal
mixed expressions	334		

SKILLS AND CONCEPTS

OBJECTIVES AND EXAMPLES

Upon completing this chapter, you should be able to:

- simplify rational expressions **(Lesson 8-1)**

$$\frac{x + y}{x^2 + 3xy + 2y^2} = \frac{\overset{1}{\cancel{x + y}}}{\underset{1}{(\cancel{x + y})}(x + 2y)} = \frac{1}{x + 2y}$$

REVIEW EXERCISES

Use these exercises to review and prepare for the chapter test.

Simplify. State the excluded values of the variables.

1. $\frac{3x^2y}{12xy^3z}$
2. $\frac{z^2 - 3z}{z - 3}$
3. $\frac{a^2 - 25}{a^2 + 3a - 10}$
4. $\frac{x^2 + 10x + 21}{x^3 + x^2 - 42x}$

- multiply rational expressions **(Lesson 8-2)**

$$\frac{1}{x^2 + x - 12} \cdot \frac{x - 3}{x + 5}$$

$$= \frac{1}{\underset{1}{(\cancel{x - 3})}(x + 4)} \cdot \frac{\overset{1}{\cancel{x - 3}}}{x + 5}$$

$$= \frac{1}{x^2 + 9x + 20}$$

Find each product. Assume that no denominator is equal to zero.

5. $\frac{7}{9} \cdot \frac{a^2}{b}$
6. $\frac{5x^2y}{8ab} \cdot \frac{12a^2b}{25x}$
7. $\frac{x^2 + x - 12}{x + 2} \cdot \frac{x + 4}{x^2 - x - 6}$
8. $\frac{b^2 + 19b + 84}{b - 3} \cdot \frac{b^2 - 9}{b^2 + 15b + 36}$

- divide rational expressions **(Lesson 8-3)**

$$\frac{y^2 - 16}{y^2 - 64} \div \frac{y + 4}{y - 8} = \frac{y^2 - 16}{y^2 - 64} \cdot \frac{y - 8}{y + 4}$$

$$= \frac{\overset{1}{(\cancel{y + 4})}(y - 4)}{(y + 8)\underset{1}{(\cancel{y - 8})}} \cdot \frac{\overset{1}{\cancel{y - 8}}}{\underset{1}{\cancel{y + 4}}}$$

$$= \frac{y - 4}{y + 8}$$

Find each quotient. Assume that no denominator is equal to zero.

9. $\frac{p^3}{2q} \div \frac{-(p^2)}{4q}$
10. $\frac{n^2}{n - 3} \div (n + 4)$
11. $\frac{7a^2b}{x^2 + x - 30} \div \frac{3a}{x^2 + 15x + 54}$
12. $\frac{m^2 + 4m - 21}{m^2 + 8m + 15} \div \frac{m^2 - 9}{m^2 + 12m + 35}$

OBJECTIVES AND EXAMPLES

- divide polynomials by binomials **(Lesson 8-4)**

$$\begin{array}{r} x^2 + x - 19 \\ x-3\overline{)x^3 - 2x^2 - 22x + 21} \\ \underline{x^3 - 3x^2} \qquad\qquad\quad \\ x^2 - 22x \qquad\; \\ \underline{x^2 - 3x} \qquad\; \\ -19x + 21 \\ \underline{-19x + 57} \\ -36 \end{array}$$

The quotient is $x^2 + x - 19 - \frac{36}{x - 3}$.

REVIEW EXERCISES

Find each quotient.

13. $(x^3 + 7x^2 + 10x - 6) \div (x + 3)$

14. $(x^4 + 3x^3 + 2x^2 - x + 6) \div (x - 2)$

15. $(2a^3 - 3a - 9) \div (a - 9)$

16. $(6a^3 - 19a^2 + 2a + 15) \div (2a - 5)$

- add and subtract rational expressions with like denominators **(Lesson 8-5)**

$$\frac{m^2}{4 + m} - \frac{16}{m + 4} = \frac{m^2}{m + 4} - \frac{16}{m + 4}$$
$$= \frac{m^2 - 16}{m + 4}$$
$$= \frac{\overset{1}{\cancel{(m + 4)}}(m - 4)}{\underset{1}{\cancel{m + 4}}}$$
$$= m - 4$$

Find each sum or difference.

17. $\frac{7}{x^2} + \frac{a}{x^2}$

18. $\frac{a^2}{a^2 - b^2} + \frac{-(b^2)}{a^2 - b^2}$

19. $\frac{2x}{x - 3} - \frac{6}{x - 3}$

20. $\frac{x}{x^2 - 1} + \frac{1}{x^2 - 1}$

- add and subtract rational expressions with unlike denominators **(Lesson 8-7)**

$$\frac{m - 1}{m + 1} + \frac{4}{2m + 5}$$
$$= \frac{2m + 5}{2m + 5} \cdot \frac{m - 1}{m + 1} + \frac{4}{2m + 5} \cdot \frac{m + 1}{m + 1}$$
$$= \frac{2m^2 + 3m - 5}{(2m + 5)(m + 1)} + \frac{4m + 4}{(2m + 5)(m + 1)}$$
$$= \frac{2m^2 + 7m - 1}{2m^2 + 7m + 5}$$

Find each sum or difference.

21. $\frac{2}{x - y} + \frac{x}{y - x}$

22. $\frac{x}{x + 3} - \frac{5}{x - 2}$

23. $\frac{2x + 3}{x^2 - 4} + \frac{6}{x + 2}$

24. $\frac{m - n}{m^2 + 2mn + n^2} - \frac{m + n}{m - n}$

- simplify mixed expressions and complex fractions **(Lesson 8-8)**

$$\frac{\frac{x - 3}{x + 5}}{\frac{x + 5}{x}} = \frac{x - 3}{x + 5} \cdot \frac{x}{x + 5}$$
$$= \frac{x^2 - 3x}{x^2 + 10x + 25}$$

Simplify.

25. $\dfrac{\frac{x^2}{y^3}}{\frac{3x}{9y^2}}$

26. $\dfrac{\frac{a^2 - 13a + 40}{a^2 - 4a - 32}}{\frac{a - 5}{a + 7}}$

27. $\dfrac{x - \frac{35}{x + 2}}{x + \frac{42}{x + 13}}$

OBJECTIVES AND EXAMPLES

- solve rational equations **(Lesson 8-9)**

$$\frac{3}{x} + \frac{1}{x-5} = \frac{1}{2x}$$
$$2x(x-5)\left(\frac{3}{x} + \frac{1}{x-5}\right) = \left(\frac{1}{2x}\right) 2x(x-5)$$
$$6(x-5) + 2x = x - 5$$
$$7x = 25$$
$$x = \frac{25}{7}$$

REVIEW EXERCISES

Solve each equation.

28. $\frac{4x}{3} + \frac{7}{2} = \frac{7x}{12}$

29. $\frac{1}{h+1} + \frac{2}{3} = \frac{2h+5}{h-1}$

30. $\frac{3x+2}{x^2+7x+6} = \frac{1}{x+6} + \frac{4}{x+1}$

31. $\frac{3m-2}{2m^2-5m-3} - \frac{2}{2m+1} = \frac{4}{m-3}$

- solve formulas for a specified variable **(Lesson 8-10)**

Solve $\frac{1}{f} = \frac{1}{a} + \frac{1}{b}$ for a.

$$abf\left(\frac{1}{f}\right) = \left(\frac{1}{a} + \frac{1}{b}\right) abf$$
$$ab = bf + af$$
$$bf = a(b - f)$$
$$\frac{bf}{b-f} = a$$

Solve each equation for *n*.

32. $\frac{1}{2}n + b = n$

33. $\frac{n}{x} = \frac{y}{r}$

34. $\frac{n}{a} + \frac{b}{c} = d$

35. $\frac{a}{c} = n + bn$

APPLICATIONS AND CONNECTIONS

Geometry: Find the measure of the area of each rectangle in simplest form. (Lesson 8-2)

36.

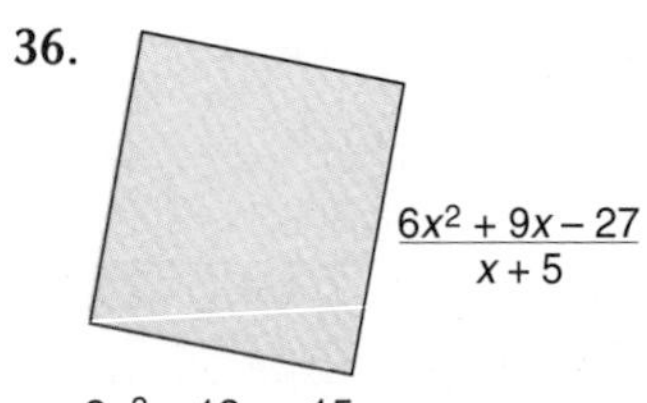

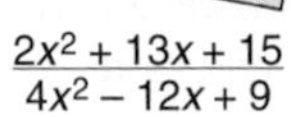

37.

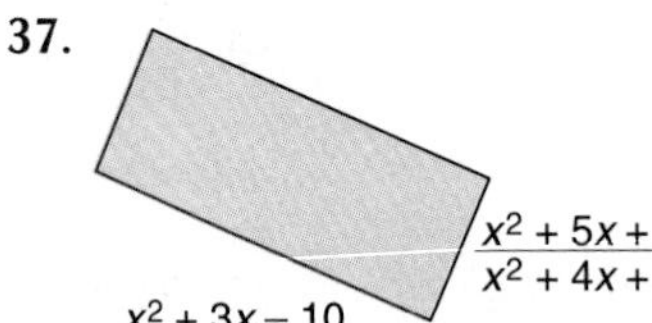

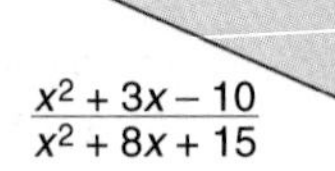

38.

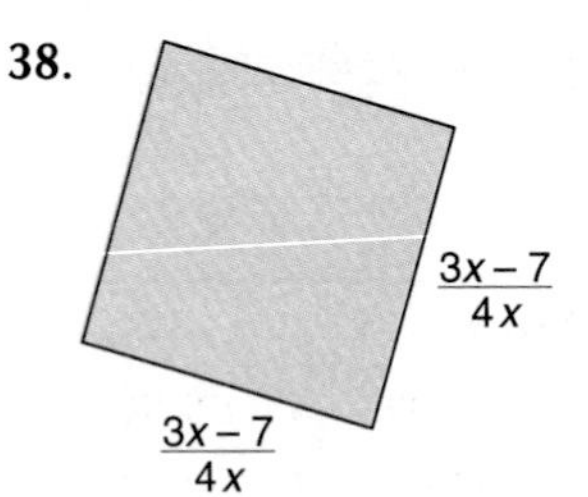

39. Solve by listing the possibilities.

Which whole numbers less than 100 are divisible by both 3 and 11? **(Lesson 8-6)**

40. **Electronics** Assume that $R_1 = 4$ ohms and $R_2 = 6$ ohms. What is the total resistance of the conductor if R_1 and R_2 are:

a. connected in series?

b. connected in parallel? **(Lesson 8-10)**

CHAPTER 8 TEST

Simplify.

1. $\dfrac{\frac{5}{9}}{\frac{2}{3}}$

2. $\dfrac{21x^2y}{28ax}$

3. $\dfrac{x^2 + 7x - 18}{x^2 + 12x + 27}$

4. $\dfrac{x^2 - x - 56}{x^2 + x - 42}$

5. $\dfrac{7x^2 - 28}{5x^3 - 20x}$

6. $\dfrac{2x^2 - 5x - 3}{x^2 + 2x - 15}$

Perform the indicated operations.

7. $\dfrac{3x}{x + 3} + \dfrac{5x}{x + 3}$

8. $\dfrac{2x}{x - 7} - \dfrac{14}{x - 7}$

9. $\dfrac{2x}{x + 7} + \dfrac{4}{x + 4}$

10. $\dfrac{2a + 1}{2a - 3} + \dfrac{a - 3}{3a + 2}$

11. $\dfrac{x + 5}{x + 2} + 6$

12. $\dfrac{x - 2}{x - 8} + x + 5$

13. $\dfrac{3x + 2}{4x + 1} + \dfrac{7}{x}$

14. $\dfrac{x}{x + 1} + \dfrac{1}{x + 1}$

15. $\dfrac{3x - 8}{x + 4} + \dfrac{9}{x + 1}$

16. $\dfrac{x^2 + 4x - 32}{x + 5} \cdot \dfrac{x - 3}{x^2 - 7x + 12}$

17. $\dfrac{3x^2 + 2x - 8}{x^2 - 4} \div \dfrac{6x^2 + 13x - 28}{2x^2 - 3x - 35}$

18. $\dfrac{4x^2 + 11x + 6}{x^2 - x - 6} \div \dfrac{x^2 + 8x + 16}{x^2 + x - 12}$

19. $\dfrac{3x^2 + 5x - 28}{x^2 - 3x - 28} \cdot \dfrac{x^2 - 8x + 7}{3x - 7}$

20. $\dfrac{x - \frac{24}{x + 5}}{x - \frac{72}{x - 1}}$

21. $\dfrac{\frac{x^2 - x - 6}{x^2 + 2x - 15}}{\frac{x^2 - 2x - 8}{x^2 + x - 20}}$

22. $\dfrac{\frac{2}{3m} + \frac{3}{m^2}}{\frac{2}{5m} + \frac{5}{m}}$

Solve each equation.

23. $\dfrac{y + 3}{6} = \dfrac{y + 2}{12} - \dfrac{2}{5}$

24. $\dfrac{x + 1}{x} + \dfrac{6}{x} = x + 7$

25. $\dfrac{4m}{m - 3} + \dfrac{6}{3 - m} = m$

26. $\dfrac{-2b - 9}{b^2 + 7b + 12} = \dfrac{b}{b + 3} + \dfrac{2}{b + 4}$

27. $\dfrac{1}{y - 4} - \dfrac{2}{y - 8} = \dfrac{-1}{y + 6}$

28. $\dfrac{m + 3}{m - 1} + \dfrac{m + 1}{m - 3} = \dfrac{22}{3}$

Solve each formula for the variable indicated.

29. $F = G\left(\dfrac{Mm}{d^2}\right)$, for G

30. $\dfrac{1}{R_T} = \dfrac{1}{R_1} + \dfrac{1}{R_2}$, for R_2

Solve.

31. **Work** Willie can do a job in 6 days. Myra can do the same job in $4\frac{1}{2}$ days. If they work together, how long will it take to complete the job?

32. **Motion** The top speed of a boat in still water is 5 miles per hour. At this speed, a 21-mile trip downstream takes the same amount of time as a 9-mile trip upstream. Find the rate of the current.

33. **Electronics** Three appliances are connected in parallel: a lamp of resistance 120 ohms, a toaster of resistance 20 ohms, and an iron of resistance 12 ohms. Find the total resistance.

Bonus Complete.

$$\frac{1}{x - y} + \frac{2}{y - x} + \frac{3}{x - y} + \frac{4}{y - x} + \frac{5}{x - y} + \frac{6}{y - x} + \frac{7}{x - y} + \frac{8}{y - x} = \frac{?}{x - y}$$

CHAPTER OBJECTIVES

In this chapter, you will:

- Identify the domain and range of a relation.
- Graph linear equations and inequalities in the coordinate plane.
- Solve linear equations for a given domain.
- Calculate functional values.
- Represent data by using graphs.

Functions and Graphs

APPLICATION IN AUTOMOBILE SAFETY

One thing driver's education teachers always emphasize is to keep a safe distance behind the car in front of you so there is enough time to stop. But what exactly is a "safe distance?" For each situation, a safe distance will depend upon the *total stopping distance* for that situation. Total stopping distance is how far your car travels from the time you decide to apply the brakes until the car stops.

Automobile manufacturers have yet to build a car that can "stop on a dime." Cars all follow one basic rule: the faster they are traveling, the farther they go before stopping once their brakes are applied.

One factor in total stopping distance is called *braking distance.* This is how far the car travels from the time you apply the brakes until the time it actually stops. The road surface, the car's condition (especially its tires), the weather, and the car's speed all affect braking distance. The other factor in total stopping distance is called *thinking distance.* This is how far the car travels from the time you decide to apply the brakes until the time you actually apply them. The driver's mental and physical condition as well as the car's speed affect thinking distance. As the chart and graph below indicate, thinking distance and braking distance are both affected by speed. We can say that each of these distances is a *function* of speed.

ALGEBRA IN ACTION

Based on the information in the chart and graph below, what would be the approximate thinking distance, braking distance, and total stopping distance for a car traveling 65 miles per hour?

Original Speed (mph)	Thinking Distance (ft)	Braking Distance (ft)	Total Stopping Distance (ft)
15	16	12	28
25	27	34	61
35	38	67	105
45	49	111	160
55	60	165	225

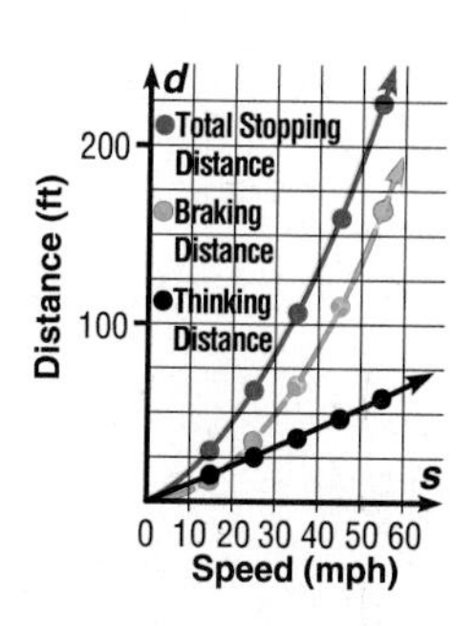

9-1 Ordered Pairs

Objective After studying this lesson, you should be able to:

- graph ordered pairs on a coordinate plane.

Application

Ms. Miyashiro makes seating charts for her classes. Her classroom has 5 rows of desks with 6 desks in each row. She assigns desks by using two numbers. The first number is for the row of the desk, and the second number is for the desk in that row. Ms. Miyashiro assigns desk (3, 2) to Gene and desk (2, 3) to Lori. This means Gene will sit in the third row at the second desk, and Lori will sit in the second row at the third desk.

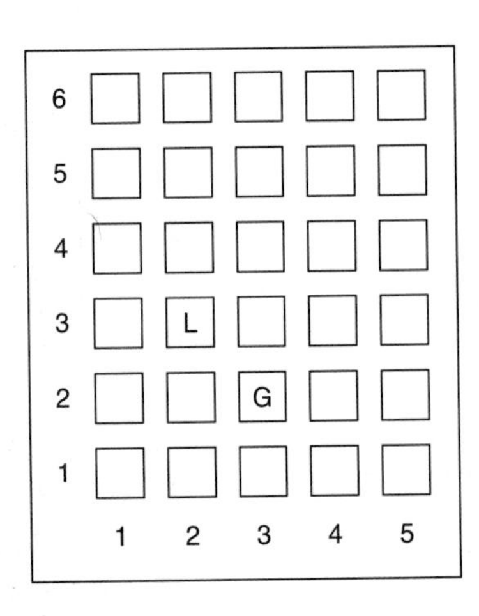

Notice that (3, 2) and (2, 3) do *not* name the same desk in the seating chart. (3, 2) and (2, 3) are called **ordered pairs** because the order in which the pair of numbers are written is important.

Perpendicular lines are lines that meet to form 90° angles.

Ordered pairs are used to locate points in a plane. The points are in reference to two *perpendicular* number lines. The point at which the number lines intersect is called the **origin.** The ordered pair that names the origin is (0, 0). The two number lines are called the **x-axis** and the **y-axis.** The plane that contains the x- and the y-axis is called the **coordinate plane.**

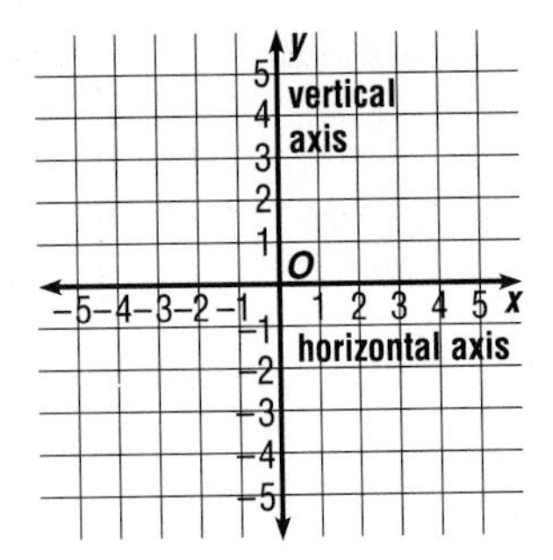

A point is often referred to by just a letter. Thus, A can be used to mean point A.

The first number, or **coordinate,** of an ordered pair corresponds to a number on the x-axis. The second number of the ordered pair corresponds to a number on the y-axis. To find the ordered pair for point A, shown at the right, think of a horizontal line and a vertical line passing through point A. Notice where these lines intersect each axis. The number on the x-axis that corresponds to A is -3. The number on the y-axis that corresponds to A is 4. Thus, the ordered pair for point A is $(-3, 4)$. The first coordinate, -3, is called the **x-coordinate** of point A. The second coordinate, 4, is called the **y-coordinate** of point A.

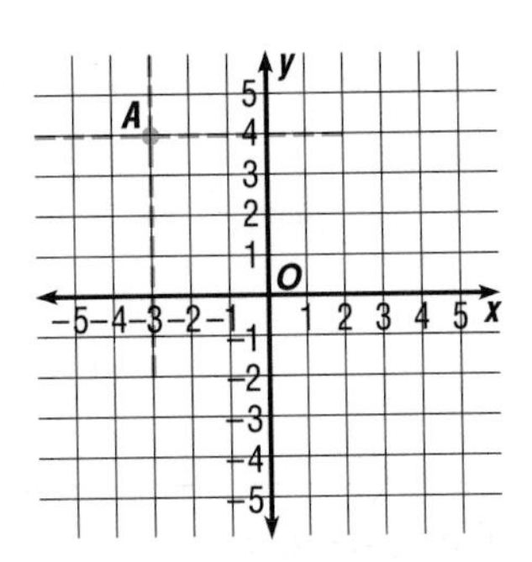

Example 1

Write the ordered pairs that name points R, S, T, and U.

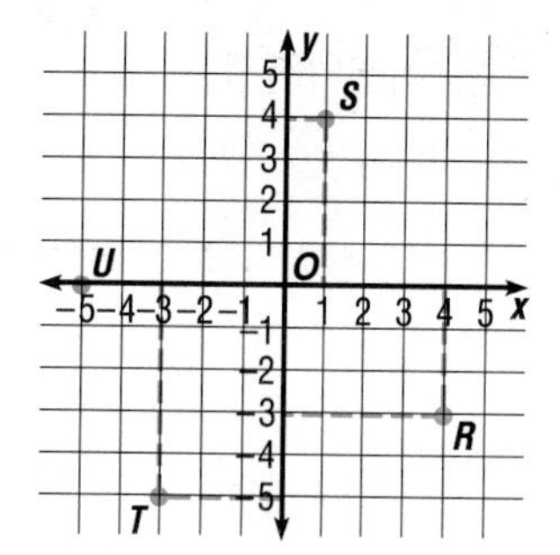

Think of a horizontal and a vertical line passing through each point.

Point R: The x-coordinate is 4 and the y-coordinate is -3. Thus, the ordered pair is $(4, -3)$.

Point S: The x-coordinate is 1 and the y-coordinate is 4. Thus, the ordered pair is $(1, 4)$.

The ordered pair for point T is $(-3, -5)$, and the ordered pair for point U is $(-5, 0)$. *Why?*

R(4, 3) also means that the location of point R is (4, 3).

Sometimes, a point is named by both a letter and its ordered pair. For example, $R(4, 3)$ means that the point R is named by the ordered pair $(4, 3)$.

To **graph** an ordered pair means to draw a dot at the point that corresponds to the ordered pair. This is sometimes called *plotting* the point. When graphing an ordered pair, start at the origin. The x-coordinate indicates the number of units to move left or right. The y-coordinate indicates the number of units to move up or down.

Example 2

Plot the following points on a coordinate plane.

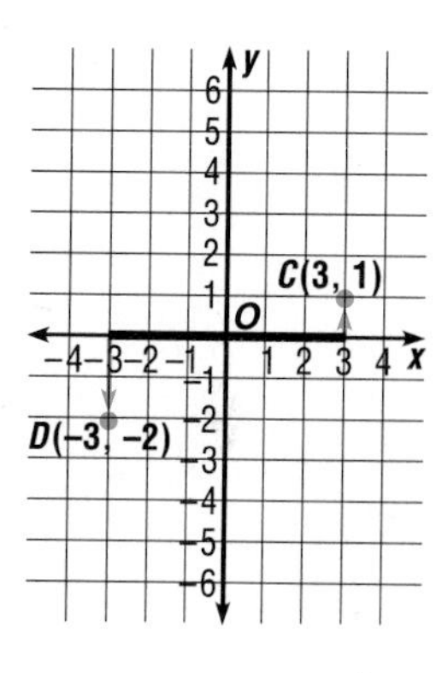

a. **$C(3, 1)$**

Start at the origin, O. Move 3 units to the right. Then move 1 unit up and draw a dot. Label this dot with the letter C. *Check that 3 on the x-axis and 1 on the y-axis correspond to C.*

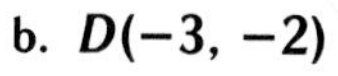

b. **$D(-3, -2)$**

Start at the origin, O. Move 3 units to the left, since the x-coordinate is -3. Then move 2 units down, since the y-coordinate is -2, and draw a dot. Label this dot with the letter D. *Check that -3 on the x-axis and -2 on the y-axis correspond to D.*

The correspondence between points in the plane and ordered pairs of real numbers is given by the following property.

Completeness Property for Points in the Plane	1. **Exactly one point in the plane is named by a given ordered pair of numbers.** 2. **Exactly one ordered pair of numbers names a given point in the plane.**

Example 3

On the map at the right, letters and numbers are used to form ordered pairs that name sectors on the map. Name all the sectors that the Scioto River passes through.

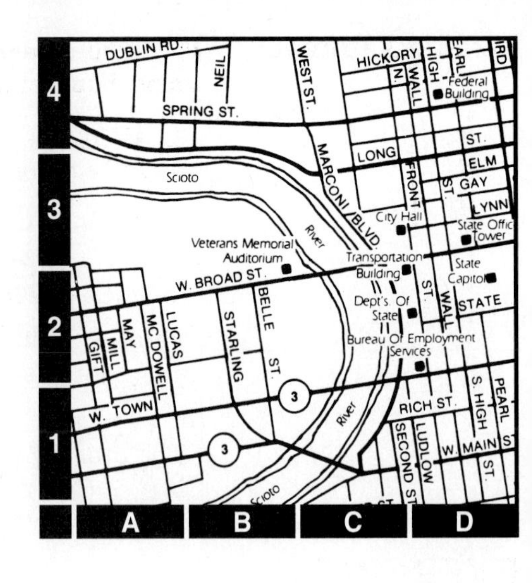

The Scioto River first appears at the bottom of the map in sectors (B, 1) and (C, 1). It then travels up through sectors (C, 2) and (C, 3) and left through sector (B, 3) before exiting the map through sectors (A, 3) and (A, 4).

The x-axis and the y-axis separate the coordinate plane into four regions, called **quadrants.** The quadrants are numbered as shown at the right. Notice that the axes are not located in any of the quadrants. *Why?*

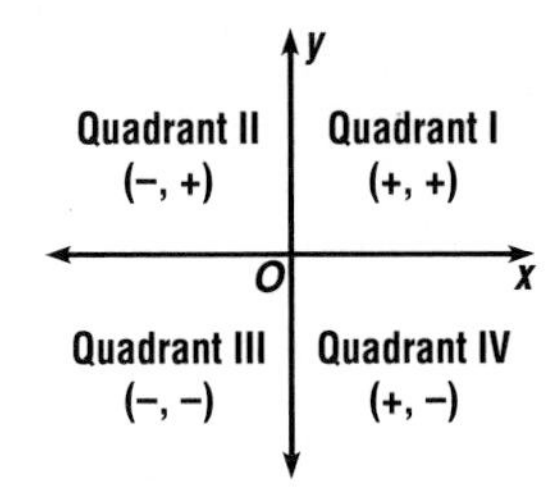

Example 4

Name the quadrant in which each point is located.

a. $A(5, -4)$

Since the x-coordinate of $A(5, -4)$ is positive and the y-coordinate is negative, point A is located in Quadrant IV.

b. $B(-2, -7)$

Since the x-coordinate of $B(-2, -7)$ is negative and the y-coordinate is negative, point B is located in Quadrant III.

c. $C(2, 0)$

Since point C lies on the x-axis, it is not located in a quadrant.

CHECKING FOR UNDERSTANDING

Communicating Mathematics

Read and study the lesson to answer each question.

1. Draw a coordinate plane. Label the origin, the x-axis, the y-axis, and quadrants I, II, III, and IV.
2. Explain the difference between the ordered pairs (5, 3) and (3, 5).
3. State the completeness property for points in a plane in your own words.
4. Name three ordered pairs whose graphs are not located in one of the four quadrants.

Guided Practice

Write the ordered pair for each point shown at the right.

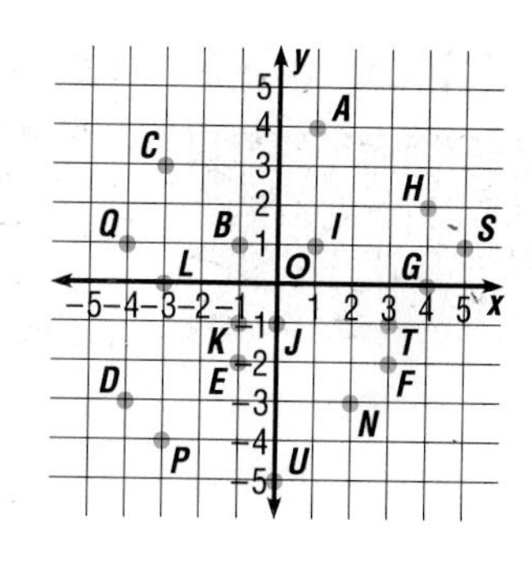

5. A
6. C
7. E
8. G
9. T
10. U

Name the quadrant in which the point named by each ordered pair is located.

11. $(5, 2)$
12. $(-3, -1)$
13. $(-2, 3)$
14. $(6, 0)$
15. $(0, -2)$
16. $(4, -3)$

EXERCISES

Practice

Write the ordered pair for each point shown above.

17. B
18. D
19. F
20. H
21. I
22. J
23. K
24. L
25. N
26. P
27. Q
28. S

If the graph of $P(x, y)$ satisfies the given conditions, name the quadrant in which point P is located.

29. $x > 0, y < 0$
30. $x < 0, y < 0$
31. $x < 0, y > 0$
32. $x = 0, y > 0$
33. $x = 1, y < 0$
34. $x = -1, y < 0$
35. $x < 0, y = 3$
36. $x > 0, y = 3$
37. $x > 0, y = 0$

Graph each point.

38. $A(5, -2)$
39. $B(3, 5)$
40. $C(-6, 0)$
41. $D(-3, 4)$
42. $E(-3, -3)$
43. $F(-5, 1)$
44. $G(2, -1)$
45. $H(4, 0)$
46. $I(3, -4)$
47. $J(-3, 0)$
48. $K(-5, -2)$
49. $L(4, 2)$
50. $M(0, 3)$
51. $N(1, 4)$
52. $P(2, -3)$
53. $Q(0, -2)$

If x and y are integers, graph all ordered pairs that satisfy the given conditions.

54. $-3 \le x \le 2$ and $-1 \le y < 3$
55. $|x| < 3$ and $|y| \le 2$

Graph each point. Then connect the points in alphabetical order and identify the figure.

56. $A(4, 6)$, $B(3, 7)$, $C(2, 6)$, $D(1, 7)$, $E(0, 6)$, $F(1, 8)$, $G(4, 9)$, $H(7, 8)$, $I(8, 6)$, $J(7, 7)$, $K(6, 6)$, $L(5, 7)$, $M(4, 6)$, $N(4, 0)$, $P(5, 0)$, $Q(5, 1)$
57. $A(-3, 0.5)$, $B(1, 0)$, $C(4, 0)$, $D(7, -4)$, $E(8, -4)$, $F(7, 0)$, $G(10, 0)$, $H(11, -2)$, $I(12, -2)$, $J(11.5, 0)$, $K(11.5, 1)$, $L(12, 3)$, $M(11, 3)$, $N(10, 1)$, $P(7, 1)$, $Q(8, 5)$, $R(7, 5)$, $S(4, 1)$, $T(1, 1)$, $U(-3, 0.5)$

Critical Thinking

Describe the possible locations, in terms of quadrants or axes, for the graph of (x, y) if x and y satisfy the given conditions.

58. $xy > 0$

59. $xy < 0$

60. $xy = 0$

Applications: Cartography

Refer to the map at the right to answer each question.

61. What city is in sector (A, 4)?

62. In what sector is the city of Garland?

63. In what 4 sectors is Love Field?

64. What highway goes from sector (A, 1) to sector (F, 3)?

65. Name all the sectors that Interstate Highway 30 passes through.

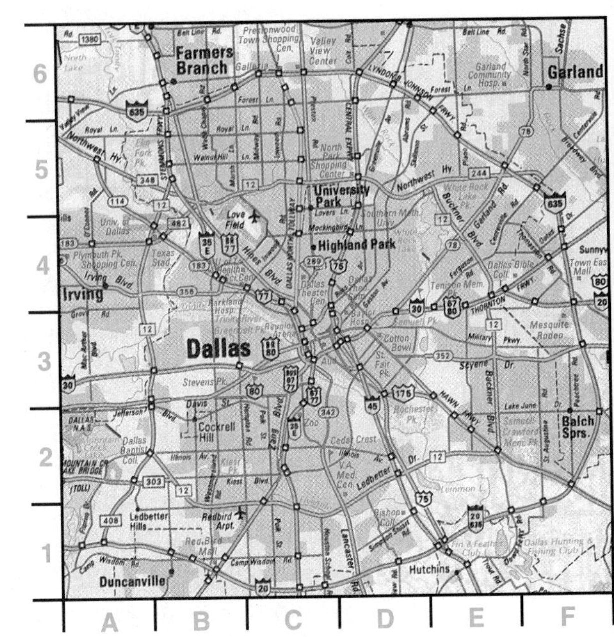

Mixed Review

66. Simplify $2x^2 + 3y + z^2 + 8x^2$. **(Lesson 1-6)**

67. **Entertainment** A theater was filled to 75% of capacity. How many of the 720 seats were filled? **(Lesson 4-2)**

68. Find $(5a-b)^2$. **(Lesson 6-9)**

69. Graph the solution set of $3 - 2p \geq -3$ or $3 - 2p \leq 7$. **(Lesson 5-5)**

70. Evaluate $\frac{8.5 \times 10^{-3}}{1.7 \times 10^{-8}}$. Express the result in scientific notation and decimal notation. **(Lesson 6-4)**

71. Factor $4x^2 - 64y^2$ completely. **(Lesson 7-6)**

72. Simplify $\frac{z^2 + 16z + 39}{z^2 + 9z + 18} \cdot \frac{z + 5}{z^2 + 18z + 65}$. **(Lesson 8-2)**

Journal

Draw a graph to represent where everyone sits in your classroom. Describe how you created your graph and name the point where your seat is located.

9-2 Relations

Objectives

After studying this lesson, you should be able to:

- identify the domain, range, and inverse of a relation, and
- show relations as sets of ordered pairs and mappings.

Application

The U.S. National debt in 1890 was 1.1 billion dollars.

As part of a project for his U.S. History class, Ramón researched the growth of the national debt since 1965. He organized his research in a table, as shown at the right.

OUR NATIONAL DEBT:
YOUR Family share $
THE NATIONAL DEBT CLOCK

National Debt	
Year	Billions of Dollars
1965	322
1970	381
1975	542
1980	909
1985	1817
1990	2601

Ramón could also have shown this data using a set of ordered pairs. Each first coordinate would be the year, and each second coordinate would be the national debt, in billions of dollars.

$$\{(1965, 322), (1970, 381), (1975, 542), (1980, 909), (1985, 1817), (1990, 2601)\}$$

A **relation** is a set of ordered pairs, like the set shown above. The set of first coordinates of the ordered pairs is called the **domain** of the relation. The set of second coordinates is called the **range** of the relation.

Definition of the Domain and Range of a Relation

The domain of a relation is the set of all first coordinates from the ordered pairs. The range of the relation is the set of all second coordinates from the ordered pairs.

For the relation given above, the domain is {1965, 1970, 1975, 1980, 1985, 1990}, and the range is {322, 381, 542, 909, 1817, 2601}.

A relation can also be shown using a table, a mapping, or a graph. A **mapping** illustrates how each element of the domain is paired with an element in the range. For example, the relation $\{(2, 2), (-2, 3), (0, -1)\}$ can be shown in each of the following ways.

Table

x	y
2	2
−2	3
0	−1

Mapping

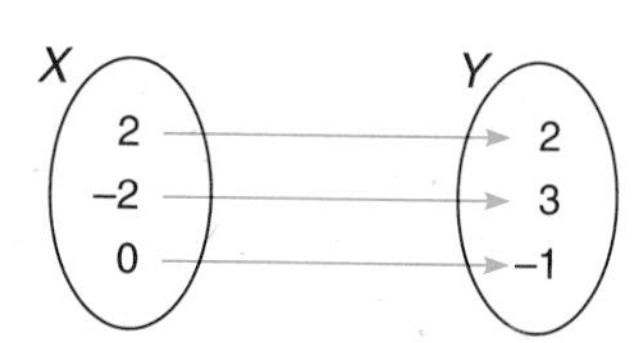

Graph

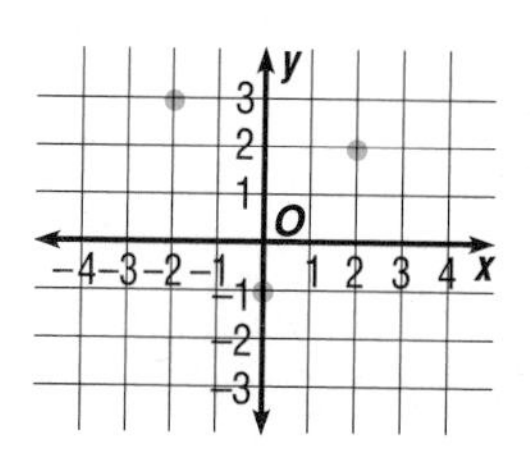

Example 1

Express the relation shown in the table below as a set of ordered pairs. Then determine its domain and range.

x	y
0	5
2	3
1	−4
−3	3
−1	−2

The set of ordered pairs for the relation is $\{(0, 5), (2, 3), (1, -4), (-3, 3), (-1, -2)\}$.

The domain is the set of all first coordinates: $\{0, 2, 1, -3, -1\}$.

The range is the set of all second coordinates: $\{5, 3, -4, -2\}$.

Example 2

Express the relation shown in the graph below as a set of ordered pairs. Determine its domain and range. Then show the relation using a mapping.

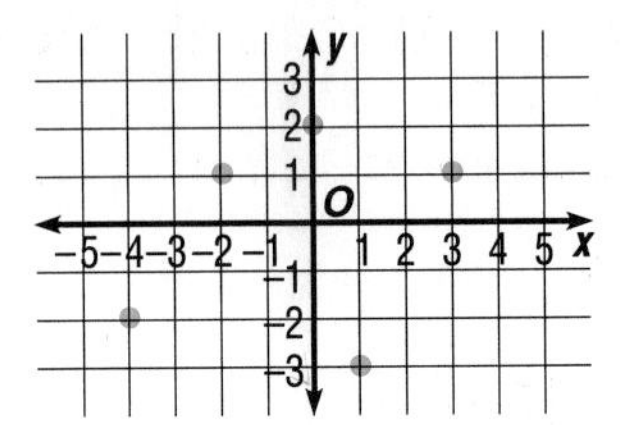

The set of ordered pairs for the relation is $\{(3, 1), (0, 2), (-2, 1), (-4, -2), (1, -3)\}$.

The domain is $\{3, 0, -2, -4, 1\}$.

The range is $\{1, 2, -2, -3\}$.

In this relation, 3 maps to 1, 0 maps to 2, −2 maps to 1, −4 maps to −2, and 1 maps to −3. This mapping is shown at the right.

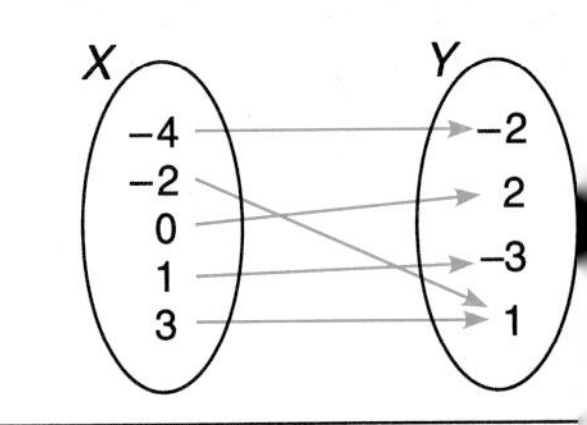

The **inverse** of any relation is obtained by switching the coordinates in each ordered pair of the relation. Thus, the inverse of the relation $\{(2, 2), (-2, 3), (0, -1)\}$ is the relation $\{(2, 2), (3, -2), (-1, 0)\}$. Notice that the domain of the relation becomes the range of the inverse and the range of the relation becomes the domain of the inverse.

Definition of the Inverse of a Relation

Relation Q is the inverse of relation S if and only if for every ordered pair (a, b) in S, there is an ordered pair (b, a) in Q.

Example 3

Express the relation shown in the mapping below as a set of ordered pairs. Write the inverse of this relation. Then determine the domain and range of the inverse.

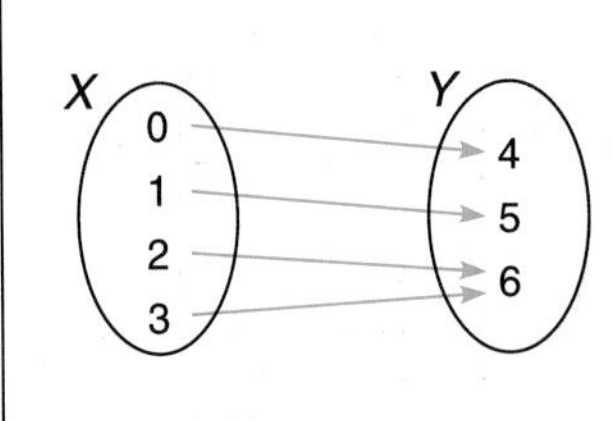

The set of ordered pairs for the relation is $\{(0, 4), (1, 5), (2, 6), (3, 6)\}$.

The inverse of the relation is $\{(4, 0), (5, 1), (6, 2), (6, 3)\}$.

The domain of the inverse is $\{4, 5, 6\}$.

The range of the inverse is $\{0, 1, 2, 3\}$.

Example 4

CONNECTION

Probability

Write a relation to show the 36 different possible outcomes when two six-sided dice are tossed. Then use this relation to help determine how many ways the faces of the dice can show a sum of 7.

Let the first coordinate of each ordered pair be the outcome on the first die. Let the second coordinate be the outcome on the second die. Each die can land in six ways. Thus, the relation can be written as follows.

{(1, 1), (1, 2), (1, 3), (1, 4), (1, 5), (1, 6),
(2, 1), (2, 2), (2, 3), (2, 4), (2, 5), (2, 6),
(3, 1), (3, 2), (3, 3), (3, 4), (3, 5), (3, 6),
(4, 1), (4, 2), (4, 3), (4, 4), (4, 5), (4, 6),
(5, 1), (5, 2), (5, 3), (5, 4), (5, 5), (5, 6),
(6, 1), (6, 2), (6, 3), (6, 4), (6, 5), (6, 6)}

The ordered pairs in this relation that show a sum of 7 are (1, 6), (2, 5), (3, 4), (4, 3), (5, 2), and (6, 1). Thus, there are 6 ways that the faces of the dice can show a sum of 7.

CHECKING FOR UNDERSTANDING

Communicating Mathematics

Read and study the lesson to answer each question.

1. What is the inverse of the national debt relation?
2. What are four different ways to show a relation?
3. What is the difference between the domain and the range of a relation?

Guided Practice

State the domain and the range of each relation.

4. {(0, 2), (1, −2), (2, 4)}
5. {(5, 2), (0, 0), (−9, −1)}
6. {(−4, 2), (−2, 0), (0, 2), (2, 4)}
7. {(7, 5), (−2, −3), (4, 0), (5, −7), (−9, 2)}
8. {(3.1, −1), (−4.7, 3.9), (2.4, −3.6), (−9, 12.12)}
9. $\left\{\left(\frac{1}{2}, \frac{1}{4}\right), \left(1\frac{1}{2}, -\frac{2}{3}\right), \left(-3, \frac{2}{5}\right), \left(-5\frac{1}{4}, -6\frac{2}{7}\right)\right\}$

Express each relation shown in each table as a set of ordered pairs. Then state the domain, range, and inverse of the relation.

10.

x	y
1	5
2	7
3	9
4	11

11.

x	y
1	3
2	2
4	9
6	5

12.

x	y
1	4
3	−2
4	4
6	−2

EXERCISES

Practice

Express the relation shown in each mapping as a set of ordered pairs.

13. 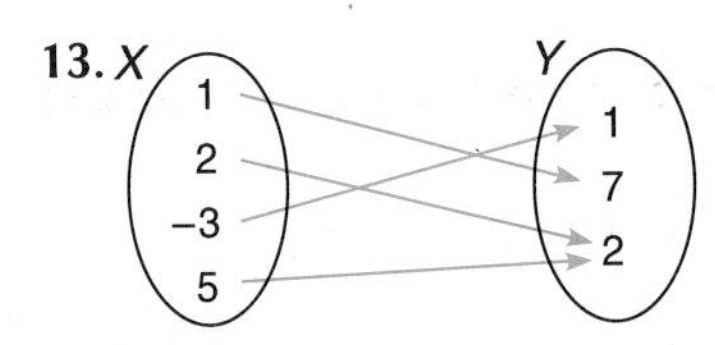

14. 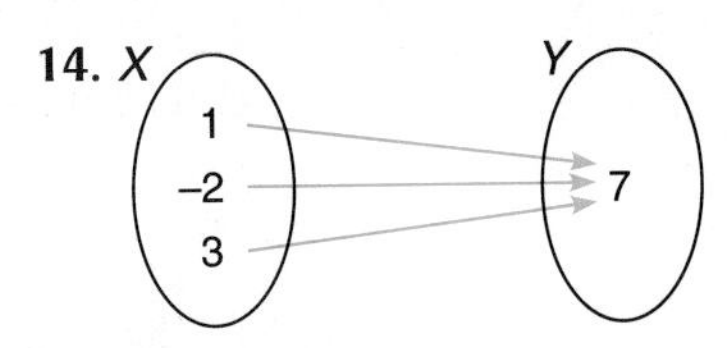

15. 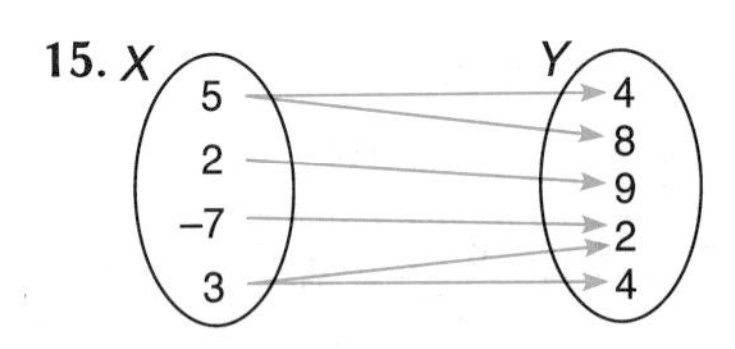

16. 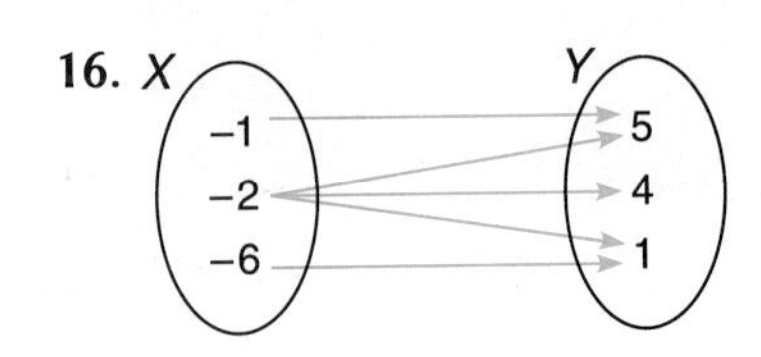

Draw a mapping for the relation shown in each table. Then express the relation and its inverse as sets of ordered pairs.

17.

x	y
1	3
2	4
3	5
4	6
5	7

18. 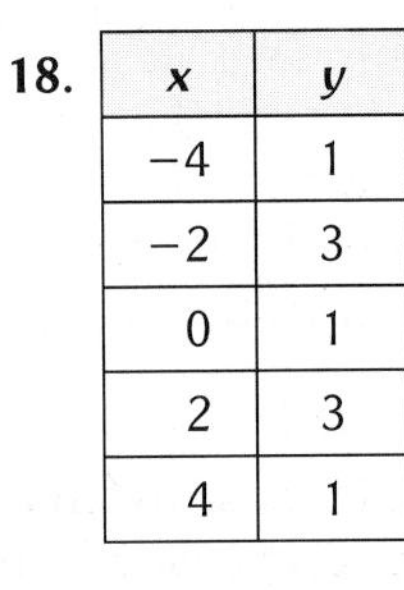

x	y
−4	1
−2	3
0	1
2	3
4	1

19. 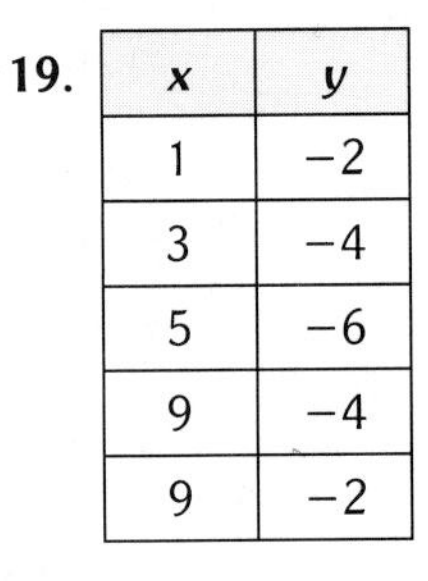

x	y
1	−2
3	−4
5	−6
9	−4
9	−2

20.

x	y
1	3
2	5
1	−7
2	9
3	3

Express the relation shown in each graph as a set of ordered pairs. Then state the domain and range of the relation.

21. 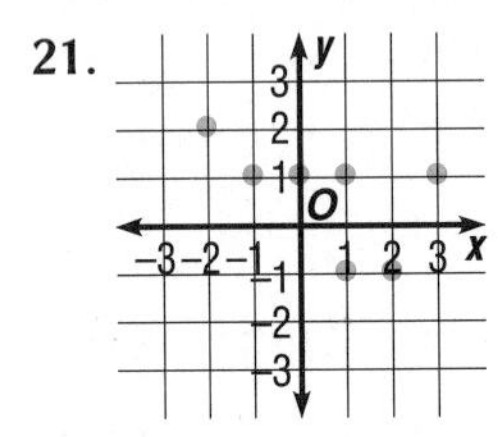

22. 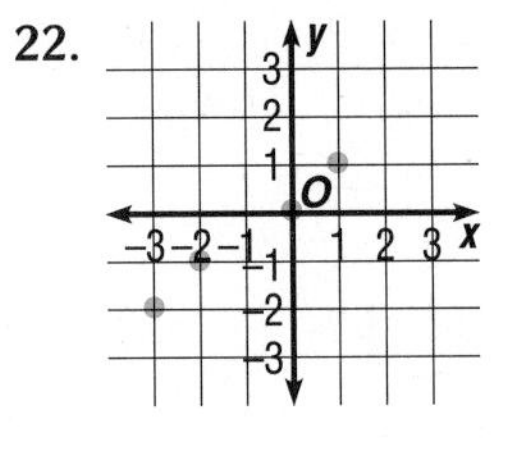

23. 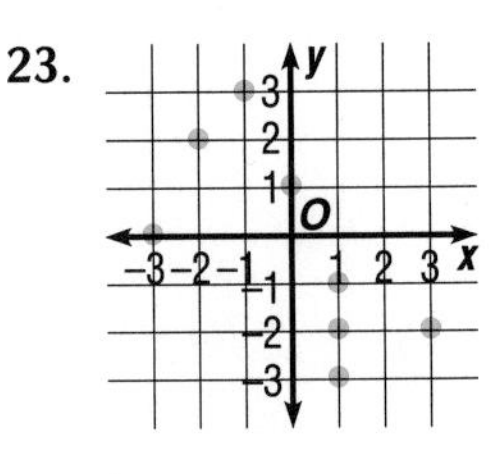

24. 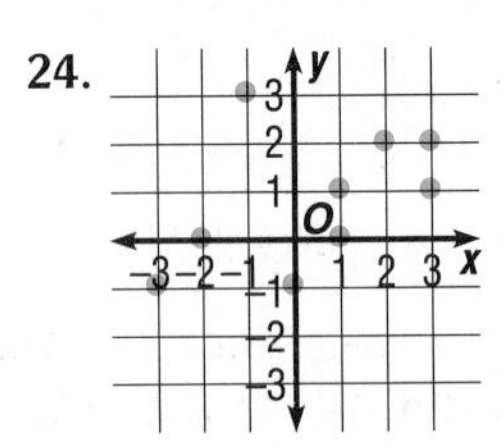

25. 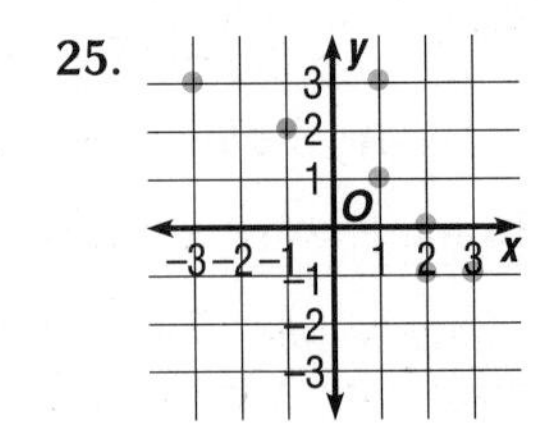

26. 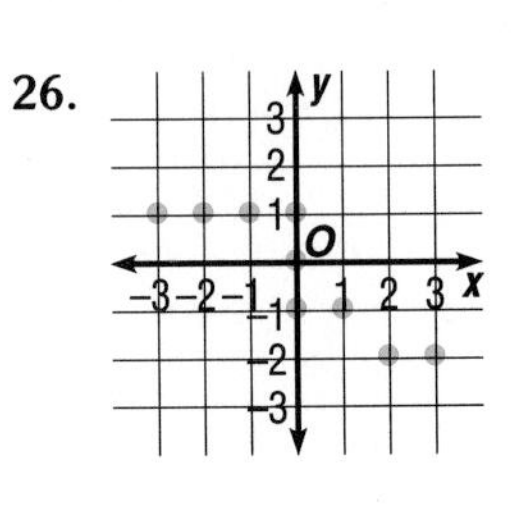

27–32. Write the inverse of the relation in Exercises 21–26.

Use the relation in Example 4 to complete the following.

33. Determine how many ways the faces of two dice can show each sum. For example, there are three ways to show a sum of 10: (5, 5), (4, 6), and (6, 4).

34. Do you notice a pattern in the answers to Exercise 33?

Draw a mapping for each relation.

35. {(0, 1)}

36. {(2, 3), (3, 2)}

37. {(1, 3), (2, 3), (2, 1), (3, 2)}

38. {(−6, 0), (−1, 2), (−3, 4)}

39. {(4, −3), (6, −4), (5, −3), (6, 0)}

40. {(1, 3), (2, 7), (4, 1), (−3, 3), (3, 3)}

41. {(4, 2), (7, 5), (−3, 2), (8, −9), (8, 2)}

Critical Thinking

42. Graph the relation {(5, 5), (1, 3), (2, −1), (4, 0)}. Connect these points in the order given. Then show the inverse of this relation as a graph and connect those points in the same order. What conclusions can be made about the two graphs?

Applications

43. **Statistics** Fifteen students participated in a free-throw shooting contest. The number of free throws made by each student is as follows: 3, 3, 4, 5, 4, 1, 2, 3, 4, 2, 2, 4, 5, 4, 4. Show this as the graph of a relation where the first coordinate is the number of free throws made and the second coordinate is the number of students who made that many free throws. Then connect the points from left to right. *This graph is called a line graph.*

44. **Food** Lisa buys a dozen doughnuts for a morning meeting. She knows that the staff prefers glazed doughnuts over cake doughnuts. If she buys at least twice as many glazed doughnuts as cake doughnuts, write a relation to show the different possibilities. (*Hint:* Let the domain be the number of glazed doughnuts.)

Mixed Review

45. Solve $2(5 - 8n) = -4(3 + 4n)$. **(Lesson 3-6)**

46. **Chemistry** A chemist has 800 mL of a solution that is 25% acid. How much pure (100%) acid must be added to the 25% solution to obtain a 40% acid solution? **(Lesson 4-6)**

47. Simplify $7a^2b^2(a^4 - 5a^2b + 6b^2)$. **(Lesson 6-7)**

48. Factor $144n^2 + 168n + 49$. **(Lesson 7-7)**

49. **Electronics** Three coils with resistances of 4 ohms, 6 ohms, and 15 ohms are connected in parallel. What is the total resistance? Use the formula $\frac{1}{R_T} = \frac{1}{R_1} + \frac{1}{R_2} + \frac{1}{R_3}$. **(Lesson 8-10)**

50. Graph the points $A(6, 2)$, $B(-3, 6)$, and $C(-5, -4)$. **(Lesson 9-1)**

9-3 Equations as Relations

Objectives

After studying this lesson, you should be able to:

- solve linear equations for a specific variable, and
- solve linear equations for a given domain.

Application

FYI...

Most American car horns beep in the key of F.

Anne Thompson is saving money to buy a used car for \$1900. She already has \$500 in her savings account. She plans to add \$80 each week from the money she earns working at Readalot Bookstore. Anne uses a chart to help determine when she will have enough money in savings to buy the car. Let x represent the number of weeks and let y represent her total savings in dollars. Then $500 + 80x$ represents her savings after x weeks.

x (weeks)	$500 + 80x$	y (savings)	(x, y)
5	500 + 80(5)	900	(5, 900)
10	500 + 80(10)	1300	(10, 1300)
12	500 + 80(12)	1460	(12, 1460)
15	500 + 80(15)	1700	(15, 1700)
18	500 + 80(18)	1940	(18, 1940)
20	500 + 80(20)	2100	(20, 2100)

Based on this chart, Anne knows that in 18 weeks her total savings will be over \$1900.

The equation $y = 500 + 80x$ describes Anne's total savings (y) after any number of weeks (x). Each of the ordered pairs listed in the chart above is a *solution* of the equation $y = 500 + 80x$.

Definition of the Solution of an Equation in Two Variables

If a true statement results when the numbers in an ordered pair are substituted into an equation in two variables, then the ordered pair is a solution of the equation.

Since the solutions of an equation in two variables are ordered pairs, such an equation describes a relation. The set of values of x is the domain of the relation. The set of corresponding values of y is the range of the relation.

Example 1

Solve $y = 2x + 3$ if the domain is $\{-5, -3, -1, 0, 1, 3, 5, 7, 9\}$.

Make a table. The values of x come from the domain. Substitute each value of x into the equation to determine the corresponding value of y.

x	$2x + 3$	y	(x, y)
−5	2(−5) + 3	−7	(−5, −7)
−3	2(−3) + 3	−3	(−3, −3)
−1	2(−1) + 3	1	(−1, 1)
0	2(0) + 3	3	(0, 3)
1	2(1) + 3	5	(1, 5)
3	2(3) + 3	9	(3, 9)
5	2(5) + 3	13	(5, 13)
7	2(7) + 3	17	(7, 17)
9	2(9) + 3	21	(9, 21)

The solution set is $\{(-5, -7), (-3, -3), (-1, 1), (0, 3), (1, 5), (3, 9), (5, 13), (7, 17), (9, 21)\}$.

The solutions of an equation in two variables are usually easier to determine when the equation is solved for one of the variables.

Example 2

Solve $3y + 6x = 12$ if the domain is $\{-4, -3, -2, 2, 3, 4\}$.

First solve the equation for y in terms of x.

$$3y + 6x = 12$$

$$3y = 12 - 6x \quad \textit{Subtract 6x from each side.}$$

$$\frac{3y}{3} = \frac{12 - 6x}{3} \quad \textit{Divide each side by 3.}$$

$$y = 4 - 2x$$

Now substitute each value of x from the domain to determine the corresponding values of y.

x	$4 - 2x$	y	(x, y)
−4	4 − 2(−4)	12	(−4, 12)
−3	4 − 2(−3)	10	(−3, 10)
−2	4 − 2(−2)	8	(−2, 8)
2	4 − 2(2)	0	(2, 0)
3	4 − 2(3)	−2	(3, −2)
4	4 − 2(4)	−4	(4, −4)

The solution set is $\{(-4, 12), (-3, 10), (-2, 8), (2, 0), (3, -2), (4, -4)\}$.

In this textbook, when variables other than x and y are used in an equation, you may assume that the values of the variable that comes first alphabetically are from the domain.

Example 3

Solve $3a + 2b = 11$ if the domain is {−3, 0, 1, 2, 5}.

Assume that the values of a come from the domain. Therefore, the equation should be solved for b in terms of a.

$$3a + 2b = 11$$
$$2b = 11 - 3a \quad \textit{Subtract 3a from each side.}$$
$$b = \frac{11 - 3a}{2} \quad \textit{Divide each side by 2.}$$

Now substitute the values of a from the domain to determine the corresponding values of b.

a	$\frac{11 - 3a}{2}$	b	(a, b)
−3	$\frac{11 - 3(-3)}{2}$	10	(−3, 10)
0	$\frac{11 - 3(0)}{2}$	$\frac{11}{2}$	$\left(0, \frac{11}{2}\right)$
1	$\frac{11 - 3(1)}{2}$	4	(1, 4)
2	$\frac{11 - 3(2)}{2}$	$\frac{5}{2}$	$\left(2, \frac{5}{2}\right)$
5	$\frac{11 - 3(5)}{2}$	−2	(5, −2)

The solution set is $\left\{(-3, 10), \left(0, \frac{11}{2}\right), (1, 4), \left(2, \frac{5}{2}\right), (5, -2)\right\}$.

Example 4

Ms. Shell works as a sales representative for Stokes Electronics. She receives a salary of $1800 per month plus a 6% commission on monthly sales over her target. Her sales in July will be $800, $1300, or $2000 over target, depending on when her orders are filled by the company distribution center. Will her total income for July be more than $1850?

Let s = Ms. Shell's sales for July, and let t = her total income. Then $t = 1800 + 0.06s$ where the domain is {800, 1300, 2000}.

s	$1800 + 0.06s$	t	(s, t)
800	1800 + 0.06(800)	1848	(800, 1848)
1300	1800 + 0.06(1300)	1878	(1300, 1878)
2000	1800 + 0.06(2000)	1920	(2000, 1920)

Her total income for July will be more than $1850 only if her sales are $1300 or $2000, and not $800.

CHECKING FOR UNDERSTANDING

Communicating Mathematics

Read and study the lesson to answer each question.

1. Refer to the application at the beginning of the lesson. How can Anne determine during which week she will have saved $1900?
2. Why does the equation $y = 2x + 1$ describe a relation?
3. If the domain of the equation $m = n + 1$ is $\{0, 1\}$, what is the range?

Guided Practice

Copy and complete each table for the given equation.

4. $y = 4x - 3$

x	y	(x, y)
-3		
-2		
-1		
0		
2		
4		

5. $n = \frac{2m + 5}{3}$

m	n	(m, n)
-4		
-2		
0		
1		
3		

Which of the ordered pairs given are solutions of the equation?

6. $3x + y = 8$ a. $(2, 2)$ b. $(3, 1)$ c. $(4, -4)$ d. $(8, 0)$
7. $2x + 3y = 11$ a. $(3, 1)$ b. $(1, 3)$ c. $(-2, 5)$ d. $(4, -1)$
8. $2m - 5n = 1$ a. $(-2, -1)$ b. $(2, 1)$ c. $(7, 3)$ d. $(-7, -3)$

Solve each equation for the variable indicated.

9. $x + y = 5$ for y
10. $3x + y = 7$ for y
11. $b - 5a = 3$ for b
12. $4m + n = 7$ for n

EXERCISES

Practice

Which of the ordered pairs given are solutions of the equation?

13. $3r = 8s - 4$ a. $\left(\frac{2}{3}, \frac{3}{4}\right)$ b. $\left(0, \frac{1}{2}\right)$ c. $(4, 2)$ d. $(2, 4)$
14. $3y = x + 7$ a. $(2, 4)$ b. $(2, -1)$ c. $(2, 3)$ d. $(-1, 2)$
15. $4x = 8 - 2y$ a. $(2, 0)$ b. $(0, 2)$ c. $(0.5, -3)$ d. $(1, -2)$

Solve each equation for the variable indicated.

16. $8x + 2y = 6$ for y
17. $6x + 3y = 12$ for y
18. $4a + 3b = 7$ for b
19. $6r + 5s = 2$ for s
20. $6x = 3y + 2$ for y
21. $3a = 7b + 8$ for b
22. $4p = 7 - 2r$ for r
23. $-4 = 5n - 7r$ for r

Solve each equation if the domain is $\{-2, -1, 0, 2, 5\}$.

24. $y = 3x$
25. $y = 2x + 1$
26. $x + y = 7$
27. $x - y = 4$
28. $5x + y = 4$
29. $2a + 3b = 13$
30. $4r + 3s = 16$
31. $2t = 3 - 5s$
32. $5a - b = -3$
33. $3x = 5 + 2y$
34. $5a = 8 - 4b$
35. $6b - a = 32$

Solve each equation if the range is $\{-3, -1, 0, 2, 3\}$.

36. $y = 2x$
37. $y = 5x + 1$
38. $2a + b = 4$
39. $4r + 3s = 13$
40. $5b = 8 - 4a$
41. $6m - n = -3$

Critical Thinking

Find the domain of each relation if the range is $\{0, 4, 36\}$.

42. $y = x^2$
43. $y = |2x| - 4$
44. $y = |2x - 4|$

Applications

45. **Gardening** Paquita has only 16 m of fencing to use for enclosing her rectangular garden. She wants the garden to have the largest possible area given that the sides have integral lengths. What are the dimensions of her garden? (*Hint:* $P = 2\ell + 2w$ and $A = \ell w$.)

46. **Physics** A ball is thrown upward with an initial velocity of 96 feet per second. The height h, in feet, of the ball above the ground after t seconds is given by the equation $h = 96t - 16t^2$. Make a table of values for this equation to determine when the ball will reach its maximum height and when the ball will hit the ground.

Mixed Review

47. Solve $5s - 6.5 < -13.4 + 4s$. **(Lesson 5-1)**

48. Find $(2xy + 6xy^2 + y^2) - (3x^2y + 2xy + 3y^2)$. **(Lesson 6-6)**

49. Factor $42abc - 12a^2b^2 + 3a^2c^2$. **(Lesson 7-2)**

50. Solve $5m^2 + 5m = 0$. **(Lesson 7-9)**

51. **Work** A swimming pool can be filled by one pipe in 12 hours and by another pipe in 4 hours. How long will it take to fill the pool if the water flows through both pipes? **(Lesson 8-9)**

52. Determine the domain, range, and inverse of the relation $\{(8, 1), (4, 2), (6, -4), (5, -3), (6, 0)\}$. **(Lesson 9-2)**

9-4 Graphing Linear Relations

Objective

After studying this lesson, you should be able to:

- graph linear equations on a coordinate plane.

Application

Every April, Maxine's Pie Shop has a $1 off sale. The following chart shows the prices for selected pies.

Pie	Original Price	Sale Price
Banana Cream	$6.00	$5.00
Lemon Meringue	$7.00	$6.00
Strawberry	$9.00	$8.00
Pecan	$10.00	$9.00

Let x represent the original price of a pie and let y represent the sale price. Then the equation $y = x - 1$ describes the sale price for any pie. For the chart above, the domain is {6, 7, 9, 10} and the solution set is {(6, 5), (7, 6), (9, 8), (10, 9)}. The solutions can also be shown in a graph like the one at the right.

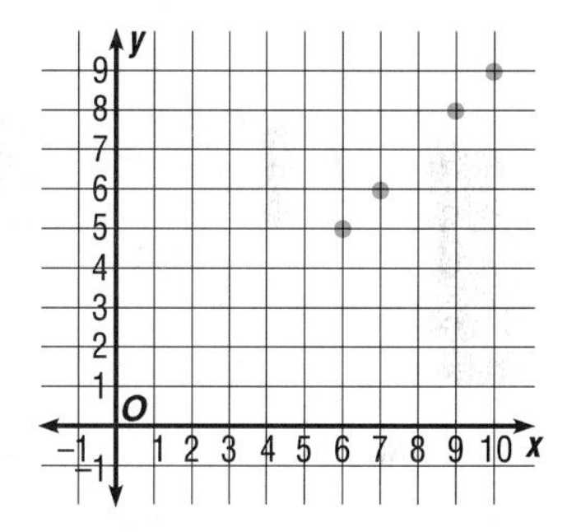

Remember, the values of the first coordinates in the ordered pairs of a relation represent its domain.

If the domain of $y = x - 1$ is the set of all real numbers, then an infinite number of ordered pairs are solutions of the equation. Suppose you draw a line connecting the points in the graph at the right. The graph of every solution of $y = x - 1$ lies on this line. The coordinates of any point on this line satisfy the equation. Hence the line is called the *graph* of $y = x - 1$.

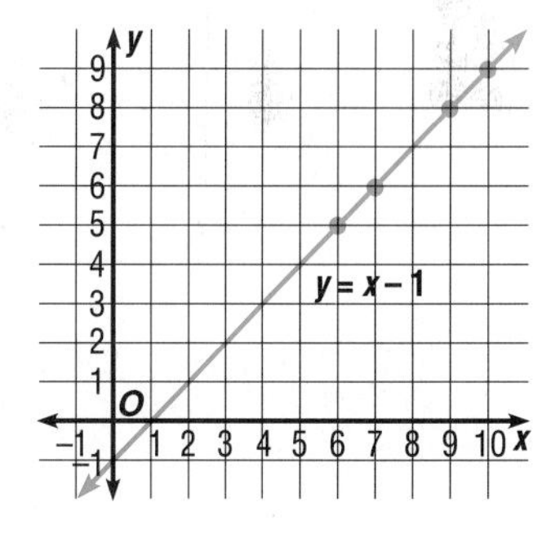

The equation $y = x - 1$ is equivalent to equations like $-y = 1 - x$ and $x - y = 1$. All of these equations have the same graph. An equation whose graph is a straight line is called a **linear equation.**

Definition of a Linear Equation	**A linear equation is an equation that can be written in the form $Ax + By = C$, where A, B, and C are any numbers, and A and B are not both zero.**

Example 1 **Determine whether each equation is a linear equation.**

a. $2x = 8 + y$

An equivalent form of this equation is $2x - y = 8$. Therefore, this is a linear equation with $A = 2$, $B = -1$, and $C = 8$.

b. $3x + y^2 = 7$

The exponent of all variables in a linear equation must be 1. Therefore, this is *not* a linear equation.

c. $y = 7$

An equivalent form of this equation is $0x + y = 7$. Therefore, this is a linear equation with $A = 0$, $B = 1$, and $C = 7$.

Example 2 **Draw the graph of $y = 2x - 1$.**

An equivalent form of this equation is $2x - y = 1$. Thus, it is a linear equation. Set up a table of values for x and y. Then graph the ordered pairs and connect the points with a line.

x	$2x - 1$	y	(x, y)
-2	$2(-2) - 1$	-5	$(-2, -5)$
-1	$2(-1) - 1$	-3	$(-1, -3)$
0	$2(0) - 1$	-1	$(0, -1)$
1	$2(1) - 1$	1	$(1, 1)$
2	$2(2) - 1$	3	$(2, 3)$

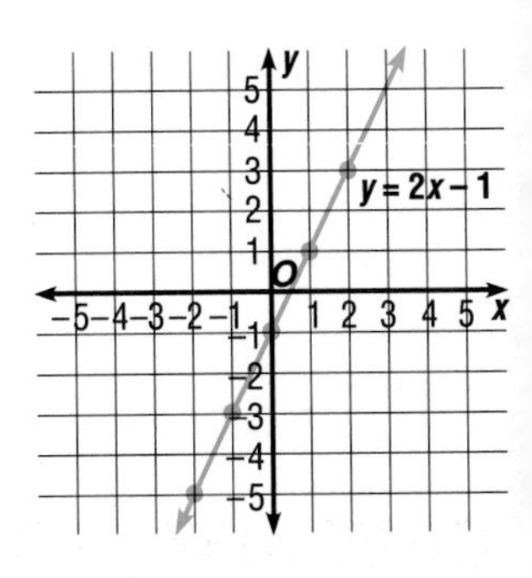

Usually, values of x such as 0 and integers near 0 are chosen.

Example 3 **Draw the graph of $3x + 2y = 4$.**

Solve the equation for y.

$$3x + 2y = 4$$
$$2y = 4 - 3x$$
$$y = \frac{4 - 3x}{2}$$

x	$\frac{4 - 3x}{2}$	y	(x, y)
-2	$\frac{4 - 3(-2)}{2}$	5	$(-2, 5)$
-1	$\frac{4 - 3(-1)}{2}$	$\frac{7}{2}$	$\left(-1, \frac{7}{2}\right)$
0	$\frac{4 - 3(0)}{2}$	2	$(0, 2)$
1	$\frac{4 - 3(1)}{2}$	$\frac{1}{2}$	$\left(1, \frac{1}{2}\right)$
2	$\frac{4 - 3(2)}{2}$	-1	$(2, -1)$

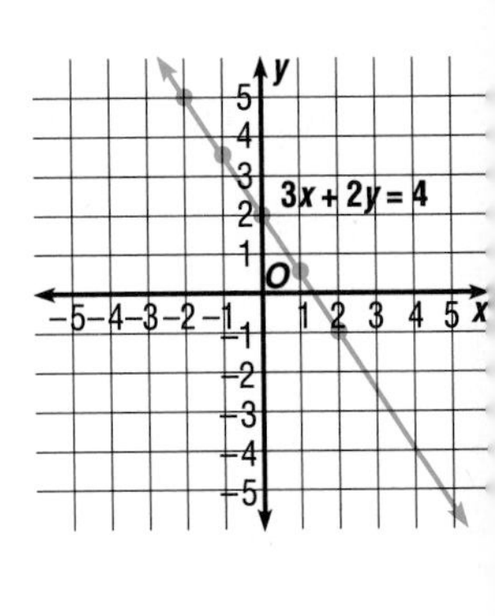

Example 4

During May of 1990, southwest Missouri received 30 inches of rain. The water level of a flooded river dropped by 3 inches per hour after cresting at 3 feet above its normal level. Let x represent the number of hours and y represent the water's height above the normal level. Then the equation $y = 36 - 3x$ describes the water's height above its normal level for any number of hours. Draw the graph of this equation. Then use the graph to estimate when the river will return to its normal level.

Set up a table of values for x and y. Then graph the ordered pairs and connect them with a line.

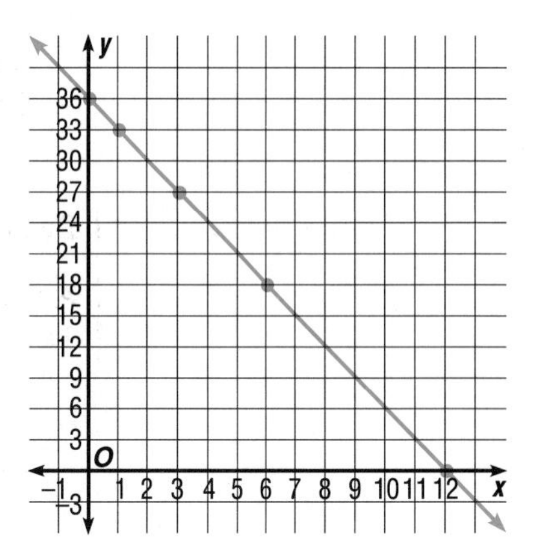

x	$36 - 3x$	y	(x, y)
0	$36 - 3(0)$	36	(0, 36)
1	$36 - 3(1)$	33	(1, 33)
3	$36 - 3(3)$	27	(3, 27)
6	$36 - 3(6)$	18	(6, 18)

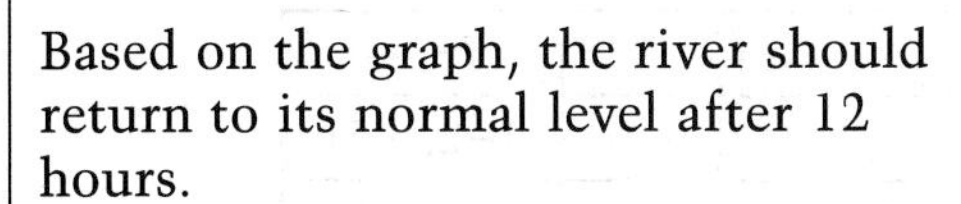

Why is it not necessary to choose negative values for x?

Based on the graph, the river should return to its normal level after 12 hours.

CHECKING FOR UNDERSTANDING

Communicating Mathematics

Read and study the lesson to answer each question.

1. What are the values of A, B, and C for the linear equation $3x = 2y$?
2. Describe the graph of a linear equation with $A = 0$.
3. Describe the graph of a linear equation with $B = 0$.
4. The graph of a linear equation with $C = 0$ always passes through what point?

Guided Practice

Determine whether each equation is a linear equation.

5. $5x + 2y = 7$ **6.** $3x^2 + 2y = 4$ **7.** $x + \frac{1}{y} = 7$

8. $\frac{1}{x} - \frac{1}{y} = \frac{1}{2}$ **9.** $\frac{3}{5}x - \frac{2}{3}y = 5$ **10.** $x = y^2$

11. $\frac{3}{5x} = y$ **12.** $\frac{3}{5}x = y$ **13.** $\frac{y}{2} = 2$

Solve each equation for y. Then graph each equation.

14. $6x + 7 = -14y$ **15.** $8x - y = 16$ **16.** $x + 5y = 16$

EXERCISES

Practice

Solve each equation for y.

17. $3x + 4y = 12$ **18.** $4x - \frac{3}{8}y = 1$ **19.** $\frac{1}{2}x - \frac{2}{3}y = 10$

Determine whether each equation is a linear equation.

20. $xy = 2$ **21.** $3x = 2y$ **22.** $x^2 = y^2$

23. $2x - 3 = y^2$ **24.** $3m - 2n = 8$ **25.** $3m - 2n = 0$

26. $4x^2 - 3x = y$ **27.** $2m + 5m = 7n$ **28.** $8a - 7b = 2a - 5$

Graph each equation.

29. $3m + n = 4$ **30.** $2x - y = 8$ **31.** $b = 5a - 7$

32. $y = 3x + 1$ **33.** $4x + 3y = 12$ **34.** $2x + 7y = 9$

35. $3x - 2y = 12$ **36.** $\frac{1}{2}x + y = 8$ **37.** $x + \frac{1}{3}y = 6$

38. $x = -\frac{5}{2}$ **39.** $y = \frac{4}{3}$ **40.** $\frac{3}{5}x = 6$

41. $-\frac{3}{4}y = 6$ **42.** $\frac{3}{4}x + \frac{1}{2}y = 6$ **43.** $\frac{4}{3}x - \frac{3}{4}y = 1$

Find a linear equation that represents each relation.

44. $\{(0, 0), (1, 1), (2, 2), (-1, -1)\}$

45. $\{(0, 0), (1, -3), (2, -6), (-1, 3)\}$

46. $\{(0, -1), (1, 1), (2, 3), (3, 5), (-1, -3)\}$

Critical Thinking

47. Graph the four equations given below on the same coordinate plane. Then compare the graphs.

$y = x - 1$ $y = 2x - 1$ $y = 3x - 1$ $y = 4x - 1$

Applications

48. Science As a thunderstorm approaches, you see lightning as it occurs, but you hear the accompanying sound of thunder a short time afterward. The distance y, in miles, that sound travels in t seconds is given by the equation $y = 0.21t$.

a. Draw a graph of this equation.

b. Use the graph to estimate how long it will take you to hear the thunder from a storm that is 3 miles away.

49. Business Creative Catering charges a basic fee of \$100 to cater a banquet plus an additional charge of \$3 per person invited to the banquet. The total charge t, in dollars, for a banquet where p people are invited is given by the equation $t = 3p + 100$.

a. Draw a graph of this equation.

b. Use the graph to determine how many people were invited to a banquet if Creative Catering charged \$850.

50. **Temperature** The relationship between Celsius temperature (C) and Fahrenheit temperature (F) is given by the formula $F = \frac{9}{5}C + 32$.

a. Draw a graph of this equation.

b. Use the graph to determine at what point the values for C and F are equal.

Mixed Review

51. Arrange the terms of $12n^4xy^3 - 4n^2x^3 + 2nxy^5$ so that the powers of y are in descending order. **(Lesson 6-5)**

Simplify each expression. (Lessons 6-3, 8-1, and 8-4)

52. $\frac{(-m)^3n^{-4}}{(3m^2n)^{-2}}$

53. $\frac{k^2 - 1}{k^2 + 2k + 1}$

54. $\frac{20x^3 + 19x^2 + 49}{4x + 7}$

55. **Gardening** A rectangular garden is 5 meters wide and 16 meters long. When the length and width are increased by the same amount in order to plant more vegetables, the area is increased by 72 square meters. What are the dimensions of the new garden? **(Lesson 7-10)**

56. Solve $3n = 10 - 4m$ if the domain is $\{-3, -1, 0, 1, 4\}$. **(Lesson 9-3)**

MID-CHAPTER REVIEW

Graph each point. (Lesson 9-1)

1. $A(-5, 7)$
2. $B(5, -7)$
3. $C(5, 0)$
4. $D(-2, -6)$

Express each relation as a set of ordered pairs. Then state the domain, the range, and the inverse of the relation. (Lesson 9-2)

5.

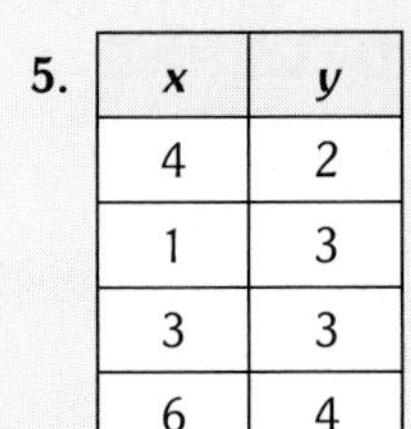

x	y
4	2
1	3
3	3
6	4

6.

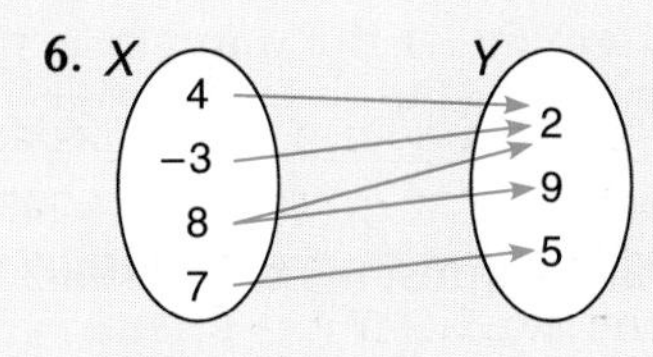

7.

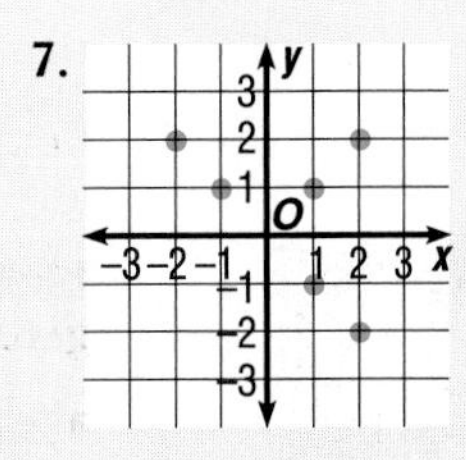

Solve each equation if the domain is $\{-5, -2, 0, 1, 3\}$. (Lesson 9-3)

8. $y = 5x + 3$
9. $2x - y = 7$
10. $3s - 4r = 16$
11. $6x = 11 + 5y$

Determine whether each equation is a linear equation. If it is, then graph the equation. (Lesson 9-4)

12. $3x - y = 4$
13. $b = 7a - 5$
14. $m = 5n^2 - 7$
15. $2x - 3y = 12$

9-5 Functions

Objectives

After studying this lesson, you should be able to:

- determine whether a given relation is a function, and
- calculate functional values for a given function.

Connection

FYI...

The ancient Greeks made dice from the ankle bones and shoulder blades of sheep.

In Example 4 on page 361, you found the 36 possible outcomes when two dice are tossed and those outcomes where the faces of the dice show a sum of 7. The graphs of these relations are shown below.

Possible outcomes when two six-sided dice are tossed

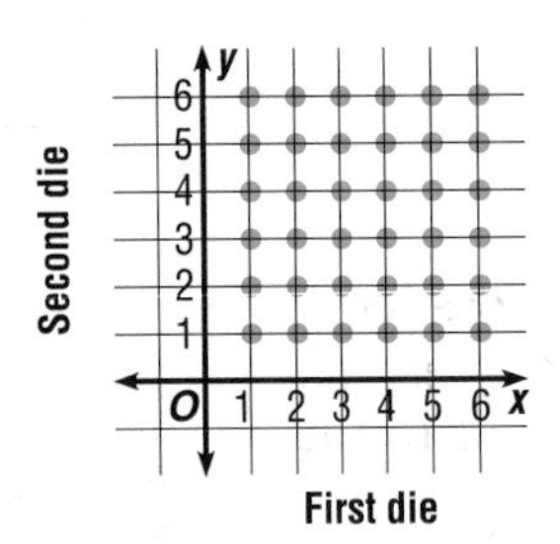

Those outcomes where the faces of the dice show a sum of 7

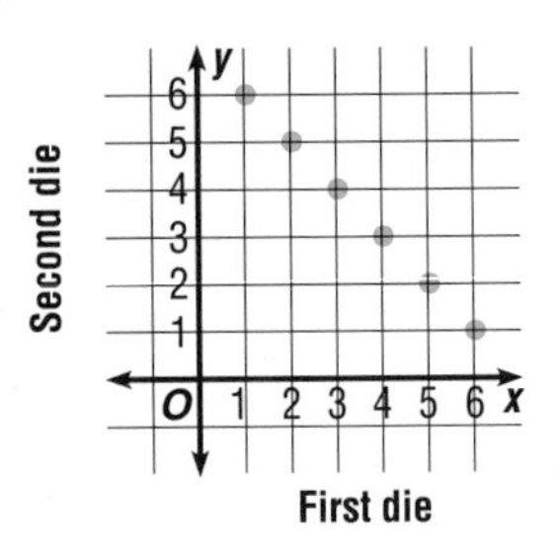

In the first relation, for each value of x, there are six different values of y: 1, 2, 3, 4, 5, and 6. In the second relation, for each value of x, there is *exactly* one value of y. The second relation is called a **function.** A function is a relation in which each element of the domain is paired with *exactly* one element of the range.

Example 1

Is {(5, −2), (3, 2), (4, −1), (−2, 2)} a function?

Since each element of the domain is paired with exactly one element of the range, this relation is a function.

Example 2

Which mapping represents a function?

a.

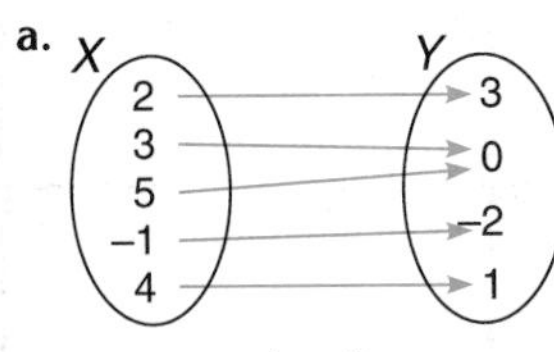

a function

The mapping in **a** represents a function since, for each element of the domain, there is *only one* corresponding element in the range.

b.

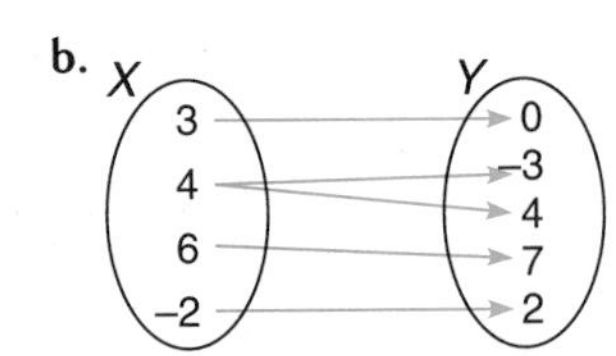

not a function

The mapping in **b** does not represent a function since the element 4 in the domain maps to two elements, −3 and 4, in the range.

Example 3 **Is the relation represented by the equation $x + 2y = 8$ a function?**

Substitute a value for x in the equation. What is the corresponding value of y? Is there more than one value for y? For example, if x is 2, then y is 3 *and* that is the only value of y that will satisfy the equation. If you try other values of x, you will see that there is always only one corresponding value of y. Therefore, the equation $x + 2y = 8$ represents a function.

For equations like $x + 2y = 8$, it may not be easy to determine whether there is an element of the domain that is paired with more than one element of the range. Often, it is simpler to look at the graph of the relation. Suppose you graph $x + 2y = 8$.

Solve $x + 2y = 8$ for y.

$$x + 2y = 8$$
$$2y = 8 - x$$
$$y = \frac{8 - x}{2}$$

Make a table of values and graph the equation.

x	2	0	8
y	3	4	0

Now place your pencil at the left of the graph to represent a vertical line. Slowly move the pencil to the right across the graph.

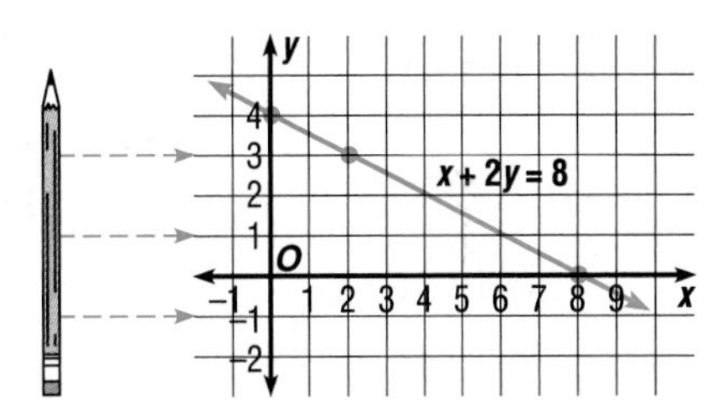

For each value of x, this vertical line passes through no more than one point on the graph. This is true for *every* function.

Vertical Line Test for a Function

If any vertical line passes through no more than one point of the graph of a relation, then the relation is a function.

Example 4 **Use the vertical line test to determine if each relation is a function.**

a. **b.** **c.** **d.**

The relations in **a** and **c** are functions since any vertical line passes through no more than one point of the graph of the relation. The relation in **b** is *not* a function, since a vertical line near the y-axis will pass through *three* points. The relation in **d** is not a function since a vertical line passing through point *P* will also intersect the horizontal line.

Letters other than f are also used for names of functions.

Equations that represent functions can be written in a form called **functional notation.** The equation $y = 2x + 1$ can be written in the form $f(x) = 2x + 1$. The symbol $f(x)$ is read "f of x" and represents the value in the range of the function that corresponds to the value of x in the domain. For example, $f(3)$ is the element in the range that corresponds to the element $x = 3$ in the domain. We say $f(3)$ is the **functional value** of f for $x = 3$.

The ordered pair (3, f(3)) is a solution of the function f.

You can determine a functional value by substituting the given value for x into the equation. For example, if $f(x) = 2x + 1$ and $x = 3$, then $f(3) = 2(3) + 1$ or 7.

Example 5

If $f(x) = 3x - 7$, find each of the following.

a. **$f(2)$**

$$f(2) = 3(2) - 7 = 6 - 7 = -1$$

b. **$f(5)$**

$$f(5) = 3(5) - 7 = 15 - 7 = 8$$

c. **$f(-3)$**

$$f(-3) = 3(-3) - 7 = -9 - 7 = -16$$

Example 6

If $g(x) = x^2 - 2x + 1$, find each of the following.

a. **$g(6a)$**

$$g(6a) = (6a)^2 - 2(6a) + 1 = 36a^2 - 12a + 1$$

Substitute 6a for x.

b. **$6[g(a)]$**

$$6[g(a)] = 6[(a)^2 - 2(a) + 1] = 6a^2 - 12a + 6$$

6[g(a)] means 6 times the value of g(a). Notice that $g(6a) \neq 6[g(a)]$.

Example 7

A rocket is launched with an initial velocity of 100 meters per second. Its height above the ground after t seconds is given by the formula $h(t) = 100t - 4.9t^2$, where $h(t)$ represents the height in meters. What is the height of this rocket 15 seconds after it is launched?

The height of the rocket after 15 seconds is $h(15)$.

$$h(t) = 100t - 4.9t^2$$
$$h(15) = 100(15) - 4.9(15)^2 = 1500 - 4.9(225) = 1500 - 1102.5 \text{ or } 397.5$$

Replace t with 15.

The rocket is 397.5 meters high after 15 seconds.

CHECKING FOR UNDERSTANDING

Communicating Mathematics

Read and study the lesson to answer each question.

1. Does a linear equation always represent a function? Explain.
2. Explain the vertical line test for a function in your own words.
3. Given a function $g(x)$, how can you determine $g(1)$?
4. Give an example of a function whose inverse is *not* a function.

Guided Practice

Determine whether each relation is a function.

5. $\{(5, 4), (-2, 3), (5, 3)\}$
6. $\{(6, 3), (5, -2), (2, 3)\}$
7. $5a^2 - 7 = b$
8. $3s^2 + 2t^2 = 7$

Use the vertical line test to determine if each relation is a function.

9. 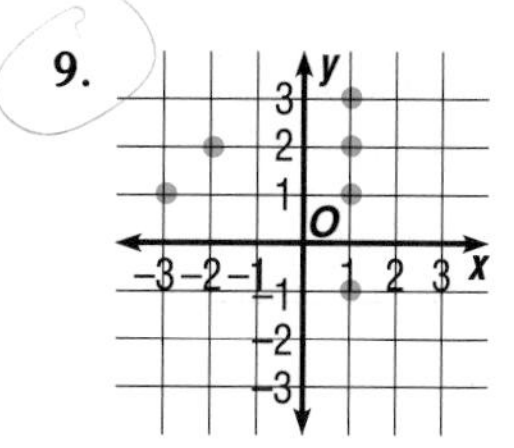

10. 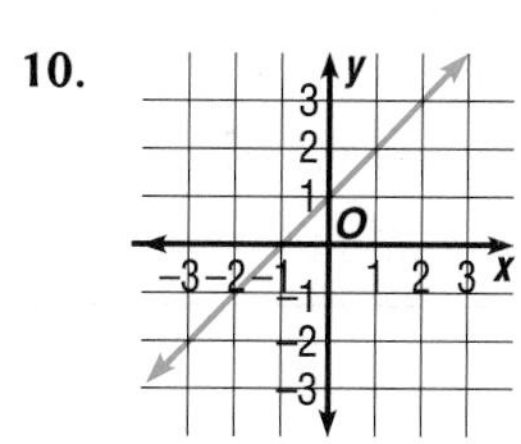

11. 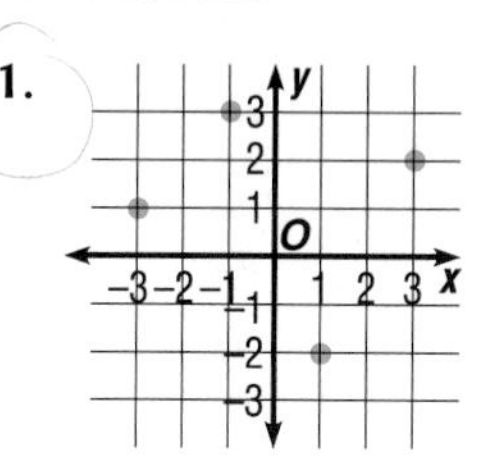

Given $g(x) = 2x - 1$, determine each value.

12. $g(2)$
13. $g(-4)$
14. $g(0)$
15. $g\left(\frac{1}{2}\right)$

EXERCISES

Practice

Determine whether each relation is a function.

16. 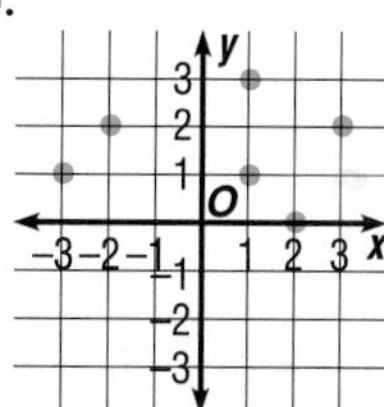

17. 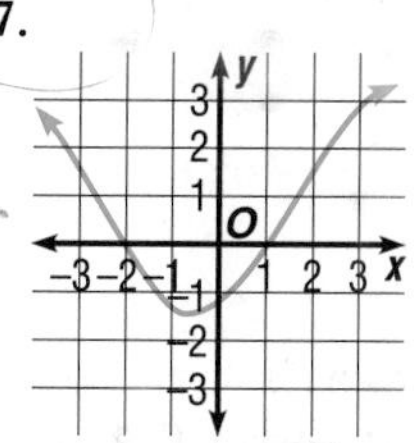

18. 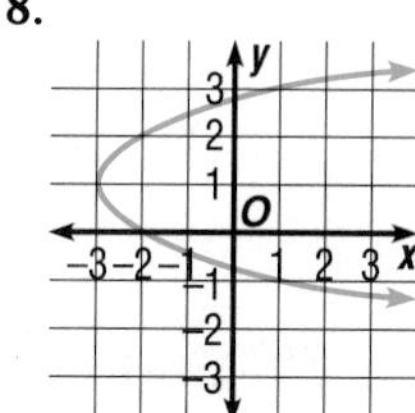

Determine whether each relation is a function. Then state the inverse of the relation and determine whether it is a function.

19. $\{(3, 1), (5, 1), (7, 1)\}$
20. $\{(1, 3), (1, 5), (1, 7)\}$
21. $\{(-2, 4), (1, 3), (5, 2), (1, 4)\}$
22. $\{(6, -1), (1, 4), (2, 3), (6, 1)\}$
23. $\{(5, 4), (-6, 5), (4, 5), (0, 4)\}$
24. $\{(3, -2), (4, 7), (-2, 5), (4, 5)\}$

Determine whether each relation is a function.

25. $3x + 5y = 7$
26. $y = 2$
27. $4x - 7y = 3$
28. $x^2 + y = 11$
29. $x + y^2 = 11$
30. $x = -3$
31. $1 = yx$
32. $x^2 - y^2 = 3$

Given $f(x) = 3x - 5$ and $g(x) = x^2 - x$, determine each value.

33. $f(-3)$ 34. $g(3)$ 35. $g\left(\frac{1}{3}\right)$ 36. $f\left(\frac{2}{3}\right)$

37. $f(5.5)$ 38. $3[f(5)]$ 39. $2[g(-2)]$ 40. $g(0.5)$

41. $f(4a)$ 42. $g(4b)$ 43. $3[f(2n)]$ 44. $2[f(3n)]$

45. $3[g(2m)]$ 46. $2[g(3m)]$ 47. $f(a + 3)$ 48. $g(b - 3)$

Critical Thinking

The inverse of a relation can be found by interchanging the variables. For example, the inverse of the relation $y = 3x - 1$ is $x = 3y - 1$. Determine if the inverse of each relation is a function.

49. $3x + 2y = 1$ 50. $y = -3x^2$ 51. $3x^2 - y^2 = 3$ 52. $x + 3y^2 = 3$

Applications

53. **Car Rental** The cost of a one-day car rental from Rossi Rentals is given by the formula $C(m) = 31 + 0.13m$, where m is the number of miles that the car is driven, \$0.13 is the cost per mile driven, and $C(m)$ is the total cost. If Sheila drove a distance of 110 miles and back in one day, what is the cost of the car rental?

54. **Business** The owner of Clean City Car Wash found that if n cars were washed in a day, the average daily profit $P(n)$, in dollars, was given by the formula $P(n) = -0.027n^2 + 8n - 280$. Find values of $P(n)$ for various values of n to determine the least number of cars that must be washed each day for Clean City Car Wash to make a profit.

Computer

55. The BASIC program at the right uses a FOR/NEXT loop to generate several ordered pairs that could be used to graph the equation $y = 2x - 1$.

```
10 FOR X = -2 TO 2
20 LET Y = 2*X-1
30 PRINT "(";X;",";Y;")"
40 NEXT X
50 END
```

a. Modify line 20 to generate ordered pairs for $y = x + 2$, $y = 2x + 2$, $y = 3x + 2$, $y = 10x + 2$, and $y = 0.5x + 2$. Graph the five equations on the same axes. At what point do the graphs intersect the y-axis?

b. Modify line 20 to generate ordered pairs for $y = -x + 2$, $y = -2x + 2$, $y = -3x + 2$, $y = -10x + 2$, and $y = -0.5x + 2$. Graph the five equations on the same axes. How do these equations differ from those in exercise **a**?

Mixed Review

56. Simplify $6(3x + 7x) + 7(-4x + 8x)$. **(Lesson 2-6)**

57. **Construction** A rectangular pool is 10 meters longer than it is wide. A deck 4 meters wide is built around the pool. If the total area of the deck is 592 m², what are the dimensions of the pool? **(Lesson 6-8)**

58. Factor $7n^2 - 22n + 3$. **(Lesson 7-5)**

59. Simplify $\frac{-3}{5 - a} + \frac{5}{a^2 - 25}$. **(Lesson 8-7)**

60. Graph $5x + 3y = 8$. **(Lesson 9-4)**

9-6 Graphing Inequalities in Two Variables

Objective

After studying this lesson, you should be able to:

- graph inequalities in the coordinate plane.

Application

Mr. Harris is taking his drama class to see the play *Hansel and Gretel* at the Granada Theatre. Tickets for the play cost either \$15 or \$20. If he plans to spend no more than \$240 on tickets, how many of each ticket can Mr. Harris purchase?

Let x = the number of \$15 tickets purchased, and let y = the number of \$20 tickets purchased. Then the following inequality can be used to represent this problem.

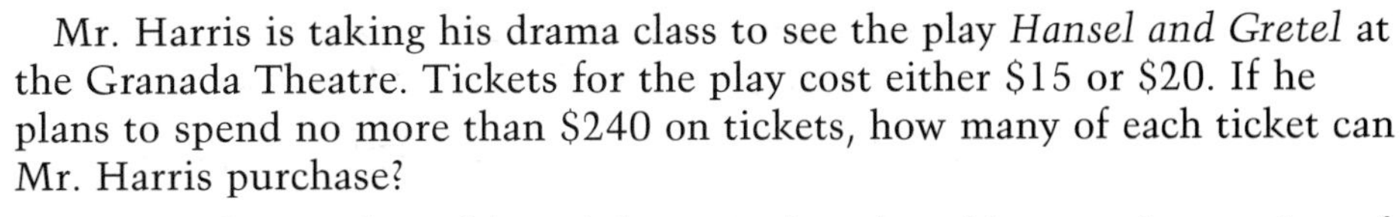

Cost of \$15 tickets · plus · cost of \$20 tickets · is no more than · \$240.

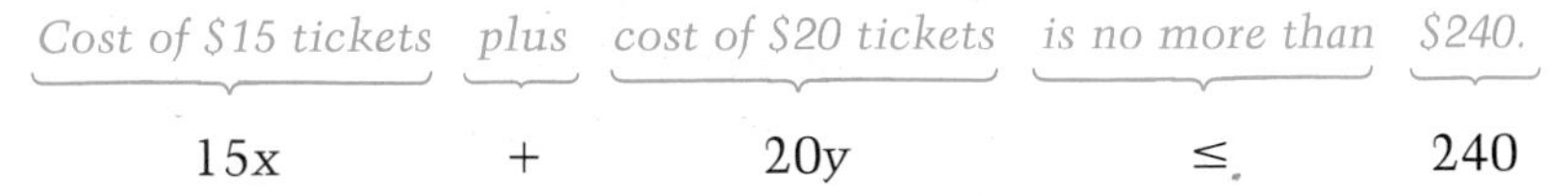

$$15x \quad + \quad 20y \quad \leq \quad 240$$

There are an infinite number of ordered pairs that are solutions to this inequality. The easiest way to show all of these solutions is to draw a *graph* of the inequality. Before doing this, let's consider some simpler inequalities. *The problem above is solved in Example 2.*

Suppose you want to draw the graph of the inequality $x < 3$. First, you need to draw the graph of the equation $x = 3$. This graph is a line that separates the coordinate plane into two regions. In the figure at the right, one of the regions is shaded blue and the other is shaded yellow. Each region is called a **half-plane.** The line $x = 3$ is called the **boundary,** or **edge,** for each half-plane.

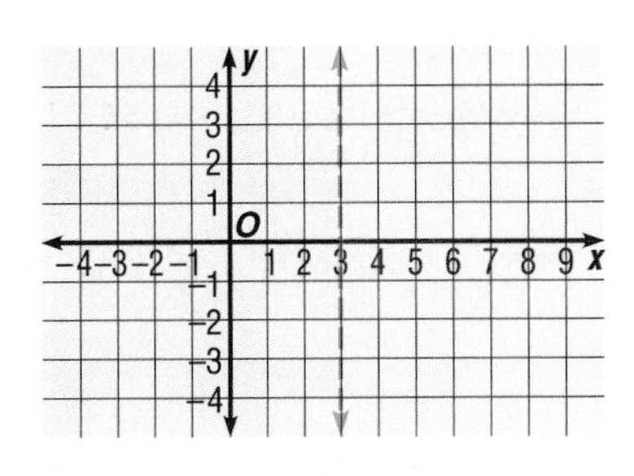

The line x = 3 means the line that is the graph of the equation x = 3.

To determine which half-plane $x < 3$ describes, choose a value of x that appears in the blue region, like 0. Since $0 < 3$, the blue region is described by $x < 3$. The x-coordinate of *every* point in that region is less than 3. The inequality $x > 3$ describes the yellow region. The points on the line $x = 3$ are in neither of these two regions.

Consider the graph of $y > x + 1$. The boundary line for this graph is the line $y = x + 1$. Since the boundary is *not* part of the graph, it is shown as a dashed line on the graph. To determine which half-plane is the graph of $y > x + 1$, test a point *not* on the boundary. For example, you can test the origin, (0, 0). Since $0 > 0 + 1$ is false, (0, 0) is *not* a solution of $y > x + 1$. Thus, the graph is all points in the half-plane that does *not* contain (0, 0). This graph is called an **open half-plane** since the boundary is not part of the graph.

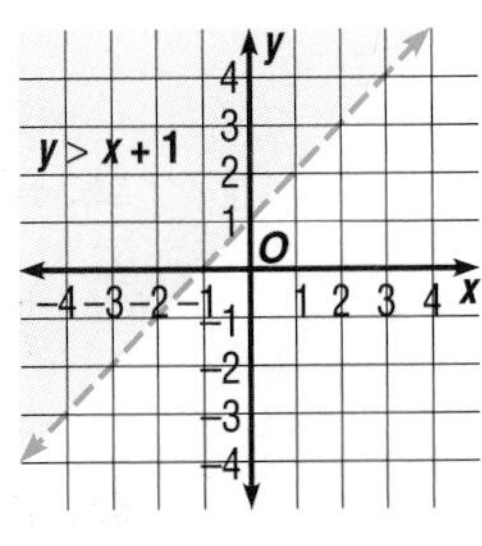

What happens if you test (0, 4)?

Now consider the graph of $y \leq x + 1$. The boundary line for this graph is also the line $y = x + 1$. Since the inequality $y \leq x + 1$ means $y < x + 1$ or $y = x + 1$, this boundary *is* part of the graph. Therefore, the boundary is shown as a solid line on the graph.

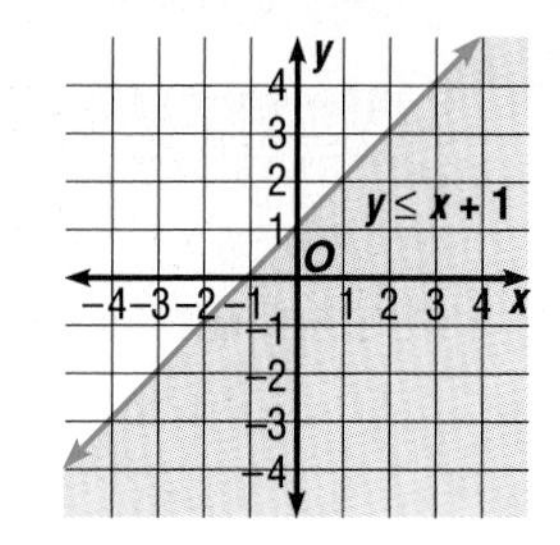

The origin, (0, 0), is a part of the graph of $y \leq x + 1$ since $0 \leq 0 + 1$ is true. Thus, the graph is all points in the half-plane that contains the origin and the line $y = x + 1$. This graph is called a **closed half-plane.**

Example 1

Graph $y > -4x - 3$.

Graph the equation $y = -4x - 3$. Draw it as a dashed line since this boundary is *not* part of the graph. The origin, (0, 0), is part of the graph since $0 > -4(0) - 3$ is true. Thus, the graph is all points in the half-plane that contains the origin.

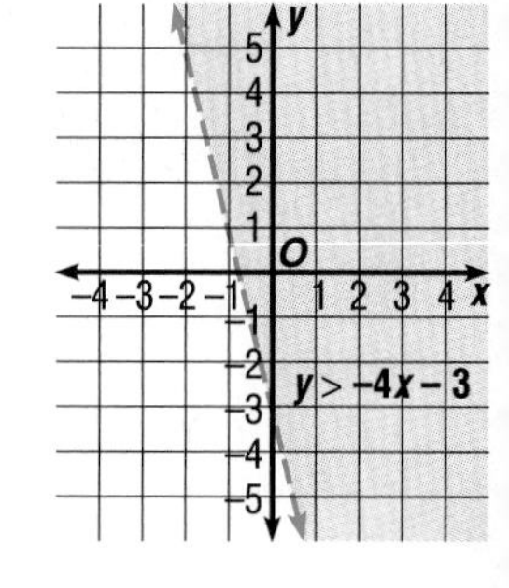

Check: Test a point on the other side of the boundary, say (−2, −2). Since $-2 > -4(-2) - 3$, or $-2 > 5$, is false, (−2, −2) is *not* part of the graph.

Example 2

Graph $15x + 20y \leq 240$ to answer the application at the beginning of the lesson. How many of each ticket can Mr. Harris purchase?

First solve for y in terms of x.

$$15x + 20y \leq 240$$
$$20y \leq 240 - 15x$$
$$y \leq 12 - \frac{3}{4}x$$

Then graph $y = 12 - \frac{3}{4}x$ as a solid line since the boundary is part of the graph. The origin is part of the graph since $15(0) + 20(0) \leq 240$ is true. Thus, the graph is all points in the half-plane that contains the origin.

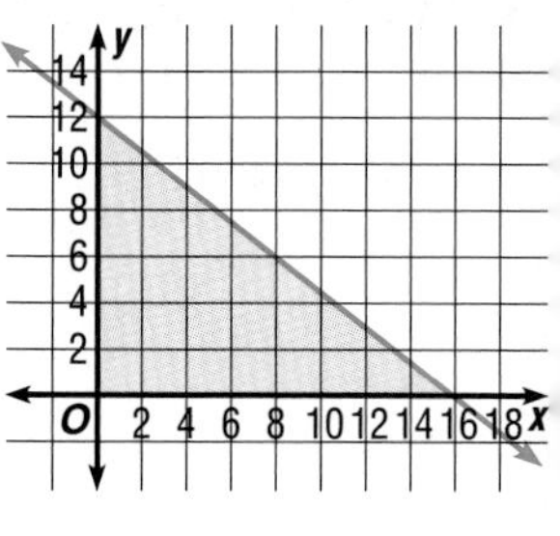

Mr. Harris cannot buy fractional or negative numbers of tickets. So, any point in the shaded region whose x- and y-coordinates are whole numbers is a possible solution. For example, (5, 8) is a solution. This corresponds to buying five \$15 tickets and eight \$20 tickets for a total cost of 15(5) + 20(8) or \$235.

Example 3

Jessica is selling two types of school notebooks. She knows that one type of notebook is much more popular with students than the other. She prices the popular notebook so that she will make \$3 profit on each sale. To help sell the other notebook, she prices it cheaply and expects to lose \$1 on each sale. How many of each type of notebook can she sell and not lose any money?

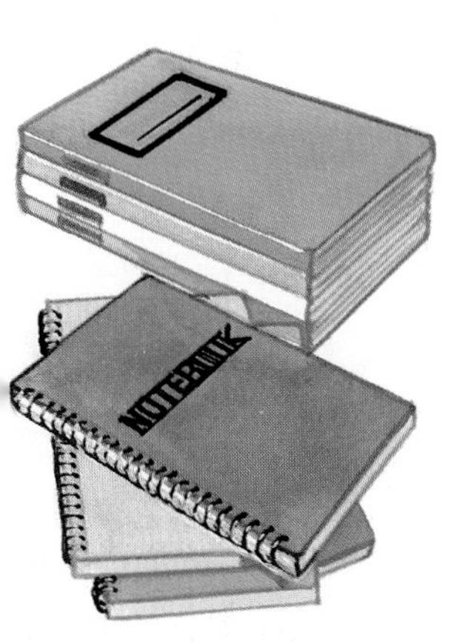

Let x = the number of the more popular notebooks sold.
Let y = the number of the other notebooks sold.
Then $3x$ = the profit from the sales of the more popular notebook, and $-1y$ = the profit (or loss) from the sales of the other notebook.

Write an inequality to represent the problem. Then graph this inequality to determine all of the possible solutions.

Profit from popular notebook	*plus*	*profit from other notebook*	*must be at least*	*0.*
$3x$	$+$	$-1y$	$\geq$	0

$$3x - y \geq 0$$
$$3x \geq y$$

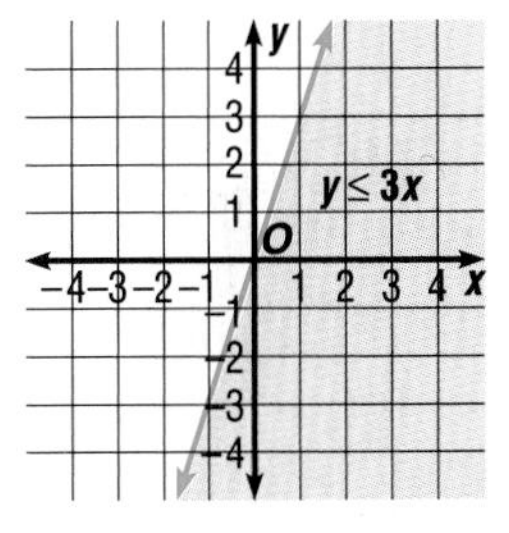

Graph the equation $3x = y$. Draw it as a solid line since this boundary is part of the graph. Since the origin is on the boundary, you must test another point, say (1, 1). Since $3(1) \geq 1$ is true, the graph of $3x \geq y$ is the half-plane that contains the point (1, 1).

Any point in the shaded region above whose x-coordinate and y-coordinate are *whole* numbers is a possible solution. For example, since the point (3, 5) is in this region, Jessica can sell 3 of the popular notebook and 5 of the other, giving her a total profit of $3(3) + (-1)(5)$ or \$4.

CHECKING FOR UNDERSTANDING

Communicating Mathematics

Complete.

1. The graph of a linear equation separates the coordinate plane into two $\underline{\quad ? \quad}$.
2. For a(n) $\underline{\quad ? \quad}$ half-plane, the boundary is part of the graph.
3. If the coordinates of a point satisfy an inequality, then the graph of the inequality is the half-plane that $\underline{\quad ? \quad}$ the point.

Guided Practice

State whether the boundary is included in the graph of each inequality.

4. $2x + y \geq 3$
5. $3x - 2y \leq 1$
6. $5x - 2 > 3y$

Determine which of the ordered pairs are solutions to the inequality.

7. $x + 2y \geq 3$	a. $(-2, 2)$	b. $(4, -1)$	c. $(3, 1)$
8. $2x - 3y \leq 1$	a. $(2, 1)$	b. $(5, -1)$	c. $(1, 1)$
9. $3x + 4y < 7$	a. $(1, 1)$	b. $(2, -1)$	c. $(-2, 4)$
10. $4y - 8 \geq 0$	a. $(0, 2)$	b. $(2, 5)$	c. $(-2, 0)$
11. $-2x < 8 - y$	a. $(5, 10)$	b. $(3, 6)$	c. $(-4, 0)$
12. $5x + 2 > 3y$	a. $(-2, -3)$	b. $(2, 3)$	c. $(4, 7)$

Use the point (0, 0) to determine which half-plane is the graph of each inequality.

13. $x < 4$

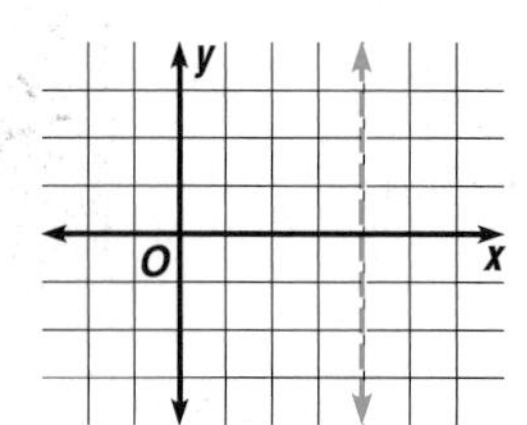

14. $y > -2$

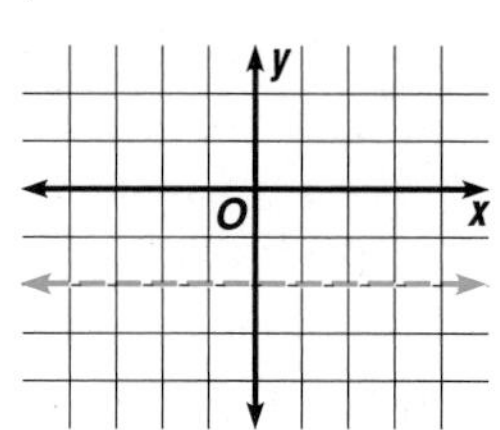

15. $3m + n > 4$

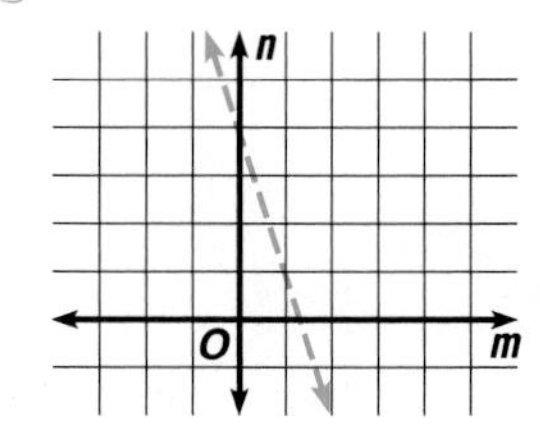

EXERCISES

Practice

Determine which half-plane is the graph of each inequality.

16. $2x - y < 6$

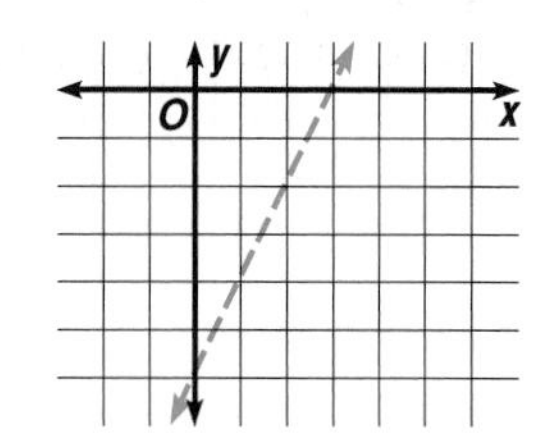

17. $5a - b < 5$

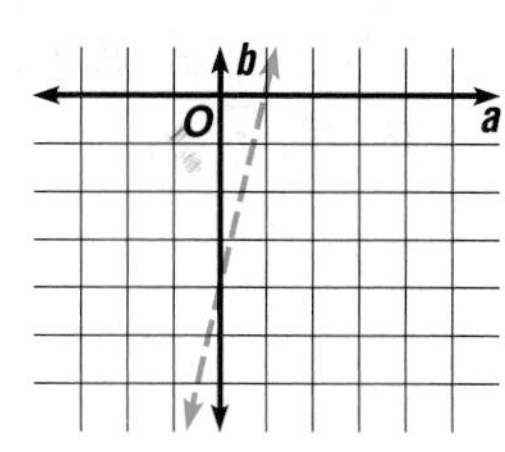

18. $3x < 2y$

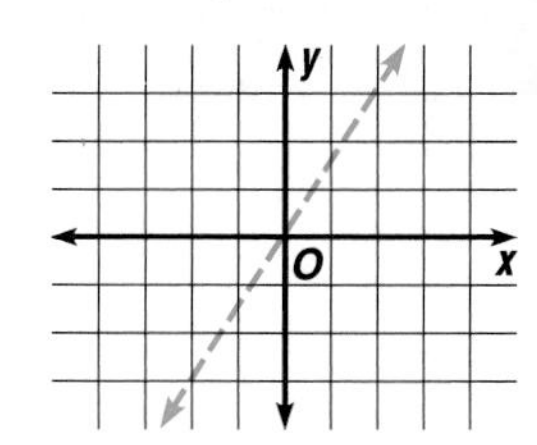

Graph each inequality.

19. $y > 3$
20. $y \leq -2$
21. $x \geq -1$
22. $x < 4$
23. $x + y > 1$
24. $x + y < 2$
25. $x + y < -2$
26. $x + y > -4$
27. $y > x - 1$
28. $y \leq x + 1$
29. $y \leq 3x - 1$
30. $y > 4x - 1$
31. $2x - y < 1$
32. $3x + y > 1$
33. $2x + 3y \geq -2$
34. $3y - 2x \leq 2$
35. $x - 2y < 4$
36. $4y + x < 16$
37. $x < y$
38. $2x > 3y$
39. $-x < -y$
40. $-y > x$
41. $2x < -y$
42. $y > |x|$
43. $|y| \geq 2$
44. $y > 2$ and $x < 3$
45. $y > 2$ or $y < 1$
46. $y \leq -x$ or $x \geq -3$
47. $3y \geq x$ and $y < 0$

Critical Thinking

48. What compound inequality is described by the graph at the right? Find an inequality that also describes this graph.

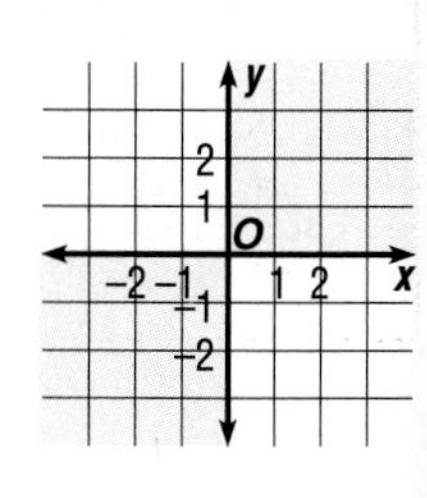

Applications

49. **Sales** Tickets for Southside High School's annual talent show are \$3 for adults and \$2 for students. In order to cover the expenses for the show, a total of \$600 must be made from ticket sales.
    a. Use a graph to determine how many of each type of ticket must be sold to cover the expenses.
    b. List three of the solutions.

50. **Manufacturing** A furniture factory can produce a table in 30 minutes and a chair in 12 minutes.
    a. Use a graph to determine how many of each can be produced during an 8-hour shift.
    b. List three possible solutions where at least 30 chairs are produced.

Mixed Review

51. Solve $2(x - 4) - 10x \leq 0$. **(Lesson 5-3)**

Journal
Write a paragraph to explain to another person how you graph an inequality.

Simplify each expression. (Lessons 6-2, 8-5, and 8-8)

52. $(r^2xy)(-2r^3x)^2$

53. $\frac{12n}{3n - 2} + \frac{8}{2 - 3n}$

54. $\dfrac{\frac{x + y}{a + b}}{\frac{x^2 - y^2}{b^2 - a^2}}$

55. Factor $15a^2 - 28bc - 21ab + 20ac$. **(Lesson 7-3)**

56. **Physics** A ball is dropped from a 100-meter tower. Its height $h(t)$, in meters, after t seconds is given by the formula $h(t) = 100 - 4.9t^2$. Will the ball have hit the ground after 4 seconds? Why or why not? **(Lesson 9-6)**

HISTORY CONNECTION

René Descartes

René Descartes (1596–1650) was one of the greatest scientists of the seventeenth century. It was he who used two perpendicular number lines to identify points in a plane. This rectangular coordinate system is called the *Cartesian coordinate system,* in honor of Descartes. The connecting of algebra and geometry through graphing is called *analytic geometry.* Although Descartes is considered to be the founder of analytic geometry, he did not consider himself a mathematician. He is often called the "father of modern philosophy," his great love.

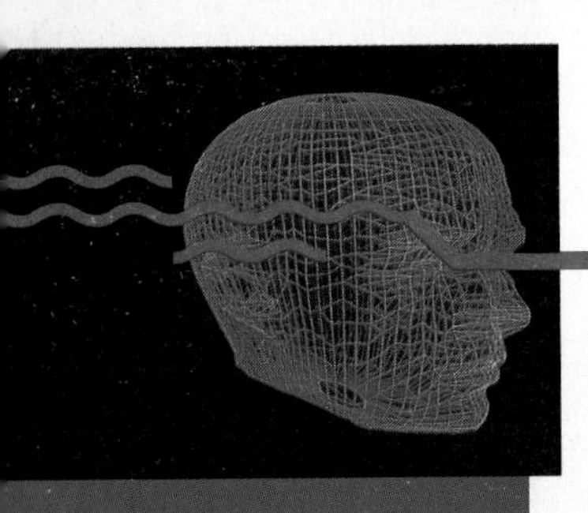

Technology

Graphing Relations

BASIC
▶ **Graphing calculators**
Graphing software
Spreadsheets

You can use a graphing calculator to graph linear relations. To graph a line on a graphing calculator, you first need to set the range. Press the Range key. Make sure it is set for the standard viewing window of Xmin = −10, Xmax = 10, Xscl = 1, Ymin = −10, Ymax = 10, Yscl = 1. To change the range on the TI-81, enter the following.

RANGE (-) 10 ENTER 10 ENTER 1 ENTER (-) 10 ENTER 10 ENTER 1 ENTER

To change the range on the Casio, enter the following.

Range (-) 10 EXE 10 EXE 1 EXE (-) 10 EXE 10 EXE 1 EXE

Be sure to use the (-) key, not the − key, when typing in a negative value. After setting the range, enter the function and graph it.

Example 1 **Graph $y = 2x + 3$.**

For the TI-81

Y = CLEAR 2 X|T + 3 GRAPH

For the Casio

Graph 2 Alpha x + 3 EXE

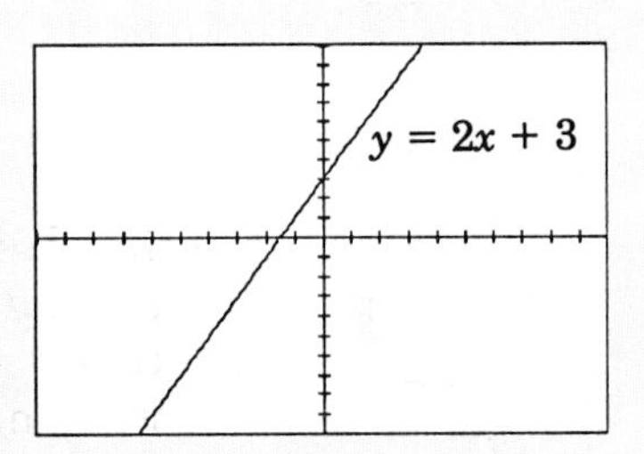

When you graph a line, you should be able to see the points at which the line crosses the x- and y-axes. When this is the case, we will call the graph a *complete graph.* Some examples follow. The range values are given below each graph.

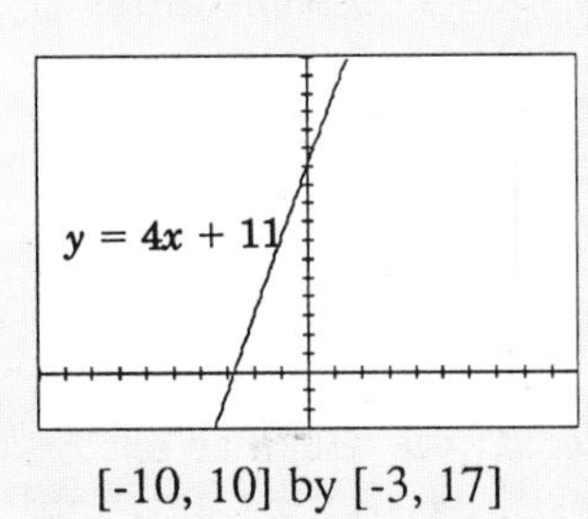

[-10, 10] by [-3, 17]

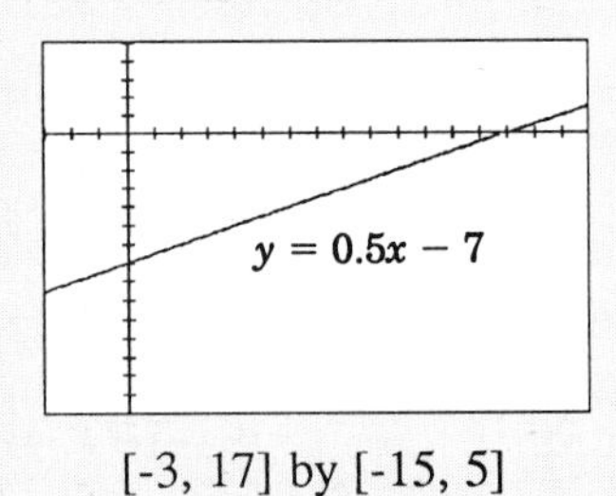

[-3, 17] by [-15, 5]

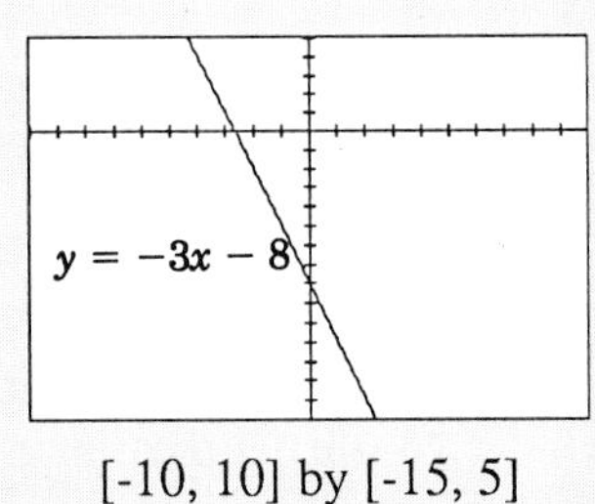

[-10, 10] by [-15, 5]

EXERCISES

Graph each equation in the standard viewing window. State whether the graph is a complete graph.

1. $y = 5x + 6$
2. $y = -10x - 7$
3. $y = 0.01x + 4$
4. $y = 5$
5. $y = 10x + 15$
6. $y = 3x - 5$

9-7 Finding Equations from Relations

Objective

After studying this lesson, you should be able to:

- write an equation to represent a relation, given a chart of values.

Application

Video Village sells blank videocassettes in packages of 5. You can make a chart to show the relationship between the number of videocassettes and the number of packages. Let y represent the number of videocassettes and let x represent the number of packages.

x	1	2	3	4	5	6
y	5	10	15	20	25	30

This relation can also be shown using an equation. Study the differences between successive values of x and y.

	+1	+1	+1	+1	+1	
x	1	2	3	4	5	6
y	5	10	15	20	25	30
	+5	+5	+5	+5	+5	

Notice that the differences of the y-values are exactly five times the differences of the corresponding x-values. This pattern suggests the relation $y = 5x$. You can check to see if this equation is correct by substituting values of x into the equation. For example, if $x = 2$, then $y = 5(2)$ or 10, as given in the chart.

Example 1

Write an equation for the relation given in the chart at the right.

a	1	2	3	4	5	6
b	1	4	7	10	13	16

Find the differences between successive values of a and b.

	+1	+1	+1	+1	+1	
a	1	2	3	4	5	6
b	1	4	7	10	13	16
	+3	+3	+3	+3	+3	

Notice that the differences of the b-values are 3 times the differences of the corresponding a-values. This pattern suggests the relation $b = 3a$. However, if $a = 1$, then $b = 3(1)$ or 3, not 1, as given in the chart. Thus, the equation $b = 3a$ does *not* describe the relation. To obtain 1 for b when a is 1, you need to subtract 2. This suggests the relation $b = 3a - 2$. Check this by using other values from the chart.

Check: If $a = 2$, then $b = 3(2) - 2$ or 4. ✓
If $a = 5$, then $b = 3(5) - 2$ or 13. ✓

Example 2

Write an equation for the relation given in the chart at the right.

m	1	2	4	5	6	9
n	9	6	0	−3	−6	−15

Find the differences.

+1 +2 +1 +1 +3

m	1	2	4	5	6	9
n	9	6	0	−3	−6	−15

−3 −6 −3 −3 −9

Notice that the values for m are increasing, but the values for n are decreasing.

The differences suggest the relation $n = -3m$. However, if $m = 1$, then $n = -3(1)$ or -3, not 9, as given in the chart. Thus, $n = -3m$ does *not* describe the relation.

To obtain 9 for n when m is 1, you need to add 12. This suggests the relation $n = -3m + 12$, or $n = 12 - 3m$.

Check: If $m = 4$, then $n = 12 - 3(4)$ or 0. ✓
If $m = 9$, then $n = 12 - 3(9)$ or -15. ✓

Example 3

McGaffin Plumbing has a standard charge for every housecall it makes. In addition, there is an hourly rate. Suppose the charges are \$60 for a one-hour job, \$95 for a two-hour job, and \$130 for a three-hour job. Write an equation to describe the relationship between the number of hours worked and the charge. Then use this equation to find the standard charge for each job and the hourly rate.

Let x = the number of hours worked, and let y = the total charge. Make a chart for this relation. Then find the differences.

+1 +1

x	1	2	3
y	60	95	130

+35 +35

The differences suggest the relation $y = 35x$. However, if $x = 1$, then $y = 35(1)$ or 35, not 60, as given in the chart. To obtain 60 for y when $x = 1$, you need to add 25.

Thus, the equation that describes this relation is $y = 35x + 25$.

Check: If $x = 2$, then $y = 35(2) + 25$ or 95. ✓
If $x = 3$, then $y = 35(3) + 25$ or 130. ✓

In the equation $y = 35x + 25$, x represents the number of hours worked. Thus, the hourly charge for each plumber is \$35. This means that the standard charge for each job is \$25.

CHECKING FOR UNDERSTANDING

Communicating Mathematics

Read and study the lesson to answer each question.

1. When you study successive differences of a relation shown in a chart, you are applying what problem-solving strategy to find an equation that describes the relation?
2. How can you determine whether an equation represents a relation that is given in a chart?
3. Find a linear equation that has both (1, 1) and (0, 3) as solutions. Is this the only linear equation that has these two solutions?

Guided Practice

In Exercises 4–7, match each equation with a relation in charts a-d.

4. $y = 2x + 6$
5. $y = 4x + 3$
6. $x - 2y = 10$
7. $2x + 3y = 3$

a.

x	2	4	6	8	10
y	−4	−3	−2	−1	0

b.

x	3	6	9	12	15
y	−1	−3	−5	−7	−9

c.

x	3	4	5	6	7
y	12	14	16	18	20

d.

x	1	2	3	4	5
y	7	11	15	19	23

Complete.

8. 5, 7, 9, 11, _?_, _?_
9. 12, 9, 6, 3, _?_, _?_
10. −1, −4, −7, −10, _?_, _?_
11. 5, _?_, 15, _?_, 25, 30

EXERCISES

Practice

Complete.

12. 3, 3.5, 4, 4.5, _?_, _?_
13. 11.75, 11.5, _?_, 11, _?_
14. −8, _?_, −8.4, −8.6, _?_
15. 1, 2, 4, 8, _?_, _?_
16. 27, 9, 3, _?_, _?_
17. 8, _?_, 2, −1, _?_

Write an equation for the relation shown in each chart. Then copy and complete each chart.

18.

x	1	2	3	4	5
y	4	8	12		

19.

m	−3	−2	−1	0	1
n	−5	−3	−1		

20.

a	−2	−1	0	1	2
b	−3	1	5		

21.

a	−4	−2	0	2	4
b	−13	−5	3		

22.

x	1	2	3	4	5	6	7
y	14	13	12				

23.

m	−2	−1	0	1	2	3	4
n	13	12	11	10			

Write an equation for the relation shown in each chart. Then copy and complete each chart.

24.

a	-5	-3	-1	1	2	4	7
b	28	18	8				

25.

x	-4	-2	0	2	4	6	8
y	26	22	18	14			

26.

r	-4	-2	0	2	4	6	8
s	-1	0	1				

27.

c	6	12	18	24	30	36	42
d		2	4			10	

Write an equation to represent each relation.

28. $\{(-2, 4), (-1, 1), (0, 0), (1, 1), (2, 4)\}$

29. $\{(1, 2), (2, 9), (3, 28), (4, 65), (5, 126)\}$

30. $\{(-6, 4), (-3, 8), (1, -24), (2, -12), (6, -4)\}$

31. $\{(-2, 11), (-1, 14), (0, 15), (1, 14), (2, 11)\}$

32. $\{(-4, 3), (-2, 12), (-1, 48), (1, 48), (2, 12)\}$

Critical Thinking

33. How can you determine whether there is *one* linear equation that has $(-2, -3)$, $(0, 5)$, and $(4, 22)$ as solutions?

Applications

34. **Geology** The underground temperature of rocks varies with their depth below the surface. The temperature at the surface is about 20°C. At a depth of 2 km, the temperature is about 90°C, and at a depth of 10 km, the temperature is about 370°C.

   a. Write an equation to describe this relationship.

   b. Use the equation to predict the temperature at a depth of 13 km.

35. **Sales** For intrastate long distance phone calls, a telephone company charges \$1.72 for a 4-minute call, \$2.40 for a 6-minute call, and \$5.46 for a 15-minute call.

   a. Write an equation to describe this relationship.

   b. Use the equation to determine the charge for a 1-minute call and the charge per minute.

Mixed Review

36. **Finance** Beatriz invested \$12,200, part at 5% annual interest and the remainder at 6.5% annual interest. After one year, the total value of her investment, including interest, was \$12,870. How much money did she invest at each rate? **(Lesson 4-4)**

37. Find the GCF of $24x^2y$, $28mnx$, and $36nx$. **(Lesson 7-1)**

38. Factor $3x^3 + 24x^2y - 99xy^2$. **(Lesson 7-8)**

39. Simplify $\frac{x^2 + 2x - 15}{x^2 - x - 30} \div \frac{x^2 - 3x - 18}{x^2 - 2x - 24}$. **(Lesson 8-3)**

40. Graph $y \geq -4x + 1$. **(Lesson 9-6)**

9-8 Problem-Solving Strategy: Use a Graph

EXPLORE
PLAN
SOLVE
EXAMINE

Objective After studying this lesson, you should be able to:

- solve problems by using bar graphs and line graphs.

Connection Statistical graphs are often used to present data and show relationships between sets of data. Two examples are shown below.

Bar graphs show how specific quantities compare. The bar graph at the right illustrates the relationship between average household income and the level of education achieved by the head of the household.

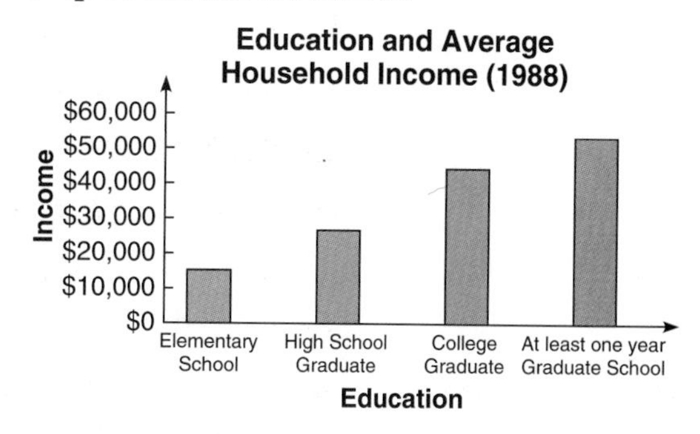

Line graphs show trends or changes. The line graph at the right shows the growth of the world's population.

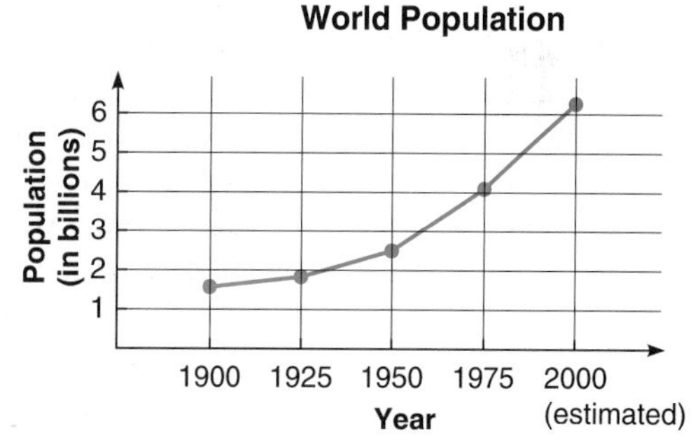

To fully understand the information presented in a graph, ask yourself the following questions.

1. What information does the title give you about the data?
2. What variable is represented along each axis?
3. What units are used along each axis?

Apply each of the three questions to the graph below.

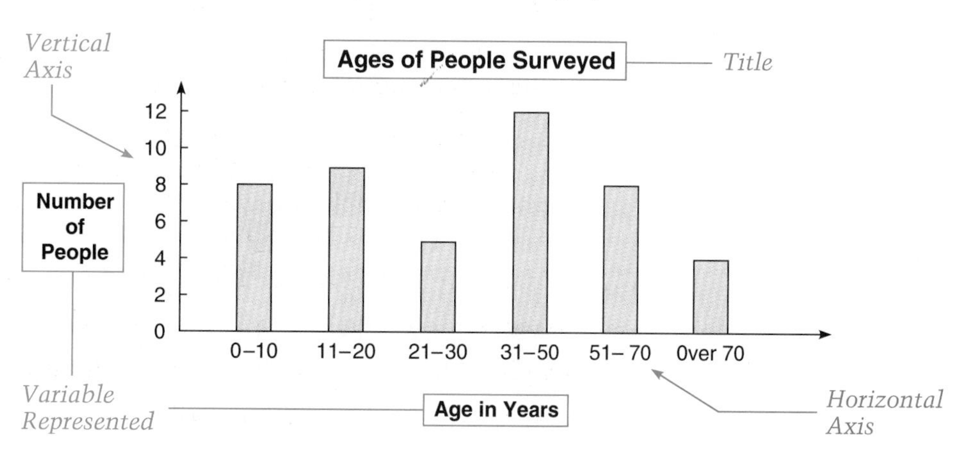

The title indicates that the graph illustrates the number of people in each age group who were surveyed. The vertical axis shows the number of people in each age group. Each unit on the vertical axis represents two people. The horizontal axis shows age groups in years. Notice that the units used for the axes do not have to be the same.

Some graphs can be misleading, as shown in the example below.

Example

Greg received scores of 72, 74, 75, 77, and 78 on his first five algebra exams. The two graphs below were made to show how his test scores have improved. Do they show the same results?

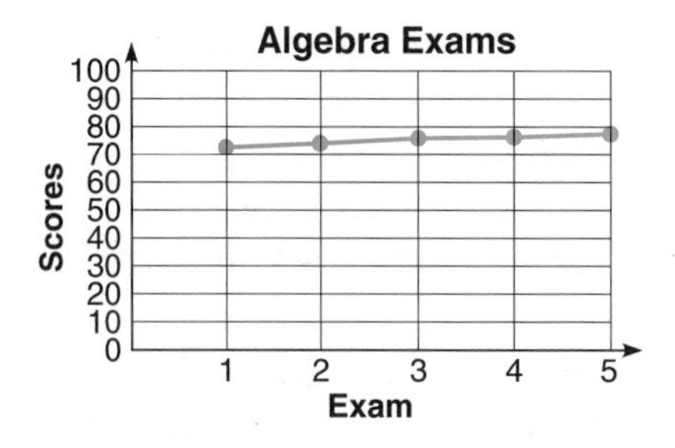

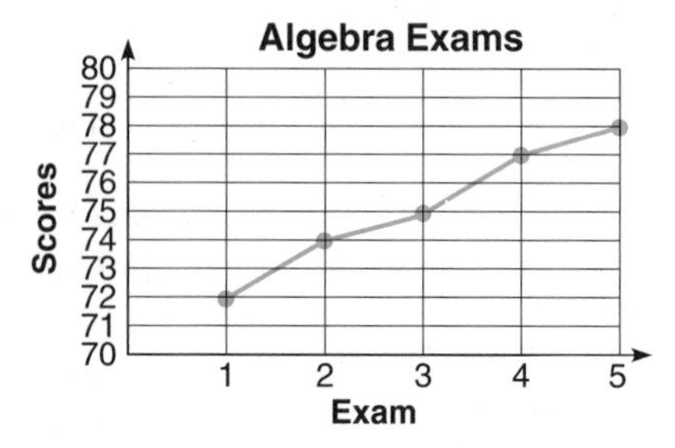

Both graphs use the same data but appear to show different results. In the graph on the left, Greg's scores show slight improvement. In the graph on the right, Greg's scores appear to show great improvement. While both graphs are correct, the graph on the right is misleading.

CHECKING FOR UNDERSTANDING

Communicating Mathematics

Read and study the lesson to answer each question.

1. What do bar graphs illustrate best?
2. What do line graphs illustrate best?

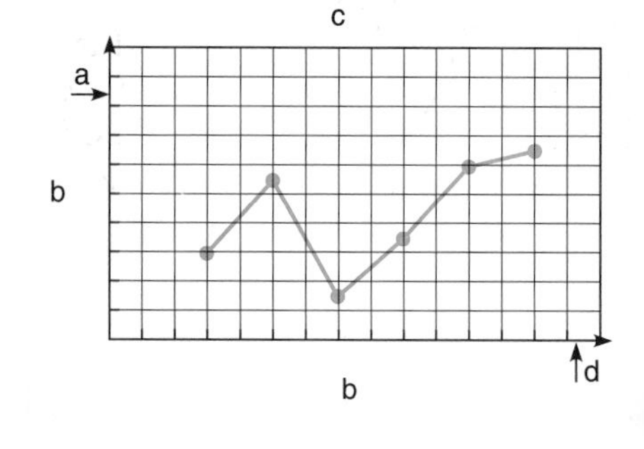

Identify each part labeled.

3. a
4. b
5. c
6. d

Guided Practice

The graphs below show the results of a survey on favorite restaurants.

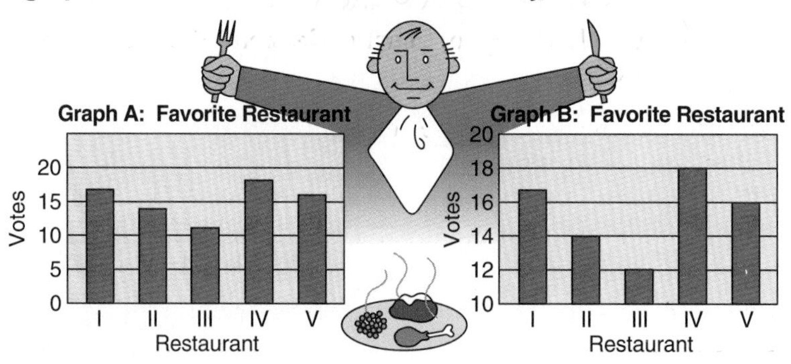

7. Do Graphs A and B display the same data?
8. In Graph B, the bar for Restaurant II is twice as high as the bar for Restaurant III. Does this mean that there are twice as many votes?
9. Why would Restaurant III prefer Graph A?
10. What causes the difference in voting to appear greater in Graph B?
11. Which graph better represents the result?

EXERCISES

Practice

Strategies

Look for a pattern.
Solve a simpler problem.
Act it out.
Guess and check.
Draw a diagram.
Make a chart.
Work backwards.

Portfolio

Select one of the assignments from this chapter that you found especially challenging and place it in your portfolio.

Solve. Use any strategy.

12. Lesharo has ten boxes. Four of his boxes contain pencils, five contain pens, and two contain both pens and pencils. How many of Lesharo's boxes are empty?
13. Numero Uno says, "I am thinking of a three-digit number. If the digits of my number are added together and the result is cubed, the answer will be the original number. What is my number?"
14. Paula spends $\frac{1}{3}$ of her monthly income on rent and $\frac{1}{4}$ of the remainder on food. She spends $\frac{1}{6}$ of what's left on clothes and $\frac{1}{5}$ of the remainder on entertainment. After she saves $\frac{1}{2}$ of what's left and makes a \$200 car payment, she has \$40 left. What is her monthly income?

15. Al is two or three years younger than Betty, who is two or three times as old as Carmelita. But Al is two or three times as old as Dwayne, who is two or three years younger than Carmelita. If they are all older than three years old, how old are they?

COOPERATIVE LEARNING ACTIVITY

Work in groups of four. Each person must understand the solution and be able to explain it to any person in the class.

Find the solution to the puzzle. The solution is unique.

Across

1. a square
4. a cube
5. a multiple of a square

Down

1. a Fibonacci number
2. a square
3. a perfect number

1	2	3
4		
5		

CHAPTER 9

SUMMARY AND REVIEW

VOCABULARY

Upon completing this chapter, you should be familiar with the following terms:

boundary	**379**	**359**	mapping
coordinate plane	**354**	**354**	ordered pair
domain	**359**	**354**	origin
function	**374**	**356**	quadrant
functional notation	**376**	**359**	range
functional value	**376**	**359**	relation
graph	**355**	**354**	x-axis
half-plane	**379**	**354**	x-coordinate
inverse of a relation	**360**	**354**	y-axis
linear equation	**369**	**354**	y-coordinate

SKILLS AND CONCEPTS

OBJECTIVES AND EXAMPLES

Upon completing this chapter, you should be able to:

- graph ordered pairs on a coordinate plane **(Lesson 9-1)**

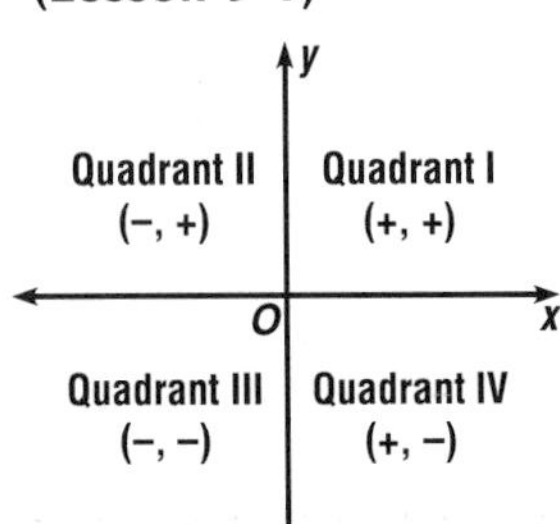

REVIEW EXERCISES

Use these exercises to review and prepare for the chapter test.

Graph each point.

1. A(6, −5)
2. B(−4, 0)
3. C(−2, −3)
4. D(3, 5)

If the graph of P(*x*, *y*) satisfies the given condition, name the quadrant in which point P is located.

5. $x > 0, y > 0$
6. $x < 0, y = -3$
7. $x = -2, y > 0$
8. $x < -3, y = 0$

- identify the domain, range, and inverse of a relation **(Lesson 9-2)**

Determine the domain, range, and inverse of {(3, 5), (2, 6), (5, 6)}.

The domain is {3, 2, 5}.
The range is {5, 6}.
The inverse is {(5, 3), (6, 2), (6, 5)}.

Determine the domain, range, and inverse of each relation.

9. {(4, 1), (4, −2), (4, 6), (4, −1)}
10. {(−3, 5), (−3, 6), (4, 5), (4, 6)}
11. {(−2, 1), (−5, 1), (−7, 1)}
12. {(−3, 1), (−2, 0), (−1, 1), (0, 2)}

OBJECTIVES AND EXAMPLES	REVIEW EXERCISES

- solve linear equations for a specific variable **(Lesson 9-3)**

Solve for y.

$$2x - 5y = 9$$
$$-5y = 9 - 2x$$
$$y = \frac{9 - 2x}{-5}$$

Solve each equation for y in terms of x.

13. $3x + y = 7$
14. $4x - 3y = 9$
15. $x + 6y = 12$

- solve linear equations for a given domain **(Lesson 9-3)**

Solve $2x + y = 7$ if the domain is $\{-2, 0\}$. $2x + y = 7 \rightarrow y = 7 - 2x$

x	$7 - 2x$	y	(x, y)
-2	$7 - 2(-2)$	11	$(-2, 11)$
0	$7 - 2(0)$	7	$(0, 7)$

Solve each equation if the domain is $\{-4, -2, 0, 2, 4\}$.

16. $y = 4x + 5$
17. $x - y = 9$
18. $3x + 2y = 9$
19. $4x - 3y = 0$

- graph linear equations on a coordinate plane **(Lesson 9-4)**

Graph $y = 3x - 5$.

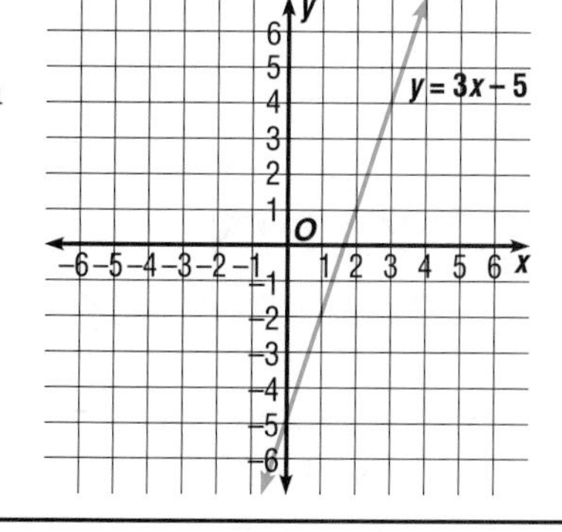

Graph each equation.

20. $x + 5y = 4$
21. $2x - 3y = 6$
22. $5x + 2y = 10$
23. $\frac{1}{2}x + \frac{1}{3}y = 3$

- determine whether a given relation is a function **(Lesson 9-5)**

Is $\{(3, 2), (5, 3), (4, 3), (5, 2)\}$ a function?

Because there are 2 values of y for one value of x, 5, the relation is *not* a function.

Determine whether each relation is a function.

24. $\{(3, 8), (9, 3), (-3, 8), (5, 3)\}$
25. $x - y^2 = 4$
26. $xy = 6$
27. $3x - 4y = 7$

- calculate functional values for a given function **(Lesson 9-5)**

Given $g(x) = 2x - 1$, find $g(-6)$.

$$g(-6) = 2(-6) - 1$$
$$= -12 - 1$$
$$= -13$$

Given $g(x) = x^2 - x + 1$, determine each value.

28. $g(2)$
29. $g(-1)$
30. $g\left(\frac{1}{2}\right)$
31. $g(a + 1)$
32. $g(-2a)$

OBJECTIVES AND EXAMPLES

- graph inequalities in the coordinate plane **(Lesson 9-6)**

Graph $2x + 7y < 9$.

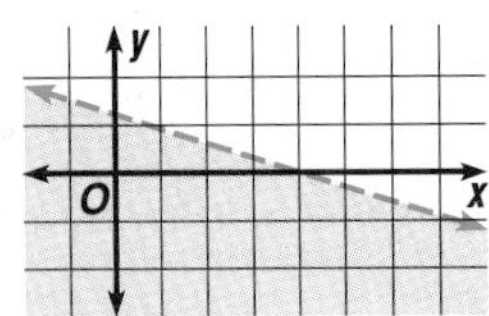

REVIEW EXERCISES

Graph each inequality.

33. $x + 2y > 5$
34. $4x - y \le 8$
35. $3x - 2y < 6$
36. $\frac{1}{2}y \ge x + 4$

- write an equation to represent a relation, given a chart of values. **(Lesson 9-7)**

Write an equation for the relation given in the chart below.

a	−2	−1	0	1	2
b	20	17	14	11	8

+1 +1 +1 +1 (a values)

−3 −3 −3 −3 (b values)

The equation is $b = 14 - 3a$.

Write an equation for the relation given in each chart.

37.

x	0	1	2	3	4
y	5	8	11	14	17

38.

x	2	4	5	7	10
y	−2	0	1	3	6

APPLICATIONS AND CONNECTIONS

39. **Physics** A ball is dropped from a 324-foot tower. Its height h, in feet, after t seconds is given by the formula $h = 324 - 16t^2$. Make a table of values for this equation to determine how many seconds it takes for the ball to hit the ground.

40. **Sales** At Recordville, Emilio earns a weekly salary of \$150 plus \$0.30 for each record over 100 that he sells each week. If Emilio sells r records in a week, then his total weekly salary is $C(r) = 150 + 0.30(r - 100)$ for $r > 100$. If he earned \$225 last week, how many records did he sell?

41. **Statistics** Eric's scores on the first three of five 100-point tests were 83, 72, and 73. Eric wants to have an average of at least 75 on all the tests.
    a. Use a graph to determine what scores Eric can receive on the last two tests.
    b. List three possible solutions where both of his last two scores are no more than 75.

42. **Car Rental** The cost of a one-day car rental from A-1 Car Rental is \$41 if you drive 100 miles, \$51.80 if you drive 160 miles, and \$63.50 if you drive 225 miles.
    a. Write an equation to describe this relationship.
    b. Use the equation to determine the per-mile charge.

CHAPTER 9 TEST

1. Graph the points $A(-4, -1)$ and $B(5, -9)$.
2. Name the quadrant in which the point $C(-3, 5)$ is located.

Express each relation as a set of ordered pairs. Then state the domain, range, and inverse.

3. 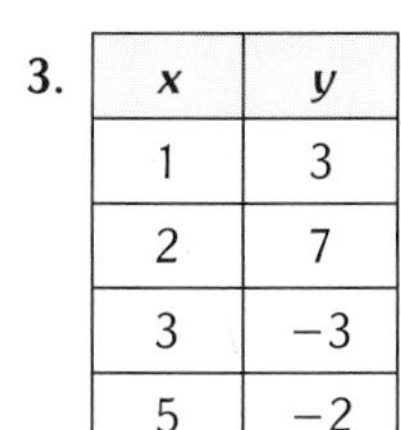

x	y
1	3
2	7
3	−3
5	−2

4. 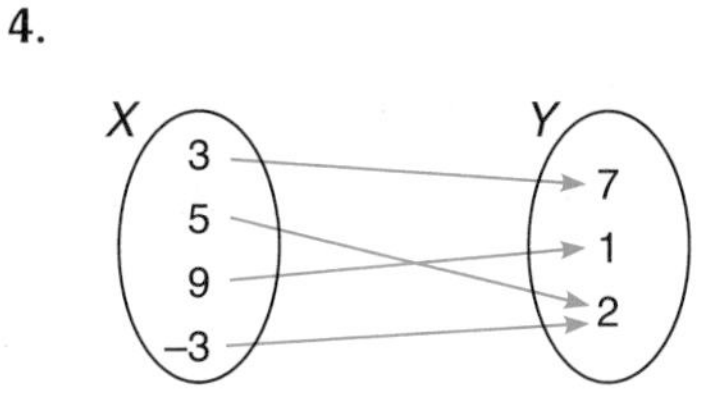

5. 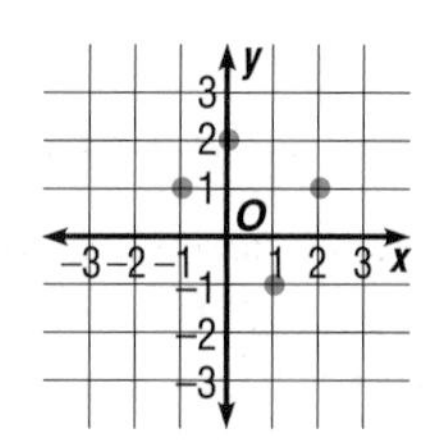

Solve each equation if the domain is {−2, −1, 0, 1, 3}.

6. $y = 3x + 10$
7. $2x - 5y = 4$
8. $2y - 8 = x$

Determine whether each equation is a linear equation.

9. $5x = 17 - 4x$
10. $y = x^2 - 4$
11. $y = xy + 1$

Determine whether each relation is a function.

12. $\{(2, 4), (3, 2), (4, 6), (5, 4)\}$
13. $8y = 7 + 3x$
14. $2x = 9$
15. Given $f(x) = 2x - 3$, determine the value of $f(-3)$, $f(7)$, and $f(0)$.

Graph each equation or inequality.

16. $x - 2y = 8$
17. $4x + 3y = 12$
18. $3x - 2y < 6$
19. $y \geq 5x + 1$
20. $4x + 2y > 9$
21. $5x - 2y = 8$

Write an equation for the relation given in each chart.

22.

x	1	2	3	4	7
y	3	8	13	18	33

23.

x	1	3	5	7	9	11	13
y	5	17	29	41	53	65	77

Solve.

24. **Physics** If a ball is thrown upward with an initial velocity of 72 meters per second, its height $h(t)$, in meters, after t seconds is given by the formula $h(t) = 72t - 4.9t^2$. Will the ball have hit the ground after 15 seconds?

25. **Sales** When you use Jay's Taxi Service, a two-mile trip costs \$6.30, a five-mile trip costs \$11.25, and a ten-mile trip costs \$19.50. Write an equation to describe this relationship and use it to find the cost of a one-mile trip.

Bonus If $f(a) = (2a + 3)^2$ and $g(a) = \sqrt{a} - 5$, for what value(s) of a does $g[f(a)] = 8$?

CHAPTER 9 College Entrance Exam Preview

The questions on these pages involve comparing two quantities, one in Column A and one in Column B. In certain questions, information related to one or both quantities is centered above them. All variables used stand for real numbers.

Directions:
Write A if the quantity in Column A is greater. Write B if the quantity in Column B is greater. Write C if the quantities are equal. Write D if there is not enough information to determine the relationship.

	Column A	Column B
	$x \geq 13$	
1.	$\frac{37}{3}$	x
2.	$\frac{0.4}{20}$	0.002
3.	$10 + 5 \div 4 - 3$	$10 - 5 \div 4 + 3$
	$b > c + 1$	
4.	b	c
	$b < c + 1$	
5.	b	c
	$r < 0 < s$	
6.	r^2	$\frac{s}{2}$
7.	a	$\lvert a \rvert$

	Column A	Column B
	$a = b$	
8.	$-5(a - b)$	$7(3b - 3a)$
9.	$\frac{3}{4}(5 + 1)\left(\frac{6 - 6}{2}\right)$	$\frac{1}{6}(1 + 3)(12 \div 2)$
10.	25% of $\frac{4}{7}$	$\frac{4}{7}$ of $\frac{1}{4}$
11.	$0.02 \div 0.2$	$0.2 \div 0.02$
12.	The value of $2y + 7$ when $y = -2$.	The value of $2x - 8$ when $x = 6$.
13.	The number indicated by A on the number line above.	The number indicated by B on the number line above.
14.	the original price of the CD player	the new price of the CD player
15.	ad	bc
16.	x	y
17.	The number of which 6 is 20%.	10% of 310

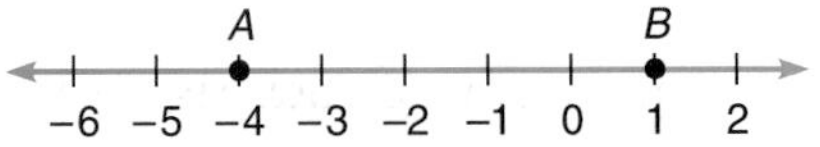

Item 13 number line (above the item 13 columns): points A at -4 and B at 1; ticks labeled $-6, -5, -4, -3, -2, -1, 0, 1, 2$.

Item 14 statement: The price of a CD player increased by 20% and then decreased by 20%.

Item 15 given: $a:b = c:d$

Item 16 given: $3x - 3y = 24$

Column A	Column B

18. Point P with coordinates (x, y) is in the second quadrant.

Column A	Column B
x	y

19.

Column A	Column B
the greatest prime factor of 1540	the greatest prime factor of 1530

20.

Column A	Column B
the number of prime numbers between 1 and 10	the number of prime numbers between 10 and 20

21. a, b, and c are consecutive odd integers.

Column A	Column B
$a + c$	$2b$

22. The perimeter of each figure is P.

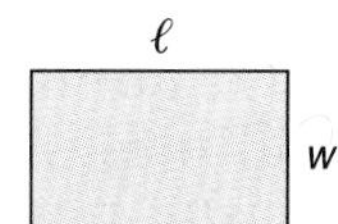

s

Column A	Column B
the area of the rectangle	the area of the square

23. If $y = 2$, then $x = 8$.

Column A	Column B
The value of x when $y = 5$ if y varies directly as x.	The value of x when $y = 5$ if y varies inversely as x.

24. A portion of \$10,000 was invested at 6% interest and the balance at 8% interest. The total interest was \$700.

Column A	Column B
the amount invested at 6%	the amount invested at 8%

TEST TAKING TIP

Treat the two expressions given as the two sides of an inequality. Add, subtract, multiply and divide by the same terms until you can more easily compare the two columns. Do not change the direction of the inequality by multiplying or dividing by negative numbers.

Substitute values for the unknown or unknowns. Be sure to use many types and combinations of numbers. Use positive and negative numbers, fractions and mixed numbers. Do not make assumptions.

Perform any indicated mathematical operations. Change the common information.

If the use of As and Bs in both the column names and answer choices is confusing, change the column names to another pair of letters or numbers, such as x and y or I and II.

Column A	Column B

$x < 0$

	Column A	Column B
25.	$x - 5$	$5 - x$
26.	a	a^2

$f(x) = x^2 + 2x - 5$

	Column A	Column B
27.	$f(3)$	$f(-3)$

CHAPTER OBJECTIVES

In this chapter, you will:

- Write an equation of a line given the coordinates of two points on the line.
- Determine the slope, *x*-intercept, and *y*-intercept of a line.
- Graph linear equations using the slope and *y*-intercept.
- Write an equation of a line that is parallel to or perpendicular to a given line.
- Represent data by using graphs.

Chapter 10 Graphing Linear Equations

Does the word *slopes* bring to mind pictures of skiers going down steep snow-covered hills? If so, then you have a pretty good idea of the mathematical meaning of the term *slope.*

APPLICATION IN SKIING

When we talk about the *slope* of a hillside, we are referring to the steepness of that hillside. Similarly, when we talk about *slope* in mathematics, we are usually referring to the steepness of a line drawn in the coordinate plane. In each case, *slope* is used to describe steepness.

Imagine you are a first-time skier facing your first run down the beginner's ski slope or "bunny hill." You certainly do not want this hill to be too steep for your level of expertise. That is, you hope the hill rises very little vertically compared to the horizontal distance, or run, from the top to the bottom of the hill. Perhaps in a few days, after you have had some lessons and practice, you might find the beginner's slope is too slow. Then you could move over to the intermediate slope where the hill rises a greater vertical distance compared to its horizontal run. In other words, you now are skiing on a steeper slope.

ALGEBRA IN ACTION

In the graph below, lines are used to represent the steepness of the beginner, intermediate, and expert slopes at a ski resort. Notice from the graph that the beginner slope rises 2 feet for every 10 horizontal feet. How many feet do the intermediate slope and the expert slope rise for every 10 horizontal feet?

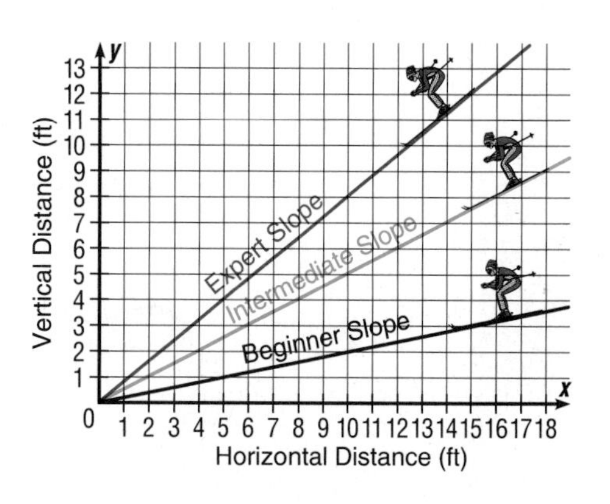

10-1 Slope of a Line

Objective

After studying this lesson, you should be able to:

- find the slope of a line, given the coordinates of two points on the line.

Application

The steepest streets in the United States are in San Francisco with grades of 31.5% or a rise of 3.15 feet for every 10 feet.

Do you ever recall seeing a sign along the highway like the one at the right? These signs are designed to inform the driver that there is a steep hill ahead. If a hill has a *grade* of 6%, this means that for every 100 feet of horizontal change, there is a vertical change of 6 feet.

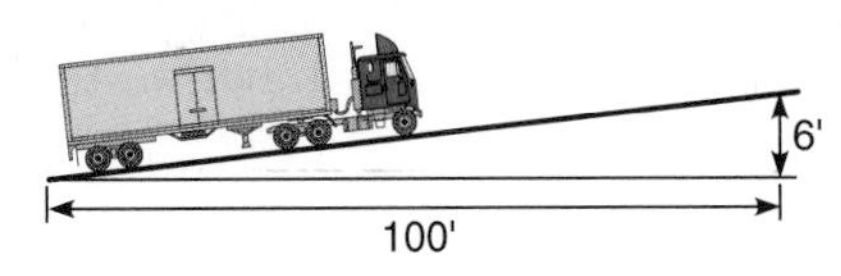

$$\text{grade} = \frac{\text{vertical change}}{\text{horizontal change}} = \frac{6}{100} \text{ or } 6\%$$

Example 1

How many feet does a road with a 6% grade drop in 3 miles?

Using 0.06 for 6%, the grade can be expressed as follows.

$0.06 = \frac{v}{h}$ *v represents the vertical change. h represents the horizontal change.*

$v = 0.06h$

$v = 0.06(3)$ *The horizontal change is 3 miles.*

$v = 0.18$ *The vertical change is 0.18 miles.*

Since 1 mile equals 5280 feet, 0.18 miles equals 0.18(5280) or 950.4 feet. Thus, a road with a 6% grade drops 950.4 feet in 3 miles.

Sometimes the vertical distance is referred to as the *rise,* and the horizontal distance is referred to as the *run.* The ratio of *rise to run* is called **slope.** The slope of a line describes its steepness, or rate of change.

On the graph below, the line passes through the origin, (0, 0), and (4, 3). The change in y or rise is 3, while the change in x or run is 4. Therefore, the slope of this line is $\frac{3}{4}$.

$$\text{slope} = \frac{\text{rise}}{\text{run}} = \frac{\text{change in } y}{\text{change in } x}$$

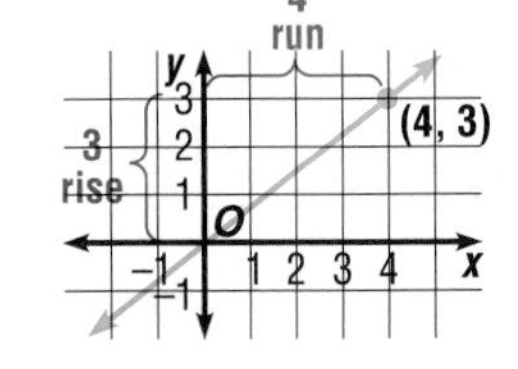

Definition of Slope	**The slope m of a line is the ratio of the change in y to the corresponding change in x.** $$\text{slope} = \frac{\text{change in } y}{\text{change in } x} \quad \text{or} \quad m = \frac{\text{change in } y}{\text{change in } x}$$

Example 2

Determine the slope of each line.

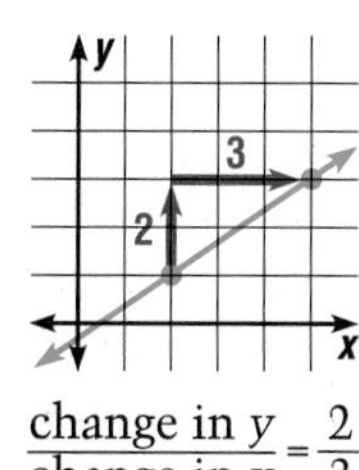

$\frac{\text{change in } y}{\text{change in } x} = \frac{2}{3}$

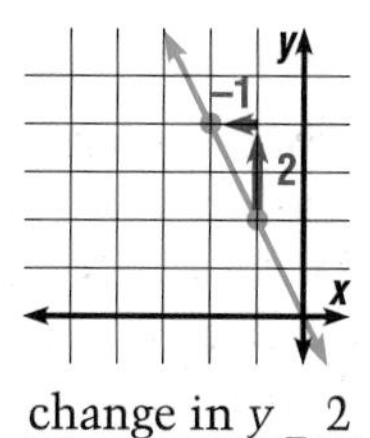

$\frac{\text{change in } y}{\text{change in } x} = \frac{2}{-1}$
$= -2$

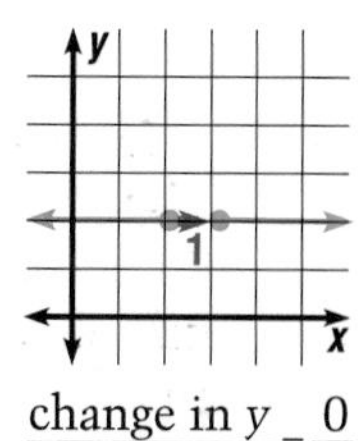

$\frac{\text{change in } y}{\text{change in } x} = \frac{0}{1}$
$= 0$

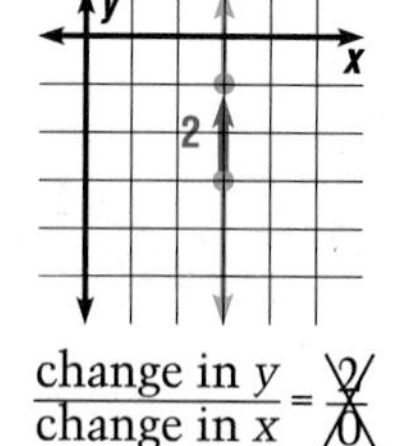

$\frac{\text{change in } y}{\text{change in } x} = \frac{2}{0}$ (crossed out)
The slope is undefined.

Notice that in Example 2, the line extending from lower left to upper right has a positive slope. The line extending from upper left to lower right has a negative slope. The slope of the horizontal line is zero. For a vertical line, the change in x would be zero. Since division by zero is not defined, the slope of a vertical line is undefined. We say that a vertical line *has no slope.*

Example 3

Determine the slope of the line containing the points with the coordinates listed in the table.

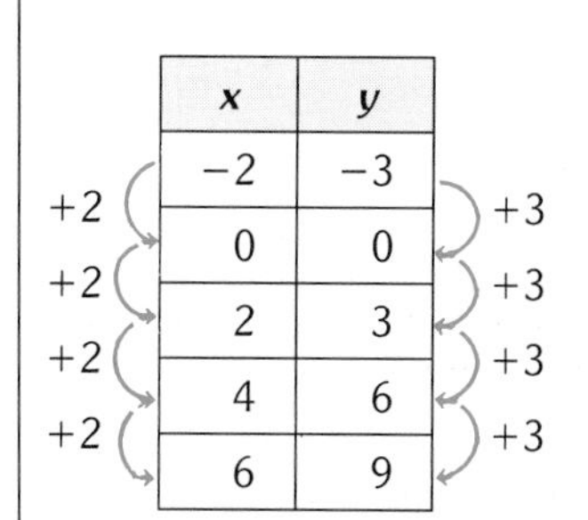

x	y
−2	−3
0	0
2	3
4	6
6	9

Notice that y increases 3 units for each 2 units that x increases.

$$\text{slope} = \frac{\text{change in } y}{\text{change in } x} = \frac{3}{2}$$

These examples suggest that the slope of a nonvertical line can be determined from the coordinates of any two points on the line.

Determining Slope Given Two Points	**Given the coordinates of two points on a line, (x_1, y_1) and (x_2, y_2), the slope, m, can be found as follows.** $$m = \frac{y_2 - y_1}{x_2 - x_1}, \text{ where } x_2 \neq x_1.$$

y_2 is read "y sub 2." The 2 is called a subscript.

Example 4 **Determine the slope of the line passing through (3, −9) and (4, −12).**

$$m = \frac{y_2 - y_1}{x_2 - x_1}$$

$$= \frac{-12 - (-9)}{4 - 3} \quad (x_1, y_1) = (3, -9) \text{ and } (x_2, y_2) = (4, -12)$$

$$= \frac{-3}{1}$$

$$= -3$$

The slope is −3.

In Example 4, the difference of the y-coordinates was expressed as $-12 - (-9)$. Suppose $-9 - (-12)$ had been used as the change in y-coordinates and $3 - 4$ had been used as the change in x-coordinates. Since $\frac{-9 - (-12)}{3 - 4}$ is also equal to −3, it does not matter which point is chosen to be (x_1, y_1). However, the coordinates of both points must be used in the same order.

Example 5 **Determine the value of r so the line through $(r, 4)$ and $(9, -2)$ has a slope of $-\frac{3}{2}$.**

$$m = \frac{y_2 - y_1}{x_2 - x_1}$$

$$-\frac{3}{2} = \frac{-2 - 4}{9 - r} \quad \textit{Replace each variable with its appropriate value.}$$

$$-\frac{3}{2} = \frac{-6}{9 - r}$$

$$-3(9 - r) = -6(2) \quad \textit{Means-extremes property}$$

$$-27 + 3r = -12 \quad \textit{Solve for r.}$$

$$3r = 15$$

$$r = 5$$

CHECKING FOR UNDERSTANDING

Communicating Mathematics

Read and study the lesson to answer each question.

1. What is meant by a road sign stating that the grade of the road is 5%?
2. Explain the meaning of the slope of a line.
3. Draw a graph of a line having each of the following.
   a. a positive slope
   b. a negative slope
   c. zero slope
   d. no slope

Guided Practice

For each table, state the change in y and the change in x. Then determine the slope of the line passing through the points with the coordinates listed.

4.

x	y
0	0
1	1
2	2
3	3
4	4

5.

x	y
−2	2
−1	1
0	0
1	−1
2	−2

6.

x	y
−2	−8
−1	−4
0	0
1	4
2	8

7.

x	y
−6	8
−3	4
0	0
3	−4
6	−8

Draw a line through the given point that has the given slope.

8. $(2, 4), m = \frac{1}{3}$ **9.** $(4, -1), m = -\frac{2}{5}$ **10.** $(-3, 4), m = 4$

11. $(-1, 5), m = -2$ **12.** $(-3, -3), m = \frac{4}{3}$ **13.** $(2, -4), m = 0$

EXERCISES

Practice

Determine the slope of each line named below.

14. a

15. b

16. c

17. d

18. e

19. f

20. g

21. h

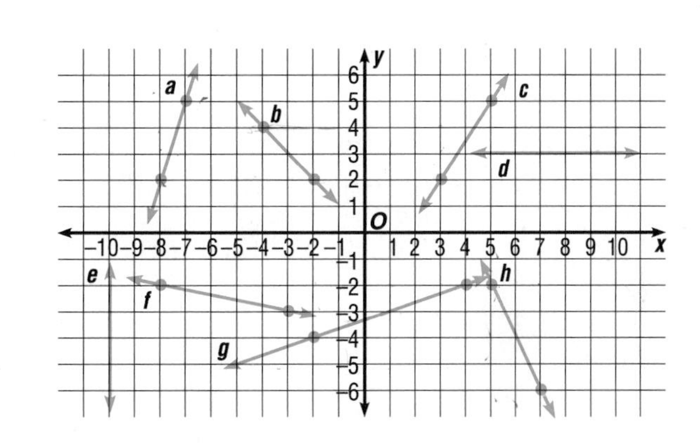

Determine the slope of the line passing through each pair of points.

22. $(3, 4), (4, 6)$ **23.** $(-3, 6), (-5, 9)$

24. $(-1, 11), (-5, 4)$ **25.** $(7, -4), (9, -1)$

26. $(18, -4), (6, -10)$ **27.** $(14, 3), (-11, 3)$

28. $(-4, -6), (-3, -8)$ **29.** $(0, 0), (0.5, 0.25)$

30. $\left(\frac{3}{4}, 1\right), \left(\frac{3}{4}, -1\right)$ **31.** $\left(3\frac{1}{2}, 5\frac{1}{4}\right), \left(2\frac{1}{2}, 6\right)$

Determine the value of r so the line passing through each pair of points has the given slope.

32. $(9, r), (6, 3), m = -\frac{1}{3}$

33. $(r, 4), (7, 3), m = \frac{3}{4}$

34. $(4, -7), (-2, r), m = \frac{8}{3}$

35. $(6, -2), (r, -6), m = -4$

36. $(r, 7), (11, r), m = -\frac{1}{5}$

37. $(4, r), (r, 2), m = -\frac{5}{3}$

38. $(9, r), (6, 2), m = r$

39. $(8, r^2), (3, -6), m = r$

Critical Thinking

40. Look at lines a, c, and g on the previous page. These lines have positive slopes. What is the general direction of lines with positive slopes?

41. Look at lines b, f, and h on the previous page. These lines have negative slopes. What is the general direction of lines with negative slopes?

Applications

42. Construction A ramp installed to give handicapped people access to a building has a 3-foot rise and a 36-foot run. What is the slope of the ramp?

43. Driving The eastern entrance of the Eisenhower Tunnel in Colorado is at an elevation of 11,080 feet. The tunnel is 8941 feet long and has an upgrade of 0.895% toward the western end. What is the elevation of the western end of the tunnel?

44. Carpentry When building a stairway, a carpenter considers the ratio of riser to tread. Write a ratio to describe the steepness of the stairs.

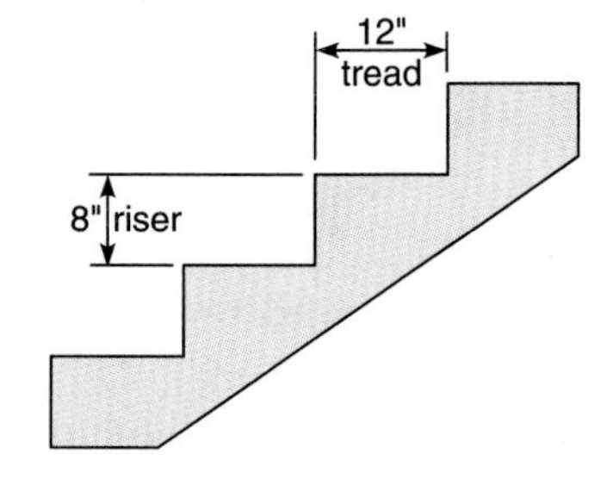

45. Aviation An airplane passing over Albuquerque at an elevation of 33,000 feet begins its descent to land at Santa Fe, 50 miles away. If the elevation of Santa Fe is 7000 feet, what should be the approximate slope of descent, expressed as a percent?

Mixed Review

46. Name the property illustrated by $8 + (12 + 34) = (8 + 12) + 34$. **(Lesson 1-6)**

47. Find the degree of $2xy^2z + 5xyz^5 + x^4$. **(Lesson 6-5)**

48. Factor $5a^3 + 3a^2b - 5ab^2 - 3b^3$. **(Lesson 7-3)**

49. Find $\frac{6}{x} - \frac{5}{x^2}$. **(Lesson 8-7)**

50. Complete: 1, 6, 11, __?__, __?__, 26, __?__. **(Lesson 9-7)**

10-2 Point-Slope and Standard Forms of Linear Equations

Objectives

After studying this lesson, you should be able to:

- write a linear equation in standard form given the coordinates of a point on the line and the slope of the line, and
- write a linear equation in standard form given the coordinates of two points on the line.

Application

> **FYI . . .**
>
> The book *Men of Good Will,* first published in 1933, consists of 4949 pages and is estimated to have 2,070,000 words.

Seth is reading a book for a book report. He decides to avoid a last-minute rush by reading 2 chapters each day. A graph representing his plan is shown at the right. By the end of the first day, Seth should have read 2 chapters, so one point on the graph has coordinates (1, 2). Since he plans to read 2 chapters in 1 day, the slope is $\frac{2}{1}$ or 2.

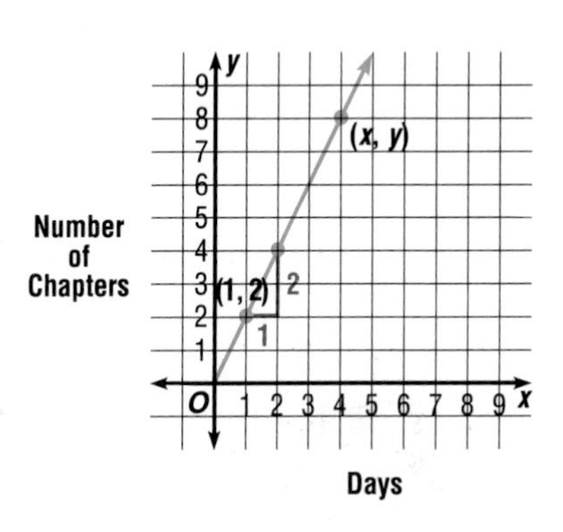

Let (x, y) represent the coordinates of any other point on the line. The equation for the slope of a line can be used to find an equation that describes Seth's plan. Replace (x_1, y_1) with (1, 2) and (x_2, y_2) with any other point (x, y). We know that the slope is 2, so replace m with 2.

$$\frac{y_2 - y_1}{x_2 - x_1} = m$$

$$\frac{y - 2}{x - 1} = 2 \qquad \text{Replace } y_2 \text{ with } y,\ y_1 \text{ with } 2,\ x_2 \text{ with } x,\ x_1 \text{ with } 1, \text{ and } m \text{ with } 2.$$

$$y - 2 = 2(x - 1) \qquad \text{Multiply each side by } (x - 1).$$

This linear equation is said to be in **point-slope form.**

Point-Slope Form

For a given point (x_1, y_1) on a nonvertical line with slope m, the point-slope form of a linear equation is as follows.

$$y - y_1 = m(x - x_1)$$

In general, you can write an equation in point-slope form for the graph of any nonvertical line. If you know the slope of a line and the coordinates of one point on the line, you can write an equation of the line.

Example 1

Write the point-slope form of an equation of the line passing through (2, −4) and having a slope of $\frac{2}{3}$.

$$y - y_1 = m(x - x_1)$$

$$y - (-4) = \frac{2}{3}(x - 2) \quad \textit{Replace } x_1 \textit{ with 2, } y_1 \textit{ with } -4\textit{, and } m \textit{ with } \frac{2}{3}.$$

$$y + 4 = \frac{2}{3}(x - 2)$$

An equation of the line is $y + 4 = \frac{2}{3}(x - 2)$.

Any linear equation can be expressed in the form $Ax + By = C$ where A, B, and C are integers and A and B are not both zero. This is called the **standard form.** An equation that is written in point-slope form can be written in standard form.

Example 2

Write $y + 4 = \frac{3}{4}(x - 2)$ in standard form.

$$y + 4 = \frac{3}{4}(x - 2)$$

$$4(y + 4) = 3(x - 2) \quad \textit{Multiply each side by 4 to eliminate the fraction.}$$

$$4y + 16 = 3x - 6 \quad \textit{Distributive property}$$

$$-3x + 4y = -22$$

$$3x - 4y = 22 \quad \textit{Multiply each side by } -1 \textit{ to get a positive coefficient for x.}$$

$3x - 4y = 22$ is in standard form.

You can also find an equation of a line if you know the coordinates of two points on the line.

Example 3

Alejandra is selling tickets to the music program at school. Mr. Foster says that he has enough money to buy 9 student tickets and 2 adult tickets or 3 student tickets and 6 adult tickets. Write an equation in standard form that represents the cost of each ticket and the amount of money that Mr. Foster has.

EXPLORE Let x = the number of student tickets and let y = the number of adult tickets. In $Ax + By = C$, A is the price in dollars of a student ticket, B is the price in dollars of an adult ticket, and C is the amount of money Mr. Foster has. The two points (9, 2) and (3, 6) lie on the graph of this equation.

PLAN Use (9, 2) and (3, 6) to find the slope. Then use the point-slope form to write the equation in standard form.

SOLVE

$$m = \frac{y_2 - y_1}{x_2 - x_1}$$

$$= \frac{6 - 2}{3 - 9}$$ *Replace y_2 with 6, y_1 with 2, x_2 with 3, and x_1 with 9.*

$$= \frac{4}{-6} \text{ or } -\frac{2}{3}$$

Now use the point-slope form. The slope is $-\frac{2}{3}$. Either (9, 2) or (3, 6) can be substituted for (x_1, y_1).

$y - y_1 = m(x - x_1)$ *Let $(x_1, y_1) = (9, 2)$.*

$y - 2 = -\frac{2}{3}(x - 9)$ *Point-slope form*

$3(y - 2) = -2(x - 9)$ *Multiply by 3 to eliminate the fraction.*

$3y - 6 = -2x + 18$

$2x + 3y = 24$ *This equation is in standard form.*

The equation in standard form is $2x + 3y = 24$. So, Mr. Foster has \$24, if the student tickets are \$2 each and the adult tickets are \$3 each.

EXAMINE Check the solution by replacing (x, y) with (9, 2) and (3, 6).

$2x + 3y = 24$	$2x + 3y = 24$
$2(9) + 3(2) \stackrel{?}{=} 24$	$2(3) + 3(6) \stackrel{?}{=} 24$
$18 + 6 \stackrel{?}{=} 24$	$6 + 18 \stackrel{?}{=} 24$
$24 = 24$ ✓	$24 = 24$ ✓

The solution checks.

A horizontal line has a slope of zero. The point-slope form can be used to write equations of horizontal lines. For example, the equation of a line through (5, 2) and (−9, 2) is $y - 2 = 0(x - 5)$ or $y = 2$.

A vertical line has no slope. Therefore, the point-slope form cannot be used for vertical lines. The equation of a line through (3, 5) and (3, −9) is $x = 3$ since the x-coordinate of every point on the line is equal to 3.

CHECKING FOR UNDERSTANDING

Communicating Mathematics

Read and study the lesson to answer each question.

1. In the point-slope form of a linear equation, what do x_1 and y_1 represent?
2. What is the standard form of a linear equation?
3. Describe the graph of the equation $y = c$, where c is any number.
4. Describe the graph of the equation $x = c$, where c is any number.

Guided Practice

State the slope and a point through which the line for each linear equation passes.

5. $y - 2 = 3(x - 5)$
6. $y - (-5) = -2(x + 1)$
7. $y + 6 = -\frac{3}{2}(x + 5)$
8. $2(x - 3) = y + \frac{3}{2}$
9. $y = 3$
10. $x = 1$

Write each equation in standard form.

11. $y - 3 = 2\left(x + \frac{3}{2}\right)$
12. $y + 5 = -3\left(x - \frac{1}{3}\right)$
13. $y + 1 = \frac{2}{3}(x + 2)$
14. $y + \frac{3}{2} = \frac{1}{2}(x + 4)$

EXERCISES

Practice

Write the standard form of an equation of the line passing through the given point and having the given slope.

15. $(5, 4), -\frac{2}{3}$
16. $(-6, -3), -\frac{1}{2}$
17. $(9, 1), \frac{2}{3}$
18. $(4, -3), 2$
19. $(-2, 4), -3$
20. $(6, -2), \frac{4}{3}$
21. $(-2, 6), 0$
22. $(1, 3)$, none

Write the standard form of an equation of the line passing through each pair of points.

23. $(5, 4), (6, 3)$
24. $(6, 1), (7, -4)$
25. $(4, -2), (4, 8)$
26. $(4, -2), (8, -3)$
27. $(-6, 1), (-8, 2)$
28. $(5, 3), (-6, 3)$
29. $(-5, 1), (6, -2)$
30. $(-8, 2), (-1, -2)$
31. $(0.75, 1), (2, 0.5)$
32. $(-8, 0.5), (9, 0.5)$
33. $\left(2\frac{1}{2}, \frac{1}{3}\right), \left(\frac{3}{4}, 1\frac{1}{2}\right)$
34. $(-2, 7), \left(-2, \frac{16}{3}\right)$

Critical Thinking

35. A line contains the points (9, 1) and (5, 5). Write a convincing argument that the same line intersects the x-axis at (10, 0).

Applications

36. **Shipping** The number of books shipped in one box is limited by the weight of the books. Six algebra books and 18 geometry books can be shipped together. Eleven algebra and 14 geometry books can also be shipped together. Write an equation in standard form that relates the number of algebra and geometry books that can be shipped in one box.

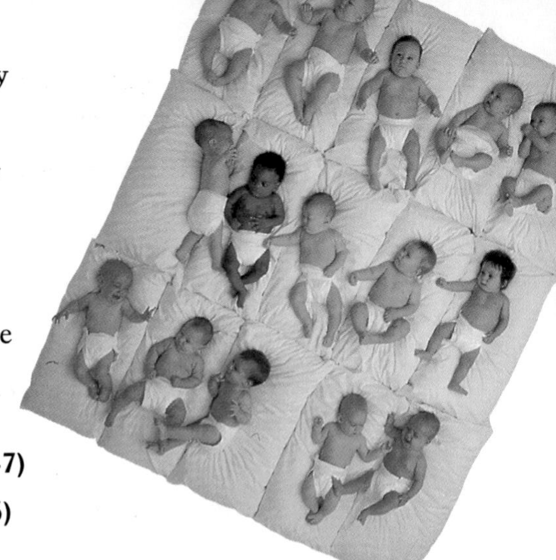

37. **Work** Kyung earned some money by baby-sitting during spring vacation. He decides to allow himself to spend \$5 of the money each week. After 3 weeks, he has \$75 left. Write an equation in standard form that relates the money left with the number of weeks that have passed. (*Hint:* The slope is -5.)

Mixed Review

38. Multiply $7a(8a + 11)$. **(Lesson 6-7)**
39. Factor $x^2 - 9x + 14$. **(Lesson 7-5)**
40. Find $\frac{y^2 + 3y^3}{y^2 - 4} \cdot \frac{2y + y^2}{y + 4y^2 + 3y^3}$. **(Lesson 8-2)**
41. Name the quadrant for the graph of (x, y), given that $x > 0$ and $y > 0$. **(Lesson 9-1)**
42. **National Landmarks** At the Royal Gorge in Colorado, an inclined railway takes visitors down to the Arkansas River. If the grade is 100% and the vertical drop is 1015 feet, what is the horizontal change of the railway? **(Lesson 10-1)**

Journal

Explain how the steepness of a line on a graph is related to the slope of the equation of the line. Include examples in your explanation.

HISTORY CONNECTION

Benjamin Banneker

Benjamin Banneker (1731–1806), a free African-American man, was a self-taught genius in the areas of mathematics, astronomy, and surveying. He was often referred to as the "sable (Black) genius." At the age of 20, Banneker obtained a book on geometry and, armed with a compass and ruler, designed and built the first clock ever built in the United States. It kept perfect time for 40 years. Banneker corresponded often with Thomas Jefferson, and it was on Jefferson's recommendation that President Washington appointed Banneker as the assistant surveyor on the team responsible for designing and building the new capital at Washington, D.C. Long before the project was completed, Pierre L'Enfant, the French engineer in charge of the project, resigned and returned to France with all of the plans and maps. But having worked closely with L'Enfant, Banneker was able to reproduce all of the plans from memory in only two days. The nation's capital stands today as a monument to Banneker's genius.

10-3 Slope-Intercept Form of Linear Equations

Objectives

After studying this lesson, you should be able to:

- write an equation in slope-intercept form given the slope and y-intercept,
- determine the slope and y-intercept of a graph, and
- determine the x- and y-intercepts of a graph.

Application

The average American consumes about 73.4 pounds of beef and 62.7 pounds of chicken each year.

Larry is taking home economics. As a class project, he is cooking dinner for some friends. Larry plans to spend \$15 on beef and/or chicken. If beef costs \$5 per pound and chicken costs \$3 per pound, the graph at the right shows all the possible combinations of beef and chicken he can buy. If he buys no beef, he can buy 5 pounds of chicken. If he buys no chicken, he can buy 3 pounds of beef. The number 5 on the x-axis and the number 3 on the y-axis are called the **intercepts.**

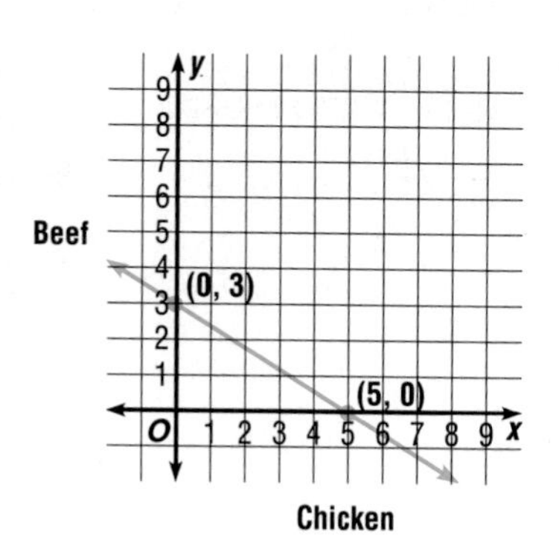

The x-coordinate of the point where a line crosses the x-axis is called the **x-intercept** of the line. The line graphed on the coordinate plane above crosses the x-axis at (5, 0). Therefore the x-intercept is 5. Note that the corresponding y-coordinate is 0. Similarly, the y-coordinate of the point where the line crosses the y-axis is called the **y-intercept** of the line. The line above crosses the y-axis at (0, 3). Therefore, the y-intercept is 3. Note that the corresponding x-coordinate is 0.

Consider the graph at the right. The line with slope m crosses the y-axis at $(0, b)$. You can write an equation for this line using the point-slope form. Let $(x_1, y_1) = (0, b)$.

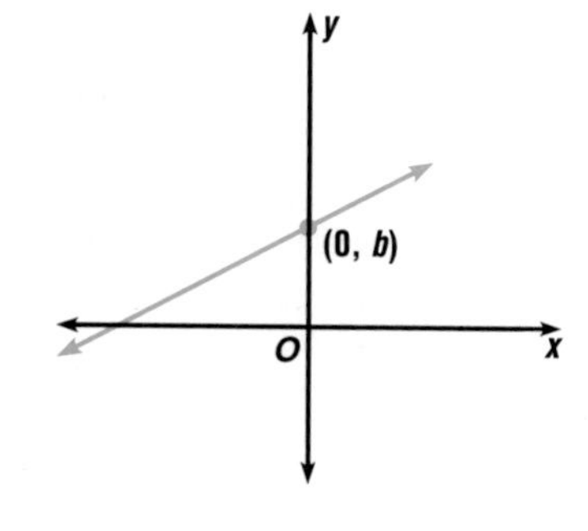

$y - y_1 = m(x - x_1)$ *Point-slope form*

$y - b = m(x - 0)$ *Replace y_1 with b. Replace x_1 with 0.*

$y = mx + b$

(↑ m: *slope*; ↑ b: *y-intercept*)

Slope-Intercept Form	Given the y-intercept b and slope m of a line, the slope-intercept form of an equation of the line is $y = mx + b$.

Example 1

State the slope and y-intercept of the graph of $y = 5x - 3$.

$$y = 5x + (-3)$$
$$\downarrow \qquad \downarrow$$
$$y = mx + \quad b$$

Since $m = 5$, the slope is 5. Since b is -3, the y-intercept is -3.

Example 2

Mini and her family are taking a trip to see her grandmother. After traveling a while, she begins to keep track of their time and mileage. Let d represent the total distance traveled in miles and let t represent the time in hours. Suppose $d = 50t + 83$ represents the total distance traveled.

a. How far had Mini's family traveled before she began keeping track of their mileage?

b. What is their average rate of speed?

a. The distance traveled would be the value of d when $t = 0$.

$d = 50t + 83$

$d = 50(0) + 83$ *Replace t with 0.*

$d = 83$ They had traveled 83 miles.

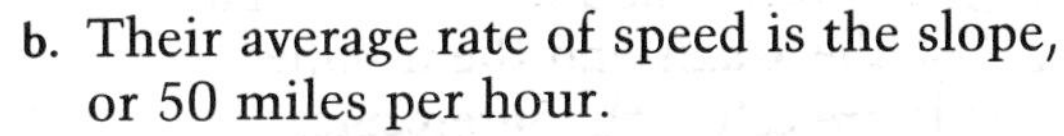

b. Their average rate of speed is the slope, or 50 miles per hour.

Lab Activity

You can learn how to use a graphing calculator to investigate the slope of a line in Lab 11 on page A18.

Compare the slope-intercept form of a linear equation with the standard form, $Ax + By = C$. Solve for y.

$Ax + By = C$

$By = -Ax + C$ *Assume $B \neq 0$.*

$y = -\frac{Ax}{B} + \frac{C}{B}$ *This equation is in slope-intercept form.*

$\uparrow$ slope $\quad \uparrow$ y-intercept

If the equation is given in standard form and B is *not* zero, the slope of the line is $-\frac{A}{B}$ and the y-intercept is $\frac{C}{B}$.

Example 3

State the slope and y-intercept of the graph of $3x + 2y = 12$.

In the equation $3x + 2y = 12$, $A = 3$, $B = 2$, and $C = 12$.

slope: $-\frac{A}{B} = -\frac{3}{2}$ $\qquad$ y-intercept: $\frac{C}{B} = \frac{12}{2}$ or 6

Check: Write the equation in slope-intercept form.

$3x + 2y = 12$

$2y = -3x + 12$

$y = -\frac{3}{2}x + 6$

The slope is $-\frac{3}{2}$ and the y-intercept is 6. The solution checks.

Recall that the x-coordinate of the ordered pair for the y-intercept is 0 and the y-coordinate of the ordered pair for the x-intercept is 0. You can use these facts to find the x- and y-intercepts of the graph of a linear equation.

Example 4

Determine the x- and y-intercepts of the graph of $4x - 5y = 10$.

To find the x-intercept, let $y = 0$.

$$\begin{aligned} 4x - 5y &= 10 \\ 4x - 5(0) &= 10 \quad \textit{Let } y = 0. \\ 4x &= 10 \\ x &= \frac{5}{2} \end{aligned}$$

The x-intercept is $\frac{5}{2}$. The graph crosses the x-axis at $\left(\frac{5}{2}, 0\right)$.

To find the y-intercept, let $x = 0$.

$$\begin{aligned} 4x - 5y &= 10 \\ 4(0) - 5y &= 10 \quad \textit{Let } x = 0. \\ -5y &= 10 \\ y &= -2 \end{aligned}$$

The y-intercept is -2. The graph crosses the y-axis at $(0, -2)$.

The y-intercept of the graph of an equation in the form $Ax + By = C$ is $\frac{C}{B}$. To find a formula for the x-intercept, let $y = 0$.

$$\begin{aligned} Ax + By &= C \\ Ax + B(0) &= C \quad \textit{Let } y = 0. \\ Ax &= C \\ x &= \frac{C}{A} \end{aligned}$$

The x-intercept is $\frac{C}{A}$, $A \neq 0$. Therefore, for the graph of $2x - 3y = 8$, the x-intercept is $\frac{8}{2}$ or 4, and the y-intercept is $-\frac{8}{3}$.

CHECKING FOR UNDERSTANDING

Communicating Mathematics

Read and study the lesson to answer each question.

1. Does the graph of every linear equation have an x-intercept?
2. How can you find the x-intercept using the equation $y = mx + b$?
3. Write the equation of a line that has no slope.
4. Write the equation of a line that has no y-intercept.

Guided Practice

State the slope and y-intercept of the graph of each equation.

5. $y = 5x + 3$
6. $y = 3x - 7$
7. $y = \frac{1}{3}x$
8. $y = \frac{3}{5}x - \frac{1}{4}$
9. $2x + 3y = 5$
10. $-x + 4y = 3$
11. $y - 6x = 5$
12. $3y - 8x = 2$
13. $5y = -8x - 2$

Write an equation in slope-intercept form of the line with the given slope and y-intercept.

14. $m = 3, b = 1$
15. $m = -3, b = 5$
16. $m = 4, b = -2$
17. $m = \frac{1}{2}, b = 5$
18. $m = -3.1, b = 0.6$
19. $m = 0, b = 14$

EXERCISES

Practice

Determine the x- and y-intercepts of the graph of each equation.

20. $3x + 2y = 6$
21. $5x + y = 10$
22. $3x + 4y = 24$
23. $2x - 7y = 28$
24. $2x + 5y = -11$
25. $x = -2$
26. $\frac{3}{4}x - 2y = 7$
27. $3y = 12$
28. $1.8x - 2.5y = 5.4$

Determine the slope and y-intercept of the graph of each equation. Then write each equation in slope-intercept form.

29. $2x + 5y = 10$
30. $5x - y = 15$
31. $7x + 4y = 8$
32. $5x - 4y = 11$
33. $12x + 9y = 15$
34. $13x - 11y = 22$
35. $2x + \frac{1}{3}y = 5$
36. $3x - \frac{1}{4}y = 6$
37. $\frac{2}{3}x + \frac{1}{6}y = 2$
38. $3x = 2y - 7$
39. $5x = 8 - 2y$
40. $8y = 4x + 12$
41. $1.1x - 0.2y = 3.2$
42. $3(x - 7) = 2y + 5x + 8$
43. $y - 3x = 6(y + 7x) + 10$
44. $4y + x = 9 - 3(2y - 2x)$
45. $\frac{4}{5}(2x - y) = 6x + \frac{2}{5}y - 10$
46. $\frac{3}{2}(4x + 9y) = 4(7x - \frac{1}{2}y)$
47. Write an equation in slope-intercept form of the line with slope $\frac{4}{5}$ and y-intercept the same as the line whose equation is $3x + 4y = 6$.
48. Write an equation in slope-intercept form of the line with slope $-\frac{3}{5}$ and y-intercept the same as the line whose equation is $7x - 3y = 12$.

49. Write an equation in slope-intercept form of the line with y-intercept 12 and slope the same as the line whose equation is $2x - 5y - 10 = 0$.

50. Write an equation in slope-intercept form of the line with y-intercept -0.65 and slope the same as the line whose equation is $\frac{1}{2}x - \frac{3}{4}y = 6$.

Critical Thinking

51. Find the coordinates of a point on the graph of $3x - 4y = -20$, if the y-coordinate is twice the x-coordinate.

52. Find the coordinates of a point on the graph of $4x - y = -2$, if the y-coordinate is three times the x-coordinate.

53. Find the coordinates of a point on the graph of $x + 2y = 11$, if the y-coordinate is 5 less than the x-coordinate.

54. Find the coordinates of a point on the graph of $7x + 3y = 2$, if the y-coordinate is 4 more than the x-coordinate.

Applications

55. **Finance** To save for a new bicycle, Cesár begins a savings plan. Cesár's savings are described by the equation $s = 5w + 56$, where s represents his total savings in dollars and w represents the number of weeks since the start of the savings plan.

 a. How much money had Cesár already saved when the plan started?

 b. How much does Cesár save each week?

56. **Education** In order to "curve" a set of test scores, a teacher uses the equation $g = 2.5p + 10$, where g is the curved test score and p is the number of problems answered correctly.

 a. How many extra points does each student receive for the test?

 b. How many points is each problem worth?

Mixed Review

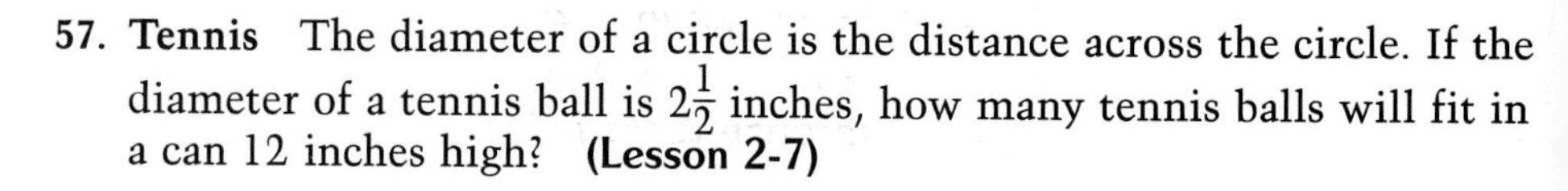

57. **Tennis** The diameter of a circle is the distance across the circle. If the diameter of a tennis ball is $2\frac{1}{2}$ inches, how many tennis balls will fit in a can 12 inches high? **(Lesson 2-7)**

58. Simplify $(3x^4)(-2x^4y^3)$. **(Lesson 6-2)**

59. Factor $m^4 + 12m^2n^2 + 36n^4$. **(Lesson 7-7)**

60. Find $\frac{y^2}{x + 2} \div \frac{y}{x + 2}$. **(Lesson 8-4)**

61. Given $g(x) = 2x - 1$, find $g\left(\frac{5}{2}\right)$. **(Lesson 9-5)**

62. Write the standard form of an equation of the line passing through (5, 7) and having slope 0. **(Lesson 10-2)**

10-4 Graphing Linear Equations

Objective

After studying this lesson, you should be able to:

- graph linear equations using the x- and y-intercepts or the slope and y-intercept.

Application

FYI ...

A crab tagged in the Red Sea walked the 101.5 miles to the Mediterranean Sea in 29 years, for an average speed of 3.5 miles per year!

Nicki is starting a training program to shape up for summer. She plans to walk and/or jog a total of 6 miles each day. Let x represent the distance she walks and let y represent the distance she jogs. Then each day $x + y$ must equal 6.

$$x + y = 6$$

For this equation, both the x-intercept and the y-intercept are 6. A simple method of graphing this equation is to graph (6, 0) and (0, 6). Then draw the line that passes through these points.

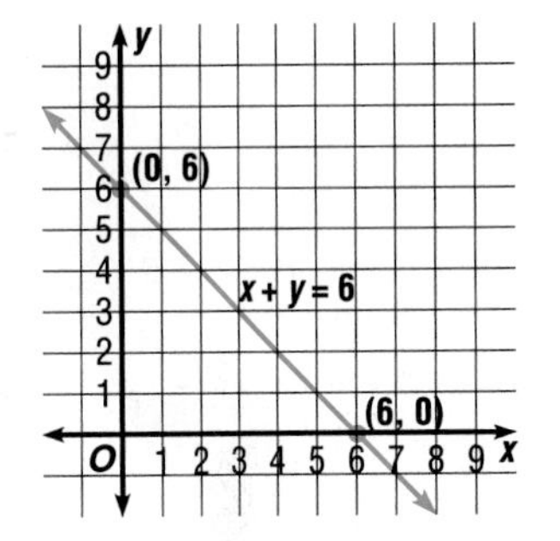

In terms of Nicki's training program, what does any point on the line represent?

Example 1

Graph $3x - 2y = 12$ using the x- and y-intercepts.

To find the x-intercept, let $y = 0$.

$$3x - 2y = 12$$
$$3x - 2(0) = 12$$
$$3x = 12$$
$$x = 4$$

To find the y-intercept, let $x = 0$.

$$3x - 2y = 12$$
$$3(0) - 2y = 12$$
$$-2y = 12$$
$$y = -6$$

Graph (4, 0) and (0, −6). Then draw the line that passes through these points.

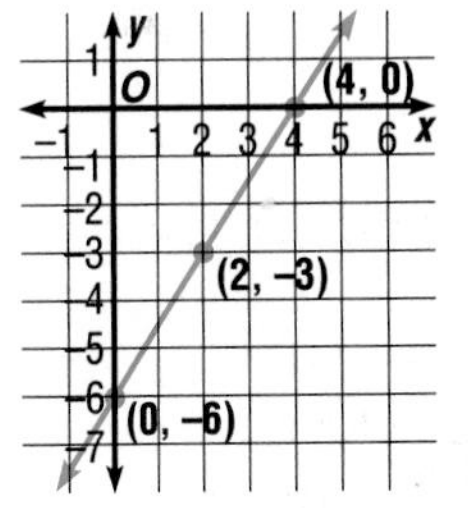

To check, choose some other point on the line and determine whether it is a solution of $3x - 2y = 12$. Try (2, −3).

$$3x - 2y = 12$$
$$3(2) - 2(-3) \stackrel{?}{=} 12$$
$$6 + 6 \stackrel{?}{=} 12$$
$$12 = 12 \checkmark$$

When $A = 0$, the graph is horizontal. When $B = 0$, the graph is vertical. When $C = 0$, the graph passes through the origin.

If an equation is in standard form and A, B, or C is zero, the graph of the equation has at most one intercept. In these cases, you will need to find some other ordered pairs that satisfy the equation. If the equation is in slope-intercept form, then it is convenient to use the slope and y-intercept to draw the graph of the equation.

Example 2

The cost of a long-distance call using a certain long-distance carrier is 50¢ plus 75¢ per minute. The equation $C = 0.75t + 0.50$, where t represents the time in minutes, represents the total cost of any call. Make a graph that can be used to find the cost of a call of any length.

The y-intercept is 0.50 or $\frac{1}{2}$ so the graph passes through $\left(0, \frac{1}{2}\right)$.

The slope is 0.75 or $\frac{3}{4}$. $\quad \frac{3}{4} = \frac{\text{change in y}}{\text{change in x}}$

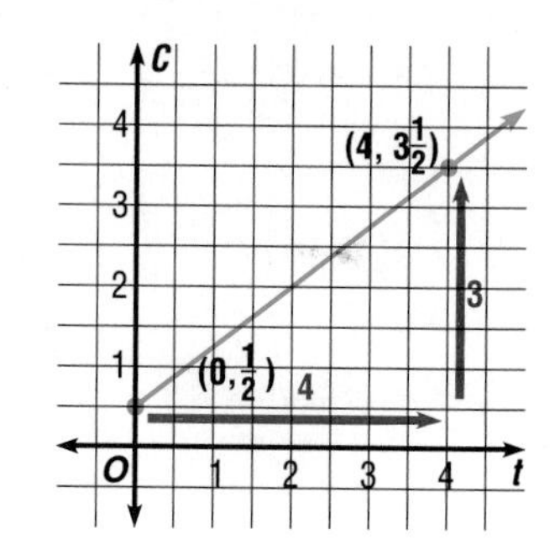

Starting at $\left(0, \frac{1}{2}\right)$, go to the right 4 units and up 3 units. This will be the point $\left(4, 3\frac{1}{2}\right)$. Then draw the line through $\left(0, \frac{1}{2}\right)$ and $\left(4, 3\frac{1}{2}\right)$. *What does the ordered pair $\left(4, 3\frac{1}{2}\right)$ represent in terms of a long-distance call?*

To check, substitute the points $\left(0, \frac{1}{2}\right)$ and $\left(4, 3\frac{1}{2}\right)$ or (0, 0.5) and (4, 3.5) in the equation $C = 0.75t + 0.50$ to determine whether they are solutions of the equation.

If the equation is in standard form, it may be more convenient to use the slope and y-intercept to draw the graph. Recall that for an equation in standard form, $-\frac{A}{B}$ is the slope and $\frac{C}{B}$ is the y-intercept. Consider $-4x + 3y = 12$. The slope of its graph, $-\frac{A}{B}$, is $\frac{4}{3}$. The y-intercept of its graph, $\frac{C}{B}$, is $\frac{12}{3}$ or 4. To draw the graph, use the slope and y-intercept, as in Example 2. To check, choose another point on the line and determine whether it is a solution of $-4x + 3y = 12$.

CHECKING FOR UNDERSTANDING

Communicating Mathematics

Read and study the lesson to answer each question.

1. How many points are needed to draw the graph of a linear equation?
2. In the equation $Ax + By = C$, suppose either A or B equals zero. What can you say about the graph of the equation?

Guided Practice

State the x- and y-intercepts for the graph of each equation. Then graph each equation using the intercepts.

3. $4x + y = 8$
4. $3x + 4y = 6$
5. $2x - y = 8$
6. $3x - 2y = 6$
7. $7x + 2y = 10$
8. $5x - \frac{1}{2}y = 2$

Given the slope of a line and a point on a line, graph the line and name two other points on the line.

9. $m = 2, (0, 1)$
10. $m = -7, (0, 4)$
11. $m = 0, (3, 2)$
12. $m = \frac{4}{3}, (0, 1)$
13. $m = \frac{1}{2}, (5, 2)$
14. $m = -\frac{2}{3}, (-3, 2)$

EXERCISES

Practice

Graph each equation using the x- and y-intercepts.

15. $6x - 3y = 6$
16. $4x + 5y = 20$
17. $5x - y = -10$
18. $2x + 5y = -10$
19. $7x - 2y = -7$
20. $y = 6x - 9$
21. $x + \frac{1}{2}y = 4$
22. $x = 8y - 4$
23. $\frac{2}{3}y = \frac{1}{2}x + 6$

Graph each equation using the slope and y-intercept.

24. $y = \frac{2}{3}x + 3$
25. $y = \frac{3}{4}x + 4$
26. $y = -\frac{3}{4}x + 4$
27. $-4x + y = 6$
28. $-2x + y = 3$
29. $3y - 7 = 2x$
30. $y = -\frac{3}{5}x - 1$
31. $y = \frac{3}{2}x - 5$
32. $\frac{3}{4}x + \frac{1}{2}y = 4$

Critical Thinking

33. Compare the equations and graphs of Exercises 24 and 28. What is true about their y-intercepts? What is their point of intersection?
34. Compare the equations and graphs of Exercises 25 and 26. What is true about their slopes? What is the relationship between the lines?
35. Compare the equations and graphs of Exercises 24 and 29. What is true about their slopes? What is the relationship between the lines?
36. Compare the equations and graphs of Exercises 29 and 32. What is true about their slopes? What is the relationship between the lines?

Applications

37. **Aviation** The equation $a = 25{,}000 - 1500t$, where t is the time in minutes and a is the altitude in feet, represents the steady descent of a certain airplane.
    a. Graph the equation.
    b. How long will it be before the airplane reaches the ground?
38. **Manufacturing** The Comfort Furniture Company plans to spend \$300,000 on the manufacture of chairs and sofas. The cost of manufacturing a chair is \$150. The cost of manufacturing a sofa is \$300. The accountant of the company knows that the equation $150c + 300s = 300{,}000$ represents all of the possibilities for the manufacture of these items. Make a graph that she can use to present this information to the president of the company.

Mixed Review

39. **National Landmarks** The Statue of Liberty and the pedestal on which it stands are 302 feet tall altogether. The pedestal is 2 feet shorter than the statue. How tall is the statue? **(Lesson 3-5)**

40. Multiply: $(3y + 2)(y - 3)$. **(Lesson 6-8)**

41. Find the GCF of 16, 24, 30. **(Lesson 7-1)**

42. Solve: $\frac{3a}{2} + \frac{5}{4} = \frac{5a}{2}$. **(Lesson 8-9)**

43. Complete: 1, 11, 21, 31, $\underline{\ ?\ }$, $\underline{\ ?\ }$, $\underline{\ ?\ }$. **(Lesson 9-8)**

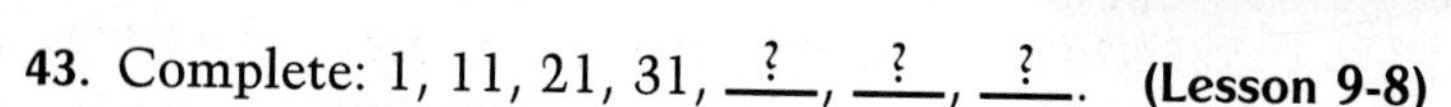

44. Determine the x- and y-intercepts of the graph of $2x + y = 6$. **(Lesson 10-3)**

MID-CHAPTER REVIEW

Determine the slope of the line passing through each pair of points. (Lesson 10-1)

1. (5, 2), (7, 6)
2. (−2, 1), (−6, 3)
3. $\left(\frac{3}{4}, \frac{1}{2}\right), \left(\frac{1}{2}, \frac{3}{4}\right)$

Write the standard form of an equation of the line passing through the given point and having the given slope. (Lesson 10-2)

4. (8, 2), $\frac{3}{4}$
5. (−6, 1), $\frac{3}{2}$
6. (−2, 1), undefined

Write the standard form of an equation of the line passing through each pair of points. (Lesson 10-2)

7. (9, 1), (8, 2)
8. (5, −4), (5, 5)
9. $\left(\frac{1}{2}, \frac{3}{4}\right), \left(\frac{2}{3}, \frac{4}{5}\right)$

Determine the *x*- and *y*-intercepts of the graph of each equation. Then graph each equation. (Lesson 10-3, 10-4)

10. $5x + 3y = 15$
11. $3x - 2y = -5$
12. $3x + \frac{1}{2}y = 8$

Determine the slope and *y*-intercept of the graph of each equation. Then write each equation in slope-intercept form and graph the equation. (Lesson 10-3, 10-4)

13. $3x + 4y = 12$
14. $10x - 14y = 21$
15. $5(x + 2) = y - 6x - 4$

10-5 Writing Slope-Intercept Equations of Lines

Objectives

After studying this lesson, you should be able to:

- write a linear equation in slope-intercept form given the slope of a line and the coordinates of a point on the line, and
- write a linear equation in slope-intercept form given the coordinates of two points on the line.

Application

FYI . . .

The world's population increases by 155 people every minute.

The present population of Cedarville is 55,000. If the population increases by 600 people each year, the equation $y = 600x + 55{,}000$ can be used to find the population x years from now. Notice that 55,000 (the present population) is the y-intercept and 600 (the growth per year) is the slope.

In the problem above, the slope and the y-intercept were used to write an equation. Other information can also be used to write an equation for a line. In fact, given any one of the three types of information below about a line, you can write an equation for that line.

1. the slope and a point on the line
2. two points on the line
3. the x- and y-intercepts

Example 1

Write an equation of the line whose slope is 3 that passes through (4, −2).

$y = mx + b$	*Use slope-intercept form.*
$y = 3x + b$	*The slope is 3.*
$-2 = 3(4) + b$	*Substitute 4 for x and −2 for y.*
$-2 = 12 + b$	*Solve for b.*
$-14 = b$	

The slope-intercept form of the equation of the line is $y = 3x + (-14)$ or $y = 3x - 14$. In standard form, the equation is $3x - y = 14$.

Example 2 illustrates a procedure that can be used to write an equation of a line when two points on the line are known.

Example 2

The Super Duper Toy Company has introduced an exciting new toy called a Gizmo Gadget. At the end of the first month, the company had manufactured 1200 Gizmo Gadgets. At the end of the fifth month, it had manufactured a total of 5800 Gizmo Gadgets. Assume that the planned production of this toy can be represented by a straight line. Write an equation for the graph used to show stockholders the planned production of Gizmo Gadgets.

The ordered pair (1, 1200) means that after 1 month, 1200 Gizmo Gadgets had been manufactured. Therefore, the line passes through the points (1, 1200) and (5, 5800). Use this information to find the slope.

$$\begin{aligned} m &= \frac{y_2 - y_1}{x_2 - x_1} \\ &= \frac{5800 - 1200}{5 - 1} \\ &= \frac{4600}{4} \\ &= 1150 \qquad \text{The slope is 1150.} \end{aligned}$$

Now, substitute the coordinates of either point into the equation $y = mx + b$ and solve for b.

$$\begin{aligned} y &= 1150x + b \\ 1200 &= 1150(1) + b \qquad (x, y) = (1, 1200) \\ 50 &= b \end{aligned}$$

The y-intercept is 50, so the equation is $y = 1150x + 50$, where x is the number of months and y is the number of Gizmo Gadgets. In standard form, the equation is $1150x - y = -50$.

Example 3

Write an equation of the line that passes through (7.6, 10.8) and (12.2, 93.7). Round values to the nearest thousandth.

The [( ] and [ )] keys on a calculator can be used to find the slope using the equation $m = \frac{y_2 - y_1}{x_2 - x_1}$.

Enter: [( ] 93.7 [−] 10.8 [ )] [÷] [( ] 12.2 [−] 7.6 [ )] [=] [STO] 18.021739

Rounded to the nearest thousandth, the slope is 18.022. In slope-intercept form, $y = 18.022x + b$. To determine the y-intercept, solve for b. Thus, $b = y - 18.022x$. Then let $(x, y) = (7.6, 10.8)$.

Enter: 10.8 [−] [RCL] [×] 7.6 [=] −126.16522

Rounded to the nearest thousandth, the y-intercept is −126.165. Therefore, the slope-intercept form of the equation is $y = 18.022x - 126.165$. In standard form, the equation is $18.022x - y = 126.165$.

You can also use the slope and y-intercept obtained from the graph to write an equation of a line.

Example 4

Write an equation for line PQ whose graph is shown below.

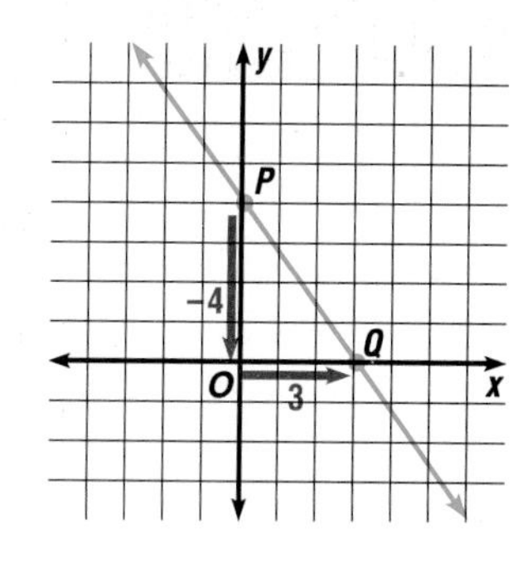

First determine the slope. Start at P. The y-coordinate *decreases by 4* as you move from P to Q. The x-coordinate *increases by* 3 as you move from P to Q.

$$\text{slope} = \frac{\text{change in } y}{\text{change in } x} = \frac{-4}{3} = -\frac{4}{3}$$

The line intersects the y-axis at (0, 4). Thus, the y-intercept is 4. Now substitute these values into the slope-intercept form.

$$y = mx + 4$$

$$y = -\frac{4}{3}x + 4 \qquad m = -\frac{4}{3} \text{ and } b = 4$$

The equation for line PQ is $y = -\frac{4}{3}x + 4$.
In standard form, the equation is $4x + 3y = 12$.

CHECKING FOR UNDERSTANDING

Communicating Mathematics

Read and study the lesson to answer each question.

1. List the information that can be used to find the equation of a line.
2. Given the coordinates of one point on a line and the slope of the line, what is the first thing that you should do in order to find an equation of the line?
3. Given the coordinates of two points, what is the first thing that you should do in order to find an equation of the line passing through these points?

Guided Practice

Given an equation of a line and a point on the line, determine the y-intercept, b.

4. $y = 3x + b$, (2, 1)
5. $y = -2x + b$, (6, 2)
6. $y = -\frac{2}{3}x + b$, (−6, 5)
7. $y = \frac{5}{6}x + b$, (3, −1)

State the slope and y-intercept for each line. Then write an equation of the line in slope-intercept form.

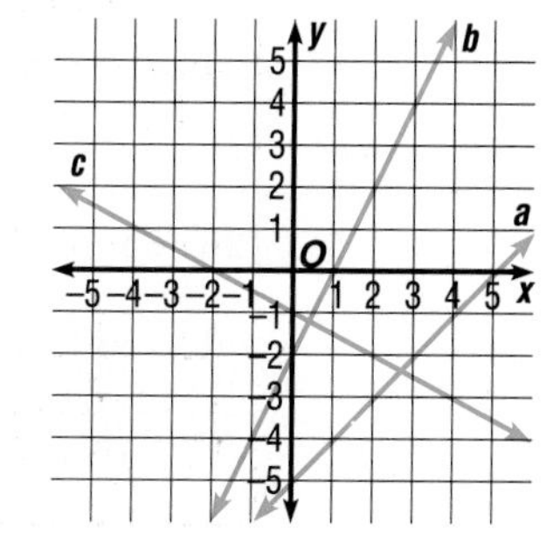

8. a

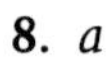

9. b
10. c

EXERCISES

Practice

State the slope and y-intercept for each line. Then write an equation of the line in slope-intercept form.

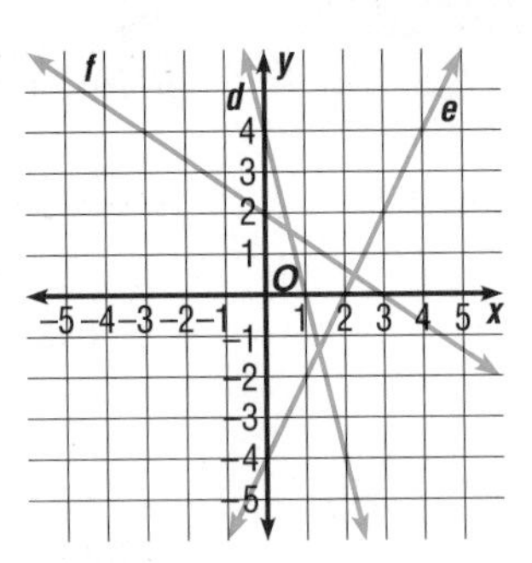

11. d

12. e

13. f

Write an equation in slope-intercept form of the line having the given slope that passes through the given point.

14. 3; (5, −2)
15. $\frac{2}{3}$; (−1, 0)
16. −5; (5, 4)
17. $\frac{3}{4}$; (−2, −4)
18. $-\frac{5}{3}$; (−3, −5)
19. $\frac{1}{4}$; (0, 8)

Journal

Write a paragraph telling what you learned today. Be sure to give examples.

Write an equation in slope-intercept form of the line that passes through each pair of points.

20. (−1, 7), (8, −2)
21. (6, 0), (0, 4)
22. (8, −1), (7, −1)
23. (1, 0), (0, 1)
24. (5, 7), (−1, 6)
25. (−6, 2), (3, −5)

Use a calculator to write an equation in slope-intercept form of the line that passes through each pair of points. Round values to the nearest thousandth.

26. (4.67, 5.235), (0.25, −1.5)
27. (−3.2, 7.198), (12.34, −0.8)
28. (0.4, 2.63), (6.25, 12.05)
29. (−2.1, −4.08), (−0.2, −7.11)
30. (6.27, −0.001), (4.33, 1.33)
31. (18.2, 1.008), (−4.3, −11.5)

Critical Thinking

32. The x-intercept of a line is s, and the y-intercept is t. Write the equation of the line.

Applications

33. **Business** Velma's Housecleaning Service charges \$65 for a three-hour job and \$115 for a six-hour job. Define variables and write an equation that Velma can use to determine what amount to bill customers for a job of any length.

34. **Skydiving** Kevin Byrd jumps from an airplane flying at 6400 feet. The wind carries him to a landing spot 580 feet away. Define variables and write an equation that represents his path of descent.

Mixed Review

35. Write 82,100,000 in scientific notation. **(Lesson 6-4)**

36. Factor $36x^2 - 81y^4$. **(Lesson 7-6)**

37. Find $\frac{a^2 + 3a - 10}{a^2 + 8a + 15} \div \frac{a^2 - 6a + 8}{12 + a - a^2}$. **(Lesson 8-3)**

38. Solve $y = 5x - 3$ if the domain is $\{-2, -1, 0, 2, 5\}$. **(Lesson 9-3)**

39. Graph $2x + 10y = 5$ using the x- and y-intercepts. **(Lesson 10-4)**

10-6 Parallel and Perpendicular Lines

Objective

After studying this lesson, you should be able to:

- write an equation of a line that passes through a given point and is parallel or perpendicular to the graph of a given equation.

Application

Isabel starts riding her bicycle at the rate of 10 miles per hour. At the same time, Heather starts riding her bike in the same direction from a point 5 miles north of Isabel. Heather also rides at a rate of 10 miles per hour. If the girls continue to ride at the same rate of speed, will Isabel ever catch up with Heather?

The equation $y = 10x$ represents Isabel's position at any given time x. The equation $y = 10x + 5$ represents Heather's position. The graph at the right represents the bike rides. Because $10x$ is never equal to $10x + 5$, Isabel will never catch up with Heather and the graphs of these two equations will never intersect. These lines are **parallel.**

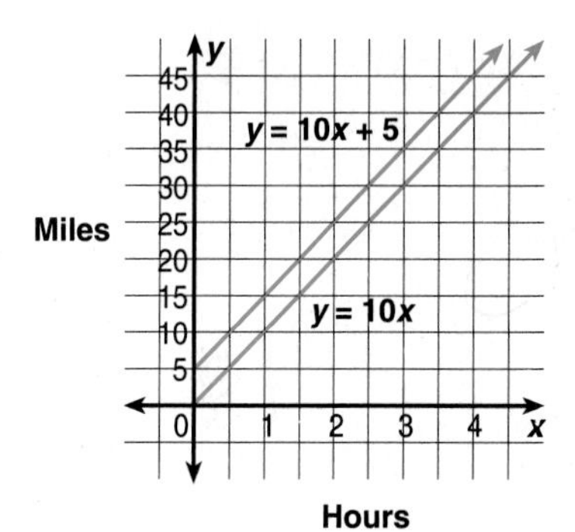

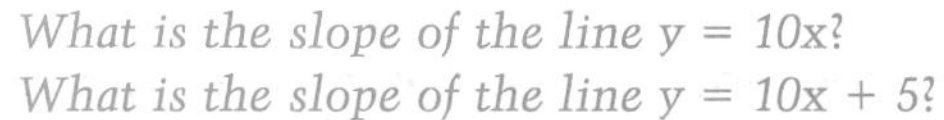

What is the slope of the line $y = 10x$?
What is the slope of the line $y = 10x + 5$?

Definition of Parallel Lines	**If two lines have the same slope, then they are parallel. All vertical lines are parallel.**

A *quadrilateral* is a four-sided figure. A *parallelogram* is a quadrilateral with two sets of parallel sides.

Example 1

Determine if the quadrilateral shown at the right is a parallelogram.

Since parallel lines have the same slope, find and compare the slope of the lines containing each side. Use $m = \frac{y_2 - y_1}{x_2 - x_1}$.

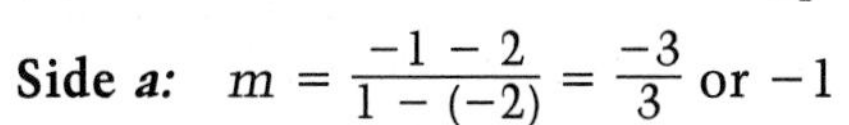

Side a: $m = \frac{-1 - 2}{1 - (-2)} = \frac{-3}{3}$ or -1

Side b: $m = \frac{-1 - 2}{1 - 7} = \frac{-3}{-6}$ or $\frac{1}{2}$

Side c: $m = \frac{2 - 5}{7 - 4} = \frac{-3}{3}$ or -1 *Lines a and c have slope -1.*

Side d: $m = \frac{5 - 2}{4 - (-2)} = \frac{3}{6}$ or $\frac{1}{2}$ *Lines b and d have slope $\frac{1}{2}$.*

The lines containing the opposite sides of the quadrilateral are parallel. Therefore, the quadrilateral is a parallelogram.

Example 2

Find an equation of the line that passes through (4, −2) and is parallel to the graph of $5x - 2y = 6$. Use slope-intercept form.

The slope of the graph of $5x - 2y = 6$ is $\frac{5}{2}$. $-\frac{A}{B} = \frac{5}{2}$

Therefore, the slope-intercept form of an equation whose graph is parallel to the graph of $5x - 2y = 6$ is $y = \frac{5}{2}x + b$. *Why?*

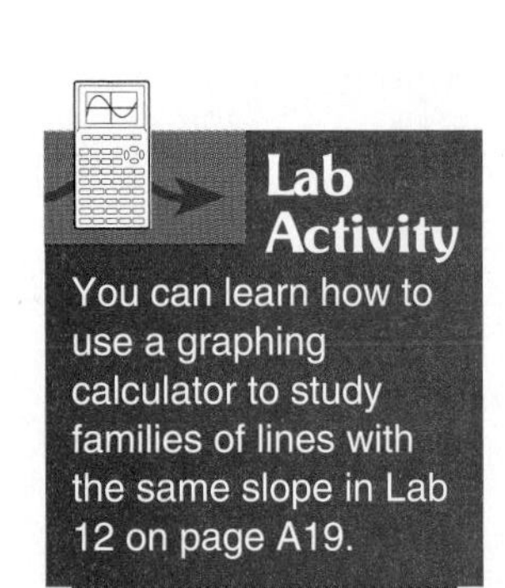

Lab Activity

You can learn how to use a graphing calculator to study families of lines with the same slope in Lab 12 on page A19.

Now, substitute (4, −2) into the equation above and solve for b.

$$y = \frac{5}{2}x + b$$

$$-2 = \frac{5}{2}(4) + b \quad \text{Let } (x, y) = (4, -2).$$

$$-2 = 10 + b$$

$$-12 = b \quad \text{The y-intercept is } -12.$$

An equation of the line is $y = \frac{5}{2}x - 12$.

ℓ′ is read "ℓ prime."

Line ℓ is shown in the figure below at left. Suppose line ℓ is rotated counterclockwise 90° to form line ℓ′. After the rotation, line ℓ′ is **perpendicular** to line ℓ.

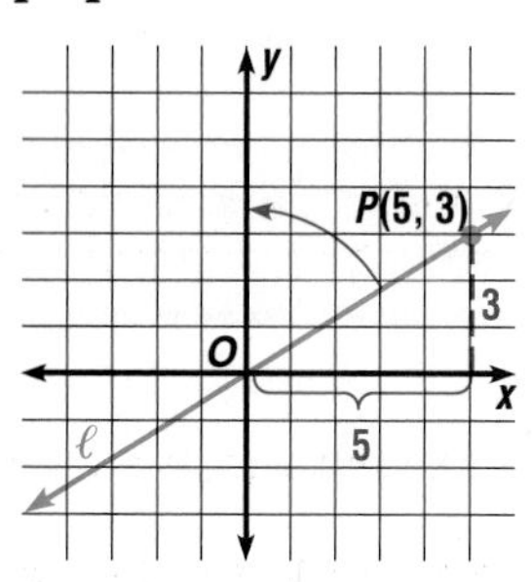

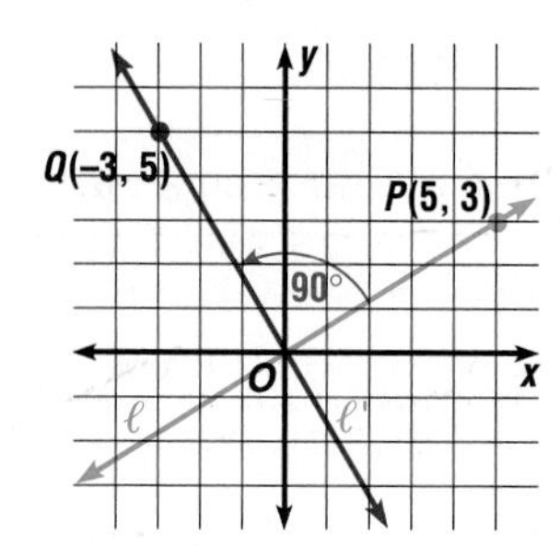

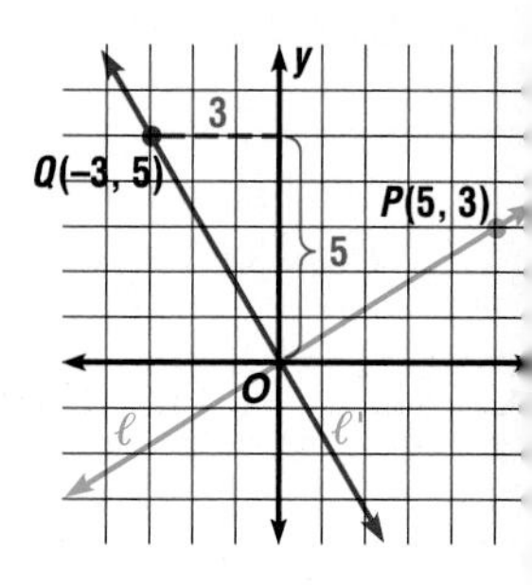

slope of ℓ

$$\frac{y_2 - y_1}{x_2 - x_1} = \frac{0 - 3}{0 - 5} = \frac{-3}{-5} = \frac{3}{5}$$

slope of ℓ′

$$\frac{y_2 - y_1}{x_2 - x_1} = \frac{0 - 5}{0 - (-3)} = \frac{-5}{3} = -\frac{5}{3}$$

Compare the slopes. What is their product?

Definition of Perpendicular Lines	**If the product of the slopes of two lines is −1, then the lines are perpendicular. In a plane, vertical lines are perpendicular to horizontal lines.**

Example 3

Show that the lines whose equations are $7x + 3y = 4$ and $3x - 7y = 1$ are perpendicular.

The slope of the graph of $7x + 3y = 4$ is $-\frac{7}{3}$. $\quad -\frac{A}{B} = -\frac{7}{3}$

The slope of the graph of $3x - 7y = 1$ is $\frac{3}{7}$. $\quad -\frac{A}{B} = -\frac{3}{-7} = \frac{3}{7}$

Since $-\frac{7}{3} \cdot \frac{3}{7} = -1$, the lines are perpendicular.

Example 4

Find an equation of the line that passes through $(-2, 7)$ and is perpendicular to the line whose equation is $2x - 5y = 3$. Use slope-intercept form.

The slope of the graph of $2x - 5y = 3$ is $\frac{2}{5}$. $\quad -\frac{A}{B} = -\frac{2}{-5} = \frac{2}{5}$

The slope of the line perpendicular to that line is $-\frac{5}{2}$.

Thus, the equation in slope-intercept form of the line perpendicular to the line whose equation is $2x - 5y = 3$ is $y = -\frac{5}{2}x + b$. Substitute $(-2, 7)$ into the equation and solve for b.

$$y = -\frac{5}{2}x + b$$

$$7 = -\frac{5}{2}(-2) + b \quad \text{Let } (x, y) = (-2, 7).$$

$$7 = 5 + b$$

$$2 = b$$

In slope-intercept form the equation is $y = -\frac{5}{2}x + 2$.

CHECKING FOR UNDERSTANDING

Communicating Mathematics

Read and study the lesson to complete the following.

1. Describe the relationship between the slopes of two parallel lines.
2. Describe the relationship between the slopes of two perpendicular lines.
3. Write some equations of lines that are perpendicular to a line that has an undefined slope.
4. Write some equations of lines that are parallel to a line that has an undefined slope.

Guided Practice

State the slopes of the lines parallel to and perpendicular to the graph of each equation.

5. $5x - y = 7$
6. $3x + 4y = 2$
7. $2x - 3y = 7$
8. $7x + y = 4$
9. $x = 7$
10. $y = 4x + 2$
11. $3y = 2x + 5$
12. $y = -4$
13. $3x = 4 - 3y$

EXERCISES

Practice

Write an equation of the line that is parallel to the graph of each equation and passes through the given point. Use slope-intercept form.

14. $y = -\frac{3}{5}x + 4$; (0, −1)
15. $y = \frac{3}{4}x - 1$; (0, 0)
16. $6x + y = 4$; (−2, 3)
17. $2x + 3y = 1$; (4, 2)
18. $5x - 2y = 7$; (0, 4)
19. $4x - 3y = 2$; (4, 0)
20. $y = -\frac{1}{3}x + 7$; (2, −5)
21. $x = y$; (7, −2)

Write an equation of the line that is perpendicular to the graph of each equation and passes through the given point. Use slope-intercept form.

22. $5x - 3y = 7$; (8, −2)
23. $3x + 8y = 4$; (0, 4)
24. $y = 3x - 2$; (6, −1)
25. $y = -3x + 7$; (−3, 1)
26. $y = 5x - 3$; (0, −1)
27. $y = \frac{2}{3}x + 1$; (−3, 0)
28. $5x + 9y = 3$; (0, 0)
29. $y = 2x - 7$; (4, −6)

Critical Thinking

30. The graphs of $5x + 8y = 3$ and $5x + 8y = 6$ are parallel lines. Find the equation of the line that is parallel to both lines and lies midway between them.

Applications

To solve each problem, write two equations and determine if the graphs of the equations are parallel. Assume that there were no price increases or decreases during the period.

31. **Sales** On Tuesday, Joey's Pizza sold 52 pizzas and 28 gallons of soda for $518. On Wednesday, Joey's Pizza sold 39 pizzas and 21 gallons of soda. Could their total sales have been $396?

32. **Sales** Lisa operates a lemonade stand. During the month of July, Lisa sold 1200 lemonades and 500 fruit punches. During the month of August, Lisa sold 1800 lemonades and 750 fruit punches. If Lisa made $1650 during July, is it possible that she made $2475 during August?

Mixed Review

33. **Geography** There is three times as much water as land on Earth's surface. What percent of Earth is covered by water? **(Lesson 4-2)**

34. Multiply $(4x^3 - 3y^2)^2$. **(Lesson 6-9)**

35. Find the number that each digit represents. (*Hint:* $T \neq 0$.) **(Lesson 7-4)**

```
  TWENTY
  TWENTY
+ THIRTY
--------
 SEVENTY
```

36. Simplify $\frac{x^2 - 49}{x^2 - 2x - 35}$. **(Lesson 8-1)**

37. Is {(−3, 3), (−2, 2), (−1, 1), (0, 0)} a function? **(Lesson 9-6)**

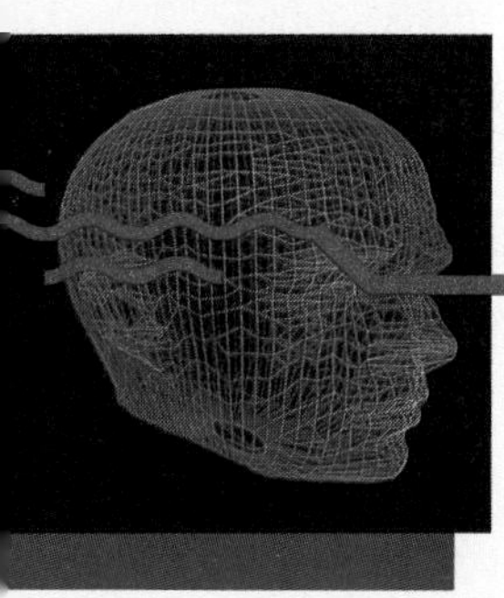

Technology

Graphing Linear Equations

BASIC
Graphing calculators
▶ **Graphing software**
Spreadsheets

The *Mathematics Exploration Toolkit* can be used to graph lines. The following CALC commands will be used.

CLEAR F (clr f) GRAPH (gra) SCALE (sca)

The CLEAR F command removes any previous graphs from the graphing window. The GRAPH command graphs the most recent equation in the expression window. The SCALE command sets limits on the x- and y-axes.

To set up the graphing window, enter clr f. Then enter sca 10. This sets limits on the axes at -10 to 10 for x and y. Only two points are needed to graph a line. So, use the command gra 2.

Example

Graph $x + y = 4$ and $x + y = -4$ on the same set of axes.

Enter	Result
$x + y = 4$	$x + y = 4$
gra 2	graphs the line
$x + y = -4$	$x + y = -4$
gra 2	graphs the line

Notice that since CLEAR was not used, the lines were graphed on the same axes. The lines have the same slope but different x- and y-intercepts.

EXERCISES

Graph each set of equations on the same set of axes.

1a. $2x + 3y = -12$ **b.** $2x + 3y = -6$ **c.** $2x + 3y = 0$
d. $2x + 3y = 6$ **e.** $2x + 3y = 12$
f. As the value of C increases, what is the effect on the slope? the x-intercept? the y-intercept?

2a. $-4x + 3y = 12$ **b.** $-2x + 3y = 12$ **c.** $-x + 3y = 12$
d. $x + 3y = 12$ **e.** $2x + 3y = 12$ **f.** $4x + 3y = 12$
g. As the value of A increases, what is the effect on the slope? the x-intercept? the y-intercept?

3a. $2x - 6y = 12$ **b.** $2x - 3y = 12$ **c.** $2x - y = 12$
d. $2x + y = 12$ **e.** $2x + 3y = 12$ **f.** $2x + 6y = 12$
g. As the value of B increases, what is the effect on the slope? the x-intercept? the y-intercept?

10-7 Midpoint of a Line Segment

Objective After studying this lesson, you should be able to:

- find the coordinates of the midpoint of a line segment in the coordinate plane given the coordinates of the endpoints.

The **midpoint** of a line segment is the point on that segment that separates it into two segments of equal length.

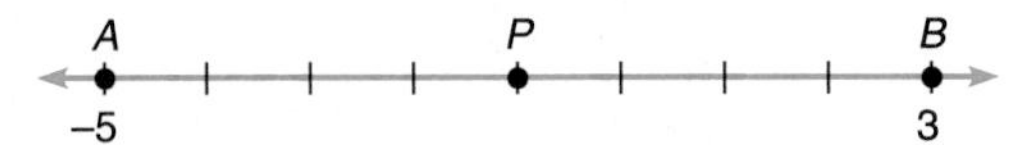

The coordinate of the midpoint, *P*, of line segment *AB* shown on the number line above can be found as follows. Find the average of the coordinates by adding the coordinates and dividing by 2.

$$P = \frac{-5 + 3}{2} = \frac{-2}{2} = -1$$

The coordinate of the midpoint is -1.

Midpoint on a Number Line

The coordinate of the midpoint, *P*, of two points, x_1 and x_2, on a number line is

$$P = \frac{x_1 + x_2}{2}.$$

This method can be extended to find the coordinates of the midpoint of a line segment in the coordinate plane.

Midpoint of a Line Segment

The coordinates of the midpoint (x, y) of a line segment whose endpoints are at (x_1, y_1) and (x_2, y_2) are

$$(x, y) = \left(\frac{x_1 + x_2}{2}, \frac{y_1 + y_2}{2}\right).$$

Example 1

The endpoints of a diameter of a circle are $(-1, 2)$ and $(3, -3)$. Find the coordinates of the center of the circle.

The diameter of a circle is a line segment. The center of a circle is the midpoint of the diameter.

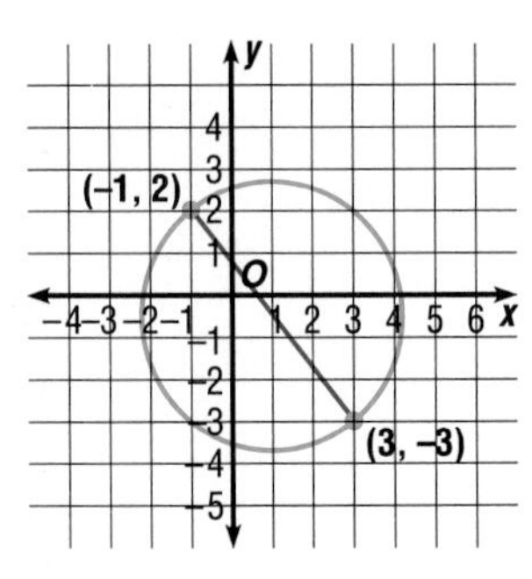

$$(x, y) = \left(\frac{x_1 + x_2}{2}, \frac{y_1 + y_2}{2}\right)$$

$$= \left(\frac{-1 + 3}{2}, \frac{2 + (-3)}{2}\right)$$

$$= \left(1, -\frac{1}{2}\right)$$

The center of the circle is at $\left(1, -\frac{1}{2}\right)$.

Example 2 **If one endpoint of a line segment is (2, 8) and the midpoint is (−1, 4), find the coordinates of the other endpoint.**

$$(x, y) = \left(\frac{x_1 + x_2}{2}, \frac{y_1 + y_2}{2}\right)$$

$$(-1, 4) = \left(\frac{2 + x_2}{2}, \frac{8 + y_2}{2}\right)$$ *Substitute (−1, 4) for (x, y).* *Substitute (2, 8) for (x_1, y_1).*

Now separate this into two equations. The x-coordinates are equal and the y-coordinates are equal.

$-1 = \frac{2 + x_2}{2}$ *Solve for x_2.*

$-2 = 2 + x_2$

$-4 = x_2$

$4 = \frac{8 + y_2}{2}$ *Solve for y_2.*

$8 = 8 + y_2$

$0 = y_2$

Thus, the coordinates of the other endpoint are (−4, 0).

CHECKING FOR UNDERSTANDING

Communicating Mathematics

Read and study the lesson to complete the following.

1. Explain how to find the coordinates of a point midway between two points on a number line.
2. Explain how to find the coordinates of the midpoint of a line segment on the coordinate plane.

Guided Practice

State the coordinate of the point midway between each pair of points on a number line.

3. 4 and 8
4. −2 and 6
5. −3 and 13
6. −4 and −10
7. −3 and 6
8. −10 and 15

State the coordinates of the midpoint of each line segment named below.

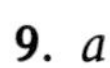

9. *a*
10. *b*
11. *c*
12. *d*
13. *e*
14. *f*
15. *g*
16. *h*

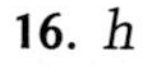

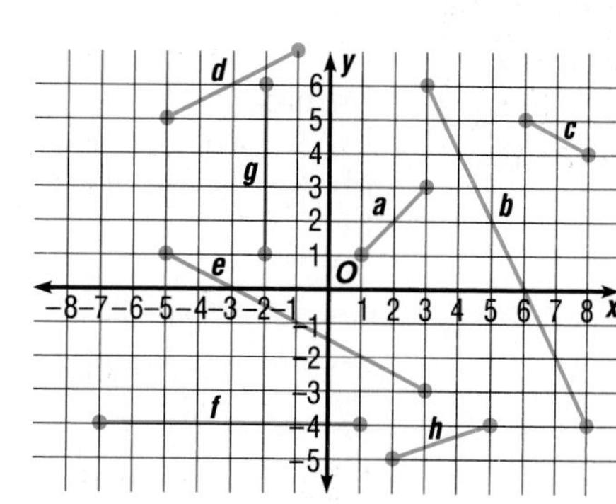

EXERCISES

Practice

Find the coordinates of the midpoint of the line segment whose endpoints are given.

17. $(8, 4), (12, 2)$
18. $(9, 5), (17, 3)$
19. $(17, 9), (11, -3)$
20. $(19, -3), (11, 5)$
21. $(4, 2), (8, -6)$
22. $(-6, 5), (8, -11)$
23. $(5, -2), (7, 3)$
24. $(-11, 6), (13, 4)$
25. $(9, 10), (-8, 4)$
26. $(x, y), (a, b)$
27. $(2x, 3y), (6x, y)$
28. $\left(\frac{5}{6}, \frac{1}{3}\right), \left(\frac{1}{6}, \frac{1}{3}\right)$

If *P* is the midpoint of line segment *AB*, find the coordinates of the missing point *A*, *B*, or *P*.

29. $A(3, 5), P(11, 7)$
30. $A(3, 5), P(5, -7)$
31. $A(5, 9), B(-7, 3)$
32. $B(11, -4), P(3, 8)$
33. $B(5, 3), P(9, 7)$
34. $A(11, -6), B(5, -9)$
35. $P(5, -9), A(4, -11)$
36. $P(3, 9), B(-4, 1)$
37. $A(4, -7), B(-8, 1)$
38. $A(7, 4), P(9, -3)$
39. $P(3, -5), A(-3, 8)$
40. $P(5, 6), B(5, 7)$

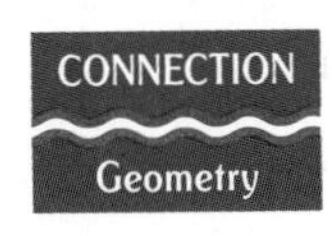

41. The two endpoints of a diameter of a circle are $(8, -2)$ and $(4, -6)$. Find the coordinates of the center.
42. The center of a circle is $(3, -2)$ and one endpoint of a diameter is $(8, 3)$. Find the other endpoint of the diameter.

Find the coordinates of *P* on line segment *AB* if *P* is one fourth of the distance from *A* to *B*.

43. $A(8, 4), B(12, 12)$
44. $A(-3, 9), B(5, 1)$
45. $A(-3, 2), B(5, 4)$
46. $A(2, -6), B(9, 5)$

Critical Thinking

For quadrilateral *ABCD*, determine whether the diagonals of *ABCD* bisect each other. Justify your answer.

47. $A(-2, 6), B(2, 11), C(3, 8), D(-1, 3)$
48. $A(11, 6), B(1, -2), C(-2, 4), D(3, 8)$

Computer

49. The BASIC program at the right tests if three points are on the same line. In the program, the points are (A, B), (C, D), and (E, F). If the slope of the line from (A, B) to (C, D) equals the slope of the line from (C, D) to (E, F), then the three points are on the same line. Points on the same line are said to be *collinear*.

```
10 PRINT "ENTER THE COORDINATES
   OF THREE POINTS."
20 INPUT A,B,C,D,E,F
30 PRINT "(";A;",";B;")","(";C;
   ",";D;")","(";E;",";F;")"
40 IF (D-B)/(C-A)=(F-D)/(E-C)
   THEN 70
50 PRINT "ARE NOT COLLINEAR."
60 GOTO 10
70 PRINT "ARE COLLINEAR."
80 GOTO 10
90 END
```

Use the program to determine whether each set of points is collinear.

a. $(1, 5)$, $(16, 14)$, $(-4, 2)$

b. $(-2, -3)$, $(2, 1)$, $(5, 6)$

c. $(-459, -80)$, $(865, 163)$, $(54, 1)$

d. $(5, 14)$, $(-5, 10)$, $(-6, 8)$

50. Modify the program to print the slope for each segment if the points are collinear.

Applications

The map coordinates of certain cities are given below.

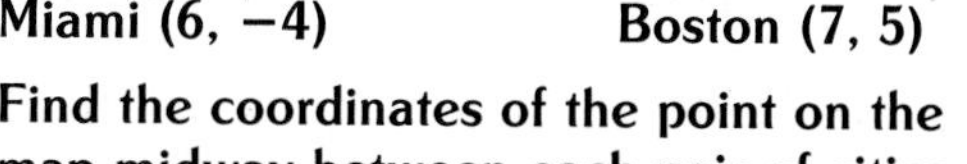

Los Angeles $(-6, -1)$ **Dallas $(1, -3)$**
Chicago $(3, 3)$ **Atlanta $(5, -1)$**
Miami $(6, -4)$ **Boston $(7, 5)$**

Find the coordinates of the point on the map midway between each pair of cities.

51. Los Angeles and Boston

52. Atlanta and Dallas

53. Chicago and Miami

54. Boston and Atlanta

55. Dallas and Boston

56. Miami and Dallas

57. Chicago and Los Angeles

58. Boston and Chicago

Mixed Review

59. Solve $3y + 7 \leq 4y + 8$. **(Lesson 5-3)**

60. Subtract $(7y + 9x) - (6x + 5y)$. **(Lesson 6-6)**

61. Factor $5x^2 + 20y^2$. **(Lesson 7-2)**

62. **Driving** The toll for the Overbrook Bridge is 50¢ per car. The machines in the exact change lanes accept any type of coin except pennies and half dollars. In how many different ways can a driver pay the toll in the exact change lane? **(Lesson 8-6)**

63. Draw a mapping for the relation $\{(1, 3), (2, 5), (8, 2), (5, -3)\}$. **(Lesson 9-2)**

EXPLORE
PLAN
SOLVE
EXAMINE

10-8 Problem-Solving Strategy: Use a Graph

Objective After studying this lesson, you should be able to:

- solve problems by using pictographs, circle graphs, and comparative graphs.

Connection In Chapter 9, you studied how bar graphs and line graphs are used to represent data. Three other types of graphs are shown below.

Pictographs use pictures or illustrations to show how specific quantities compare. The pictograph at the right shows the average motor fuel consumption in the United States. *Why do you think fuel consumption dropped so much in the 1970s?*

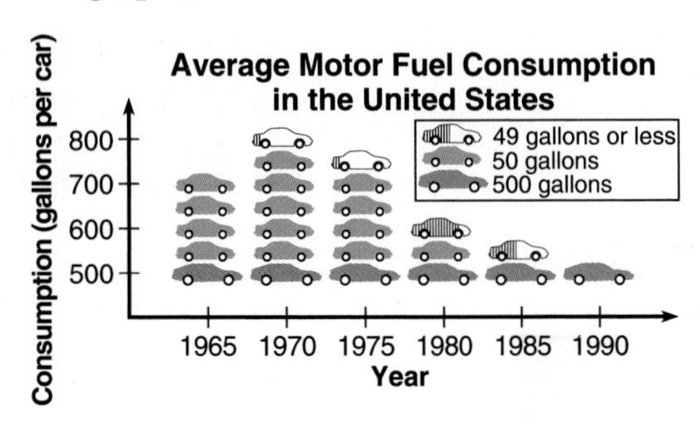

Circle graphs show how parts are related to the whole. The circle graph at the right shows how often Americans eat at fast-food restaurants each week.

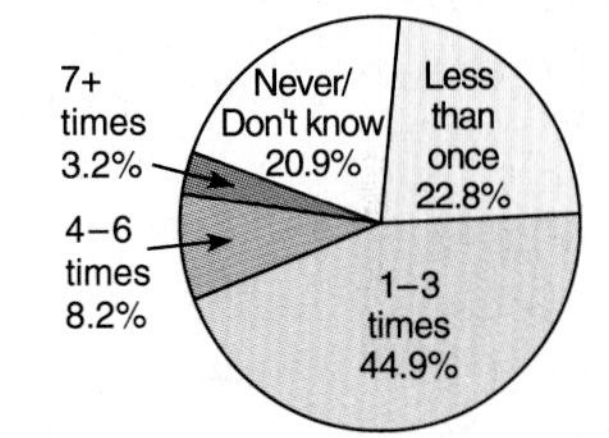

Comparative graphs show trends. These graphs are often used to compare results in two or more similar groups. The graph at the right compares the college enrollments of men and women.

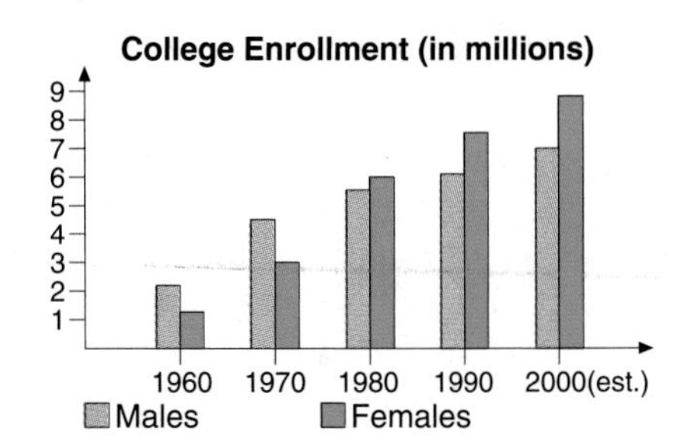

Portfolio

Select an item from your work that shows your creativity and place it in your portfolio.

CHECKING FOR UNDERSTANDING

Communicating Mathematics

Match.

1. Shows how parts are related to the whole.
2. Shows trends or changes of a single quantity.
3. Shows how specific quantities compare.
4. Uses pictures or illustrations to show how specific quantities compare.
5. Compares results in two or more similar groups.

a. bar graph
b. circle graph
c. comparative graph
d. line graph
e. pictograph

Guided Practice

Use the graphs below to answer each question.

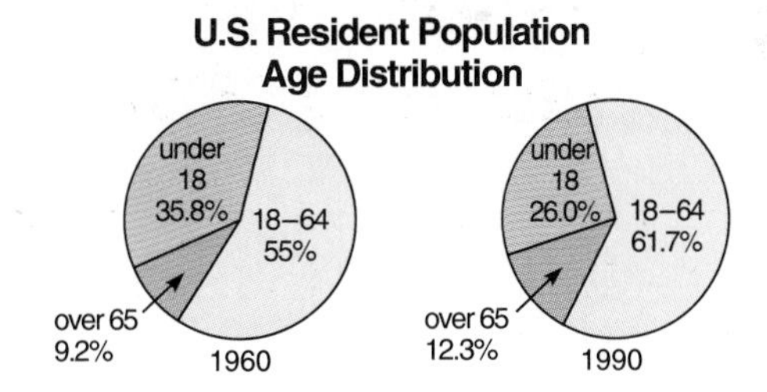

6. Did the percentage of people under 18 increase or decrease from 1960 to 1990?
7. Which age group had the least number of people in both years?
8. If the population in 1960 was 200 million and in 1990 it was 250 million, how many more people were in the 18–64 age group in 1990 than in 1960?

EXERCISES

Practice

Solve. Use any strategy.

Strategies

- Look for a pattern.
- Solve a simpler problem.
- Act it out.
- Guess and check.
- Draw a diagram.
- Make a chart.
- Work backwards.

9. Three couples received a total of $5400 in income tax refunds. Altogether, the wives received $2400. Ana received $200 more than Daisy, and Betty received $200 more than Ana. Carl received half as much as his wife, Ed received the same amount as his wife, and Frank received twice as much as his wife. Who is married to whom?
10. **Number Theory** The *persistence* of a number is the number of times you can multiply the digits until you get a one-digit product. For example, 34 has a persistence of 2 since 34 → 12 → 2 (2 steps). If 0 is the least number with a persistence of 0, and 10 is the least number with a persistence of 1, find the least numbers that have persistences of 2, 3, and 4.

COOPERATIVE LEARNING ACTIVITY

Work in groups of three. Each person must understand the solution and be able to explain it to any person in the class.

The map of Hokeyville is shown at the right. Darnell plans to visit Karen once a day until he has tried every route between his house and Karen's house.

If Darnell goes only east or north, how many routes can he take? (*Hint:* Find the number of different routes to each point on the grid.)

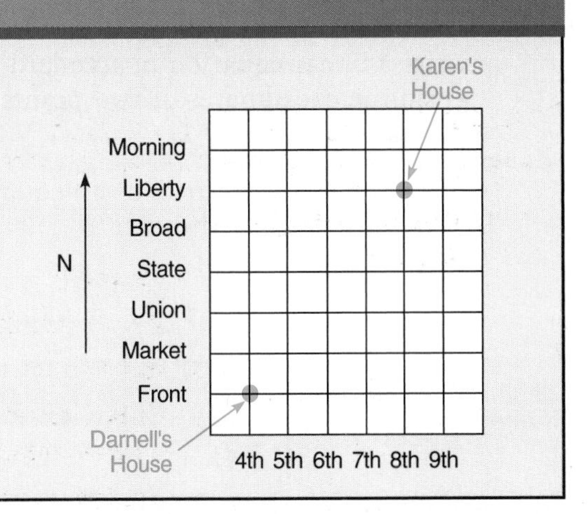

CHAPTER 10

SUMMARY AND REVIEW

VOCABULARY

Upon completing this chapter, you should be familiar with the following terms:

intercepts	**410**	**400**	slope
midpoint	**428**	**406**	standard form
parallel	**423**	**410**	x-intercept
perpendicular	**424**	**410**	y-intercept
point-slope form	**405**		

SKILLS AND CONCEPTS

OBJECTIVES AND EXAMPLES

Upon completing this chapter, you should be able to:

- find the slope of a line, given the coordinates of two points on the line **(Lesson 10-1)**

Determine the slope of the line passing through (−3, 5) and (4, 5).

$$m = \frac{y_2 - y_1}{x_2 - x_1} = \frac{5 - 5}{4 - (-3)} = 0$$

The slope is 0.

REVIEW EXERCISES

Use these exercises to review and prepare for the chapter test.

Determine the slope of the line passing through each pair of points.

1. (8, 3), (2, 5)
2. (−2, 5), (−2, 9)
3. (−3, −5), (9, −1)
4. (−3, 6), (−8, 4)
5. (11, −1), (14, −6)

- write a linear equation in standard form given the coordinates of two points on the line **(Lesson 10-2)**

Write the standard form of an equation of the line passing through (8, 1) and (−3, 5).

$$m = \frac{y_2 - y_1}{x_2 - x_1} = \frac{5 - 1}{-3 - 8} = \frac{4}{11}$$

$$y - y_1 = m(x - x_1)$$
$$y - 1 = \frac{4}{11}(x - 8)$$
$$11y - 11 = 4x - 32$$
$$-4x + 11y = -21$$

Write the standard form of an equation of the line passing through each pair of points.

6. (−2, 5), (9, 5)
7. (0, 5), (−2, 0)
8. (−3, 0), (0, −6)
9. (4, 2), (−7, 2)
10. $\left(-2, \frac{2}{3}\right), \left(-2, \frac{2}{7}\right)$

OBJECTIVES AND EXAMPLES

- determine the slope and y-intercept of a graph **(Lesson 10-3)**

Determine the slope and y-intercept of the graph of $3x - 2y = 7$.

$$3x - 2y = 7$$
$$-2y = -3x + 7$$
$$y = \frac{3}{2}x - \frac{7}{2}$$

The slope is $\frac{3}{2}$; the y-intercept is $-\frac{7}{2}$.

REVIEW EXERCISES

Determine the slope and y-intercept of the graph of each equation.

11. $y = \frac{1}{4}x + 3$

12. $8x + y = 4$

13. $x = 2y - 7$

14. $14x + 20y = 10$

15. $\frac{1}{2}x + \frac{1}{4}y = 3$

- determine the x- and y-intercepts of a graph **(Lesson 10-3)**

Determine the x- and y-intercepts of $2x + 5y = 10$.

$$2x + 5(0) = 10 \qquad 2(0) + 5y = 10$$
$$2x = 10 \qquad 5y = 10$$
$$x = 5 \qquad y = 2$$

x-intercept = 5, y-intercept = 2

Determine the x- and y-intercepts of the graph of each equation.

16. $3x + 4y = 15$

17. $8x + y = 4$

18. $6x + 2y = 3$

19. $\frac{1}{2}x - \frac{3}{2}y = 4$

20. $2.2x + 0.5y = 1.1$

- graph linear equations using the x- and y-intercepts or the slope and y-intercept **(Lesson 10-4)**

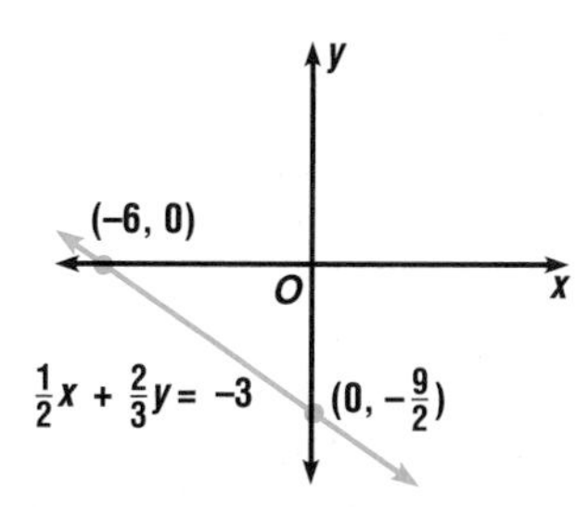

Graph each equation using the x- and y-intercepts.

21. $3x - y = 9$

22. $5x + 2y = 12$

Graph each equation using the slope and y-intercept.

23. $y = \frac{2}{3}x + 4$

24. $y = -\frac{3}{2}x - 6$

- write a linear equation in slope-intercept form **(Lesson 10-5)**

Write an equation in slope-intercept form of the line passing through $(4, -2)$ with slope 2.

$$y = mx + b$$
$$-2 = 2(4) + b$$
$$-10 = b$$

The slope-intercept form is $y = 2x - 10$.

Write an equation of the line satisfying the given conditions. Use the slope-intercept form.

25. slope = 4, passes through $(6, -2)$

26. passes through $(9, 5)$ and $(-3, -4)$

27. passes through $(2, 2)$, y-intercept 7

28. slope = $-\frac{3}{5}$, y-intercept = 3

OBJECTIVES AND EXAMPLES	REVIEW EXERCISES

- write an equation of a line that passes through a given point and is parallel or perpendicular to the graph of a given equation **(Lesson 10-6)**

Write an equation of the line that is perpendicular to the graph of $2x + y = 6$ and passes through $(2, 3)$.

$$2x + y = 6 \qquad y = \frac{1}{2}x + b$$
$$y = -2x + 6 \qquad 3 = \frac{1}{2}(2) + b$$
$$m = \frac{1}{2} \qquad 2 = b$$

The equation is $y = \frac{1}{2}x + 2$.

Write an equation of the line that passes through a given point and is parallel to the graph of each equation.

29. $4x - y = 7$; $(2, -1)$

30. $3x + 9y = 1$; $(3, 0)$

Write an equation of the line that passes through a given point and is perpendicular to the graph of each equation.

31. $2x - 7y = 1$; $(-4, 0)$

32. $8x - 3y = 7$; $(4, 5)$

- find the coordinates of the midpoint of a line segment in the coordinate plane given the coordinates of the endpoints **(Lesson 10-7)**

The endpoints of a segment are (11, 4) and (9, 2). Find its midpoint.

$$(x, y) = \left(\frac{x_1 + x_2}{2}, \frac{y_1 + y_2}{2}\right)$$
$$= \left(\frac{11 + 9}{2}, \frac{4 + 2}{2}\right)$$
$$= (10, 3)$$

Find the coordinates of the midpoint of line segment *AB*.

33. $A(3, 5)$, $B(9, -3)$

34. $A(14, 4)$, $B(2, 0)$

35. $A(-6, 6)$, $B(8, -11)$

36. $A(2, 7)$, $B(8, 4)$

37. $A(2, 5)$, $B(4, 1)$

APPLICATIONS AND CONNECTIONS

38. **Skiing** A course for cross-country skiing is regulated so that the slope of any hill cannot be greater than 0.33. Suppose a hill rises 60 meters over a horizontal distance of 250 meters. Does the hill meet the requirement? **(Lesson 10-1)**

39. **Travel** Jon Erlanger is taking a long trip. In the first 2 hours he drives 80 miles. After that, he averages 45 miles per hour. Write an equation in slope-intercept form relating distance traveled and time. **(Lesson 10-2)**

40. **Entertainment** Carolyn Parks owns stock in Star Gazer Motion Picture Company. Every other week, she graphs the closing value of a share of the stock. What is the midpoint between the highest and lowest values of the stock? **(Lessons 10-7 and 10-8)**

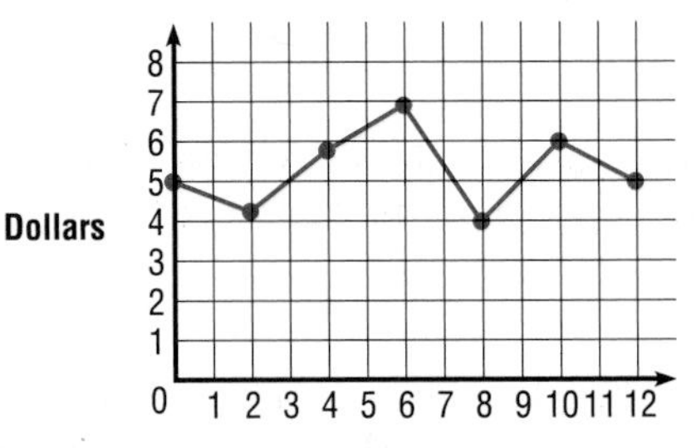

CHAPTER 10 TEST

Determine the slope of the line passing through each pair of points.

1. $(9, 2), (3, -4)$
2. $(8, 3), (8, 1)$

Determine the slope and y-intercept of the graph of each equation.

3. $x - 8y = 3$
4. $3x - 2y = 9$
5. $\frac{1}{2}x + \frac{3}{4}y = 2$
6. $y = 7$

Write an equation in standard form of the line satisfying the given conditions.

7. passes through $(2, 5)$ and $(8, -3)$
8. passes through $(-2, -1)$ and $(6, -4)$
9. has slope of 2 and y-intercept $= 3$
10. has y-intercept $= -4$ and passes through $(5, -3)$
11. slope $= \frac{3}{4}$ and passes through $(6, -2)$

Write an equation in slope-intercept form of the line satisfying the given conditions.

12. passes through $(4, -2)$ and the origin
13. passes through $(-2, -5)$ and $(8, -3)$
14. passes through $(6, 4)$ with y-intercept $= -2$
15. slope $= -\frac{2}{3}$ and y-intercept $= 5$
16. slope $= 6$ and passes through $(-3, -4)$
17. parallel to $6x - y = 7$ and passes through $(-2, 8)$
18. parallel to $3x + 7y = 4$ and passes through $(5, -2)$
19. perpendicular to $5x - 3y = 9$ and passes through the origin
20. perpendicular to $x + 3y = 7$ and passes through $(5, 2)$

Find the coordinates of the midpoint of the segment whose endpoints are given.

21. $(9, 3), (3, 6)$
22. $(-2, -7), (6, -5)$

Graph each equation.

23. $4x - 3y = 24$
24. $2x + 7y = 16$

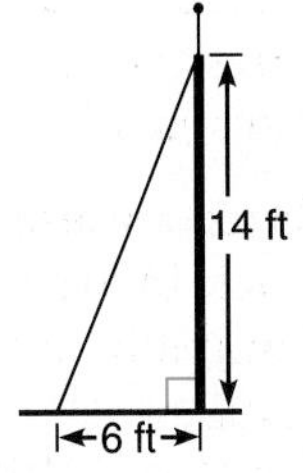

25. **Construction** Roberto drew a sketch of the antenna and guy wire support in his back yard. What is the slope of the guy wire?

Bonus The diameter of a circle has the endpoints $(8, 6)$ and $(2, -2)$. Find the center of the circle and at least two other points on the circle.

CHAPTER OBJECTIVES

In this chapter, you will:

- Solve systems of equations and inequalities by graphing.
- Solve systems of equations algebraically.
- Solve problems after checking for hidden assumptions.

Systems of Open Sentences

APPLICATION IN LITERATURE

Charles Lutwidge Dodgson (1832–1898) is better known as Lewis Carroll, the author of *Alice in Wonderland* and *Through the Looking Glass.* Dodgson, an English mathematician, weaved logic, mathematics, and science into his children's tales in ways that often appeared nonsensical. Look at the passage below. What is the hidden assumption?

> ". . . Let me see: four times five is twelve, and four times six is thirteen, and four times seven is—oh dear! I shall never get to twenty at that rate."

If you continue the progression described above, you should assume that it ends at 4 times 12, since multiplication tables taught during the time of Mr. Dodgson ended with twelves. Based on the examples in the passage, 4 times 12 must be 19, which is why the speaker exclaimed that twenty could not be reached "at that rate."

"Contrariwise," continued Tweedledee, "if it was so, it might be; and if it were so, it would be; but as it isn't, it ain't. That's logic."

Dodgson also used straightforward mathematical problems in his books. Consider this passage from *Through the Looking Glass.*

> "Tweedledum said to Tweedledee: 'The sum of your weight and twice mine is 361 pounds.' Tweedledee said to Tweedledum: 'Contrariwise, the sum of your weight and twice mine is 362 pounds.'"

Do you know how much each one weighs?

ALGEBRA IN ACTION

What two equations could you use to represent the statements made by Tweedledum and Tweedledee? Can you think of some methods that you could use to find numbers that are solutions to *both* equations?

11-1

Problem-Solving Strategy: Hidden Assumptions

EXPLORE
PLAN
SOLVE
EXAMINE

Objective

After studying this lesson, you should be able to:

- solve problems after checking for hidden assumptions.

Sometimes we make problem solving more difficult than it is by incorporating hidden assumptions into the process. You can avoid this by asking yourself the following questions.

1. What *exactly* does the problem say?
2. What does the problem *not* say?
3. What am I assuming that isn't found in the problem?
4. Does my answer make sense?

Ask yourself these questions as you work through the problem below.

Example

FYI · · ·

The Hilton Hotel in Las Vegas is the largest hotel in the United States. It has 3174 rooms.

Two women agree to share a hotel room and pay a total of $65 when they check in. Later, the clerk realizes that he should have applied a corporate discount and charged them $60. He sends a bellhop up with the $5 refund for the room. The bellhop doesn't carry coins, so he returns only $4 to the women and keeps $1. This means that each woman has paid $30.50 as her share of the room, or a total of $61. The bellhop kept the other $1 for a total of $62. What happened to the other $3?

In this problem, an incorrect relationship is suggested. Can you discover this relationship? Let's change the amount of the discount and compare the results to those given.

Amount Clerk Returns	Amount Returned to Women	Amount Kept by Bellhop	Room Cost Per Person	Total Room Cost Plus Bellhop's "Tip"
$5	$4	$1	$30.50	$61 + $1 = $62
$7	$6	$1	$29.50	$59 + $1 = $60
$9	$8	$1	$28.50	$57 + $1 = $58

If you go by the amounts in the last column, we could ask what happened to the $3, $5, or $7. But perhaps you now realize that there is no reason why the total room charge to the women plus the bellhop's "tip" should equal $65. That is the hidden assumption in this problem.

CHECKING FOR UNDERSTANDING

Communicating Mathematics

Read and study the lesson to answer each question.

1. What *exactly* did the problem in the example ask?
2. What did the problem *not* say?
3. What relationship does exist between the numbers in this problem?

Guided Practice

Solve each problem. Try to eliminate any hidden assumptions.

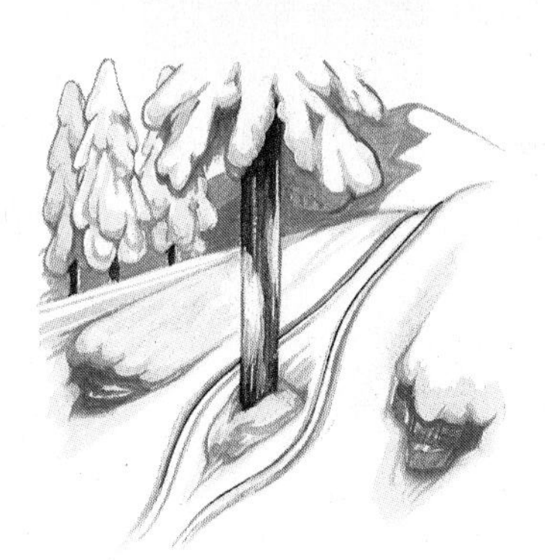

4. As Barry prepared to ski down a hill, he noticed that there were some strange ski tracks nearby. One track went around a tree on the left side. The other track went around on the right side. Give five reasonable explanations as to how this could have happened.

EXERCISES

Practice

Solve. Use any strategy.

Strategies

Look for a pattern.
Solve a simpler problem.
Act it out.
Guess and check.
Draw a diagram.
Make a chart.
Work backwards.

5. At Walnut Bluff High School, there are 2100 students. Three percent of the students wear one ring. Of the other 97 percent, half wear two rings, and half wear no rings. How many rings are worn by the students at Walnut Bluff?
6. In how many ways can you receive change for a quarter if at least one coin is a dime?
7. The MB Construction Company is building a new apartment complex that will contain 1037 units. If the apartment numbers are 1, 2, 3, and so on, how many single digits will be needed for the numbers on the apartment doors in the complex?
8. A firefighter spraying water on a fire stood on the middle rung of a ladder. The smoke lessened, so he moved up 3 rungs. When it got too hot, he backed down 5 rungs. Later, he moved up 7 rungs and stayed until the fire was out. Then he climbed the remaining 6 rungs and went into the building. How many rungs did the ladder have?

COOPERATIVE LEARNING ACTIVITY

Work in groups of four. Each person must understand the solution and be able to explain it to any person in the class.

Chef Cook's four best recipes appear on different pages of her cookbook, *Chef Cook Cooks,* which has recipes on pages 5 through 420. The page numbers of the four recipes have no repeated digits. The page number of the onion soup recipe is a divisor of the other three page numbers. The chicken salad page number is composed partially of consecutive digits. It is also more than twice the beef wellington page number, which is exactly five times the devil's food cake page number.

On what pages can Chef Cook find her four favorite recipes?

11-2 Graphing Systems of Equations

Objectives

After studying this lesson, you should be able to:

- solve systems of equations by graphing, and
- determine whether a system of equations has one solution, no solution, or infinitely many solutions by graphing.

Application

The Southside High School basketball team held a raffle to earn money for a trip to an out-of-state basketball tournament. Tickets for the raffle were sold for either \$5 or \$2. The team sold a total of 500 tickets for a total of \$1450. How many \$5 tickets and how many \$2 tickets did the team sell?

To solve this problem, let x = the number of \$5 tickets, and let y = the number of \$2 tickets. Then 5x = the amount of money collected from the sale of the \$5 tickets and $2y$ = the amount of money collected from the sale of the \$2 tickets. You can write two equations to represent this situation.

Number of \$5 tickets	*plus*	*number of \$2 tickets*	*is 500.*
x	$+$	y	$= 500$

Money from \$5 tickets	*plus*	*money from \$2 tickets*	*is \$1450.*
$5x$	$+$	$2y$	$= 1450$

The equations $x + y = 500$ and $5x + 2y = 1450$ together are called a **system of equations.** The solution to this problem is the ordered pair of numbers that satisfies both of these equations.

One method for solving a system of equations is to graph the equations on the same coordinate plane. The coordinates of the point where the graphs intersect is the solution.

Example 1

Graph the equations $x + y = 500$ and $5x + 2y = 1450$ on the same coordinate plane. Then find the solution of the system of equations to answer the problem given above.

Kareem Abdul-Jabbar scored a professional career record 37,639 points from 1970 through 1988.

The graphs intersect at the point (150, 350).

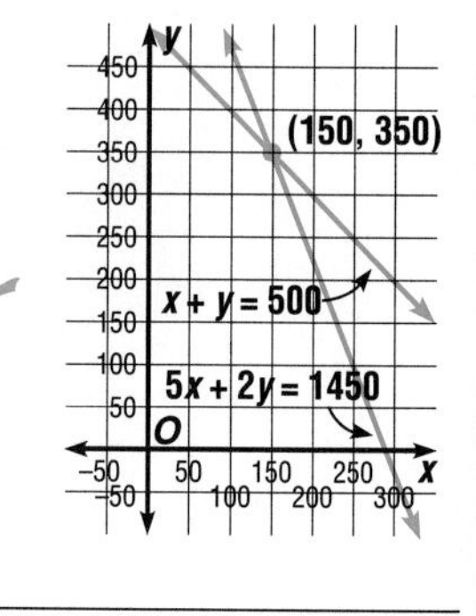

Check:

$$x + y = 500 \qquad\qquad 5x + 2y = 1450$$
$$150 + 350 \stackrel{?}{=} 500 \qquad\qquad 5(150) + 2(350) \stackrel{?}{=} 1450$$
$$500 = 500 \;\checkmark \qquad\qquad 1450 = 1450 \;\checkmark$$

The solution of the system of equations $x + y = 500$ and $5x + 2y = 1450$ is (150, 350). Therefore, the basketball team sold 150 of the \$5 tickets and 350 of the \$2 tickets.

Example 2

Graph the equations $y = x - 4$ and $x + \frac{1}{2}y = \frac{5}{2}$ to find the solution of the system of equations.

The graphs intersect at the point $(3, -1)$.

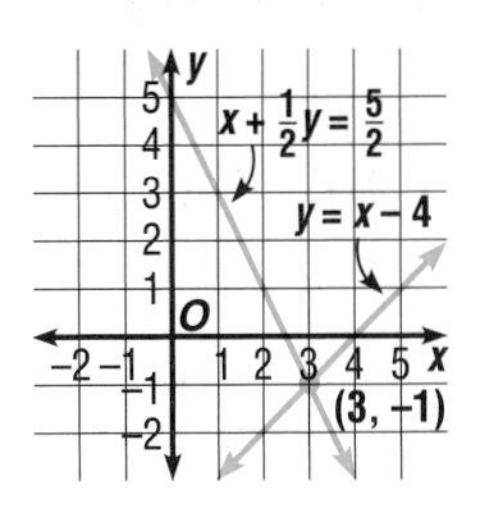

Check:

$y = x - 4$	$x + \frac{1}{2}y = \frac{5}{2}$
$-1 \stackrel{?}{=} 3 - 4$	$3 + \frac{1}{2}(-1) \stackrel{?}{=} \frac{5}{2}$
$-1 = -1$ ✓	$\frac{5}{2} = \frac{5}{2}$ ✓

The solution is $(3, -1)$.

A system of equations that has exactly one solution is said to be consistent and independent.

A system of two linear equations has exactly one ordered pair as its solution when the graphs of the equations intersect at exactly one point. It is also possible for the two graphs to be parallel lines or to be on the same line. When the graphs are parallel lines, then the system of equations does not have a solution. When the graphs are the same line, the system of equations has infinitely many solutions.

Example 3

Graph the equations $x + y = 3$ and $x + y = 4$. Then determine the number of solutions to the system of equations.

The graphs of the equations are parallel lines. Since they do not intersect, there is no solution to this system of equations. Notice that the two lines have the same slope but different y-intercepts.

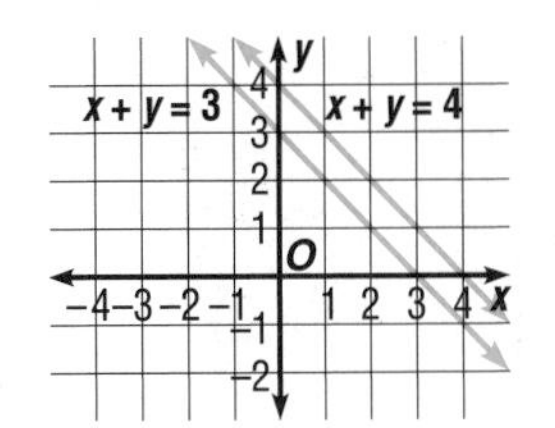

A system of equations that has no solution is said to be inconsistent.

Example 4

Graph the equations $2x + y = 3$ and $4x + 2y = 6$. Then determine the number of solutions to the system of equations.

Each equation has the same graph. Thus, any ordered pair on the graph will satisfy both equations. Therefore, there are infinitely many solutions to this system of equations. Notice that the graphs have the same slope and intercepts.

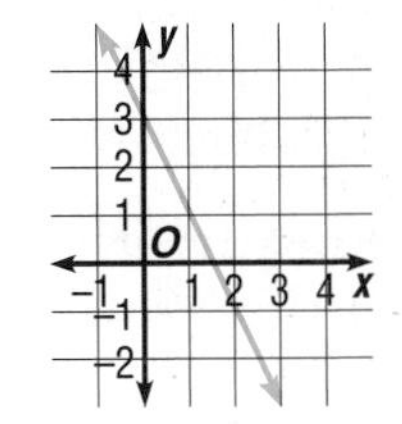

A system of equations that has infinitely many solutions is said to be consistent and dependent.

The chart at the top of the next page summarizes the possible solutions to systems of linear equations.

Description of Graph	Number of Solutions	Special Terminology
intersecting lines	exactly one	consistent and independent
parallel lines	none	inconsistent
same line	infinitely many	consistent and dependent

Example 5

The points $A(-8, -3)$, $B(2, 1)$, $C(1, -6)$, and $D(-4, -8)$ are vertices of a quadrilateral. What is the point of intersection of the two diagonals?

Draw quadrilateral *ABCD* with diagonals *AC* and *BD*. The diagonals appear to intersect at the point $(-2, -5)$. To check this, find the equations of lines *AC* and *BD* and verify that $(-2, -5)$ is a solution of both equations. First, find the slope of each line using $m = \frac{y_2 - y_1}{x_2 - x_1}$.

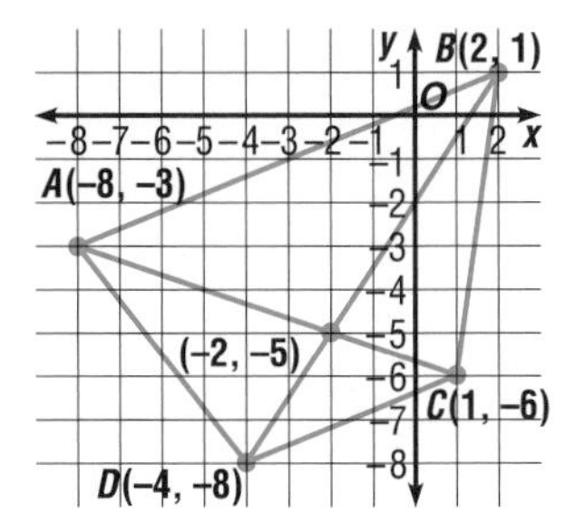

Equation for $\overline{AC}$

$$m = \frac{-3 - (-6)}{-8 - 1}$$

$$= \frac{3}{-9} \text{ or } -\frac{1}{3}$$

Equation for $\overline{BD}$

$$m = \frac{1 - (-8)}{2 - (-4)}$$

$$= \frac{9}{6} \text{ or } \frac{3}{2}$$

Then use the slope-intercept form, $y = mx + b$.

$y = -\frac{1}{3}x + b$	$y = \frac{3}{2}x + b$
$-6 = -\frac{1}{3}(1) + b$	$1 = \frac{3}{2}(2) + b$
$-\frac{17}{3} = b$	$-2 = b$
$y = -\frac{1}{3}x - \frac{17}{3}$	$y = \frac{3}{2}x - 2$

Check that $(-2, -5)$ is a solution to both equations.

$-5 \stackrel{?}{=} -\frac{1}{3}(-2) - \frac{17}{3}$ $\qquad$ $-5 \stackrel{?}{=} \frac{3}{2}(-2) - 2$

$-5 = -5$ ✓ $\qquad$ $-5 = -5$ ✓ $\qquad$ The solution checks.

CHECKING FOR UNDERSTANDING

Communicating Mathematics

Complete.

1. If a system of two linear equations has no solution, then the graphs of the two equations must be ___?___ lines.
2. If a system of two linear equations has exactly one solution, then the graphs of the two equations must be ___?___ lines.
3. If a system of two linear equations has infinitely many solutions, then the lines have the same slope and ___?___.
4. Write a system of linear equations that has (2, 1) as its only solution.

Guided Practice

State the ordered pair for the point of intersection of each pair of lines.

5. a and b
6. a and c
7. a and d
8. b and c
9. b and d
10. c and d

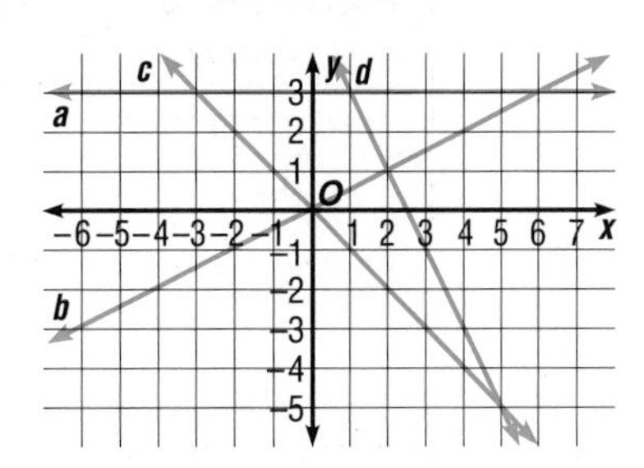

State the slope and y-intercept of the graph of each equation. Then determine whether the system of equations has one solution, no solution, or infinitely many solutions.

11. $x + y = 6$
    $x - y = 2$
12. $x + y = 6$
    $3x + 3y = 3$
13. $x + 2y = 5$
    $3x - 15 = -6y$
14. $2x + 3y = 5$
    $-6x + 15 = 9y$
15. $3x - 8y = 4$
    $6x - 42 = 16y$
16. $y = -3x$
    $6y - x = -38$

EXERCISES

Practice

Use the graphs at the right to determine whether each system has one solution, no solution, or infinitely many solutions. If the system has one solution, name it.

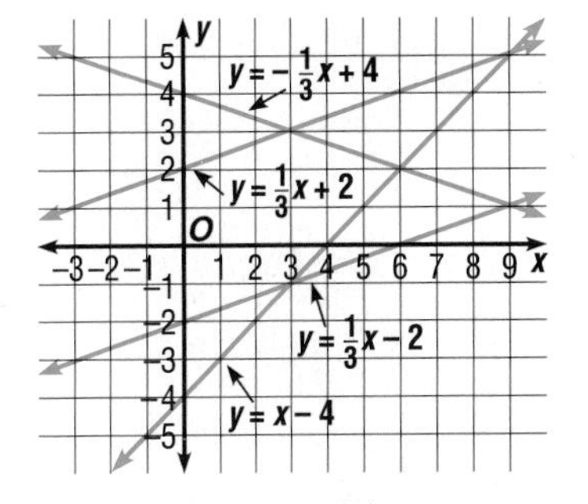

17. $x + 3y = 12$
    $x - 3y = -6$
18. $x - 3y = 6$
    $x - y = 4$
19. $x - 3y = 6$
    $x + 3y = 12$
20. $x - 3y = -6$
    $x - 3y = 6$
21. $x + 3y = 12$
    $x - y = 4$
22. $x - y = 4$
    $x - 3y = -6$

Graph each system of equations. Then determine whether the system has one solution, no solution, or infinitely many solutions. If the system has one solution, name it.

23. $y = -x$
    $y = 2x$
24. $x + y = 8$
    $x - y = 2$
25. $y = -3x$
    $4x + y = 2$
26. $x + 2y = 5$
    $2x + 4y = 2$
27. $2x + 3y = 4$
    $-4x - 6y = -8$
28. $3x - y = 4$
    $6x + 2y = -8$
29. $x - y = 2$
    $3y + 2x = 9$
30. $y = x + 3$
    $3y + x = 5$
31. $2x + 3y = -17$
    $y = x - 4$
32. $x + 2y = 0$
    $y + 3 = -x$
33. $3x + y = -8$
    $x + 6y = 3$
34. $x + 2y = 6$
    $2y - 8 = -x$

Graph each system of equations. Then determine whether the system has one solution, no solution, or infinitely many solutions. If the system has one solution, name it.

35. $x + 2y = -9$
$x - y = 6$

36. $4x + 3y = 24$
$10x - 16y = -34$

37. $12x - y = -21$
$\frac{1}{2}x + \frac{2}{3}y = -3$

38. $\frac{1}{2}x + \frac{1}{3}y = 6$
$y = \frac{1}{2}x + 2$

39. $\frac{2}{3}x + \frac{1}{4}y = 4$
$x = -\frac{3}{8}y + 6$

40. $3.4x + 6.3y = 4.4$
$2.1x + 3.7y = 3.1$

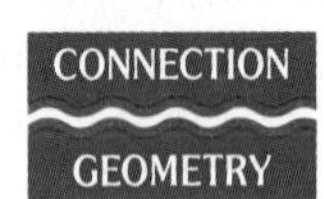

CONNECTION GEOMETRY

41. The graphs of the equations $2x + y = 8$, $-x + 2y = 6$, and $7x + y = 3$ contain the sides of a triangle. Find the coordinates of the vertices of the triangle.

42. The graphs of the equations $y = 2$, $3x + 2y = 1$, and $3x - 4y = -29$ contain the sides of a triangle. Find the measure of the area of the triangle. (*Hint:* Use the formula $A = \frac{1}{2}bh$.)

Critical Thinking

43. The solution to the system of equations $Ax + y = 5$ and $Ax + By = 7$ is $(-1, 2)$. What are the values of A and B?

Applications

44. Gardening The perimeter of a rectangular garden is 40 meters. The length of the garden is 1 meter less than twice its width. What are the dimensions of this garden?

45. Sales A cab ride costs \$1.70 plus \$0.10 per tenth of a mile traveled if you use the Blue Cab Company. The cost is \$1.55 plus \$0.15 per tenth of a mile traveled if you use the Red Cab Company. For what distance will the cab rides cost the same?

Journal

Describe the three different possibilities that may occur when graphing a system of two linear equations. What types of solutions occur with each possibility?

46. Ballooning A hot-air balloon is 10 meters above the ground rising at a rate of 15 meters per minute. Another balloon is 150 meters above the ground descending at a rate of 20 meters per minute.

a. After how long will the balloons be at the same height?

b. What is that height?

Mixed Review

47. Replace $\underline{?}$ with $<$, $>$, or $=$ to make the sentence $\frac{7}{8} \underline{?} \frac{29}{33}$ true. **(Lesson 2-3)**.

48. Simplify $(m^2n)(am)(an^2)$. **(Lesson 6-2)**

49. Boating The rate of the current of a river is 5 miles per hour. A boat travels downstream 78 miles and returns in 32 hours. What is the rate of the boat in still water? **(Lesson 8-7)**

50. Geometry The center of a circle is at $(-2, 3)$. One endpoint of a diameter is at $(3, 9)$. Find the other endpoint of this diameter. **(Lesson 10-8)**

11-3 Substitution

Objective After studying this lesson, you should be able to:

- solve systems of equations by the substitution method.

Application

Kiyo wants to build a rectangular corral for her horse using picket fencing. To make the best use of the available land, the length of the corral must be three times its width. If Kiyo has 210 meters of fencing and wants to use it all, what are the dimensions of her corral?

Let y = the length of the corral and let x = its width. You know that the length must be three times the width. Also, the perimeter must be 210 meters since Kiyo has that much fencing. This information can be described by the following system of equations.

$$y = 3x$$
$$2y + 2x = 210$$

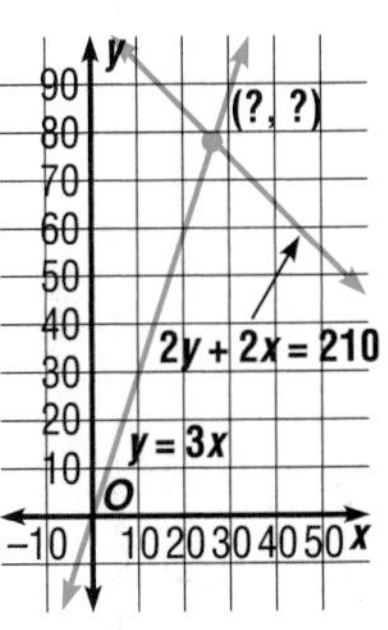

You could try to solve this system of equations by graphing, as shown at the right. Notice that the *exact* coordinates of the point where the lines intersect cannot be easily determined from this graph. The exact solution of this system of equations can be found by using algebraic methods. One such method is called **substitution.**

In the system $y = 3x$ and $2y + 2x = 210$, you know that y is equal to $3x$. Since y must have the same value in *both* equations, you can substitute $3x$ for y in the equation $2y + 2x = 210$.

$$2y + 2x = 210$$
$$2(3x) + 2x = 210 \quad \text{Substitute } 3x \text{ for } y \text{ so the equation will have only one variable.}$$
$$8x = 210$$
$$x = 26.25$$

Now find the value of y by substituting 26.25 for x in $y = 3x$.

$$y = 3x$$
$$y = 3(26.25) \quad \text{You could also substitute 26.25 for } x \text{ in } 2y + 2x = 210.$$
$$y = 78.75$$

Check:

$$y = 3x \qquad\qquad 2y + 2x = 210$$
$$78.75 \stackrel{?}{=} 3(26.25) \qquad\qquad 2(78.75) + 2(26.25) \stackrel{?}{=} 210$$
$$78.75 = 78.75 \ \checkmark \qquad\qquad 210 = 210 \ \checkmark$$

The solution of the system of equations is (26.25, 78.75). Therefore, the dimensions of Kiyo's corral are 78.75 meters by 26.25 meters.

FYI . . .

The world's horse population is estimated to be around 75 million.

Example 1

Use substitution to solve the system of equations $x + 6y = 1$ and $3x - 10y = 31$.

Solve the first equation for x since the coefficient of x is 1.

$$x + 6y = 1$$
$$x = 1 - 6y$$

Next find the value of y by substituting $1 - 6y$ for x in the second equation.

$$3x - 10y = 31$$
$$3(1 - 6y) - 10y = 31$$
$$3 - 18y - 10y = 31$$
$$-28y = 28$$
$$y = -1$$

Then substitute -1 for y in either of the original equations and find the value of x. *Choose the equation that is easier for you to solve.*

$$x + 6y = 1$$
$$x + 6(-1) = 1$$
$$x - 6 = 1$$
$$x = 7$$

The solution of this system is $(7, -1)$. *Check this result.*

Example 2

Use substitution to solve the system of equations $\frac{3}{2}x - y = 3$ and $3x - 2y = 12$.

Solve the first equation for y since the coefficient of y is -1.

$$\frac{3}{2}x - y = 3$$
$$-y = -\frac{3}{2}x + 3$$
$$y = \frac{3}{2}x - 3$$

Then find the value of x by substituting $\frac{3}{2}x - 3$ for y in the second equation.

$$3x - 2y = 12$$
$$3x - 2\left(\frac{3}{2}x - 3\right) = 12$$
$$3x - 3x + 6 = 12$$
$$6 = 12$$

The statement $6 = 12$ is false. This means that there are no ordered pairs that are solutions to both equations. Compare the slope-intercept forms of the equations, which are $y = \frac{3}{2}x - 3$ and $y = \frac{3}{2}x - 6$. Notice that these lines have the same slope but different y-intercepts. Thus, the lines are parallel and the system of equations has no solution.

Example 3

A metal alloy is 25% copper. Another metal alloy is 50% copper. How much of each alloy should be used to make 1000 grams of a metal alloy that is 45% copper?

EXPLORE Let a = the number of grams of the 25% copper alloy.
Let b = the number of grams of the 50% copper alloy.

	25% Copper	50% Copper	45% Copper
Total Grams	a	b	1000
Grams of Copper	$0.25a$	$0.50b$	0.45(1000)

PLAN The system of equations is $a + b = 1000$ and $0.25a + 0.50b = 0.45(1000)$. Use substitution to solve this system.

SOLVE Since $a + b = 1000$, $a = 1000 - b$.

$$0.25a + 0.50b = 0.45(1000)$$
$$0.25(1000 - b) + 0.50b = 450 \quad \text{Substitute } 1000 - b \text{ for } a.$$
$$250 - 0.25b + 0.50b = 450 \quad \text{Solve for } b.$$
$$0.25b = 200$$
$$b = 800$$

$$a + b = 1000$$
$$a + 800 = 1000 \quad \text{Substitute 800 for } b.$$
$$a = 200 \quad \text{Solve for } a.$$

200 grams of the 25% copper alloy and 800 grams of the 50% copper alloy should be used.

EXAMINE The 45% copper alloy contains $0.25(200) + 0.50(800) = 50 + 400$, or 450 grams of copper. Since $0.45(1000) = 450$, the answer is correct.

Example 4

CONNECTION
Number Theory

The sum of the digits of a two-digit number is 9. The number is 6 times the units digit. Find the number.

Let t = the tens digit and let u = the units digit of the number. Then any two-digit number can be represented as $10t + u$.

Since the sum of the digits is 9, one equation is $t + u = 9$. Since the number is 6 times its unit digit, another equation is $10t + u = 6u$. Use substitution to solve this system.

Since $t + u = 9$, $t = 9 - u$.

$$10t + u = 6u$$
$$10(9 - u) + u = 6u \quad \text{Substitute } 9 - u \text{ for } t.$$
$$90 - 10u + u = 6u \quad \text{Solve for } u.$$
$$90 = 15u$$
$$6 = u$$

$$t + u = 9 \quad \text{Substitute 6 for } u.$$
$$t + 6 = 9$$
$$t = 3 \quad \text{Solve for } t.$$

The number is $10(3) + 6$, or 36. *Check this result.*

CHECKING FOR UNDERSTANDING

Communicating Mathematics

Read and study the lesson to answer each question.

1. To solve the system of equations $y = -5x + 1$ and $2x + 3y = 7$, why can you substitute $-5x + 1$ for y in the second equation?
2. In Example 3, what is the system of equations if a 30% copper alloy and a 70% copper alloy is used to make 1000 grams of the 45% copper alloy?
3. How many two-digit whole numbers are there such that the sum of the digits of the number is 9?
4. Explain why any 2-digit number can be represented by $10t + u$, where t is the tens digit and u is the ones digit.

Guided Practice

Solve each equation for x. Then, solve each equation for y.

5. $x + y = 5$
6. $2x + y = 3$
7. $2x + 3y = 6$
8. $3y - \frac{1}{2}x = 7$
9. $0.75x + 6 = -0.8y$
10. $\frac{2}{3}x - \frac{4}{5}y = 3$

For each system of equations, use the first equation to make a substitution in the second equation. Then solve the second equation.

11. $y = 3 + 2x$
    $x + y = 7$
12. $y = 7 - x$
    $2x - y = 8$
13. $y = x$
    $5x = 12y$
14. $x = 5 - y$
    $3y = 3x + 1$
15. $3x = -18 + 2y$
    $x + 3y = 4$
16. $2x = 3 - y$
    $2y = 12 - x$

EXERCISES

Practice

Use substitution to solve each system of equations. If the system *does not* have exactly one solution, state whether it has no solution or infinitely many solutions.

17. $y = 3x$
    $x + 2y = -21$
18. $y = 2x$
    $x + 2y = 8$
19. $x = 2y$
    $4x + 2y = 15$
20. $3x + y = 6$
    $y + 2 = x$
21. $2x - y = -4$
    $-3x + y = -9$
22. $x = 3 - 2y$
    $2x + 4y = 6$
23. $2x + 3y = 5$
    $4x - 9y = 9$
24. $x - 3y = 3$
    $2x + 9y = 11$
25. $9x + 6y = 14$
    $3x + 2y = 11$
26. $3x + y = 2$
    $4x - 2y = 1$
27. $3x + 5y = 2x$
    $x + 3y = y$
28. $x - 2y = 5$
    $3x - 5y = 8$
29. $2x + 3 = 3y$
    $4x - 3y = 3$
30. $3x - 2y = -3$
    $25x + 10y = 215$
31. $0.5x - 2y = 17$
    $2x + y = 104$
32. $3x + 2y = 18$
    $-\frac{1}{4}x - \frac{2}{3}y = -3$
33. $8x + 6y = 44$
    $\frac{1}{4}x - 2y = -3$
34. $0.3x + 0.2y = 0.5$
    $0.5x - 0.3y = 0.2$

Use a system of equations and substitution to solve each problem.

35. A two-digit number is 6 times its units digit. Find the number if the sum of its digits is 6.

36. A two-digit number is 2 more than 8 times the sum of its digits. Find the number if its tens digit is 6 more than its units digit.

37. A two-digit number is 7 times its unit digit. If 18 is added to the number, its digits are reversed. Find the number.

Use substitution to solve each system of equations. Write each solution as an ordered triple of the form (x, y, z).

38. $x + y + z = -54$
$x = -6y$
$z = 14y$

39. $2x + 3y - z = 17$
$y = -3z - 7$
$2x = z + 2$

40. $12x - y + 7z = 99$
$x + 2z = 2$
$y + 3z = 9$

Critical Thinking

41. If 36 is subtracted from a certain two-digit positive integer, then its digits are reversed. Find all integers for which this is true.

Applications

42. **Sports** Eric is preparing to run in the Bay Marathon. One day, he ran and walked a total of 16 miles. If he ran one mile more than twice as far as he walked, how many miles did he run?

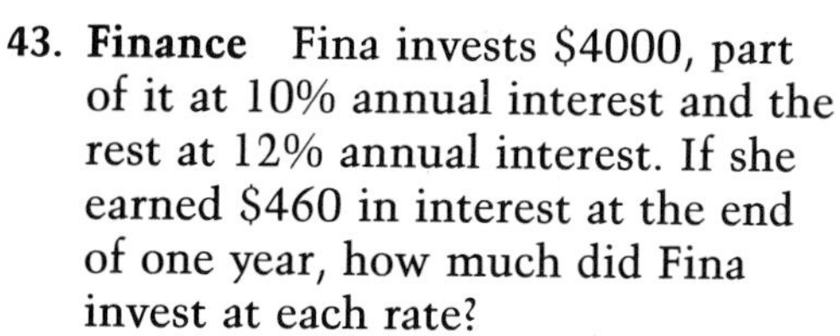

Journal

Look at Example 2 on page 85. Show how you could solve this problem using two equations.

43. **Finance** Fina invests \$4000, part of it at 10% annual interest and the rest at 12% annual interest. If she earned \$460 in interest at the end of one year, how much did Fina invest at each rate?

44. **Chemistry** MX Labs needs to make 500 gallons of a 34% acid solution. The only solutions available are 25% acid and 50% acid. How many gallons of each solution should be mixed to make the 34% solution?

45. **Sales** Musicville had a sale on 500 assorted cassette tapes. Some tapes were sold for \$10 while the rest were sold for \$8. After all of the tapes were sold, the average price of a tape was \$9.50. How many \$8 tapes were sold?

Mixed Review

46. Solve $|13 - 2y| < 9$ and graph its solution set. **(Lesson 5-6)**

47. Factor $x^3 + 2x^2 - 4x - 8$. **(Lesson 7-8)**

48. **Sales** Amy works at Savemore Shoes. Her daily income is described by the formula $f(s) = 25 + 0.15s$, where s is the amount of her total sales for the day. If Amy earned \$94 on Monday, what were her total sales that day? **(Lesson 9-6)**

49. Write the standard form of the equation of the line passing through $(-5, 1)$ and $(6, -2)$. **(Lesson 10-2)**

50. Graph the equations $5x - 3y = 12$ and $2x - 5y = 1$. Then state the solution of the system of equations. **(Lesson 11-2)**

11-4 Elimination Using Addition and Subtraction

Objective

After studying this lesson, you should be able to:

- solve systems of equations by the elimination method using addition or subtraction.

Application

FYI . . .

The world's largest amusement park is Disney World, located on 28,000 acres in Central Florida.

Winona and Larry each had a birthday party at Water World last weekend. The cost of admission to Water World was \$137.50 for the 13 children and 2 adults at Winona's party. The admission was \$103.50 for the 9 children and 2 adults at Larry's party. What was the price of admission to Water World for an adult and for a child?

Let a = the price of admission for one adult, and let c = the price of admission for one child. Then the information in this problem can be represented by the following system of equations.

$$2a + 13c = 137.50$$
$$2a + 9c = 103.50$$

You could solve this system by first solving either of the equations for either a or c and then using substitution. However, a simpler method is to subtract one equation from the other since the coefficients of the variable a are the same. This method is called **elimination** because the subtraction eliminates one of the variables.

$$\begin{array}{r} 2a + 13c = 137.50 \\ (-)\ 2a + 9c = 103.50 \\ \hline 4c = 34 \\ c = 8.5 \end{array}$$

Write the equations in column form and subtract.

Notice that the variable a is eliminated.

Now substitute 8.5 for c in either equation and find the value of a.

$$2a + 9c = 103.50$$
$$2a + 9(8.5) = 103.50 \quad \textit{Substitute 8.5 for c.}$$
$$2a + 76.50 = 103.50$$
$$2a = 27$$
$$a = 13.5$$

Check:

$$2a + 13c = 137.50$$
$$2(13.5) + 13(8.5) \stackrel{?}{=} 137.50$$
$$137.5 = 137.50 \ \checkmark$$

$$2a + 9c = 103.50$$
$$2(13.5) + 9(8.5) \stackrel{?}{=} 103.50$$
$$103.5 = 103.50 \ \checkmark$$

The solution of this system of equations is (13.5, 8.5). Thus, the cost of admission to Water World is \$13.50 for an adult and \$8.50 for a child.

In some systems of equations, the coefficients of terms containing the same variable are additive inverses. For these systems, the elimination method can be applied by adding the equations.

Example 1

Use elimination to solve the system of equations $x - 4y = 6$ and $3x + 4y = 10$.

Since the coefficients of the y-terms, -4 and 4, are additive inverses, you can solve the system by adding the equations.

$$\begin{array}{rl} x - 4y = 6 & \\ (+)\ 3x + 4y = 10 & \\ \hline 4x \quad = 16 & \\ x = 4 & \end{array}$$

Write the equations in column form and add.

Notice that the variable y is eliminated.

Now substitute 4 for x in either equation and find the value of y.

$$\begin{aligned} x - 4y &= 6 \\ 4 - 4y &= 6 \\ -4y &= 2 \\ y &= -\frac{1}{2} \end{aligned}$$

The solution of the system is $\left(4, -\frac{1}{2}\right)$. *Check this result.*

Example 2

The sum of two numbers is 42. Their difference is 6. Find the numbers.

Let x = the greater number.
Let y = the lesser number.

Since the sum of the numbers is 42, one equation is $x + y = 42$. Since the difference of the numbers is 6, another equation is $x - y = 6$. Use elimination to solve this system.

$$\begin{array}{r} x + y = 42 \\ (+)\ x - y = 6 \\ \hline 2x \quad = 48 \\ x = 24 \end{array}$$

Since the coefficients of the y-terms are additive inverses, use elimination by addition.

$$\begin{aligned} x + y &= 42 \\ 24 + y &= 42 && \textit{Substitute 24 for x.} \\ y &= 18 && \textit{Solve for y.} \end{aligned}$$

The numbers are 24 and 18. *Check this result.*

A boat is rowed 24 miles downstream in 4 hours. In order to make the return trip upstream in the same amount of time, the rate of the boat in still water was doubled. Find the rate of the current and the rate of the boat in still water on the downstream trip.

EXPLORE Let b = the rate of the boat in still water on the downstream trip. Let c = the rate of the current. Then $2b$ = the rate of the boat in still water on the upstream trip.

Use the formula rate × time = distance, or $rt = d$.

	r	t	d	$rt = d$
Downstream	$b + c$	4	24	$4(b + c) = 24$
Upstream	$2b - c$	4	24	$4(2b - c) = 24$

PLAN

$4(b + c) = 24 \rightarrow b + c = 6$ *Divide each side of*

$4(2b - c) = 24 \rightarrow 2b - c = 6$ *both equations by 4.*

SOLVE

$$\begin{array}{r} b + c = 6 \\ (+)\ 2b - c = 6 \\ \hline 3b \quad = 12 \\ b = 4 \end{array}$$

Since the coefficients of the c-terms are additive inverses, use elimination by addition.

$b + c = 6$ *Substitute 4 for b.*

$4 + c = 6$ *Solve for c.*

$c = 2$

The rate of the current is 2 miles per hour and the rate of the boat in still water on the downstream trip is 4 miles per hour.

EXAMINE The rate of the boat on the downstream trip is 4 + 2 or 6 miles per hour. Its rate on the upstream trip is 2(4) − 2 or 6 miles per hour. Since these rates are the same, the answer is correct.

CHECKING FOR UNDERSTANDING

Communicating Mathematics

Read and study the lesson to answer each question.

1. When is it easier to solve a system of equations by elimination using subtraction?
2. When is it easier to solve a system of equations by elimination using addition?
3. In Example 2, would you get the same answer if you let y = the greater number and x = the lesser number? Explain.
4. What is the result when you add $2x - 5y = 23$ and $-2x + 5y = 12$? What does the result tell you about the system of equations?

Guided Practice

State whether addition, subtraction, both, or neither could be used to solve each system of equations. Then solve the system.

5. $3a + b = 6$
   $4a + b = 7$
6. $m + 3n = 5$
   $n + 2m = 3$
7. $3x + y = 12$
   $3y - 3x = 6$
8. $5x + y = 9$
   $y - 5x = 7$

For each system of equations, first eliminate y. Then solve the system.

9. $x + y = -3$
   $2x + y = 6$
10. $x + y = 6$
    $2x - y = 6$
11. $5x - 2y = 23$
    $5x + 2y = 17$
12. $2x = 4 - 3y$
    $3y - x = 11$

EXERCISES

Practice

Use elimination to solve each system of equations.

13. $x + y = 7$
    $x - y = 9$
14. $r - s = -5$
    $r + s = 25$
15. $2x - y = 32$
    $2x + y = 60$
16. $x - y = 3$
    $y + x = 3$
17. $-n + m = 6$
    $m + n = 5$
18. $x + y = 8$
    $2x - y = 6$
19. $x + 2y = 8$
    $3x + 2y = 6$
20. $3x + 1 = -7y$
    $6x + 7y = 0$
21. $3x = 13 - y$
    $2x - y = 2$
22. $2x - 3y = -4$
    $x = 7 - 3y$
23. $5s + 4t = 12$
    $3s = 4 + 4t$
24. $12x - 9y = 114$
    $7y + 12x = 82$
25. $4x - \frac{1}{3}y = 8$
    $5x + \frac{1}{3}y = 6$
26. $\frac{3}{4}x + \frac{1}{5}y = 5$
    $\frac{3}{4}x - \frac{1}{5}y = -5$
27. $\frac{2}{3}x - \frac{1}{2}y = 14$
    $\frac{5}{6}x - \frac{1}{2}y = 18$
28. $9x + 2y = 26$
    $1.5x - 2y = 13$
29. $3x + 0.2y = 7$
    $3x = 0.4y + 4$
30. $0.6m - 0.2n = 0.9$
    $0.3m = 0.9 - 0.2n$

Use a system of equations and elimination to solve each problem.

31. Find two numbers whose sum is 64 and whose difference is 42.
32. The units digit of a two-digit number exceeds twice the tens digit by 1. Find the number if the sum of its digits is 7.

Use elimination twice to solve each system of equations. Write each answer as an ordered triple of the form (x, y, z).

33. $x - y + 2z = 8$
    $2x + y + z = 13$
    $4x - 3z = 7$
34. $-3x + y - z = 6$
    $3x - 2y + 2z = -9$
    $-y - 3z = 1$
35. $5x - 2y + z = 0$
    $2x - y + z = -3$
    $3x + 4y = 18$

36. **Statistics** The mean of two numbers is 28. Find the numbers if 3 times one of the numbers equals half the other number.

Critical Thinking

37. Find the values of A and B if $(11, -5)$ is the only solution to the system of equations $Ax + By = 7$ and $Ax + (1 - 2B)y = 47$.

Applications

38. **Conferences** Last year, 2713 teachers attended a technology conference. If there were 163 more men than women at the conference, how many men and how many women attended?

39. **Testing** Jerrod received a total score of 1340 on the Scholastic Aptitude Test (SAT). His math score was 400 points less than twice his verbal score. What was his math score and his verbal score?

40. **Uniform Motion** In still water, a speedboat travels 5 times faster than the current of the river. If the speedboat can travel 48 miles upstream and then back in 5 hours, find the rate of the current.

Mixed Review

41. Translate *the sum of* y *and the cube of* n *is equal to twice* x into an equation. **(Lesson 1-7)**

42. Solve $12m^2 + 3 = -20m$. Check your solution. **(Lesson 7-10)**

43. Determine the domain, range, and inverse for $\{(5, 1), (-3, 2), (4, 2), (5, 0), (2, 2)\}$. **(Lesson 9-2)**

44. **Aviation** An airplane passing over Sacramento at an elevation of 37,000 feet begins its descent to land at Reno, 140 miles away. If the elevation of Reno is 4500 feet, what should be the approximate slope of descent? **(Lesson 10-1)**

45. Use substitution to solve the system of equations $y = -2x + 8$ and $3x - y = 17$. **(Lesson 11-3)**

MID-CHAPTER REVIEW

1. 376 teams are playing in a single-elimination basketball tournament. If a team loses one game, it is out of the tournament. How many games must be played before the winning team can be determined? **(Lesson 11-1)**

Graph each system of equations. Then determine if the system has one solution, no solution, or infinitely many solutions. If the system has one solution, name it. (Lesson 11-2)

2. $x - y = 3$; $3x + y = 1$
3. $2x - 3y = 7$; $3y = 7 + 2x$
4. $4x + y = 12$; $x = 3 - \frac{1}{4}y$
5. $2x - y = 3$; $\frac{2}{3}x = y - 1$

Use substitution to solve each system of equations. (Lesson 11-3)

6. $y = 5x$; $x + 2y = 22$
7. $x = 2y + 3$; $3x + 4y = -1$
8. $2y - x = -5$; $4y - 3x = -1$
9. $3x + 2y = 18$; $\frac{1}{4}x + \frac{2}{3}y = 3$

10. **Uniform Motion** Odina walks from her house to a friend's house in 1 hour. She can travel the same distance on her bicycle in 15 minutes. If she rides 6 miles per hour faster than she can walk, what is her speed on the bicycle? **(Lesson 11-3)**

11-5 Elimination Using Multiplication

Objective

After studying this lesson, you should be able to:

- solve systems of equations by the elimination method using multiplication and addition.

Application

A-1 Car Rental rents compact cars for a fixed amount per day plus a fixed amount for each mile driven. Benito Sanchez rented a car from A-1 for 6 days, drove it 550 miles, and spent \$337. Lisa McGuire rented the same car for 3 days, drove it 350 miles, and spent \$185. What are the charge per day and charge per mile driven, excluding gas, insurance, and taxes?

Let d = the charge per day for renting the car, and let m = the charge per mile driven. Then the information in this problem can be represented by the following system of equations.

$$6d + 550m = 337$$
$$3d + 350m = 185$$

Neither of the variables in this system can be eliminated by simply adding or subtracting the equations. Substitution could be used but the computations would not be easy. A simpler method is to multiply one of the equations by some number so that adding or subtracting eliminates one of the variables. For this system, multiply the second equation by −2 and add.

$$\begin{aligned} 6d + 550m &= 337 \\ 3d + 350m &= 185 \end{aligned} \quad \xrightarrow{\text{Multiply by } -2.} \quad \begin{aligned} 6d + 550m &= 337 \\ (+)\ -6d - 700m &= -370 \\ \hline -150m &= -33 \\ m &= 0.22 \end{aligned}$$

$$\begin{aligned} 6d + 550m &= 337 \\ 6d + 550(0.22) &= 337 && \textit{Substitute 0.22 for m.} \\ 6d + 121 &= 337 && \textit{Solve for d.} \\ 6d &= 216 \\ d &= 36 \end{aligned}$$

Check:

$$\begin{aligned} 6d + 550m &= 337 \\ 6(36) + 550(0.22) &\stackrel{?}{=} 337 \\ 337 &= 337 \ \checkmark \end{aligned} \qquad \begin{aligned} 3d + 350m &= 185 \\ 3(36) + 350(0.22) &\stackrel{?}{=} 185 \\ 185 &= 185 \ \checkmark \end{aligned}$$

The solution to the system is (36, 0.22). Thus, the charge per day is \$36 and the charge per mile driven is \$0.22 for this compact car.

For some systems of equations, it is necessary to multiply *each* equation by a different number in order to solve the system by elimination. This can be accomplished in several ways, depending on which of the variables you choose to eliminate.

Example 1

Use elimination to solve the system of equations $3x + 4y = -25$ and $2x - 3y = 6$ in two different ways.

Method 1 You can eliminate the variable x by multiplying the first equation by 2 and the second equation by −3 and then adding the resulting equations.

$3x + 4y = -25$ → Multiply by 2. → $6x + 8y = -50$

$2x - 3y = 6$ → Multiply by −3. → $(+)\ -6x + 9y = -18$

$$17y = -68$$
$$y = -4$$

Now find x using one of the original equations.

$$3x + 4y = -25$$
$$3x + 4(-4) = -25 \quad \textit{Substitute } -4 \textit{ for } y.$$
$$3x - 16 = -25 \quad \textit{Solve for } x.$$
$$3x = -9$$
$$x = -3$$

The solution of the system is $(-3, -4)$.

Method 2 You can also solve this system by eliminating the variable y. Multiply the first equation by 3 and the second equation by 4. Then add.

$3x + 4y = -25$ → Multiply by 3. → $9x + 12y = -75$

$2x - 3y = 6$ → Multiply by 4. → $(+)\ 8x - 12y = 24$

$$17x = -51$$
$$x = -3$$

Now find y.

$$3x + 4y = -25$$
$$3(-3) + 4y = -25 \quad \textit{Substitute } -3 \textit{ for } x.$$
$$-9 + 4y = -25 \quad \textit{Solve for } y.$$
$$4y = -16$$
$$y = -4$$

The solution is $(-3, -4)$, which matches the result obtained above.

Example 2

A bank teller accidentally reversed the digits in the amount of a check and overpaid a customer by $36. If the sum of the digits in the two-digit amount was 10, what was the actual amount of the check?

EXPLORE Let t = the tens digit of the amount of the check.
Let u = the units digit.
The actual amount of the check can be represented by $10t + u$.
The amount paid by the teller can be represented by $10u + t$. *Why?*

PLAN Since the sum of the digits is 10, one equation is $t + u = 10$. Since the teller overpaid the customer by $36, another equation is $(10u + t) - (10t + u) = 36$, or $-9t + 9u = 36$.

SOLVE

$$t + u = 10 \quad \xrightarrow{\text{Multiply by 9.}} \quad 9t + 9u = 90$$

$$-9t + 9u = 36 \qquad\qquad (+)\ -9t + 9u = 36$$

$$18u = 126$$

$$u = 7$$

$$t + u = 10$$

$t + 7 = 10$ *Substitute 7 for u.*

$t = 3$ *Solve for t.*

The actual amount of the check was $10(3) + 7$, or \$37.
Check this result.

Example 3

A coal barge on the Ohio River travels 24 miles upstream in 3 hours. The return trip takes the barge only 2 hours. Find the rate of the barge in still water.

EXPLORE Let b = the rate of the barge in still water.
Let c = the rate of the current.

PLAN Use the formula $rt = d$ to write a system of equations. Then solve the system to find the value of b.

	r	t	d	rt = d
Downstream	$b + c$	2	24	$2b + 2c = 24$
Upstream	$b - c$	3	24	$3b - 3c = 24$

SOLVE

$3b - 3c = 24$ → Multiply by 2. → $6b - 6c = 48$

$2b + 2c = 24$ → Multiply by 3. → $(+)\ 6b + 6c = 72$

$$12b = 120$$

$$b = 10$$

The rate of the barge in still water is 10 miles per hour.
Find the value of c for this system and then check the solution.

CHECKING FOR UNDERSTANDING

Communicating Mathematics

Read and study the lesson to answer each question.

1. Are $5x - 7y = 3$ and $-15x + 21y = -9$ equivalent equations?
2. When using elimination to solve a system of equations, why might you need to multiply each equation by a different number?
3. Write a system of equations where you can eliminate the variable y by multiplying one equation by 3 and then adding the equations.

Guided Practice Explain the steps you would follow to eliminate the variable x in each system of equations. Then solve the system.

4. $x + 2y = 5$
   $3x + y = 7$
5. $4x + y = 8$
   $x - 7y = 2$
6. $y + x = 9$
   $2y - x = 1$
7. $2x + y = 6$
   $3x - 7y = 9$

For each system of equations, use multiplication and then addition to eliminate the variable y. Then solve the system.

8. $x + 8y = 3$
   $4x - 2y = 7$
9. $4x - y = 4$
   $x + 2y = 3$
10. $3y - 8x = 9$
    $y - x = 2$
11. $5y - 4x = 2$
    $2y + x = 6$

EXERCISES

Practice Use elimination to solve each system of equations.

12. $x - 5y = 0$
    $2x - 3y = 7$
13. $x + 4y = 30$
    $2x - y = -6$
14. $9x + 8y = 7$
    $18x - 15y = 14$
15. $-5x + 8y = 21$
    $10x + 3y = 15$
16. $5x + 3y = 12$
    $4x - 5y = 17$
17. $4x + 3y = 19$
    $3x - 4y = 8$
18. $7x + 2y = 3(x + 16)$
    $x + 16 = 5y + 3x$
19. $2x - y = 36$
    $3x - 0.5y = 26$
20. $x - 0.5y = 1$
    $0.4x + y = -2$
21. $\frac{1}{3}x - y = -1$
    $\frac{1}{5}x - \frac{2}{5}y = -1$
22. $\frac{1}{2}x - \frac{2}{3}y = \frac{7}{3}$
    $\frac{3}{2}x + 2y = -25$
23. $\frac{2x + y}{3} = 15$
    $\frac{3x - y}{5} = 1$
24. $x + y = 600$
    $0.06x + 0.08y = 46$
25. $x + y = 20$
    $0.4x + 0.15y = 4$
26. $0.25(x + 4y) = 3.5$
    $0.5x - 0.25y = 1$

Use a system of equations and elimination to solve each problem.

27. The sum of the digits of a two-digit number is 7. If the digits are reversed, the new number is 3 less than 4 times the original number. Find the number.

28. The ratio of the tens digit to the units digit of a two-digit number is 1:4. If the digits are reversed, the sum of the new number and the original number is 110. Find the number.

Use elimination to solve each system of equations.

29. $\frac{2}{x + 7} - \frac{1}{y - 3} = 0$
    $\frac{1}{x - 5} - \frac{3}{y + 6} = 0$
30. $\frac{1}{x} + \frac{1}{y} = 7$
    $\frac{2}{x} + \frac{3}{y} = 16$
31. $\frac{1}{x + y} = 2$
    $\frac{1}{x - y} = \frac{1}{y}$

Critical Thinking

32. Explain how the elimination or substitution methods show that a system of equations is inconsistent or that a system of equations is consistent and dependent.

Applications

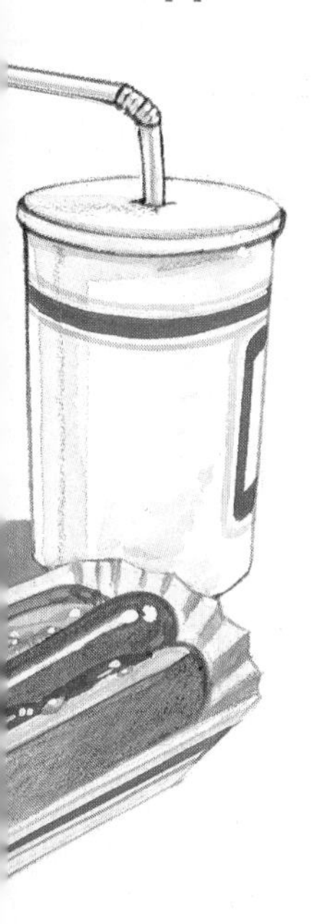

33. **Uniform Motion** A riverboat traveled 48 miles downstream in 2 hours. The return trip upstream took 2 hours and 40 minutes.
    a. Find the rate of the riverboat in still water.
    b. Find the rate of the current.

34. **Sales** The concession stand sells hot dogs and soda during Beck High School football games. John bought 6 hot dogs and 4 sodas and paid $6.70. Jessica bought 4 hot dogs and 3 sodas and paid $4.65.
    a. What is the price of a hot dog?
    b. What is the price of a soda?

35. **Uniform Motion** Traveling against the wind, a plane flies 2100 miles from Chicago to San Diego in 4 hours and 40 minutes. The return trip, traveling with a wind that is twice as fast, takes 4 hours. Find the rate of the plane in still air.

36. **Sales** The Beach Resort is offering two weekend specials. One includes a 2-night stay with 3 meals and costs $195. The other includes a 3-night stay with 5 meals and costs $300.
    a. What is the cost of a 1-night stay?
    b. What is the cost per meal?

Computer

37. The BASIC program at the right finds the solution of the following system of equations.

$$ax + by = c$$
$$dx + ey = f$$

The formulas for the solution of this system are as follows.

$$x = \frac{ce - bf}{ae - bd},\ y = \frac{af - cd}{ae - bd}$$

Use the program to solve each system.

```
10 PRINT "ENTER THE COEFFICIENTS."
20 INPUT A,B,C,D,E,F
30 IF A*E-B*D=0 THEN 80
40 LET X = (C*E-B*F)/(A*E-B*D)
50 LET Y = (A*F-C*D)/(A*E-B*D)
60 PRINT "(";X;",";Y;") IS A
    SOLUTION."
70 GOTO 10
80 IF C*E-B*F=0 OR A*F-C*D=0
    THEN 110
90 PRINT "NO SOLUTION."
100 GOTO 10
110 PRINT "INFINITE NUMBER OF
    SOLUTIONS."
120 GOTO 10
130 END
```

a. $5x + 5y = 16$
   $2x + 2y = 5$

b. $7x - 3y = 5$
   $14x - 6y = 10$

c. $x - 2y = 5$
   $3x - 5y = 8$

d. $6x + 3y = 0$
   $4x + 2y = 0$

Mixed Review

38. Solve $\frac{1}{3}a + 3 = \frac{1}{2}a$. Check your solution. **(Lesson 3-7)**

39. **Gardening** Ms. Salgado has a rectangular garden that is 12 feet longer than it is wide. In order to surround the garden with railroad ties, she must reduce each side of the garden by 1 foot. In all, she reduced the area of the garden by 55 square feet. What were the original dimensions of her garden? **(Lesson 6-8)**

40. Divide $\frac{m + 4}{m^2 + 4m + 4}$ by $\frac{m^2 - 16}{4m + 8}$. **(Lesson 8-3)**

41. Use elimination to solve the system of equations $x - 2y = 12$ and $-3x + 2y = 16$. **(Lesson 11-4)**

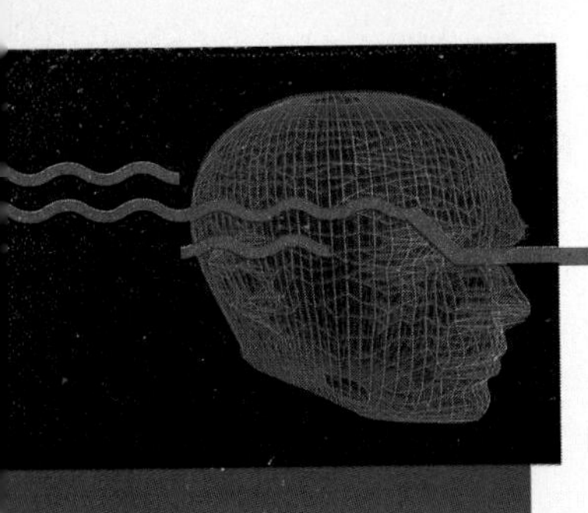

Technology

Solving Systems of Equations

BASIC
▶ **Graphing calculators**
Graphing software
Spreadsheets

Your graphing calculator can help you determine if a solution of a system of equations exists and, if it does, the calculator can help you find it. Using the trace function, you can move to the intersection of two lines and read the x- and y-values of the point of intersection.

Example **Graph the system $3x + y = 8$ and $x + y = 4$. Then find its solution.**

For the TI-81

Set your calculator for the standard viewing window of [−10, 10] by [−10, 10]. Then enter the equations and graph. You must first solve each equation for y.

$3x + y = 8 \rightarrow y = -3x + 8$

$x + y = 4 \rightarrow y = -x + 4$

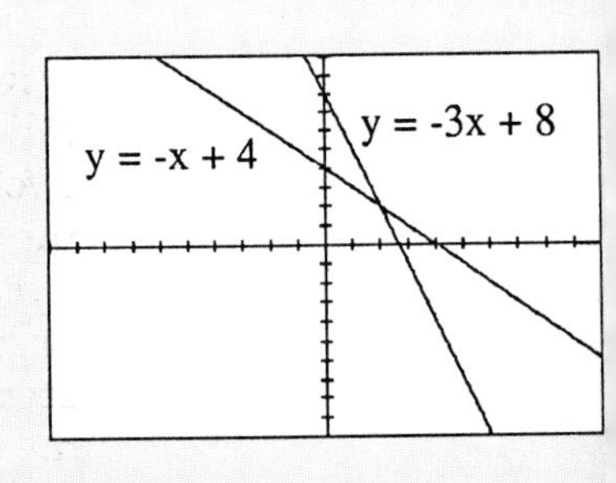

For the Casio

Set your calculator for the default viewing window. Then enter the equations and graph.

Graph (-) 3 ALPHA X + 8 :

Graph (-) ALPHA X + 4 EXE

Trace along the graph to the point of intersection. Use the TRACE and arrow keys. On the Casio, use the X↔Y key to show the y-value. The solution is (2, 2).

EXERCISES

Graph each pair of equations and use the trace key to find the solution. For the TI-81, use the viewing window [−5, 5] by [−5, 5]. For the Casio, use the viewing window [−4.7, 4.7] by [−3.1, 3.1].

1. $y = 3x - 3$
   $y = -3x + 3$
2. $y = 2$
   $y = 3x + 5$
3. $y = 2x - 4$
   $y = -4x + 14$
4. $y = x$
   $y = 4x + 6$
5. $y = -0.8x + 1.28$
   $y = 0.4x - 1.24$
6. $y = 0.4x + 0.6$
   $y = -0.1x + 0.3$

11-6 Graphing Systems of Inequalities

Objective

After studying this lesson, you should be able to:

- solve systems of inequalities by graphing.

Consider the system of inequalities shown below.

$$y \geq x + 2$$
$$y \leq -2x - 1$$

The solution of the system is the set of all ordered pairs that satisfy *both* inequalities. This solution can be determined by graphing each inequality in the same coordinate plane as shown below.

Refer to Lesson 9-7 to review graphing inequalities in two variables.

Recall that the graph of each inequality is called a *half-plane.* The intersection of the two half-planes represents the solution to the system of inequalities. This solution is a region that contains the graphs of an infinite number of ordered pairs. The graphs of $y = x + 2$ and $y = -2x - 1$ are the boundaries of the region and are included in the graph of the system.

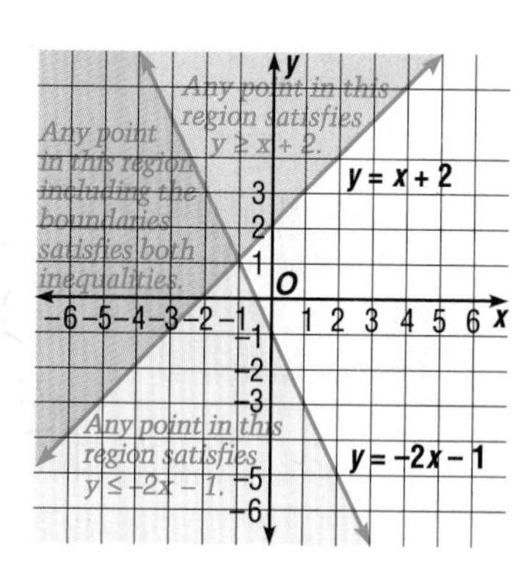

Example 1

Solve each system of inequalities by graphing.

a. $y > x - 3$ **and** $y \leq -1$

The solution is the ordered pairs in the intersection of the graphs of $y > x - 3$ and $y \leq -1$. This region is shaded in green at the right. The graphs of $y = -1$ and $y = x - 3$ are the boundaries of this region. The graph of $y = x - 3$ is a dashed line and is *not* included in the graph of the system.

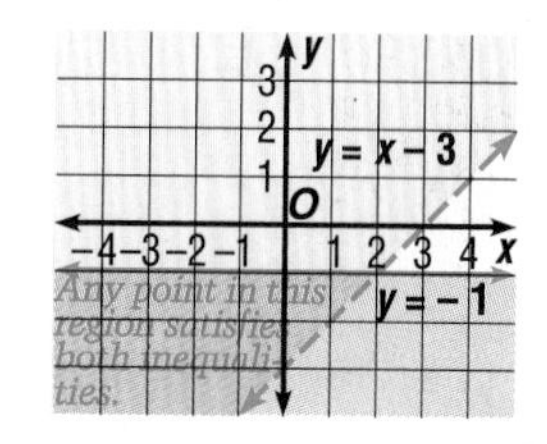

b. $x - 2y \leq -4$ **and** $4y < 2x - 4$

The graphs of $x - 2y = -4$ and $4y = 2x - 4$ are parallel lines. Because the two regions have no points in common, the system of inequalities has no solutions.

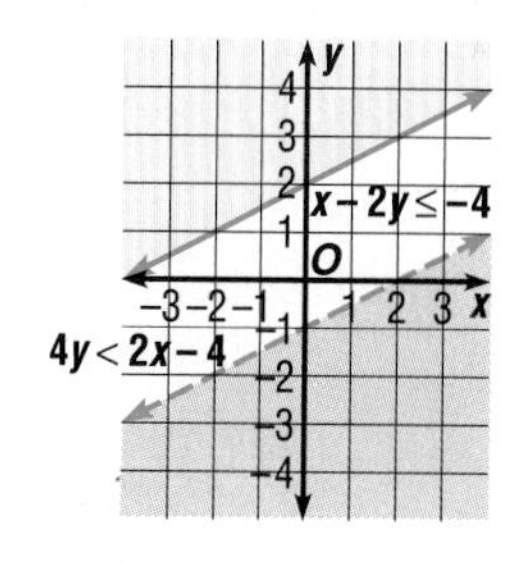

The solution to an inequality that contains an absolute value expression can also be determined by graphing. To do this, write the absolute value inequality as a system of inequalities that does not contain absolute value expressions.

Example 2

Solve the inequality $y \geq |x|$ by graphing.

The inequality $y \geq |x|$ is equivalent to the system of inequalities $y \geq x$ and $y \geq -x$. The solution is the set of all ordered pairs whose graphs are in the intersection of the graphs of these two inequalities. This region is shown in green at the right. The parts of the graphs of $y = x$ and $y = -x$ that are boundaries of this region are included in the graph of the system.

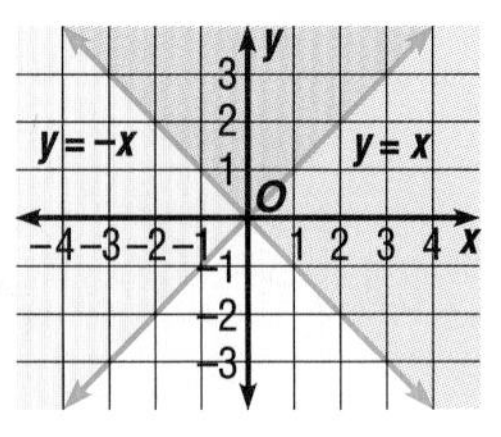

Example 3

To ensure a growing season of sufficient length, Mr. Hobson has at most 16 days left to plant his corn and soybean crops. He can plant corn at a rate of 10 acres per day and soybeans at a rate of 15 acres per day. If he has at most 200 acres available, how many acres of each type of crop can he plant?

Let c = the number of days that corn will be planted.
Let s = the number of days that soybeans will be planted.
Since both c and s represent a number of days, neither can be a negative number. Thus, $c \geq 0$ and $s \geq 0$.

Then the following system of inequalities can be used to represent the conditions of this problem.

$c + s \leq 16$
$10c + 15s \leq 200$, where $c \geq 0$ and $s \geq 0$

The solution is the set of all ordered pairs whose graphs are in the intersection of the graphs of these inequalities. This region is shown in green at the right. Only the portion of the region in the first quadrant is used since $c \geq 0$ and $s \geq 0$.

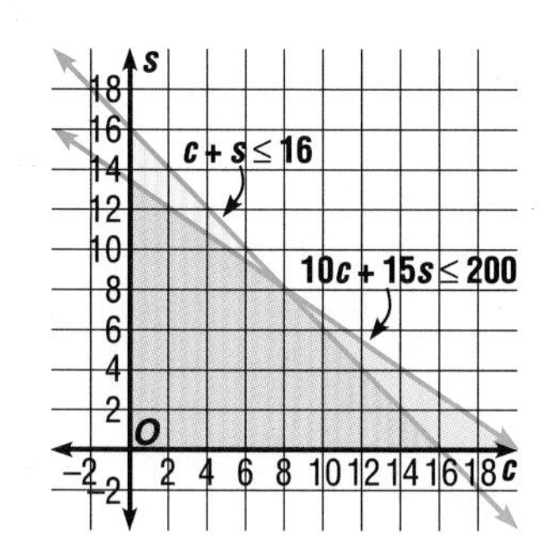

Any point in this region is a possible solution. For example, since (7, 8) is a point in the region, Mr. Hobson could plant corn for 7 days and soybeans for 8 days. In this case, he would use 15 days to plant 10(7) or 70 acres of corn and 15(8) or 120 acres of soybeans.

CHECKING FOR UNDERSTANDING

Communicating Mathematics

Read and study the lesson to answer each question.

1. If a system of inequalities has a solution, how many ordered pairs will be contained in the region that is the solution?
2. What are the equations of the four lines that are the boundaries of the region in Example 3?
3. Write a system of inequalities that has no solutions.

Guided Practice

State whether each ordered pair is a solution of the system of inequalities $x \leq 3$ and $y > 6$.

4. (3, 7)
5. (−2, 6)
6. (7, 8)
7. (0, 8)

State which region in the graph at the right is the solution of each system of inequalities.

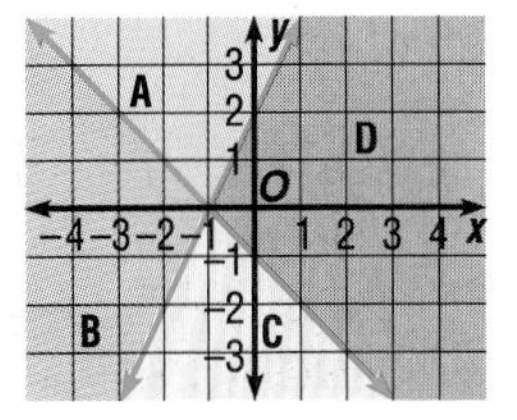

8. $y \geq 2x + 2$, $y \leq -x - 1$
9. $y \geq 2x + 2$, $y \geq -x - 1$
10. $y \leq 2x + 2$, $y \leq -x - 1$
11. $y \leq 2x + 2$, $y \geq -x - 1$

EXERCISES

Practice

Solve each system of inequalities by graphing.

12. $x > 3$, $y < 6$
13. $y > 0$, $x \leq 0$
14. $y > 2$, $y > -x + 2$
15. $y < -2$, $y - x > 1$
16. $x \leq 2$, $y - 3 \geq 5$
17. $x \geq 1$, $y + x \leq 3$
18. $|y| < x$
19. $|x| \geq y$
20. $y \geq 2x + 1$, $y \leq -x + 1$
21. $y \leq x + 3$, $y \geq x + 2$
22. $y \geq 3x$, $3y \leq 5x$
23. $y \geq x - 3$, $y \geq -x - 1$
24. $y - x < 1$, $y - x > 3$
25. $2y + x < 4$, $3x - y > 6$
26. $y + 2 < x$, $2y - 3 > 2x$
27. $x + 2y \leq 7$, $3x - 4y < 1$
28. $|y| + 1 < x$
29. $|y| > x + 3$
30. $|y - 4| > x$
31. $|2y + 4| \leq x$

Determine whether the point of intersection of the two boundaries is part of the solution set for each system of inequalities.

32. $y < -x$, $x \geq 6$
33. $y + 4 \leq 8$, $2y \leq x$
34. $x + 3y \geq 4$, $2x - y < 5$

Write a system of inequalities for each graph.

35. 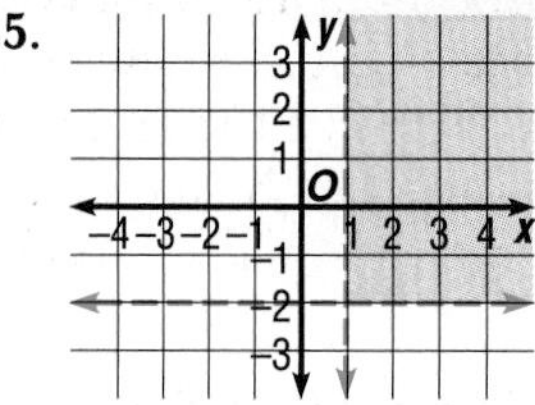

36. 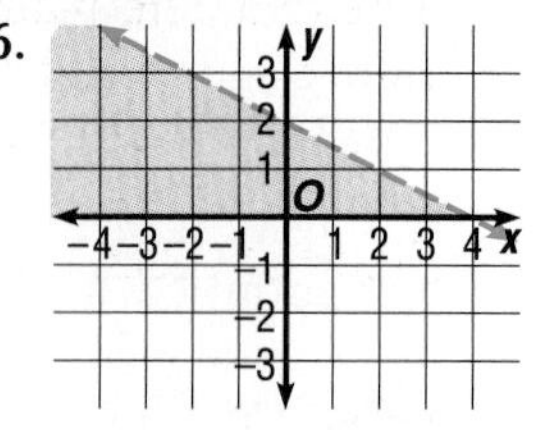

37. 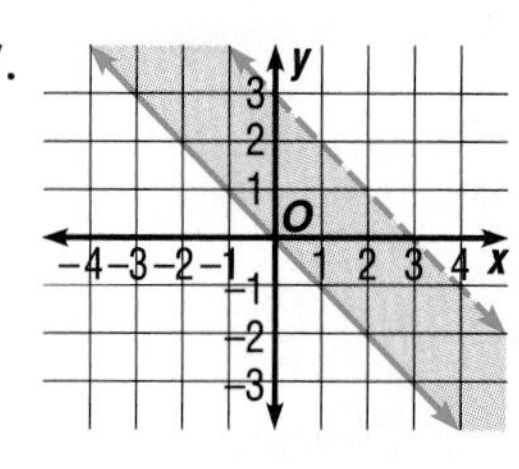

38. 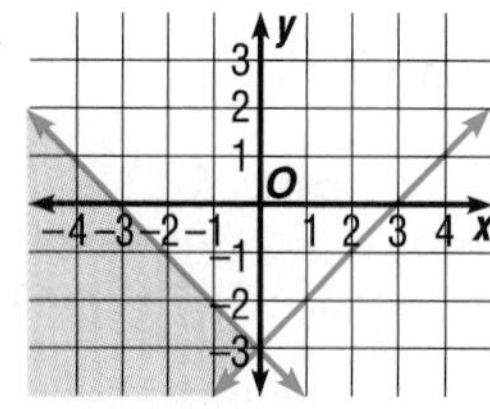

39. 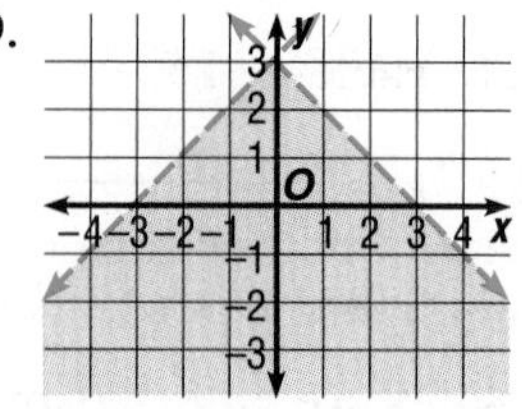

40. 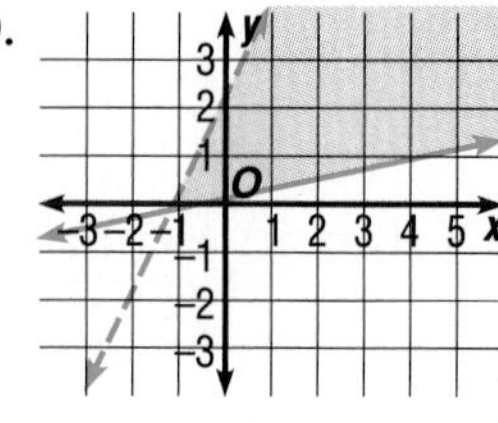

41. 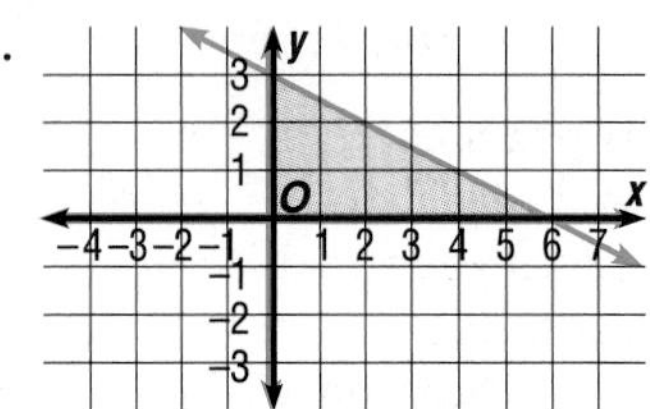

42. 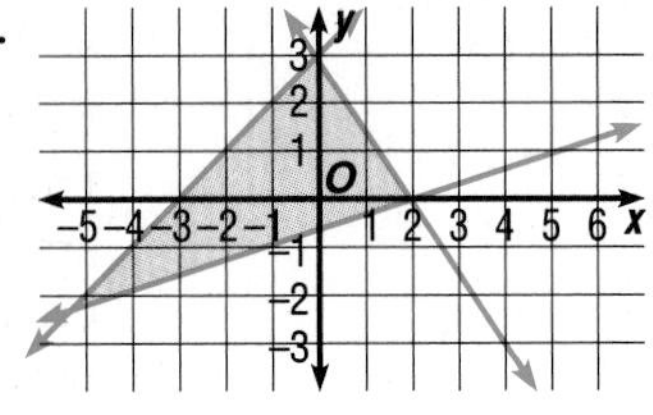

Solve each system of inequalities by graphing.

43. $x - y \leq 3$
    $x + y \leq 1$
    $y \geq 0$

44. $x + 4y < 4$
    $5x - 8y < -8$
    $3x - 2y \geq -16$

45. $x < 3$
    $5y < x$
    $x + 3y < 9$
    $2x - y < -9$

Critical Thinking

46. Write an absolute value inequality for the graph at the right. (*Hint:* Find the equations for the four boundaries of this region.)

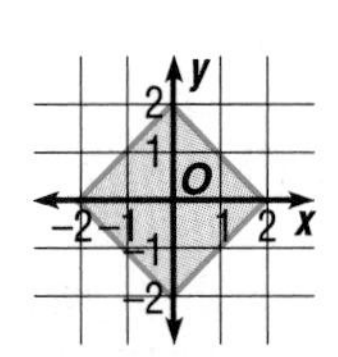

Applications

Graph a system of inequalities to solve each problem.

47. **Sales** Abby plans to spend at most \$24 to buy cashews and peanuts for her Fourth of July party. The Nut Shoppe sells peanuts for \$3 a pound and cashews for \$5 a pound. If Abby needs to have at least 5 pounds of nuts for the party, how many of each type can she buy? List three possible solutions.

48. **Sales** The Debate Team needed to earn at least \$500 to attend a national competition. The members decided to hold a raffle. They sold tickets at a price of \$2 for students and \$3 for everyone else. If they were allowed to sell at most 200 raffle tickets, how many of each type of ticket must the members sell to make enough money for the trip? List three possible solutions.

49. **Painting** A painter has exactly 32 units of yellow dye and 54 units of blue dye. She plans to mix the dyes to make two shades of green. Each gallon of the lighter shade of green requires 4 units of yellow dye and 1 unit of blue dye. Each gallon of the darker shade of green requires 1 unit of yellow dye and 6 units of blue dye. How many gallons of each shade of green can she mix? List three possible solutions.

Mixed Review

50. **Sales** The price of a suit was marked down $37.50. If the original price was $125, what was the rate of discount? **(Lesson 4-2)**

51. Factor $-16 + 9x^2$. **(Lesson 7-6)**

52. Add $\frac{x}{x^2 + 2x + 1} + \frac{1}{x + 1}$. **(Lesson 8-7)**

53. Solve $4y = 3 + 2x$ if the domain is $\{-2, -1, 0, 1, 3\}$. **(Lesson 9-3)**

Portfolio

Review the items in your portfolio. Make a table of contents of the items, noting why each item was chosen. Replace any items that are no longer appropriate.

54. Write an equation in slope-intercept form for the line that passes through $(-4, 3)$ and $(6, -5)$. **(Lesson 10-5)**

55. **Sales** Joey sold 30 peaches from his fruit stand for a total of $7.50. He sold small ones for 20 cents each and large ones for 35 cents each. How many of each kind did he sell? **(Lesson 11-5)**

HISTORY CONNECTION

The *K'iu-ch'ang Suan-shu*

The *K'iu-ch'ang Suan-shu,* or *Arithmetic in Nine Sections,* is the greatest of the ancient Chinese books on mathematics. A mathematician named Ch'ang Ts'ang is thought to have collected and edited these writings of the ancients around 213 B.C. The titles of its nine chapters are as follows.

1. **Squaring the Farm,** surveying
2. **Calculating the Cereals,** percentage and proportions
3. **Calculating the Shares,** partnership and the rule of three
4. **Finding Length,** sides of figures and square and cube roots
5. **Finding Volumes**
6. **Allegation,** motion problems
7. **Excess and Deficiency,** rule of false position
8. **Equations,** systems of linear equations
9. **Right Triangles,** Pythagorean Theorem, trigonometry

This work indicates that Chinese mathematicians were among the pioneers in establishing the early science of mathematics.

CHAPTER 11 SUMMARY AND REVIEW

VOCABULARY

Upon completing this chapter, you should be familiar with the following terms:

system of equations **442**
substitution **447**
452 elimination
463 system of inequalities

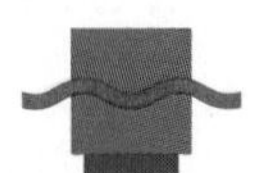

SKILLS AND CONCEPTS

OBJECTIVES AND EXAMPLES

Upon completing this chapter, you should be able to:

- solve systems of equations by graphing **(Lesson 11-2)**

Graph $y = x$ and $y = 2 - x$. Then find the solution.

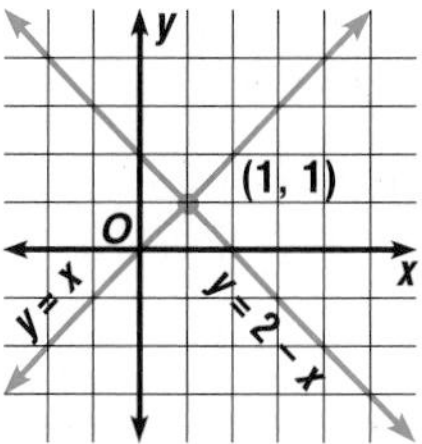

REVIEW EXERCISES

Use these exercises to review and prepare for the chapter test.

Graph each system of equations. Then find the solution to the system of equations.

1. $x + y = 6$
   $x - y = 2$
2. $y = 2x - 7$
   $x + y = 11$
3. $5x - 3y = 11$
   $2x + 3y = -25$

- determine whether a system of equations has one solution, no solution, or infinitely many solutions by graphing **(Lesson 11-2)**

Graph $3x + y = -4$ and $6x + 2y = -8$. Then determine the number of solutions.

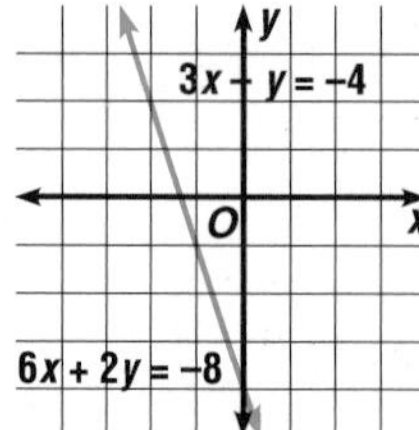

There are infinitely many solutions.

Graph each pair of equations. Then determine whether the system of equations has one solution, no solution, or infinitely many solutions. If the system has one solution, state it.

4. $x - y = 9$
   $x + y = 11$
5. $9x + 2 = 3y$
   $y - 3x = 8$
6. $2x - 3y = 4$
   $6y = 4x - 8$
7. $3x - y = 8$
   $3x = 4 - y$

OBJECTIVES AND EXAMPLES

- solve systems of equations by the substitution method **(Lesson 11-3)**

Use substitution to solve the system of equations $y = x - 1$ and $4x - y = 19$.

$$\begin{aligned} 4x - y &= 19 \\ 4x - (x - 1) &= 19 \\ 3x + 1 &= 19 \\ 3x &= 18 \\ x &= 6 \end{aligned} \qquad \begin{aligned} y &= x - 1 \\ &= 6 - 1 \\ &= 5 \end{aligned}$$

The solution is (6, 5).

REVIEW EXERCISES

Use substitution to solve each system of equations.

8. $x = 2y$
$x + y = 6$

9. $2m + n = 1$
$m - n = 8$

10. $3a - 2b = -4$
$3a + b = 2$

11. $3x - y = 1$
$2x + 4y = 3$

- solve systems of equations by the elimination method using addition or subtraction **(Lesson 11-4)**

Use elimination to solve the system of equations $2m - n = 4$ and $m + n = 2$.

$$\begin{aligned} 2m - n &= 4 \\ (+) \quad m + n &= 2 \\ \hline 3m \qquad &= 6 \\ m &= 2 \end{aligned} \qquad \begin{aligned} m + n &= 2 \\ 2 + n &= 2 \\ n &= 0 \end{aligned}$$

The solution is (2, 0).

Use elimination to solve each system of equations.

12. $x + 2y = 6$
$x - 3y = -4$

13. $2m - n = 5$
$2m + n = 3$

14. $3x - y = 11$
$x + y = 5$

15. $2s + 6r = 32$
$6r - 9s = 21$

- solve systems of equations by the elimination method using multiplication and addition **(Lesson 11-5)**

Use elimination to solve the system of equations $3x - 4y = 7$ and $2x + y = 1$.

$$\begin{aligned} 3x - 4y &= 7 \\ 2x + y &= 1 \end{aligned} \quad \xrightarrow{\times 4} \quad \begin{aligned} 3x - 4y &= 7 \\ (+)\ 8x + 4y &= 4 \\ \hline 11x \qquad &= 11 \\ x &= 1 \end{aligned}$$

$$\begin{aligned} 2x + y &= 1 \\ 2(1) + y &= 1 \\ y &= -1 \end{aligned}$$

The solution is (1, −1).

Use elimination to solve each system of equations.

16. $x - 2y = 5$
$3x - 5y = 8$

17. $2x + 3y = 8$
$x - y = 2$

18. $6x + 7y = 5$
$2x - 3y = 7$

19. $5m + 2n = -8$
$4m + 3n = 2$

OBJECTIVES AND EXAMPLES	REVIEW EXERCISES

- solve systems of inequalities by graphing **(Lesson 11-6)**

Solve the system of inequalities $x \geq -3$ and $y \leq x + 2$ by graphing.

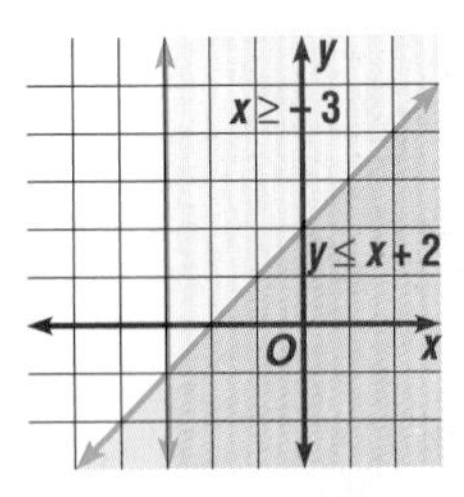

Solve each system of inequalities by graphing.

20. $y > -x - 1$
$y \leq 2x + 1$

21. $2r + s < 9$
$r + 11s < -6$

22. $|x + 2| \geq y$

APPLICATIONS AND CONNECTIONS

Solve the problem below. Try to eliminate any hidden assumptions. (Lesson 11-1)

23. Copy the array of dots at the right onto a piece of paper. Without lifting your pencil from the paper, draw four straight line segments through the 9 dots.

Use a system of equations to solve each problem.

24. Number Theory A two-digit number is 7 times its units digit. If 18 is added to the number, its digits are reversed. Find the original number. **(Lesson 11-3)**

25. Geometry The difference between the length and width of a rectangle is 7 cm. Find the dimensions of the rectangle if its perimeter is 50 cm. **(Lesson 11-4)**

26. Uniform Motion Two trains start toward each other on parallel tracks at the same time from towns 450 miles apart. One train travels 6 miles per hour faster than the other train. What is the rate of each train if they meet in 5 hours? **(Lesson 11-3)**

27. Sales Mr. Ayala bought 5 shirts and 3 ties at the Clothes Outlet for \$102. Mr. Gilmore bought 8 shirts and 3 ties for \$147. What is the price of a shirt and the price of a tie at the Clothes Outlet? **(Lesson 11-4)**

28. Finance Last year, Jodi invested \$10,000, part at 6% annual interest and the rest at 8% annual interest. If she received \$760 in interest at the end of the year, how much did she invest at each rate? **(Lesson 11-5)**

29. Uniform Motion A speedboat travels 60 miles with the current in 3 hours. The return trip against the current takes 4 hours. What is the rate of the speedboat in still water and the rate of the current? **(Lesson 11-5)**

CHAPTER 11

TEST

Graph each system of equations. Then determine whether the system has one solution, no solution, or infinitely many solutions. If the system has one solution, name it.

1. $y = x + 2$
$y = 2x + 7$

2. $x + 2y = 11$
$x = 14 - 2y$

3. $2x + 5y = 16$
$5x - 2y = 11$

4. $3x + y = 5$
$2y - 10 = -6x$

Use substitution or elimination to solve each system of equations.

5. $y = 7 - x$
$x - y = -3$

6. $x = 2y - 7$
$y - 3x = -9$

7. $x + y = 8$
$x - y = 2$

8. $3x - y = 11$
$x + 2y = -36$

9. $3x + y = 10$
$3x - 2y = 16$

10. $5x - 3y = 12$
$-2x + 3y = -3$

11. $2x + 5y = 12$
$x - 6y = -11$

12. $x + y = 6$
$3x - 3y = 13$

13. $3x + \frac{1}{3}y = 10$
$2x - \frac{5}{3}y = 35$

14. $8x - 6y = 14$
$6x - 9y = 15$

Solve each system of inequalities by graphing.

15. $y \leq -3$
$y > -x + 2$

16. $x \leq 2y$
$2x + 3y \leq 7$

17. Number Theory The units digit of a two-digit number exceeds twice the tens digit by 1. Find the number if the sum of its digits is 10.

18. Uniform Motion Marisel rode her bicycle against the wind for 1 hour and traveled 15 km. She returned the same distance with the wind in 36 minutes. What was the rate of the wind?

19. Sales Mr. Salvatore mixed nuts that cost \$3.90 per pound with nuts that cost \$4.30 per pound. He now has a mixture of 50 pounds of nuts that costs \$4.20 per pound. How many pounds of each type of nut did he use?

20. Automobiles A gas station attendant is making 1000 gallons of antifreeze that is 48% alcohol. He has some antifreeze that is 40% alcohol and some that is 60% alcohol. How much of each type of antifreeze should he use?

Bonus

Number Theory The sum of the digits of a three-digit number is 20. The tens digit exceeds twice the units digit by 1. The hundreds digit is one less than twice the units digit. Find the number.

CHAPTER OBJECTIVES

In this chapter, you will:

- Find exact and approximate values for square roots.
- Simplify radical expressions.
- Solve radical equations.
- Solve problems that can be represented by radical equations.
- Solve problems by using tables.

CHAPTER 12

Radical Expressions

APPLICATION IN NATURE

The shell of a chambered nautilus is one of the many examples of mathematics occurring in nature's designs. This shell takes on its spiral shape because of the nautilus's growth pattern and the formation of small chambers of increasing size to accommodate this growth. The growth pattern of the nautilus can be studied by looking at the size and shape of the shell's spiral.

Have you ever noticed that Mother Nature is a very skilled designer? Just look around and you can find many examples of mathematics in nature's designs.

One method for studying the spiral is to use a set of right triangles winding around a point to represent the shape of the spiral. These right triangles are drawn so that the sides used to represent the spiral all have the same length, one unit, as shown in the figure below. Notice that each triangle has the same apex (top point). The longest side of each right triangle, called the *hypotenuse*, is also one of the shorter sides of the next triangle in the set. This design is called a "spiral of square roots" because of the relationship between the lengths of the sides of the right triangles.

The shell of the chambered nautilus is just one example of naturally occurring spirals. The next time you see a daisy, a sunflower, or a pineapple, see if you can find the spirals that are a part of their design.

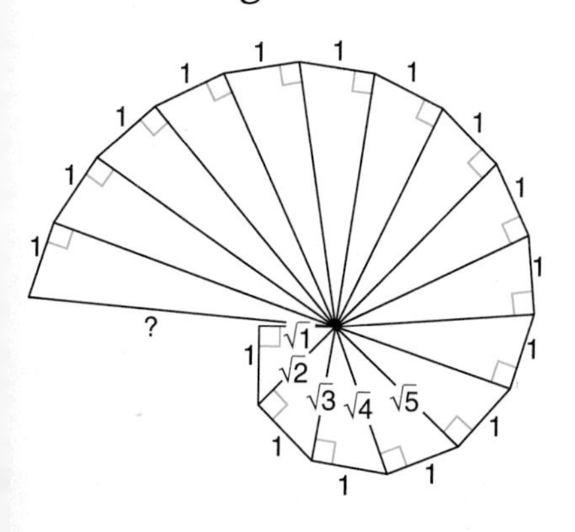

ALGEBRA IN ACTION

Look at the triangle with sides 1, $\sqrt{1}$, and $\sqrt{2}$ in the figure at the left. Notice that $1^2 + (\sqrt{1})^2 = (\sqrt{2})^2$ since $(\sqrt{1})^2 = 1$ and $(\sqrt{2})^2 = 2$. Similarly, in the next few triangles, $1^2 + (\sqrt{2})^2 = (\sqrt{3})^2$, $1^2 + (\sqrt{3})^2 = (\sqrt{4})^2$, and $1^2 + (\sqrt{4})^2 = (\sqrt{5})^2$. This relationship between the measures of the sides of the right triangles is the basis for the *Pythagorean Theorem.* If you continue in this manner, what will be the measure of the side labeled ? in the figure?

EXPLORE
PLAN
SOLVE
EXAMINE

12-1 Problem-Solving Strategy: Use a Table

Objective After studying this lesson, you should be able to:

- solve problems by using a table.

One important problem-solving strategy that you have already used is to make a table or chart. In some situations, the information from a problem has already been organized in a table. In order to solve the problem, you must be able to read and interpret this information.

Example

FYI...

Women received the right to vote with the passage of the 19th Amendment to the Constitution, August 26, 1920.

The Guessright Poll Company set up a survey to determine the voter's preference for mayor during various months prior to the election. A sample of 1000 people from the city's population of 550,000 was taken every other month. The results of the surveys are shown in the chart below. Based on this information, determine the following.

a. Which candidate gained the most votes from April to October?

b. In which month was the difference in votes between the two leading candidates the least?

Preference for Mayor

Candidate	Month			
	April	June	Aug.	Oct.
Alvarez	223	286	294	347
Lewis	167	202	387	399
Rinehart	287	268	168	157
Others	209	152	64	31
Undecided	114	92	87	66

a. To determine which candidate gained the most votes from April to October, look at the totals in those columns of the table. Notice that only candidates Alvarez and Lewis had more votes in October than in April. Alvarez had 347 − 223 or 124 more votes. Lewis had 399 − 167 or 232 more votes. Therefore, Lewis gained the most votes from April to October.

b. To determine in which month the difference in votes between the leading candidates was the least, first look at each column to determine the leading candidates. Then, find each difference and compare.

April		**June**		**August**		**October**	
Rinehart	287	Alvarez	286	Lewis	387	Lewis	399
Alvarez	−223	Rinehart	−268	Alvarez	−294	Alvarez	−347
Difference	64		18		93		52

Therefore, June was the month in which the difference in votes between the two leading candidates was the least.

In mathematics, numerical values, especially those involving approximations, are often presented in tables or charts. For example, on page 644 of this text, there is a table of squares and approximate *square roots* of integers from 1 to 100. You may need to use the information in this table to solve problems in this chapter.

CHECKING FOR UNDERSTANDING

Communicating Mathematics

Read and study the lesson to answer each question.

1. In the example on page 474, during which month was the difference in votes between the two leading candidates the greatest?
2. Name another problem-solving strategy that can be used with the strategy described in this lesson.

Guided Practice

3. The table below shows the normal precipitation, in inches, for selected cities in the United States. Use this information to answer the following questions.

City	Jan.	Feb.	Mar.	Apr.	May	Jun.	July	Aug.	Sep.	Oct.	Nov.	Dec.
Albuquerque, NM	0.30	0.39	0.47	0.48	0.53	0.50	1.39	1.34	0.77	0.79	0.29	0.52
Boston, MA	3.69	3.54	4.01	3.49	3.47	3.19	2.74	3.46	3.16	3.02	4.51	4.24
Chicago, IL	1.85	1.59	2.73	3.75	3.41	3.95	4.09	3.14	3.00	2.62	2.20	2.11
Houston, TX	3.57	3.54	2.68	3.54	5.10	4.52	4.12	4.35	4.65	4.05	4.03	4.04
Mobile, AL	4.71	4.76	7.07	5.59	4.52	6.09	8.86	6.93	6.59	2.55	3.39	5.92
San Francisco, CA	4.37	3.04	2.54	1.59	0.41	0.13	0.01	0.03	0.16	0.98	2.29	3.98

a. Which city had the most precipitation in November?
b. What is the driest month in Houston?
c. What is the greatest monthly precipitation for any of the six cities? What month is it? What city is it?
d. What is the least difference between the greatest and the least monthly precipitation for any one month? What is the month?

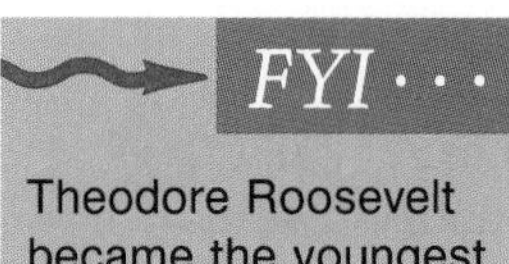

FYI · · ·

Theodore Roosevelt became the youngest man ever to become president of the United States when President William McKinley was assassinated in September of 1901. Roosevelt was 42.

4. Each number below represents the age, in years, of U. S. presidents on their first inaugurations.

57 61 57 57 58 57 61 54 68 51 49 64 50 48
65 52 56 46 54 49 50 47 55 55 54 42 51 56
55 51 54 51 60 62 43 55 56 61 52 69 64 46

a. Organize this information into a table with the headings *Age at First Inauguration* and *Number of Times.* Then use the information in this table to answer questions b-e.
b. How many different ages are given?
c. What was the difference in age between the youngest and oldest president?
d. Which age(s) occurred most frequently?
e. How many presidents were in their 60s on their first inauguration?

EXERCISES

Practice

Strategies

Look for a pattern.
Solve a simpler problem.
Act it out.
Guess and check.
Draw a diagram.
Make a chart.
Work backwards.

Solve. Use any strategy.

5. A vending machine dispenses products that each cost 50¢. It accepts quarters, dimes, and nickels only. How many different combinations of coins must the machine be programmed to accept?

6. To determine a grade point average, four points are given for an A, three for a B, two for a C, one for a D, and zero for an F. If Greg has a total of 13 points for 5 classes, what combinations of grades could he receive?

7. What is the next number in the sequence below? (*Hint:* This sequence involves powers of 3.)

   0, 1, 8, 7, 4, 5, 6, ?

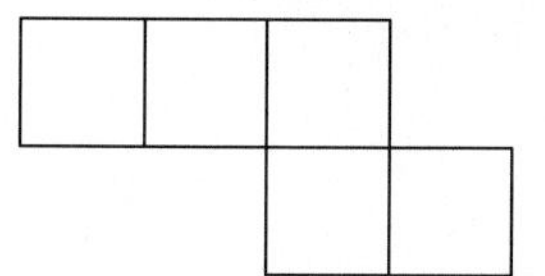

8. In the figure at the right, exactly two line segments can be moved to make four squares. Which two segments should be moved?

9. A woman was born in the 19th century. In a certain year, the square of her age was equal to the year. How old was she in 1885?

10. In a six-team baseball tournament, the teams must play each other exactly one time. How many games will be played in this tournament?

COOPERATIVE LEARNING ACTIVITY

Work in groups of three. Each person in the group should understand the solution and be able to explain it to any person in the class.

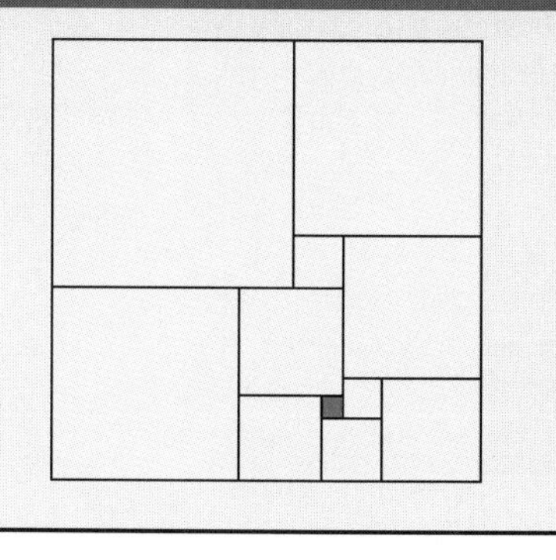

The large rectangle, shown in blue at the right, has been divided into eleven squares of various sizes. The sides of the smallest square, shown in red, are 9 cm long. What are the dimensions of the large rectangle?

12-2 Square Roots

Objectives

After studying this lesson, you should be able to:

- simplify rational square roots, and
- find approximate values for square roots.

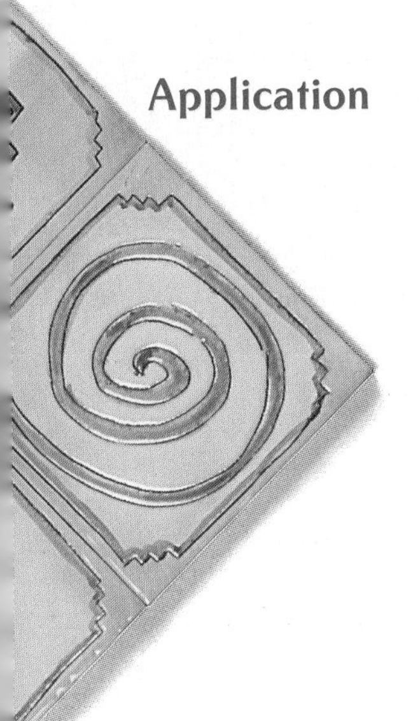

Application

The Cutright Tile Co. has donated 144 tiles to the Southside Community Center to be used to cover the floor of a square patio. If each of the tiles is 18 inches by 18 inches, what are the dimensions of the largest patio that can be covered?

To solve this problem, determine how many tiles will be on each side of this patio. Find a number whose square is 144. Since $12^2 = 144$, the square patio will have 12 tiles on each side. Each side of this patio will be 12(18) or 216 inches long. Thus, the dimensions of the largest square patio are 216 inches by 216 inches or 18 feet by 18 feet.

In the problem, you needed a number whose square was 144. Recall that **squaring** a number means using that number as a factor two times.

$8^2 = 8 \cdot 8 = 64$ — *8^2 is read "eight squared" and means eight is used as a factor two times.*

$(-8)^2 = (-8)(-8) = 64$ — *-8 is used as a factor two times.*

Finding a square root of 64 is the same as finding a number whose square is 64.

The opposite of squaring a number is finding its **square root.** To find a square root of 64, you must find two *equal* factors whose product is 64.

$$x^2 = x \cdot x = 64$$

Since 8 times 8 is 64, one square root of 64 is 8. Since -8 times -8 is also 64, another square root of 64 is -8. In the problem above, 12 is a square root of 144. Another square root of 144 is -12, but it was not considered as a solution to the tile problem. *Why?*

Definition of Square Root	**If $x^2 = y$, then x is a square root of y.**

The square root of a negative number is not defined for the sets of numbers covered thus far in this text.

An expression like $\sqrt{64}$ is called a **radical expression.** The symbol $\sqrt{\ }$ is a **radical sign.** It indicates the *nonnegative* or **principal** square root of the expression under the radical sign. The expression under the radical sign is called the **radicand.**

radical sign → $\sqrt{64}$ ← radicand

$\sqrt{64} = 8$ — $\sqrt{64}$ indicates the *principal* square root of 64.

$-\sqrt{64} = -8$ — $-\sqrt{64}$ indicates the *negative* square root of 64.

$\pm\sqrt{64} = \pm 8$ — $\pm\sqrt{64}$ indicates *both* square roots of 64.
This is read "plus or minus the square root of 64."

$\pm$ means positive or negative.

Example 1

Find each square root.

a. $\sqrt{81}$

Since $9^2 = 81$, you know that $\sqrt{81} = 9$.

b. $\pm\sqrt{0.09}$

Since $(0.3)^2 = 0.09$, you know that $\pm\sqrt{0.09} = \pm 0.3$.

You may sometimes need to use prime factorization to find a square root of a number.

Example 2

Find each square root.

a. $-\sqrt{576}$

$$576 = 2^6 \cdot 3^2 \quad \text{Find the prime factorization of 576.}$$
$$= (2^3 \cdot 3)^2$$
$$= 24^2$$

Since $24^2 = 576$, you know that $-\sqrt{576} = -24$.

b. $\sqrt{\frac{256}{2025}}$

$$\frac{256}{2025} = \frac{2^8}{3^4 \cdot 5^2} \quad \text{Find the prime factorizations of 256 and 2025.}$$
$$= \left(\frac{2^4}{3^2 \cdot 5}\right)^2$$
$$= \left(\frac{16}{45}\right)^2$$

Since $\left(\frac{16}{45}\right)^2 = \frac{256}{2025}$, you know that $\sqrt{\frac{256}{2025}} = \frac{16}{45}$.

When will the square root of a whole number not be an integer?

Some calculators have a *square root key* labeled $\sqrt{\ }$ or $\sqrt{x}$. When you press this key, the number in the display is replaced by its principal square root. If a principal square root is *not* a whole number, then most calculators will round the result and display as many decimal places as they can handle.

Example 3

Find $\sqrt{2209}$ and $\pm\sqrt{1236}$. If the principal square root is *not* a whole number, round the result to the nearest hundredth.

a. Enter: 2209 [$\sqrt{x}$] 47

Therefore, $\sqrt{2209} = 47$.

b. Enter: 1236 [$\sqrt{x}$] 35.1567917

Therefore, $\pm\sqrt{1236}$ to the nearest hundredth is ± 35.16.

To check this result, compare $(35.16)^2$ to 1236.

In Example 3, you could also write $\pm\sqrt{1236} \approx \pm 35.16$. The symbol $\approx$ means *is approximately equal to.*

Example 4

In order to allow for the proper flow of water, the opening of a circular pipe from a hot water heater must have an area of 0.785 square inches. Use the formula for the area of a circle, $A = \pi r^2$, to find the radius of the opening of the pipe. Use 3.14 for π.

$A = \pi r^2$

$0.785 \approx 3.14r^2$ *Substitute 0.785 for A and 3.14 for π.*

$\frac{0.785}{3.14} \approx \frac{3.14r^2}{3.14}$ *Divide each side by 3.14.*

$0.25 \approx r^2$

$0.5 \approx r$ *Since $0.25 = (0.5)^2$, 0.5 is the square root of 0.25.*

Therefore, the radius of the opening of the pipe should be about 0.5 inches.

CHECKING FOR UNDERSTANDING

Communicating Mathematics

Read and study the lesson to answer each question.

1. What are the radicand and principal square root of $\sqrt{25}$?
2. Why are both 2 and −2 square roots of 4?
3. Choose any positive number and enter it on a calculator. Then press the square root key, followed by the x^2 key. What is the result? Why?
4. Choose any negative number and enter it on a calculator. Then press the square root key. What is the result? Why?

Guided Practice

State the square of each number.

5. 12 6. −20 7. 0.3 8. $\frac{4}{7}$ 9. $-\frac{11}{4}$

Simplify.

10. $\sqrt{121}$ 11. $-\sqrt{81}$ 12. $\pm\sqrt{\frac{81}{64}}$ 13. $\sqrt{0.0016}$

Use a calculator to find each square root. Round answers to the nearest hundredth.

14. $\sqrt{85}$ 15. $-\sqrt{149}$ 16. $\pm\sqrt{206}$ 17. $\sqrt{60.3}$

EXERCISES

Practice

Find the principal square root of each number.

18. 169 19. 256 20. $\frac{81}{121}$ 21. $\frac{36}{196}$

22. $\frac{400}{225}$ 23. 0.0025 24. 0.0289 25. 3.24

Simplify.

26. $\pm\sqrt{144}$
27. $-\sqrt{100}$
28. $\sqrt{529}$
29. $\sqrt{484}$
30. $-\sqrt{676}$
31. $\pm\sqrt{961}$
32. $\pm\sqrt{1764}$
33. $\sqrt{2025}$
34. $\sqrt{0.0729}$
35. $-\sqrt{10.24}$
36. $-\sqrt{\frac{169}{121}}$
37. $\pm\sqrt{\frac{144}{1521}}$

Use a calculator to find each square root. Round answers to the nearest thousandth.

38. $\sqrt{115.7}$
39. $-\sqrt{175.6}$
40. $\pm\sqrt{155.1}$
41. $\sqrt{531.4}$
42. $-\sqrt{0.61}$
43. $\sqrt{2.314}$
44. $\sqrt{0.00462}$
45. $\pm\sqrt{0.00932}$

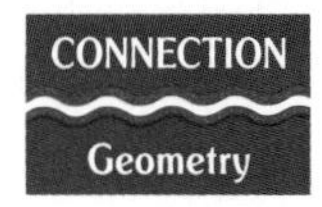

Find the length of the side of each square. The area is given. Round answers to the nearest hundredth.

46. 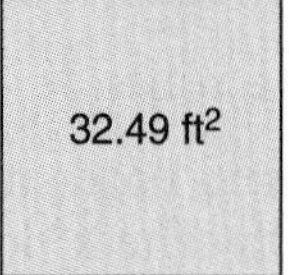

47. 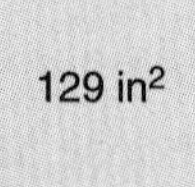

48. 1400 cm^2

Simplify.

49. $\sqrt{\sqrt{81}}$
50. $\sqrt{\sqrt{625}}$
51. $\sqrt{\sqrt{\sqrt{256}}}$

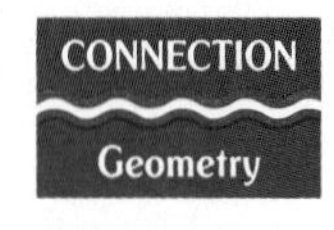

52. The volume of a rectangular solid is 100 cm^3. Its height is the product of its length and width. If the base of the solid is a square, find the dimensions of the solid. Use a calculator as needed. Round decimal answers to the nearest hundredth.

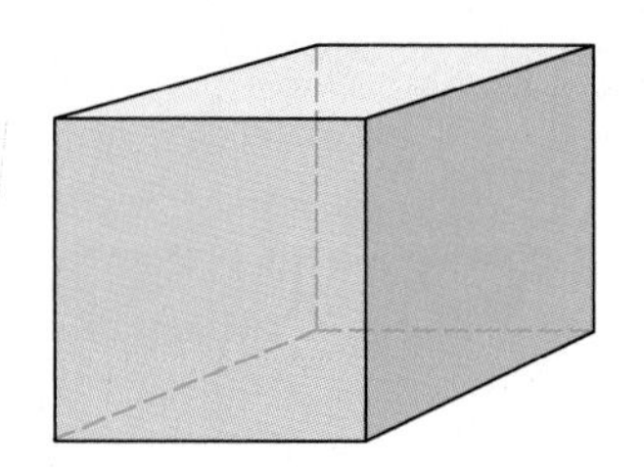

Critical Thinking

53. Choose any negative number and enter it on a calculator. Then press the x^2 key, followed by the square root key. What is the result? Why?

Applications

54. **Law Enforcement** The approximate speed s, in miles per hour, of a car traveling on a dry concrete road if it skidded d feet after its brakes were applied, is given by the formula $s = \sqrt{24d}$. What was the approximate speed of a car that skidded 150 feet on a dry concrete road after the brakes were applied?

Journal

Suppose a square has an area of 1500 square feet. Tell how you could find between what two whole numbers the measure of its side lies.

55. **Physics** For an object moving in a circular path, its acceleration, a, toward the center is given by the formula $a = \frac{v^2}{r}$. In this formula, v is the velocity of the object and r is the radius of the circular path. Find the velocity, in meters per second, of an object moving in a circular path with radius 20 meters, if its acceleration toward the center is 17 meters per second squared. Round your answer to the nearest tenth of a unit.

56. **Electricity** The power P, in watts, of a circuit is given by the formula $P = I^2R$. In this formula, I is the current in amperes and R is the resistance in ohms. Find the current in a circuit that produces 1200 watts of power if the resistance is 5 ohms. Round your answer to the nearest tenth of an ampere.

Mixed Review

57. Eight is 20% of what number? **(Lesson 4-2)**

58. Factor $20a^2c^2 + 60a^2c + 45a^2$. **(Lesson 7-7)**

59. **Aviation** An airplane can fly at a rate of 600 mph in calm air. It can fly 2413 miles with the wind in the same time it can fly 2147 miles against the wind. What is the speed of the wind? **(Lesson 8-9)**

60. Graph $4x - 3y = 24$. **(Lesson 9-4)**

61. Write an equation in slope-intercept form of the line that passes through (6, 3) and (−2, 4). **(Lesson 10-5)**

62. Solve the system $y - x > 1$ and $y + 2x \leq 10$ by graphing. **(Lesson 11-6)**

HISTORY CONNECTION

Jaime Escalante

Jaime Escalante is the former Garfield High School calculus teacher on whom the movie *Stand and Deliver* is based. Escalante left a comfortable career in industry to teach mathematics in an economically depressed area of Los Angeles. When he joined the staff at Garfield, his junior class students were having trouble with fractions and percentages. Before the end of their senior year, 18 of these students would pass the most difficult mathematics exam given to high-school seniors, the National Advanced Placement Calculus Exam. Escalante believed that "math is the great equalizer," so he showed his students the misery into which a lack of education would lead them and raised their expectations of what was ahead of them. As a result, his students went on to colleges and universities, many with college credit and scholarships.

12-3 The Pythagorean Theorem

Objective After studying this lesson, you should be able to:

- use the Pythagorean Theorem.

Application

> **FYI . . .**
>
> Before 1845, the bases on a baseball field were arranged in the shape of a U. Before 1859, umpires sat in a padded rocking chair behind the catcher.

A baseball scout uses many different tests to determine whether or not to draft a particular player. One test for catchers is to see how quickly they can throw a ball from home plate to second base. The scout must know the distance between the two bases in case a player cannot be tested on a baseball diamond. This distance can be found by separating the baseball diamond into two right triangles, as shown at the left.

The side opposite the right angle in a right triangle is called the **hypotenuse.** This side is *always* the longest side of a right triangle. The other two sides are called the **legs** of the right triangle.

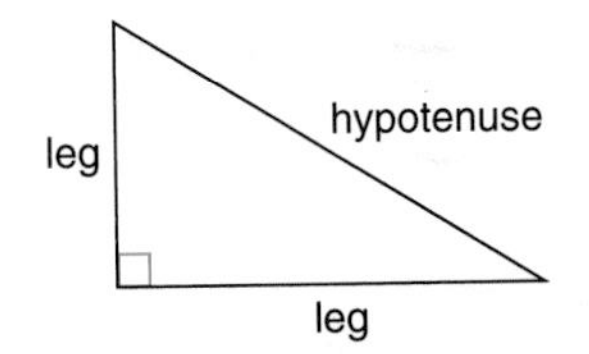

For the baseball diamond above, the distance from home plate to second base is equal to the length of the hypotenuse of the right triangle. The distance from home plate to first base and the distance from first base to second base are the lengths of the legs. On a baseball diamond, the distance from one base to the next is 90 feet.

To find the length of the hypotenuse, given the lengths of the legs, you can use a formula developed by the Greek mathematician, Pythagoras.

The Pythagorean Theorem

In a right triangle, if a and b are the measures of the legs and c is the measure of the hypotenuse, then

$$c^2 = a^2 + b^2.$$

For the baseball diamond, $a = 90$ and $b = 90$. You can use the Pythagorean Theorem to find the distance from home plate to second base.

second base
?
90 ft
home plate
90 ft
first base

$$c^2 = a^2 + b^2$$
$$c^2 = 90^2 + 90^2 \qquad a = 90 \text{ and } b = 90$$
$$c^2 = 8100 + 8100$$
$$c^2 = 16{,}200$$
$$c = \sqrt{16{,}200}$$
$$c \approx 127$$

Use a calculator to approximate $\sqrt{16{,}200}$ to the nearest unit.

The distance from home plate to second base is approximately 127 feet.

A geometric model can also be used to illustrate the Pythagorean Theorem, as shown below.

Lab Activity

You can learn how to use a geoboard and the Pythagorean theorem to construct a square with a given area in Lab 13 on page A20.

The area of this square is 25 square units.

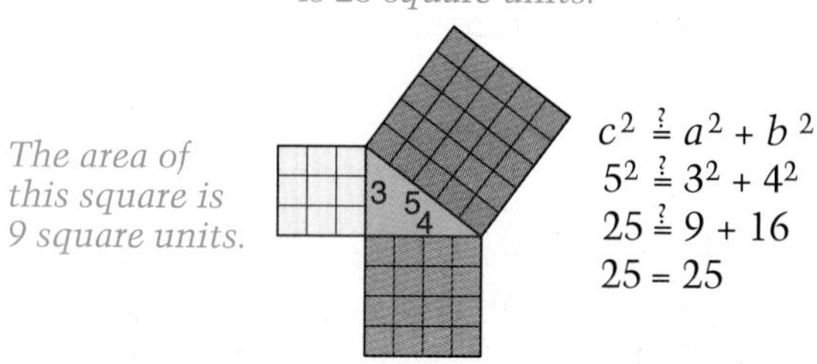

The area of this square is 16 square units.

You can use the Pythagorean Theorem to find the length of any side of a right triangle when the lengths of the other two sides are known.

Example 1 **Find the length of the hypotenuse of a right triangle if $a = 15$ and $b = 8$.**

$c^2 = a^2 + b^2$

$c^2 = 15^2 + 8^2$ *$a = 15$ and $b = 8$*

$c^2 = 225 + 64$

$c^2 = 289$

$c = \pm\sqrt{289}$

$c = 17$ *Disregard -17. Why?*

The length of the hypotenuse is 17 units.

Example 2 **Find the length of the other leg of a right triangle, to the nearest hundredth, if $a = 6$ and $c = 14$.**

$c^2 = a^2 + b^2$

$14^2 = 6^2 + b^2$ *$a = 6$ and $c = 14$*

$196 = 36 + b^2$

$b^2 = 160$

$b = \sqrt{160}$ *Use a calculator to approximate $\sqrt{160}$ to the nearest hundredth.*

$b \approx 12.65$

The length of the leg, to the nearest hundredth, is 12.65 units.

The following statement, which is based on the Pythagorean Theorem, can be used to determine whether a triangle is a right triangle.

If c is the measure of the longest side of a triangle and $c^2 \neq a^2 + b^2$, then the triangle is *not* a right triangle.

Example 3

The measures of the sides of a triangle are 5, 7, and 9. Determine whether this triangle is a right triangle.

Since the measure of the longest side is 9, let $c = 9$, $a = 5$, and $b = 7$. Then determine whether $c^2 = a^2 + b^2$.

$9^2 \stackrel{?}{=} 5^2 + 7^2$

$81 \stackrel{?}{=} 25 + 49$

$81 \neq 74$ Since $c^2 \neq a^2 + b^2$, the triangle is *not* a right triangle.

Example 4

The walls of the Downtown Recreation Center are being covered with paneling. The doorway into one room is 0.9 m wide and 2.5 m high. What is the longest rectangular panel that can be taken through this doorway?

EXPLORE The panel needs to be taken through the doorway diagonally. Let c = the length of the diagonal, as shown at the right.

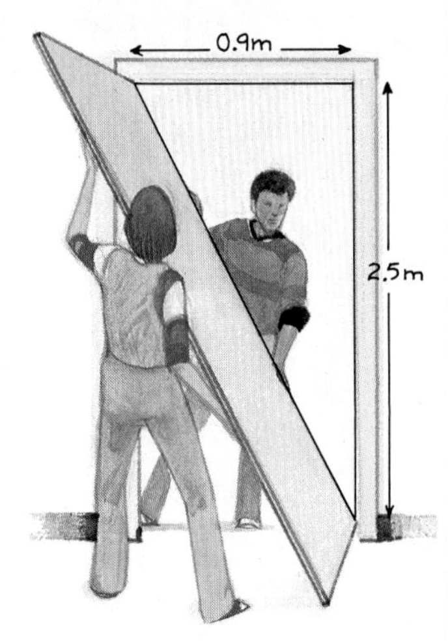

PLAN Use the Pythagorean Theorem to find c. Let $a = 0.9$ and $b = 2.5$.

$c^2 = (0.9)^2 + (2.5)^2$

SOLVE

$c^2 = 0.81 + 6.25$

$c^2 = 7.06$

$c = \sqrt{7.06}$ *Use a calculator to approximate $\sqrt{7.06}$.*

$c \approx 2.66$

The diagonal is about 2.66 meters long. Since space is needed to slide the panel through the doorway, the panel should be slightly less than 2.66 meters long.

EXAMINE Check the solution by substituting 2.66 for c in the Pythagorean Theorem.

$c^2 = a^2 + b^2$

$(2.66)^2 \stackrel{?}{=} (0.9)^2 + (2.5)^2$

$7.0756 \approx 7.06$ ✓

CHECKING FOR UNDERSTANDING

Communicating Mathematics

Read and study the lesson to answer each question.

1. What is the name of the longest side of a right triangle?
2. State the Pythagorean Theorem in your own words.
3. If you know the lengths of the three sides of a triangle, how can you determine if the triangle is a right triangle?

Guided Practice

State whether each sentence is *true* or *false*.

4. $3^2 + 4^2 = 5^2$
5. $9^2 + 10^2 = 11^2$
6. $6^2 + 8^2 = 9^2$

Solve each equation. Assume each variable represents a positive number.

7. $6^2 + 8^2 = c^2$
8. $5^2 + 12^2 = c^2$
9. $a^2 + 15^2 = 17^2$
10. $a^2 + 24^2 = 25^2$
11. $12^2 + b^2 = 20^2$
12. $10^2 + b^2 = 26^2$

If c is the measure of the hypotenuse of a right triangle, find each missing measure.

13. $a = 9, b = 12, c = ?$
14. $a = \sqrt{7}, b = \sqrt{9}, c = ?$
15. $b = \sqrt{30}, c = \sqrt{34}, a = ?$
16. $a = \sqrt{11}, c = \sqrt{47}, b = ?$

EXERCISES

Practice

If c is the measure of the hypotenuse of a right triangle, find each missing measure. Round answers to the nearest hundredth.

17. $a = 16, b = 30, c = ?$
18. $a = 11, c = 61, b = ?$
19. $b = 21, c = 29, a = ?$
20. $a = \sqrt{13}, b = 6, c = ?$
21. $a = \sqrt{11}, c = 6, b = ?$
22. $b = 13, c = \sqrt{233}, a = ?$
23. $a = 6, b = 3, c = ?$
24. $a = 4, b = \sqrt{11}, c = ?$
25. $a = 15, c = \sqrt{253}, b = ?$
26. $b = \sqrt{77}, c = 12, a = ?$
27. $b = 10, c = 11, a = ?$
28. $a = 12, c = 17, b = ?$

The measures of the sides of a triangle are given. Determine whether each triangle is a right triangle.

29. 9, 16, 20
30. 9, 40, 41
31. 45, 60, 75
32. 12, 11, 15
33. 18, $\sqrt{24}$, 30
34. 15, $\sqrt{31}$, 16

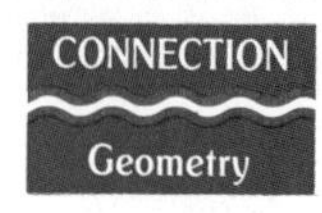

For each problem, make a drawing. Then use an equation to solve the problem. Round answers to the nearest hundredth.

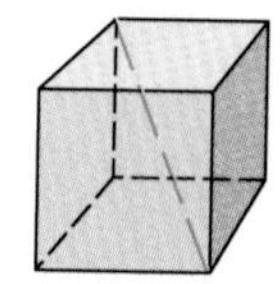

35. Find the length of the diagonal of a square if its area is 98 cm^2.
36. Find the length of the diagonal of a cube if each side of the cube is 5 inches long.
37. The area of a rectangle is 40 square meters. Find the length of a diagonal of the rectangle if its length is 2 meters less than twice its width.
38. Exactly two right triangles will be formed when what figure is cut along one of its diagonals?

Critical Thinking

39. Mary hikes 7 km north, 5 km west, and then 5 km north again. How far is Mary from the starting point of her hike?

Applications

For Exercises 40–42, make a drawing. Then use an equation to solve the problem. Round answers to the nearest hundredth.

40. **Baseball** The person on third base fields the ball directly on the third-base line 20 feet beyond third base. How far must she throw the ball to reach first base? (*Hint:* The distance between bases is 90 feet.)

41. **Construction** James wants to position a 16-foot ladder so that the top of the ladder is 15 feet above the base of an outside wall of his house. How far should the bottom of the ladder be placed from the base of the wall?

42. **Sailing** A rope from the top of a mast on a sailboat is attached to a point 2 meters from the base of the mast. If the rope is 8 meters long, how high is the mast?

43. **Construction** A wire is run from the top of a 52-meter tower to the top of a 12-meter tower and then to the base of the 52-meter tower. If the two towers are 9 meters apart, how much wire is needed?

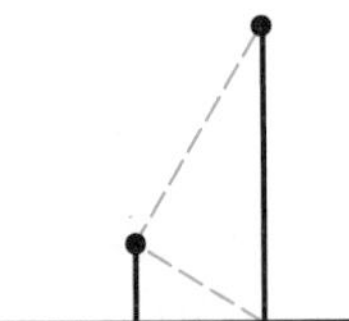

Mixed Review

44. **Statistics** Find the mean for the following set of data: 1, 14, 4, 8, 8, 7.35, 2.9, 12.75. **(Lesson 2-7)**

45. Find $\frac{2x^2 + 11x + 15}{2x^2 - 5x - 3} \div \frac{x^2 + 7x + 12}{x^2 + x - 12}$. **(Lesson 8-3)**

46. Graph $x = \frac{3}{4}y + 6$ using the x- and y-intercepts. **(Lesson 10-4)**

47. **Chemistry** One solution is 50% glycol and another is 30% glycol. How much of each should be mixed to make a 100-gallon solution that is 45% glycol? **(Lesson 11-3)**

48. Find $\pm\sqrt{4356}$. **(Lesson 12-2)**

APPLICATION

Escape Velocity

The minimum velocity v that a spacecraft must have to escape the gravitational force of a planet can be calculated using the formula $v = \sqrt{\frac{2GM}{r}}$, where G is a gravitational constant, M is the mass of the planet, and r is the radius of the planet.

Earth's mass is 5.98×10^{24} kg, its radius is 6.37×10^6 m, and G = 6.67×10^{-11} N-m²/kg². Compute the escape velocity for Earth.

$$v = \sqrt{\frac{2GM}{r}} \qquad G = 6.67 \times 10^{-11},\ M = 5.98 \times 10^{24},\ r = 6.37 \times 10^6$$

$$= \sqrt{\frac{2(6.67 \times 10^{-11})(5.98 \times 10^{24})}{6.37 \times 10^6}}$$

$$= \sqrt{\frac{7.97732 \times 10^{14}}{6.37 \times 10^6}}$$

$$= \sqrt{1.2523265 \times 10^8} \approx 1.12 \times 10^4$$

The escape velocity for Earth is about 1.12×10^4 meters per second.

12-4 Real Numbers

Objective

After studying this lesson, you should be able to:

- identify irrational numbers.

Application

FYI...

The symbol for a radical probably comes from an old European lowercase letter *r*, an abbreviation for the Latin word *radix*, meaning root.

Drew is building a storage shed in his backyard. The shed needs to have 120 square feet of floor space in order to store all of his equipment. Is it possible for Drew to build the shed with a square floor that has an area of *exactly* 120 square feet?

In order to build the storage shed with a square floor, Drew needs to know the length of each side of the square. Let s = the measure of the side of the square floor.

$s^2 = 120$ *The floor is to have an area of 120 ft².*

$s = \sqrt{120}$ *Use a calculator to find $\sqrt{120}$.*

$s = 10.9544511\ldots$

If the area of the square floor is to be exactly 120 square feet, then the length of each side must be exactly 10.9544511 . . . feet! Based on your knowledge of measurement and measuring devices, do you think that it is possible for Drew to build this floor?

The number $\sqrt{120} = 10.9544511\ldots$ is not a repeating or terminating decimal. Therefore, it is *not* a member of any of the sets of numbers you have encountered so far. These sets of numbers are listed below.

Natural numbers, **N**	$\{1, 2, 3, 4, \ldots\}$	*Natural numbers are also called counting numbers.*
Whole numbers, **W**	$\{0, 1, 2, 3, 4, \ldots\}$	
Integers, **Z**	$\{\ldots, -2, -1, 0, 1, 2, 3, 4, \ldots\}$	
Rational numbers, **Q**	$\left\{\begin{array}{l}\text{all numbers that can be expressed in the} \\ \text{form } \frac{a}{b}\text{, where } a \text{ and } b \text{ are integers and } b \neq 0\end{array}\right\}$	

Recall that repeating or terminating decimals name rational numbers since they can be represented as quotients of integers. The square roots of *perfect squares* also name rational numbers. For example, $\sqrt{0.16}$ names a rational number since it is equivalent to the rational number 0.4.

Numbers such as $\sqrt{120}$ are the square roots of numbers that are *not* perfect squares. Some other examples are shown below. What do you notice about these numbers?

None of these decimals terminate.

$\sqrt{2} = 1.414213\ldots$ $\sqrt{3} = 1.732050\ldots$ $\sqrt{7} = 2.645751\ldots$

These numbers are *not* rational numbers since they are not repeating or terminating decimals. They are called **irrational numbers.** The set of irrational numbers is often denoted by the capital letter **I.**

Definition of Irrational Numbers

Irrational numbers are numbers that *cannot* be expressed in the form $\frac{a}{b}$ where a and b are integers and $b \neq 0$.

Example 1

Name the set or sets of numbers to which each number belongs.

a. **0.8888 . . .** This repeating decimal is a rational number since it is equivalent to $\frac{8}{9}$. This number can also be expressed as $0.\overline{8}$.

b. **3.141592 . . .** This decimal is an irrational number since it does not repeat or terminate. Decimal approximations are often used for such numbers. $\pi \approx 3.14$

c. $-\sqrt{9}$ Since $-\sqrt{9} = -3$, this number is an integer and a rational number.

Lab Activity

You can learn how to use base-ten blocks to estimate the square root of a number in Lab 14 on page A21.

You have graphed rational numbers on number lines. Yet, if you graphed *all* of the rational numbers, the number line would still not be complete. The irrational numbers complete the number line. The set of irrational numbers together with the set of rational numbers form the set of **real numbers, R.** The graph of all real numbers is the entire number line.

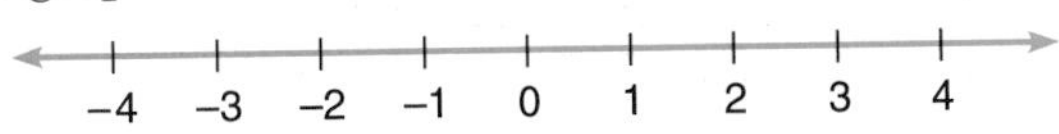

This is illustrated by the completeness property.

Completeness Property for Points on the Number Line

Each real number corresponds to exactly one point on the number line. Each point on the number line corresponds to exactly one real number.

You can use a calculator, computer, or table of square roots, like the one on page 644, to find approximate square roots. These values can be used to approximate the graphs of square roots.

Example 2

Use a calculator to find an approximate value for $\sqrt{54}$. Then graph $\sqrt{54}$ on the number line.

Enter: 54 $\sqrt{x}$ 7.34846923

An approximate value for $\sqrt{54}$ is 7.348.

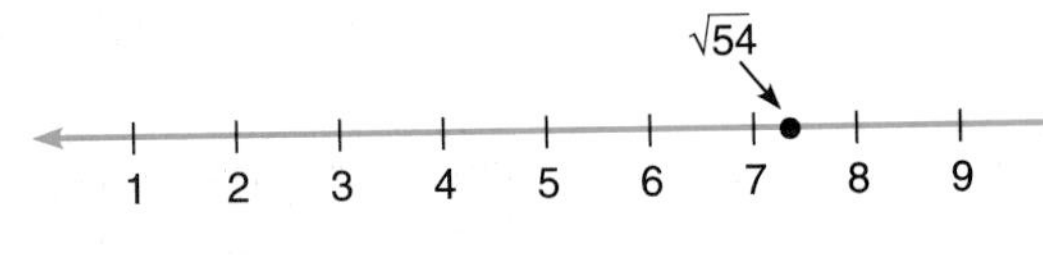

Example 3

Is $\sqrt{8556}$ an irrational number? If it is, then name two consecutive integers between which its graph lies on the number line.

The value of $\sqrt{8556}$ is approximately 92.498649. It is irrational. This value can be found by using a calculator, computer, or table of square roots. The graph of $\sqrt{8556}$ lies between 92 and 93 on the number line.

Example 4

The voltage, V, in a circuit is given by the formula $V = \sqrt{PR}$. In this formula, V is in volts, P is the power in watts, and R is the resistance in ohms. An electrician has a circuit that produces 1800 watts of power. She wants the voltage in the circuit to be at most 110 volts. Should she design the circuit with a resistance of 6.4 ohms or 6.7 ohms?

For this circuit, $P = 1800$ and $V = 110$. To determine which resistance to use, you can use a calculator to evaluate $\sqrt{1800R}$ for $R = 6.4$ and for $R = 6.7$ to find out which one will produce at most 110 volts.

Evaluate $\sqrt{1800R}$ for $R = 6.4$.

Enter: 1800 [×] 6.4 [=] [√x] 107.331263

Evaluate $\sqrt{1800R}$ for $R = 6.7$.

Enter: 1800 [×] 6.7 [=] [√x] 109.818031

She can design the circuit with *either* resistance. If she wants the voltage in the circuit to be as close to 110 volts as possible, then she should design the circuit with a resistance of 6.7 ohms.

CHECKING FOR UNDERSTANDING

Communicating Mathematics

Read and study the lesson to answer each question.

1. What must be true about the measure of the area of a square if the measure of its side is a natural number?
2. Explain the difference between the rational and irrational numbers. Are there any numbers that are both rational *and* irrational?
3. The graph of which set of numbers is the entire number line?

Determine whether each statement is *true* or *false*.

4. Every integer is also a real number.
5. Every rational number is also an irrational number.
6. Every real number is also a rational number.
7. Every natural number is also a whole number.

Guided Practice

Name the set or sets of numbers to which each real number belongs. Use N for natural numbers, W for whole numbers, Z for integers, Q for rational numbers, and I for irrational numbers.

8. $-\frac{1}{2}$
9. $\frac{6}{3}$
10. 0
11. 0.3333 . . .
12. $\sqrt{11}$
13. $\sqrt{36}$
14. 0.6125
15. 0.53694 . . .

Find an approximation, to the nearest hundredth, for each square root.

16. $\sqrt{11}$
17. $\sqrt{40}$
18. $\sqrt{91}$
19. $-\sqrt{89}$

EXERCISES

Practice

State whether each decimal represents a rational or an irrational number.

20. 1.23123412 . . .
21. 0.4444 . . .
22. 4.3434343 . . .
23. 4.34334333 . . .
24. 7.6567876 . . .
25. $1.24\overline{37}$

Find an approximation, to the nearest hundredth, for each square root. Then graph the square root on a number line.

26. $\sqrt{7}$
27. $\sqrt{20}$
28. $-\sqrt{50}$
29. $-\sqrt{66}$
30. $\sqrt{84}$
31. $-\sqrt{31}$
32. $-\sqrt{98}$
33. $\sqrt{107}$

Determine whether each number is rational or irrational. If it is irrational, find two consecutive integers between which its graph lies on the number line.

34. $\sqrt{6436}$
35. $\sqrt{9025}$
36. $\sqrt{3840}$
37. $\sqrt{7511}$

Find the area of each rectangle. Round answers to the nearest hundredth. (*Hint:* Use the Pythagorean Theorem.)

38.

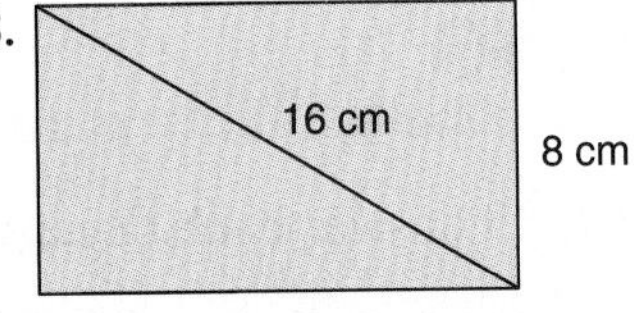

39.

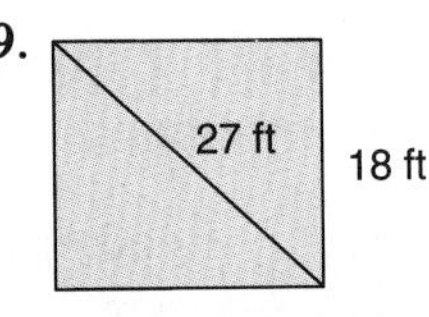

Solve each problem. Round answers to the nearest tenth.

40. The area of a square is 549 square inches. How many additional square inches of area would result in a new square whose sides have integral lengths?
41. The length of a rectangle is three times its width. What are the dimensions of the rectangle if its area is 186 ft^2?
42. A square is inscribed in a circle as shown at the right. If the radius of the circle is 3 cm, find the perimeter of the square.
43. What is the area of the shaded region in the figure at the right? Use 3.14 for π.

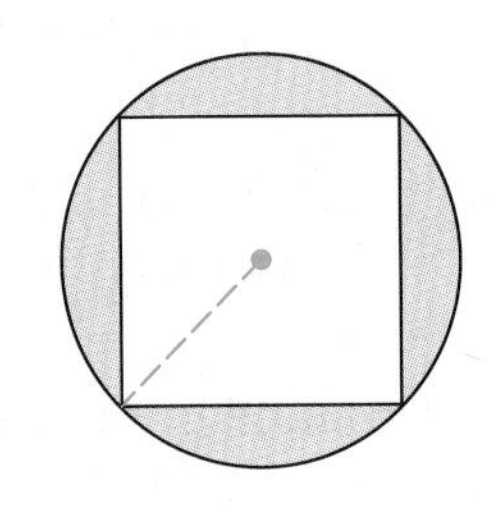

Critical Thinking

Find all numbers of the form $\sqrt{n}$ such that n is a natural number and the graph of $\sqrt{n}$ lies between each pair of numbers on the number line.

44. 3 and 4
45. 5.25 and 5.5

Applications

46. **Electricity** A circuit is designed with two resistance settings R, 4.6 ohms and 5.2 ohms, and two power settings P, 1200 watts and 1500 watts. Which settings can be used so that the voltage of the circuit is between 75 volts and 85 volts? Use $V = \sqrt{PR}$.

47. **Weather** The time, t, in hours, that a storm will last is given by the formula $t = \sqrt{\frac{d^3}{216}}$, where d is the diameter of the storm in miles. Suppose the umpires at a baseball game declared a rain delay at 10:00 P.M. The storm causing the rain delay has a diameter of 12 miles. If it takes 20 minutes to get the field ready after the rain has stopped, can the game be restarted before 1:00 A.M.?

Mixed Review

48. Multiply $(3x - 0.5)(3x + 0.5)$. **(Lesson 6-9)**

49. **Physics** Solve the motion formula $F = \frac{GMn}{d^2}$ for d. **(Lesson 8-10)**

50. Determine the slope and y-intercept of the graph of $7x - 3y = 10$. **(Lesson 10-3)**

51. Use elimination to solve the system of equations $3x + 4y = 7$ and $3x - 4y = 8$. **(Lesson 11-4)**

52. **Construction** Lorena needs to run a wire from the top of a telephone pole to a stake on the ground 10 meters from the base of the pole. Lorena only has 16 meters of wire to use. Does she have enough wire if the pole is 14 meters tall? **(Lesson 12-3)**

MID-CHAPTER REVIEW

The following high temperatures were recorded in Cleveland during a cold spell that lasted 30 days. (Lesson 12-1)

29°F	26°F	17°F	12°F	5°F	4°F	25°F	17°F	23°F	18°F
2°F	12°F	27°F	16°F	27°F	16°F	30°F	6°F	16°F	5°F
0°F	5°F	29°F	18°F	16°F	22°F	29°F	8°F	23°F	24°F

1. Organize this information into a table with headings *Temperature in Degrees Fahrenheit* and *Number of Days.* Then use the table to answer the following questions.

2. Which high temperature occurs the greatest number of times?

3. How many days was the high temperature in the range 10–19?

Simplify. (Lesson 12-2)

4. $-\sqrt{441}$

5. $\sqrt{10.89}$

6. $\pm\sqrt{0.0841}$

7. $\sqrt{\frac{576}{729}}$

If c is the measure of the hypotenuse of a right triangle, find each missing measure. (Lesson 12-3)

8. $a = 21, b = 28, c = ?$

9. $a = 0.5, c = 1.3, b = ?$

10. $b = \sqrt{17}, c = 9, a = ?$

11. Can the measures of the sides of a right triangle be 24, 30, and 36? **(Lesson 12-3)**

12-5 Simplifying Square Roots

Objectives

After studying this lesson, you should be able to:

- simplify square roots, and
- simplify radical expressions that contain variables.

A radical expression is in *simplest form* if the radicand has no perfect square factors other than one. The following property can be used to simplify square roots.

Product Property of Square Roots	**For any numbers a and b, where $a \geq 0$ and $b \geq 0$, $\sqrt{ab} = \sqrt{a} \cdot \sqrt{b}$.**

The product property of square roots and prime factorization can be used to simplify radical expressions in which the radicand is not a perfect square.

Example 1

Simplify $\sqrt{375}$.

$\sqrt{375} = \sqrt{3} \cdot \sqrt{5} \cdot \sqrt{5} \cdot \sqrt{5}$ — *Prime factorization of 375*

$= \sqrt{3} \cdot \sqrt{5} \cdot \sqrt{5^2}$ — *Product property of square roots*

$= \sqrt{3 \cdot 5} \cdot 5$

$= 5\sqrt{15}$

When finding the principal square root of an expression containing variables, be sure that the result is not negative. Consider the expression $\sqrt{x^2}$. Its simplest form is not x since, for example, $\sqrt{(-3)^2} \neq -3$. For radical expressions like $\sqrt{x^2}$, use absolute value to ensure nonnegative results.

$$\sqrt{x^2} = |x| \qquad \sqrt{x^3} = x\sqrt{x} \qquad \sqrt{x^4} = x^2 \qquad \sqrt{x^6} = |x^3|$$

For $\sqrt{x^3}$, absolute value is not used since x cannot be negative. If x were negative, then x^3 would be negative and $\sqrt{x^3}$ would not be defined. Why is absolute value not used for $\sqrt{x^4}$?

Example 2

Simplify $\sqrt{200m^2y^3}$.

$\sqrt{200m^2y^3} = \sqrt{2 \cdot 2 \cdot 2 \cdot 5 \cdot 5 \cdot m \cdot m \cdot y \cdot y \cdot y}$ — *Prime factorization*

$= \sqrt{2} \cdot \sqrt{2^2} \cdot \sqrt{5^2} \cdot \sqrt{m^2} \cdot \sqrt{y} \cdot \sqrt{y^2}$ — *Product property of squar roots*

$= \sqrt{2} \cdot 2 \cdot 5 \cdot |m| \cdot \sqrt{y} \cdot y$ — *The absolute value of m ensures a nonnegative result. Why is the absolut value not indicated for y?*

$= 10|m|y\sqrt{2y}$

The product property can also be used to multiply square roots.

Example 3 **Simplify $\sqrt{10} \cdot \sqrt{20}$.**

$$\begin{aligned} \sqrt{10} \cdot \sqrt{20} &= \sqrt{10 \cdot 20} && \textit{Product property of square roots} \\ &= \sqrt{10^2 \cdot 2} \\ &= \sqrt{10^2} \cdot \sqrt{2} \text{ or } 10\sqrt{2} \end{aligned}$$

You can divide square roots and simplify radical expressions that involve division by using the quotient property of square roots.

Quotient Property of Square Roots

For any numbers a and b, where $a \geq 0$ and $b > 0$,

$$\sqrt{\frac{a}{b}} = \frac{\sqrt{a}}{\sqrt{b}}.$$

A fraction containing radicals is in simplest form if no radicals are left in the denominator.

Example 4 **Simplify $\frac{\sqrt{72}}{\sqrt{6}}$.**

$$\begin{aligned} \frac{\sqrt{72}}{\sqrt{6}} &= \sqrt{\frac{72}{6}} && \textit{Quotient property of square roots} \\ &= \sqrt{12} \\ &= \sqrt{2^2 \cdot 3} && \textit{Prime factorization} \\ &= \sqrt{2^2} \cdot \sqrt{3} \text{ or } 2\sqrt{3} \end{aligned}$$

The next example illustrates a method for simplifying radical expressions called **rationalizing the denominator.** This method may be used to eliminate radicals from the denominator of a fraction.

Example 5 **Simplify $\frac{\sqrt{32}}{\sqrt{3}}$.**

$$\begin{aligned} \frac{\sqrt{32}}{\sqrt{3}} &= \frac{\sqrt{32}}{\sqrt{3}} \cdot \frac{\sqrt{3}}{\sqrt{3}} && \textit{Notice that } \frac{\sqrt{3}}{\sqrt{3}} = 1. \\ &= \frac{\sqrt{32 \cdot 3}}{\sqrt{3 \cdot 3}} \\ &= \frac{\sqrt{16 \cdot 2 \cdot 3}}{\sqrt{3^2}} \\ &= \frac{\sqrt{16} \cdot \sqrt{2} \cdot \sqrt{3}}{3} && \textit{Do you see why } \frac{\sqrt{3}}{\sqrt{3}} \textit{ was used?} \\ &= \frac{4\sqrt{6}}{3} \end{aligned}$$

Example 6

The ratio of the measures of the legs of a right triangle is 2:1. Find the ratio of the measure of the shorter leg to the measure of the hypotenuse of the triangle.

Let a = the measure of the shorter leg of the right triangle.
Let b = the measure of the longer leg.
Then $\frac{b}{a} = \frac{2}{1}$ or $b = 2a$.

By the Pythagorean Theorem, $c^2 = a^2 + b^2$. Therefore, $c = \sqrt{a^2 + b^2}$ and the ratio of the shorter leg to the hypotenuse is $\frac{a}{\sqrt{a^2 + b^2}}$.

$$\frac{a}{\sqrt{a^2 + b^2}} = \frac{a}{\sqrt{a^2 + (2a)^2}} \quad \textit{Substitute 2a for b.}$$

$$= \frac{a}{\sqrt{5a^2}} \quad a^2 + (2a)^2 = a^2 + 4a^2 = 5a^2$$

$$= \frac{a}{\sqrt{5} \cdot \sqrt{a^2}}$$

$$= \frac{a}{\sqrt{5} \cdot a} \cdot \frac{\sqrt{5}}{\sqrt{5}} \quad \textit{a must be positive, so absolute value does not need to be indicated for a.}$$

$$= \frac{\sqrt{5}}{5}$$

The ratio of the measure of the shorter leg to the measure of the hypotenuse is $\frac{\sqrt{5}}{5}$ to 1 or about 0.45 to 1.

Binomials of the form $a\sqrt{b} + c\sqrt{d}$ and $a\sqrt{b} - c\sqrt{d}$ are **conjugates** of each other. For example, $8 + \sqrt{2}$ and $8 - \sqrt{2}$ are conjugates. Conjugates are useful for simplifying radical expressions because their product is always a rational number with no radicals.

$$(\sqrt{2})^2 = \sqrt{2} \cdot \sqrt{2} = \sqrt{2 \cdot 2} = \sqrt{2^2} = 2$$

$$(8 + \sqrt{2})(8 - \sqrt{2}) = 8^2 - (\sqrt{2})^2 \quad \textit{Use the pattern } (a - b)(a + b) = a^2 - b^2 \textit{ to simplify the product.}$$

$$= 64 - 2$$

$$= 62$$

Conjugates are often used to rationalize the denominators of fractions containing square roots.

Example 7

Simplify $\frac{3}{3 - \sqrt{5}}$.

To rationalize the denominator, multiply both the numerator and denominator of the fraction by the conjugate of $3 - \sqrt{5}$, which is $3 + \sqrt{5}$

$$\frac{3}{3 - \sqrt{5}} = \frac{3}{3 - \sqrt{5}} \cdot \frac{3 + \sqrt{5}}{3 + \sqrt{5}} \quad \textit{Notice that } \frac{3 + \sqrt{5}}{3 + \sqrt{5}} = 1.$$

$$= \frac{3(3) + 3\sqrt{5}}{3^2 - (\sqrt{5})^2} \quad \textit{Use the distributive property to multiply numerators. Use the pattern } (a - b)(a + b) = a^2 - b^2 \textit{ to multiply denominators.}$$

$$= \frac{9 + 3\sqrt{5}}{9 - 5}$$

$$= \frac{9 + 3\sqrt{5}}{4}$$

When simplifying radical expressions, check the following conditions to determine if you have the expression in simplest form.

Simplified Form for Radicals

A radical expression is in simplest form when the following three conditions have been met.
1. No radicands have perfect square factors other than one.
2. No radicands contain fractions.
3. No radicals appear in the denominator of a fraction.

CHECKING FOR UNDERSTANDING

Communicating Mathematics

Read and study the lesson to answer each question.

1. State the product property of square roots in your own words.
2. Why are absolute values sometimes needed when simplifying radical expressions containing variables?
3. What do you do when you rationalize a denominator?
4. What is the conjugate of $5 - 9\sqrt{2}$?

Guided Practice

Simplify.

5. $\sqrt{20}$ 6. $\sqrt{18}$ 7. $\sqrt{48}$ 8. $\frac{\sqrt{42}}{\sqrt{6}}$ 9. $\frac{\sqrt{20}}{\sqrt{5}}$

State the conjugate of each expression. Then multiply the expression by its conjugate.

10. $3 + \sqrt{2}$ 11. $\sqrt{5} - 7$ 12. $\sqrt{3} - \sqrt{7}$ 13. $2\sqrt{8} + 3\sqrt{5}$

State the fraction by which each expression should be multiplied to rationalize the denominator.

14. $\frac{3}{\sqrt{5}}$ 15. $\frac{2\sqrt{3}}{\sqrt{8}}$ 16. $\sqrt{\frac{8}{7}}$ 17. $\frac{2\sqrt{5}}{4 - \sqrt{3}}$

EXERCISES

Practice

Simplify. Use absolute value symbols when necessary.

18. $\sqrt{75}$ 19. $\sqrt{45}$ 20. $\sqrt{80}$ 21. $\sqrt{72}$
22. $\sqrt{98}$ 23. $\sqrt{280}$ 24. $\sqrt{500}$ 25. $\sqrt{1000}$
26. $\frac{\sqrt{7}}{\sqrt{3}}$ 27. $\frac{\sqrt{5}}{\sqrt{10}}$ 28. $\sqrt{\frac{3}{7}}$ 29. $\sqrt{\frac{11}{32}}$
30. $\sqrt{10} \cdot \sqrt{30}$ 31. $2\sqrt{5} \cdot \sqrt{5}$ 32. $5\sqrt{10} \cdot 3\sqrt{10}$ 33. $7\sqrt{30} \cdot 2\sqrt{6}$
34. $\sqrt{\frac{2}{3}} \cdot \sqrt{\frac{5}{2}}$ 35. $\sqrt{\frac{1}{6}} \cdot \sqrt{\frac{6}{11}}$ 36. $\sqrt{32x^2}$ 37. $\sqrt{40b^4}$

Simplify. Use absolute value symbols when necessary.

38. $\sqrt{54a^2b^2}$
39. $\sqrt{80x^2y^3}$
40. $\sqrt{60m^2y^4}$
41. $\sqrt{147x^5y^7}$
42. $\sqrt{\frac{b}{6}}$
43. $\sqrt{\frac{27}{r^2}}$
44. $\sqrt{\frac{5n^5}{4m^5}}$
45. $\frac{\sqrt{9x^5y}}{\sqrt{12x^2y^6}}$
46. $\frac{1}{6 + \sqrt{3}}$
47. $\frac{10}{\sqrt{5} - 9}$
48. $\frac{12}{\sqrt{6} - \sqrt{5}}$
49. $\frac{9b}{6 + \sqrt{b}}$
50. $\frac{2\sqrt{5}}{-4 + \sqrt{8}}$
51. $\frac{2\sqrt{7}}{3\sqrt{5} + 5\sqrt{3}}$
52. $\frac{3\sqrt{2} - \sqrt{7}}{2\sqrt{3} - 5\sqrt{2}}$
53. $\frac{\sqrt{a} - \sqrt{b}}{\sqrt{a} + \sqrt{b}}$

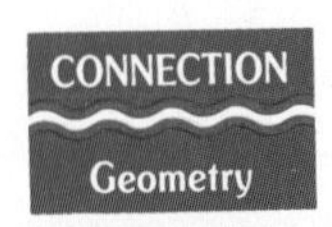

54. Find the length of the diagonal of a square whose area is $48s^3$ square feet.

Critical Thinking

55. Is the sentence $\sqrt{a \cdot b} = \sqrt{a} \cdot \sqrt{b}$ true for negative numbers? Why or why not?

Applications

56. **Physics** The period of a pendulum is the time in seconds that it takes the pendulum to make one complete swing back and forth. The formula for the period P of a pendulum is $P = 2\pi\sqrt{\frac{\ell}{32}}$, where ℓ is the length of the pendulum in feet. Suppose a clock sounds one "tick" after each complete swing back and forth of a 2-foot-long pendulum. How many ticks would the clock sound in one minute? Use 3.14 for π and round to the nearest whole number.

57. **Travel** Lois rode her bike due west for 2 hours at a constant speed. She then rode due north for 3 hours at twice that speed.
    a. Write an expression, in simplest form, to represent how far Lois is from the starting point of her bike ride. Let s = the speed at which Lois rode due west.
    b. Find this distance, to the nearest tenth of a mile, if her starting speed was 5 miles per hour.

Mixed Review

58. Write as an equation: *the square of the sum of a and b is equal to the product of the squares of a and b.* **(Lesson 1-7)**

59. Factor $12r^2 - 16rs - 11s^2$. **(Lesson 7-5)**

60. Is {(5, 4), (6, 1), (−2, 3), (0, 3)} a function? Is the inverse of this relation a function? **(Lesson 9-5)**

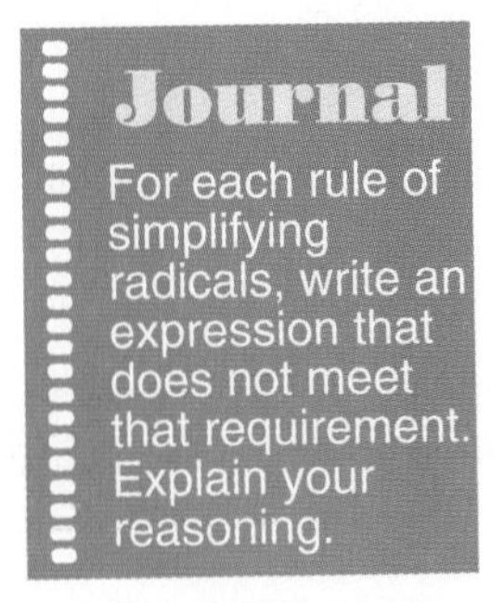

61. **Geometry** If one endpoint of a line segment is at (−4, 11) and the midpoint is at (8, 3), find the coordinates of the other endpoint. **(Lesson 10-8)**

62. Graph the equations $x + y = 5$ and $x - 2y = -4$. Then find the solution to the system of equations. **(Lesson 11-2)**

63. **Electricity** The resistance R of a power circuit is 4.5 ohms. How much current I, in amperes, can the circuit generate if it can produce at most 1500 watts of power P? Use $I^2R = P$. **(Lesson 12-4)**

12-6 Adding and Subtracting Radical Expressions

Objective After studying this lesson, you should be able to:

- simplify radical expressions involving addition and subtraction.

Radical expressions in which the radicands are alike can be added or subtracted in the same way that monomials are added or subtracted.

Monomials	Radical Expressions
$3x + 2x = (3 + 2)x = 5x$	$3\sqrt{2} + 2\sqrt{2} = (3 + 2)\sqrt{2} = 5\sqrt{2}$
$7y - 4y = (7 - 4)y = 3y$	$7\sqrt{5} - 4\sqrt{5} = (7 - 4)\sqrt{5} = 3\sqrt{5}$

Notice that the distributive property was used to simplify each radical expression.

Example 1

Simplify $3\sqrt{11} + 6\sqrt{11} - 2\sqrt{11}$.

$$3\sqrt{11} + 6\sqrt{11} - 2\sqrt{11} = (3 + 6 - 2)\sqrt{11}$$
$$= 7\sqrt{11}$$

Example 2

Simplify $9\sqrt{7} - 4\sqrt{2} + 3\sqrt{2} + 5\sqrt{7}$.

$$9\sqrt{7} - 4\sqrt{2} + 3\sqrt{2} + 5\sqrt{7} = 9\sqrt{7} + 5\sqrt{7} - 4\sqrt{2} + 3\sqrt{2}$$ *Commutative property*
$$= (9 + 5)\sqrt{7} + (-4 + 3)\sqrt{2}$$
$$= 14\sqrt{7} - \sqrt{2}$$

Example 3

CONNECTION Geometry

Find the exact measure of the perimeter of the rectangle.

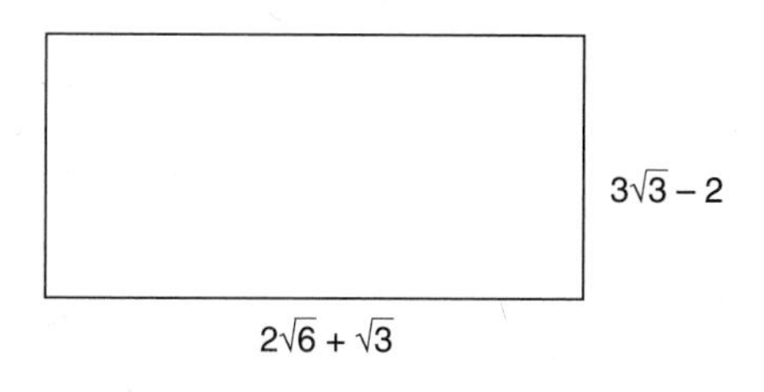

$$p = 2\ell + 2w$$
$$= 2(2\sqrt{6} + \sqrt{3}) + 2(3\sqrt{3} - 2)$$
$$= 4\sqrt{6} + 2\sqrt{3} + 6\sqrt{3} - 4$$
$$= 4\sqrt{6} + (2 + 6)\sqrt{3} - 4$$
$$= 4\sqrt{6} + 8\sqrt{3} - 4$$

The exact measure of the perimeter is $4\sqrt{6} + 8\sqrt{3} - 4$.

In Example 2, the expression $14\sqrt{7} - \sqrt{2}$ cannot be simplified further because the radicands are different, there are no common factors, and each radicand is in simplest form. The same is true for the expression $4\sqrt{6} + 8\sqrt{3} - 4$ in Example 3.

If each radical in a radical expression is not in simplest form, simplify them first. Then use the distributive property, whenever possible, to further simplify the expression.

Example 4

Simplify $7\sqrt{98} + 5\sqrt{32} - 2\sqrt{75}$.

$$\begin{aligned} 7\sqrt{98} + 5\sqrt{32} - 2\sqrt{75} &= 7\sqrt{7^2 \cdot 2} + 5\sqrt{4^2 \cdot 2} - 2\sqrt{5^2 \cdot 3} \\ &= 7(\sqrt{7^2} \cdot \sqrt{2}) + 5(\sqrt{4^2} \cdot \sqrt{2}) - 2(\sqrt{5^2} \cdot \sqrt{3}) \\ &= 7(7\sqrt{2}) + 5(4\sqrt{2}) - 2(5\sqrt{3}) \\ &= 49\sqrt{2} + 20\sqrt{2} - 10\sqrt{3} \\ &= 69\sqrt{2} - 10\sqrt{3} \quad \textit{Distributive property} \end{aligned}$$

Example 5

Find a decimal approximation for the area of the rectangle. Then find the exact measure of the area. Compare the decimal approximation of the exact measure to the original decimal approximation for the area to verify your results.

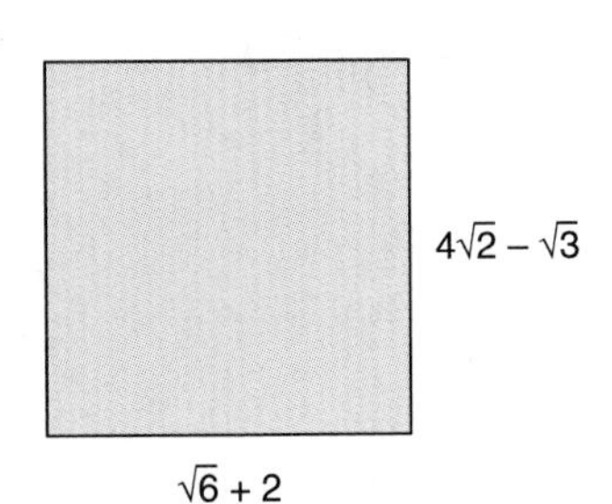

$$\begin{aligned} A &= \ell w \\ &= (4\sqrt{2} - \sqrt{3})(\sqrt{6} + 2) \end{aligned}$$

Use a calculator to find a decimal approximation for the area.

Enter: (4 × 2 √x − 3 √x) × (6 √x + 2) =

Display: 17.4633727

The measure of the area is approximately 17.46.

Simplify $(4\sqrt{2} - \sqrt{3})(\sqrt{6} + 2)$ to find the exact measure of the area.

$$\begin{aligned} (4\sqrt{2} - \sqrt{3})(\sqrt{6} + 2) &= 4\sqrt{2} \cdot \sqrt{6} + 4\sqrt{2} \cdot 2 - \sqrt{3} \cdot \sqrt{6} - \sqrt{3} \cdot 2 \quad \textit{Use FOIL.} \\ &= 4\sqrt{12} + 8\sqrt{2} - \sqrt{18} - 2\sqrt{3} \\ &= 4\sqrt{2^2 \cdot 3} + 8\sqrt{2} - \sqrt{3^2 \cdot 2} - 2\sqrt{3} \\ &= 4(2\sqrt{3}) + 8\sqrt{2} - 3\sqrt{2} - 2\sqrt{3} \\ &= 8\sqrt{3} + 5\sqrt{2} - 2\sqrt{3} \\ &= 6\sqrt{3} + 5\sqrt{2} \end{aligned}$$

The exact measure of the area is $6\sqrt{3} + 5\sqrt{2}$.

Now, use a calculator to find a decimal approximation for the measure of the area.

Enter: 6 × 3 √x + 5 × 2 √x =

Display: 17.4633727

Since the approximations are equal, the results have been verified.

CHECKING FOR UNDERSTANDING

Communicating Mathematics

Read and study the lesson to answer each question.

1. What property do you use to simplify the sum or difference of radicals?
2. Why should you simplify each radical in a radical expression before adding or subtracting?
3. How can you use a calculator to verify that two radical expressions are equal?

Guided Practice

Name the expressions in each group that have the same radicand.

4. $5\sqrt{3}$, $4\sqrt{6}$, $3\sqrt{3}$
5. $4\sqrt{2}$, $7\sqrt{2}$, $2\sqrt{7}$
6. $2\sqrt{10}$, $-5\sqrt{10}$, $10\sqrt{5}$

Name the expressions in each group that will have the same radicand after each expression is written in simplest form.

7. $5\sqrt{14}$, $-3\sqrt{7}$, $2\sqrt{28}$
8. $3\sqrt{20}$, $3\sqrt{5}$, $5\sqrt{6}$
9. $\sqrt{18}$, $\sqrt{24}$, $\sqrt{12}$, $\sqrt{28}$
10. $3\sqrt{32}$, $2\sqrt{48}$, $\sqrt{50}$, $7\sqrt{200}$

Simplify.

11. $8\sqrt{6} + 3\sqrt{6}$
12. $4\sqrt{3} - 7\sqrt{3}$
13. $3\sqrt{5} - 5\sqrt{3}$
14. $25\sqrt{13} + \sqrt{13}$
15. $18\sqrt{2x} + 3\sqrt{2x}$
16. $3\sqrt{5m} - 5\sqrt{5m}$

EXERCISES

Practice

Simplify. Then use a calculator to verify your answer.

17. $4\sqrt{3} + 7\sqrt{3} - 2\sqrt{3}$
18. $2\sqrt{11} - 6\sqrt{11} - 3\sqrt{11}$
19. $5\sqrt{5} + 3\sqrt{5} - 18\sqrt{5}$
20. $\sqrt{6} + 2\sqrt{2} + \sqrt{10}$
21. $8\sqrt{3} - 2\sqrt{2} + 3\sqrt{2} + 5\sqrt{3}$
22. $4\sqrt{6} + \sqrt{7} - 6\sqrt{2} + 4\sqrt{7}$
23. $2\sqrt{3} + \sqrt{12}$
24. $3\sqrt{7} - 2\sqrt{28}$
25. $2\sqrt{50} - 3\sqrt{32}$
26. $3\sqrt{27} + 5\sqrt{48}$
27. $\sqrt{18} + \sqrt{108} + \sqrt{50}$
28. $2\sqrt{20} - 3\sqrt{24} - \sqrt{180}$
29. $\sqrt{7} + \sqrt{\frac{1}{7}}$
30. $\sqrt{10} - \sqrt{\frac{2}{5}}$
31. $3\sqrt{3} - \sqrt{45} + 3\sqrt{\frac{1}{3}}$
32. $6\sqrt{\frac{7}{4}} + 3\sqrt{28} - 10\sqrt{\frac{1}{7}}$

Find the exact measures of the perimeter and area, in simplest form, for each rectangle.

33.

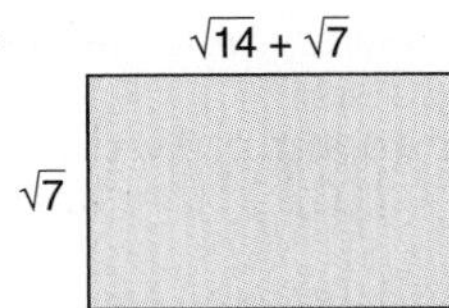

34. $4\sqrt{7} - 2\sqrt{12}$; $\sqrt{3}$

35. $\sqrt{8} + \sqrt{27}$; $\sqrt{3} - \sqrt{2}$

Simplify.

36. $(\sqrt{14} + \sqrt{35})(\sqrt{5} - \sqrt{2})$

37. $(2\sqrt{10} + 3\sqrt{15})(3\sqrt{3} - 2\sqrt{2})$

38. $(\sqrt{6} + \sqrt{8})(\sqrt{24} + \sqrt{2})$

39. $(2\sqrt{10} - 3)(3\sqrt{5} + 5\sqrt{2})$

Critical Thinking

40. Is the sum of two irrational numbers always an irrational number? Explain.

41. Is the sentence $\sqrt{a + b} = \sqrt{a} + \sqrt{b}$ ever true?

Applications

42. Construction Akikta wants to build a wooden border around a square bulletin board that has an area of 44 square feet. The wood for the border can only be purchased in lengths that are an integral number of feet. What is the least number of feet of wood that Akikta can purchase to build the border?

43. Construction A wire is stretched from the top of a 10-foot pole to a stake in the ground and then to the base of the pole. If a total of 18 feet of wire is needed, how far is the stake from the pole? (*Hint:* In the figure, $a + b = 18$. So, $b = 18 - a$.)

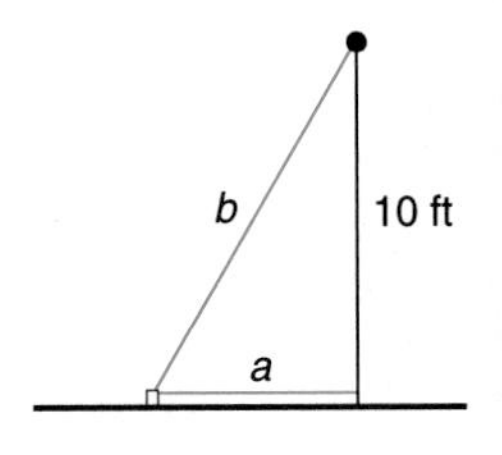

Mixed Review

44. Solve $\frac{2n + 1}{7} \geq \frac{n + 4}{5}$. **(Lesson 5-3)**

45. Simplify $\frac{k^2 + 2k - 3}{k^2 + 6k + 5} \cdot \frac{k^2 - 1}{3k + 9}$. **(Lesson 8-7)**

46. Car Rental The cost of a one-day car rental from Best Car Rental is \$42.20 if you drive 110 miles, \$56.50 if you drive 175 miles, and \$62 if you drive 200 miles. Write an equation to describe this relationship and use it to determine the per-mile charge for a one-day rental. **(Lesson 9-8)**

47. Write an equation for the line that is perpendicular to the graph of $x - 4y = 16$ and that passes through $(-1, 1)$. **(Lesson 10-6)**

48. Simplify $\sqrt{1944}$. **(Lesson 12-5)**

12-7 Radical Equations

Objective

After studying this lesson, you should be able to:

- solve radical equations.

Application

FYI · · ·

In 1905, the Bosco Company of Akron, Ohio, marketed a "Collapsible Rubber Automobile Driver" to scare thieves away from parked cars.

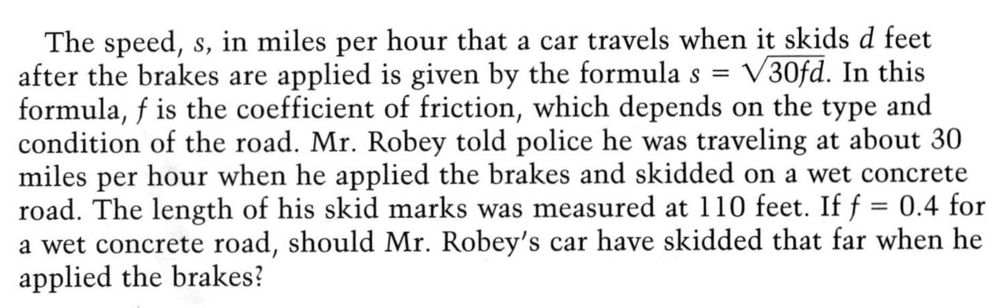

The speed, s, in miles per hour that a car travels when it skids d feet after the brakes are applied is given by the formula $s = \sqrt{30fd}$. In this formula, f is the coefficient of friction, which depends on the type and condition of the road. Mr. Robey told police he was traveling at about 30 miles per hour when he applied the brakes and skidded on a wet concrete road. The length of his skid marks was measured at 110 feet. If $f = 0.4$ for a wet concrete road, should Mr. Robey's car have skidded that far when he applied the brakes?

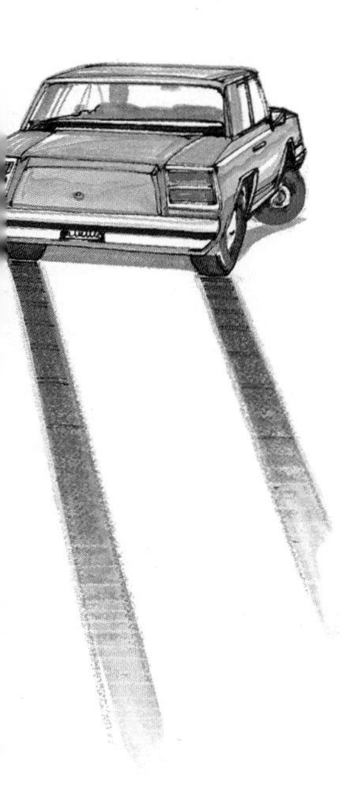

Use the formula $s = \sqrt{30fd}$ to find d when $s = 30$ and $f = 0.4$.

$$s = \sqrt{30fd}$$

$$30 = \sqrt{30(0.4)d} \quad \text{Substitute 30 for } s \text{ and 0.4 for } f.$$

$$30 = \sqrt{12d}$$

Equations like $30 = \sqrt{12d}$ that contain radicals with variables in the radicand are called **radical equations.** To solve these equations, first isolate the radical on one side of the equation. Then square each side of the equation to eliminate the radical.

$$30 = \sqrt{12d}$$

$$30^2 = (\sqrt{12d})^2 \quad \text{Square each side to eliminate the radical.}$$

$$900 = 12d$$

$$d = 75$$

Check: $30 = \sqrt{12d}$

$$30 \stackrel{?}{=} \sqrt{12(75)}$$

$$30 \stackrel{?}{=} \sqrt{900}$$

$$30 = 30 \checkmark$$

At 30 miles per hour, Mr. Robey's car should have skidded 75 feet after the brakes were applied and not 110 feet. Therefore, he was traveling faster than 30 miles per hour.

Consider the equation $x = 3$.

$$x = 3$$

$$x^2 = 9 \quad \text{Square each side.}$$

The solutions are $x = 3$ *and* $x = -3$. Squaring each side of an equation does not necessarily produce results that satisfy the *original* equation. In this case, the solution -3 for $x^2 = 9$ is not a solution of the original equation $x = 3$. Therefore, you must check *all* solutions when you solve radical equations.

Example 1

Solve $\sqrt{3x - 14} + x = 6$ and check.

$$\sqrt{3x - 14} + x = 6$$

$$\sqrt{3x - 14} = 6 - x$$ *Isolate the radical by subtracting x from each side.*

$$3x - 14 = (6 - x)^2$$ *Square each side of the equation.*

$$3x - 14 = 36 - 12x + x^2$$ *Recall that $(a - b)^2 = a^2 - 2ab + b^2$.*

$$0 = x^2 - 15x + 50$$

$$0 = (x - 5)(x - 10)$$ *Factor.*

$x - 5 = 0$ or $x - 10 = 0$ *Zero product property*

$x = 5$ $\quad$ $x = 10$

Check: $\sqrt{3x - 14} + x = 6$

$\sqrt{3(5) - 14} + 5 \stackrel{?}{=} 6$ or $\sqrt{3(10) - 14} + 10 \stackrel{?}{=} 6$

$\sqrt{15 - 14} + 5 \stackrel{?}{=} 6$ $\quad$ $\sqrt{30 - 14} + 10 \stackrel{?}{=} 6$

$\sqrt{1} + 5 \stackrel{?}{=} 6$ $\quad$ $\sqrt{16} + 10 \stackrel{?}{=} 6$

$1 + 5 = 6$ ✓ $\quad$ $4 + 10 \neq 6$

Notice that 10 does *not* satisfy the original equation. Therefore, 5 is the *only* solution of $\sqrt{3x - 14} + x = 6$.

Example 2

The *geometric mean* of two numbers is a square root of their product. Thus, the geometric mean of numbers *a* and *b* is $\pm\sqrt{ab}$. Find two numbers that have a geometric mean of 6 given that one number is 5 more than the other.

EXPLORE Let x = the lesser number.
Then x + 5 = the greater number.

PLAN

$$\pm\sqrt{x(x + 5)} = 6$$ *The geometric mean of the numbers is 6.*

$$x(x + 5) = 36$$ *Square each side.*

SOLVE

$$x^2 + 5x = 36$$

$$x^2 + 5x - 36 = 0$$

$$(x + 9)(x - 4) = 0$$ *Factor.*

$x + 9 = 0$ or $x - 4 = 0$ *Zero product property*

$x = -9$ $\quad$ $x = 4$

If $x = -9$, then $x + 5 = -4$. If $x = 4$, then $x + 5 = 9$. Thus, the numbers are -9 and -4, or 4 and 9.

EXAMINE $\sqrt{-9(-4)} \stackrel{?}{=} 6$ $\quad$ $\sqrt{4(9)} \stackrel{?}{=} 6$

$\sqrt{36} = 6$ ✓ $\quad$ $\sqrt{36} = 6$ ✓

CHECKING FOR UNDERSTANDING

Communicating Mathematics

Read and study the lesson to answer each question.

1. When solving a radical equation, what is the first thing you should do?
2. *True* or *false:* Squaring each side of an equation always results in an equivalent equation.
3. Write a radical equation that has *no* solution.
4. Write an expression for the geometric mean of 5 and x.

Guided Practice

Square each side of the following equations.

5. $\sqrt{x} = 5$
6. $\sqrt{y + 1} = 3$
7. $11 = \sqrt{2a - 5}$

Solve each equation. Check the solutions.

8. $\sqrt{y} = 3$
9. $\sqrt{m} = -2$
10. $-\sqrt{a} = -8$
11. $\sqrt{5x} = 5$
12. $\sqrt{-3a} = 6$
13. $\sqrt{x - 3} = 6$

Find the geometric mean of each pair of numbers.

14. 4, 9
15. 5, 20
16. 7, 10
17. 4, 8

EXERCISES

Practice

Solve each equation. Check the solutions.

18. $\sqrt{r} = 3\sqrt{5}$
19. $4\sqrt{7} = \sqrt{-m}$
20. $\sqrt{b} - 5 = 0$
21. $\sqrt{2d} + 1 = 0$
22. $5 - \sqrt{3x} = 1$
23. $2 + 3\sqrt{m} = 13$
24. $\sqrt{4x + 1} = 3$
25. $\sqrt{8s + 1} - 5 = 0$
26. $\sqrt{3b - 5} + 6 = 2$
27. $\sqrt{\frac{x}{4}} = 6$
28. $\sqrt{\frac{5k}{7}} - 8 = 2$
29. $5\sqrt{\frac{4a}{3}} - 2 = 0$
30. $4\sqrt{3m^2 - 15} = 4$
31. $\sqrt{2z^2 - 121} = z$
32. $\sqrt{5x^2 - 7} = 2x$
33. $\sqrt{x + 2} = x - 4$
34. $\sqrt{1 - 2x} = 1 + x$
35. $4 + \sqrt{x - 2} = x$

36. The geometric mean of a certain number and 4 is 26. Find the number.
37. Find two numbers with a geometric mean of $\sqrt{24}$ given that one number is 2 more than the other.
38. Find two numbers with a geometric mean of 12 given that one number is 11 less than three times the other.

Solve each equation. Check the solution. (*Hint:* You will need to square each side twice.)

39. $\sqrt{x + 16} = \sqrt{x} + 4$ 40. $6 - \sqrt{x} = \sqrt{x - 12}$ 41. $\sqrt{x + 5} = 5 + \sqrt{x}$

Solve each system of equations.

42. $3\sqrt{x} - 5\sqrt{y} = 9$
$2\sqrt{x} + 5\sqrt{y} = 6$

43. $-4\sqrt{a} + 6\sqrt{b} = 3$
$-3\sqrt{a} + 3\sqrt{b} = 1$

44. $m = 4n$
$\sqrt{m} - 5\sqrt{n} = -6$

Critical Thinking

45. Find two numbers such that the square root of their sum is 5 and the square root of their product is 12.

Applications

46. **Science** The horizontal distance you can see is related to your height above the ground by the formula $V = 3.5\sqrt{h}$. In this formula, V is the distance in kilometers that you can see horizontally and h is your height in meters above the ground. If you look out the window of an airplane on a cloudless day and can see for a distance of about 315 km, what is the altitude of the plane?

47. **Physics** The time, in seconds, that it takes an object, initially at rest, to fall a distance of s meters is given by the formula $t = \sqrt{\frac{2s}{g}}$. In this formula, g is the acceleration due to gravity in meters per second squared. On the moon, a rock falls 7.2 meters in 3 seconds. What is the acceleration due to gravity on the moon?

48. **Recreation** The rangers at an aid station received a distress call from a group camping 60 miles east and 10 miles south of the station. A jeep sent to the campsite travels directly east for some number of miles and then turns and heads directly to the campsite. If the jeep traveled a total of 66 miles to get to the campsite, for how many miles did it travel due east?

60 mi
10 mi

Mixed Review

49. Solve $8 = \frac{11t - 10}{7}$. **(Lesson 3-5)**

50. **Gardening** Mr. Schultz doubled the area of his rectangular garden by adding a strip of new soil of uniform width along each of the sides. If the dimensions of the original garden were 10 feet by 15 feet, how wide a strip of new soil did he add? **(Lesson 7-10)**

51. Solve $6x + 5y = 11$ if the domain is $\{-4, -2, 0, 1, 5\}$. **(Lesson 9-3)**

52. Determine the value of r so that the line that passes through $(r, 4)$ and $(3, -r)$ has a slope of $2r$. **(Lesson 10-1)**

53. Use substitution to solve the system of equations $y = -2x + 10$ and $2x + 3y = 6$. **(Lesson 11-3)**

54. **Geometry** The measures of the sides of a triangle are $\sqrt{363}$, $\sqrt{27}$, and $2\sqrt{108}$. Find the measure of its perimeter. **(Lesson 12-6)**

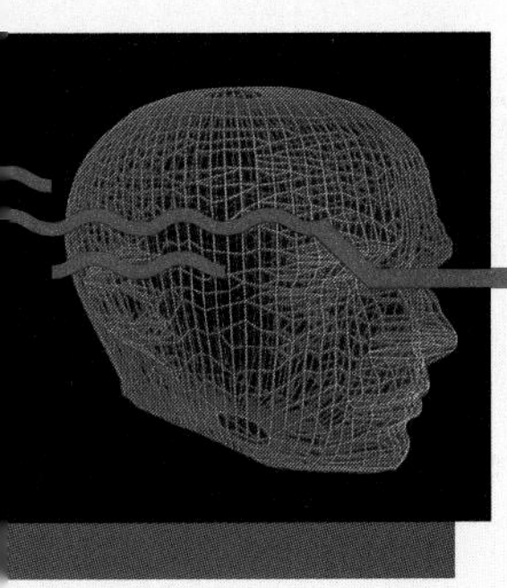

Technology

Solving Radical Equations

BASIC
Graphing calculators
▶ Graphing software
Spreadsheets

You can use the *Mathematics Exploration Toolkit* to solve equations involving square roots. The steps are as follows. First, isolate the radical on one side of the equation. Then raise both sides to the second power to remove the radical. The CALC commands below will be used.

ADD (add)	SUBTRACT (sub)	MULTIPLY (mult)
DIVIDE (div)	FACTOR (fac)	RAISETO (rai)
SIMPLIFY (simp)	STORE (sto)	SUBSTITUTE (subs)

To enter the square root symbol, type a &.

Example

Solve $\sqrt{x - 1} = x - 3$.

Enter	Result
&(x − 1) = x − 3	$\sqrt{x - 1} = x - 3$
sto a	saves the equation as *a*
rai 2	$(\sqrt{x - 1})^2 = (x - 3)^2$
simp	$x - 1 = x^2 - 6x + 9$
sub x−1	$x - 1 - (x - 1) = x^2 - 6x + 9 - (x - 1)$
simp	$0 = x^2 - 7x + 10$
fac	$0 = (x - 5)(x - 2)$

By inspection, the solutions are x = 5 or x = 2.

Check:	
a	$\sqrt{x - 1} = x - 3$
subs 5 x	$\sqrt{5 - 1} = 5 - 3$
simp	$2 = 2$ ✓
a	$\sqrt{x - 1} = x - 3$
subs 2 x	$\sqrt{2 - 1} = 2 - 3$
simp	$1 = -1$

The only solution is 5.

EXERCISES

Use CALC to solve each equation. Check all solutions. Record the steps used to solve each equation.

1. $\sqrt{x} + 4 = 1$

2. $3 + \sqrt{2x} = 7$

3. $\sqrt{3x - 8} = 5$

4. $x - \sqrt{x + 1} = 1$

5. $\sqrt{6 - x} = 4 - x$

6. $\sqrt{2x + 6} = \sqrt{3x - 9}$

12-8 The Distance Formula

Objective After studying this lesson, you should be able to:

- find the distance between two points in the coordinate plane.

Application The Beck Corporation is having a fiber optic cable system installed between two new offices. Becktower I is 4 miles east and 5 miles north of Beck Central. Becktower II is 5 miles west and 2 miles north of Beck Central. How many miles of cable will be needed to connect the new offices?

To help solve this problem, the engineer in charge of the project drew a map on a grid, like the one shown at the right. The sides of each small square in the grid represent a distance of 1 mile. Since the distances are given from Beck Central, she placed it at the origin of the grid. Thus, Becktower I is located at the point (4, 5) and Becktower II is located at the point $(-5, 2)$. The amount of cable needed is the length of the segment joining these two points. Let d = the length of this side.

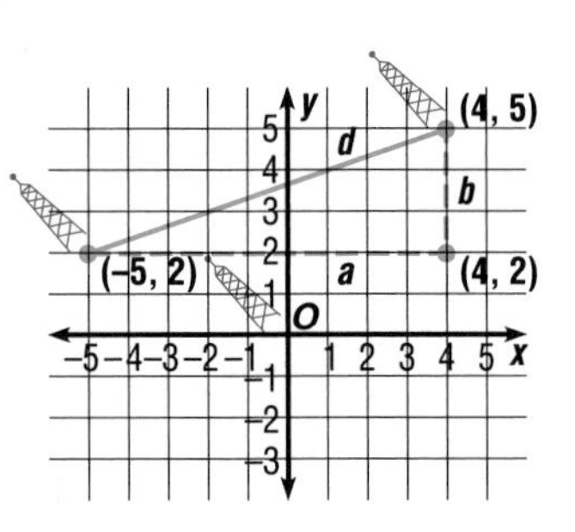

In the illustration above, notice that a right triangle can be formed by drawing lines parallel to the axes from points $(-5, 2)$ and $(4, 5)$. These lines intersect at the point (4, 2). The measure, a, of the side that has $(-5, 2)$ as an endpoint is the difference of the x-coordinates, $4 - (-5)$, or 9. The measure, b, of the side that has (4, 5) as an endpoint is the difference of the y-coordinates, $5 - 2$, or 3.

Now the Pythagorean Theorem can be used to find d, the distance from Becktower I to Becktower II. This is also the distance between the points $(-5, 2)$ and $(4, 5)$.

$c^2 = a^2 + b^2$ *Pythagorean Theorem*

$d^2 = 9^2 + 3^2$ *Substitute 9 for a, 3 for b, and d for c.*

$d^2 = 81 + 9$

$d^2 = 90$

$d = \sqrt{90}$ *In simplest form, $\sqrt{90} = 3\sqrt{10}$.*

$d \approx 9.49$

About 9.49 miles of cable will be needed to connect the two new offices.

The method used for finding the distance between $(-5, 2)$ and $(4, 5)$ can also be used to find the distance between *any* two points in the coordinate plane. The result can be described by the following formula.

The Distance Formula

The distance, d, between any two points with coordinates (x_1, y_1) and (x_2, y_2) is given by the following formula.

$$d = \sqrt{(x_2 - x_1)^2 + (y_2 - y_1)^2}$$

Example 1

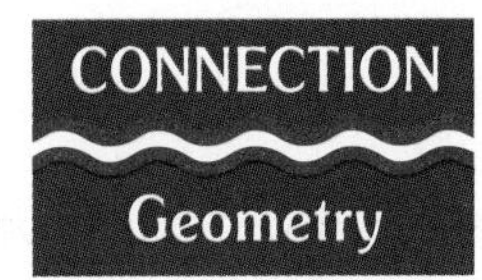

If the diagonals of a trapezoid have the same length, then the trapezoid is isosceles. Find the lengths of the diagonals of the trapezoid with vertices $A(-2, 2)$, $B(10, 6)$, $C(9, 8)$ and $D(0, 5)$ to determine if it is isosceles.

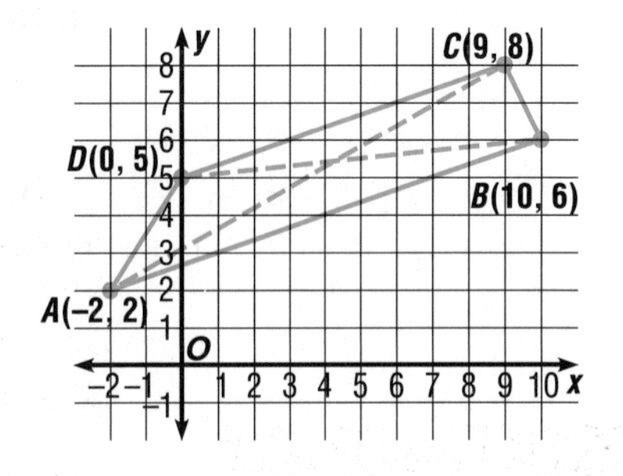

The diagonals of the trapezoid are segments AC and BD. Use the distance formula to compare the lengths of the segments.

AC is the measure of $\overline{AC}$ and BD is the measure of $\overline{BD}$.

For AC, let $x_2 = 9$, $x_1 = -2$, $y_2 = 8$, and $y_1 = 2$.

$$\begin{aligned} d &= \sqrt{(x_2 - x_1)^2 + (y_2 - y_1)^2} \\ AC &= \sqrt{[9 - (-2)]^2 + (8 - 2)^2} \\ &= \sqrt{11^2 + 6^2} \\ &= \sqrt{121 + 36} \\ &= \sqrt{157} \end{aligned}$$

For BD, let $x_2 = 0$, $x_1 = 10$, $y_2 = 5$, and $y_1 = 6$.

$$\begin{aligned} d &= \sqrt{(x_2 - x_1)^2 + (y_2 - y_1)^2} \\ BD &= \sqrt{(0 - 10)^2 + (5 - 6)^2} \\ &= \sqrt{(-10)^2 + (-1)^2} \\ &= \sqrt{100 + 1} \\ &= \sqrt{101} \end{aligned}$$

Since $\sqrt{157} \neq \sqrt{101}$, trapezoid $ABCD$ is not isosceles.

Example 2

Find the value of a if the distance between points $(5, a)$ and $(7, -3)$ is $\sqrt{85}$ units.

$$\begin{aligned} d &= \sqrt{(x_2 - x_1)^2 + (y_2 - y_1)^2} && \text{Let } x_2 = 7,\ x_1 = 5,\ y_2 = -3, \text{ and } y_1 = a. \\ \sqrt{85} &= \sqrt{(7 - 5)^2 + (-3 - a)^2} \\ \sqrt{85} &= \sqrt{2^2 + (-3 - a)^2} \\ \sqrt{85} &= \sqrt{4 + 9 + 6a + a^2} \\ \sqrt{85} &= \sqrt{a^2 + 6a + 13} \\ 85 &= a^2 + 6a + 13 && \text{Square each side.} \\ 0 &= a^2 + 6a - 72 \\ 0 &= (a + 12)(a - 6) && \text{Factor.} \end{aligned}$$

$a + 12 = 0$ or $a - 6 = 0$

$a = -12$ $\quad a = 6$

You can check these answers by substituting -12 and 6 for a in the equation $\sqrt{85} = \sqrt{(7 - 5)^2 + (-3 - a)^2}$.

The value of a is -12 or 6.

CHECKING FOR UNDERSTANDING

Communicating Mathematics

Read and study the lesson to answer each question.

1. Explain how the distance formula is actually an application of the Pythagorean Theorem.
2. When finding the distance between (18, 8) and (5, 7), do you have to choose 18 for x_1? Explain.
3. Explain how you can find the distance between (10, 3) and (2, 3) without using the distance formula.
4. Explain how you can find the distance between $(-1, 5)$ and $(-1, -2)$ without using the distance formula.

Guided Practice

Find the distance between each pair of points whose coordinates are given.

5. (3, 4), (6, 8)
6. $(-4, 2)$, (4, 17)
7. (3, 7), $(-2, -5)$
8. $(5, -1)$, (11, 7)

EXERCISES

Practice

Find the distance between each pair of points whose coordinates are given. Express answers in simplest form and as decimal approximations rounded to the nearest hundredth.

9. (2, 2), $(5, -1)$
10. $(-8, -4)$, $(-3, -8)$
11. (5, 4), $(-3, 8)$
12. (2, 7), $(10, -4)$
13. $(9, -2)$, $(3, -6)$
14. (4, 2), $\left(6, -\frac{2}{3}\right)$
15. $\left(6, -\frac{2}{7}\right)$, $\left(5, \frac{3}{7}\right)$
16. $\left(\frac{4}{5}, -1\right)$, $\left(2, -\frac{1}{2}\right)$
17. $(2\sqrt{5}, 9)$, $(4\sqrt{5}, 3)$
18. $(3\sqrt{2}, 7)(5\sqrt{2}, 9)$

Find the value of *a* if the points whose coordinates are given are the indicated distance apart.

19. $(4, 7)$, $(a, 3)$; $d = 5$
20. $(-3, a)$, $(5, 2)$; $d = 17$
21. $(a, 5)$, $(-7, 3)$; $d = \sqrt{29}$
22. $(4, -2)$, $(-5, a)$; $d = \sqrt{130}$

Find the lengths of the diagonals of each trapezoid with the given vertices to determine whether it is isosceles.

23. (0, 0), (7, 0), (7, 4), (1, 4)
24. (1, 1), (5, 9), (2, 8), (0, 4)
25. Find the perimeter of the triangle with vertices $A(2, -1)$, $B(-2, 2)$, and $C(-6, 14)$.

Find the distance between each pair of points whose coordinates are given. Express answers in simplest form.

26. $(\sqrt{8}, \sqrt{3})$, $(\sqrt{3}, -\sqrt{8})$
27. $(-3\sqrt{6}, \sqrt{10})$, $(2\sqrt{5}, 6\sqrt{3})$

28. Find the value of a if the distance between points $(a, 5)$ and $(3, -2a)$ is 5 units.

Critical Thinking

29. Show that the triangle with vertices $(3, -2)$, $(-3, 7)$, and $(-9, 3)$ is a right triangle.

Applications

30. **Telecommunications** Refer to the application at the beginning of this lesson. Suppose the Beck Corporation decides to build a distribution center 12 miles south of Beck Central. How many additional miles of cable will be needed to connect Beck Central, Becktower I, and Becktower II to the distribution center?

31. **Telecommunications** In order to set long distance rates, phone companies first superimpose an imaginary coordinate grid over the United States. Then the location of each exchange is represented by an ordered pair on the grid. The units on this grid are approximately equal to 0.316 miles. So, a distance of 3 units on the grid equals an actual distance of about 3(0.316) or 0.948 miles. Suppose the exchanges in two cities are (1583, 5622) and (7878, 9213). Find the actual distance between these cities, to the nearest mile.

Portfolio

Place your favorite word problem from this chapter in your portfolio with a note explaining why it is your favorite.

Computer

32. Use the BASIC program at the right to find the distance between each pair of points whose coordinates are given.

    a. $(5, -1)$, $(11, 7)$

    b. $(12, -2)$, $(-3, 5)$

    c. $(0.67, -4)$, $(3, -2)$

```
10 PRINT "ENTER COORDINATES OF
   POINT (X1, Y1)."
20 INPUT X1, Y1
30 PRINT "ENTER COORDINATES OF
   POINT (X2, Y2)."
40 INPUT X2, Y2
50 D = SQR((X2-X1)*(X2-X1) +
   (Y2-Y1)*(Y2-Y1))
60 PRINT "DISTANCE FROM (";X1;", ";
   Y1;") TO (";X2;", ";Y2;") IS ";
   D;" UNITS."
```

Mixed Review

33. Simplify $\frac{4m^2 - 6m - 4}{2m^2 - 8m + 8}$. **(Lesson 8-1)**

34. If $f(x) = 3x - 5$ and $g(x) = x^2 - x$, find $f[g(-2)]$. **(Lesson 9-5)**

35. Write the standard form of an equation of the line passing through $(8, 3)$ and $(5, -1)$. **(Lesson 10-2)**

36. **Aviation** Flying with the wind, a plane travels 300 miles in 40 minutes. Flying against the wind, it travels 300 miles in 45 minutes. Find the air speed of the plane. **(Lesson 11-5)**

37. Solve $\sqrt{x^2 + 3} = 3 - x$. **(Lesson 12-7)**

CHAPTER 12

SUMMARY AND REVIEW

VOCABULARY

Upon completing this chapter, you should be familiar with the following terms:

conjugate	**494**	**477**	radical expression
hypotenuse	**482**	**477**	radical sign
irrational numbers	**487**	**477**	radicand
legs	**482**	**493**	rationalizing the denominator
principal square root	**477**	**488**	real numbers
Pythagorean Theorem	**482**	**477**	square root
radical equation	**501**		

SKILLS AND CONCEPTS

OBJECTIVES AND EXAMPLES

Upon completing this chapter you should be able to:

- simplify rational square roots **(Lesson 12-2)**

$\sqrt{1225} = \sqrt{5^2 \cdot 7^2}$
$= \sqrt{(5 \cdot 7)^2}$
$= \sqrt{35^2}$ or 35

REVIEW EXERCISES

Use these exercises to review and prepare for the chapter test.

Simplify.

1. $\sqrt{169}$
2. $-\sqrt{784}$
3. $-\sqrt{0.0289}$
4. $\pm\sqrt{\frac{196}{225}}$

- find approximate values for square roots **(Lesson 12-2)**

Find $\sqrt{167.3}$ to the nearest hundredth.

Using a calculator, $\sqrt{167.3} \approx 12.93$.

Use a calculator to find each square root. Round answers to the nearest hundredth.

5. $-\sqrt{61.7}$
6. $\sqrt{191.6}$
7. $\pm\sqrt{23.04}$
8. $\pm\sqrt{10.028}$

- use the Pythagorean Theorem **(Lesson 12-3)**

Find the measure of the hypotenuse of a right triangle if $a = 15$ and $b = 20$.

$c^2 = a^2 + b^2$
$c^2 = 15^2 + 20^2$
$c^2 = 225 + 400$
$c^2 = 625$
$c = 25$

a c b

Use the Pythagorean Theorem to find each missing measure to the nearest hundredth.

9. $a = 30, b = 16, c = ?$
10. $a = 6, b = 10, c = ?$
11. $a = 10, c = 15, b = ?$
12. The measures of the sides of a triangle are 20, 21, and 29. Determine whether this triangle is a right triangle.

OBJECTIVES AND EXAMPLES

REVIEW EXERCISES

- identify irrational numbers **(Lesson 12-4)**

Identify $\sqrt{21.16}$ and $-17.121121112\ldots$ as rational or irrational numbers.

Since $\sqrt{21.16} = 4.6$, it is a rational number. Since $-17.121121112\ldots$ does not repeat or terminate, it is an irrational number.

Determine whether each number is rational or irrational.

13. $\sqrt{31.25}$

14. $\sqrt{0.1296}$

15. $-8.10101010\ldots$

16. $4.21222324\ldots$

- simplify square roots **(Lesson 12-5)**

$$\begin{aligned}\sqrt{450} &= \sqrt{2 \cdot 3 \cdot 3 \cdot 5 \cdot 5}\\ &= \sqrt{2} \cdot \sqrt{3^2} \cdot \sqrt{5^2}\\ &= \sqrt{2} \cdot 3 \cdot 5 \text{ or } 15\sqrt{2}\end{aligned}$$

$$\begin{aligned}\frac{\sqrt{12}}{\sqrt{5}} &= \frac{\sqrt{12}}{\sqrt{5}} \cdot \frac{\sqrt{5}}{\sqrt{5}}\\ &= \frac{\sqrt{2 \cdot 2 \cdot 3 \cdot 5}}{\sqrt{5 \cdot 5}}\\ &= \frac{\sqrt{2^2} \cdot \sqrt{3} \cdot \sqrt{5}}{\sqrt{5^2}} \text{ or } \frac{2\sqrt{15}}{5}\end{aligned}$$

Simplify.

17. $\sqrt{108}$

18. $\sqrt{720}$

19. $2\sqrt{6} - \sqrt{48}$

20. $\frac{\sqrt{5}}{\sqrt{55}}$

21. $\sqrt{\frac{20}{7}}$

22. $\frac{9}{3 + \sqrt{2}}$

- simplify radical expressions that contain variables **(Lesson 12-5)**

$$\begin{aligned}\sqrt{343x^2y^3} &= \sqrt{7 \cdot 7^2 \cdot x^2 \cdot y \cdot y^2}\\ &= \sqrt{7^2} \cdot \sqrt{x^2} \cdot \sqrt{y^2} \cdot \sqrt{7} \cdot \sqrt{y}\\ &= 7|x|y\sqrt{7y}\end{aligned}$$

Simplify. Use absolute value symbols when necessary.

23. $\sqrt{96x^4}$

24. $\sqrt{44a^2b^5}$

25. $\sqrt{\frac{60}{y^2}}$

26. $\frac{\sqrt{3a^3b^4}}{\sqrt{8ab^{10}}}$

- simplify radical expressions involving addition and subtraction **(Lesson 12-6)**

$$\begin{aligned}&\sqrt{6} - \sqrt{54} + 3\sqrt{12} + 5\sqrt{3}\\ &= \sqrt{6} - \sqrt{3^2 \cdot 6} + 3\sqrt{2^2 \cdot 3} + 5\sqrt{3}\\ &= \sqrt{6} - \sqrt{3^2} \cdot \sqrt{6} + 3(\sqrt{2^2} \cdot \sqrt{3}) + 5\sqrt{3}\\ &= \sqrt{6} - 3\sqrt{6} + 3(2\sqrt{3}) + 5\sqrt{3}\\ &= -2\sqrt{6} + 11\sqrt{3}\end{aligned}$$

Simplify.

27. $2\sqrt{13} + 8\sqrt{15} - 3\sqrt{15} + 3\sqrt{13}$

28. $4\sqrt{27} + 6\sqrt{48}$

29. $5\sqrt{18} - 3\sqrt{112} - 3\sqrt{98}$

30. $\sqrt{8} + \sqrt{\frac{1}{8}}$

- solve radical equations **(Lesson 12-7)**

$$\begin{aligned}\sqrt{5 - 4x} &= 13\\ 5 - 4x &= 169\\ -4x &= 164\\ x &= -41\end{aligned}$$

Solve each equation. Check the solutions.

31. $\sqrt{3x} = 6$

32. $\sqrt{7x - 1} = 5$

33. $\sqrt{\frac{4a}{3}} - 2 = 0$

34. $\sqrt{x + 4} = x - 8$

OBJECTIVES AND EXAMPLES	REVIEW EXERCISES

- find the distance between two points in the coordinate plane **(Lesson 12-8)**

Find the distance between the points $(-5, 1)$ and $(1, 5)$.

$$d = \sqrt{(x_2 - x_1)^2 + (y_2 - y_1)^2}$$
$$= \sqrt{(-5 - 1)^2 + (1 - 5)^2}$$
$$= \sqrt{(-6)^2 + (-4)^2}$$
$$= \sqrt{36 + 16} \text{ or } \sqrt{52} \approx 7.21$$

Find the distance between each pair of points whose coordinates are given.

35. $(9, -2), (1, 13)$

36. $(4, 2), (7, -9)$

37. Find the value of a if the distance between the points $(5, -2)$ and $(a, -3)$ is $\sqrt{170}$ units.

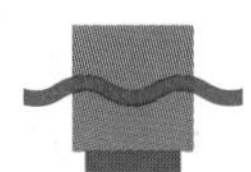

APPLICATIONS AND CONNECTIONS

Use the information in the table shown below to answer each question. (Lesson 12-1)

Number of Hits by Home Run Champions							
Year	1960	1965	1970	1975	1980	1985	1990
National League	41	52	45	38	48	37	40
American League	40	32	44	36	41	40	51

38. How many times did a home run champion hit less than 40 home runs?

39. In what year(s) did the American League champion hit more home runs than the National League champion?

Solve each problem. Round answers to the nearest hundredth.

40. Geometry The measure of the area of a square is 108. What is the measure of a side of this square, rounded to the nearest hundredth? **(Lesson 12-2)**

41. Geometry The length of a rectangle is 1.1 cm and its width is 6.0 cm. Find the length of a diagonal of the rectangle. **(Lesson 12-3)**

42. Number Theory Find two numbers with geometric mean 18 if one number is 3 more than twice the other. **(Lesson 12-7)**

43. Geometry Find the perimeter of the triangle with vertices $A(0, 0)$, $B(-3, 4)$, and $C(6, 8)$. **(Lesson 12-8)**

44. Nature An 18-foot tall tree is broken by the wind. The top of the tree falls and touches the ground 12 feet from its base. How many feet from the base of the tree did the break occur? **(Lesson 12-3)**

45. Law Enforcement Lina told the police officer that she was traveling at 55 mph when she applied the brakes and skidded. The skid marks at the scene were 240 feet long. Should Lina's car have skidded that far if it was traveling at 55 mph? Use the formula $s = \sqrt{15d}$. **(Lesson 12-7)**

CHAPTER 12 TEST

The table at the right shows American consumption of fresh fruits per capita in pounds. Use this information to answer the following questions.

Fruit	1965	1970	1975	1980	1985
Bananas	17.9	17.4	17.7	20.8	23.4
Apples	15.7	16.3	18.3	18.3	16.7
Grapes	3.8	2.5	2.9	3.3	6.6

1. What is the consistently most-popular fruit?
2. In what year did the most-popular fruit change?
3. Why do you think grapes have such low consumption?

Name the set or sets of numbers to which each of the following numbers belongs. Use N for natural numbers, W for whole numbers, Z for integers, Q for rational numbers, and I for irrational numbers.

4. $\sqrt{16}$
5. $-\sqrt{18}$
6. $4.565656\ldots$
7. $\frac{3}{8}$

Use the Pythagorean Theorem to find each missing measure to the nearest hundredth.

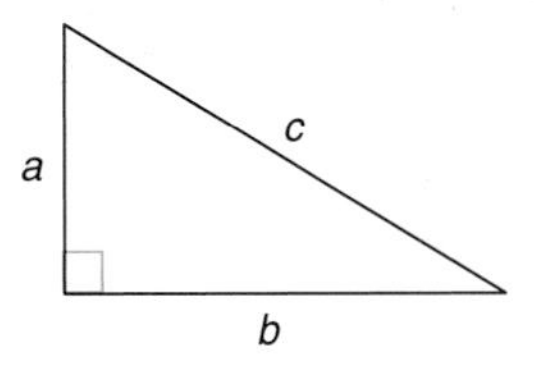

8. $a = 8, b = 10, c = ?$
9. $a = 12, c = 20, b = ?$
10. $a = b, c = 12, b = ?$

Simplify. Use absolute value symbols when necessary.

11. $\sqrt{40}$
12. $\sqrt{72} \cdot \sqrt{48}$
13. $\sqrt{54x^4y}$
14. $\sqrt{45xy^3}$
15. $\sqrt{\frac{32}{25}}$
16. $\sqrt{\frac{3x^2}{4n^3}}$
17. $\frac{7}{7 + \sqrt{5}}$
18. $3\sqrt{50} - 2\sqrt{8}$
19. $\sqrt{6} + \sqrt{\frac{2}{3}}$
20. $2\sqrt{27} + \sqrt{63} - 4\sqrt{3}$
21. $(4 + \sqrt{5})(4 - \sqrt{5})$
22. $\sqrt{2}(\sqrt{18} + 4\sqrt{3})$

Solve each equation. Check the solutions.

23. $\sqrt{t} + 5 = 3$
24. $\sqrt{5x^2 - 9} = 2x$
25. $\sqrt{4x + 1} = 5$
26. $\sqrt{4x - 3} = 6 - x$

Find the distance between each pair of points whose coordinates are given.

27. $(4, 7), (4, -2)$
28. $(-9, 2), (3, -3)$
29. $(-1, 1), (1, -5)$

$2\sqrt{32} - 3\sqrt{6}$

$\sqrt{6}$

30. **Geometry** Find the measures of the perimeter and the area, in simplest form, for the rectangle shown at the right.
31. Find the value of a if the distance between the points $(8, 1), (5, a)$ is 5 units.
32. **Geometry** The length of a rectangle is 4 times its width. Find the dimensions of the rectangle if its area is 224 cm^2.
33. **Construction** Khoa wants a 12-foot ladder to reach a window 10 feet from the base of a wall. How far out from the base of the wall should he position the bottom of the ladder?

Bonus
The diagonal of a cube is 96 cm long. What is the volume of the cube?

CHAPTER 12 College Entrance Exam Preview

The test questions on these pages deal with coordinates and geometry. The figures shown may not be drawn to scale.

Directions: Choose the best answer. Write A, B, C, or D.

1. The midpoint of $\overline{AB}$ is M. If the coordinates of A are $(-3, 2)$ and the coordinates of M are $(-1, 5)$, what are the coordinates of B?

   (A) (1, 10) (B) (1, 8)
   (C) (0, 7) (D) (−5, 8)

2. 

   The length of $\overline{ST}$ is

   (A) 5 (B) $4\frac{1}{2}$ (C) $5\frac{1}{2}$ (D) 6

3. What is the area of the shaded triangle in square units?

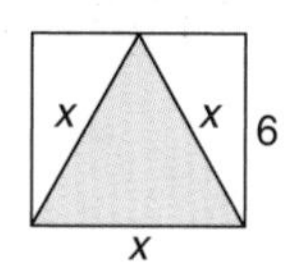

   (A) 12 (B) $12\sqrt{3}$
   (C) $24\sqrt{3}$ (D) 24

4. Seven squares of the same size form a rectangle when placed side by side. The perimeter of the rectangle is 496. What is the area of each square?

   (A) 72 ft^2 (B) 324 ft^2
   (C) 900 ft^2 (D) 961 ft^2

5. What is the total length of fencing needed to enclose a rectangular area 46 feet by 34 feet?

   (A) 26 yards 1 foot
   (B) $26\frac{2}{3}$ yards
   (C) 52 yards 2 feet
   (D) $53\frac{1}{3}$ yards

6. The number of degrees in the smaller angle formed by the hands of a clock at 12:15 is

   (A) 120 (B) $82\frac{1}{2}$ (C) $92\frac{1}{2}$ (D) 90

7. The length of each side of a square is $\frac{3}{5}x + 1$. The perimeter of the square is

   (A) $\frac{12x + 20}{5}$ (B) $\frac{12x + 4}{5}$
   (C) $\frac{3x + 4}{5}$ (D) $\frac{3}{5}x + 16$

8. If a line passes through point (0, 2) and has a slope of 4, what is the equation of the line?

   (A) $x = 2y + 4$ (B) $x = 4y + 2$
   (C) $y = 4x + 2$ (D) $y = 2x + 4$

9. The larger circle has a diameter of b. The area of the shaded ring in square units is

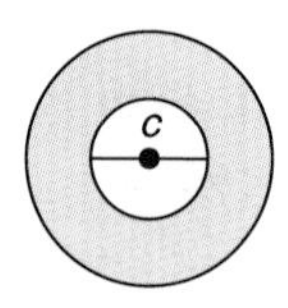

   (A) $b^2 - c^2$ (B) $\pi b^2 - \pi c^2$
   (C) $\frac{1}{4}\pi(b^2 - c^2)$ (D) $\frac{1}{2}(b^2 - c^2)$

10. If a circle of radius 10 meters has its radius decreased by 2 meters, by what percent is its area decreased?

    (A) 20% (B) 40% (C) 80% (D) 36%

11. What is the value of x if the area of the triangle is $\frac{1}{4}$ the area of the square?

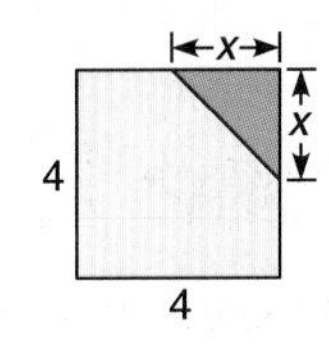

(A) $\sqrt{2}$ (B) $2\sqrt{2}$ (C) 4 (D) 8

12. The slope of the line passing through the points (2, −5) and (−4, 7) is

(A) −2 (B) −1 (C) $\frac{-1}{2}$ (D) 2

13. The coordinates of the x-intercept of the line with equation $3x + 4y = 12$ is

(A) (0, 3) (B) (3, 0)
(C) (0, 4) (D) (4, 0)

14. The area of the triangle formed by the x-axis, the y-axis and the line $y = 2x - 3$ is

(A) $\frac{9}{4}$ (B) $\frac{9}{2}$ (C) 9 (D) 36

15. The area of the square is 64 and the perimeter of each of the two congruent triangles is 20. What is the perimeter of the figure?

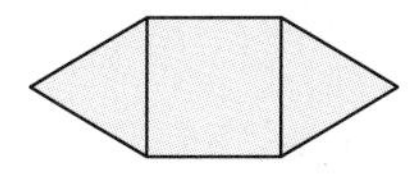

(A) 40 (B) 48 (C) 56 (D) 104

16. The ratio of the area of a circle to its circumference is

(A) $\frac{2}{r}$ (B) $\frac{r}{2}$ (C) π (D) $\frac{\pi}{2r}$

17. Find the slope of the line that is perpendicular to the line $2x - 3y = 12$.

(A) $\frac{2}{3}$ (B) $-\frac{2}{3}$ (C) $\frac{3}{2}$ (D) $-\frac{3}{2}$

TEST TAKING TIP

Although most of the basic formulas that you need to answer the questions on a standardized test are usually given to you in the test booklet, it is a good idea to review the formulas beforehand.

The area of a circle with radius r is πr^2.

The area of a triangle is the product of $\frac{1}{2}$ the base and the height.

The perimeter of a rectangle is twice the sum of its length and width.

The circumference of a circle with radius r is $2\pi r$.

The coordinates of the midpoint, (x, y) of a line segment whose endpoints are at (x_1, y_1) and (x_2, y_2) are $(x, y) = \left(\frac{x_1 + x_2}{2}, \frac{y_1 + y_2}{2}\right)$.

18. What are the coordinates of the intersection of the lines whose equations are $x - 2y = 6$ and $3x + y = 4$?

(A) (2, −2) (B) (2, 2)
(C) (−2, 2) (D) (−2, −2)

CHAPTER OBJECTIVES

In this chapter, you will:

- Graph quadratic functions.
- Solve quadratic equations.
- Determine the nature of the roots of a quadratic equation.
- Solve problems that can be represented by quadratic equations.
- Solve problems by identifying subgoals.

CHAPTER 13

Quadratics

The suspension bridge is one of the oldest engineering forms. As early as the 4th century A.D., these bridges were built by using vines for cables and placing the roadway directly on the cables.

APPLICATION IN CONSTRUCTION

The longest bridge in the world is the Humber suspension bridge in Hull, England. Its main span is 1410 meters long, which is over 860 meters longer than the longest non-suspension bridge, the Quebec Railway cantilever bridge in Canada. In fact, there are 22 suspension bridges that are longer than the Quebec Railway bridge. Why can suspension bridges be built so much longer than other types of bridges? The answer lies in the shape of the cables connecting the towers on these bridges.

When you hang a cable between two supports and then attach weights at equal intervals along the cable, the curve of the cable has the shape of a *parabola.* The cable between two towers of a suspension bridge has a parabolic shape since the shorter cables attached to this cable are equally spaced between the towers. The weight of the bridge exerts tension on the cable, which in turn exerts an inward pull on the towers holding the cable. This method of distributing the bridge's weight allows for a greater distance between the supporting towers.

ALGEBRA IN ACTION

Another shape often used in construction is the *catenary,* which looks very much like a parabola. A cable has the shape of a catenary when it is hung loosely between two supports and *no* weights are attached.

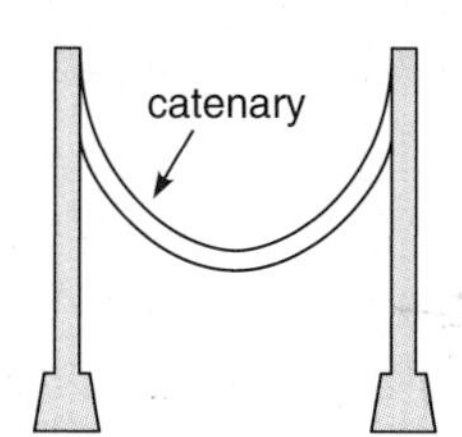

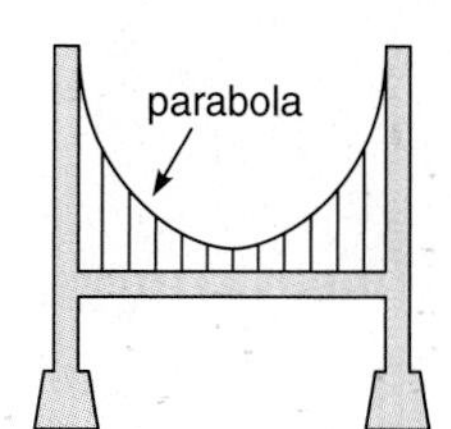

Research the following structures to see if they have the shape of a catenary or the shape of a parabola.

a. Memorial Arch

b. Golden Gate Bridge

13-1 Graphing Quadratic Functions

Objectives

After studying this lesson, you should be able to:

- find the equation of the axis of symmetry and the coordinates of the vertex of the graph of a quadratic function, and
- graph quadratic functions.

Application

The path of an object when it is thrown or dropped is called the *trajectory* of the object. A bouncing object, like the tennis ball in the photo, will have a trajectory in a general shape called a **parabola.**

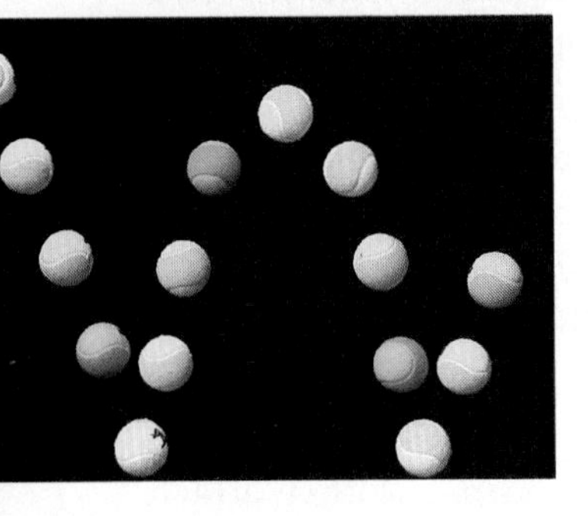

The path of this tennis ball, between its first and second bounces, is given by the equation $y = 6x - 0.5x^2$. In this equation, y is the height of the ball in inches, and x is the horizontal distance of the ball in inches, from the spot of its first bounce. A graph of this equation is shown at the right.

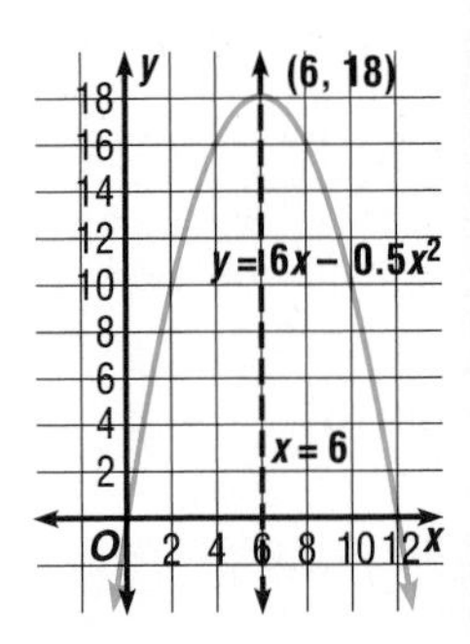

Equations such as $y = 6x - 0.5x^2$ and $y = x^2 - 4x + 1$ describe a type of function known as a **quadratic function.**

Definition of Quadratic Function

A quadratic function is a function that can be described by an equation of the form $y = ax^2 + bx + c$, where $a \neq 0$.

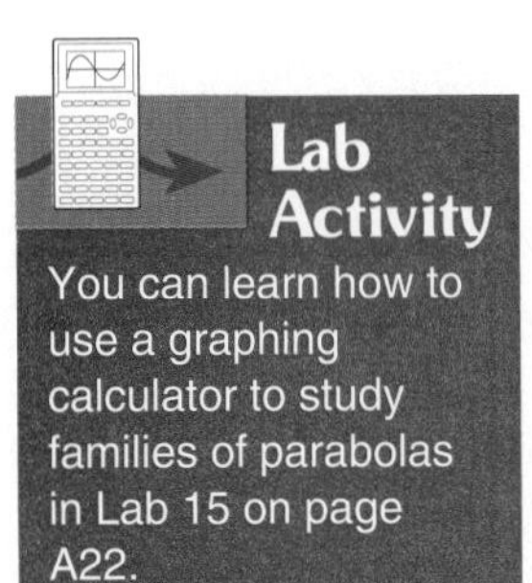

Graphs of quadratic functions have certain common characteristics. For instance, they all have the general shape of a parabola. The table and graph below can be used to illustrate other common characteristics of quadratic functions. Notice the matching values in the y-column of the table.

x	$x^2 - 4x + 1$	y
−1	$(-1)^2 - 4(-1) + 1$	6
0	$0^2 - 4(0) + 1$	1
1	$1^2 - 4(1) + 1$	−2
2	$2^2 - 4(2) + 1$	−3
3	$3^2 - 4(3) + 1$	−2
4	$4^2 - 4(4) + 1$	1
5	$5^2 - 4(5) + 1$	6

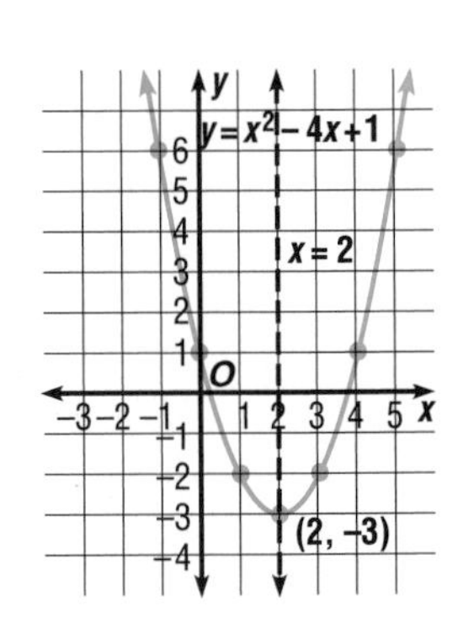

Notice that in the y-column of the table, -3 does not have a matching value. Also, notice that -3 is the y-coordinate of the lowest point of the graph. The point $(2, -3)$ is the *lowest point,* or **minimum point,** of the graph of $y = x^2 - 4x + 1$. For the graph of $y = 6x - 0.5x^2$ on page 518, the point $(6, 18)$ is the *highest point,* or **maximum point.** The maximum point or minimum point of a parabola is also called the **vertex** of the parabola.

The graph of a quadratic function will have a minimum point or a maximum point but not both.

The vertical line containing the vertex of a parabola is called the **axis of symmetry** for the graph. Thus, the equation of the axis of symmetry for the graph of $y = x^2 - 4x + 1$ is $x = 2$.

In general, the equation of the axis of symmetry for the graph of a quadratic function can be found by using the following rule.

Equation of the Axis of Symmetry

The equation of the axis of symmetry for the graph of $y = ax^2 + bx + c$, where $a \neq 0$, is $x = -\frac{b}{2a}$.

Example 1

Find the equation of the axis of symmetry and the coordinates of the vertex of the graph of $y = x^2 - x - 6$. Then use the information to draw the graph.

First, find the equation of the axis of symmetry.

$$x = -\frac{b}{2a}$$

$$= -\left(\frac{-1}{2 \cdot 1}\right) \quad \textit{For } y = x^2 - x - 6,\ a = 1 \textit{ and } b = -1.$$

$$= \frac{1}{2}$$

Next, find the vertex. Since the equation of the axis of symmetry is $x = \frac{1}{2}$, the x-coordinate of the vertex must be $\frac{1}{2}$. You can find the y-coordinate by substituting $\frac{1}{2}$ for x in $y = x^2 - x - 6$.

$$y = \left(\frac{1}{2}\right)^2 - \frac{1}{2} - 6$$

$$= \frac{1}{4} - \frac{1}{2} - 6 \text{ or } -\frac{25}{4}$$

The point $\left(\frac{1}{2}, -\frac{25}{4}\right)$ is the vertex of the graph. *This point is a minimum.*

Finally, construct a table. For values of x, be sure to choose some integers greater than $\frac{1}{2}$ and some less than $\frac{1}{2}$. This insures that points on each side of the axis of symmetry are plotted.

x	$x^2 - x - 6$	y
−2	$(-2)^2 - (-2) - 6$	0
−1	$(-1)^2 - (-1) - 6$	−4
0	$0^2 - 0 - 6$	−6
1	$1^2 - 1 - 6$	−6
2	$2^2 - 2 - 6$	−4
3	$3^2 - 3 - 6$	0

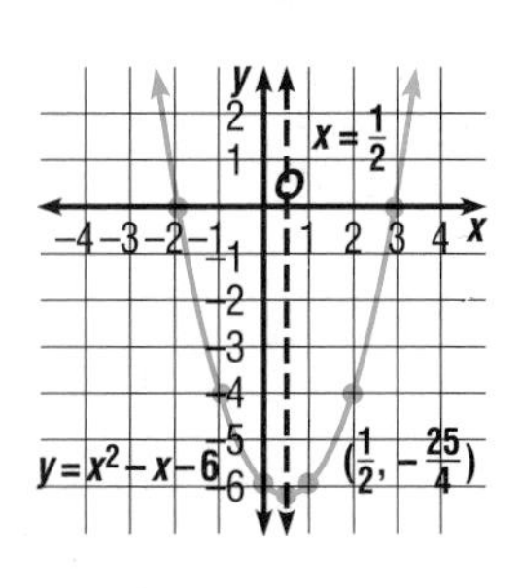

Look at the graph in Example 1. If this graph is folded along its axis of symmetry, the two halves of the graph would coincide. In other words, the parabola is *symmetric* with respect to the axis of symmetry. This is true for the graph of any quadratic function.

In general, the graph of a quadratic function will have a minimum point and open *upward* when the coefficients of y and x^2 have the same sign. The graph will have a maximum point and open *downward* when the coefficients of y and x^2 have opposite signs. Also, the vertex of the graph *always* lies on the axis of symmetry.

Example 2

A movie theater has seats for 1200 people and has been filled to capacity for almost every showing. Tickets currently cost $5.00 and the owner wants to increase this price. She estimates that for each $0.50 increase in the ticket price, 100 fewer people will attend each showing. Based on this estimate, what ticket price will maximize her income?

EXPLORE Let x = the number of $0.50 price increases.
Then, $5.00 + 0.50x$ = the ticket price, and
$1200 - 100x$ = the number of tickets sold for one showing.
Finally, let y = the income from one showing.

PLAN $\underbrace{\text{Income}}$ $\underbrace{\text{is}}$ $\underbrace{\text{number of tickets sold}}$ $\underbrace{\text{times}}$ $\underbrace{\text{ticket price.}}$

$$\begin{aligned} y &= (1200 - 100x) \times (5.00 + 0.50x) \\ &= 6000 + 600x - 500x - 50x^2 \\ &= -50x^2 + 100x + 6000 \end{aligned}$$

SOLVE Notice that the result is a quadratic function. Since the coefficients of y and x^2 have opposite signs, the graph of $y = -50x^2 + 100x + 6000$ has a *maximum* point. The x-coordinate of this maximum point will indicate the number of $0.50 price increases needed to maximize the income.

Since the vertex of the graph lies on the axis of symmetry, its x-coordinate is given by the equation of the axis of symmetry.

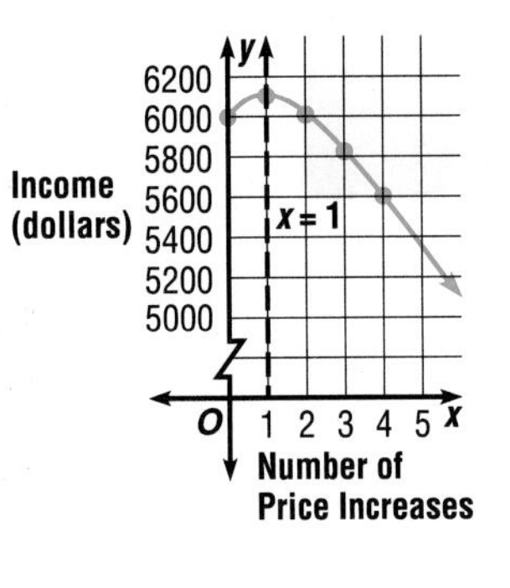

$$\begin{aligned} x &= -\frac{b}{2a} \\ &= -\left(\frac{100}{2(-50)}\right) \qquad a = -50,\ b = 100 \\ &= 1 \end{aligned}$$

The equation of the axis of symmetry is $x = 1$.

The income is maximized when the owner makes one $0.50 price increase. Thus, the ticket price should be $5.00 + 0.50(1)$ or $5.50. *Examine this solution by trying other values of x in the income equation to see if the income can be greater than [1200 − 100(1)] · [5.00 + 0.50(1)] or $6050.*

CHECKING FOR UNDERSTANDING

Communicating Mathematics

Complete.

1. The graphs of all quadratic functions have a general shape called a __?__.
2. The vertex of a parabola is the __?__ point or the __?__ point of the graph.
3. A parabola is __?__ with respect to its axis of symmetry.
4. The graph of $y = ax^2 + 3x + 5$ has a maximum point if a is __?__.

Guided Practice

State whether the graph of each quadratic function opens upward or downward.

5. $y = x^2 - 1$
6. $y = -x^2 + x + 1$
7. $y = -5x^2 - 3x + 2$
8. $y = 2x^2 + 5x - 2$

Find the equation of the axis of symmetry of the graph of each quadratic function.

9. $y = x^2 + x + 3$
10. $y = -x^2 + 4x + 5$
11. $y = 3x^2 + 6x + 16$

Find the coordinates of the vertex of the graph of each quadratic equation.

12. $y = x^2 + 6x + 8$
13. $y = -x^2 + 3x$
14. $y = 5x^2 - 20x + 37$

EXERCISES

Practice

Find the equation of the axis of symmetry and the coordinates of the vertex of the graph of each quadratic function.

15. $y = -x^2 + 5x + 6$
16. $y = x^2 - 4x + 13$
17. $y = x^2 + 2x$
18. $y = -3x^2 + 4$
19. $y = 3x^2 + 24x + 80$
20. $y = -4x^2 + 8x + 13$

Find the equation of the axis of symmetry and the coordinates of the vertex of the graph of each quadratic function. Then draw the graph.

21. $y = x^2 - 4x - 5$
22. $y = -x^2 + 4x + 5$
23. $y = -x^2 + 6x + 5$
24. $y = x^2 - x - 6$
25. $y = x^2 - 3$
26. $y = -x^2 + 7$
27. $y = 2x^2 + 3$
28. $y = \frac{1}{2}x^2 + 3x + \frac{9}{2}$
29. $y = \frac{1}{4}x^2 - 4x + \frac{15}{4}$
30. $y = -3x^2 - 6x + 4$
31. $y = -1(x - 2)^2 + 1$
32. $y = 3(x + 1)^2 - 20$

The axis of symmetry for a parabola is the *y*-axis. Given that each point below lies on the graph, find another point that also lies on the graph.

33. $(1, 1)$
34. $(-3, 17)$
35. $(-4, 0)$

Match each graph with its equation.

36. $y = x^2 + 2$

37. $y = -x^2 + 2$

38. $y = x^2 - 2$

a. y (0, 2) O x

b. y O x (0, −2)

c. 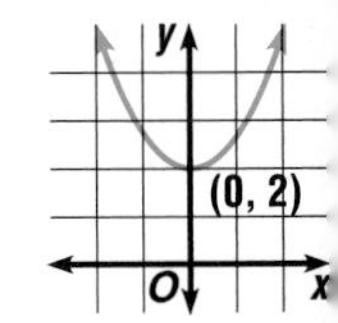

39. The x-intercepts of the graph of a quadratic function are -5 and 2. What is the equation of the axis of symmetry for this graph?

40. The vertex of the graph of a quadratic function is (6, 8). If one x-intercept of the graph is -1, what is the other x-intercept?

41. The function $y = \frac{x(x-1)}{2}$ describes the number of handshakes, y, that occur when x different people all shake each other's hands. Graph this function.

Critical Thinking

42. Draw the graphs of $y = \frac{1}{2}x^2$, $y = x^2$, $y = 2x^2$, and $y = 3x^2$ on the same set of axes. Then compare the graphs.

Applications

43. **Physics** The height h, in feet, that a certain arrow will reach t seconds after being shot directly upward is given by the formula $h = 112t - 16t^2$. What is the maximum height for this arrow?

44. **Finance** A bus company transports 900 people a day between Morse Rd. and High St. A one-way bus fare is \$1.00. The owners estimate that for each \$0.10 price increase, 60 passengers will be lost. What should the one-way fare be in order to maximize their income?

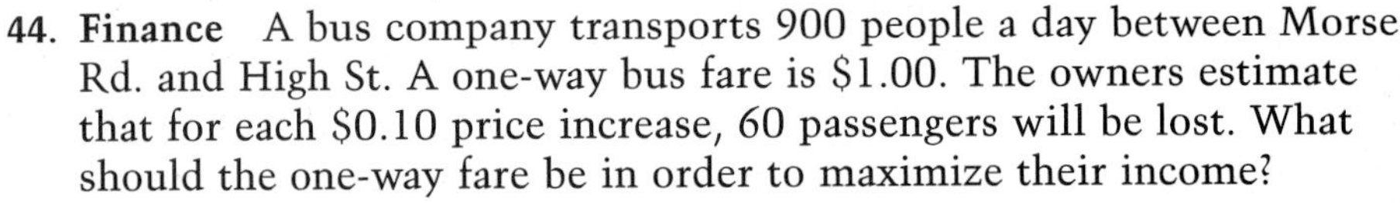

45. **Agriculture** At the beginning of the harvest season, a farmer has 20,000 pounds of potatoes for which the selling price is \$9.50 per 100-pound lot. For each week that he waits there will be an additional 1000 pounds to sell, but the selling price will drop by \$0.25 per 100 pounds per week. How many weeks should the farmer wait before selling his crop in order to have the greatest possible income?

Mixed Review

46. **Consumerism** Julio is offered two payment plans when he buys a sofa. Under one plan, he pays \$400 down and x dollars a month for 9 months. Under the other, he pays no money down and x + 25 dollars a month for 12 months. How much did Julio pay for the sofa? **(Lesson 3-6)**

47. Graph $3x - 2y \leq 6$. **(Lesson 9-6)**

48. Write an equation of the line that is parallel to the graph of $-5x + 6y = 12$ and passes through $(-2, -2)$. **(Lesson 10-6)**

49. Use elimination to solve the system of equations $2x + y = 5$ and $-2x + 3y = 7$. **(Lesson 11-4)**

50. Find two values for a if the distance between points (8, 1) and (5, a) is 5 units. **(Lesson 12-8)**

13-2 Solving Quadratic Equations by Graphing

Objective

After studying this lesson, you should be able to:

- find the roots of a quadratic equation by graphing.

Application

FYI . . .

The Chinese first used rockets and firelances in battle in 1100 A.D.

Tonio launched a toy rocket from a point 50 meters above ground level. The height above ground level h, in meters, of the rocket after t seconds is given by the formula $h = 50 + 45t - 5t^2$. How many seconds after the launch will his rocket hit the ground?

When Tonio's rocket hits the ground, the value of h will be 0. Thus, you can solve this problem by solving the quadratic equation $0 = 50 + 45t - 5t^2$. This equation can be solved by using factoring.

$$0 = 50 + 45t - 5t^2$$
$$0 = 5(10 + 9t - t^2)$$
$$0 = 5(10 - t)(1 + t)$$

$$10 - t = 0 \quad \text{or} \quad 1 + t = 0 \qquad \textit{Zero product property}$$
$$10 = t \qquad\qquad t = -1$$

Since t represents time, t cannot be negative. Therefore, $t = -1$ is not a solution. Tonio's rocket will hit the ground 10 seconds after launch.

The solutions of the equation $0 = 50 + 45t - 5t^2$ are also called the **roots** of the equation. Notice that the roots, 10 and -1, of this equation are the x-intercepts of the graph of the *related quadratic function*, $h = 50 + 45t - 5t^2$.

x	$50 + 45t - 5t^2$	y
-1	$50 + 45(-1) - 5(-1)^2$	0
0	$50 + 45(0) - 5(0)^2$	50
2	$50 + 45(2) - 5(2)^2$	120
4	$50 + 45(4) - 5(4)^2$	150
5	$50 + 45(5) - 5(5)^2$	150
7	$50 + 45(7) - 5(7)^2$	120
9	$50 + 45(9) - 5(9)^2$	50
10	$50 + 45(10) - 5(10)^2$	0

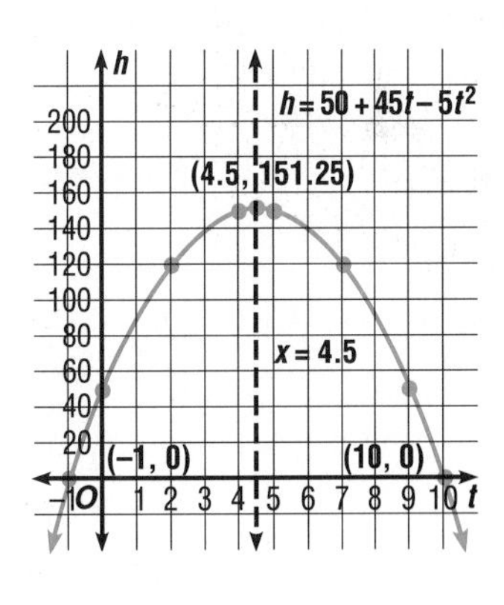

Real roots are roots that are real numbers.

In general, the real roots of any quadratic equation of the form $ax^2 + bx + c = 0$ are the x-intercepts of the graph of the related function $y = ax^2 + bx + c$.

You can always solve a quadratic equation by graphing the related function. However, you may only be able to find approximations of the roots by graphing.

Example 1

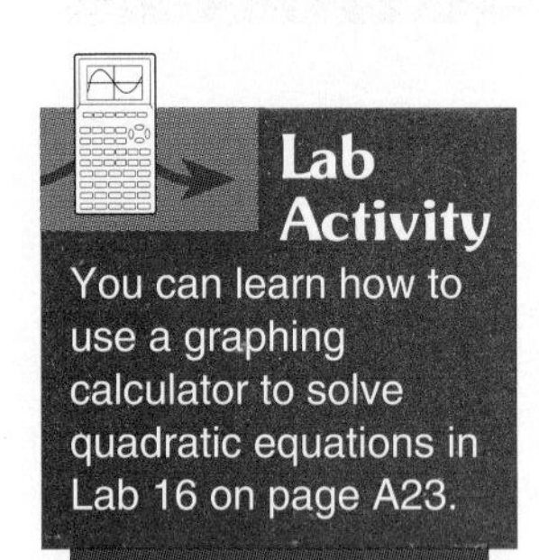

Find the roots of $x^2 + 6x + 6 = 0$ by graphing the related function. If exact roots cannot be found, state the consecutive integers between which the roots are located.

The graph of the related function $y = x^2 + 6x + 6$ has a minimum point and opens upward since y and x^2 have the same sign.

$$\text{axis of symmetry: } x = -\frac{b}{2a}$$

$$= -\frac{6}{2(1)} \text{ or } -3 \qquad \textit{For } y = x^2 + 6x + 6,\ a = 1 \textit{ and } b = 6.$$

The equation of the axis of symmetry is $x = -3$. *Be sure to find the coordinates of the minimum point by finding the value of y when x is −3.*

x	$x^2 + 6x + 6$	y
−5	$(-5)^2 + 6(-5) + 6$	1
−4	$(-4)^2 + 6(-4) + 6$	−2
−3	$(-3)^2 + 6(-3) + 6$	−3
−2	$(-2)^2 + 6(-2) + 6$	−2
−1	$(-1)^2 + 6(-1) + 6$	1

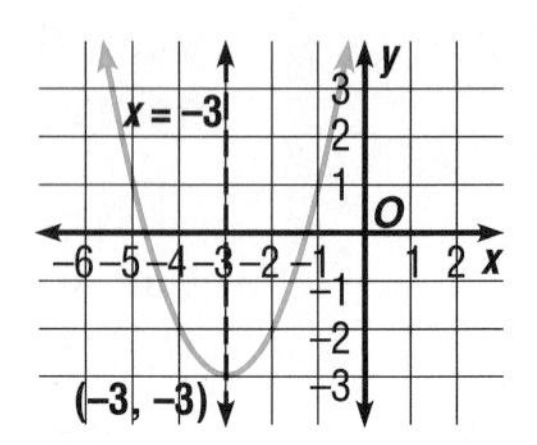

From the graph you can see that one root is between −5 and −4, and the other root is between −2 and −1.

Example 2

Find two real numbers whose sum is 4 and whose product is 5.

EXPLORE Let x = one of the numbers.
Then $4 - x$ = the other number.

PLAN Since the product of the two numbers is 5, you know that $x(4 - x) = 5$.

$$x(4 - x) = 5$$
$$4x - x^2 = 5$$
$$0 = x^2 - 4x + 5$$

SOLVE You can solve $0 = x^2 - 4x + 5$ by graphing the related function $y = x^2 - 4x + 5$. This graph opens upward and the equation of its axis of symmetry is $x = -\frac{-4}{2(1)}$, or $x = 2$.

x	$x^2 - 4x + 5$	y
0	$0^2 - 4(0) + 5$	5
1	$1^2 - 4(1) + 5$	2
2	$2^2 - 4(2) + 5$	1
3	$3^2 - 4(3) + 5$	2
4	$4^2 - 4(4) + 5$	5

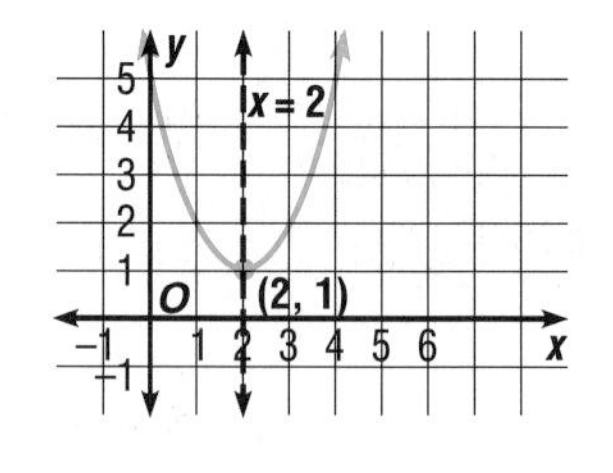

The graph has no x-intercepts since it does not cross the x-axis. This means the equation $x^2 - 4x + 5 = 0$ has no real roots. Thus, it is *not* possible for two real numbers to have a sum of 4 and a product of 5. *Examine this solution.*

CHECKING FOR UNDERSTANDING

Communicating Mathematics

Read and study the lesson to answer each question.

1. What is the related function for the equation $x^2 - 5x + 6 = 0$?

2. How can you find the real roots of a quadratic equation given the graph of its related function?

3. Can you always find the *exact* roots of a quadratic equation by graphing its related function?

Guided Practice

State the roots of each quadratic equation whose related function is graphed below.

4. 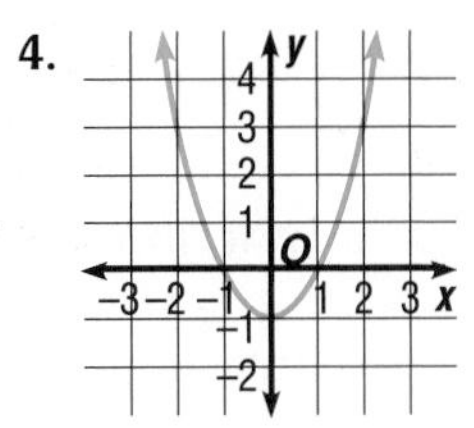

5. 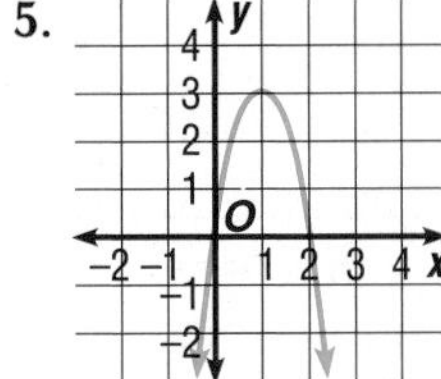

6. 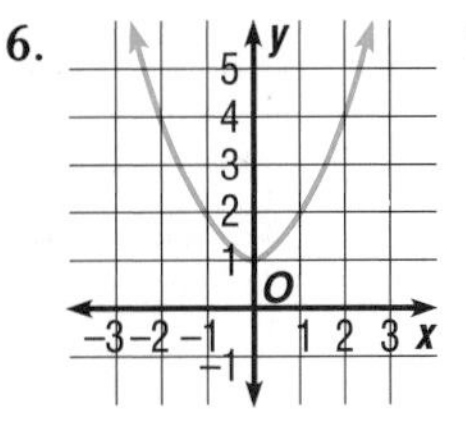

7. 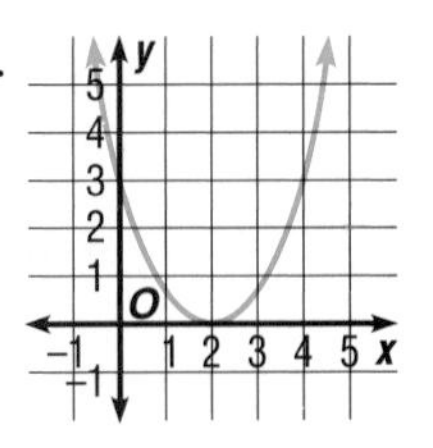

Find the roots of each equation by graphing its related function. If exact roots cannot be found, state the consecutive integers between which the roots are located.

8. $x^2 - x - 12 = 0$
9. $x^2 + 7x + 12 = 0$
10. $x^2 - 9 = 0$

EXERCISES

Practice

Find the roots of each equation by graphing its related function. If exact roots cannot be found, state the consecutive integers between which the roots are located.

11. $x^2 - 10x = -21$
12. $x^2 + 4x = 12$
13. $x^2 - 2x + 2 = 0$
14. $x^2 - 4x + 1 = 0$
15. $x^2 - 8x + 16 = 0$
16. $3x^2 + 2x + 4 = 0$
17. $x^2 + 6x = -7$
18. $6x^2 - 13x = 15$
19. $4x^2 = 35 - 4x$

20. Find two real numbers whose sum is 18 and whose product is 81.

21. Find two real numbers whose difference is 6 and whose product is 91.

Match each equation with the graph of its related function.

22. $x^2 - 3x + 2 = 0$
23. $x^2 + x - 2 = 0$
24. $x^2 - x - 2 = 0$

a.

b. 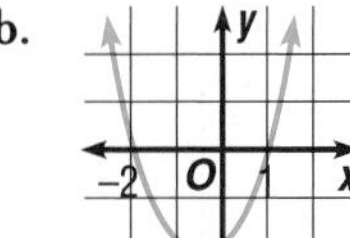

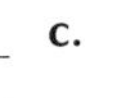

c.

The roots of a quadratic equation are given. Graph the related quadratic function if it has the indicated maximum or minimum point.

25. roots: 2, 6
minimum point: $(4, -2)$

26. roots: 0, -6
maximum point: $(-3, 4)$

27. root: -5
maximum point: $(-5, 0)$

28. roots: no real roots
minimum point: $(-5, 1)$

29. roots: no real roots
maximum point: $(3, -2)$

30. root: 2
minimum point: $(2, 0)$

31. roots: $-6 < x < -5, -5, < x < -4$
minimum point: $(-5, -1)$

32. roots: $-4 < x < -3, 1 < x < 2$
maximum point: $(-1, 6)$

Locate the *y*-intercepts of each quadratic relation by graphing.

33. $x = 2y^2 - 8y + 7$

34. $x = y^2 - 2y + 3$

35. $x = -\frac{3}{4}y^2 - 6y - 9$

Critical Thinking

36. Suppose the value of a quadratic function is negative when $x = 1$ and positive when $x = 2$. Explain why it is reasonable to assume that the related equation has a root between 1 and 2.

Applications

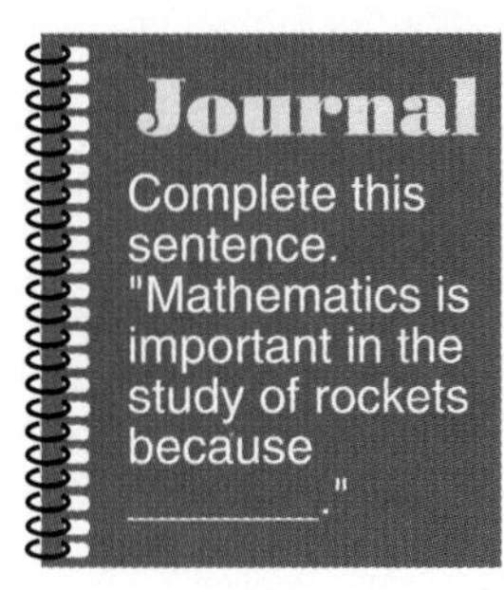

37. Diving Wendy is diving from a 10-meter platform. Her height h in meters above the water when she is x meters away from the platform is given by the formula $h = -x^2 + 2x + 10$. Approximately how far away from the platform is she when she enters the water?

38. Physics The height h in feet of a ball t seconds after being tossed upward is given by the formula $h = 84t - 16t^2$.

a. After how many seconds will the ball reach a height of 80 feet for the *second* time?

b. After how many seconds will it hit the ground?

c. What is its maximum height?

Mixed Review

39. Statistics What is the average of $2b - 4$, $b + 5$, and $3b + 8$? **(Lesson 2-7)**

40. Write an equation for the relation $\{(-2, 0), (-1, -3), (0, -4), (1, -3), (2, 0)\}$. **(Lesson 9-7)**

41. Determine the slope, y-intercept, and x-intercept of the graph of $2x - 3y = 13$. **(Lesson 10-3)**

42. Simplify $\sqrt{720}$. **(Lesson 12-5)**

43. Construction Muturi has 120 meters of fence to make a rectangular pen for his rabbits. If a shed is used as one side of the pen, what would be the maximum area for the pen? **(Lesson 13-1)**

13-3 Problem-Solving Strategy: Identify Subgoals

EXPLORE
PLAN
SOLVE
EXAMINE

Objective After studying this lesson, you should be able to:

- solve problems by identifying subgoals.

Sometimes finding the solution to a problem requires several steps. An important strategy for solving such problems is to *identify subgoals.* This strategy involves taking steps that will either produce part of the solution or make the problem easier to solve.

Example 1

Funtime Toys uses robots to assemble their Speedy cars. Three of the robots can assemble 10 cars in 21 minutes. If all the robots assemble cars at the same rate, how many cars can 14 robots assemble in 90 minutes?

Finding an equation to represent this problem will be easier if you develop the equation in steps rather than trying to write one directly from the given information.

FYI . . . The word *robot* is from the Czech word *robota* for *work.*

Step 1 Determine the rates at which the cars are assembled when 3 robots are working and when 14 robots are working. These two rates can then be used to determine the rate of assembly when 1 robot is working.

$$\text{rate of assembly} = \frac{\text{number of cars assembled}}{\text{time needed to assemble cars}}$$

You know that 3 robots can assemble 10 cars in 21 minutes. Let x = the number of cars that 14 robots can assemble in 90 minutes.

$$\text{rate of assembly for 3 robots} = \frac{10}{21} \qquad \text{rate of assembly for 14 robots} = \frac{x}{90}$$

$$\text{rate of assembly for 1 robot} = \frac{10}{21} \div 3 \qquad \text{rate of assembly for 1 robot} = \frac{x}{90} \div 14$$

$$= \frac{10}{63} \qquad\qquad = \frac{x}{1260}$$

Step 2 Since all of the robots assemble cars at the same rate, the two rates of assembly for 1 robot must be equal.

$$\frac{10}{63} = \frac{x}{1260}$$

$$12600 = 63x \quad \textit{Means-extremes property}$$

$$200 = x$$

Therefore, 14 robots can assemble 200 cars in 90 minutes.

Step 3 Let's check our solution. There are about 5 times as many robots working for about 4 times as long. Hence, they should assemble about $5 \cdot 4$ or 20 times as many cars. Since $10 \cdot 20 = 200$, the solution checks.

Example 2

How many pairs of unit fractions have a sum of $\frac{1}{2}$?

A unit fraction has 1 as its numerator and a positive integer as its denominator. The unit fractions are $\frac{1}{2}, \frac{1}{3}, \frac{1}{4}, \frac{1}{5}, \ldots$.

Suppose one of the fractions in the pair is $\frac{1}{3}$. You can use subtraction to find the other fraction. Since $\frac{1}{2} - \frac{1}{3} = \frac{1}{6}$, it follows that $\frac{1}{6} + \frac{1}{3} = \frac{1}{2}$. Therefore, the pair $\frac{1}{6}$ and $\frac{1}{3}$ is one solution.

Suppose one fraction in the pair is $\frac{1}{4}$. Since $\frac{1}{2} - \frac{1}{4} = \frac{1}{4}$, $\frac{1}{4} + \frac{1}{4} = \frac{1}{2}$. The pair $\frac{1}{4}$ and $\frac{1}{4}$ is a second solution.

Suppose one fraction in the pair is $\frac{1}{5}$. Since $\frac{1}{2} - \frac{1}{5} = \frac{3}{10}$ and $\frac{3}{10}$ is *not* a unit fraction, there is no unit fraction that can be added to $\frac{1}{5}$ to get $\frac{1}{2}$.

Are there any other pairs of unit fractions whose sum is $\frac{1}{2}$? Is so, one of the fractions must be greater than $\frac{1}{4}$ and the other less than $\frac{1}{4}$. Why?

Since $\frac{1}{3}$ and $\frac{1}{2}$ are the *only* unit fractions greater than $\frac{1}{4}$, the two solutions given above are the only pairs of unit fractions whose sum is $\frac{1}{2}$.

CHECKING FOR UNDERSTANDING

Communicating Mathematics

Read and study the lesson to answer each question.

1. The strategy described in this lesson involves taking steps that will result in one of two outcomes. What are these outcomes?
2. Name another strategy that can be used with the strategy described in this lesson.

Guided Practice

Strategies

- Look for a pattern.
- Solve a simpler problem.
- Act it out.
- Guess and check.
- Draw a diagram.
- Make a chart.
- Work backwards.

Solve each problem by identifying subgoals.

3. Two dogs need to eat 3 pounds of dog food per week for proper nutrition. What is the maximum number of dogs that can eat 90 pounds of dog food for 6 weeks?
4. How many pairs of unit fractions have a sum of $\frac{1}{6}$?
5. How many whole numbers less than 200 have digits whose sum is 10?
6. **Number Theory** *Palindromes* are words or numbers that read the same backward or forward. For example, 11, 686, and 1881 are palindromes. How many numbers between 10 and 1000 are palindromes?

EXERCISES

Practice **Solve. Use any strategy.**

7. Suppose the scoring in football is simplified to 7 points for a touchdown and 3 points for a field goal. What scores are impossible to achieve?

8. Mr. Apple, Mr. Pear, Miss Peach, and Mrs. Berry are eating apples, pears, peaches, and berries, although none of them are eating the same fruit as their last name. Neither Mr. Pear nor Mrs. Berry is eating apples. If Mr. Apple and Mrs. Berry are eating pieces of fruit that have different first initials, who is eating what?

9. The three co-captains from West High School meet with the four co-captains from East High School for the coin toss. If each co-captain shakes hands with the referee and each opposing co-captain, how many handshakes occur?

10. What is the next symbol in the following sequence?

    ?

11. Babe Ruth hit 714 home runs over 22 seasons, from 1914 to 1935. He averaged 102 games per season. During what season would he have hit his 714th home run if he had averaged 162 games per season and hit home runs at the same rate?

12. Jim is 17 times as old as his brother was when Jim was as old as his brother was when Jim was as old as his brother is now. If their father is less than 90 years old, how old are Jim and his brother?

COOPERATIVE LEARNING ACTIVITY

Work in groups of four. Each person in the group should understand the solution and be able to explain it to any person in the class.

The large hexagon, shown in blue at the right, has been divided into 16 regions of various shapes and sizes. How many rectangles and how many triangles can you find in this figure? (*Hint:* Number each of the regions. Then use the numbers to name the rectangles and triangles.)

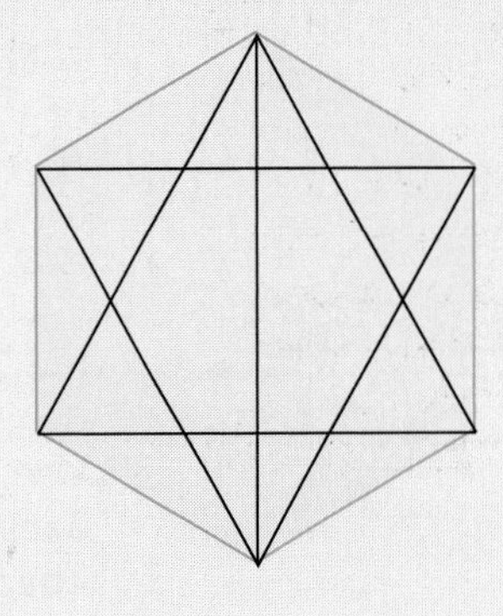

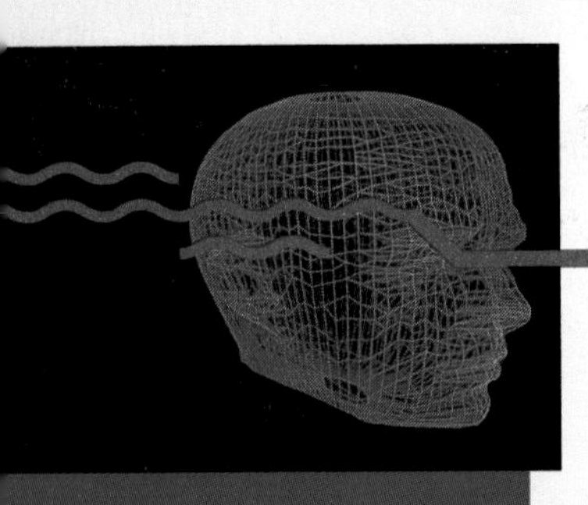

Technology

Solving Quadratic Equations

BASIC
▶ **Graphing calculators**
Graphing software
Spreadsheets

A graphing calculator is a powerful tool for studying functions. When a group of functions is graphed on the same axes, you can compare the graphs to make generalizations about types of functions. Then you can predict what a function will look like by studying the equation.

Example **Graph $y = x^2$, $y = 3x^2$, and $y = (x + 2)^2 + 5$ on the same set of axes. Use a range of [−8, 6] by [−4, 20].**

For the TI-81

First, set the range.

RANGE (-) 8 ENTER 6 ENTER 1 ENTER (-) 4 ENTER 20 ENTER 1 ENTER

Enter $y = x^2$. Y= CLEAR X|T x^2 ENTER

Enter $y = 3x^2$. 

Enter $y = (x + 2)^2 + 5$ and graph.

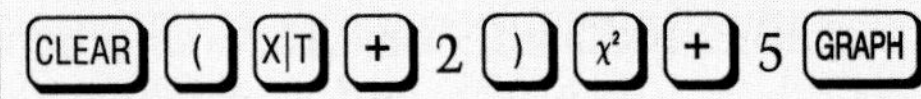

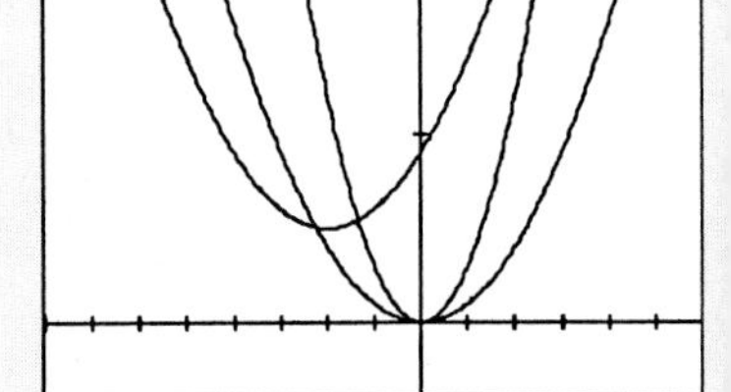

For the Casio

First, set the range.

RANGE (-) 8 EXE 6 EXE 1 EXE (-) 4 EXE 20 EXE 1 EXE

Graph $y = x^2$. Graph ALPHA × x^2 EXE

Graph $y = 3x^2$. Graph 3 ALPHA × x^2

Graph $(x + 2)^2 + 5$. Graph (ALPHA × + 2) x^2 + 5 EXE

Notice that the three graphs look very much alike. However, the graph of $y = 3x^2$ is narrower than the graph of $y = x^2$. The graph of $y = (x + 2)^2 + 5$ is translated 2 units to the left and 5 units upward.

EXERCISES

Use a graphing calculator to graph each pair of equations on the same set of axes. Then compare the graph of the second equation to the graph of the first equation.

1. $y = x^2$; $y = x^2 + 3$
2. $y = x^2$; $y = (x - 10)^2$
3. $y = x^2$; $y = 0.25x^2$
4. $y = x^2$; $y = (x + 4)^2 - 8$

13-4 Solving Quadratic Equations by Completing the Square

Objective After studying this lesson, you should be able to:

- solve quadratic equations by completing the square.

Application

A photograph of a nautilus shell sold for $115,000 in April 1989.

Carlos Rodriguez is framing a picture. He begins with a square piece of matting and then cuts out a 6-inch by 6-inch square to accommodate the picture. If the area of the remaining matting is 28 square inches, how large was the original square piece of matting?

Let x = the length of a side of the original square piece of matting. The area of the cutout is 36 square inches. Then, $x^2 - 36 =$ the area of the matting after the 6-inch-by-6-inch square is cut out.

Since the area of the remaining matting is 28 square inches, you can solve the equation $x^2 - 36 = 28$ to find x. One method for solving this equation is shown below.

$x^2 - 36 = 28$

$x^2 = 64$ *Add 36 to each side.*

$x = \pm\sqrt{64}$ *Find the square root of each side.*

$x = \pm 8$

Since -8 inches is not a reasonable solution, the length of a side of the original square piece of matting must be 8 inches. *Check this result.*

The method used for solving the equation $x^2 - 36 = 28$ can also be used to solve equations such as $x^2 - 4x + 4 = 3$.

Example 1 **Solve $x^2 - 4x + 4 = 3$.**

$x^2 - 4x + 4 = 3$

$(x - 2)^2 = 3$ *$x^2 - 4x + 4$ is a perfect square trinomial.*

$\sqrt{(x - 2)^2} = \sqrt{3}$ *Find the square root of each side.*

$|x - 2| = \sqrt{3}$

$x - 2 = \pm\sqrt{3}$ *Why is this true?*

$x = 2 \pm\sqrt{3}$ *Add 2 to each side.*

The solution set is $\{2 + \sqrt{3}, 2 - \sqrt{3}\}$.

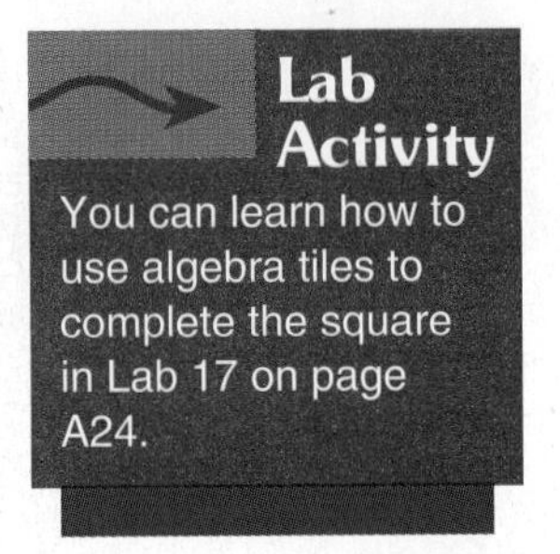

To use the method shown in Example 1, the quadratic expression on one side of the equation must be a perfect square. To make it a perfect square, a method called **completing the square** may be used.

Consider the pattern for squaring a binomial such as $x + 6$.

$$(x + 6)^2 = x^2 + 2(6)(x) + 6^2$$
$$= x^2 + 12x + 36$$
$$\left(\frac{12}{2}\right)^2 \rightarrow 6^2$$

Notice that one-half of 12 is 6 and 6^2 is 36.

To complete the square for a quadratic expression of the form $x^2 + bx$, you can follow the steps below.

Step 1 Find one-half of b, the coefficient of x.
Step 2 Square the result of Step 1.
Step 3 Add the result of Step 2 to $x^2 + bx$.

Example 2

Find the value of c that makes $x^2 + 14x + c$ a perfect square.

Step 1	Find one-half of 14.	$\frac{14}{2} = 7$
Step 2	Square the result of Step 1.	$7^2 = 49$
Step 3	Add the result of Step 2 to $x^2 + 14x$.	$x^2 + 14x + 49$

Thus, $c = 49$. Notice that $x^2 + 14x + 49 = (x + 7)^2$.

Example 3

Solve $x^2 + 6x - 16 = 0$ by completing the square.

$x^2 + 6x - 16 = 0$ — *Notice that $x^2 + 6x - 16$ is not a perfect square.*
$x^2 + 6x = 16$ — *Add 16 to each side. Then complete the square.*
$x^2 + 6x + 9 = 16 + 9$ — *Since $\left(\frac{6}{2}\right)^2 = 9$, add 9 to each side.*
$(x + 3)^2 = 25$ — *Factor $x^2 + 6x + 9$.*
$x + 3 = \pm 5$ — *Find the square root of each side.*
$x = -3 \pm 5$ — *Subtract 3 from each side.*

$x = -3 + 5$ or $x = -3 - 5$
$= 2$ $\quad = -8$

The solution set is $\{2, -8\}$. *Check this result.*

This method for solving quadratic equations cannot be used unless the coefficient of the first term is 1. To solve a quadratic equation where this coefficient is not 1, divide each term by the coefficient, as shown in Example 4.

Example 4

The Cartland Company wants to redesign the main conference room at its corporate headquarters. Because of the existing design of the building, the maximum length of the redesigned rectangular conference room can be 9 meters less than twice its maximum width. Find the dimensions, to the nearest tenth of a meter, for the largest possible conference room if its area is to be 390 square meters.

EXPLORE Let w = the maximum width of the redesigned conference room. Then $2w - 9$ = the maximum length of the conference room.

PLAN $\underbrace{\text{Length}}\ \underbrace{\text{times}}\ \underbrace{\text{width}}\ \underbrace{\text{equals}}\ \underbrace{\text{area.}}$

$$(2w \cdot 9) \quad \times \quad w \quad = \quad 390$$

SOLVE

$$2w^2 - 9w = 390$$

$$w^2 - \frac{9}{2}w = 195 \qquad \textit{Divide each side by 2.}$$

$$w^2 - \frac{9}{2}w + \frac{81}{16} = 195 + \frac{81}{16} \qquad \textit{Complete the square.}$$

$$\left(w - \frac{9}{4}\right)^2 = \frac{3201}{16}$$

$$w - \frac{9}{4} = \pm\frac{\sqrt{3201}}{4}$$

$$w = \frac{9 \pm \sqrt{3201}}{4}$$

The roots of $(2w - 9)w = 390$ are $\frac{9 + \sqrt{3201}}{4}$ and $\frac{9 - \sqrt{3201}}{4}$.

You can use a calculator to find decimal approximations for these numbers.

Enter: (9 + 3201 √x) ÷ 4 = 16.3943452

Enter: (9 − 3201 √x) ÷ 4 = −11.8943452

The roots of the equation are approximately 16.4 and −11.9.

Since −11.9 meters is not a reasonable solution, the width must be 16.4 meters. Therefore, the dimensions of the redesigned conference room, rounded to the nearest tenth of a meter, are 16.4 meters by 2(16.4) − 9 or 23.8 meters.

EXAMINE Since $23.8 \times 16.4 = 390.32$, the answer seems reasonable.

CHECKING FOR UNDERSTANDING

Communicating Mathematics

Read and study the lesson to answer each question.

1. What is the value of $\sqrt{x^2}$?
2. What are the three steps used to complete the square for the expression $x^2 + bx$?
3. If you were solving the equation $3x^2 - 8x = 9$ by completing the square, what should be your first step?
4. Which method for solving a quadratic equation always produces an *exact* solution, graphing or completing the square?

Guided Practice

State whether each trinomial is a perfect square.

5. $b^2 + 4b + 3$
6. $m^2 - 10m + 25$
7. $r^2 - 8r - 16$
8. $d^2 + 11d + 121$
9. $h^2 - 13h + \frac{169}{4}$
10. $4x^2 + 12x + 9$

Find the value of c that makes each trinomial a perfect square.

11. $x^2 + 8x + c$
12. $a^2 - 6a + c$
13. $m^2 + 7m + c$

Solve each equation by completing the square.

14. $y^2 + 4y + 3 = 0$
15. $n^2 - 8n + 7 = 0$
16. $t^2 - 4t = 21$

EXERCISES

Practice

Find the value of c that makes each trinomial a perfect square.

17. $x^2 - 7x + c$
18. $a^2 + 5a + c$
19. $9x^2 - 18x + c$

Solve each equation by completing the square. Leave irrational roots in simplest radical form.

20. $r^2 + 14r - 10 = 5$
21. $y^2 + 7y + 10 = -2$
22. $x^2 - 5x + 2 = -2$
23. $4x^2 - 20x + 25 = 0$
24. $z^2 - 4z = 2$
25. $b^2 + 4 = 6b$
26. $y^2 - 8y = 4$
27. $x^2 - 10x = 23$
28. $2d^2 + 3d - 20 = 0$
29. $a^2 - \frac{7}{2}a + \frac{3}{2} = 0$
30. $\frac{1}{2}q^2 - \frac{5}{4}q - 3 = 0$
31. $0.3x^2 + 0.1x = 0.2$
32. $r^2 + 0.25r = 0.5$
33. $2x^2 - 5x + 1 = 0$
34. $3x^2 - 7x - 3 = 0$
35. **Geometry** Find the dimensions of a rectangle whose perimeter is 37 yards and whose area is 78 square yards.

Find the value of c that makes each trinomial a perfect square.

36. $x^2 + cx + 64$

37. $4x^2 + cx + 225$

38. $cx^2 + 28x + 49$

39. $cx^2 - 18x + 36$

Solve each equation by completing the square. Leave irrational roots in simplest radical form.

40. $x^2 + 4x + c = 0$

41. $x^2 + bx + c = 0$

42. $x^2 + 4bx + b^2 = 0$

Critical Thinking

43. Geometry Write the quadratic function $y = x^2 - 4x + 7$ in the form $y = (x - h)^2 + k$. Then graph the function. What is the relationship of the point (h, k) to the graph?

Application

44. Construction The rectangular penguin pond at the Bay Park Zoo is 12 meters long by 8 meters wide. The zoo wants to double the area of the pond by increasing the length and width by the same amount. By how much should the length and width be increased?

45. Photography Sheila is placing a photograph behind a 12-inch-by-12-inch piece of matting. The photograph is to be positioned so that the matting is twice as wide at the top and bottom as it is at the sides. If the area of the photograph is to be 54 square inches, what are its dimensions?

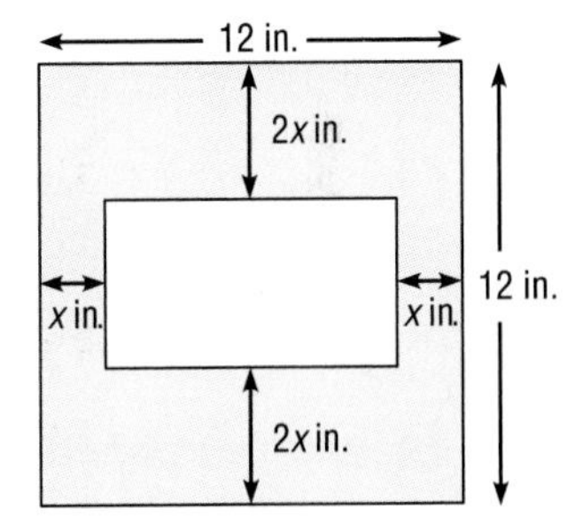

46. Travel Two trains left the same station at the same time. One was traveling due north at a speed that was 10 mph faster than the other train, which was traveling due east. After one hour, the trains were 71 miles apart. How fast, to the nearest mile per hour, was each train traveling? (*Hint:* Use the Pythagorean Theorem)

Mixed Review

47. Consumerism Denise Williams has budgeted \$150 to have business cards printed. A card printer charges \$11 to set up each job and an additional \$6 per box of 100 cards printed. What is the greatest number of cards Ms. Williams can have printed? **(Lesson 5-3)**

48. Number Theory The tens digit of a two-digit number exceeds twice its units digit by 1. If the digits are reversed, the number is 4 more than 3 times the sum of the digits. Find the original number. **(Lesson 11-3)**

49. Simplify $5\sqrt{72} + 2\sqrt{20} - 3\sqrt{5}$. **(Lesson 12-6)**

50. Locate the roots of $4x^2 - 12x + 3 = 0$ between two consecutive integers by graphing the related function. **(Lesson 13-2)**

51. Determine how many whole numbers less than 1000 have digits whose sum is 10. **(Lesson 13-3)**

13-5 Solving Quadratic Equations Using the Quadratic Formula

Objective

After studying this lesson, you should be able to:

- solve quadratic equations by using the quadratic formula.

Application

FYI...

Captain Eugene A. Cernan and Dr. Harrison H. Schmitt spent 74 hours 59 minutes on the moon in December 1972.

The height H of an object t seconds after it is propelled upward is given by the formula $H = -\frac{1}{2}gt^2 + vt + h$. In this formula, v is the initial upward velocity of the object, h is its initial height, and g is the acceleration due to gravity. Suppose an astronaut on the moon throws a baseball with an initial upward velocity of 10 meters per second while letting go of the ball 2 meters above the ground. How much longer will the ball stay in the air than a baseball thrown on Earth under the exact same conditions? On Earth, g is 9.8 meters per second squared and on the moon, g is 1.6 meters per second squared.

On both Earth and the moon, $v = 10$ and $h = 2$. The value of H is 0 when the baseball hits the ground. Therefore, the two equations below can be used to describe this situation.

Baseball thrown on the moon	**Baseball thrown on Earth**
$0 = -\frac{1}{2}(1.6)t^2 + 10t + 2$	$0 = -\frac{1}{2}(9.8)t^2 + 10t + 2$
$0 = -0.8t^2 + 10t + 2$	$0 = -4.9t^2 + 10t + 2$

You could solve these equations by graphing the related functions, but that will produce only a rough approximation of the actual solution. You could also try solving the equations by completing the square, but that would not be a simple task. Another alternative is to develop a general formula for solving *any* quadratic equation. To do this, you begin with the general form, $ax^2 + bx + c = 0$, where $a \neq 0$.

$$ax^2 + bx + c = 0$$

$$x^2 + \frac{b}{a}x + \frac{c}{a} = 0 \qquad \textit{Divide by a so the coefficient of } x^2 \textit{ becomes 1.}$$

$$x^2 + \frac{b}{a}x = -\frac{c}{a} \qquad \textit{Subtract } \tfrac{c}{a} \textit{ from each side.}$$

Now complete the square.

$$x^2 + \frac{b}{a}x + \left(\frac{b}{2a}\right)^2 = -\frac{c}{a} + \left(\frac{b}{2a}\right)^2$$

$$\left(x + \frac{b}{2a}\right)^2 = -\frac{c}{a} + \frac{b^2}{4a^2} \qquad \textit{Factor the left side.}$$

$$\left(x + \frac{b}{2a}\right)^2 = \frac{b^2 - 4ac}{4a^2} \qquad \textit{Simplify the right side.}$$

Finally, find the square root of each side and solve for x.

$$x + \frac{b}{2a} = \pm\sqrt{\frac{b^2 - 4ac}{4a^2}}$$

$$x + \frac{b}{2a} = \frac{\pm\sqrt{b^2 - 4ac}}{2a} \qquad \textit{Simplify the square root on the right side.}$$

$$x = \frac{\pm\sqrt{b^2 - 4ac}}{2a} - \frac{b}{2a} \qquad \textit{Subtract } \frac{b}{2a} \textit{ from each side.}$$

$$x = \frac{-b \pm \sqrt{b^2 - 4ac}}{2a} \qquad \textit{The result is an expression for x.}$$

This result is called the **quadratic formula** and can be used to solve *any* quadratic equation.

The Quadratic Formula

The roots of a quadratic equation of the form $ax^2 + bx + c = 0$, where $a \neq 0$, are given by the formula

$$x = \frac{-b \pm \sqrt{b^2 - 4ac}}{2a}.$$

In order for the roots of $ax^2 + bx + c = 0$ to be real numbers, the value of $b^2 - 4ac$ must be nonnegative. If $b^2 - 4ac$ is negative, then $\sqrt{b^2 - 4ac}$ is not defined and the equation has no real roots.

Example 1

Use the quadratic formula to solve $x^2 - 8x - 4 = 0$.

$$x = \frac{-b \pm \sqrt{b^2 - 4ac}}{2a}$$

$$= \frac{-(-8) \pm \sqrt{(-8)^2 - 4(1)(-4)}}{2(1)} \qquad a = 1, b = -8, c = -4$$

$$= \frac{8 \pm \sqrt{64 + 16}}{2}$$

$$= \frac{8 \pm \sqrt{80}}{2} \qquad \sqrt{80} = \sqrt{16 \cdot 5} \text{ or } 4\sqrt{5}$$

$$x = \frac{8 + 4\sqrt{5}}{2} \quad \text{or} \quad x = \frac{8 - 4\sqrt{5}}{2}$$

$$= 4 + 2\sqrt{5} \qquad = 4 - 2\sqrt{5} \qquad \textit{Check this result.}$$

The solution set is $\{4 + 2\sqrt{5}, 4 - 2\sqrt{5}\}$.

Example 2

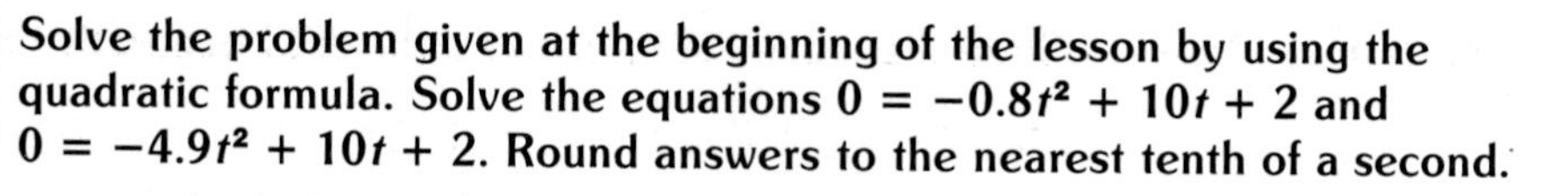

Solve the problem given at the beginning of the lesson by using the quadratic formula. Solve the equations $0 = -0.8t^2 + 10t + 2$ and $0 = -4.9t^2 + 10t + 2$. Round answers to the nearest tenth of a second.

You can use a calculator to help with the computations when using the quadratic formula.

Moon For $0 = -0.8t^2 + 10t + 2$, $a = -0.8$, $b = 10$ and $c = 2$. First, evaluate $\sqrt{b^2 - 4ac}$, or $\sqrt{10^2 - 4(-0.8)(2)}$ and store the value.

Enter: 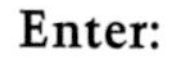 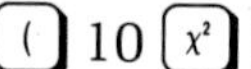 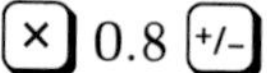  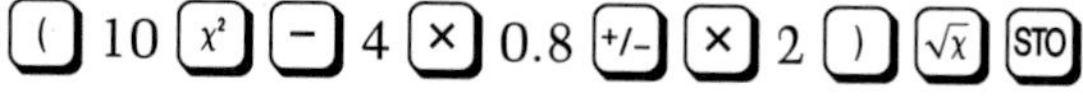

(10 x² − 4 × 0.8 +/- × 2) √x STO

Display:  10.3150376

Next, find the value of $\frac{-b + \sqrt{b^2 - 4ac}}{2a}$. *Remember, you have stored the value of* $\sqrt{b^2 - 4ac}$.

Enter: 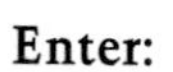  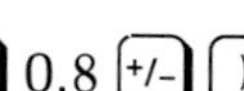 

(10 +/- + RCL) ÷ (2 × 0.8 +/-) =

Display: −0.19689848

Finally, find the value of $\frac{-b - \sqrt{b^2 - 4ac}}{2a}$.

Enter:  

(10 +/- − RCL) ÷ (2 × 0.8 +/-) =

Display: 12.6968985

The decimal approximations for the roots of $0 = -0.8t^2 + 10t + 2$ are -0.2 and 12.7. Thus, the ball will stay in the air for about 12.7 seconds when thrown on the moon. *−0.2 seconds is not an acceptable answer.*

Earth By following the steps used above, you will find that the decimal approximations for the roots of the equation $0 = -4.9t^2 + 10t + 2$ are -0.2 and 2.2. Thus, the ball will stay in the air for about 2.2 seconds when thrown on Earth. *Again, −0.2 seconds is not an acceptable answer.*

Therefore, the ball thrown on the moon will stay in the air for $12.7 - 2.2$ or about 10.5 seconds longer than the ball thrown on Earth.

CHECKING FOR UNDERSTANDING

Communicating Mathematics

Read and study the lesson to answer each question.

1. What method for solving quadratic equations was used to develop the quadratic formula?
2. What coefficient of a quadratic equation cannot equal 0 in the quadratic formula?
3. *True* or *false*: If $b^2 > 4ac$, then the quadratic equation $ax^2 + bx + c = 0$ has no real roots.
4. What are the two roots of the quadratic equation $ax^2 + bx + c = 0$?

Guided Practice

State the values of *a*, *b*, and *c* for each quadratic equation.

5. $x^2 + 7x + 6 = 0$
6. $2t^2 - t - 15 = 0$
7. $2y^2 + 3 = -7y$
8. $2x^2 = 98$
9. $4a^2 + 8a = 0$
10. $3k^2 + 11k = 4$

Find the value of $b^2 - 4ac$ for each quadratic equation.

11. $x^2 + 5x - 6 = 0$
12. $y^2 - 7y - 8 = 0$
13. $m^2 - 2m = 35$
14. $4n^2 - 20n = 0$
15. $5t^2 = 125$
16. $3x^2 + 14x = 5$

Solve each equation.

17. $m^2 + 4m + 2 = 0$
18. $-4x^2 + 8x = -3$
19. $3k^2 + 2 = -8k$

EXERCISES

Practice

Solve each equation.

20. $x^2 + 7x + 6 = 0$
21. $r^2 + 10r + 9 = 0$
22. $-a^2 + 5a - 6 = 0$
23. $y^2 - 25 = 0$
24. $-2d^2 + 8d + 3 = 3$
25. $2r^2 + r - 15 = 0$
26. $3n^2 - 2n = 1$
27. $2y^2 + 3 = -7y$
28. $8t^2 + 10t + 3 = 0$
29. $z^2 - 13z - 32 = 0$
30. $y^2 - \frac{3y}{5} + \frac{2}{25} = 0$
31. $3x^2 - \frac{5}{4}x - \frac{1}{2} = 0$
32. $24x^2 - 2x - 15 = 0$
33. $21a^2 + 5a - 6 = 0$
34. $2x^2 = 0.7x + 0.3$
35. $-r^2 - 6r + 3 = 0$
36. $4b^2 + 20b + 23 = 0$
37. $-4y^2 + 13 = -16y$

Approximate the *x*-intercepts of the graph of each function to the nearest tenth. (*Hint:* the *x*-intercepts are the roots of the related equation.)

38. $y = x^2 - 6x + 1$
39. $y = 4x^2 + 8x - 1$
40. $y = 2x^2 - x - 2$
41. $y = 3x^2 - 5x + 1$
42. $y = 3.2x^2 - 5.6x - 7.1$
43. $y = 1.9x^2 + 6.5x + 2.7$

Write a quadratic equation that has the given roots.

44. $-1 \pm \sqrt{3}$
45. $\frac{-3 \pm \sqrt{5}}{2}$
46. $\frac{1 \pm \sqrt{33}}{4}$
47. $\frac{4 \pm \sqrt{7}}{3}$

Critical Thinking

48. If a quadratic equation can be solved by factoring, what can you say about the value of $b^2 - 4ac$?
49. If a quadratic equation has exactly one real root, what can you say about the value of $b^2 - 4ac$?

Application

50. **Physics** Rafael is tossing rocks off the edge of a 10-meter-high cliff. He throws one rock with an initial upward velocity of 15 meters per second. Use the formula $H = -4.9t^2 + vt + h$ to answer each question. Round your answers to the nearest tenth of a second.
    a. When will the rock reach a height of 25 meters above the ground?
    b. When will the rock return to the height from which it was thrown?
    c. When will the rock hit the ground?

51. **Finance** Elise received \$100 from her father on her birthday and placed it in a savings account. After 3 years, she has $100[(1 + r)^2 + (1 + r) + 1]$ dollars in the account, where r is the annual interest rate. Find this rate, to the nearest tenth of a percent, if Elise has \$325 in the account after 3 years.

Mixed Review

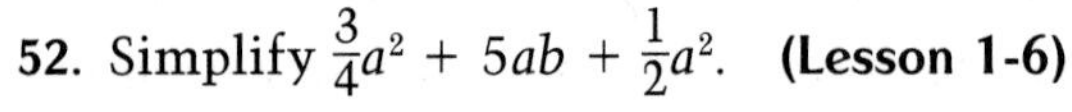

52. Simplify $\frac{3}{4}a^2 + 5ab + \frac{1}{2}a^2$. **(Lesson 1-6)**

53. Simplify $\frac{a^2}{a^2 - b^2} + \frac{a}{(a - b)^2}$. **(Lesson 8-7)**

54. Graph $\frac{5}{2}x + \frac{2}{3}y = 1$. **(Lesson 9-4)**

55. Write the standard form of the equation of the line passing through $(4, -1)$ and $(2, 5)$. **(Lesson 10-2)**

56. **Physics** The time t, in seconds, it takes an object to drop d feet is given by the formula $4t = \sqrt{d}$. Jessica and Lu-Chan each dropped a stone at the same time, but Jessica dropped hers from a spot higher than Lu-Chan's. Lu-Chan's stone hit the ground 1 second before Jessica's. If Jessica's stone dropped 112 feet farther than Lu-Chan's, how long did it take her stone to hit the ground? **(Lesson 12-7)**

57. Solve $2x^2 - 6x - 5 = 0$ by completing the square. **(Lesson 13-4)**

Journal

Write a paragraph to explain how you can use a calculator to solve a quadratic equation.

MID-CHAPTER REVIEW

Find an equation of the axis of symmetry and the coordinates of the vertex of the graph of each quadratic function. (Lesson 13-1)

1. $y = x^2 - x - 12$
2. $y = -2x^2 - 9$
3. $y = -3x^2 - 6x + 5$

Find the roots of each equation by graphing its related function. If exact roots cannot be found, state the consecutive integers between which the roots are located. (Lesson 13-2)

4. $x^2 + 6x + 10 = 0$
5. $x^2 - 2x - 1 = 0$
6. $x^2 - 5x - 6 = 0$

Solve each equation by completing the square. (Lesson 13-4)

7. $x^2 - 6x + 7 = 0$
8. $2b^2 - b - 7 = 14$

Solve each equation by using the quadratic formula. (Lesson 13-5)

9. $y^2 + 8y + 15 = 0$
10. $p^2 + 5p + 3 = 0$

11. Three painters can paint 4 houses in 5 days. How long would it take 5 painters to paint 18 houses if they all worked at the same rate? **(Lesson 13-3)**

13-6 Using the Discriminant

Objective

After studying this lesson, you should be able to:

- evaluate the discriminant of a quadratic equation to determine the nature of the roots of the equation.

Application

Trevor and Mineku are mountain climbing on Sopher Peak. In order to climb one of the cliffs, Trevor must throw his grappling hook up onto a ledge that is 15 meters above him. If Trevor can throw the grappling hook with an initial velocity of at most 13 meters per second, can he throw the hook onto the ledge?

In Lesson 13-5, you learned that the height H of an object t seconds after it is propelled upward is given by the formula $H = -\frac{1}{2}gt^2 + vt + h$. For this problem, $v = 13$ and $g = 9.8$, since Trevor is on Earth. Since the ledge is 15 meters above Trevor, we can let $H = 15$ and $h = 0$. Thus, this problem can be represented by the following equation.

$$15 = -\frac{1}{2}(9.8)t^2 + 13t + 0$$

$$0 = -4.9t^2 + 13t - 15$$

To determine if Trevor will be able to throw the grappling hook onto the ledge, we can use the quadratic formula to solve this equation for t.

$$t = \frac{-b \pm \sqrt{b^2 - 4ac}}{2a} \qquad a = -4.9,\ b = 13,\ c = -15$$

$$= \frac{-13 \pm \sqrt{13^2 - 4(-4.9)(-15)}}{2(-4.9)}$$

$$= \frac{-13 \pm \sqrt{169 - 294}}{-9.8}$$

$$= \frac{-13 \pm \sqrt{-125}}{-9.8}$$

Notice that the number under the radical sign is negative. This indicates that the equation $0 = -4.9t^2 + 13t - 15$ has *no* real roots. Therefore, Trevor cannot throw the grappling hook up onto the ledge.

In the quadratic formula, the expression under the radical sign, $b^2 - 4ac$, is called the **discriminant.** The value of the discriminant for a quadratic equation can give you information about the nature of the roots of the equation. In particular, the value of the discriminant is used to determine the number of real roots for a quadratic equation. Recall that the real roots of a quadratic equation are the x-intercepts of the graph of its related function.

There are three cases to consider: when the value of the discriminant is positive, when it is zero, and when it is negative.

Case 1: Positive Discriminant

Solve $x^2 + 3x - 2 = 0$.

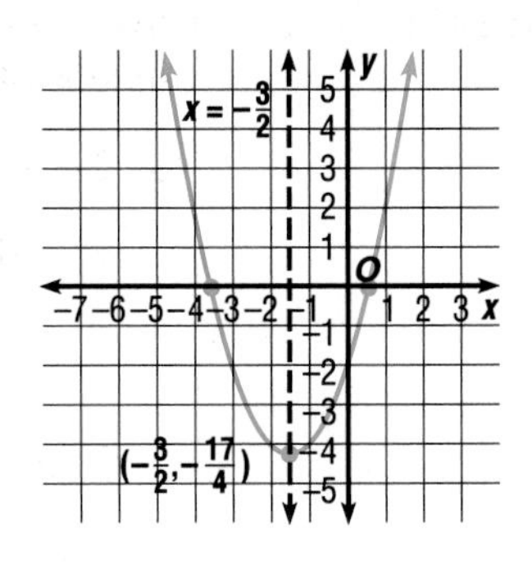

$$x = \frac{-b \pm \sqrt{b^2 - 4ac}}{2a}$$

$$= \frac{-3 \pm \sqrt{3^2 - 4(1)(-2)}}{2(1)} \quad a = 1, b = 3, c = -2$$

$$= \frac{-3 \pm \sqrt{9 + 8}}{2}$$

$$= \frac{-3 \pm \sqrt{17}}{2}$$

Notice that the graph of the related function $y = x^2 + 3x - 2$ has two distinct x-intercepts.

$x \approx -3.56$ or $x \approx 0.56$

In this case, $b^2 - 4ac > 0$ and there are two distinct real roots.

Case 2: Discriminant of Zero

Solve $x^2 - 8x + 16 = 0$.

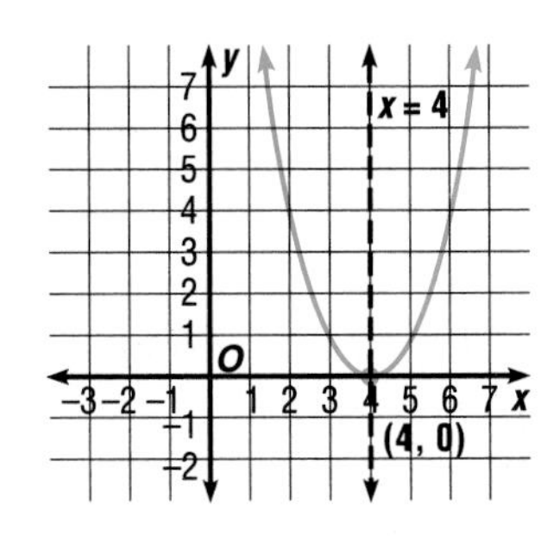

$$x = \frac{-b \pm \sqrt{b^2 - 4ac}}{2a}$$

$$= \frac{-(-8) \pm \sqrt{(-8)^2 - 4(1)(16)}}{2(1)}$$

$$= \frac{8 \pm \sqrt{64 - 64}}{2}$$

$$= \frac{8 \pm 0}{2} \text{ or } 4$$

Notice that the graph of the related function $y = x^2 - 8x + 16$ has one distinct x-intercept.

In this case, $b^2 - 4ac = 0$ and there is exactly one distinct real root.
Notice that the x-intercept is the vertex of the parabola.

Case 3: Negative Discriminant

Solve $x^2 + 6x + 10 = 0$.

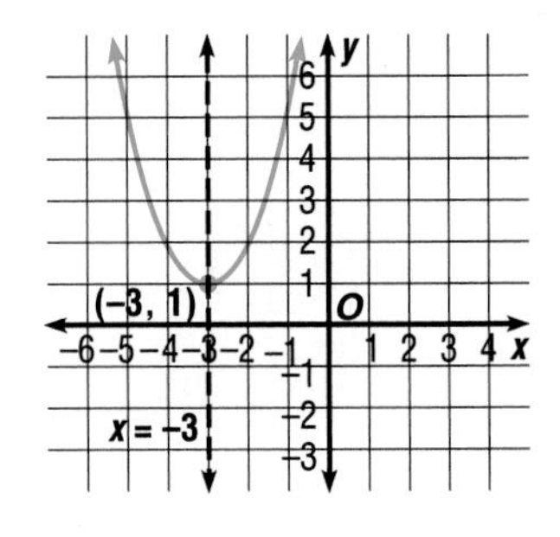

$$x = \frac{-b \pm \sqrt{b^2 - 4ac}}{2a}$$

$$= \frac{-6 \pm \sqrt{6^2 - 4(1)(10)}}{2(1)}$$

$$= \frac{-6 \pm \sqrt{36 - 40}}{2}$$

$$= \frac{-6 \pm \sqrt{-4}}{2}$$

Notice that the graph of the related function $y = x^2 + 6x + 10$ has no x-intercepts.

In this case, $b^2 - 4ac < 0$ and there are no real roots since *no* real number can be the square root of a negative number.

The relationship between the value of the discriminant and the nature of the roots of a quadratic equation can be summarized as follows.

Nature of Roots of a Quadratic Equation

Discriminant	Nature of Roots
$b^2 - 4ac > 0$	two distinct real roots
$b^2 - 4ac = 0$	one distinct real root
$b^2 - 4ac < 0$	no real roots

Example 1

State the value of the discriminant of each equation. Then determine the nature of the roots of the equation.

a. $\mathbf{2x^2 + 10x + 11 = 0}$

$$\begin{aligned} b^2 - 4ac &= (10)^2 - 4(2)(11) \quad && a = 2, b = 10, c = 11 \\ &= 100 - 88 \\ &= 12 \end{aligned}$$

Since $b^2 - 4ac > 0$, then $2x^2 + 10x + 11 = 0$ has two distinct real roots.

b. $\mathbf{3x^2 + 4x + 2 = 0}$

$$\begin{aligned} b^2 - 4ac &= (4)^2 - 4(3)(2) \quad && a = 3, b = 4, c = 2 \\ &= 16 - 24 \\ &= -8 \end{aligned}$$

Since $b^2 - 4ac < 0$, then $3x^2 + 4x + 2 = 0$ has no real roots.

Example 2

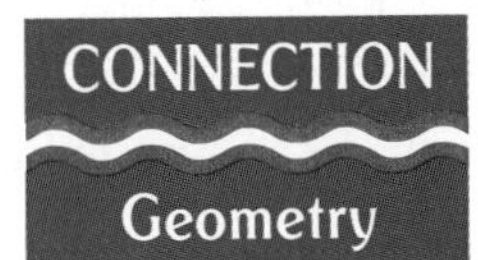

Can a rectangle with a perimeter of 42 cm have an area of 110.25 cm²?

EXPLORE Let ℓ = the length of the rectangle.
Let w = the width of the rectangle.
To solve this problem, you need to determine if there are values of ℓ and w such that $2\ell + 2w = 42$ and $\ell w = 110.25$.

PLAN First solve $2\ell + 2w = 42$ for ℓ. Then substitute the expression for ℓ into $\ell w = 110.25$. A quadratic equation results that represents the area of a rectangle whose perimeter is 42 cm.

SOLVE

$$\begin{aligned} 2\ell + 2w &= 42 \\ \ell + w &= 21 \quad && \text{Divide each side by 2.} \\ \ell &= 21 - w \end{aligned}$$

$$\begin{aligned} (21 - w)w &= 110.25 \quad && \text{Substitute } 21 - w \text{ for } \ell. \\ 21w - w^2 &= 110.25 \\ 0 &= w^2 - 21w + 110.25 \\ 0 &= 4w^2 - 84w + 441 \quad && \text{Multiply each side by 4.} \end{aligned}$$

You could solve the equation $0 = 4w^2 - 84w + 441$. However, you *only* need to determine if the equation has *any* real roots. You can do this by evaluating the determinant.

$$\begin{aligned} b^2 - 4ac &= (-84)^2 - 4(4)(441) \quad && a = 4, b = -84, c = 441 \\ &= 7056 - 7056 \\ &= 0 \end{aligned}$$

Since $b^2 - 4ac = 0$, then $0 = 4w^2 - 84w + 441$ has one distinct real root. Thus, there is a rectangle with a perimeter of 42 cm and an area of 110.25 cm². *Examine this solution.*

CHECKING FOR UNDERSTANDING

Communicating Mathematics

Complete.

1. In the quadratic formula, the expression __?__ is the discriminant.
2. A quadratic equation has two real roots when the discriminant is __?__.
3. If the discriminant of a quadratic equation is 0, then the __?__ of the graph of the related function is its x-intercept.
4. If a quadratic equation has two *irrational* roots, then the discriminant of the equation __?__ a perfect square.

Guided Practice

State the value of the discriminant for each equation. Then determine the nature of the roots of the equation.

5. $x^2 + 3x - 4 = 0$
6. $m^2 + 5m - 6 = 0$
7. $s^2 + 8s + 16 = 0$
8. $2z^2 + 7z + 50 = 0$
9. $3x^2 + x + 1 = 0$
10. $2a^2 - 2a - 1 = 0$

EXERCISES

Practice

State the value of the discriminant for each equation. Then determine the nature of the roots of the equation.

11. $y^2 + 3y + 1 = 0$
12. $x^2 - 1.2x = 0$
13. $4a^2 + 10a = -6.25$
14. $\frac{4}{3}n^2 + 4n + 3 = 0$
15. $\frac{3}{2}m^2 + m = -\frac{7}{2}$
16. $2r^2 = \frac{1}{2}r - \frac{2}{3}$

Determine the nature of the roots of each equation. Then find all real roots. Express irrational roots in simplest radical form and in decimal form rounded to the nearest hundredth.

17. $y^2 - 4y + 1 = 0$
18. $k^2 + 6k + 10 = 0$
19. $r^2 + 4r - 12 = 0$
20. $h^2 - 16h + 64 = 0$
21. $2x^2 + 3x + 1 = 0$
22. $3y^2 + y - 1 = 0$
23. $6r^2 - 5r = 7$
24. $8p^2 + 1 = -7p$
25. $9y^2 = 6y - 1$
26. $0.3a^2 + 0.8a = -0.4$
27. $x^2 - \frac{5}{3}x = \frac{-2}{3}$
28. $\frac{1}{3}x^2 + 13\frac{1}{3} = 4x$
29. $5c^2 - 7c = 1$
30. $15a^2 + 2a + 16 = 0$
31. $11z^2 = z + 3$

Determine the number of x-intercepts of the graph of each function *without* graphing the function.

32. $y = x^2 + 5x + 3$
33. $y = x^2 + 4x + 7$
34. $y = 7x^2 - 3x - 1$
35. $y = 0.6x^2 + x - 1.8$
36. $y = \frac{3}{2}x^2 + 2x + \frac{5}{4}$
37. $y = 4x^2 - \frac{4}{3}x + \frac{1}{9}$

38. **Geometry** Can a rectangle with a perimeter of 56 meters have an area of 200 square meters?

Determine the values of k so that each quadratic equation has the indicated number of real roots.

39. $x^2 + kx + 36 = 0$; 1
40. $x^2 + 8x + k = 0$; 2
41. $kx^2 + 5x = 1$; 0

Critical Thinking

42. In the quadratic equation $ax^2 + bx + c = 0$, if $ac < 0$, what must be true about the nature of the roots of the equation?

Applications

43. **Physics** The height h, in feet, of a certain rocket t seconds after blast-off is given by the formula $h = -16t^2 + 2320t + 125$.
   a. Approximately how long after blast-off will this rocket reach a height of 84,225 feet?
   b. Is this the maximum height of the rocket?

44. **Consumerism** A grocer sells 50 loaves of bread a day at $1.15 per loaf. The grocer estimates that for each $0.05 price increase, 2 fewer loaves of bread will be sold each day. What price can she charge in order to have a daily income of $65?

Computer

45. This BASIC program uses the quadratic discriminant to determine the nature of the roots of any quadratic equation. Real roots are computed by means of the quadratic formula.

```
10 INPUT "ENTER THE COEFFICIENTS
     OF AX^2 + BX + C = 0.";A,B,C
20 LET D = B^2-4*A*C
30 IF D < 0 THEN 110
40 LET X1 = (-B+SQR(D))/(2*A)
50 LET X2 = (-B-SQR(D))/(2*A)
60 IF D = 0 THEN 90
70 PRINT "THE TWO REAL DISTINCT
     ROOTS ARE ";X1;" AND ";X2;"."
80 GOTO 120
90 PRINT "THE ONE REAL DISTINCT
     ROOT IS ";X1;"."
100 GOTO 120
110 PRINT "THERE ARE NO REAL
     ROOTS."
120 END
```

Use the program to find and determine the roots of each quadratic equation.

a. $3x^2 - 2x + 1 = 0$
b. $4x^2 + 4x + 1 = 0$
c. $7x^2 + 2x - 5 = 0$
d. $x^2 - 11x + 10 = 0$
e. $2x^2 + x + 1 = 0$
f. $x^2 - 6x = 0$

Mixed Review

46. **Sales** Ruth paid $19.61 for a videotape. This included 6% sales tax. What was the cost of the videotape before taxes? **(Lesson 4-4)**

47. Is the inverse of $\{(-1, 3), (-1, 0), (2, -1)\}$ a function? **(Lesson 9-5)**

48. **Business** Jake's Garage charges $83 for a two-hour repair job and $185 for a five-hour repair job. Define variables and write an equation that Jake can use to bill customers for repair jobs of any length of time. **(Lesson 10-5)**

49. Solve the system $y < 3x$ and $x + 2y \geq -21$ by graphing. **(Lesson 11-6)**

50. Simplify $\sqrt{59.29}$. **(Lesson 12-2)**

51. **Geometry** The perimeter of a rectangle is 8 m and its area is 2 m². Find its dimensions, to the nearest tenth of a meter. **(Lesson 13-5)**

13-7 Application: Solving Quadratic Equations

Objective After studying this lesson, you should be able to:

- solve problems that can be represented by quadratic equations.

You have studied a variety of methods for solving quadratic equations. The table below summarizes these methods.

Method	Can be Used	Comments
graphing	always	Not always exact; use only when an approximate solution is sufficient.
factoring	sometimes	Use if constant term is 0 or factors are easily determined.
completing the square	always	Useful for equations of the form $x^2 + bx + c = 0$ where b is an even number.
quadratic formula	always	Other methods may be easier to use, but this method always gives exact values.

You can use these methods to solve problems that can be represented by quadratic equations.

Example 1

CONNECTION Geometry

The perimeter of a rectangle is 60 cm. Find the dimensions of the rectangle if its area is 221 cm².

EXPLORE Let ℓ = the measure of the length of the rectangle.
Let w = the measure of its width.
Since the perimeter of the rectangle is 60 cm, $2\ell + 2w = 60$.
Since the area of the rectangle is 221 cm², $221 = \ell w$.

PLAN First solve $2\ell + 2w = 60$ for ℓ.

$$2\ell + 2w = 60$$
$$\ell + w = 30 \quad \textit{Divide each side by 2.}$$
$$\ell = 30 - w$$

SOLVE Then substitute $30 - w$ for ℓ in $221 = \ell w$ and solve.

$$221 = (30 - w)w$$
$$221 = 30w - w^2$$
$$w^2 - 30w + 221 = 0$$
$$(w - 13)(w - 17) = 0 \quad \textit{Factor.}$$
$$w - 13 = 0 \quad \text{or} \quad w - 17 = 0$$
$$w = 13 \qquad\qquad w = 17$$

If $w = 13$, then $\ell = 30 - 13$ or 17.
If $w = 17$, then $\ell = 30 - 17$ or 13.
Thus, the dimensions of the rectangle are 13 cm and 17 cm.

EXAMINE If the dimension of the rectangle are 13 cm and 17 cm, then its perimeter is $2(13) + 2(17)$ or 60 cm and its area is $13(17)$ or 221 cm^2. Thus, the solution is correct.

Example 2

A pan is to be formed by cutting 2 cm-by-2 cm squares from each corner of a square piece of sheet metal and then folding the sides. If the volume of the pan is to be 441 cm², what are the dimensions of the original piece of sheet metal?

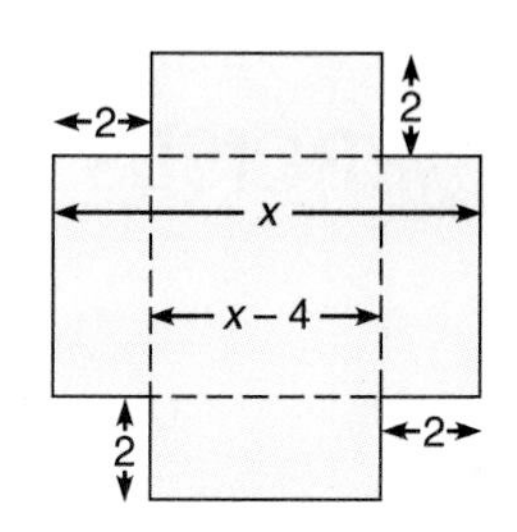

Let x = the measure of a side of the square piece of sheet metal. Then $x - 4$ = the measure of a side of the square base of the pan.

You can use the formula for the volume of a rectangular solid, $\ell wh = V$, to write an equation to represent the problem.

$$(x - 4)(x - 4)(2) = 441 \quad \ell = x - 4, w = x - 4, h = 2, V = 441$$
$$2x^2 - 16x + 32 = 441$$
$$2x^2 - 16x - 409 = 0$$

$$x = \frac{16 \pm \sqrt{(-16)^2 - 4(2)(-409)}}{2(2)}$$ *Use the quadratic formula with $a = 2$, $b = -16$, and $c = -409$.*

$$= \frac{16 \pm \sqrt{256 + 3272}}{4}$$

$$x = \frac{16 + \sqrt{3528}}{4} \quad \text{or} \quad x = \frac{16 - \sqrt{3528}}{4}$$ *Use a calculator to approximate each root to the nearest tenth.*

$$\approx 18.8 \qquad \approx -10.8$$

The length of each side of the piece of sheet metal should be about 18.8 cm since the length cannot be −10.8 cm. The volume of this pan is $(18.8 - 4)(18.8 - 4)(2)$ or 438.08 cm^3. Thus, the answer appears reasonable.

CHECKING FOR UNDERSTANDING

Communicating Mathematics

Read and study the lesson to answer each question.

1. Name four methods for solving quadratic equations.
2. Which of the four methods does not always find an exact solution?
3. In Example 2, what would be the dimensions of the original piece of sheet metal if the squares measured 1 cm on a side instead of 2 cm?

Guided Practice

State the method that is most appropriate for solving each equation.

4. $x^2 - 12x + 27 = 0$
5. $2m^2 + 19m + 9 = 0$
6. $a^2 - 12a - 4 = 0$
7. $3r^2 - 7r = 5$
8. $3x^2 - 2x = 5$
9. $y^2 + 4y = 9$

Solve each quadratic equation by an appropriate method.

10. $x^2 - 8x - 20 = 0$
11. $y^2 + 10y - 2 = 0$
12. $r^2 + 13r = -42$
13. $3x^2 - 7x - 6 = 0$
14. $2a^2 + 4a + 1 = 0$
15. $3z^2 - 7z = 3$

EXERCISES

Practice

Solve each quadratic equation by an appropriate method. Express irrational roots in simplest radical form and in decimal form rounded to the nearest hundredth.

16. $3h^2 - 5h - 2 = 0$
17. $2k^2 + k - 5 = 0$
18. $2y^2 - 4y + 3 = 0$
19. $3z^2 = 5z - 1$
20. $2m^2 + 4m = 5$
21. $x^2 - 1.1x = 0.6$
22. $x^2 - \frac{17}{20}x + \frac{3}{20} = 0$
23. $\frac{1}{4}y^2 = y + \frac{1}{2}$
24. $0.7a^2 - 2.8a = 7$

25. Find two integers whose sum is 12 and whose squares differ by 24.
26. The sum of a number and its reciprocal is $\frac{10}{3}$. Find the number.

Applications

Solve each problem. Approximate any irrational solutions to the nearest tenth.

27. **Construction** A box is to be formed from a rectangular piece of sheet metal by cutting squares measuring 5 inches on a side and then folding the sides. The piece of sheet metal is twice as long as it is wide. If the volume of the box is to be 1760 in^3, what are the dimensions of the original piece of sheet metal?

28. **Gardening** Gene Knight has a rectangular flower garden that measures 15 meters by 20 meters. He wishes to build a concrete walk of uniform width around the garden. His budget for the project allows him to buy enough concrete to cover an area of 74 square meters. How wide can he build the walk?

29. **Physics** Janice tosses a softball directly upward with an initial velocity of 100 feet per second. Use the formula $h = vt - 16t^2$ to answer each question.
    a. When will the ball reach a height of 84 feet?
    b. When will the ball hit the ground?

30. **Framing** Miguel wants to make a square picture frame that is 5 cm wide. The frame is to be designed so that the area available for a picture is two-thirds of the total area of the picture and the frame. What will be the dimensions of this frame?

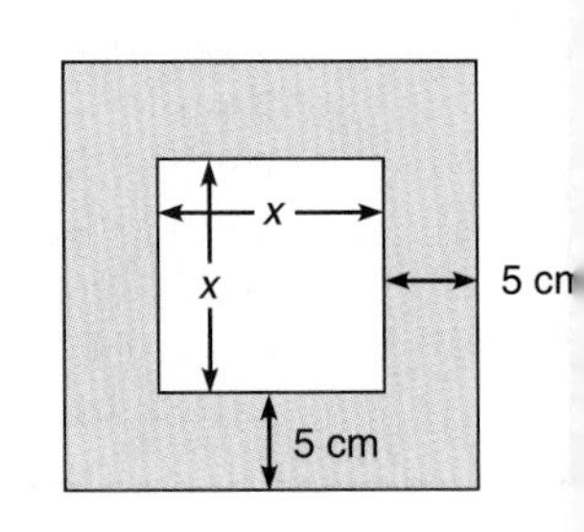

31. **Mowing** Lisa, Nicole, and Amber are to mow a rectangular lawn that measures 100 feet by 120 feet. Lisa is going to mow one-third of the lawn by mowing a strip of uniform width around the outer edge of the lawn. What are the dimensions of the lawn still to be mowed?

32. **Sales** The Computer Club bought a \$68 graphics software package for club use. If there had been 1 more student in the club, then cost per member to buy the software would have been \$0.25 less. How many students are in the club?

Critical Thinking

33. Write a proportion to represent the following statement: *The ratio of 1 to a positive number is equal to the ratio of the number to 1 minus the number.* Then solve the proportion to find the number. Express your answer in simplest radical form and in decimal form rounded to the nearest thousandth. This number is called the golden ratio.

Mixed Review

34. **Geometry** The area of a square is $4x^2 - 28x + 49$ cm². What is the value of x if the perimeter of the square is 60 cm? **(Lesson 7-10)**

35. **Work** Pipe A can fill a tank in 4 hours and pipe B can fill the tank in 8 hours. With the tank empty, pipe A is turned on, and one hour later, pipe B is turned on. How long will pipe B run before the tank is full? Use $rt = w$. **(Lesson 8-9)**

36. Use substitution to solve the system of equations $5x + 6y = 74$ and $2x - 3y = 8$. **(Lesson 11-3)**

37. **Geometry** The measures of the sides of a triangle are 0.9, 1.6, and 2.0. Is this triangle a right triangle? **(Lesson 12-3)**

38. Find the discriminant of $5b^2 = 1 + 6b$. Determine the nature of the roots of the equation. **(Lesson 13-6)**

Portfolio

Select some of your work from this chapter that shows how you used a calculator or computer. Place it in your portfolio.

HISTORY CONNECTION

Muhammed ibn Musa al Khwarizmi

Just where did this thing called "algebra" come from? Around 825 A.D., an Arab mathematician, Al-Khwarizmi, wrote about a practical form of mathematics for everyday people, for use in trade and commerce, law, surveying, and construction. The word *algebra* is taken from the title of his work, *Hisab al-jabr w'al muqabalah,* which mainly dealt with solving linear equations. The title has been translated as *The Science of Restoring and Cancelling.* This probably refers to the process of adding a number to each side of an equation (restoring) and dividing each side of an equation by a number (cancelling). Al-Khwarizmi also solved quadratic equations by completing the square. Although he was not the first person to do this, his work became very important in shaping the way we solve these types of equations today. For these reasons, Al-Khwarizmi is often called "The Father of Algebra."

13-8 The Sum and Product of Roots

Objectives After studying this lesson, you should be able to:

- find the sum and product of the roots of a quadratic equation, and
- write a quadratic equation given its roots.

Engineers, scientists, and mathematicians are often asked to find an equation that can be used to describe a certain situation or problem. Sometimes, the results of the problem are known and an equation can be found that fits these results. Finding this equation is just like finding an equation given the roots of the equation.

Suppose the roots of a quadratic equation are known to be -3 and 8.

If $x = -3$, then $x + 3 = 0$. If $x = 8$, then $x - 8 = 0$.

The quadratic equation $(x + 3)(x - 8) = 0$, or $x^2 - 5x - 24 = 0$, has roots -3 and 8. *Why?*

Suppose you find the sum and the product of the roots of this quadratic equation.

Sum of roots: $-3 + 8 = 5$ **Product of roots:** $-3 \cdot 8 = -24$

Now look at the equation again. What do you notice?

product of roots — *The product of the roots is the constant term.*

$x^2 - 5x - 24 = 0$

opposite of sum of roots — *The opposite of the sum of the roots is the coefficient of x.*

This relationship between the roots and coefficients is true for all quadratic equations. Suppose the two roots are r_1 and r_2. Then the quadratic equation is $(x - r_1)(x - r_2) = 0$. Multiplying the binomials results in the equation $x^2 - (r_1 + r_2)x + r_1r_2 = 0$. Notice that the coefficient of x, $-(r_1 + r_2)$, is the opposite of the sum of the roots. Also, the constant term, r_1r_2, is the product of the roots.

Example 1 **Use the sum and the product of roots to find a quadratic equation whose roots are $1 + \sqrt{5}$ and $1 - \sqrt{5}$.**

Opposite of the sum of roots

$$= -[(1 + \sqrt{5}) + (1 - \sqrt{5})]$$
$$= -[(1 + 1) + (\sqrt{5} - \sqrt{5})]$$
$$= -2$$

Product of roots

$$= (1 + \sqrt{5})(1 - \sqrt{5})$$
$$= 1 - \sqrt{5} + \sqrt{5} - 5$$
$$= -4$$

The opposite of the sum of the roots, −2, is the coefficient of x. The product of the roots, −4, is the constant term. Thus, a quadratic equation whose roots are $1 + \sqrt{5}$ and $1 - \sqrt{5}$ is $x^2 - 2x - 4 = 0$.

Now consider the quadratic equation of the form $ax^2 + bx + c = 0$, where $a \neq 0$. Dividing each side by a results in the equation $x^2 + \frac{b}{a}x + \frac{c}{a} = 0$. By comparing this equation to the equation $x^2 - (r_1 + r_2)x + r_1r_2 = 0$, the following rule may be derived.

Sum and Product of Roots of a Quadratic Equation

For a quadratic equation of the form $ax^2 + bx + c = 0$, where $a \neq 0$, the sum of the roots of the equation is $-\frac{b}{a}$ and the product of the roots of the equation is $\frac{c}{a}$.

Example 2

Are $\frac{3}{2}$ and $-\frac{4}{3}$ roots of $6x^2 - x - 12 = 0$?

If $\frac{3}{2}$ and $-\frac{4}{3}$ are the roots, then their sum must be equal to $-\frac{b}{a}$ and their product must be equal to $\frac{c}{a}$.

Sum of Roots

$$\frac{3}{2} + \left(-\frac{4}{3}\right) = \frac{9}{6} - \frac{8}{6}$$
$$= \frac{1}{6}$$

Product of Roots

$$\frac{3}{2}\left(-\frac{4}{3}\right) = -2$$

Since $-\frac{b}{a} = -\left(\frac{-1}{6}\right)$ or $\frac{1}{6}$ and $\frac{c}{a} = \frac{-12}{6}$ or -2, $\frac{3}{2}$ and $-\frac{4}{3}$ are roots of the equation.

Example 3

A research engineer proposes that the height h, in feet, of a rocket t seconds after blast-off is given by the formula $h = -16t^2 + 256t + 3124$. If the altitude of the rocket was 4500 feet after 5.5 seconds and after 10.5 seconds, is the proposed equation correct?

For the proposed equation to be correct, 5.5 and 10.5 must be the roots of the equation $4500 = -16t^2 + 256t + 3124$, or $16t^2 - 256t + 1376 = 0$. Rather than solving this equation to determine if 5.5 and 10.5 are the roots, you can check their sum and their product.

For $16t^2 - 256t + 1376 = 0$, $-\frac{b}{a} = \frac{-(-256)}{16}$ or 16 and $\frac{c}{a} = \frac{1376}{16}$ or 86.

Sum of Roots

$5.5 + 10.5 = 16$

Product of Roots

$5.5(10.5) = 57.75$

Since $57.75 \neq 86$, 5.5 and 10.5 are *not* roots of the proposed equation. Therefore, the proposed equation is not correct.

CHECKING FOR UNDERSTANDING

Communicating Mathematics

Read and study the lesson to answer each question.

1. For the equation $x^2 + bx + c = 0$, how is b related to the roots of the equation?
2. For the equation $x^2 + bx + c = 0$, how is c related to the roots of the equation?
3. If r and s are roots of a quadratic equation, what is the factored form of the equation?

Guided Practice

State the sum and the product of the roots of each equation.

4. $x^2 - 5x + 6 = 0$
5. $3m^2 + 6m - 3 = 0$
6. $4t^2 + 8t + 3 = 0$
7. $6a^2 - 13a = 15$

State whether each pair of numbers are the roots of each equation.

8. $-6, 3; y^2 + 3y - 18 = 0$
9. $2, 3; n^2 + 5n - 6 = 0$
10. $-1, 7; b^2 - 8b = -7$
11. $\frac{-1}{3}, \frac{1}{2}; x^2 - \frac{1}{6}x - \frac{1}{6} = 0$
12. $1 + \sqrt{7}, 1 - \sqrt{7}; r^2 - 2r - 6 = 0$
13. $4 + \sqrt{3}, 4 - \sqrt{3}; x^2 + 8x = -1$

Write a quadratic equation having the given roots.

14. 5, 2
15. 1, -6
16. $\frac{2}{3}$, 7
17. 0.3, -0.6

EXERCISES

Practice

Find the sum and the product of the roots of each equation.

18. $y^2 + 15y + 54 = 0$
19. $a^2 - 5a - 24 = 0$
20. $6x^2 + 31x + 35 =$
21. $z^2 + \frac{13}{2}z - \frac{9}{4} = 0$
22. $\frac{1}{2}m^2 - \frac{3}{2}m + 4 = 0$
23. $2c^2 - \frac{2}{3}c = \frac{1}{6}$
24. $12x^2 - 4x = 9$
25. $6k^2 - 0.02 = 0.4k$
26. $2y^2 - 6 = -y\sqrt{2}$

Write a quadratic equation having the given roots.

27. 4, 7
28. 6, -5
29. 1, -10
30. $-2, -17$
31. $\frac{5}{2}$, 2
32. $\frac{-3}{4}$, 8
33. $\frac{2}{3}, \frac{-3}{2}$
34. $-1.4, -2.2$
35. $\sqrt{2}, -\sqrt{2}$
36. $\sqrt{3}, \sqrt{3}$
37. $2 + \sqrt{3}, 2 - \sqrt{3}$
38. $-4 + \sqrt{10}, -4 - \sqrt{10}$
39. $3 + \sqrt{5}, 5 + \sqrt{3}$
40. $\frac{2 - \sqrt{11}}{3}, \frac{2 + \sqrt{11}}{3}$

Find the value of k such that indicated number is a root of the given equation.

41. $3;\ x^2 + kx - 21 = 0$

42. $1;\ x^2 + kx + 5 = 0$

43. $-5;\ x^2 + 12x + k = 0$

44. $3;\ x^2 + 6x - k = 0$

45. Write the general form of the quadratic equation with roots q and r.

Critical Thinking

Show that the following statements are true for the equation $ax^2 + bx + c = 0$. (*Hint:* Use the roots from the quadratic formula.)

46. The sum of the roots is $-\frac{b}{a}$.

47. The product of the roots is $\frac{c}{a}$.

Applications

48. **Physics** A rocket is launched from a height of 50 feet. The rocket reaches a height of 122 feet at 1.5 seconds and at 3 seconds. Was the initial velocity of this rocket 72 feet per second or 80 feet per second? Use the formula $H = -16t^2 + vt + h$.

49. **Finance** A theater has seats for 500 people. To increase income, the owner plans to raise the price of tickets which is now $6.00. She estimates that if p is the number of $0.50 price increases, then the expression $3000 + 100p - 12.5p^2$ represents her income. If the maximum income actually occurs when the ticket price is $8.00, is this expression for the income correct?

Mixed Review

50. Find $\left(\frac{5}{7}a^2 - \frac{3}{4}a + \frac{1}{2}\right) - \left(\frac{3}{7}a^2 + \frac{1}{2}a - \frac{1}{2}\right)$. **(Lesson 6-6)**

51. Simplify $\frac{n^2 - 4}{n^2 - 4n - 12} \div (n - 2)$. **(Lesson 8-4)**

52. Draw a mapping of the relation shown at the right. Then write the relation and its inverse as sets of ordered pairs. **(Lesson 9-2)**

x	2	6	7	8
y	−2	5	3	−2

53. **Uniform Motion** When Neil was 2 miles upstream from camp on a canoe trip, he passed a log floating downstream with the current. He paddled upstream for 1 more hour and then returned to camp just as the log arrived. What was the rate of the current? **(Lesson 11-5)**

54. Solve $\sqrt{3x - 5} + x = 11$ and check. **(Lesson 12-7)**

55. **Physics** The height h, in meters, that Andreina is above the water t seconds after beginning her dive is given by the formula $h = -5t^2 + 6t + 10$. After how many seconds will Andreina hit the water, rounded to the nearest tenth of a second? **(Lesson 13-7)**

CHAPTER 13

SUMMARY AND REVIEW

VOCABULARY

Upon completing this chapter, you should be familiar with the following terms:

axis of symmetry	519	518	parabola
completing the square	532	537	quadratic formula
discriminant	541	518	quadratic function
maximum point	519	523	root
minimum point	519	519	vertex

SKILLS AND CONCEPTS

OBJECTIVES AND EXAMPLES

Upon completing this chapter you should be able to:

- find the equation of the axis of symmetry and the coordinates of the vertex of the graph of a quadratic function **(Lesson 13-1)**

For the graph of $y = x^2 - 8x + 12$, the equation of the axis of symmetry is $x = -\frac{-8}{2(1)}$ or $x = 4$.
Since $4^2 - 8(4) + 12 = -4$, the graph has a minimum point at $(4, -4)$.

REVIEW EXERCISES

Use these exercises to review and prepare for the chapter test.

Find the equation of the axis of symmetry and the coordinates of the vertex of the graph of each quadratic function.

1. $y = x^2 - 3x - 4$
2. $y = -x^2 + 6x + 16$
3. $y = -2x^2 + 9x - 9$
4. $y = 3x^2 + 6x - 17$

- graph a quadratic function **(Lesson 13-1)**

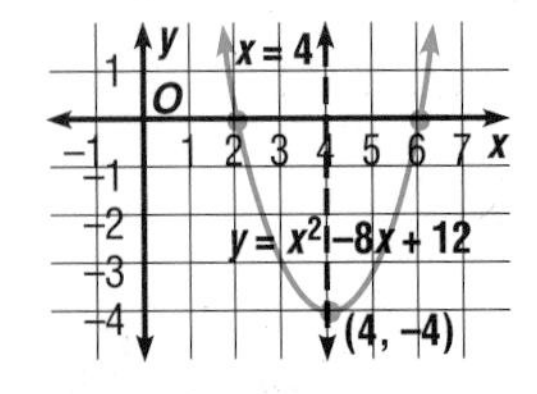

Using the results from Exercises 1–4, graph each quadratic function.

5. $y = x^2 - 3x - 4$
6. $y = -x^2 + 6x + 16$
7. $y = -2x^2 + 9x - 9$
8. $y = 3x^2 + 6x - 17$

- find the roots of a quadratic equation by graphing **(Lesson 13-2)**

Based on the graph of $y = x^2 - 8x + 12$ shown above, the roots of the equation $x^2 - 8x + 12 = 0$ are 2 and 6.

Find the roots of each equation by graphing its related function.

9. $x^2 - x - 12 = 0$
10. $x^2 + 6x + 9 = 0$
11. $x^2 + 4x - 3 = 0$
12. $2x^2 - 5x + 4 = 0$

OBJECTIVES AND EXAMPLES

- solve quadratic equations by completing the square **(Lesson 13-4)**

$$y^2 + 6y + 2 = 0$$
$$y^2 + 6y = -2$$
$$y^2 + 6y + 9 = -2 + 9$$
$$(y + 3)^2 = 7$$
$$y + 3 = \pm\sqrt{7}$$
$$y = -3 \pm \sqrt{7}$$

The roots are $-3 + \sqrt{7}$ and $-3 - \sqrt{7}$.

REVIEW EXERCISES

Find the value of c that makes each trinomial a perfect square.

13. $x^2 + 8x + c$

14. $r^2 - 5r + c$

Solve each equation by completing the square.

15. $x^2 - 16x + 32 = 0$

16. $m^2 - 7m = 5$

17. $4a^2 + 16a + 15 = 0$

- solve quadratic equations by using the quadratic formula **(Lesson 13-5)**

Solve $2x^2 + 7x - 15 = 0$.

$$x = \frac{-7 \pm \sqrt{(7)^2 - 4(2)(-15)}}{2(2)}$$
$$= \frac{-7 \pm \sqrt{169}}{4}$$
$$x = \frac{-7 + 13}{4} \quad \text{or} \quad x = \frac{-7 - 13}{4}$$
$$= \frac{3}{2} \qquad\qquad = -5$$

Solve each equation by using the quadratic formula.

18. $x^2 - 8x = 20$

19. $5b^2 + 9b + 3 = 0$

20. $9k^2 - 1 = 12k$

21. $2m^2 = \frac{17}{6}m - 1$

22. $3s^2 - 7s - 2 = 0$

- evaluate the discriminant of a quadratic equation to determine the nature of the roots of the equation **(Lesson 13-6)**

The discriminant of $3x^2 - 8x - 40 = 0$ is $(-8)^2 - 4(3)(-40)$ or 544. Thus, the equation has 2 real roots.

Use the discriminant to determine the nature of the roots of each equation.

23. $9k^2 - 13k + 4 = 0$

24. $7x^2 - 6x + 5 = 0$

25. $9a^2 + 25 = 30a$

26. $4p^2 + 4p = 15$

- solve problems that can be represented by quadratic equations **(Lesson 13-7)**

A rectangle has a perimeter of 38 in. and an area of 84 in^2. Find the dimensions of this rectangle.

Since $2\ell + 2w = 38$, $2\ell = 38 - 2w$ or $\ell = 19 - w$. Since $\ell w = 84$, $(19 - w)w = 84$ or $0 = w^2 - 19w + 84$. The roots of this equation are 7 and 12. Therefore, the dimensions are 7 in. and 12 in.

Solve.

27. Find two integers whose sum is 21 and whose product is 90.

28. The length of a rectangular garden is 8 feet more than its width. A walkway 3 feet wide surrounds the outside of the garden. If the total area of the walkway alone is 288 square feet, what are the dimensions of the garden?

OBJECTIVES AND EXAMPLES	REVIEW EXERCISES

- find the sum and product of the roots of a quadratic equation **(Lesson 13-8)**

For the equation $3x^2 + 8x + 9 = 0$, the sum of the roots is $-\frac{b}{a} = \frac{-8}{3}$, and the product of the roots is $\frac{c}{a} = \frac{9}{3}$ or 3.

Find the sum and product of the roots of each equation.

29. $y^2 + 8y - 14 = 0$

30. $4a^2 - 6a + 11 = 0$

31. $2x^2 - x = 6$

- write a quadratic equation given its roots **(Lesson 13-8)**

Find a quadratic equation whose roots are 2 and $\frac{5}{3}$.

Sum of roots: $2 + \frac{5}{3} = \frac{11}{3}$ $\quad -\frac{b}{a} = \frac{11}{3}$

Product of roots: $2 \cdot \frac{5}{3} = \frac{10}{3}$ $\quad \frac{c}{a} = \frac{10}{3}$

The equation is $3x^2 - 11x + 10 = 0$.

Find a quadratic equation having the given roots.

32. $1, -8$

33. $\frac{3}{2}, -4$

34. $3 + \sqrt{5}, 3 - \sqrt{5}$

APPLICATIONS AND CONNECTIONS

35. Number Theory How many numbers between 1000 and 10,000 are palindromes? **(Lesson 13-3)**

36. Geometry The area of a certain square is one-half the area of the rectangle formed if the length of one side of the square is increased by 2 cm and the length of an adjacent side is increased by 3 cm. What are the dimensions of the square? **(Lesson 13-7)**

37. Sales A helicopter service transports passengers to an island during vacation season. Each day, 500 people are transported for a round trip fare of \$20. The owner has decided to increase the fare. A survey has shown that for each \$1 increase in fare, 20 less people will use the service. What fare will maximize the owner's income? **(Lesson 13-1)**

38. Physics The height h, in feet, of a rocket t seconds after blast-off is given by the formula $h = 1440t - 16t^2$.

a. After how many seconds will the rocket reach a height of 25,000 ft?

b. After how many seconds will the rocket reach a height of 35,000 ft?

c. After how many seconds will the rocket hit the ground? **(Lesson 13-6)**

CHAPTER 13 TEST

Find the equation of the axis of symmetry and the coordinates of the vertex of the graph of each quadratic function.

1. $y = 4x^2 - 8x - 17$
2. $y = -3x^2 + 12x + 34$

Find the roots of each equation by graphing its related function. If exact roots cannot be found, state the consecutive integers between which the roots are located.

3. $x^2 + x - 2 = 0$
4. $x^2 - 8x + 11 = 0$
5. Solve $m^2 - 8m - 4 = 0$ by completing the square.
6. Solve $2k^2 - 9k + 8 = 0$ by using the quadratic formula.

Solve each quadratic equation by an appropriate method. Express irrational roots in simplest radical form and in decimal form rounded to the nearest hundredth.

7. $2x^2 - 5x - 12 = 0$
8. $m^2 + 18m + 75 = 0$
9. $3y^2 - 2y - 4 = 0$
10. $3k^2 + 2k = 5$
11. $6n^2 + 7n = 20$
12. $2x^2 - 10 = 3x$
13. $7a^2 + \frac{23}{3}a + 2 = 0$
14. $x^2 - 4.4x + 4.2 = 0$

Use the discriminant to determine the nature of the roots of each equation.

15. $3y^2 - y - 10 = 0$
16. $4b^2 + 12b + 9 = 0$
17. $3m^2 - 9m + 7 = 0$
18. $y^2 + y\sqrt{3} - 5 = 0$
19. Find the sum and the product of the roots of $3m^2 - 15m + 41 = 0$.

Find a quadratic equation having the given roots.

20. $-2, \frac{5}{4}$
21. $6 + \sqrt{3}, 6 - \sqrt{3}$
22. **Geometry** A rectangle has a perimeter of 44 cm and area of 105 cm^2. Find its dimensions.
23. **Number Theory** Find two real numbers whose sum is 22 and whose product is 125.
24. **Construction** A rectangular piece of sheet metal is 3 times as long as it is wide. Squares measuring 2 cm on a side are cut from each corner and the sides are folded up in order to make a box. If the volume of the box needs to be 512 cm^3, what must be the dimensions of the piece of sheet metal?
25. **Diving** Greg is diving off a 3-meter springboard. His height h, in meters, above the water when he is x meters away from the board is given by the formula $y = -x^2 + 3x + 3$. What is the maximum height Greg will reach on a dive? About how far away from the board is he when he enters the water?

Bonus Find the coordinates of the vertex of the graph of the equation $y = ax^2 + bx + c$, where $a \neq 0$.

CHAPTER OBJECTIVES

In this chapter, you will:

- Display data on line plots, stem-and-leaf plots, and box-and-whisker plots.
- Find the mean, median, mode, range, and interquartile range of a set of data.
- Draw scatter plots for sets of data.
- Find the probability and odds of events.

CHAPTER 14

Statistics and Probability

The study of probability arose from our desire to estimate the chance that the outcome of an event, like the toss of a coin, the roll of a die, or the birth of a child, will occur in a certain way.

APPLICATION IN PROBABILITY

Probability is a way to measure the chances that something will occur in relation to the possible alternatives. For example, the probability that a woman gives birth to a boy is $\frac{1}{2}$. But this probability is *not* a guarantee. A couple might have four children and all are boys, or they might have six children and all are girls.

Now you might think that a couple with six girls would not expect to have another girl if they decided to have a seventh child. In fact, the probability that the seventh child is a girl is still $\frac{1}{2}$, since the gender of this child is not affected by the gender of the previous six children.

What if you know that a family has seven children and six of them are girls? Is the probability that the seventh child is a girl still $\frac{1}{2}$? There are eight possibilities: all seven children are girls, or either the first, second, third, fourth, fifth, sixth, or seventh child is a boy. Since in only one of these cases the seventh child is a girl, the probability is $\frac{1}{8}$.

ALGEBRA IN ACTION

Complete the chart below to find each probability for the given conditions and number of children.

Conditions	Number of Children					
	2	3	4	5	6	7
All children are girls given that you know all but the last child are girls.	$\frac{1}{2}$	$\frac{1}{2}$	?	?	?	$\frac{1}{2}$
All children are girls given that you know all but one child are girls.	$\frac{1}{3}$	$\frac{1}{4}$	?	?	?	$\frac{1}{8}$
All children are girls.	$\frac{1}{4}$	$\frac{1}{8}$	?	?	?	$\frac{1}{128}$

14-1 Statistics and Line Plots

Objectives

After studying this lesson, you should be able to:

- interpret numerical data from a table, and
- display and interpret statistical data on a line plot.

The study of statistics is sometimes called data analysis.

Each day when you read newspapers or magazines, watch television, or listen to the radio, you are bombarded with numerical information about the national economy, sports, politics, and so on. Interpreting this numerical information, or **data,** is important to your understanding of the world around you. A branch of mathematics called **statistics** helps provide you with methods for collecting, organizing, and interpreting data.

Statistical data can be organized and presented in numerous ways. One of the most common ways is to use a table or chart. The chart at the right shows the hourly wages earned by the principal wage earner in ten families.

Using tables or charts like the one at the right should enable you to more easily analyze the given data.

Family	Hourly Wage
A	$8.00
B	$10.50
C	$20.25
D	$9.40
E	$11.00
F	$13.75
G	$8.50
H	$10.50
I	$9.00
J	$11.00

Example 1

Use the information in the chart above to answer each question.

a. What are the maximum and minimum hourly wages of the principal wage earner for the ten families?

The principal wage earner in Family C makes $20.25 per hour. This is the maximum hourly wage of all the families.
The principal wage earner in Family A makes $8.00 per hour. This is the minimum hourly wage of the ten families.

b. What percent of the families have a principal wage earner that makes less than $10.00 per hour?

The principal wage earners in Family A ($8.00), Family D ($9.40), Family G ($8.50), and Family I ($9.00) each make less than $10.00 per hour. Thus, 4 out of 10, or 40%, of the families have a principal wage earner that makes less than $10.00 per hour.

In some instances, statistical data can be presented on a number line. Numerical information displayed on a number line is called a **line plot.** For example, the data in the table above can be presented in a line plot.

Note that some data values are located between integer values on the number line.

From Example 1, you know that the data in the chart range from \$8.00 per hour to \$20.25 per hour. In order to represent each hourly wage on a number line, the scale used must include these values. A "w" is used to represent each hourly wage. When more than one "w" has the same location on the number line, additional "w"s are placed one above the other. A line plot for the hourly wages is shown below.

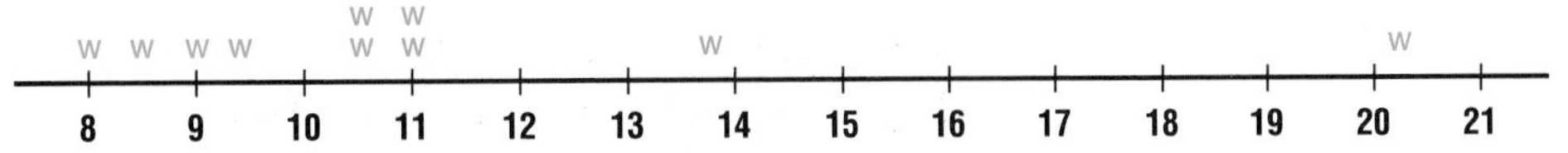

Example 2

APPLICATION

Transportation

The number of passengers arriving at and departing from the ten busiest airports in the United States for a one-year period are listed below.

Airport	Passengers	Airport	Passengers
Chicago (O'Hare)	59,130,007	San Francisco	29,939,835
Dallas/Ft. Worth	47,579,046	Denver	27,568,003
Los Angeles	44,967,221	Miami	23,385,010
Atlanta	43,312,285	New York (LGA)	23,158,317
New York (JFK)	30,323,077	Honolulu	22,617,340

The busiest foreign airport was Heathrow in London with 39,905,200 arriving and departing passengers in 1989.

a. Make a line plot of the data.

The numbers in the table are too large to represent easily on a number line. Change each number to represent every 1,000,000 passengers arriving and departing, and round to the nearest whole number. For example, since $59{,}130{,}007 = 59.130007 \times 1{,}000{,}000$, you would plot an "x" at 59 to represent O'Hare.

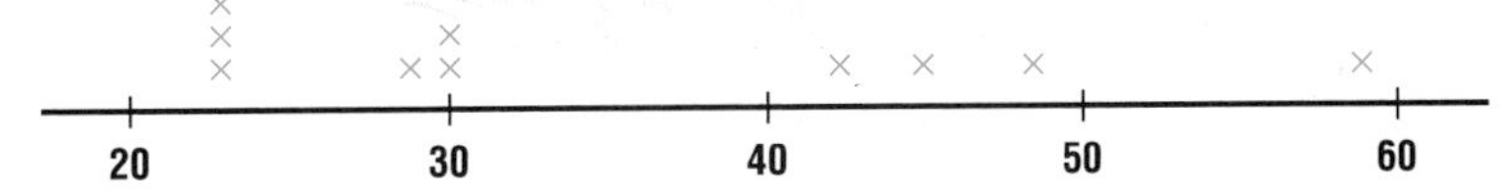

b. Determine how many airports had approximately 25,000,000 passengers arrive and depart during the year.

Four airports, Denver, Miami, New York (LGA), and Honolulu, had approximately 25,000,000 passengers arrive and depart during the year.

c. Determine if any airport had a much greater number of passengers arrive and depart than the other airports.

Based on the line plot, it appears that O'Hare Airport had a much greater number of passengers arriving and departing.

The data in the table in Example 2 were collected by checking airport records. Data can also be collected by taking actual measurements, by conducting surveys or polls, or by using questionnaires.

When you study and analyze data to draw conclusions, it is important that you know how the data were obtained. For example, would you want to draw conclusions about the favorite subject at your entire school based on the results of a survey of seniors only? Why or why not?

CHECKING FOR UNDERSTANDING

Communicating Mathematics

Read and study the lesson to answer each question.

1. What is the name for the numerical information used in statistics?
2. Why do you think tables and charts are useful for presenting statistical data?
3. What are some other ways you could survey some of the students at your school in order to draw conclusions about the favorite subject at your school?

Guided Practice

State the scale you would use to make a line plot for the following data.

4. 4.2, 5.3, 7.6, 9.6, 7.3, 6.7
5. 30, 30, 40, 40, 10, 20, 50, 40
6. 123, 234, 789, 456, 111, 420, 397, 334
7. 7895, 3785, 9987, 1672, 4444, 6754, 5550, 3197, 7965, 10,615
8. The following table lists the percent of 18-year-olds with high school diplomas for each of 9 years.

Year	1950	1955	1960	1965	1970	1975	1980	1985	1990
Percent	56	65	72	71	77	74	72	74	72

   a. What was the lowest percent of 18-year-olds with high school diplomas?
   b. In what year did the highest percent occur?
   c. In how many years was the percent with diplomas less than 74%?
   d. During how many of the 5-year intervals did the percent increase?
   e. What was the greatest decrease over a 5-year interval for the percent with diplomas?
   f. Make a line plot of the data.

EXERCISES

Practice

9. Use the line plot at the right to answer each question.
   a. What was the highest score on the test?
   b. What was the lowest score on the test?
   c. How many students took the test?
   d. How many students scored in the 40s?
   e. What score was received by the most students?

Class Scores on a Science Test

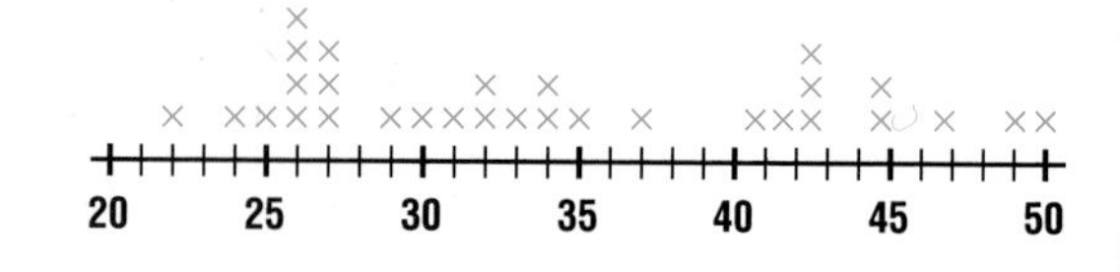

10. The high temperatures in degrees Fahrenheit for 50 cities in 1992 are listed below.

94 103 82 90 95 94 96 103 108 90 98 101 96
95 103 92 92 107 91 98 99 94 98 102 99 91
97 97 100 99 96 100 104 94 118 95 92 110 96
98 106 100 92 91 102 94 97 96 87 91

a. Organize the data into a table with the headings **High Temperature** and **Number of Cities.**

b. What was the highest high temperature for any of the cities?

c. What was the lowest high temperature for any of the cities?

d. Which temperature(s) occurred most frequently?

e. Which temperature(s) between 90°F and 110°F are not in the table?

f. How many cities had a high temperature of 100°F?

g. How many cities had a high temperature of at least 100°F?

h. How many cities had a high temperature of at most 95°F?

11. Each number below represents the age of a United States president on his first inauguration.

57 61 57 57 58 57 61
54 68 51 49 64 50 48
65 52 56 46 54 49 50
47 55 55 54 42 51 56
55 51 54 51 60 62 43
55 56 61 52 69 64 46

a. Make a line plot of the ages of United States presidents on their first inauguration.

b. How many presidents' ages are listed?

c. Do any of the ages appear clustered? If so, which ones?

12. The players with the most runs batted in (RBI) for the National League (1969–1992) are listed below.

a. Make a line plot of the data.

b. What was the greatest number of RBIs during a season?

c. What was the least number of RBIs during a season?

d. What was the most frequent number of RBIs during a season?

e. How many of the players had from 119 to 129 RBIs?

Year	Name	RBI	Year	Name	RBI
1969	Willie McCovey	126	1982	Dale Murphy	109
1970	Johnny Bench	148		Al Oliver	
1971	Joe Torre	137	1983	Dale Murphy	121
1972	Johnny Bench	125	1984	Mike Schmidt	106
1973	Willie Stargell	119		Gary Carter	
1974	Johnny Bench	129	1985	Dave Parker	125
1975	Greg Luzinski	120	1986	Mike Schmidt	119
1976	George Foster	121	1987	Andre Dawson	137
1977	George Foster	149	1988	Will Clark	109
1978	George Foster	120	1989	Kevin Mitchell	125
1979	Dave Winfield	118	1990	Matt Williams	122
1980	Mike Schmidt	121	1991	Howard Johnson	117
1981	Mike Schmidt	91	1992	Darren Daulton	109

13. Ms. Lee and Mr. Jebson each asked 15 students from their algebra classes how many hours of television they watched last week. The results are shown in the line plot at the right. Use this plot to answer each question.

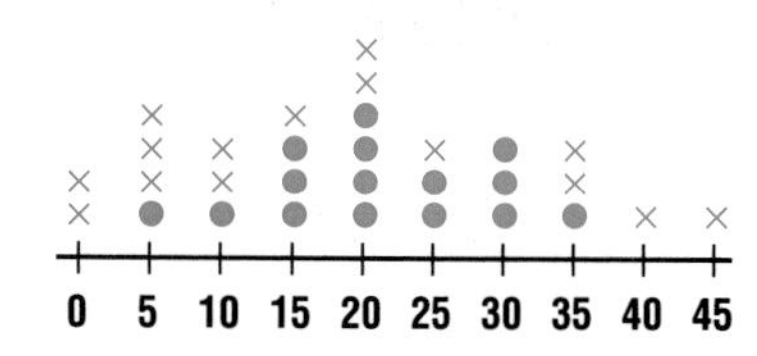

a. Which group of 15 students watched the most television?

b. Does the pattern for the number of hours watching television appear to be the same for both groups? Explain.

Critical Thinking

14. Use the line plot for Exercise 13. Find the average number of hours that the students in Ms. Lee's class and Mr. Jebson's class watched television. Do the values support your answer to Exercise 13?

Applications

15. **Meteorology** Make a line plot of the data about the high temperatures for 50 cities in 1989 used in Exercise 10. Which temperatures occurred exactly four times?

16. **History** Refer to the data on the ages of presidents used for Exercise 11. Organize the data into a table with the headings **Age on First Inauguration** and **Number of Times.** Under **Age on First Inauguration,** group the data by fives, starting with 41–45. In which of these five-year interval(s) did the most ages occur?

17. **Consumerism** The cost per cup, in cents, of 28 different liquid laundry detergents are listed below. Make a line plot of the data. Then determine how many detergents cost at most 17¢ per cup.

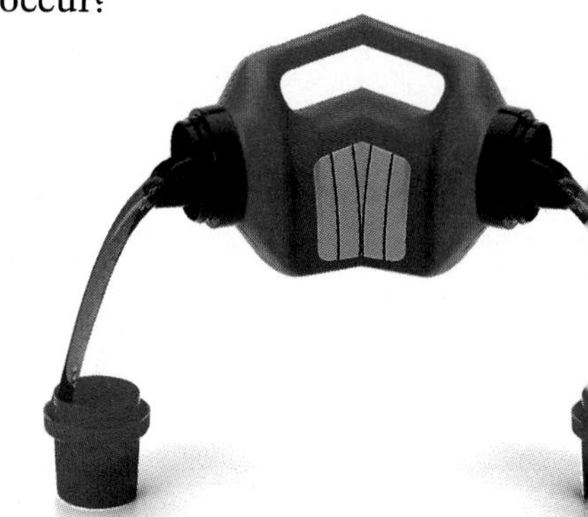

28	17	16	18	19	21	26
15	19	19	16	14	21	12
26	17	30	17	13	18	14
22	10	12	19	9	15	12

Mixed Review

18. **Travel** Paula wants to reach Urbanville at 10 A.M. If she drives at 36 miles per hour, she would reach Urbanville at 11 A.M. But if she drives at 54 miles per hour, she would arrive at 9 A.M. At what speed should she drive to reach Urbanville exactly at 10 A.M.? **(Lesson 4-7)**

19. Given $f(x) = x^2 - 2x$ and $g(x) = 1 - 3x$, find $g[f(a)]$. **(Lesson 9-5)**

20. **Manufacturing** At the end of January, Don's Sporting Equipment had manufactured 450 tennis racket covers. At the end of May, the company had manufactured a total of 3250 tennis racket covers. Assuming that the planned production of tennis racket covers can be represented by a straight line, determine how many will be manufactured by the end of the year. **(Lesson 10-5)**

21. Find the geometric mean of 12 and 27. **(Lesson 12-7)**

22. Find a quadratic equation whose roots are $\frac{7}{3}$ and -3. **(Lesson 13-8)**

14-2 Stem-and-Leaf Plots

Objective After studying this lesson, you should be able to:

- display and interpret data on a stem-and-leaf plot.

Application Mr. Juarez wants to study the distribution of the scores for a 100-point unit exam given in his first-period biology class. The scores of the 35 students in the class are listed below.

82 77 49 84 44 98 93 71 76 65 89 95 78 69 89 64
88 54 96 87 92 80 44 85 93 89 55 62 79 90 86 75
74 99 62

He can organize and display the scores in a compact way using a **stem-and-leaf plot.**

A stem may have one or more digits. A leaf always has just one digit.

In a stem-and-leaf plot, the greatest common place value of the data is used to form the *stems.* The numbers in the next greatest common place-value position are then used to form the *leaves.* In the list above, the greatest place value is tens. Thus, the number 82 would have stem 8 and leaf 2.

To make the stem-and-leaf plot, first make a vertical list of the stems. Since the test scores range from 44 to 99, the stems range from 4 to 9. Then, plot each number by placing the units digit (leaf) to the right of its correct stem. Thus, the score 82 is plotted by placing leaf 2 to the right of stem 8. The complete stem-and-leaf plot is shown at the right.

Stem	Leaf
4	9 4 4
5	4 5
6	5 9 4 2 2
7	7 1 6 8 9 5 4
8	2 4 9 9 8 7 0 5 9 6
9	8 3 5 6 2 3 0 9

8|2 represents a score of 82.

A second stem-and-leaf plot can be made to arrange the leaves in numerical order from least to greatest as shown at the right. This will make it easier for Mr. Juarez to analyze the data.

Stem	Leaf
4	4 4 9
5	4 5
6	2 2 4 5 9
7	1 4 5 6 7 8 9
8	0 2 4 5 6 7 8 9 9 9
9	0 2 3 3 5 6 8 9

Example 1 **Use the information in the stem-and-leaf plots above to answer each question.**

a. What were the highest and lowest scores on the test? 99 and 44

b. Which test score occurred most frequently? 89 (3 times)

c. In which 10-point interval did the most students score?
80–89 (10 students)

d. How many students received a score of 70 or better? 25 students

Sometimes the data for a stem-and-leaf plot are numbers that have more than two digits. Before plotting these numbers, they may need to be rounded or *truncated* to determine each stem and leaf. Suppose you wanted to plot 356 using the hundreds digit for the stem.

Rounded

Round 356 to 360. Thus, you would plot 356 using stem 3 and leaf 6. What would be the stem and leaf of 499? *5 and 0*

Truncated

To truncate means to cut off, so truncate 356 as 350. Thus, you would plot 356 using stem 3 and leaf 5. What would be the stem and leaf of 499? *4 and 9*

A *back-to-back stem-and-leaf plot* is sometimes used to compare two sets of data or rounded and truncated values of the same set of data. In a back-to-back plot, the same stem is used for the leaves of both plots.

Example 2

The average annual pay for workers in selected states are listed below. Make a back-to-back stem-and-leaf plot of the average annual pay comparing rounded values and truncated values. Then answer each question.

State	Avg. Annual Pay	State	Avg. Annual Pay
Alaska	$28,033	Michigan	$24,193
California	$24,126	Minnesota	$21,481
Colorado	$21,472	New Jersey	$25,748
Connecticut	$26,234	New York	$26,347
Delaware	$21,977	Ohio	$21,501
Illinois	$23,608	Pennsylvania	$21,485
Maryland	$22,515	Texas	$21,130
Massachusetts	$24,143	Virginia	$21,053

Since the data range from $21,053 to $28,033, the stems range from 21 to 28 for both plots.

Rounded	Stem	Truncated
5 5 5 5 1 1	21	0 1 4 4 4 5 9
5 0	22	5
6	23	6
2 1 1	24	1 1 1
7	25	7
3 2	26	2 3
	27	
0	28	0

Using rounded data, 28|0 represents $27,950 to $28,049.

Using truncated data, 28|0 represents $28,000 to $28,099.

a. What does 21|5 represent in each plot?
It represents $21,450 to $21,549 for rounded data and $21,500 to $21,599 for truncated data.

b. What is the difference between the highest and lowest average annual pay? about $6900 for rounded data and $7000 for truncated data

c. Did more of the states have average annual pay above or below $25,000? below $25,000

d. Does there appear to be any significant difference between the two stem-and-leaf plots? no

CHECKING FOR UNDERSTANDING

Communicating Mathematics

Read and study the lesson to answer each question.

1. Suppose a student in Mr. Juarez's biology class received a score of 100. How would this score be represented in the stem-and-leaf plot?
2. If the number 53.7 is plotted using stem 5 and leaf 3, how would the number 99.8 be plotted?
3. If the number 6373 is plotted using stem 63 and leaf 7, how would the number 6498 be plotted?
4. In what situations is a back-to-back stem-and-leaf plot used?

Guided Practice

Write the stems that would be used for a plot of each set of data.

5. 23, 46, 54, 13, 28, 54, 37
6. 8, 13, 21, 73, 101, 49, 52, 33
7. 24.5, 26.1, 25.8, 19.1, 23.2, 24.1, 23.7, 22.9
8. 236, 450, 748, 254, 755, 347, 97, 396

Suppose two digits of each number are to be used in a stem-and-leaf plot. Find the rounded and the truncated values for each number.

9. 456,876
10. 34,591
11. 4321
12. 1,234,567

13. Use the stem-and-leaf plot below to answer each question.
    a. What was the highest temperature recorded?
    b. What was the lowest temperature recorded?
    c. On how many days was the high temperature in the 70s?
    d. What temperature(s) occurred most frequently?

Daily High Temperature in April

Stem	Leaf
8	0 2 2 5
7	0 3 3 4 5 7 7 7 8 9 9
6	1 2 3 5 5 6 7 8 8 9
5	4 7 8 8
4	9

4|9 = 49°F

EXERCISES

Applications

14. **Football** The stem-and-leaf plot below gives the number of catches of the NFL's leading pass receiver for each season through 1990.
    a. What was the greatest number of catches during a season?
    b. What was the least number of catches during a season?
    c. How many seasons are listed?
    d. What number of catches occurred most frequently?
    e. How many times did the leading pass receiver have at least 90 catches?

Stem	Leaf
6	0 1 2 6 7
7	1 1 1 2 2 3 3 3 4 5 8
8	0 2 6 8 8 9
9	0 1 2 2 5
10	0 0 1 6

9|2 represents 92 catches.

15. **School** Each number below represents the age of a student in Ms. Nichols' evening calculus class at DeAnzio Community College.

22 17 25 24 19 27 33 16 35 26 20 18 24
33 18 19 48 36 19 23 55 18 18 19 27 18
19 25 17 32 19 45 19 20 30

a. Make a stem-and-leaf plot of the data.
b. How many people attend Ms. Nichols' class?
c. What is the difference in ages between the oldest and youngest person in class?
d. What is the most common age for a student in the class?
e. Which age group is most widely represented in the class?
f. How many of Ms. Nichols' students are older than 25 years old?

16. **Work** The stem-and-leaf plot below gives the average weekly earnings in 1990 for various occupations.

a. What were the highest weekly earnings?
b. What were the lowest weekly earnings?
c. How many occupations have weekly earnings of at least $500?
d. What range does 5|1 represent?
e. Have the leaf values in this data been truncated or rounded?

Stem	Leaf
2	6 7
3	1 2 8 9
4	1 3 6 8 8 8 9
5	0 1 6
6	
7	1 2 4

3|8 represents $380–$389.

17. **Auto Racing** Each number below represents the average speed, in miles per hour, of the winning car in the last 31 Indianapolis 500s.

139 139 140 143 147 151 144
151 153 157 156 158 163 159
159 149 149 161 161 159 143
139 162 162 164 153 171 162
145 168 186

a. Make a stem-and-leaf plot of this data.
b. In what speed range were most of the winning cars?

18. **Stock Market** Each number below represents shares of CCXY stock traded each day over a 28-day period.

924 832 154 523 932 128 324 643 329 364 293
843 734 231 435 276 832 876 924 833 435 555
498 349 467 856 647 165

a. Make a stem-and-leaf plot of the data. Round numbers of shares to the nearest ten.
b. What is the difference between the greatest and least number of shares traded during this period?
c. How many times were less than 400 shares of stock traded?
d. What 10-share range of stocks traded occurred most frequently?

19. **Housing** The following table lists the median price, in thousands of dollars, of existing single-family homes in various cities in the U.S. for 1990.

City	Price	City	Price	City	Price
Atlanta	85	Honolulu	290	Omaha	65
Baltimore	104	Indianapolis	73	Orlando	80
Boston	182	Kansas City	74	Philadelphia	98
Chicago	112	Knoxville	74	Phoenix	82
Cleveland	77	Las Vegas	89	Pittsburgh	69
Dallas	94	Los Angeles	212	St. Louis	83
Denver	85	Milwaukee	82	San Francisco	262
Detroit	76	New Orleans	68	Seattle	136
Hartford	157	New York	174	Washington, D.C.	145

a. Make a stem-and-leaf plot of the data.

b. What is the difference in price of homes in the most and least expensive cities?

c. In what price range are most of the homes for these cities?

20. **Education** The table below lists the average Scholastic Aptitude Test (SAT) scores for males and females for certain years between 1967 and 1989.

Year	1967	1970	1975	1977	1980	1982	1985	1987	1989
Males	977	968	932	928	919	924	936	935	934
Females	935	926	880	872	863	864	877	878	875

a. Make a back-to-back stem-and-leaf plot of the SAT scores for males and females.

b. What information about the relationship between the SAT scores of males and females does it present?

Critical Thinking

21. The height, in inches, of the students in a health class are 65, 63, 68, 66, 72, 61, 62, 63, 59, 58, 61, 74, 65, 63, 71, 60, 62, 63, 71, 70, 59, 66, 61, 62, 68, 69, 64, 63, 70, 61, 68, and 67. If the stems 5, 6, and 7 are used in a stem-and-leaf plot, the distribution of heights would not be easy to analyze. How could you change the stems so the data is displayed in a way that is easier to analyze?

Mixed Review

22. **Statistics** Carita's bowling scores for the first four games of a five-game series are $b + 2$, $b + 3$, $b - 2$, and $b - 1$. What must her score be for the last game to have an average of $b + 2$? **(Lesson 3-7)**

23. Simplify $\frac{12x^4 + 12x^3 - 9x^2}{12x^3 + 18x^2 - 12x}$. **(Lesson 8-1)**

24. **Travel** While driving to Fullerton, Mrs. Sumner travels at an average speed of 40 mph. On the return trip, she travels at an average speed of 56 mph and saves two hours of travel time. How far does Mrs. Sumner live from Fullerton? **(Lesson 11-5)**

25. Find the distance between the points (10, 8) and (2, −3). **(Lesson 12-8)**

26. Solve $\frac{1}{2}t^2 - 2t - \frac{3}{2} = 0$ by completing the square. **(Lesson 13-4)**

27. Make a line plot of the data on the heights of the students in health class given in Exercise 21. What was the most common height for the students in this class? **(Lesson 14-1)**

14-3 Measures of Central Tendency

Objective After studying this lesson, you should be able to:

- calculate and interpret the mean, median, and mode of a set of data.

Application As part of its grand opening celebration, Zumpone's World of Sports is giving away two tickets to next year's Super Bowl to the person who correctly guesses the winning score of this year's Super Bowl.

In order to make the best possible guess, Jason has decided to study the results of previous Super Bowls. He lists the winning scores of the first 25 Super Bowls in a stem-and-leaf plot as shown at the right. Based on this information, what should be Jason's guess for the winning score? *This problem will be solved in Example 5.*

Stem	Leaf
1	4 6 6 6
2	0 0 1 3 4 4 6 7 7 7
3	1 2 3 5 5 8 8 9
4	2 6
5	5

2|1 represents 21 points.

In analyzing statistical data, it is useful to have a number that describes a set of data. In the problem above, Jason wants to find a number that best describes all of the data. Numbers known as *measures of central tendency* are often used to describe sets of data since they represent centralized or *middle* values of the data. Three measures of central tendency are called the **mean, median,** and **mode.**

The mean of a set of data is a number that represents an average of the numbers in the set.

Definition of Mean **The mean of a set of data is the sum of the numbers in the set divided by the number of numbers in the set.**

Example 1

The high temperatures for a week during January in Milwaukee were 19°, 21°, 18°, 17°, 18°, 22°, 46°. Find the mean high temperature for that week.

$$\text{mean} = \frac{19 + 21 + 18 + 17 + 18 + 22 + 46}{7}$$ *The mean is the sum of the 7 numbers divided by 7.*

$$= \frac{161}{7}$$

$$= 23$$

The mean, or average, high temperature for the week was 23°.

Notice in Example 1 that the mean high temperature, 23°, is greater than all of the daily high temperatures except one, 46°. Thus, 23° does not appear to be the best number to use to describe this set of data. Extremely low or high values, like 46°, affect the mean a great deal. In such cases, the mean becomes less representative of the values in a set of data.

The second measure of central tendency is the median.

Definition of Median

The median of a set of data is the middle number when the numbers in the set are arranged in numerical order.

Example 2

Find the median high temperature for the high temperatures given in Example 1.

First arrange the temperatures in order from least to greatest.

17°   18°   18°   19°   21°   22°   46°

Since there are seven temperatures, the middle one is the fourth value, 19°. Thus, the median high temperature for the week is 19°. *The median is not affected by the extremely high temperature, 46°.*

If a set of data contains an even number of elements, then the median of the set is the value halfway between the two middle elements.

Example 3

FYI . . .

The major league record for the highest batting average in a season is held by Rogers Hornsby who hit 0.424 for St. Louis in 1924.

The batting averages for 10 players on a baseball team are 0.234, 0.253, 0.312, 0.333, 0.286, 0.240, 0.183, 0.222, 0.297, and 0.275. Find the median batting average for these players.

Arrange the batting averages in ascending order.

0.183 0.222 0.234 0.240 0.253 0.275 0.286 0.297 0.312 0.333

Since there are an even number of batting averages, 10, the median is halfway between the two middle values, 0.253 and 0.275.

$$\frac{0.253 + 0.275}{2} = 0.264$$ *Find the mean of the two middle values.*

The median batting average for the 10 players is 0.264.

Notice in Examples 2 and 3 that the number of values greater than the median is the same as the number of values less than the median.

The third measure of central tendency is the mode.

Definition of Mode

The mode of a set of data is the number that occurs most often in the set.

Example 4

Find the mode of the high temperatures given in Example 1.

In the set of temperatures 19°, 21°, 18°, 17°, 18°, 22°, 46°, the temperature 18° occurs twice. Thus, 18° is the mode of the high temperatures for the week. *The mode is not affected by the extremely high temperature.*

Sets of data with two modes are called bimodal.

If no number in a set of data occurs more often than the other numbers, then the set has no mode. It is also possible for a set of data to have more than one mode. For example, the set of data {2, 3, 3, 4, 6, 6} has two modes, 3 and 6.

Based on the results of Examples 1, 2, and 4, the set of data {19°, 21°, 18°, 17°, 18°, 22°, 46°} has mean 23°, median 19°, and mode 18°. These examples show that the mean, median, and mode are not always the same value.

Example 5

Refer to the problem presented at the beginning of the lesson. Find the median, mode, and mean for the winning scores to determine what Jason's guess should be for the winning score this year.

The stem-and-leaf plot for the 25 scores is shown below.

Stem	Leaf
1	4 6 6 6
2	0 0 1 3 4 4 6 7 7 7
3	1 2 3 5 5 8 8 9
4	2 6
5	5

2|1 represents 21 points.

Median Since there are 25 values, the median is the 13th value. Counting from the top down, you will find that this value is 27. Thus, the median score is 27.

Mode Note that for stem 1, there are three leaves with a value of 6. Also, for stem 2, there are three leaves with a value of 7. Thus, the scores have two modes, 16 and 27.

Mean Add the 25 scores and then divide by 25. Since the sum of the 25 scores is 725, the mean score is $\frac{725}{25}$ or 29.

Since the median and one of the modes of the scores is 27, it appears that a good guess for the winning score is 27 points.

CHECKING FOR UNDERSTANDING

Communicating Mathematics

Complete.

1. Measures of central tendency represent __?__ values of a set of data.
2. If the numbers in a set of data are arranged in numerical order, then the __?__ of a set is the middle number.
3. Extremely high or low values affect the __?__ of a set of data.
4. If all the numbers in a set of data occur the same number of times, then the set has no __?__.

Guided Practice

Find the mean, median, and mode for each set of data.

5. 4, 6, 12, 5, 8
6. 9, 9, 9, 9, 8
7. 7, 19, 9, 4, 7, 2
8. 300, 34, 40, 50, 60
9. 23, 23, 23, 12, 12, 12
10. 10, 3, 17, 1, 8, 6, 12, 15

EXERCISES

Practice

Find the median and mode of the data shown in each line plot.

11.

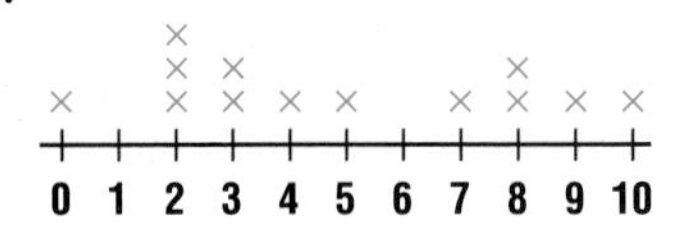

12.

30 35 40 45 50 55 60 65

Find the median and mode of the data shown in each stem-and-leaf plot.

13.

Stem	Leaf
7	3 5
8	2 2 4
9	0 4 7 9
10	5 8
11	4 6

9|4 = 94

14.

Stem	Leaf
5	3 6 8
6	5 8
7	0 3 7 7 9
8	1 4 8 8 9
9	9

6|8 = 68

15.

Stem	Leaf
19	3 5 5
20	2 2 5 8
21	5 8 8 9 9 9
22	0 1 7 8 9

21|5 = 215

List six numbers that satisfy each set of conditions.

16. The mean is 50, the median is 40, and the mode is 20.

17. The mean is 70, the median is 75, and the modes are 65 and 100.

18. The mean of a set of ten numbers is 5. When the greatest number in the set is eliminated, the mean of the new set of numbers is 4. What number was eliminated from the original set of numbers?

Applications

19. Football Michael Anderson of the West High Bears averages 137.6 yards rushing per game for the first five games of the season. He rushes for 155 yards in the sixth game. What is his new rushing average?

20. Advertising A magazine ad shows 5 videotape cameras for sale. The prices given are \$499, \$895, \$679, \$1195, and \$1400. Find the mean and median price of the cameras.

21. Basketball In a girl's basketball game between Lincoln High School and Taft High School, the Taft players' individual scores were 12, 4, 5, 3, 11, 23, 4, 6, 7, and 8. Find the mean, median, and mode of the individual points.

22. School Patty's scores on the 25-point quizzes in her English class are 20, 21, 18, 21, 22, 22, 24, 21, 20, 19, and 23. Find the mean, median, and mode of her quiz scores.

23. Sports One of the events in the Winter Olympics is the Men's 500-meter Speed Skating. The winning times for this event are shown at the right. Find the mean, median, and mode of the times.

Year	Time(s)	Year	Time(s)
1932	43.4	1968	40.3
1936	43.4	1972	39.44
1948	43.1	1976	39.17
1952	43.2	1980	38.03
1956	40.2	1984	38.19
1960	40.2	1988	36.45
1964	40.1	1992	37.14

24. **Geography** The areas in square miles of the 20 largest natural U. S. lakes are given below. Find the mean, median, and mode of these areas.

31,700	1697	242	700	22,300	451	374	432	23,000	207
1000	1361	315	625	9,910	215	458	360	7,550	435

25. **Work** Each number below represents the number of days that each employee of the Simpson Corporation was absent during 1990. Find the mean, median, and mode of the number of days.

0	10	8	5	8	9	3	3	2	9	7	0	4	2	4	6	2
9	13	3	1	5	5	7	2	6	5	3	4	7	1	1	5	3
3	1	4	5	1	2											

26. **Diving** Participants in diving events have each of their dives rated by 7 judges using a scale from 0 to 10. A diver's score is computed by eliminating the highest and lowest ratings of the judges and then finding the mean of the remaining five scores. This rating is then multiplied by a number that represents the difficulty of the dive attempted. Wendy received ratings of 8.2, 9.0, 7.3, 8.2, 7.7, 8.6, and 8.3 on a dive with difficulty 3.3. What was her score?

27. **Business** Of the 42 employees at Speedy Pizza, sixteen make $4.75 an hour, four earn $5.50 an hour, three earn $6.85 an hour, six earn $4.85 an hour, and thirteen earn $5.25 an hour. Find the mean, median, and mode of the hourly wages.

Critical Thinking

28. The annual salaries of the 12 employees at CompuSoftware Inc. are $38,500, $34,000, $27,500, $38,500, $63,500, $125,000, $31,500, $30,000, $38,500, $31,500, $92,500, and $31,000.
    a. If you are the personnel director, would you quote the median, mean, or mode as the "average" salary of the employees when interviewing a job applicant? Why?
    b. If the employees were trying to justify a pay raise to the management, would they quote the median, mean, or mode as their "average" salary? Why?

Mixed Review

29. **Number Theory** Draw a Venn diagram to find the intersection of {letters in the word *statistics*} and {letters in the words *data analysis*}. **(Lesson 2-1)**

30. Factor $24x^2 - 61xy + 35y^2$. **(Lesson 7-8)**

31. Write an equation in standard form of the line with slope of -3 that passes through (3, 7). **(Lesson 10-2)**

32. **Physics** An object is fired upward from the top of tower. Its height h, in feet, above the ground t seconds after firing is given by the formula $h = -16t^2 + 96t + 125$.
    a. How long after firing does the object reach its maximum height?
    b. What is this maximum height? **(Lesson 13-1)**

33. Make a stem-and-leaf plot for the winning times in the Men's 500-meter Speed Skating event using the data given in Exercise 23. **(Lesson 14-2)**

14-4 Measures of Variation

Objective After studying this lesson, you should be able to:

- calculate and interpret the range, quartiles, and the interquartile range of a set of data.

Application Pacquita Colón and Larry Nielson are two candidates for promotion to manager of sales at Fitright Shoes. In order to determine who should be promoted, the owner, Mr. Tarsel, looked at each person's quarterly sales record for the last two years.

Quarterly Sales (thousands of dollars)								
Ms. Colón	30.8	29.9	30.0	31.0	30.1	30.5	30.7	31.0
Mr. Nielson	31.0	28.1	30.2	33.2	31.8	29.8	28.9	31.0

After studying the data, Mr. Tarsel found that the mean of the quarterly sales was \$30,500, the median was \$30,600, and the mode was \$31,000 for both Ms. Colón and Mr. Nielson. If he was to decide between the two, Mr. Tarsel needed to find more numbers to describe this data.

The example above shows that measures of central tendency may not give an accurate enough description of a set of data. Often, *measures of variation* are also used to help describe the *spread* of the data. One of the most commonly used measures of variation is the **range.**

Definition of Range **The range of a set of data is the difference between the greatest and the least values of the set.**

Example 1 **Use the information in the table above to determine the range in the quarterly sales for Ms. Colón and Mr. Nielson during the last two years.**

Ms. Colón's greatest quarterly sales were \$31,000 and her least were \$29,900. Therefore, the range is \$31,000 − \$29,900 or \$1100.

Mr. Nielson's greatest quarterly sales were \$33,200 and his least were \$28,100. Therefore, the range is \$33,200 − \$28,100 or \$5100.

Based on this analysis, Ms. Colón's sales are much more consistent, a quality Mr. Tarsel values. Therefore, Ms. Colón is promoted.

Another commonly used measure of variation is called the **interquartile range.** In a set of data, the *quartiles* are values that divide the data into four equal parts. The median of a set of data divides the data in half. The **upper quartile (UQ)** divides the upper half into two equal parts. The **lower quartile (LQ)** divides the lower half into two equal parts. The difference between the upper and lower quartile is the interquartile range.

Definition of Interquartile Range	**The difference between the upper quartile and the lower quartile of a set of data is called the interquartile range. It represents the middle half, or 50%, of the data in the set.**

Example 2

The Birch Corporation held its annual golf tournament for its employees. The scores for 18 holes were 88, 91, 102, 80, 115, 99, 101, 103, 139, 105, 99, 95, 76, 105, and 112. Find the median, upper and lower quartiles, and the interquartile range for these scores.

First, order the 15 scores. Then, find the median.

76 80 88 91 95 99 99 101 102 103 105 105 112 115 139

(101 ↑ median)

The lower quartile is the median of the lower half of the data and the upper quartile is the median of the upper half.

76 80 88 91 95 99 99 101 102 103 105 105 112 115 139

(91 ↑ lower quartile; 101 ↑ median; 105 ↑ upper quartile)

The interquartile range is 105 − 91 or 14. Therefore, the middle half, or 50%, of the golf scores vary by 14.

An outlier will not affect the median, LQ, or UQ, but it will affect the mean.

In Example 2, one score, 139, is much greater than the others. In a set of data, a value that is much higher or much lower than the rest of the data is called an **outlier.** An outlier is defined as any element of the set of data that is at least 1.5 interquartile ranges above the upper quartile or below the lower quartile.

$(1.5)(14) = 21$
$105 + 21 = 126$
$91 - 21 = 70$

To determine if 139 or any of the other scores from Example 2 is an outlier, first multiply 1.5 times the interquartile range, 14. Then add this product to the UQ, 105, and subtract it from the LQ, 91. These values give the boundaries for determining outliers. Since $139 > 126$, the score 139 is an outlier. This is the only outlier. Why?

Example 3

The stem-and-leaf plot at the right represents the number of shares of the 20 most active stocks that were bought and sold on the New York Stock Exchange during 1989.

Stem	Leaf
1	[2 2 7
2	3 [3 3] 4 4 5 6] [6 8 8 9
3	[0 1] 4 6
4	0 6]

4|0 represents 400,000,000 shares.

a. The brackets group the values in the lower half and the values in the upper half. What do the boxes contain?
values used to find the lower and upper quartiles

b. Find the interquartile range.
The median is 26. LQ $= \frac{23 + 23}{2}$ or 23 and UQ $= \frac{30 + 31}{2}$ or 30.5. The interquartile range is 30.5 − 23 or 7.5.

c. Find any outliers.

$30.5 + (1.5)(7.5) = 30.5 + 11.25$
$= 41.75$

$23 - (1.5)(7.5) = 23 - 11.25$
$= 11.75$

Since $46 > 41.75$, 46 is an outlier.

CHECKING FOR UNDERSTANDING

Communicating Mathematics

Read and study the lesson to answer each question.

1. The range is the difference between which two values in a set of data?
2. Quartiles divide a set of data into how many equal parts?
3. What measure of central tendency is affected by an outlier?
4. Which would you expect to have a greater interquartile range, the set consisting of the weights of each student in your school or the set consisting of the heights of each of these students?

Guided Practice

Find the range for each set of data.

5. 12, 17, 16, 23, 18
6. 56, 45, 37, 43, 10, 34
7. 77, 78, 68, 96, 99, 84, 65
8. 30, 90, 40, 70, 50, 100, 80, 60
9. 3, 3.2, 6, 45, 7, 26, 2, 3.4, 4, 5.3, 5, 78, 8, 1, 5

Find the median and upper and lower quartiles for each set of data.

10. 12, 17, 16, 23, 18
11. 56, 45, 37, 43, 10, 34
12. 77, 78, 68, 96, 99, 84, 65
13. 30, 90, 40, 70, 50, 100, 80, 60
14. 3, 3.2, 6, 45, 7, 26, 1, 3.4, 4, 5.3, 5, 78, 8, 21, 5

EXERCISES

Practice

Find the range, median, upper and lower quartiles, and interquartile range for each set of data.

15. 85, 77, 58, 69, 62, 73, 55, 82, 67, 77, 59, 92, 75, 69, 76
16. 1050, 1175, 835, 1075, 1025, 1145, 1100, 1125, 975, 1005, 1125, 1095, 1075, 1055
17. 211, 225, 205, 207, 208, 213, 180, 200, 210, 229, 199, 206, 212, 208, 220, 197, 211, 204, 206, 212

18.

Stem	Leaf
0	0 2 3
1	1 7 9
2	2 3 5 6
3	3 4 4 5 9
4	0 7 8 8

2|2 = 22

19.

Stem	Leaf
7	3 4 7 8
8	0 0 3 5 7
9	4 6 8
10	0 1 8
11	1 9

9|4 = 9.4

20.

Stem	Leaf
25	0 3 7 9
26	1 3 4 5 5 6
27	1 5 6 6 9
28	1 2 3 5 8
29	2 5 6 9

27|5 = 2750

Critical Thinking

21. Give an example of a set of 11 numbers that has range 60, median 50, and interquartile range 15. Does this set have an outlier?

Applications

22. **School** Art and Gina's test scores in algebra are given below.

    Art: 87, 54, 78, 97, 65, 82, 75, 68, 82, 73, 66, 75
    Gina: 70, 80, 57, 100, 73, 74, 65, 77, 91, 69, 71, 76

    a. Find the range and interquartile range for each set of scores.

    b. Identify any outliers.

    c. Which one had the more consistent scores?

23. **Foods** The stem-and-leaf plot at the right represents the cost per cup of various brands of coffee.

    a. Find the range and interquartile range for the costs.

    b. Identify any outliers.

Stem	Leaf
0	6 6 6 6 8 9 9 9 9
1	0 2 3 4 5 7 7 8 8
2	4 8 9
3	0 2

1|2 = $0.12

24. **Entertainment** The total rental sales of the all-time top 20 movies, as of January 1992, are listed below.

Rental Sales (in millions of dollars)

Title	Total Rentals	Title	Total Rentals
Back to the Future	105.5	Indiana Jones and the Last Crusade	109.0
Batman	150.5	Indiana Jones and the Temple of Doom	115.5
Beverly Hills Cop	108.0	Jaws	129.5
The Empire Strikes Back	141.6	Raiders of the Lost Ark	115.6
E.T.	228.6	Rain Man	86.8
The Exorcist	89.0	Return of the Jedi	168.0
Ghost	98.2	Star Wars	193.5
Ghostbusters	132.7	Terminator 2	112.0
The Godfather	86.3	Tootsie	94.4
Grease	96.3		
Home Alone	140.6		

   a. Find the range, quartiles, and interquartile range for the rentals.

   b. Identify any outliers.

Mixed Review

25. Solve $-5 < 4 - 3x < 13$. Graph the solution set. **(Lesson 5-5)**

26. Write the inverse of the relation $\{(-3, 3), (2, -2), (-1, 1), (0, 0), (1, -1)\}$. **(Lesson 9-2)**

27. **Geometry** The radius r, in inches, of a sphere with surface area S in^2 is given by the formula $r = \frac{1}{2}\sqrt{\frac{S}{\pi}}$, where $\pi \approx \frac{22}{7}$. Find the radius of a sphere with surface area 440 in^2. **(Lesson 12-1)**

28. Solve $35x^2 - 11x = 6$ using the quadratic formula. **(Lesson 13-5)**

29. **Consumerism** The prices of six different models of printers in a computer store are $299, $369, $525, $359, $228, and $398. Find the mean and median prices for the printers. **(Lesson 14-3)**

14-5 Box-and-Whisker Plots

Objective After studying this lesson, you should be able to:

- display and interpret data on a box-and-whisker plot.

Box-and-whisker plots are sometimes called box plots.

When analyzing a set of data, it is often helpful to draw a graphical representation of the data. One such graph, which shows the quartiles and *extreme* values of the data, is called a **box-and-whisker plot.**

In Lesson 14-2, we discussed the following 35 scores for an exam given to Mr. Juarez's first-period biology class.

82, 77, 49, 84, 44, 98, 93, 71, 76, 64, 89, 95, 78, 69, 89, 65, 88, 54, 96, 87, 92, 80, 44, 85, 93, 89, 55, 62, 79, 90, 86, 75, 74, 99, 62

Suppose Mr. Juarez wants to display this data in a box-and-whisker plot. Since the quartiles must be determined before drawing a box-and-whisker plot, he must first arrange the data in numerical order or make a stem-and-leaf plot. The stem-and-leaf plot that we made in Lesson 14-2 is shown below.

Stem	Leaf
4	[4 4 9
5	4 5
6	2 2 4 [5] 9
7	1 4 5 6 7 8 9]
8	0 [2 4 5 6 7 8 9 9 [9]
9	0 2 3 3 5 6 8 9]

8|2 represents a score of 82.

The median for this set of data is 80. The lower quartile is 65 and the upper quartile is 89. The *extreme values* are the least value (LV), 44, and the greatest value (GV), 99.

The number line may be drawn horizontally or vertically.

To make the box-and-whisker plot, first draw a number line. Assign a scale to the number line that includes the extreme values. Plot dots to represent the extreme values (LV and GV), the upper and lower quartiles (LQ and GQ), and the median (M).

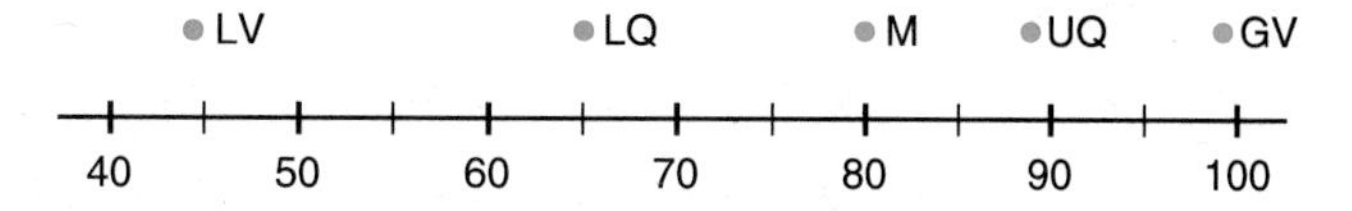

Draw a box around the interquartile range. Mark the median by a vertical line through its point in the box. The median line will not always divide the box into equal parts. Draw a segment from the lower quartile to the least value and one from the upper quartile to the greatest value. These segments are the *whiskers* of the plot.

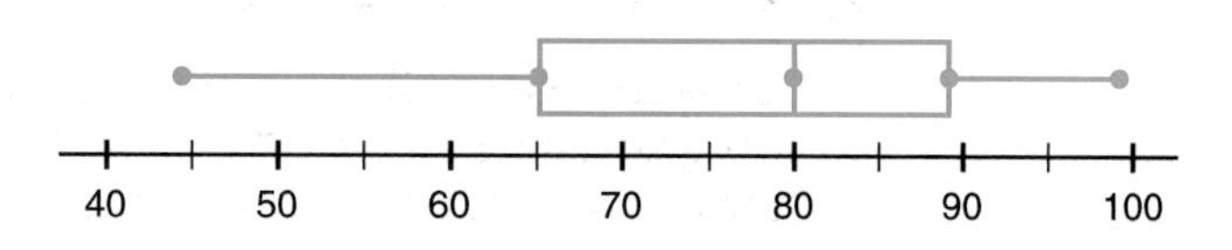

Even though the whiskers are different lengths, each whisker contains at least one-fourth of the data while the box contains at least one-half of the data.

Now check for outliers. The interquartile range of these scores is 89 − 65 or 24.

$$65 - (1.5)(24) = 65 - 36 = 29 \qquad 89 + (1.5)(24) = 89 + 36 = 125$$

Since none of the test scores is above 125 or below 29, there are no outliers.

If a set of data contains outliers, then a box-and-whisker plot can be altered to show them. This is shown in the following example.

Example

Twelve members of the Beck High School Pep Club are selling programs at the football game. The number of programs sold by each person is listed below.

George	**51**	**Anthony**	**27**	**Kendra**	**55**	**Eddie**	**54**
Vinette	**69**	**Marlene**	**60**	**Carmen**	**39**	**Jason**	**46**
Tomás	**46**	**Kohana**	**53**	**Danny**	**81**	**Nashoba**	**23**

a. Make a box-and-whisker plot of this data.

First arrange the data in numerical order.

[23 27 | 39 46 | 46 51] [53 54 | 55 60 | 69 81]

The extreme values are 23 and 81. The median is $\frac{51 + 53}{2}$ or 52. The lower quartile is $\frac{39 + 46}{2}$ or 42.5. The upper quartile is $\frac{55 + 60}{2}$ or 57.5. Thus, the interquartile range is 57.5 − 42.5 or 15.

Now check for outliers.

$$57.5 + (1.5)(15) = 57.5 + 22.5 = 80 \qquad 42.5 - (1.5)(15) = 42.5 - 22.5 = 20$$

Since 81 > 80, the value 81 is an outlier.

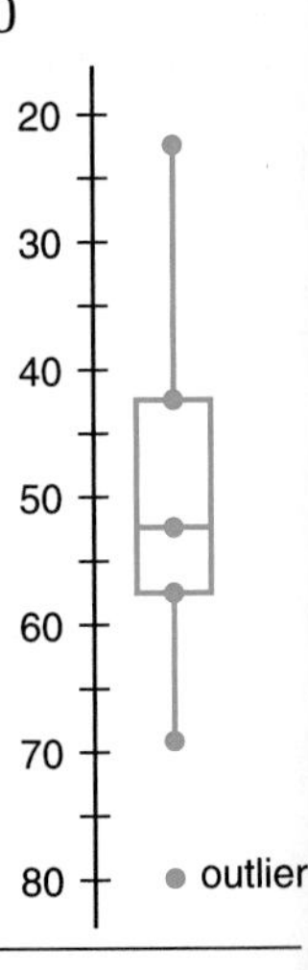

The box-and-whisker plot for this set of data is shown at the right. A point is plotted for 69 since it is the last value that is not an outlier. The whisker is drawn to this point as shown. Outliers are plotted as isolated points.

b. Analyze the box-and-whisker plot to determine if any of the members did an exceptional job selling programs.

Based on this plot, Danny did an exceptionally good job selling programs.

CHECKING FOR UNDERSTANDING

Communicating Mathematics

Read and study the lesson to answer each question.

1. Explain how to determine the scale of the number line in a box-and-whisker plot.
2. The two whiskers in a box-and-whisker plot connect what values?
3. What percent of the data is included in the box of a box-and-whisker plot?
4. What information about a set of data can you determine from its box-and-whisker plot?

Guided Practice

Answer each question for the indicated box-and-whisker plot.

5. What percent of the data is between 120 and 130?
6. What is the median?
7. What is the least value in this set of data?
8. Between what two values of the data is the middle 50% of the data?

80 90 100 110 120 130

Exercises 5-8

9. What is the upper quartile?
10. What is the greatest number in this set of data?
11. What percent of the data is between 50 and 90?
12. Why does this plot have only one whisker?

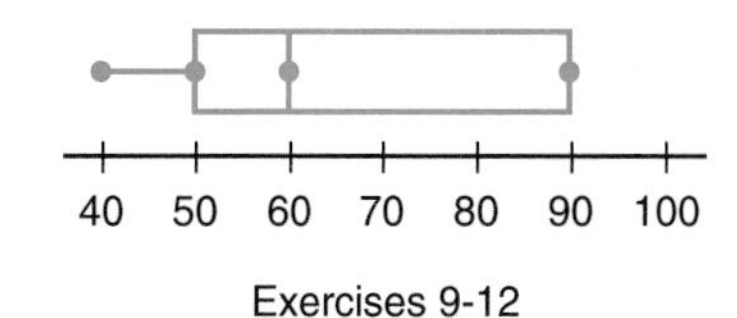

Exercises 9-12

EXERCISES

Practice

Compare box-and-whisker plots X and Y to answer each question.

13. Which plot has the lesser median?
14. Which plot has the greater range?
15. Which plot has the lesser interquartile range?

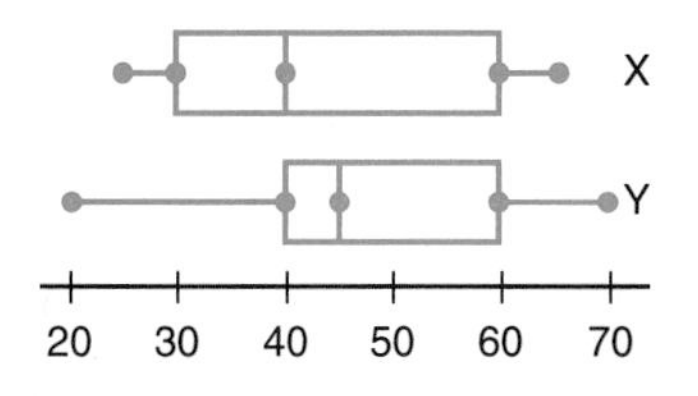

Applications

16. **Travel** Speeds of the fastest train runs in the U.S. and Canada are given below in miles per hour. Make a box-and-whisker plot of the data.

93.5	82.5	89.3	83.8	81.8	86.8
90.8	84.9	95.0	83.1	83.2	88.2

17. **Weather** The following high temperatures were recorded during a two-week cold spell in St. Louis. Make a box-and-whisker plot of the temperatures.

20° 2° 12° 5° 4° 16° 17° 7° 6° 16° 5° 0° 5° 30°

18. **Basketball** The numbers below represent the 20 highest points-scored-per-game averages for a season in the NBA from 1947 to 1990. Make a box-and-whisker plot of this data.

35.0 33.5 32.5 37.9 31.2 38.4 34.5 34.7 32.9 44.8
31.7 37.1 36.5 34.0 50.4 32.3 33.6 34.8 33.1 35.6

19. **Baseball** The stem-and-leaf plot at the right shows the number of home runs hit by the home run leaders in the National League in 1990. Make a box-and-whisker plot of the data.

Stem	Leaf
2	2 3 3 4 4 4 4 5 5 6 7 7 8
3	2 2 3 3 5 7
4	0

3|3 = 33

20. **Consumerism** Ten light bulbs were purchased from each of two different manufacturers, A and B. The bulbs were tested to determine how many hours of light they would provide. The results are given below.

A: 290, 300, 497, 395, 450, 360, 740, 500, 520, 370
B: 400, 410, 460, 450, 350, 495, 375, 405, 485, 520

a. Make a box-and-whisker plot comparing the data.

b. Based on this plot, from which manufacturer would you buy your light bulbs? Explain.

21. **Health** The table below shows the number of years of life expected for men and women born in various years from 1920 to 1989. Make a box-and-whisker plot for the men's data and for the women's data. Then compare the plots.

Year	Men	Women	Year	Men	Women	Year	Men	Women
1920	53.6	54.6	1960	66.6	73.1	1980	70.0	77.5
1930	58.1	61.6	1965	66.8	73.7	1985	71.2	78.2
1940	60.8	65.2	1970	67.1	74.7	1989	71.8	78.5
1950	65.6	71.1	1975	68.8	76.6			

Critical Thinking

22. Name a set of data with ten numbers for which its box-and-whisker plot would have no whiskers.

Mixed Review

23. Solve $29 - 3a = 2(3a - 4) + 3$. **(Lesson 6-7)**

24. Graph $2x - y = 6$. **(Lesson 9-4)**

25. Solve the system $x + y = 4$ and $3x - 5y = 60$. **(Lesson 11-3)**

26. Simplify $\frac{1}{7 - \sqrt{3}}$. **(Lesson 12-5)**

27. Find the discriminant and determine the nature of the roots of $3n^2 - n - 5 = 0$. **(Lesson 13-6)**

28. Use the information from Exercise 18 to find the range, quartiles and interquartile range of the 20 highest points-scored-per-game averages in the NBA. Identify any outliers. **(Lesson 14-4)**

14-6 Scatter Plots

Objective

After studying this lesson, you should be able to:

- graph and interpret pairs of numbers on a scatter plot.

Application

A branch of statistics called *regression analysis* deals with finding functions that approximate the relationship between sets of data.

Each student in Ms. Suber's sociology class was given the assignment of conducting a survey and interpreting the results. Jenelle decided to conduct a survey to determine whether there was a relationship between the amount of time spent studying for a test and the grade earned on the test. To do this, she asked 15 classmates in her history class to record the amount of time that they studied for an upcoming test. Then she recorded their test scores. She organized the results in the table shown at the right.

Student	Study Time	Test Score
Justin	60 min	92
Penny	55 min	79
Doug	10 min	65
Brad	75 min	87
Carmen	120 min	98
Yoshica	90 min	95
Greg	110 min	75
Allison	45 min	73
Montega	30 min	77
Shelley	95 min	94
Emilio	60 min	83
Cecilia	15 min	68
Latoya	35 min	79
Ben	70 min	78
Hanna	25 min	97

Based on this information, can she conclude that there is a relationship between study time and test scores?

In order to determine if there is a relationship, Jenelle needs to compare study times and test scores. One way to do this is to display the information in a graph called a **scatter plot.** In a scatter plot, the two sets of data are plotted as ordered pairs in the coordinate plane. Let the horizontal axis represent the study time, and let the vertical axis represent the test score. Then plot the data. For example, the information about Justin is plotted as the point (60, 92).

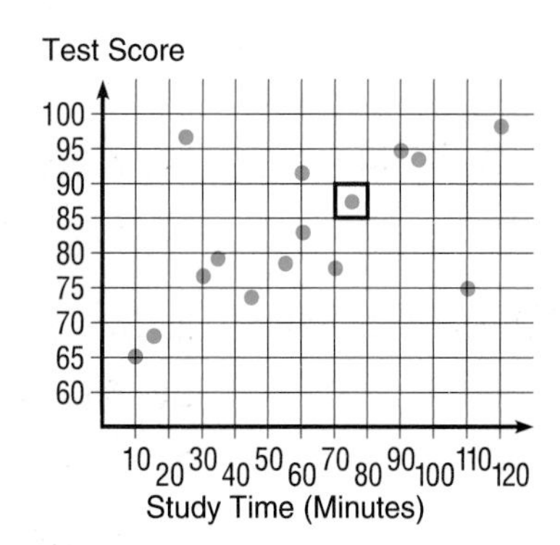

a. Describe the point with the box around it.

It represents a study time of 75 minutes with a resulting test score of 87.

b. Who studied quite a bit and had a fairly low test score?

Greg

c. Who did not study much and had a fairly high test score?

Hanna

d. Does the scatter plot show a relationship between study time and test scores?

In general, the scatter plot seems to show that a higher grade is directly related to the amount of time spent studying.

When data is displayed in a scatter plot, usually the purpose is to determine if there is a pattern, or association, between the variables on the graph. If an association does exist, it may be *negative* or *positive*. In the application on the previous page, if you could draw a line that is suggested by the points in the scatter plot, it would have a positive slope. Thus, the association between study time and test scores is said to be positive.

Example 1

The scatter plot below compares the number of hours per week people watched television to the number of hours per week they spent doing some physical activity. Does there appear to be any association between time spent watching television and physical activity?

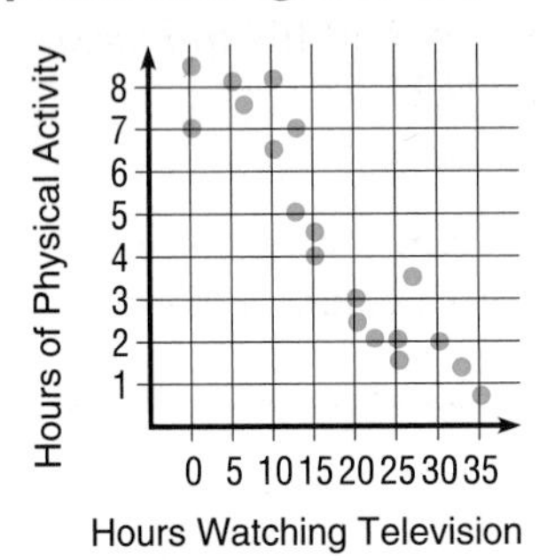

The plot seems to indicate that the greater the amount of time spent watching television, the less the amount of time spent doing physical activity. Thus, there appears to be a *negative* association between watching television and physical activity.

Example 2

Each week, Ms. Suber gives a 50-point quiz in her sociology class. The scatter plots below compare the week the quizzes were taken to the quiz scores of two students, Mike and Melanie. Does there appear to be any association between the week they took the quiz and their quiz score?

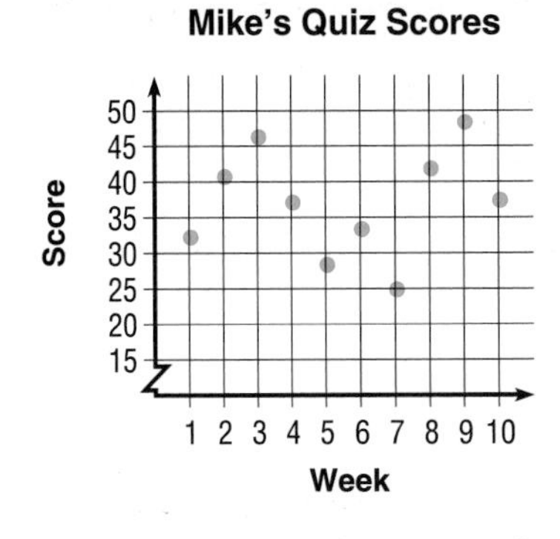

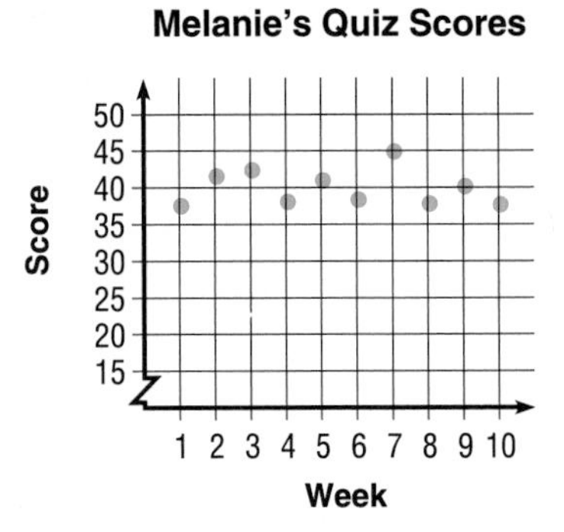

In the scatter plot on the left, the points are very spread out. There appears to be no association between the week Mike took the quiz and his score. In the scatter plot on the right, there appears to be an association between the week Melanie took the quiz and her score. However, it is impossible to tell whether this association is positive or negative.

Even when a scatter plot shows no association between two sets of data, the plot can still be used to answer questions about the data. For example, you can use the scatter plot in Example 2 to determine how many weeks Mike scored over 40 points.

CHECKING FOR UNDERSTANDING

Communicating Mathematics

Read and study the lesson to answer each question.

1. Describe how to draw a scatter plot for two sets of data.
2. How do you determine from a scatter plot whether there appears to be any association between the two sets of data plotted?
3. *True* or *false:* If two sets of data do not show a positive association in a scatter plot, then they will show a negative association.
4. In Example 2, suppose there was a positive association between the week Mike took the quiz and his quiz score. What would this tell you about Mike's quiz scores?

Guided Practice

Determine whether a scatter plot of the data for the following would show positive, negative, or no association between the variables.

5. amount of money earned and spent
6. heights of sons and fathers
7. speed of car and miles per gallon
8. age of car and its value
9. your height and month of birth
10. playing time and points scored

EXERCISES

Practice

Which graphs show an association between the variables? If there is an association, is it positive, negative, or not possible to tell?

11.

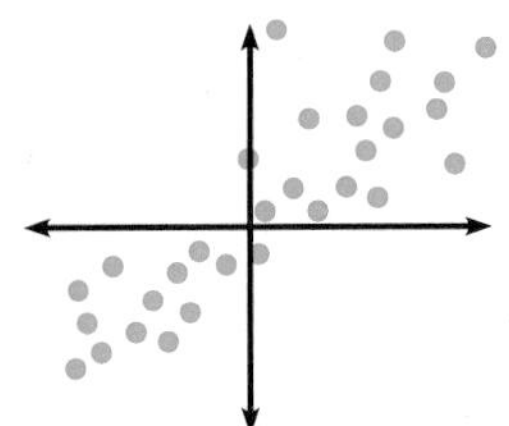

12.

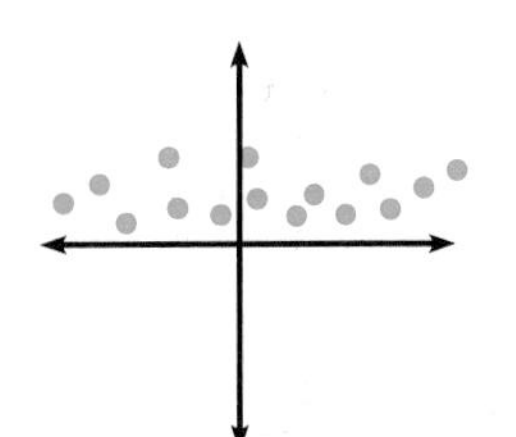

13.

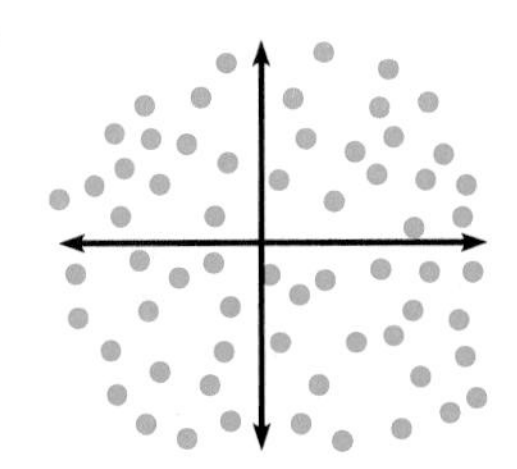

14. Collect data from your classmates on the number of hours per week they watch television and the number of hours per week they spend studying. Make a scatter plot of the data.

Applications

15. **Finance** The table below shows the annual income and the number of years of college education for eleven people.

Income (thousands)	\$23	\$20	\$25	\$47	\$19	\$48	\$35	\$10	\$39	\$26	\$36
College Education (years)	3	2	4	6	2.5	7.5	6.5	1	5.5	4.5	4

a. Make a scatter plot of the data.
b. Based on this plot, how does the number of years of college affect income?

16. **Weather** The maximum and minimum monthly temperatures in July of ten cities are given in the table below.

a. Make a scatter plot of the data.

b. Which city has the highest maximum and minimum temperatures?

c. Which city has the lowest minimum and maximum temperatures?

d. Is there a cluster of points in the plot? If so, explain why.

City	Maximum	Minimum
Hartford, CT	85	62
Baltimore, MD	87	67
Boston, MA	82	65
Portland, ME	79	57
Concord, NH	83	56
Albany, NY	83	60
New York, NY	84	69
Burlington, VT	81	59
Norfolk, VA	90	70
Huntington, WV	86	65

17. **Hockey** The number of points scored and assists made by members of the 1990 Stanley Cup Champion Edmonton Oilers is given in the table.

a. Make a scatter plot of the data.

b. Does the scatter plot show a positive or negative association?

c. Could you predict the number of assists a player would have if you were given the number of goals for that player?

d. Do you think that the position of a player affects the number of points scored by that player?

e. Do you think that the amount of time played affects the number of points scored by a player?

Player	Goals	Assists
Messier	45	84
Kurri	33	60
Anderson	34	38
Klima	30	33
Tikkanen	30	33
Simpson	29	32
MacTarish	21	22
Smith	7	34
Lowe	7	26
Murphy	10	19
Lamb	12	16
Gelinas	17	8
Gregg	4	20
Muni	5	12

18. **Golf** The earnings of the leading money winners on the PGA and LPGA golf tours from 1978–1989 are given in the table below.

Year	PGA Earnings (in thousands)	LPGA Earnings (in thousands)	Year	PGA Earnings (in thousands)	LPGA Earnings (in thousands)
1978	362	190	1984	476	267
1979	463	216	1985	542	416
1980	531	231	1986	653	492
1981	376	207	1987	926	466
1982	446	310	1988	1148	347
1983	427	291	1989	1395	654

a. Make a scatter plot of each set of data, using the year as the first variable.

b. What do the scatter plots indicate?

19. **School** The table below shows the typing speeds of 12 students.

Typing Speed (wpm)	33	45	46	20	40	30	38	22	52	44	42	55
Experience (weeks)	4	7	8	1	6	3	5	2	9	6	7	10

a. Make a scatter plot of this data.

b. Draw a line that "fits" the data in the scatter plot. Find an equation of the line.

c. Use the equation to predict the typing speed of a student after a 12-week course.

Critical Thinking

20. Use the data from Exercise 18 to make a scatter plot comparing LPGA and PGA earnings from 1978 to 1989. What does the association between the variables in this scatter plot indicate?

Mixed Review

21. Write an algebraic expression for *25 less than six times the square of a number.* **(Lesson 1-1)**

22. **Aviation** The top flying speed of an open cockpit biplane is 120 mph. At this speed, a 420-mile trip flying with the wind takes the same amount of time as a 300-mile trip flying against the wind. What is the speed of the wind? **(Lesson 8-9)**

23. Graph $5x + 2 = 7y$. **(Lesson 10-4)**

24. **Geometry** Find the dimensions of a rectangle with a perimeter of 15.4 cm and an area of 14.4 cm^2? **(Lesson 13-7)**

25. Use the data from Exercise 18 to make a box-and-whisker plot for the earnings of the leading money winner on the PGA tour from 1978 to 1989. **(Lesson 14-5)**

Journal

Write a few sentences to tell how changing the scale on either axis might make the scatter plot appear differently.

MID-CHAPTER REVIEW

The table below shows the normal seasonal precipitation, in inches, for selected cities in the U.S. Use this data to complete Exercises 1–6.

Season	Albuquerque	Boston	Chicago	Denver	Houston	Kansas City	New Orleans	San Francisco
Winter	1.3	11.8	5.5	2.4	9.2	4.1	14.9	10.5
Spring	1.4	10.1	11.0	5.9	12.9	10.2	14.2	1.9
Summer	3.7	9.8	10.5	4.6	11.9	10.0	18.6	0.3
Fall	1.8	12.5	6.5	2.6	10.8	4.8	12.1	7.1
Totals	8.2	44.2	33.5	15.5	44.8	29.1	59.8	19.8

1. What are the greatest and the least seasonal precipitation for any of the cities? **(Lesson 14-1)**
2. Make a line plot of the winter precipitation for the eight cities. **(Lesson 14-1)**
3. Make a stem-and-leaf plot of the fall precipitation for the eight cities. **(Lesson 14-2)**
4. Find the mean, median, and mode of the spring precipitation for the eight cities. **(Lesson 14-3)**
5. Find the range, quartiles, and interquartile range of the summer precipitation in the eight cities. Identify any outliers. **(Lesson 14-4)**
6. Make a box-and-whisker plot of the total precipitation for the eight cities. **(Lesson 14-5)**
7. Make a scatter plot of the following data. **(Lesson 14-6)**

Miles Driven	200	322	250	290	310	135	60	150	180	70	315	175
Fuel Used (gallons)	7.5	14	11	10	10	5	2.3	5	6.2	3	11	6.5

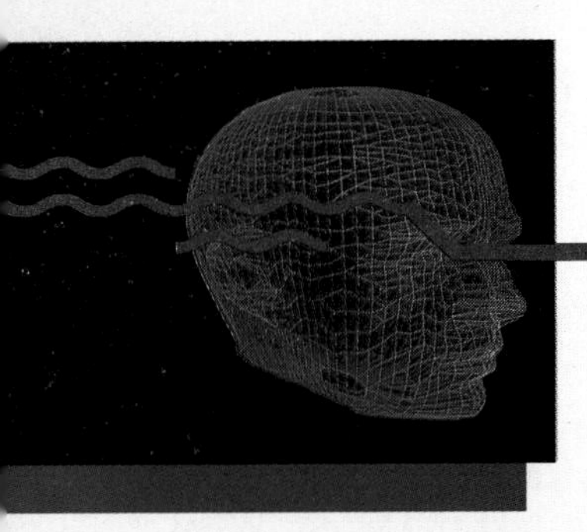

Technology

Regression Lines

BASIC
▶ Graphing calculators
Graphing software
Spreadsheets

Is the number of police officers related to the number of crimes? Is the amount of time spent studying related to test scores? Each of these relationships involves two variables. Therefore, they can be denoted with ordered pairs of the form (x, y). The ordered pairs of data can be graphed on a coordinate system as points. If the points lie on the same line or close to a line, they are said to be *linearly related.* A **regression line** is a line that best fits data that are linearly related.

Example **Graph the ordered pairs of data in the table and draw a regression line.**

x	−8	−6	2	4	−1
y	−4	3	−1	5	9

	T1-81	*Casio*
1. Set the range values as desired.	(Possible choice: [−10, 10] by [−5, 15].)	
2. Clear the graphics screen.	2nd DRAW 1 ENTER	Cls EXE
3. Set the mode and clear the statistical memories.	2nd STAT ◀ 2 ENTER 2nd STAT ◀ ENTER	SHIFT MODE ÷ SHIFT Scl EXE
4. Enter the data.	(-) 8 ENTER (-) 4 ENTER (-) 6 ENTER 3 ENTER 2 ENTER (-) 1 ENTER 4 ENTER 5 ENTER (-) 1 ENTER 9 ENTER	(-) 8 SHIFT , (-) 4 DT (-) 6 SHIFT , 3 DT 2 SHIFT , (-) 1 DT 4 SHIFT , 5 DT (-) 1 SHIFT , 9 DT
5. Graph the regression line and scatter plot.	2nd STAT ▶ 2 ENTER 2nd STAT 2 ENTER Y= VARS ▶ ▶ 4 ENTER GRAPH 2nd STAT ▶ 2 ENTER	GRAPH SHIFT LINE 1 EXE

EXERCISES

Graph each set of data and draw a regression line.

1.

x	−2	−1	2	3	4	7
y	1	−3	−2	5	2	6

2.

x	−3	−3	−2	−1	1	2
y	6	3	5	1	−1	−1

14-7 Probability and Odds

Objectives

After studying this lesson, you should be able to:

- find the probability of a simple event, and
- find the odds of a simple event.

Application

Denise is one of 32 students entered in the Paradise Valley Chess Club's annual tournament. Of the 32 students, 7 are freshmen, 6 are sophomores, 9 are juniors, and 10 are seniors. If Denise is a junior, how likely is it that her first opponent is also a junior?

The probability of an event may be written as a percent, a fraction, or a decimal.

We do not know who Denise's first opponent will be. When we are uncertain about the occurrence of an event, we can measure the chances of its happening with **probability.** The probability of an event is a ratio that tells how likely it is that an event will take place. The numerator is the number of favorable outcomes and the denominator is the number of possible outcomes. For example, suppose you want to know the probability of getting a 2 on one roll of a die. When you roll a die, there are six possible outcomes. Of these outcomes, only one is favorable, a 2. Therefore, the probability is $\frac{1}{6}$. We write $P(2)$ to represent *the probability of getting a 2 on one roll of a die.*

Definition of Probability

$$P(\text{event}) = \frac{\text{number of favorable outcomes}}{\text{number of possible outcomes}}$$

Example 1

Determine the probability that Denise's first opponent in the chess tournament is a junior to answer the problem presented above.

Since there are 32 students, Denise has 31 possible opponents. Since she is a junior, there are only 8 juniors left to be her first opponent.

$$P(\text{junior}) = \frac{\text{number of juniors}}{\text{total number of opponents}} \quad \frac{\textit{number of favorable outcomes}}{\textit{number of possible outcomes}}$$

$$= \frac{8}{31}$$

The probability that Denise's first opponent is a junior is $\frac{8}{31}$.

Example 2

Diego has a collection of tapes that he plays regularly. He has 10 rock tapes, 5 jazz tapes, and 8 country tapes. If Diego chooses a tape at random, what is the probability that he will choose a jazz tape?

$$P(\text{jazz tape}) = \frac{\text{number of jazz tapes}}{\text{total number of tapes}} \quad \frac{\textit{number of favorable outcomes}}{\textit{number of possible outcomes}}$$

$$= \frac{5}{23}$$

The probability that Diego chooses a jazz tape is $\frac{5}{23}$.

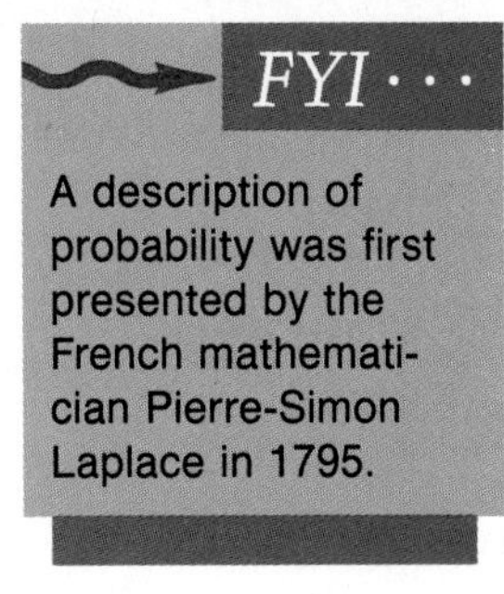

A description of probability was first presented by the French mathematician Pierre-Simon Laplace in 1795.

Some outcomes have an equal chance of occurring. We say that such outcomes are **equally likely.**

What are the possible values for the probability of an event? Consider the following cases.

1. Suppose it is impossible for an event to occur. This means that there are no favorable outcomes. Thus, the probability of an impossible event must be 0.
2. Suppose an event is certain to occur. This means that every possible outcome is a favorable outcome. Thus, the probability of a certain event must be 1.
3. Suppose an event is neither impossible nor certain. This means that the number of favorable outcomes is greater than 0 but less than the number of possible outcomes. Thus, the probability of this event must be greater than 0 and less than 1.

Based on these three cases, we can conclude that the probability of any event is always a value between 0 and 1, inclusive. This can be expressed as $0 \leq P(\text{event}) \leq 1$.

Another way to measure the chance of an event's occurring is with **odds.** The odds of an event is the ratio that compares the number of ways an event can occur to the number of ways it *cannot* occur.

Definition of Odds

The odds of an event occurring is the ratio of the number of ways the event can occur (successes) to the number of ways the event cannot occur (failures).

Odds = number of successes : number of failures

Example 3

The Southside Youth Center is having a raffle to make money for a new gymnasium. The Ryder family bought 10 raffle tickets, two for each family member. If 1000 raffle tickets are sold, what are the odds that a member of the Ryder family will win the raffle?

The Ryder family has 10 of the 1000 tickets. Thus, there are 1000 − 10 or 990 tickets that will not be winning tickets for the Ryder family.

$$\text{Odds of winning} = \frac{\text{number of chances of}}{\text{drawing winning ticket}} : \frac{\text{number of chances of}}{\text{drawing other tickets}}$$

$= 10{:}990$ or $1{:}99$ *This is read "1 to 99."*

Example 4

The Channel 8 weather forecaster states that the probability of rain tomorrow is 40% or $\frac{2}{5}$. Find the odds that it will not rain tomorrow.

If the probability of rain tomorrow is $\frac{2}{5}$, then the number of failures (rain) is 2, while the total number of outcomes is 5. This means that the number of successes (no rain) must be 5 − 2 or 3.

Odds of no rain = number of successes : number of failures

$= 3{:}2$ *This is read "3 to 2."*

CHECKING FOR UNDERSTANDING

Communicating Mathematics

Read and study the lesson to answer each question.

1. Give three examples of events that have *equally likely* outcomes.
2. Suppose you roll a die. Give examples of an impossible event and a certain event.
3. If the odds that an event will occur are 3:5, what are the odds that the event will *not* occur?
4. Nick says that the probability of getting heads in one toss of a coin is 1 out of 2. Felicia says that the odds of getting heads in one toss of a coin are 1 to 1. Who is correct?

Guided Practice

The probability of an event can be graphed on a number line like the one shown below. Copy the number line and then graph the probability of each event.

0		$\frac{1}{2}$		1
Impossible	Unlikely	50-50 Chance	Likely	Certain

5. It will rain today.
6. You will pass your next test.
7. This is an algebra book.
8. A coin will land tails up.
9. Today is Friday.
10. You will go skiing tomorrow.

Determine if each event described below has equally likely outcomes.

11. tossing a fair coin
12. passing a test
13. winning a golf game
14. rolling a die
15. winning a lottery
16. choosing a soft drink at random from a machine

EXERCISES

Practice

Find the probability of each outcome if a die is rolled.

17. a 3
18. a number less than 1
19. an even number
20. a number greater than 1

Find the odds of each outcome if a die is rolled.

21. a multiple of 3
22. a number greater than 3
23. not a 2
24. a number less than 5

A card is selected at random from a deck of 52 cards.

25. What is the probability of selecting a black card?
26. What is the probability of selecting a king?
27. What are the odds of selecting a club?
28. What are the odds of *not* selecting a red 7?
29. What is the probability of selecting a club *or* an ace?

30. If the probability that an event will occur is $\frac{2}{3}$, what are the odds that the event will occur?

31. If the probability that an event will occur is $\frac{3}{7}$, what are the odds that the event will *not* occur?

32. If the odds that an event will occur are 8:5, what is the probability that the event will occur?

33. If the odds that an event does *not* occur are 9:14, what is the probability that the event does occur?

34. The number of males and females enrolled in each grade at Oak Grove High School is given in the table below. You are conducting a survey on reading habits for the school newspaper by selecting students at random.

    a. What is the probability that a student chosen is a female?

Grade	9th	10th	11th	12th
Male	130	150	100	120
Female	150	100	110	175

    b. What is the probability that a student chosen is a male in the 9th grade?
    c. What are the odds that a student chosen is in the 10th grade?
    d. If a student from the 11th grade is selected, is it more likely that the student is a male or a female?

35. What is the probability that a number chosen at random from the domain {0, 1, 2, 3, 4, 5, 6, 7, 8, 9} will satisfy the inequality $2x + 3 < 17$?

36. If a three-digit number is selected at random from the set of all three-digit numbers, what are the odds that all three of the digits will be prime numbers?

Critical Thinking

37. A butterfly lands on one of the six squares of the T-shaped figure shown and then randomly moves to an adjacent square. What is the probability that the butterfly ends up on the red square?

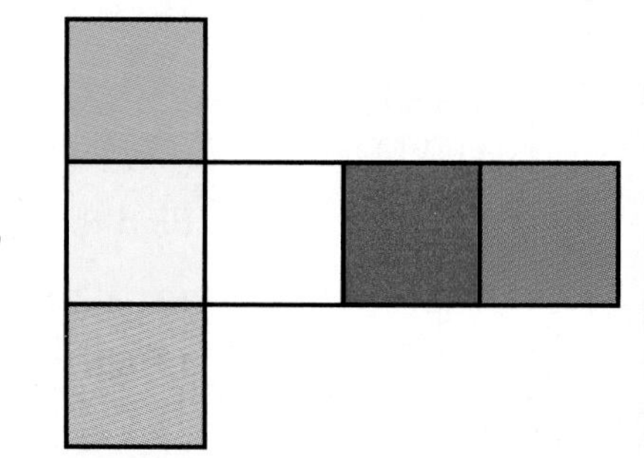

Applications

38. **Entertainment** A reception is scheduled for an open area of Spring Grove Park, and a tent has been ordered in case of rain. Weather reports indicate a 50% chance of rain on the day of the reception. Whether or not it rains, there is a 50% chance the tent will not arrive in time for the reception due to traffic problems. What is the probability the people at the reception will get wet?

39. **Consumerism** A shipment of 100 clocks was just received by The Clock Shoppe. There is a 4% probability that one of the clocks was damaged during shipment, even though the package does not give any indication that the clock is damaged. If Alison buys one of these clocks, what are the odds that she is buying a damaged clock?

40. **Entertainment** The freshman class of Perry High School is planning an end-of-the-year dance. The dance committee decides to award door prizes. They want the odds of winning a prize to be 1:5. If they plan for 180 tickets to be sold for the dance, how many prizes will be needed?

Mixed Review

41. **Finance** Dolores earned \$340 in 4 days by mowing lawns and doing yardwork. At that rate, how long will it take her to earn \$935? **(Lesson 4-8)**
42. Solve the inequality $|x - 1| < y$ by graphing. **(Lesson 11-6)**
43. **Recreation** Frank is competing in a kite-flying contest. At the time of the judging, he lets out 250 feet of string. The judge standing directly below Frank's kite is 70 feet away from Frank. How high is Frank's kite? **(Lesson 12-3)**
44. Locate the real roots of $x^2 - 8x + 18 = 0$ by graphing. **(Lesson 13-2)**
45. The table below shows the heights and weights of each of 12 players on a pro basketball team. **(Lesson 14-6)**

Height (inches)	75	82	75	74	80	80	75	79	80	78	76	81
Weight (pounds)	180	235	184	185	230	205	185	230	221	195	205	215

a. Make a scatter plot of this data.
b. Describe the association between height and weight.

HISTORY CONNECTION

Graunt and Bernoulli

Statistical methods have been developed in only the last 350 years, largely as a result of researchers in other fields needing better methods for analyzing data. John Graunt, for example, collected records of births and deaths in London. He published his findings in *Natural and Political Observations* in 1662. Graunt, a merchant, was one of the first to study population in this way.

Probability has also only recently been studied by mathematicians. Jacob Bernoulli (1654–1705) is considered to be the founder of probability. He wrote *The Art of Conjecturing,* which was published in 1713, eight years after his death. This book showed how probability theory can be applied to a great number of collected statistical data.

Jacob Bernoulli

14-8 Empirical Probability

Objective After studying this lesson, you should be able to:

- conduct and interpret probability experiments.

Application Mike wanted to know what was the most popular color for cars in his town. He decided to conduct an experiment by looking at cars at two intersections. The number of cars of each color was counted and recorded in the table below.

Color	Intersection 1	Intersection 2	Total
Blue	21	19	40
Brown	16	19	35
Red	12	15	27
Black	9	15	24
White	8	7	15
Green	4	5	9
Totals	70	80	150

Using the results of his experiment, Mike computed the following probabilities.

$$P(\text{red car}) = \frac{\text{number of red cars}}{\text{total number of cars}} = \frac{27}{150} \text{ or } 18\%$$

$P(\text{brown car}) = \frac{35}{150}$ or $23.\overline{3}\%$ $\quad P(\text{blue car}) = \frac{40}{150}$ or $26.\overline{6}\%$

$P(\text{black car}) = \frac{24}{150}$ or 16% $\quad P(\text{green car}) = \frac{9}{150}$ or 6%

$P(\text{white car}) = \frac{15}{150}$ or 10%

Based on these results, which color car would you most likely see in Mike's town? Which color car would you least likely see?

Probability calculated by making observations or conducting experiments, like the one Mike did, is called **empirical probability.** Empirical probabilities are not exact, since the results may vary when the experiment is repeated. For example, Mike's result might have been different if he had chosen two different intersections or even if he had counted cars at the same intersections on a different day. One way to make an empirical probability more accurate is to gather a lot of data.

As the number of trials in an experiment increases, the empirical probability will get closer to the probability that is expected. For example, suppose you are tossing a fair coin to determine the probability of getting

A coin or die is called fair if each of its outcomes are equally likely.

a tail. Then the more times you toss the coin, the closer your empirical probability should get to the expected probability of $\frac{1}{2}$.

One important use of empirical probability involves making predictions about a large group of people based on the results of a poll or survey. This technique, called *sampling,* is used when it is impractical or impossible to question every member of the group.

Example

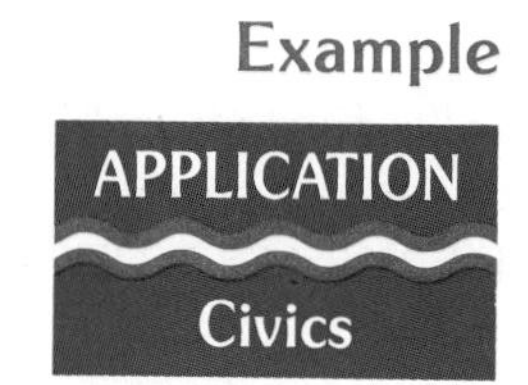

Mrs. Roberts gave her civics class the assignment of determining the candidate for student body president preferred by students in the upcoming election. The class randomly chose 150 students and asked who they preferred, Jane or Troy. Use the results of the survey, shown at the right, to answer each question.

Preference for Student Body President		
Students	**Jane**	**Troy**
Male	38	42
Female	54	16

a. **If a female student is chosen at random, what is the probability that she prefers Jane?**

The poll included 54 + 16 or 70 females. Since 54 of 70 females preferred Jane, the probability is $\frac{54}{70}$ or about 0.77.

b. **A student selected at random prefers Jane. What is the probability that this student is male?**

In the poll, 38 + 54 or 92 students preferred Jane. Since 38 of the 92 were male, the probability is $\frac{38}{92}$ or about 0.41.

c. **If a student is chosen at random, what is the probability that the student favors Troy?**

Of the 150 students polled, 42 + 16 or 58 students preferred Troy. Thus, the probability is $\frac{58}{150}$ or about 0.39.

CHECKING FOR UNDERSTANDING

Communicating Mathematics

Read and study the lesson to answer each question.

1. What is empirical probability?
2. If you toss a fair coin ten times, should the result always be 5 heads and 5 tails? Why?
3. Suppose you roll a die ten times and get a 6 all ten times. Can you say this die is *not* fair? Why or why not?

Guided Practice

4. Ten students in Mr. Howard's algebra class believe that a coin is not fair. To test this assumption, each one tossed the coin 30 times and recorded results as shown in the table below.

Student	1	2	3	4	5	6	7	8	9	10
Number of Heads	21	22	18	20	21	21	19	22	18	19
Number of Tails	9	8	12	10	9	9	11	8	12	11

a. Find *P*(heads) for each of the ten students.
b. Based on this data, does the coin appear to be fair?
c. Give an estimate for *P*(head) for this coin. (*Hint:* you could use the mean, median, or mode of this data.)
d. How can you get a better estimate of *P*(heads) for this coin?

EXERCISES

Practice

5. Roll two dice at least 100 times.
   a. Record the sums of the faces.
   b. What is the probability of a sum of 3?
   c. What is the probability of a sum of 7?
   d. What is the probability of a sum less than 5?
   e. What is the probability of a sum greater than 6?
   f. Which sum occurred most often? Is this what you would have expected? Why or why not?
   g. Which sum occurred least often? Is this what you would have expected? Why or why not?
6. Toss three coins at least 100 times.
   a. Record the number of heads and tails.
   b. What is the probability of tossing three heads?
   c. What is the probability of tossing exactly one tail?
   d. What is the probability of tossing at least two tails?
   e. What is the probability of tossing at most two heads?
   f. What result occurred most often? Is this what you would have expected? Why or why not?
   g. What result occurred least often? Is this what you would have expected? Why or why not?
7. From a deck of 52 cards, remove the ace to 10 from one suit. Shuffle the 42 cards and then deal out one card. Record the result, replace the card, and shuffle again. Repeat the experiment 100 times.
   a. What is the probability of getting the same card twice in a row?
   b. What is the probability of getting the same card three times in a row?
   c. What is the probability of getting an ace, then a two, then a three, and so on until the tenth card drawn is a ten?

Critical Thinking

8. In a survey done by Yogurt and Stuff, 415 of the 700 people interviewed said they liked blueberry frozen yogurt, 269 people said they liked chocolate swirl frozen yogurt, and 124 people said they liked both flavors. Based on this survey, what is the probability that a person dislikes both of these flavors?

Applications

9. **Baseball** Of the 12 National League baseball stadiums, 6 have natural grass. Of the 14 American League baseball stadiums, 10 have natural grass. If a game on television is being played at a stadium with natural grass, what are the odds that it is an American League stadium?

10. **Employment** The table below gives the employment status of the students at Monroe High School. Based on this data, find the probability that a student selected at random who works after school is a senior and the probability that a senior selected at random works after school. Should these probabilities be equal?

Employment	Freshman	Sophomore	Junior	Senior
Does not work	115	76	45	28
Works after school	45	54	65	72

Computer

11. The BASIC program at the right simulates a random experiment in which an outcome is chosen (with replacement) from among equally likely outcomes. First enter the number of outcomes. Then enter the number of trials to be run. In the example below, there were 5 equally likely outcomes. The experiment was run for 200 trials.

OUTCOME	FREQUENCY
1	37
2	41
3	39
4	44
5	39

```
10 INPUT "ENTER THE NUMBER
   OF EQUALLY LIKELY
   OUTCOMES: ";N
20 INPUT "ENTER THE NUMBER
   TRIALS: ";K
30 FOR X = 1 TO K
40 LET P = INT(N * RND(1) + 1)
50 LET F(P) = F(P) + 1
60 NEXT X
70 PRINT
80 PRINT "OUTCOME",
   "FREQUENCY"
90 FOR I = 1 TO N
100 PRINT I, F(I)
110 NEXT I
120 END
```

a. Use the program to simulate tossing a coin for 50 trials and for 400 trials.

b. Use the program to simulate rolling a die for 100 trials and for 500 trials.

Mixed Review

12. Write an equation for the relation {(−4, 3), (−2, 2), (0, 1), (2, 0), (4, −1). **(Lesson 9-7)**

13. Determine the value of r so that the line passing through $(2, r)$ and $(5, 1)$ has a slope of $-\frac{5}{3}$. **(Lesson 10-1)**

14. **Number Theory** If the digits of a two-digit positive integer are reversed, the result is 6 less than twice the original number. Find all such integers for which this is true. **(Lesson 11-5)**

15. Simplify $8\sqrt{50} + 5\sqrt{72} - 2\sqrt{98}$. **(Lesson 12-6)**

16. If a card is selected at random from a deck of 52 cards, what are the odds that it is not a face card? **(Lesson 14-7)**

EXPLORE
PLAN
SOLVE
EXAMINE

14-9 Problem-Solving Strategy: Solve a Simpler Problem

Objective

After studying this lesson, you should be able to:

- solve problems by first solving a simpler but related problem.

An important strategy for solving complicated or unfamiliar problems is to solve a simpler problem. This strategy involves setting aside the original problem and solving a simpler or more familiar case. The same methods that are used to solve the simpler problem can then be applied to the original problem.

Example

Find the sum of the first 1000 natural numbers.

EXPLORE The first 1000 natural numbers are the numbers 1 through 1000, inclusive. Thus, we want to find the following sum.

$$S = 1 + 2 + 3 + \ldots + 998 + 999 + 1000$$

PLAN This problem can be solved by actually adding all of the numbers, but this would be very tedious, even if a calculator is used. Let's consider a simpler version of the problem: finding the sum of the first ten natural numbers.

$$S = 1 + 2 + 3 + 4 + 5 + 6 + 7 + 8 + 9 + 10$$

Notice that pairs of addends have a sum of 11.

$$S = 10 + 9 + 8 + 7 + 6 + 5 + 4 + 3 + 2 + 1$$

11
11
11
11
11

Each sum is 11. Since there are $10 \div 2$ or 5 such sums, the sum of the first ten natural numbers is $5 \cdot 11$ or 55.

SOLVE Now use this method to solve the original problem.

$$S = 1 + 2 + 3 + \ldots + 998 + 999 + 1000$$

1001
1001

Each sum is 1001. Since there are $1000 \div 2$ or 500 such sums, the sum of the first 1000 natural numbers is $500 \cdot 1001$ or 500,500.

CHECKING FOR UNDERSTANDING

Communicating Mathematics

Read and study the lesson to answer each question.

1. Why do you use the solve-a-simpler-problem strategy?
2. What other problem-solving strategy was also used in the example?

Guided Practice

Solve each problem by solving a simpler problem.

3. A total of 3001 digits was used to print all of the page numbers (beginning with 1) of the Northern College yearbook. How many pages are in this yearbook?
4. How many line segments are needed to connect 1001 points if each pair of points must be connected by a line segment?

EXERCISES

Practice

Solve. Use any strategy.

5. While hiking with his dog, cat, and bird, Jesse came to a rope bridge. He can only carry one animal at a time across the bridge. He cannot leave the dog and cat together or the cat and bird together since they will fight. Can he get all the animals across the bridge without having any fights among the animals? If so, explain how.
6. Numero Uno says, "I'm thinking of a three-digit number that is both a perfect square and a perfect cube. What is my number?"
7. Alice bought a dozen peaches and pears for $3.78. A peach cost 3¢ more than a pear. What is the cost of each type of fruit? How many of each type did she buy?
8. Complete.
   1, 1, 2, 4, 7, 13, 24, ___?___, ___?___, ___?___, 274, 504
9. If a hen and a half can lay an egg and a half in a day and a half, how many hens are needed to lay a dozen eggs in one day?

COOPERATIVE LEARNING ACTIVITY

Work in groups of five. Each person in the group should understand the solution and be able to explain it to any person in the class.

Design a pair of dice with the following restrictions.

1. The dice must be shaped like regular dice, with 6 sides.
2. You must be able to get a sum of 1, 2, 3, 4, 5, 6, 7, 8, 9, 10, 11, or 12 when the dice are rolled.
3. Each of the sums must have the same probability of occurring.

Describe your dice.

14-10 Compound Events

Objective After studying this lesson, you should be able to:

- find the probability of a compound event.

Application Gwen has three skirts that she wears to school: 1 blue, 1 yellow, and 1 red. She also has 4 blouses that she can wear with any of the skirts: 1 yellow, 1 white, 1 striped, and 1 tan. Maria owns the same blue skirt and white blouse as Gwen, and she is wearing that outfit today. If Gwen chose one of her skirt-blouse outfits at random, what is the probability she and Maria are wearing the same outfit?

In order to calculate this probability, you need to know how many different skirt-blouse outfits are possible for Gwen. One method for finding all of the outfits is by using a *tree diagram.*

Notice that the number of possible outfits is the product of the number of skirts (3) and the number of blouses (4).

The colored lines in the tree diagram show the blue skirt, white blouse outfit.

In the tree diagram shown, each skirt color is given at the left. The four blouse possibilities branch from each skirt color. Then, each of the possible outfits is shown at the right. Since there are 12 possible outcomes, Gwen has 12 possible outfits to wear.

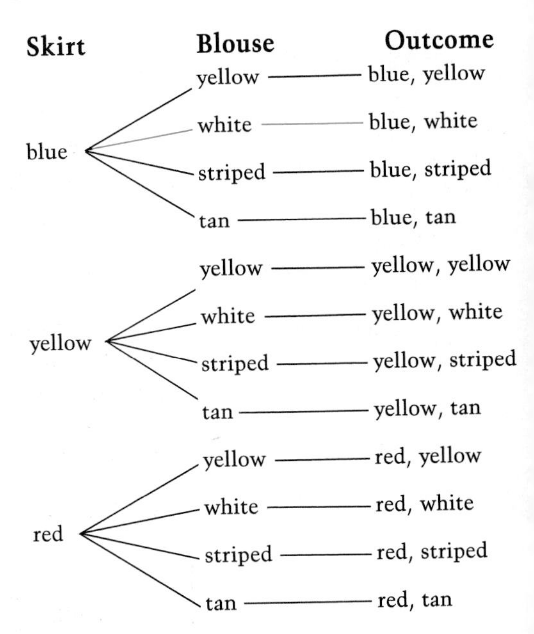

Since Gwen chose her outfit randomly, you can assume each skirt-blouse outfit is equally likely. Therefore, the probability that Maria and Gwen are wearing the same blue skirt and white blouse is $\frac{1}{12}$ or about 0.083.

The problem above is an example of finding the probability of a **compound event.** A compound event consists of two or more *simple events.* Gwen's choice of a skirt is a simple event, and her choice of a blouse is a simple event. Gwen's choice of a skirt and a blouse is a compound event.

Example 1 **Use the tree diagram above to answer each question.**

a. What is the probability Gwen and Maria are wearing the same skirt?

Since 4 of the 12 outfits use the blue skirt, the probability is $\frac{4}{12}$ or about 0.33.

b. What is the probability Gwen and Maria are wearing the same blouse?

Since 3 of the 12 outfits use the white blouse, the probability is $\frac{3}{12}$ or 0.25.

Example 2

Sean and Aaron have enough money to order a large pizza with 3 different toppings. Their four favorite toppings are pepperoni (P), mushrooms (M), olives (O), and red peppers (R). After a few minutes of arguing, they decide that one topping must be pepperoni. If they choose the other 2 toppings at random, what is the probability they will also choose olives?

1st Choice	2nd Choice	3rd Choice	Outcomes
pepperoni	mushrooms	olives	PMO ✓
		red peppers	PMR
	olives	mushrooms	POM ✓
		red peppers	POR ✓
	red peppers	mushrooms	PRM
		olives	PRO ✓

There are six ways Sean and Aaron can choose the other 2 toppings. In four cases, the pizza chosen will have olives as a topping. Thus, the probability they will also choose olives is $\frac{4}{6}$ or about 0.67.

Example 3

Ralph had three questions remaining in the true-false section of his biology final. Because he was almost out of time, Ralph had to guess the answer to each of these questions. If the answers are true, false, true, what is the probability that Ralph answered at least two of the last three questions correctly?

1st Question	2nd Question	3rd Question	Outcome
T	T	T	TTT ✓
		F	TTF
	F	T	TFT ✓
		F	TFF ✓
F	T	T	FTT
		F	FTF
	F	T	FFT ✓
		F	FFF

There are eight possible outcomes. In four cases, Ralph answers two or three of the questions correctly. Therefore, the probability that he answers at least two of the last three questions correctly is $\frac{4}{8}$ or 0.5.

CHECKING FOR UNDERSTANDING

Communicating Mathematics

Read and study the lesson to answer each question.

1. Refer to the problem at the beginning of the lesson. How many possible outfits does Gwen have if she buys **a.** a beige skirt, **b.** a beige blouse, or **c.** a beige skirt and beige blouse?

2. What is the difference between a simple event and a compound event?

Guided Practice

3. An automobile dealer has cars available with the combinations of colors, engines, and transmissions indicated in the tree diagram shown. A car is selected at random.

   a. What is the probability of selecting a car with manual transmission?

   b. What is the probability of selecting a blue car with manual transmission?

   c. What is the probability of selecting a car with a 4-cylinder engine and a manual transmission?

   d. What is the probability of selecting a blue car with a 6-cylinder engine and an automatic transmission?

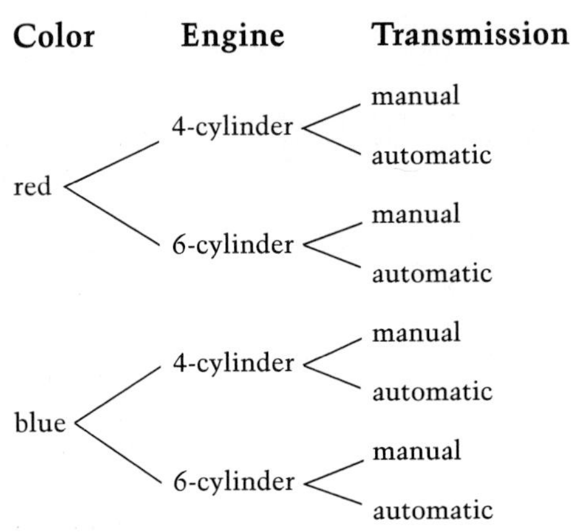

EXERCISES

Practice

4. Draw a tree diagram to show the possibilities for boys and girls in a family of 4 children. Assume that the probabilities for girls and boys being born are the same.

   a. What is the probability the family has exactly 4 girls?

   b. What is the probability the family has 2 boys and 2 girls in any order?

5. Use a tree diagram to find the probability of getting at least one tail when four fair coins are tossed.

6. Compare and contrast the tree diagram used for Exercise 5 with the one used for Exercise 4.

7. Box A contains one blue and one green marble. Box B contains one green and one red marble. Box C contains one white and one green marble. A marble is drawn at random from each box.

   a. What is the probability that all of the marbles are green?

   b. What is the probability that exactly two marbles are green?

   c. What is the probability that at least one marble is not green?

Portfolio

Select an item from your work in this chapter that shows your creativity and place it in your portfolio.

8. With each shrimp, salmon, or crab dinner at the Seafood Palace, you may have soup or salad. With shrimp, you may have broccoli or a baked potato. With salmon, you may have rice or broccoli. With crab, you may have rice, broccoli, or a potato. If all combinations are equally likely, find the probability of an order containing each item.
   a. salmon
   b. soup
   c. rice
   d. shrimp and rice
   e. salad and broccoli
   f. crab, soup, and rice

9. Two men and three women are each waiting for a job interview. There is only enough time to interview two people before lunch.
   a. What is the probability that both people are women?
   b. What is the probability at least one person is a woman?
   c. Which is more likely, one of the people is a woman and the other is a man, or both people are either men or women?

Critical Thinking

10. After each football game, 80% of the students at the game go to pick up pizza while the others go home. After picking up the pizza, 40% of the students go home while the rest go to a friend's house. If Kelly goes to the game, what is the probability she will go to a friend's house that night?

Applications

11. **Entertainment** In order to raise money for a trip to the opera, the music club has set up a lottery using two-digit numbers. The first digit will be a numeral from 1 to 4. The second digit will be a numeral from 3 to 8. One of the digits in Trudy's lottery number is 3, but she can't remember which digit. If only one two-digit lottery number is drawn, what is the probability that Trudy will win?

12. **Law** A three-judge panel is being used to settle a dispute. Both sides in the dispute have decided that a majority decision will be upheld. If each judge will render a favorable decision based on the evidence presented two-thirds of the time, what is the probability that the correct side will win the dispute?

Mixed Review

13. **Finance** A selling price of $145,000 for a home included a 6.5% commission for the real estate agent. How much money did the owners receive from this sale? **(Lesson 4-4)**

14. Simplify $\frac{4}{5-p} - \frac{3}{p-5}$. **(Lesson 8-7)**

15. **Construction** Janet's family room has an area rug that is 9 feet by 12 feet. A strip of floor of equal width is uncovered along all edges of the rug. Janet wants to place ceramic tiles on this part of the floor. If the area of the uncovered floor is 270 ft^2, what is the width of the space to be tiled? **(Lesson 13-5)**

16. Toss a coin 50 times. Then determine P(heads) for this coin. Are the results what you expected? If not, does this imply that the coin is not fair? Explain. **(Lesson 14-8)**

17. Solve by solving a simpler problem. **(Lesson 14-9)**

    Determine the number of diagonals of a convex polygon with 100 sides.

CHAPTER 14

SUMMARY AND REVIEW

VOCABULARY

Upon completing this chapter, you should be familiar with the following terms:

box-and-whisker plot	**579**	**570**	mode
compound event	**600**	**589**	probability
data	**560**	**590**	odds
empirical probability	**594**	**576**	outlier
equally likely outcomes	**590**	**575**	range
interquartile range	**575**	**583**	scatter plot
line plot	**560**	**560**	statistics
lower quartile	**575**	**565**	stem-and-leaf plot
mean	**570**	**575**	upper quartile
median	**570**		

SKILLS AND CONCEPTS

OBJECTIVES AND EXAMPLES

Upon completing this chapter you should be able to:

- interpret numerical data from a table or chart **(Lesson 14-1)**

Company	Sales	Income
Exxon	87	3.5
Ford Motor	97	3.8
General Electric	55	3.9
General Motors	127	4.2
IBM	63	3.8
Mobil	60	1.8
Philip Morris	39	2.9

Which one had the greatest sales? GM

REVIEW EXERCISES

Use these exercises to review and prepare for the chapter test.

Use the data in the table at the left to answer each question. Sales and income are given in billions of dollars.

1. Which company had the least sales?
2. Which company had the least income?
3. For each company, determine income as a percent of sales (income ÷ sales). Which one had the greatest income as a percent of sales? Was it the company with the greatest income?

- display and interpret statistical data on a line plot **(Lesson 14-1)**

Use the data in the table above to make a line plot of sales.

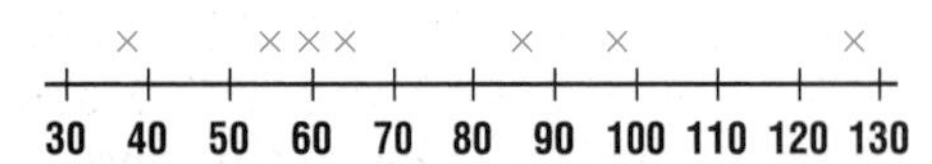

Make a line plot for each set of data.

4. 78, 74, 86, 88, 99, 63, 85, 85, 85
5. 134, 167, 137, 138, 120, 134, 145, 155, 152, 159, 164, 135, 144, 156

OBJECTIVES AND EXAMPLES	REVIEW EXERCISES

- display and interpret data on a stem-and-leaf plot **(Lesson 14-2)**

Use the data in the table on the previous page to make a stem-and-leaf plot of income.

Stem	Leaf
1	8
2	9
3	5 8 8 9
4	2

2|9 represents 2.9 billion dollars.

6. The prices for 15 different types of athletic shoes are listed below. Make a stem-and-leaf plot of the prices. How many of the prices are between \$40 and \$50 inclusive?

\$42 \$44 \$76 \$56 \$78 \$62 \$65 \$69
\$55 \$66 \$42 \$83 \$50 \$40 \$41

7. Make a back-to-back stem-and-leaf plot of the following data comparing rounded and truncated values.

4267, 5679, 3623, 6791, 3471, 3124, 5629, 4444, 3812, 5814, 4967

- calculate and interpret the mean, median, and mode of a set of data **(Lesson 14-3)**

Find the mean, median, and mode of 4, 5, 6, 6, 10, 9, 14, 16, 16, 16, 30.

mean: The sum of the 11 values is 132.

$\frac{132}{11} = 12$

median: 6th value = 9

mode: 16

8. Find the mean, median, and mode of 6.8, 8.4, 6.2, 5.7, 5.6, 7.1, 9.9, 1.5, 7.1, 5.4, 3.4.

9. The stem-and-leaf plot below shows points scored by losing teams in the first 25 Super Bowls. Find the mean, median, and mode of the scores.

Stem	Leaf
0	3 6 7 7 7 7 9
1	0 0 0 0 0 0 3 4 4 6 6 7 7 9 9
2	0 1
3	1

3|1 represents 31 points.

- calculate and interpret the range, quartiles, and the interquartile range of a set of data **(Lesson 14-4)**

Find the range, upper and lower quartiles, and interquartile range of the data for Lesson 14-3.

range: 30 − 4 = 26
lower quartile: 3rd value = 6
upper quartile: 9th value = 16
interquartile range: 16 − 6 = 10

10. Use the data in Exercise 9 to find the range, quartiles, and interquartile range of points scored by losing teams in the first 25 Super Bowls.

11. The average annual snowfall, in inches, for 12 northeastern cities are listed below. Find the interquartile range for this data. Identify any outliers.

111.5	70.7	59.8	68.6	63.8	254.8
64.3	82.3	91.7	88.9	110.5	77.1

- display and interpret data on a box-and-whisker plot **(Lesson 14-5)**

Make a box-and-whisker plot of the data for Lesson 14-3.

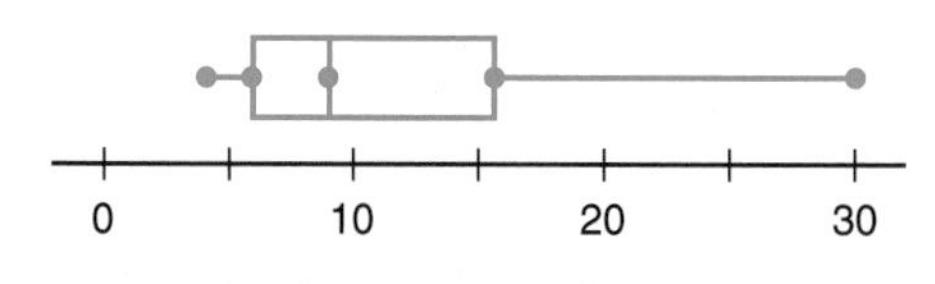

12. Use the data in Exercise 9 to make a box-and-whisker plot of points scored by losing teams in the first 25 Super Bowls.

13. The number of calories in a serving of french fries at 13 restaurants are 250, 240, 220, 348, 199, 200, 125, 230, 274, 239, 212, 240, and 327. Make a box-and-whisker plot of this data.

OBJECTIVES AND EXAMPLES	REVIEW EXERCISES

- Graph and interpret pairs of numbers on a scatter plot **(Lesson 14-6)**

This scatter plot shows a negative association since the line suggested by the points would have a negative slope.

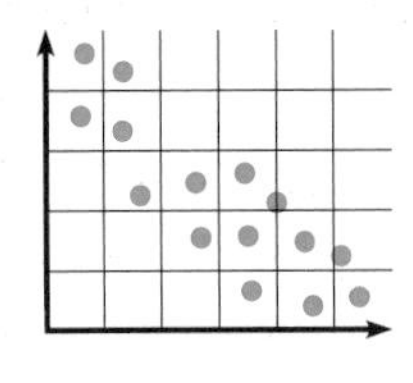

14. Draw a scatter plot of the data on income and sales for the top U.S. companies given in the table on page 604. Let the horizontal axis be sales. What type of association, if any, is shown between sales and income?

- find the probability of a simple event **(Lesson 14-7)**

If a soccer team consists of 8 seniors, 7 juniors, 3 sophomores, and 2 freshmen, then the probability that a player chosen at random is a senior is $\frac{8}{20}$ or 0.4.

15. For the soccer team listed at the left, find the probability that a player chosen at random is not a junior or a freshman.

16. If one of the first 100 positive integers is selected at random, what is the probability that the number is prime?

- find the odds of a simple event **(Lesson 14-7)**

On the soccer team given above, the odds that a player chosen at random is *not* a junior are 13:7.

17. If the odds that an event occurs are 9:4, what is the probability that the event occurs?

18. What are the odds that one of the 30 people at Terry's birthday party will not win one of the 8 door prizes?

- conduct and interpret probability experiments **(Lesson 14-8)**

Two dice are rolled 100 times. If the sum 11 appears 8 times, then $P(\text{sum of } 11) = \frac{8}{100}$ or 0.08.

Toss two coins 100 times. Record the results and use them to answer each question.

19. What is the probability of tossing exactly one head?

20. What is the probability of tossing at least one head?

- find the probability of a compound event **(Lesson 14-10)**

Marcus guesses on two multiple-choice questions on a quiz. If each question has four possible answers, then the probability that he answers both questions incorrectly is $\frac{9}{16}$ or 0.5625.

21. Matthew has 2 brown and 4 black socks in his dresser. While dressing one morning, he pulled out 2 socks without looking. What is the probability that he chose a matching pair?

22. Each day, Angie takes one of 4 ferries from Lehigh to Port City. Find the probability that Angie takes a different ferry on two consecutive days.

---

23. Solve by solving a simpler problem. **(Lesson 14-9)**
There are 208 baseball teams in the state playoff. If a team loses, it is eliminated and plays no more games. How many games will be needed to decide the state champion?

CHAPTER 14 TEST

The numbers below represent the mean number of students per teacher in the 50 states in 1988.

18.7 17.0 18.2 15.7 22.7 17.8 13.1 16.4 13.3 17.1 18.5 21.1 16.0
20.6 17.1 17.8 15.8 15.2 17.8 18.2 14.6 16.8 13.7 19.8 17.0 14.6
18.4 15.9 15.8 15.0 20.3 16.2 13.6 18.5 14.9 17.5 15.4 17.6
16.5 18.4 15.9 14.6 17.2 15.4 19.3 24.5 13.6 16.1 20.4 15.1

1. Make a stem-and-leaf plot of the ratios.
2. How many of the ratios were in the 17s?
3. Find the median and the mode of the ratios.
4. What is the range of the ratios?
5. Find the upper and lower quartiles and the interquartile range of the ratios.
6. Identify any outliers in the ratios.

The table at the right shows the shots attempted and the shots made for 10 players on the Patriots basketball team during the first five games of the season.

Basketball Player	A	B	C	D	E	F	G	H	I	J
Shots Attempted	60	25	35	12	80	4	15	42	11	22
Shots Made	25	10	15	4	36	1	6	16	4	8

7. Make a line plot of the shots attempted.
8. What are the greatest number and the least number of shots made?
9. Find the mean number of shots attempted and shots made.
10. Find the quartiles and interquartile range of the shots attempted.
11. Make a box-and-whisker plot of the shots attempted.
12. Draw a scatter plot of the data. Let the horizontal axis be *Shots Attempted.* Does this plot show positive, negative, or no association?

During a 20-song sequence on a popular radio station, 8 soft rock, 7 hard rock, and 5 rap songs are played at random. You tune in to the station.

13. What is the probability that a hard rock song is playing?
14. What are the odds that a rap song is playing?

The table at the right shows how registered voters voted on Issue 12 in the 4th District.

Vote	Yes	No	Did Not Vote
Men	165	140	185
Women	205	105	200

15. If a voter is chosen at random, what is the probability that the person did not vote?
16. If a man who voted is chosen at random, what are the odds that he voted yes?
17. If a person who voted no is chosen at random, what is the probability that the person is a woman?
18. If the odds that it will rain are 3:7, find the probability of rain.
19. If the probability of rain is 0.45, find the odds that it will not rain.

Bonus If a couple has four children, are they more likely to have two boys and two girls or three of one gender and one of the other?

CHAPTER OBJECTIVES

In this chapter, you will:

- Find the complement and supplement of an angle.
- Use proportions to find the missing measures for similar triangles.
- Find the sine, cosine, and tangent of an acute angle of a right triangle given the measure of its sides.
- Use trigonometric ratios to solve problems.
- Solve problems by making a model.

15 Trigonometry

APPLICATION IN OPTICS

Light will change direction, or bend, as it moves from one substance to another, such as from air to water. This bending of light at the boundary of two substances is called *refraction.* The amount of refraction depends on the angle at which the light falls on the boundary of the two substances.

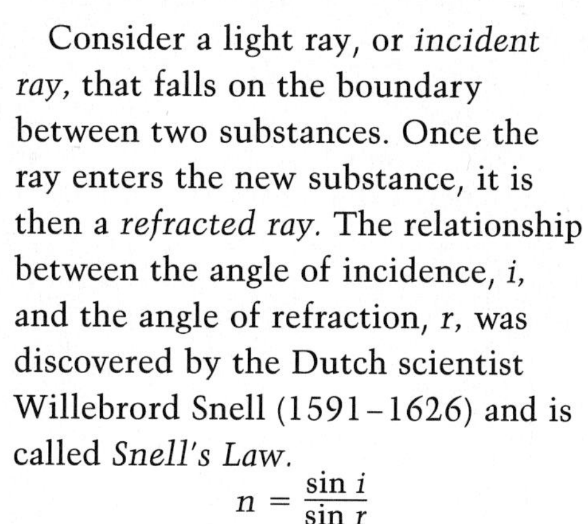

Consider a light ray, or *incident ray,* that falls on the boundary between two substances. Once the ray enters the new substance, it is then a *refracted ray.* The relationship between the angle of incidence, i, and the angle of refraction, r, was discovered by the Dutch scientist Willebrord Snell (1591–1626) and is called *Snell's Law.*

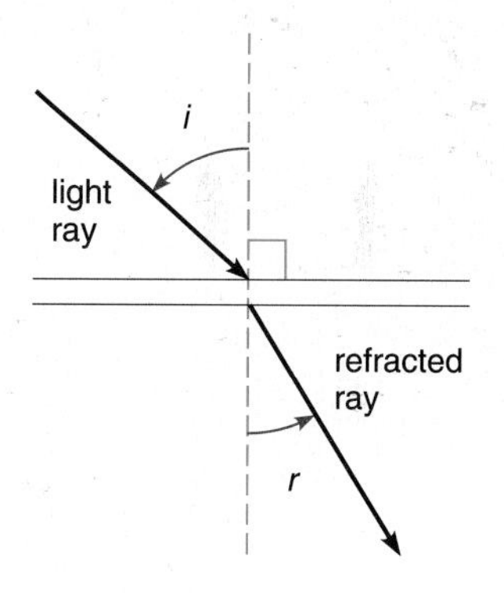

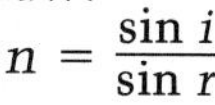

$$n = \frac{\sin i}{\sin r}$$

In this equation, n is called the *index of refraction* and is constant for any light ray traveling from one substance to another. The abbreviation *sin* represents the *sine* of an angle, a ratio that you will learn about in this chapter.

When you place part of your leg in a pool, why does it look like the leg "bends" at the surface of the water? The answer is that light traveling from the water to your eyes "bends" when it hits the surface of the water.

ALGEBRA IN ACTION

Copy and complete the chart below. Which has the smallest index of refraction?

Substances	sin i	sin r	Index of Refraction
air to water	0.8660	0.6508	?
air to glass	0.5000	0.3107	?
air to diamond	0.7071	0.2924	?
water to glass	0.2588	0.2135	?
water to diamond	0.7760	0.4210	?
glass to diamond	0.9659	0.6428	?

15-1 Angles and Triangles

Objectives After studying this lesson, you should be able to:

- find the complement and the supplement of an angle, and
- find the measure of the third angle of a triangle given the measure of the other two angles.

Application On fourth down in a football game, the team with the ball usually punts. How far will the football go? The distance depends not only on the strength of the punter, but also on the angle at which the ball is kicked. If the angle is too small, the ball will fall short. If the angle is too large, the ball will go very high but not far.

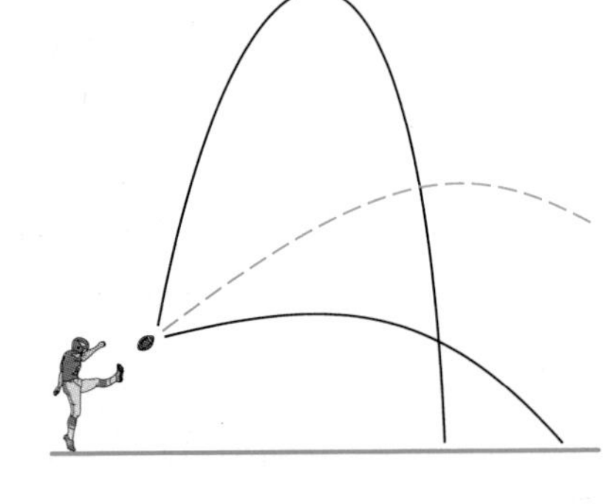

In this lesson, you will learn about different angle measures and their applications to practical problems.

A *protractor* can be used to measure angles as shown below.

The symbol for angle is ∠.

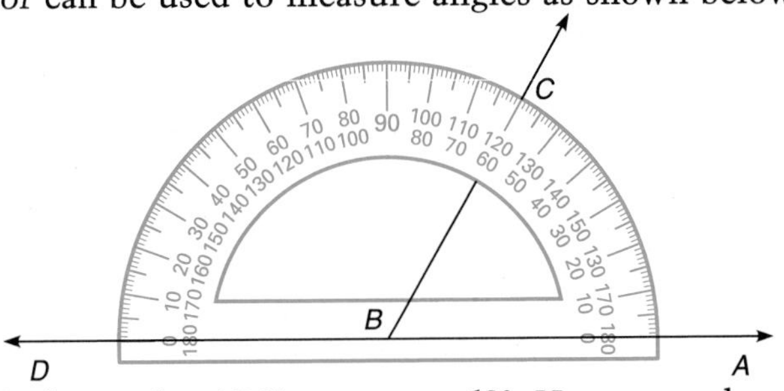

Angle *ABC* (denoted ∠*ABC*) measures 60°. However, where ray *BC* intersects the curve of the protractor, there are two readings, 60° and 120°. The measure of ∠*DBC* is 120°. What is the sum of the measures of ∠*ABC* and ∠*DBC*?

Supplementary Angles **Two angles are supplementary if the sum of their measures is 180°.**

Example 1

The measure of an angle is four times the measure of its supplement. Find the measure of each angle.

Let x = the lesser measure.
Then 4x = the greater measure.

$$x + 4x = 180 \quad \textit{The sum of the measures is 180°.}$$
$$5x = 180$$
$$x = 36$$

4x° x°

The measures are 36° and 4 · 36° or 144°. *Check this result.*

Complementary Angles

Two angles are complementary if the sum of their measures is 90°.

Example 2

The measure of an angle is 26° greater than its complement. Find the measure of each angle.

Let x = the lesser measure.
Then x + 26 = the greater measure.

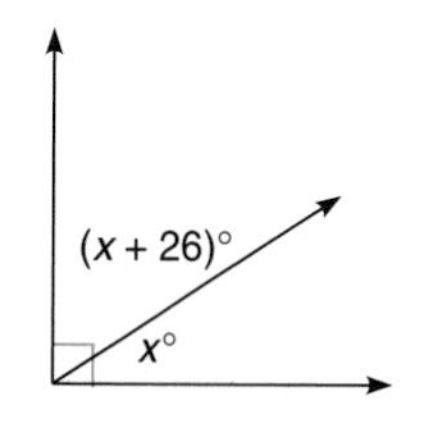

$$x + (x + 26) = 90 \quad \textit{The sum of the measures is 90°.}$$
$$2x + 26 = 90$$
$$2x = 64$$
$$x = 32$$

The measures are 32° and 32° + 26° or 58°. *Check this result.*

What is the sum of the measures of the three angles of a triangle? Use a protractor to measure the angles of each triangle below. Then find the sum of the measures of each triangle.

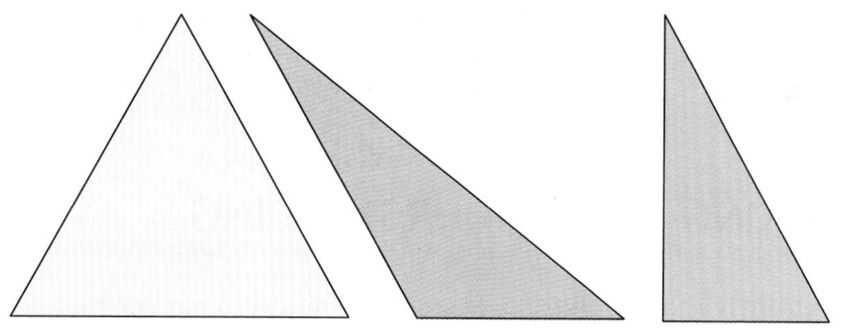

What did you discover? In each case, your sum should be approximately 180°.

Sum of the Angles of a Triangle

The sum of the measures of the angles in any triangle is 180°.

In an equilateral triangle, the angles have the same measure. We say that the angles are **congruent.** The sides of an equilateral triangle are also congruent.

Example 3

What is the measure of the angles of an equilateral triangle?

Let x = the measure of each angle.

$$x + x + x = 180$$
$$3x = 180$$
$$x = 60$$

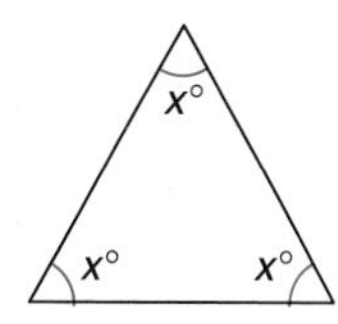

Each angle measures 60°.

Example 4 **In an isosceles triangle, at least two angles have the same measure. What are the measures of the base angles of an isosceles triangle in which the vertex angle measures 40°?**

In an isosceles triangle, the base angles are equal.
Let x = the measure of each base angle.

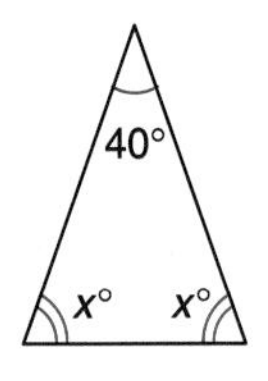

$$x + x + 40 = 180$$
$$2x + 40 = 180$$
$$2x = 140$$
$$x = 70$$

The base angles each measure 70°.

Example 5 **The measures of the angles of a triangular vegetable garden are given as $x°$, $3x°$, and $4x°$. What are the measures of each angle?**

The sum of the measures of the angles of a triangle is 180°.

$$x + 3x + 4x = 180$$
$$8x = 180$$
$$x = 22.5$$

The measures are 22.5°, 3(22.5°) or 67.5°, and 4(22.5°) or 90°.

CHECKING FOR UNDERSTANDING

Communicating Mathematics

Read and study the lesson to answer each question.

1. The words *compliment* and *complement* sound very much alike. What are their meanings? Which word has the mathematical meaning?
2. Three types of triangles are right, obtuse, and acute. A right triangle has one angle with measure equal to 90°, an obtuse triangle has one angle with measure greater than 90°, and an acute triangle has all angles with measures less than 90°. In the diagram on page 611, identify each triangle as right, obtuse, or acute.
3. Can a right triangle have two 90° angles? Explain.

Guided Practice

Find the complement of each angle measure.

4. 42°
5. 13°
6. 45°
7. 24°
8. 11°
9. 76°
10. $3x°$
11. $(2x + 40)°$
12. $(x - 7)°$

Find the supplement of each angle measure.

13. 130°
14. 65°
15. 87°
16. 90°
17. 32°
18. 156°
19. $y°$
20. $6m°$
21. $(x - 20)°$

EXERCISES

Practice

Find both the complement and the supplement of each angle measure.

22. $42°$	**23.** $87°$	**24.** $125°$	**25.** $160°$
26. $90°$	**27.** $68°$	**28.** $21°$	**29.** $174°$
30. $99°$	**31.** $a°$	**32.** $3y°$	**33.** $(x + 30)°$
34. $(x - 38)°$	**35.** $5x°$	**36.** $(90 - x)°$	**37.** $(180 - y)°$

Find the measure of the third angle of each triangle in which the measures of two angles of the triangle are given.

38. $16°, 42°$	**39.** $40°, 70°$	**40.** $50°, 45°$
41. $90°, 30°$	**42.** $89°, 90°$	**43.** $63°, 12°$
44. $43°, 118°$	**45.** $4°, 38°$	**46.** $x°, y°$
47. $x°, (x + 20)°$	**48.** $y°, (y - 10)°$	**49.** $m°, (2m + 1)°$

50. One of the congruent angles of an isosceles triangle measures 37°. Find the measures of the other angles.

51. The measures of the angles of a certain triangle are consecutive even integers. Find their measures.

52. An angle measures 38° less than its complement. Find the measures of the two angles.

53. One of the angles of a triangle measures 53°. Another angle measures 37°. What is the measure of the third angle?

54. One angle of a triangle measures 10° more than the second. The measure of the third angle is twice the sum of the first two angles. Find the measure of each angle.

55. One of the two complementary angles measures 30° more than three times the other. Find the measure of each angle.

56. Find the measure of an angle that is 10° more than its complement.

57. Find the measure of an angle that is 30° less than its supplement.

58. Find the measure of an angle that is one-half the measure of its complement.

59. Find the measure of an angle that is one-half the measure of its supplement.

60. The measures of the angles of a triangle are given as $x°$, $2x°$, and $3x°$. What are the measures of each angle?

61. The measures of the angles of a triangle are given as $x°$, $(x + 5)°$, and $(2x + 3)°$. What are the measures of each angle?

62. The measures of the angles of a triangle are given as $6x°$, $(x - 3)°$, and $(3x + 7)°$. What are the measures of each angle?

Critical Thinking

63. Using the fact that the sum of the measures of the angles of any triangle is 180°, determine the sum of the measures of the interior angles of an octagon. Justify your answer.

Applications

64. **Football** A punter for Darien High School punts a football into the air at an angle of 40° from the horizontal (the ground). At what angle measure from the vertical is the ball punted?

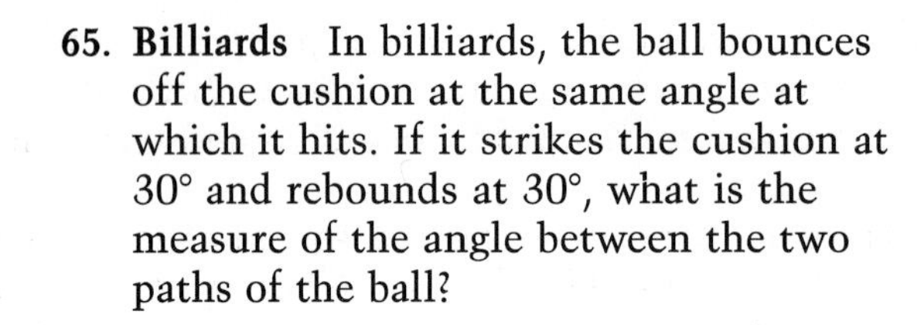

65. **Billiards** In billiards, the ball bounces off the cushion at the same angle at which it hits. If it strikes the cushion at 30° and rebounds at 30°, what is the measure of the angle between the two paths of the ball?

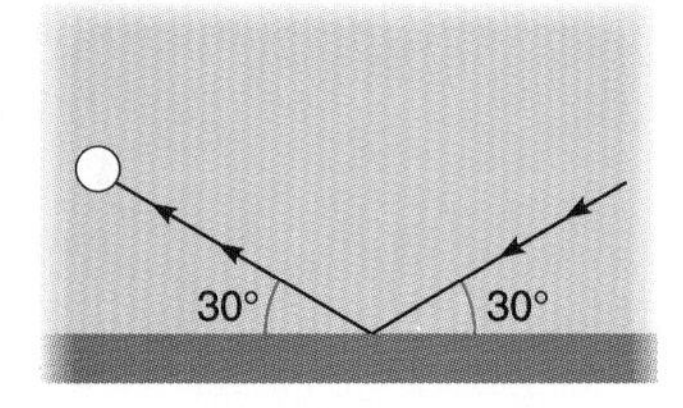

66. **Mountain Climbing** *Rapelling* is a technique used by climbers to make a difficult descent. Climbers back to the edge and spring off. To avoid slipping, the climber's legs should remain perpendicular to the side of the ledge. The ledge at the right is 50° off the vertical. What is the measure of the climber's angle with the vertical?

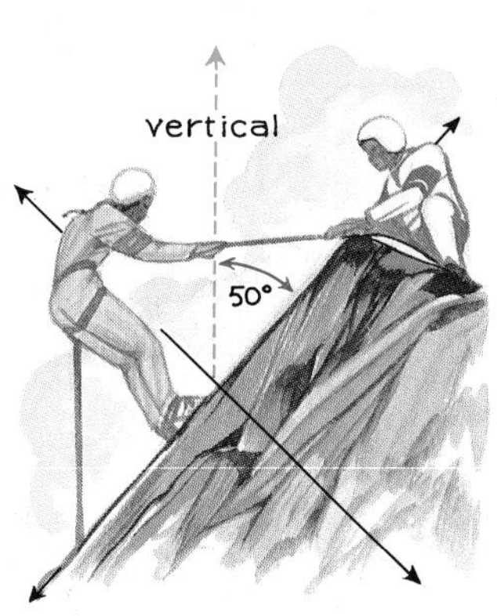

Mixed Review

67. Write *R is the product of a and m decreased by z* as an equation. **(Lesson 1-7)**

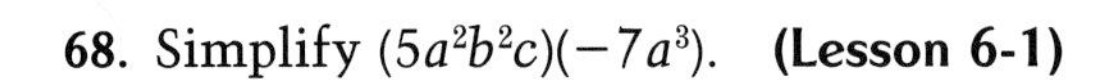

68. Simplify $(5a^2b^2c)(-7a^3)$. **(Lesson 6-1)**

Journal

Make up a problem in which you must use an equation to find the measures of the angles of a triangle. Then solve the problem, explaining each step.

69. Write a system of inequalities for the graph at the right. **(Lesson 11-6)**

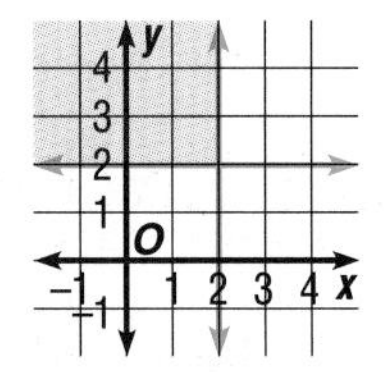

70. Find the distance between (−3, 5) and (2, 7). **(Lesson 12-8)**

71. Find the equation of the axis of symmetry and the coordinates of the maximum or minimum point of the graph of $y = -5x^2 + 15x + 23$. **(Lesson 13-1)**

72. Find the sum and product of the roots of $b^2 + 12b - 28 = 0$. **(Lesson 13-8)**

73. Bill, Raul, and Joe are in a bicycle race. If each boy has an equal chance of winning, find the probability that Raul finishes last. **(Lesson 14-10)**

15-2 Problem-Solving Strategy: Make a Model

EXPLORE
PLAN
SOLVE
EXAMINE

Objective After studying this lesson, you should be able to:

- solve problems by making models.

Models are often useful in solving problems. A model can be a simple sketch, a precise scale drawing, or a three-dimensional object. In this chapter, you will make models of angles and triangular shapes.

Example

A landscaping plan suggests that 6 square concrete tiles be placed in a garden. At least one side of each tile is to be matched evenly with a side of another tile. If the tiles are one foot long on each side and the perimeter of the area to be covered by the tiles is 14 feet, what are some possible arrangements of the tiles?

This problem can be solved by cutting 6 congruent squares from a piece of paper. Arrange these squares until you find a pattern with a perimeter of 14 units. Some possible patterns are shown below.

perimeter = 10 units

perimeter = 12 units

perimeter = 14 units

perimeter = 12 units

perimeter = 12 units

perimeter = 14 units

Two possible configurations are and .

CHECKING FOR UNDERSTANDING

Communicating Mathematics

Read and study the lesson to answer each question.

1. What other models could you use to solve the tile problem?
2. The following diagrams are all variations of the same pattern.

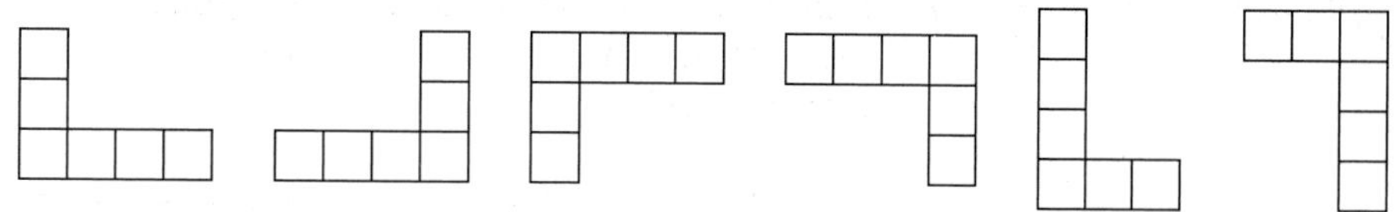

Can you find a pattern with a perimeter of 14 units that is different than the ones already given?

3. How could an interior decorator use models to design ways to arrange furniture in a room?

Guided Practice **Solve. Make a model.**

4. Two tiles are put together to form a pentagon. One tile is in the shape of a square, and the other is in the shape of an equilateral triangle. If one side of the triangle is 10 centimeters long, what is the perimeter of the pentagon?
5. A painting is 12 inches by 20 inches. The painting is bordered by a mat that is 3 inches wide. The frame around the mat is 2 inches wide. What is the area of the picture, including the frame and mat?
6. In the Centerville Community Softball League tournament, the losing team from each game is out of the tournament and the winning team goes on to play another winning team until a first-place team is determined. If 16 teams enter the tournament, how many games will be played?

EXERCISES

Practice **Solve. Use any strategy.**

7. Hong and Dora each have a bag of marbles. If Hong gave Dora one marble, they would each have the same number of marbles. However, if Dora would give Hong one marble, Hong would have twice as many marbles as Dora. How many marbles does each one have?
8. The six faces of a wooden cube are painted orange, and then the cube is cut into 64 one-inch cubes.
   a. How many of the smaller cubes have exactly 3 painted faces?
   b. How many of the smaller cubes have exactly 2 painted faces?
   c. How many have exactly 1 painted face?
   d. How many have no painted faces?
9. Mr. Zerman went shopping. In the first store he spent half of his money plus an additional dollar. At the second store he spent half of his remaining money plus another dollar. This pattern continued until he left the fifth store with no money left. How much money did he have before he started shopping?

COOPERATIVE LEARNING ACTIVITY

Work in groups of three. Each person must understand the solution and be able to explain it to any person in the class.

There are 8 small boxes identical in color, size, and shape. Seven of the boxes each contain an inexpensive costume ring and therefore have the same weight. One box contains a valuable diamond ring and weighs slightly more than the other boxes. You cannot open the boxes, but you can use a balance scale. How can you be sure to pick the box with the diamond ring, if you can only use the balance scale two times?

15-3 30°–60° Right Triangles

Objective

After studying this lesson, you should be able to:

- find the measures of the sides of a 30°–60° right triangle given the measure of one side.

Application

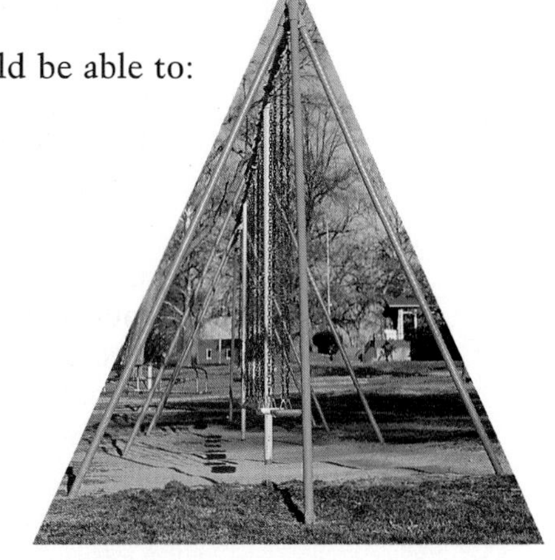

The two poles at the end of the swing set at the right form a triangle with the ground. You can use a special right triangle to find the height of the swing set.

C

A D B

Examine the equilateral triangle shown at the left. What is the measure of $\angle B$?

Line segment CD (denoted $\overline{CD}$) can be drawn perpendicular to $\overline{AB}$. $\overline{CD}$ bisects $\angle ACB$. What is the measure of $\angle DCB$?

Triangle CDB is a special triangle called a **30°–60° right triangle.**

The hypotenuse of a right triangle is the side opposite the right angle.

$\overline{CD}$ bisects $\overline{AB}$. Since the measure of $\overline{DB}$ is one-half the measure of $\overline{AB}$, it is also one-half the measure of the hypotenuse, $\overline{BC}$. Why?

In a 30°–60° right triangle, the measure of the hypotenuse is twice the measure of the side opposite the 30° angle.

Lower case letters are often used to designate the measures of the sides of a triangle. For example, the measure of the side opposite angle R is r.

Example 1

For each 30°–60° right triangle, find the indicated measures.

a. Find the measure of $\overline{MK}$ in $\triangle RMK$.

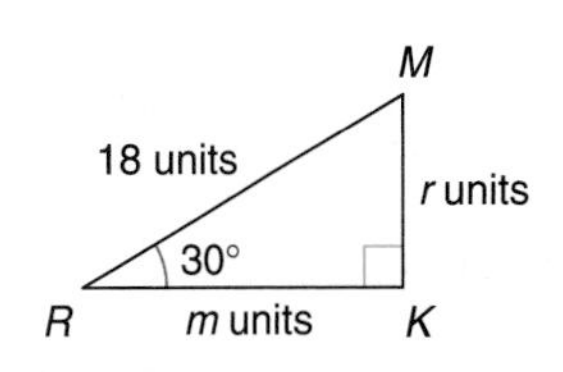

$r = \frac{1}{2}k$ *The measure of the hypotenuse is k.*

$= \frac{1}{2}(18)$

$= 9$

Thus, $\overline{MK}$ is 9 units long.

b. Find the measure of $\overline{PQ}$ in $\triangle PQR$.

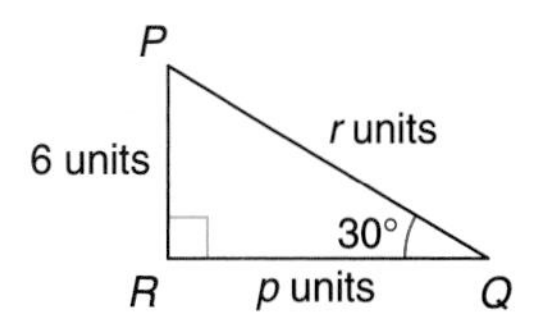

$r = 2q$ *The measure of the side opposite the 30° angle is 6.*

$= 2(6)$

$= 12$

Thus, $\overline{PQ}$ is 12 units long.

In triangle ABC (denoted $\triangle ABC$), a represents the measure of the side opposite the 30° angle. The Pythagorean Theorem can be used to find b, the measure of the side opposite the 60° angle.

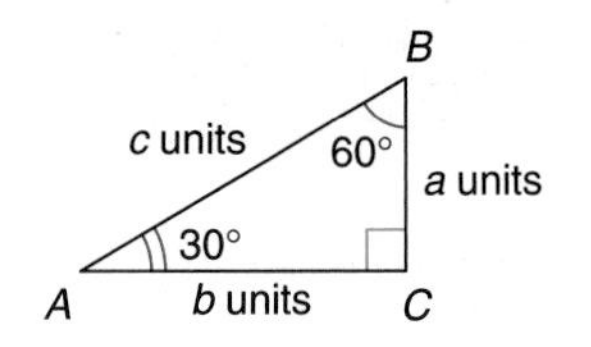

$a^2 + b^2 = c^2$ *Pythagorean Theorem*

$a^2 + b^2 = (2a)^2$ *Replace c with 2a.*

$a^2 + b^2 = 4a^2$

$b^2 = 4a^2 - a^2$ *Solve for b.*

$b^2 = 3a^2$

$b = a\sqrt{3}$

The measures of the sides of a 30°–60° right triangle are summarized as follows.

30°–60° Right Triangle

In a 30°–60° right triangle, if a is the measure of the side opposite the 30° angle, then $2a$ is the measure of the hypotenuse, and $a\sqrt{3}$ is the measure of the side opposite the 60° angle.

Example 2

For each 30°–60° right triangle, find the length of the indicated side.

a. Find the measure of $\overline{RT}$ in $\triangle RST$.

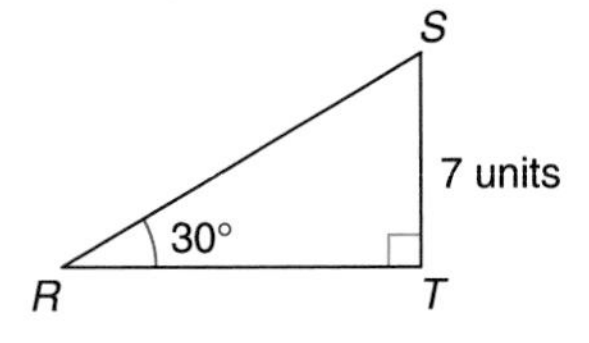

$s = r\sqrt{3}$ *The measure of the side*

$= 7\sqrt{3}$ *opposite the 30° angle is r.*

≈ 12.124

Thus, $\overline{RT}$ is approximately 12.124 units long.

b. Find the measures of $\overline{XY}$ and $\overline{XZ}$ in $\triangle XYZ$.

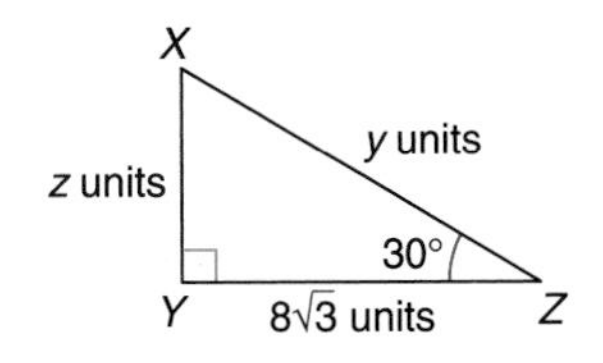

$x = z\sqrt{3}$

$8\sqrt{3} = z\sqrt{3}$

$8 = z$ *Divide each side by $\sqrt{3}$.*

Since $y = 2z$, then $y = 2(8)$ or 16. Thus, $\overline{XY}$ is 8 units long, and $\overline{XZ}$ is 16 units long.

Example 3

Suppose each pole holding up the swing set on page 617 is 10 feet long. What is the height of the swing set?

Draw a model of the problem. Look at one of the right triangles. The hypotenuse is 10 feet long, and the side opposite the 30° angle is 5 feet long. The height is the length of the side opposite the 60° angle.

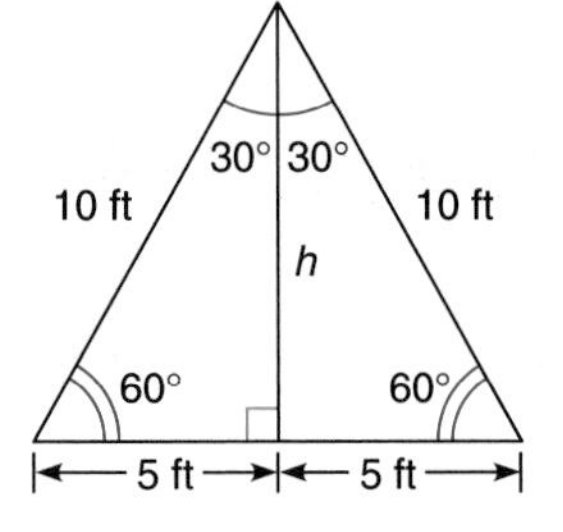

$h = 5\sqrt{3}$

≈ 8.66

The height of the swing is about 8.66 feet, or 8 ft 8 in.

Example 4

Trusses are also used in many kinds of machinery, such as cranes and lifts, and in aircraft wings and fuselages.

In construction, a truss is often used to support the weight of a roof. An architect designs a building so that the rafters form an angle of 30° with the span. If the rafters are 14 feet long, how high is the vertical support and how wide is the span?

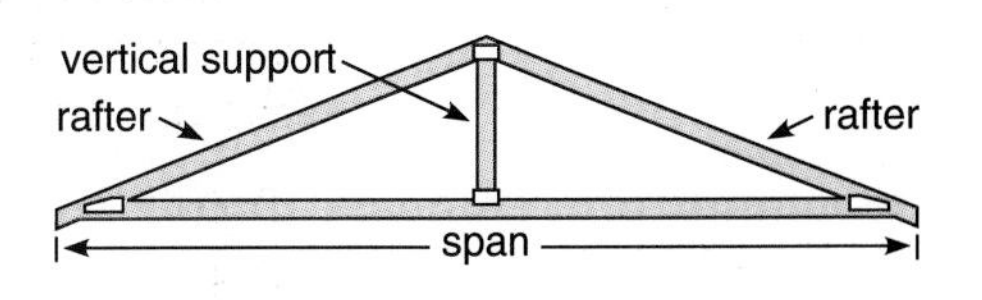

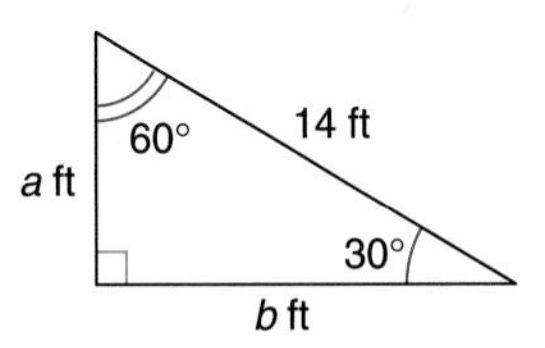

The truss forms two congruent right triangles. Draw a model of one triangle to find the measures.

Let a be the measure of the side opposite the 30° angle.
Let b be the measure of the side opposite the 60° angle.

$a = \frac{1}{2}(14)$ $\quad\quad$ $b = 7\sqrt{3}$

$= 7$ $\quad\quad$ ≈ 12.124

The vertical support is 7 feet. Since b represents half the span, the span is about 2(12.124) or 24.25 feet.

CHECKING FOR UNDERSTANDING

Communicating Mathematics

Read and study the lesson to answer each question.

1. In a 30°–60° right triangle, the ratio of the measures of the sides is $1{:}\sqrt{3}{:}2$. Will the ratio stay the same no matter how large the 30°–60° right triangle is? Justify your answer.
2. What is the length of the other leg of the triangle at the right?
3. What is the length of the hypotenuse of the triangle at the right?

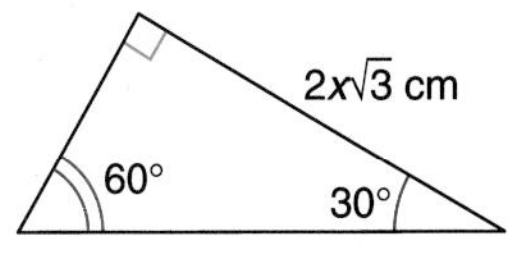

Guided Practice

The length of the hypotenuse of a 30°–60° right triangle is given. Find the length of the side opposite the 30° angle in each triangle.

4. 8 m
5. 16 cm
6. 13 mm
7. 9 mi
8. $4\frac{1}{2}$ in.
9. $3\frac{3}{8}$ in.
10. 16.36 m
11. 4.63 cm

The length of the side opposite the 60° angle in a 30°–60° right triangle is given. Find the length of the other two sides.

12. $4\sqrt{3}$ ft
13. $2\sqrt{3}$ cm
14. $8\sqrt{3}$ m
15. $\sqrt{3}$ mi

EXERCISES

Practice

The length of the side opposite the 30° angle in a 30°–60° right triangle is given. Find the length of the hypotenuse in each triangle.

16. 7 m
17. 6.2 cm
18. 4.35 mm
19. $4\frac{1}{2}$ mi
20. $6\frac{3}{8}$ in.
21. 13 m
22. 3.86 cm
23. $7\frac{3}{4}$ in.

Find the missing length in each triangle.

24.

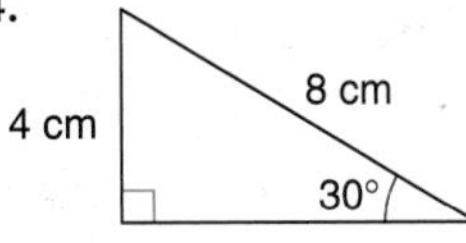

25.

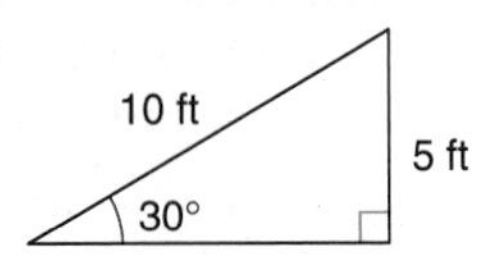

26.

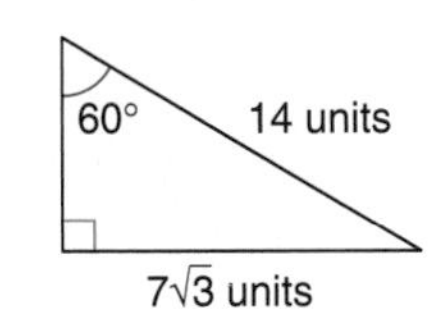

27.

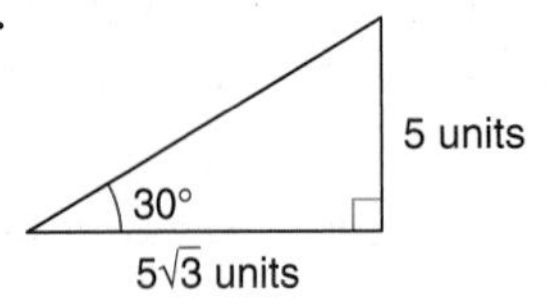

28.

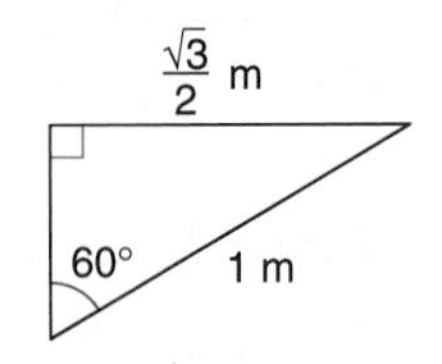

29.

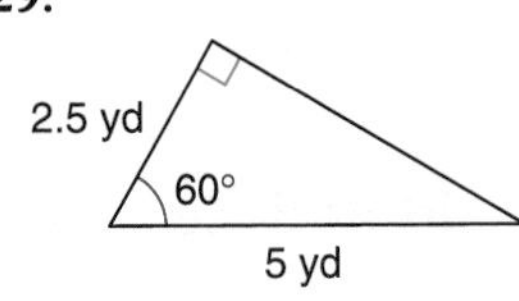

Find the missing lengths for each 30°–60° right triangle described below.

	Hypotenuse	Side Opposite 30° Angle	Side Opposite 60° Angle
30.	6 m	?	?
31.	4.75 mm	?	?
32.	$3\frac{1}{2}$ in.	?	?
33.	?	8 cm	?
34.	?	6.5 m	?
35.	?	$3\frac{1}{4}$ in.	?
36.	?	?	$2\sqrt{3}$ m
37.	?	?	$3.5\sqrt{3}$ cm

38. In $\triangle ABC$ below, $\overline{AC}$ is 10 meters long, AB is $3\sqrt{3}$ meters long, and $\angle A$ measures 30°. Find the length of $\overline{BC}$.

39. In $\triangle PQR$ below, $\overline{PS}$ is $8\sqrt{3}$ yards long. Find the perimeter of $\triangle PQR$.

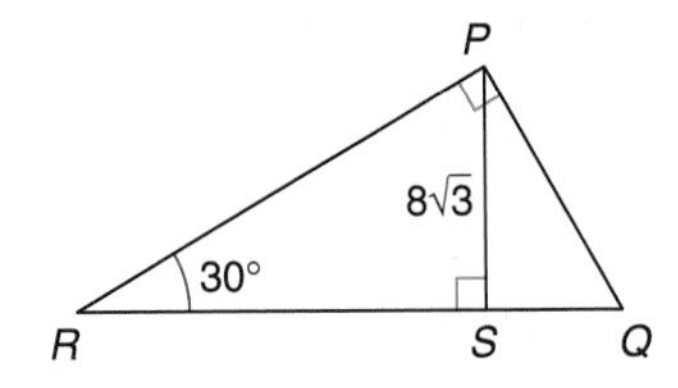

40. Find the perimeter of the quadrilateral shown at the right.

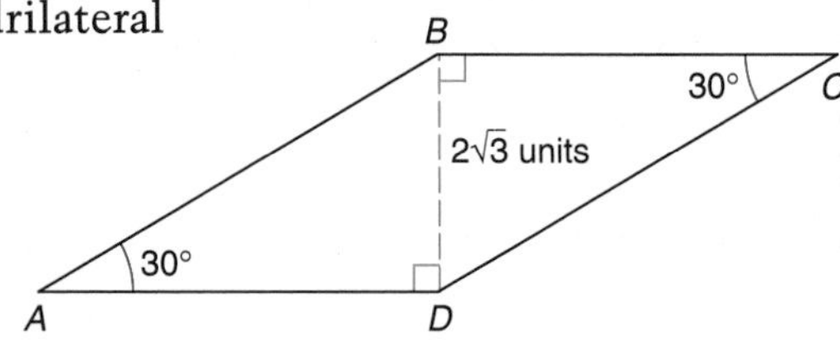

Critical Thinking

41. In rectangle $JKLM$ shown at the right, the measure of $\angle NJK$ is 30° and $\overline{KN}$ is 5 meters long. Find the perimeter of rectangle $JKLM$.

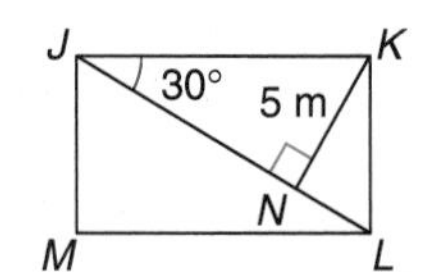

Applications

42. **Architecture** An A-frame house is one that looks like the letter A when viewed from the front. An architect is designing such a house. The sides of the building form an angle of 60° with the floor (base). If the house is to be 30 feet wide at the base, how high will it be?

43. **Construction** Rhonda wants to get a refrigerator into her house. She cannot lift it up the steps leading to her back door, so she decides to build a ramp. She decides that the ramp should make a 30° angle with the ground. If her back door is 2 feet above the ground, how long should Rhonda cut the boards to make the ramp?

Mixed Review

44. Find $-37.12 + 42.18 + (-12.6)$. **(Lesson 2-5)**

45. Factor $3y^2 + 5y - 2$. **(Lesson 7-5)**

46. Solve $2x - y = 3x$ and $2x + y = 3y$ by using substitution. **(Lesson 11-3)**

47. Simplify $-\sqrt{\frac{289}{100}}$. **(Lesson 12-2)**

48. Solve $2t^2 - t = 4$. **(Lesson 13-5)**

49. What is the outlier in the box-and-whisker plot at the right? **(Lesson 14-5)**

30 40 50 60 70 80 90 100

50. Find the complement of 17°. **(Lesson 15-1)**

MID-CHAPTER REVIEW

Find both the complement and supplement of each angle measure. (Lesson 15-1)

1. 85°
2. 127°
3. 65°
4. 108°
5. $x°$
6. $(3x + 5)°$

The length of the side opposite the 60° angle in a 30°–60° right triangle is given. Find the lengths of the other two sides. (Lesson 15-3)

7. $3\sqrt{3}$ m
8. $7\sqrt{3}$ yd
9. $9\sqrt{3}$ mm
10. $\frac{\sqrt{3}}{3}$ in.

11. Solve by making a model. **(Lesson 15-2)**

The bowling pin machine at Abbey Lanes is broken, making the pins point in the wrong direction. Can you make them face the opposite direction by moving just three pins? Explain. (*Hint:* All of the pins are exactly alike.)

15-4 Similar Triangles

Objective

After studying this lesson, you should be able to:

- find the unknown measures of the sides of two similar triangles.

Application

Janet is an ichnologist. She studies dinosaur tracks to estimate their speeds and weights. She has taken a picture of a fossilized trail and wants to develop it.

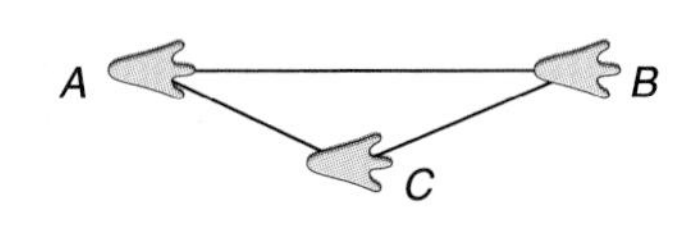

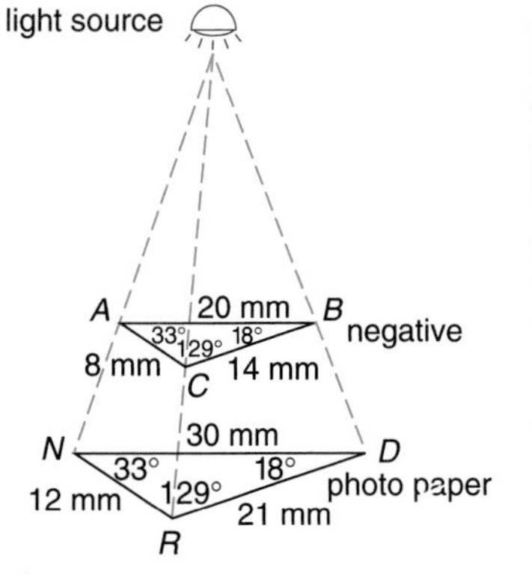

$\triangle ABC$ is on a photo negative. $\triangle NDR$ is its image on the photo paper below. An angle and its image are called **corresponding angles.** The sides opposite corresponding angles are called **corresponding sides.**

corresponding angles	**corresponding sides**
$\angle A$ and $\angle N$	$\overline{BC}$ and $\overline{DR}$
$\angle B$ and $\angle D$	$\overline{AC}$ and $\overline{NR}$
$\angle C$ and $\angle R$	$\overline{AB}$ and $\overline{ND}$

Two figures are **similar** if they have the same shape but not necessarily the same size. If corresponding angles of two triangles have equal measures, the triangles are similar. The two triangles in the example are similar. We write $\triangle ABC \sim \triangle NDR$. The order of the letters indicates the angles that correspond.

Compare the measures of the corresponding sides. Note that BC means the measure of $\overline{BC}$, DR means the measure of $\overline{DR}$, and so on.

$$\frac{BC}{DR} = \frac{14}{21} = \frac{2}{3} \qquad \frac{AC}{NR} = \frac{8}{12} = \frac{2}{3} \qquad \frac{AB}{ND} = \frac{20}{30} = \frac{2}{3}$$

When the measures of the corresponding sides form equal ratios, the measures are said to be **proportional.**

Similar Triangles

If two triangles are similar, the measures of their corresponding sides are proportional, and the measures of their corresponding angles are equal.

Example 1

If a tree 6 feet tall casts a shadow 4 feet long, how high is a flagpole that casts a shadow 18 feet long?

$\triangle JKL$ is similar to $\triangle PQR$.

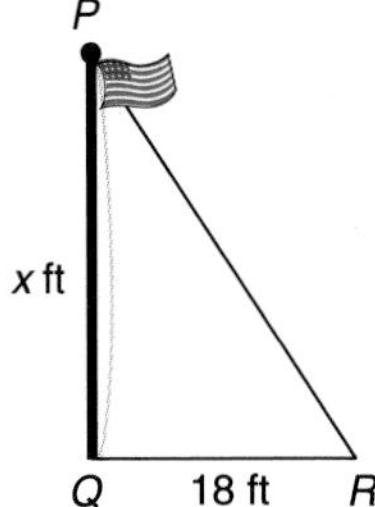

$\frac{JK}{PQ} = \frac{KL}{QR}$ *Corresponding sides of similar triangles are proportional.*

$\frac{6}{x} = \frac{4}{18}$

$4x = 6(18)$ *Cross multiply.*

$4x = 108$

$x = 27$ The flagpole is 27 feet high.

Similar triangles do not have to be positioned in the same way.

Example 2

Find the missing measures of the sides.

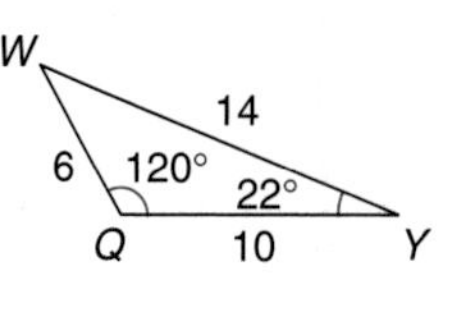

The measure of $\angle V = 180° - (120° + 38°)$ or 22°.
The measure of $\angle W = 180° - (120° + 22°)$ or 38°.

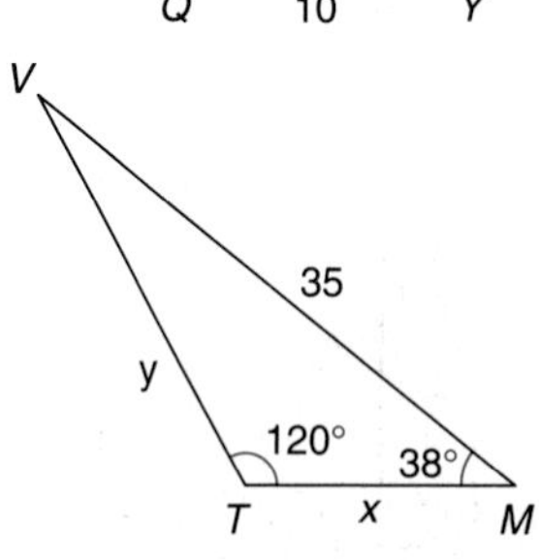

Since the corresponding angles have equal measures, $\triangle VMT \sim \triangle YWQ$. This means that the lengths of the corresponding sides are proportional.

$$\frac{VM}{YW} = \frac{MT}{WQ} \qquad \frac{VM}{YW} = \frac{VT}{YQ}$$

$$\frac{35}{14} = \frac{x}{6} \qquad \frac{35}{14} = \frac{y}{10}$$

$$14x = 35(6) \qquad 14y = 35(10)$$

$$14x = 210 \qquad 14y = 350$$

$$x = 15 \qquad y = 25$$

The missing measures are 15 and 25.

Example 3

FYI . . .

The record for the longest consecutive run in billiards is 625 shots, set by Michael Eufemia on February 2, 1960.

Eddie is playing billiards on a table like the one shown at the right. If he can make this next shot, he figures he will have no trouble winning. He wants to strike the cue ball at *D*, bank it at *C*, and hit another ball at the mouth of pocket *A*. Use similar triangles to find where Eddie's cue ball should strike the rail.

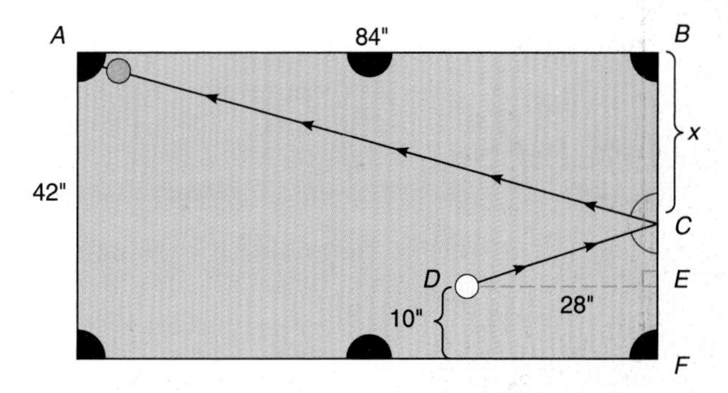

The ball bounces off the cushion at the same angle at which it hits. Since $\angle DCE$ has the same measure as $\angle ACB$, $\triangle ABC \sim \triangle DEC$.

$\frac{BC}{EC} = \frac{AB}{DE}$ *$\overline{BC}$, $\overline{EC}$ and $\overline{AB}$, $\overline{DE}$ are pairs of corresponding sides.*

$\frac{x}{32 - x} = \frac{84}{28}$ *BC = x, EC = 32 − x, AB = 84, DE = 28*

$\frac{x}{32 - x} = \frac{3}{1}$ *Simplify.*

$x = 3(32 - x)$ *Cross multiply.*

$x = 96 - 3x$ *Distributive property*

$4x = 96$

$x = 24$

Eddie should aim the cue ball 24 inches below pocket *B*.

CHECKING FOR UNDERSTANDING

Communicating Mathematics

Read and study the lesson to answer each question.

1. In Example 2, name the pairs of corresponding angles.
2. If two triangles are similar, list two things you know about the triangles.

Guided Practice

If $\triangle BIG \sim \triangle RED$, complete each of the following.

3. List the corresponding angles.
4. List the corresponding sides.
5. List three proportions that correspond to these triangles.

For each pair of similar triangles, name the triangle that is similar to $\triangle ABC$. Make sure you have the letters in the correct order.

6.

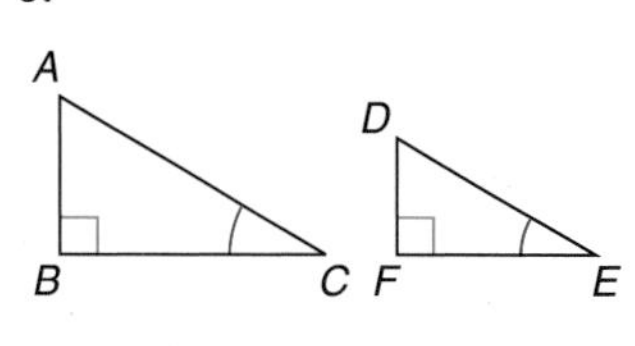

7.

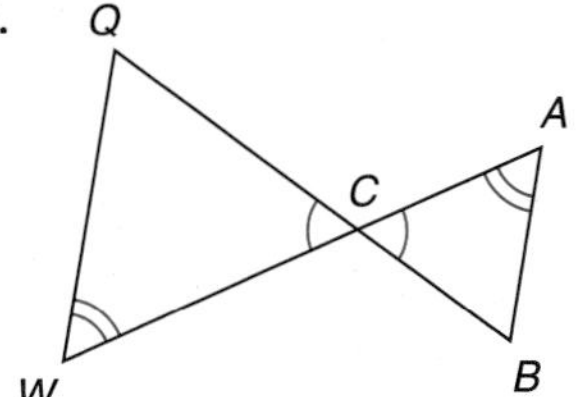

8.

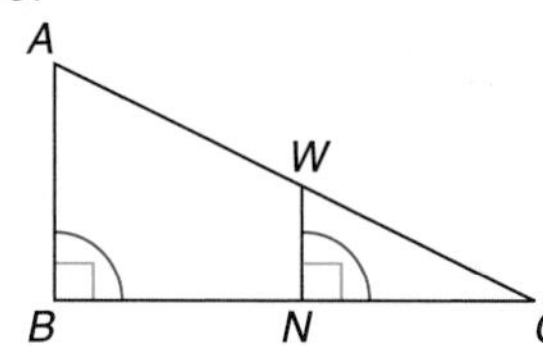

EXERCISES

Practice

Determine if each pair of triangles is similar.

9.

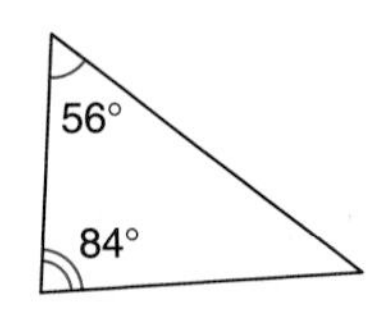

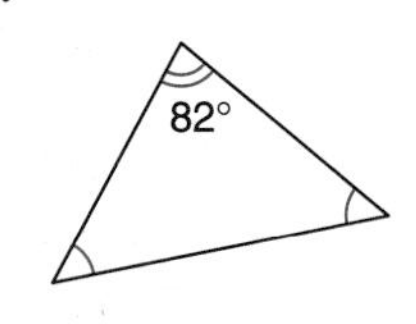

10.

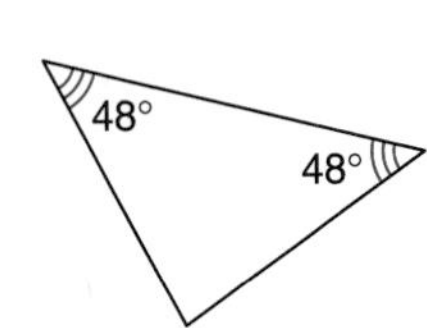

$\triangle ABC$ and $\triangle DEF$ are similar. For each set of measures given, find the measures of the remaining sides.

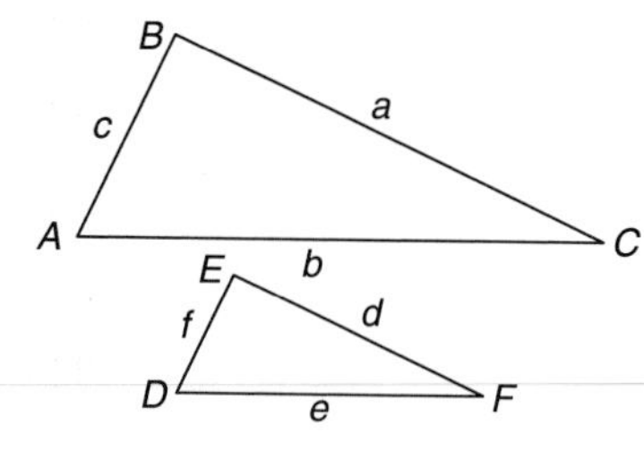

11. $a = 5, d = 7, f = 6, e = 5$
12. $c = 11, f = 6, d = 5, e = 4$
13. $b = 4.5, d = 2.1, e = 3.4, f = 3.2$
14. $a = 16, c = 12, b = 13, e = 7$
15. $a = 17, b = 15, c = 10, f = 6$
16. $c = 18, f = 12, d = 18, e = 16$
17. $a = 4\frac{1}{4}, b = 5\frac{1}{2}, e = 2\frac{3}{4}, f = 1\frac{3}{4}$
18. $c = 7\frac{1}{2}, f = 5, a = 10\frac{1}{2}, b = 15$
19. $c = 5, f = 2.5, a = 12.6, e = 8.1$

Critical Thinking

20. Find the value of x.

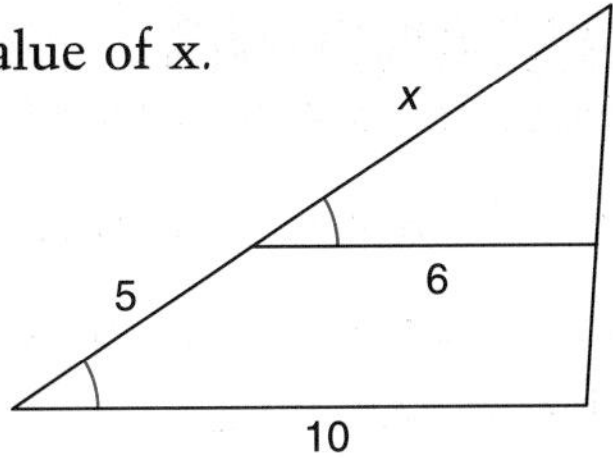

21. $\triangle ABC$ and $\triangle RST$ are similar. If the ratio of AC to RT is 2 to 3, what is the ratio of the area of $\triangle ABC$ to the area of $\triangle RST$?

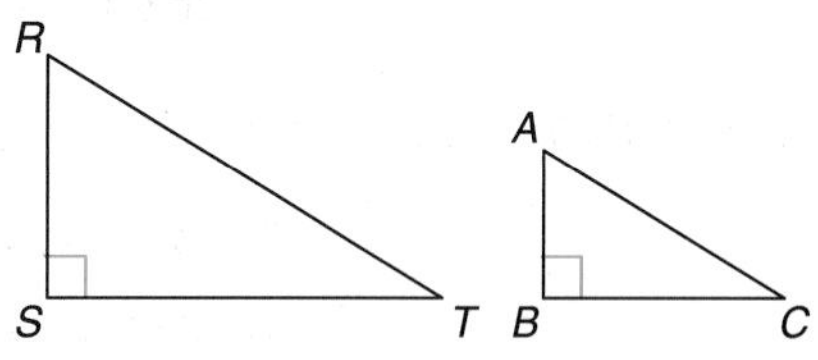

Applications

22. **Construction** In building a roof, a 5-foot support is to be placed at point B as shown on the diagram. Find the length of the support that is to be placed at point A.

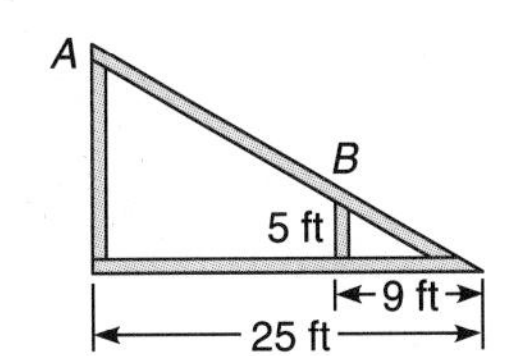

> **Journal**
> Explain how similar triangles are related to what you learned about proportions in Lesson 4-1. Include an example.

23. **Surveying** $\triangle ABC$ is similar to $\triangle EDC$. Find the distance across the lake from point A to point B.

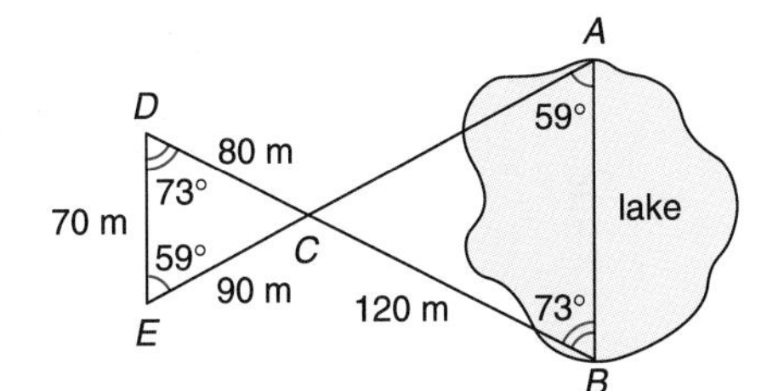

24. **Cinematography** When a movie theater is being designed, the size of the screen depends on the placement of the projector lens. Suppose a theater requires a screen 14 feet wide (DE). The distance from the projector to the screen (AC) is 90 feet. The film is 0.87 inches wide (FG). Find the focal length (AB) of the lens needed for this theater to the nearest tenth of an inch.

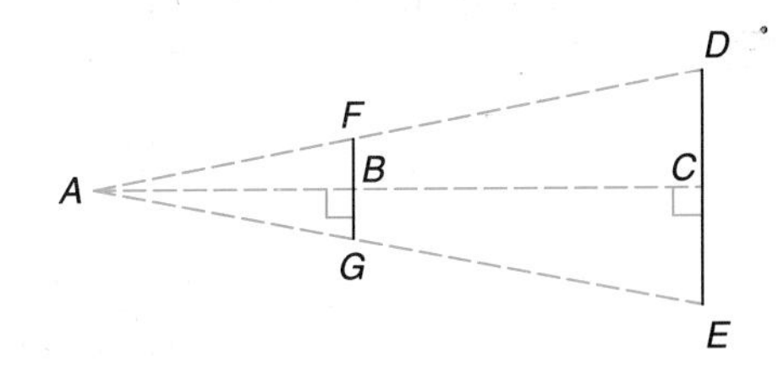

Mixed Review

25. Solve $13 - 8x = 5x + 2$. **(Lesson 3-6)**

26. Find $\frac{6}{a^2 - 2ab + b^2} - \frac{6}{a - b}$. **(Lesson 8-7)**

27. Solve the system $2x + y = 8$ and $x - y = 3$. **(Lesson 11-4)**

28. Simplify $\sqrt{48} - \sqrt{12} + \sqrt{300}$. **(Lesson 12-6)**

29. Solve $x^2 - 4x + 1 = 0$ by completing the square. **(Lesson 13-4)**

30. **Statistics** Olivia swims the 50-yard freestyle for the Wachung High School swim team. Her times in the last six meets were 26.89 seconds, 26.27 seconds, 25.18 seconds, 25.63 seconds, 27.16 seconds, and 27.18 seconds. Find the mean and median of her times. **(Lesson 14-3)**

31. The length of the side opposite the 60° angle in a 30°–60° right triangle is $5\sqrt{3}$ cm. Find the length of the other two sides. **(Lesson 15-3)**

15-5 Trigonometric Ratios

Objectives After studying this lesson, you should be able to:

- compute the sine, cosine, and tangent of an acute angle of a right triangle given the measures of its sides, and
- find the measure of an acute angle of a right triangle given a trigonometric value or the lengths of two of the sides.

If enough is known about a right triangle, certain ratios can be used to find the measures of the remaining parts of the triangle. These ratios are called **trigonometric ratios.**

A typical right triangle is shown at the right.

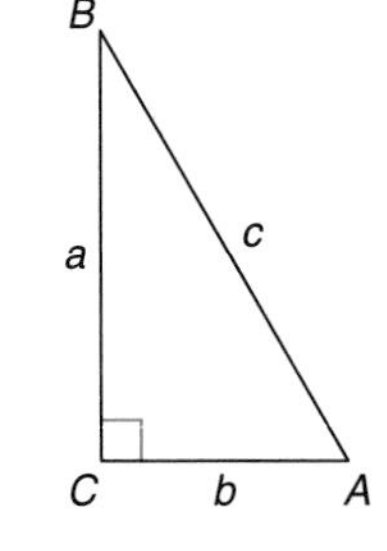

a is the measure of the side *opposite* $\angle A$, $\overline{BC}$.
b is the measure of the side *opposite* $\angle B$, $\overline{AC}$.
c is the measure of the side *opposite* $\angle C$, $\overline{AB}$.

a is *adjacent* to $\angle B$ and $\angle C$.
b is *adjacent* to $\angle A$ and $\angle C$.
c is *adjacent* to $\angle A$ and $\angle B$.

Recall that the side opposite $\angle C$, the right angle, is called the **hypotenuse.** The other two sides are called **legs.**

Three common trigonometric ratios are defined as follows.

Definition of Trigonometric Ratios

$$\text{sine of } \angle A = \frac{\text{measure of side opposite } \angle A}{\text{measure of hypotenuse}}$$

$$\sin A = \frac{a}{c}$$

$$\text{cosine of } \angle A = \frac{\text{measure of side adjacent to } \angle A}{\text{measure of hypotenuse}}$$

$$\cos A = \frac{b}{c}$$

$$\text{tangent of } \angle A = \frac{\text{measure of side opposite } \angle A}{\text{measure of side adjacent to } \angle A}$$

$$\tan A = \frac{a}{b}$$

Notice that sine, cosine, and tangent are abbreviated as sin, cos, and tan, respectively.

Example 1

Find the sine, cosine, and tangent of each acute angle. Round your answers to the nearest thousandth.

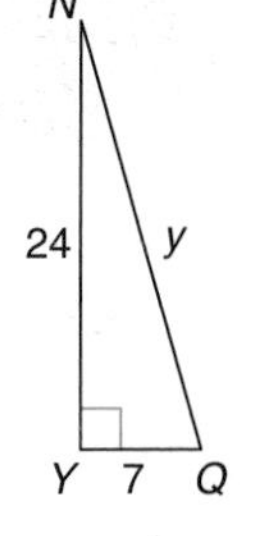

Use the Pythagorean Theorem to find the value of y.

$$7^2 + 24^2 = y^2$$
$$49 + 576 = y^2$$
$$625 = y^2$$
$$25 = y$$

$\sin N = \dfrac{\text{opposite leg}}{\text{hypotenuse}}$ | $\sin Q = \dfrac{\text{opposite leg}}{\text{hypotenuse}}$

$\sin N = \dfrac{7}{25}$ or 0.280 | $\sin Q = \dfrac{24}{25}$ or 0.960

$\cos N = \dfrac{\text{adjacent leg}}{\text{hypotenuse}}$ | $\cos Q = \dfrac{\text{adjacent leg}}{\text{hypotenuse}}$

$\cos N = \dfrac{24}{25}$ or 0.960 | $\cos Q = \dfrac{7}{25}$ or 0.280

$\tan N = \dfrac{\text{opposite leg}}{\text{adjacent leg}}$ | $\tan Q = \dfrac{\text{opposite leg}}{\text{adjacent leg}}$

$\tan N = \dfrac{7}{24}$ or about 0.292 | $\tan Q = \dfrac{24}{7}$ or about 3.429

Consider triangles MTP and DFE. Since corresponding angles have the same measure, $\triangle MTP \sim \triangle DFE$. In similar triangles, the corresponding sides are proportional.

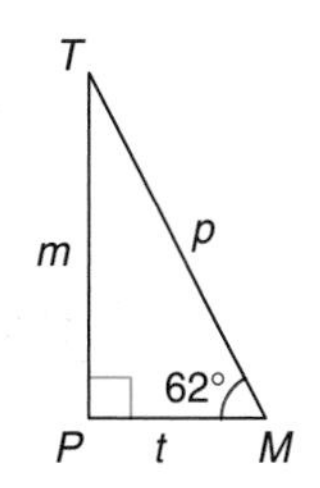

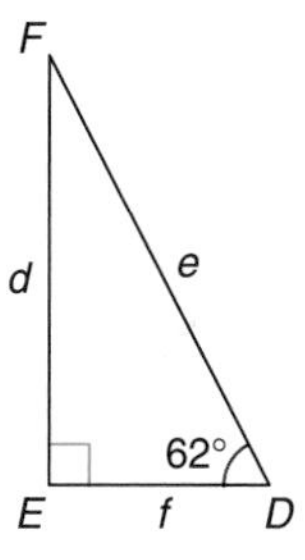

$$\frac{d}{m} = \frac{e}{p}$$

$$\frac{m}{e} \cdot \frac{d}{m} = \frac{m}{e} \cdot \frac{e}{p} \quad \textit{Multiply each side by } \frac{m}{e}.$$

$$\frac{d}{e} = \frac{m}{p}$$

$$\sin D = \sin M$$

FYI ... The first table of values for the sine and cosine was calculated by the Greek mathematician Hipparchus during the 2nd century B.C.

In general, the sine of a 62° angle of a right triangle will be the same number no matter how big or small the triangle is. A similar result holds for cosine and tangent.

Values of trigonometric functions can be found using the table on page 646 or by using a calculator.

Example 2

Find the value of sin 62° to the nearest ten thousandth.

Use a calculator.

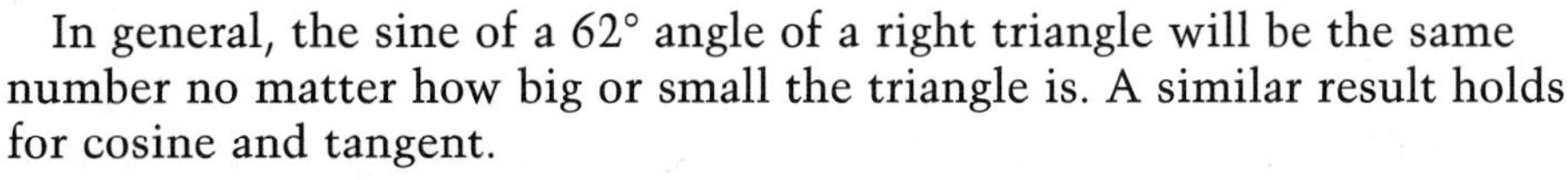

Enter: 62 [SIN] 0.8829476 *The calculator must be in degree mode.*

Rounded to the nearest ten thousandth, $\sin 62° \approx 0.8829$.

Example 3

Find the measure of $\angle H$ to the nearest degree.

Use the cosine since the lengths of the side adjacent to $\angle H$ and the hypotenuse are given.

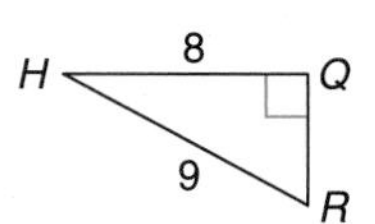

$$\cos H = \frac{\text{adjacent leg}}{\text{hypotenuse}}$$
$$= \frac{8}{9}$$

Use a calculator.

Enter: 8 [÷] 9 [=] [INV] [COS] 27.266044 *The inverse key "undoes" the original function.*

To the nearest degree, the measure of $\angle H$ is 27°.

Example 4

The steepest grade of any standard railway system in the world is in France between Chedde and Servoz. The slope of the track is $\frac{1}{11}$. Find the measure of the angle formed by the track between Chedde and Servoz with the horizontal.

Make a model.

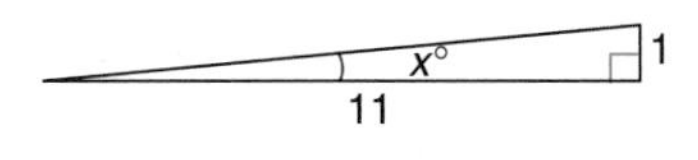

$$\text{slope} = \frac{\text{rise}}{\text{run}} = \frac{1}{11}$$

Since the length of the opposite and adjacent sides are known, use the tangent.

$$\tan x° = \frac{\text{opposite leg}}{\text{adjacent leg}}$$
$$= \frac{1}{11}$$
$$= 0.090909$$ *Use a calculator to find x.*
$$x \approx 5.1944$$

To the nearest degree, the measure of the angle is 5°.

Example 5

When a beam of light in air enters water, the beam is bent. According to Snell's Law, $\frac{\sin y°}{\sin x°} = 0.752$. If $x = 30$, find y to the nearest degree.

$$\frac{\sin y°}{\sin 30°} = 0.752$$ *Use a calculator to find sin 30°.*
$$\frac{\sin y°}{0.5} = 0.752$$
$$\sin y° = (0.5)(0.752)$$
$$\sin y° = 0.376$$ *Use a calculator to find y.*
$$y \approx 22.086$$

To the nearest degree, the measure of the angle is 22°.

CHECKING FOR UNDERSTANDING

Communicating Mathematics

Read and study the lesson to answer each question.

1. What is the measure of the leg opposite $\angle Y$?
2. What is the measure of the leg adjacent to $\angle Y$?
3. What trigonometric ratio or ratios involve the measure of the hypotenuse?

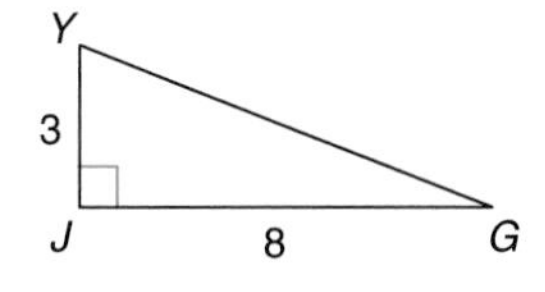

4. What trigonometric ratio or ratios do *not* involve the measure of the hypotenuse?

Guided Practice

Using $\triangle NTK$ express each trigonometric ratio as a fraction.

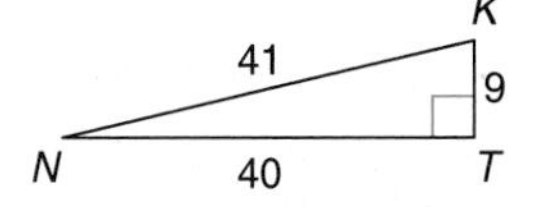

5. $\sin N$
6. $\sin K$
7. $\cos N$
8. $\cos K$
9. $\tan N$
10. $\tan K$
11. Are $\angle N$ and $\angle K$ complementary angles?
12. What is true about the sine of an angle and the cosine of that angle's complement?

Use a calculator to find the value of each trigonometric ratio to the nearest ten thousandth.

13. $\cos 25°$
14. $\tan 31°$
15. $\sin 71°$
16. $\cos 64°$
17. $\tan 9°$
18. $\sin 2°$

EXERCISES

Practice

For each triangle, find $\sin N$, $\cos N$, and $\tan N$ to the nearest thousandth.

19.

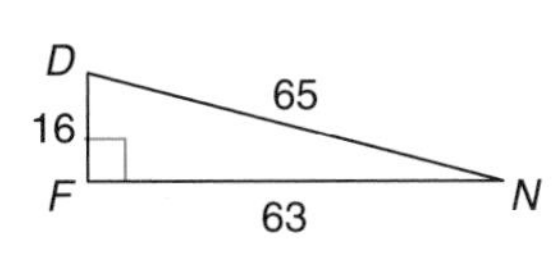

20.

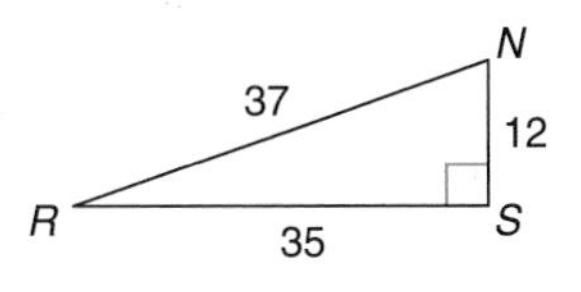

21.

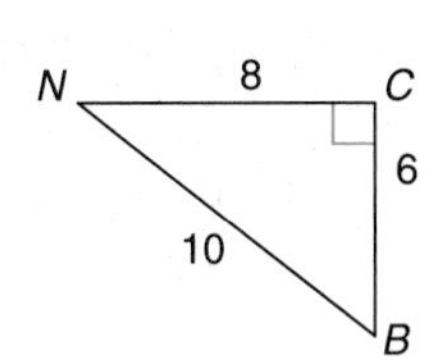

22.

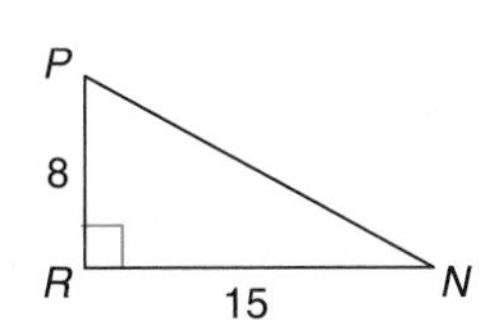

23.

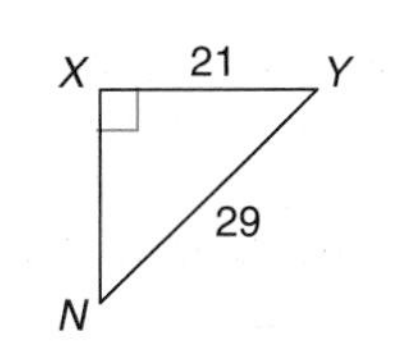

24.

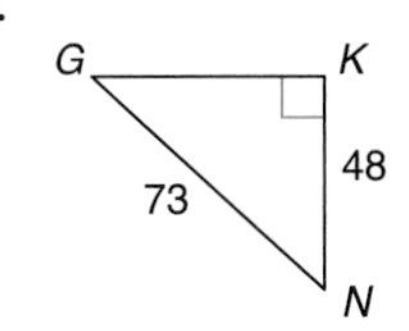

Use a calculator to find the measure of each angle to the nearest degree.

25. $\sin A = 0.2756$ **26.** $\cos B = 0.8480$ **27.** $\cos W = 0.2598$

28. $\sin N = 0.6124$ **29.** $\tan V = 0.956$ **30.** $\tan Q = 7.84$

For each triangle, find the measure of the marked acute angle to the nearest degree.

31.

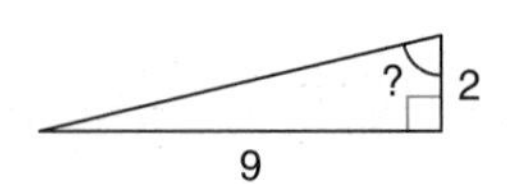

32.

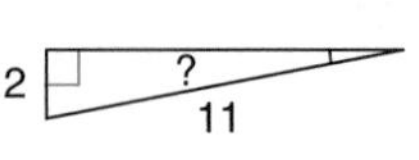

33.

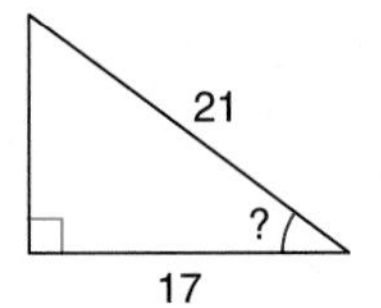

34.

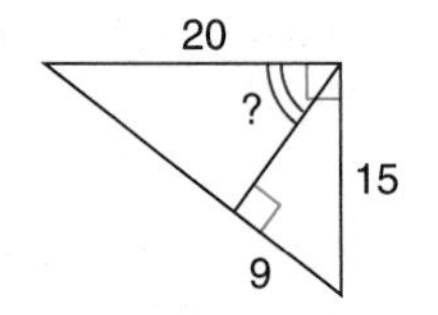

35.

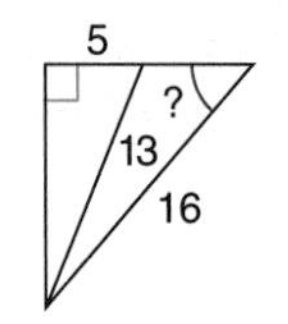

36.

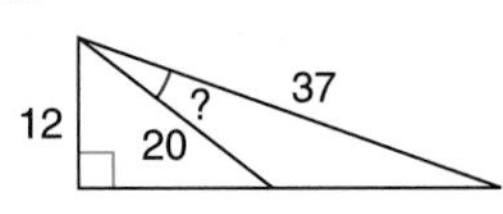

37. The formula area $= \frac{1}{2}bc \sin A$ can be used to find the area of any triangle. Find the area of the triangle at the right to the nearest tenth of a square centimeter if $b = 16$, $c = 9$, and $\angle A$ is a 36° angle.

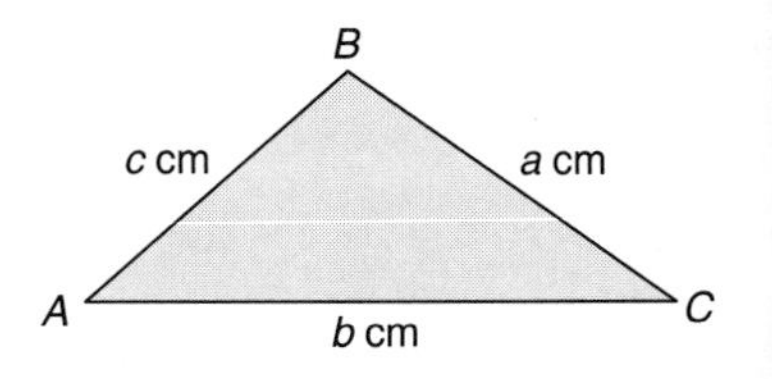

Critical Thinking

38. Use the triangle at the right to determine which of the following statements are true. Justify your answers.

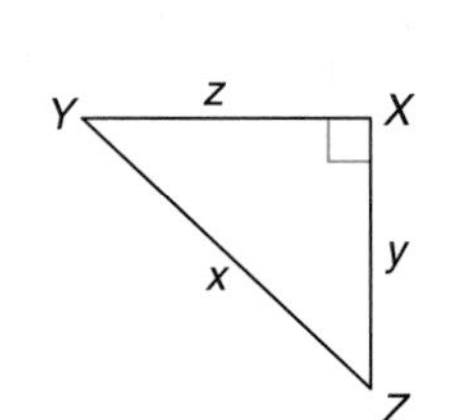

a. $\sin Z = \cos Y$

b. $\cos Y = \dfrac{1}{\sin Y}$

c. $\tan Z = \dfrac{\cos Z}{\sin Z}$

d. $\tan Z = \dfrac{\sin Z}{\cos Z}$

e. $\sin Y = (\tan Y)(\cos Y)$

Applications

39. Physics When an object is fired into the air, its path forms a parabola. The range of the object is given by $R = \frac{v^2}{g} \sin (2x)°$, where v is the velocity of the object when fired in feet/second, x is the angle at which the object is fired, and g is 32 feet/second². Find the range to the nearest foot of an object fired at 40 feet per second on an angle of 35°.

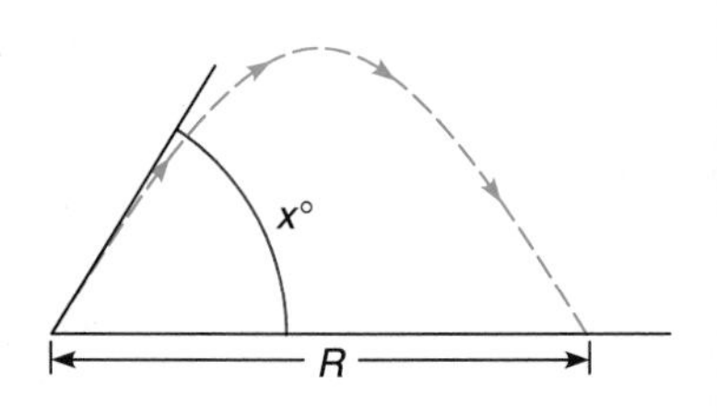

40. **Meteorology** The mean temperature in Chicago on the nth day of the year can be estimated using the formula

$$t = 25.5 \sin\left[\frac{360}{365}(n - 106)\right]^\circ + 50,$$

where t is the mean temperature for the day in degrees Fahrenheit. Find the mean temperature on October 1 (the 274th day of the year) to the nearest degree.

41. **Physics** Refer to the diagram in Example 5. The following equation is also true.

$$\frac{\sin y^\circ}{\sin x^\circ} = \frac{\text{the speed of light in water}}{\text{the speed of light in air}}$$

The speed of light in air is about 3×10^8 meters/second. If $x = 40$, find y to the nearest degree using Snell's Law. Find the speed of light in water to two significant digits.

Computer

42. Given the lengths of the legs of a right triangle, this BASIC program computes the length of the hypotenuse, the measures of the acute angles, and the values of the trigonometric ratios for each acute angle.

```
10 INPUT "ENTER THE LENGTHS OF
   THE LEGS: ";L1,L2
20 LET P = 3.1415927
30 LET H = SQR(L1^2 + L2^2)
40 LET A1 = ATN(L1/L2) * 180/P
50 LET A2 = ATN(L2/L1) * 180/P
60 PRINT "SIDES: ";L1;", ";L2;", ";H
70 PRINT "ANGLES: ";A1;", ";A2;", 90"
80 PRINT "SIN(";A1;") = ";L1/H
90 PRINT "SIN(";A2;") = ";L2/H
100 PRINT "COS(";A1;") = ";L2/H
110 PRINT "COS(";A2;") = ";L1/H
120 PRINT "TAN(";A1;") = ";L1/L2
130 PRINT "TAN(";A2;") = ";L2/L1
140 END
```

Given the lengths of the legs of each right triangle, use the program to find the length of the hypotenuse, the angle measures, and the trigonometric ratios of the acute angles.

a. 3, 4   b. 5, 12   c. 20, 40   d. 125, 100

Mixed Review

43. Nineteen is what percent of 76? **(Lesson 4-1)**

44. State the domain and range of $\{(1, 1), (2, 4), (-2, 4), (3, 9)\}$. **(Lesson 9-2)**

45. Solve the system $3x - 2y = 10$ and $x + y = 0$ by graphing. **(Lesson 11-1)**

46. Simplify $\sqrt{120a^3b}$. **(Lesson 12-5)**

47. **Statistics** The Goodwill Games were first held in Moscow in July 1985. The total number of medals won by the top eleven countries are 241, 28, 31, 18, 6, 10, 6, 142, 9, 11, and 6. Find the interquartile range. **(Lesson 14-4)**

15-6 Solving Right Triangles

Objectives After studying this lesson, you should be able to:

- use trigonometric ratios to solve verbal problems, and
- use trigonometric ratios to solve right triangles.

Solving a triangle means to find all of its missing measures.

Example 1

Solve $\triangle PQR$.

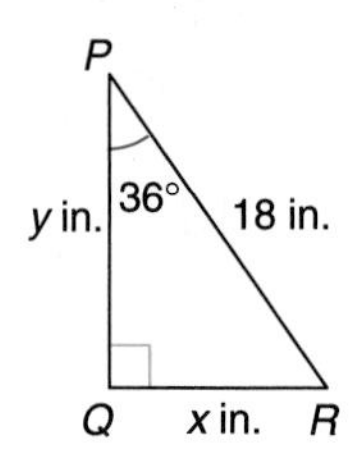

The measure of $\angle R$ is $90° - 36°$ or $54°$.

$\sin 36° = \frac{x}{18}$

$0.5878 \approx \frac{x}{18}$

$10.6 \approx x$

Thus, $\overline{QR}$ is about 10.6 inches long.

$\cos 36° = \frac{y}{18}$

$0.8090 \approx \frac{y}{18}$

$14.6 \approx y$

Thus, $\overline{PQ}$ is about 14.6 inches long.

When using trigonometric ratios to solve problems, you will often deal with angles of elevation and depression. An **angle of elevation** is formed by a horizontal line and another line of sight above it. An **angle of depression** is formed by a horizontal line and another line of sight below it.

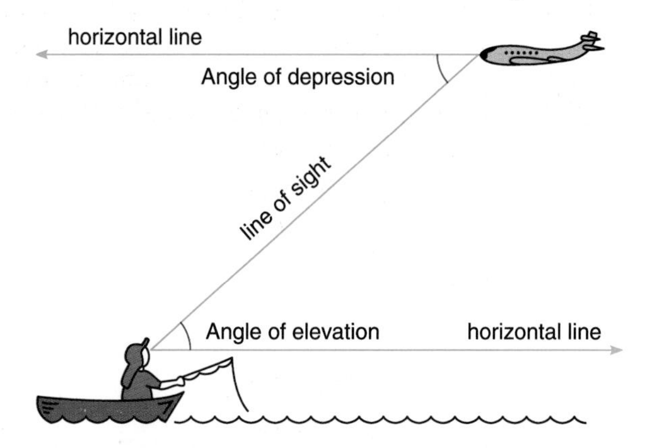

These concepts are used with trigonometric ratios to find missing measures of right triangles. Sometimes these measures are difficult to obtain directly.

Example 2

The National Champion Elm Tree in Louisville, Kansas is 97 feet tall. As you are walking in your neighborhood, you notice a tall elm tree. Could this tree be taller than the National Champion? You walk 50 feet from the base of the tree. From this position, the angle of elevation to the top of the tree is 65°. Find the height of the tree to the nearest foot.

First, make a model. Notice that 50 feet is the length of the leg adjacent to the 65° angle and h is the length of the leg opposite the 65° angle. The tangent ratio should be used to solve this problem.

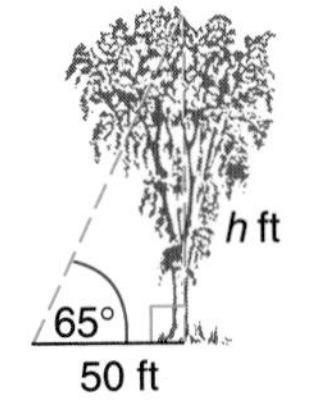

$$\tan 65° = \frac{\text{opposite leg}}{\text{adjacent leg}}$$

$$\tan 65° = \frac{h}{50}$$

$$2.1445 \approx \frac{h}{50}$$

$$2.1445(50) \approx h$$

$107.225 \approx h$ The tree is about 107 feet tall.

Example 3

Jack is fishing for salmon. He has 60 yards of line in the water, and the angle of depression to the lure is 20°. Find the lowest possible depth of the lure.

$$\sin 20° = \frac{\text{opposite leg}}{\text{hypotenuse}}$$

$$0.3420 \approx \frac{d}{60}$$

$$60\ (0.3420) \approx d$$

$$20.52 \approx d$$

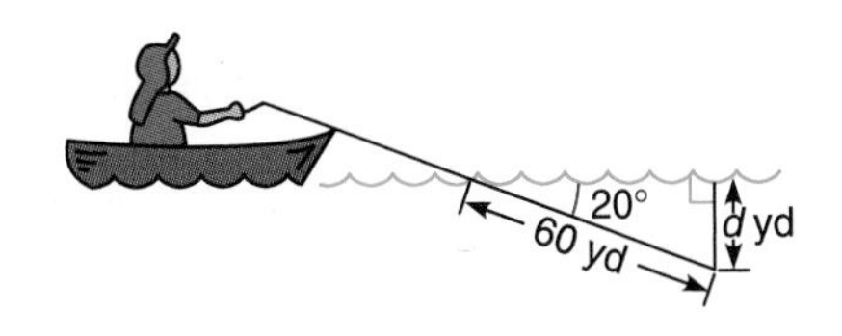

The lure could be at most about 21 yards below the surface.

Example 4

Alexis Martin has been hired to roof houses during the summer. She needs to purchase an extension ladder that reaches at least 24 feet off the ground. Ladder manufacturers recommend the angle formed by the ladder and the ground be no more than 75°. What is the shortest ladder she could buy to reach 24 feet safely?

$$\sin 75° = \frac{\text{opposite leg}}{\text{hypotenuse}}$$

$$0.9659 \approx \frac{24}{c}$$

$$0.9659c \approx 24$$

$$c \approx \frac{24}{0.9659}$$

$$c \approx 24.847$$

To the nearest foot, the ladder should be at least 25 feet long.

CHECKING FOR UNDERSTANDING

Communicating Mathematics

Read and study the lesson to answer each question.

1. In Example 1, once you know x is about 10.6, how could you find y without using trigonometry?
2. In all four examples you were given a right angle and the measures of two other parts of the triangle. Can you think of a right triangle in which you were given two other parts and could not solve the triangle?

Guided Practice

Name the angles of elevation and depression in each drawing.

3. 4. 5.

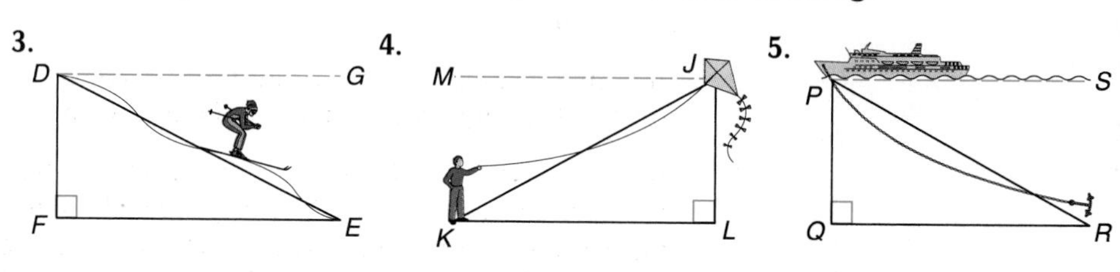

State which trigonometric ratio you could use to find x.

6.

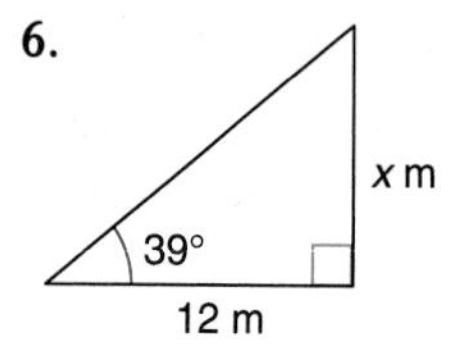

7.

46°
x ft
15 ft

8.

12 ft
x°
13 ft

9. Find the value of x in Exercise 6 to the nearest tenth.
10. Find the value of x in Exercise 8 to the nearest degree.

EXERCISES

Practice

Solve each triangle. Give side lengths to the nearest tenth and angle measures to the nearest degree.

11.

A
21°
13 in.
C
B

12.

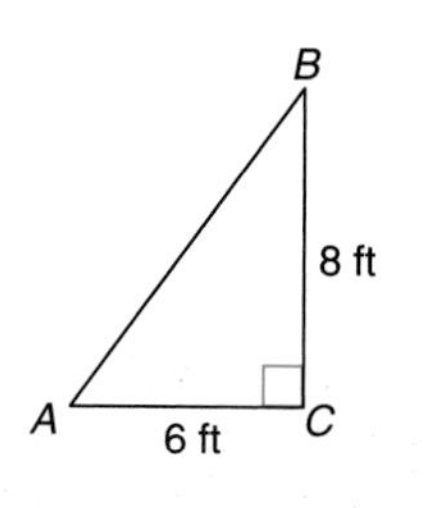

13.

A
C
70°
9 cm
B

14.

A
16 m
60°
C
B

15.

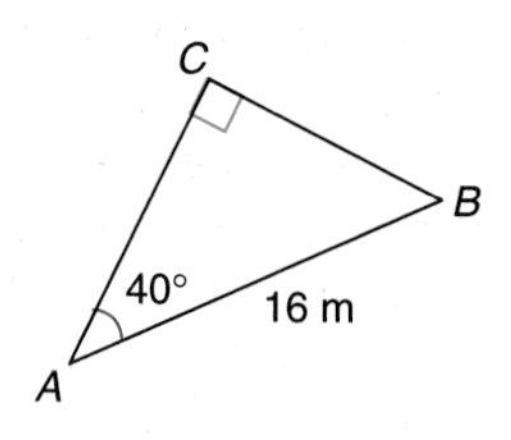

16.

A
35°
C
B
7 km

Use trigonometric ratios to solve each problem.

17. At a point 200 feet from the base of a flagpole, the angle of elevation is 62°. Find the height of the flagpole and the distance from the point to the top.

18. A guy wire is fastened to a TV tower 40 feet above the ground and forms an angle of 52° with the tower. How long is the wire?

19. A chimney casts a shadow 75 feet long when the angle of elevation of the sun is 41°. How tall is the chimney?

20. In a parking garage, there are 20 feet between each level. Each ramp to a level is 130 feet long. Find the measure of the angle of elevation of each ramp.

21. Find the area of a right triangle in which one acute angle measures 25° and the leg opposite that angle is 40 cm long.

22. From the top of a 70-meter lighthouse, an airplane was observed that was directly over a ship. The angle of elevation of the plane was 18°, while the angle of depression of the ship was 25°. Find the distance from the ship to the foot of the lighthouse and the height of the plane.

23. A weather balloon is directly above a tree. The angle of elevation is 60° when you are 100 meters from the tree. How high is the balloon?

24. A train in the mountains rises 15 feet for every 250 feet it moves along the track. Find the angle of elevation of the track.

Critical Thinking

25. Find the equation of this line in slope-intercept form.

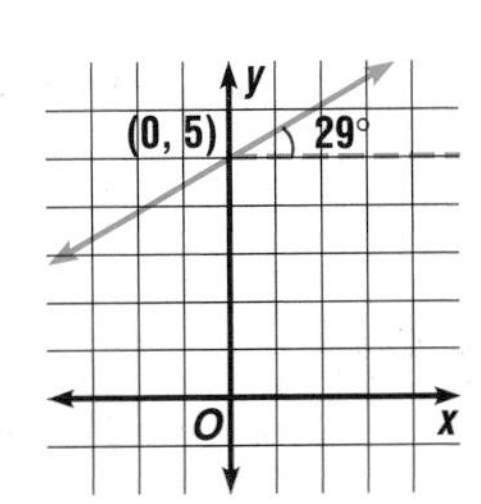

26. Solve the triangle. (*Hint:* Draw an altitude from *B*.)

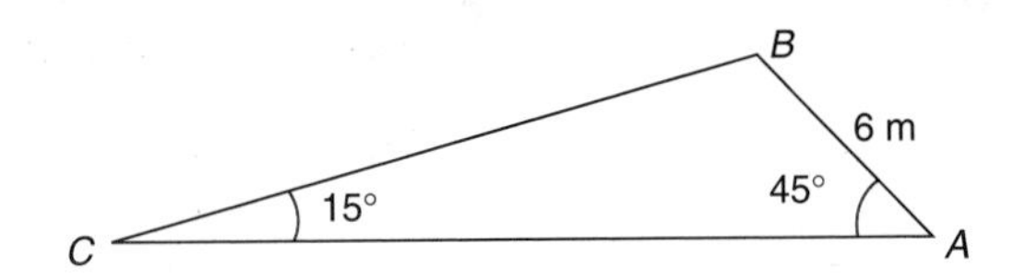

Applications

27. **Space Travel** An astronaut in a spacecraft h miles from Earth measures the angle ($\angle ABD$) formed by the lines of sight to the Earth's horizon to be 80°. Use right triangle ABC to find h to the nearest mile.

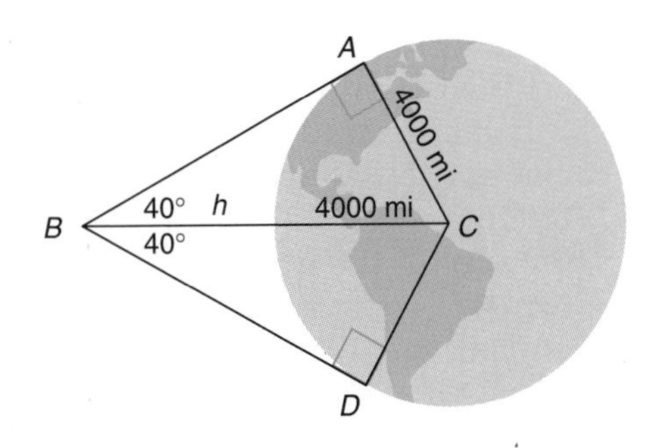

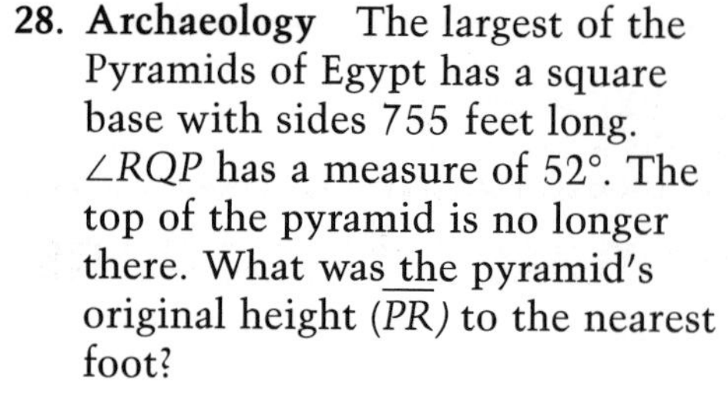

28. **Archaeology** The largest of the Pyramids of Egypt has a square base with sides 755 feet long. $\angle RQP$ has a measure of 52°. The top of the pyramid is no longer there. What was the pyramid's original height ($\overline{PR}$) to the nearest foot?

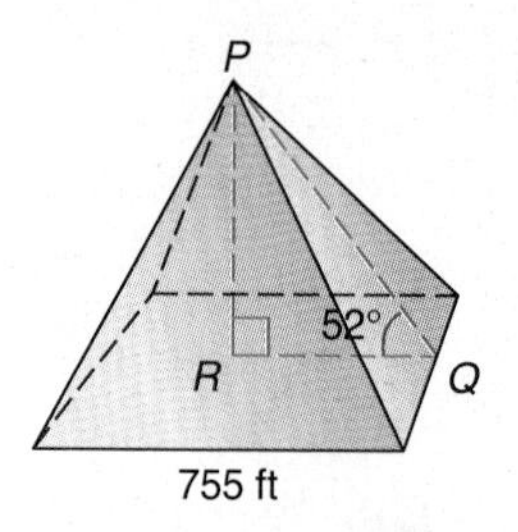

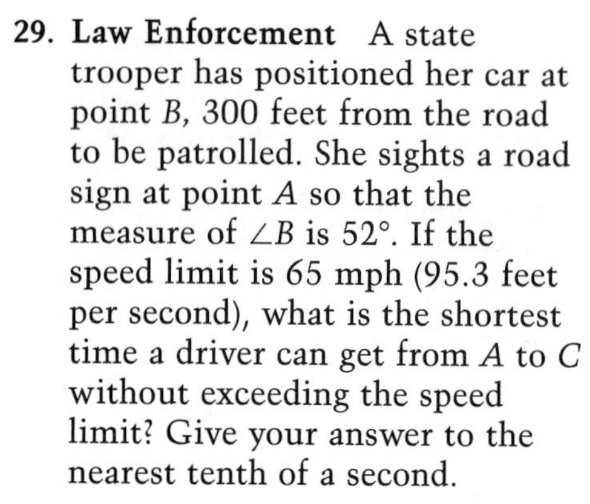

29. **Law Enforcement** A state trooper has positioned her car at point B, 300 feet from the road to be patrolled. She sights a road sign at point A so that the measure of $\angle B$ is 52°. If the speed limit is 65 mph (95.3 feet per second), what is the shortest time a driver can get from A to C without exceeding the speed limit? Give your answer to the nearest tenth of a second.

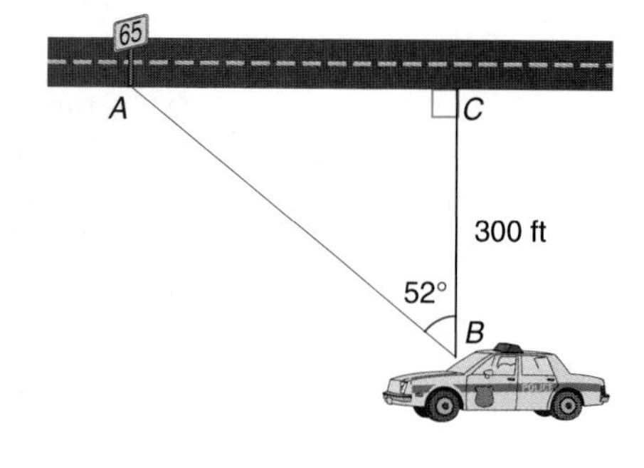

Portfolio

Review the items in your portfolio. Make a table of contents of the items, noting why each item was chosen. Replace any items that are no longer appropriate.

30. **Navigation** How far will a submarine travel when going to a depth of 300 feet if its course has an angle of depression of 25°?

Mixed Review

31. Solve $0.1x < 0.2x - 8$. **(Lesson 5-3)**
32. What is the slope of the line through $(12, -3)$ and $(14, -7)$? **(Lesson 10-1)**
33. Solve the system $2x + y = 3(x - 5)$ and $x + 5 = 4y + 2x$. **(Lesson 11-5)**
34. Solve $\sqrt{2x + 7} = 5$. **(Lesson 12-7)**
35. How many real roots does $2p^2 - p - 3 = 0$ have? **(Lesson 13-6)**
36. Find P (a number divisible by 4) on a roll of a die. **(Lesson 14-7)**
37. For what angle are the sine and cosine equal? **(Lesson 15-5)**

HISTORY CONNECTION

Pythagoras

About 540 B.C., Pythagoras of Samos (585–507 B.C.) founded a school for his followers, called the Pythagoreans, in what is now Italy. The Pythagorean Theorem had been discovered a thousand years earlier by the Babylonians, but the Pythagorean school is credited with proving it. Along with studying numbers, Pythagoras taught his disciples to worship them as well. The number one, they argued, was the generator of all numbers and not considered to be an odd number, all even numbers were considered to be feminine, and all odd numbers masculine. Five was the number for marriage, since it was the sum of the first even number and the first odd number. The holiest number of all was ten, for it was the number of the universe. Philolaus, a Pythagorean, once wrote that the number ten was "great, all-powerful and all-producing, the beginning and the guide of the divine as of the terrestrial life."

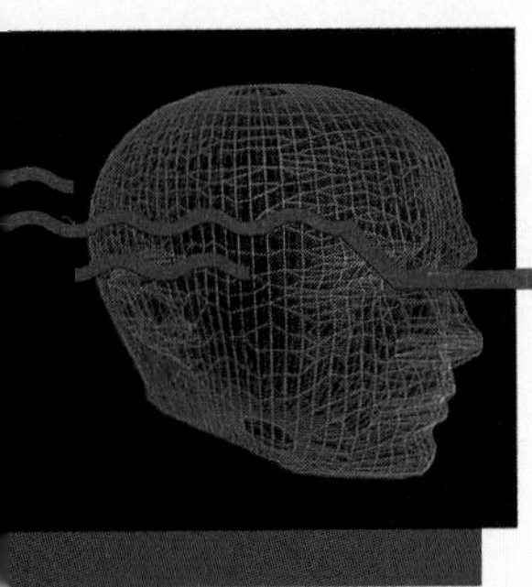

Technology

Trigonometric Functions

BASIC
Graphing calculators
▶ Graphing software
Spreadsheets

A computer can calculate values of trigonometric functions very rapidly. Trigonometric functions usually available in BASIC include SIN, COS, and TAN. The program at the right uses the SIN function to print a table of sines.

The sine function generates ordered pairs. To graph the function, use the horizontal axis for the degree values and the vertical axis for the sine. After plotting the points, connect them with a smooth curve as shown below.

```
10 PRINT "ANGLE", "SINE
20 FOR D = 0 TO 360 STEP 30
30 LET R = D*3.1416/180
40 LET R1 = INT(SIN(R)*1000+.5)/1000
50 PRINT D, R1
60 NEXT D
70 END

RUN
ANGLE        SINE
0            0
30           .5
60           .866
90           1
120          .866
150          .5
180          0
210          -.5
240          -.866
270          -1
300          -.866
330          -.5
360          0
```

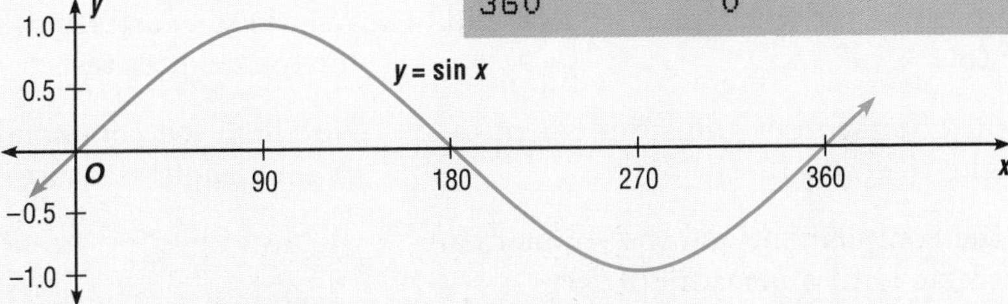

EXERCISES

1. Modify the program to print a table of cosines.

State whether the value of each trigonometric ratio is *positive* or *negative*.

2. sin 30°
3. cos 150°
4. sin 330°
5. cos 60°

State which is greater.

6. cos 30° or cos 90°
7. sin 0° or sin 90°

Find the values of x for which each equation is true.

8. $\cos x = 1$
9. $\sin x = 1$
10. $\sin x = 0$
11. $\cos x = -1$

CHAPTER 15

SUMMARY AND REVIEW

VOCABULARY

Upon completing this chapter, you should be familiar with the following terms:

angle of depression	632	626	tangent
angle of elevation	632	617	30°–60° right triangle
complementary angles	611	622	similar triangles
corresponding angles	622	626	sine
corresponding sides	622	610	supplementary angles
cosine	626	626	trigonometric ratios

SKILLS AND CONCEPTS

OBJECTIVES AND EXAMPLES

Upon completing this chapter, you should be able to:

- find the complement and supplement of an angle **(Lesson 15-1)**

Find the complement and the supplement of an angle with a measure of 28°.

complement: 90° − 28° or 62°
supplement: 180° − 28° or 152°

REVIEW EXERCISES

Use these exercises to review and prepare for the chapter test.

Find both the complement and supplement of each angle measure.

1. 66°

2. 62°

3. 148°

4. $y°$

- find the measure of the third angle of a triangle given the measures of the other two angles **(Lesson 15-1)**

The measures of two angles of a triangle are 38° and 41°. Find the measure of the third angle.

180° − (38° + 41°) or 101°

Find the measure of the third angle of each triangle in which the measure of two angles of the triangle are given.

5. 16°, 72°

6. 41°, 121°

7. 37°, 90°

8. $y°$, $x°$

OBJECTIVES AND EXAMPLES	REVIEW EXERCISES

- find the measures of the sides of a 30°–60° right triangle given the measure of one side **(Lesson 15-3)**

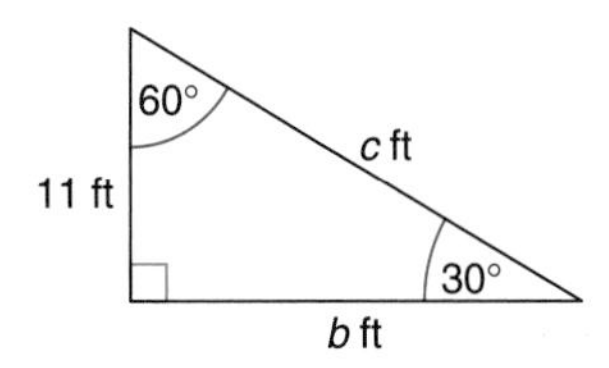

$b = 11\sqrt{3}$ $c = 22$

Find the missing lengths for each 30°–60° right triangle described below. Express irrational answers in simplest radical form.

	Hypotenuse	Side Opposite 30° Angle	Side Opposite 60° Angle
9.	8 cm	?	?
10.	4.25 cm	?	?
11.	?	3 in.	?
12.	?	?	$2\sqrt{3}$ in.

- find the unknown measures of the sides of two similar triangles given the measures of four of the sides **(Lesson 15-4)**

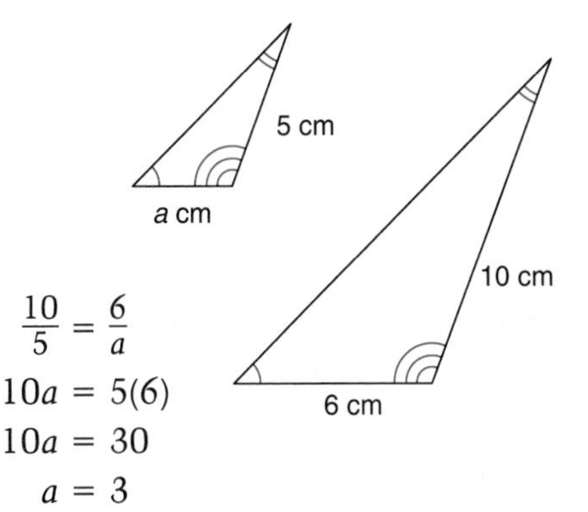

$\frac{10}{5} = \frac{6}{a}$
$10a = 5(6)$
$10a = 30$
$a = 3$

$\triangle ABC$ and $\triangle DEF$ are similar. For each set of measures, find the missing measures.

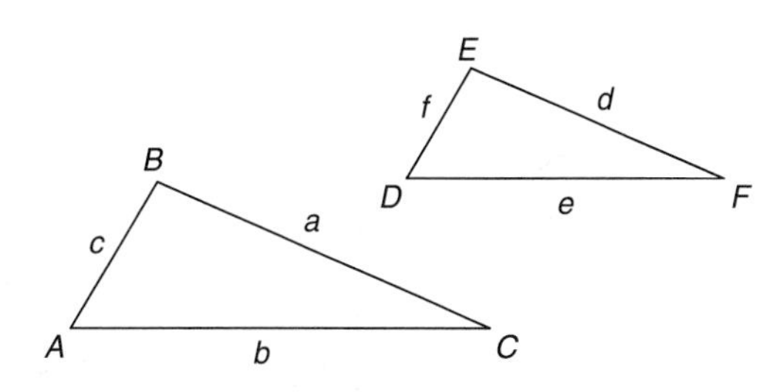

13. $a = 5, d = 11, f = 6, e = 14$
14. $c = 16, b = 12, a = 10, f = 9$
15. $a = 8, c = 10, b = 6, f = 12$
16. $c = 12, f = 9, a = 8, e = 11$

- compute the sine, cosine, and tangent of an acute angle of a right triangle given the measures of its sides **(Lesson 15-5)**

$\sin A = \frac{\text{measure of leg opposite } \angle A}{\text{measure of hypotenuse}}$

$\cos A = \frac{\text{measure of leg adjacent to } \angle A}{\text{measure of hypotenuse}}$

$\tan A = \frac{\text{measure of leg opposite } \angle A}{\text{measure of leg adjacent to } \angle A}$

For $\triangle ABC$, express each trigonometric ratio as a fraction in simplest form.

17. $\sin A$
18. $\cos B$
19. $\cos A$
20. $\sin B$
21. $\tan A$
22. $\tan B$

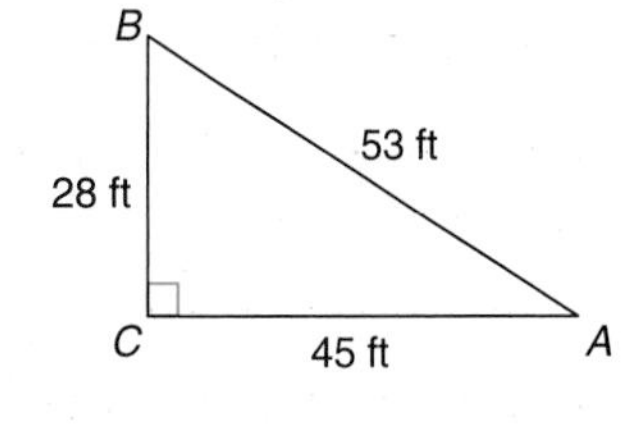

- find the measure of an acute angle of a right triangle given a trigonometric value **(Lesson 15-5)**

$\cos M = 0.3245$

Enter: 0.3245 INV COS 71.064715

The measure of $\angle M$ is about 71°

Use a calculator to find the measure of each angle to the nearest degree.

23. $\tan M = 0.8043$
24. $\sin T = 0.1212$
25. $\tan Q = 5.9080$
26. $\cos F = 0.7443$

OBJECTIVES AND EXAMPLES

- use trigonometric ratios to solve right triangles **(Lesson 15-6)**

$m\angle B = 40°$ and $c = 6$

$m\angle a = 180° - (90° + 40°)$ or $50°$

$\cos 40° = \frac{a}{6}$ $\qquad$ $\sin 40° = \frac{b}{6}$

$0.7660 \approx \frac{a}{6}$ $\qquad$ $0.6428 \approx \frac{b}{6}$

$0.7660(6) \approx a$ $\qquad$ $0.6428(6) \approx b$

$4.596 \approx a$ $\qquad$ $3.857 \approx b$

REVIEW EXERCISES

Solve each triangle.

27.

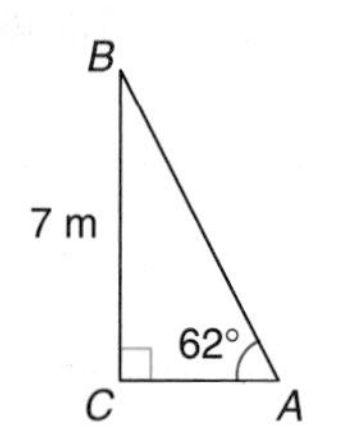

28.

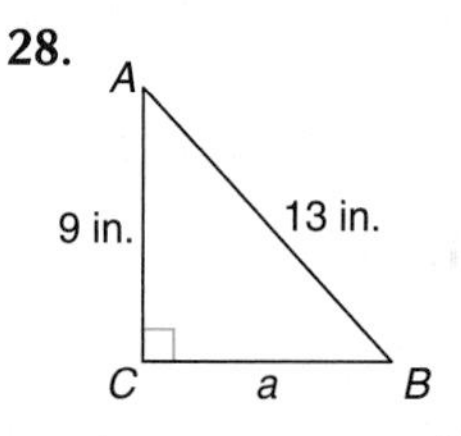

APPLICATIONS AND CONNECTIONS

Solve by making models. (Lesson 15-2)

29. Sam is making his family tree. He plans to include himself, his parents, his grandparents, his great grandparents, and his great-great grandparents. If Sam does not include any stepparents or stepgrandparents, how many people will he include in his family tree?

30. A rectangular swimming pool is 4 meters by 10 meters. A walkway is 1 meter wide and goes around the pool. Find the area of the walkway.

Solve.

31. Find the area of an equilateral triangle if the length of each side is 12 feet. **(Lesson 15-3)**

32. Without using a calculator or a table, find sin 30°. **(Lesson 15-3)**

33. Broadcasting A radio tower casts a shadow 120 meters long when the angle of elevation of the sun is 41°. How tall is the tower? **(Lesson 15-5)**

x m
41°
120 m

34. Forest Management From the top of an observation tower 50 meters high, a forest ranger spots a deer at an angle of depression of 28°. How far is the deer from the base of the tower? **(Lesson 15-6)**

35. Navigation An airplane, at an altitude of 2000 feet, is directly over a power plant. The navigator finds the angle of depression of the airport to be 19°. How far is the plane from the airport? How far is the power plant from the airport? **(Lesson 15-6)**

36. Construction A roof is constructed as shown in the diagram below. Find the pitch (angle of elevation) of the roof. **(Lesson 15-5)**

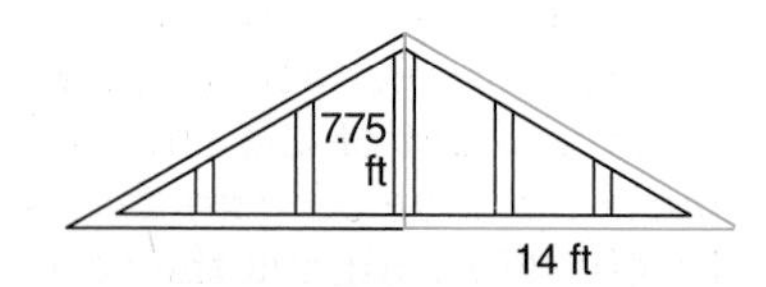

CHAPTER 15 TEST

Find both the complement and the supplement of each angle measure.

1. $28°$
2. $69°$
3. $(y + 20)°$

Find the measure of the third angle of each triangle in which the measures of two angles of the triangle are given.

4. $16°, 47°$
5. $89°, 66°$
6. $45°, 120°$

Find the missing lengths for each 30°–60° right triangle described below. Express irrational answers in simplest radical form and in decimal form to the nearest thousandth.

	Hypotenuse	Side Opposite 30° Angle	Side Opposite 60° Angle
7.	17 in.	?	?
8.	?	8 ft	?
9.	?	?	$9\sqrt{3}$ m

$\triangle ABC$ and $\triangle JKH$ are similar. For each set of measures, find the missing lengths.

10. $c = 20, h = 15, k = 16, j = 12$
11. $c = 12, b = 13, a = 6, h = 10$
12. $k = 5, c = 6.5, b = 7.5, a = 4.5$
13. $h = 1\frac{1}{2}, c = 4\frac{1}{2}, k = 2\frac{1}{4}, a = 3$

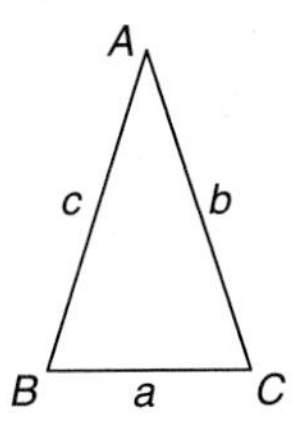

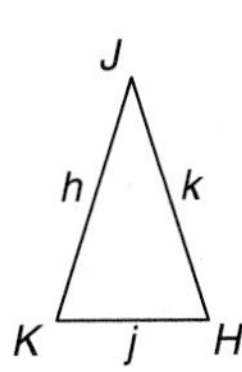

Solve each triangle.

14.
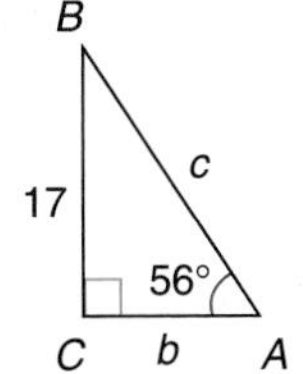

15.
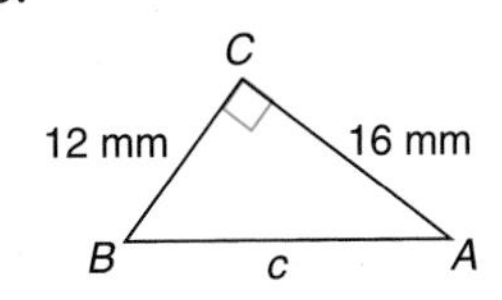

16.
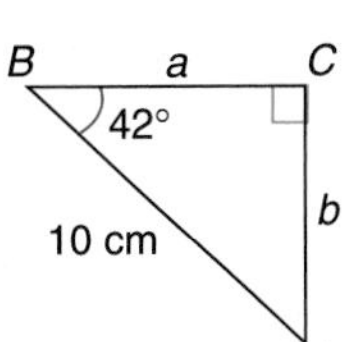

17.
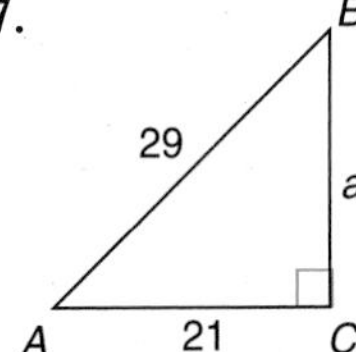

Solve.

18. **Recreation** A kite is flying at the end of a 300-foot string. Assuming the string is straight and forms an angle of 58° with the ground, how high is the kite?

19. **Navigation** A plane is 1000 feet above the ground. The angle of depression of the landing strip is 20°. Find the diagonal distance between the plane and the landing strip.

20. A 6-foot pole casts a 4-foot shadow. How tall is a tree that casts a 50-foot shadow?

Bonus Two buildings are separated by an alley. Joe is looking out of a window 60 feet above the ground in one building. He observes the measurement of the angle of depression of the base of the second building to be 50° and that of the angle of elevation of the top to be 40°. How high is the second building?

CHAPTER 15

College Entrance Exam Preview

The test questions on these pages deal with expression and equations.

Directions: Choose the best answer. Write A, B, C, or D.

1. What is the average of $70 - c$, $70 + 2c$, and $45 - c$?

   (A) 75 (B) $61\frac{2}{3} + \frac{2}{3}c$

   (C) $61\frac{2}{3}$ (D) $75 + c$

2. If $(1 - 2 + 3 - 4 + 5 - 6) = -3$, then $2(1 - 2 + 3 - 4 + 5 - 6) =$

   (A) -6 (B) 6 (C) 3 (D) -3

3. If $ab + 6 = 5abc$, then $b =$

   (A) $\frac{a+6}{5c}$ (B) $\frac{6}{a-5c}$

   (C) $\frac{-6}{5c-a}$ (D) $\frac{6}{5ac-a}$

4. If $23(66 + x) = 2300$, then $x =$

   (A) 1185 (B) 728

   (C) 34 (D) 44

5. If $3y + 2$ is an odd integer, what is the next consecutive odd integer?

   (A) $3y + 4$ (B) $5y + 2$

   (C) $5y + 4$ (D) $y + 2$

6. $\frac{3(1.8 - 2.6) - (1.8 - 2.6)}{2} =$

   (A) 0.8 (B) -0.8

   (C) -1.6 (D) -3.4

7. If $\frac{x}{8} + 3 = 1$, the value of $\frac{x}{2}$ is

   (A) -32 (B) -16 (C) -8 (D) 4

8. If $2m - n = 8$ and $m + p = 14$, what is the value of n in terms of p?

   (A) $2p - 20$ (B) $10 - p$

   (C) $20 - 2p$ (D) $p - 10$

9. $-3(a - b) =$

   (A) $3(b - a)$ (B) $3(a \div b)$

   (C) $3(-b) + (-a)$ (D) $3ab$

10. If $x = 1$, $y = -2$, and $z = 2$, then $\frac{x^2y}{(x-z)^2} =$

    (A) -2 (B) 2 (C) $\frac{-1}{3}$ (D) $\frac{1}{3}$

11. If $1 + \frac{c}{12} = 2\frac{3}{4}$, then $c =$

    (A) 33 (B) 32 (C) 21 (D) 12

12. A number added to one-third of itself results in a sum of 40. What is the number?

    (A) 32 (B) 30 (C) 10 (D) 27

13. How many dollars do you have if you have n nickels, d dimes, and k quarters?

    (A) $\frac{5n + 25k + 10d}{100}$

    (B) $\frac{25n + 10d + 4k}{100}$

    (C) $20n + 10d + 4k$

    (D) $\frac{20n + 10d + 4k}{100}$

14. If x, y, and z are three consecutive integers and $x > y > z$, then $(x - y)(x - z)(y - z) =$

    (A) 2 (B) -2 (C) -16 (D) 16

15. Emily, who is 24 years old, is three times as old as Barb. Barb is four years younger than twice Pedro's age. How old is Pedro?

(A) 2 (B) 12 (C) 10 (D) 6

16. If 20x cartons fill $\frac{x}{5}$ trucks, how many trucks are needed to hold 400 cartons?

(A) 20 (B) 8000
(C) 100 (D) 4

17. If $\frac{x}{2}$, $\frac{x}{5}$, and $\frac{x}{7}$ are whole numbers, x may be

(A) 20 (B) 35 (C) 50 (D) 70

18. If $\frac{1}{b - d} = 4$, then $d =$

(A) $b - \frac{1}{4}$ (B) $4b - 1$
(C) $\frac{b + 1}{4}$ (D) $b + 4$

19. If $\frac{2a}{5b} = 6$, then $\frac{2a - 5b}{5b} =$

(A) 6 (B) $\frac{2}{5}$ (C) 15 (D) 5

20. If $3\frac{1}{5}c = 2\frac{1}{2}b$ and $c \neq 0$, then $\frac{b}{c} =$

(A) $\frac{25}{32}$ (B) $\frac{32}{25}$ (C) $\frac{7}{8}$ (D) $\frac{11}{10}$

21. Carrie's bowling scores for four games are $b + 2$, $b + 3$, $b - 2$, and $b - 1$. What must her score be on her fifth game to average $b + 2$?

(A) $b + 8$ (B) b
(C) $b - 2$ (D) $b + 5$

TEST TAKING TIP

Guessing can improve your score if you make educated guesses. First, find out if there is a penalty for incorrect answers. If there is no penalty, a guess can only increase your score, or at worse, leave your score the same. If there is a penalty, try to eliminate enough choices to make the probability of a correct guess greater than the penalty for an incorrect answer. That is, if there is a quarter point penalty for an incorrect response and you have a choice of three responses, the probability of making a correct guess is 1 out of 3, which is better than the penalty.

22. If $abc = 4$ and $b = c$, then $a =$

(A) c^2 (B) $\frac{1}{4c}$ (C) $\frac{4}{c^2}$ (D) $\frac{1}{c^2}$

23. The sum of six integers is what percent of the average of six integers?

(A) 0.001% (B) 2%
(C) 10% (D) 600%

24. Which of the following is an irrational number?

(A) 0.123 . . . (B) $\sqrt{169}$
(C) $\sqrt{-8}$ (D) π

25. If r and s are roots of $x^2 + bx + c = 0$, then $r + s =$

(A) 0 (B) $-b$ (C) c (D) $b - c$

APPENDIX: USING TABLES

A table of squares and approximate square roots and a table of values of trigonometric functions are provided for use in case a scientific calculator is not available. This guide will show you how to use the tables to find squares, square roots, and values of trigonometric functions.

Example 1

Find the square and square root of 51.

Read across the row labeled 51.

SQUARES AND APPROXIMATE SQUARE ROOTS

n	n^2	$\sqrt{n}$	n	n^2	$\sqrt{n}$
1	1	1.000	51	2601	7.141
2	4	1.414	52	2704	7.211
3	9	1.732	53	2809	7.280
4	16	2.000	54	2916	7.348
5	25	2.236	55	3025	7.416

The n^2 column shows that the square of 51 is 2601.
The $\sqrt{n}$ column shows that the square root of 51 to the nearest thousandth is 7.141.

Example 2

Find the sine, cosine, and tangent of 37° to the nearest ten thousandth.

Read across the row labeled 37°.

TRIGONOMETRIC RATIOS

Angle	sin	cos	tan	Angle	sin	cos	tan
36°	0.5878	0.8090	0.7265	81°	0.9877	0.1564	6.3138
37°	0.6018	0.7986	0.7536	82°	0.9903	0.1392	7.1154
38°	0.6157	0.7880	0.7813	83°	0.9925	0.1219	8.1443
39°	0.6293	0.7771	0.8098	84°	0.9945	0.1045	9.5144
40°	0.6428	0.7660	0.8391	85°	0.9962	0.0872	11.4301

The sin column shows that the sine of 37° is 0.6018.
The cos column shows that the cosine of 37° is 0.7986.
The tan column shows that the tangent of 37° is 0.7536.

SQUARES AND APPROXIMATE SQUARE ROOTS

n	n^2	$\sqrt{n}$	n	n^2	$\sqrt{n}$
1	1	1.000	51	2601	7.141
2	4	1.414	52	2704	7.211
3	9	1.732	53	2809	7.280
4	16	2.000	54	2916	7.348
5	25	2.236	55	3025	7.416
6	36	2.449	56	3136	7.483
7	49	2.646	57	3249	7.550
8	64	2.828	58	3364	7.616
9	81	3.000	59	3481	7.681
10	100	3.162	60	3600	7.746
11	121	3.317	61	3721	7.810
12	144	3.464	62	3844	7.874
13	169	3.606	63	3969	7.937
14	196	3.742	64	4096	8.000
15	225	3.873	65	4225	8.062
16	256	4.000	66	4356	8.124
17	289	4.123	67	4489	8.185
18	324	4.243	68	4624	8.246
19	361	4.359	69	4761	8.307
20	400	4.472	70	4900	8.367
21	441	4.583	71	5041	8.426
22	484	4.690	72	5184	8.485
23	529	4.796	73	5329	8.544
24	576	4.899	74	5476	8.602
25	625	5.000	75	5625	8.660
26	676	5.099	76	5776	8.718
27	729	5.196	77	5929	8.775
28	784	5.292	78	6084	8.832
29	841	5.385	79	6241	8.888
30	900	5.477	80	6400	8.944
31	961	5.568	81	6561	9.000
32	1024	5.657	82	6724	9.055
33	1089	5.745	83	6889	9.110
34	1156	5.831	84	7056	9.165
35	1225	5.916	85	7225	9.220
36	1296	6.000	86	7396	9.274
37	1369	6.083	87	7569	9.327
38	1444	6.164	88	7744	9.381
39	1521	6.245	89	7921	9.434
40	1600	6.325	90	8100	9.487
41	1681	6.403	91	8281	9.539
42	1764	6.481	92	8464	9.592
43	1849	6.557	93	8649	9.644
44	1936	6.633	94	8836	9.695
45	2025	6.708	95	9025	9.747
46	2116	6.782	96	9216	9.798
47	2209	6.856	97	9409	9.849
48	2304	6.928	98	9604	9.899
49	2401	7.000	99	9801	9.950
50	2500	7.071	100	10000	10.000

TRIGONOMETRIC RATIOS

Angle	sin	cos	tan	Angle	sin	cos	tan
0°	0.0000	1.0000	0.0000	45°	0.7071	0.7071	1.0000
1°	0.0175	0.9998	0.0175	46°	0.7193	0.6947	1.0355
2°	0.0349	0.9994	0.0349	47°	0.7314	0.6820	1.0724
3°	0.0523	0.9986	0.0524	48°	0.7431	0.6691	1.1106
4°	0.0698	0.9976	0.0699	49°	0.7547	0.6561	1.1504
5°	0.0872	0.9962	0.0875	50°	0.7660	0.6428	1.1918
6°	0.1045	0.9945	0.1051	51°	0.7771	0.6293	1.2349
7°	0.1219	0.9925	0.1228	52°	0.7880	0.6157	1.2799
8°	0.1392	0.9903	0.1405	53°	0.7986	0.6018	1.3270
9°	0.1564	0.9877	0.1584	54°	0.8090	0.5878	1.3764
10°	0.1736	0.9848	0.1763	55°	0.8192	0.5736	1.4281
11°	0.1908	0.9816	0.1944	56°	0.8290	0.5592	1.4826
12°	0.2079	0.9781	0.2126	57°	0.8387	0.5446	1.5399
13°	0.2250	0.9744	0.2309	58°	0.8480	0.5299	1.6003
14°	0.2419	0.9703	0.2493	59°	0.8572	0.5150	1.6643
15°	0.2588	0.9659	0.2679	60°	0.8660	0.5000	1.7321
16°	0.2756	0.9613	0.2867	61°	0.8746	0.4848	1.8040
17°	0.2924	0.9563	0.3057	62°	0.8829	0.4695	1.8807
18°	0.3090	0.9511	0.3249	63°	0.8910	0.4540	1.9626
19°	0.3256	0.9455	0.3443	64°	0.8988	0.4384	2.0503
20°	0.3420	0.9397	0.3640	65°	0.9063	0.4226	2.1445
21°	0.3584	0.9336	0.3839	66°	0.9135	0.4067	2.2460
22°	0.3746	0.9272	0.4040	67°	0.9205	0.3907	2.3559
23°	0.3907	0.9205	0.4245	68°	0.9272	0.3746	2.4751
24°	0.4067	0.9135	0.4452	69°	0.9336	0.3584	2.6051
25°	0.4226	0.9063	0.4663	70°	0.9397	0.3420	2.7475
26°	0.4384	0.8988	0.4877	71°	0.9455	0.3256	2.9042
27°	0.4540	0.8910	0.5095	72°	0.9511	0.3090	3.0777
28°	0.4695	0.8829	0.5317	73°	0.9563	0.2924	3.2709
29°	0.4848	0.8746	0.5543	74°	0.9613	0.2756	3.4874
30°	0.5000	0.8660	0.5774	75°	0.9659	0.2588	3.7321
31°	0.5150	0.8572	0.6009	76°	0.9703	0.2419	4.0108
32°	0.5299	0.8480	0.6249	77°	0.9744	0.2250	4.3315
33°	0.5446	0.8387	0.6494	78°	0.9781	0.2079	4.7046
34°	0.5592	0.8290	0.6745	79°	0.9816	0.1908	5.1446
35°	0.5736	0.8192	0.7002	80°	0.9848	0.1736	5.6713
36°	0.5878	0.8090	0.7265	81°	0.9877	0.1564	6.3138
37°	0.6018	0.7986	0.7536	82°	0.9903	0.1392	7.1154
38°	0.6157	0.7880	0.7813	83°	0.9925	0.1219	8.1443
39°	0.6293	0.7771	0.8098	84°	0.9945	0.1045	9.5144
40°	0.6428	0.7660	0.8391	85°	0.9962	0.0872	11.4301
41°	0.6561	0.7547	0.8693	86°	0.9976	0.0698	14.3007
42°	0.6691	0.7431	0.9004	87°	0.9986	0.0523	19.0811
43°	0.6820	0.7314	0.9325	88°	0.9994	0.0349	28.6363
44°	0.6947	0.7193	0.9657	89°	0.9998	0.0175	57.2900
45°	0.7071	0.7071	1.0000	90°	1.0000	0.0000	∞

ALGEBRAIC SKILLS REVIEW

Integer Equations: Addition and Subtraction

Solve each equation.

1. $-5 + (-8) = x$
2. $-7 + 4 = y$
3. $-4 + 8 = t$
4. $9 + (-2) = a$
5. $6 + 6 = b$
6. $3 + (-8) = m$
7. $-7 + (-9) = v$
8. $-5 + 5 = z$
9. $-19 + 43 = c$
10. $51 + (-26) = w$
11. $-37 + (-48) = d$
12. $-93 + 44 = e$
13. $67 + (-82) = n$
14. $28 + 46 = f$
15. $29 + (-37) = s$
16. $-94 + (-58) = g$
17. $-18 + 63 = p$
18. $28 + (-52) = j$
19. $77 + 57 = r$
20. $47 + (-29) = x$
21. $-18 + 26 = a$
22. $-65 + (-75) = k$
23. $21 + (-47) = h$
24. $-15 + 52 = q$
25. $y = -13 + (-98)$
26. $u = -5 + 82$
27. $s = -47 + 26 + (-18)$
28. $a = -71 + (-85) + (-16)$
29. $41 + 57 + (-32) = m$
30. $82 + (-14) + (-35) = c$
31. $-4 - (-2) = t$
32. $5 - (-6) = p$
33. $-9 - 3 = x$
34. $5 - (-5) = k$
35. $-3 - (-8) = m$
36. $3 - 9 = w$
37. $-6 - 8 = w$
38. $0 - 6 = v$
39. $j = -10 - (-4)$
40. $-23 - 45 = a$
41. $28 - (-14) = z$
42. $-53 - (-61) = f$
43. $c = -16 - 47$
44. $90 - 43 = g$
45. $71 - (-47) = q$
46. $-99 - (-26) = s$
47. $38 - (-19) = t$
48. $-20 - (-92) = j$
49. $18 - 47 = y$
50. $h = -15 - (-81)$
51. $-42 - 63 = b$
52. $-84 - 47 = r$
53. $42 - (-47) = d$
54. $y = -19 - (-63)$
55. $16 - (-84) = n$
56. $42 - (-26) = k$
57. $-52 - (-33) = x$
58. $-35 - 86 = a$
59. $v = -8 - (-47)$
60. $33 - 51 = t$
61. $-2 + g = 7$
62. $9 + s = -5$
63. $-7 + k = -2$
64. $-4 + y = -9$
65. $m + 6 = 2$
66. $t + (-4) = 10$
67. $h - (-2) = 6$
68. $v - 7 = -4$
69. $a - (-6) = -5$
70. $r - (-3) = -8$
71. $j - (-8) = 5$
72. $x - 8 = -9$
73. $-2 - x = -8$
74. $14 = -48 + b$
75. $c + (-26) = 45$
76. $z - (-57) = -39$
77. $d + (-44) = -61$
78. $n - 38 = -19$
79. $-77 = w + 23$
80. $e - (-26) = 41$
81. $p - 47 = 22$
82. $-63 - f = -82$
83. $87 = t + (-14)$
84. $q + (-53) = 27$

Integer Equations: Multiplication and Division

Solve each equation.

1. $x = (-8)(-4)$
2. $(-3)5 = t$
3. $(-7)(-2) = a$
4. $(-9)8 = b$
5. $6(-5) = v$
6. $k = 8(6)$
7. $14(-26) = s$
8. $(-46)(-25) = g$
9. $(-71)(-20) = y$
10. $(-42)66 = h$
11. $(-97)47 = w$
12. $53(-32) = c$
13. $19(-46) = x$
14. $(-82)0 = e$
15. $72(43) = m$
16. $(-18)(-18) = d$
17. $24(-29) = u$
18. $f = (-39)45$
19. $(-76)(-34) = s$
20. $(-81)(-18) = q$
21. $(-65)28 = t$
22. $71(-38) = p$
23. $j = 49(-92)$
24. $36(24) = a$
25. $(-42)78 = z$
26. $(-54)(-77) = r$
27. $n = (-6)(-127)(-4)$
28. $(13)(-12)(95) = w$
29. $(-1)(45)(-45) = v$
30. $(-3)(61)(99) = y$
31. $72 \div (-8) = g$
32. $-64 \div 8 = b$
33. $-45 \div (-9) = y$
34. $56 \div (-7) = z$
35. $42 \div 6 = e$
36. $-24 \div (-6) = m$
37. $992 \div (-32) = a$
38. $-4428 \div 54 = k$
39. $x = -600 \div (-24)$
40. $1472 \div (-64) = p$
41. $-564 \div (-47) = h$
42. $-504 \div 14 = j$
43. $-2201 \div 71 = r$
44. $1512 \div (-28) = n$
45. $765 \div (-85) = q$
46. $-1591 \div (-37) = f$
47. $s = -1080 \div 36$
48. $3432 \div (-52) = v$
49. $2730 \div 78 = k$
50. $-3936 \div 96 = c$
51. $-1476 \div 41 = z$
52. $1496 \div (-22) = a$
53. $2646 \div (-63) = t$
54. $w = -4730 \div (-55)$
55. $-1092 \div (-26) = x$
56. $-2700 \div (-75) = e$
57. $1127 \div 49 = y$
58. $d = 1900 \div (-38)$
59. $-845 \div 13 = w$
60. $-1596 \div (-42) = a$
61. $-5p = 35$
62. $7g = -49$
63. $-3x = -24$
64. $a \div (-6) = -2$
65. $m \div (-8) = 8$
66. $q \div 9 = -3$
67. $41j = 1476$
68. $62y = -2356$
69. $b \div (-21) = 13$
70. $-33n = -1815$
71. $k \div 46 = -41$
72. $w \div 17 = 24$
73. $c \div (-59) = -7$
74. $-56h = 1792$
75. $-42z = 1512$
76. $j \div (-27) = 27$
77. $89s = -712$
78. $-18v = -1044$
79. $d \div (-34) = -43$
80. $f \div 14 = -63$
81. $45t = 810$
82. $-74w = 1554$
83. $-49e = -2058$
84. $r \div (-16) = -77$
85. $x \div (-26) = 47$
86. $-23 = t \div 44$
87. $-962 = -37g$
88. $-3040 = 95k$
89. $84 = x \div 97$
90. $-108 = m \div (-12)$

Fraction Equations: Addition and Subtraction

Solve each equation and express answers in simplest form.

1. $\frac{3}{11} + \frac{6}{11} = x$
2. $\frac{4}{7} + \frac{5}{7} = a$
3. $\frac{5}{9} - \frac{2}{9} = t$
4. $\frac{17}{18} - \frac{5}{18} = w$
5. $\frac{1}{3} + \frac{2}{9} = b$
6. $\frac{1}{2} - \frac{1}{3} = v$
7. $\frac{3}{4} - \frac{9}{16} = s$
8. $\frac{2}{3} + \frac{8}{15} = r$
9. $\frac{5}{6} - \frac{3}{4} = d$
10. $\frac{4}{9} + \frac{1}{6} = c$
11. $m = \frac{7}{9} + \frac{3}{8}$
12. $\frac{11}{12} - \frac{7}{10} = j$
13. $\frac{5}{6} - \frac{5}{12} = p$
14. $4\frac{2}{3} + 1\frac{8}{15} = k$
15. $5\frac{1}{2} - 2\frac{1}{3} = w$
16. $8\frac{1}{12} - 5\frac{5}{12} = e$
17. $7 - 1\frac{4}{9} = h$
18. $n = \frac{3}{16} + \frac{7}{12}$
19. $4\frac{1}{2} - 2\frac{2}{3} = q$
20. $7\frac{1}{12} - 4\frac{5}{8} = x$
21. $11\frac{5}{6} + 9\frac{7}{15} = f$
22. $y = 9\frac{2}{7} - 5\frac{5}{6}$
23. $\frac{1}{4} + \frac{5}{6} + \frac{7}{12} = c$
24. $\frac{5}{6} + \frac{2}{9} + \frac{3}{4} = z$
25. $-\frac{2}{13} + \left(-\frac{3}{13}\right) = t$
26. $-\frac{11}{18} + \frac{17}{18} = f$
27. $-\frac{9}{10} - \frac{7}{10} = n$
28. $-\frac{7}{11} - \left(-\frac{3}{11}\right) = a$
29. $\frac{1}{12} - \left(-\frac{7}{12}\right) = w$
30. $\frac{17}{21} + \left(-\frac{10}{21}\right) = g$
31. $\frac{1}{4} + \left(-\frac{2}{3}\right) = p$
32. $b = -\frac{1}{6} - \frac{8}{9}$
33. $\frac{1}{3} - \frac{5}{6} = m$
34. $-\frac{1}{2} + \left(-\frac{3}{5}\right) = a$
35. $\frac{3}{7} + (-5) = s$
36. $-\frac{5}{9} - 2 = k$
37. $t = 1\frac{1}{2} - \left(-\frac{3}{4}\right)$
38. $-\frac{3}{8} + \frac{4}{7} = c$
39. $\frac{3}{5} - \left(-3\frac{1}{4}\right) = v$
40. $-8\frac{7}{8} - \left(-4\frac{5}{12}\right) = r$
41. $-3\frac{1}{6} + 5\frac{1}{15} = d$
42. $-1\frac{8}{9} + \left(-5\frac{7}{12}\right) = h$
43. $7\frac{5}{6} + \left(-8\frac{7}{8}\right) = j$
44. $-3\frac{1}{2} - 4\frac{5}{9} = e$
45. $q = \frac{11}{16} - 12$
46. $-5\frac{11}{20} + 4\frac{7}{12} = z$
47. $-1\frac{1}{12} - \left(-\frac{2}{3}\right) = w$
48. $-4\frac{16}{21} + \left(-7\frac{5}{9}\right) = y$
49. $\frac{3}{13} + p = \frac{10}{13}$
50. $e + \frac{4}{15} = \frac{13}{15}$
51. $\frac{2}{5} + n = \frac{2}{3}$
52. $j - \frac{5}{18} = \frac{17}{18}$
53. $r - \frac{1}{4} = \frac{5}{16}$
54. $b - \frac{1}{2} = \frac{2}{5}$
55. $s + \frac{2}{7} = 2$
56. $\frac{7}{10} - a = \frac{1}{2}$
57. $1\frac{5}{6} + x = 2\frac{1}{4}$
58. $4\frac{1}{4} = w + 2\frac{1}{3}$
59. $d - 1\frac{5}{7} = 6\frac{1}{4}$
60. $h - \frac{3}{4} = 2\frac{5}{8}$
61. $t - \frac{2}{3} = 1\frac{5}{8}$
62. $g + \frac{5}{6} = \frac{4}{9}$
63. $q - \frac{7}{10} = -\frac{11}{15}$
64. $-\frac{3}{7} + c = \frac{1}{2}$
65. $-\frac{3}{4} = v + \left(-\frac{1}{8}\right)$
66. $f - \left(-\frac{1}{8}\right) = \frac{3}{10}$
67. $m - \left(-1\frac{3}{8}\right) = -2\frac{1}{2}$
68. $-6\frac{5}{6} + y = 7\frac{7}{15}$
69. $7\frac{1}{6} - z = -5\frac{2}{3}$
70. $-2\frac{1}{3} + w = -5\frac{5}{6}$
71. $-6\frac{1}{7} + k = -\frac{4}{21}$
72. $-4\frac{5}{12} = t - \left(-10\frac{1}{36}\right)$

Fraction Equations: Multiplication and Division

Solve each equation and express answers in simplest form.

1. $\frac{1}{7}\left(\frac{1}{3}\right) = x$
2. $\frac{2}{3}\left(\frac{1}{5}\right) = v$
3. $\frac{5}{6}\left(\frac{3}{10}\right) = y$
4. $2 \div \frac{1}{3} = j$
5. $\frac{5}{6} \div \frac{1}{6} = f$
6. $\frac{1}{4} \div \frac{5}{8} = c$
7. $\frac{2}{3}(9) = b$
8. $\frac{5}{18}\left(\frac{3}{10}\right) = r$
9. $\frac{8}{15} \div \frac{1}{10} = k$
10. $g = \frac{1}{2} \div 8$
11. $\frac{3}{14} \div \frac{2}{7} = y$
12. $\frac{7}{12}\left(\frac{4}{5}\right) = w$
13. $\frac{6}{13} \div \frac{5}{7} = z$
14. $\frac{24}{25}\left(\frac{15}{32}\right) = a$
15. $\frac{7}{10}\left(\frac{5}{28}\right) = w$
16. $2\frac{2}{3}\left(\frac{4}{5}\right) = n$
17. $t = \frac{7}{8}\left(4\frac{1}{4}\right)$
18. $1\frac{3}{4} \div \frac{7}{12} = e$
19. $1 \div 2\frac{3}{5} = d$
20. $2\frac{1}{10}\left(4\frac{2}{7}\right) = q$
21. $1\frac{3}{5} \div 11\frac{1}{5} = s$
22. $3\frac{1}{8}\left(2\frac{4}{5}\right)\left(\frac{5}{7}\right) = p$
23. $3\frac{2}{3}\left(\frac{1}{8}\right)\left(1\frac{1}{11}\right) = h$
24. $m = 3\frac{1}{21} \div 1\frac{21}{35}$
25. $\frac{1}{5}\left(-\frac{1}{8}\right) = p$
26. $-\frac{2}{9}\left(\frac{1}{3}\right) = w$
27. $-\frac{5}{8}\left(-\frac{4}{5}\right) = c$
28. $-3 \div \frac{1}{2} = y$
29. $-\frac{7}{9} \div \left(-\frac{1}{9}\right) = r$
30. $\frac{2}{3} \div \left(-\frac{4}{9}\right) = t$
31. $-\frac{2}{5}(-10) = m$
32. $\frac{2}{3} \div \left(-\frac{7}{9}\right) = h$
33. $-\frac{9}{15}\left(\frac{5}{9}\right) = s$
34. $-\frac{9}{14} \div \left(-\frac{3}{7}\right) = a$
35. $\frac{7}{16} \div \left(-\frac{7}{11}\right) = z$
36. $j = -\frac{4}{5}(30)$
37. $-7 \div 4 = q$
38. $-4\frac{9}{10}\left(-1\frac{5}{21}\right) = b$
39. $-5\frac{3}{5} \div 4\frac{1}{5} = g$
40. $-5\frac{3}{5} \div \left(-4\frac{1}{5}\right) = e$
41. $v = 6\frac{1}{4}\left(-1\frac{7}{15}\right)$
42. $3\frac{1}{3}\left(-4\frac{1}{2}\right) = w$
43. $-2\left(1\frac{5}{18}\right) = k$
44. $4\frac{2}{5} \div \left(-\frac{11}{15}\right) = p$
45. $-2\frac{5}{8} \div 7\frac{1}{2} = x$
46. $5 \div (-11) = n$
47. $d = -2\frac{3}{10}\left(-\frac{5}{12}\right)$
48. $-9\frac{1}{3}\left(-3\frac{3}{4}\right) = f$
49. $\frac{1}{3}a = 5$
50. $\frac{4}{7}k = 4$
51. $\frac{2}{5}x = \frac{4}{7}$
52. $w \div 5 = 3$
53. $w \div \frac{1}{4} = \frac{3}{8}$
54. $c \div \frac{3}{10} = \frac{1}{2}$
55. $\frac{7}{11}t = \frac{4}{5}$
56. $h \div \frac{1}{8} = \frac{4}{11}$
57. $z \div 6 = \frac{5}{12}$
58. $1\frac{1}{2}d = \frac{6}{7}$
59. $\frac{10}{33} = b \div 4\frac{2}{5}$
60. $2\frac{1}{6}j = 5\frac{1}{5}$
61. $s \div 2\frac{1}{6} = 2\frac{2}{5}$
62. $1\frac{3}{24}g = 3\frac{1}{8}$
63. $3 = 1\frac{7}{11}q$
64. $n \div \frac{2}{3} = -\frac{4}{9}$
65. $-1\frac{3}{4}p = -\frac{5}{8}$
66. $v \div \left(-\frac{7}{11}\right) = 1\frac{2}{7}$
67. $-1\frac{3}{5} = e \div \left(-3\frac{1}{5}\right)$
68. $-\frac{5}{9}r = 7\frac{1}{2}$
69. $3\frac{4}{7}x = -3\frac{3}{4}$
70. $a \div 3\frac{2}{7} = -8\frac{3}{4}$
71. $-2\frac{4}{7}m = -3\frac{3}{8}$
72. $f \div \left(-3\frac{1}{8}\right) = -3\frac{2}{5}$

Decimal Equations: Addition and Subtraction

Solve each equation.

1. $0.53 + 0.26 = x$
2. $14.756 + 0.185 = k$
3. $0.711 - 0.158 = z$
4. $12.01 - 0.83 = s$
5. $0.4 + 0.86 = n$
6. $1.4 - 0.12 = a$
7. $57.5 + 7.94 = m$
8. $10.04 - 0.18 = f$
9. $5 - 1.63 = r$
10. $5.92 + 7.3 = b$
11. $12 + 9.6 = y$
12. $28.05 - 9.95 = c$
13. $0.2 + 6.51 + 2.03 = y$
14. $4.4 + 30.6 + 11.2 = z$
15. $0.007 + 3 + 10.02 = h$
16. $w = 20.13 - 12.5$
17. $2.3 - 0.846 = t$
18. $11 - 1.1 = p$
19. $6.2 + 5.54 + 13.66 = g$
20. $a = 412 - 0.007$
21. $101.12 + 9.099 = s$
22. $66.4 - 5.288 = d$
23. $84.083 - 17 = m$
24. $q = 0.046 + 5.8 + 11.37$
25. $8 - 3.49 = n$
26. $8.77 + 0.3 + 52.9 = x$
27. $14.7 - 5.8364 = e$
28. $66.68421 - 18.465 = v$
29. $y = 0.0013 + 2.881$
30. $127.11 + 48 + 0.143 = u$
31. $-0.47 + 0.62 = h$
32. $-4.5 + (-12.8) = x$
33. $-1.7 + 0.24 = p$
34. $-6.831 - (-2.648) = c$
35. $-4.23 - 2.47 = b$
36. $2.64 - (-5.9) = k$
37. $10 + (-0.43) = r$
38. $6.7 - (-0.64) = v$
39. $-6.71 - (-8) = e$
40. $14.14 + (-1.4) = a$
41. $1.2 - 6.73 = j$
42. $-9.7 + (-0.86) = d$
43. $-7 - 4.63 = w$
44. $-0.17 - (-14.6) = g$
45. $m = 1.8 + (-14.14)$
46. $5.003 + (-0.47) = f$
47. $0.88 - 42 = s$
48. $-6.2 + (-27.47) = j$
49. $n = -1.4962 + 2.118$
50. $2.4 - (-1.736) = q$
51. $4.16 + (-5.909) = t$
52. $17 + (-0.45) = w$
53. $10 - 13.463 = a$
54. $f = -82.007 - 3.218$
55. $-11.264 + (-8.2) = z$
56. $-56 + 2.783 = s$
57. $-0.682 - (-0.81) = y$
58. $r = -23 + 4.093$
59. $2.08 - (-0.094) = t$
60. $-51.34 + (-5.1346) = x$
61. $2.2 + a = 11.4$
62. $h + 1.83 = 8.42$
63. $c + 5.4 = -11.33$
64. $m - 0.41 = 0.85$
65. $p - 1.1 = 14.9$
66. $r - 0.76 = -3.2$
67. $t + (-6.47) = -22.3$
68. $-6.11 + b = 14.321$
69. $k - (-4) = 7.9$
70. $k - 99.7 = -46.88$
71. $w + (-17.8) = -5.63$
72. $-5 = y - 22.7$
73. $13.475 + d = 4.09$
74. $-5 - q = 1.19$
75. $-3.214 + f = -16.04$
76. $-88.9 = s - 6.21$
77. $2 + e = 1.008$
78. $n + (-4.361) = 59.78$
79. $4.8 - j = -5.834$
80. $w - 0.73 = -1.8$
81. $-8 = g - (-4.821)$
82. $-2.315 + x = -15$
83. $m + (-1.4) = 0.07$
84. $v - 5.234 = -1.051$
85. $7.1 = v - (-0.62)$
86. $s + 6.4 = -0.11$
87. $t - (-46.1) = -3.673$
88. $k + (-1.604) = -0.45$
89. $81.6 + p = -6.73$
90. $-0.1448 - z = -2.6$

Decimal Equations: Multiplication and Division

Solve each equation.

1. $46(0.5) = e$
2. $108(0.9) = b$
3. $g = 6.47(39)$
4. $0.04(197) = f$
5. $r = 67(5.892)$
6. $2.8(4.27) = d$
7. $0.061(5.5) = m$
8. $0.62(0.13) = c$
9. $4.007(1.95) = q$
10. $6.25 \div 5 = w$
11. $t = 91.8 \div 27$
12. $7.31 \div 43 = h$
13. $5.91 \div 0.3 = a$
14. $167.5 \div 2.5 = k$
15. $4.7208 \div 0.84 = v$
16. $p = 278.1 \div 6.18$
17. $30{,}176 \div 9.43 = n$
18. $0.1001 \div 0.77 = j$
19. $2.11(0.059) = w$
20. $s = 0.4484 \div 1.18$
21. $0.0062(84.7) = x$
22. $0.03912 \div 1.63 = z$
23. $230.4 \div 0.072 = m$
24. $w = 59.8(100.23)$
25. $v = 432 \div 9.6$
26. $0.008(0.0045) = x$
27. $1.21(0.47)(9.3) = s$
28. $0.0418 \div 0.19 = x$
29. $0.032(13)(2.6) = t$
30. $0.0001926 \div 0.00321 = y$
31. $-5(0.2) = x$
32. $-1.7(-44) = f$
33. $72(1.01) = c$
34. $627(-0.14) = a$
35. $-2.3(7.81) = n$
36. $r = -1.02(-4.4)$
37. $57.6 \div (-12) = b$
38. $160.8 \div 24 = h$
39. $-16.38 \div (-0.7) = t$
40. $m = -15.54 \div 2.1$
41. $-0.405 \div (-0.27) = a$
42. $-598 \div 0.13 = p$
43. $0.45(-0.0016) = k$
44. $y = -0.002052 \div 0.054$
45. $6.7284 \div 1.08 = d$
46. $-0.0066(-91.8) = w$
47. $455 \div (-1.82) = q$
48. $-0.905(0.208) = g$
49. $-2.4827 \div (-6.71) = e$
50. $0.153 \div (-0.017) = z$
51. $j = -462.1(0.0094)$
52. $56.1(2.3) = y$
53. $0.07553 \div 0.0083 = v$
54. $-1.7(-0.121) = s$
55. $t = -0.6612 \div (-0.114)$
56. $-0.026(45.1) = x$
57. $59(-0.00042) = w$
58. $7.93(-5.036) = c$
59. $9.2397 \div 1.9 = t$
60. $-0.000101 \div 0.001 = m$
61. $7c = 4.2$
62. $37p = 81.4$
63. $57k = 0.1824$
64. $1.5m = 9.9$
65. $1.296 = 0.48d$
66. $0.0022b = 0.1958$
67. $t \div 110 = 2.8$
68. $x \div 71 = 0.33$
69. $r \div 0.85 = 10$
70. $h \div 1.98 = 6.7$
71. $a \div 0.002 = 0.109$
72. $n \div 40.6 = 0.021$
73. $100.8x = 9374.4$
74. $2.61 = f \div 9.5$
75. $1.7118 = 0.317e$
76. $0.0603g = 0.0043416$
77. $w \div 0.0412 = 60$
78. $q \div 1.07 = 0.088$
79. $5j = -32.15$
80. $-1.2v = 112.8$
81. $-0.013s = -0.00923$
82. $w \div (-2) = -2.48$
83. $z \div 2.8 = -6.2$
84. $a \div (-0.53) = -0.034$
85. $k \div (-0.013) = -0.7$
86. $-4.63t = -125.473$
87. $7.9y = 1583.16$
88. $6.05p = -1573$
89. $g \div 9.9 = 12$
90. $x \div (-0.063) = 0.015$

Forms of Real Numbers

Write each fraction in simplest form.

1. $\frac{13}{26}$ **2.** $\frac{9}{12}$ **3.** $-\frac{36}{42}$ **4.** $\frac{5}{60}$

5. $-\frac{24}{32}$ **6.** $-\frac{10}{35}$ **7.** $\frac{54}{63}$ **8.** $-\frac{45}{60}$

9. $\frac{48}{84}$ **10.** $-\frac{28}{42}$ **11.** $-\frac{72}{96}$ **12.** $\frac{75}{105}$

13. $-\frac{16}{100}$ **14.** $\frac{24}{60}$ **15.** $\frac{15}{27}$ **16.** $-\frac{99}{111}$

17. $\frac{126}{700}$ **18.** $-\frac{198}{462}$ **19.** $-\frac{84}{1080}$ **20.** $-\frac{525}{1155}$

Write each fraction as a decimal.

21. $\frac{1}{4}$ **22.** $-\frac{3}{10}$ **23.** $\frac{1}{50}$ **24.** $\frac{2}{3}$

25. $-\frac{1}{9}$ **26.** $-\frac{16}{25}$ **27.** $-\frac{9}{20}$ **28.** $\frac{1}{11}$

29. $\frac{5}{9}$ **30.** $-\frac{5}{8}$ **31.** $\frac{43}{100}$ **32.** $-\frac{5}{6}$

33. $-\frac{7}{11}$ **34.** $\frac{3}{7}$ **35.** $\frac{4}{5}$ **36.** $-\frac{7}{12}$

37. $-\frac{15}{16}$ **38.** $-\frac{8}{15}$ **39.** $\frac{1}{6}$ **40.** $-\frac{11}{32}$

41. $\frac{9}{11}$ **42.** $\frac{11}{16}$ **43.** $-\frac{11}{15}$ **44.** $\frac{124}{125}$

Write each mixed numeral as a decimal.

45. $-5\frac{1}{2}$ **46.** $14\frac{17}{100}$ **47.** $6\frac{3}{25}$ **48.** $-7\frac{1}{3}$

49. $4\frac{3}{25}$ **50.** $-20\frac{2}{9}$ **51.** $12\frac{3}{4}$ **52.** $-10\frac{5}{6}$

53. $-1\frac{4}{9}$ **54.** $-9\frac{16}{50}$ **55.** $-2\frac{2}{11}$ **56.** $13\frac{13}{40}$

57. $3\frac{5}{12}$ **58.** $-8\frac{5}{7}$ **59.** $2\frac{3}{5}$ **60.** $11\frac{1}{12}$

61. $-44\frac{3}{8}$ **62.** $19\frac{8}{15}$ **63.** $-67\frac{7}{10}$ **64.** $5\frac{3}{16}$

65. $78\frac{2}{9}$ **66.** $-108\frac{1}{20}$ **67.** $-51\frac{6}{7}$ **68.** $8\frac{1}{15}$

Write each decimal as a fraction in simplest form.

69. 0.3 **70.** 0.14 **71.** 0.013 **72.** -1.25

73. 4.2 **74.** -20.05 **75.** $0.\overline{3}$ **76.** -14.50

77. $-12.\overline{7}$ **78.** 6.125 **79.** -8.6 **80.** $-8.\overline{6}$

81. 23.15 **82.** -0.37 **83.** -33.85 **84.** 1.16

85. -2.27 **86.** 16.75 **87.** -5.375 **88.** 4.26

89. 7.1875 **90.** -9.45 **91.** $5.2\overline{6}$ **92.** -0.324

Percents

Write each decimal as a percent.

1. 0.71
2. 0.4
3. 0.835
4. 1.05
5. 0.009
6. 0.27
7. 2.5
8. 0.706

Write each fraction as a percent.

9. $\frac{31}{100}$
10. $\frac{1}{2}$
11. $\frac{4}{5}$
12. $\frac{3}{10}$
13. $\frac{1}{8}$
14. $\frac{5}{4}$
15. $\frac{2}{3}$
16. $\frac{4}{11}$

Write each percent as a decimal.

17. 14%
18. 10%
19. 450%
20. 6%
21. 27.5%
22. 4.2%
23. 190.5%
24. 0.3%

Write each percent as a fraction in simplest form.

25. 17%
26. 40%
27. 8%
28. 75%
29. 0.9%
30. 2.5%
31. 45.6%
32. 1.05%

Solve.

33. 10% of 70 is ___.
34. 20% of 35 is ___.
35. 4% of 250 is ___.
36. 255% of 160 is ___.
37. 115% of 24 is ___.
38. 130% of 60 is ___.
39. ___ is 3.7% of 300.
40. ___ is 22.5% of 260.
41. ___ is 52.6% of 150.
42. 5 is ___% of 20.
43. 3 is ___% of 10.
44. 17 is ___% of 68.
45. ___% of 500 is 55.
46. ___% of 96 is 12.
47. ___% of 81 is 27.
48. 40% of ___ is 12.
49. 10% of ___ is 16.
50. 65% of ___ is 26.
51. 25 is $33\frac{1}{3}$% of ___.
52. 54 is 108% of ___.
53. 1.28 is 16% of ___.
54. 6.3% of 400 is ___.
55. ___% of 40 is 25.
56. 68% of ___ is 85.
57. 44% of ___ is 37.4.
58. 53% of 62 is ___.
59. ___% of 16 is 56.
60. ___ is 235% of 270.
61. ___ is 5.8% of 45.
62. 28 is ___% of 21.
63. 2.5% of ___ is 1.
64. ___% of 20 is 26.2.
65. 420% of ___ is 336.
66. ___% of 45 is 30.
67. $87\frac{1}{2}$% of ___ is 14.
68. $66\frac{2}{3}$% of 81 is ___.
69. ___ is 12.4% of 15.
70. 135 is 675% of ___.
71. 45 is ___% of 36.
72. 14.5% of 18 is ___.
73. 180.5% of 200 is ___.
74. 98.1 is ___% of 90.
75. ___% of 85 is 102.
76. 3% of ___ is 18.
77. 44% of ___ is 37.4.
78. ___% of 170 is 153.
79. 738 is 72% of ___.
80. $266\frac{2}{3}$% of 561 is ___.

Evaluating Expressions

Evaluate each expression if $a = 3$, $b = 5$, $c = 12$, and $d = 9$.

1. $8 + c$
2. $d - 4$
3. $b \cdot d$
4. $c \div a$
5. $a + c + d$
6. $c - d$
7. $a \cdot b \cdot c$
8. $d - a$
9. $a \cdot c$
10. $\frac{d}{a}$
11. $\frac{13 + c}{b}$
12. $\frac{a + d}{4}$

Evaluate each expression if $e = 2$, $f = 5$, $g = 6$, and $h = 10$.

13. $8g$
14. fh
15. g^2
16. e^5
17. $7f^2$
18. g^2h^3
19. $\frac{h^4}{f^2}$
20. $3e^2g$
21. $8g^2f^2$
22. $e^4f^2h^3$
23. $20e^3f^3g$
24. $\frac{3e^2f^3}{g}$

Evaluate each expression if $x = 3$, $j = 4$, $k = 9$, and $m = 20$.

25. $k^2 - 4k + 6$
26. $(m + j) \div 3$
27. $x^3j^2 - 4m$
28. $(xj + k) \div x$
29. $5j^2 \div m + k^2$
30. $j^3 + mk + 4x^4$
31. $(j^3 + m)k - 4x$
32. $(xj)^2 + km^2$
33. $k^3 + m \div j - 5j^2$
34. $(5 + j)^2 \div k + m^2$
35. $(m - k)^3 \div (2j + 3)$
36. $(x^4 + k)m^2 - kx$

Evaluate each expression if $n = -1$, $p = 6$, $q = -8$, $r = 15$, and $s = -24$.

37. $pr + 2q$
38. $pn^4 + s$
39. $r^2 - q + 5s$
40. $pq^2 \div ns$
41. $(p + 2q)n - r$
42. $(p + q)^5n^5 + s$
43. $pr^2 + ns - 6q$
44. $p(r^2 + ns) - 6q$
45. $(q + r + s)p + n$
46. $\frac{4(p^2 + q^2)}{2q} - s$
47. $(p + n)^3 + \left(\frac{s}{p}\right)q$
48. $[(r + s)q]p$

The formula for the total surface area of a rectangular solid is $T = 2\ell w + 2wh + 2\ell h$, where T is the total surface area of the solid, ℓ is its length, w is its width, and h is its height. Find the total surface area of each rectangular solid.

49. $\ell = 8, w = 5, h = 14$
50. $\ell = 4, w = 2.5, h = 3$
51. $\ell = 7, w = 7, h = 16$
52. $\ell = 14, w = 17, h = 11$
53. $\ell = 21, w = 18, h = 6$
54. $\ell = 3.7, w = 1.2, h = 3.5$

The formula to change Fahrenheit degrees to Celsius degrees is $C = \frac{5}{9}(F - 32)$, where C is the temperature in Celsius degrees and F is the temperature in Fahrenheit degrees. Change each temperature in Fahrenheit degrees to Celsius degrees.

55. 86°F
56. 5°F
57. −13°F
58. 41°F
59. −40°F
60. 23°F
61. −58°F
62. 374°F

Inequalities

Replace each ▒ with >, <, or = to make each sentence true.

1. 9 ▒ 12
2. 14 ▒ 7
3. -3 ▒ 0
4. -7 ▒ -3
5. -5 ▒ 3
6. $7 + 8$ ▒ 15
7. $-8 + 5$ ▒ -6
8. $-24 \div (-8)$ ▒ -3
9. $-3 - (-9)$ ▒ -12
10. -9 ▒ $-3 + (-4)$
11. $-6 \cdot 3$ ▒ -18
12. 10 ▒ $36 \div 4$
13. 8 ▒ $-40 \div 5$
14. $5 - (-4)$ ▒ 9
15. 27 ▒ $4 \cdot 7$
16. $4 - 7$ ▒ 3

Solve each inequality.

17. $x + 4 < 10$
18. $a + 7 \geq 15$
19. $g + 5 > -8$
20. $c + 9 \leq 3$
21. $z - 4 > 20$
22. $h - (-7) > -2$
23. $m - 14 \leq -9$
24. $d - (-3) < 13$
25. $\frac{g}{-8} < 4$
26. $\frac{w}{3} > -12$
27. $\frac{p}{5} < 8$
28. $\frac{t}{-4} \geq -10$
29. $7b \geq -49$
30. $-5j < -60$
31. $-8f < 48$
32. $-2 + 9n \leq 10n$
33. $-5e + 9 > 24$
34. $3y - 4 > -37$
35. $7s - 12 < 13$
36. $-6v - 3 \geq -33$
37. $-2k + 12 < 30$
38. $-2x + 1 < 16 - x$
39. $15t - 4 > 11t - 16$
40. $13 - y \leq 29 + 2y$
41. $5q + 7 \leq 3(q + 1)$
42. $2(w + 4) \geq 7(w - 1)$
43. $-4t - 5 > 2t + 13$
44. $9m + 7 < 2(4m - 1)$
45. $3\left(a + \frac{2}{3}\right) \geq a - 1$
46. $3(3y + 1) < 13y - 8$
47. $2 + x < -5$ or $2 + x > 5$
48. $-4 + t > -5$ or $-4 + t < 7$
49. $3 \leq 2g + 7$ and $2g + 7 \leq 15$
50. $7 - 3s < 13$ and $7s < 3s + 12$
51. $2x + 1 < -3$ or $3x - 2 > 4$
52. $2v - 2 \leq 3v$ and $4v - 1 \geq 3v$
53. $3b - 4 \leq 7b + 12$ and $8b - 7 \leq 25$
54. $-9 < 2z + 7 < 10$
55. $5m - 8 \geq 10 - m$ or $5m + 11 < -9$
56. $12c - 4 \leq 5c + 10$ or $-4c - 1 \leq c + 24$
57. $2h - 2 \leq 3h \leq 4h - 1$
58. $3p + 6 < 8 - p$ and $5p + 8 \geq p + 6$
59. $4a + 3 < 3 - 5a$ or $a - 1 \geq -a$
60. $d - 4 < 5d + 14 < 3d + 26$
61. $2r + 8 > 16 - 2r$ and $7r + 21 < r - 9$
62. $-4j + 3 < j + 22$ and $j - 3 < 2j - 15$
63. $3n \neq 9$ and $6n - 5 \leq 2n + 7$
64. $7e \neq -21$ and $5e + 8 \geq e + 6$
65. $2(q - 4) \leq 3(q + 2)$ or $q - 8 \leq 4 - q$
66. $\frac{1}{2}w + 5 \geq w + 2 \geq \frac{1}{2}w + 9$
67. $|g + 6| > 8$
68. $|t - 5| \leq 3$
69. $|a + 5| \geq 0$
70. $|y - 9| < 19$
71. $|2m - 5| > 13$
72. $|14 - w| \geq 20$
73. $|3p + 5| \leq 23$
74. $|6b - 12| \leq 36$
75. $|25 - 3x| < 5$
76. $|7 + 8x| > 39$
77. $|4c + 5| \geq 25$
78. $|4 - 5s| > 46$

Polynomials: Addition and Subtraction

Find each sum.

1. $(3x - 4y) + (8x + 6y)$
2. $(12b + 2a) + (7b - 13a)$
3. $(7m - 8n) + (4m - 5n)$
4. $(5x^2 + 3x) + (4x^2 + 2x)$
5. $(-6s - 11t) + (5s - 6t)$
6. $(-14g - h) + (-8g + 5h)$
7. $(4p - 7q) + (5q - 8p)$
8. $(5y^2 - 7y) + (7y - 3y^2)$
9. $(9b^3 - 3b^2) + (12b^2 + 4b)$
10. $(2r + 8s) + (-3s - 9t)$
11. $(2a^2 + 4a + 5) + (2a^2 - 10a + 6)$
12. $(7x - 2y - 5z) + (x + 7y - 8z)$
13. $(-3m + 9mn - 5n) + (14m - 2n - 5mn)$
14. $(5x + 8y + 3z) + (-6z + 6y)$
15. $(6 - 4g - 9h) + (12g - 4h - 6j)$
16. $(x^2 - 4x + 8) + (12 + 7x - 4x^2)$
17. $(-7t^2 + 4ts - 6s^2) + (3s^2 - 12ts - 5t^2)$
18. $(7g + 8h - 9) + (-g - 3h - 6k)$
19. $(8a^2 - 4ab - 3b^2 + a - 4b) + (3a^2 + 6ab - 9b^2 + 7a + 9b)$
20. $(-3v + 14w - 12x - 13y + 6z - 8) + (16 - 6v + 2x - 5y - 2z)$
21. $(3y^2 - 7y + 6) + (3 - 2y^2 - 5y) + (y^2 - 8y - 12)$
22. $(4a^2 - 10b^2 + 7c^2) + (2c^2 - 5a^2 + 2b) + (7b^2 - 7c^2 + 7a)$
23. $(5x^2 + 3) + (4 - 7x - 9x^2) + (2x - 3x^2 - 5) + (2x - 6)$
24. $(9p - 13p^2) + (7p^2 + 5q^2) + (-6p - 12q) + (3q - 8q^2)$

Find each difference.

25. $(5g + 3h) - (2g + 7h)$
26. $(2e - 5f) - (7e - f)$
27. $(-3m + 8n) - (6m - 4n)$
28. $(6a^2 - 9a) - (4a^2 + 2a)$
29. $(-11k + 6) - (-6k - 8)$
30. $(9y^2 - 4y) - (-6y^2 - 8y)$
31. $(-r - 3s) - (2s - 5r)$
32. $(13c^2 - 4c) - (5c - 12c^2)$
33. $(g^3 - 2g^2) - (5g^2 - 7)$
34. $(7a + 4b) - (7b - 6c)$
35. $(z^2 + 6z - 8) - (4z^2 - 7z - 5)$
36. $(6v - 12w - 2x) - (2v + 8w - 10x)$
37. $(6a^2 - 7ab - 4b^2) - (6b^2 + 2a^2 + 5ab)$
38. $(3r - 7t) - (2t + 2s + 9r)$
39. $(7ax^2 + 2ax - 4a) - (5ax - 2ax^2 + 8a)$
40. $(h^3 + 4h^2 - 7h) - (3h^2 - 7h - 8)$
41. $(4d + 3e - 8f) - (-3d + 10e - 5f + 6)$
42. $(-3z^2 + 4x^2 - 8y^2) - (7x^2 - 14z^2 - 12)$
43. $(2b^2 + 7b - 2) - (2b^2 + 3b - 16)$
44. $(15j^4k^2 - 7j^2k + 8) - (8j^2k + 11)$
45. $(9x^2 - 11xy - 3y^2) - (12y^2 + x^2 - 16xy)$
46. $(17z^4 - 5z^2 + 3z) - (4z^4 + 2z^3 + 3z)$
47. $(-4p - 7q - 3t) - (-8t - 5q - 8p)$
48. $(-14h + 16j - 7k) - (-3j + 5h - 6k - 3)$
49. $(14a + 9b - 2x - 11y + 4z) - (8a + 8b + 6x - 5y - 7z)$
50. $(7m^2 - 3mn + 4n^2 - 2m - 8n) - (4m^2 + 3m - 4n^2 - 13n + 4mn)$

Polynomials: Multiplication and Division

Find each product.

1. $t^3 \cdot t^6$
2. $g^5 \cdot g^9$
3. $(3x^2y)(-5x^3y^8)$
4. $(7p^4q^7r^2)(4p^5r^7)$
5. $(2a^2b)(-b^2c^3)(-8ab^2c^4)$
6. $(e^4f^6g)^5$
7. $(-2m^6n^2)^6$
8. $(-3h^2k^3)^3(5hj^6k^8)^2$
9. $(v^4w)^6(-1v^3w^2)^8$
10. $5y(y^2 - 3y + 6)$
11. $-ab(3b^2 + 4ab - 6a^2)$
12. $4st^2(-4s^2t^3 + 7s^5 - 3st^3)$
13. $(d + 2)(d + 3)$
14. $(z + 7)(z - 4)$
15. $(m - 5)(m - 8)$
16. $(2x - 5)(x + 6)$
17. $(7a - 4)(2a - 5)$
18. $(t + 7)^2$
19. $(q - 4h)^2$
20. $(w - 12)(w + 12)$
21. $(2b + 4d)(2b - 4d)$
22. $(4x + y)(2x - 3y)$
23. $(7v + 3)(v + 4)$
24. $(4e + 3)(4e + 3)$
25. $(7s - 8)(3s - 2)$
26. $(5b - 6)(5b + 6)$
27. $(4g + 3h)(2g - 5h)$
28. $(5c - 2d)^2$
29. $(10x + 11y)(10x - 11y)$
30. $(12r - 4s)(5r + 8s)$

Simplify.

31. $\frac{24x^5}{8x^2}$
32. $\frac{s^7t^4}{s^5}$
33. $\frac{-9h^2k^4}{18h^5j^3k^4}$
34. $\frac{3m^7n^2p^4}{9m^2np^3}$
35. $\frac{9a^2b^7c^3}{12a^5b^4c}$
36. $\frac{-15xy^5z^7}{-10x^4y^6z^4}$
37. $\frac{-5w^4v^2 - 3w^3v}{w^3}$
38. $\frac{8g^4h^4 + 4g^3h^3}{2gh^2}$
39. $\frac{x^2 - 2x - 15}{x - 3}$
40. $\frac{q^2 - 10q + 24}{q - 4}$
41. $\frac{2j^2 + 10j + 12}{j + 3}$
42. $\frac{12d^2 + d - 6}{3d - 2}$
43. $\frac{4s^3 + 4s^2 - 9s - 18}{2s + 3}$
44. $\frac{z^3 - 27}{z - 3}$
45. $\frac{6n^3 + 7n^2 - 29n + 12}{3n - 4}$
46. $\frac{3e^2 + 3e - 80}{e - 5}$
47. $\frac{-6k^2 + 3k + 12}{2k + 3}$
48. $\frac{t^3 + 3t}{t - 1}$
49. $\frac{4g^3 - 4g - 20}{2g - 4}$
50. $\frac{9a^3 - 3a^2 + 7a + 2}{3a + 1}$
51. $\frac{8c^3 - 2c^2 + 2c - 4}{2c - 2}$
52. $\frac{8y^3 - 1}{2y + 1}$
53. $\frac{12m^3 + m^2 - 20}{4m - 5}$
54. $\frac{5b^3 - 2b^2 + 7b + 4}{5b + 3}$

Factoring

Find the prime factorization of each integer. Write each negative integer as the product of −1 and its prime factors.

1. 35 **2.** 12 **3.** 72 **4.** 64

5. −75 **6.** 70 **7.** 85 **8.** −92

9. −117 **10.** −114 **11.** 243 **12.** −360

13. 405 **14.** 605 **15.** −5292 **16.** 5076

Factor.

17. $10g + 35h$

18. $t^3s^2 - t^2$

19. $15a^2b - 24a^5b^2$

20. $18c^4d - 30c^3e$

21. $36m^4n^2p + 12m^5n^3p^2$

22. $6g^5h^3 - 12g^4h^6k - 18g^6h^5$

23. $p^2 - q^2$

24. $144x^2 - 49y^2$

25. $75r^2 - 48$

26. $64v^2 - 100w^4$

27. $g^2 + 4g + 4$

28. $t^2 - 22t + 121$

29. $9n^2 - 36nm + 36m^2$

30. $2a^2b^2 + 4ab^2c^2 + 2b^2c^4$

31. $g^2 - 14g + 48$

32. $z^2 + 15z + 36$

33. $12 - 13b + b^2$

34. $x^2 + 17xy + 16y^2$

35. $g^2 - 4g - 32$

36. $h^2 + 12h - 28$

37. $s^2 - 13st - 30t^2$

38. $3a^2 + 11a + 10$

39. $6y^2 + 2y - 20$

40. $12j^2 - 34j - 20$

41. $24a^2 - 57ax + 18x^2$

42. $2sx - 4tx + 2sy - 4ty$

43. $8ac - 2ad + 4bc - bd$

44. $2e^2g + 2fg + 4e^2h + 4fh$

45. $5x^3 - 2x^2y - 5xy^2 + 2y^3$

46. $4p^2 + 12pr + 9r^2$

47. $169 - 16t^2$

48. $b^2 - 11b - 42$

49. $30g^2h - 15g^3$

50. $3b^2 - 13bd + 4d^2$

51. $a^2x - 2a^2y - 5x + 10y$

52. $s^2 + 30s + 225$

53. $18v^2 + 42v + 12$

54. $4k^2 + 2k - 12$

55. $5z^3 - 8z^2 - 21z$

56. $5g^2 - 20h^2$

57. $30x^2 - 125x + 70$

58. $a^2b^2 - b^2 + a^2 - 1$

59. $8t^4 + 56t^3 + 98t^2$

60. $3p^3q^2 + 27pq^2$

61. $a^2c^2 + b^2c^2 - 4a^2d^2 - 4b^2d^2$

62. $36m^3n - 90m^2n^2 + 36mn^2$

63. $4x^2z^2 + 7xyz^2 - 36y^2z^2$

64. $4g^2j^2 - 25h^2j^2 - 4g^2 + 25h^2$

Algebraic Fractions

Simplify.

1. $\frac{48a^2b^5c}{32a^7b^2c^3}$
2. $\frac{-28x^3y^4z^5}{42xyz^2}$
3. $\frac{k+3}{4k^2+7k-15}$
4. $\frac{t^2-s^2}{5t^2-2st-3s^2}$
5. $\frac{6g^2-19g+15}{12g^2-6g-18}$
6. $\frac{2d^2+4d-6}{d^4-10d^2+9}$

Find each product or quotient.

7. $\frac{5m^2n}{12a^2} \cdot \frac{18an}{30m^4}$
8. $\frac{25g^7h}{28t^3} \cdot \frac{42s^2t^3}{5g^5h^2}$
9. $\frac{6a+4b}{36} \cdot \frac{45}{3a+2b}$
10. $\frac{x^2y}{18z} \div \frac{2yz}{3x^2}$
11. $\frac{p^2}{14qr^3} \div \frac{2r^2p}{7q}$
12. $\frac{3d}{2d^2-3d} \div \frac{9}{2d-3}$
13. $\frac{t^2-2t-15}{t-5} \cdot \frac{t+5}{t+3}$
14. $\frac{5e-f}{5e+f} \div (25e^2-f^2)$
15. $\frac{8}{6c^2+17c+10} \div \frac{6}{c+2}$
16. $\frac{3v^2-27}{15v} \cdot \frac{v^2}{v+3}$
17. $\frac{3k^2-10k+3}{5} \div \frac{3k-1}{15k}$
18. $\frac{3g^2+15g}{4} \cdot \frac{g^2}{g+5}$
19. $\frac{x^2-2x-15}{2x^2-7x-15} \cdot \frac{4x^2-4x-15}{2x^2+x-15}$
20. $\frac{a^2-16}{3a^2-13a+4} \div \frac{3a^2+11a-4}{a^3}$

Find each sum or difference.

21. $\frac{j+4}{3} + \frac{j-7}{3}$
22. $\frac{15n}{5n+3} + \frac{9}{5n+3}$
23. $\frac{2a-3}{b} - \frac{a-5}{b}$
24. $\frac{25}{5-g} - \frac{g^2}{5-g}$
25. $\frac{s}{t^2} - \frac{r}{3t}$
26. $\frac{7}{ab} + \frac{4}{bc}$
27. $\frac{2}{2p+3} + \frac{p}{3p+2}$
28. $\frac{x}{x+y} - \frac{5}{y}$
29. $\frac{c}{c^2-4c} - \frac{5c}{c-4}$
30. $\frac{t+10}{t^2-100} + \frac{1}{t-10}$
31. $\frac{1}{g^2-6gh+9h^2} - \frac{3}{g-3h}$
32. $\frac{x}{x+2} + \frac{x^2+3x}{x^2+5x+6}$
33. $\frac{3d}{d^2-3d-10} + \frac{d+1}{d^2-8d+15}$
34. $\frac{k+2}{k^2-8k+16} - \frac{k+3}{k^2+k-20}$

Solving Equations

Solve each equation.

1. $2x - 5 = 3$

2. $4t + 5 = 37$

3. $7a + 6 = -36$

4. $47 = -8g + 7$

5. $-3c - 9 = -24$

6. $5k - 7 = -52$

7. $5s + 4s = -72$

8. $6(y - 5) = 18$

9. $-21 = 7(p - 10)$

10. $2m + 5 - 6m = 25$

11. $3z - 1 = 23 - 3z$

12. $5b + 12 = 3b - 6$

13. $\frac{e}{5} + 6 = -2$

14. $\frac{d}{4} - 8 = -5$

15. $\frac{p + 10}{3} = 4$

16. $\frac{h - 7}{6} = 1$

17. $\frac{5f + 1}{8} = -3$

18. $\frac{4n - 8}{-2} = 12$

19. $\frac{2a}{7} + 9 = 3$

20. $\frac{-3t - 4}{2} = 8$

21. $\frac{6v - 9}{3} = v$

22. $|s - 4| = 7$

23. $|5g + 8| = 33$

24. $|16 - 3b| = 22$

25. $t(t - 4) = 0$

26. $4p(p + 7) = 0$

27. $7m(2m - 12) = 0$

28. $(x - 5)(x + 8) = 0$

29. $(3s + 6)(2s - 7) = 0$

30. $(4g + 5)(2g + 10) = 0$

31. $h^2 + 4h = 0$

32. $3b^2 - 3b = 0$

33. $m^2 - 16 = 0$

34. $w^2 + w - 30 = 0$

35. $2c^2 - 14c + 24 = 0$

36. $6p^2 + 10p - 24 = 0$

37. $5n^2 = 15n$

38. $4x^2 + 20x + 25 = 0$

39. $3y^2 = 75$

40. $\frac{a}{3} - \frac{a}{4} = 3$

41. $\frac{k}{6} + \frac{2k}{3} = -\frac{5}{2}$

42. $\frac{r + 3}{r} + \frac{r - 12}{r} = 5$

43. $\frac{2y}{y - 4} - \frac{3}{5} = 3$

44. $\frac{2t}{t + 3} + \frac{3}{t} = 2$

45. $\frac{2g}{2g + 1} - \frac{1}{2g - 1} = 1$

46. $\frac{8n^2}{2n^2 - 5n - 3} = 4$

47. $5 - \frac{5x}{x - 4} = \frac{2}{x^2 - 4x}$

48. $\frac{2}{e + 1} - \frac{3}{e + 2} = 0$

49. $\frac{1}{2d - 1} - \frac{12d}{6d^2 + d - 2} = \frac{-4}{3d + 2}$

50. $\frac{4z}{2z + 1} - \frac{6 - z}{2z^2 - 5z - 3} = 2$

51. $\frac{b + 1}{b - 2} - \frac{3b}{3b + 2} = \frac{20}{3b^2 - 4b - 4}$

52. $\frac{2s}{9s^2 - 3s - 2} + \frac{2}{3s + 1} = \frac{4}{3s - 2}$

53. $\frac{m}{20} = \frac{9}{15}$

54. $\frac{12}{21} = \frac{20}{f}$

55. $\frac{4}{9} = \frac{16}{t + 8}$

56. $\frac{4}{14} = \frac{2h - 1}{21}$

57. $\frac{2 + c}{c - 5} = \frac{8}{9}$

58. $\frac{2y + 4}{y - 3} = \frac{2}{3}$

Radicals

Simplify.

1. $\sqrt{25}$
2. $-\sqrt{64}$
3. $\pm\sqrt{576}$
4. $\sqrt{900}$
5. $\pm\sqrt{0.01}$
6. $\sqrt{1.44}$
7. $-\sqrt{0.0016}$
8. $\sqrt{4.84}$
9. $\sqrt{\frac{36}{121}}$
10. $-\sqrt{\frac{16}{36}}$
11. $\sqrt{\frac{81}{49}}$
12. $\pm\sqrt{\frac{64}{100}}$
13. $\sqrt{75}$
14. $\sqrt{20}$
15. $\sqrt{162}$
16. $\sqrt{700}$
17. $\sqrt{4x^4y^3}$
18. $\sqrt{12ts^3}$
19. $\sqrt{175m^4n^6}$
20. $\sqrt{99a^3b^7}$
21. $\sqrt{\frac{54}{g^2}}$
22. $\sqrt{\frac{32c^5}{9d^2}}$
23. $\sqrt{\frac{27p^4}{3p^2}}$
24. $\sqrt{\frac{243y^7}{3y^4}}$
25. $\sqrt{7}\,\sqrt{3}$
26. $\sqrt{3}\,\sqrt{15}$
27. $6\sqrt{2}\,\sqrt{3}$
28. $5\sqrt{6} \cdot 2\sqrt{3}$
29. $\sqrt{5}(\sqrt{3} + \sqrt{10})$
30. $\sqrt{2}(\sqrt{6} + \sqrt{32})$
31. $\sqrt{5}\,\sqrt{t}$
32. $\sqrt{18}\,\sqrt{g^3}$
33. $\sqrt{12k}\,\sqrt{3k^5}$
34. $\sqrt{15m^2}\,\sqrt{6n^3}$
35. $\frac{\sqrt{3}}{\sqrt{5}}$
36. $\sqrt{\frac{2}{7}}$
37. $\frac{3\sqrt{6}}{\sqrt{2}}$
38. $\sqrt{\frac{x}{8}}$
39. $\sqrt{\frac{t^2}{3}}$
40. $\sqrt{\frac{20}{a}}$
41. $(5 + \sqrt{3})(5 - \sqrt{3})$
42. $(\sqrt{17} + \sqrt{11})(\sqrt{17} - \sqrt{11})$
43. $(2\sqrt{5} + \sqrt{7})(2\sqrt{5} - \sqrt{7})$
44. $\frac{1}{3 + \sqrt{5}}$
45. $\frac{2}{\sqrt{3} - 5}$
46. $\frac{12}{\sqrt{8} - \sqrt{6}}$
47. $\frac{10}{\sqrt{13} + \sqrt{7}}$
48. $\frac{14}{3\sqrt{2} + \sqrt{5}}$
49. $\frac{\sqrt{3}}{\sqrt{3} - 5}$
50. $\frac{\sqrt{6}}{7 - 2\sqrt{3}}$
51. $\frac{5\sqrt{10}}{3\sqrt{3} + 2\sqrt{5}}$
52. $6\sqrt{13} + 7\sqrt{13}$
53. $9\sqrt{15} - 4\sqrt{15}$
54. $2\sqrt{11} - 8\sqrt{11}$
55. $2\sqrt{12} + 5\sqrt{3}$
56. $2\sqrt{27} - 4\sqrt{12}$
57. $4\sqrt{8} - 3\sqrt{5}$
58. $8\sqrt{32} + 4\sqrt{50}$
59. $6\sqrt{20} + \sqrt{45}$
60. $2\sqrt{63} + 8\sqrt{45} - 6\sqrt{28}$
61. $10\sqrt{\frac{1}{5}} - \sqrt{45} - 12\sqrt{\frac{5}{9}}$
62. $3\sqrt{\frac{1}{3}} - 9\sqrt{\frac{1}{12}} + \sqrt{243}$

Solve and check.

63. $\sqrt{t} = 10$
64. $\sqrt{3g} = 6$
65. $\sqrt{y} - 2 = 0$
66. $5 + \sqrt{a} = 9$
67. $\sqrt{2k} - 4 = 8$
68. $\sqrt{5y + 4} = 7$
69. $\sqrt{10x^2 - 5} = 3x$
70. $\sqrt{2a^2 - 144} = a$
71. $\sqrt{b^2 + 16} + 2b = 5b$
72. $\sqrt{m + 2} + m = 4$
73. $\sqrt{3 - 2c} + 3 = 2c$

Use the quadratic formula to solve each equation.

74. $s^2 + 8s + 7 = 0$
75. $d^2 - 14d + 24 = 0$
76. $3h^2 = 27$
77. $n^2 - 3n + 1 = 0$
78. $2z^2 + 5z - 1 = 0$
79. $3w^2 - 8w + 2 = 0$
80. $3f^2 + 2f = 6$
81. $2r^2 - r - 3 = 0$
82. $x^2 - 9x = 5$

Equations in Two Variables

Write an equation in slope-intercept form for each given slope and y-intercept.

1. $m = 2, b = 5$
2. $m = -4, b = 1$
3. $m = \frac{1}{2}, b = -3$
4. $m = -1, b = -6$
5. $m = \frac{3}{2}, b = 1$
6. $m = 5, b = \frac{1}{4}$
7. $m = \frac{2}{5}, b = -4$
8. $m = -\frac{3}{4}, b = \frac{1}{2}$

Write each equation in slope-intercept form.

9. $-3x + y = 2$
10. $6x + y = -5$
11. $2x - y = -3$
12. $-x - y = 4$
13. $2x + 5y = 10$
14. $x - 4y = -2$
15. $-9x + 3y = -18$
16. $4x + 7y = 3$

Write each equation in standard form.

17. $y = 3x + 6$
18. $y = -4x + 1$
19. $y = \frac{2}{3}x - 7$
20. $y = \frac{1}{4}x + \frac{1}{2}$
21. $y = -\frac{5}{3}x - \frac{1}{3}$
22. $y = 2x - \frac{1}{2}$
23. $\frac{5}{6}y = \frac{1}{4}x + \frac{2}{3}$
24. $\frac{1}{5}x = \frac{7}{10}y - \frac{3}{4}$

Write the equation of the line with each x-intercept and y-intercept. Use slope-intercept form.

25. x-intercept $= 2$; y-intercept $= -1$
26. x-intercept $= -3$; y-intercept $= -2$
27. x-intercept $= 1$; y-intercept $= 5$
28. x-intercept $= -\frac{1}{2}$; y-intercept $= 3$

Write the equation of the line that passes through each pair of points. Use slope-intercept form.

29. $(1, 3), (2, 7)$
30. $(4, 8), (2, 4)$
31. $(-2, 3), (3, 1)$
32. $(0, 0), (-2, -3)$
33. $(5, 1), (3, -2)$
34. $(2, -1), (5, -4)$
35. $(8, 6), (10, 3)$
36. $(-4, -1), (-1, -7)$

Solve each system of equations.

37. $y = 3x$; $4x + 2y = 30$
38. $a = -2b$; $3a + 5b = 21$
39. $n = m + 4$; $3m + 2n = 19$
40. $h = k - 7$; $2h - 5k = -2$
41. $s + 2t = 6$; $3s - 2t = 2$
42. $c + 2d = 10$; $-c + d = 2$
43. $3v + 5w = -16$; $3v - 2w = -2$
44. $e - 5f = 12$; $3e - 5f = 6$
45. $-3p + 2q = 10$; $-2p - q = -5$
46. $2a + 5b = 13$; $4a - 3b = -13$
47. $5s + 3t = 4$; $-4s + 5t = -18$
48. $2g - 7h = 9$; $-3g + 4h = 6$
49. $2c - 6d = -16$; $5c + 7d = -18$
50. $6m - 3n = -9$; $-8m + 2n = 4$
51. $3x - 5y = 8$; $4x - 7y = 10$
52. $9a - 3b = 5$; $a + b = 1$

Find the equation of the axis of symmetry and the maximum or minimum point for the graph of each quadratic function.

53. $y = -x^2 + 2x - 3$
54. $y = x^2 - 4x - 4$
55. $y = 3x^2 + 6x + 3$
56. $y = 2x^2 + 12x$
57. $y = x^2 - 6x + 5$
58. $y = 4x^2 - 1$
59. $y = -2x^2 - 2x + 4$
60. $y = \frac{1}{2}x^2 + 4x + \frac{1}{4}$
61. $y = 6x^2 - 12x - 4$

GLOSSARY

A

absolute value The absolute value of a number is the number of units that it is from zero on the number line. (55)

addition property for inequalities For all numbers a, b, and c,
1. if $a > b$, then $a + c > b + c$, and
2. if $a < b$, then $a + c < b + c$. (176)

addition property of equality For any numbers a, b, and c, if $a = b$, then $a + c = b + c$. (94)

additive identity The number 0 is the additive identity since the sum of any number and 0 is equal to the number. (22)

additive inverse Two numbers are additive inverses if their sum is zero. The additive inverse, or opposite, of a is $-a$. (56)

algebraic expression An expression consisting of one or more numbers and variables along with one or more arithmetic operations. (8)

angle of depression An angle of depression is formed by a line of sight along the horizontal and another line of sight below it. (632)

angle of elevation An angle of elevation is formed by a line of sight along the horizontal and another line of sight above it. (632)

associative property of addition For any numbers a, b, and c,
$(a + b) + c = a + (b + c)$. (31)

associative property of multiplication For any numbers a, b, and c,
$(ab)c = a(bc)$. (31)

axis of symmetry A straight line with respect to which a figure is symmetric. (519)

B

base In an expression of the form x^n, the base is x. (9)
In the proportion $\frac{17}{25} = \frac{r}{100}$, the base is 25. (138)

binomial A polynomial with exactly two terms. (226)

boundary A line that separates a graph into half-planes. (379)

box-and-whisker plot In a box-and-whisker plot, the quartiles and extreme value of a set of data are displayed using a number line. (579)

closed half-plane A half-plane that includes the boundary. (380)

coefficient The numerical part of a term. (28)

commutative property of addition For any numbers a and b,
$$a + b = b + a. \quad (31)$$

commutative property of multiplication For any numbers a and b, $ab = ba$. (31)

comparison property For any two numbers a and b, exactly one of the following sentences is true.
$a < b \qquad a = b \qquad a > b$ (60)

comparison property for rational numbers For any rational numbers $\frac{a}{b}$ and $\frac{c}{d}$, with $b > 0$ and $d > 0$,
1. if $\frac{a}{b} < \frac{c}{d}$, then $ad < bc$, and
2. if $ad < bc$, then $\frac{a}{b} < \frac{c}{d}$. (65)

complementary angles Two angles are complementary if the sum of their measures is 90°. (611)

completeness property for points in the plane When plotting points, the following is true. (355)

1. Exactly one point in the plane is named by a given ordered pair of numbers.
2. Exactly one ordered pair of numbers names a given point in the plane.

completing the square Completing the square is a method of solving a quadratic equation where a perfect square trinomial is formed on one side of the equation. (532)

complex fraction If a fraction has one or more fractions in the numerator or denominator, it is called a complex fraction. (81)

composite number Any positive integer, except 1, that is not prime. (256)

compound event A compound event consists of two or more simple events. (600)

compound inequalities Two inequalities connected by *and* or *or*. (194)

compound interest The amount of interest paid or earned on the original principal plus the accumulated interest. (145)

compound sentence Two sentences connected by *and* or *or*. (185)

conjugates Two binomials of the form $a\sqrt{b} + c\sqrt{d}$ and $a\sqrt{b} - c\sqrt{d}$. (494)

conjunction A compound sentence where the statements are connected using *and*. (185)

consecutive even integers Numbers given when beginning with an even integer and counting by two's. (112)

consecutive numbers Numbers in counting order. (112)

consecutive odd integers Numbers given when beginning with an odd integer and counting by two's. (112)

consistent A system of equations is consistent and independent if it has one ordered pair as its solution. A system of equations is consistent and dependent if it has infinitely many ordered pairs as its solution. (443)

constant A monomial that does not contain variables. (213)

constant of variation In the direct variation equation $y = kx$, k is called the constant of variation. (162)

coordinate The coordinate of a point is the number that corresponds to it on the number line. (51)

coordinate plane The number plane formed by two perpendicular number lines that intersect at their zero points. (354)

corresponding angles In similar triangles, the measures of corresponding angles are equal. (622)

corresponding sides In similar triangles, the sides opposite corresponding angles are called corresponding sides. The measures of corresponding sides are proportional. (622)

cosine In a right triangle, the cosine of angle A =

$$\frac{\text{measure of side adjacent to angle } A}{\text{measure of hypotenuse}}. \quad (626)$$

cross product In the proportion $\frac{a}{b} = \frac{c}{d}$, the cross products are $a \times d$ and $b \times c$. (65)

data Numerical information. (560)

decimal notation A way of expressing numbers using a base ten system. 483.26 is expressed in decimal notation. (221)

degree The degree of a monomial is the sum of the exponents of its variables. The degree of a nonzero constant is 0. The degree of a polynomial is the greatest of the degrees of its terms. (227)

density property Between every pair of distinct rational numbers, there is another rational number. (137)(66)

difference of squares Two perfect squares separated by a subtraction sign. $a^2 - b^2$ (244)

direct variation A direct variation is described by an equation of the form $y = kx$, where k is not zero. (162)

discriminant In the quadratic formula, the expression $b^2 - 4ac$ is called the discriminant. (541)

disjoint sets Two sets that have no members in common. (51)

disjunction A compound sentence where the statements are connected using *or*. (185)

distance formula The distance between any two points (x_1, y_1) and (x_2, y_2) is given by the formula
$d = \sqrt{(x_2 - x_1)^2 + (y_2 - y_1)^2}$. (506)

distributive property For any numbers a, b, and c:
1. $a(b + c) = ab + ac$ and
   $(b + c)a = ba + ca$.
2. $a(b - c) = ab - ac$ and
   $(b - c)a = ba - ca$. (26)

division property for inequalities For all numbers a, b, and c, with $c \neq 0$,
1. if c is positive and $a < b$, then $\frac{a}{c} < \frac{b}{c}$, and if c is positive and $a > b$, then $\frac{a}{c} > \frac{b}{c}$.
2. If c is negative and $a < b$, then $\frac{a}{c} > \frac{b}{c}$, and if c is negative and $a > b$, then $\frac{a}{c} < \frac{b}{c}$. (182)

division property of equality For any numbers a, b, and c, with $c \neq 0$, if $a = b$, then $\frac{a}{c} = \frac{b}{c}$. (104)

domain The domain of a relation is the set of all first components from each ordered pair. (359)

edge A line that separates a graph into half-planes. (379)

element One of the members of a set. (19)

elimination method A method for solving systems of equations in which the equations are added or subtracted to eliminate one of the variables. Multiplication of one or both equations may occur before the equations are added or subtracted. (452)

empty set A set with no elements. (19)

equally likely events Events that have an equal chance of occurring. (589)

equals sign The equals sign, =, between two expressions indicates that if the sentence is true, the expressions name the same number. (19)

equation A mathematical sentence that contains the equals sign. (19)

equivalent equations Equations that have the same solution. (94)

evaluate To find the value of an expression when the values of the variables are known. (13)

excluded value A value excluded from the domain of a variable because that value substituted for the variable would result in a denominator of zero. (306)

experimental probability Probability calculated by performing experiments. (594)

exponent A number used to tell how many times a number is used as a factor. In an expression of the form x^n, the exponent is n. (9)

extremes *See* proportion. (134)

factor In a multiplication expression, the quantities being multiplied are called factors. (8)

FOIL method for multiplying binomials To multiply two binomials, find the sum of the products of
F the first terms,
O the outer terms,
I the inner terms, and
L the last terms. (238)

formula An equation that states a rule for the relationship between certain quantities. (36)

function A function is a relation in which each element of the domain is paired with exactly one element of the range. (374)

functional notation The functional notation of the equation $y = x + 5$ is
$f(x) = x + 5$. (376)

functional value The symbol $f(3)$ represents the functional value of f for $x = 3$. (376)

graph To graph a set of numbers means to locate the points by those numbers on a number line. (51)
To graph an ordered pair means to draw a dot at the point on a coordinate plane that corresponds to the ordered pair. (335)

greatest common factor (GCF) The GCF of two or more integers is the greatest factor that is common to each of the integers. (257)

grouping symbols Symbols used to clarify or change the order of operations in an expression. Parentheses, brackets, and the fraction bar are grouping symbols. (13)

half-plane The region of a graph on one side of a boundary. (379)

hypotenuse The side opposite the right angle in a right triangle. (482)

identity An equation that is true for every value of the variable. (118)

inconsistent A system of equations is inconsistent if it has no solution. (443)

inequality Any sentence containing $<$, $>$, $\neq$, $\leq$, or $\geq$. (60)

integers (Z) The set of numbers $\{\ldots, -3, -2, -1, 0, 1, 2, 3, \ldots\}$. (50)

intercept An intercept is a point where a graph crosses the x-axis or y-axis. (410)

interquartile range The difference between the upper quartile and the lower quartile of a set of data is called the interquartile range. It represents the middle half of the data in the set. (575)

intersection The intersection of two sets A and B ($A \cap B$) is the set of elements common to both A and B. (51)

inverse The inverse of any relation is obtained by switching the coordinates in each ordered pair of the relation. (360)

inverse operations Operations that undo each other, such as multiplication and division. (80)

inverse variation An inverse variation is described by an equation of the form $xy = k$, where k is not zero. (166)

irrational numbers (I) Numbers that cannot be expressed in the form $\frac{a}{b}$, where a and b are integers, $b \neq 0$. (487)

least common denominator (LCD) The least common multiple of the denominators of two or more fractions. (330)

least common multiple (LCM) The LCM of two or more integers is the least positive integer that is divisible by each of the integers. (329)

legs The adjacent sides of the right angle of a right triangle. (482)

like terms Terms that contain the same variables, with corresponding variables raised to the same power. (27)

linear equation A linear equation is an equation that may be written in the form $Ax + By = C$, where A, B, and C are any numbers and A and B are not both 0. (369)

line plot Numerical information displayed on a number line. (560)

lower quartile The lower quartile divides the lower half of the set of data into two equal parts. (575)

maximum point The highest point on the graph of a curve, such as a parabola, which opens down. (518)

mean The mean of a set of data is the sum of the elements in the set, divided by the number of elements in the set. (80)

means *See* proportion. (134)

median The median is the middle number of a set of data when the numbers are arranged in numerical order. (570)

midpoint The midpoint of a line segment is the point that is halfway between the endpoints of the segment. (428)

minimum point The lowest point on the graph of a curve, such as a parabola which opens up. (518)

mixed expression Algebraic expression which contain monomials and rational expressions. (334)

mode The mode is the number that occurs most often in a set of data. (570)

monomial A number, a variable, or a product of numbers and variables. (213)

multiplication property for inequalities For all numbers a, b, and c,
1. *if* c is positive and $a < b$, then $ac < bc$, and if c is positive and $a > b$, then $ac > bc$.
2. If c is negative and $a < b$, then $ac > bc$, and if c is negative and $a > b$, then $ac < bc$. (181)

multiplication property of equality For any numbers a, b, and c, with $c \neq 0$, if $a = b$, then $ac = bc$. (103)

multiplicative identity The number 1 is the multiplicative identity since the product of any number and 1 is equal to the number. (22)

multiplicative inverse Two numbers are multiplicative inverses if their product is 1. The multiplicative inverse, or reciprocal, of a is $\frac{1}{a}$. (81)(84)

multiplicative property of zero For any number a, $a \cdot 0 = 0$. (23)

natural numbers (N) The set of numbers $\{1, 2, 3, \ldots\}$. (51)

negative number A number that is graphed on the negative side of the number line. (50)

null set A set with no elements. (19)

number line A line with equal distances marked off to represent numbers. (50)

number theory The study of numbers. (112)

numerical coefficient The numerical part of a term. (28)(16)

odds The odds of an event occurring is the ratio of the number of ways the event can occur (successes) to the number of ways the event cannot occur (failures). (590)

open half-plane A half-plane that does not include the boundary. (379)

open sentence A sentence containing a symbol(s) to be replaced in order to determine if the sentence is true or false. (18)

opposite The opposite of a number is its additive inverse. (56)

ordered pair In mathematics, pairs of numbers used to locate points in the plane. (354)

order of operations
1. Simplify the expressions inside grouping symbols.
2. Evaluate all powers.
3. Then do all multiplications and divisions from left to right.
4. Then do all additions and subtractions from left to right. (13)

origin The point of intersection of the two axes of the coordinate plane. (354)

outlier Any element of a set of data that is at least 1.5 interquartile ranges above the upper quartile or below the lower quartile. (576)

parabola The general shape of the graph of a quadratic function. (518)

parallel lines Lines that have the same slope are parallel. All vertical lines are parallel. (423)

percent Per hundred, or hundredths. (138)

percentage A number which is compared to another number (base) in the percent proportion. (138)

percent of decrease The ratio of an amount of decrease to the previous amount, expressed as a percent. (146)

percent of increase The ratio of an amount of increase to the previous amount, expressed as a percent. (146)

percent proportion

$$\frac{\text{Percentage}}{\text{Base}} = \text{Rate or } \frac{P}{B} = \frac{r}{100}. \quad (139)$$

perfect square trinomial A perfect square trinomial is a trinomial of the form $(a + b)^2 = a^2 + 2ab + b^2$ or $(a - b)^2 = a^2 - 2ab + b^2$. (281)

perpendicular lines Two lines are perpendicular if the product of their slopes is -1. In a plane, vertical lines are perpendicular to horizontal lines. (424)

point-slope form For a given point (x, y) on a nonvertical line with slope m, the point-slope form of a linear equation is $y - y_1 = m(x - x_1)$. (405)

polynomial An expression that can be written as a sum of monomials. (226)

power An expression of the form x^n. (10)

prime factorization The expression of a composite number as the product of its prime factors. (256)

prime number An integer, greater than 1, whose only positive factors are 1 and itself. (256)

prime polynomial A polynomial that cannot be written as a product of two or more polynomials. (273)

principal square root The nonnegative square root of the expression. (477)

probability The ratio that tells how likely it is that an event will take place.

$$P(\text{event}) = \frac{\text{number of favorable outcomes}}{\text{number of possible outcomes}}$$

(589)

product The result of multiplication. (8)

product property of square roots For any numbers a and b, where $a \geq 0$ and $b \geq 0$, $\sqrt{ab} = \sqrt{a} \cdot \sqrt{b}$. (492)

properties Algebraic statements that are true for any number. (22)

proportion An equation of the form $\frac{a}{b} = \frac{c}{d}$ which states that two ratios are equal. The first and fourth terms (a and d) are called the extremes. The second and third terms (b and c) are called the means. (134)

Pythagorean Theorem In a right triangle, if a and b are the measures of the legs, and c is the measure of the hypotenuse, then $c^2 = a^2 + b^2$. (482)

quadrant One of the four regions into which two perpendicular number lines separate the plane. (256)

quadratic formula The solutions of a quadratic equation of the form $ax^2 + bx + c = 0$, where $a \neq 0$, are given by $x = \frac{-b \pm \sqrt{b^2 - 4ac}}{2a}$. (537)

quadratic function A quadratic function is a function described by an equation of the form $y = ax^2 + bx + c$, where $a \neq 0$. (518)

quotient property of square roots For any numbers a and b, where $a \geq 0$ and $b > 0$, $\sqrt{\frac{a}{b}} = \frac{\sqrt{a}}{\sqrt{b}}$. (493)

radical equations Equations containing radicals with variables in the radicand. (501)

radical expression An expression of the form $\sqrt{a}$. (477)

radical sign The symbol $\sqrt{}$ indicating the principal or nonnegative square root. (477)

radicand The expression under the radical sign. (477)

range The range of a relation is the set of all second components from each ordered pair. (359)
The difference between the greatest and the least values of a set of data. (575)

rate A ratio of two measurements having different units of measure. (135)
In the percent proportion, the rate is the fraction with a denominator of 100. (138)

ratio A comparison of two numbers by division. The ratio of a to b is $\frac{a}{b}$. (134)

rational equations Equations containing rational expressions. (338)

rational expression An algebraic fraction whose numerator and denominator are polynomials. (306)

rationalizing the denominator Rationalizing the denominator is a method used to eliminate radicals from the denominator of a fraction. (493)

rational numbers (Q) Numbers that can be expressed in the form $\frac{a}{b}$, where a and b are integers, $b \neq 0$. (61)

real numbers (R) Irrational numbers together with rational numbers form the set of real numbers. (51)

reciprocal The reciprocal of a number is its multiplicative inverse. (81)

reflexive property of equality For any number a, $a = a$. (23)

relation A set of ordered pairs. (359)

replacement set The set of numbers for which replacements for a variable may be chosen. (19)

right triangle A triangle that has a 90° angle. (612)

root of an equation A solution of the equation. (523)

scale A ratio that compares the size of a model to the actual size of the object being modeled. (135)

scatter plot In a scatter plot, two sets of data are plotted as ordered pairs in the coordinate plane. (583)

scientific notation A number is expressed in scientific notation when it is in the form $a \times 10^n$, where $1 \leq a < 10$ and n is an integer. (221)

set A collection of objects or numbers. (19)

set-builder notation A notation used to describe the members of a set. For example, $\{y|y < 17\}$ represents the set of all numbers y such that y is less than 17. (177)

similar triangles If two triangles are similar, the measures of their corresponding angles are equal and the measures of their corresponding sides are proportional. (622)

simple interest The amount paid or earned for the use of money for a unit of time.
$I = prt$ (142)

simplest form of an expression An expression in simplest form has no like terms and no parentheses. (27)

simplify To simplify an expression involving monomials, write an equivalent expression that has positive exponents and no powers of powers. Each base should appear only once and all fractions should be in simplest form. (218)

sine In a right triangle, the sine of angle

A is $\frac{\text{measure of leg opposite angle } A}{\text{measure of hypotenuse}}$.

(626)

slope The slope of a line is the ratio of the change in y to the corresponding change in x.

$$\text{slope} = \frac{\text{change in } y}{\text{change in } x} \quad (400)$$

slope-intercept form The slope-intercept form of the equation of a line is $y = mx + b$. The slope of the line is m, and the y-intercept is b. (410)

solution A replacement for the variable in an open sentence which results in a true sentence. (18)

solution set The set of all replacements for the variable in an open sentence which make the sentence true. (19)

solve To solve an open sentence means to find all the solutions. (18)

spreadsheets Computer programs especially designed to create charts involving many calculations. (40)

square root If $x^2 = y$, then x is a square root of y. (477)

squaring Squaring a number means using that number as a factor two times. (477)

standard form of linear equation A linear equation in standard form is $Ax + By = C$ where A, B, and C are integers, and A and B are not both zero. (406)

statement Any sentence that is either true or false. (18)

statistics A branch of mathematics which provides methods for collecting, organizing, and interpreting data. (560)

stem-and-leaf plot In a stem-and-leaf plot, each piece of data is separated into two numbers that are used to form the stem and leaf. The data are organized into two columns. The column on the left is the stem and the column on the right is the leaf. (565)

subset A set that is made from the elements of another set. (19)

substitution method A method for solving systems of equations. One variable is expressed in terms of the other variable in one equation. Then the expression is substituted into the other equation. (447)

substitution property of equality For any numbers a and b, if $a = b$ then a may be replaced by b. (23)

subtraction property for inequalities For all numbers a, b, and c,
1. if $a > b$, then $a - c > b - c$, and
2. if $a < b$, then $a - c < b - c$. (176)

subtraction property of equality For any numbers a, b, and c, if $a = b$, then $a - c = b - c$. (99)

supplementary angles Two angles are supplementary if the sum of their measures is 180°. (610)

symmetric property of equality For any numbers a and b, if $a = b$ then $b = a$. (23)

system of equations A set of equations with the same variables. (442)

system of inequalities A set of inequalities with the same variables. (463)

tangent In a right triangle, the tangent of angle A =

$$\frac{\text{measure of side opposite angle } A}{\text{measure of side adjacent to angle } A}. \quad (626)$$

term A number, a variable, or a *product* or *quotient* of numbers and variables. The terms of an expression are separated by the symbols + and −. (27)

transitive property of equality For any numbers a, b, and c, if $a = b$ and $b = c$, then $a = c$. (23)

transitive property of order For all numbers a, b, and c,
1. if $a < b$ and $b < c$, then $a < c$, and
2. if $a > b$ and $b > c$, then $a > c$. (61)

tree diagram A diagram used to show all of the possibilities. (326)

trigonometric ratios Ratios in a right triangle that involve the measures of the sides and the measures of the angles. (626)

trinomial A polynomial having exactly three terms. (226)

uniform motion When an object moves at a constant speed, or rate, it is said to be in uniform motion. (158)

union The union of two sets A and B ($A \cup B$) is the set of all elements contained either in A or in B or in both. (51)

unit cost The cost of one unit of something. (66)

upper quartile The upper quartile divides the upper half of the set of data into two equal parts. (575)

variable In a mathematical sentence, a variable is a symbol used to represent an unspecified number. (8)

Venn diagram A diagram using circles or ovals inside a rectangle to show relationships of sets. (51)

vertex The vertex of a parabola is the maximum or minimum point of the parabola. (518)

vertical line test If any vertical line drawn on the graph of a relation passes through no more than one point of its graph, then the relation is a function. (375)

whole numbers (W) The set of numbers $\{0, 1, 2, 3, \ldots\}$. (50)

x-axis The horizontal number line which helps to form the coordinate plane. (354)

x-coordinate The first component of an ordered pair. (354)

x-intercept The value of x when y is 0. (410)

***y*-axis** The vertical number line which helps to form the coordinate plane. (354)

***y*-coordinate** The second component of an ordered pair. (354)

***y*-intercept** The value of y when x is 0. (410)

zero product property For all numbers a and b, if $ab = 0$, then $a = 0$, $b = 0$, or both a and b equal 0. (290)

SELECTED ANSWERS

CHAPTER 1 AN INTRODUCTION TO ALGEBRA

Pages 11–12 Lesson 1-1

5. $7x$ **7.** $a + 19$ **9.** b^3 **11.** 5^3 **13.** $2m^3$ **15.** 16 **17.** 10,000 **19.** m minus 1 **21.** n to the fourth power **23.** 8 times y squared **25.** $x + 17$ **27.** $2x^3$ **29.** $\frac{1x^2}{2}$ or $\frac{x^2}{2}$ **31.** $94 + 2x$ **33.** 15 m^2 **35.** 56,800.236 **37.** 873,324 **39.** $a + b - ab$ **43.** 14,280 square feet

Pages 15–17 Lesson 1-2

7. 2 **9.** 316 **11.** 31 **13.** 75 **15.** 26 **17.** 14 **19.** 20 **21.** 408 **23.** 6 **25.** 16 **27.** 12.5 mm **29.** 14 in. **31.** 12.56 ft **33.** 18 **35.** 13.2 **37.** 413 **39.** $\frac{11}{18}$; $0.6\overline{1}$ **41.** $2(a + b)$; 7 **43.** $b^2 + c$; $\frac{1}{4}$; 0.25 **47.** 13 year old: $R = 144.9$, 14 year old: $R = 144.2$, 15 year old: $R = 143.5$ **49.** $w - 7$ **50.** a number y raised to the fifth power **51.** 4^4 **52.** $2n - 25$ **53.** 24 ft^2

Pages 20–21 Lesson 1-3

7. false **9.** true **11.** 2 **13.** Foster, Winters, Gell, Rath, or Gordon **15.** 11 **17.** 9 **19.** The capital of the U. S. is not Houston; true. **21.** Birds do not have wings; false. **23.** {1}, {2}, {1, 2}, Ø **25.** 5 **27.** 11.05 **29.** $\frac{7}{13}$ **31.** $\frac{7}{4}$ **33.** {6, 7, 8} **35.** {5} **37.** {4, 5, 6, 7, 8} **41.** \$614.54 **43.** h^3 **44.** $\frac{1}{2}a^2b^3$ **45.** 15 **46.** $\frac{11}{13}$ **47.** 25.7

Pages 24–25 Lesson 1-4

5. symmetric (=) **7.** substitution (=) **9.** 0 **11.** 7 **13.** 1 **15.** symmetric (=) **17.** multiplicative identity **19.** reflexive (=) **21.** substitution (=) **23.** multiplicative property of zero **25.** transitive (=) **27.** 3 **29.** 7 **31.** 15 **33.** 4(20) + 7; 87 years **35.** \$36.15 **36.** a number x squared **37.** 39 **38.** true **39.** 20

Page 25 Mid-Chapter Review

1. n^3 **2.** $n^2 + 7$ **3.** Divide 8 by 2. Multiply 2 by 6. Then add. **4.** Square 3. Subtract 3. Then multiply by 3. **5.** Add 8 and 6. Divide by 2. Then add 2. **6.** $\frac{26}{3}$ **7.** 15 **8.** 2 **9.** 6.28 **10.** 4 **11.** 4 **12.** {4, 5, 6, 7, 8} **13.** {4, 5, 6} **14.** {8} **15.** multiplicative property of zero **16.** symmetric (=)

Pages 29–30 Lesson 1-5

3. 5 **5.** $\frac{1}{3}$ **7.** $6bc$, bc **9.** $4xy$, $5xy$ **11.** $5x$ **13.** in simplest form **15.** 645 **17.** $24x + 56$ **19.** $18a$ **21.** $15am - 12$ **23.** $22y^2 + 3$ **25.** in simplest form **27.** $3x + 4y$ **29.** $30a + 6b$ **31.** $14x + 14$ **33.** $8.827xy^3 - 0.012y^3$ **35.** $6336x^2$ **39a.** 172.4 cm **b.** 179.6 cm **41.** 3^2a^3 or $9a^3$ **42.** 7 **43.** 14.13 **44.** symmetric (=) **45.** 4

Pages 33–35 Lesson 1-6

3. associative (+) **5.** distributive **7.** commutative (×) **9.** associative (×) **11.** commutative (+) **13a.** commutative (+) **b.** associative (+) **c.** distributive **d.** substitution (=) **15.** commutative (+) **17.** associative (+) **19.** commutative (×) **21.** additive identity **23.** commutative (+) **25.** distributive **27.** multiplicative identity **29.** $12a + 6b$ **31.** $5x + 10y$ **33.** $3a + 13b + 2c$ **35.** $14x + 3y$ **37.** $\frac{3}{4} + \frac{5}{3}x + \frac{4}{3}y$ **39.** $3.1x + 1.54$ **43a.** $7\frac{7}{8}$ in. × 11 in. × 3 in. **c.** 6 ways **45.** 39 inches or 3 feet 3 inches **47.** $2x^2$ **48.** 10.69 **49.** 25.7 **50.** 15 **51.** $\frac{4}{5}$ **52.** $23a + 42$ **53.** substitution (=) **54.** distributive

Pages 38–39 Lesson 1-7

3. $A = s^2$ **5.** $P = 4s$ **7.** 236 **9.** 584 **11.** 499 **13.** $2x + y^2 = z$ **15.** $(x + a)^2 = m$ **17.** $(abc)^2 = k$ **19.** $29 - xy = z$ **21.** 330 **23.** $33\frac{3}{4}$ **25.** 63.2 **27.** 13.68 **29.** 9 **31.** 2335 **33.** 18 **35.** 428 **37.** $A = s^2 + \frac{\pi}{2}s^2$ **41.** 19,800 ft **43.** 500 seconds or 8 minutes 20 seconds **44.** 6 **45.** $1\frac{1}{9}$ **46.** b^3, $4b^3$ **47.** $37a + 23b$ **48.** commutative (×)

Pages 42–43 Lesson 1-8

3a. \$1 bills **b.** 7 **c.** \$267 **d.** none **e.** \$157 **f.** end of the day **g.** $5n$ dollars **5a.** does not say **b.** 7¢ **c.** \$7.18 **d.** more **e.** $(n - 7)$¢ or $(359 - n)$¢ **7a.** no **b.** yes **c.** rock **d.** 13 **e.** $n + 3$ **9. a.** 48 **b.** 72 **c.** no

Pages 44–46 Chapter 1 Summary and Review

1. $8y$ **3.** a^4 **5.** 320 **7.** 4 **9.** 2.2 **11.** additive identity **13.** symmetric (=) **15.** $2a + 3b$ **17.** $9r + 7s$ **19.** associative (×) **21.** commutative (×) **23.** $\frac{3a^2}{4} + \frac{5ab}{3}$ **25.** $c = (2x)^3$ **27.** 48 cm **29.** Ostriches can fly; false **31.** less **33.** 59¢ **35.** $c = 3 \cdot 111$; 333 calories **37.** 9 bricks

CHAPTER 2 RATIONAL NUMBERS

Pages 52-54 Lesson 2-1

5. -7 **7.** -10 **9.** 2 **11.** -4 **13.** $\{3, 6, 9, 12\}$
15. $\{4, 6, 8, 10, 12\}$ **17.** $\{3, 4, 5, 6, 9, 12\}$
19. $\{3, 4, 5, 6, 8, 10, 12\}$ **21.** $-3 + 5 = 2$
23. $-1 + (-4) = -5$ **25.** $\{-3, -2, -1, 0\}$
27. $\{-1, 0, 1, 2, 3, 4, \ldots\}$

29.
-2 -1 0 1 2 3 4 5 6

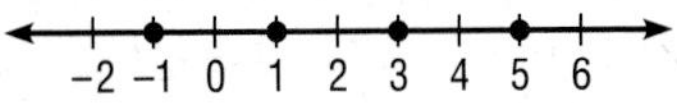

33.
-5 -4 -3 -2 -1 0 1 2 3

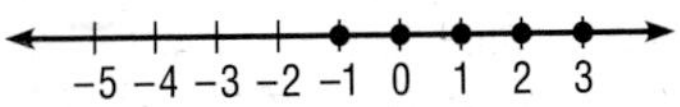

35. 13 **37.** -20 **39.** -5 **41.** 0 **43.** 6 **45.** -5
47. 5 **49.** -22 **51.** $\emptyset$ **53.** $\{d, i, c, t, o, n, a, r, y, l\}$
57. 16 yd loss **59.** $\frac{3}{4}xy^5$ **60.** 3 **61.** $55y^2$
62. $x + a^2 = n$

Pages 58-59 Lesson 2-2

5. $-8, 8$ **7.** $0, 0$ **9.** + **11.** 5 **13.** -4 **15.** -6
17. 40 **19.** 13 **21.** -5 **23.** 5 **25.** -29 **27.** $31m$
29. 22 **31.** 18 **33.** $33b$ **35.** -26 **37.** 8 **39.** 9
41. 0 **43.** 2 **45.** 2 **47.** -30 **49.** 926 **51.** -275
55. $s = 30 - 6 + 15$; \$39 **57a.** $2, -3, -8, \ldots$
b. $-15, -5, 5, \ldots$ **c.** $-5, -7, -9, \ldots$ **d.** The sum of the steps between the addends is equal to the step between the sums. **58.** 1.32 **59.** 6 **60.** $3\frac{1}{4}x + \frac{3}{4}y$
61. associative (+)

62.
-4 -3 -2 -1 0 1 2 3 4

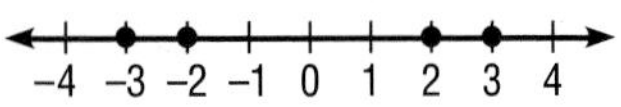

63. 9

Pages 63-64 Lesson 2-3

5. F **7.** F **9.** F **11.** $\{4, 5, 6, \ldots\}$
13. $\{0, 1, 2, 3, 4, 5, 7, \ldots\}$ **15.** no **17.** yes
19. $x \neq -3$ **21.** $x < -3$ **23.** $x \leq 2$

25.
-4 -3 -2 -1 0 1 2 3

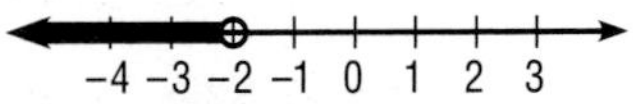

29.
-15 -14 -13 -12 -11 -10 -9 -8

33. $y \geq -5$ **35.** $b < 0$ **37.** $>$ **39.** $<$ **41.** $=$ **43.** $>$

45.
Real Numbers
Rational Numbers
Integers
Whole Numbers
Natural Numbers

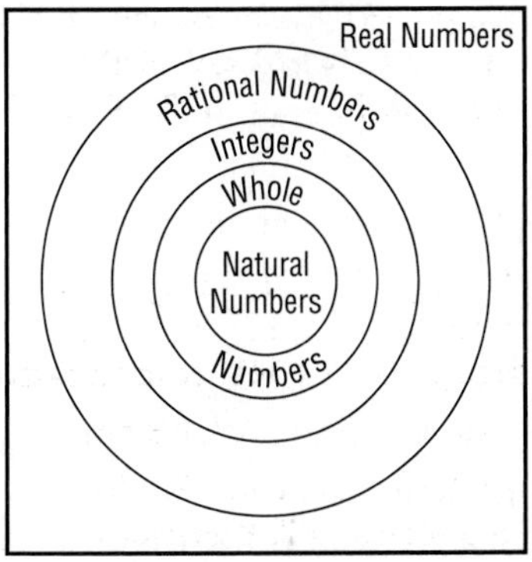

49. 4800 residents **51.** 1955 **52.** 9.5
53. $\frac{m^2n}{2}$, $5m^2n$ **54.** 81, 243, 729
55. $4 + (-6) = -2$ **56.** 4 **57.** -20

Pages 67-68 Lesson 2-4

3. $\frac{4}{5}$ **5.** $\frac{10}{11}$ **7.** $\frac{6}{5}$ **9.** a 28-ounce can for 97¢ **11.** $<$
13. $>$ **15.** $=$ **17.** $\frac{20}{27}, \frac{19}{24}, \frac{17}{21}$ **19.** $\frac{9}{43}, \frac{3}{14}, \frac{5}{23}$
21. $\frac{79}{56}$; 1.411 **23.** $\frac{119}{180}$; $0.66\overline{1}$ **25.** 0 **29.** three liters of soda for \$2.25 **31.** a dozen oranges for \$1.59
33. a 22-ounce bottle for \$1.09 **34.** $r + 7$ or $7 + r$
35. multiplicative identity **36.** commutative (+)
37. $>$ **39.** $<$

Pages 71-73 Lesson 2-5

3. 4 **5.** 18 **7.** $\frac{4}{21}$ **9.** -153.8 **11.** $\frac{5}{14}$ **13.** -2
15. $-\frac{1}{6}$ **17.** -14.7 **19.** $\frac{1}{14}$ **21.** $\frac{4}{9}$ **23.** -0.3005
25. $-\frac{5}{24}$ **27.** -13.1 **29.** $-8\frac{5}{8}$ **31.** $-9m$ **33.** $2.2k$
35. 8.9 **37.** -5.5 **39.** $-\frac{5}{8}$ **41.** $-28y$ **43.** $\frac{5}{6}$
45. $\frac{7}{32}$ **47.** 0.32 **53.** $2\frac{1}{2}$ points **55.** +30 yards
57. $\frac{1}{8}$ **58.** +650 **59.** -25 **60.** $36 + (-11) = f$; 25th floor

61.
-6 -5 -4 -3 -2 -1 0 1 2 3

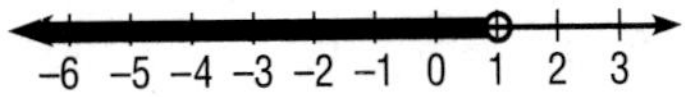

62. a 184-gram can of peanuts for 91¢

Page 73 Mid-Chapter Review

1.
-3 -2 -1 0 1 2 3 4

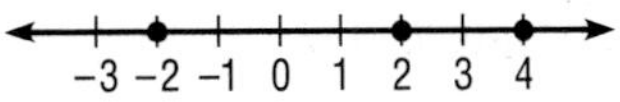

4. -24 **5.** -70 **6.** -54 **7.** -18 **8.** -11 **9.** -82
10. $-\frac{19}{13}$ **11.** $-\frac{1}{6}$ **12.** -8.774 **13.** $=$ **14.** $>$
15. $>$ **16.** $>$ **17.** $=$ **18.** $<$ **19.** $+6\frac{3}{8}$ **20.** Brand X

Pages 77-78 Lesson 2-6

5. +16 **7.** +24 **9.** 60 **11.** 30 **13.** $\frac{7}{24}$
15. 0.00879 **17.** $-\frac{6}{5}$ **19.** 114.1482 **21.** -24
23. -6 **25.** 9 **27.** -332 **29.** -10 **31.** $98xy$
33. -68.416 **35.** $\frac{10}{7}$ **37.** $-21x$ **39.** $5.1x - 7.6y$
47. \$7.13 **49.** $6x^2 + 5x$ **50.** 3, 6, 9 **51.** 286
52. $\{-4, -3, -2, \ldots\}$ **53.** $\frac{1}{3}$ **54.** $82y$

Pages 83-84 Lesson 2-7

5. $\frac{1}{3}$ **7.** none **9.** $\frac{3}{2}$ **11.** $\frac{4}{13}$ **13.** $-\frac{11}{71}$ **15.** -6
17. $5b$ **19.** $a + 3$ **21.** -5 **23.** $-45n$ **25.** $-7a$
27. -2 **29.** $-\frac{1}{12}$ **31.** $\frac{153}{10}$ **33.** $-\frac{35}{2}$ **35.** $a + 4$
37. $x + 2y$ **39.** $7\frac{1}{5}$ or 7.2 **41.** $1\frac{2}{5}$ or 1.4 **43.** $<$

45. < 47. < 53. 1543 airplanes 55. 51,173 slices 56. $b = x - m^3$ 57. 5 58. true 59. a 1-pound package of lunch meat for $1.98
60. $36m$ 61. $-\frac{7}{2}$

Pages 86-87 Lesson 2-8

5. $49 - n$ 7. $n - 8$ 9. 5.65n$ 11. Let k = Kimiko's age now; $k - 27 = 21$. 13. Let x = number of years for a tree to become $33\frac{1}{2}$ feet tall; $17 + 1\frac{1}{2}x = 33\frac{1}{2}$. 15. Let t = number of tapes; $t + \left(\frac{1}{2}t + 4\right) = 31$. 17. Let q = number of quarters; $q + (q + 4) + (q + 4 - 7) = 28$. 19. Twice Quincy's age 7 years ago was 58. How old is Quincy now?
21. Ramón's car weighs 250 pounds more than Seth's car. The sum of the weights of both cars is 7140 pounds. How much does each car weigh?
23. Reggie is 31 cm shorter than Soto. The sum of Soto's height and twice Reggie's height is 502 cm. How tall is Soto?

Page 88-90 Chapter 2 Summary and Review

1.
−3 −2 −1 0 1 2 3 4 5

3. 8 5. −4 7. −22 9. 4 11. 6 13. −15

15.
−5 −4 −3 −2 −1 0 1 2

17. < 19. > 21. > 23. $\frac{71}{120}$ 25. −1 27. 12.37
29. −99 31. $-\frac{3}{7}$ 33. $-5a - 12b$ 35. $-9b$
37. $-3a - 6$ 39. {1, 3, 5} 41. $38 43. −69
45. 1.25 liters of soda for $1.31 47a. $\frac{5}{12}$; $\frac{7}{12}$ b. 16 karats 49. Let n = the number; $3n - 21 = 57$.

CHAPTER 3 EQUATIONS

Pages 97-98 Lesson 3-1

5. −13 7. 9 9. 13 11. −17 13. −12 15. 20
17. 1.4 19. −24 21. 58 23. 15 25. −10
27. −6 29. 21 31. −4 33. −2.8 35. 2.32
37. −32 39. 122 41. −32 43. 85 meters 45. 63 cars 47. 339.223 48. 31 49. −28.9 50. $-a - 5$
51. Let d = dimes; $d + (d + 8) + 15 = 51$.

Pages 101-102 Lesson 3-2

3. 16 5. 5 7. −3 9. $m - 8$ 11. $z - 31$ 13. −9
15. −23 17. −75 19. −0.8 21. −3 23. −21
25. 34 27. 24 29. −7 31. $\frac{5}{4}$; 1.25 33. −32
35. −26 37. −101 41. 126.1 seconds 43. 443 cattle 44. false

45. 

46. A possible answer is 0. 47. −26 48. 43
49. −16

Pages 106-107 Lesson 3-3

5. 3 7. $\frac{4}{3}$ 9. $-\frac{1}{8}$ 11. 4 13. 4 15. −8 17. 7
19. −9 21. $\frac{40}{9}$; $4.\overline{4}$ 23. $-\frac{11}{3}$; $-3.\overline{6}$ 25. 0.188
27. −23 29. 48 31. 70 33. −1885 35. −25
37. $\frac{250}{3}$; $83.\overline{3}$ 39. $\frac{110}{13}$; 8.462 41. $\frac{14}{9}$; $1.\overline{5}$ 43. −8
45. 13 47. −65.3 49. 15 51. $7\frac{1}{2}$ 53. 23 55. $-\frac{1}{3}$; $-0.\overline{3}$ 59. 20 cans 60. $4 + 80x + 32y$ 61. 4, −7, −18 62. −3 63. −19

Page 107 Mid-Chapter Review

1. −1.9 2. 52 3. −25 4. −15 5. −60 6. −7
7. 104 8. −16 9. 6 10. $115.62 11. 6.25 kg

Pages 109-110 Lesson 3-4

3. 80 pounds 5. $50 7. In the figure below, all sums are 12.

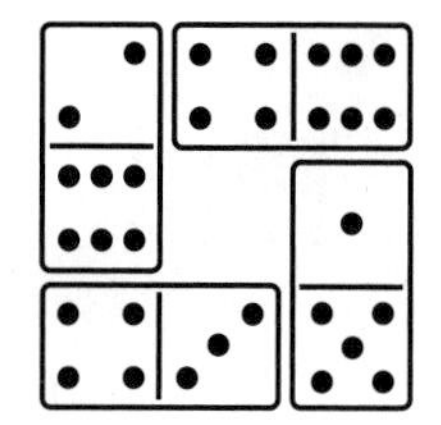

9. 981 11. top row: three-letter words; middle row: four-letter words; bottom row: five-letter words; the words in each row are in alphabetical order.
13. 15 15. Kevin, Kim, Maria, Marquita, and Paul

Pages 113-115 Lesson 3-5

5. −1 7. $-42\frac{1}{4}$; −42.25 9. 6, 8, 10
11. $x + (x + 1) = 17$ 13. $x + (x + 2) = -36$
15. 3 17. 5 19. $-\frac{25}{3}$; $-8.\overline{3}$ 21. 16 23. 4
25. 136 27. −38 29. −104 31. $\frac{65}{7}$; 9.286 33. 28, 29, 30 35. no solution 37. 30, 32, 34, 36 39. 46 customers 41. 88; no 43a. They are multiples of 3 and odd. b. They are multiples of 3 and even.
c. They are not divisible by three. 44. substitution (=) 45. 11 46. $\frac{0.9}{5}$ 47. −36 48. 7 49. 32 students

Pages 118-120 Lesson 3-6

5. Add 10 to each side. Add $3y$ to each side. Divide each side by 11. 7. Multiply $x - 3$ by −7. Subtract 21 from each side. Divide each side by −7. 9. 2
11. −2 13. 65 yd × 120 yd 15. $-\frac{1}{2}$; −0.5 17. $\frac{7}{8}$;

0.875 **19.** 5.6 **21.** −3 **23.** identity **25.** no solution **27.** 8 **29.** 3 **31.** 42 **33.** identity **35.** 43 **37.** 26, 28, 30 **39.** 38 vans **41.** 65 yd by 120 yd **43.** $a = m + n^2$ **44.** $\frac{1}{6}$ **45.** 0.00879 **46.** 12.4 **47.** −9 **48.** $-\frac{20}{3}$; $-6.\overline{6}$ **49.** 3

Pages 123–125 Lesson 3-7

5. 100; $817y = 420 - 370y$ **7.** 20; $8x = 140 - 15x$ **9.** 49 **11.** $\frac{8}{5}$; 1.6 **13.** $\frac{11}{6}$; $1.8\overline{3}$ **15.** −17 **17.** $\frac{3}{2}$; 1.5 **19.** 4 **21.** −2 **23.** −1 **25.** $3c - a$ **27.** $\frac{3z + 2y}{e}$ **29.** $\frac{am - z}{a + n}$ **31.** $\frac{5}{3}(b - a)$ **33.** 120 **35.** 10 **37.** 18 and 30 or 150 and 162 **39.** 260 m **41a.** identity **b.** $\frac{5}{2}$; 2.5 **c.** 2 **d.** −5 **e.** no solution **42.** > **43.** $-\frac{1}{22}$ **44.** Yvette has 17 less pennies than nickels. Yvette has a total of 63 nickels and pennies. How many of each type of coin does she have? **45.** 3 **46.** −3

Page 126–128 Chapter 3 Summary and Review

1. 53 **3.** 6 **5.** 27 **7.** 77 **9.** −24 **11.** 38 **13.** −26 **15.** 55 **17.** 8 **19.** −6 **21.** −16 **23.** −16 **25.** 10 **27.** −18 **29.** 69 **31.** 7 **33.** 3 **35.** 43 **37.** 3 **39.** −0.5 **41.** $\frac{3b - 10a}{5}$ **43.** $\frac{3}{2}b$ **45.** 5.1 **47.** 900 mL **49.** 25, 27, 29 **51.** 15.5 m × 78.5 m **53.** 36 ft, 38 ft, 40 ft, 42 ft

CHAPTER 4 APPLICATIONS OF RATIONAL NUMBERS

Pages 136–137 Lesson 4-1

5. $\frac{3}{11}$ **7.** $\frac{2}{1}$ **9.** $\frac{4}{1}$ **11.** 6 **13.** 14 **15.** $\frac{28}{3}$; $9.\overline{3}$ **17.** −2 **19.** $\frac{58}{3}$; $19.\overline{3}$ **21.** 23 **23.** $\frac{52}{41}$; 1.268 **25.** $-\frac{149}{6}$; $-24.8\overline{3}$ **27.** 2.28 **29.** 1.251 **31.** 63.368 **35.** 1127 deer **37.** 267.9 km **39.** 53 **40.** $14x + 14y$ **41.** $18px - 15bg$ **42.** 2 cups **43.** 2

Pages 140–141 Lesson 4-2

3. 31% **5.** 4% **7.** $37\frac{1}{2}$%; 37.5% **9.** 24 **11.** 60 **13.** 40% **15.** $12\frac{1}{2}$%; 12.5% **17.** 32 **19.** 25% **21.** 80 **23.** $242.80 **25.** $50,000 **27.** $6.24 **31.** 40 questions **33.** 7520 people **35.** $2m + n^2 = y$ **36.** 52 **37.** $-\frac{17}{16}$ **38.** 18 weeks

Pages 143–145 Lesson 4-3

3. $480 **5.** $3\frac{1}{2}$ years **7.** 12% **9.** $2400 **11.** $6400 at 8%, $3600 at 12% **13.** $3800 **15.** $3000 **17.** 11.5% **19.** $18,000 **22.** 15 **23.** < **24.** A possible answer is 0. **25.** 88 gallons in X, 40 gallons in Y **26.** 140

Pages 148–149 Lesson 4-4

5. D; $6; 6% **7.** $172; $28 **9.** $47.89; 25.5% **11.** 10 **13.** 28 **15.** 39 **17.** 20% **19.** $3.00 **21.** $303.60 **23.** $41.18 **27.** $88.50 **29.** $307.80 **30.** $23a + 16b$ **31.** $-7b$ **32.** $\frac{19}{2}$ **33.** 30% **34.** 4 years

Pages 152–153 Lesson 4-5

3. 4 hours **5.** 3:30 P.M. **7.** left: 7, middle: 3, right: 2, bottom: 5 **9.** 4

Pages 155–157 Lesson 4-6

3. $0.25x + 0.10(x + 8) = 2.55$ **5.** $0.10(5) + 1.00n = 0.40(5 + n)$ **7.** 5 adults, 16 children **9.** 15 pounds **11.** 15 pounds **13.** 3.2 quarts **15.** 20 liters **17.** Let y = number of yards Diego gained in both games; $134 + (134 - 17) = y$. **18.** 13 **19.** 3.2 pounds **20.** $41.95

Page 157 Mid-Chapter Review

1. 15 **2.** 14 **3.** 10.29 **4.** 30 **5.** 40 **6.** $12\frac{1}{2}$% **7.** $800 **8.** $121.36 **9.** $40.81 **10.** $9.12

Pages 159–161 Lesson 4-7

3a. 120 mi **b.** 180 mi **c.** 10 mi **d.** $\frac{2}{3}k$ mi **5a.** 9 hours **b.** 12 hours **c.** $\frac{360}{x}$ hours **7.** $3\frac{1}{2}$ hours **9.** $2\frac{1}{2}$ hours **11.** 11:30 A.M. **13.** 46 mph **15.** 240 km **19.** $w - 5$ **20.** multiplicative property of zero **21.** 8 **22.** $12,500 at each rate

Pages 164–165 Lesson 4-8

3. 3 **5.** 7 **7.** 28 **9.** −5 **11.** 6 **13.** $26\frac{1}{4}$ **15.** $\frac{1}{5}$, $y = \frac{1}{5}x$ **17.** $-\frac{2}{3}$, $y = -\frac{2}{3}x$ **19.** $\frac{4}{5}$, $y = \frac{4}{5}x$ **21.** $\frac{567}{8}$ **25.** $13.33 **27.** $3\frac{1}{2}$ ft³ **28.** $\frac{6}{5}$ **29.** $11.66 **30.** 2^{17} or 131,072 **31.** 5 pounds **32.** 5 hours

Pages 168–169 Lesson 4-9

5. direct, 3.14 **7.** direct, $\frac{1}{5}$ **9.** inverse, 14 **11.** inverse, 7 **13.** Emilio **15.** 48 **17.** 99 **19.** 6.075 **23.** $21\frac{1}{3}$ m^3 **25.** 720 cycles per second **27.** $8\frac{1}{43}$ feet; about 8 ft $\frac{1}{4}$ in. **29.** −15 **30.** 67 **31.** 16.6% **32.** 1 liter **33.** $467.50

Page 170–172 Chapter 4 Summary and Review

1. 18 **3.** 7 **5.** 16 **7.** 48 **9.** 87.5% **11.** 0.1881 **13.** 12.5% **15.** $189.86 **17.** $0.09x + 0.04 = 0.06(x + 1)$; 0.67 gal **19a.** 40 mph **b.** $\frac{240}{t}$ mph **21.** 15 **23.** $\frac{75}{7}$; 10.714 **25.** 21 **27.** turkey -

12:30, potatoes - 3:45, yams - 3:30, green beans - 4:10, cranberry sauce - 4:55, gravy - 4:50, rolls - 4:45, jello - 12:45

CHAPTER 5 INEQUALITIES

Pages 178–180 Lesson 5-1

5. 17 **7.** 14.5 **9.** 1 **11.** −8 **13.** $\{r|r < -5\}$ **15.** $\{x|x \le 13\}$ **17.** $\{a|a < 9\}$ **19.** $\{r|r \ge 42\}$ **21.** $\{w|w \le 0\}$ **23.** $\{d|d < -7\}$ **25.** $\{x|x > -3\}$ **27.** $\{z|z < 23\}$ **29.** $\{x|x \ge 3\}$ **31.** $\{b|b \ge -1\}$ **33.** $\{s|s < -4.9\}$ **35.** $\left\{x|x \ge \frac{1}{5}\right\}$ **37.** $\{z|z \ge -1.654\}$ **39.** $\{x|x \ge -1\}$ **41.** $\{x|x < 8\}$ **43.** $\{x|x < 7\}$ **45.** 12 **47.** −5 **49.** 177 points **51.** any score less than 9.9 **53.** $1.11y + 0.06$

54.
−5 −4 −3 −2 −1 0 1 2 3 4 5

55. $-\frac{81}{4}$ **56.** $31.00 **57.** 32

Pages 183–185 Lesson 5-2

5. 4; no **7.** $\frac{1}{10}$; no **9.** −10; yes **11.** $\{x|x < -6\}$ **13.** $\left\{d|d \ge -\frac{5}{2}\right\}$ **15.** $\{x|x < 6\}$ **17.** $\{z|z \le 9\}$ **19.** $\{t|t > -36\}$ **21.** $\left\{c|c \ge -\frac{3}{2}\right\}$ **23.** $\{x|x > 0.6\}$ **25.** $\{a|a > 150\}$ **27.** $\{m|m \ge -33\}$ **29.** $\left\{y|y \le \frac{256}{3}\right\}$ **31.** $\left\{z|z < -\frac{7}{10}\right\}$ **33.** $d < 15$ **35.** $d < 144$ **37.** $\{x|x < 36\}$ **39.** $\{x|x \le 8\}$ **41.** $\{x|x > 30\}$ **43.** 48 **45.** < **49.** at least 37.5 pounds **51.** 37 shares **53.** 12 ounces of orange juice at $1.69 **54.** −9 **55.** $\{y|y < -6\}$ **56.** 26 **57.** $\frac{3}{5}$ **58.** $1\frac{11}{13}$

Pages 188–190 Lesson 5-3

5. $\{x|x > 5\}$ **7.** $\{n|n \ge 23\}$ **9.** $\{y|y \le -1\}$ **11.** $\{x|x > 2\}$ **13.** $\{a|a \ge -6\}$ **15.** $\{m|m \le 6\}$ **17.** $\{z|z \ge -48\}$ **19.** $\{d|d > -125\}$ **21.** $\{w|w < -6.5\}$ **23.** $\{r|r > 8\}$ **25.** $\{y|y \ge 15\}$ **27.** $\{g|g < -5\}$ **29.** $\{c|c > 2\}$ **31.** $\{x|x \le 12\}$ **33.** $\{x|x < 17\}$ **35.** 38 and 40 **37.** 7 and 9, 5 and 7, 3 and 5, 1 and 3 **39.** $y > \frac{9}{4}$ **41.** $c \le -4$ **43.** at least 90 points **45.** more than 35 hours **46.** Let x = number of points Megan scored in both games; $12 + (12 + 4) = x$. **47.** 20 **48.** $403.33 **49.** $\{z|z < 0.08\}$

Page 190 Mid-Chapter Review

1. $\{a|a < 8\}$ **2.** $\{s|s \ge -15\}$ **3.** $\{z|z < -3\}$ **4.** $\{b|b \ge 12\}$ **5.** $\left\{k|k < -\frac{18}{7}\right\}$ **6.** $\left\{a|a \le -\frac{65}{32}\right\}$ **7.** $\{x|x > -2\}$ **8.** $\left\{d|d \ge \frac{13}{10}\right\}$ **9.** $\left\{n|n > -\frac{5}{12}\right\}$ **10.** $\{n|n \ge 12\}$ **11.** $\{n|n < 120\}$ **12.** $\{n|n \le 42\}$ **13.** $\{n|n \le 1\}$ **14.** 63 newspapers or more **15.** 33 or less

Page 193 Lesson 5-4

3. 25 thumbtacks **5.** 1-Darryl, 2-Adrienne, 3-Allison, 4-Mr. Crawford, 5-Don, 6-Benito, 7-Chumani, 8-Belinda **7.** $1^2 = 1$, $(25)^2 = 625$, or $(76)^2 = 5776$

Pages 197–198 Lesson 5-5

5. $0 \le m < 9$ **7.** $-\frac{4}{5} < z < \frac{2}{3}$

9.
−8 −7 −6 −5 −4 −3 −2 −1 0 1

13.
−5 −4 −3 −2 −1 0 1 2 3 4

17.
−8 −7 −6 −5 −4 −3 −2 −1 0

21.
−4 −3 −2 −1 0 1 2 3 4 Ø

25. {all numbers} **27.** Ø **29.** $\{t|t < -2 \text{ or } t > -1\}$ **31.** $\{x|x \le 4\}$ **33.** $\left\{x|x < \frac{3}{2}\right\}$ **35.** $x < -3$ or $x > 3$ **37.** $-3 \le x < 5$ **39.** $1 < n < 3$ **41.** $x < -\frac{5}{3}$, $x \ne 0$ **43.** $-4 < x < 1$ **47.** between $131,250 and $181,250 **49.** −7 **50.** 2.5 **51.** 4 adult, 12 children **52.** $\{r|r < -6.6\}$ **53.** $\{x|x \ge -1\}$

Pages 202–203 Lesson 5-6

7. two **9.** and **11.** distance from 0 to y is more than 3 units; $y > 3$ or $y < -3$; $\{y|y > 3 \text{ or } y < -3\}$ **13.** distance from 0 to $x - 12$ is less than 9 units; $x - 12 < 9$ and $x - 12 > -9$; $\{x|3 < x < 21\}$ **15.** distance from 0 to $7 - r$ is 4 units; $7 - r = 4$ or $7 - r = -4$; $\{3, 11\}$ **17.** $|s - 90| < 6$ **19.** $\{x|-3 < x < 5\}$ **21.** $\{a|a \le -9 \text{ or } a \ge -7\}$ **23.** $\{y|y \le -4 \text{ or } y \ge 22\}$ **25.** $\{z|z = -1 \text{ or } z = 15\}$ **27.** $\{x|0 < x < 8\}$ **29.** $\{w|w < -10 \text{ or } w > 8\}$ **31.** {all numbers} **33.** Ø **35.** $\left\{t|t \le -\frac{5}{3} \text{ or } t \ge 2\right\}$ **37.** $|x| = 1$ **39.** $|x| \ge 2$ **41.** $|x + 1| < 3$ **43.** $\{-2, -1, 0, 1, 2\}$ **45.** $2a + 1$ **49.** between −259°C and −255°C **51.** 1.96 cm to 2.04 cm, inclusive **53.** $\frac{7n - 1}{3}$ **54.** 80 **55.** $6350 at 8% and $3650 at 10% **56.** $333\frac{1}{3}\%$ **57.** 21 line segments **58.** between 83 and 99, inclusive

Pages 204–206 Chapter 5 Summary and Review

1. $\{n|n < 13\}$ **3.** $\{a|a \ge -5.5\}$ **5.** $\{x|x > 7\}$ **7.** $\{n|n \ge 5\}$ **9.** $\{x|x \le -4\}$ **11.** $\{t|t > 1.2\}$ **13.** $\left\{k|k \ge \frac{1}{5}\right\}$ **15.** $\{n|n < -22\}$ **17.** $\{x|x < 6\}$ **19.** $\left\{y|y \le -\frac{9}{2}\right\}$ **21.** $\{z|z \le 20\}$ **23.** $\{n|n \le -5\}$

25.
−5 −4 −3 −2 −1 0 1 2 3

29. $-2 \leq x < 3$ **31.** $\{n|-4 \leq n \leq 6\}$ **33.** $\left\{p|p < -2 \text{ or } p > \frac{5}{2}\right\}$ **35.** 32, 33, 34 **37.** Wednesday **39.** Jill: \$208, Sung: \$104 **41.** 17 to 19 books

CHAPTER 6 POLYNOMIALS

Pages 211-212 Lesson 6-1

3. 15,624 **5.** 21^2; 15^2 **7.** 1, 4, 9, 16, 25 **9.** 29 days **11.** You could never say this for the first time. The first time you say it, you would be lying. Therefore, only a Falsite could say it.

Pages 215-216 Lesson 6-2

5. no **7.** yes **9.** no **11.** p^{14} **13.** x^8 **15.** a^6 **17.** a^{13} **19.** m^4n^3 **21.** $(-4)^4$ or 256 **23.** r^7t^8 **25.** $-20x^5y$ **27.** $-343z^3$ **29.** $21y^7z$ **31.** $6x^4y^4z^4$ **33.** $a^{12}x^8$ **35.** $a^2b^2c^2$ **37.** $9a^2y^6$ **39.** $\frac{1}{8}x^3y^6$ **41.** $-54a^3b^6$ **43.** $90y^{10}$ **45.** $-6a^3b^5c^5$ **47.** $-30x^9y^3$ **51.** No, you will have \$6908.22. **52.** $-1, -\frac{1}{2}, -\frac{1}{4}$ **53.** 1 daughter **54.** $\{x|x \leq 4\}$ **55.** $\{x|x \geq 3\}$ **56.** 15 rectangles

Pages 219-220 Lesson 6-3

5. $-\frac{1}{8}$ **7.** $-\frac{1}{8}$ **9.** $\frac{1}{64}$ **11.** 81 **13.** $\frac{1}{b^2c}$ **15.** $\frac{5}{n^3}$ **17.** r^4 **19.** an **21.** r^3 **23.** m^6 **25.** b **27.** $3b$ **29.** y^3 **31.** $-s^6$ **33.** $-\frac{4b^3}{c^2}$ **35.** $-4y^4z^2$ **37.** $\frac{9ab^6}{5c^6}$ **39.** $\frac{s^3}{r^3}$ **41.** 1 **43.** $\frac{m}{7nr^2}$ **51.** \$4597.87 **53.** 12 extra-large eggs weighing 27 ounces for \$1.09 **54.** 2 **55.** $5\frac{7}{10}$ feet or 5 feet $8\frac{2}{5}$ in. **56.** Sam **57.** a^7

Pages 223-225 Lesson 6-4

5. 5000; 5.79×10^7 **7.** 12,760; 1.4959×10^8 **9.** 143,200; 7.782×10^8 **11.** 51,800; 2.87×10^9 **13.** 3000; 5.9×10^9 **15.** 2.4×10^5 **17.** 4.296×10^{-3} **19.** 3.17×10^{-9} **21.** 2.84×10^5 **23.** 3×10^2; 300 **25.** 6×10^2; 600 **27.** 5.5×10^{-9}, 0.0000000055 **29.** 6×10^{-3}; 0.006 **31.** 6.51×10^3; 6510 **33.** 7.8×10^8; 780,000,000 **35.** 2.1×10^{-1}, 0.21 **37.** 4×10^8; 400,000,000 **43.** 3000 calories **45.** \$886.38 **47a.** 1.416E + 7 **b.** 884,736 **c.** 5.706E + 14 **d.** 33,092 **48.** $-10a + 5b$ **49.** 8 legs **50.** 70 **51.** $\{q|q < 8\}$ **52.** $4a^6b^9$ **53.** $4a^7$

Pages 228-229 Lesson 6-5

5. not a polynomial **7.** yes, monomial **9.** yes, trinomial **11.** 0 **13.** 2 **15.** none **17.** 4 **19.** 29 **21.** 7 **23.** 4 **25.** 4 **27.** 3 **29.** 3 **31.** 7 **33.** $a^3 + 5ax + 2x^2$ **35.** $3xy^3 - 2x^2y + x^3$ **37.** $5b + \frac{2}{3}bx + b^3x^2$ **39.** $-3x^3 + 5x^2 + 2x + 7$ **41.** $7ax^3 + 11x^2 - 3x + 2a$ **43.** $\frac{1}{5}x^5 - 8a^3x^3 + \frac{2}{3}x^2 + 7a^3x$ **45.** yes **47.** 211 eggs

48. 2 3 4 5 6 7 8

49. -7 **50.** \$4500 at 6.2%, \$8000 at 8.6% **51.** $\left\{p|p < -\frac{3}{2}\right\}$ **52.** 4.235×10^4 **53.** 6.28×10^{-6}

Pages 229 Mid-Chapter Review

1. 7 **2.** b^9 **3.** x^4y^4 **4.** $-6n^5y^7$ **5.** 15,625 **6.** $-48x^3y^2$ **7.** $9a^4b^{10}$ **8.** n^3 **9.** $-12ab^4$ **10.** $\frac{125}{r^3s^3}$ **11.** 2.85×10^7 **12.** 5×10^{-6} **13.** $\frac{1}{4}x - \frac{2}{5}s^4x^2 + \frac{1}{3}s^2x^3 + 4x^4$ **14.** $p^4 + 21p^2x + 3px^3$

Pages 232-233 Lesson 6-6

3. $-3x - 2y$ **5.** $-x^2 - 3x - 7$ **7.** $3ab^2 - 5a^2b + b^3$ **9.** $5m$ and $-3m$, $4mn$ and $-mn$, $2n$ and $8n$ **11.** $8a^2b$ and $16a^2b$, $11b^2$ and $-2b^2$ **13.** $7ax^2 - 5a^2x - 7a^3 + 4$ **15.** $9a - 3b - 4c + 16d$ **17.** $x^2y^2 - 5xy - 10$ **19.** $6m^2n^2 + 8mn - 28$ **21.** $7x + 14y$ **23.** $7x - 4y$ **25.** $13m + 3n$ **27.** $-2 - 6a$ **29.** $7ax^2 + 3a^2x - 5ax + 2x$ **31.** $x + y$ **33.** $5x^2 - 23x - 23$ **39.** 184 **40.** 2.1 **41.** 21 sports **42.** 8 hours, 4 hours **43.** $\{m|m < -7\}$ **44.** 3

Pages 236-237 Lesson 6-7

5. $24m^2 + 21m$ **7.** $10x - 6$ **9.** $35a^2b^3 + 7ab^3$ **11.** $7a^4 + 21a^3 - a + 25$ **13.** 0 **15.** $-24x - 15$ **17.** $15b^2 + 24b$ **19.** $2.2a^2 + 7.7a$ **21.** $15s^3t + 6s^2t^2$ **23.** $10a^4 - 14a^3 + 4a$ **25.** $40y^4 + 35y^3 - 15y^2$ **27.** $15x^4y - 35x^3y^2 + 5x^2y^3$ **29.** $-32x^2y^2 - 56x^2y + 112xy^3$ **31.** $-\frac{1}{4}ab^4 + \frac{1}{3}ab^3 - \frac{3}{4}ab^2$ **33.** $15t^2 + t$ **35.** $x^2 + 6x$ **37.** $50.6t^2 + 21t - 102$ **39.** $-3m^3 + 41m^2 - 14m - 16$ **41.** $21a^3 - 6a^2 - 46a + 28$ **43.** $\frac{29}{3}$ **45.** $\frac{58}{33}$ **47.** $\frac{1}{3}$ **49.** $-\frac{3}{2}$ **51.** 17 **55.** 4.5 ft **56.** -2 **57.** $-\frac{1}{5}$ **58.** $-\frac{14}{27}$ **59.** $7\frac{1}{2}\%$ **60.** $\{x|x \leq 3 \text{ or } x \geq 11\}$ **61.** $6a + 9b$

Pages 241-242 Lesson 6-8

3. $8x$ **5.** $22x$ **7.** $-b$ **9.** $a^2 + 10a + 21$ **11.** $x^2 + 7x - 44$ **13.** $c^2 + 10c + 16$ **15.** $y^2 - 4y - 21$ **17.** $8a^2 + 2a - 3$ **19.** $10x^2 + 19xy + 6y^2$ **21.** $40q^2 + qr - 6r^2$ **23.** $10r^2 - 0.1r - 0.03$ **25.** $18x^2 - \frac{1}{8}$ **27.** $2x^3 + 15x^2 - 11x - 9$ **29.** $12x^3 + 11x^2y - 11xy^2 - 3y^3$ **31.** $6x^3 - 5x^2 + 8x + 55$ **33.** $-24a^3 - 34a^2 + 19a + 15$ **35.** $20x^4 - 9x^3 + 73x^2 - 39x + 99$ **37.** $2x^4 - 17x^3 + 23x^2 + 30x - 24$ **39.** $-6x^4 - 5x^3 + 7x^2 + 71x + 40$ **45.** 21 ft by 47 ft **47.** 20, 35, 65 **48.** \$11.66 **49.** $\{x|x < 14\}$ **50.** b^5 **51.** $4a^2 + 10a$

Pages 245–246 Lesson 6-9

5. $a^2 - 6ab + 9b^2$ **7.** $9x^2 - 12xy + 4y^2$ **9.** $25a^2 - 9b^2$ **11.** $4a^2 - 4ab + b^2$ **13.** $16x^2 - 72xy + 81y^2$ **15.** $25x^2 + 60xy + 36y^2$ **17.** $25 - 10x + x^2$ **19.** $1.21x^2 + 2.2xy + y^2$ **21.** $x^6 - 10x^3y^2 + 25y^4$ **23.** $64a^2 - 4b^2$ **25.** $64x^4 - 9y^2$ **27.** $2x^3 + 5x^2 - 8x - 20$ **29.** $x^4 - 25x^2 + 144$ **31.** $8x^3 - 36x^2y + 54xy^2 - 27y^3$ **33.** $16x^4 - 32x^3y + 24x^2y^2 - 8xy^3 + y^4$ **35.** $a^5 - 5a^4b + 10a^3b^2 - 10a^2b^3 + 5ab^4 - b^5$ **37.** 8 miles **39.** $\frac{3}{8}$ **40.** 110 at 95¢, 154 at \$1.25 **41.** false

Pages 248–250 Chapter 6 Summary and Review

1. y^7 **3.** $20a^5x^5$ **5.** $576x^5y^2$ **7.** $-\frac{1}{2}m^4n^8$ **9.** y^4 **11.** $3b^3$ **13.** $\frac{a^4}{2b}$ **15.** 2.4×10^5 **17.** 3.14×10^{-4} **19.** 6×10^{11} **21.** 6×10^{-1} **23.** 2 **25.** 4 **27.** $3x^4 + x^2 - x - 5$ **29.** $-3x^3 + x^2 - 5x + 5$ **31.** $16m^2n^2 - 2mn + 11$ **33.** $12a^3b - 28ab^3$ **35.** $10x^2 - 19x + 63$ **37.** $r^2 + 4r - 21$ **39.** $4x^2 + 13x - 12$ **41.** $x^3 + x^2 - 27x + 28$ **43.** $x^2 - 36$ **45.** $16x^2 + 56x + 49$ **47.** $64x^2 - 80x + 25$ **49.** 2; 3; Fibonacci sequence **51.** $x^2 - 4$

CHAPTER 7 FACTORING

Pages 258–260 Lesson 7-1

5. prime **7.** composite; $2^3 \cdot 3$ **9.** 4 **11.** 5 **13.** 5 **15.** 1 **17.** $3 \cdot 7$ **19.** $2^2 \cdot 3 \cdot 5$ **21.** $3^2 \cdot 7$ **23.** $2^4 \cdot 7$ **25.** $2^4 \cdot 19$ **27.** $2^2 \cdot 3 \cdot 5^2$ **29.** $-1 \cdot 2 \cdot 2 \cdot 2 \cdot 2 \cdot 2 \cdot 2$ **31.** $-1 \cdot 2 \cdot 2 \cdot 2 \cdot 2 \cdot 3 \cdot 5$ **33.** $2 \cdot 7 \cdot 7 \cdot a \cdot a \cdot b$ **35.** $2 \cdot 2 \cdot 3 \cdot 3 \cdot 3 \cdot 7 \cdot m \cdot m \cdot m \cdot n \cdot n \cdot n$ **37.** 4 **39.** 5 **41.** 5 **43.** 1 **45.** $7y^2$ **47.** $4ab$ **49.** $10n^2$ **51.** $4a$ **53.** 5 **55.** 6 **57.** 2 **59.** 3 **61.** $5a$ **63.** $6y^2$ **65.** $-4x^2y^3$ **67.** $6a^3b^2$ **71.** 5 rows of 20 plants; 10 rows of 10 plants; 20 rows of 5 plants; 25 rows of 4 plants; 50 rows of 2 plants **73.** 60 × 25 or 1500 squares of sod **75a.** 16 **b.** 1 **c.** 12 **76.** −64 **77.** $x = 2d - r$ **78.** 36 pounds **79.** $\left\{x \mid x \geq -\frac{8}{5}\right\}$ **80.** $\left\{y \mid y \geq \frac{9}{2} \text{ or } y \leq \frac{1}{2}\right\}$ **81.** $-40a^3 - 64a^2 + 24a$ **82.** $25r^2 - 70rs + 49s^2$

Pages 263–264 Lesson 7-2

5. 2 **7.** 1 **9.** $7m$ **11.** 1 **13.** r^2 **15.** $x(x^4y - 1)$ **17.** $3c^2d(1 - 2d)$ **19.** $2x^2$ **21.** $5ab$ **23.** $8y(2y + 1)$ **25.** $2mn(7n + 1)$ **27.** $y^3(15x + y)$ **29.** $12pq(3pq - 1)$ **31.** $2mn(m^2n - 8mn^2 + 4)$ **33.** $7abc(4abc + 3ac - 2)$ **35.** $a(1 + ab + a^2b^3)$ **37.** $A = 3(4a + b + 12)$ **39.** $A = 8(a + b + c + 2d + 8)$ **41.** $y(y^4 + 5y^3 + 3y + 2)$ **43.** $\frac{1}{5}ab(4a - 3b - 1)$

47. $(3x - 10)$ shares **48.** $x \leq -3$ **49.** \$1599 **50.** $\frac{61}{9}$ **51.** 3×10^2, 300 **52.** $-3x^5 + 2mx^4 + 6x^3 + 4m^5$ **53.** $2ab$

Pages 267–268 Lesson 7-3

5. $(t + s)(t - s)$ **7.** $(8m + 1)(x + y)$ **9.** $3x + 2y$ **11.** $2y - 5$ **13.** $(a - c)(y - b)$ **15.** $(5a + 2b) \cdot (1 + 2a)$ **17.** $x^2 + 1$ **19.** $5k - 7p$ **21.** $(2m + r) \cdot (3x - 2)$ **23.** $(m - b)(3y + a)$ **25.** $(a + 1)(a - 2b)$ **27.** $(m^2 + p^2)(3 - 5p)$ **29.** $(2x - 5y)(2a - 7b)$ **31.** $a(x + ax - 1 - 2a)$ **33.** $(2x^2 - 5y^2)(x - y)$ **35.** $4z + 3m$ ft by $z - 6$ ft **37.** $(4x + 3y)(a + b)$ **39.** $(7x + 3m - 4)(a + b)$ **41.** $(x - 3y - z)(2a + b)$ **45.** $(s + 8)(2s - 3)$ **46.** {2, 3} **47.** \$595.99 **48.** 61 **49.** $6x^2 - x - 40$ **50.** $2x^2(5x^2 - 3xy - 4y^2)$

Pages 269–270 Lesson 7-4

3. $4 \cdot (5 - 2) + 7 = 19$

5.

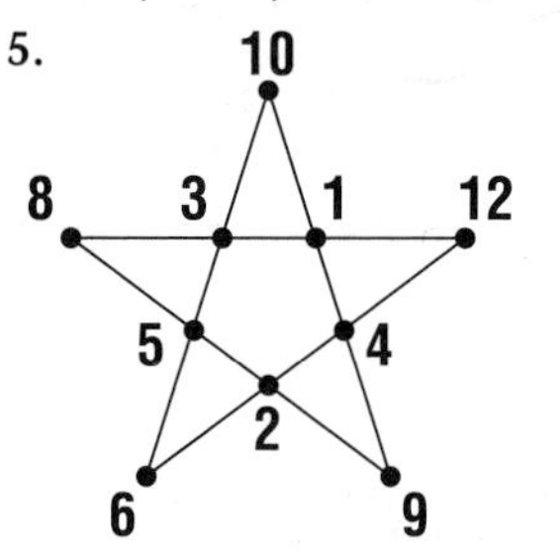

7. 631, 542 **9.** $3 \times 54 = 162$ **11.** July 9 and September 7

Pages 274–275 Lesson 7-5

5. 2, 9 **7.** 2, 7 **9.** −2, −6 **11.** 10 **13.** 9 **15.** $(x + 5)(x - 3)$ **17.** $(b + 5)(b + 7)$ **19.** −3 **21.** $2z$, 3 **23.** $(y + 3)(y + 9)$ **25.** $(c + 3)(c - 1)$ **27.** $(x - 8)(x + 3)$ **29.** $(7a + 1)(a + 3)$ **31.** $(3y - 4)(2y - 1)$ **33.** $(2x + 3)(x - 4)$ **35.** prime **37.** $(9 - y)(4 - y)$ **39.** $(a + 3b)(a - b)$ **41.** $(3s + 2t)(s - 4t)$ **43.** prime **45.** 7, −7, 11, −11 **47.** 12, −12 **49.** 1, −1, 11, −11, 19, −19, 41, −41 **51.** $4x^2(5x - 7)(2x - 3)$ **53.** $(a - 3)(a - 4) \cdot (2x + 3y)$ **57.** 60 m by 80 m **58.** \$4300 at 5%, \$7400 at 7% **59.** $\{n \mid n > 14.2\}$ **60.** $12x^6y^8$ **61.** $x^2 - 15x + 10$ **62.** $(5a + 2b)(3 + 2a)$ **63.** 60 guests

Pages 278–280 Lesson 7-6

5. yes **7.** yes **9.** yes **11.** d **13.** a **15.** $(x - 7)(x + 7)$ **17.** $(x - 6y)(x + 6y)$ **19.** $(4a - 3b)(4a + 3b)$ **21.** prime **23.** $2(z - 7) \cdot (z + 7)$ **25.** $2(2x - 3)(2x + 3)$ **27.** $(5y - 7z^2) \cdot (5y + 7z^2)$ **29.** $(0.1n - 1.3r)(0.1n + 1.3r)$ **31.** prime **33.** $(7x - 4)(7x + 4)$ **35.** $\left(\frac{1}{4}x - 5z\right)\left(\frac{1}{4}x + 5z\right)$ **37.** $\left(\frac{1}{2}n - 4\right)\left(\frac{1}{2}n + 4\right)$ **39.** prime **41.** 5 **43.** $3a - 2$, $3a + 2$, $a + 2$ **45.** $a - 5b$, $a + 5b$, $5a + 3b$

47. $(2 - a)(2 + a)(4 + a^2)$ **49.** $3x(2x - y)(2x + y)(4x^2 + y^2)$ **51.** $(x - 1) \cdot (x + 1)(x^2 + 1)(x^4 + 1)$ **57.** 12 in. by 12 in. **59.** $\frac{1}{6}$ **60.** 168 miles **61.** $\{x|-9 < x < 2\}$ **62.** $-8ab$ **63.** 7 **64.** $(8x - 5)(2x + 3)$

Page 280 Mid-Chapter Review

1. $13a^2$ **2.** $4ac^2$ **3.** $55m^2n$ **4.** $(y - 1)(y - 7)$ **5.** $5m^2n(5m + 3n)$ **6.** $(r + 6)(r - 3)$ **7.** prime **8.** $5a(a^2 + 9a - 3)$ **9.** $(2p + 3)(3p - 1)$ **10.** $8(k + 3z)(k - 3z)$ **11.** $7ab^2(-a - 11a^2 + 11b)$ **12.** $(5x - 14)(x - 1)$ **13.** $(y + 2)(y - 2)(m + n)$ **14.** 41,312,432 or 23,421,314 **15.** $4x^2 + 36x - 115\ m^2$

Pages 283–284 Lesson 7-7

5. 7 **7.** $8b$ **9.** $6x$ **11.** no **13.** yes, $(b - 7)^2$ **15.** no **17.** yes, $(p + 6)^2$ **19.** no **21.** yes, $(2a - 5)^2$ **23.** $(n - 4)^2$ **25.** $(2k - 1)^2$ **27.** prime **29.** $(1 - 5z)^2$ **31.** $2(5x + 2)^2$ **33.** $(7m - 9)^2$ **35.** $(5x - 12)^2$ **37.** $(m + 8n)^2$ **39.** $(2x + z^2)^2$ **41.** $\left(\frac{1}{2}a + 3\right)^2$ **43.** yes; $11y + 1$ **45.** no **47.** 4 **49.** $4b^2$ **51.** 40 or -40 **53.** $(a + 2 - 3b)(a + 2 + 3b)$ **55.** $(m - k + 3)(m + k - 3)$ **57.** $(m - 2)(a + 3)^2$ **59.** $6y + 26$ **61.** 21 in. by 21 in. **63.** 37.5 **64.** 15.152 gallons **65.** 7, 8, or 9 **66.** $9x^2 - 4$ **67.** $16t^2 + 34t - 15$ **68.** $5(x - 4y^2)(x + 4y^2)$ **69.** $15x^2y(xy^3 - 2z)$

Pages 288–289 Lesson 7-8

5. greatest common factor **7.** perfect square trinomial **9.** trinomial that has two binomial factors **11.** $3(x^2 + 5)$ **13.** $(a - 3b)(a + 3b)$ **15.** $(a + 4)^2$ **17.** $6a(2a + 3y^2)$ **19.** $3(y - 7)(y + 7)$ **21.** $m(m + 3)^2$ **23.** $(3r + 2)(2r + 3)$ **25.** $(m^2 - p)(m^2 + p)$ **27.** $3x(x - 3)(x + 3)$ **29.** $2(5n + 1)(2n + 3)$ **31.** $(2a + 3b)^2$ **33.** $3t(3t - 2)(t + 8)$ **35.** $m(mn + 7)(mn - 7)$ **37.** $0.7(y - 2)(y - 3)$ **39.** $\frac{1}{12}(4m - 1)(3m + 2)$ **41.** $(x^2 + 8)(xy + 2)$ **43.** $(y^2 + z^2)(x + 1)(x - 1)$ **45.** $(x + y + a - b)(x + y - a + b)$ **47.** $x(x - 1)$ **49.** $(x^2 - 3y)(x + 3)^2$ **51.** $x - 3y$, $x + 3y$, $xy + 7$ **53.** 23 feet by 35 feet **55.** 29.4% **56.** $21.60 or less **57.** 28 games **58.** $-0.008x^9y^3$ **59.** ab^3 **60.** $12s - 28$

Pages 293–294 Lesson 7-9

5. $3r = 0$ or $r - 4 = 0$ **7.** $x - 6 = 0$ or $x + 4 = 0$ **9.** $4x - 7 = 0$ or $3x + 5 = 0$ **11.** $\{0, -4\}$ **13.** $\{0, 2\}$ **15.** $\left\{0, \frac{1}{2}\right\}$ **17.** $\{0, -6\}$ **19.** $\{3, 5\}$ **17.** $\{-1, -3\}$ **23.** $\left\{\frac{5}{3}\right\}$ **25.** $\{0, -36\}$ **27.** $\{0, -9\}$ **29.** $\{0, -8\}$ **31.** $\{0, -5\}$ **33.** $\{0, 2\}$ **35.** $\left\{0, \frac{1}{6}\right\}$ **37.** 16 years old **39.** 0, 1; -3, -2 **41.** 9 ft by 9 ft **43.** 12 seconds **45.** 125 tickets **46.** $\{s|s \leq -6\}$ **47.** $8mn^2 + 3mn - n - 3n^3$ **48.** $2^2 \cdot 3^3 \cdot 5^2$ **49.** 15, -15, 6, -6, 0 **50.** $2(3c + 7d)(2c - 3d)$

Pages 297–299 Lesson 7-10

5. $\left\{-2, -\frac{1}{2}\right\}$ **7.** $\{4, -4\}$ **9.** $\{0, -1, -28\}$ **11.** Let x = one of the integers; $x(15 - x) = 44$. **13.** Let x = the amount the length and width should be increased; $(7 + x)(4 + x) = 28 + 26$. **15.** $\{-8, 7\}$ **17.** $\{8, -8\}$ **19.** $\{12\}$ **21.** $\{9, -2\}$ **23.** $\left\{\frac{5}{3}, -7\right\}$ **25.** $\left\{\frac{5}{2}, -\frac{7}{3}\right\}$ **27.** $\left\{0, \frac{1}{5}, -7\right\}$ **29.** $\{-4, -5\}$ **31.** $\left\{-4, \frac{2}{3}\right\}$ **33.** 10, 12; -12, -10 **35.** 8, 11; -8, -11 **37.** 6, 8; -12, -10 **39.** $\left\{2, -\frac{1}{3}, \frac{1}{3}\right\}$ **41.** $\left\{(p, z)|p = \frac{1}{4} \text{ or } z = -3\right\}$ **45.** 18 yd by 13 yd **47.** 1.5 km **49.** 20 seconds **50.** 4133 points **51.** -6 **52.** 16 **53.** $\{t|t > 2\}$ **54.** $56c^3d^3 - 8c^2d^3 + 8cd^4$ **55.** 16 or -16 **56.** 8 seconds

Pages 300–302 Chapter 7 Summary and Review

1. 5 **3.** 2 **5.** $4n$ **7.** mnp **9.** $6(x^2y + 2xy + 1)$ **11.** $2a(13b + 9c + 16a)$ **13.** $\frac{3}{5}(a - b + 2c)$ **15.** $(3a + 5b)(8m - 3n)$ **17.** $(4k - p^2)(4k^2 - 7p)$ **19.** $(y + 3)(y + 4)$ **21.** $(b - 1)(b + 6)$ **23.** $(3a - 7)(a - 2)$ **25.** $(a - b)(a - 9b)$ **27.** $(8m - 3n)(7m - 9n)$ **29.** $(5 - 3y)(5 + 3y)$ **31.** $2y(y - 8)(y + 8)$ **33.** $(9x^2 + 4)(3x - 2)(3x + 2)$ **35.** $(4x - 1)^2$ **37.** $2(4n - 5)^2$ **39.** $\left(y - \frac{3}{4}z^2\right)^2$ **41.** $(7y + 2)(4y - 3)$ **43.** $m(3m + 5)(2m - 3)$ **45.** $2r(r^2 - 9r + 15)$ **47.** $(x + 7)(x - 7)(m + b)$ **49.** $\{0, 5\}$ **51.** $\left\{0, \frac{9}{2}\right\}$ **53.** $\left\{0, \frac{3}{2}\right\}$ **55.** $\{7, -7\}$ **57.** $\left\{-\frac{2}{5}\right\}$ **59.** $\left\{0, -\frac{4}{3}, -\frac{7}{2}\right\}$ **61.** 5 of the 24 and 3 of the 36 **63.** $16x$ **65.** 9, 11; -11, -9

CHAPTER 8 RATIONAL EXPRESSIONS

Pages 308–310 Lesson 8-1

5. $6y$; $x \neq 0, y \neq 0$ **7.** x: $x \neq 0, y \neq 2$ **9.** $a - b$; $a \neq b$ **11.** $\frac{1}{3x}$; $x \neq 0$ **13.** $\frac{19a}{21b}$; $a \neq 0, b \neq 0$ **15.** $\frac{1}{2}$; $m \neq -5$ **17.** $\frac{1}{y - 4}$; $y \neq 4, y \neq -4$ **19.** $-\frac{1}{w + 4}$; $w \neq \frac{2}{3}, w \neq -4$ **21.** $\frac{c - 2}{c + 2}$; $c \neq -2$ **23.** m; $m \neq 2$ **25.** r^2; $r \neq 1$ **27.** $\frac{m^2}{2m - 1}$; $m \neq \frac{1}{2}, m \neq 0$ **29.** $\frac{2}{1 - 2y}$; $y \neq \frac{1}{2}$; $y \neq 0$ **31.** $\frac{a^2b}{3a + 7b^2}$; $a \neq -\frac{7b^2}{3}, a \neq 0, b \neq 0$ **33.** $\frac{1}{x + y}$; $x \neq -y$ **35.** $\frac{6x}{x + 4}$; $x \neq -4$ **37.** $\frac{2}{x + 3}$; $x \neq 7, x \neq -3$

39. $\frac{5}{3}$; $x \neq -1$ 41. $\frac{2k+5}{2k-5}$; $k \neq \frac{5}{2}$ 43. $\frac{1}{(b+3)(b+2)}$; $b \neq 2, b \neq 3, b \neq -2, b \neq -3$ 45. $\frac{n-2}{n(n-6)}$; $n \neq 0$, $n \neq 6$ 47. -1; $x \neq 3$ 49. $\frac{x^2+4}{x^2-4}$; $x \neq 2, x \neq -2$
51. 36 seconds 53. about 9.5°F 55. $\frac{9}{17}$
56. 6 years 57. $\left\{q \middle| q \leq -\frac{27}{2}\right\}$ 58. 7.124
59. $-20x^7$ 60. $3ab + 2b^3$ 61. $\{-6, 6\}$

Pages 313-314 Lesson 8-2

3. $\frac{3b}{4a}$ 5. $\frac{b}{d}$ 7. 2 9. $\frac{a^2}{bd}$ 11. $\frac{10ac}{3b^2}$ 13. $\frac{4m^4}{15n^2}$ 15. 15
17. 6 19. $\frac{3}{2}$ 21. $4a - 4b$ 23. 7 25. $\frac{x+y}{x+1}$
27. $\frac{3k}{k-3}$ 29. $\frac{-xy-x^2}{y}$ 31. $\frac{9m^2}{n}$ 33. $\frac{1}{3}$
35. $\frac{x+5}{x^2-12x+35}$ 37. $-\frac{x}{xy^2-y}$ 39. $\frac{y^2}{y^2-y-2}$
41. $\frac{2y^2+11y-21}{12y^3+19y^2+5y}$ 43. $\frac{-2(a+b)(y+1)}{y}$
47. 742.5 ft^3 49. 9, 25, 2500 50. $\{h|h < 450\}$
51. 7.6×10^6 52. $15x^2 + 32x - 7$
53. $6(y+2x)(y-2x)$ 54. $\frac{x}{x^2-4}$

Pages 317-318 Lesson 8-3

3. $\frac{2}{m}$ 5. $-\frac{3n}{8}$ 7. $\frac{5}{2m^2}$ 9. $\frac{x-y}{x+y}$ 11. $\frac{-y}{5y+25}$ 13. $\frac{a^4}{b^4}$
15. a 17. $\frac{b+3}{4b}$ 19. $2a$ 21. $-\frac{x}{7}$ 23. $\frac{ax+bx}{2}$
25. $\frac{3t+12}{2w-6}$ 27. $\frac{x+3}{x}$ 29. $\frac{2m-3}{3m-2}$ 31. $\frac{x+5}{x+7}$
33. $\frac{4x^2+10x-24}{2x^2+11x-6}$ 35. \$8.50 37. 12.5%
39. \$4500 40. $x < -20$ 41. $9m^4 + 12m^2n + 4n^2$
42. $\frac{1}{4}x(2x-a)$ 43. $2m + 4$

Pages 320-321 Lesson 8-4

3. a 5. $4m^2$ 7. x^2 9. $x + 4$ 11. $a + 5$
13. $c + 3 + \frac{9}{c+9}$ 15. $r - 5$ 17. $2x + 3$
19. $x^2 + 2x - 3$ 21. $t^2 + 4t - 1$
23. $3c - 2 + \frac{4}{9c-2}$ 25. $3n^2 - 2n + 3 + \frac{3}{2n+3}$
27. $3s^2 + 2s - 3 - \frac{1}{s+2}$ 29. $5t^2 - 3t - 2$
31. $8x^2 - 9$ 35. 9 football fields
37a. $3x + 11 + \frac{24}{x-2}$ b. $3x + 1 + \frac{4}{x+2}$
c. $x^2 + 4x + 4$ d. $x^2 + 4$
e. $2x^3 + 3x^2 + x + 4 + \frac{5}{x-1}$ f. $2x^3 - x + 2$
38. $9\frac{3}{5}$ ft 39. 9 days 40. $\frac{9yz^5}{x^3}$ 41. $(p+5)(p+5)$
42. $\frac{y^2}{a^2}$

Pages 324-325 Lesson 8-5

5. $\frac{7}{a}$ 7. $-\frac{1}{z}$ 9. $-\frac{1}{16}$ 11. $\frac{k}{m}$ 13. $\frac{a}{4}$ 15. -1
17. $\frac{2m+3}{5}$ 19. 1 21. $\frac{-y}{b+6}$ 23. $\frac{2y}{y-2}$ 25. 0
27. $\frac{r^2+s^2}{r-s}$ 29. $m + n$ 31. 4 33. $a + b$ 35. $\frac{x+1}{x-1}$
37. 0 39. $\frac{1}{x+1}$ 41. $-\frac{1}{t+1}$ 43. $P = -2x - 4$
47. \$1,506,167,664 48. 12:00 noon

49. $-\frac{1}{2} < r < \frac{8}{3}$ 50. $56m + 16$ 51. $2^3 \cdot 5^3$
52. $m^2 - 2m - 9 + \frac{-4}{m-4}$

Page 325 Mid-Chapter Review

1. $\frac{1}{a+1}$; $a \neq -1, a \neq 7$ 2. $\frac{1}{y+3}$; $y \neq 3, y \neq -3$
3. $\frac{4y-1}{8y-1}$; $y \neq \frac{1}{8}, y \neq -2$ 4. $\frac{y-2}{y-1}$ 5. $\frac{m^2+16}{m^2+8m+16}$
6. $\frac{x}{x^2+8x+16}$ 7. b 8. $\frac{1}{q(y-2)}$ 9. $\frac{2m+2n}{3m^2-3n}$
10. $2m - 1$ 11. $t^2 - 4t - 9 + \frac{16}{3t+1}$

Pages 327-328 Lesson 8-6

3. 24, 48 5. 27 7. \$180,000 9. Since Ed was freshly shaved and neatly trimmed and there are only two barbers, we can assume that Floyd did it. Therefore, the visitor went to Floyd's. 11. 5 people 13a. It is also happy. b. 1, 7, 10, 13, 19, 23, 28, 31, 32, 44, 49, 68, 70, 79, 82, 86, 91, 94, 97

Pages 331-332 Lesson 8-7

3. a^2 5. $120a^2b^3$ 7. $392ax^3y$ 9. $n(m+n)$
11. $\frac{m^2-2mn+n^2-1}{m^2-n^2}$ 13. $\frac{13t}{21}$ 15. $\frac{2}{a}$ 17. $\frac{5z+6x}{xyz}$
19. $\frac{2s+t^2+3t}{st}$ 21. $\frac{2a+3}{a+3}$ 23. $\frac{y^2+12y+25}{y^2-25}$
25. $\frac{2(x^3-1)}{x^2-1}$ or $\frac{2(x^2-x+1)}{x-1}$ 27. $\frac{-7y-39}{y^2-9}$
29. $\frac{-x^3+x^2+3xy-3y}{9xy^2}$ or $\frac{(3y-x^2)(x-1)}{9xy^2}$
31. $\frac{2x^3+5x^2-3x}{(2x-3)(2x+3)^2}$ 33. $\frac{x^2+6x-11}{x^2-2x-3}$
35. $\frac{9m+6}{(m+2)^2(m+1)}$ 37. $\frac{x^2+12x+2}{(x-1)^2(x+4)}$ 39a. 360; 2; 180; 360 b. 768; 16; 48; 768 c. The product of the numbers is equal to the product of their GCF and LCM. d. Divide the product of the two numbers by their LCM. 41. eldest: 9 cows; second: 6 cows; youngest 2 cows 43. 3 tables of 5 people and 8 tables of 8 people or 11 tables of 5 people and 3 tables of 8 people 45. 16 46. 160 balls 47. 7
48. \$3.16, \$1.50, \$1.25, \$1.20 49. 1 50. 6 ways

Pages 336-337 Lesson 8-8

3. $\frac{4x+2}{x}$ 5. $\frac{2m^2+m+4}{m}$ 7. $\frac{b^3-2b^2+2}{b-2}$ 9. $\frac{14}{19}$
11. $\frac{y^3(x+4)}{x^2(y-2)}$ 13. $\frac{a-b}{x-y}$ 15. $\frac{x+y}{x-y}$ 17. $\frac{x+1}{x-2}$
19. $\frac{1}{y+4}$ 21. $\frac{a+2}{a+3}$ 23. $\frac{(x+3)(x-1)}{(x-2)(x+4)}$
25. $\frac{8x^2-27y^2}{x^2-4y^2}$ 27. $\frac{x+1}{x+5}$ 29. 1 31. 1888 33. 45 pounds 34. $\frac{6}{9}, \frac{57}{99}, \frac{253}{999}, \frac{6001}{9999}$ 35. $12(a+1)(a-1)$
36. $\frac{15bx+3a}{21x^2}$

Pages 341-342 Lesson 8-9

3. 6 5. 8 7. $r^2 - 1$ 9. $(k+5)(k+3)$
11. $3(2x+1)(2x-1)$ 13a. $\frac{1}{n}$ b. $\frac{4}{n}$ c. $\frac{x}{n}$ 15. -3

17. $-\frac{9}{4}$ **19.** 5 **21.** $\frac{1}{4}$ **23.** $-\frac{3}{2}$ **25.** $\frac{1}{2}$ **27.** 20, 10 **29.** -3 **31.** 2 **33.** $\frac{27}{7}$ **35.** 7 **37.** 3, 1 **41.** 30 hours **43.** 12 mph **45.** 30 mph **47.** 4.355 minutes or about 4 minutes 21 seconds **48.** 8.8 ft **49.** $6a + 9b$ **50.** $(a + 6)^2$ **51.** $\frac{xy + x}{x}$

Pages 345-347 Lesson 8-10

5. 3.429 ohms **7.** 4 ohms **9.** 8 ohms, 4 ohms **11.** $t = \frac{v}{a}$ **13.** $v = \frac{2s - at^2}{2t}$ **15.** $M = \frac{Fd^2}{Gm}$ **17.** $p = \frac{A}{1 + rt}$ **19.** $P = \frac{36{,}500}{IR + 365}$ **21.** $y = \frac{r}{2a + 0.5}$ **23.** $R = \frac{H}{0.24I^2t}$ **25.** $R_1 = \frac{R_T R_2}{R_2 - R_T}$ **27.** $n = \frac{IR}{E - Ir}$ **29.** $m = \frac{y - b}{x}$ **31.** $y_2 = mx_2 - mx_1 + y_1$ **33.** $R = P - DQ$ **35.** $2.91\overline{6}$ ohms **37.** $7.\overline{6}$ ohms **39.** 96 ohms **41.** 12.632 ohms **44.** 15°C **45.** $\left\{x \middle| -\frac{9}{2} \le x \le \frac{5}{2}\right\}$ **46.** \$8405.68 **47.** 50 hours or about 2 days **48.** $2(m + 4n)(m - 4n)$ **49.** 60 minutes

Pages 348-350 Chapter 8 Summary and Review

1. $\frac{x}{4y^2z}$ **3.** $\frac{a - 5}{a - 2}$ **5.** $\frac{7a^2}{9b}$ **7.** $\frac{(x + 4)^2}{(x + 2)^2}$ **9.** $-2p$ **11.** $\frac{7ab(x + 9)}{3(x - 5)}$ **13.** $x^2 + 4x - 2$ **15.** $2a^2 + 18a + 159 + \frac{1422}{a - 9}$ **17.** $\frac{7 + a}{x^2}$ **19.** 2 **21.** $\frac{2 - x}{x - y}$ **23.** $\frac{8x - 9}{x^2 - 4}$ **25.** $\frac{3x}{y}$ **27.** $\frac{(x - 5)(x + 13)}{(x + 2)(x + 6)}$ or $\frac{x^2 + 8x - 65}{x^2 + 8x + 12}$ **29.** $-2, -\frac{5}{2}$ **31.** 0 **33.** $\frac{xy}{r}$ **35.** $\frac{a}{c + cb}$ **37.** $\frac{x - 2}{x + 2}$ **39.** 33, 66, 99

CHAPTER 9 FUNCTIONS AND GRAPHS

Pages 357-358 Lesson 9-1

5. (1, 4) **7.** $(-1, -2)$ **9.** $(3, -1)$ **11.** I **13.** II **15.** none **17.** $(-1, 1)$ **19.** $(3, -2)$ **21.** (1, 1) **23.** $(-1, -1)$ **25.** $(2, -3)$ **27.** $(-4, 1)$ **29.** IV **31.** II **33.** IV **35.** II **37.** none

55.

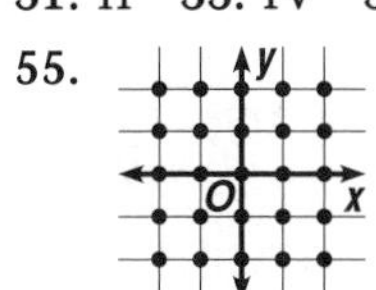

57.

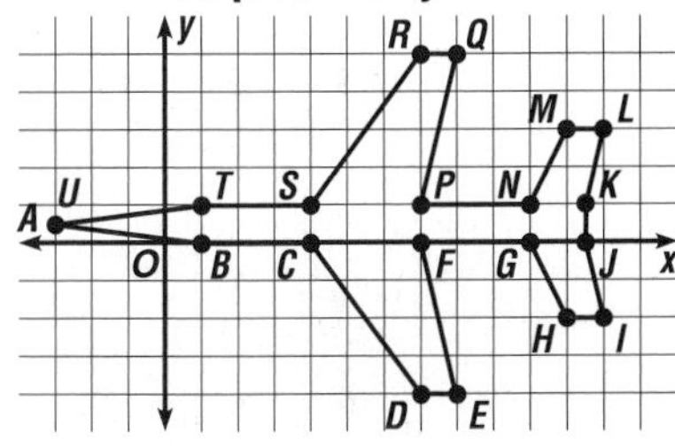

61. Irving **63.** (B, 4), (B, 5), (C, 4), (C, 5) **65.** (A, 3), (B, 3), (C, 3), (D, 3), (E, 3), (E, 4), (F, 4) **66.** $10x^2 + 3y + z^2$ **67.** 540 seats **68.** $25a^2 - 10ab + b^2$

69. −2 −1 0 1 2 3 4 5 6

70. 5×10^5; 500,000 **71.** $4(x + 4y)(x - 4y)$ **72.** $\frac{1}{z + 6}$

Pages 361-363 Lesson 9-2

5. $\{5, 0, -9\}$; $\{2, 0, -1\}$ **7.** $\{7, -2, 4, 5, -9\}$; $\{5, -3, 0, -7, 2\}$ **9.** $\left\{\frac{1}{2}, 1\frac{1}{2}, -3, -5\frac{1}{4}\right\}$; $\left\{\frac{1}{4}, -\frac{2}{3}, \frac{2}{5}, -6\frac{2}{7}\right\}$ **11.** $\{(1, 3), (2, 2), (4, 9), (6, 5)\}$; $\{1, 2, 4, 6\}$; $\{3, 2, 9, 5\}$; $\{(3, 1), (2, 2), (9, 4), (5, 6)\}$ **13.** $\{(1, 7), (2, 2), (-3, 1), (5, 2)\}$ **15.** $\{(5, 4), (5, 8), (2, 9), (-7, 2), (3, 2), (3, 4)\}$ **17.** $\{(1, 3), (2, 4), (3, 5), (4, 6), (5, 7)\}$; $\{(3, 1), (4, 2), (5, 3), (6, 4), (7, 5)\}$ **19.** $\{(1, -2), (3, -4), (5, -6), (9, -4), (9, -2)\}$; $\{(-2, 1), (-4, 3), (-6, 5), (-4, 9), (-2, 9)\}$ **21.** $\{(-2, 2), (-1, 1), (0, 1), (1, 1), (1, -1), (2, -1), (3, 1)\}$; $\{-2, -1, 0, 1, 2, 3\}$; $\{2, 1, -1\}$ **23.** $\{(-3, 0), (-2, 2), (-1, 3), (0, 1), (1, -1), (1, -2), (1, -3), (3, -2)\}$; $\{-3, -2, -1, 0, 1, 3\}$; $\{0, 2, 3, 1, -1, -2, -3\}$ **25.** $\{(-3, 3), (-1, 2), (1, 1), (1, 3), (2, 0), (2, -1), (3, -1)\}$; $\{-3, -1, 1, 2, 3\}$; $\{3, 2, 1, 0, -1\}$ **27.** $\{(2, -2), (1, -1), (1, 0), (1, 1), (-1, 1), (-1, 2), (1, 3)\}$ **29.** $\{(0, -3), (2, -2), (3, -1), (1, 0), (-1, 1), (-2, 1), (-3, 1), (-2, 3)\}$ **31.** $\{(3, -3), (2, -1), (1, 1), (3, 1), (0, 2), (-1, 2), (-1, 3)\}$ **33.** sum of 2: 1; sum of 3: 2; sum of 4: 3; sum of 5: 4; sum of 6: 5; sum of 7: 6; sum of 8: 5; sum of 9: 4; sum of 10: 3; sum of 11: 2; sum of 12: 1

35.

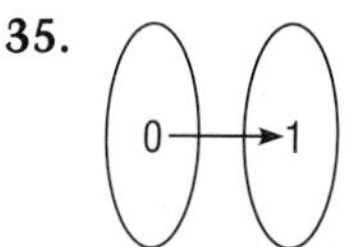

39.

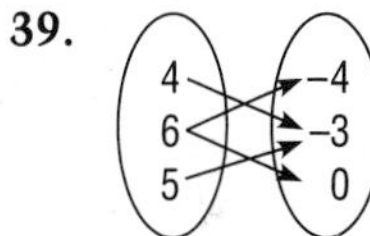

43.

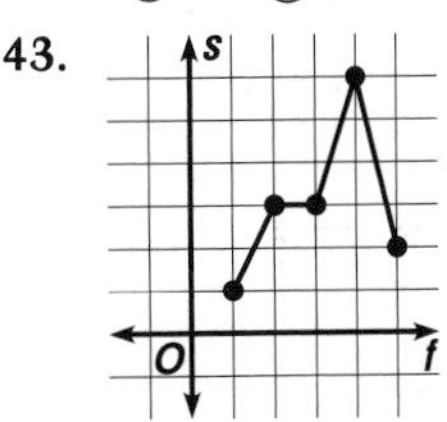

45. no solution **46.** 200 mL **47.** $7a^6b^2 - 35a^4b^3 + 42a^2b^4$ **48.** $(12n + 7)^2$ **49.** $2\frac{2}{29}$ ohms

50.

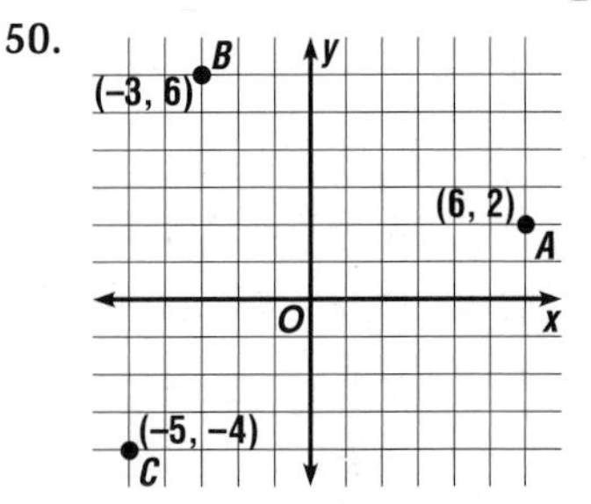

Pages 367-368 Lesson 9-3

5. $-1, (-4, -1); \frac{1}{3}, \left(-2, \frac{1}{3}\right); \frac{5}{3}, \left(0, \frac{5}{3}\right); \frac{7}{3}, \left(1, \frac{7}{3}\right); \frac{11}{3}, \left(3, \frac{11}{3}\right)$ **7.** b, c **9.** $y = 5 - x$ **11.** $b = 3 + 5a$ **13.** a, b, c **15.** a **17.** $y = 4 - 2x$ **19.** $s = \frac{2 - 6r}{5}$ **21.** $b = \frac{3a - 8}{7}$ **23.** $r = \frac{4 + 5n}{7}$ **25.** $\{(-2, -3), (-1, -1), (0, 1), (2, 5), (5, 11)\}$ **27.** $\{(-2, -6), (-1, -5), (0, -4), (2, -2), (5, 1)\}$ **29.** $\left\{\left(-2, \frac{17}{3}\right), (-1, 5), \left(0, \frac{13}{3}\right), (2, 3), (5, 1)\right\}$ **31.** $\left\{\left(-2, \frac{13}{2}\right), (-1, 4), \left(0, \frac{3}{2}\right), \left(2, -\frac{7}{2}\right), (5, -11)\right\}$ **33.** $\left\{\left(-2, -\frac{11}{2}\right), (-1, -4), \left(0, -\frac{5}{2}\right), \left(2, \frac{1}{2}\right), (5, 5)\right\}$ **35.** $\left\{(-2, 5), \left(-1, \frac{31}{6}\right), \left(0, \frac{16}{3}\right), \left(2, \frac{17}{3}\right), \left(5, \frac{37}{6}\right)\right\}$ **37.** $\left\{\left(-\frac{4}{5}, -3\right), \left(-\frac{2}{5}, -1\right), \left(-\frac{1}{5}, 0\right), \left(\frac{1}{5}, 2\right), \left(\frac{2}{5}, 3\right)\right\}$ **39.** $\left\{\left(\frac{11}{2}, -3\right), (4, -1), \left(\frac{13}{4}, 0\right), \left(\frac{7}{4}, 2\right), (1, 3)\right\}$ **41.** $\left\{(-1, -3), \left(-\frac{2}{3}, -1\right), \left(-\frac{1}{2}, 0\right), \left(-\frac{1}{6}, 2\right), (0, 3)\right\}$ **45.** 4 m by 4 m **47.** $\{s|s < -6.9\}$ **48.** $-3x^2y + 6xy^2 - 2y^2$ **49.** $3a(14bc - 4ab^2 + ac^2)$ **50.** $\{-1, 0\}$ **51.** 3 hours **52.** $\{8, 4, 6, 5\}$; $\{1, 2, -4, -3, 0\}$; $\{(1, 8), (2, 4), (-4, 6), (-3, 5), (0, 6)\}$

Pages 371-373 Lesson 9-4

5. yes **7.** no **9.** yes **11.** no **13.** yes **15.** $y = 8x - 16$ **17.** $y = \frac{12 - 3x}{4}$ **19.** $y = \frac{3}{4}x - 15$ **21.** yes **23.** no **25.** yes **27.** yes

29.

33.

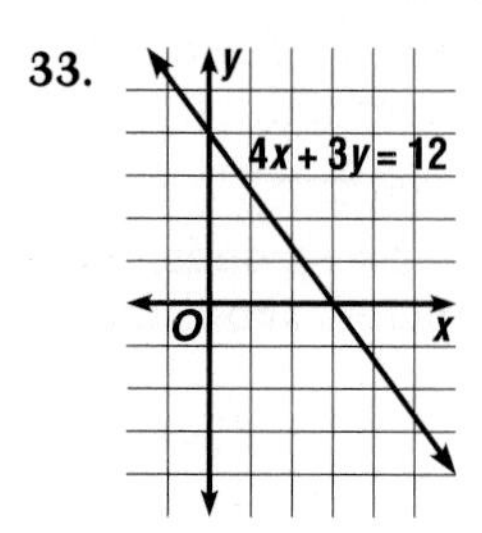

37.

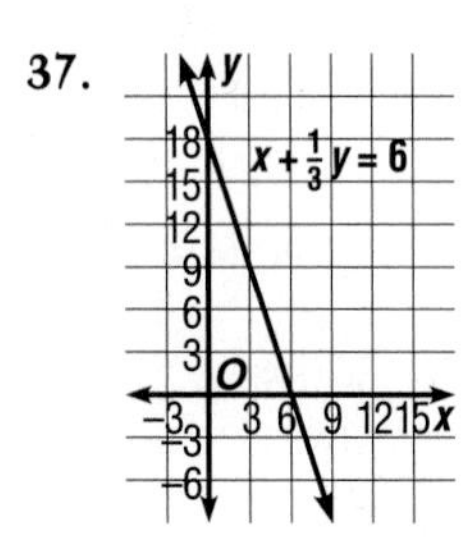

41.

45. $y = -3x$ or $3x + y = 0$

49a.

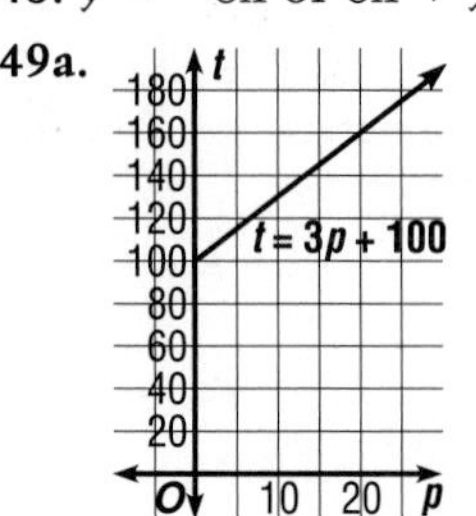

49b. 250 people **51.** $2nxy^5 + 12n^4xy^3 - 4n^2x^3$ **52.** $-\frac{9m^7}{n^2}$ **53.** $\frac{k - 1}{k + 1}$ **54.** $5x^2 - 4x + 7$ **55.** 8 m by 19 m **56.** $\left\{\left(-3, \frac{22}{3}\right), \left(-1, \frac{14}{3}\right), \left(0, \frac{10}{3}\right), (1, 2), (4, -2)\right\}$

Page 373 Mid-Chapter Review

1-4.

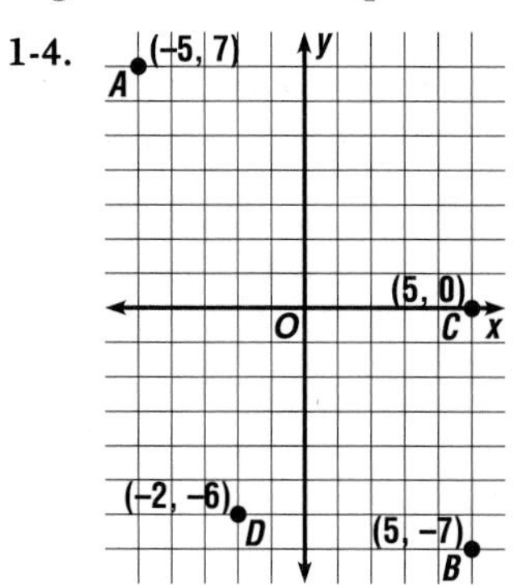

5. $\{(4, 2), (1, 3), (3, 3), (6, 4)\}$; $\{4, 1, 3, 6\}$; $\{2, 3, 4\}$; $\{(2, 4), (3, 1), (3, 3), (4, 6)\}$ **6.** $\{(4, 2), (-3, 2), (8, 2), (8, 9), (7, 5)\}$; $\{4, -3, 8, 7\}$; $\{2, 9, 5\}$; $\{(2, 4), (2, -3), (2, 8), (9, 8), (5, 7)\}$ **7.** $\{(-2, 2), (-1, 1), (1, 1), (1, -1), (2, 2), (2, -2)\}$; $\{-2, -1, 1, 2\}$; $\{2, 1, -1, -2\}$; $\{(2, -2), (1, -1), (1, 1), (-1, 1), (2, 2), (-2, 2)\}$ **8.** $\{(-5, -22), (-2, -7), (0, 3), (1, 8), (3, 18)\}$ **9.** $\{(-5, -17), (-2, -11), (0, -7), (1, -5), (3, -1)\}$ **10.** $\left\{\left(-5, -\frac{4}{3}\right), \left(-2, \frac{8}{3}\right), \left(0, \frac{16}{3}\right), \left(1, \frac{20}{3}\right), \left(3, \frac{28}{3}\right)\right\}$ **11.** $\left\{\left(-5, -\frac{41}{5}\right), \left(-2, -\frac{23}{5}\right), \left(0, -\frac{11}{5}\right), (1, -1), \left(3, \frac{7}{5}\right)\right\}$ **12.** yes **13.** yes

14. no **15.** yes

Pages 377-378 Lesson 9-5

5. no **7.** yes **9.** no **11.** yes **13.** -9 **15.** 0 **17.** yes **19.** yes; {(1, 3), (1, 5), (1, 7)}; no **21.** no; {(4, −2), (3, 1), (2, 5), (4, 1)}; no **23.** yes; {(4, 5), (5, −6), (5, 4), (4, 0)}; no **25.** yes **27.** yes **29.** no **31.** yes **33.** -14 **35.** $-\frac{2}{9}$ **37.** 11.5 **39.** 12 **41.** $12a - 5$ **43.** $18n - 15$ **45.** $12m^2 - 6m$ **47.** $3a + 4$ **53.** $59.60 **55a.** (0, 2) **b.** The coefficients of x are the additive inverses of those in Exercise **a.** **56.** 88x **57.** 28 m by 38 m **58.** $(7n - 1)(n - 3)$ **59.** $\frac{3a + 20}{a^2 - 25}$

60.

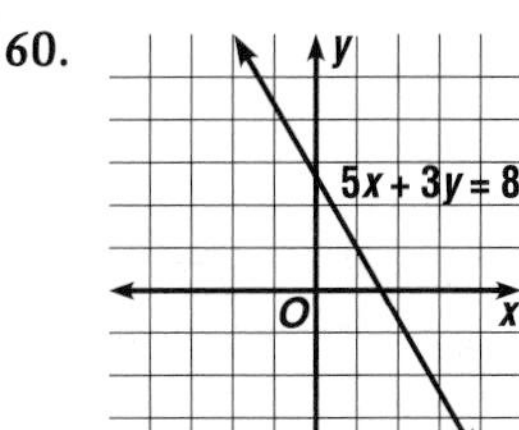

Pages 381-383 Lesson 9-6

5. yes **7.** c **9.** b **11.** a, b **13.** half-plane to left of line **15.** half-plane to right of line **17.** half-plane to left of line

19.

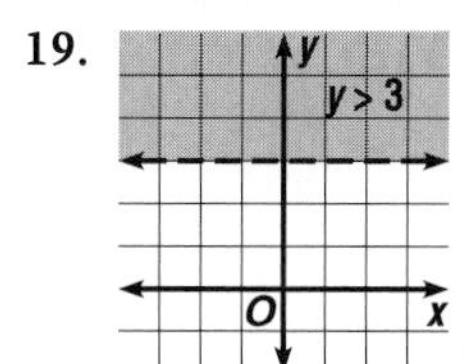

23.

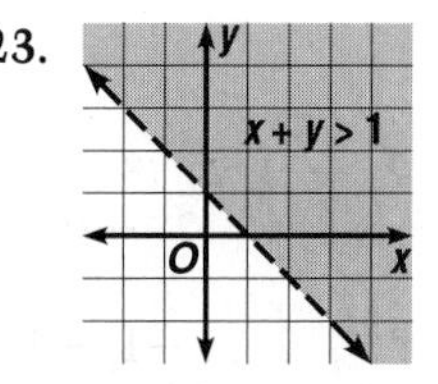

27.

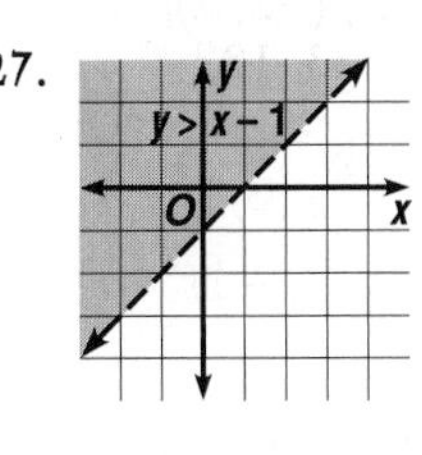

31.

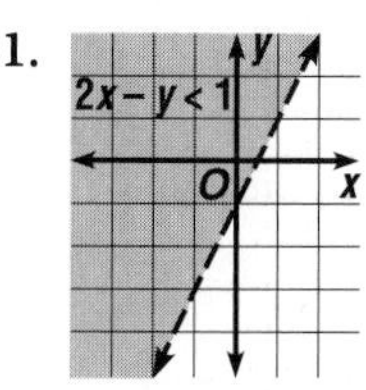

35.

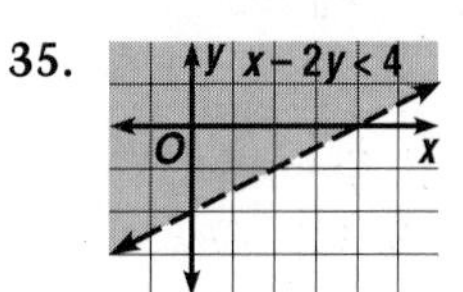

39.

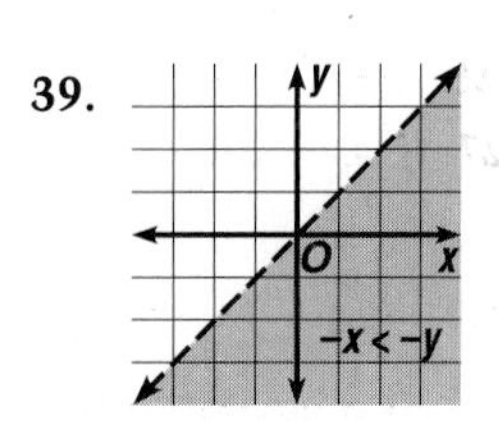

43.

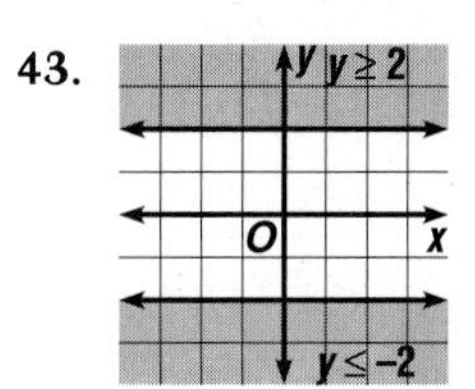

47.

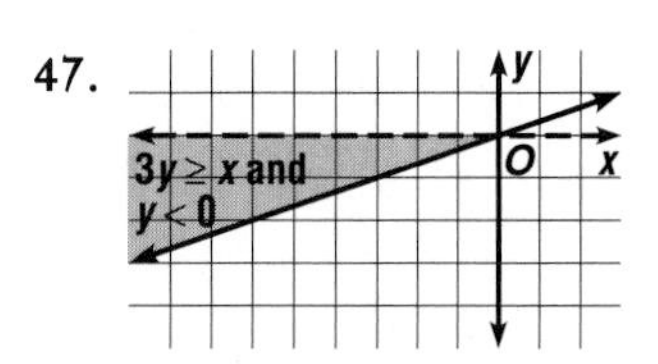

49a.

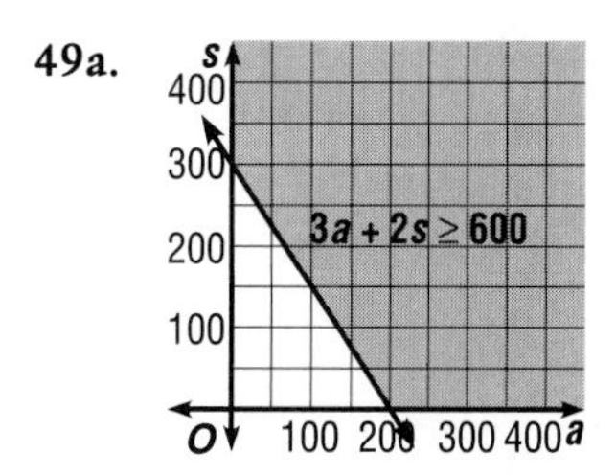

49b. Answers will vary; typical answers are 200 adult and 0 student tickets, 0 adult and 300 student tickets, and 100 adult and 200 student tickets. **51.** $\{x|x \geq -1\}$ **52.** $4r^8x^3y$ **53.** 4 **54.** $\frac{b - a}{x - y}$ **55.** $(3a + 4c)(5a - 7b)$ **56.** No; after 4 seconds, the ball will be $100 - 4.9(4)^2$ or 21.6 meters above the ground.

Pages 387-388 Lesson 9-7

5. d **7.** b **9.** 0, -3 **11.** 10, 20 **13.** 11.25, 10.75 **15.** 16, 32 **17.** $-4, \frac{1}{2}$ **19.** $n = 2m + 1$; 1, 3 **21.** $b = 4a + 3$; 11, 19 **23.** $n = 11 - m$; 9, 8, 7 **25.** $y = 18 - 2x$; 10, 6, 2 **27.** $d = \frac{1}{3}c - 2$; 6, 8, 12 **29.** $y = x^3 + 1$ **31.** $y = 15 - x^2$ **35. a.** $y = \$0.34x + \$0.36$ **b.** $0.70; $0.34 **36.** $8200 at 5%, $4000 at 6.5% **37.** 4x **38.** $3x(x + 11y)(x - 3y)$ **39.** $\frac{x^2 + x - 12}{x^2 - 3x - 18}$

40.

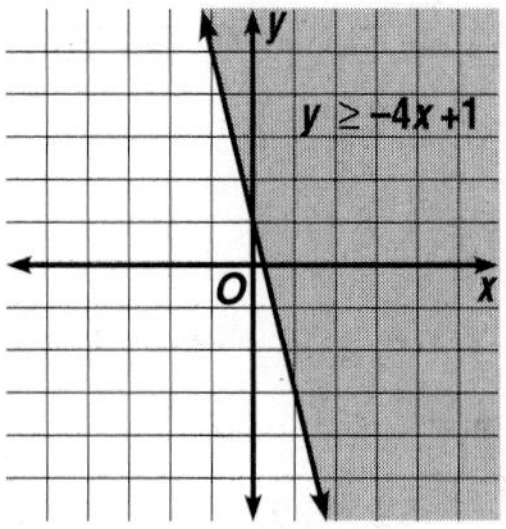

Pages 390-391 Lesson 9-8

7. yes **9.** In Graph A, Restaurant III does not appear as unpopular as it does in Graph B. **11.** Graph A **13.** 512 **15.** Al, 12 years old; Betty, 14 years old; Carmelita, 7 years old; Dwayne, 4 years old

Pages 392-394 Chapter 9 Summary and Review

5. I **7.** II **9.** {4}; {1, −2, 6, −1}; {(1, 4), (−2, 4), (6, 4), (−1, 4)} **11.** {−2, −5 −7}; {1}; {(1, −2), (1, −5), (1, −7)} **13.** $y = 7 - 3x$ **15.** $y = \frac{12 - x}{6}$
17. {(−4, −13), (−2, −11), (0, −9), (2, −7), (4, −5)}
19. $\left\{\left(-4, -\frac{16}{3}\right), \left(-2, -\frac{8}{3}\right), (0, 0), \left(2, \frac{8}{3}\right), \left(4, \frac{16}{3}\right)\right\}$

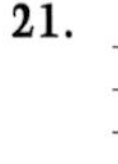

21.

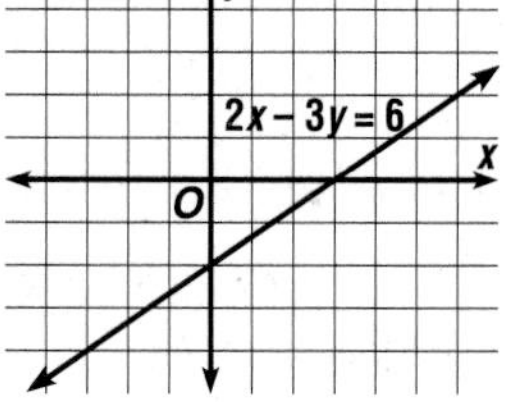

25. no **27.** yes **29.** 3 **31.** $a^2 + a + 1$

33.

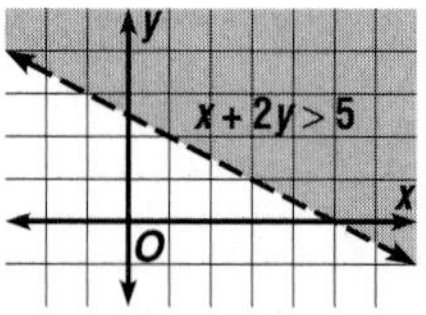

37. $y = 3x + 5$ **39.** 4.5 seconds

41a.

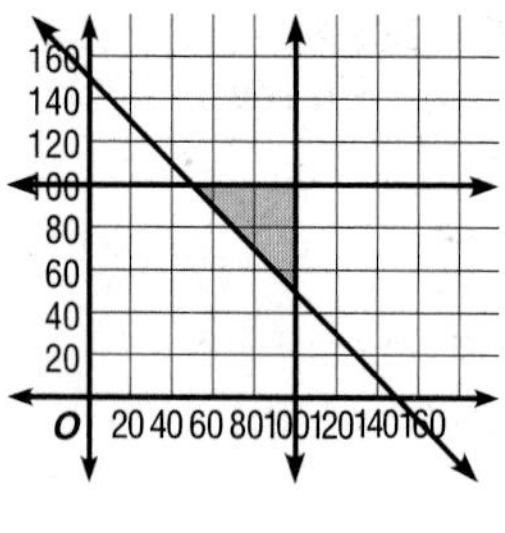

b. Answers will vary; typical answers are 74 and 73, 80 and 67, and 100 and 47.

CHAPTER 10 GRAPHING LINEAR EQUATIONS

Pages 403-404 Lesson 10-1

5. $-1; 1; \frac{^-1}{1} = -1$ **7.** $-4; 3; \frac{^-4}{3} = -\frac{4}{3}$

9.

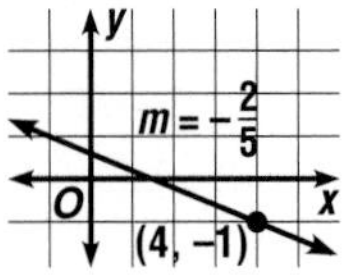

15. −1 **17.** 0 **19.** $-\frac{1}{5}$ **21.** −2 **23.** $-\frac{3}{2}$ **25.** $\frac{3}{2}$
27. 0 **29.** 0.5 **31.** $-\frac{3}{4}$ **33.** $\frac{25}{3}$ **35.** 7 **37.** 7
39. 2, 3 **43.** 11,160 feet **45.** 9.8%
46. associative (+) **47.** 7
48. $(5a + 3b)(a + b)(a - b)$ **49.** $\frac{6x - 5}{x^2}$
50. 16, 21, 31

Pages 408-409 Lesson 10-2

5. 3; (5, 2) **7.** $-\frac{3}{2}$; (−5, −6) **9.** 0; (0, 3)
11. $2x - y = -6$ **13.** $2x - 3y = -1$
15. $2x + 3y = 22$ **17.** $2x - 3y = 15$
19. $3x + y = -2$ **21.** $y = 6$ **23.** $x + y = 9$
25. $x = 4$ **27.** $x + 2y = -4$ **29.** $3x + 11y = -4$
31. $4x + 10y = 13$ **33.** $2x + 3y = 6$
37. $5x + y = 90$, where x = the number of weeks that have passed and y = the amount of money he has left **38.** $56a^2 + 77a$ **39.** $(x - 7)(x - 2)$
40. $\frac{y^2}{y^2 - y - 2}$ **41.** Quadrant I **42.** 1015 feet

Pages 413-414 Lesson 10-3

5. 5, 3 **7.** $\frac{1}{3}$, 0 **9.** $-\frac{2}{3}, \frac{5}{3}$ **11.** 6, 5 **13.** $-\frac{8}{5}, -\frac{2}{5}$
15. $y = -3x + 5$ **17.** $y = \frac{1}{2}x + 5$ **19.** $y = 14$
21. 2, 10 **23.** 14, −4 **25.** −2, none **27.** none, 4
29. $-\frac{2}{5}$, 2; $y = -\frac{2}{5}x + 2$ **31.** $-\frac{7}{4}$, 2; $y = -\frac{7}{4}x + 2$
33. $-\frac{4}{3}, \frac{5}{3}$; $y = -\frac{4}{3}x + \frac{5}{3}$ **35.** −6, 15; $y = -6x + 15$
37. −4, 12; $y = -4x + 12$ **39.** $-\frac{5}{2}$, 4; $y = -\frac{5}{2}x + 4$
41. $\frac{11}{2}$, −16; $y = \frac{11}{2}x - 16$ **43.** −9, −2; $y = -9x - 2$ **45.** $-\frac{11}{3}, \frac{25}{3}$; $y = -\frac{11}{3}x + \frac{25}{3}$
47. $y = \frac{4}{5}x + \frac{3}{2}$ **49.** $y = \frac{2}{5}x + 12$ **55. a.** $56 **b.** $5
57. 4 tennis balls **58.** $-6x^8y^3$ **59.** $(m^2 + 6n^2)^2$
60. y **61.** 4 **62.** $y = 7$

Pages 417-418 Lesson 10-4

3. 2, 8 **5.** 4, −8 **7.** $\frac{10}{7}$, 5 **9.** Possible answers are (1, 3) and (2, 5). **11.** Possible answers are (4, 2) and (2, 2). **13.** Possible answers are (7, 3) and (9, 4).

15. 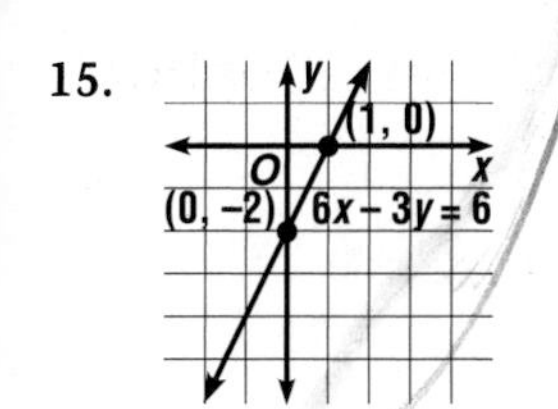

19. 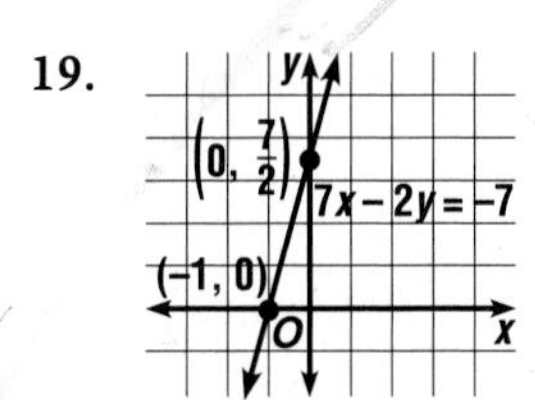

23. 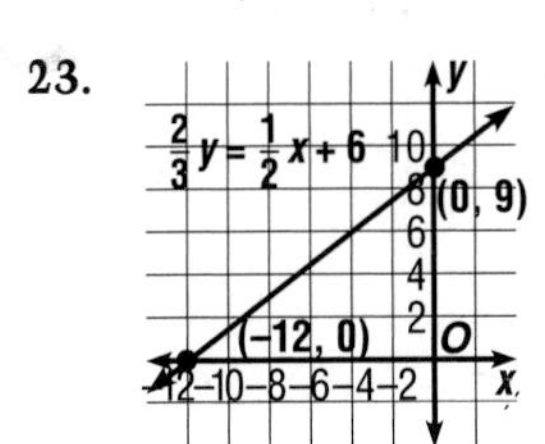

27. 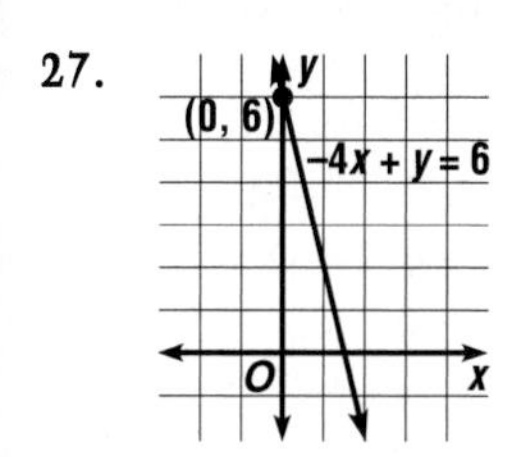

31. 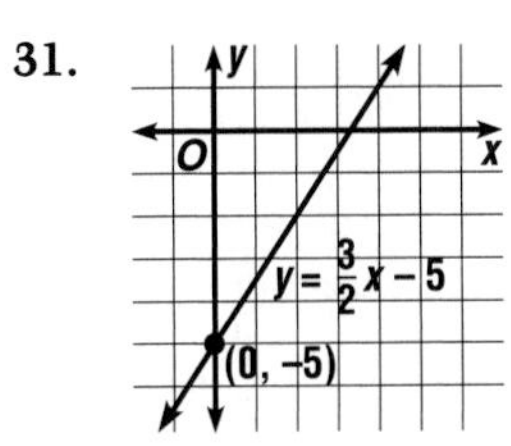

37a. 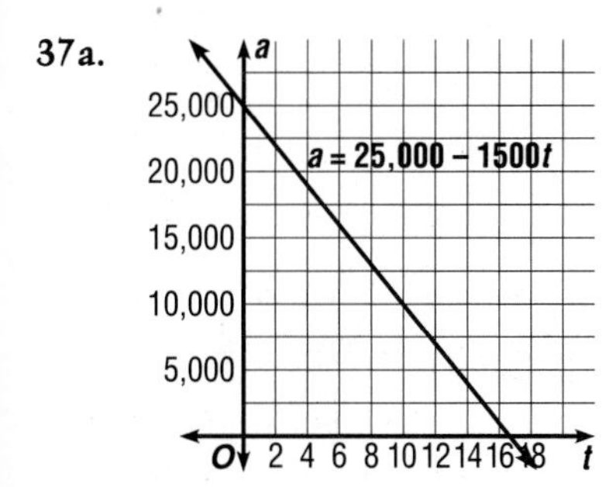

37b. $16\frac{2}{3}$ minutes or 16 minutes 40 seconds
39. 152 feet **40.** $3y^2 - 7y - 6$ **41.** 2 **42.** $\frac{5}{4}$
43. 41, 51, 61 **44.** 3, 6

Page 418 Mid-Chapter Review

1. 2 **2.** $-\frac{1}{2}$ **3.** −1 **4.** $3x - 4y = 16$
5. $3x - 2y = -20$ **6.** $x = -2$ **7.** $x + y = 10$
8. $x = 5$ **9.** $3x - 10y = -6$ **10.** 3, 5 **11.** $-\frac{5}{3}, \frac{5}{2}$
12. $\frac{8}{3}$, 16 **13.** $-\frac{3}{4}$, 3; $y = -\frac{3}{4}x + 3$ **14.** $\frac{5}{7}, -\frac{3}{2}$; $y = \frac{5}{7}x - \frac{3}{2}$ **15.** 11, 14; $y = 11x + 14$

Pages 421-422 Lesson 10-5

5. 14 **7.** $-\frac{7}{2}$ **9.** 2, −2, $y = 2x - 2$ **11.** −4, 4, $y = -4x + 4$ **13.** $-\frac{2}{3}$, 2, $y = -\frac{2}{3}x + 2$
15. $y = \frac{2}{3}x + \frac{2}{3}$ **17.** $y = \frac{3}{4}x - \frac{5}{2}$ **19.** $y = \frac{1}{4}x + 8$
21. $y = -\frac{2}{3}x + 4$ **23.** $y = -x + 1$
25. $y = -\frac{7}{9}x - \frac{8}{3}$ **27.** $y = -0.515x + 5.551$
29. $y = -1.595x - 7.429$ **31.** $y = 0.556x - 9.110$
33. $d = \frac{50}{3}h + 15$, where h = the number of hours worked and d = the total charge **35.** 8.21×10^7
36. $9(2x - 3y^2)(2x + 3y^2)$ **37.** −1 **38.** {(−2, −13), (−1, −8), (0, −3), (2, 7), (5, 22)}
39.

Pages 425-426 Lesson 10-6

5. 5, $-\frac{1}{5}$ **7.** $\frac{2}{3}, -\frac{3}{2}$ **9.** undefined, 0 **11.** $\frac{2}{3}, -\frac{3}{2}$
13. −1, 1 **15.** $y = \frac{3}{4}x$ **17.** $y = -\frac{2}{3}x + \frac{14}{3}$
19. $y = \frac{4}{3}x - \frac{16}{3}$ **21.** $y = x - 9$ **23.** $y = \frac{8}{3}x + 4$
25. $y = \frac{1}{3}x + 2$ **27.** $y = -\frac{3}{2}x - \frac{9}{2}$ **29.** $y = -\frac{1}{2}x - 4$
31. no **33.** 75% **34.** $16x^6 - 24x^3y^2 + 9y^4$
35. T = 5, W = 4, E = 6, Y = 0, H = 9, I = 3, S = 1, V = 8, and either N = 7 and R = 2 or N = 2 and R = 7 **36.** $\frac{x + 7}{x + 5}$ **37.** yes

Pages 429-431 Lesson 10-7

3. 6 **5.** 5 **7.** $\frac{3}{2}$ **9.** (2, 2) **11.** (7, 4.5)
13. (−1, −1) **15.** (−2, 3.5) **17.** (10, 3) **19.** (14, 3)
21. (6, −2) **23.** $\left(6, \frac{1}{2}\right)$ **25.** $\left(\frac{1}{2}, 7\right)$ **27.** $(4x, 2y)$
29. $B(19, 9)$ **31.** $P(-1, 6)$ **33.** $A(13, 11)$
35. $B(6, -7)$ **37.** $P(-2, -3)$ **39.** $B(9, -18)$
41. (6, −4) **43.** (9, 6) **45.** $\left(-1, \frac{5}{2}\right)$ **49. a.** yes **b.** no **c.** no **d.** no **51.** $\left(\frac{1}{2}, 2\right)$ **53.** $\left(\frac{9}{2}, -\frac{1}{2}\right)$
55. (4, 1) **57.** $\left(-\frac{3}{2}, 1\right)$ **59.** $\{y \mid y \geq -1\}$
60. $3x + 2y$ **61.** $5(x^2 + 4y^2)$ **62.** 10 ways
63.

Page 433 Lesson 10-8

7. over 65 **9.** Ana and Carl, Betty and Frank, Daisy and Ed

Pages 434-436 Chapter 10 Summary and Review

1. $-\frac{1}{3}$ **3.** $\frac{1}{3}$ **5.** $-\frac{5}{3}$ **7.** $5x - 2y = -10$ **9.** $y = 2$
11. $\frac{1}{4}$, 3 **13.** $\frac{1}{2}$, $\frac{7}{2}$ **15.** −2, 12 **17.** $\frac{1}{2}$, 4 **19.** 8, $-\frac{8}{3}$

21.

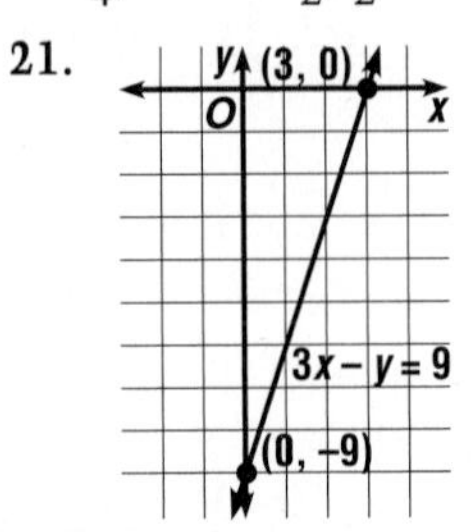

25. $y = 4x - 26$ **27.** $y = -\frac{5}{2}x + 7$ **29.** $y = 4x - 9$
31. $y = -\frac{7}{2}x - 14$ **33.** (6, 1) **35.** $\left(1, -\frac{5}{2}\right)$
37. (3, 3) **39.** $45x - 10$, where x = time, y = distance, and $x \geq 2$

CHAPTER 11 SYSTEMS OF OPEN SENTENCES

Page 441 Lesson 11-1

5. 2100 rings **7.** 3041 digits

Pages 445-446 Lesson 11-2

5. (6, 3) **7.** (1, 3) **9.** (2, 1) **11.** $m = -1, b = 6$; $m = 1, b = -2$; one solution **13.** $m = -\frac{1}{2}, b = \frac{5}{2}$; $m = -\frac{1}{2}, b = \frac{5}{2}$; infinitely many solutions
15. $m = \frac{3}{8}, b = -\frac{1}{2}$; $m = \frac{3}{8}, b = -\frac{21}{8}$; no solution
17. one, (3, 3) **19.** one, (9, 1) **21.** one, (6, 2)
23. (0, 0) **25.** (2, −6) **27.** infinitely many solutions **29.** (3, 1) **31.** (−1, −5) **33.** (−3, 1)
35. (1, −5) **37.** (−2, −3) **39.** infinitely many solutions **41.** (0, 3), (2, 4), (−1, 10) **45.** 0.3 mile
47. < **48.** $a^2m^3n^3$ **49.** 8 mph **50.** (−7, −3)

Pages 450-451 Lesson 11-3

5. $x = 5 - y$; $y = 5 - x$ **7.** $x = 3 - \frac{3}{2}y$; $y = 2 - \frac{2}{3}x$
9. $x = -\frac{0.8}{0.75}y - 8$; $y = -\frac{0.75}{0.8}x - 7.5$
11. $x + (3 + 2x) = 7$; $x = \frac{4}{3}$ **13.** $5x = 12x$; $x = 0$
15. $\left(-6 + \frac{2}{3}y\right) + 3y = 4$; $y = \frac{30}{11}$ **17.** (−3, −9)
19. $\left(3, \frac{3}{2}\right)$ **21.** (13, 30) **23.** $\left(\frac{12}{5}, \frac{1}{15}\right)$
25. no solution **27.** (0, 0) **29.** (3, 3) **31.** (50, 4)
33. (4, 2) **35.** 24 **37.** 35 **39.** (−1, 5, −4)
43. \$1000 at 10% and \$3000 at 12%
45. 125 tapes **46.** $\{y|2 < y < 11\}$ **47.** $(x + 2)^2(x - 2)$
48. \$460 **49.** $3x + 11y = -4$ **50.** (3, 1)

Pages 455-456 Lesson 11-4

5. subtraction, (1, 3) **7.** addition, $\left(\frac{5}{2}, \frac{9}{2}\right)$
9. $-x = -9$; (9, −12) **11.** $10x = 40$; $\left(4, -\frac{3}{2}\right)$
13. (8, −1) **15.** (23, 14) **17.** $\left(\frac{11}{2}, -\frac{1}{2}\right)$ **19.** $\left(-1, \frac{9}{2}\right)$
21. (3, 4) **23.** $\left(2, \frac{1}{2}\right)$ **25.** $\left(\frac{14}{9}, -\frac{16}{3}\right)$ **27.** (24, 4)
29. (2, 5) **31.** 11, 53 **33.** (4, 2, 3) **35.** (2, 3, −4)
39. math: 760 points, verbal: 580 points
41. $y + n^3 = 2x$ **42.** $-\frac{1}{6}, -\frac{3}{2}$ **43.** {5, −3, 4, 2}; {1, 2, 0}; {(1, 5), (2, −3), (2, 4), (0, 5), (2, 2)}
44. ≈232 **45.** (5, −2)

Page 456 Mid-Chapter Review

1. 375 games **2.** (1, −2) **3.** no solution
4. infinitely many **5.** (3, 3) **6.** (2, 10) **7.** (1, −1)
8. (−9, −7) **9.** (4, 3) **10.** 8 mph

Pages 460-461 Lesson 11-5

5. Multiply the second equation by −4, then add; (2, 0). **7.** Multiply the first equation by 3, multiply the second equation by −2, then add; (3, 0).
9. Multiply the first equation by 2, then add; $\left(\frac{11}{9}, \frac{8}{9}\right)$. **11.** Multiply the first equation by 2, multiply the second equation by −5, then add; (2, 2). **13.** $\left(\frac{2}{3}, \frac{22}{3}\right)$ **15.** $\left(\frac{3}{5}, 3\right)$ **17.** (4, 1)
19. (4, −28) **21.** (−9, −2) **23.** (10, 25)
25. (4, 16) **27.** 16 **29.** (11, 12) **31.** $\left(\frac{1}{3}, \frac{1}{6}\right)$
33a. 21 mph **b.** 3 mph **35.** 475 mph
37a. no solution **b.** infinite number of solutions
c. (−9, −7) **d.** infinite number of solutions
38. 18 **39.** 34 ft by 22 ft **40.** $\frac{4}{(m + 2)(m - 4)}$
41. (−14, −13)

Pages 465-467 Lesson 11-6

5. no **7.** yes **9.** A **11.** D

13.

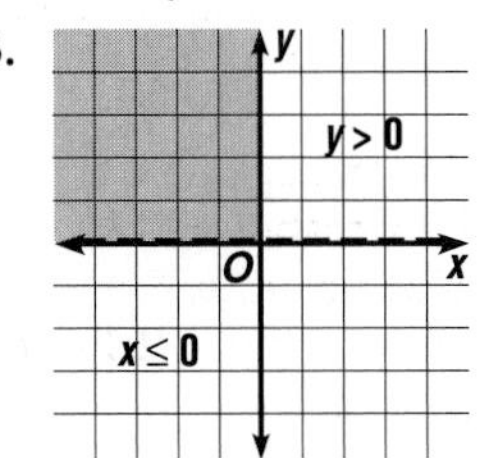

17.

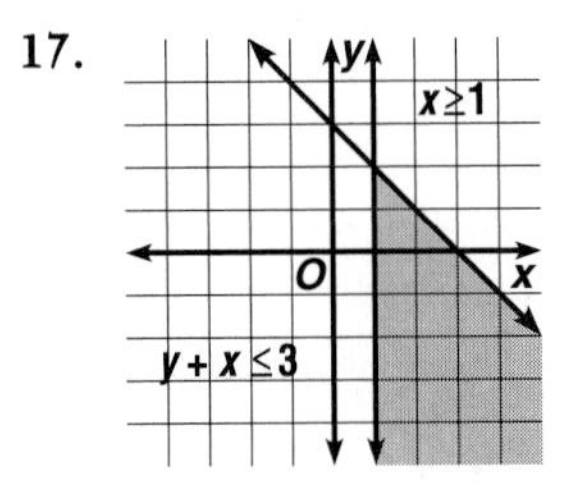

21.

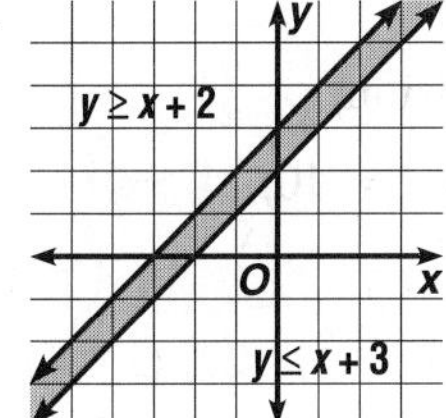

25.

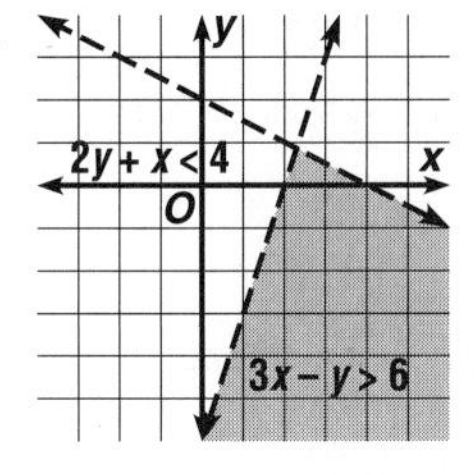

29.

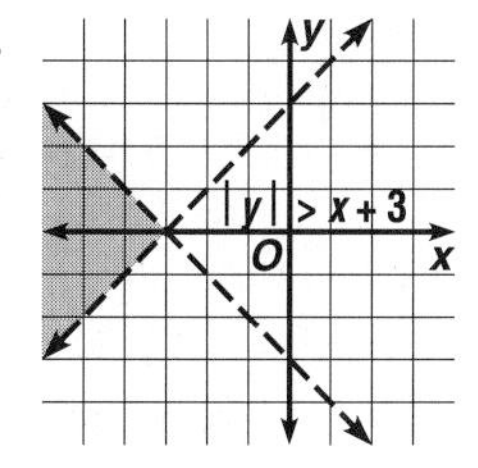

33. yes 35. $x > 1, y > -2$ 37. $y \geq -x, y < 3 - x$ 39. $y < 3 - x, y < 3 + x$ 41. $x \geq 0, y \geq 0, x + 2y \leq 6$

43.

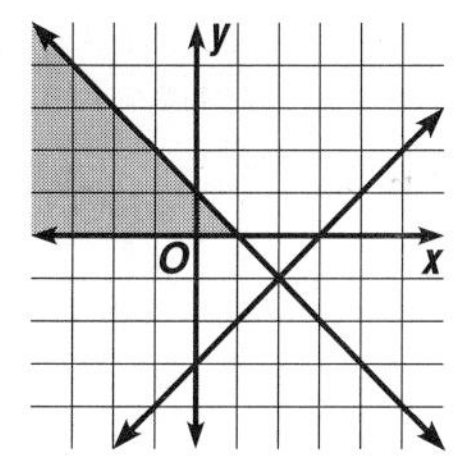

47.

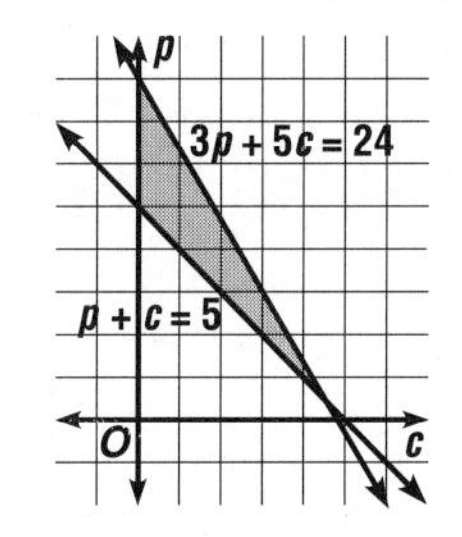

Answers will vary. Possible solutions: 1 lb of cashews, 6 lb of peanuts; 3 lb of cashews, 3 lb of peanuts; 4 lb of cashews, 1 lb of peanuts
49. Answers will vary. Possible solutions: 2 light, 8 dark; 6 light, 8 dark; 7 light, 4 dark **50.** 30%
51. $(4 + 3x)(-4 + 3x)$ **52.** $\frac{2x + 1}{(x + 1)^2}$ **53.** $\left\{\left(-2, -\frac{1}{4}\right), \left(-1, \frac{1}{4}\right), \left(0, \frac{3}{4}\right), \left(1, \frac{5}{4}\right), \left(3, \frac{9}{4}\right)\right\}$ **54.** $y = -\frac{4}{5}x - \frac{1}{5}$
55. 20 small and 10 large

Pages 468-470 Chapter 11 Summary and Review

1. (4, 2) **3.** (−2, −7) **5.** no solution
7. one, (2, −2) **9.** (3, −5) **11.** $\left(\frac{1}{2}, \frac{1}{2}\right)$ **13.** (2, −1)
15. (5, 1) **17.** $\left(\frac{14}{5}, \frac{4}{5}\right)$ **19.** (−4, 6)

21.

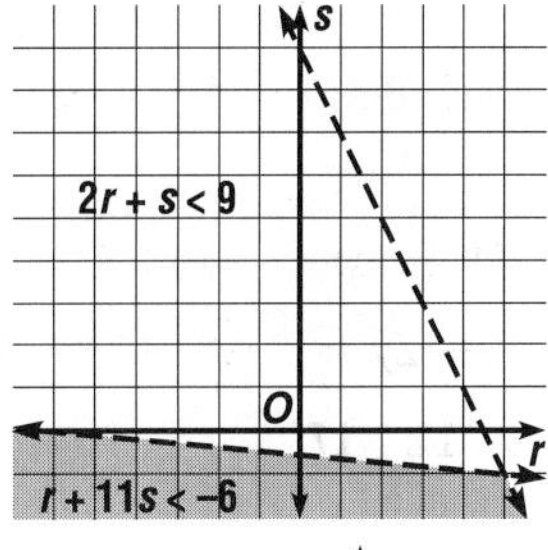

23.

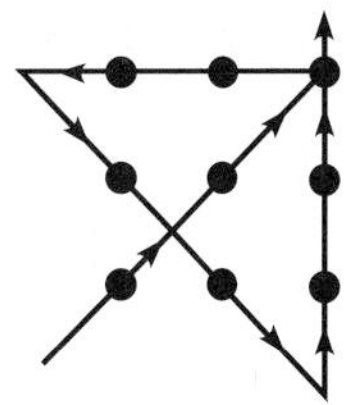

25. 16 cm by 9 cm **27.** \$15, \$9 **29.** $\frac{35}{2}$ mph, $\frac{5}{2}$ mph

CHAPTER 12 RADICAL EXPRESSIONS

Pages 475-476 Lesson 12-1

3a. Boston **b.** March **c.** 8.86 in., July, Mobile **d.** 3.26 in., October **5.** 10 **7.** 3, for the units digit of 7^3 **9.** 79 years old

Pages 479-481 Lesson 12-2

5. 144 **7.** 0.09 **9.** $\frac{121}{16}$ **11.** −9 **13.** 0.04
15. −12.21 **17.** 7.77 **19.** 16 **21.** $\frac{6}{14}$ or $\frac{3}{7}$
23. 0.05 **25.** 1.8 **27.** −10 **29.** 22 **31.** ±31
33. 45 **35.** −3.2 **37.** $\pm\frac{12}{39}$ **39.** −13.251
41. 23.052 **43.** 1.521 **45.** ±0.097 **47.** 11.36 in.
49. 3 **51.** 2 **55.** 18.4 meters per second **57.** 40
58. $5a^2(2c + 3)^2$ **59.** 35 mph

60.

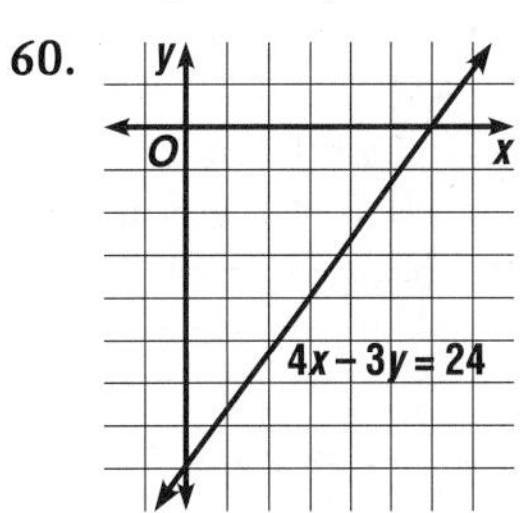

61. $y = -\frac{1}{8}x + \frac{15}{4}$

62.

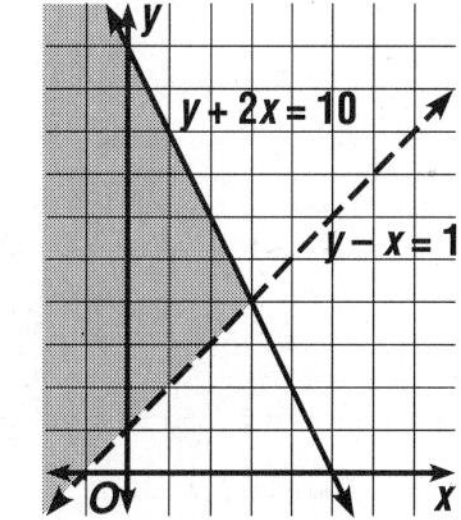

Pages 484–486 Lesson 12-3

5. false **7.** 10 **9.** 8 **11.** 16 **13.** 15 **15.** 2 **17.** 34 **19.** 20 **21.** 5 **23.** $\sqrt{45}$ or about 6.71 **25.** $\sqrt{28}$ or about 5.29 **27.** $\sqrt{21}$ or about 4.58 **29.** no **31.** yes **33.** no **35.** 14 cm **37.** $\sqrt{89}$ or about 9.43 m **41.** $\sqrt{31}$ or about 5.57 ft **43.** 56 m **44.** 7.25 **45.** $\frac{2x+5}{2x+1}$

46.

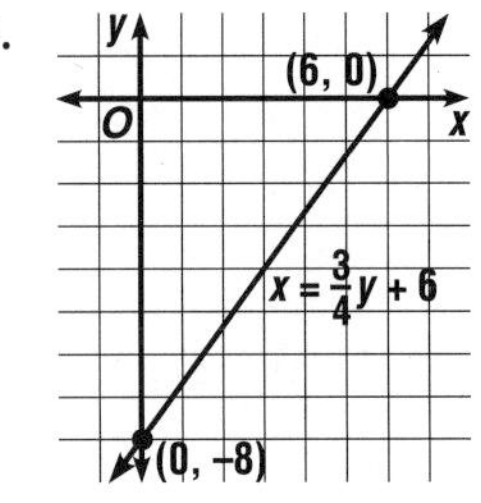

47. 75 gallons of 50% solution, 25 gallons of 30% solution **48.** ±66

Pages 489–491 Lesson 12-4

9. N, W, Z, Q **11.** Q **13.** N, W, Z, Q **15.** I **17.** 6.32 **19.** −9.43 **21.** Q **23.** I **25.** Q **27.** 4.47 **29.** −8.12 **31.** −5.57 **33.** 10.34 **35.** Q; 95 **37.** I; 86, 87 **39.** 362.24 ft^2 **41.** 7.9 ft by 23.7 ft **43.** 10.3 cm^2 **47.** No, since the rain will end at about 12:50 A.M. **48.** $9x^2 - 0.25$ **49.** $d = \pm\sqrt{\frac{GMn}{F}}$ **50.** $m = \frac{7}{3}$, $b = -\frac{10}{3}$ **51.** $\left(\frac{5}{2}, -\frac{1}{8}\right)$ **52.** No, she needs about 17.2 meters of wire.

Page 491 Mid-Chapter Review

1.

Temperature in Degrees Fahrenheit	Number of Days	Temperature in Degrees Fahrenheit	Number of Days
0	1	16	4
1	0	17	2
2	1	18	2
3	0	19	0
4	1	20	0
5	3	21	0
6	1	22	1
7	0	23	2
8	1	24	1
9	0	25	1
10	0	26	1
11	0	27	2
12	2	28	0
13	0	29	3
14	0	30	1
15	0		

2. 16°F **3.** 10 days **4.** −21 **5.** 3.3 **6.** ±0.29 **7.** $\frac{8}{9}$ **8.** 35 **9.** 1.2 **10.** 8 **11.** no

Pages 495–496 Lesson 12-5

5. $2\sqrt{5}$ **7.** $4\sqrt{3}$ **9.** 2 **11.** $\sqrt{5} + 7$; −44 **13.** $2\sqrt{8} - 3\sqrt{5}$; −13 **15.** $\frac{\sqrt{2}}{\sqrt{2}}$ **17.** $\frac{4+\sqrt{3}}{4+\sqrt{3}}$ **19.** $3\sqrt{5}$ **21.** $6\sqrt{2}$ **23.** $2\sqrt{70}$ **25.** $10\sqrt{10}$ **27.** $\frac{\sqrt{2}}{2}$ **29.** $\frac{\sqrt{22}}{8}$ **31.** 10 **33.** $84\sqrt{5}$ **35.** $\frac{\sqrt{11}}{11}$ **37.** $2b^2\sqrt{10}$ **39.** $4|x|y\sqrt{5y}$ **41.** $7x^2y^3\sqrt{3xy}$ **43.** $\frac{3\sqrt{3}}{|r|}$ **45.** $\frac{|x|\sqrt{3xy}}{2|y^3|}$ **47.** $\frac{5\sqrt{5}+45}{-38}$ **49.** $\frac{54b - 9b\sqrt{b}}{36 - b}$ **51.** $\frac{3\sqrt{35} - 5\sqrt{21}}{-15}$ **53.** $\frac{a - 2\sqrt{ab} + b}{a - b}$ **57a.** $2s\sqrt{10}$ miles **b.** 31.62 miles **58.** $(a+b)^2 = a^2b^2$ **59.** $(2r + s)(6r - 11s)$ **60.** yes; no **61.** (20, −5) **62.** (2, 3) **63.** about 18.26 amperes

Pages 499–500 Lesson 12-6

5. $4\sqrt{2}, 7\sqrt{2}$ **7.** $-3\sqrt{7}, 2\sqrt{28}$ **9.** none **11.** $11\sqrt{6}$ **13.** in simplest form **15.** $21\sqrt{2x}$ **17.** $9\sqrt{3}$, 15.59 **19.** $-10\sqrt{5}$, −22.36 **21.** $13\sqrt{3} + \sqrt{2}$, 23.93 **23.** $4\sqrt{3}$, 6.93 **25.** $-2\sqrt{2}$, −2.83 **27.** $8\sqrt{2} + 6\sqrt{3}$, 21.71 **29.** $\frac{8\sqrt{7}}{7}$, 3.02 **31.** $4\sqrt{3} - 3\sqrt{5}$, 0.22 **33.** $4\sqrt{7} + 2\sqrt{14}$, $7 + 7\sqrt{7}$ **35.** $2\sqrt{2} + 8\sqrt{3}$, $5 - \sqrt{6}$ **37.** $19\sqrt{5}$ **39.** $15\sqrt{2} + 11\sqrt{5}$ **43.** $6.\bar{2}$ or $6\frac{2}{9}$ feet **44.** $n \geq \frac{23}{3}$ **45.** $\frac{k^2 - 2k + 1}{3k + 15}$ **46.** \$0.22 per mile **47.** $y = -4x - 3$ **48.** $18\sqrt{6}$

Pages 503–504 Lesson 12-7

5. $x = 25$ **7.** $121 = 2a - 5$ **9.** no real solution **11.** 5 **13.** 39 **15.** 10 **17.** $4\sqrt{2}$ **19.** −112 **21.** no real solution **23.** $\frac{121}{9}$ **25.** 3 **27.** 144 **29.** $\frac{3}{25}$ **31.** 11 **33.** 7 **35.** 6 **37.** −6, −4 or 4, 6 **39.** 0 **41.** no real solution **43.** $\left(\frac{1}{4}, \frac{25}{36}\right)$ **47.** 1.6 meters per second squared **49.** 6 **50.** 2.5 feet **51.** $\left\{(-4, 7), \left(-2, \frac{23}{5}\right), \left(0, \frac{11}{5}\right), (1, 1), \left(5, -\frac{19}{5}\right)\right\}$ **52.** $\left(-\frac{1}{2}, 4\right)$ **53.** (6, −2) **54.** $26\sqrt{3}$ or about 45.03

Pages 508–509 Lesson 12-8

5. 5 **7.** 13 **9.** $3\sqrt{2}$ or 4.24 **11.** $4\sqrt{5}$ or 8.94 **13.** $2\sqrt{13}$ or 7.21 **15.** $\frac{\sqrt{74}}{7}$ or 1.23 **17.** $2\sqrt{14}$ or 7.48 **19.** 7 or 1 **21.** −2 or −12

23. The distance between (0, 0) and (7, 4) is $\sqrt{65}$ units. The distance between (7, 0) and (1, 4) is $\sqrt{52}$ units. Thus, the trapezoid is not isosceles.
25. $22 + 4\sqrt{10}$ or about 34.65 units **27.** $8\sqrt{3}$
31. 2290 miles **33.** $\frac{2m + 1}{m - 2}$ **34.** 13
35. $4x - 3y = 23$ **36.** 425 mph **37.** 1

Pages 510–512 Chapter 12 Summary and Review

1. 13 **3.** −0.17 **5.** −7.85 **7.** ±4.8 **9.** 34
11. $\sqrt{125}$ or about 11.18 **13.** irrational
15. rational **17.** $6\sqrt{3}$ **19.** $2\sqrt{6} - 4\sqrt{3}$ **21.** $\frac{2\sqrt{35}}{7}$
23. $4x^2\sqrt{6}$ **25.** $\frac{2\sqrt{15}}{|y|}$ **27.** $5\sqrt{13} + 5\sqrt{15}$ or about 37.39 **29.** $-6\sqrt{2} - 12\sqrt{7}$ or about −40.23
31. 12 **33.** 3 **35.** 17 **37.** 18 or −8
39. 1985, 1990 **41.** 6.1 cm **43.** 24.85 units
45. No; it should skid about 201.7 feet.

CHAPTER 13 QUADRATICS

Pages 521–522 Lesson 13-1

5. up **7.** down **9.** $x = -\frac{1}{2}$ **11.** $x = -1$ **13.** $\left(\frac{3}{2}, \frac{9}{4}\right)$
15. $x = \frac{5}{2}$; $\left(\frac{5}{2}, \frac{49}{4}\right)$ **17.** $x = -1$; $(-1, -1)$
19. $x = -4$; $(-4, 32)$
21. $x = 2$; $(2, -9)$

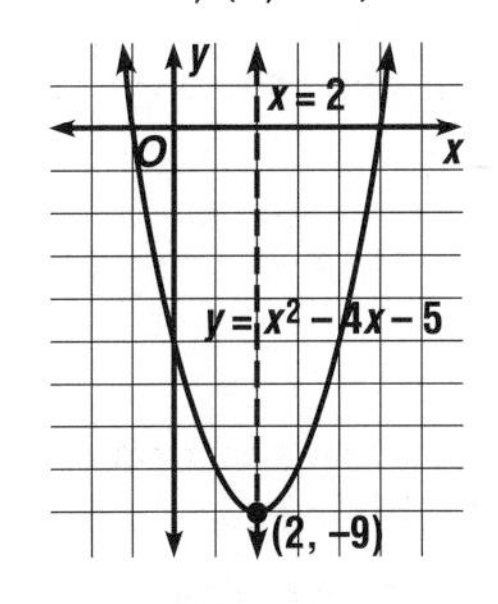

23. $x = 3$; $(3, 14)$
25. $x = 0$; $(0, -3)$

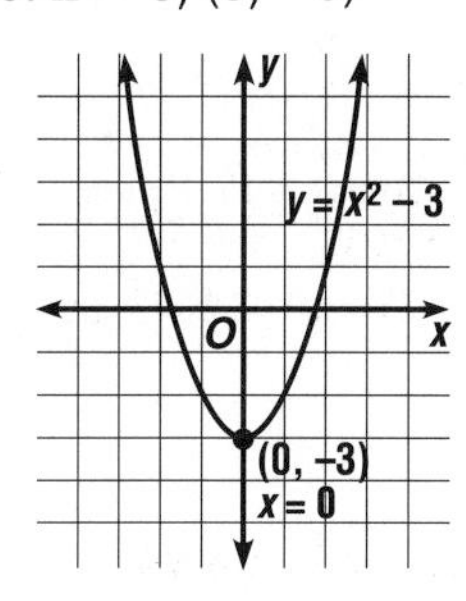

27. $x = 0$; $(0, 3)$
29. $x = 8$; $\left(8, -\frac{49}{4}\right)$

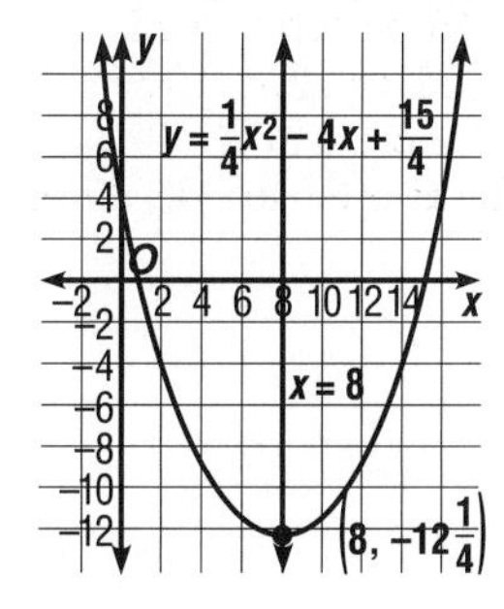

31. $x = 2$; $(2, 1)$ **33.** $(-1, 1)$ **35.** $(4, 0)$ **37.** a
39. $x = -\frac{3}{2}$

41.

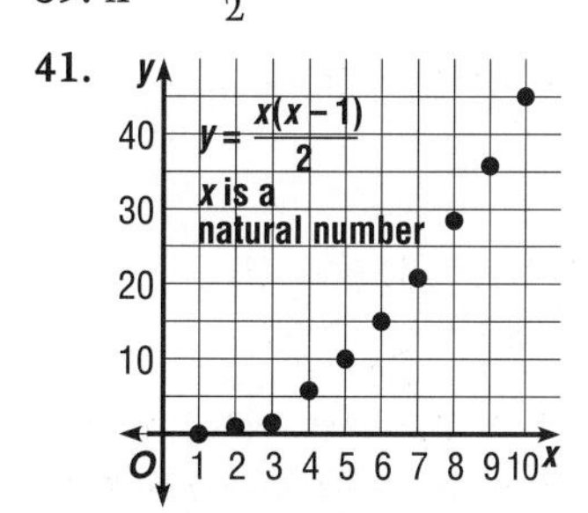

43. 196 ft **44.** 9 weeks **46.** $700

47.

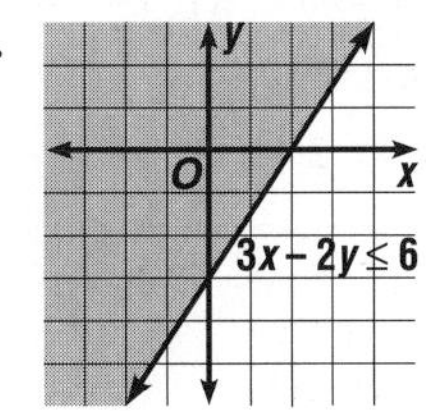

48. $-5x + 6y = -2$ **49.** (1, 3) **50.** 5 or −3

Pages 525–526 Lesson 13-2

5. 0, 2 **7.** 2 **9.** −3, −4 **11.** 3, 7 **13.** no real roots **15.** 4 **17.** $-5 < x < -4$, $-2 < x < -1$
19. $2 < x < 3$, $-4 < x < -3$ **21.** −13, −7 or 7, 13
23. b

25.

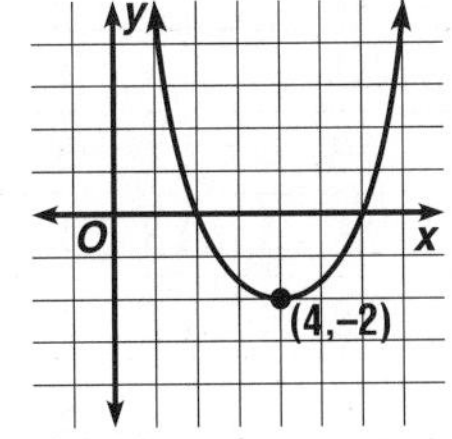

33. $1 < y < 2$, $2 < y < 3$ **35.** −2, −6 **37.** between 4 meters and 5 meters **39.** $2b + 3$ **40.** $y = x^2 - 4$
41. $\frac{2}{3}$; $-\frac{13}{3}$; $\frac{13}{2}$ **42.** $12\sqrt{5}$ **43.** 1800 m²

Pages 528–529 Lesson 13-3

3. 10 dogs **5.** 19 numbers **7.** 1, 2, 4, 5, 8, 11 **9.** 19 handshakes **11.** his 13th season

Pages 534–535 Lesson 13-4

5. no **7.** no **9.** yes **11.** 16 **13.** $\frac{49}{4}$ **15.** 1, 7 **17.** $\frac{49}{4}$ **19.** 9 **21.** −3, −4 **23.** $\frac{5}{2}$ **25.** $3 \pm \sqrt{5}$ **27.** $5 \pm 4\sqrt{3}$ **29.** 3, $\frac{1}{2}$ **31.** $\frac{2}{3}$, −1 **33.** $\frac{5 \pm \sqrt{17}}{4}$ **35.** 6.5 yd by 12 yd **37.** 60 or −60 **39.** $\frac{9}{4}$ **41.** $\frac{-b \pm \sqrt{b^2 - 4c}}{2}$ **45.** 9 in. by 6 in. **47.** 2300 cards **48.** 52 **49.** $30\sqrt{2} + \sqrt{5}$ **50.** $0 < x < 1$, $2 < x < 3$ **51.** 63

Pages 539–540 Lesson 13-5

5. 1, 7, 6 **7.** 2, 7, 3 **9.** 4, 8, 0 **11.** 49 **13.** 144 **15.** 2500 **17.** $-2 \pm \sqrt{2}$; −0.56, −3.41 **19.** $\frac{-4 \pm \sqrt{10}}{3}$; −0.28, −2.39 **21.** −1, −9 **23.** 5, −5 **25.** $\frac{5}{2}$, −3 **27.** $-\frac{1}{2}$, −3 **29.** $\frac{13 \pm 3\sqrt{33}}{2}$; 15.12, −2.12 **31.** $\frac{2}{3}$, $-\frac{1}{4}$ **33.** $\frac{3}{7}$, $-\frac{2}{3}$ **35.** $-3 \pm 2\sqrt{3}$; 0.46, −6.46 **37.** $\frac{4 \pm \sqrt{29}}{2}$; 4.69, −0.69 **39.** −2.1, 0.1 **41.** 0.2, 1.4 **43.** −0.5, −2.9 **45.** $x^2 + 3x + 1 = 0$ **47.** $3x^2 - 8x + 3 = 0$ **51.** 8.1% **52.** $\frac{5}{4}a^2 + 5ab$ **53.** $\frac{a^3 - a^2b + a^2 + ab}{(a + b)(a - b)^2}$

54.

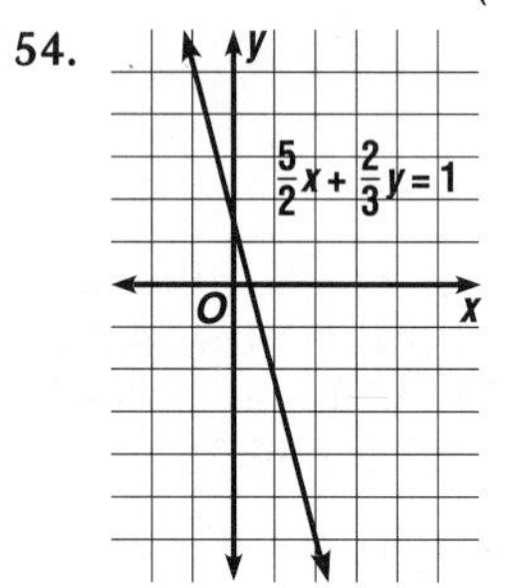

55. $3x + y = 11$ **56.** 4 seconds **57.** $\frac{3 \pm \sqrt{19}}{2}$

Page 540 Mid-Chapter Review

1. $x = \frac{1}{2}$; $\left(\frac{1}{2}, -\frac{49}{4}\right)$ **2.** $x = 0$; (0, −9) **3.** $x = -1$; (−1, 8) **4.** no real roots **5.** $-1 < x < 0$, $2 < x < 3$ **6.** 6, −1 **7.** $3 \pm \sqrt{2}$ **8.** $\frac{7}{2}$, −3 **9.** −5, −3 **10.** $\frac{-5 \pm \sqrt{13}}{2}$ **11.** 13.5 days

Pages 544–545 Lesson 13-6

5. 25; 2 real roots **7.** 0; 1 real root **9.** −11; no real roots **11.** 5; 2 real roots **13.** 0; 1 real root **15.** −20; no real roots **17.** $2 \pm \sqrt{3}$; 3.73, 0.27 **19.** 2, −6 **21.** $-\frac{1}{2}$, −1 **23.** $\frac{5 \pm \sqrt{193}}{12}$; 1.57, −0.74 **25.** $\frac{1}{3}$ **27.** 1, $\frac{2}{3}$ **29.** $\frac{7 \pm \sqrt{69}}{10}$; 1.53, −0.13 **31.** $\frac{1 \pm \sqrt{133}}{22}$; 0.57, −0.48 **33.** 0 **35.** 2 **37.** 1 **39.** 12, −12 **41.** $k < -\frac{25}{4}$ **43a.** 72.5 seconds **b.** yes **45a.** no real roots **b.** 0.5 **c.** 0.7142857, −1 **d.** 10, 1 **e.** no real roots **f.** 6, 0 **46.** $18.50 **47.** yes **48.** $c = 34h + 15$, where h = the number of hours for the repair job and c = the total charge

49.

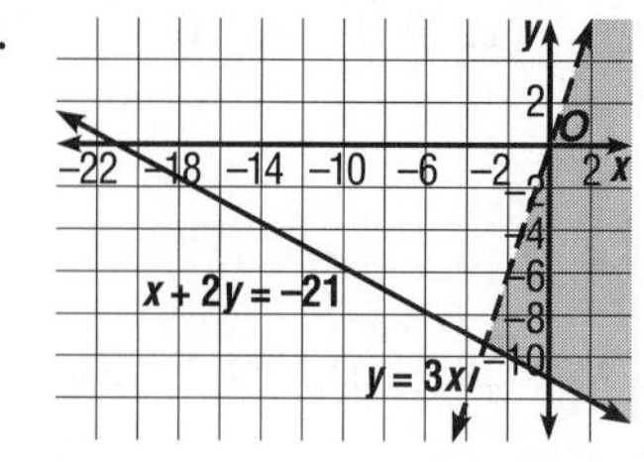

50. 7.7 **51.** 3.4 meters by 0.6 meters

Pages 548–549 Lesson 13-7

5. factoring or quadratic formula **7.** quadratic formula **9.** quadratic formula or completing the square **11.** $-5 \pm 3\sqrt{3}$; 0.20; −10.20 **13.** $-\frac{2}{3}$, 3 **15.** $\frac{7 \pm \sqrt{85}}{6}$; 2.70, −0.37 **17.** $\frac{-1 \pm \sqrt{41}}{4}$; 1.35, −1.85 **19.** $\frac{5 \pm \sqrt{13}}{6}$; 1.43, 0.23 **21.** 1.5, −0.4 **23.** $2 \pm \sqrt{6}$; 4.45, −0.45 **25.** 5, 7 **27.** 21 in. by 42 in. **29a.** 1 second, 5.25 seconds **b.** 6.25 seconds **31.** 80 ft by 100 ft **34.** 11 **35.** 2 hours **36.** (10, 4) **37.** no **38.** 56; 2 real roots

Pages 552–553 Lesson 13-8

5. −2, −1 **7.** $\frac{13}{6}$, $-\frac{5}{2}$ **9.** no **11.** yes **13.** no **15.** $x^2 + 5x - 6 = 0$ **17.** $x^2 + 0.3x - 0.18 = 0$ **19.** 5, −24 **21.** $-\frac{13}{2}$, $-\frac{9}{4}$ **23.** $\frac{1}{3}$, $-\frac{1}{12}$ **25.** $\frac{1}{15}$, $-\frac{1}{300}$ **27.** $x^2 - 11x + 28 = 0$ **29.** $x^2 + 9x - 10 = 0$ **31.** $2x^2 - 9x + 10 = 0$ **33.** $6x^2 + 5x - 6 = 0$ **35.** $x^2 - 2 = 0$ **37.** $x^2 - 4x + 1 = 0$ **39.** $x^2 - (8 + \sqrt{3} + \sqrt{5})x + (15 + 5\sqrt{5} + 3\sqrt{3} + \sqrt{15}) = 0$ **41.** 4 **43.** 35 **45.** $x^2 - (q + r)x + qr = 0$ **49.** yes **50.** $\frac{2}{7}a^2 - \frac{5}{4}a + 1$ **51.** $\frac{1}{n - 6}$

52.

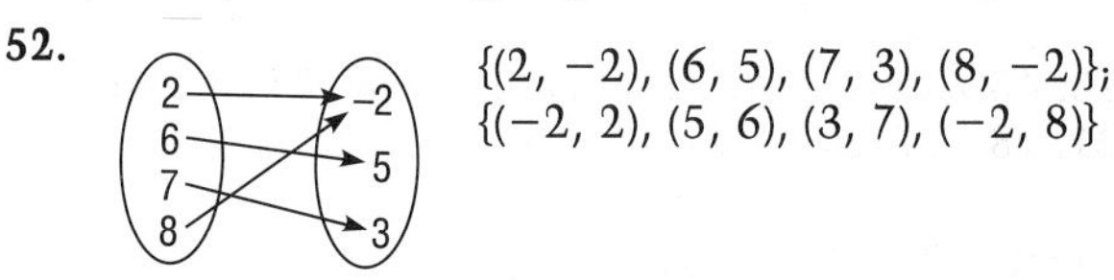

53. 1 mph **54.** 7 **55.** 2.1 seconds

Pages 554–556 Chapter 13 Summary and Review

1. $x = \frac{3}{2}$; $\left(\frac{3}{2}, -\frac{25}{4}\right)$ **3.** $x = \frac{9}{4}$; $\left(\frac{9}{4}, \frac{9}{8}\right)$

5.

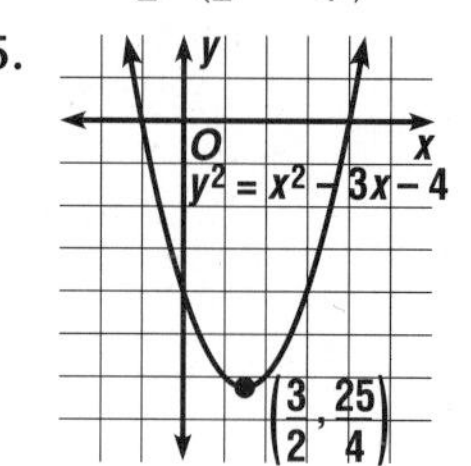

9. −3, 4 **11.** $-5 < x < -4$, $0 < x < 1$ **13.** 16 **15.** $8 \pm 4\sqrt{2}$; 13.66, 2.34 **17.** $-\frac{3}{2}, -\frac{5}{2}$ **19.** $\frac{-9 \pm \sqrt{21}}{10}$; −0.44, −1.36 **21.** $\frac{3}{4}, \frac{2}{3}$ **23.** 2 real roots **25.** 1 real root **27.** 6, 15 **29.** −8, −14 **31.** $\frac{1}{2}$, −3 **33.** $2x^2 + 5x - 12 = 0$ **35.** 90 **37.** \$22.50

CHAPTER 14 STATISTICS AND PROBABILITY

Pages 562–564 Lesson 14-1

5. from 10 to 50, intervals of 10 **7.** from 1000 to 11,000, intervals of 1000 **9a.** 50 **b.** 22 **c.** 30 students **d.** 9 students **e.** 26

11a.

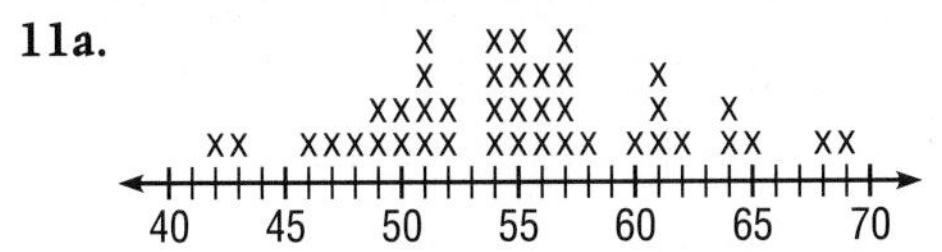

b. 42 **c.** yes; 54 years old to 57 years old **13a.** Mr. Jebson's **b.** No; the number of hours of television watched for Ms. Lee's students is more spread out while the number of hours watched for Mr. Jebson's students is more clustered between 15 and 30 hours. **15.** 91°F, 92°F, 98°F

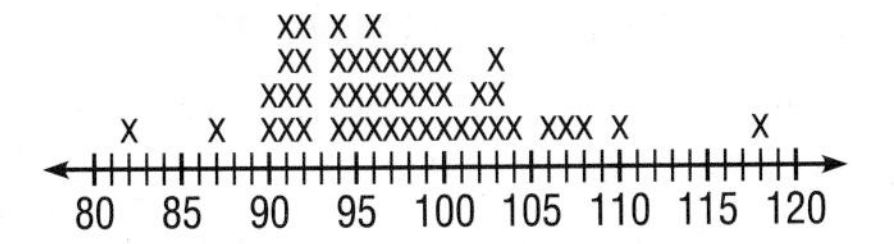

17. 15

18. 43.2 mph **19.** $3a^2 + 6a + 1$ **20.** 8150 covers **21.** 18 **22.** $3x^2 + 2x - 21 = 0$

Pages 567–569 Lesson 14-2

5. 1, 2, 3, 4, 5 **7.** 19, 20, 21, 22, 23, 24, 25, 26 **9.** 46, 45 **11.** 43, 43 **13a.** 85°F **b.** 49°F **c.** 11 days **d.** 77°F

15a.

Stem	Leaf
1	6 7 7 8 8 8 8 8 9 9 9 9 9 9 9
2	0 0 2 3 4 4 5 5 6 7 7
3	0 2 3 3 5 6
4	5 8
5	5

4\|8 = 48 years old

b. 35 people **c.** 39 years **d.** 19 years old **e.** teens **f.** 12 students

17a.

Stem	Leaf
13	9 9 9
14	0 3 3 4 5 7 9 9
15	1 1 3 3 6 7 8 9 9 9
16	1 1 2 2 2 3 4 8
17	1
18	6

15\|1 = 151 mph

b. 150-159 mph

19a.

Stem	Leaf	Stem	Leaf
6	5 8 9	14	5
7	3 4 4 6 7	15	7
8	0 2 2 3 5	17	4
	5 9	18	2
9	4 8	21	2
10	4	26	2
11	2	29	0
13	6		

9\|4 = \$94,000

b. \$225,000 **c.** \$80,000-\$89,000

22. $b + 8$ **22.** $\frac{x(2x + 3)}{2(x + 2)}$ **24.** 280 miles **25.** $\sqrt{185}$ or about 13.6 units **26.** $2 \pm \sqrt{7}$

27. 63 in.

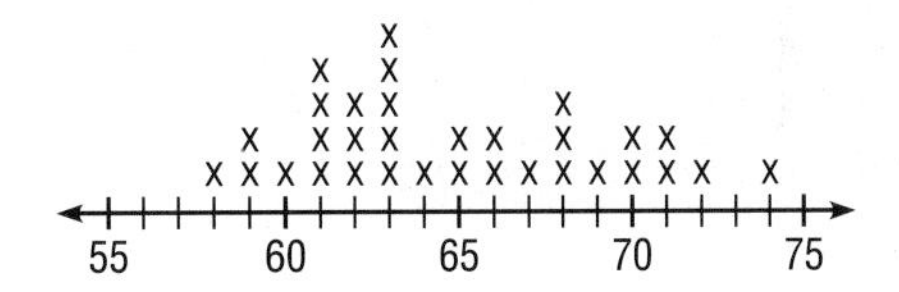

Pages 572–574 Lesson 14-3

5. 7; 6; no mode **7.** 8; 7; 7 **9.** 17.5; 17.5; 12 and 23 **11.** 4; 2 **13.** 94; 82 **15.** 218; 219 **19.** 140.5 yards **21.** 8.3; 6.5; 4 **23.** 40.17; 40.15; 40.2, 43.4 **25.** 4.45; 4; 3 and 5 **27.** \$5.14; \$4.85; \$4.75

29.

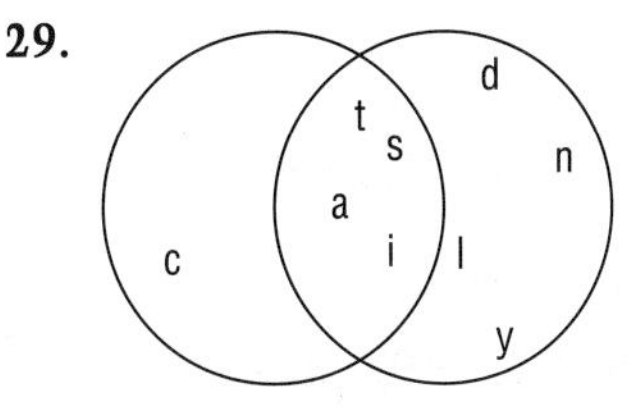

30. $(8x - 7y)(3x - 5y)$ **31.** $3x + y = 16$ **32a.** 3 seconds **b.** 269 feet

33.

Stem	Leaf (rounded)
36	5
37	
38	2 3
39	2 4
40	1 2 2 3
41	
42	
43	1 2 4 4 4

39\|2 = 39.15 to 39.24 seconds

Pages 577–578 Lesson 14-4

5. 11 **7.** 34 **9.** 77 **11.** 40; 45; 34 **13.** 65; 85; 45 **15.** 37; 73, 77, 62; 15 **17.** 49; 208, 212, 204.5; 7.5 **19.** 4.6; 8.7, 10.05, 7.9; 2.15 **23a.** $0.26, $0.09 **b.** $0.32 **25.** $\{x \mid -3 < x < 3\}$ **26.** $\{(3, -3), (-2, 2), (1, -1), (0, 0), (-1, 1)\}$ **27.** $\sqrt{35}$ or about 5.92 in. **28.** $\frac{3}{5}, -\frac{2}{7}$ **29.** $363; $364

Pages 581–582 Lesson 14-5

5. 25% **7.** 85 **9.** 90 **11.** 75% **13.** X **15.** Y

17.

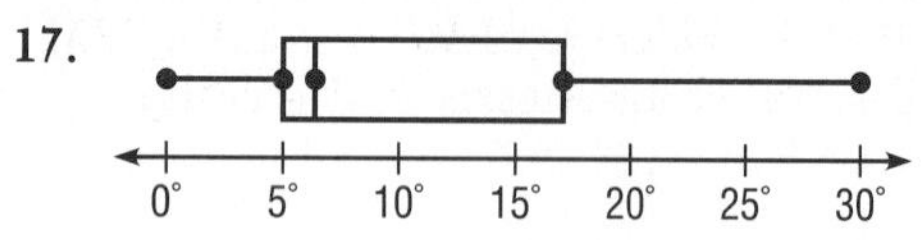

21.

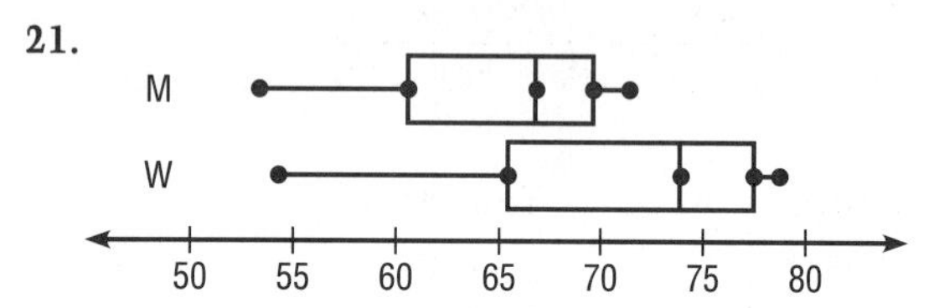

23. $\frac{34}{9}$

24.

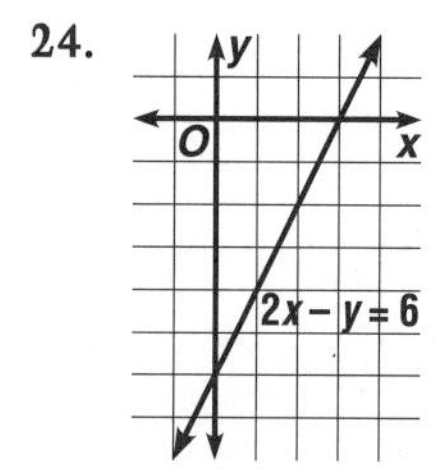

25. (10, −6) **26.** $\frac{7 + \sqrt{3}}{46}$ **27.** 61, 2 real roots **28.** 19.2; 33.0, 34.6, 36.8; 3.8; 44.8, 50.4

Pages 585–587 Lesson 14-6

5. positive **7.** negative or positive **9.** none **11.** positive **13.** none

15a.

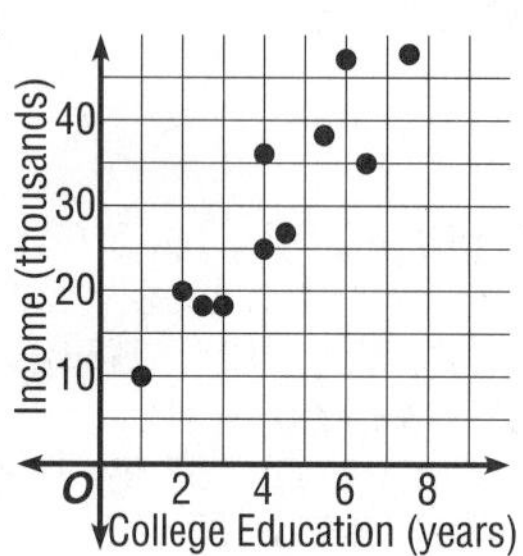

b. the more years of college education, the greater the income **17b.** positive **c.** Yes; you could estimate the number of assists by using the points in the scatter plot. **d.** Yes; some players have more opportunities to score points because of the position they play for the team. **e.** Yes, since more playing time provides a player with more opportunities to score points.

19a.

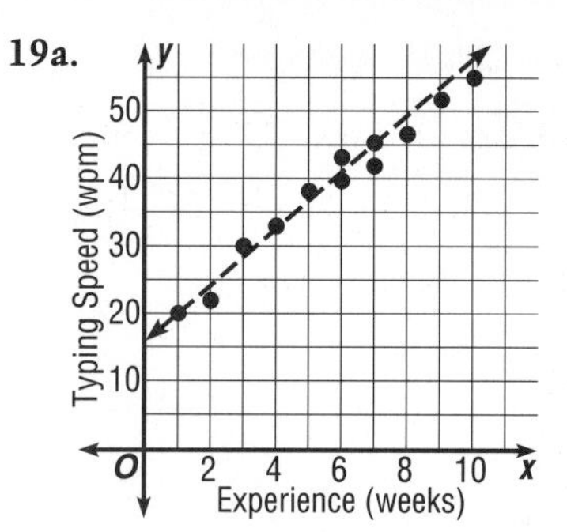

b. $y = \frac{13x}{3} + \frac{43}{3}$ **c.** about 66 words per minute **21.** $6x^2 - 25$ **22.** 20 mph

23.

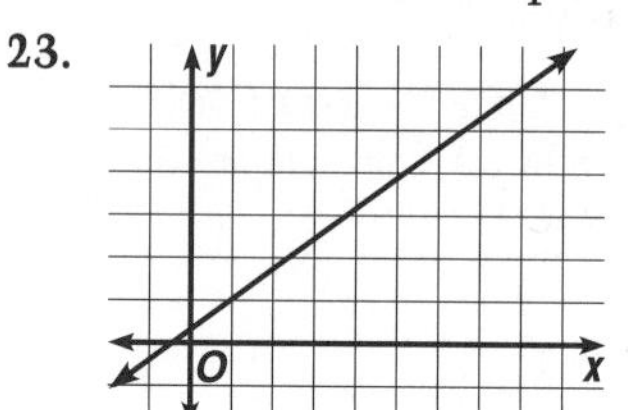

24. 3.2 cm by 4.5 cm

25.

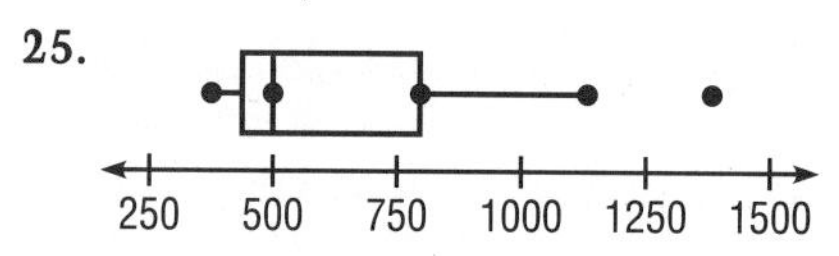

Page 587 Mid-Chapter Review

1. 18.6 in., summer, New Orleans; 0.3 in., summer, San Francisco

2.

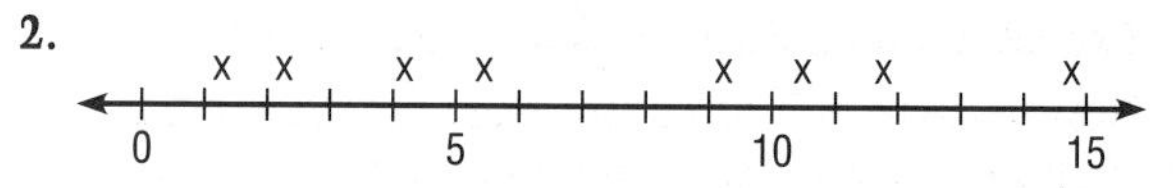

3.

Stem	Leaf
1	8
2	6
4	8
6	5
7	1
10	8
12	1 5

6|5 = 6.5 inches

4. 8.45 in.; 10.15 in.; no mode **5.** 18.3 in.; 4.15 in., 9.9 in., 11.2 in.; 7.05 in.; no outliers

6.

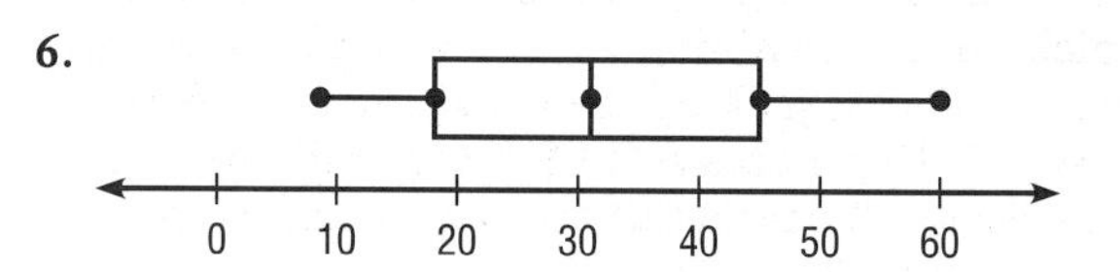

7.

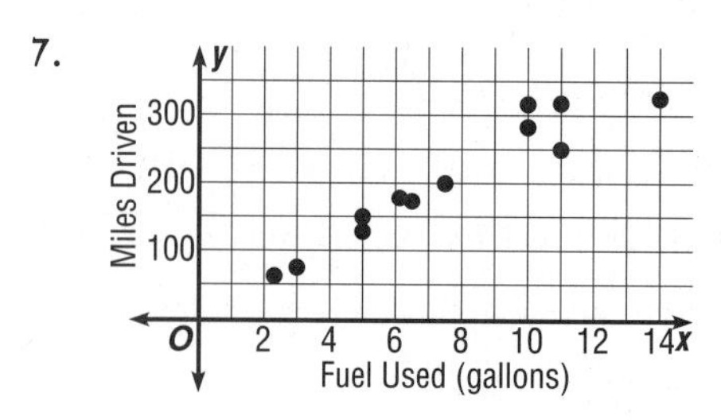

Pages 591–593 Lesson 14-7

7. 1 **9.** $\frac{1}{7}$ **11.** yes **13.** no **15.** yes **17.** $\frac{1}{6}$ **19.** $\frac{1}{2}$ **21.** 1:2 **23.** 5:1 **25.** $\frac{1}{2}$ **27.** 1:3 **29.** $\frac{4}{13}$ **31.** 4:3 **33.** $\frac{14}{23}$ **35.** 0.7 **39.** 1:24 **41.** 11 days

42.

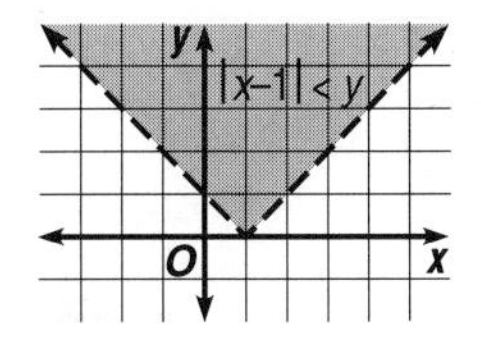

43. 240 feet **44.** no real roots

45a.

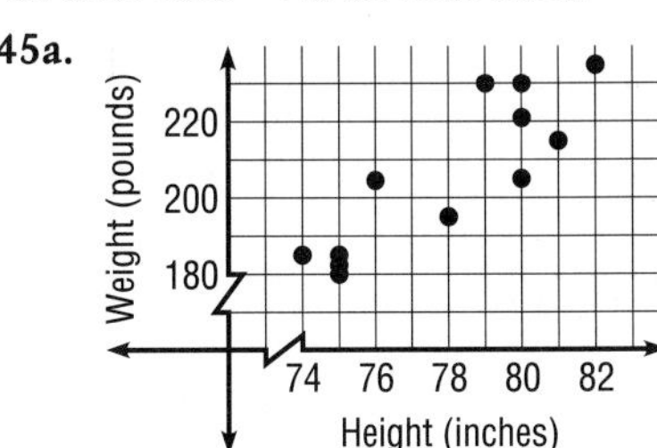

b. the taller the player, the greater the weight

Pages 596–597 Lesson 14-8

5b. Expected probability is $\frac{2}{36}$ or about 0.056.
c. Expected probability is $\frac{6}{36}$ or about 0.167.
d. Expected probability is $\frac{6}{36}$ or about 0.167.
e. Expected probability is $\frac{21}{36}$ or about 0.583.
f. Sum that should occur most often is 7. **g.** Sums that should occur least often are 2 and 12.
7a. $\frac{1}{42} \approx 0.02$ **b.** $\left(\frac{1}{42}\right)^2 \approx 0.0006$ **c.** 1.1×10^{-11}
9. 5:3 **12.** $y = 1 - \frac{1}{2}x$ **13.** 6 **14.** 24
15. $56\sqrt{2}$ or about 79.20 **16.** 10:3

Page 599 Lesson 14-9

3. 1027 pages **5.** Take the cat across first. Then go back, pick up either the dog or the bird, and take it across the bridge. Now, go back with the cat, leave it, pick up the remaining animal, and take it across. Finally, go back, pick up the cat again, and take it across. **7.** 10 peaches at \$0.32 each and 2 pears at \$0.29 each or 2 peaches at \$0.34 each and 10 pears at \$0.31 each **9.** 12 hens

Pages 602–603 Lesson 14-10

3a. 0.5 **b.** 0.25 **c.** 0.25 **d.** 0.125 **5.** $\frac{15}{16} = 0.9375$ **7a.** 0.125 **b.** 0.375 **c.** 0.875 **9a.** $\frac{3}{10} = 0.3$ **b.** $\frac{9}{10} = 0.9$ **c.** man and woman **11.** $\frac{1}{4} = 0.25$ **13.** \$135,575 **14.** $\frac{7}{5 - p}$ **15.** 4.5 ft

16. P(heads) should be approximately $\frac{1}{2}$ or 0.50. If not, then the results do not necessarily imply that the coin is not fair because of the low number of trials. More trials must be done in order to determine whether or not the coin is fair.
17. 4950 diagonals

Pages 604–606 Chapter 14 Summary and Review

1. Philip Morris **3.** Percents are rounded to the nearest tenth. Exxon: 4.0%, Ford Motor: 3.9%, General Electric: 7.1%, General Motors: 3.3%, IBM: 6.0%, Mobil: 7.4%, Philip Morris: 7.4%; Philip Morris; no

5.

120 130 140 150 160 170

7.

Rounded	Stem	Truncated
8 6 5 1	3	1 4 6 8
4 3	4	2 4 9
8 7 6 0	5	6 6 8
8	6	7

rounded data: 4|3 = 4250-4349
truncated data: 4|3 = 4300-4399

9. 12.92; 10; 10 **11.** 34.7; 254.8 in.

13.

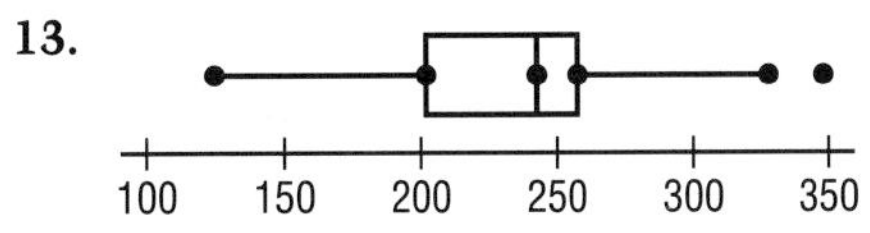

15. $\frac{11}{20} = 0.55$ **17.** $\frac{9}{13}$ or about 0.692 **19.** Expected probability is $\frac{1}{2}$ or 0.5. **21.** $\frac{7}{15} = 0.4\overline{6}$ **23.** 207 games

CHAPTER 15 TRIGONOMETRY

Pages 612–614 Lesson 15-1

5. 77° **7.** 66° **9.** 14° **11.** $(50 - 2x)°$ **13.** 50° **15.** 93° **17.** 148° **19.** $(180 - y)°$ **21.** $(200 - x)°$ **23.** 3°, 93° **25.** none, 20° **27.** 22°, 112° **29.** none, 6° **31.** $(90 - a)°$, $(180 - a)°$ **33.** $(60 - x)°$, $(150 - x)°$ **35.** $(90 - 5x)°$, $(180 - x)°$ **37.** $(y - 90)°$, $y°$ **39.** 70° **41.** 60° **43.** 105° **45.** 138° **47.** $(160 - 2x)°$ **49.** $(179 - 3m)°$ **51.** 58°, 60°, 62° **53.** 90° **55.** 15°, 75° **57.** 75° **59.** 60° **61.** 43°, 48°, 89° **65.** 120° **67.** $R = am - z$ **68.** $-35a^5b^2c$ **69.** $x \leq 2, y \geq 2$ **70.** $\sqrt{29}$ or about 5.39 units **71.** $x = \frac{3}{2}, \left(\frac{3}{2}, \frac{137}{4}\right)$ **72.** −12, −28 **73.** $\frac{1}{3} = 0.\overline{3}$

Page 616 Lesson 15-2

5. 660 in^2 **7.** Hong: 7, Dora: 5 **9.** \$62

Pages 619–621 Lesson 15-3

5. 8 cm **7.** 4.5 mi **9.** $1\frac{11}{16}$ in. **11.** 2.315 cm **13.** 2 cm, 4 cm **15.** 1 mi, 2 mi **17.** 12.4 cm **19.** 9 mi **21.** 26 m **23.** $15\frac{1}{2}$ in. **25.** $5\sqrt{3}$ or about 8.660 ft **27.** 10 units **29.** $2.5\sqrt{3}$ or about 4.330 yd **31.** 2.375 mm, $2.375\sqrt{3}$ or about 4.114 mm **33.** 16 cm, $8\sqrt{3}$ or about 13.856 cm **35.** $6\frac{1}{2}$ in., $\frac{13}{4}\sqrt{3}$ or about 5.629 in. **37.** 7 cm, 3.5 cm **39.** $48 + 16\sqrt{3}$ or about 75.71 yd **43.** 4 ft **44.** −7.54 **45.** $(3y - 1)(y + 2)$ **46.** no solution **47.** $-\frac{17}{10}$ **48.** $\frac{1 \pm \sqrt{33}}{4}$ **49.** 34 **50.** 73°

Page 621 Mid-Chapter Review

1. 5°, 95° **2.** none, 53° **3.** 25°, 115° **4.** none, 72° **5.** $(90 - x)°$, $(180 - x)°$ **6.** $(85 - 3x)°$, $(175 - 3x)°$ **7.** 3 m, 6 m **8.** 7 yd, 14 yd **9.** 9 mm, 18 mm **10.** $\frac{1}{3}$ in., $\frac{2}{3}$ in. **11.** The figure below shows the view from directly above the pins. Move the three pins as indicated in this figure.

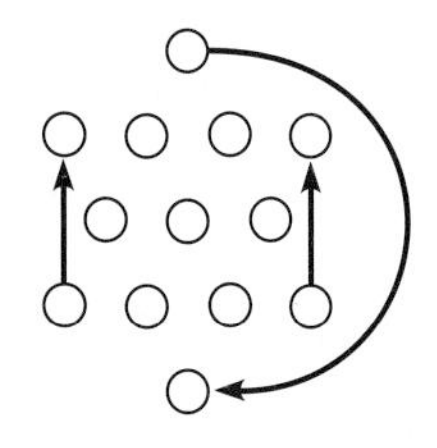

Pages 624–625 Lesson 15-4

3. $\angle B$ and $\angle R$, $\angle I$ and $\angle E$, $\angle G$ and $\angle D$ **5.** $\frac{BI}{RE} = \frac{IG}{ED}$, $\frac{BI}{RE} = \frac{BG}{RD}$, $\frac{IG}{ED} = \frac{BG}{BD}$ **7.** $\triangle WQC$ **9.** yes **11.** $b = \frac{25}{7}$, $c = \frac{30}{7}$ **13.** $a \approx 2.78$, $c \approx 4.23$ **15.** $d = \frac{51}{5}$, $e = 9$ **17.** $c = \frac{7}{2}$, $d = \frac{17}{8}$ **19.** $b = 16.2$, $d = 6.3$ **23.** 105 m **25.** $\frac{11}{13}$ **26.** $\frac{6 - 6a + 6b}{a^2 - 2ab - b^2}$ **27.** $\left(\frac{11}{3}, \frac{2}{3}\right)$ **28.** $12\sqrt{3}$ **29.** $2 \pm \sqrt{3}$ **30.** 26.385, 26.58 **31.** 5, 10

Pages 629–631 Lesson 15-5

5. $\frac{9}{41}$ **7.** $\frac{40}{41}$ **9.** $\frac{9}{40}$ **11.** yes **13.** 0.9063 **15.** 0.9455 **17.** 0.1584 **19.** $\sin N = 0.246$, $\cos N = 0.969$, $\tan N = 0.254$ **21.** $\sin N = 0.600$, $\cos N = 0.800$, $\tan N = 0.750$ **23.** $\sin N = 0.724$, $\cos N = 0.690$, $\tan N = 1.050$ **25.** 16° **27.** 75° **29.** 44° **31.** 77° **33.** 36° **35.** 49° **37.** 42.3 cm² **39.** 47 feet **41.** 29°; 2.3×10^8 meters per second **43.** 25% **44.** {1, 2, −2, 3}, {1, 4, 9} **45.** (2, −2) **46.** $2|a|\sqrt{30ab}$ **47.** 25

Pages 634–636 Lesson 15-6

3. $\angle FED$; $\angle EDG$ **5.** $\angle QRP$; $\angle SPR$ **7.** sine **9.** 9.7 **11.** $\angle B$: 69°, $\overline{AB}$: 13.9 in., $\overline{BC}$: 5.0 in. **13.** $\angle B$: 20°, $\overline{AB}$: 9.6 cm, $\overline{AC}$: 3.3 cm **15.** $\angle B$: 50°, $\overline{AC}$: 12.3 m, $\overline{BC}$: 10.3 m **17.** 376 ft, 426 ft **19.** 65 ft **21.** 1716 cm² **23.** 173.2 m **27.** 22 miles **29.** 4.0 seconds **31.** $\{x|x > 80\}$ **32.** −2 **33.** $x = 13$, $y = -2$ **34.** 9 **35.** 2 **36.** $\frac{1}{6}$ **37.** 45°

Pages 638–640 Chapter 15 Summary and Review

1. 24°, 114° **3.** none, 32° **5.** 92° **7.** 53° **9.** 4 cm, $4\sqrt{3}$ or about 6.928 cm **11.** 6 in., $3\sqrt{3}$ or about 5.196 in. **13.** $b = \frac{70}{11}$, $c = \frac{30}{11}$ **15.** $d = \frac{48}{5}$, $e = \frac{36}{5}$ **17.** $\frac{28}{53}$ **19.** $\frac{45}{53}$ **21.** $\frac{28}{45}$ **23.** 39° **25.** 80° **27.** $\angle B$: 28°, $\overline{AC}$: 3.7 m, $\overline{AB}$: 7.9 m **29.** 31 people **31.** $36\sqrt{3}$ or about 62.4 ft² **33.** about 104 m **35.** about 6143 ft, about 5808 ft

INDEX

Absolute values, 55–58, 69–70, 89, 492
  open sentences involving, 199–203, 206, 464, 466
Acute angles, 182
Acute triangles, 612
Addition
  additive identity, 22–24
  additive inverse property, 74–75
  associative property of, 31–35, 45, 70, 230
  commutative property of, 31–34, 70, 230, 497
  elimination using, 453–459, 469
  of integers, 52–59, 88
  of polynomials, 230–233, 497
  properties of, 22–24, 31–34
  of radical expressions, 497–500, 511
  of rational expressions, 322–325, 330–332, 335, 349
  of rational numbers, 69–72, 79, 89, 322
  in solving equations, 94–98, 100, 112, 126
  in solving inequalities, 176–179, 187
  symbol in BASIC, 10
  words used to indicate, 9
Addition property for inequality, 176
Addition property of equality, 94, 99–100
Additive identity property, 22–24
Additive inverse property, 74–75
Additive inverses, 56–58, 230–232, 266–267, 322, 453–454
Adjacent sides, 626–629, 633, 639
A'h-Mose, 165
Algebraic expressions, 8–17, 36
  evaluating, 10–16, 45
  rational, 306–325, 330–351
  simplifying, 27–30, 45
Al-Khwarizmi, Muhammed ibn Musa, 93, 549
Analytic geometry, 383
Angles
  acute, 182
  complementary, 611–613, 638
  corresponding, 622–624
  of depression, 632–634
  of elevation, 632–634
  measures of, 610–614, 638
  supplementary, 610, 612–613, 638
  of triangles, 233, 611–614, 638
Applications
  in automobile safety, 353
  in biology, 209
  black holes, 225
  compound interest, 145
  in construction, 517
  in consumer awareness, 133
  escape velocity, 486
  formulas, 343–347
  in history, 93
  in literature, 439
  in marine biology, 255
  meteorology, 54
  mixtures, 154–157, 171
  in nature, 473
  in optics, 609
  in physics, 305
  in probability, 559
  Punnett squares, 246
  in recreation, 49
  simple interest, 142–144
  in skiing, 399
  solving quadratic equations, 546–549
  in sports, 7
  in statistics, 175
  travel agents, 161
  uniform motion, 158–161, 171
Area, 36–40
  of circles, 37, 479
  cross-sectional, 12
  of rectangles, 37, 105, 166, 213, 215, 234, 238, 240, 243, 256, 261–265, 272, 276–277, 287, 295, 312–313, 368, 498, 546, 555
  of squares, 10, 38, 243, 245–246, 276, 282, 480
  surface, 37, 247
  of trapezoids, 38, 40, 237, 246, 343
  of triangles, 36, 39–40, 446, 630
Arrays, 242
Ascending order, 227–228
Associative properties
  of addition, 31–35, 45, 70, 230
  of multiplication, 31–35, 75–76, 214, 223
Automorphic numbers, 193
Averages
  batting, 7
  means, 80, 188, 570–576, 605
  medians, 570–581, 605
  modes, 570–575, 605
Axes
  of symmetry, 519–522, 554
  x, 354, 356
  y, 354, 356

Back-to-back stem-and-leaf plots, 566–568
Balanced numbers, 109
Banneker, Benjamin, 409
Bar graphs, 389–390, 432
Base functions, 530
Bases
  percents, 138–139
  of powers, 9, 213, 218
BASIC, 12
  DATA statements, 40
  FOR/NEXT loops, 378
  IF-THEN statements, 180
  programs, 12, 30, 40, 59, 78–79, 115, 125, 144, 150, 185, 225, 247, 260, 321, 378, 431, 461, 509, 545, 597, 631, 637
  READ statements, 40
  symbol for exponents, 12, 225
Batting averages, 7
Bernoulli, Jacob, 593
Bimodal data, 571
Binomials, 226
  conjugates, 494–495
  difference of squares, 276–280, 286, 301
  dividing by, 319–321, 349
  factoring, 261–263, 276–279, 286, 300–301
  multiplying, 238–246
Boundaries, 379–381, 463–464

Box-and-whisker plots, 579–582, 605
Boyle's Law, 169

CALC commands, 121, 191, 315, 427, 505
Calculators, 227, 533, 538
  change-sign keys, 57
  degree mode, 627
  graphing, 384, 462, 530, 588
  inverse keys, 628
  parentheses keys, 420
  percent keys, 139
  power keys, 215
  raising numbers to powers, 10
  recall keys, 420
  reciprocal keys, 84
  scientific, 15
  scientific notation, 223–224
  square root keys, 478–480, 488–489, 498
  store keys, 420
  for trigonometric ratios, 627–630, 639
Capture-recapture method, 134–135
Carroll, Lewis, 439
Cartesian coordinate system, 383
Catenaries, 517
Celsius, 36, 78, 120, 199, 347, 373
Centers, 428, 430
Central tendencies
  means, 570–576, 605
  medians, 570–581, 605
  modes, 570–575, 605
Change-sign keys, 57
Ch'ang Ts'ang, 467
Chapter reviews, 44–46, 88–90, 126–128, 170–172, 204–206, 248–250, 300–302, 348–350, 392–394, 434–436, 468–470, 510–512, 554–556, 604–606, 638–640
Chapter tests, 47, 91, 129, 173, 207, 251, 303, 351, 395, 437, 471, 513, 557, 607, 641
Charles' Law, 165
Circle graphs, 141, 432–433
Circles
  area of, 37, 479
  centers of, 428, 430
  circumferences of, 16, 38
  diameters of, 414, 428, 430
Circumferences, 16, 38
Closed half-planes, 380–381
Coefficients, 28–29, 242
  of one, 94
College entrance exams previews, 130–131, 252–253, 296–297, 514–515, 642–643
Collinear points, 431
Common denominators, 322–323
  least, 330–331
Common factors, 257–258
  greatest, 257–263, 277, 286–288, 291, 300–301
Common monomial factors, 265, 271, 286, 301
Commutative properties
  of addition, 31–34, 70, 230, 497
  of multiplication, 31–34, 75, 214, 223
Comparative graphs, 432
Comparison property, 65, 82
Comparisons of numbers, 60, 65–68, 89
Complementary angles, 611–613, 638
Complete graphs, 384
Completeness properties
  for points in the plane, 355–356
  for points on the number line, 488
Completing the square, 531–536, 555
Complex fractions, 81, 334–337, 349
Composite numbers, 256, 258
Compound events, 600–603, 606
Compound inequalities, 175, 194–198, 200–203, 205–206
Compound interest, 226–227, 325
Compound sentences, 185, 199
Computers
  BASIC, 12, 40, 225, 321, 637
  CALC commands, 121, 191, 315, 427, 505
  E notation, 225
  Mathematics Exploration Toolkit, 121, 191, 315, 427, 505
  programs, 12, 30, 40, 59, 78–79, 115, 125, 144, 150, 180, 185, 225, 247, 260, 321, 378, 431, 461, 509, 545, 597, 631, 637
  spreadsheets, 40, 247
Conjugates, 494–495
Conjunctions, 185
Connections
  geometry, 10, 14, 32, 36–37, 105, 117, 166, 182, 213, 231, 243, 262, 272, 278, 287, 312, 323, 343, 423, 428, 444, 494, 497–498, 507, 543
  history, 35, 78, 125, 165, 180, 233, 260, 310, 383, 409, 467, 481, 549, 593
  number theory, 449, 453, 502–503, 525
  probability, 361
  sequences, 76
  set theory, 51
  statistics, 188
Consecutive numbers, 112–115
  even, 112–115
  odd, 112–115
Consistent equations, 443–444
Constants, 213
  of variation, 162
Cooperative learning activities, 43, 87, 110, 153, 193, 212, 270, 328, 391, 433, 441, 476, 529, 599, 616
Coordinate planes, 354–358, 392
  origins, 354, 380
  quadrants, 356–357, 392
  x-axis, 354, 356
  y-axis, 354, 356
Coordinates, 354, 359–361
  on number lines, 51–52
  x, 354–356, 381
  y, 354–356, 381
Corresponding angles, 622–624
Corresponding sides, 622–624
Cosine, 626–632, 639–640
Counting numbers, 487
Cross products, 65, 134, 163
Cross-sectional area, 12
Cubes, 10

Data, 560
  bimodal, 571
  box-and-whisker plots of, 579–582, 605
  interquartile ranges of, 575–578, 580, 605
  line plots of, 560–564, 604
  means of, 80, 188, 570–576, 605
  medians of, 570–581, 605
  modes of, 570–575, 605
  outliers, 576–578, 580

quartiles of, 575–581, 605
ranges of, 575, 577–578, 605
scatter plots of, 583–587, 606
stem-and-leaf plots of, 565–570, 572–573, 576–579, 582, 605
DATA statements, 40
Decimal notation, 221–224, 249
Decimals
in equations, 122–125
repeating, 487–488, 511
terminating, 487, 511
Degree mode, 627
Degrees
of monomials, 227, 249
of polynomials, 227–228, 249
Denominators
common, 322–323, 330–331
least common, 330–332, 338–341, 343
rationalizing, 493–496
Density property, 66–67
Dependent equations, 443–444
Descartes, René, 383
Descending order, 227–228, 249, 282
Diameters, 414, 428, 430
Differences
of squares, 244, 276–280, 286, 301
squares of, 244–245
Diophantus, 93, 125
Direct variations, 162–165, 172
Discounts, 133, 146–150, 171
successive, 150
Discriminants, 541–545, 555
Disjoint sets, 51
Disjunctions, 185
Distance formula, 506–509, 512
Distributive property, 26–30, 32–34, 45, 57, 74–75, 82, 122, 135, 187, 230, 234, 238–239, 494, 497
factoring using, 261–264, 271, 300
Dividends, 319–320
Division
dividends, 319–320
divisors, 319–320
of monomials, 217–219
of polynomials, 319–321, 349
of powers, 217–219
quotients, 319–321
of rational expressions, 316–318, 334, 348
of rational numbers, 80–84, 99
remainders, 319–321
in solving equations, 104–107, 111, 121, 127
in solving inequalities, 182–187, 200
symbol in BASIC, 12
words used to indicate, 9
by zero, 123
Division property for inequalities, 182–183
Division property of equality, 104–105, 111
Divisors, 319–320
binomials, 319–321, 349
proper, 328
Dodgson, Charles Lutwidge, 439
Domains, 359–362, 364–366, 368–369, 376, 392

Edges, 379
Einstein, Albert, 233
Elements, 19–20
Elimination, 452–461, 469
Empirical probability, 594
Empty set, 19, 51, 195
E notation, 225
Equally likely events, 589, 591, 600
Equations, 19–22
with absolute values, 199
of axes of symmetry, 519–522, 554
checking solutions, 95–98, 100–101, 104, 112, 123, 124, 291, 296, 407, 444, 447, 452, 457, 501–504
consistent, 443–444
with decimals, 122–125
dependent, 443–444
equivalent, 94
formulas, 36–40
with fractions, 122–124, 127
identities, 118–119
inconsistent, 443–444
independent, 443–444
linear, 369–373, 375, 393, 405–427, 434–436, 442–461, 468–469
with more than one variable, 123–124, 128
percents, 139–140
point-slope form, 405–407, 410
quadratic, 290–299, 302, 523–526, 531–557
radical, 501–505, 512
rational, 338–347, 350
as relations, 364–373
from relations, 385–388, 394
slope-intercept form, 410–414, 416–422, 424–426, 435, 444
solving, 94–107, 111–129, 290–299, 302, 338–347, 350, 501–504, 512, 523–526, 531–540, 546–549, 554–555
standard form, 406–409, 411–412, 416, 420–421
systems of, 442–461, 468–469
in two variables, 364–373, 375
with variables on both sides, 116–121, 127
writing, 85–87
Equilateral triangles, 611, 617
Equivalent equations, 94
Eratosthenes, 260
Escalante, Jaime, 481
Escape velocity, 486
Even numbers, 636
consecutive, 112–115
Events
compound, 600–603, 606
equally likely, 589, 591, 600
simple, 589–592, 600, 606
Excluded values, 306–309
Experimental probability, 594–597, 606
Exponents, 9–12, 44, 213–214, 217–223, 227
negative, 218–219, 222–224, 248
symbol in BASIC, 12, 225
zero, 218–219
Expressions, 8–17, 36
evaluating, 10–16, 45
mixed, 334
radical, 477–500, 511
rational, 306–325, 330–351
simplest form of, 27–30, 45
simplifying, 218–219, 492–500, 511
Extra, 242, 318
Extremes, 134–136

Factors, 8, 256–268, 271–303, 319
of binomials, 261–263, 276–279, 286, 300–301
checking, 265–267, 271
common, 257–258
common monomial, 265, 271, 286, 301
of differences of squares, 276–280, 286, 301
using distributive property, 261–264, 271, 300

greatest common, 257–263, 277, 286–288, 291, 300–301, 307–308, 322–323
by grouping, 265–268, 287, 300
of monomials, 257–259, 300
of perfect square trinomials, 281–288, 296, 301
of polynomials, 257–259, 261–268, 271–289, 300–301
prime, 256–258, 478, 492–493
in solving equations, 290–299, 302, 546–547
of trinomials, 262–264, 271–275, 281–288, 301

Fahrenheit, 36, 54, 78, 120, 347, 373

Fibonacci sequence, 212

Figurate numbers, 153, 270

FOIL method, 238–241, 244, 265, 271–272, 277, 287, 498

Formulas, 36–40, 343–347, 350
for acceleration, 481
for area of circles, 37, 479
for area of rectangles, 105, 213, 312
for area of squares, 38
for area of trapezoids, 38, 40, 237, 246, 343
for area of triangles, 36, 40, 446, 630
for changing Celsius to Fahrenheit, 36, 120, 373
for changing Fahrenheit to Celsius, 78, 347
for circumferences, 38
for compound interest, 145, 226
distance, 506–509, 512
for distance, 39, 158–159, 161, 171, 339, 342
for earned-run average, 46
for escape velocity, 486
for focal length, 343
for height of projectile, 255, 290, 294, 299, 368, 376, 522–523, 526, 536, 540, 548, 551, 553, 556
for interest, 142–143, 145, 170
for length of storm, 491
for monthly payments, 220
for perimeters of parallelograms, 38
for perimeters of rectangles, 117, 323, 497
for perimeters of squares, 38
for periods of pendulums, 496
for pitch of roofs, 105
for power, 481
to predict time of storm, 35
quadratic, 537–547, 555
for range of projectiles, 630
for resistance, 344–345, 363
for Schwarzschild radius, 225
for simple interest, 142–143, 170
Snell's Law, 609, 628, 631
for speed, 501, 512
for surface area of rectangular solids, 37
for terms of arithmetic sequences, 76
for time to fall to Earth, 504
for visible distance, 504
for voltage, 489, 491
for volume of rectangular prisms, 32
for volume of trapezoidal prisms, 314

FOR/NEXT loop, 378

Fractions
adding, 79, 322–325, 330–332, 335, 349
algebraic, 306–325, 330–351
bars, 13–15
complex, 81, 334–337, 349
dividing, 316–318, 334, 348
in equations, 122–124, 127
history of, 310
improper, 104, 334
least common denominators, 330–332, 338–341, 343
mixed numbers, 334
multiplying, 311–316, 348
to percents, 138
simplifying, 307–309, 311, 323, 348
subtracting, 322–324, 331–332, 335, 349
unit, 165, 528

Functional notation, 376–378

Functional values, 376–378, 393

Functions, 374–378, 393
base, 530
families of, 530
functional notation, 376–378
functional values, 376–378, 393
graphs of, 375, 377, 518–526, 542, 554
greatest integer, 260
quadratic, 518–526, 542, 554
Trace, 462
vertical line test, 375, 377

Geometric means, 502–503

Geometry
analytic, 383
angles, 182, 610–614, 632–634, 638
area, 10, 36–40, 105, 166, 446, 479, 498, 546, 555, 630
circles, 16, 37–38, 414, 428, 430, 479
circumferences, 16, 38
collinear points, 431
midpoints, 428–431, 436
parallel lines, 423–426
parallelograms, 423
perimeters, 14, 16, 38, 40, 117–118, 497, 543, 555
perpendicular lines, 354, 424–426, 436
quadrilaterals, 423
rectangles, 105, 117, 166, 497–498, 543, 555
rectangular prisms, 32, 37, 247
squares, 10, 40
surface area, 37, 247
trapezoids, 38, 40, 507
triangles, 36, 39–40, 185, 446, 473, 482–486, 494, 510, 611–614, 617–641
volume, 10, 32, 247, 278–279, 547

Grade, 400, 402

Graphing calculators, 384, 462, 530, 588, 411, 424, 518, 524, A18, A19, A22, A23
standard viewing window, 384
Trace functions, 462

Graphs, 353
bar, 389–390, 432
boundaries, 379–381, 463–464
circle, 141, 432–433
comparative, 432
complete, 384
distance formula, 506–509, 512
of functions, 375, 377, 518–526, 542, 554
graphing calculators, 384, 462, 530, 588, 637
of inequalities, 195–198, 200–202, 205–206, 379–384, 394, 463–466, 470
intercepts, 443
line, 389–390, 436
line plots, 560–564, 604
of lines, 369–373, 375, 384, 393, 400–405, 410–418, 421–427, 435, 442–447, 468
misleading, 390
number lines, 50–53, 55, 62–63, 69, 88–89, 195–202, 205–206, 354
of ordered pairs, 354–360, 369–371
of parabolas, 517–526, 530, 542, 554
pictographs, 432

of points, 354–360, 369–371
of quadratic functions, 518–526, 542, 554
of relations, 359–360, 362, 369–375, 377, 379–383, 393–394
scatter plots, 583–587, 606
slopes, 399–414, 416–427, 431, 434–435, 443–444, 448, 584, 606, 628
of solution sets, 195–202, 205–206
of systems of equations, 442–447, 468
of systems of inequalities, 463–466, 470
of trigonometric functions, 637
using, 389–391, 432–433
x-intercepts, 384, 410, 412–413, 415, 417–419, 435, 523–524, 541–542, 544
y-intercepts, 384, 410–422, 424, 435
Graunt, John, 593
Greatest common factors, 257–263, 277, 286–288, 291, 300–301, 307–308, 322–323
Greatest integer function, 260
Guess and check, 269–270

H

Half-planes, 379–382, 463
closed, 380–381
open, 379
Hamilton, Sir William Rowan, 35
Happy numbers, 328
Harriot, Thomas, 180
History connections
A'h-Mose, 165
Benjamin Bannaker, 409
Diophantus, 125
Emmy Noether, 233
Eratosthenes, 260
fractions, 310
Graunt and Bernoulli, 593
Jaime Escalante, 481
K'iu-ch'ang Suan-shu, 467
Muhammed ibn Musa al Khwarizmi, 549
Pythagoras, 636
Recorde, Harriott, and Oughtred, 180
René Descartes, 383
Sir William Rowan Hamilton, 35
Sonya Kavalevskaya, 78
Horizontal lines, 401–402, 407
Hypotenuse, 473, 482–483, 485, 494, 510, 617–618, 626–629, 631, 633, 639

Identities, 118–119
additive, 22–24
multiplicative, 22–24, 28
IF-THEN statements, 180
Improper fractions, 104, 334
Inches
united, 230
Inconsistent equations, 443–444
Independent equations, 443–444
Inequalities, 60–64, 89
addition property for, 176
checking solutions, 176
compound, 175, 194–198, 200–203, 205–206
division property for, 182–183
graphs of, 195–198, 200–202, 205–206, 379–384, 394, 463–466, 470
involving absolute values, 200–203, 206, 464, 466
with more than one operation, 186–191, 196, 200, 205
multiplication property for, 181–182, 186
phrases for, 177
solving, 176–191, 195–198, 200–207
subtraction property for, 176, 186
systems of, 463–467, 470
in two variables, 379–383
Integers, 50–60, 487, 489
absolute values of, 55–58, 89
adding, 52–59, 88
consecutive even, 112–115
consecutive odd, 112–115
factoring, 256–259
subtracting, 57–58, 88
Intercepts, 410, 443
x, 410, 412–413, 415, 417–419, 435, 523–524, 541–542, 544
y, 410–422, 424, 435
Interest, 216
compound, 145, 226–227, 325
principals, 142, 145, 170
rates, 142, 145, 170
simple, 142–144, 170
Interquartile ranges, 575–578, 580, 605
Intersecting lines, 442–447
Intersections, 51–52, 194–195
Inverse keys, 628
Inverse operations, 80
Inverses, 99
additive, 266–267, 322, 453–454
multiplicative, 316
of relations, 360–363, 392
Inverse variation, 166–169, 172
product rule for, 167–168
proportions, 167, 172
Irrational numbers, 487–489, 511
Isosceles trapezoids, 507
Isosceles triangles, 612–613

Jefferson, Thomas, 409

K'iu-ch'ang Suan-shu, 467
Kovalevskaya, Sonya, 78

Labs, *see Manipulative Activities*
Least common denominators, 330–332
Least common multiples, 329–330
Legs, 482–483, 494, 626–629, 631
adjacent, 626–629, 633, 639
opposite, 626–629, 633, 639
L'Enfant, Pierre, 409
Like terms, 27–29, 230–232, 234, 239–240, 262, 323
Linear equations, 369–373, 375, 393, 405–427, 434–436
point-slope form, 405–407, 410
slope-intercept form, 410–414, 416–422, 424–426, 435, 444
standard form of, 406–409, 411–412, 416, 420–421
systems of, 442–461, 468–469
Line graphs, 389–390, 436
Line plots, 560–564, 604

Lines
graphs of, 369–373, 375, 393, 400–405, 410–418, 421–427, 435, 442–447, 468
horizontal, 401–402, 407
intersecting, 442–447
number, 354, 560–564, 579–582
parallel, 423–426, 443–444, 448
perpendicular, 354, 424–426, 436
regression, 588
slopes of, 399–414, 416–427, 431, 434–435, 443–444, 448, 584, 606, 628
vertical, 401–402, 407

Line segments
midpoints of, 428–431, 436

Logic, 18
compound sentences, 185, 199
conjunction, 185
disjunctions, 185
negations, 18
statements, 18

Lower quartiles, 575–581, 605

Magic squares, 59

Manipulative activities, 27, 57, 94, 99, 103, 112, 117, 227, 230, 235, 238, 261, 271, 411, 424, 483, 488, 518, 524, 532, A1–A24

Mappings, 359–360, 362, 374

Mathematics Exploration Toolkit, 121, 191, 315, 427, 505

Matrix, 151–152

Maximum points, 518, 520–522

McKinley, William, 475

Means
averages, 80, 188, 570–576, 605
geometric, 502–503
of proportions, 134–136

Means-extremes property of proportions, 134–135, 402, 527

Measurement
of angles, 610–611
area, 10, 36–40, 105, 166, 446, 479, 498, 546, 555, 630
circumferences, 16, 38
ohms, 344–347
perimeters, 14, 16, 38, 40, 117–118, 497, 543, 555
protractors, 610–611
surface area, 37, 247
temperature, 36–37, 54, 78, 120, 199, 347
united inches, 230
volume, 10, 32, 247, 278–279, 314, 547

Measures of central tendency, 570–575
means, 80, 188, 570–576, 605
medians, 570–581, 605
modes, 570–575, 605

Measures of variation, 575–578
interquartile ranges, 575–578, 580, 605
ranges, 575, 577–578, 605

Medians, 570–581, 605

Meteorology, 54

Mid-chapter reviews, 25, 73, 107, 157, 190, 229, 280, 325, 373, 418, 456, 491, 540, 587, 621

Midpoints, 428–431, 436

Minimum points, 518–521, 554

Misleading graphs, 390

Mixed expressions, 334

Mixed numbers, 104, 334

Mixture problems, 154–157, 171

Modes, 570–575, 605

Money
commissions, 140
consumer awareness, 133
discounts, 133, 146–150, 171
interest, 142–145, 170, 216, 226–227, 325
taxes, 146–149

Monomials, 213–220, 226–228
adding, 497
constants, 213
degrees of, 227
dividing, 217–219
factoring, 256–259, 300
greatest common factors of, 258–259, 261–262, 300
multiplying by, 234–237, 250
powers of, 214–216
simplifying, 218–219, 248
subtracting, 497

Multiples
least common, 329–330

Multiplication
associative property of, 31–35, 75–76, 214, 223
of binomials, 238–246
commutative property of, 31–34, 75, 214, 223
elimination using, 457–461, 469
factors, 8, 256–268, 271–303
FOIL method, 238–241, 244, 250, 265, 271–272, 277, 287, 498
multiplicative identity, 22–24, 28
of polynomials, 234–246, 250
of powers, 213–216, 234, 348
product, 8–9
properties of, 22–24, 31–34
of radical expressions, 498, 500
of rational expressions, 311–316, 348
of rational numbers, 74–78, 89
in solving equations, 103–107, 112, 116–117, 127
in solving inequalities, 181–186, 188, 205
symbol in BASIC, 12
words used to indicate, 9
zero product property, 290–293, 295–296, 302

Multiplication property for inequalities, 181–182, 186

Multiplication property of equality, 103, 105, 122

Multiplicative identity property, 22–24, 28

Multiplicative inverse property, 81

Multiplicative inverses, 81–82, 316

Multiplicative property of zero, 23–24, 32, 74

Nasty numbers, 289

Natural numbers, 51, 487, 489, 598

Negations, 18

Negative numbers, 50
as exponents, 218–219, 222–224, 348
factoring, 257
as slopes, 401–402, 584, 606

Noether, Emmy, 233

Null set, 19

Number lines, 50–53, 55, 354
in adding, 52–53, 55, 69
box-and-whisker plots, 579–582, 605
comparing numbers on, 61
completeness property, 488
graphing inequalities, 62–63, 89, 195–198, 200–202, 205–206
graphing sets of numbers, 51–53, 88
line plots, 560–564, 604
x-axis, 354, 356
y-axis, 354, 356

Numbers
absolute values of, 55–58, 69–70, 89
additive inverses, 56–58, 453–454
automorphic, 193
balanced, 109
comparing, 60, 65–68, 89
composite, 256, 258
consecutive, 112–115
counting, 487
decimal notation, 221, 249
even, 636
factors of, 256–259
Fibonacci, 212
figurate, 153, 270
happy, 328
integers, 50–60, 88, 487, 489
inverses, 99
irrational, 487–489, 511
mixed, 104, 334
multiplicative inverses, 81–82
nasty, 289
natural, 51, 487, 489, 598
negative, 50
odd, 636
ordered pairs, 354–374, 392–393, 442–445, 463–465, 583
palindromes, 528
perfect, 328
perfect squares, 281
persistence of, 433
prime, 256, 258, 260
prime factorization of, 256–259
rational, 61–84, 487–489, 511
real, 51, 369, 488–489, 523–524
rounded, 566–568
scientific notation, 221–225, 249
square roots of, 474, 477–513, 531–532
squares of, 477, 479, 501–502
triangular, 153
truncated, 566–567
twin prime, 259
whole, 50–52, 487, 489
Number theory, 112, 212, 328
automorphic numbers, 193
balanced numbers, 109
composite numbers, 256, 258
consecutive numbers, 112–115
figurate numbers, 153, 270
geometric means, 502–503
happy numbers, 328
nasty numbers, 289
palindromes, 528
perfect numbers, 328
persistence of numbers, 433
prime numbers, 256, 258, 260
triangular numbers, 153
twin primes, 259
Numerical coefficient, 28–29

Obtuse triangles, 612
Odd numbers, 636
consecutive, 112–115
Odds, 590–593, 606
Ohm, Georg Simon, 344
Ohms, 344–347
Open half-planes, 379
Open sentences, 18–21
with absolute values, 199–203, 206
equations, 19–22, 94–107, 111–129, 338–347, 350, 369–373, 375, 393, 405–427, 434–436
inequalities, 60–64, 89, 176–191, 194–198, 200–207, 379–383, 394
Operations
inverse, 80
order of, 13–14, 19
Opposites, 56
Opposite sides, 617–620, 626–629, 633, 639
Ordered pairs, 354–374, 392–393, 442–445, 463–465, 583
graphing, 354–360, 369–371
relations, 359–374, 392
Order of operations, 13–14, 19
Origins, 354, 380
Oughtred, William, 180
Outliers, 576–578, 580

Palindromes, 528
Parabolas, 517–526, 530, 542, 554
axes of symmetry, 519–522, 554
maximum points, 518, 520–522
minimum points, 518–521, 554
vertices, 518–522
Parallel lines, 423–426, 443–444, 448
slopes of, 423–425
Parallelograms, 423
perimeters of, 38
Parentheses, 9, 13–17, 27, 220
Parentheses keys, 420
Percentages, 138–139
Percent keys, 139
Percents, 138–150
of change, 146–150, 171
commissions, 140
of decrease, 146–150, 171
discounts, 146–150, 171
fractions to, 138
of increase, 146–149, 171
interest, 142–145, 170
mixture problems, 155–157
sales taxes, 146–149
solving using equations, 139–140
solving using proportions, 138–140, 170
Perfect numbers, 328
Perfect squares, 281, 487
trinomials, 281–288, 296, 301, 531–535
Perimeters, 14, 16, 231
of parallelograms, 38
of rectangles, 40, 117–118, 323, 325, 368, 497, 543, 555
of squares, 38
Perpendicular lines, 354, 424–426, 436
slopes of, 424–425
Persistence, 433
Philolaus, 636
Pictographs, 432
Planes
completeness property for points in, 355–356
coordinate, 354–358, 392
half, 379–382
Points, 354
collinear, 431
graphing, 354–360, 369–371
maximum, 518, 520–522
midpoints, 428–431, 436
minimum, 518–521, 554
origins, 354, 380
vertices, 518–522
Point-slope form, 405–407, 410
Polynomials, 226–246, 306
adding, 230–233, 249, 497
ascending order, 227–228
binomials, 226, 261–263, 276–279, 286, 300–301, 494–495
degrees of, 227–228
descending order, 227–228, 249, 282
dividing, 319–321, 349
factoring, 257–259, 261–268, 271–289, 300–301, 306–308, 323, 330

monomials, 213–220, 226–228, 257–259, 300
multiplying, 234–246, 250
prime, 273–274, 286
subtracting, 230–233, 249, 497
trinomials, 226–228, 262–264, 271–275, 281–288, 301, 531–535

Positive numbers
as slopes, 401–402, 584

Power keys, 215

Powers, 9–13, 44, 213–228
of monomials, 214–216
of powers, 214–216
of products, 214–216
products of, 213–216, 234, 248
quotients of, 217–219
reading, 220
symbols in BASIC
zero, 218–219

Prime factorization, 256–258
to find least common multiples, 329
to find square roots, 478
to simplify radical expressions, 492–493

Prime factors, 256–258

Prime numbers, 256, 258
sieve of Eratosthenes, 260
twin, 259

Prime polynomials, 273–274, 286

Principals, 142, 145, 170

Principal square roots, 477–480, 492

Prisms
rectangular, 32, 247
trapezoidal, 314
volume of, 32, 214, 247, 547

Probability, 198, 559, 589–597, 599–603
of compound events, 600–603, 606
experimental, 594–597, 606
of one, 590–591
of simple events, 589–592, 606
of zero, 590–591

Problem-solving strategies
check for hidden assumptions, 440–441
explore verbal problems, 41–43
guess and check, 269–270
identify subgoals, 527–529
list possibilities, 326–328
look for a pattern, 210–212
make a diagram, 192–193
make a model, 615–616
make a table or chart, 151–153
solve a simpler problem, 210–211, 598–599
use a graph, 389–391, 432–433
use a table, 474–476
work backwards, 108–110
write an equation, 85–87

Product property of square roots, 492–493, 495

Product rule for inverse variations, 167–168

Products, 8–9
cross, 65, 134, 163
factors of, 256–268, 271–303
of powers, 213–216, 234
powers of, 214–216
of roots, 550–553, 556
special, 243–246
of sums and differences, 244–245

Programs, 12, 30, 40, 59, 78–79, 115, 125, 144, 150, 185, 225, 260, 321, 378, 431, 461, 509, 545, 597, 631
DATA statements, 40
FOR/NEXT loops, 378
IF-THEN statements, 180
READ statements, 40
spreadsheets, 40, 247

Projects, extended, B1–B16

Properties, 22–35
of addition, 22–24, 31–34, 497
additive identity, 22–24
additive inverse, 74–75
associative, 70, 75–76, 214, 223, 230
commutative, 31–34, 70, 75, 214, 223, 230, 497
comparison, 65, 82
completeness, 355–356, 488
density, 66–67
distributive, 26–30, 32–34, 45, 57, 74–75, 82, 122, 135, 187, 230, 234, 238–239, 261–262, 271, 300, 494, 497
of equalities, 23–24, 94, 99–101, 103–105
for inequalities, 176, 181–182
of levers, 167, 169
means-extremes, 134–135, 402, 527
of multiplication, 22–24, 31–34
multiplicative identity, 22–24, 28
multiplicative inverse, 81
product, 492–493, 495
of proportions, 134
quotient, 493
reflexive, 23–24, 60
of square roots, 492–493
substitution, 23–24, 28, 31–33, 45, 55, 74–76, 230
symmetric, 23–24, 27, 60
transitive, 23–24, 60–61
zero product, 290–293, 295–296, 302, 306–307, 340, 502

Proportions, 134–140, 170, 622–623, 627, 639
capture-recapture method, 134–135
direct variations, 163–165, 172
extremes, 134–136
inverse variations, 167, 172
means, 134–136
means-extremes property of, 134–135, 402
percents, 138–140
scales, 135

Protractors, 610–611

Punnett squares, 246

Puzzles, 59, 115, 120

Pythagoras, 636

Pythagorean Theorem, 473, 482–486, 494, 506, 508, 510, 618, 627, 636

Quadrants, 356–357, 392

Quadratic equations, 523–526, 531–557
applications, 546–549
completing the square, 531–536, 555
discriminant, 541–545, 555
nature of roots, 541–545, 555
products of roots, 550–553, 556
quadratic formula, 537–547, 555
solving by factoring, 290–299, 302, 546–547
solving by graphing, 523–526, 554
sums of roots, 550–553, 556

Quadratic formula, 537–547, 555
discriminant, 541–545, 555

Quadratic functions, 518–526, 542, 554

Quadrilaterals, 423

Quartiles, 575–581
lower, 575–577, 579–581, 605
upper, 575–577, 579–581, 605

Quotient property of square roots, 493

Quotients, 319–321
of powers, 217–219

Radical equations, 501–505, 512
Radical expressions, 477–500, 511
  adding, 497–500, 511
  multiplying, 498, 500
  rationalizing denominators, 493–496
  simplifying, 492–500, 511
  subtracting, 497–500, 511
Radical signs, 477–480, 487
Radicands, 477, 479
Radii
  Schwarzschild, 225
Ranges
  of data, 575, 577–578, 605
  interquartile, 575–578, 580, 605
  of relations, 359–362, 364, 368, 384, 392
Rates
  of interest, 142, 145, 170
  percents, 138–140
  ratios, 135–136
  of speed, 158
  tax, 147–148
Rational equations, 338–347, 350
Rational expressions, 306–325, 330–351
  adding, 322–325, 330–332, 335, 349
  complex, 334–337, 349
  dividing, 316–318, 334, 348
  excluded values, 306–309
  least common denominators, 330–332, 338–341, 343
  mixed expressions, 334
  multiplying, 311–316, 348
  simplifying, 307–309, 311, 323, 348
  subtracting, 322–324, 331–332, 335, 349
Rationalizing denominators, 493–496
Rational numbers, 61–84, 487–489, 511
  adding, 69–72, 79, 89, 322
  comparing, 65–68, 89
  comparison property for, 65, 82
  dividing, 80–84, 99
  multiplying, 74–78, 89
  subtracting, 70–72
Ratios, 134–138, 146
  cosine, 626–632, 639–640
  percents, 138–150, 155–157
  rates, 135–136
  scales, 135
  sine, 609, 626–633, 639–640
  tangent, 626–631, 633, 639
  trigonometric, 626–636, 639–640
Reading algebra, 17, 185, 220
READ statements, 40
Real numbers, 51, 369, 488–489, 523–524
Real roots, 523–525, 537–545, 555
Recall keys, 420
Reciprocal keys, 84
Reciprocals, 81–83, 182, 316–317, 334–336
Recorde, Robert, 180
Rectangles
  area of, 37, 105, 166, 213, 215, 234, 238, 240, 243, 256, 261–265, 272, 276–277, 287, 295, 312–313, 368, 498, 546, 555
  perimeters of, 40, 117–118, 323, 325, 368, 497, 543, 555
Rectangular prisms
  surface area of, 37
  volume of, 32, 247, 278–279, 547
Reflexive property, 23–24, 60
Regression lines, 588
Relations, 359–388, 392–395
  domain of, 359–362, 364–366, 368–369, 376, 392
  equations as, 364–373
  equations from, 385–388, 394
  functions, 374–378, 393
  graphs of, 359–360, 362, 369–375, 377, 379–383, 393–394
  inverses of, 360–363, 392
  mappings, 359–360, 362, 374
  ordered pairs, 359–374, 392
  range of, 359–362, 364, 368, 384, 392
  tables, 359–362
Remainders, 319–321
Repeating decimals, 487–488, 511
Replacement sets, 19
Reviews
  chapter, 44–46, 88–90, 126–128, 170–172, 204–206, 248–250, 300–302, 348–350, 392–394, 434–436, 468–470, 510–512, 554–556, 604–606, 638–640
  mid-chapter, 25, 73, 107, 157, 190, 229, 280, 325, 373, 418, 456, 491, 540, 587, 621
Right triangles, 473, 482–486, 612
  hypotenuse, 473, 482–483, 485, 494, 510, 617–618, 626–629, 631, 633, 639
  legs, 482–483, 494, 626–629, 631, 633, 639
  Pythagorean Theorem, 473, 482–486, 494, 506, 508, 510, 618, 627, 636
  solving, 632–636, 640
  30°-60°, 617–621, 639
  trigonometric ratios, 626–636, 639–640
Roosevelt, Theodore, 475
Roots, 523–526, 533–557
  nature of, 541–545, 555
  products of, 550–553, 556
  real, 523–525, 537–545, 555
  square, 474, 477–513, 531–532
  sums of, 550–553, 556
Rotations, 424
Rounded numbers, 566–568

Sales taxes, 146–149
Samples, 175
Scales, 135
Scatter plots, 583–587, 606
Schwarzschild radius, 225
Scientific calculators, 15
Scientific notation, 221–225, 249
  E notation, 225
Sequences, 76–77
  Fibonacci, 212
Set-builder notation, 177
Sets, 19–21
  disjoint, 51
  elements of, 19–20
  empty, 19, 51, 195
  intersections of, 51–52, 194–195
  replacement, 19
  set-builder notation, 177
  solutions, 19, 176–177, 182, 186–187, 195–196, 199–201
  subsets, 19–20, 51
  unions of, 51–52, 195–196
  universal, 51
Sides
  adjacent, 626–629, 633, 639
  corresponding, 622–624
  opposite, 617–620, 626–629, 633, 639

Sieve of Eratosthenes, 260
Similar triangles, 622–625, 627, 639
Simple events, 589–592, 600, 606
Simple interest, 142–144, 170
Simplest form, 27–30
Sine, 609, 626–633, 639–640
Slope-intercept form, 410–414, 416–422, 424–426, 435, 444
Slopes, 399–414, 416–427, 431, 434–435, 443–444, 448, 628
of horizontal lines, 401–402, 407
negative, 401–402, 584, 606
of parallel lines, 423–425
of perpendicular lines, 424–425
positive, 401–402, 584
of vertical lines, 401–402, 407
Snell, Willebrord, 609
Snell's Law, 609, 628, 631
Solutions, 18–19
checking, 95–98, 100–101, 104, 112, 123–124, 176, 291, 296, 407, 444, 447, 452, 457, 501–504
of equations in two variables, 364–373
roots, 523–526, 533–557
Solution sets, 19, 176–177, 182, 186–187, 195–196, 199–201
Spreadsheets, 40, 247
Square root keys, 478–480, 488–489, 498
Square roots, 474, 477–513, 531–532
principal, 477–480, 492
product property of, 492–493, 495
quotient property of, 493
simplifying, 492–496, 511
Squares
area of, 10, 38, 243, 245–246, 276, 282, 287, 480
completing, 531–536, 555
of differences, 244–245
differences of, 244
magic, 59
of numbers, 477, 479, 501–502
perfect, 281–288, 296, 301, 487
perimeters of, 38
Punnett, 246
of sums, 243–245
Standard form, 406–409, 411–412, 416, 420–421
Standard viewing window, 384
Statements, 18
Statistics, 175, 560–587, 604–606
bar graphs, 389–390, 432
box-and-whisker plots, 579–582, 605
circle graphs, 141, 432–433
comparative graphs, 432
interquartile ranges, 575–578, 580, 605
line graphs, 389–390, 436
line plots, 560–564, 604
means, 80, 188, 570–576, 605
medians, 570–581, 605
modes, 570–575, 605
outliers, 576–578, 580
pictographs, 432
quartiles, 575–581, 605
ranges, 575, 577–578, 605
regression lines, 588
samples, 175
scatter plots, 583–587, 606
stem-and-leaf plots, 565–570, 572–573, 576–579, 582, 605
tolerance, 175
Stem-and-leaf plots, 565–570, 572–573, 576–579, 582, 605
back-to-back, 566–568
Store keys, 420
Subsets, 19–20, 51
Substitution, 447–451, 469
Substitution property, 23–24, 28, 31–33, 45, 55, 74–76, 230
Subtraction
elimination using, 452, 454–456
of integers, 57–58, 88
of polynomials, 230–233, 497
of radical expressions, 497–500, 511
of rational expressions, 322–324, 331–332, 335, 349
of rational numbers, 70–72, 322
in solving equations, 99–102, 111–112, 116, 126
in solving inequalities, 176–179, 186–188, 200–201, 204
symbol in BASIC, 10
words used to indicate, 9
Subtraction property for inequality, 176, 186
Subtraction property of equality, 99, 111
Summaries, 44–46, 88–90, 126–128, 170–172, 204–206, 248–250, 300–302, 348–350, 392–394, 434–436, 468–470, 510–512, 554–556, 604–606, 638–640
Sums
of roots, 550–553, 556
squares of, 243–246
Supplementary angles, 610, 612–613, 638
Surface area, 37, 247
Symmetric property, 23–24, 27, 60
Symmetry, 519–522, 554
Systems of equations, 442–461, 468–469
elimination, 452–461, 469
graphing, 442–447, 468
number of solutions, 443–446, 448, 450, 468
substitution, 447–451, 469
Systems of inequalities, 463–467, 470

Tangent, 626–631, 633, 639
Taxes, 146–149
Technologies
adding fractions, 79
formulas, 40
graphing linear equations, 427
graphing linear relations, 384
quadratic equations, 530
radical equations, 505
rational expressions, 315
regression lines, 588
solving equations, 121
solving inequalities, 191
successive discounts, 150
systems of equations, 462
trigonometric functions, 637
volume and surface area, 247
Temperature
Celsius, 36, 78, 120, 199, 347, 373
Fahrenheit, 36, 54, 78, 120, 347, 373
windchill factors, 54
Terminating decimals, 487, 511
Terms, 27–29
coefficients of, 28–29
like, 27–29, 230–232, 234, 239–240, 262, 323
Tests
chapter, 47, 91, 129, 173, 207, 251, 303, 351, 395, 437, 471, 513, 557, 607, 641
college entrance, 130–131, 252–253, 296–297, 514–515
vertical line, 375, 377
Theorems
Pythagorean, 473, 482–486, 494, 506, 508, 510, 618, 627, 636
unique factorization, 256
Tolerance, 175
Trace functions, 462

Trajectory, 518
Transitive property, 23-24
  of order, 60-61
Trapezoidal prisms, 314
Trapezoids
  area of, 38, 40, 237, 246, 343
  diagonals of, 507
  isosceles, 507
Tree diagrams, 326-327, 600-602
Triangles
  acute, 612
  area of, 36, 39-40, 446, 630
  equilateral, 611, 617
  isosceles, 612-613
  obtuse, 612
  perimeters of, 231
  right, 339-340, 473, 482-486, 494, 510, 612, 617-621, 626-636
  sides of, 185
  similar, 622-625, 627, 639
  sum of angles of, 233, 611-614, 638
Triangular numbers, 153
Trigonometric ratios, 626-636, 639-640
  using calculators, 627-630, 639
  cosine, 626-632, 639-640
  graphs of, 637
  sine, 609, 626-633, 639-640
  tangent, 626-631, 633, 639
Trinomials, 226-228
  factoring, 262-264, 271-275, 281-288, 301
  perfect square, 281-288, 296, 301, 531-535
Truncated numbers, 566-567
Twin primes, 259

Uniform motion problems, 158-161, 171, 339-340, 342, 454, 456, 459, 461
Unions, 51-52, 195-196
Unique Factorization Theorem, 256
Unit cost, 66
United inches, 230
Unit fractions, 165, 528
Universal sets, 51
Upper quartiles, 575-577, 579-581, 605

Variables, 8
  in BASIC, 10
Variations
  constants of, 162
  direct, 162-165, 172
  interquartile ranges, 575-578, 580, 605
  inverse, 166-169, 172
  ranges, 575, 577-578, 605
Venn diagrams, 51
Vertical lines, 401-402, 407
Vertical line test, 375, 377
Vertices, 518-522
Volume
  of cubes, 10
  of rectangular prisms, 32, 247, 278-279, 547
  of trapezoidal prisms, 314

Washington, George, 409
Whole numbers, 50-52, 487, 489
Windchill factors, 54

x-axis, 354, 356
x-coordinates, 354-356, 381
x-intercepts, 384, 410, 412-413, 415, 417-419, 435, 523-524, 541-542, 544

y-axis, 354, 356
y-coordinates, 354-356, 381
y-intercepts, 384, 410-422, 424, 435

Zero
  division by, 123
  exponent, 218-219
  multiplicative property of, 23-24, 32, 74
  opposite of, 56
  probability of, 590-591
  product property, 306-307, 340
  slope of, 401-402, 407, 434
Zero product property, 290-293, 295-296, 302, 502

PHOTO CREDITS

Cover, Bo Zaunders/The Stock Market

v(t), Rick Rickman/DUOMO, **(b),** Skip Comer; **vi,** Aaron Haupt; **vii(t),** Art Resource, **(b),** Skip Comer; **viii(t),** Mike & Carol Werner/Comstock, Inc, **(b),** Carl R. Sams II/Peter Arnold, Inc; **ix(t),** Bruce Hando/Comstock, Inc, **(b),** Tim Cairns/Cobalt; **x,** Aaron Haupt; **xi(t),** Romilly Lockyer/Image Bank, **(b),** Henry Georgi/Comstock, Inc; **xii(t),** Coco McCoy/Rainbow, **(b),** R. Michael Stuckey/Comstock, Inc; **xiii(t),** Mark Ippolito/Comstock, Inc, **(b),** David Lissy/The Photo File; **xiv,** Aaron Haupt; **xv,** Ted Mahieu/The Stock Market; **6,** Rick Rickman/DUOMO; **7,** Stephen Wilkes/The Image Bank; **8,** Official White House Photo; **10,** Elaine Shay; **12,** Kul Bhatia/Photo Researchers; **13,17,** Skip Comer; **18,** Howard Sochurck/Woodfin Camp & Assoc; **20,** Kenneth Fink/Photo Researchers; **21,** Doug Martin; **25,** Cobalt Productions; **26,27,** Skip Comer; **30,** The Bettmann Archive; **31,** John McDermott/Tony Stone Images; **35,** Ralph Wetmore/Tony Stone Images; **36,** Skip Comer; **37,** Pete Saloutos/The Stock Market; **39,** Sheila Goode-Green; **42,** Elaine Shay; **48,49,** Skip Comer; **54,** Brian Park/Tom Stack & Assoc; **56,** Mitchell Layton/DUOMO; **60,** Howard Zyrb/FPG; **64,** Bo Zaunders/The Stock Market; **65,** Randy Scheiber; **68,69,** Skip Comer; **72,** Heinz Kluetmeier/Sports Illustrated; **74,** The Bettmann Archive; **78,80,** Skip Comer; **84,** Reinhard Kunkel/The Image Bank; **85,** Elaine Shay; **87,** David Jentz/Superstock; **92,** Art Resource; **93,** Al Assid/The Stock Market; **96,** Skip Comer; **98,** Robert Garvey/The Stock Market; **99,** Skip Comer; **102,** Richard Price/Westlight; **103,** Tom Tracy/FPG; **108,** Skip Comer; **109,** George Hunter/Tony Stone Images; **110,** Aaron Haupt; **111,** Dave Wilhelm/The Stock Market; **115,** Henley & Savage/The Stock Market; **116(l),** Howard Millard/The Stock Market, **(r),** Four by Five/Superstock; **120,** Carl Purcell/FPG; **122,** Skip Comer; **123,** Elaine Shay; **132,** Skip Comer; **133,** Martin Rogers/FPG; **134,** Mack Albin; **135,** Elaine Shay; **136,** Steve Vidler/Superstock; **138,** Ben Simmons/The Stock Market; **139,** R. Dahlquist/Superstock; **141,** Ted Mahieu/The Stock Market; **144,** Skip Comer; **147,** Elaine Shay; **148,** John H. Curtis/The Stock Market; **149,** Skip Comer; **152,** Brownie Harris/The Stock Market; **154,** Elaine Shay; **155,** Skip Comer; **156,** Aaron Haupt; **160,** Barney Oldfield/FPG; **162,** NASA; **165,** British Museum/Bridgeman Art Library/Superstock; **166,** Skip Comer; **174,** Doug Watzstein/FPG; **175,** Mike and Carol Werner/Comstock, Inc; **177,** Elaine Shay; **178,** Dan Helms/DUOMO; **180,** Focus on Sports, Inc; **181,** Sheila Goode-Green; **184,** Skip Comer; **186(t),** Doug Martin, **(b),** Courtesy D. C. Comics/Movie Still Archives; **190,** Doug Martin; **198,** Skip Comer; **199,** Gerard Photography; **201,** Randy Scheiber; **203,** Art Montes de Oca/FPG; **208,** Carl R. Sams II/Peter Arnold, Inc; **209,** Animals Animals/Gary W. Griffen; **212,** Rod Planck/Photo Researchers; **216,** C. Simpson/FPG; **220,** file photo; **221,** NASA; **222,** Tom Dietrich/Tony Stone Images; **223,** NASA; **224,** Heinz Kluetmeier/Sports Illustrated; **226,** Bud Fowle; **229,** Stan Osolinski/The Stock Market; **230,** Skip Comer; **240,** Ping Amranand/Superstock; **246,** Elaine Shay; **254,** Bruce Handy/Comstock, Inc; **255,** Dan McCoy/Rainbow; **260,** Larry Hamill; **264,** Randy Duchaine/The Stock Market; **268,** David Lokey/Comstock, Inc; **270,** Tim Davis/Photo Researchers; **275,** Serraillier/Rapho/Photo Researchers; **280,** David Stoecklein/The Stock Market; **282,** Donald Johnson/The Stock Market; **289,** Clyde H. Smith/Peter Arnold, Inc; **294,** Tim Courlas; **295(l),** Doug Martin, **(r),** James Westwater; **299,** Comstock, Inc; **304,** Tim Cairns; **305,** file photo; **309,** Spencer Swanger/Tom Stack & Assoc; **314,** SV&B Productions/The Image Bank; **318,** Sharon Chester/Comstock, Inc; **321,** John Iacono/Sports Illustrated; **325,** Comstock, Inc; **326,** Brian Heston; **327,** Robert Kligge/The Stock Market; **329,** Ted Rice; **333,** Larry Hamill; **337,** Aaron Haupt; **338,** Tim Courlas; **339,** David Stoecklein/The Stock Market; **341,** Doug Martin; **342,** David Frazier; **344,** ITTC Productions/The Image Bank; **346,** Murray Alcosser/The Image Bank; **347,** Jean Anderson/The Stock Market; **352,** Romilly Lockyear/The Image Bank; **353,** David Frazier/Photo Researchers; **356,** file photo; **358(t),** file photo, **(b),** Larry Hamill; **359,** Steve Elmore/Tom Stack & Assoc; **361,363,** Elaine Shay; **364,** Cobalt Productions; **366,** Skip Comer; **368,** David Cavagnaro; **372,** Craig Hammell/The Stock Market; **376,** NASA; **379,** Ted Rice; **383,** Michael Furman/The Stock Market; **385,** Skip Comer; **388,** Tony Stone Images; **391,** Elyse Lewis/The Image Bank; **398,** A. Hurtman/Comstock, Inc; **399,** David Lokey/Comstock, Inc; **400(t),** David Frazier, **(b),** David Barnes/The Stock Market; **404,** Doug Martin; **407,** Lawrence Migdale/Photo Researchers; **409,** Michel Tcherevkoff/The Image Bank; **410,** Tim Courlas; **411,** Skip Comer; **414,** Elaine Shay; **416,** Skip Comer; **418,** Manfred Gottschalk/Tom Stack & Assoc; **419,** Aaron Haupt; **423,** Doug Martin; **426,** Ed Bock/The Stock Market; **431,** Howard Millard/The Stock Market; **438,** Coco McCoy/Rainbow; **439,** file photo; **446,** Garry McMichael/Photo Researchers; **447,** Bob Pool/Tom Stack & Assoc; **449,** Ken Frick; **451,** Andy Levin/Photo Researchers; **452,** Stanley Tess, Jr/The Stock Market; **454,** Karl Weatherly/Tony Stone Images; **459,** Michael Melford/The Image Bank; **467,** Bill Weber; **472,473,** R. Michael Stuckey/Comstock, Inc; **475,** The Bettmann Archive; **476,** Doug Martin; **482,** J. Blank/FPG; **486,** Steve Lissau; **491,** Marc Romanelli/The Image Bank; **496,** Doug Martin; **504,** Comstock, Inc; **506,** Jon Feingersh/Tom Stack & Assoc; **509,** Washnik Studio/The Stock Market; **516,** Mark Ippolito/Comstock, Inc; **517,** Albert Normundin/The Image Bank; **518,** Tim Courlas; **520,** UPI/Bettmann Newsphotos; **522,** Bruce Hands/Comstock, Inc; **523,** Doug Martin; **526,** Cheryl Blair;

527, Don E. Carroll/The Image Bank; **528,** Doug Martin; **529,** Frank Whitney/The Image Bank; **533,** Larry Dale Gordon/The Image Bank; **535,** Sharon Chester/Comstock, Inc; **537,** Gary Irving/Tony Stone Images; **538,** Al Tielemans/DUOMO; **540,** Tom Bean/The Stock Market; **541,** George D. Lepp/Comstock, Inc; **547,** Skip Comer; **549,** Lightscapes/The Stock Market; **558,** David Lissy/The Photo File; **559,** Ken Biggs/The Stock Market; **560,** Jon Feingersh/Tom Stack & Assoc; **561,** Peter Pearson/Tony Stone Images; **562,** R. Michael Stuckey/Comstock, Inc; **563,** S. Meiselas/Magnum; **564,** Skip Comer; **566(t),** Tom McHugh/Photo Researchers, **(b),** John Roberts/The Stock Market; **568,** Michael Dunn/The Stock Market; **570,** David Frazier; **573,** Doug Martin; **574,** Antonio Rosario/The Image Bank; **575,** Skip Comer; **578,** Courtesy Lucas Productions/ Movie Still Archives; **584,** Comstock, Inc; **586,** David E. Klutho/Sports Illustrated; **591,** Elaine Shay; **592,** Kevin Barry; **599,** Welzenback/The Stock Market; **601,603,** Elaine Shay; **608,** Ted Mahieu/The Stock Market; **609,** Skip Comer; **616,** Elaine Shay; **617,** Skip Comer; **622,** Mike and Carol Werner/Comstock, Inc; **631,** Rod Planck/Tom Stack & Assoc; **633,** Henry Groskinsky; **636,** Wendy Shattil/Bob Rozinski/Tom Stack & Assoc; **B1,** Gabe Palmer/The Stock Market; **B2,** White House Historical Association; **B3(tl),** Latent Image, **(tr),** Joseph DiChello, **(b),** Doug Martin; **B5,** Mak-1; **B6,** Don Smetzer/Tony Stone Images; **B7(t),** Tim Courlas, **(b),** Grandadan/Tony Stone Images; **B8,** Matt Meadows; **B9,** Mak-1; **B10,** Smithsonian Institution; **B11,** Mak-1; **B12(t),** Aaron Haupt, **(bl),** Skip Comer, **(br),** Aaron Haupt; **B13,** Hal Lynes; **B15,** Johnny Johnson; **B16(t),** Aaron Haupt, **(b),** Tim Courlas

MANIPULATIVE ACTIVITIES

for Merrill Algebra 1

UNDERSTANDING THROUGH HANDS-ON EXPERIENCES

Lab	Title	Use with lesson(s)
Lab 1	The Distributive Property	1-5
Lab 2	Adding and Subtracting Integers	2-2
Lab 3	Solving One-Step Equations	3-1, 3-2, 3-3
Lab 4	Solving Multi-Step Equations	3-5, 3-6
Lab 5	Polynomials	6-5
Lab 6	Adding and Subtracting Polynomials	6-6
Lab 7	Multiplying a Polynomial by a Monomial	6-7
Lab 8	Multiplying Polynomials	6-8
Lab 9	Factoring Using the Distributive Property	7-2
Lab 10	Factoring Trinomials	7-5
Lab 11	**Graphing Calculator:** Slope	10-3
Lab 12	**Graphing Calculator:** Intercepts	10-6
Lab 13	The Pythagorean Theorem	12-3
Lab 14	Estimating Square Roots	12-4
Lab 15	**Graphing Calculator:** Quadratic Functions	13-1
Lab 16	**Graphing Calculator:** Solving Quadratic Equations	13-2
Lab 17	Completing the Square	13-4

GLENCOE

Macmillan/McGraw-Hill

LAB 1 The Distributive Property

Use with: *Lesson 1–5, pages 26–30*
Materials: *algebra tiles, product mat*

Throughout your study of mathematics, you have used rectangles to model multiplication. For example, the figure at the right shows the multiplication $2(3 + 1)$ as a rectangle that is 2 units wide and $3 + 1$ units long. Its area is $2 \cdot 3 + 2 \cdot 1$ or 8 square units.

In this text, you will use special tiles called algebra tiles to form rectangles showing multiplication. A 1-tile is a square that is 1 unit long and 1 unit wide. Its area is 1 square unit. An x-tile is a rectangle that is 1 unit wide and x units long. Its area is x square units.

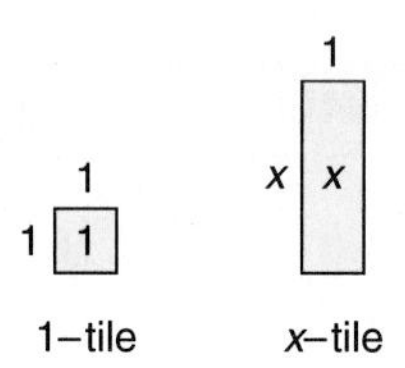

Activity: Use algebra tiles to find the product $2(x + 2)$.

- The rectangle has a width of 2 units and a length of $x + 2$ units. Use your area tiles to mark off the dimensions on a product mat.

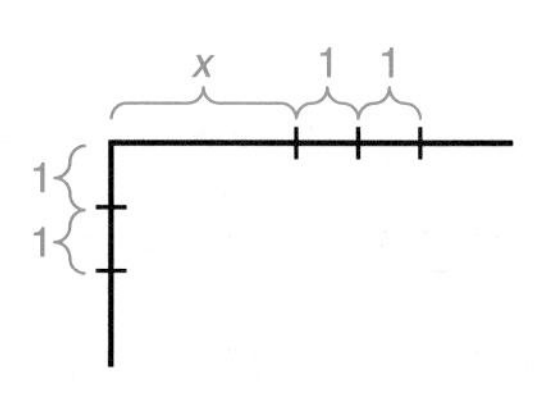

- Using the marks as a guide, make the rectangle with algebra tiles.

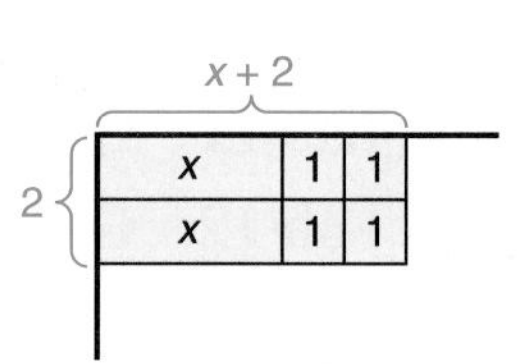

- The rectangle has 2 x-tiles and 4 1-tiles. The area of the rectangle is $x + 1 + 1 + x + 1 + 1$, or $2x + 4$. Thus, $2(x + 2) = 2x + 4$.

Model **Find each product using area tiles.**

1. $2(x + 1)$ **2.** $3(x + 2)$ **3.** $2(2x + 1)$ **4.** $2(3x + 3)$

Draw **Tell whether each statement is true or false. Justify your answer with a drawing.**

5. $3(x + 3) = 3x + 3$ **6.** $x(3 + 2) = 3x + 2x$

Write **7.** A classmate says that $3(x + 4) = 3x + 4$. How would you show your classmate that $3(x + 4) = 3x + 12$? Write your solution in paragraph form, complete with drawings.

8. Write a paragraph explaining how to find the product $10(x + 3)$ without using tiles.

LAB 2 Adding and Subtracting Integers

Use with: *Lesson 2–2, pages 55–59*
Materials: *counters, integer mat*

You can use counters to help you understand addition and subtraction of integers. In these activities, yellow counters represent positive integers, and red counters represent negative integers.

Integer Models
• A *zero-pair* is formed by pairing one positive counter with one negative counter.
• You can remove or add zero-pairs to a set because removing or adding zero does not change the value of the set.

Activity 1: **Use counters to find the sum $-3 + (-2)$.**

- Place 3 negative counters and 2 negative counters on the mat.

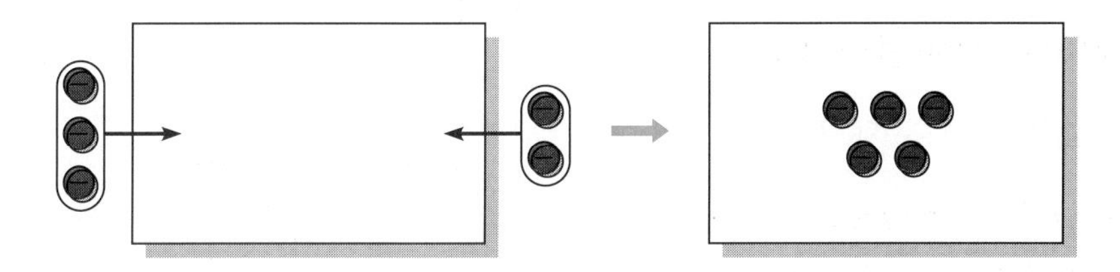

- Since there are 5 negative counters on the mat, the sum is -5. Therefore, $-3 + (-2) = -5$.

Activity 2: **Use counters to find the sum $-2 + 3$.**

- Place 2 negative counters and 3 positive counters on the mat. It is possible to remove 2 zero-pairs.

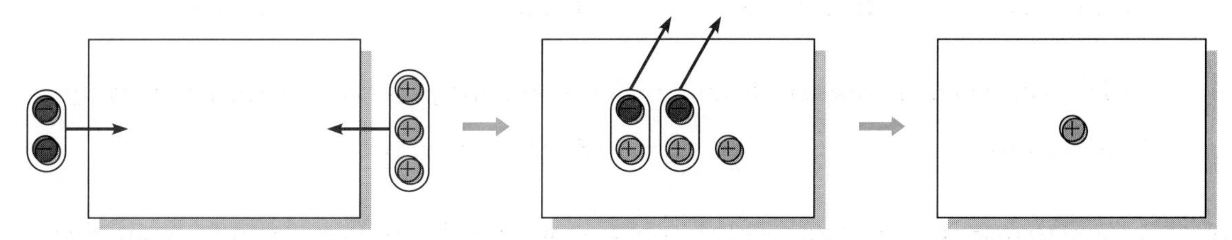

- Since 1 positive counter remains, the sum is 1. Therefore, $-2 + 3 = 1$.

Activity 3: **Use counters to find the difference $-4 - (-1)$.**

- Place 4 negative counters on the mat. Remove 1 negative counter.

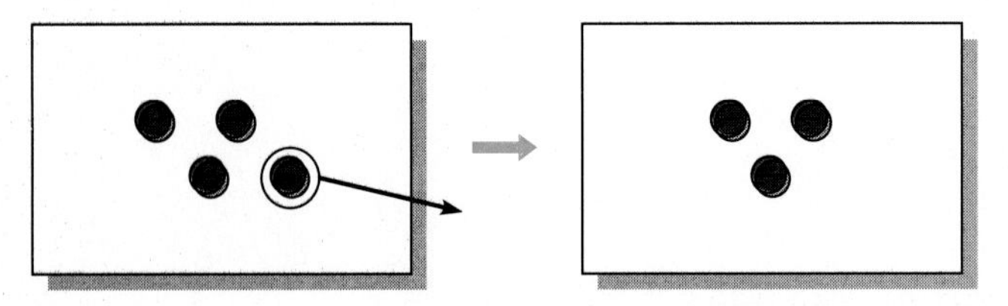

- Since 3 negative counters remain, the difference is -3.
  Therefore, $-4 - (-1) = -3$.

Activity 4: **Use counters to find the difference $3 - (-2)$.**

- Place 3 positive counters on the mat. There are no negative counters, so you can't remove 2 negatives. Add 2 zero-pairs to the mat. Remember, adding zero-pairs does not change the value of the set. Now remove 2 negative counters.

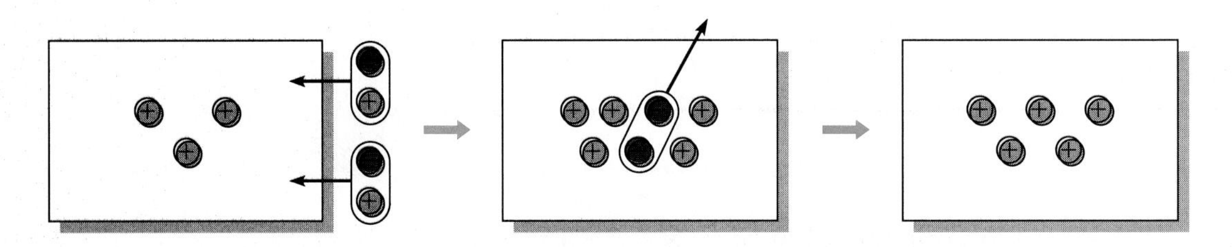

- Since 5 positive counters remain, the difference is 5.
  Therefore, $3 - (-2) = 5$.

Model **Find each sum or difference using counters.**

1. $4 + 2$ **2.** $4 + (-2)$ **3.** $-4 + 2$ **4.** $-4 + (-2)$

5. $4 - 2$ **6.** $-4 - (-2)$ **7.** $4 - (-2)$ **8.** $-4 - 2$

9. $1 - 4$ **10.** $2 - (-7)$ **11.** $-3 + 6$ **12.** $-3 + 3$

Draw **Tell whether each statement is true or false. Justify your answer with a drawing.**

13. $5 - (-2) = 3$ **14.** $-5 + 7 = 2$ **15.** $2 - 3 = -1$ **16.** $-1 - 1 = 0$

Write **17.** Write a paragraph explaining how to find the sum of two integers without using counters. Be sure to include all possibilities.

LAB 3 Solving One-Step Equations

Use with: *Lessons 3–1, 3–2, and 3–3, pages 94–107*
Materials: *cups, counters, equation mat*

You can use cups and counters as a model for solving equations. In this model, a cup represents the variable, yellow counters represent positive integers, and red counters represent negative integers. After you model the equation, the goal is to get the cup by itself on one side of the mat, by using the rules stated below.

Equation Models
• A zero-pair is formed by pairing one positive identical counter with one negative counter. • You can remove or add the same number of identical counters to each side of the equation mat. • You can remove or add zero-pairs to either side of the equation mat without changing the equation.

Activity 1: **Use an equation model to solve $x + (-3) = -5$.**

- Place 1 cup and 3 negative counters on one side of the mat. Place 5 negative counters on the other side of the mat. The two sides of the mat represent equal quantities.

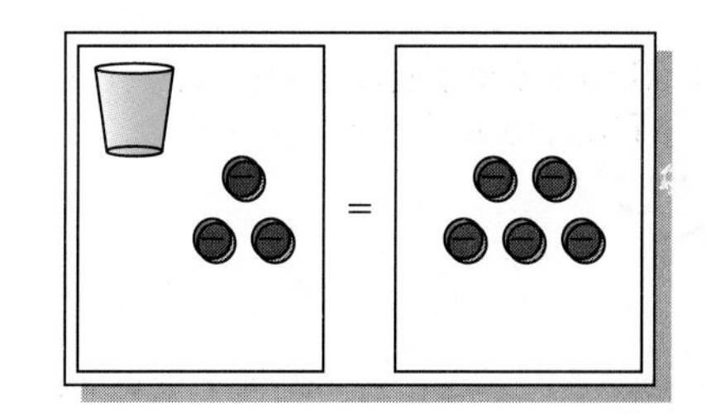

- Remove 3 negative counters from each side to get the cup by itself.

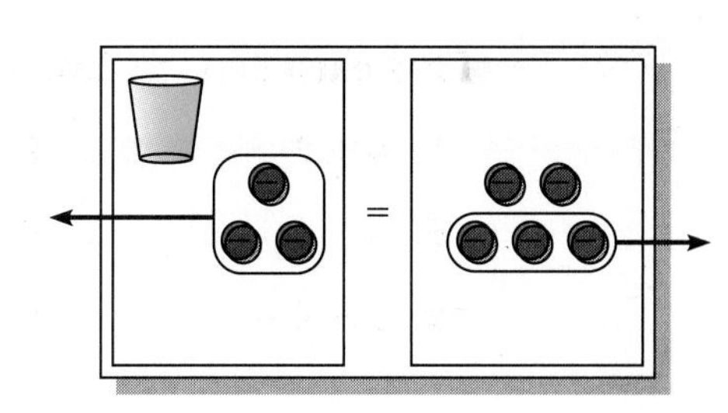

- The cup on the left side of the mat is matched with 2 negative counters. Therefore, $x = -2$.

Activity 2: **Use an equation model to solve $2p = -6$.**

- Place 2 cups on one side of the mat. Place 6 negative counters on the other side of the mat.

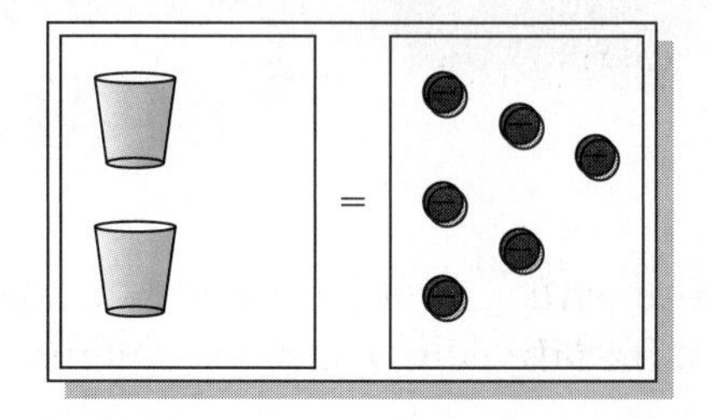

- Separate the counters into 2 equal groups to correspond to the 2 cups.

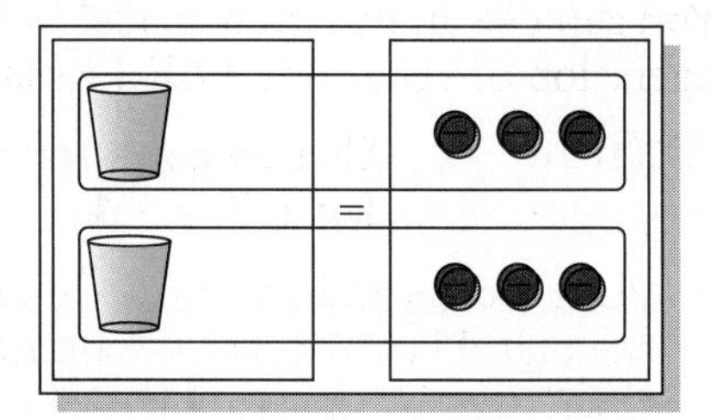

- Each cup on the left is matched with 3 negative counters. Therefore, $p = -3$.

Activity 3: **Use an equation model to solve $r - 2 = 3$.**

- Write the equation in the form $r + (-2) = 3$. Place 1 cup and 2 negative counters on one side of the mat. Place 3 positive counters on the other side of the mat. Notice that it is not possible to remove the same kind of counters from each side. Add 2 positive counters to each side.

- Group the counters to form zero-pairs. Then remove all zero-pairs.

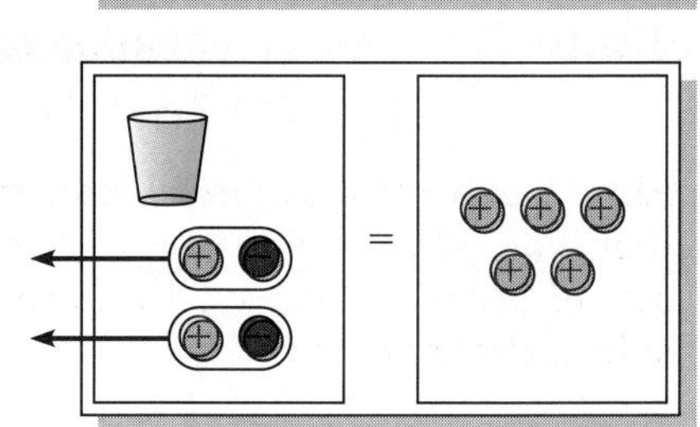

- The cup on the left is matched with 5 positive counters. Therefore, $r = 5$.

Model **Use an equation model to solve each equation.**

1. $x + 4 = 5$
2. $y + (-3) = -1$
3. $y + 7 = -4$
4. $3z = -9$
5. $m - 6 = 2$
6. $-2 = x + 6$
7. $8 = 2a$
8. $w - (-2) = 2$

Draw **Tell whether each number is a solution of the given equation. Justify your answer with a drawing.**

9. -3; $x + 5 = -2$
10. -1; $5b = -5$
11. -4; $y - 4 = -8$

Write 12. Write a paragraph explaining why you use zero-pairs to solve an equation such as $m + 5 = -8$.

LAB 4

Solving Multi-Step Equations

Use with: *Lessons 3–5 and 3–6, pages 111–120*
Materials: *cups, counters, equation mat*

You can use an equation model to solve equations with more than one operation or equations with a variable on each side.

Activity 1: Use an equation model to solve $2x + 2 = -4$.

- Place 2 cups and 2 positive counters on one side of the mat. Place 4 negative counters on the other side of the mat. Notice it is not possible to remove the same kind of counters from each side. Add 2 negative counters to each side.
- Group the counters to form zero-pairs and remove all zero-pairs. Separate the remaining counters into 2 equal groups to correspond to the 2 cups.
- Each cup is matched with 3 negative counters. Therefore, $x = -3$.

Activity 2: Use an equation model to solve $w - 3 = 2w - 1$.

- Place 1 cup and 3 negative counters on one side of the mat. Place 2 cups and 1 negative counter on the other side of the mat. Remove 1 negative counter from each side of the mat.

- Just as you can remove the same kind of counter from each side of the mat, you can remove cups from each side of the mat. In this case, remove 1 cup from each side.
- The cup on the right is matched with 2 negative counters. Therefore, $w = -2$.

Model Use an equation model to solve each equation.

1. $2x + 3 = 13$ **2.** $2y - 2 = -4$ **3.** $-4 = 3a + 2$

4. $3m - 2 = 4$ **5.** $3x + 2 = x + 6$ **6.** $3x + 7 = x + 1$

7. $3x - 2 = x + 6$ **8.** $y + 1 = 3y - 7$ **9.** $2b + 3 = b + 1$

LAB 5 Polynomials

Use with: *Lesson 6–5, pages 226–229*
Materials: *algebra tiles*

You can use algebra tiles as a model for polynomials.

Polynomial Models

- The model is based on these three tiles.

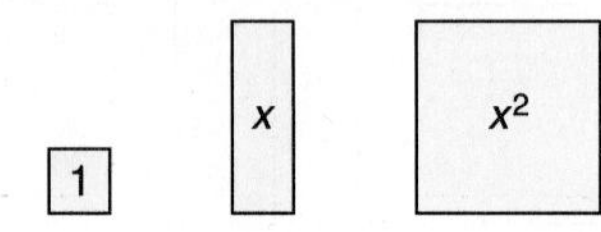

- Each tile shown above has an opposite.

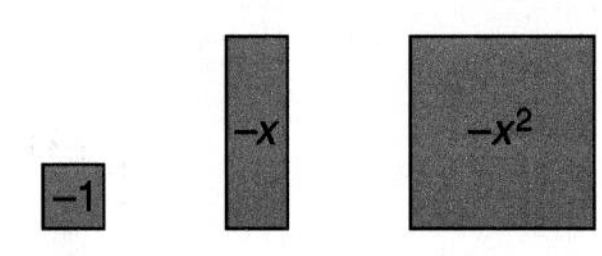

- A zero-pair is formed by pairing one tile with its opposite.

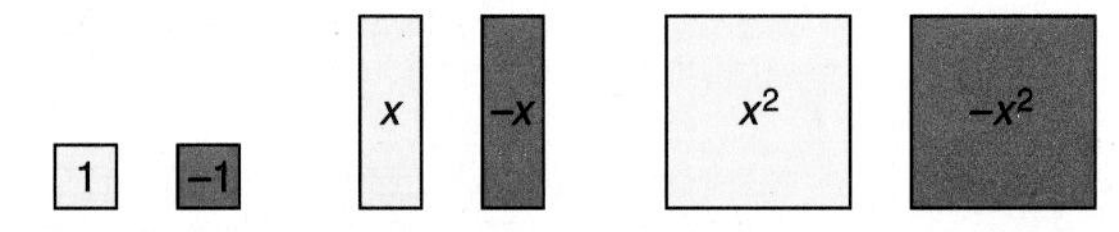

- You can remove or add zero-pairs without changing the value of the polynomial.
- *Like terms* are represented by tiles that are the same shape and size.

Activity 1: **Use algebra tiles to model each monomial or polynomial.**

a. $3x^2$

To model this expression, you need 3 yellow x^2-tiles.

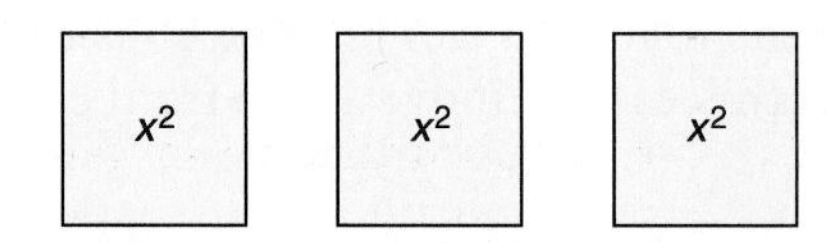

b. $x^2 - 2d$

To model this expression, you need 1 yellow x^2-tile and 2 red x-tiles.

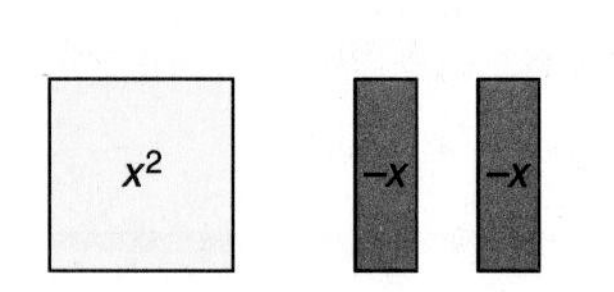

c. $2x^2 + x - 2$

To model this expression, you need 2 yellow x^2-tiles, 1 yellow x-tile, and 2 red 1-tiles.

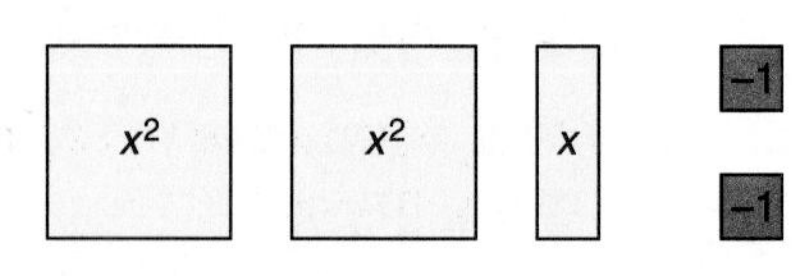

Monomials such as $3x$ and $5x$ are called *like terms* because they have the same variable to the same power. When you use algebra tiles, you can recognize like terms because they have the same size and shape. In the activities that follow, you will simplify polynomials that have like terms.

Activity 2: Use algebra tiles to simplify $2x^2 + x^2 + 2x$.

- Model the polynomial.
- Combine like terms.

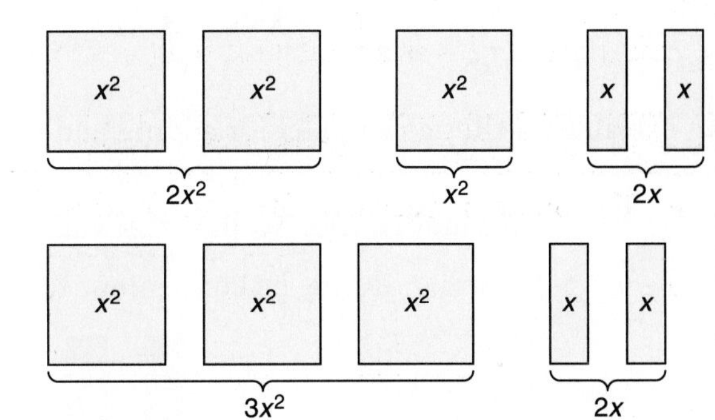

- In simplest form, $2x^2 + x^2 + 2x = 3x^2 + 2x$.

Activity 3: Use algebra tiles to simplify $3x + 2 - 5x + 1$.

- Model the polynomial.
- Rearrange the tiles so that like terms are next to each other. Form zero-pairs.
- Remove all zero-pairs.

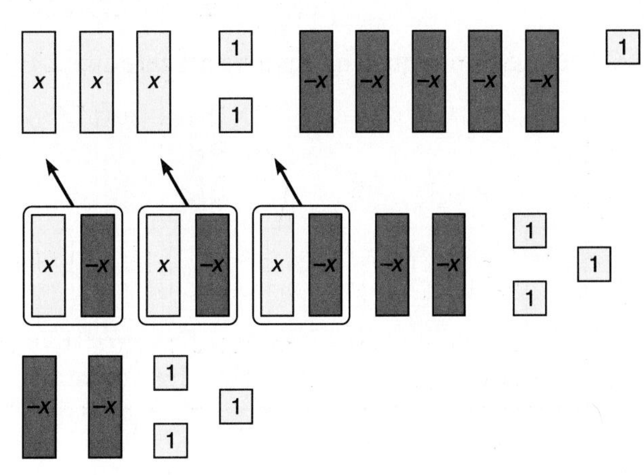

- In simplest form, $3x + 2 - 5x + 1 = -2x + 3$.

Model **Use algebra tiles to model each monomial or polynomial.**

1. $-2x^2$ **2.** $5x + 3$ **3.** $3x^2 + 2x + 6$ **4.** $x^2 - 8$

Use algebra tiles to simplify each polynomial.

5. $2x + 3 - x + x^2 - 4$ **6.** $3x^2 - 2x^2 + 3x$

7. $-x^2 + 4x + 2x + x^2$ **8.** $x^2 + 4x - 2x + 3$

Draw **Use a drawing to simplify each polynomial.**

9. $2a^2 - 3a + a^2 + 3 + 2a$ **10.** $-2m^2 - 3m^2 - 4m + 3$

Write **11.** Write a sentence to explain how subtracting polynomials is related to adding polynomials.

LAB 6 Adding and Subtracting Polynomials

Use with: *Lesson 6–6, pages 230–233*
Materials: *algebra tiles*

You can use algebra tiles as a model for adding and subtracting polynomials.

Activity 1: Use algebra tiles to find the sum $(x^2 - 2x + 4) + (2x^2 + 5x - 5)$.

- Model each polynomial. It may be convenient to arrange like terms in columns.

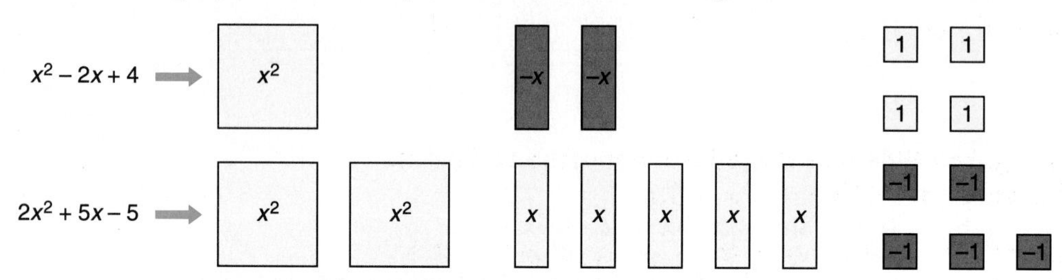

- Combine like terms and remove all zero-pairs.

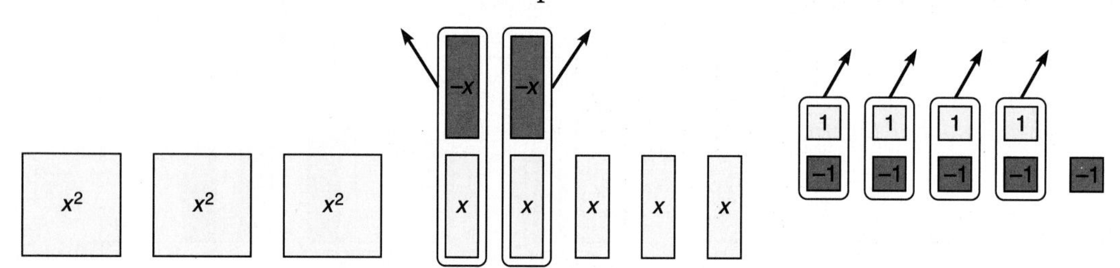

- Write the polynomial for the tiles that remain.
  Therefore, $(x^2 - 2x + 4) + (2x^2 + 5x - 5) = 3x^2 + 3x - 1$.

Activity 2: Use algebra tiles to find the difference $(3x + 5) - (-2x + 2)$.

- Model the polynomial $3x + 5$.

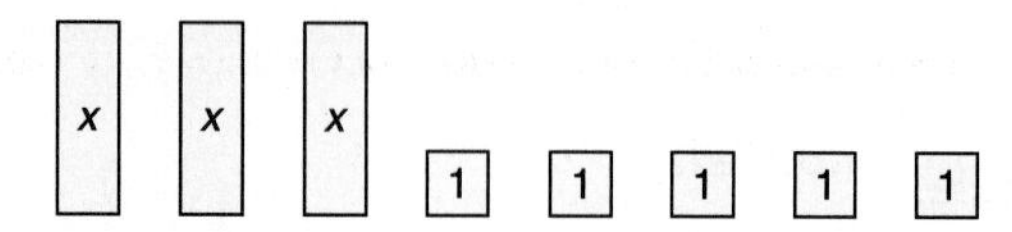

- To subtract $-2x + 2$, you must remove 2 red x-tiles and 2 yellow 1-tiles. You can remove the 1-tiles, but there are no red x-tiles. Add 2 zero-pairs of x-tiles. Then remove 2 red x-tiles.

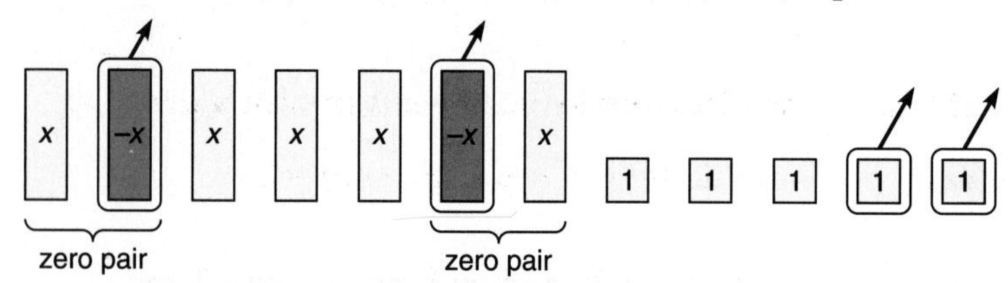

- Write the polynomial for the tiles that remain.
  Therefore, $(3x + 5) - (-2x + 2) = 5x + 3$.

Recall that you can subtract a rational number by adding its additive inverse or opposite. Similarly, you can subtract a polynomial by adding its opposite.

Activity 3: **Use algebra tiles to model the opposite of $-2x + 2$.**

- Model the expression.

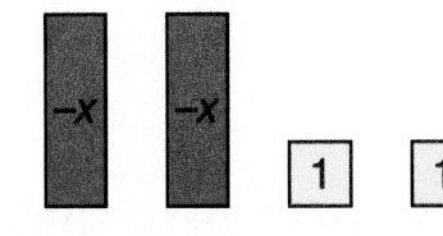

- Replace each tile with its opposite.

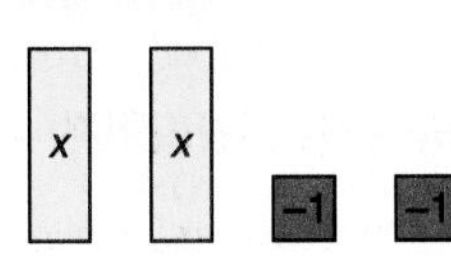

- The opposite of $-2x + 2$ is $2x - 2$.

Activity 4: **Use algebra tiles and the additive inverse to find the difference $(3x + 5) - (-2x + 2)$.**

- Model the polynomial $3x + 5$ and model the opposite of $-2x + 2$. In Activity 3, you found that the opposite of $-2x + 2$ is $2x - 2$. Combine like terms and remove all zero-pairs.

$3x + 5$ →

$-2x + 2$ additive inverse $2x - 2$ →

- Write the polynomial for the tiles that remain.
Therefore, $(3x + 5) - (-2x + 2) = 5x + 3$. Note this is the same answer as Activity 2.

Model **Use algebra tiles to find each sum or difference.**

1. $(3x^2 - 3x + 5) + (-2x^2 + 3x)$
2. $(-x^2 + 2x) + (3x - 4)$
3. $(x^2 + 2x) - (2x^2 + x)$
4. $(2x^2 - 5x - 1) - (x^2 - x - 1)$

Draw **Tell whether each statement is true or false. Justify your answer with a drawing.**

5. $(9y - 1) - (7y + 2) = 2y + 1$
6. $(a^2 - 4a) + (2a^2 - 6a) = 3a^2 - 10a$

Write 7. Find the difference $(3x + 4) - (2x - 3)$ using each method from Activity 2 and Activity 4. Write a paragraph, including drawings, that explains how zero-pairs are used in each case.

LAB 7

Multiplying a Polynomial by a Monomial

Use with: *Lesson 6–7, pages 234–237*
Materials: *algebra tiles, product mat*

You have used rectangles to model multiplication. In this activity, you will use algebra tiles to find the product of simple polynomials. The width and length of the rectangle will each represent a polynomial. So, the area of the rectangle represents the product of the polynomials.

Activity 1: Use algebra tiles to find the product $x(x + 2)$.

- The rectangle has a width of x units and a length of $x + 2$ units. Use your algebra tiles to mark off the dimensions on a product mat.

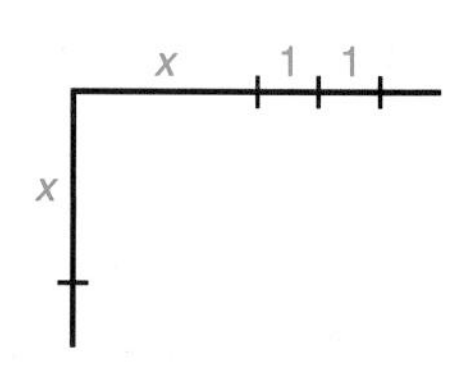

- Using the marks as a guide, make the rectangle with algebra tiles.
- The rectangle has 1 x^2-tile and 2 x-tiles. The area of the rectangle is $x^2 + 2x$. Thus, $x(x + 2) = x^2 + 2x$.

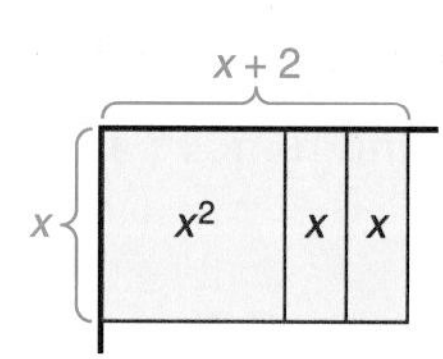

Activity 2: Use algebra tiles to find the product $x(x - 1)$.

- The rectangle has a width of x units and a length of $x - 1$ units. Mark off the dimensions on a product mat as shown at the right.

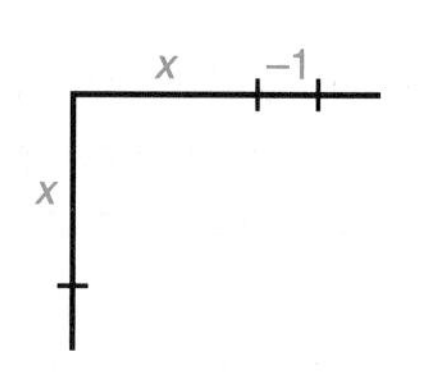

- Make the rectangle with a yellow x^2-tile and a red x-tile. The red tile is used because the length is -1.
- The area of the rectangle is $x^2 - x$. Thus, $x(x - 1) = x^2 - x$.

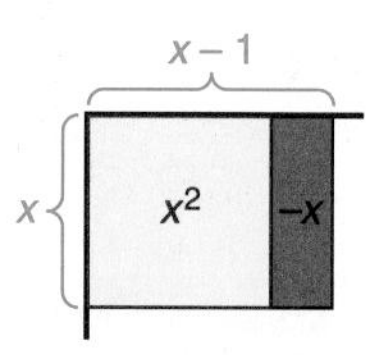

Model **Use algebra tiles to find each product.**

1. $x(x + 3)$
2. $x(x - 2)$
3. $2x(x + 1)$
4. $3x(2x - 1)$

Draw **Tell whether each statement is true or false. Justify your answer with a drawing.**

5. $x(2x + 3) = 2x^2 + 3x$
6. $2x(3x - 4) = 6x^2 - 4x$

Write

7. Suppose you have a square garden plot that measures x feet on a side. If you double the length of the plot and increase the width by 3 feet, how large will the new plot be? Write your solution in paragraph form, complete with drawings.

LAB 8 Multiplying Polynomials

Use with: *Lesson 6–8*
Materials: *algebra tiles, product mat*

You can find the product of simple binomials using algebra tiles.

Activity 1: **Use algebra tiles to find the product $(x + 1)(x + 2)$.**

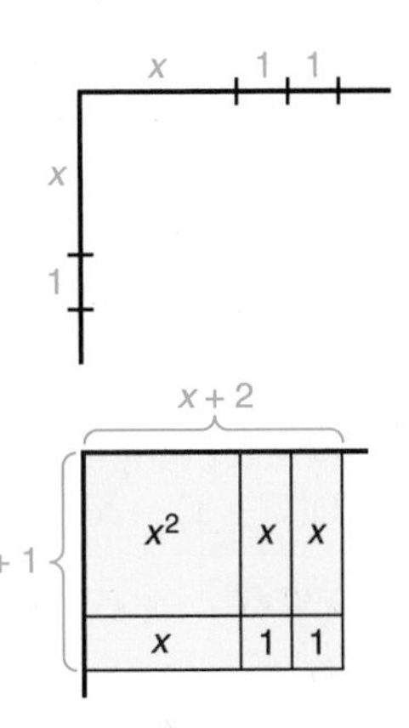

- The rectangle has a width of $x + 1$ units and a length of $x + 2$ units. Use your algebra tiles to mark off the dimensions on a product mat.
- Using the marks as a guide, make the rectangle with algebra tiles.
- The rectangle has 1 x^2-tile, 3 x-tiles, and 2 1-tiles. The area of the rectangle is $x^2 + 3x + 2$. Thus, $(x + 1)(x + 2) = x^2 + 3x + 2$.

Activity 2: **Use algebra tiles to find the product $(x + 2)(x - 1)$.**

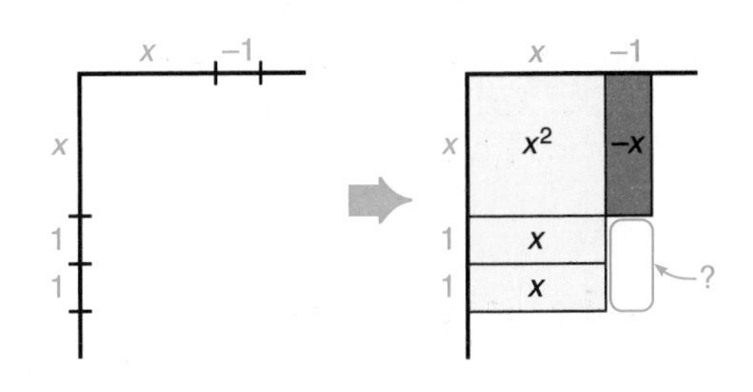

- The rectangle has a width of $x + 2$ units and a length of $x - 1$ units. Use your algebra tiles to mark off the dimensions on a product mat.
- Begin to make the rectangle as shown at the right. You need 1 yellow x^2-tile, 1 red x-tile, and 2 yellow x-tiles. Now you need to determine whether to use 2 yellow 1-tiles or 2 red 1-tiles to complete the rectangle.

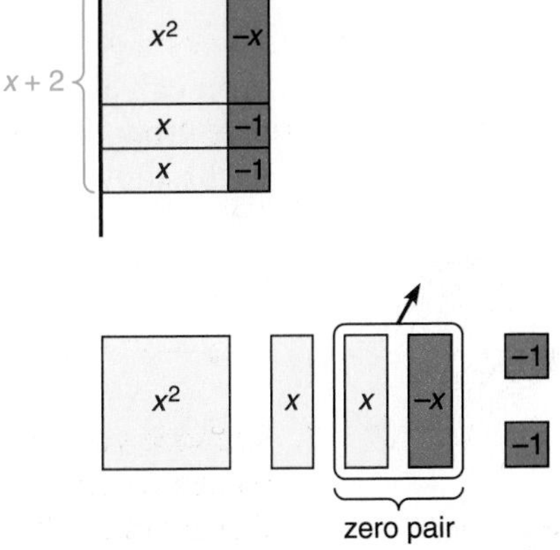

- Remember that the numbers at the top and on the side give the dimensions of the tile needed. The area of each tile is the product of -1 and 1, or -1. This is represented by a red 1-tile. Fill in the space with 2 red 1-tiles to complete a rectangle.
- You can rearrange the tiles to simplify the polynomial you have formed. Notice that a zero pair is formed by the x-tiles. In simplest form, $(x + 2)(x - 1) = x^2 + x - 2$.

Activity 3: Use algebra tiles to find the product $(x - 1)(x - 1)$.

- The rectangle has a width of $x - 1$ units and a length of $x - 1$ units. Use your algebra tiles to mark off the dimensions on a product mat.

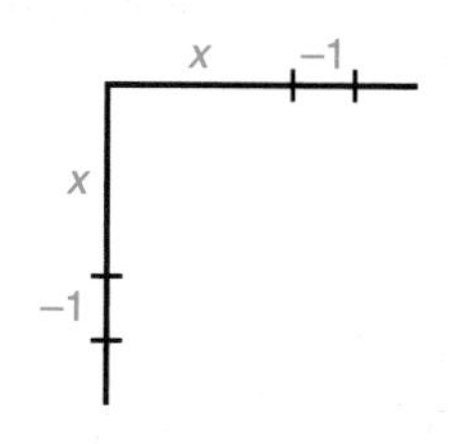

- Begin to make the rectangle as shown at the right. You need 1 yellow x^2-tile and 2 red x-tiles. Now you need to determine whether to use a yellow 1-tile or a red 1-tile.

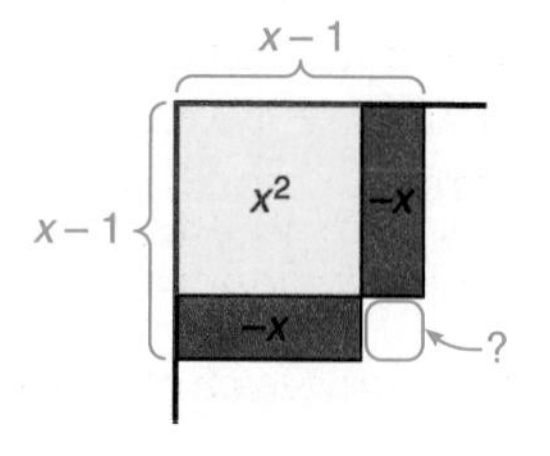

- The area of the tile is the product of -1 and -1. Since $-1 \cdot -1 = 1$, the tile needed is a yellow 1-tile.

- The area of the rectangle is $x^2 - x - x + 1$. In simplest form, $(x - 1)(x - 1) = x^2 - 2x + 1$.

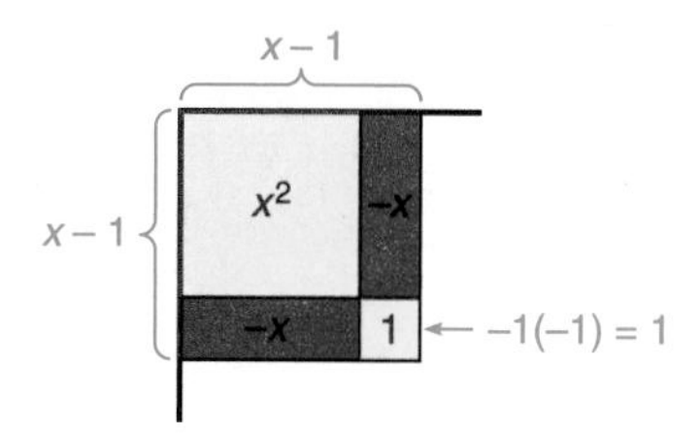

Model **Use algebra tiles to find each product.**

1. $(x + 2)(x + 2)$
2. $(x + 1)(x + 3)$
3. $(2x + 1)(x + 2)$
4. $(x + 1)(2x + 3)$
5. $(x + 2)(x - 3)$
6. $(x - 3)(x + 1)$
7. $2x + 1)(x - 2)$
8. $(x - 1)(x - 2)$
9. $(x - 2)(x - 3)$

Draw **Tell whether each statement is true or false. Justify your answer with a drawing.**

10. $(x + 3)(x + 2) = x^2 + 6$
11. $(x + 3)(x - 2) = x^2 - x - 6$
12. $(x + 2)(x - 2) = x^2 - 4$
13. $(x - 1)(x - 3) = x^2 + 4x + 3$

Write

14. You can also use the distributive property to find the product of two binomials. The figure at the right shows the model for $(x + 2)(x + 1)$ separated into four parts. Write a paragraph explaining how the model shows the use of the distributive property.

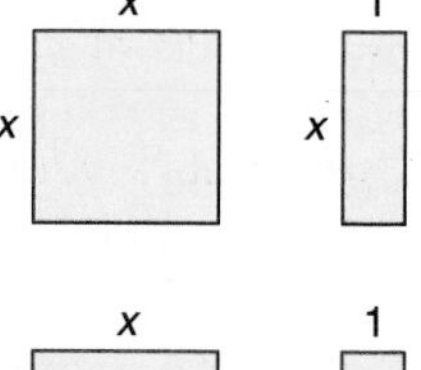

15. Write a paragraph describing the visual patterns possible in the rectangle representing the product of two binomials. Include drawings in your answer.

LAB 9 Factoring Using the Distributive Property

Use with: *Lesson 7–2, pages 261–264*
Materials: *algebra tiles, product mat*

You may recall that when two or more numbers are multiplied, these numbers are called *factors* of the product. Sometimes you know the product of polynomials and are asked to find the factors. This is called *factoring.* Using the algebra tile model, this means that you know the area of a rectangle and are asked to find the length and width.

Activity 1: Use algebra tiles to factor $3x + 6$.

- Model the polynomial $3x + 6$.
- Arrange the tiles into a rectangle.
- The rectangle has a width of 3 and a length of $x + 2$. Therefore, $3x + 6 = 3(x + 2)$.

Activity 2: Use algebra tiles to factor $x^2 - 4x$.

- Model the polynomial $x^2 - 4x$.

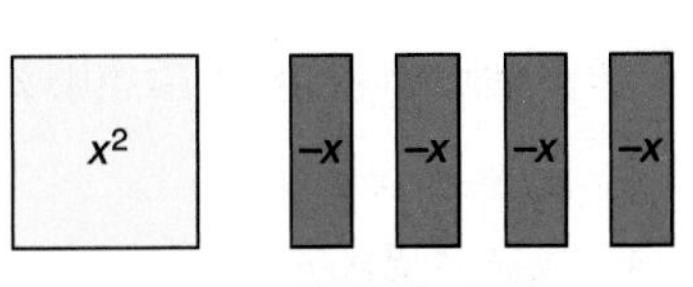

- Arrange the tiles into a rectangle.

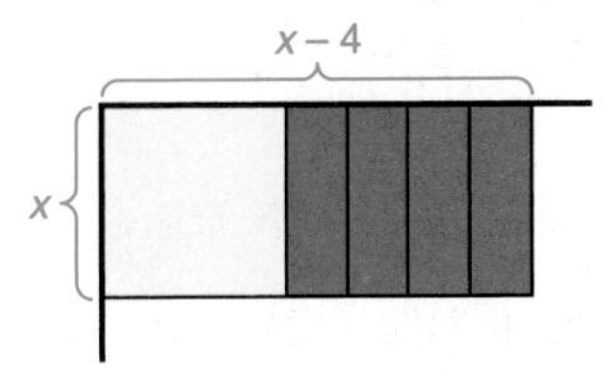

- The rectangle has a width of x and a length of $x - 4$. Therefore, $x^2 - 4x = x(x - 4)$.

Model **Use algebra tiles to factor each binomial.**

1. $5x + 10$ **2.** $4x - 14$ **3.** $4x^2 + 6x$ **4.** $6 - 6x$

Draw **Tell whether each binomial can be factored. Justify your answer with a drawing.**

5. $3x + 5$ **6.** $4 - 20x$ **7.** $3x^2 - 6x$ **8.** $5x^2 - 3$

Write **9.** Write a paragraph that explains how you can determine whether a binomial can be factored. Include an example of one binomial that can be factored and one that cannot.

LAB 10 Factoring Trinomials

***Use with:** Lesson 7–5, pages 271–275*
***Materials:** algebra tiles, product mat*

You can use algebra tiles as a model for factoring simple trinomials. If a rectangle cannot be formed to represent the trinomial, then the trinomial is not factorable.

Activity 1: Use algebra tiles to factor $x^2 + 3x + 2$.

- Model the polynomial $x^2 + 3x + 2$.

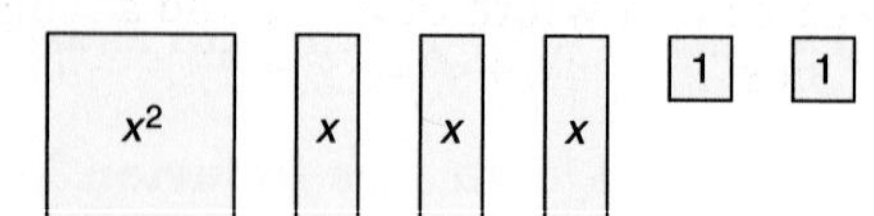

- Place the x^2-tile at the corner of the product mat. Arrange the 1-tiles into a 1-by-2 rectangular array as shown.

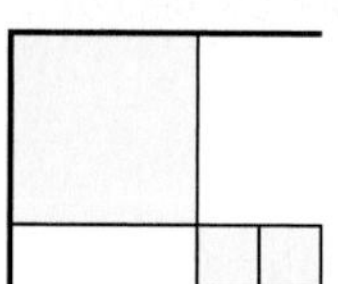

- Complete the rectangle with the x-tiles.

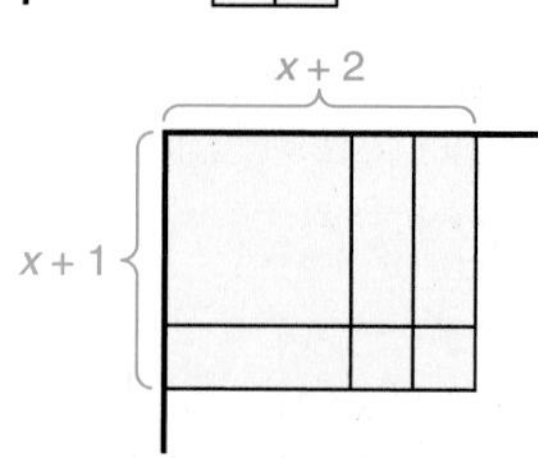

- The rectangle has a width of $x + 1$ and a length of $x + 2$. Thus, $x^2 + 3x + 2 = (x + 1)(x + 2)$.

You will need to use the guess-and-check strategy with many trinomials.

Activity 2: Use algebra tiles to factor $x^2 + 5x + 4$.

- Model the polynomial $x^2 + 5x + 4$.

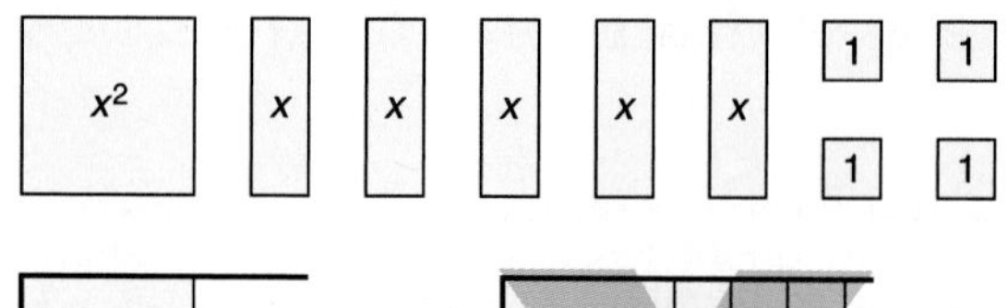

- Place the x^2-tile at the corner of the product mat. Arrange the 1-tiles into a 2-by-2 rectangular array as shown. Try to complete the rectangle. Notice there seems to be an extra x-tile.

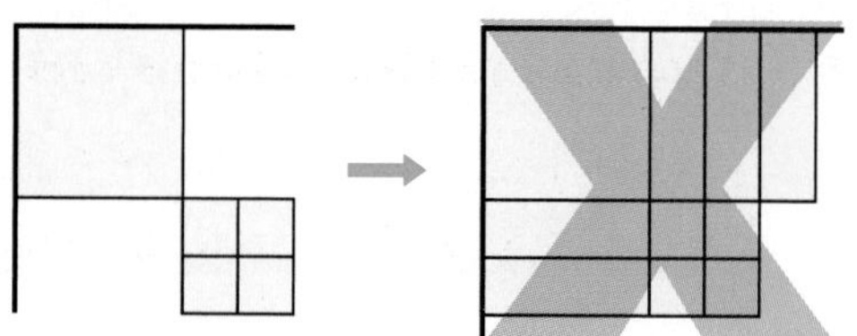

- Arrange the 1-tiles into a 1 by 4 rectangular array. This time, you can complete the rectangle with the x-tiles.

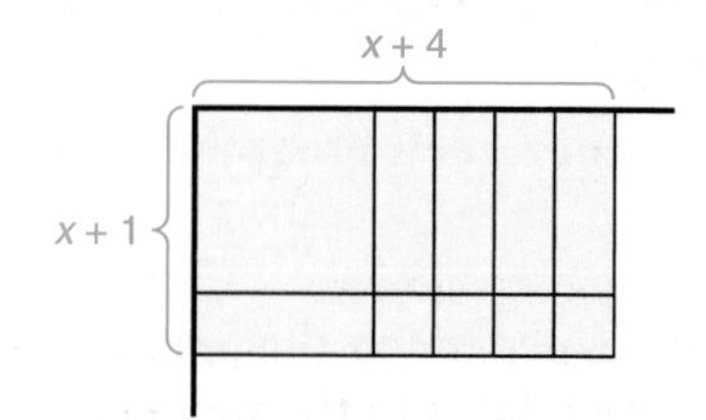

- The rectangle has a width of $x + 1$ and a length of $x + 4$. Thus, $x^2 + 5x + 4 = (x + 1)(x + 4)$.

Activity 3: Use algebra tiles to factor $x^2 - 4x + 4$.

- Model the polynomial $x^2 - 4x + 4$.

- Place the x^2-tile at the corner of the product mat. Arrange the 1-tiles into a 2-by-2 rectangular array as shown.

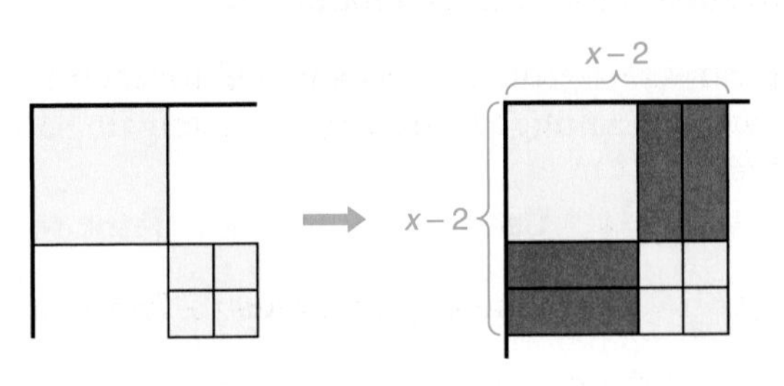

- Complete the rectangle with the x-tiles.

- The rectangle has a width of $x - 2$ and a length of $x - 2$. Thus, $x^2 - 4x + 4 = (x - 2)(x - 2)$.

Activity 4: Use algebra tiles to factor $x^2 - 2x - 3$.

- Model the polynomial $x^2 - 2x - 3$.

- Place the x^2-tile at the corner of the product mat. Arrange the 1-tiles into a 1-by-3 rectangular array as shown.

- Place the x-tiles as shown. Recall that you can add zero-pairs without changing the value of the polynomial. In this case, add a yellow x-tile and a red x-tile.

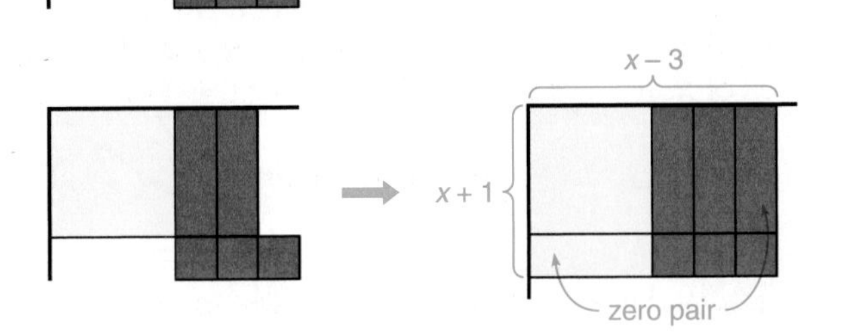

- The rectangle has a width of $x + 1$ and a length of $x - 3$. Thus, $x^2 - 2x - 3 = (x + 1)(x - 3)$.

Model **Use algebra tiles to factor each trinomial.**

1. $x^2 + 6x + 8$ **2.** $x^2 + 6x + 9$ **3.** $x^2 + 7x + 10$ **4.** $x^2 - 7x + 10$

5. $x^2 - 5x + 6$ **6.** $x^2 - 5x + 4$ **7.** $x^2 + 2x - 3$ **8.** $x^2 + x - 1$

Draw **If you cannot form a rectangle with the algebra tiles, the trinomial cannot be factored. Tell whether each trinomial can be factored. Justify your answer with a drawing.**

9. $x^2 - 8x + 15$ **10.** $x^2 + 3x - 8$ **11.** $x^2 - 5x - 24$ **12.** $x^2 + 2x + 6$

Write **13.** The expression $x^2 - 6x + 9$ is an example of a perfect square trinomial. Use algebra tiles to factor this expression. Describe the shape of the model and explain how the name relates to the model. Give examples of two other perfect square trinomials.

LAB 11 Slope

Use with: *Lesson 10–3, pages 410–414*
Materials: *graphing calculator*

You learned to use the graphing calculator to graph linear functions on page 384. You can also use a graphing calculator to investigate patterns that occur within families of graphs.

Activity 1: **Use a graphing calculator to graph $y = x$, $y = 3x$, and $y = 5x$ on the same screen. Describe the similarities and differences among the graphs.**

- Graph each of the equations. Make sure your calculator is set for the standard viewing window, $[-10, 10]$ by $[-10, 10]$.
- The graphs are all lines with a left-to-right upward slope, and all pass through the origin. The graphs have a different slope. The graph of $y = 5x$ is the steepest, and the graph of $y = x$ is the least steep.

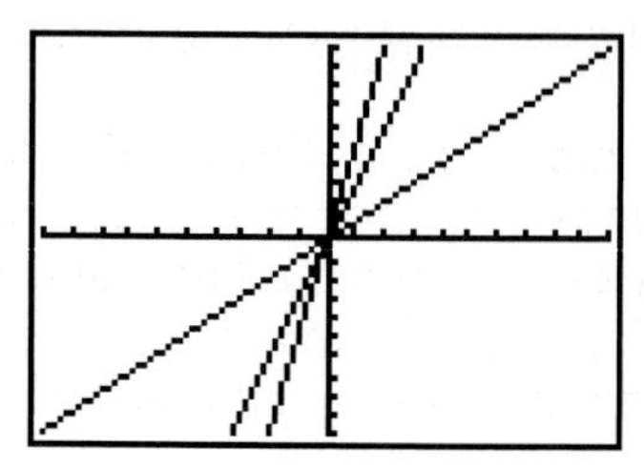

Activity 2: **For each equation, predict what its graph will look like. Then graph each equation on the same screen as the equations in Activity 1.**

a. $y = 2x$

The graph of $y = 2x$ is a line that passes through the origin. It will fall in between the graphs of $y = x$ and $y = 3x$.

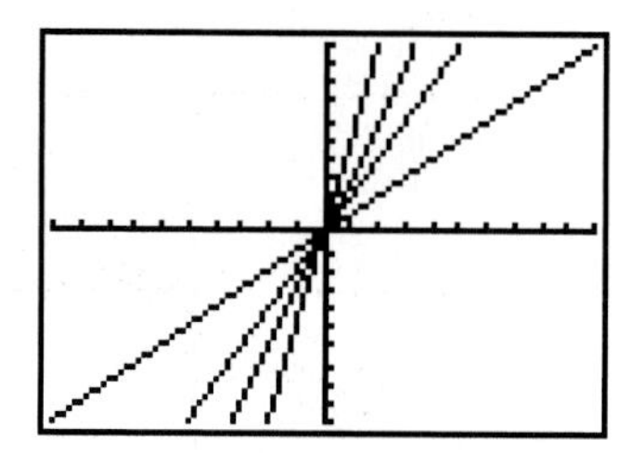

b. $y = 0.5x$

The graph of $y = 0.5x$ is a line that passes through the origin. It will fall in between the graph of $y = x$ and the x-axis.

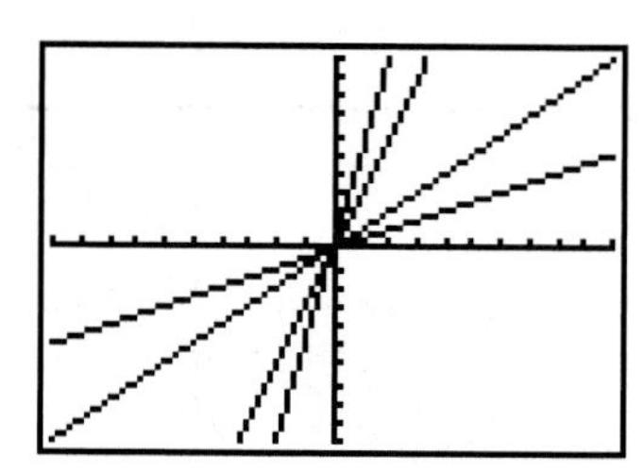

Draw

1. Graph $y = -x$, $y = -3x$, and $y = -5x$ on the same screen. Make a sketch of the graphs. Describe the similarities and differences among the graphs.
2. Graph each equation on the same screen as the equations in Exercise 1. Sketch the graphs. Describe the similarities and differences among the graphs.
   a. $y = -2x$ **b.** $y = -0.5x$

Write

3. Write a paragraph comparing the graphs of equations with a positive coefficient of x and a negative coefficient of x. Justify your answer with a drawing.
4. Write a paragraph describing how the coefficient of x affects the graph of the equation.
5. Write an equation of a line whose graph has a left-to-right downward slope.
6. Write an equation of a line whose slope is greater than the slope of $y = 10x$.
7. Write an equation of a line whose graph lies between the graphs of $y = -2x$ and $y = -3x$.

LAB 12 Intercepts

Use with: *Lesson 10–6, pages 423–426*
Materials: *graphing calculator*

Activity 1: **Use a graphing calculator to graph $y = x$, $y = x + 2$, and $y = x - 2$ on the same screen. Describe the similarities and differences among the graphs.**

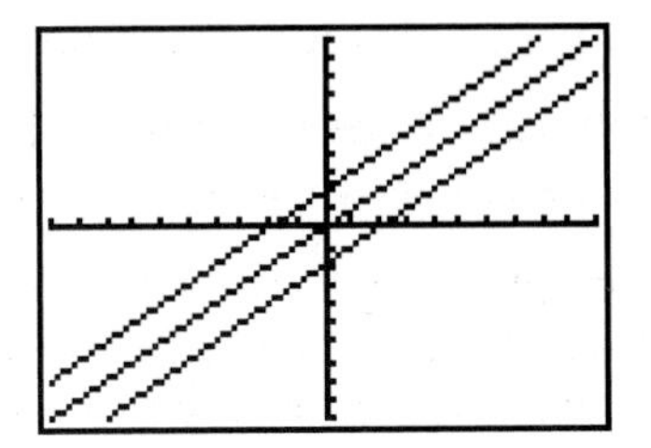

- Graph each of the equations. Make sure your calculator is set for the standard viewing screen, $[-10, 10]$ by $[-10, 10]$.

- The graphs are all lines with a left-to-right upward slope, and all lines have the same slope. The lines pass through different points on the y-axis. The graph of $y = x$ passes through the origin, (0, 0), the graph of $y = x + 2$ passes through (0, 2), and the graph of $y = x - 2$ passes through (0, -2).

Activity 2: **For each equation, predict what its graph will look like. Then graph each equation on the same screen as the equations in Activity 1.**

a. $y = x + 3$

The graph of $y = x + 3$ is a line that passes through (0, 3) with the same slope as $y = x$, $y = x + 2$, and $y = x - 2$.

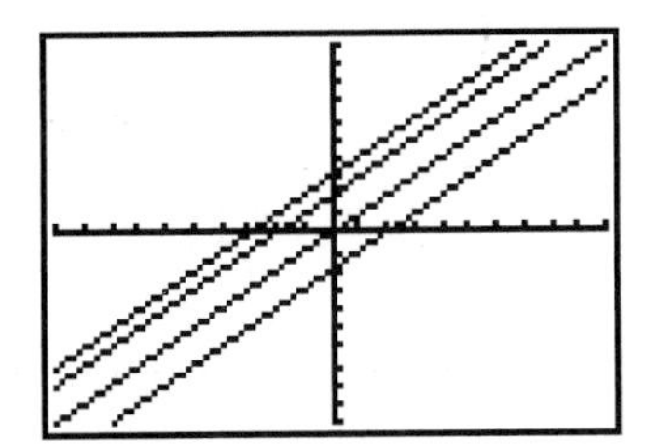

b. $y = x - 1$

The graph of $y = x - 1$ is a line that passes through (0, -1) with the same slope as $y = x$, $y = x + 2$, and $y = x - 2$.

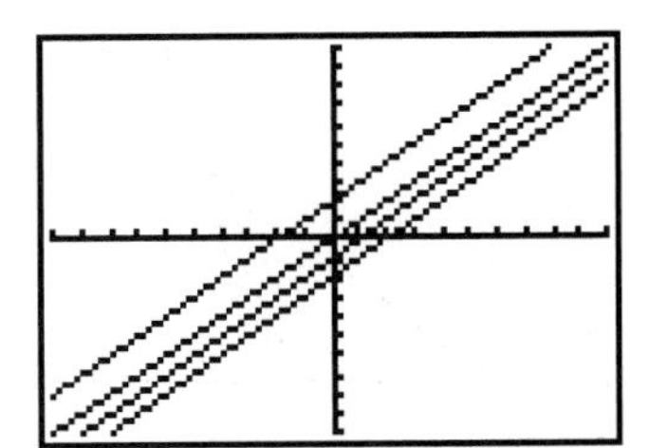

Draw

1. Graph $y = -x$, $y = -x + 2$, and $y = -x - 2$ on the same screen. Sketch the graphs. Describe the similarities and differences among the graphs.

2. For each equation, predict what its graph will look like. Then graph the equation on the same screen with those in Exercise 1. Describe the similarities and differences among the graphs.

a. $y = -x + 3$ **b.** $y = -x - 1$

Write

3. The slope-intercept form of an equation of the line is $y = mx + b$, where m is the slope and b is the y-intercept. Write a paragraph explaining how the values of m and b affect the graph of the equation. Include several drawings with your paragraph.

LAB 13 The Pythagorean Theorem

Use with: *Lesson 12–3, pages 482–486*
Materials: *geoboard, dot paper*

In this activity, you will use the Pythagorean Theorem to build squares on a geoboard or dot paper.

Activity: **Make a square with an area of 2 units.**

- Start with a right triangle like the one shown below.

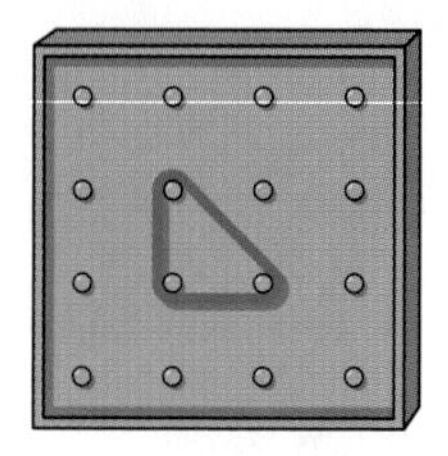

- Build squares on the two legs. Each square has an area of 1 square unit.

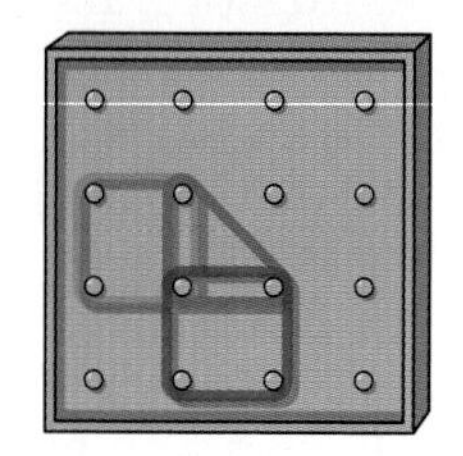

- Now build a square on the hypotenuse.

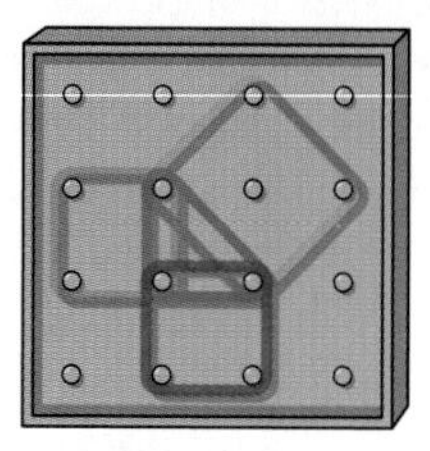

You can find the area of the square on the hypotenuse by using the Pythagorean Theorem.

$c^2 = a^2 + b^2$
$c^2 = 1^2 + 1^2$
$c^2 = 1 + 1$ or 2 The area of the square on the hypotenuse is 2 square units.

Model **Build squares on each side of the triangles shown below. Record the areas of the squares.**

1.

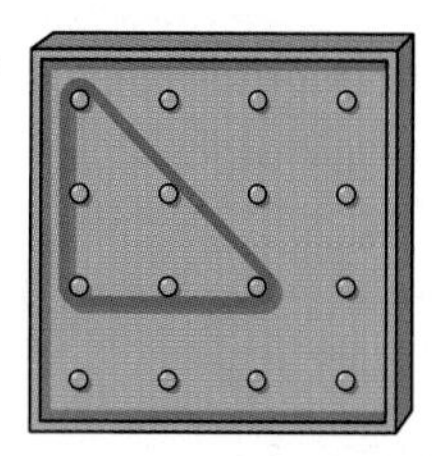

2.

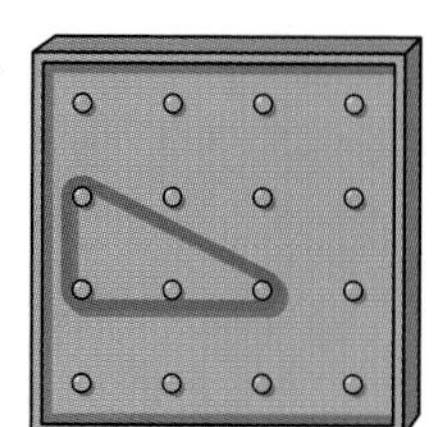

3.

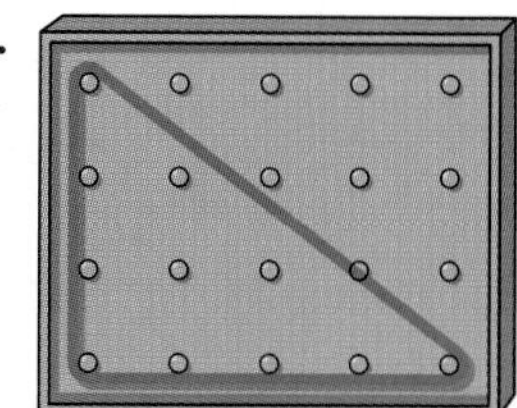

Draw **Draw a square having each area.**

4. 9 square units **5.** 4 square units **6.** 8 square units **7.** 13 square units
8. 10 square units **9.** 17 square units **10.** 32 square units **11.** 40 square units

Write

12. Write a paragraph explaining how to find the total area of the four small triangles in the drawing at the right.

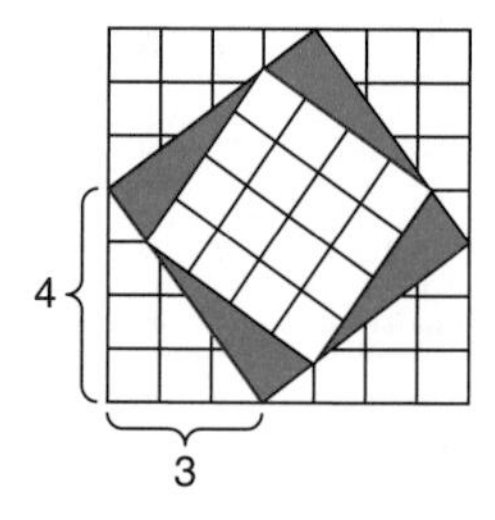

LAB 14 Estimating Square Roots

Use with: *Lesson 12–4, pages 487–491*
Materials: *base–ten tiles, product mat, graph paper*

You can use base-ten tiles to model square roots.

Activity 1: **Use base-ten tiles to find the square root of 121.**

- Model 121 with base-ten tiles.

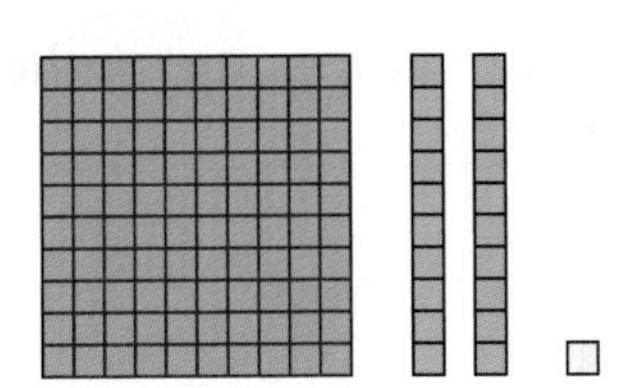

- Arrange the tiles into a square. The square root of 121 is 11 because $11^2 = 121$.

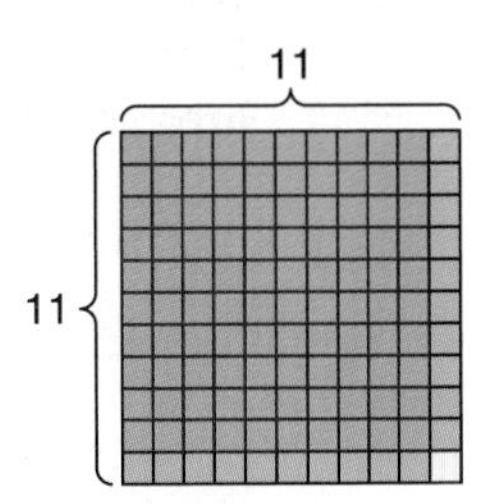

Activity 2: **Use base-ten tiles to estimate the square root of 151.**

- Model 151 with base-ten tiles.

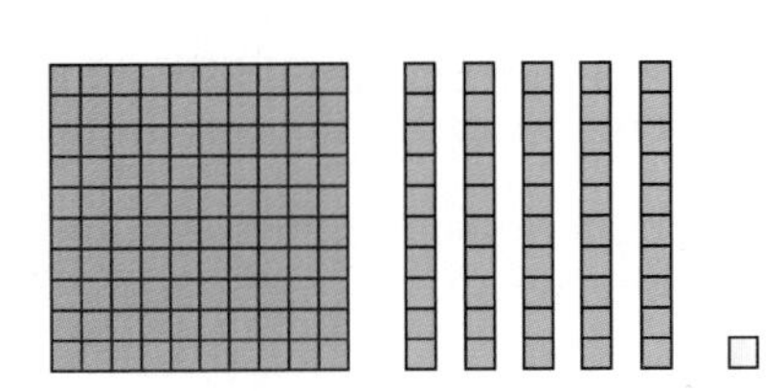

- Arrange the tiles into a square. The largest square possible has 144 tiles, with 7 left over.

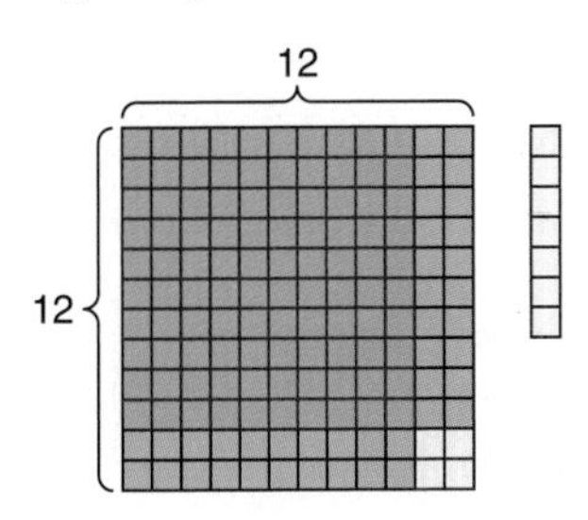

Notice that you need to trade one 10-tile for 10 1-tiles.

- Add tiles until you have the next larger square. You need to add 18 tiles.

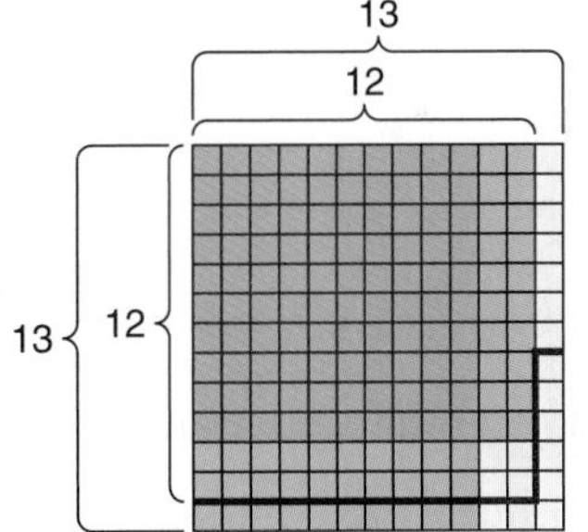

The square has 169 tiles.

Since 151 is between 144 and 169, the square root of 151 is between 12 and 13.

Model **Use base-ten tiles to estimate the square root of each number.**

1. 20 **2.** 450 **3.** 180 **4.** 200 **5.** 2

Draw **Use graph paper to estimate each square root.**

6. $\sqrt{28}$ **7.** $\sqrt{54}$ **8.** $\sqrt{125}$ **9.** $\sqrt{169}$ **10.** $\sqrt{250}$

Write **11.** If you list all of the factors of number in numerical order, the square root of the number either is the middle number or lies between the two middle numbers. How can a model show this?

LAB 15 Quadratic Functions

Use with: *Lesson 13–1, pages 518–522*
Materials: *graphing calculator*

You can use a graphing calculator to investigate patterns that occur within families of graphs.

Activity 1: **Use a graphing calculator to graph $y = x^2$, $y = x^2 + 2$, and $y = x^2 - 3$ on the same screen. Describe the similarities and differences among the graphs.**

- Graph each of the equations. Make sure your calculator is set for the standard viewing screen, $[-10, 10]$ by $[-10, 10]$.

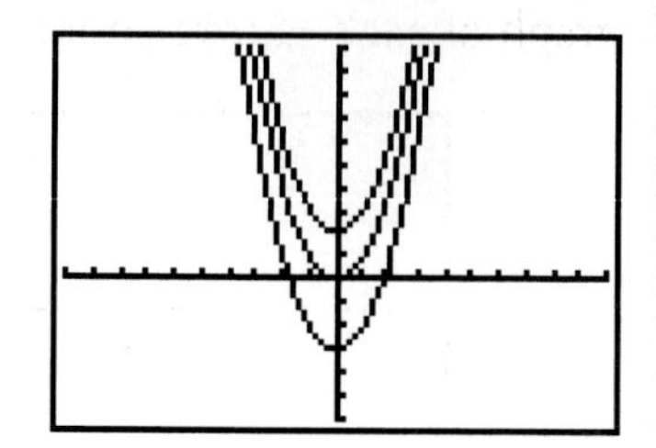

- The graphs are all the same shape, all open upward, and all have their vertex on the y-axis. The graphs pass through different points on the y-axis.

Activity 2: **Use a graphing calculator to graph $y = x^2$, $y = 2x^2$, and $y = 5x^2$ on the same screen. Describe the similarities and differences among the graphs.**

- Graph each of the equations. Make sure your calculator is set for the standard viewing screen.

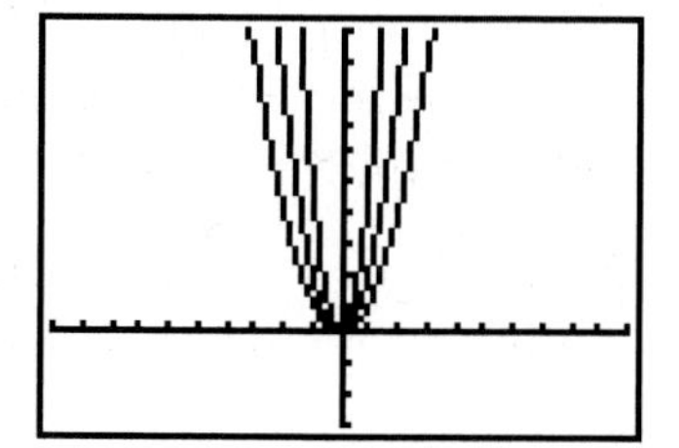

- The graphs all open upward, and all have their vertex at the origin. The graphs have different shapes. The graph of $y = 5x^2$ is the thinnest, and the graph of $y = x^2$ is the widest.

Draw **Repeat Activity 1 using the equations $y = x^2$, $y = (x + 2)^2$, and $y = (x - 3)^2$.**

1. Make a sketch of each graph.
2. Describe the similarities and differences among the graphs.
3. Compare the graph of $y = x^2 + 2$ with the graph of $y = (x + 2)^2$.

Repeat Activity 2 using the equations $y = -x^2$, $y = -2x^2$, and $y = -5x^2$.

4. Make a sketch of each graph.
5. Describe the similarities and differences among the graphs.
6. Compare the graph of $y = x^2$ with the graph of $y = -x^2$.

Write 7. The general form of a quadratic function can be written $y = a(x + h)^2 + b$. Write a paragraph explaining how the values of a, h, and b affect the graph of the function. Include several drawings with your paragraph.

LAB 16 Solving Quadratic Equations

Use with: *Lesson 13–2, pages 523–526*
Materials: *graphing calculator*

You can use a graphing calculator to solve quadratic equations graphically.

There are three possible outcomes when solving a quadratic equation. The equation will have either two real roots, one real root, or no real roots. A graph of each of these outcomes is shown below.

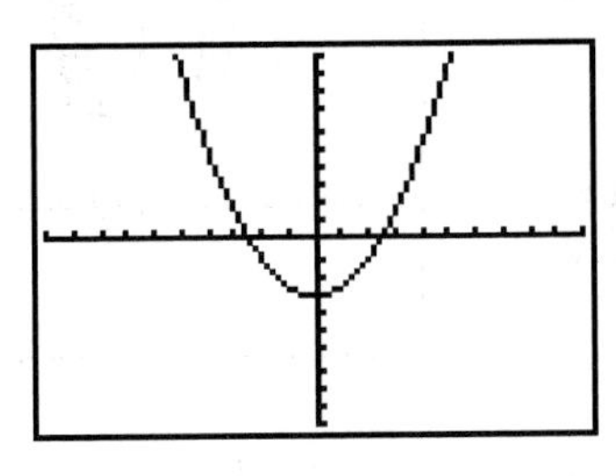

two real roots

two equal roots

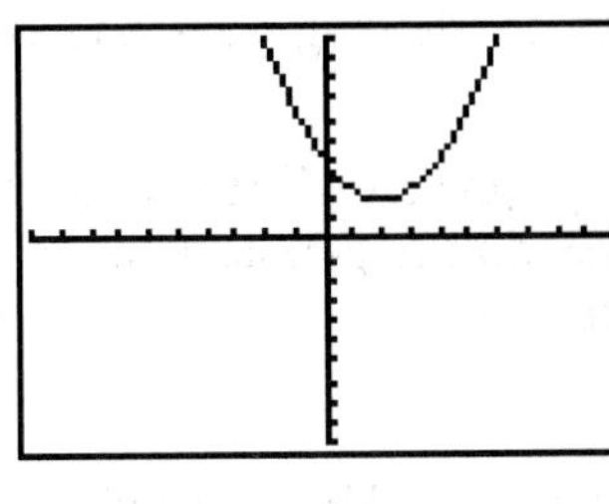

no real roots

Activity: Use a graphing calculator to solve $3x^2 + 6x - 1 = 0$.

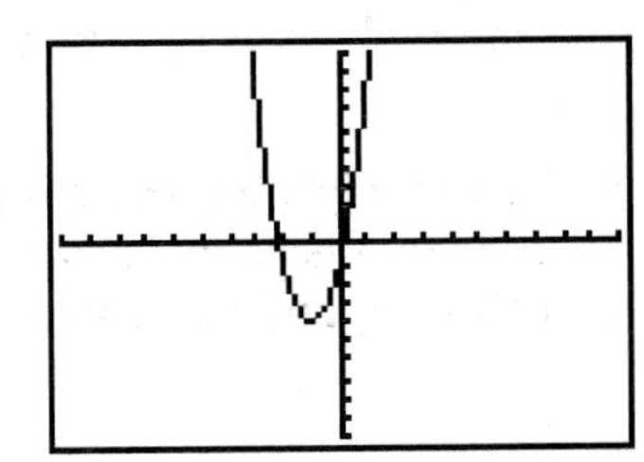

- Graph the related function, $y = 3x^2 + 6x - 1$. Make sure your calculator is set for the standard viewing screen, $[-10, 10]$ by $[-10, 10]$.
- Press TRACE and then use the arrow keys to determine estimate the x-intercepts. These x-intercepts are the roots of the equation.
- To determine the x-intercepts with greater accuracy, use the ZOOM feature. Set the cursor on the x-intercept and observe the value of x. Then zoom-in and place the cursor on the intersection point again. Any digits that are unchanged since the last trace are accurate. Repeat this process of zooming-in and checking digits until you have the number of accurate digits that you desire.
- The graph shown at the right indicates that one root of $3x^2 + 6x - 1 = 0$ is close to 0.1529. Actually, the root is approximately 0.1547. Another root is about -2.1547.

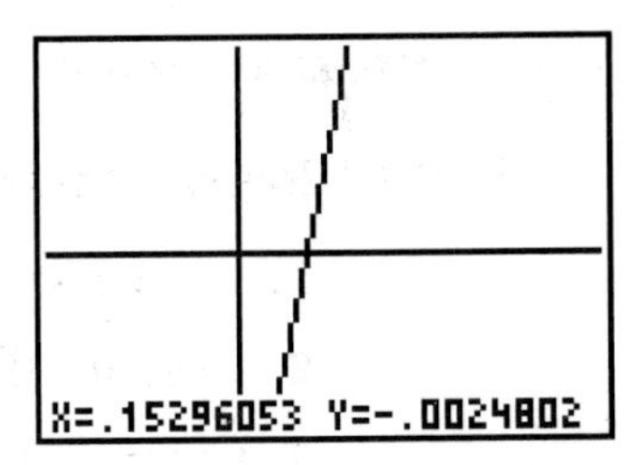

Model Use a graphing calculator to find the roots of each quadratic equation accurate to four decimal places.

1. $2x^2 - 11x - 15 = 0$

2. $0.2x^2 - 0.250x - 0.0738 = 0$

3. $1.2x^2 - 3.6x + 5.8 = 0$

4. $3x^2 + 3.08x - 1.36 = 0$

5. $x^2 + 31.54x + 229.068 = 0$

6. $35x^2 + 66x - 65 = 0$

Write **7.** Write a paragraph to explain how graphing the related function can help find the roots for a quadratic equation.

LAB 17 Completing the Square

Use with: *Lesson 13–4, pages 531–535*
Materials: *algebra tiles, equation mat*

One way to solve a quadratic equation is by *completing the square.* To use this method, the quadratic expression on one side of the equation must be a perfect square. You can use algebra tiles as a model for completing the square.

Activity: Use algebra tiles to complete the square for the equation $x^2 + 4x + 1 = 0$.

- Subtract 1 from each side of the equation.

$$x^2 + 4x + 1 - 1 = 0 - 1$$
$$x^2 + 4x = -1$$

- Model the equation $x^2 + 4x = -1$.

- Begin to arrange the x^2-tile and x-tiles into a square.

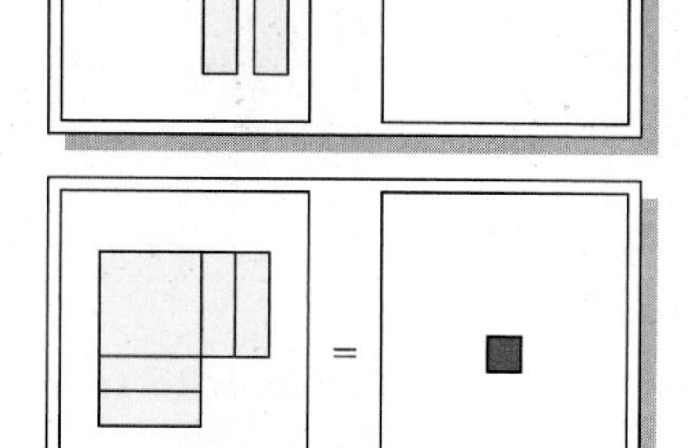

- In order to complete the square you need to add 4 1-tiles to the left side of the mat. Since you are modeling an equation, add 4 1-tiles to the right side of the mat.

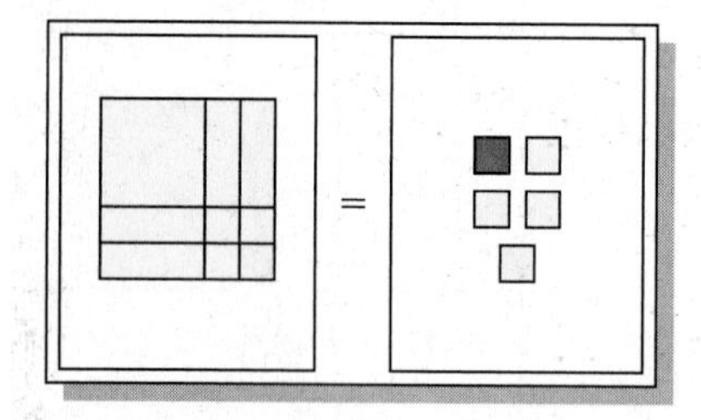

- Remove the zero-pairs on the right side of the mat. You have completed the square, and the equation is $x^2 + 4x + 4 = 3$ or $(x + 2)^2 = 3$.

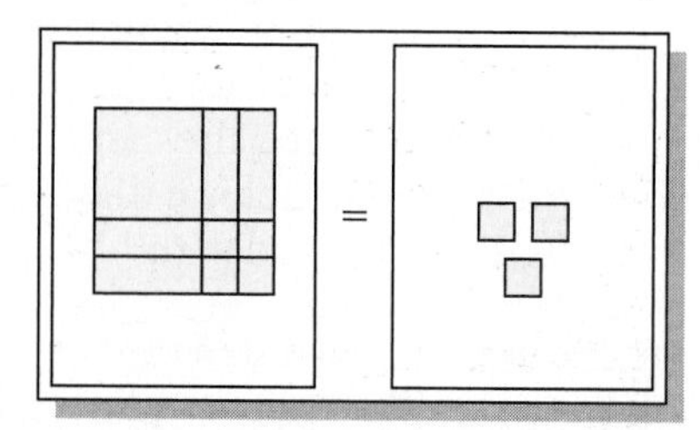

Model Use algebra tiles to complete the square for each equation.

1. $x^2 + 4x + 3 = 0$
2. $x^2 - 6x - 3 = 0$
3. $x^2 - 2x + 5 = 3$
4. $4 = x^2 + 2x - 5$
5. $0 = x^2 - 4x - 1$
6. $x^2 + 8x + 2 = 5$

Draw

7. In the equations shown above, the coefficient of x was always an even number. Sometimes you have an equation like $x^2 + 3x - 2 = 0$ in which the coefficient of x is an odd number. Complete the square by making a drawing.

Write

8. Write a paragraph explaining how you could complete the square with models without first rewriting the equation. Include a drawing.

GLENCOE

EXTENDED PROJECTS

for Merrill Algebra 1

PROJECTS

1 The Principle of Balance B2

2 Mathematics and Cryptography B6

3 Extending the Pythagorean Theorem B10

4 What's on TV? B14

PROJECT 1

The Principle of Balance

Chapter 3 defines four properties of equality. They are the addition, subtraction, multiplication, and division properties of equality. Each definition begins with an equation in *balance* .

$$a = b$$

Each property states that if something is done to one side of the equation, the equation can be kept in balance by doing the same thing to the other side of the equation.

$$a + 3 = b + 3 \qquad a - 7 = b - 7 \qquad 5a = 5b \qquad \frac{a}{2} = \frac{b}{2}$$

This is the basic principle of balance. Two quantities that are in balance will remain that way until something is done to only one of the quantities. To bring the quantities back into balance once something has been done to one side, you must do the same thing to the other side of the equation. Likewise, you could bring the two sides back into balance by undoing what was done to the first side.

The word *balance* has numerous meanings in addition to the meaning in algebra. World War I officially began on July 28, 1914, when Austria-Hungary declared war against Serbia after the June assassination of Archduke Francis Ferdinand of Austria-Hungary. Before it ended, 28 nations would become involved in this global conflict.

When President Woodrow Wilson asked Congress for a declaration of war on April 2, 1917, he said, "It is a fearful thing to lead this great peaceful people into war, into the most terrible and disastrous of all wars, civilization itself seeming to be in the balance." Wilson meant that by avoiding war, the United States might help to secure victory for an uncivilized enemy. The way to avoid such a calamity was to reverse that possibility by defeating the enemy. At the end of the war on November 11, 1918, Wilson declared that America's participation in the effort had helped bring civilization back into balance.

Investigating Balance

In this project, you will construct a device that you can use to study and demonstrate the principle of balance. Then you will investigate some of the many meanings of the word *balance* and show how the principle of balance relates to each of them.

Getting Started

Some balances are so precise that they can detect a difference in weights no greater than the weight of the period at the end of this sentence. But all balances, from the simplest to the most precise, have certain features in common. A horizontal bar is balanced on a vertical support. Pans are suspended from the ends of the bar. A pointer attached to the bar moves along a scale on the support, indicating when the pans are balanced.

Follow these steps to carry out your project.

- Design a balance that you can make with materials you can easily find.
- Build your balance. Keep in mind that the precision of the balance depends more on the care and accuracy with which you build it than it does on the materials you use.
- Outline a plan you can follow to demonstrate the principle of balance. Your plan should include demonstrations of various ways that a balanced system can be thrown out of balance and steps that can be taken at such times to return the system to a state of balance.
- Carry out your plan. Take accurate notes so that you can describe your work later.
- Discuss your results with your group. Did you learn anything about balance that surprised you? How could you improve the construction of your balance to ensure more accurate results?
- Determine how you might use your balance to represent an equation. Include ways to solve the equation using a balance. Discuss the limitations of your balance.
- Apply what you have learned to interpreting the principle of balance as it is found in the real world.

  Listed below are six everyday phrases containing the word *balance*:
  a. *balance* the budget
  b. *balance* of trade with Japan
  c. *balance* of nature
  d. the government system of checks and *balances*
  e. *balance* the tires
  f. *balance* your checkbook

  For each use of the word *balance*, carry out the following directives:

  - Research the meaning of the word.
  - Research the principles governing the working of the system suggested by the phrase.
  - Determine what it means to be "in balance" in the system.
  - Determine how the system can be thrown out of balance.
  - Determine how an out-of-balance system can be returned to a state of equilibrium.

Extensions

1. Research and report on additional meanings of the word *balance* that are not included in the above list.
2. Make a list of properties of equality not mentioned in Chapter 3. State each property.
   Example: The Squaring Property of Equality
   Statement: For any numbers a and b, if $a = b$, then $a^2 = b^2$.

Culminating Activities

Show what you have learned in this project by completing one of the following activities.

1. Write a report explaining how you built your balance, outlining the method you used to demonstrate the principle of balance, and describing the results of your research into the various meanings of the word *balance*. Supplement your report with diagrams, graphs, illustrations, or any other means that might help a reader better understand your research.
2. Create a "balance exhibit." Use displays, artwork, props, experiments, music, or any other materials to produce an exhibit for other students to view that will illustrate some of the many ways that the idea of balance can be applied.

PROJECT 2

Mathematics and Cryptography

Alan Turing (June 23, 1912-June 7, 1954) was an English mathematician who made major contributions to the development of the computer. In 1940, he began serving with the Government Code and Cypher School at Bletchley, Buckinghamshire, where he played a signigicant role in breaking the German "Enigma" codes that Germany used for all of its communications during World War II. German messages were encoded on a machine that was capable of generating more than 100,000 different letter combinations. In response, Turing's research unit created the most sophisticated decoding device that had ever been built, a machine they called "the Bombe." Early in 1941, Turing tested the machine using a German message recently intercepted by British intelligence. The Bombe responded:

FROM: ADMIRAL COMMANDING U-BOATS
ESCORT FOR U69 AND U107 WILL BE AT
POINT 2 ON MARCH 1 AT 0800 HOURS

Turing's success was instrumental in bringing about Germany's defeat in the war.

In the Pacific arena of World War II, the United States was at war with Japan. One of the United States' greatest assets during this conflict was the development of a code that the Japanese were completely unsuccessful in breaking. The American secret: the Navajo Indian language, one of the most complex languages on earth. Navajo "code talkers" translated U.S. messages into Navajo, then relayed them to other Navajos, who translated them back into English.

These examples show the importance of codes to a country's defense during wartime. With the advent of computers, even Turing's Bombe would be considered a primitive device today. High-tech devices can assign, juggle, and sort through billions of combinations of letters to encode and decode messages. No fool-proof code has yet been devised, but cryptographers—mathematicians who work with codes—are constantly searching for one.

Perhaps you will be the cryptographer to discover it.

Create a Code

In this project, you will design a code that uses mathematics to encode and decode messages. After looking at several common types of codes, you and the members of your group will devise an encoding device that is based on one or more mathematical techniques you are learning this year. You will work out a method for decoding messages and then practice coding and decoding secret messages sent using your code.

Getting Started

Follow these steps to carry out your project.

- Research common codes. Here are two to get you started.

  Letter-to-Letter This code "shifts" the alphabet a specified number of letters. Here the shift is 6 letters:

  A B C D E F G H I J K L M N O P Q R S T U V W X Y Z
  ↓
  F G H I J K L M N O P Q R S T U V W X Y Z A B C D E

  To encode the word DOG, write the letters beneath the letters D-O-G. The encoded word is ITL. The decoder must know the shift number 6, or else must try all 25 possible shifts until one yields a sensible message.

  Remainders A more sophisticated code, and a harder one to crack, uses remainders after division by 26. First choose a code multiplier, 5 for example. To encode the letter K, note that K is letter number 11. Multiply the letter number by 5 ($11 \times 5 = 55$). Divide by 26 ($55 \div 26 = 2$ R 3). The remainder 3 tells you to encode K as the 3rd letter, C.

- Work with your group to devise a code that is based on a topic you have studied this year. Some possibilities include equations, integers, formulas, patterns, polynomials, prime numbers, and factors. For example, suppose the alphabet letters consecutively corresponded to the numbers 1–26. The equation $C = \frac{2L + 48}{2} - 26n$ could be used to create a code, where C is the number corresponding to the code letter, L is the number corresponding to the original letter, and $n = 0$ or 1 so that a positive number less than 27 results.

  Keep in mind the fundamental properties of a successful code:
  1. For each letter, number, or other symbol to be encoded, the code must assign a unique symbol.
  2. There must exist a decoding system that the receiver of the message can use to translate the message.
  3. The code must be difficult to break.

- Practice sending and receiving messages with your code.

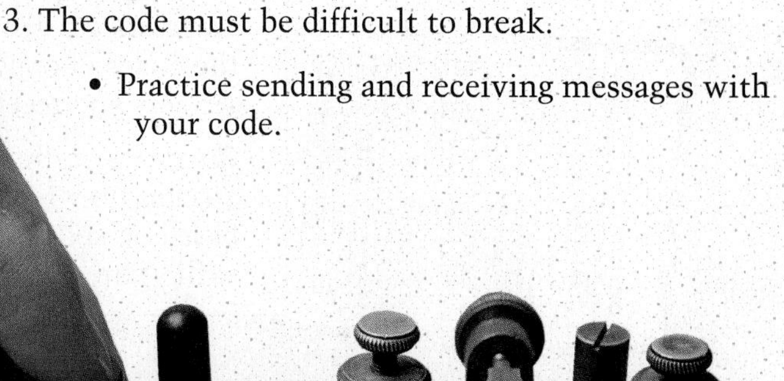

CRYPTOQUIP

PHODGGHX BLML SK

PKEWDGO

SK LSXD

Today's Crypt

Extensions

1. Research and report on the breaking of the German war code by Alan Turing's team in England.
2. Research and report on the work of the Navajo code talkers during World War II.
3. Many newspapers carry daily "cryptograms" that use letter frequencies to encode famous quotes letter-for-letter. Study the method used to solve cryptograms and then report on the method to the class.

Culminating Activities

Show what you have learned in this project by completing one of the following activities.

1. Write a manual that clearly explains how to encode and decode messages using your method. Assume that the reader is unfamiliar with common coding methods. Give several examples and include a set of encoded messages that readers can use to practice your decoding system.
2. Participate in a "Cryptogra-Fair." In the fair, you will compete against other groups that have invented codes. Divide your group into "senders" and "receivers." Your teacher will give secret messages to senders, who will code the messages and pass them both to group receivers and to senders in other groups. While receivers are decoding messages, senders will attempt to break the codes of opposing groups. For a second set of messages, senders and receivers should reverse roles.

   Before beginning, participants in the Cryptogra-Fair should discuss how it will be judged. Groups may wish to assign points based on speed and accuracy of sending and receiving, and on skill at breaking codes.

PROJECT 3

Extending the Pythagorean Theorem

The Pythagorean Theorem, discussed in Lesson 12-3, may be one of the most elegant, amazing, and useful results in all of mathematics. Mathematicians Gary Musser and William Burger have characterized the theorem as "perhaps the most spectacular result in geometry." Some mathematicians consider it a mark of honor to discover their own proofs of the theorem. In the classic 1907 book *The Pythagorean Proposition*, author Elisha Loomis gave more than 370 different proofs of the theorem, including one devised by President James Garfield. The theorem has wide applications and appears in nearly every branch of mathematics. What continues to intrigue and delight people who enjoy mathematics, even thousands of years after its discovery, is the utter simplicity of the Pythagorean relationship, combined with its far-reaching implications. No other relationship combines these two features so elegantly.

The Pythagorean Theorem states that in a right triangle, where a and b are the measures of the legs and c is the measure of the hypotenuse, then $c^2 = a^2 + b^2$.

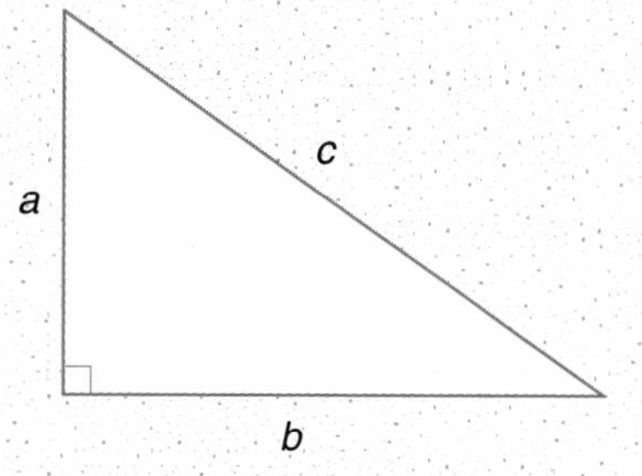

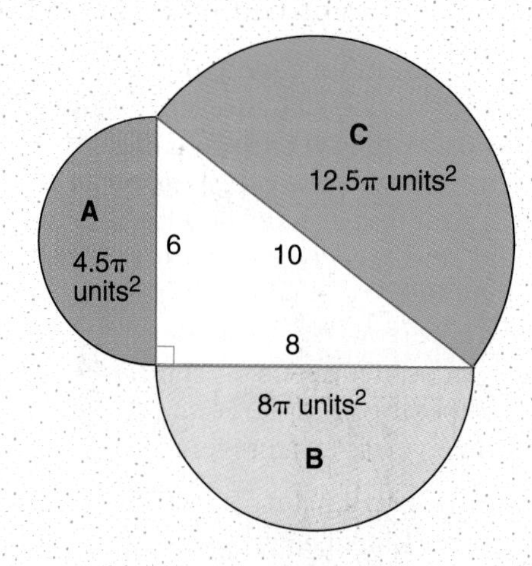

What is not so well known, and what is likely to increase your appreciation for the Pythagorean relationship considerably, is that the theorem given above is only one of many closely related Pythagorean theorems. Consider a 6–8–10 right triangle with semicircles constructed on each side.

Notice that the radius of each semicircle is half the measure of that side of the triangle. By using the formula for the area of a circle ($A = \pi r^2$), we can determine the areas for the semicircles constructed on the legs. They are $\frac{1}{2}(\pi \cdot 3^2)$, $\frac{1}{2}(\pi \cdot 4^2)$, and $\frac{1}{2}(\pi \cdot 5^2)$, or 4.5π, 8π, and 12.5π. Notice that $4.5\pi + 8\pi = 12.5\pi$.

At least in this example, we have a *Pythagorean Theorem for Semicircles:*

In a right triangle, if A and B are the areas of semicircles constructed on the legs and C is the area of a semicircle constructed on the hypotenuse, then $C = A + B$.

Is this true for all right triangles? What about the area of equilateral triangles constructed on the legs and hypotenuses of right triangles? What about pentagons? Hexagons? Is there a Pythagorean Theorem for volumes or surface areas of cubes, cylinders, or pyramids constructed on the legs and hypotenuses of right triangles? The answer to many of these questions and many others like them is: Yes!

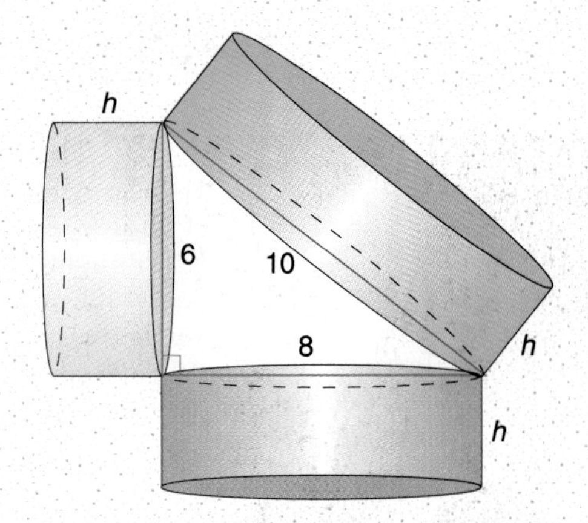

Build a PythagoLab

In this project, you will attempt to discover some of the little known but fascinating variations of the Pythagorean Theorem. You and the members of your group will choose a set of variations of the theorem that you would like to investigate. You will use calculators, paper and pencil, and 2-dimensional and 3-dimensional models of geometrical shapes to test the proposed variations and then summarize your results.

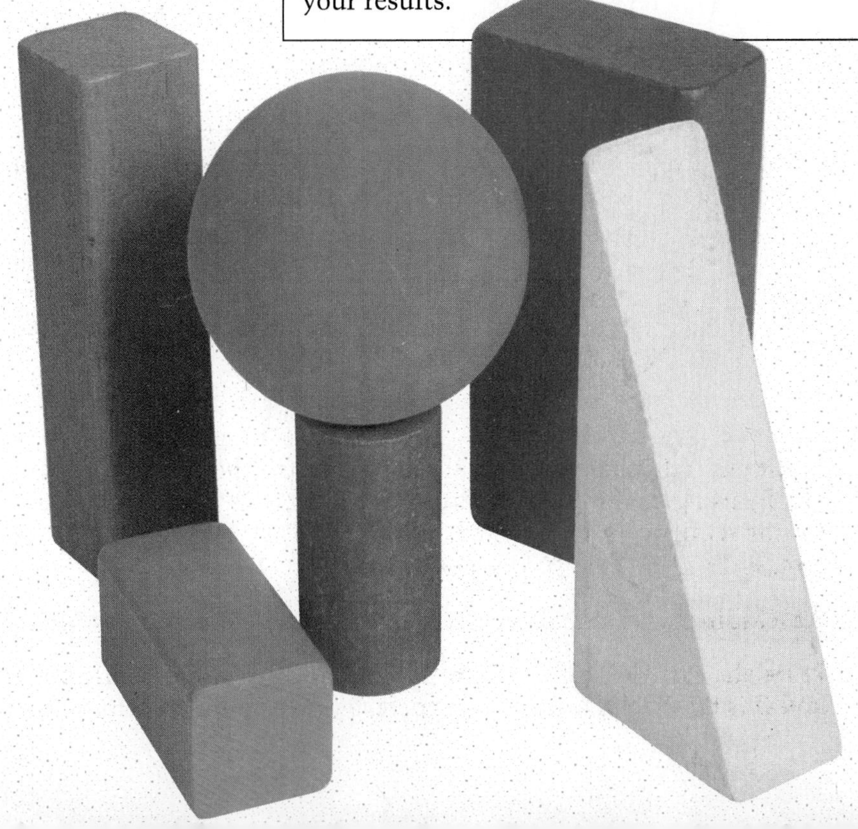

Getting Started

Follow these steps to carry out your project.

- Go over the previous discussion of the Pythagorean Theorem with the members of your group. Which variations mentioned in the discussion seem most likely to be true? Which ones are most intriguing? What other potential variations can you think of that you would like to investigate? Let your imagination run free. Ask yourself: What operations can I perform on the three sides of a right triangle such that the sum of the results of the operations performed on the legs might equal the result of the operation performed on the hypotenuse?
- Make a list of Pythagorean variations you want to investigate. Your teacher may specify the number of variations.
- Test each variation thoroughly until you are convinced that it is either true or false. In the discussion of the "semicircle" variation, only the 6–8–10 triangle was tested. Does the variation hold for other right triangles? How many should you test?
- Make models both to test variations and to illustrate those that you are able to confirm. The diagram on page 483 shows a model that illustrates the conventional Pythagorean Theorem. Use graph paper, construction paper, and other art supplies of your choosing to model 2-dimensional variations. If you test any 3-dimensional variations, build appropriate models.
- Discuss your results with your group. Do you see any patterns in your results that might enable you to say with assurance that some variations you did not test are almost certainly true?

Extensions

1. Find as many examples of practical applications of the Pythagorean Theorem as you can. You may wish to talk to science teachers, surveyors, engineers, architects, and others in professions that rely heavily on mathematics.
2. Research and report on Pythagoras, the Order of the Pythagoreans, and the discovery of the Pythagorean Theorem.
3. Research and report on Fermat's Last Theorem.
4. *Pythagorean triples* are triples of whole numbers that satisfy the Pythagorean relationship. The numbers 3, 4, and 5 form a Pythagorean triple because $3^2 + 4^2 = 5^2$. Find other examples of Pythagorean triples.
5. Write a proof or detailed explanation to verify the results you found in your PythagoLab.

Culminating Activities

Show what you have learned in this project by completing one of the following activities.

1. Put together an exhibit that summarizes your investigation into variations of the Pythagorean Theorem. The design of the exhibit is up to you, but you will probably want to include your 2- and 3-dimensional models, an explanation of the procedure that you followed, and a summary of your conclusions.
2. Create a detailed checklist that investigators could follow to test proposed variations of the Pythagorean Theorem.

PROJECT 4

What's on TV?

There were only about 100 television sets in Great Britain when the world's first television broadcasting service began in London on November 2, 1936. Regular commercial broadcasts began in the United States three years later for an audience not much larger than the one that began in London.

From those modest beginnings, television has grown in less than 60 years into one of the most influential forces in the world. Today, 98% of American households have at least one TV set. Nearly two-thirds have two or more sets. In 1950, the average set was on about $4\frac{1}{2}$ hours per day. In 1983, the average surpassed 7 hours per day for the first time, and it has remained near that level ever since. Where once newspapers and magazines provided the main sources of news and information for Americans, today television serves that function. Americans also obtain most of their information about consumer products from television commercials. In response, advertisers spend billions of dollars annually on TV advertising. Car makers alone spend more than $3 billion each year on television commercials. The cost of a 30-second advertisement on a much-watched program like the Super Bowl can run as high as $800,000.

Given the influence of television on the lives of most people, it is no surprise that pollsters are continually surveying the tastes, opinions, and viewing habits of television viewers. Perhaps you have participated in a survey. Even if you have not, you will have an opportunity to do so now.

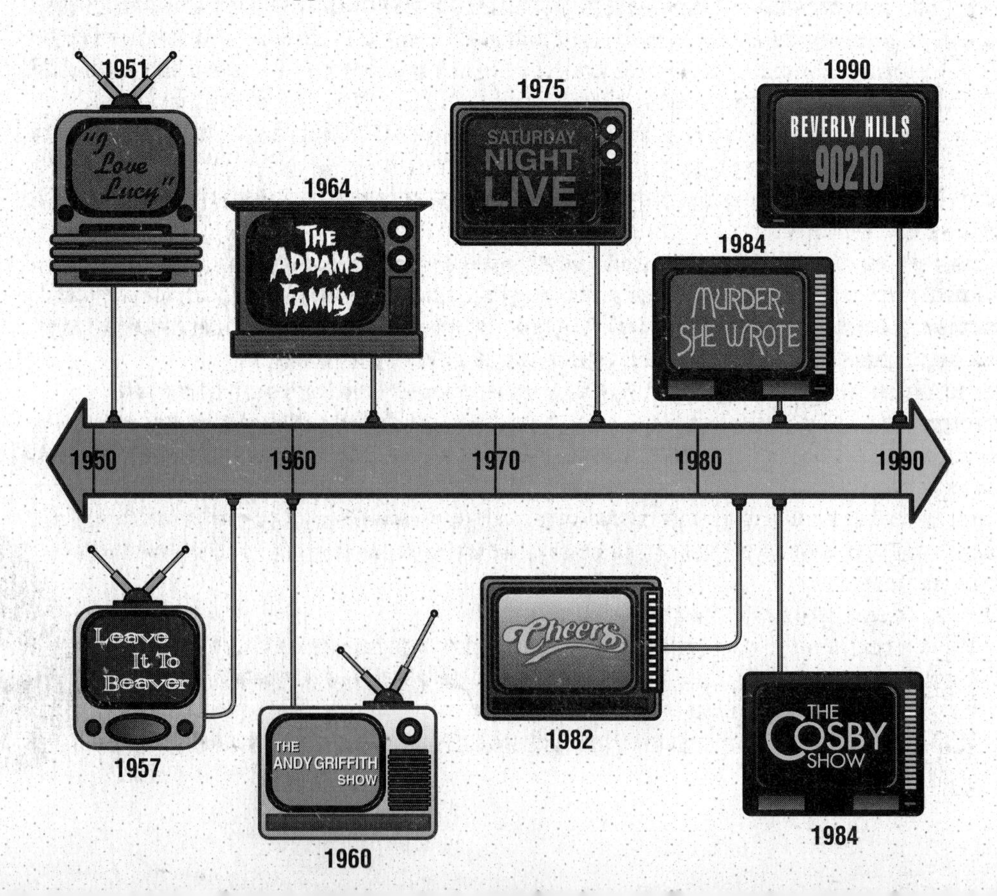

Conduct a TV Survey

In this project, you and the members of your group will decide on an issue relating to television that you would like to investigate. You will write, conduct, and analyze the results of a television survey intended to shed light on the issue you have chosen. Within certain broad limitations, the design of the survey will be left to you so that you can investigate the questions that interest you most. After you conduct your survey, you will analyze your results using some of the statistical methods discussed in Chapter 14.

Getting Started

Follow these steps to carry out your project.

- With your group, discuss the issues relating to television that interest you most. Try to move beyond a comparison of favorite programs to deeper issues involving television. Is television generally a positive influence on young people? Do people watch too much television? Is there too much violence on television? Are commercials truthful?
- After you have talked about some of these issues and identified areas of common interest, choose a broad question about television that intrigues you. You can choose one of the above questions or another one that better reflects your own interests. Be sure that the question you choose is one you can shed light on by surveying viewers.
- Create a survey consisting of at least eight questions. Each question should be designed to furnish data that will help you answer the main question your group has chosen to investigate. Spend a good deal of time on this step. Your questions should form an integrated whole aimed at revealing key viewer attitudes and habits. Try to break away from tired survey questions like "What is your favorite program?" Instead, devise creative questions that will illuminate the central issue you have chosen to explore. "How many people do you usually watch television with?" "Of the following products advertised on television, which ones would you absolutely *not* buy?"
- Think about the people you will be surveying. How many should you survey? Should they be chosen from the general population or should you target a small, precisely defined group? How will you record the responses when considering the sex, age, marital status, income, and so on of the interviewee?
- It may be necessary for people you intend to survey to keep track of certain information for a period of time—programs watched, length of time, and so on. If so, design a "viewer data sheet" and distribute it to the people you will be interviewing.
- Gather your data.
- Analyze your data. Use plots, measures of central tendency, graphs, scatter plots, regression lines drawn with graphing calculators, or any other statistical tools that seem useful.
- Discuss these questions with your group:
  1. What conclusions can we draw, based on the results of our survey?
  2. How has our survey shed light on the central question we posed at the beginning of the project?
  3. What questions has the survey raised that should be asked in a follow-up survey?

Extensions

1. Networks and sponsors closely watch the results of the Nielson ratings. A low rating often means the television show will be canceled. A high rating ensures another season of shows and often gives the network validation for charging sponsors more for the commercials that appear during the show. Contact the A.C. Nielsen Company to learn about the methods they use to survey public viewing habits. Its address is Nielsen Media Research, 1290 Avenue of the Americas, New York, NY 10104.
2. Research and report on changes in viewer tastes in television programming over the past several decades.
3. Contact an advertising agency to find out how it makes decisions on where to advertise its client's products (television, radio, newspapers, and so on).

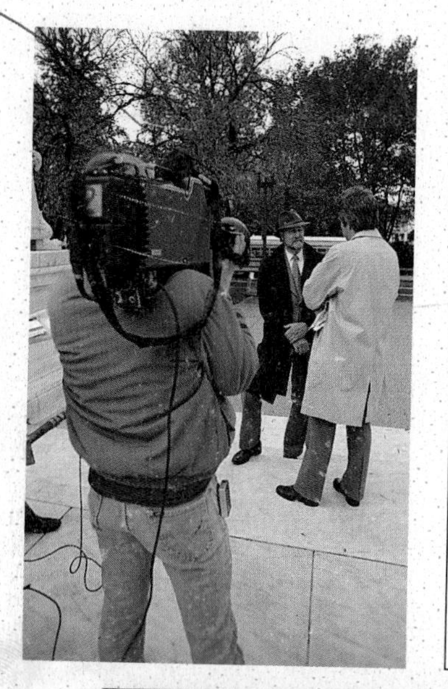

Culminating Activities

Show what you have learned in this project by completing one of the following activities.

1. Write a report summarizing the work of your group. State the question you chose to answer and describe how you decided on which survey questions to ask. Explain how you analyzed your data and detail your conclusions. Include with your report any graphs, plots, displays, or other supporting material you feel is needed to clarify your work.
2. Create a series of posters designed around the issue raised by your main question. In your display, describe your survey, detail your results, and summarize your conclusions. Include graphs, plots, photos, art work, or any other visual aids that might help viewers understand your survey.